Properties of Earth and Planetary Materials at High Pressure and Temperature

Geophysical Monograph Series

Including

IUGG Volumes

Maurice Ewing volumes

Mineral Physics Volumes

68 Microwave Remote Sensing of Sea Ice *Frank Carsey, Roger Barry, Josefino Comiso, D. Andrew Rothrock, Robert Shuchman, W. Terry Tucker, Wilford Weeks, and Dale Winebrenner (Eds.)*

69 Sea Level Changes: Determination and Effects (IUGG Volume 11) *P. L. Woodworth, D. T. Pugh, J. G. DeRonde, R. G. Warrick, and J. Hannah (Eds.)*

70 Synthesis of Results from Scientific Drilling in the Indian Ocean *Robert A. Duncan, David K. Rea, Robert B. Kidd, Ulrich von Rad, and Jeffrey K. Weissel (Eds.)*

71 Mantle Flow and Melt Generation at Mid-Ocean Ridges *Jason Phipps Morgan, Donna K. Blackman, and John M. Sinton (Eds.)*

72 Dynamics of Earth's Deep Interior and Earth Rotation (IUGG Volume 12) *Jean-Louis Le Mouël, D.E. Smylie, and Thomas Herring (Eds.)*

73 Environmental Effects on Spacecraft Positioning and Trajectories (IUGG Volume 13) *A. Vallance Jones (Ed.)*

74 Evolution of the Earth and Planets (IUGG Volume 14) *E. Takahashi, Raymond Jeanloz, and David Rubie (Eds.)*

75 Interactions Between Global Climate Subsystems: The Legacy of Hann (IUGG Volume 15) *G. A. McBean and M. Hantel (Eds.)*

76 Relating Geophysical Structures and Processes: The Jeffreys Volume (IUGG Volume 16) *K. Aki and R. Dmowska (Eds.)*

77 The Mesozoic Pacific: Geology, Tectonics, and Volcanism *Malcolm S. Pringle, William W. Sager, William V. Sliter, and Seth Stein (Eds.)*

78 Climate Change in Continental Isotopic Records *P. K. Swart, K. C. Lohmann, J. McKenzie, and S. Savin (Eds.)*

79 The Tornado: Its Structure, Dynamics, Prediction, and Hazards *C. Church, D. Burgess, C. Doswell, R. Davies-Jones (Eds.)*

80 Auroral Plasma Dynamics *R. L. Lysak (Ed.)*

81 Solar Wind Sources of Magnetospheric Ultra-Low Frequency Waves *M. J. Engebretson, K. Takahashi, and M. Scholer (Eds.)*

82 Gravimetry and Space Techniques Applied to Geodynamics and Ocean Dynamics (IUGG Volume 17) *Bob E. Schutz, Allen Anderson, Claude Froidevaux, and Michael Parke (Eds.)*

83 Nonlinear Dynamics and Predictability of Geophysical Phenomena (IUGG Volume 18) *William I. Newman, Andrei Gabrielov, and Donald L. Turcotte (Eds.)*

84 Solar System Plasmas in Space and Time *J. Burch, J. H. Waite, Jr. (Eds.)*

85 The Polar Oceans and Their Role in Shaping the Global Environment *O. M. Johannessen, R. D. Muench, and J. E. Overland (Eds.)*

86 Space Plasmas: Coupling Between Small and Medium Scale Processes *Maha Ashour-Abdalla, Tom Chang, and Paul Dusenbery (Eds.)*

87 The Upper Mesosphere and Lower Thermosphere: A Review of Experiment and Theory *R. M. Johnson and T. L. Killeen (Eds.)*

88 Active Margins and Marginal Basins of the Western Pacific *Brian Taylor and James Natland (Eds.)*

89 Natural and Anthropogenic Influences in Fluvial Geomorphology *John E. Costa, Andrew J. Miller, Kenneth W. Potter, and Peter R. Wilcock (Eds.)*

90 Physics of the Magnetopause *Paul Song, B.U.Ö. Sonnerup, and M.F. Thomsen (Eds.)*

91 Seafloor Hydrothermal Systems: Physical, Chemical, Biological, and Geological Interactions *Susan E. Humphris, Robert A. Zierenberg, Lauren S. Mullineaux, and Richard E. Thomson (Eds.)*

92 Mauna Loa Revealed: Structure, Composition, History, and Hazards *J. M. Rhodes and John P. Lockwood (Eds.)*

93 Cross-Scale Coupling in Space Plasmas *James L. Horwitz, Nagendra Singh, and James L. Burch (Eds.)*

94 Double-Diffusive Convection *Alan Brandt and H.J.S. Fernando (Eds.)*

95 Earth Processes: Reading the Isotopic Code *Asish Basu and Stan Hart (Eds.)*

96 Subduction Top to Bottom *Gray E. Bebout, David Scholl, Stephen Kirby, and John Platt (Eds.)*

97 Radiation Belts:Models and Standards *J. F. Lemaire, D. Heynderickx, and D. N. Baker (Eds.)*

98 Magnetic Storms *Bruce T. Tsurutani, Walter D. Gonzalez, Yohsuke Kamide, and John K. Arballo (Eds.)*

99 Coronal Mass Ejections *Nancy Crooker, Jo Ann Joselyn, and Joan Feynman (Eds.)*

100 Large Igneous Provinces *John J. Mahoney and Millard F. Coffin (Eds.)*

Geophysical Monograph 101

Properties of Earth and Planetary Materials at High Pressure and Temperature

Murli H. Manghnani
Takehiko Yagi
Editors

American Geophysical Union

Published under the aegis of the AGU Books Board

Library of Congress Cataloging-in-Publication Data

Properties of earth and planetary materials at high pressure and
temperature / Murli H. Manghnani, Takehiko Yagi, editors.
p. cm. --- (Geophysical Monograph ; 101)
Includes bibliographical references and index.
ISBN 0-87590-083-6
1. Mineralogy. 2. Geophysics. 3. Materials at high pressures.
4. Materials at high temperatures. I. Manghnani, M. H. (Murli H.)
Il. Yagi, Takehiko . Ill. Series.
QE364.2.H54P67 1998 97-46529
521.1--dc21 CIP

ISBN 0-87590-083-6
ISSN 0065-8448

Printed in the United States of America.

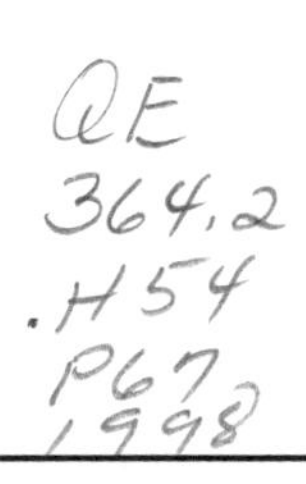

CONTENTS

PREFACE

Vital to understanding geophysical processes is knowledge of the physical and chemical properties of the Earth and planetary materials, including their variations with temperature, pressure, composition and volatile content. Information on chemical bonding and structure as well as chemical composition is necessary to understand the behavior of materials and to interpret global processes related to the Earth and planetary interiors. Properties of Earth materials can be measured either in situ via geophysical methods such as seismology or in the laboratory under appropriate high pressure-temperature environments, or they can be deduced from the atomic-level structure. Geophysicists increasingly recognize that high pressure-temperature research is vital to understanding the formation and evolution of the Earth and planetary interiors. Recent advances enable us to conduct experiments under conditions that replicate those throughout the Earth's mantle and core and in the outer portions of the interiors of the major planets. Current themes in high-pressure mineral physics are the simultaneous application of ultrahigh pressure-temperature techniques (e.g., the use of intense X ray synchrotron beam and extremely small sample volume in the laser-heated diamond-anvil), the characterization of important mineral phases under such P-T conditions, and the geophysical and geochemical interpretation of the experimental data thus obtained.

This volume presents new laboratory data as well as geophysical and geochemical field observations and their implications for describing the nature and composition of the Earth and planetary interiors. The chapters in this volume describe advances in high pressure-temperature techniques; in-situ properties under controlled P-T conditions; theoretical and planetary aspects of ultrahigh pressure research; and melts, melting and partitioning/segregation. Also included are chapters on iron and Earth's core; shock wave and equation of state measurements; phase equilibria at high pressure and temperature; rheological and electrical properties; roles of hydrogen, oxygen, and water; properties of hydrous phases; and amorphization at high temperature and pressure. These advances in state-of-the-art high pressure-temperature research and applications will interest mineral physicists, petrologists, geochemists, and planetologists.

The work included in this volume has been developed from material presented at the fifth U.S.-Japan seminar on High Pressure-Temperature Research: Properties of Earth and Planetary Materials, held in Maui, Hawaii, in 1996 (in addition to the editors, Thomas J. Ahrens, Yasuhiko Syono, Donald J. Weidner, Eiji Ito, Russell J. Hemley, and Tetsuo Irifune served on the organizing committee). These seminars, held every five years since 1976, are co-sponsored by the U.S. National Science Foundation and the Japan Society for the Promotion of Science. This series of seminars has played a significant role in promoting communications among high pressure research communities in Japan, the United States, and elsewhere.

Grateful acknowledgment is due to the Japan Society for the Promotion of Science and to the U.S. National Science Foundation for their financial support of the seminar. Financial support from the Facilities and Instrumentation Program, Division of the Earth Sciences, National Science Foundation, and from the School of Ocean and Earth Science and Technology, University of Hawaii, for the publication of this volume is also gratefully acknowledged.

The editors recognize the efforts of the many colleagues who provided constructive, critical reviews of these papers. Without their painstaking and enthusiastic efforts, the volume could not have been completed in its present good form. The editors are especially indebted to these conscientious reviewers for their time and effort. The editors express their utmost gratitude to Diane Henderson for invaluable help with the editorial work.

Murli H. Manghnani
University of Hawaii

Takehiko Yagi
University of Tokyo

Editors

Reviewers

M. Akaogi
S. Akimoto
T. J. Ahrens
H. Arashi
W. A. Bassett
R. Boehler
J. M. Brown
A. Chopelas
R. C. Cohen
T. S. Duffy
S. Endo
T. Fujii
K. Fujino
Y. Fukai
Y. Fukao
D. Grady
H. W. Green, II
J. Hama
N. Hamaya
R. J. Hemley
T. Irifune
E. Ito
I. Jackson
R. Jeanloz
M. Kanzaki
M. Kato
K. Kato
T. Katsura
T. Kawasaki
T. Kikegawa
M. Kitamura
E. Knittle
K. Kondo
M. Kruger
Y. Kudoh
K. Kusaba
K. Leinenweber
R. C. Liebermann
I. Maeda
M. H. Manghnani
H. K. Mao
Y. Matsui
C. Meade
L. C. Ming
H. Mizutani
Y. Morioka
A. Navrotsky
W. J. Nellis
E. Ohtani
J. P. Poirier
P. C. Presnall
C. T. Prewitt
S. M. Rigden
Q. Rubie
Y. Sato-Sorensen
H. Sawamoto
T. Sekine
H. Shimizu
O. Shimomura
K. Suito
Y. Syono
E. Takahashi
M. Tokonami
S. Tsuneyuki
B. Velde
D. J. Weidner
Q. Williams
G. Wolf
T. Yagi
T. Yamanka

X Ray Diffraction Measurements in a Double-Stage Multianvil Apparatus using ADC Anvils

T. Irifune[1], K. Kuroda[1], N. Nishiyama[1], T. Inoue[2], N. Funamori[3], T. Uchida[3], T. Yagi[3], W. Utsumi[4], N. Miyajima[5], K. Fujino[5], S. Urakawa[6], T. Kikegawa[7], and O. Shimomura[7]

Anvils made of ADC (Advanced Diamond Composite) have been introduced for a double-stage multianvil system. Using a hybrid system for the second stage anvils, composed of four ADC and four WC cubes, we were able to produce pressures to 28 GPa and temperatures exceeding 1500°C. In situ X ray diffraction measurements on some minerals have been successfully performed with a combination of the present high pressure system and synchrotron radiation. Only in two runs some failures of ADC anvils have been observed out of more than 10 runs so far conducted using the MAX80 and MAX90 apparatus at the Photon Factory, National Laboratory for High Energy Physics (KEK). The present system may be used on a routine basis for experiments under pressures to 30 GPa, and at temperatures approaching 2000°C within the force capacity of these apparatus.

INTRODUCTION

Application of sintered diamond anvils to multianvil apparatus has been successful in the last decade [e.g., *Ohtani et al.*, 1989; *Shimomura et al.*, 1992; *Utsumi et al.*, 1992; *Kato et al.*, 1992; *Kondo et al.*, 1993; *Funamori et al.*, 1996a,b]. Pressures greater than 30 GPa and temperatures over 1500°C are now accessible with this technique using eight sintered diamond cubes for the second stage anvils [*Irifune et al.*, 1992a; *Kondo et al.*, 1993; *Funamori et al.*, 1996a,b]. In addition, in situ X ray diffraction measurements using synchrotron radiation are also possible with sintered diamond anvils, as such anvils are relatively more transparent to X ray beams than those made of tungsten carbide (WC). Manufacturers that supply sintered diamond materials sufficiently large for anvil cubes, however, have been few, and such anvils remain quite expensive for use in routine high-pressure experiments.

A new diamond composite called ADC was invented by a research group at Australian National University [*Ringwood et al.*, 1989]. This material is produced at a pressure lower than the diamond stability field and, as a

[1]Department of Earth Sciences, Ehime University, Matsuyama 790, Japan

[2]Department of Earth and Space Sciences, SUNY at Stony Brook, Stony Brook, NY 11794, USA

[3]Institute for Solid State Physics, University of Tokyo, Roppongi, Tokyo 106, Japan

[4]Department of Synchrotron Radiation Facilities Project, Japan Atomic Energy Research Institute, SPring-8, Kamigori, Hyogo 678-12, Japan

[5]Department of Earth and Planetary Sciences, Hokkaido University, Sapporo 060, Japan

[6]Department of Earth Sciences, Okayama University, Okayama 700, Japan

[7]Photon Factory, National Laboratory for High Energy Physics, Tsukuba 305, Japan

Properties of Earth and Planetary Materials
at High Pressure and Temperature
Geophysical Monograph 101

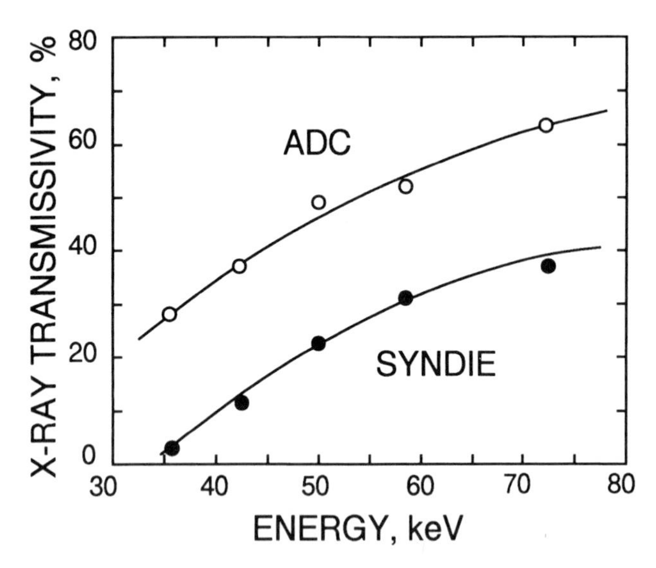

Figure 1. X ray transmissivity versus photon energy for ADC and SYNDIE cubes of 9.5 mm edge lengths (modified after *Irifune et al.*, 1992a).

result, is inexpensive relative to other competitive diamond composite materials. Moreover, *Ringwood et al.* [1989] suggested that ADC has mechanical properties comparable to those of commercially available sintered diamond materials. *Irifune et al.* 1992a] tested and compared some properties, such as electric resistivity and X ray transmissivity, of cubes made of ADC and those of a sintered diamond composite (SYNDIE), the latter having been commonly used in Japanese high-pressure research institutions. They found that ADC is highly transsparent to a wide photon energy range of synchrotron radiation (Figure 1) and suggested that this material may be used as a window for X ray measurements, although its use as an electric lead is limited due to its relatively high electric resistance.

Irifune et al. [1992a] also tested ADC cubes as second-stage anvils for a multianvil 6-8 apparatus using a DIA-type cubic press. They devised a hybrid anvil system (4 ADC and 4 WC anvils) for practical use of ADC anvils and demonstrated that this system has the potential to produce pressures over 30 GPa and temperatures to 1500°C, when a small anvil truncation (truncated edge length (TEL) = 1.0 mm) is adopted. However, they realized that manufacturing and handling of the cell assembly were extremely difficult due to the small dimensions of the assembly parts and, thus, their system with TEL = 1.0 mm was not very suitable for routine experiments. Moreover, a few ADC anvils were occasionally found cracked after high-pressure runs with this system of TEL = 1.0 mm, particularly those subjected to high temperatures [*Irifune et al.*, 1992a].

We recently modified the furnace assembly introduced by *Irifune et al.* [1992a] for routine experiments at pressures to 30 GPa and temperatures exceeding 1500°C. We also attempted to introduce the hybrid anvil system for in situ X ray diffraction measurements using synchrotron radiation at the Photon Factory of the National Laboratory for High Energy Physics (KEK). We have conducted more than 10 high pressure runs with the hybrid anvil system newly designed and optimized for such measurements and describe herein the details of the experimental techniques and the performance of the present system.

CELL ASSEMBLY AND IN SITU X RAY DIFFRACTION MEASUREMENTS

Figure 2 illustrates the cell assembly used in the present experiments, which was newly designed by modifying those of *Irifune et al.* [1992b] and *Funamori et al.* [1996a]. We used second-stage anvils of TEL = 1.5 mm. Two sheet heaters made of TiC were used and the sample was put directly into a capsule of amorphous boron cemented by epoxy resin. The TiC sheet heaters were rotated relative to each other by 90° so that electric power could be supplied via the four WC anvils of the hybrid anvil system, as

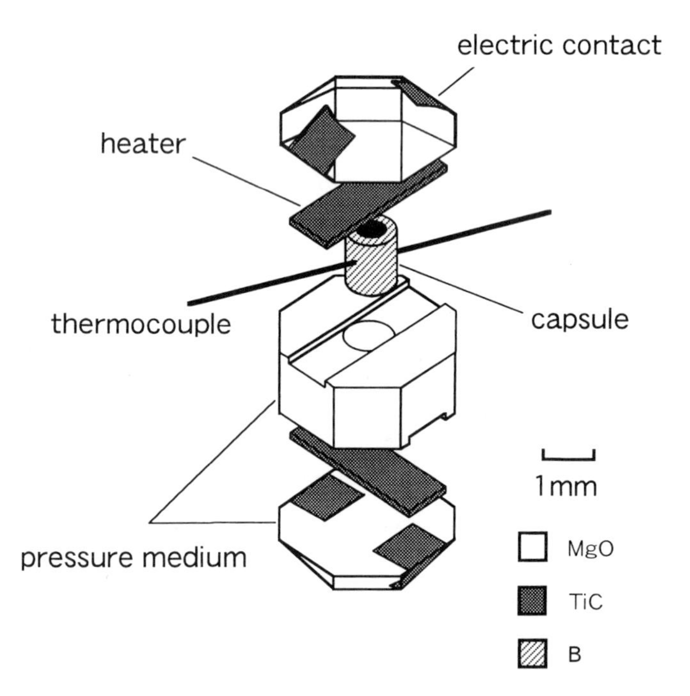

Figure 2. A schematic illustration of the cell. The upper and lower TiC sheet heaters are rotated by 90 degrees so that electric power is supplied via four WC anvils. Sample is put directly into a boron capsule, and the temperature at the central part of the sample is measured by a thermocouple.

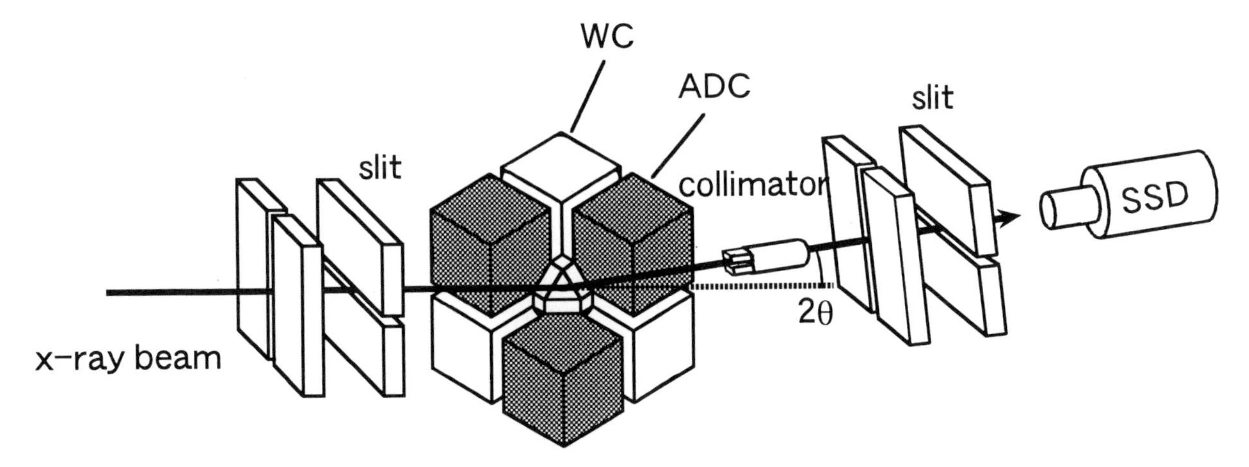

Figure 3. An illustration showing the present X ray diffraction system. The incident X ray beam is introduced between the anvil gap via vertical and horizontal slits and is directed to the sample in the middle of the pressure medium. The diffracted X ray beam, taken trough one of the ADC anvils, is collimated and detected by a solid state detector via receiving vertical and horizontal slits. Four WC anvils (one removed in this figure) are used for the electric leads to supply power to the twin heaters (see Figure 1).

illustrated in Figures 2 and 3. The temperature of the sample at the central portion of the capsule was measured by a $W_{95}Re_5$-$W_{74}Re_{26}$ thermocouple. Semi-sintered magnesia was used as the pressure medium, which was surrounded by preformed pyrophyllite gaskets 1.5 mm thick and 3.5 mm wide.

The cell assembly was pressurized in the hybrid anvil system as illustrated in Figure 3. The DIA type cubic apparatus (MAX80 and MAX90) at KEK were used to apply a load to these second stage anvils. A synchrotron X ray beam was directed to the sample through the anvil gap via horizontal (50 μm) and vertical (300 μm) slits. The diffracted beam was collected via horizontal (50 μm) and vertical (500 μm) slits at fixed diffraction angles of $2\theta = 4$ and 6°. A collimator having a narrow horizontal slit of 50 μm was also used to register the information from a limited region of the sample. An energy-dispersive method was adopted using a solid state detector and a multichannel analyzer, which was carefully calibrated using characteristic X ray radiation from some standard metals.

As there were significant temperature gradients across the capsule, the sample adjacent to the hot junction of the thermocouple was examined. Pressure was monitored by the unit-cell volume changes in Au powder, which was mixed with the sample by 1:100 in volume, using an equation of state [*Anderson et al.*, 1989]. As some of the diffraction peaks of Au overlapped with those of the starting materials and of the product phases, we selected one or two Au peaks that were not affected for accurate pressure determination. Nevertheless, it was difficult to remove this effect completely, and accordingly the uncertainty of the pressure estimation was somewhat greater than that obtained with a techniques using separate sample chambers for pressure measurements [e.g., *Funamori et al.*, 1996a].

EXPERIMENTAL RESULTS AND DISCUSSION

High Pressure Generation with the Hybrid Anvil System

Figure 4 depicts the relation between pressures and applied loads at room temperature for some runs conducted at KEK. Also shown is the covalent-metallic transition in GaP (~ 23 GPa) and the ε-γ transition in Fe-11%V (~ 26 GPa) for the same anvil system in separate runs. The slight deviation of these values from those determined by the volume changes in Au is probably because the cell assembly used in the calibration runs is different from those for the in situ X ray measurements, or it may be due to uncertainty of the transition pressures of these pressure calibration points. Further, it should also be noted that there is some uncertainty in the pressure estimation based on the volume changes in Au as discussed in *Funamori et al.* [1996a].

The pressure versus load curves for most of the present runs reasonably agree with each other, and pressures of about 28 GPa were obtained at a press load of about 300 tons. However, one run (ME-2) exhibited a different pressure generation curve as compared to those of other runs shown in Figure 4. In this particular run, temperature was raised to about 200°C at some fixed press loads in the course of increasing pressure [see *Kuroda and Irifune*, this

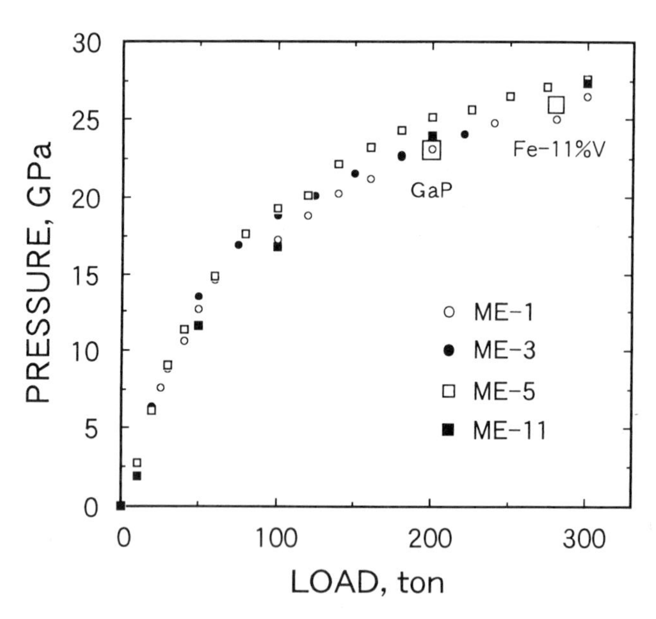

Figure 4. Some examples of the pressure-load relationship determined by molar volume changes in Au under pressure. Those based on the phase transitions in GaP (~23 GPa) and Fe-11%V (~26 GPa) are also shown for comparison.

volume]. The pressure dropped significantly upon heating to this temperature and was not recovered to the initial values when the temperature was decreased to room temperature. Thus the final pressure obtained at the maximum load (200 ton) in this run was lower than that of the "normal" runs by about 4 GPa. Similar preheating was also made in run ME-10, where the final pressure was also significantly lower than that expected from the pressure-load relation in Figure 4 (see Table 1).

Figure 4 suggests that pressures approaching 30 GPa may be produced in the present hybrid anvil system if we apply the maximum press load of 500 tons available in the MAX80 apparatus. *Funamori et al.* [1996a] reported generation of pressures up to 30 GPa using sintered diamond cubes (TEL = 2.0 mm; c.f., TEL = 1.5 mm in the present system) for all of the second stage anvils. Moreover, generation of much higher pressures up to 37 GPa at 400 ton were confirmed when they adopted the smaller truncation edge length (TEL = 1.5 mm), as adopted in our present system [*Funamori et al.*, 1996b]. They demonstrated that about 32 GPa was produced at a load of 300 ton using eight sintered-diamond anvils, which is higher than produced in our hybrid anvil system by 4 GPa. These results show that the mixing of tungsten carbide anvils with sintered diamond anvils notably decreases the efficiency of pressure generation, particularly at pressures above 25 GPa, where plastic deformation in the WC anvils becomes significant *[Irifune et al.*, 1992a,b]. Nevertheless the present hybrid anvil system has great advantage in designing the heating system and in cost performance of the experiments and provides a potential tool to pressures to 30 GPA.

The nominal pressure at a fixed ram load decreases significantly with increasing temperature to about 600°C, above which the pressure increases again and approaches the initial value [see Figure 5 of *Kuroda and Irifune*, this volume]. The decrease of pressure with increasing temperature may be due to stress relaxation in either the gasket-pressure medium system or in the sample itself. The magnitude of the apparent pressure drop with increasing temperature seems to be enhanced in the lower-pressure regime, suggesting that the stress relaxation in the gasket-pressure medium may predominate over that in the sample, at least in the pressure region below 10 GPa [*Kuroda and Irifune*, this volume]. This is probably because the gasket readily yields at high temperature and low pressures, where the normal stress acting on the gasket from the adjacent two anvils is relatively small and hence the friction between gasket and anvil surfaces is low. Such pressure drop due to the stress redistribution in the gasket an/or the sample should be canceled by thermal pressure effects with further increase in temperature.

High Temperature Generation Under Pressure

We were able to produce temperatures exceeding 1500°C for more than 30 minutes using the present heating system. However, heating became unstable in some runs using serpentine starting material at temperatures of about 700–1000°C, and these runs (ME2 and ME4, see Table 1) were forced to be terminated after a few minutes heating at these temperatures. This is probably because the heating element was contaminated by the water released from the dehydration of serpentine at these temperatures, as the sample was directly in contact with the heater. We were, however, able to continue heating to temperatures of above ~1500°C when anhydrous starting materials were used and/or only minor dehydration occurred (ME-3, ME-7, ME-9, and ME-11 of Table 1). Figure 5 shows temperature versus applied electric power for some runs performed at similar press loads of 200–300 tons. The temperature increased almost linearly with input electric power to 1000–1200°C. However, we noticed abrupt changes in the temperature-power relations at higher temperatures, as seen in Figure 5. It is most likely that the $W_{95}Re_5$-$W_{74}Re_{26}$ thermocouple reacted with the boron capsule, as we observed substantial increases in the electric resistance of the thermocouple wire at these temperatures. No obvious changes in the performance of the heaters were observed in

TABLE 1. Experimental Conditions, Results, and Anvil Damage in the Runs Conducted at KEK Using the Hybrid Anvil System

Run	Load (ton)	P_{max} (GPa)	T_{max} (°C)	Starting material	Phases identified	Anvil damages ADC	Anvil damages WC	Remarks
ME-1	300	26	1200	antigorite	PhD + ShB + (W)	0	0	
ME-2	200	20	1050	antigorite	γ + St + (W)	0	1	pressure drop due to preheating
ME-3	300	27	1500*	lizardite	PhD + ShB + (W)	0	4	
ME-4	50	13	700	antigorite	α + En + (W)	0	0	
ME-5	300	28	1200	diopside	Mg-Pv + Ca-Pv	2	4	blow-out during heating at 300 tons
ME-6	70	17	1200	antigorite	α + En + (W)	0	0	
ME-7	150	23	1700**	Au (+ MgO)	L	1	3	
ME-8	300	28	1200**	diopside	Mg-Pv + Ca-Pv	0	1	
ME-9	200	26	1700**	Au (+ MgO)	L	0	1	
ME-10	100	12	1500	orthoenstatite	En (clino-form)	0	0	pressure drop due to preheating
ME-11	300	28	2000*	diopside	Mg-Pv + Ca-Pv	0	4	blow-out on release of pressure

PhD = phase D [see *Kuroda and Irifune*, this volume], ShB = superhydrous phase B, W = water, γ = Mg_2SiO_4 spinel, St = stishovite, α = forsterite, L = liquid, En = enstatite, Mg-Pv = $MgSiO_3$ perovskite, Ca-Pv = $CaSiO_3$ perovskite.
* may be lower than the nominal values (see,text).
** temperature was estimated by the supplied electric power because of thermocouple failure.

these runs, and accordingly the nominal temperatures from the e.m.f.s of the thermocouple (Table 1) may have been somewhat overestimated, as the electric power needed for such temperatures was lower than that expected from the extrapolation of the power-temperature relations from lower temperatures. Thus the highest temperature (2000°C) we achieved in run ME-11 should be reduced to about 1600–1700°C, if we use the temperature-power relations observed in the low-temperature region. Some modifications of the cell assembly are therefore needed to make more reliable temperature measurements above 1000–1200°C. Nevertheless, the power-temperature relations suggest that temperatures exceeding 1500°C can be reached using the present heating system. Actually, *Funamori et al.* [1996b] demonstrated that a similar heating system using smaller disk heaters has potential to produce temperatures of ~1700°C. We suppose even higher temperatures approaching 2000°C may be generated in our present system, as was confirmed by quench experiments using virtually the same furnace assembly [e.g., *Irifune et al.*, 1996] as used in the present study.

Figure 6 shows the cross section of the furnace assembly after a run at 26 GPa, at 1200°C (ME-1). We noticed the presence of significant temperature gradients across the

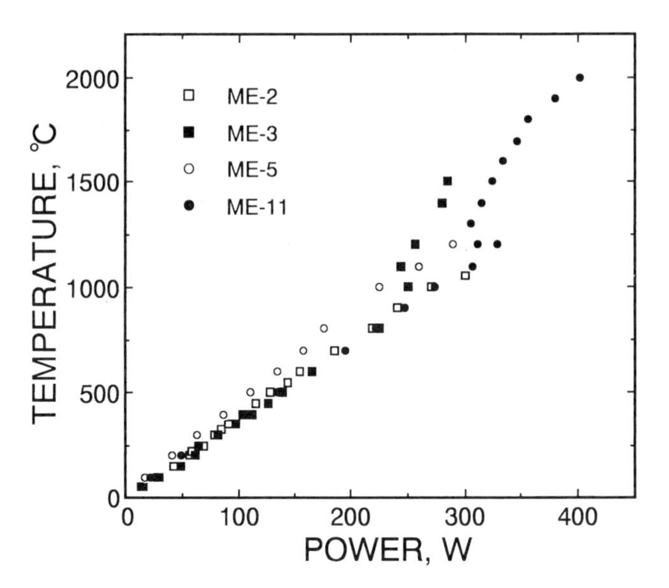

Figure 5. The temperature-power relationship of the runs conducted at press loads of 200-300 tons. Temperature changes almost linearly with increasing power supply to temperatures of 1000-1200°C, while some notable changes in this relation are observed in most runs conducted at higher temperatures (see text).

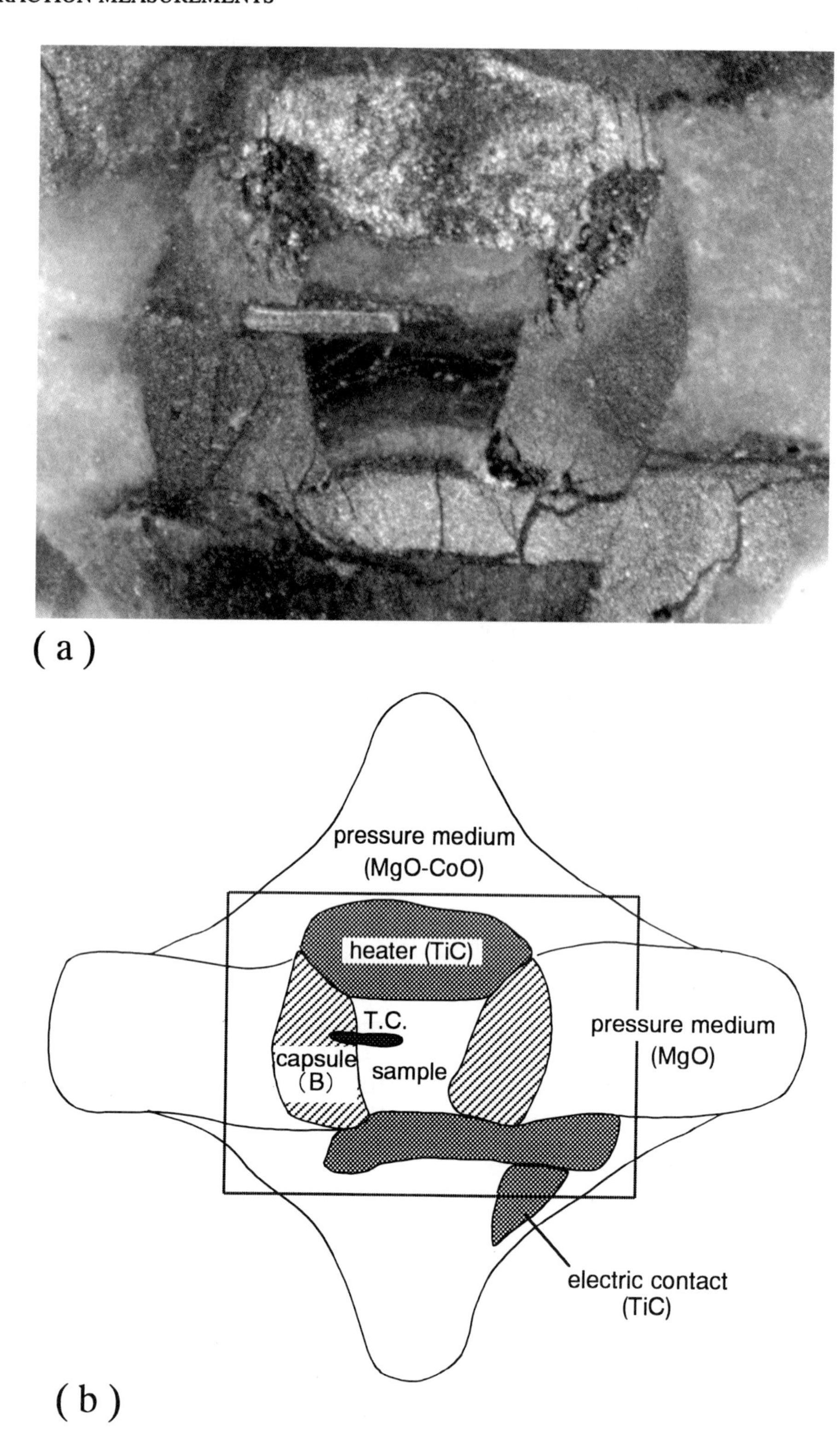

Figure 6. A photomicrograph (a) and its schematic illustration (b) of the cross section of the cell assembly after a run (ME-1). The color of the sample changes along the cylindrical capsule, as there were significant temperature gradients and the dehydration of the sample (antigorite) occurred at the hotter end parts. Only a portion of the sample was examined near the hot junction of the thermocouple using the thin X ray beam.

sample charge, as suggested by this picture. Although we did not measure the temperature gradients in the present sample, such measurements for a similar cell assembly in quench experiments suggest that the temperature at the central part of the sample, where the thermocouple was placed, is lower by about 10% than those near the sheet heaters [Irifune, unpublished data]. Careful scanning of the X ray beam by shifting the press position [*Shimomura et al.*, 1992], however, allows diffraction data from the sample very close to the hot junction of the thermocouple to be obtained. Thus we can avoid the temperature uncertainty due to the large gradient using this technique, though there remain some uncertainty in temperature due to the possible contamination of the thermocouple from the boron capsule at temperatures greater than 1000–1200°C.

In Situ X Ray Diffraction Measurements under High Pressure and High Temperature and the Performance of ADC Anvils

Irifune et al. [1992a] measured the practical X ray transmissivity of ADC as a function of photon energy using synchrotron radiation. Figure 1 compares the results on ADC and SYNDIE (produced by De Beers) cubes, both of which are used for the second-stage anvils in a double-stage multianvil system. It is demonstrated that ADC is transparent to a wider energy range of X ray beams as compared to the latter material. For instance, at photon energies below 40 keV, less than 10% of the incident beam can pass through a SYNDIE cube of about 1 cm edge length, while the corresponding ADC cube is several times more transparent. Thus, X ray diffraction data of the same quality may be collected in a much shorter duration when we adopt ADC as a window for X ray beams.

We actually obtained X ray diffraction patterns from very small volumes (about 0.5 mm wide x 1 mm long × 50 mm thick) of the sample within a few minutes using synchrotron radiation [see Figures 1-4 *of Kuroda and Irifune*, this volume]. Although we have not compared the data acquisition time necessary to obtain X ray diffraction patterns of similar quality, it is most likely that we need two or three times longer exposure time when we adopt a SYNDIE anvil instead of ADC as the X ray window [see also *Funamori et al.*, 1996a,b]. Saving data acquisition time can be a very important factor especially for in situ X ray diffraction measurements at high temperature under pressure, as the heaters tend to be unstable in the higher temperature regime and rapid data collection is essential for such experiments. In addition, efficient data collection also saves total machine time, which is a consideration for other visiting users of synchrotron facilities.

Table 1 summarizes the maximum pressure and temperature conditions and the results of our runs so far performed at KEK. We were able to collect diffraction data of sufficient quality normally in 5 minutes or so at pressures to 28 GPa and at temperatures exceeding 1500°C. Moreover, very few instances of damage to ADC anvils were recognized after 11 runs using the present hybrid anvil system, whereas WC anvils were occasionally found cracked after these runs. As can be seen from Table 1, we had two ADC anvils cracked after a blowout during heating at 1200°C at a pressure near 28 GPa, but all ADC anvils survived in most runs except one (ME-7) in which the presence of a minor crack was recognized. As the WC anvils are far cheaper than those of ADC, frequent loss of small WC anvils is not problematic for this type of experiment.

Although performance of the ADC anvils seems to be quite good as seen in Table 1, we previously noticed that the ADC cubes occasionally cracked after some quench runs, especially when we had a blowout during heating under pressure [e.g., *Irifune et al.*, 1992a]. Moreover, other users of ADC [e.g., *Kato et al.*, 1995] point out that ADC cubes are generally more fragile than SYNDIE cubes when they are mixed in use as the second stage anvils for the MA8 system. Nevertheless, one of the present authors (N. Funamori) recently found that the ADC anvils persist to pressures as high as 32 GPa, even if the ADC anvils are used in the mixed configuration. Our present study also suggests that ADC anvils may be repeatedly used without any major failures when they are properly used. Because of its high X ray transmissivity and cost performance, ADC is suitable for anvils for in situ X ray diffraction measurements under pressures to about 30 GPa. The present hybrid anvil system is one of the promising methods for the practical use of this material in such experiments on a routine basis.

Acknowledgments. One of the authors (T.I.) thanks P. Willis of RSES, ANU, for supplying the ADC cubes. We are indebted to M. Mizobuchi, M. Miyashita, M. Isshiki, N. Kubo,.and Y. Yamasaki for assistance of our in situ X ray diffraction measurements at KEK. We acknowledg invaluable comments from K. Kusaba, K Leinenweber, and an aonymouse reviewer. The present study is supported by Grant-in-Aid for Scientific Research from the Ministry of Education, Science and Culture of Japan, Inoue Science Foundation, and also by Japan Society for Promotion of Sciences (JSPS).

REFERENCES

Anderson, O. L., D. G. Isaak, and S. Yamamoto, Anharmonicity and the equation of stttate for gold, *J. Appl. Phys.*, *65*, 1534-1543, 1989.

Funamori, N., T. Yagi, W. Utsumi, T. Kondo, T. Uchida, and M. Funamori, Thermoelastic properties of $MgSiO_3$ perovskite determined by in situ X ray observations up to 30 GPa and 2000K, *J. Geophys. Res., 101*, 8257-8269, 1996a.

Funamori, N., T. Yagi, and T. Uchida, High-pressure and high-temperature in situe X ray diffraction study of iron to above 30 GPa using MA8-type apparatus, *Geophys. Res. Lett.*, 23, 953-956, 1996b.

Irifune, T., W. Utsumi, and T. Yagi, Use of a new diamond composite for multianvil high-pressure apparatus, *Proc. Jpn. Acad., 68B*, 161-166, 1992a.

Irifune, T., Y. Adachi, K. Fujino, E. Ohtani, A. Yoneda, and H. Sawamoto, A performance test for WC anvils for multianvil apparatus and phase transformations in some aluminous minerals up to 28 GPa, in High Pressure Research in *Mineral Physics: Application to Earth and Planetary Sciences, Geophys. Monogr. Ser.*, vol. 67, edited by Y. Syono and M. H. Manghnani, pp. 43-50, Terra Scientific Publishing, Tokyo/AGU, Washington, D.C., 1992b.

Irifune, T., T. Koizumi, and J. Ando, An experimental study of the garnet-perovskite transformation in the system $MgSiO_3$-$Mg_3Al_2Si_3O_{12}$, *Phys. Earth Planet. Inter., 96*, 147-157, 1996.

Kato, T., E. Ohtani, N. Kamaya, O. Shimomura, and T. Kikegawa, Double-stage multi-anvil system with a sintered diamond anvil for X ray diffraction experiment at high pressures and temperatures, in *High Pressure Research in Mineral Physics: Application to Earth and Planetary Sciences, Geophys. Monogr. Ser.*, vol. 67, edited by Y. Syono and M. H. Manghnani, pp. 33-36, Terra Scientific Publishing, Tokyo/AGU, Washington, D.C., 1992.

Kato, T., E. Ohtani, H. Morishima, D. Yamazaki, A. Suzuki, M. Suto, T. Kubo, T. Kikegawa, and O. Shimomura, In situ X ray observation of high-pressure phase transitions of $MgSiO_3$ and thermal expansion of $MgSiO_3$ perovskite at 25 GPa by double-stage multianvil system, *J. Geophys. Res., 100*, 20,475-20,481, 1995.

Kondo, T., H. Sawamoto, A. Yoneda, M. Kato, A. Matsumuro, and T. Yagi, Ultrahigh-pressure and high-temperature generation by use of the MA8 system with sintered diamond anvils, *High Temp.-High Press., 25*, 105-112, 1993.

Ohtani, E., N. Kagawa, O. Shimomura, M. Togaya, K. Suito, A. Onodera, H. Sawamoto, A. Yoneda, S. Tanaka, W. Utsumi, E. Ito, A. Matsumuro, and T. Kikegawa, High-pressure generation by a multiple anvil system with sintered diamond anvils, *Rev. Sci. Instrum., 60*, 922-925, 1989.

Ringwood, A. E., A. Major, P. Willis, and W. O. Hibberson, Diamond composite tools, Annual Report 1989, Research School of Earth Sciences, ANU, pp. 38-43, 1989.

Shimomura, O., W. Utsumi, T. Taniguchi, T. Kikegawa and T. Nagashima, A new high pressure and high temperature apparatus with the sintered diamond anvil for synchrotron radiation use, *in High Pressure Research in Mineral Physics: Application to Earth and Planetary Sciences, Geophys. Monogr.* Ser., vol. 67, edited by Y. Syono and M. H. Manghnani, pp. 3-12, Terra Scientific Publishing, Tokyo/ AGU, Washington, D.C., 1992.

Utsumi, W., T. Yagi, K. Leinenweber, O. Shimomura and T. Taniguchi, High pressure and high temperature generation using sintered diamond anvils, in *High Pressure Research in Mineral Physics: Application to Earth and Planetary Sciences, Geophys. Monogr. Ser.*, vol. 67, edited by Y. Syono and M. H. Manghnani, pp. 37-42, Terra Scientific Publishing, Tokyo/AGU, Washington, D.C., 1992.

T. Irifune, K. Kuroda, and N. Nishiyama, Department of Earth Sciences, Ehime University, Matsuyama 790, Japan

T. Inoue, Department of Earth and Space Sciences, SUNY at Stony Brook, Stony Brook, NY 11794, USA

N. Funamori, T. Uchida, and T. Yagi, Institute for Solid State Physics, University of Tokyo, Roppongi, Tokyo 106, Japan

W. Utsumi, Department of Synchrotron Radiation Facilities Project, Japan Atomic Energy Research Institute, SPring-8, Kamigori, Hyogo 678-12, Japan

N. Miyajima, and K. Fujino, Department of Earth and Planetary Sciences, Hokkaido University, Sapporo 060, Japan

S. Urukawa, Department of Earth Sciences, Okayama University, Okayama 700, Japan

T. Kikegawa and O. Shimomura, Photon Factory, National Laboratory for High Energy Physics, Tsukuba 305, Japan

Single-Crystal Elasticity of the α and β of Mg_2SiO_4 Polymorphs at High Pressure

Chang-Sheng Zha[1], Thomas S. Duffy[2], Robert T. Downs[1], Ho-kwang Mao[1], Russell J. Hemley[1,3], and Donald J. Weidner[4]

The full set of single-crystal elastic moduli of forsterite (α-Mg_2SiO_4) andwadsleyite (β-Mg_2SiO_4) were measured at pressures of 3-16 GPa and 0-14 GPa, respectively. For forsterite, the pressure derivatives of the bulk (K_0's) and shear (G_0') modulus are 4.2 ± 0.2 and 1.4 ± 0.1, respectively. For wadsleyite, the corresponding pressure derivatives are 4.3 ± 0.2 and 1.4 ± 0.2. These values are much lower than those reported in earlier, low-pressure studies for both materials. At the pressure of the 410-km seismic discontinuity, the room-temperature velocity increase across the α-β transition is 9.8% for compressional waves and 12.4% for shear waves. The results are consistent with an olivine fraction in the upper mantle of 30-50%.

INTRODUCTION

The elastic stiffness coefficients provide fundamental insight into the nature of atomic forces in solids. While the atomic structures of a vast range of substances have been explored at high pressure, the complete set of elastic stiffnesses are known only for a few, usually simple, materials. X ray diffraction experiments at high pressure can constrain the bulk modulus [e.g., *Knittle*, 1995], but these measurements typically provide no information on material response to shear deformation. Such information for Earth materials are critical for constraining mantle composition and structure. Knowledge of the individual elastic moduli are necessary for calculating acoustic wave velocities and their orientation dependence. Comparison of laboratory measurements of acoustic velocities with seismic velocities in the mantle has long been recognized as the most direct means to determine the mineralogy of the deep inaccessible regions of planetary interiors. Seismological studies employing stacking techniques are providing increasingly detailed models of impedance contrasts and velocity gradients in and near the mantle transition zone [*Shearer*, 1996]. In this region, the elastic moduli of the forsterite (α), wadsleyite (β), and spinel (γ) polymorphs of Mg_2SiO_4 are the most critical quantities. Phase equilibria data suggest that the seismic discontinuity near 410 km depth (about 13.8 GPa) is related to the α-β transformation in $(Mg,Fe)_2SiO_4$. The possible existence of a weaker discontinuity near 520 km depth (about 17.9 GPa) [*Shearer*, 1996] may be due to the β-γ transition. Elasticity data on the Mg_2SiO_4 polymorphs at high pressure place constraints

[1]Geophysical Laboratory and Center for High-Pressure Research, Carnegie Institution of Washington, 5251 Broad Branch Rd, NW, Washington, DC 20015

[2]Consortium for Advanced Radiation Sources, The University of Chicago, 5640 S. Ellis Ave, Chicago, IL 60637

[3]Also at: Laboratoire de Sciences de la Terre, Ecole Normale Superieure de Lyon, 69364 Lyon, France

[4]Department of Earth and Space Sciences and Center for High-Pressure Research, University at Stony Brook, Stony Brook, NY 11974 USA

Properties of Earth and Planetary Materials
at High Pressure and Temperature
Geophysical Monograph 101

on the total olivine fraction of the upper mantle at these depths [e.g., *Duffy et al.*, 1995].

Recently there have been a number of developments in experimental methods for acoustic velocity and elastic constant determination. At ambient pressure, there have been advances in resonant ultrasound spectroscopy [*Maynard,* 1996] and in direct imaging of acoustic velocity surfaces [*Wolfe and Hauser*, 1995]. There has also been considerable progress in the study of sound velocity and elasticity at high pressure. In the diamond anvil cell, Brillouin scattering [*Shimizu and Sasaki*, 1992; *Zha et al.*, 1993] and impulsive stimulated scattering [*Zaug et al.*, 1993] have been used to obtain single-crystal elastic constants to pressure as high as 24 GPa. Ultrasonic sound velocity measurements in the large volume press have been pioneered recently using both single crystals to 6 GPa [*Yoneda and Morioka,* 1992] and polycrystals to 12 GPa [*Li et al.*, 1996]. A recently developed gigahertz ultrasonic interferometer has also been used with the large volume press [*Chen et al.*, 1996] to obtain selected single-crystal constants at simultaneous high pressure and temperature.

EXPERIMENTAL TECHNIQUE

Brillouin scattering is the interaction of light (photons) with thermally excited long-wavelength lattice vibrations (phonons) in a crystal. Propagating acoustic waves produce fluctuations in the refractive index from which light is scattered. Since the fluctuations are moving at the acoustic velocity, the scattered light is shifted in frequency by the Doppler effect. Brillouin spectroscopy is similar to the more familiar Raman spectroscopy except that the frequency shifts in Brillouin scattering are very small (~1-2 cm^{-1}) and the signal intensity is weaker. The experimental challenge is to extract the weak Brillouin signal from the intense, elastically scattered light.

The geometry for Brillouin scattering in the diamond cell is shown in Figure 1. Light from a frequency-stabilized Ar laser is directed into the sample through one of the diamond anvils. The scattered radiation passes through the other anvil in a symmetric arrangement. A diamond anvil cell with a short piston-cylinder and a large optical opening was used in these experiments. The Brillouin scattered light is collected, spatially filtered, and passed through a tandem Fabry-Perot interferometer which eliminates overlap of successive interference orders. The frequency spectrum is recorded with a multichannel scaler. Further experimental details can be found in *Zha et al.* [1993; 1996].

In the platelet scattering geometry, light enters through one sample face and exits through another parallel face (Figure 1). The acoustic wave propagation direction is then perpendicular to the axis of the cell and coplanar with incident and scattered light. The equation relating the acoustic velocity, V, to the measured frequency shift, $\delta\nu$, for this geometry is

$$V = \delta\nu \, \lambda_0 \, (2 \sin \theta)^{-1} \quad (1)$$

where λ_0 is the incident laser wavelength, and θ is the angle between incident or scattered light and the diamond cell axis at the outer diamond surface. The refractive index is not needed to determine the velocity in this case. By rotating the diamond cell around its axis, the acoustic velocity distribution within the sample plane can be completely characterized.

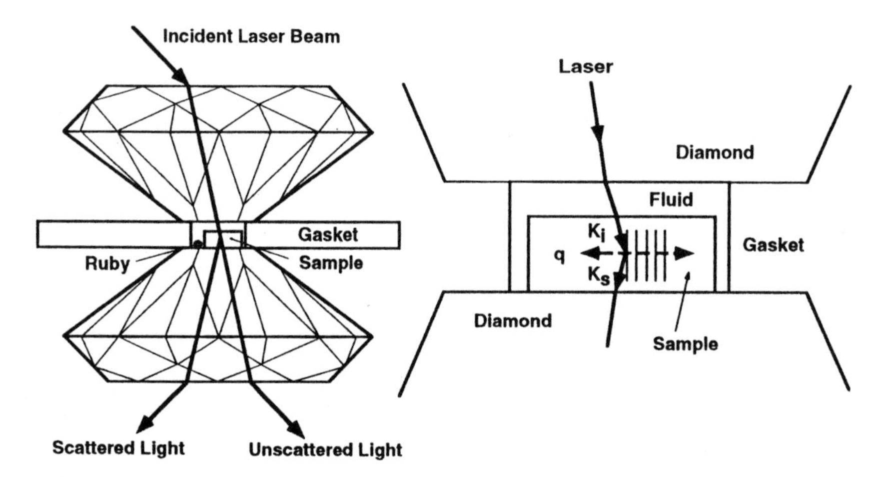

Figure 1. Brillouin scattering in the diamond anvil cell. Wavevectors for the incident photon (k_i), scattered photon (k_s), and phonon (q) are shown at the right. In the platelet geometry, light enters and exits the sample from opposite, parallel faces, and the acoustic wave propagates in the plane of the sample.

Single-crystal samples of Mg_2SiO_4 in the forsterite (α) and wadsleyite (β) structures were examined in this study. Electron microprobe analysis confirmed that both samples were pure end-members. For β-Mg_2SiO_4, electron microscopy also revealed the presence of submicron inclusions. No peaks attributable to hydroxyl bonds were observed by Raman spectroscopy. Several pieces of each material were polished to flat, parallel plates. Sample thicknesses ranged from 18 to 50 μm. The samples were oriented such that the polished plane intersected the three crystal axes at equal angles (within a few degrees). X ray diffraction was performed on both materials at ambient and selected high pressures. For forsterite, the cell volume determined by single-crystal X ray diffraction at ambient pressure was 290.22(9) $Å^3$, while for wadsleyite, the volume was 535.8(2) $Å^3$. These are consistent with previously reported values [*Smyth and McCormick*, 1995].

At low pressures, argon was used as a pressure transmitting medium. A 4:1 mixture of methanol-ethanol was used at intermediate pressures, and helium was the medium at the highest pressures. The choice of pressure medium was dictated by the need to establish nearly hydrostatic pressure conditions within the cell and to avoid overlap of Brillouin peaks from the sample and the pressure medium. At high pressures, it was found that the Brillouin peaks of the wadsleyite sample nearly overlapped the shear wave peaks from the diamond anvils. Since the volume of diamond is much larger than the sample volume, the Brillouin signal from the diamond can overwhelm the weaker sample signal. To overcome this problem, it was necessary to improve the spatial filtering by inserting cylindrical lenses in front and back of the cell to correct the astigmatic aberration introduced by the inclined positioned diamond cell. This was successful in reducing the intensity of the diamond peak by more than 90% for a 30-μm thick sample. It was sufficient to allow the sample Brillouin peaks to be measured over the pressure range studied here.

RESULTS

Sound velocities were generally measured at 10° intervals in the platelet plane at each pressure for both samples. In each direction, three acoustic velocities were determined: one quasi-longitudinal and two quasi-transverse. Figure 2 shows the complete set of acoustic velocities for wadsleyite in different loadings. A total of 144 directions and 400 individual acoustic velocities was measured. The average data collection time for a single direction is about 3 hours (but can vary considerably), so roughly a total data collection time of about 18 days (432 hours) was required. A comparable number of measurements was carried out on the forsterite sample.

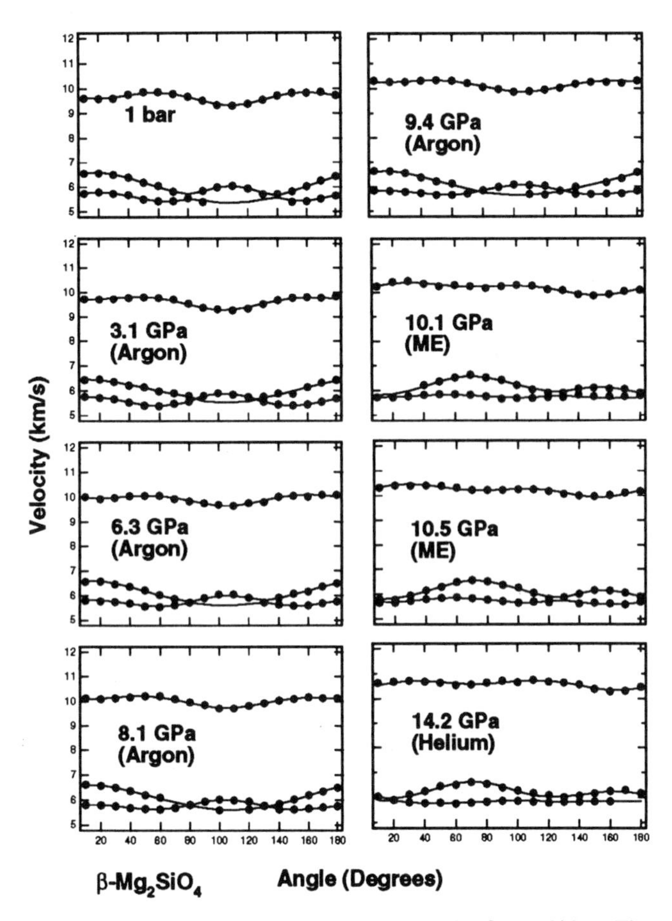

Figure 2. Acoustic velocity measurements for β-Mg_2SiO_4 The symbols are the experimental data and the solid lines are calculated using the best fitting elastic constants. The pressure and pressure-transmitting medium are indicated in each panel. The 10.1, 10.5 and 14.2 GPa are measured from second sample with different orientations. The angle is relative to an arbitrary marking on the cell. ME - 4:1 methanol-ethanol mixture.

The propagation of acoustic waves in anisotropic solids is governed by Christoffel's equation [*Auld*, 1973]. Using Cardan's solution of the cubic equations, this can be written as follows [*Rokhlin and Wang*, 1992]:

$$\rho V_j^2 = -a/3 + 2\sqrt{\frac{-p}{3}}\cos\left(\frac{\psi + 2\pi j}{3}\right), j = 0, 1, 2, \quad (2)$$

where ρ is the density, V_j are the acoustic velocities, and the other parameters are

$$\psi = \arccos\{-q/[2(p/3)]^{3/2}\}, \quad (3)$$

$$p = a^2/3 - b, \quad (4)$$

$$q = c - ab/3 + 2(a/3)^3, \quad (5)$$

$$a = - G_{ii}, \quad (6)$$

$$b = -(G^2_{12} + G^2_{13} + G^2_{23} - G_{11}G_{22} - G_{11}G_{33} - G_{22}G_{33}) \quad (7)$$

$$c = -(G_{11}G_{22}G_{33} + 2G_{12}G_{13}G_{23} - G_{11}G^2_{23} - G_{22}G^2_{13} - G_{33}G^2_{12}), \quad (8)$$

$$G_{im} = C_{ijlm} n_j n_l, \quad (9)$$

Summation over repeated indices is implied; $\vec{n}$ [vectors n_1, n_2, n_3...] specifies the acoustic wave propagation direction. C_{ijlm} is the tensor of elastic constants which can also be written in the contracted Voigt notation, Cij, where i,j = 1,2,...,6 [*Nye*, 1985]. Both forsterite and wadsleyite are of orthorhombic symmetry and are thus characterized by nine independent elastic moduli. Three of these (C_{11}, C_{22}, and C_{33}) are the longitudinal elastic moduli, C_{44}, C_{55}, C_{66} are the shear elastic moduli, and the remaining three (C_{12}, C_{13}, and C_{23}) are the off-diagonal moduli.

The acoustic velocity data were inverted using Equation (2) by nonlinear least squares to yield the elastic stiffness coefficients and three Eulerian angles which relate the laboratory and crystallographic reference frames [*Shimizu and Sasaki*, 1992; *Zha et al.*, 1996]. Acoustic velocity distributions for the best fitting elasticity and orientation parameters are shown by the solid curves in Figure 2 for

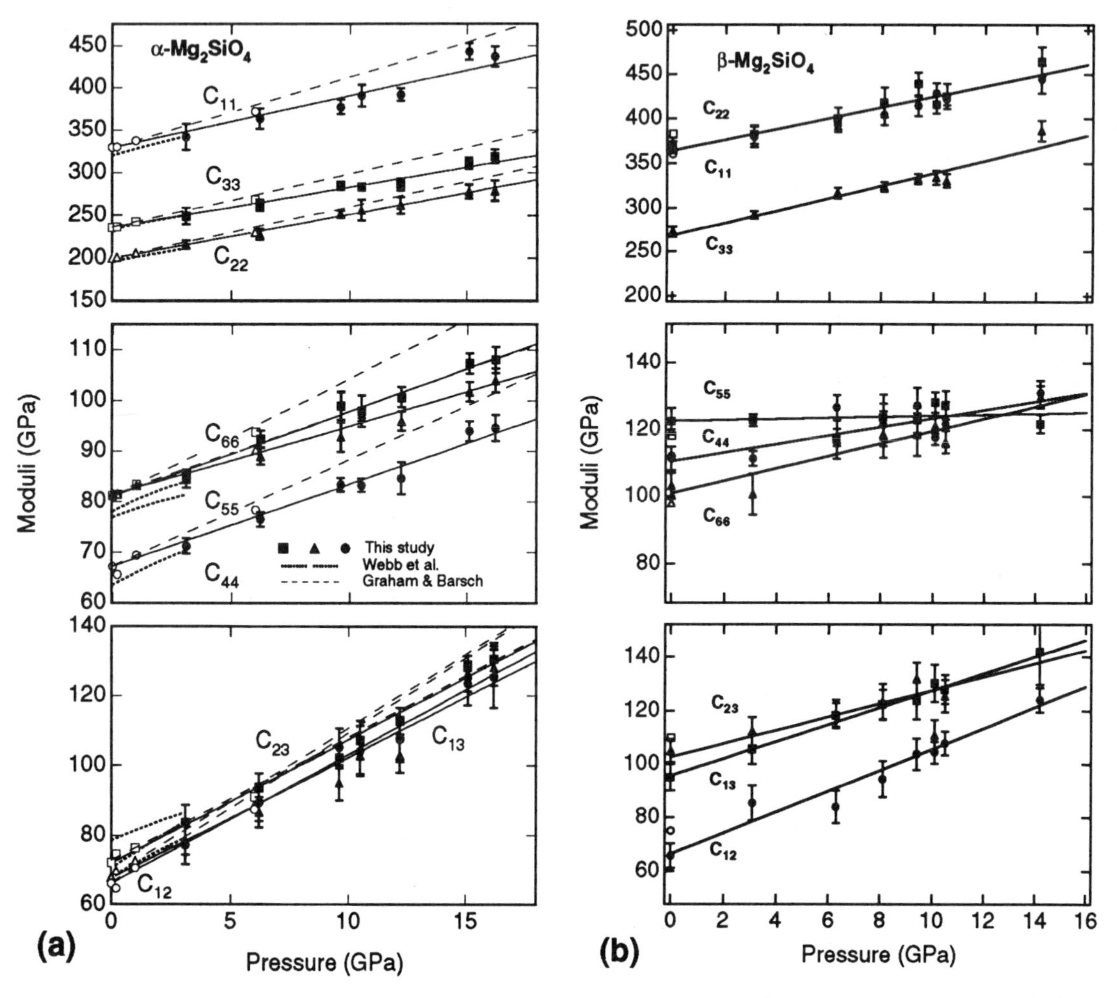

Figure 3. Elastic moduli as a function of pressure for (a) α-Mg_2SiO_4 and (b) β-Mg_2SiO_4. The filled symbols with error bars (2σ) are the present data points, and the solid lines are weighted least squares fits to the data. In (b), ambient pressure data of *Sawamoto et al.* [1984] are shown as open symbols. In (a), the dashed lines show extrapolation of 1-GPa results of *Graham and Barsch* [1969]. The doted lines show the results of *Webb* [1989]. In the middle panel, the curve for C_{55} of *Graham and Barsch* overlays that of C_{66} from this study. Ambient pressure data of *Isaak et al.* [1989] were used in determining the least squares fits.

wadsleyite. The root-mean-square deviation between measured and calculated acoustic velocities ranges from 20 m/s at 8.1 GPa to 42 m/s at 14.2 GPa. For forsterite, the RMS deviations were between 24 and 56 m/s [*Zha et al.*, 1996].

The pressure dependence of the elastic moduli of forsterite and wadsleyite is shown in Figures 3a and b. In all cases, a linear dependence of the moduli on pressure is obtained within the resolution of the data. For wadsleyite, the ambient pressure results of *Sawamoto et al.* [1984] are shown for comparison. Table 1 shows all elastic constants at different pressures for wadsleyite. There are no other single-crystal elasticity data at high pressure of which we are aware for this material. For forsterite, the present results are compared with extrapolations of previous single-crystal ultrasonic elasticity data to 1 GPa [*Graham and Barsch*, 1969]. For all the individual moduli, our measured values lie significantly below the values extrapolated from low-pressure.

Established velocity averaging techniques (Voigt-Reuss-Hill) were used to compute bounds on the elastic properties of a randomly oriented polycrystalline aggregate from the single-crystal properties. The resulting bulk and shear moduli for the two polymorphs are shown in Figure 4. Values for the aggregate moduli and their pressure derivatives were determined by fitting the experimental data to finite strain expressions for the bulk and shear moduli [*Davies*, 1974; *Davies and Dziewonski*, 1975] (Table 2). For forsterite, the pressure derivatives of the bulk and shear modulus are much lower than those found in early, low-pressure ultrasonic studies which reported pressure derivatives of the bulk modulus between 5.0 and 5.4, and pressure derivatives of the shear modulus of 1.8 [*Kumazawa and Anderson,* 1969; *Graham and Barsch*, 1969]. More recent high-pressure ultrasonic studies of forsterite single-crystals [*Yoneda and Morioka*, 1992] and polycrystalline aggregates [*Li et al.*, 1996] in the large-volume press give results consistent with ours. Recent work on a forsterite sample containing 10 mol% Fe using the impulsive stimulated scattering technique [*Zaug et al.*, 1993] is also generally consistent with the results reported here, except that no nonlinear behavior of the shear modulus is observed in our study of the iron-free end-member.

For wadsleyite, our ambient-pressure aggregate elastic moduli (Table 2) are comparable to those reported by *Sawamoto et al.* [1984] (K_0 = 174 GPa, G_0 = 114 GPa). Pressure derivatives for this material were first reported by *Gwanmesia et al.* [1990] from measurements on poly-

TABLE 1. Elastic Moduli of Wadsleyite (all in GPa)

P	C_{11}	C_{22}	C_{33}	C_{44}	C_{55}	C_{66}	C_{12}	C_{13}	C_{23}
0.0	370.47	367.74	272.43	111.20	122.48	103.05	65.59	95.20	105.14
	±7.84	±6.50	±5.82	±3.58	±4.00	±3.86	±4.54	±5.18	±4.36
3.1	379.28	382.01	292.21	111.24	122.71	100.70	85.39	105.53	112.39
	±11.46	±10.40	±4.46	±2.16	±1.84	±6.10	±6.62	±5.44	±5.28
6.3	393.35	399.90	316.69	126.59	117.53	116.96	83.91	118.23	119.10
	±7.90	±12.30	±6.06	±3.68	±2.58	±5.76	±6.14	±4.62	±4.98
8.1	404.22	418.33	323.99	121.37	123.58	118.46	94.42	122.31	123.37
	±11.22	±16.78	±5.42	±6.36	±6.74	±6.98	±6.82	±5.48	±6.70
9.4	414.42	439.78	333.08	127.13	122.77	118.43	103.89	123.78	131.81
	±11.16	±13.04	±5.76	±5.32	±3.80	±6.40	±5.90	±6.80	±5.92
10.1	428.83	415.97	333.91	117.55	128.19	121.09	104.48	130.21	110.77
	±11.30	±9.86	±7.90	±2.22	±3.08	±2.40	±4.10	±6.90	±5.80
10.5	421.89	425.66	330.73	121.65	127.21	116.16	107.78	127.90	125.39
	±6.72	±14.22	±7.70	±6.76	±4.32	±3.32	±4.38	±5.42	±5.98
14.2	444.47	464.89	386.83	130.94	121.67	129.71	123.93	141.95	151.85
	±15.86	±16.13	±11.30	±3.72	±2.76	±3.30	±4.56	±12.18	±19.60

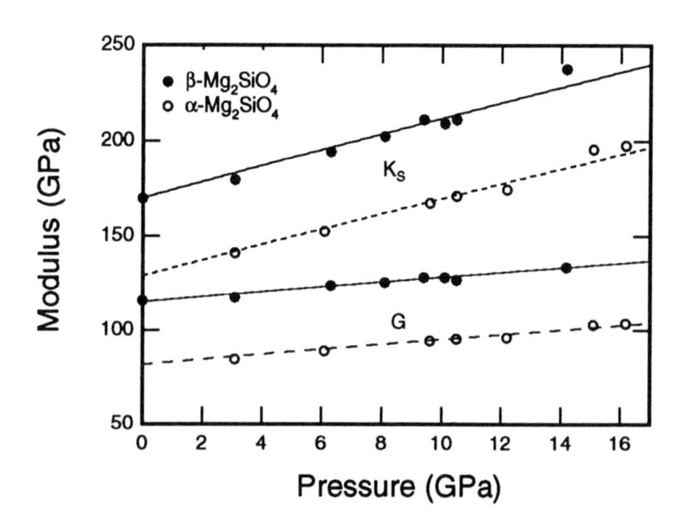

Figure 4. Bulk and shear moduli of α-Mg_2SiO_4 (open symbols) and β-Mg_2SiO_4(filled symbols). Lines are third order finite strain fits to the experimental data. Ambient pressure data of *Isaak et al.* [1989] were used in fitting the α-Mg_2SiO_4 data.

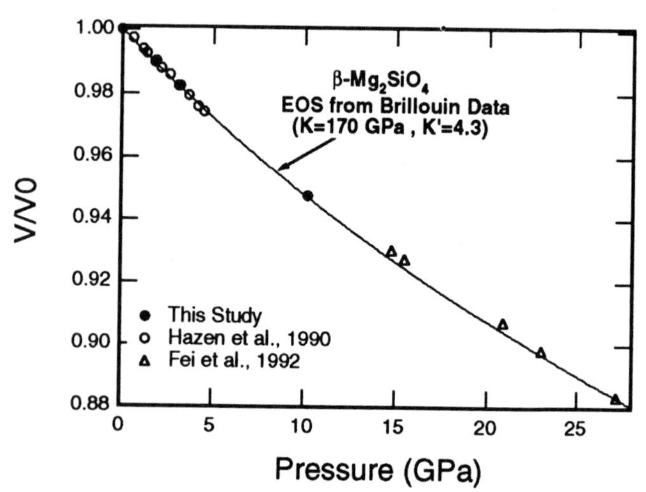

Figure 5. Compression data for β-$(Mg,Fe)_2SiO_4$. The data *of Fei et al.* [1992] were for a sample containing 16 mol% Fe; other data are for β-Mg_2SiO_4. The solid curve is a third-order equation of state from the present Brillouin data.

crystalline aggregates to 3 GPa. The values obtained in that study (K_0' = 4.8±0.1 and G_0' = 1.8±0.1) are considerably higher than those obtained here. Very recently, *Li et al.* [1996] using a similar approach to *Gwanmesia et al.* have measured pressure derivatives of β-Mg_2SiO_4 to 12 GPa. When fit to finite strain expressions, the resulting pressure derivatives (K_0' = 4.5 and G_0' = 1.6) are intermediate between our single-crystal results and the work of *Gwanmesia et al.* [1990]. The use of polycrystalline samples, which retain some porosity even at very high pressure, may be responsible for the higher pressure derivatives obtained in those studies.

A pressure-volume equation of state can be constructed from the Brillouin data by correcting the bulk modulus and its pressure derivatives from adiabatic to isothermal conditions. The correction is small and can be reliably estimated. Figure 5 compares the Brillouin equation of state for wadsleyite to X ray diffraction data from this and other studies. The good agreement obtained demonstrates the self-consistency of our results in that the densities predicted by our data are nearly identical to those used in the solution of Equation (2) above. Similar results were obtained for the forsterite samples [*Downs et al.*, 1996; *Zha et al.*, 1996].

TABLE 2. Aggregate Elastic Properties of Mg_2SiO_4 Polymorphs

Property	α-Mg_2SiO_4	β-Mg_2SiO_4
K_{0S} (GPa)	129[a]	170 (2)
G_0 (GPa)	79 [a]	115 (2)
K'_{0S}	4.2 (2)	4.3 (2)
G'_0	1.4 (1)	1.4 (2)

[a] *Isaak et al.* [1989]

DISCUSSION

This is the first study in which the elasticity of both the α and β polymorphs of Mg_2SiO_4 has been measured at pressures greater than that of the 410-km discontinuity in the mantle. Our primary finding is that the pressure dependencies of the aggregate moduli are nearly the same for both phases. This is consistent with values used in some earlier studies (e.g., *Duffy and Anderson*, 1989; *Gwanmesia et al.*, 1990), but the magnitudes of the derivatives are much lower than previously believed. Aggregate compressional and shear wave velocities are shown as a function of pressure in Figure 6. The velocity contrast between the α and β polymorphs is 12.3% and 14.2% at ambient pressure for compressional (P) and shear (S) waves, respectively, and decreases to 9.8% for P-waves and 12.4% for S-waves at the pressure of the 410-km discontinuity (~ 13.8 GPa). In comparison, the velocity seismic contrast across the discontinuity is 4-5%, although a recent study using stacks of long period records obtained an S-velocity jump at the discontinuity of only 3.4% [*Shearer*, 1996].

Figure 7 compares calculated sound velocities in the $(Mg,Fe)_2SiO_4$ polymorphs at high temperature to seismic velocity profiles for the upper mantle. The high-temperature mineral velocities were calculated along a 1600-K adiabat using the method of *Duffy et al.* [1995].

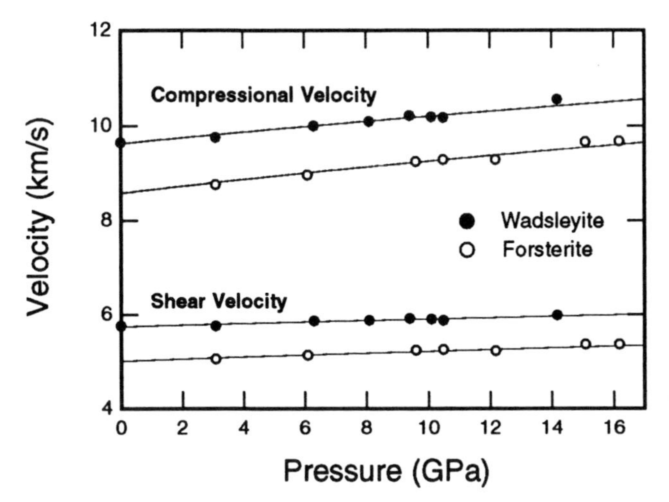

Figure 6. Aggregate compressional and shear velocity of wadsleyite and forsterite. Open and filled symbols are experimental data, solid lines are third-order finite strain fits to the data.

The only difference in this calculation is that our new measured values of the aggregate elastic properties of the β-phase have been used. The major uncertainty in calculating the high-temperature sound velocities is that the temperature derivative of the shear modulus ($\partial G/\partial T$) of β-Mg_2SiO_4 is unknown. Following *Duffy et al.* [1995], we adopt a range of values for this parameter between -0.014 GPa/K and -0.024 GPa/K, based on an evaluation of systematic trends in the larger mineral elasticity database. The lower estimate of the magnitude of the $\partial G/\partial T$ would be appropriate if the

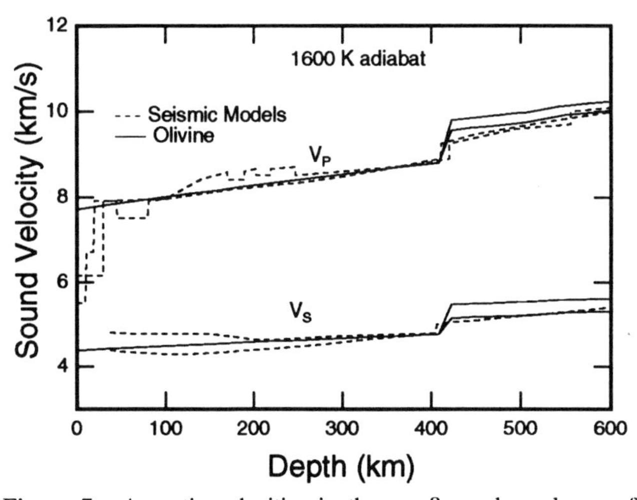

Figure 7. Acoustic velocities in the α-, β-, and γ- phases of $(Mg_{0.9},Fe_{0.1})_2SiO_4$ along a 1600-K adiabat (solid lines) compared with seismic velocity profiles for the upper mantle (dashed lines). The lower solid curve (at depths > 410 km) is for $\partial G/\partial T$ = -0.024 GPa/K. The upper solid curve is for $\partial G/\partial T$ = -0.014 GPa/K. The seismic models are from *Grand and Helmberger* [1984], *Walck et al.* [1984], and *Mechie et al.* [1993].

shear modulus temperature sensitivities of the α and β phases are the same. The larger value holds if the shear modulus of the β-phase is more sensitive to temperature than typical silicates are and has a value comparable to MgO and Al_2O_3.

In the case of the lower temperature sensitivity, the velocity increase at 410 km under mantle conditions is consistent with olivine fractions of 30% (for shear waves) and 40% (for compressional waves) (Figure 7). Using the higher value for the shear modulus temperature sensitivity of β-Mg_2SiO_4, the amount of olivine consistent with the seismic discontinuity becomes 50% and 51% for compressional and shear waves, respectively. These values are slightly larger than those obtained by *Duffy et al.* [1995] because the pressure dependence of the aggregate elastic moduli of β-Mg_2SiO_4 measured here is lower than the values reported in the earlier work of *Gwanmesia et al.* [1990]. Note that the comparison with the recent long-period sdudy of *Shearer* [1996] would reduce the olivine content contributing at this depth.

Acknowledgments. We thank H. Yoder for providing the forsterite specimen and R. Liebermann and B. Li for providing preprints of papers. This work was supported by the Center for High Pressuer Research, National Science Foundation subcontract No. 431-3860A.

REFERENCES

Auld, B. A., *Acoustic Waves and Fields in Solids*, vol. 1, Wiley, New York, 1973.

Chen, G., H. A. Spetzler, I. C. Getting, and A. Yoneda, Selected elastic moduli and their temperature derivatives for olivine and garnet with different Mg/(Mg+Fe) contents - results from GHZ ultrasonic interferometry, *Geophys. Res. Lett.*, *23*, 5-8, 1996.

Davies, G. F., Effective elastic moduli under hydrostatic stress - 1. Quasi-harmonic theory, *J. Phys. Chem. Solids*, *35*, 1513-1520, 1974.

Davies, G. F., and A. M. Dziewonski, Homogeneity and constitution of the Earth's lower mantle and outer core, *Phys. Earth Planet. Inter.*, *10*, 336-343, 1975.

Downs, R. T., C. S. Zha, T. S. Duffy, and L. W. Finger, The equation of state of forsterite to 17.2 GPa and effects of pressure media, *Am. Mineral.*, *81*, 51-55, 1996.

Duffy, T. S., and D. L. Anderson, Seismic velocities in mantle minerals and the mineralogy of the upper mantle, *J. Geophys. Res.*, *94*, 1895-1912, 1989.

Duffy, T. S., C. S. Zha, R. T. Downs, H. K. Mao, and R. J. Hemley, Elasticity of forsterite to 16 GPa and the composition of the upper mantle, *Nature*, *378*, 170-173, 1995.

Fei, Y., H. K. Mao, J. Shu, G. Parthasarathy, and W. A. Bassett, Simultaneous high-P, high-T X ray diffraction study of β-$(Mg,Fe)_2SiO_4$ to 26 GPa and 900 K, *J. Geophys. Res.*, *97*, 4489-4495, 1992.

Graham, E. K., and G. R. Barsch, Elastic constants of single-crystal forsterite as a function of temperature and pressure, *J. Geophys. Res.*, *74*, 5949-5959, 1969.

Grand, S. P., and D. V. Helmberger, Upper mantle shear structure of North America, *Geophys. J. R. astr. Soc.*, *76*, 399-438, 1984.

Gwanmesia, G. D., S. Rigden, I. Jackson, and R. C. Liebermann, Pressure dependence of elastic wave velocity in β-Mg_2SiO_4 and the composition of the Earth's mantle, *Science*, *250*, 794-797, 1990.

Hazen, R. M., J. Zhang, and J. Ko, Effects of Fe/Mg on the compressibility of synthetic wadsleyite: β-$(Mg_{1-x}Fe_x)_2SiO_4$ (x ² 0.25), *Phys. Chem. Minerals*, *17*, 416-419, 1990.

Isaak, D. G., O. L. Anderson, T. Goto, and I. Suzuki, Elasticity of single-crystal forsterite measured to 1700 K, *J. Geophys. Res.*, *94*, 5895-5906, 1989.

Knittle, E., Static compression measurements of equations of state, in *Mineral Physics and Crystallography: A Handbook of Physical Constants*, edited by T. J. Ahrens, pp. 98-142, AGU, Washington D. C., 1995.

Kumazawa, M., and O. L. Anderson, Elastic moduli, pressure derivatives, and temperature derivatives of single-crystal olivine and single-crystal forsterite, *J. Geophys. Res.*, *74*, 5961-5972, 1969.

Li, B., G. D. Gwanmesia, and R. C. Liebermann, Sound velocities of olivine and beta polymorphs of Mg_2SiO_4 at Earth's transition zone pressures, *Geophys. Res. Lett.*, *23*, 2259-2262, 1996.

Maynard, J., Resonant ultrasound spectroscopy, *Physics Today*, 26-31, Jan. 1996.

Mechie, J., A. V. Egorkin, K. Fuchs, T. Ryberg, L. Solodilov, and F. Wenzel, P-wave mantle velocity structure beneath northern Eurasia from long-range recordings along profile quartz, *Phys. Earth Planet. Int.*, *79*, 269-186, 1993.

Nye, J. F., *Physical Properties of Crystals*, Oxford Press, Oxford, 1985.

Rokhlin, S. I., and W. Wang, Double through-transmission bulk wave method for ultrasonic phase velocity measurement and determination of elastic constants of composite materials, *J. Acoust. Soc. Am.*, *91*, 3303-3312, 1992.

Sawamoto, H., D. J. Weidner, S. Sasaki, and M. Kumazawa, single-crystal elastic properties of the modified spinel (Beta) phase of magnesium orthosilicate, *Science*, *224*, 749-751, 1984.

Shearer, P. M., Transition zone velocity gradients and the 520-km discontinuity, *J. Geophys. Res.*, *101*, 3053-3066, 1996.

Shimizu, H., and S. Sasaki, High-pressure Brillouin studies and elastic properties of single-crystal H_2S grown in a diamond cell, *Science*, *257*, 514-516, 1992.

Smyth, J. R., and T. C. McCormick, Crystallographic data for minerals, in *Mineral Physics and Crystallography: A Handbook of Physical Constants*, edited by T. J. Ahrens, pp.1-17, AGU, Washington D. C., 1995.

Walck, M. C., The P-wave upper mantle structure beneath an active spreading center: The Gulf of California, *Geophys. J. r. astro. Soc.*, *76*, 697-723, 1984.

Webb S. L., The elasticity of the upper mantle orthosilicates olivine and garnet to 3 GPa, *Phys Chem Minerals*, *16*, 684-692, 1989.

Wolfe, J. P., and M. R. Hauser, Acoustic wavefront imaging, *Ann. Physik*, *4*, 99-126, 1995.

Yoneda, A., and M. Morioka, Pressure derivatives of elastic constants of single crystal forsterite, in *High-Pressure Research: Application to Earth and Planetary Sciences*, *Geophys. Monogr. Ser.*, vol. 67, edited by Y. Syono and M. H. Manghnani, pp. 207-214, Terra Scientific, Tokyo/AGU, Washington, D. C., 1992.

Zaug, J. M., E. H. Abramson, J. M. Brown, and L. J. Slutsky, Sound velocities in olivine at Earth mantle pressures, *Science*, *260*, 1487-1489, 1993.

Zha, C. S., T. S. Duffy, H. K. Mao, and R. J. Hemley, Elasticity of hydrogen to 24 GPa from single-crystal Brillouin scattering and synchrotron x-ray diffraction, *Phys. Rev. B*, *48*, 9246-9255, 1993.

Zha, C. S., T. S. Duffy, R. T. Downs, H. K. Mao, and R. J. Hemley, Sound velocity and elasticity of single-crystal forsterite to 16 GPa, *J. Geophys. Res.*, *101*, B8, 17535-17545, 1996.

Robert T. Downs, Geophysical Laboratory and Center for High-Pressure Research, Carnegie Institution of Washington, 5251 Broad Branch Rd., NW, Washington, DC 20015.

Thomas S. Duffy and Russell J. Hemley, Consortium for Advanced Radiation Sources, The University of Chicago, 5640 S. Ellis Ave., Chicago, IL 60637.

Russell J. Hemley, also at: Laboratoire de Sciences de la Terre, Ecole Normale Superieure de Lyon, 69364 Lyon, France.

Ho-kwang Mao, Geophysical Laboratory and Center for High-Pressure Research, Carnegie Institution of Washington, 5251 Broad Branch Rd., NW, Washington, DC 20015.

Donald J. Weidner, Department of Earth and Space Sciences and Center for High-Pressure Research, State University of New York at Stony Brook, Stony Brook, NY 11974.

Chang-Sheng Zha, Geophysical Laboratory and Center for High-Pressure Research, Carnegie Institution of Washington, 5251 Broad Branch Rd., NW, Washington, DC 20015.

Temperature Distribution in the Laser-Heated Diamond Cell

Michael Manga

Department of Geological Sciences, University of Oregon, Eugene

Raymond Jeanloz

Department of Geology and Geophysics, University of California, Berkeley

The relative importance of different heat transfer mechanisms is evaluated, and thermal conduction is found to dominate under most conditions achieved in the diamond anvil cell. The temperature distribution is calculated for metals and dielectrics heated by a laser. A comparison of calculated and measured temperature profiles across laser-heated perovskite samples indicates that the thermal conductivity is inversely proportional to temperature, as expected for dielectrics in which thermal conduction is due to phonon transport. A model is developed which relates laser power and sample temperature and permits the determination of thermal conductivity and absorption at high pressures and temperatures. The bias introduced by axial temperature gradients on measurements is negligible for experiments in which a metal foil is heated, but may lead to underestimating melting temperatures by 10% for dielectric samples at 4000 K.

1. INTRODUCTION

Despite widespread use of laser-heated diamond cells in high pressure research, there has been little theoretical modeling of the heat transfer involved or of its effect on the inference of material properties at elevated pressures and temperatures. Previous modeling of the temperature distribution has been for a dielectric sample with constant material properties: *Bodea and Jeanloz* [1989] solved the heat conduction equation in the sample, gasket and diamonds, but did not relate model calculations to observations; *Heinz and Jeanloz* [1987a,b] assumed a Gaussian temperature distribution and related the temperature measured by a spectroradiometer to the actual sample temperature.

Here we develop models for the temperature distribution in the laser-heated diamond cell, focusing on two experimental configurations. In the first case (Figure 1a), a thin opaque sample (e.g., metal foil) is surrounded by an inert, transparent medium (e.g., a dielectric such as Al_2O_3 or Ar), and heated by the focused laser. In the second case (Figure 1b), a relatively transparent (dielectric) sample is heated by the laser beam, and energy is absorbed throughout the thickness of the sample [e.g., *Heinz and Jeanloz*, 1987b; *Bodea and Jeanloz*, 1989]. The first experimental configuration is used to study metals [e.g., *Williams et al.*, 1991; *Boehler*, 1993; *Saxena et al.*, 1994], as well as the dielectrics which insulate and isolate the metal foil from the diamonds [e.g., *Shen and Lazor*, 1995].

Configurations intermediate between those summarized in Figure 1 can also arise, for example, if the sample is a strongly absorbing dielectric. Nevertheless, by considering the two endmembers shown in the figure, it is possible to deduce the temperature distribution for intermediate cases.

Properties of Earth and Planetary Materials
at High Pressure and Temperature
Geophysical Monograph 101

a)

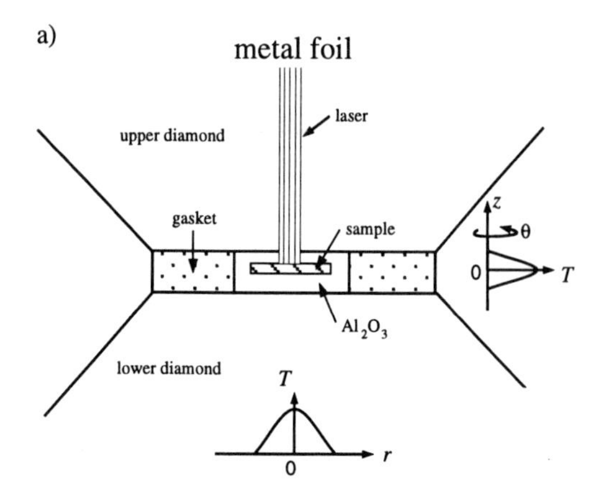

b)

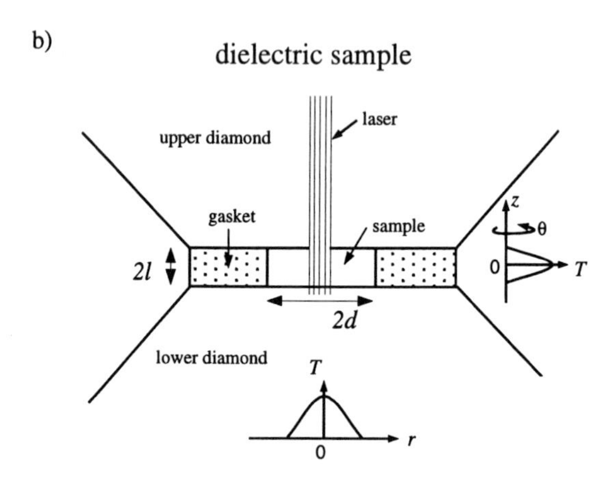

Figure 1. Schematic illustration of (a) a metal-foil surrounded by a dielectric medium (e.g., Al_2O_3) and (b) a dielectric sample in the laser-heated diamond anvil cell. Characteristic dimensions in the axial (z, parallel to laser beam) and radial (r, perpendicular to laser beam) directions are given by a sample thickness $2l \approx 20$ μm and radius $2d \approx 200$ μm, respectively.

2. MECHANISMS OF HEAT TRANSFER IN THE DIAMOND CELL

Heat can be transported by radiation, conduction, and convection. Here we consider the relative importance of these three modes of heat transfer and show that conductive heat transfer dominates for temperatures less than 6000 K. Our discussion in this section focuses on the case of a metal foil insulated by an optically thin dielectric medium (Figure 1a), but can also be extended to laser-heated dielectric samples (Figure 1b).

2.1 Radiation

Consider the metal foil to be blackbody at 4000 K radiating into a dielectric medium with index of refraction n: assuming an emissivity $\epsilon = 1$ yields the maximum value of radiative flux,

$$Q_{\text{radiative}} = n^2 \epsilon \sigma T^4 = 3.3 \times 10^7 \text{ W/m}^2, \qquad (1)$$

where $\sigma = 5.67 \times 10^{-8}$ W/m^2 is the Stefan-Boltzmann constant and T is temperature, and we assume $n = 1.5$. For comparison, the conductive heat transfer through the insulating (dielectric) material is

$$Q_{\text{conductive}} = k \nabla T = 4 \times 10^8 \text{ W/m}^2, \qquad (2)$$

assuming that the insulating layer is 10 μm thick and has a (low) thermal conductivity k of 1 W/mK, and that the diamonds are close to room temperature [*Bodea and Jeanloz*, 1989]. Argon, sometimes used as the insulating medium, has a low thermal conductivity in the gas phase but is a condensed fluid or solid at pressures typical of diamond cell experiments and is expected to have a thermal conductivity similar to that of other liquids, $O(1)$ W/mK, at these conditions [*Powell and Childs*, 1972]. Because the absorption mean-free path for dielectrics is typically $10^{-2} - 100$ cm [*Fukao et al.*, 1968] and thus much greater than the thickness of the insulating dielectrics, it is reasonable to assume these layers are optically thin.

The relative importance of radiative and conductive transport involves a tradeoff between temperature and the lengthscale over which temperature varies, accord-

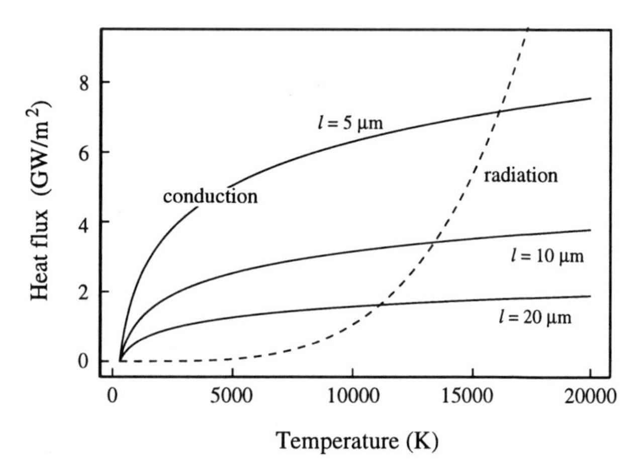

Figure 2. Contribution to the heat flux from radiation (dashed curve) and conduction (solid curves) for a heated metal foil embedded between two layers of Al_2O_3. The conducted heat flux is given by equation (29), with the thickness of the Al_2O_3 layers around the sample assumed to be $l = 5$, 10 or 20 μm. The thermal conductivity is given by $k_{Al_2O_3} = aT^{-1}$ with $a = 9000$ W/m, and we assume an index of refraction of 1.5. The metal foil is assumed to emit radiation as a blackbody, i.e., $\epsilon = 1$, thus yielding a typical upper bound for heat transfer by radiation.

ing to equations (1) and (2). Thus, in the diamond cell, conductive heat transfer dominates due to the small length scales within the experimental apparatus, of order 10 μm (l) along the axial direction (z). It is for exactly the same reasons, small lengthscales, that conduction dominates over radiation in the measurement of Hugoniot temperatures [*Grover and Urtiew*, 1974].

A more accurate estimate of $Q_{\text{conductive}}$ is derived later in the paper for a more realistic temperature-dependent thermal conductivity, $k = a/T$. Comparison with equation (1) yields the same result, that conduction dominates over radiation.

In Figure 2, for example, we compare the contributions of radiation and conduction to heat transport, choosing the largest possible emissivity, $\epsilon = 1$ (i.e., a blackbody), so that the dashed curve in Figure 2 is an upper bound on radiative transfer. For the dielectric, we assume $n = 1.5$ and $k = aT^{-1} = 9000T^{-1}$ W/mK, which corresponds to Al_2O_3 at room pressure and is typical of other oxides. Since a increases as pressure increases, the solid curves in Figure 2 should be lower bounds on conductive transport for samples insulated with Al_2O_3. Even for thick layers of Al_2O_3, it is evident that conductive transport dominates over radiation for T less than about 10,000 K.

For the case of an optically thin dielectric sample, the integrated emissivity across the sample will be $\ll 1$ (which follows from the samples being optically thin) and conduction will also dominate. As we show later, the measured temperature distributions are not compatible with the effective thermal conductivity being dominated by a radiative contribution that scales as T^3 [*Siegel and Howell*, 1968].

2.2 Conduction

Heat transfer by conduction satisfies the thermal diffusion equation

$$\rho C \frac{\partial T}{\partial t} = \nabla \cdot k \nabla T + A \tag{3}$$

where ρ, C, and k are the density, specific heat, and thermal conductivity for the region of interest, and A is the absorbed power density and describes heating by the laser. The unsteady part of equation (3), $\rho C \partial T/\partial t$, can typically be ignored because the characteristic diffusive timescale is short relative to experimental durations: $O(\rho C l^2/k) \approx 10^{-4}$ s, where l is a characteristic dimension of the sample (we use the half-thickness l rather than the radius of the sample d, because $l \ll d$ in most diamond cell experiments and it is the axial dimension l that determines the maximum conducted heat flux out of the sample area: see Figure 1). For comparison, measurements are made on timescales of $10^{-2} - 10^{2}$ s.

Since the power distribution P across the laser beam in TEM_{00} mode is axisymmetric, $P \propto \exp[-(r/R)^2]$ where r is the radial position and R is the beam radius, the temperature distribution should also be axisymmetric. The absorption of the sample is given by

$$A = A_0 e^{-\Lambda z} e^{-(r/R)^2} \tag{4}$$

where $1/\Lambda$ is the optical absorption length, and z is the distance within the material the beam has penetrated. The distinction made in the Introduction between optically thin (e.g., dielectric) and optically thick (e.g., metallic) samples is quantified through this relation: these two limits correspond, respectively, to $\Lambda l \gg 1$ and $\Lambda l \ll 1$ where l is the sample thickness as before.

Equation (3), which applies in each region of the experimental apparatus, can therefore be rewritten in cylindrical coordinates as

$$\frac{1}{r}\frac{\partial}{\partial r}\left[rk\frac{\partial T}{\partial r}\right] + \frac{\partial}{\partial z}\left[k\frac{\partial T}{\partial z}\right] = -A. \tag{5}$$

Equation (5), as we show later, can be simplied further for each of the experimental configurations illustrated in Figure 1.

2.3 Convection

Convection is sometimes observed in diamond cell experiments, for example, within the argon medium used to insulate the sample or within the melted region at the center of the heated sample. In fact, the observation of fluid flow is among the criteria used for inferring that melting has indeed occurred [*Jeanloz and Kavner*, 1996].

It has be argued that convection should not take place because the Rayleigh number, Ra, is much less than the critical Rayleigh number for Rayleigh-Benard convection, Ra_{cr}, i.e.,

$$Ra = \frac{\rho g \alpha \Delta T D^3}{\kappa \mu} < 10^{-5} \ll Ra_{cr} \approx 10^3 \tag{6}$$

where ρ is density, g is gravitational acceleration, ΔT is the temperature difference across the fluid, α is thermal expansion, D is the dimension of the convecting region, κ is thermal diffusivity, and μ is viscosity. The upper bound on Ra is calculated assuming a $\Delta T = 2000$ K,

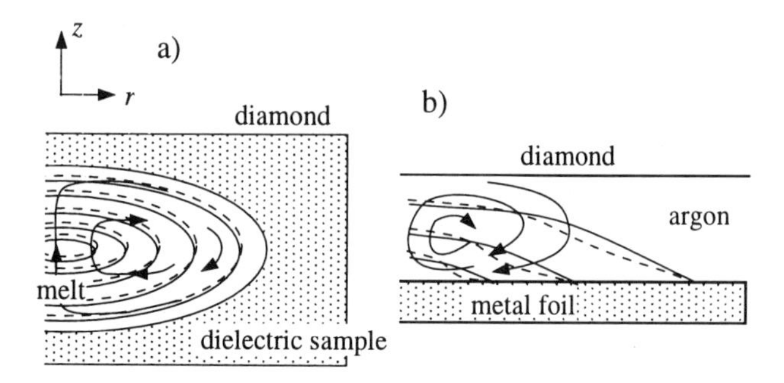

Figure 3. Schematic illustration of isotherms (solid and dashed lines) and velocities (arrows) in (a) the molten spot within a crystalline sample and (b) a fluid sample or pressure medium. The solid curves are isotherms in the absence of fluid flow, and the dashed curves illustrate the distortion of isotherms due to the flow.

$\rho = 5000$ kg/m^3, $D = 10\mu$m, $\kappa = 10^{-5}$ m^2/s, $\alpha = 10^{-5}$ ^{o}C^{-1}, and a low viscosity (similar to that of water) of $\mu = 10^{-3}$ Pa s.

This argument only considers vertical temperature gradients, however. Because of the radial dependence of laser power, temperature varies laterally across the sample (Figure 1). In this case, free convection can occur as a result of the lateral temperature gradients [e.g., *Tritton*, 1988, chapter 4]. We have assumed that the laser beam which defines the axial direction is vertical, that is, parallel to the gravitational acceleration. Nevertheless, the temperature distribution is similarly three-dimensional in other orientations of the laser beam and diamond cell, so that fluid motion can generally be induced by lateral temperature variations across the sample.

A schematic illustration of the temperature distribution and induced fluid flow are shown in Figure 3. The solid lines represent isotherms in the absence of flow, and the dashed lines show the position of the same isotherms accounting for advection by the flow. The distortion of isotherms due to flow is negligible because advective heat transport is many orders of magnitude less than conductive transport, as we now show.

We assume that the Reynolds number, which characterizes the relative importance inertial and viscous forces, is

$$Re = \frac{\rho U D}{\mu} \ll 1, \qquad (7)$$

where U is velocity, a reasonble assumption because D is so small. Thus, we can estimate U by balancing thermally induced buoyancy and viscous resistance forces,

$$U = O\left(\frac{\rho g \alpha \Delta T D^2}{\mu}\right) < 10^{-5}\text{m/s}. \qquad (8)$$

This upper bound on U is calculated with the same values of material properties used to estimate the upper bound on Ra in equation (6).

The relative importance of advective heat transport compared with heat conduction is characterized by the Peclet number,

$$Pe = \frac{UD}{\kappa} < 10^{-5}. \qquad (9)$$

As equation (9) clearly shows, although convection can occur in the laser-heated diamond cell it has a negligible effect on heat transport. Again, the small length scale (D) is a dominant factor in equations (6)-(9).

3. TEMPERATURE DISTRIBUTION WITHIN LASER-HEATED DIELECTRIC SAMPLES

For samples at temperatures less than about 6000 K, conduction is the dominant mechanism of heat transfer in the laser-heated diamond cell. The temperature distribution along the axial direction through the sample, i.e., in a direction parallel to the laser beam, is sketched in Figure 4a for a dielectric being heated at high pressures (the case of metals is considered in §4, and illustrated in Figure 4b). Due to the large size and high thermal conductivity of the diamond anvils, the diamonds are effective heat sinks and the diamond-dielectric interface is close to room temperature [*Bodea and Jeanloz*, 1989].

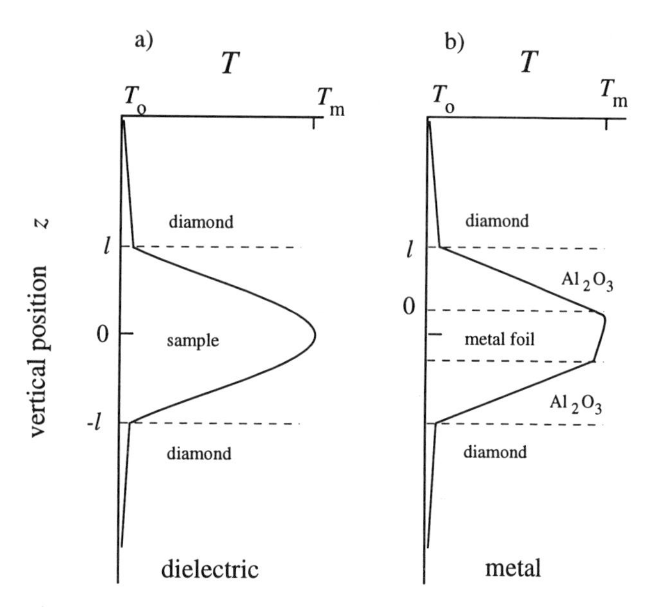

Figure 4. chematic illustration of the axial temperature distribution in the laser-heated diamond cell for a (a) dielectric sample and (b) heated metal foil.

If the radial dimension of the sample d is much greater than the thickness of the sample l (Figure 1), then the terms in equation (5) involving radial gradients are much smaller than the terms involving gradients in the axial direction, and can be neglected. Thus, equation (5) can be be approximated as a one-dimensional conduction equation,

$$\frac{d}{dz}\left[k(T)\frac{dT}{dz}\right] = -A, \tag{10}$$

where the separation of the axial from the radial temperature variation is an approximation good to $O(l/d)^2$. In the diamond cell, $(l/d)^2$ is typically less than 0.1.

The absorbed power density A can often be taken as constant throughout the depth of dielectric samples because the absorption mean-free path, $1/\Lambda$ in equation (4), for dielectrics is typically $10^{-2} - 100$ cm, which is much greater than the sample thickness [e.g., *Fukao et al.*, 1968]. For more strongly absorbing dielectric samples, our analysis can be modified by setting $A = A(z)$ in equation (10). In the limit of a fully absorbing sample (e.g., true blackbody), the sample is usually surrounded by a dielectric (non-absorbing) medium in order to avoid damaging the diamond anvils through chemical reaction with the hot sample. Therefore, this last case is identical to that of metal-foil samples treated in the following section (Figure 4b).

Due to the symmetry of the temperature distribution in the sample,

$$\partial T/\partial z = 0 \text{ at } z = 0. \tag{11}$$

The sample-diamond interface is taken to be at room temperature due to the high thermal conductivity of the diamonds, i.e.,

$$T = T_0 \text{ at } z = \pm l. \tag{12}$$

This last boundary condition can be modified for the small but finite heating of the anvils, as illustrated schematically in Figure 4, but three-dimensional calculations which account for heat transfer in the sample, gasket and diamonds confirm that the sample-diamond interface is indeed close to room temperature [*Bodea and Jeanloz*, 1989].

Making the change of variables,

$$s = \int k(T)dT, \tag{13}$$

with integration constants chosen to satisfy boundary conditions (11-12), equation (10) has the solution

$$s(z) = \frac{Al^2}{2}\left[1 - \left(\frac{z}{l}\right)^2\right]. \tag{14}$$

In dielectrics, heat is carried by phonons, and the thermal conductivity is due to the coupling of phonon modes. At temperatures greater than a few hundred kelvin, thermal conductivity is inversely proportional to T [e.g., *Peierls*, 1955; *Berman*, 1976],

$$k = \frac{a}{T}. \tag{15}$$

Accounting for the temperature dependence of thermal conductivity,

$$T(z) = T_0 \exp\left\{\frac{Al^2}{2a}\left[1 - \left(\frac{z}{l}\right)^2\right]\right\}. \tag{16}$$

Equation (16) is valid if $A = A(r)$, and a solution to equation (10) can also be found if $A = A(r, z)$.

3.1 Radial temperature distribution

Due to the geometry of the experiment, Figure 1, only the radial distribution of temperature is measured (i.e., temperature variations perpendicular to the laser beam). We can extend the one-dimensional solution to describe the radial temperature distribution, by accounting for the radial variation of the laser power, $A(r)$, in equation (10). As we show in the following section, the measured temperature is very nearly the maximum temperature in the middle of the sample at $z = 0$.

Accounting for the temperature-dependence of thermal conductivity, described by equation (15), the radial temperature distribution is

$$T(r) = T_0\left(\frac{T_m}{T_0}\right)^{\exp[-(r/R)^2]}, \tag{17}$$

where T_m is the peak temperature in the middle of the hotspot. The terms omitted in equation (10) are $O(l/d)^2$ smaller than the terms retained, so that solutions, e.g. equations (16) and (17), are accurate to $O(l/d)^2$. Since l/d is typically less than 0.1, temperatures calculated based on the one-dimensional approximation should be accurate to within a few percent. Indeed, the temperature distribution calculated from equation (10) for $k =$ constant is nearly identical to that calculated in a fully three-dimensional model [*Bodea and Jeanloz*, 1989].

In Figure 5 we show the theoretically predicted temperature profile, assuming a beam radius $R = 16\mu$m

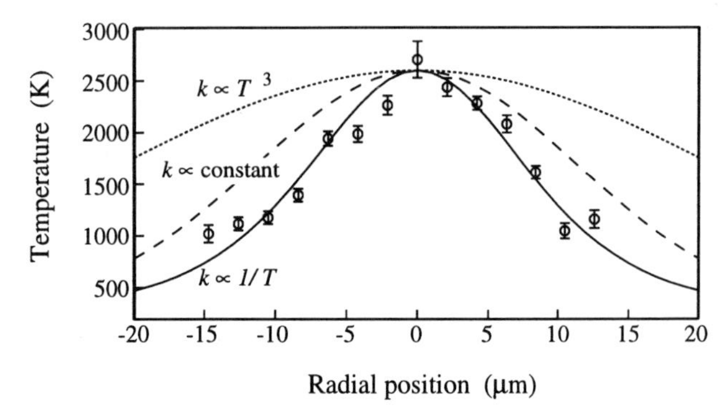

Figure 5. Calculated temperature distribution for $k \propto 1/T$ (solid curve), k = constant (dashed curve) and $k \propto T^3$ (dotted curve); the first and last of these correspond to heat transport by only phonons and radiation, respectively. Data points are for $(Mg_{0.86},Fe_{0.14})SiO_3$ perovskite at 56 GPa [*Li et al.*, 1996].

[*Williams et al.*, 1991], and a measured temperature profile for $(Mg_{0.86},Fe_{0.14})SiO_3$ perovskite at 56 GPa [*Li et al.*, 1996]. For comparison, temperature profiles are also shown for $k \propto T^3$, which accounts for a radiative contribution to the effective thermal conductivity in the diffusion limit [*Siegel and Howell*, 1992], and for k = constant.

The excellent agreement between the measured profile and the model prediction for $k \propto 1/T$ suggests that the inverse relation between thermal conductivity and temperature is valid for silicate perovskite at elevated pressure and $T > 1000$ K. Room-pressure measurements of the thermal conductivity of dielectrics often show that the $1/T$ dependence of thermal conductivity breaks down at $T \approx 1000 - 1500$ K, with the thermal conductivity increasing for greater temperatures [e.g., *Touloukian and Ho*, 1971]. In such cases, however, the deviation from a $1/T$ behavior at large T is thought to be due to radiative transfer influencing the measurements; radiation depends strongly on sample configuration and other experimental details, and can be considered a bias in the measurements [*McQuarrie*, 1954]. Therefore, the fact that our measurements empirically support the $1/T$ dependence of thermal conductivity (Figure 5) reinforces our conclusion that heat transfer is dominated by conduction in the diamond cell. We note that the temperature dependence of absorption will further modulate the temperature distribution so that our inference that $k \propto 1/T$ is not unique [e.g., *Li et al.*, 1996].

The large difference between the three theoretical profiles in Figure 5 highlights the importance of correctly accounting for the temperature dependence of thermal conduction. If we had instead assumed k = constant, then $T(r) = T_m \exp[-(r/R)^2]$, the distribution obtained numerically by *Bodea and Jeanloz* [1989] and shown with a dashed curve in Figure 5. Instead, the actual temperature distribution deviates from this Gaussian form.

3.2 Relationship Between Measured and Actual Sample Temperatures

A variety of spectroradiometric techniques have been developed to measure temperatures inside the diamond cell [e.g., *Heinz and Jeanloz*, 1987a; *Boehler et al.*, 1990; *Lazor et al.*, 1993; *Jeanloz and Kavner*, 1996]. Due to the geometry of the experiment, only the radial distribution of temperature is measured. Thus, for an optically thin sample, the spectroradiometer measures a depth-averaged intensity

$$I(\lambda) = \int_{-l}^{l} L_p[\lambda, T(z)]dz \tag{18}$$

where λ is the wavelength of the light. L_p is the intensity described by Planck's law

$$L_p(\lambda, T) = \frac{2\pi c^2 h\epsilon}{\lambda^5 (e^{hc/\lambda k_b T} - 1)} \tag{19}$$

where h, k_b and c are Planck's constant, Boltzmann's constant and speed of light, respectively.

We solve equation (18) numerically, assuming ϵ = constant (greybody approximation), a thermal conductivity described by equation (15), and $T(z)$ given by equation (16). A more detailed study of the effect of axial temperature gradients on measured temperatures is presented by *Manga and Jeanloz* [1996].

Figure 6 summarizes the relationship between the apparent temperature, corresponding to the integrated light intensities measured by the spectroradiometer at wavelengths between 600 and 900 nm, and the peak sample temperature. At 2000 and 4000 K, respectively, the measured temperature is about 4% and 8% lower than the actual peak temperature, for example. Thus, the measured temperature distribution shown in Figure 5 is relatively unaffected by depth averaging: the measurements are within 100 K of the true peak temperatures within the sample.

4. TEMPERATURE DISTRIBUTION FOR HEATED METAL FOILS

For the case of a heated metal foil, laser energy that is not reflected is fully absorbed by the foil (absorption length scale $1/\Lambda \sim O(10)$ nm), and is conducted axially through the insulating dielectric layers and radially through the metal foil. The total heat flux conducted radially through a sample with radial dimension d and thickness L is

$$Q_{\text{radial}} \approx 2\pi d L k_m (T_m - T_0)/d \tag{20}$$

where T_m is the temperature of the metal foil at the center of the hotspot, T_0 is the ambient temperature, and k_m is the thermal conductivity of the metal. The heat conducted axially from the heated foil through dielectric layers with thickness l is

$$Q_{\text{axial}} \approx 2\pi R^2 k_d (T_m - T_0)/l, \tag{21}$$

where πR^2 is approximately the area of the hot spot and k_d is the thermal conductivity of the dielectric.

We can identify two limits, corresponding to the cases in which either radial or axial heat conduction dominates. If

$$\frac{Q_{\text{radial}}}{Q_{\text{axial}}} = \frac{k_m l L}{k_d R^2} \ll 1 \quad \text{(thin film)} \tag{22}$$

then axial conduction dominates. Since this condition is more likely to hold when the film thicknesses are small, i.e., $(l, L) \ll R$, we refer to this condition as the thin-film limit. If

$$\frac{Q_{\text{radial}}}{Q_{\text{axial}}} \gg 1 \quad \text{(thick film)} \tag{23}$$

then radial conduction dominates, and we refer to this condition as the thick-film limit. Because the thermal conductivity of metals is typically much greater than that of dielectrics, the thick-film limit applies for $l, L \approx R$.

For a typical experiment with $L \approx 3$ μm [*Boehler* et al., 1990], $l \approx 10$ μm, $R \approx 20$ μm, and $k_m/k_d \approx 10$ [*Manga and Jeanloz*, 1997] (iron compared with Al_2O_3), we find

$$\frac{Q_{\text{radial}}}{Q_{\text{axial}}} \approx 1/6. \tag{24}$$

The preceding analysis assumes that the radial extent of the metal foil is large, so that the edge of the foil is at the ambient temperature (typically 300 K). However, if the radial extent of the foil is not much greater than the beam radius, and the foil is surrounded on all sides by an insulating dielectric, radial temperature gradients can be significantly reduced, and heat transfer by conduction occurs primarily in the axial direction. Temperature measurements for metal foils of limited width confirm that radial temperature gradients can be small [e.g., *Boehler et al.*, 1990; *Shen and Lazor*, 1995]. Unlike the case of dielectric samples, we cannot calculate the radial temperature distribution without knowledge of the thickness and radial extent of both the metal foil and the insulating layers.

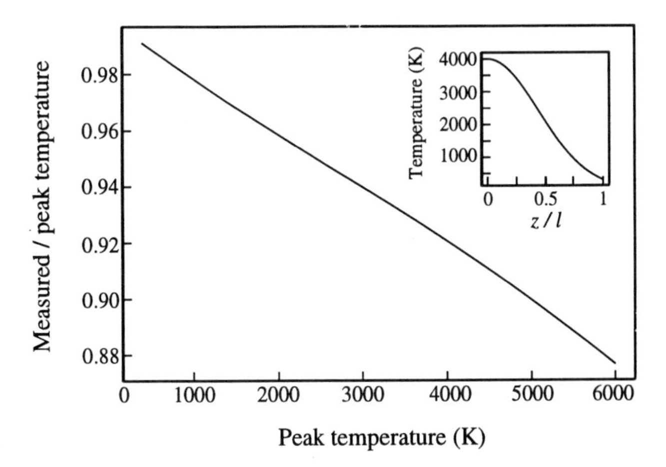

Figure 6. Relationship between the actual peak temperature in an optically thin (dielectric) sample and the temperature measured by a spectroradiometer. The inset shows the temperature distribution for $k \propto 1/T$. The results shown here are independent of sample thickness, l.

4.1 Axial Temperature Distribution: Thin Films

If heat transfer is dominated by axial conduction through the insulating dielectric, then the temperature distribution within the dielectric layer satisfies the one-dimensional steady heat conduction equation

$$\frac{d}{dz}\left[k(T)\frac{dT}{dz}\right] = 0, \tag{25}$$

with boundary conditions

$$T = T_0 \text{ at } z = l \quad \text{and} \quad T = T_m \text{ at } z = 0, \tag{26}$$

where l is the thickness of each insulating layer, and T_m is the temperature of the metal foil. Since the absorption of the laser by the foil is much greater than absorption by the dielectric, $A \approx 0$ within the insulation, see equation (4).

Assuming the thermal conductivity of the insulating layer is inversely proportional to temperature, equation (15), we obtain the temperature distribution

$$T(z) = T_m \left(\frac{T_o}{T_m}\right)^{z/l}. \tag{27}$$

Since the laser in TEM_{00} mode has a Gaussian distribution, the absorbed energy flux at the center of the hotspot is

$$Q_{\text{laser}} \approx \frac{P_0 \beta(T)}{\pi R^2}, \tag{28}$$

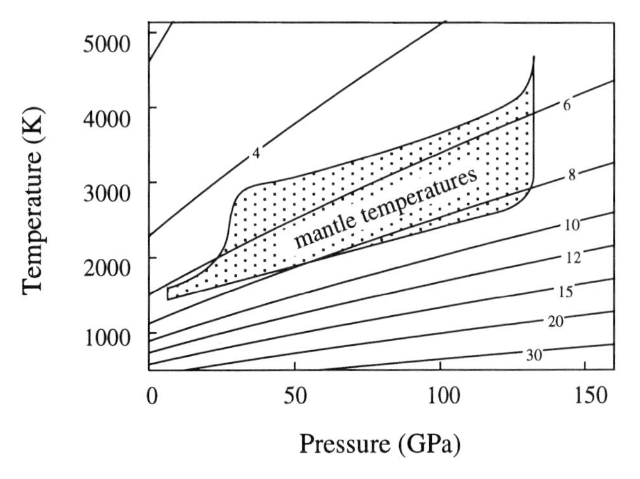

Figure 7. Average thermal conductivity of MgO and Al_2O_3 in W/mK obtained from data collected by *Saxena et al.* [1994] in the diamond cell. Typical mantle temperatures are denoted by the hatched region.

where P_0 is the laser power and β is the temperature dependent absorption of the metal (this is actually an effective absorption coefficient that includes the effect of sample reflectivity; because reflectivity is often strongly influenced by extrinsic effects, such as the detailed character of the sample surface, we cannot separate the effects of intrinsic sample absorption from those of reflectivity in the present analysis).

The heat flux conducted through the dielectric layer is

$$Q_{\text{conducted}} = -k\frac{dT}{dz} = \frac{a}{l}\ln\frac{T_m}{T_0}. \quad (29)$$

From conservation of energy, we can equate the absorbed and conducted heat fluxes, and determine the relationship between the laser power, P_0, and foil temperature, T_m.

4.2 Determining Thermal Conductivity From Diamond Cell Experiments

Equations (28) and (29), combined with the measured relationship between T_m and P_0, as illustrated in Figure 3 of *Saxena et al.* [1994] and Figures 2-3 of *Shen and Lazor* [1995], can be used to determine the temperature dependence of absorption, $\beta(T)$, and the pressure dependence of thermal conductivity, $k(P)$.

For example, using the data of temperature versus laser power presented in Figure 3 of *Saxena et al.* [1994], we can determine the average high-pressure and high-temperature thermal conductivity of the MgO and Al_2O_3 used to insulate the iron foil. The calculations involve several assumptions: *(i)* $(\partial \ln k/\partial \ln \rho)_T$ is constant [*Roufosse and Jeanloz*, 1983]; *(ii)* the Murnaghan equation-of-state is used to relate compression and pressure; *(iii)* k is inversely proportional to temperature, compatible with our results for (Mg,Fe)SiO_3 perovskite (Figure 5); and *(iv)* the absorption of the iron foil is a function of only temperature. Details of the analysis are discussed elsewhere [*Manga and Jeanloz*, 1997]; results are shown in Figure 7. For comparison, typical mantle temperatures are also shown. The thermal conductivities of MgO and Al_2O_3 should be representative of other lower mantle minerals and imply a thermal conductivity in the lower mantle of $5 - 9$ W/mK.

5. SUMMARY

We have considered quantitatively various heat transfer mechanisms in the laser-heated diamond cell, and developed models which relate measured temperatures and temperature distributions to thermophysical material properties. The temperature distribution within the laser-heated diamond cell is governed by the steady-state thermal diffusion equation: radiative and convective heat transfer can be neglected. The temperature distribution within heated dielectric (optically thin) samples can be calculated; a comparison of experimental measurements for perovskite with calculated temperatures indicates that thermal conductivity is inversely proportional to temperature, as expected for nonmetals. For the case of heated metal foils, the relationship between laser power and measured temperatures can sometimes be used to determine the pressure dependence of the thermal conductivity of the insulating nonmetal.

Acknowledgments. This work was supported by NSF, NASA and the Miller Institute for Basic Research in Science. L. Dubrovinsky, S.K. Saxena and an anonymous reviewer are thanked for reviews. A. Kavner, S. Morris and H.A. Stone are thanked for discussions.

REFERENCES

Berman, B. R., *Thermal Conduction in Solids*, Oxford University Press, New York, 1976.

Bodea, S., and R. Jeanloz, Model calculations of the temperature distribution in the laser-heated diamond cell, *J. Appl. Phys., 65*, 4688-4692, 1989.

Boehler, R., Temperature in the Earth's core from melting-point measurements of iron at high pressures, *Nature, 363*, 534-536, 1993.

Boehler, R., N. vonBargen, and A. Chopelas, Melting, thermal expansion, and phase transitions of iron at high pressures, *J. Geophys. Res., 95*, 21,731-21,736, 1990.

Fukao, Y., H. Mizutani, and S. Uyeda, Optical absorption spectra at high temperatures and radiative thermal conductivity of olivines, *Phys. Earth Planet. Int., 1*, 57-62, 1968.

Grover, R., and P. A. Urtiew, Thermal relaxation at interfaces following shock compression, *J. Appl. Phys., 45*, 146-152, 1974.

Heinz, D.L., and R. Jeanloz, Temperature measurements in the laser-heated diamond cell, in *High-Pressure Research in Mineral Physics, Geophys. Monogr. Ser.*, vol. 39, edited by M. Manghnani and Y. Syono, Terra Scientific Publishing, Tokyo, and AGU, Washington, DC, 113-127, 1987a.

Heinz, D.L., and R. Jeanloz, Measurement of the melting curve of (Mg,Fe)SiO_3 at lower mantle conditions and its geophysical implications, *J. Geophys. Res., 92*, 11,437-11,444, 1987b.

Jeanloz, R., and D.L. Heinz, Experiments at high temperature and pressure: Laser heating through the diamond cell, *J. Physique, 45*, 83-92, 1984.

Jeanloz, R., and A. Kavner, Melting criteria and imaging spectroradiometry in laser-heated diamond-cell experiments, *Phil. Trans. Roy. Soc. Lond., 354*, 1279-1305, 1996.

Lazor, P., G. Shen, and S.K. Saxena, Laser-heated diamond anvil cell experiments at high pressure - Melting curve of nickel up to 700 kbar, *Phys. Chem. Minerals, 20*, 86-90, 1993.

Li, X., M. Manga, J.N. Nguyen, and R. Jeanloz, Temperature distribution in the laser-heated diamond cell with external heating, and implications for the thermal conductivity of perovskite, *Geophys. Res. Lett., 23*, 3775-3778, 1996.

Manga, M., and R. Jeanloz, Vertical temperature gradients in the laser-heated diamond cell, *Geophys. Res. Lett., 23*, 1845-1848, 1996.

Manga, M., and R. Jeanloz, Thermal conductivity of corundum and periclase and implications for the lower mantle, *J. Geophys. Res., 102*, 2999-3008, 1997.

McQuarrie, M., Thermal conductivity: VII, Analysis of variation of conductivity with temperature for Al_2O_3, BeO and MgO, *J. Am. Ceram. Soc., 37*, 91-96, 1954.

Peierls, R., *Quantum Theory of Solids*, Oxford University Press, Oxford, 1955.

Powell, R. L., and G. E. Childs, Thermal conductivity, in *American Institute of Physics Handbook*, edited by D. E. Grey, McGraw-Hill, New York, 4-142-4-162, 1972.

Roufosse, M.C., and R. Jeanloz, Thermal conductivity of minerals at high pressure: The effect of phase transitions, *J. Geophys. Res., 88*, 7399-7409, 1983.

Saxena, S.K., G. Shen, and P. Lazor, Temperatures in Earth's core based on melting and phase transformation experiments on iron, *Science, 264*, 405-407, 1994.

Seigel, R., and J.R. Howell, *Thermal Radiation Heat Transfer*, Hemisphere Publishing Corp., Washington, DC., 1992.

Shen, G., and P. Lazor, Measurement of melting temperatures of some minerals under lower mantle pressures, *J. Geophys. Res., 100*, 17,699-17,713, 1995.

Tritton, D.J., *Physical Fluid Dynamics*, second edition, Oxford University Press, 1988.

Touloukian, Y.S., and C.Y. Ho, *Thermophysical Properties of Matter*, volume 2, MacMillan, 1971.

Williams, Q., E. Knittle, and R. Jeanloz, A high-pressure melting curve of iron: A technical discussion, *J. Geophys. Res., 96*, 2171-2184, 1991.

M. Manga, Department of Geological Sciences, University of Oregon, Eugene, OR 97403. (email: manga@newberry.uoregon.edu)

R. Jeanloz, Department of Geology and Geophysics, University of California, Berkeley, CA 94720.

X Ray Diffraction with a Double Hot-Plate Laser-Heated Diamond Cell

Ho-kwang Mao, Guoyin Shen,[1] and Russell J. Hemley

Geophysical Laboratory and Center for High Pressure Research, Carnegie Institution of Washington, Washington, D. C.

Thomas S. Duffy[2]

Consortium for Advanced Radiation Sources, The University of Chicago, Chicago, IL

The laser-heated diamond cell has been improved with the integration of in-situ X ray microprobe and double hot-plate heating techniques. A multimode YAG laser provides a flat-top power distribution at the focal spot. A hot-plate configuration is created where the heat generation and temperature measurement are concentrated at the planar interface of an opaque sample and transparent medium. The heating laser is split into two beams that pass through the opposed diamond anvils to heat the sample simultaneously from both sides. The temperatures of the two sides are measured separately with an imaging spectrograph and CCD and equalized by controlling the ratio of beam splitting. The axial temperature gradient in the sample layer is eliminated within the cavity of the two parallel hot plates. Uniform temperatures of 3000 ($\pm$ 20) K have been achieved in high-pressure samples of 15-μm diameter $\times$ 10-μm thickness. X ray microprobe beam sizes down to 3.5 $\times$ 7 μm (i.e., significantly smaller than the laser heating spot) were used for in situ characterization of the samples under high *P-T* conditions. The technique has been used to study phase relations and melting of iron.

1. INTRODUCTION

Accurate determination of phase diagrams and *P-V-T* equations of state of materials at conditions of planetary deep interiors has been a long-sought goal of experimental geophysics [*Mao and Hemley*, 1996]. Although the conditions at the Earth's core can be achieved with a laser-heated diamond cell [*Boehler*, 1996; *Jeanloz and Kavner*, 1996; *Jephcoat and Besedin*, 1996; *Lazor and Saxena*, 1996] the results are often perceived as ambiguous or represent disequilibrium conditions. While the samples absorbing the laser beam are heated to thousands of degrees, other parts of the high-pressure apparatus remain cool and undisturbed [*Bodea and Jeanloz*, 1989]. The large temperature gradient and the small heating spot are essential for achieving extreme conditions, but they are detrimental to defining precise temperatures. Thermal pressure generated by localized heating adds large uncertainty and inhomogeneity to the pressure, which is normally measured in a calibrant in an unheated region adjacent to the sample. Temperature gradients also cause non-equilibrium chemical diffusion and differentiation in an originally homogeneous sample [*Campbell et al.*, 1992].

[1]Present Address: Consortium for Advanced Radiation Sources, The University of Chicago, 5640 S. Ellis Ave., Chicago, IL 60637

[2]Present Address: Department of Geosciences, Princeton University, Princeton, NJ 08544

Properties of Earth and Planetary Materials
at High Pressure and Temperature
Geophysical Monograph 101

Characterization of laser-heated diamond-cell samples is typically carried out after quenching to ambient temperature at high pressures [*Bassett and Ming*, 1972; *Liu*, 1975; *Mao et al.*, 1977]. Measurement techniques used with in situ studies at high *P-T* often suffer from non-unique interpretations. For example, melting of iron has been determined on the basis of visual criteria or temperature-laser power correlation [*Williams and Jeanloz*, 1991; *Boehler*, 1993; *Saxena et al.*, 1993; *Jeanloz and Kavner*, 1996]. In principle, in situ X ray diffraction during laser heating provides definitive information on crystal structure and density. However, only a handful of examples have been reported to date and these are considered largely reconnaissance studies [*Boehler et al.*, 1990; *Meade et al.*, 1995; *Saxena et al.*, 1995; *Serghiou et al.*, 1995; *Yoo et al.*, 1995; *Fiquet et al.*, 1996]. X ray diffraction studies could be improved by maximizing the heating spot size, X ray intensity, spatial resolution, and diffraction accuracy, and minimizing the *P* gradients, *T* gradients, *T* fluctuations, sample differentiation, and X ray sampling time. Because of the interdependence of these factors, an integrated approach for constraining the overall experimental uncertainty is essential.

2. TEMPERATURE, PRESSURE, AND SAMPLE

2.1. *Double Hot Plates*

During laser heating, a steady-state temperature distribution is reached in diamond cell samples as heat is generated by absorption of the laser and dissipated by flowing down temperature gradient. Severe temperature gradients in the sample exist along the path of the laser beam (i.e., axially) [*Manga and Jeanloz*, 1996]. The diamond anvils, with their characteristic high thermal conductivity, dissipate the heat efficiently and constrain the low-temperature boundary condition. In a semi-transparent dielectric sample (absorbance $A < 1$), the laser is partially absorbed as it passes through the sample, resulting in a peak temperature near the middle and lower temperatures on both sides of the sample. For an opaque sample ($A >> 1$), the laser beam is blocked and absorbed at the sample-medium interface. Heat is generated only at the interface and is conducted to the opposite side of the sample, resulting in a monotonically decreasing temperature through the thickness of the sample. The axial temperature difference in the sample can be reduced by minimizing the sample thickness. However, a very thin sample is undesirable because it also reduces the signal for in situ measurement.

In the present study, the axial temperature gradient in the sample is eliminated by adopting a "double hot-plate" configuration (Figure 1) [*Shen et al.*, 1996]. An opaque sample is sandwiched between two layers of transparent media. Two coaxial but opposing laser beams are focused on opposite sides of the sample. The power of the two beams is adjusted so that both sample-medium interfaces are heated to the same temperature. Acting like planar heat sources, the two "hot plates" eliminate the axial temperature gradient in the sample between the plates and leave a steep gradient in the transparent media outside. The double-side laser-heating geometry, although not necessarily in the hot-plate configuration, has been previously reported by *R. Boehler* [1993, personal communication].

2.2. *Radial Temperature Gradient*

The radial temperature distribution across a heated spot corresponds closely to the power distribution of the laser beam, being only slightly modified by radial heat conduction. The commonly used TEM_{00} laser mode has a small beam diameter and low divergence that can be focused to generate a Gaussian-like power distribution; this is undesirable because the spot size at the peak temperature is a very small portion of the Gaussian. For example, in a conventional experiment of heating an iron sample at 50 GPa to 3000 K by a 15W TEM_{00} YAG laser, the diameter of the spot at 95% maximum temperature (2700 K) is only ~5 μm, but that at 50% maximum (1500 K) is 20 μm. Enlarging the area in the sample at the peak

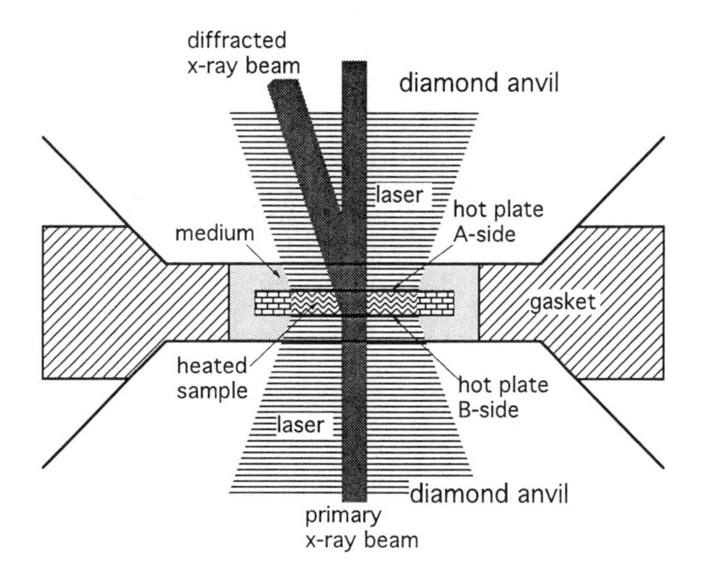

Figure 1. Double hot plate geometry.

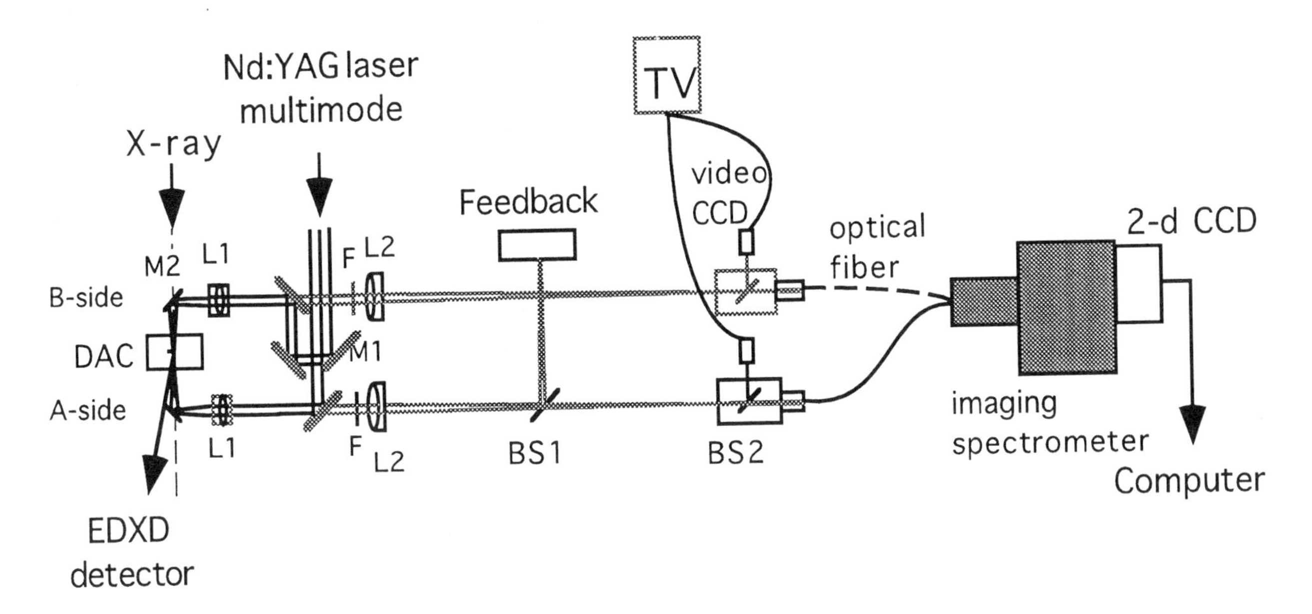

Figure 2. Schematic of the optical system for simultaneous double hot-plate laser-heating and X ray diffraction.

temperature requires increasing the laser power and altering the beam profile to a flat-top (i.e., boxcar) shape. Typical TEM_{00} YAG (or YLF) lasers are limited to ~30 W. The CO_2 laser has been used to deliver 100-200 W and to provide a larger heating spot [*Boehler and Chopelas*, 1992; *Yagi and Susaki*, 1992].

We explored several methods to achieve a flat-top power distribution with high peak temperature. (1) We used a 100-W multimode cw YAG laser which naturally provides a favorable flat-top power distribution. The multimode laser has sufficient power to be split into two beams for double hot-plate heating. Experimental results obtained by this method are presented in this paper. (2) We tested focusing the laser beam onto an optical fiber, which scrambles the mode shape and provides an output with a uniform power distribution across the fiber diameter. The output is refocused into the diamond cell to produce a hot spot with a flat-top power distribution (full-width-95%-maximum of 50 μm). (3) We have also directed the output of two separate YAG lasers (one adjusted for TEM_{00}, the other TEM_{01}) onto each side of the hot plates, which also doubles the total power.

2.3. *Coaxial Optical System*

A schematic diagram of the optical system is shown in Figure 2. A multimode Nd:YAG laser beam is split into two paths: A, reflected from mirror M1, and B, which bypassing M1. The two beams are focused by lens L1 and reflected by beryllium mirrors M2, which reflect visible-to-infrared radiation and transmit incident and diffracted high-energy X radiation. Optical images are monitored from both sides (A and B) of the diamond cell by 2 video CCD (TV) and are adjusted for coaxial alignment of the two heating spots to a spatial accuracy of ±2 μm. Thermal radiation from both sides is focused into optical fibers and dispersed by an imaging spectrometer onto a two-dimensional CCD for temperature measurements. Temperatures of A and B sides are equalized by adjusting the M1 mirror position, which controls the ratio of the laser beam splitting. The laser beams, thermal radiation, and optical images are all focused to the diamond surface with normal incidence to minimize astigmatism.

2.4. *Temperature Determination, Stabilization, and Distribution*

The temperature of the laser-heated sample is determined from its incandescence spectrum. For a transparent or semi-transparent sample, the spectrum is the summation of emission from the sample and media through the entire axial temperature gradient. Corrections and assumptions are applied to deconvolute the peak temperature from the combined spectra [*Boehler*, 1996; *Jeanloz and Kavner*, 1996]. The uncertainty is minimized in the present configuration with an opaque sample (absorbance $A >> 1$ at 400–900 nm) and transparent medium ($A < 0.01$ at 400–900 nm). The transparent medium, in which the steep temperature gradient is located, has a very low emissivity. Its thermal emission is

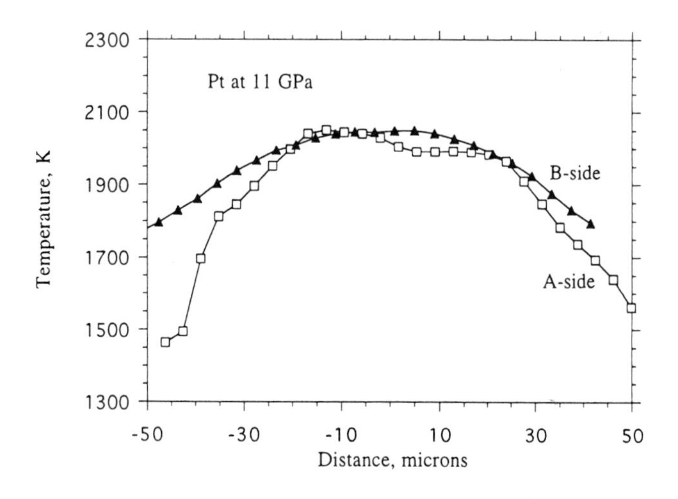

Figure 3. Temperature distribution of A and B sides of a Pt foil at 11 GPa.

insignificant in comparison to that of the highly-emissive, opaque hot plates, and has little effect on the temperature measurement at the opaque sample-medium interface. The spectra are collected with an imaging spectrograph and area (CCD) detector. The radial temperature distribution is mapped with optical microprobes of 3-μm spatial resolution. Figure 3 shows a representative measurement at 11 GPa and 2000 K in which the temperature is uniform to within ±0.5% in a 15-μm area, or ±3% within a 50-μm area, on both sides A and B.

Temperature varies with the fluctuation in laser power and the change of sample-medium absorption characteristics. Variations are reduced with feedback control of the heating power [*Boehler and Chopelas*, 1991]. However, the use of a polarizer to attenuate TEM_{00} laser power [*Heinz et al.*, 1991] is ineffective for the multimode laser used in the present system. We regulate the temperature by an electronic feedback circuit monitoring thermal radiation from the sample and adjusting the electrical power of the laser (Figure 2). The precision of temperature stabilization depends upon the temperature, pressure, and heating time. Figure 4 shows temperatures stabilized to ±0.5% at 2000 K and 11 GPa for 300 s and to ±3% at 2950 K and 30 GPa for 550 s, both are sufficient time for X ray diffraction studies.

2.5. *Symmetrical Diamond Cell*

In the Mao-Bell type of piston-cylinder diamond cell commonly used for experiments in the 100-GPa pressure range [*Mao et al.*, 1994], the sample is located near the cylinder end. The asymmetric arrangement severely restricts the optical access on the piston side and is unsuitable for the double hot-plate configuration. A new type of diamond cell with symmetrical access has been designed and used up to 220 GPa at room temperature. In the new device (Figure 5), the piston-cylinder is sufficiently long (length : diameter = 1) for mechanical stability at ultrahigh pressures. The sample is at a symmetrical position with 60° conical openings on both piston and cylinder sides for double hot-plate heating and measurements.

2.6. *Samples and media*

Opaque materials are studied with the double laser heating of the sample sandwiched between two layers of transparent material. The thickness of medium layers affects the heat dissipation and temperature distribution. Plastic flow of an initially uniform medium may result in an uneven thickness that reduces the size of the uniform-temperature region on increasing pressures. In Figure 3, for example, the 3% variation in temperature of sides A and B over the 50-μm area is due to a kink in the distribution caused by uneven thickness of the medium on side A. We have experimented with Pt, Re, W, Ir, Os, Fe, Si, Al, B, Be, FeO, and FeS as the opaque sample and sapphire, MgO, NaCl, LiF, Ne, and Ar as the transparent media. Because chemical reactivity between medium and sample may change at high pressures and temperatures, it is prudent to conduct redundant studies of the same sample with different media to test their chemical compatibility.

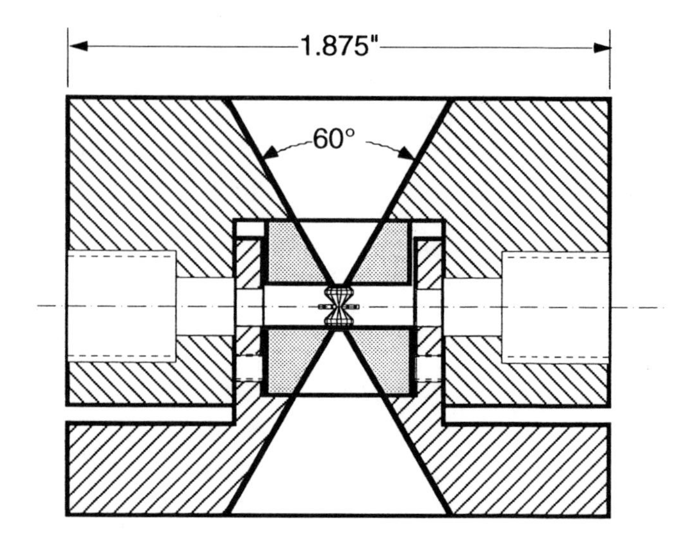

Figure 4. Temporal temperature variations with the feedback stabilizer at 11 GPa - 2000 K and 30 GPa - 3000 K.

For transparent samples (e.g., $MgSiO_3$) which do not naturally absorb the laser and have a black-body emission spectrum, we have experimented with three methods for fabricating the opaque layer: (1) by mixing the sample with metallic powders, (2) by sputtering with a metallic coating, or (3) by sandwiching between thin metallic foils. The third method, although requiring complicated 5-layer fabrication (medium-metal-sample-metal-medium), more reliably produces an opaque interface for the double hot-plate configuration.

3. X RAY DIFFRACTION MICROPROBE

3.1. *X ray Microbeam Size*

A glancing angle mirror system [*Yang et al.*, 1995] has been developed for focusing polychromatic synchrotron X ray beams to a microscopic spot size. The system consists of two bent Kirkpatrick-Baez (K-B) mirrors made of platinum-coated glass plates that reflect 90% of incident X ray photons up to the cut-off energy of 70 keV at 1 mrad glancing angle. In trial experiments, the synchrotron X ray beam was first focused in the vertical plane by the first mirror (260 mm focal distance) and in the horizontal plane by the second mirror (150 mm focal distance). The mirrors intercepted a 50-μm (h) × 70-μm (v) beam and focused it to 7-μm (h) × 3.5-μm (v) (Figure 6), resulting in a hundred fold increase in unit-area intensity. In routine experiments, the beam size was always ≤ 10 × 10 μm. The size of the focused X ray beam at the sample is therefore significantly smaller than the double laser-heated spot. With improved spatial and temporal resolution, energy

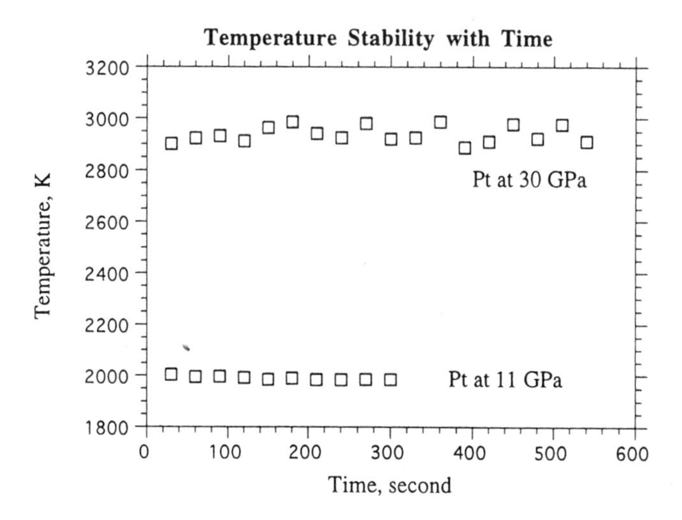

Figure 5. Symmetrical diamond cell for double-sided laser heating.

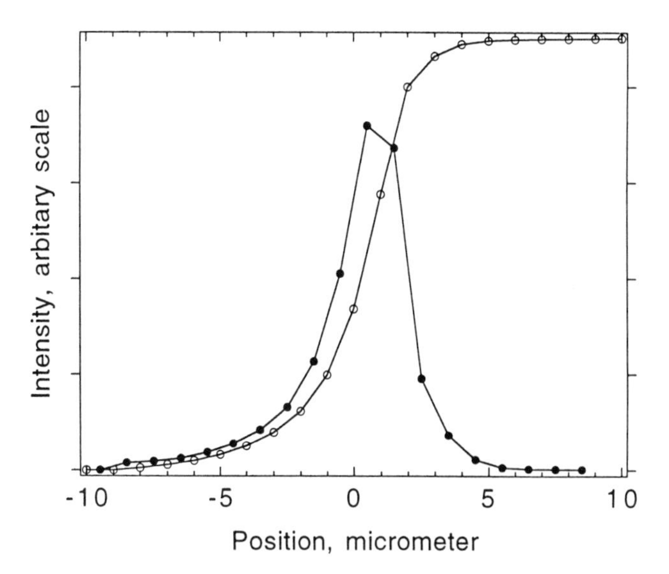

Figure 6. Step-scanning a sharp edge (for blocking the beam) across the focused X ray microbeam in the vertical direction. S-shaped curve (opened circles) shows the intensity measurement. The FWHM of the derivation (bell-shaped curve and solid circles) gives an approximate measure of the focused beam size.

dispersive X ray diffraction (EDXD) from diamond cell samples can be obtained at comparatively uniform and constant *P-T* (±0.5% in *T*)

3.2. *X ray Microbeam Position*

The position of the X ray microbeam must be coordinated with the laser-heating spot. For coarse alignment, a two-dimensional step scan of the diamond cell across the X ray microbeam provides a transmission radiographic image showing the details of the diamond culet, gasket hole, and sample shape. The heating laser spot and the details of the sample chamber are also viewed with an optical microscope and a video camera. The X ray image is coordinated with the optical image to ±5 μm accuracy. Finer alignment to ±2 μm accuracy is achieved by tuning the X ray beam to maximize the fluorescence of a 5-μm grain of gold marker in the sample chamber. With its position known, the X ray microbeam can be placed precisely at the center of laser spot where the temperature is maximum, and gradient is minimum.

3.3. *Preferred Orientation and Coarse Crystallinity*

The extremely small and intense X ray beam ensures *P-T* uniformity within the probed region, but greatly reduces the number of crystallites sampled. Crystal growth within the laser-heated zone further compounds the problem of

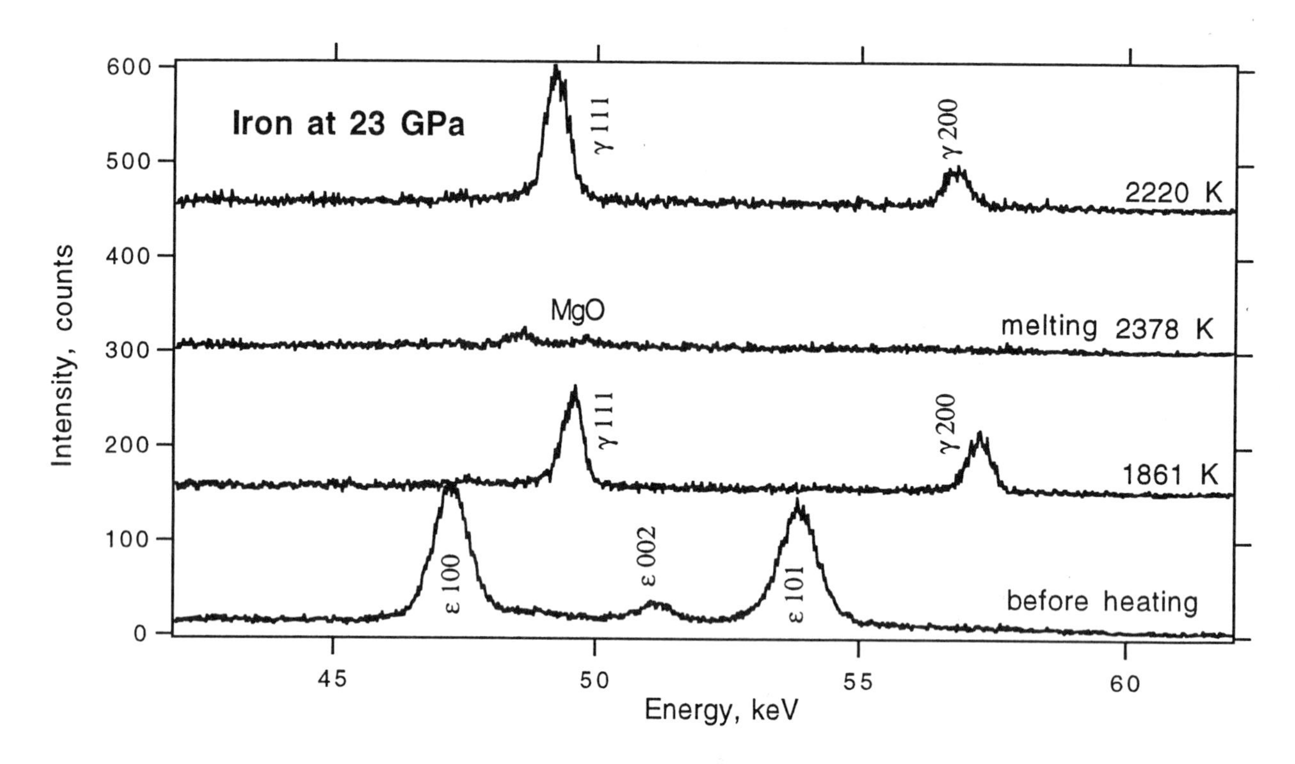

Figure 7. EDXD pattern of iron at 23 GPa as a function of the heating cycle; $2\theta = 7.166°$; $E(\text{keV})d(\text{Å}) = 99.198$.

preferred orientation and coarse crystallinity; this can lead to unreliable peak intensities in the polycrystalline EDXD pattern and even missing diffraction peaks. We have resolved this problem by rotating the diamond cell around its axis; the entire 2θ cone is sampled and averaged at each revolution (typically 50–100 rpm). The diamond cell is centered in the rotation stage (±3 μm) so that the laser beam, X ray beam, and diamond axis are all coaxial with the rotation stage. Consequently, the heating spot position and the sample temperature is constant during rotation. The rotation is crucial for high *P-T* melting studies where recrystallization is rapid near the liquidus.

4. EXAMPLE—PHASE TRANSITIONS AND MELTING OF IRON

Understanding phase transitions and melting of iron is central to models of the Earth's core, and thus the high *P-T* behavior of iron has been the focus of numerous static compression studies. However, the reported melting curves spread over a wide temperature range as a result of the aforementioned experimental difficulties [*Boehler*, 1996; *Jeanloz and Kavner*, 1996; *Jephcoat and Besedin*, 1996; *Lazor and Saxena*, 1996]. We applied the present technique to study iron under high *P-T* conditions. Thin iron foils sandwiched between Al_2O_3, MgO, or NaCl were compressed in the new type of diamond cell and heated with the double hot-plate method and a multimode YAG laser. X ray diffraction was performed at the superconducting wiggler beamline X17B1 at the National Synchrotron Light Source.

Representative diffraction patterns are shown in Figure 7. At 23 GPa and room temperature, iron crystallizes in hexagonal-close-packed (hcp) structure (ε-Fe). As the temperature was raised to 1861 (±50) K, the hcp diffraction lines disappeared completely, and new lines corresponding to the face-centered-cubic (fcc) phase (γ-Fe) appeared. Above 2378 (±50) K, the diffraction peaks of γ-Fe disappeared, and MgO diffraction peaks, which are much weaker that of crystalline iron, became observable. Iron peaks reappeared when temperature was lowered to 2220 (±50) K, bracketing the melting temperature. The completeness of the transition indicates that the X ray beam sampled a comparatively uniform *P-T* region, a major improvement over previous experiments [e.g., *Yoo et al.*, 1995]. Melt and crystalline phases of iron observed in distinct *P-T* regions are shown in Figure 8. Although a systematic study is needed to complete the determination of the phase diagram, the present data indicate that the melting curve is lower than that of *Williams and Jeanloz* [1991] but higher than that of *Boehler et al.* [1990], *Shen et al.* [1993], and *Saxena et al.* [1993]. The liquidus phase

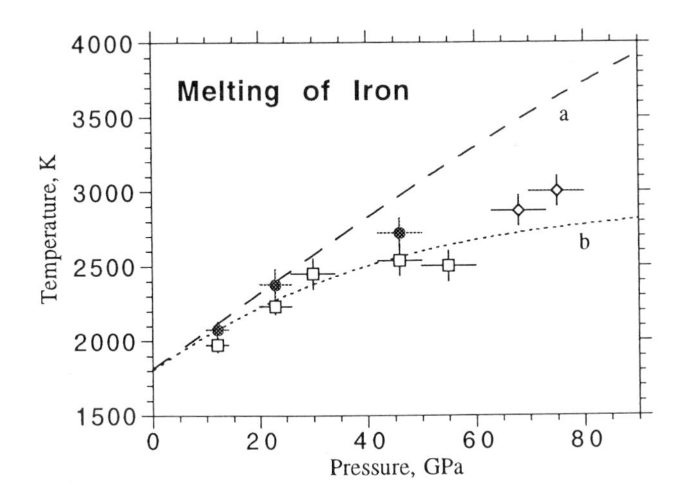

Figure 8. Results of high-*P*-*T* X ray diffraction of iron; open square, fcc pattern; open diamond, hcp pattern; solid circle, no diffraction peaks; dashed line (a), melting curve from *Williams and Jeanloz* [1991]; dotted line (b), melting curve taken from the combined studies of *Boehler et al.* [1990], *Shen et al.* [1993], and *Saxena et al.* [1993].

changes from γ at 55 GPa to ε at 68 GPa, resulting in a γ-ε-liquid triple point near 60 GPa [*Yoo et al.*, 1995]. The new data demonstrate the feasibility of performing high-precision phase equilibria studies and the prospect for measurements such as equations of state, at the *P-T* conditions of the deep mantle and core with the double hot-plate laser-heated diamond cell.

Acknowledgments. This work is supported by the National Science Foundation. We thank the National Synchrotron Light Source for beam time at X17B1.

REFERENCES

Bassett, W. A., and L. C. Ming, Disproportionation of Fe_2SiO_4 to $2FeO + SiO_2$ at pressure up to 250 kilobars and temperatures up to 3000°C, *Phys. Earth Planet. Interiors*, *6*, 154-160, 1972.

Bodea, S., and R. Jeanloz, Model calculations of the temperature distribution in the laser-heated diamond cell., *J. Appl. Phys.*, *65*, 4688-4692, 1989.

Boehler, R., Temperatures in the Earth's core from melting-point measurements of iron at high static pressures, *Nature*, *363*, 534-536, 1993.

Boehler, R., Melting of mantle and core materials at very high pressures, *Phil. Trans. R. Soc. Lond. A*, *354*, 1265-1278, 1996.

Boehler, R., and A. Chopelas, A new approach to laser heating in high pressure mineral physics, *Geophys. Res. Lett.*, *18*, 1147-1150, 1991.

Boehler, R., and A. Chopelas, Phase transition in a 500 kbar - 3000 K gas apparatus, *High-Pressure Research: Application to Earth and Planetary Sciences*, edited by Y. Syono and M. H. Manghnani, pp. 55-60, Terra Scientific Publishing Co., Tokyo/AGU, Washington, D. C. 1992.

Boehler, R., N. von Bargen, and A. Chopelas, Melting, thermal expansion, and phase transitions of iron at high pressures, *J. Geophys. Res.*, *95*, 21,731-21,736, 1990.

Campbell, A. J., D. L. Heinz, and A. M. Davis, Material transport in laser-heated diamond anvil cell melting experiments, *Geophys. Res. Lett.*, *19*, 1061-1064, 1992.

Fiquet, G., D. Andrault, J. P. Itié, P. Gillet, and P. Richet, X-ray diffraction in a laser-heating diamond-anvil cell, *Advanced Materials '96 - New Trends in High Pressure Research*, pp. 153-158, Nat. Inst. Res. Inorganic Mat., Tsukuba, Japan, 1996.

Heinz, D. L., J. S. Sweeney, and P. Miller, A laser heating system that stabilizes and controls the temperature: Diamond anvil cell applications, *Rev. Sci. Instrum.*, *62*, 1568-, 1991.

Jeanloz, R., and A. Kavner, Melting criteria and imaging spectroradiometry in laser heated diamond-cell experiments, *Phil. Trans. R. Soc. Lond. A*, *354*, 1279-1305, 1996.

Jephcoat, A. P., and S. P. Besedin, Temperature measurement and melting determination in laser-heated diamond-anvil cells, *Phil. Trans. R. Soc. Lond. A*, *354*, 1333-1360, 1996.

Lazor, P., and S. K. Saxena, Discussion comment on melting criteria and imaging spectroradiometry in laser-heated diamond-cell experiments (by R. Jeanloz & A. Kavner), *Phil. Trans. R. Soc. Lond. A*, *354*, 1307 -1313, 1996.

Liu, L. G., Post-oxide phases of olivine and pyroxene and mineralogy of the mantle, *Nature*, *258*, 510-512, 1975.

Manga, M., and R. Jeanloz, Axial temperature gradients in dielectric samples in the laser-heated diamond cell, *Geophys. Res. Lett.*, *23*, 1845-1848, 1996.

Mao, H. K., and R. J. Hemley, Experimental studies of Earth deep interior: Accuracy and versatility of diamond cells, *Phil. Trans. R. Soc. Lond. A*, *354*, 1-18, 1996.

Mao, H. K., R. J. Hemley, and A. L. Mao, Recent design of ultrahigh-pressure diamond cell, *High Pressure Science and Technology --1993*, vol. 2, edited by S. C. Schmidt, J. W. Shaner, G. A. Samara and M. Ross, pp. 1613-1616, AIP Press, New York, 1994.

Mao, H. K., T. Yagi, and P. M. Bell, Mineralogy of the earth's deep mantle: quenching experiments on mineral compositions at high pressure and temperature, *Carnegie Inst. Washington Yearb.*, *76*, 502-504, 1977.

Meade, C., H. K. Mao and J. Hu, High-temperature phase transition and dissociation of $(Mg,Fe)SiO_3$ perovskite at lower mantle pressures, *Science*, *268*, 1743-1745, 1995.

Saxena, S. K., L. S. Dubrovinsky, P. Haggkvist, Y. Cerenius, G. Shen and H. K. Mao, Synchrotron X-ray study of iron at high pressure and temperature, *Science*, *269*, 1703-1704, 1995.

Saxena, S. K., G. Shen, and P. Lazor, Experimental evidence for a new iron phase and implications for Earth's core, *Science*, *260*, 1312-1314, 1993.

Serghiou, G., A. Zerr, L. Chudinovskikh, and R. Boehler, The coesite-stishovite transition in a laser-heated diamond cell, *Geophys. Res. Lett.*, *22*, 441-444, 1995.

Shen, G., P. Lazor, and S. K. Saxena, Melting of wüstite and iron up to pressures of 600 kbar, *Phys. Chem. Mineral.*, *20*, 91-96, 1993.

Shen, G., H. K. Mao, and R. J. Hemley, Laser-heating diamond-anvil cell technique: Double-sided heating with multimode Nd:YAG laser, *Advanced Materials '96 - New Trends in High Pressure Research*, pp. 149-152, Nat. Inst. Res. Inorganic Mat., Tsukuba, Japan, 1996.

Williams, Q., and R. Jeanloz, The high pressure melting curve of iron: A technical discussion, *J. Geophys. Res.*, *96*, 2171-2184, 1991.

Yagi, T., and J. Susaki, A laser heating system for diamond anvil using CO_2 laser, *High-Pressure Research: Application to Earth and Planetary Sciences*, edited by Y. Syono and M. H. Manghnani, pp. 51-54, Terra Scientific, Tokyo/AGU, Washington, D. C. 1992.

Yang, B. X., M. Rivers, W. Schildkamp, and P. J. Eng, GeoCARS microfocusing Kirkpatrick-Baez mirror bender development, *Rev. Sci. Instrum.*, *66*, 2278-2280, 1995.

Yoo, C. S., J. Akella, A. J. Campbell, H. K. Mao, and R. J. Hemley, Phase diagram of iron by in situ x-ray diffraction: implications for the Earth's core, *Science*, *270*, 1473-1475, 1995.

H. K. Mao, G. Shen, and R. J. Hemley, Geophysical Laboratory and Center for High Pressure Research, Carnegie Institution of Washington, 5251 Broad Branch Rd. NW, Washington, D. C. 20015

T. S. Duffy, Consortium for Advanced Radiation Sources, The University of Chicago, 5640 S. Ellis Ave., Chicago, IL 60637

Performance of Tapered Anvils in a DIA-Type, Cubic-Anvil, High-Pressure Apparatus for X Ray Diffraction Studies

Yanbin Wang[1]

Center for High Pressure Research (CHiPR) and Department of Earth and Space Sciences, State University of New York at Stony Brook

Ivan C. Getting

CHiPR and Cooperative Institute for Research in Environmental Science (CIRES), University of Colorado, Boulder

Donald J. Weidner and Michael T. Vaughan

CHiPR and Department of Earth and Space Sciences, State University of New York at Stony Brook

A modified DIA anvil geometry is demonstrated to be able to generate at least 30% higher pressures without sacrificing sample volume. This tapered WC anvil system has less pressure-versus-load hysteresis, is more flexible in varying pressures, and has much lower rate of blowouts and higher thermocouple survival rate. Stress analyses indicate that tapering significantly reduces normal stress in the gasket near the tip of the anvil, thereby making the stress distribution inside the anvil more homogeneous. The same approach may be applicable to sintered diamond anvils.

1. INTRODUCTION

In high-pressure and high-temperature X ray diffraction experiments, one of the most commonly used large-volume, high-pressure devices is the so-called DIA-type, cubic-anvil apparatus. Originally developed in Japan, this apparatus consists of six anvils, each with a square truncation anvil tip, forming a cubic cavity in which the pressure medium is compressed (Figure 1). Two of the anvils are fixed opposite to each other on the lower and upper guide blocks which are driven by a uniaxial hydraulic ram. The other four anvils are horizontally located on thrust blocks that transform a uniaxial force into six components acting along three orthogonal directions. The travel of the anvils is synchronized so that the sample assembly cube is compressed isotropically. This geometry has four vertical anvil gaps, convenient for incident and exiting X rays, as well as thermocouple leads. The DIA apparatus currently in operation with synchrotron sources are MAX-80 and MAX-90 at the Photon Factory, Tsukuba, Japan [*Shimomura et al.*, 1985, 1992], and SAM-85 at State University of New

[1]Now at Consortium for Advanced Radiation Sources, The University of Chicago, 5640 South Ellis Avenue, Chicago, IL 60637.

Properties of Earth and Planetary Materials
at High Pressure and Temperature
Geophysical Monograph 101

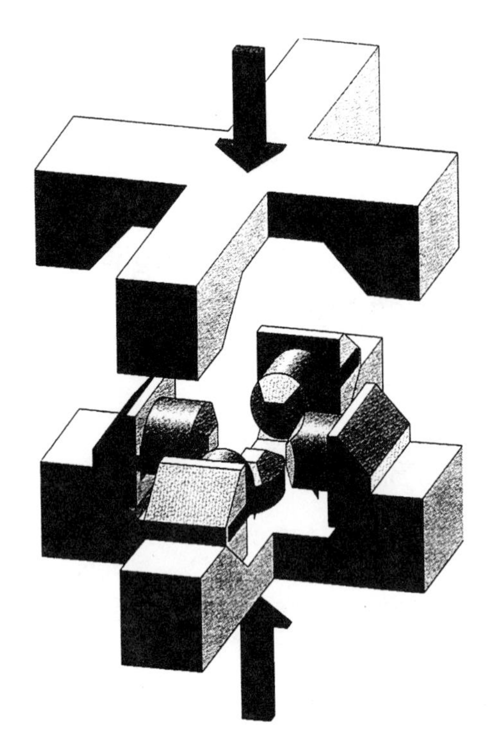

Figure 1. Conceptual diagram of the DIA apparatus. Only the guide blocks and anvils are shown. Uniaxial ram load is represented by the two vertical arrows.

York at Stony Brook and the National Synchrotron Light Source, Brookhaven National Laboratory, USA [*Weidner et al.*, 1992].

The maximum attainable pressure in the DIA system depends on the truncation size on the anvils. In order to reach high pressures, one is forced to sacrifice volume of the sample assembly, therefore losing pressure and temperature homogeneity enjoyed by the larger cell assemblies. In addition, as thermal loss becomes more important in the smaller assemblies, anvils are heated to higher temperatures, thereby limiting temperature capabilities of the system. With polycrystalline sintered diamond (PSD) as the anvil material, pressure and temperature conditions that have been reached are 1500°C and 15 GPa for anvils with 4 × 4 mm truncation tips [*e.g., Weidner et al.*, 1992], and about 1000°C and 20 GPa for 3 × 3 mm [*Utsumi et al.*, 1992]. Pressures are considerably lower when WC anvils are used; generally, 4 × 4 mm anvils are limited to pressures below 10 GPa with a 6-mm edge length cubic sample assembly.

In this paper, we present recent modifications on the DIA anvil geometry to achieve higher pressures without losing sample volume, thereby maintaining the pressure and temperature homogeneity in the sample. We found that a slight taper on the anvil flanks improves the pressure efficiency significantly. In the original 4 × 4 mm anvil system, not only a ~ 30% higher pressure can be gained with a 4° taper, but also the entire performance is greatly improved.

2. TYPICAL DEFORMATION PATTERN IN WC ANVILS

The DIA anvils are cylinders with four flat flanks inclined at an angle of 45° from the cylindrical axis, truncated to form a square shaped anvil tip. Figure 2A illustrates typical plastic deformation in WC anvils in the DIA after high-pressure experiments. Characteristics of the deformation are depression in the middle of the truncation, dipping at the anvil flank near the tip, and bulging on the flanks further away from the anvil tip. This deformation pattern suggests exceedingly high normal stresses on the anvil flanks near the anvil tip, probably even higher than the center pressure.

In order to achieve high pressures, gaskets have been used to provide certain lateral support for the anvils to minimize plastic deformation or brittle fracture. Some pressure gain can be achieved when the properties and geometry of the gaskets are carefully selected. However, anvil failures are frequent and large load-pressure hysteresis is observed. In addition, failures of thermocouples become very frequent and intolerable.

Post-mortem analysis on the failed anvils reveals that typical failures are Mode I cracks: Most anvils are split in planes nearly perpendicular to the truncation face, suggesting an inadequate lateral support. Furthermore, the deformation pattern shown in Figure 2A indicates that failure must be a complex combination of brittle fracture and plastic flow.

3. NORMAL STRESS DISTRIBUTION IN GASKETS

Analyses of stress distribution in simple gasket systems, such as the Bridgman and diamond anvils, have been carried out by many authors (e.g., see papers cited by *Kamarad*, 1980). In these cases the stress distribution can be considered as cylindrically symmetric with the following form:

$$\sigma_n = \sigma_0 \exp[2\mu(a - r)/h] \qquad (1)$$

where σ_n is the normal stress acting on the anvil, σ_0 is yield stress of the gasket material, μ the coefficient of friction, h the thickness of the gasket disc, and $r \in (0, a)$, where a is the radius of the flat face of the anvil.

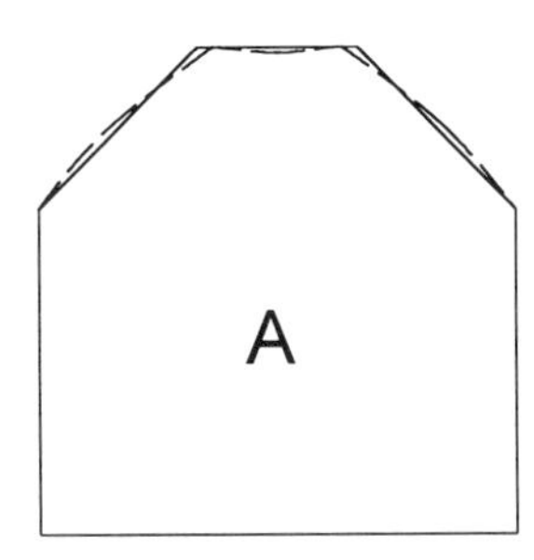

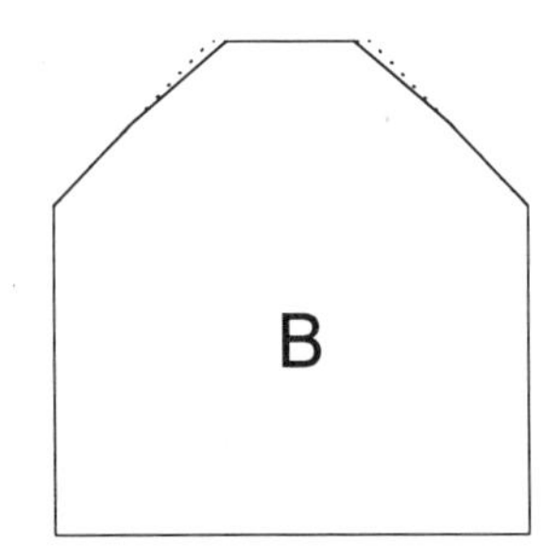

Figure 2. Schematic illustration of the DIA anvil geometry and the modified, tapered geometry. A: Conventional straight anvil geometry (solid outline) and typical deformation after high pressure experiments (dashed lines). B: Comparison of tapered anvil geometry (solid contour) with straight anvil (dashed contour).

Kamarad [1980] modified equation (1) to describe stress distribution in more complex gasket systems such as the belt or girdle apparatus. It was assumed that the normal stress acting on the anvil flanks can be described with an equation similar to (1), i.e.,

$$\sigma_n(l) = \sigma_0 \exp[2\mu^* (l_2 - l)/t] \qquad (2)$$

where $l \in (l_1, l_2)$ is measured along the anvil surface, which is inclined at an angle α from the symmetrical axis of the anvil; l_1 is the starting point of the gasket at the anvil tip, $l_2 - l_1$ the width of the gasket. The thickness of the gasket is t and effective friction coefficient $\mu^* = \mu / \sin^2\alpha$.

Equation (2) can be further used to describe more complex systems such as the DIA, with an empirical geometric correction factor. Here we will use (2) directly to examine some of the features qualitatively for the DIA system using anvils with 4-mm truncation size, $\alpha = 45°$, and $l_1 = 2.828$ mm. Assuming that the gasket material (most commonly used is pyrophyllite) has a friction coefficient of $\mu = 0.1$ and yield strength of 0.1 GPa, Figure 3 illustrates the normal stress distribution in a gasket with a thickness of 0.5 mm and width of 5 mm (typical gasket geometric parameters for the 4 × 4 straight anvil system under high pressure). This qualitative example (solid curve) indicates that, close to the anvil tip, normal stresses are extremely high and may cause the anvil to yield, thereby resulting a deformation profile as observed in the DIA.

One way to reduce normal stress is to increase the thickness of gasket near the tip of the anvil. As an example, we show in Figure 3 a case where the thickness of the gasket is 1 mm at the tip of the anvil and decreases to 0.5 mm at 2.5 mm down the anvil flank (a set of parameters similar to the tapered anvils discussed below). This slight change in gasket thickness results in a dramatic decrease in normal stress near the anvil tip (dashed curve). The force required in the gasket is proportional to the area under the $\sigma_n(l)$ curve. This reduction in normal stress will significantly decrease the total force required in generating the maximum pressure in the cell, thus improving force efficiency.

4. THE TAPERED ANVIL DESIGN AND PERFORMANCE

To reduce the high normal stress component on the anvil flank, we decided to taper the anvils by a small angle, the geometry of the tapered anvils is shown in Figure 2B. Two taper angles, 4° and 8°, were tested, corresponding to ultimate anvil tip sizes of 3.5 × 3.5 mm and 3 × 3 mm, respectively. The tapered region on the anvil flank extends to 2 mm from the edge of the tip. As the gap between the two adjacent anvils increases with decreasing distance from the anvil tips, normal stress near the anvil tip is expected to be reduced considerably. Tests were carried out with a cell assembly of about 6 mm in edge length, the same cube size as in the 4 × 4 mm straight anvil system. All of the tests

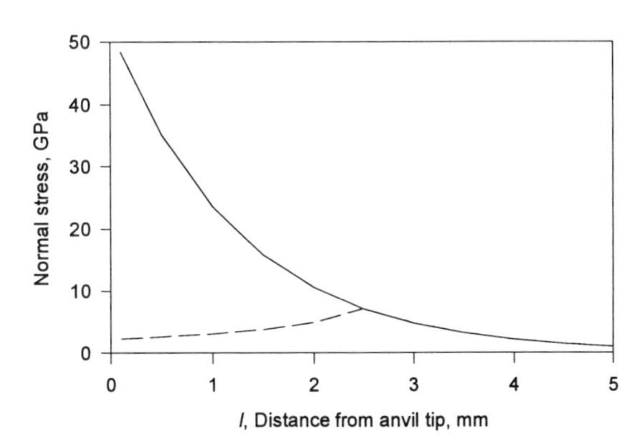

Figure 3. Distribution of normal stress in the gasket as predicted by equation (2). Solid curve corresponds to a uniform thickness (0.5 mm) gasket and dashed curve is for gasket that has a thickness of 1 mm at $l = 0$ and 0.5 mm at $l > 2.5$ mm.

were performed at room temperature, with NaCl as the pressure standard. The resultant pressure-versus-load curves are plotted in Figure 4, which also compares straight WC and sintered diamond anvils.

Also tested were effects of cell dimension on pressure generation. Several linear cell dimensions, ranging from 5.8 to 6.2 mm, were tried out and the resultant pressures varied only slightly (by about 0.1 - 0.2 GPa). The optimal system, found by trial-and-error, is 4° taper angle with 6 mm boron-epoxy cell. Figure 4 shows the performance of this system. A few points are worthy of notice: (1) There is at least a 30% gain in pressure capability when compared with conventional straight WC anvils; (2) The overall pressure efficiency is similar to that of sintered diamond anvils (with gaskets), and yet the cost of tapered WC anvils is only ~5% that of the diamonds; (3) There is very little hysteresis in the load versus pressure behavior, thus allowing much greater pressure variation in a single experiment. The higher apparent pressures on decompression between 70 and 130 ton ram load are most likely due to the presence of macroscopic deviatoric stress that changes sign when ram direction is reversed, resulting in an underestimate and overestimate in pressure, respectively, during compression and decompression [see analyses by *Weidner et al.*, 1992]; (4) There have been very few blowouts, which were much more frequent in the conventional system; and (5) The rate of failure of thermocouples is also markedly lower in the tapered system. Overall, the tapered system increased efficiency in synchrotron X ray experiments by a factor of about 5. It is worthwhile to note that in the tapered anvil system the size of the cell assembly remains the same as that used in the 4 mm straight anvil system. Volume is not sacrificed.

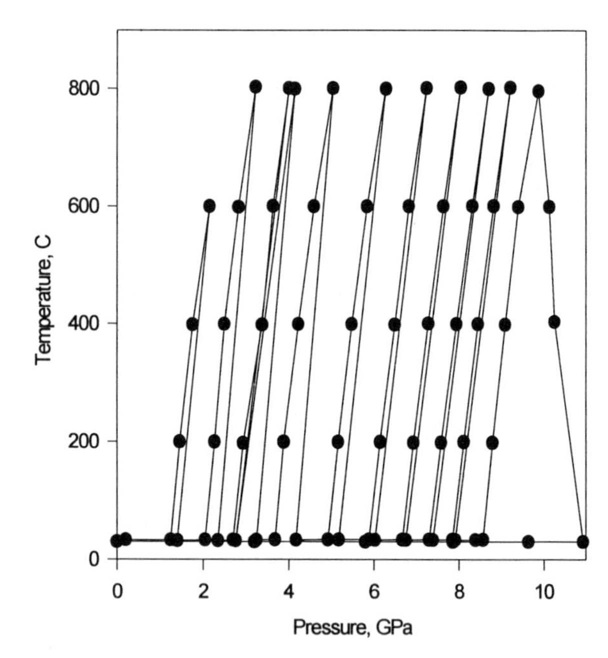

Figure 5. Pressure-temperature path for a P-V-T equation-of-state experiment using the tapered anvils. Pressure is first increased at room temperature to the maximum, temperature is then raised. Pressure decreases on cooling at a fixed ram load due to the loss of thermal pressure. At a lower ram load, pressure increases on heating and regain the thermal pressure (note similar slopes on both cooling and heating). A wide P, T range is covered in a single experiment.

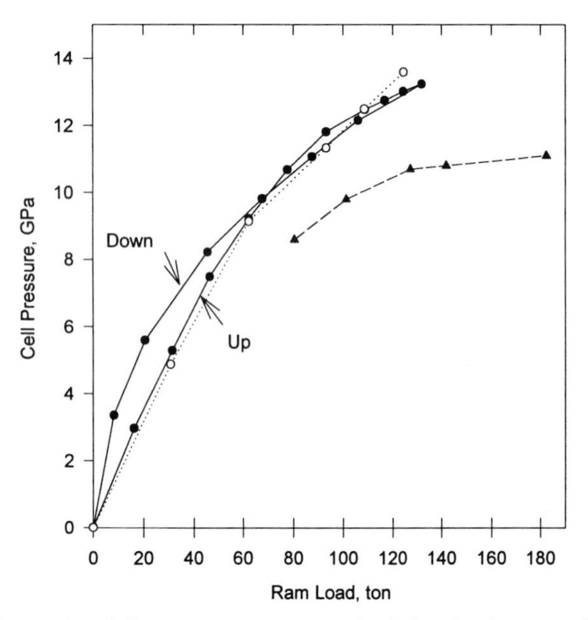

Figure 4. Cell pressure versus ram load for the 4-mm anvils tapered by 4° to a final tip dimension of 3.5 mm (solid circles). Note small hysteresis in pressure. Loading (Up) and unloading (Down) curves are indicated. Solid triangles show a typical curve for 4 mm straight WC anvils with square cross sectioned (2×2 mm) preformed pyrophyllite gaskets. Note the change in slope at ~120 ton; pressures remained virtually unchanged at higher loads and the anvils failed at about 180 ton load. Open circles are from a typical run using 4-mm straight sintered diamond anvils with a similar gasket system. The tapered anvils give a similar pressure efficiency as the diamonds.

The absence of hysteresis in the pressure-versus-load relation for the tapered anvils may be due to the fact that there are two length-scale changes during pressure generation: the change in overall cell dimension, denoted here as $\Delta L/L$ (where L ~ 6 mm is the edge length of the cubic pressure medium) and the change in thickness of the gaskets, $\Delta l/l$, where l is the order of 0.5 mm for straight anvils and varies in the tapered anvil case from ~0.5 mm (on the flank of the anvil, outside the tapered area) to ~1 mm (near the tip). For straight anvils, during decompression the cell pressure remains virtually unchanged until the ram load has been reduced by more than 50%. This suggests that the first load being decreased is mostly supported by the gaskets, where Δl is small, though $\Delta l/l$ may be large. This is consistent with the solid curve in Figure 3, which predicts large normal stress on the gasketed area. When the cell pressure begins to decrease, there is no more support from the gaskets. Blowouts are likely to occur. In addition, the high normal stress is unfavorable for the thermocouple; indeed, failure rate for the thermocouple is much higher in the straight anvil cases during compression.

In the tapered anvil system, $\Delta l/l$ varies continuously, because the length scale l varies. Thus, there is always some support from the gasket during most of the decompression process. The stress field around the thermocouple is not as severe as in the straight anvil case because the normal stress near the gasket-cube interface is low. Overall, the system performs much better both in terms of thermocouple survival and the ability of varying pressure. Figure 5 shows an example of equation-of-state experiments performed in the tapered-anvil system. Between zero and the target pressure, we are now able to carry out many temperature excursions to map out the entire pressure - temperature space in a single experiment.

5. CONCLUSIONS

A tapered anvil geometry is designed for WC anvils to reach higher pressures comparable to conventional straight sintered diamond anvils. Analyses on stress distribution in the gaskets indicate that the tapered anvils are under significantly lower normal stress near the anvil tip, resulting in marked increase in pressure efficiency. Compared with the conventional straight anvils, at least 30% more pressure can be generated with a 4° taper. The tapered anvils have much lower risk of blowouts and have very little load-versus-pressure hysteresis. They allow measurements covering the entire pressure and temperature space offered by the DIA apparatus with fewer runs. The overall improvement in efficiency in performing DIA experiments is increased several fold.

Acknowledgments. The authors wish to thank all the users of SAM-85 for assistance and helpful input and D. Walker and H. Green for useful review. Work supported by NSF grant to Center for High Pressure Research EAR 89-20239 and partly supported by EAR95-26634. MPI No.191.

REFERENCES

Kamarad, J., Pressure distribution in gaskets of high pressure devices, *Rev. Sci. Instrum.*, *51*, 848-849, 1980.

Shimomura, O., W. Utsumi, T. Taniguchi, T. Kikegawa, and T. Nagashima, A new high pressure and high temperature apparatus with sintered diamond anvils for synchrotron radiation use, in *High-Pressure Research: Application to Earth and Planetary Sciences*, *Geophys. Monogr. Ser.*, vol. 67, edited by Y. Syono and M. H. Manghnani, pp 3-11, Terra Scientific Publishing, Tokyo/ AGU, Washington, D. C., 1992.

Shimomura, O., S. Yamaoka, T. Yagi, M. Wakatsuki, K. Tsuji, O. Fukunaga, H. Kawamura, K. Aoki, and S. Akimoto, Multi-anvil type high pressure apparatus for synchrotron radiation, Solid State Physics under Pressure, in *Recent Advance with Anvil Devices*, edited by S. Minomura, pp 351-356, KTK/Reidel, Tokyo/Dortrecht, 1985.

Utsumi, W., T. Yagi, K. Leinenweber, O. Shimomura, and T. Taniguchi, High pressure and high temperature generation using sintered diamond anvils, in *High-Pressure Research: Application to Earth and Planetary Sciences*, *Geophys. Monogr. Ser.*, vol. 67, edited by Y. Syono and M. H. Manghnani, pp 37-42, Terra Scientific Publishing, Tokyo/ AGU, Washington, D. C., 1992.

Weidner, D. J., M. T. Vaughan, J. Ko, Y. Wang, X. Liu, A. Yeganeh-Haeri, R. E. Pacalo, and Y. Zhao, Characterization of stress, pressure, and temperature in SAM85, a DIA type high pressure apparatus, in *High-Pressure Research: Application to Earth and Planetary Sciences*, *Geophys. Monogr. Ser.*, vol. 67, edited by Y. Syono and M. H. Manghnani, pp 13-17, Terra Scientific Publishing, Tokyo/ AGU, Washington, D. C., 1992.

Weidner, D. J., Y. Wang, and M. T. Vaughan, Yield strength at high pressure and temperature, *Geophys. Res. Lett.*, *21*, 753-756, 1994.

I. C. Getting, Center for High Pressure Research (CHiPR) and Cooperative Institute for Research in Environmental Science (CIRES), University of Colorado, Boulder, CO 80309-0216.

Y. Wang, CHiPR, Department of Earth and Space Sciences, State University of New York at Stony Brook, Stony Brook, NY 11794-2100 and now at Consortium for Advanced Radiation Sources, The University of Chicago, 5640 South Ellis Avenue, Chicago, IL 60637.

D. J. Weidner and M. T. Vaughan, CHiPR, Department of Earth and Space Sciences, State University of New York at Stony Brook, Stony Brook, NY 11794-2100.

Sound Velocity Measurements at Mantle Transition Zone Conditions of Pressure and Temperature Using Ultrasonic Interferometry in a Multianvil Apparatus

Baosheng Li, Ganglin Chen, Gabriel D. Gwanmesia[1], and Robert C. Liebermann[2]

Center for High Pressure Research and Mineral Physics Institute, State University of New York at Stony Brook*

Technological developments using a 1000-ton uniaxial split-cylinder apparatus enable elastic wave velocities to be measured using millimeter-sized polycrystalline and single crystal specimens by ultrasonic interferometry. Measurements on polycrystalline Lucalox alumina to 10 GPa at room T show very good agreement with extrapolations of previous data at low pressures (<1GPa). Specimens recovered from the high pressure experiments are undamaged and re-usable; travel times measured during successive pressure cycles are reproducible within 0.5%. Sound velocities of the olivine (α)and beta-phase (β) polymorphs of Mg_2SiO_4 are reported to P>12 GPa at room T on polycrystalline specimens previously hot-pressed in a multianvil apparatus. The new velocity data for olivine are lower by 1% for P waves and 2% for S wave at transition zone pressures than finite-strain extrapolations of low pressure data, but exhibit good agreement to 10 GPa with recent Brillouin scattering and ultrasonic data for single crystals. The velocity data for the beta phase at 12 GPa agree within 1% with finite strain extrapolations of previous data to 3 GPa for similar specimens. Linear fittings to K_S and G vs. pressure yield $dK_S/dP = 4.4$ and $dG/dP = 1.3$ for olivine and $dK_S/dP = 4.2$ and $dG/dP = 1.5$ for the beta phase. The longitudinal modulus C_{22} of San Carlos olivine and the shear modulus C_{55} of San Carlos olivine and forsterite were measured to pressures above 10 GPa at room T. Examination of the specimens recovered from ~10 GPa revealed no internal cracks and no increase in the dislocation density. Good agreement is observed for the C_{22} of San Carlos olivine and C_{55} of forsterite as compared to previous studies. The new ultrasonic data indicate that C_{55} of San Carlos olivine remains linear with pressure to 13.5 GPa and does not exhibit the pronounced curvature reported earlier. Preliminary results on polycrystalline forsterite demonstrate the feasibility of performing ultrasonic velocity measurements to simultaneous pressure of 10 GPa and temperature of 1000 °C. A specially designed P-T path in pressurization/heating and decompression/cooling ensures the integrity of the sample at these conditions.These technological advances make it now possible to measure acoustic velocities at the P and T conditions of the Earth's mantle transition zone.

Properties of Earth and Planetary Materials
at High Pressure and Temperature
Geophysical Monograph 101

INTRODUCTION

Seismic studies of the Earth's interior lead to velocity-depth profiles that feature distinct gradients and

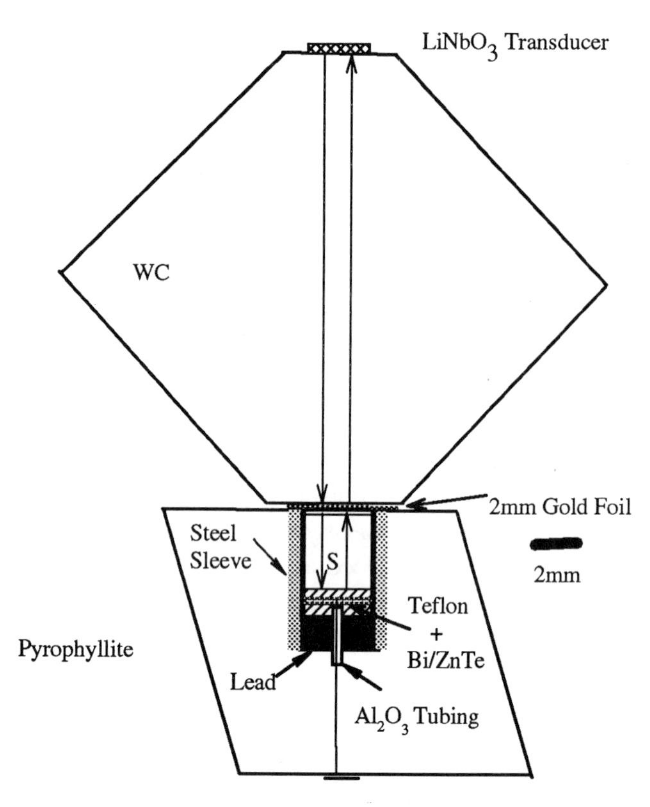

Figure 1. A cross section of the WC buffer rod and the pyrophyllite octahedral cell assembly for the room temperature acoustic experiments.

discontinuities in the upper mantle, transition zone, and lower mantle of the Earth. These profiles vary from region to region in the upper mantle due to lateral heterogeneities, but these variations largely vanish at depths of the transition zone (400 - 700 km) and below [e.g., *Nolet et al., 1994*]. In the absence of direct sampling of the deep interior of the Earth, different compositional and mineralogical models for the Earth's mantle have been proposed based on evidence from petrology, geochemistry, and seismology. Measurements of the sound velocity rock-forming minerals and their high pressure phases at mantle conditions of pressure and temperature are necessary to interpret the seismic profiles in terms of chemical composition and mineralogy.

The techniques of ultrasonic interferometry have been used extensively to determine the elastic properties of ceramics and rock-forming minerals since they were developed by McSkimin [1950]. The high resolution of these techniques make possible very accurate measurements of compressional and shear wave velocities (and thus the bulk and shear moduli) for single crystal or high-density polycrystals fabricated by hot-pressing at ambient or elevated pressures [e.g., *Liebermann et al., 1975*]. Because of the geological interest in elastic properties at mantle conditions, many investigators have attempted to extend the experimental conditions for these velocity measurements.

In the 1960's, studies on oxides and other minerals using ultrasonic interferometry measurements and gas apparatus reached P ~ 1GPa and/or 500°C under hydrostatic stress conditions [e.g., *Kumazawa and Anderson, 1969; Schreiber and Anderson, 1966; Gieske and Barsch, 1968; Spetzler et al., 1970; Wang and Simmons, 1972*]. Using a liquid-medium piston-cylinder apparatus, velocities in both single crystal and polycrystalline specimens have been measured to 3 GPa at room temperature for upper mantle minerals [*Jackson and Niesler, 1982; Niesler and Jackson, 1989; Webb, 1989*]. A modification to the piston cylinder apparatus enabled Niesler and Fisher [1992] to determine sound velocities up to 2 GPa and 600 °C by installing an internal heater.

With the development of large volume multianvil apparatus, pressures representative of the transition zone and uppermost part of the lower mantle have become accessible in large sample volumes. Kinoshita et al. [1979] and Fujisawa and Ito [1984, 1985] have explored the possibility of doing ultrasonic interferometry in cubic anvil apparatus and uniaxial split-sphere apparatus [see also Suito et al., 1992]. A hybrid liquid-solid assembly has been used in such apparatus by Yoneda [1990] to measure the pressure derivatives of the elastic moduli for single crystal MgO and $MgAl_2O_4$ to 8 GPa at room temperature. However, these developments in ultrasonic interferometry at high pressure still fall short of achieving transition zone conditions. Recently, velocity measurements at transition zone pressures at room temperature have been reported using ultrasonic interferometry, Brillouin and impulsive stimulated scattering techniques [see *Li et al., 1994, 1995a, b, 1996b, c; Duffy et al., 1995; Zha et al., 1996; Zaug et al., 1992, 1993*]. Subsequently, ultrasonic measurements have been made at simultaneous high P and T conditions typical of the transition zone in the Bayreuth [*Knoche et al., 1995*] and Stony Brook [*Li et al., 1995b*] laboratories.

The purpose of this paper are to: (1) demonstrate the feasibility of performing velocity measurements in a multianvil high pressure apparatus using ultrasonic interferometry and millimeter-sized specimens; (2) report our new data for high-quality polycrystals alumina and the olivine and beta phases of Mg_2SiO_4, and for single crystals of forsterite and San Carlos olivine to P > 12 GPa at room temperature; and (3) describe pilot studies of the elasticity of polycrystalline olivine at P ~ 10 GPa and T~ 1000 °C.

EXPERIMENTAL PROCEDURES

Multianvil Apparatus Cell Assembly

The implementation of ultrasonic measurements using MA-8 multianvil high pressure apparatus (USCA-1000) has been described in detail in our early studies [*Li et al., 1996b*]. Briefly, diagonally opposite corners of one cube are truncated to yield lapped surfaces on which the transducer and sample are mounted. This cube thus serves as the buffer rod to transmit the acoustic signals to and from the sample. Figure 1 is a cross section of the octahedral cell assembly and the buffer rod cube designed to perform acoustic velocity measurements. The sample is positioned flush with the surface of the octahedron. It is surrounded by lead on the sides (0.3-mm wall thickness) and bottom (2.0-mm-thick disk), thereby providing a pseudohydrostatic pressure medium which protects the sample from cracking at high pressures. A teflon disk (~2 mm thick) is placed at the end of the sample to enhance the ultrasonic signals by increasing the mismatch between the sample and the backup material (Figure 1). To prevent the lead from extruding to the gasket area between the cubes while applying pressure, the sample and surrounding lead are inserted into a steel sleeve (6.0 mm long, OD 4.4 mm, ID 3.4 mm). Pressure sensors (Bi and/or ZnTe) are placed in a tiny hole in the teflon disk next to the sample as shown in Figure 1. Therefore, the pressure scale is obtained in each individual run through observation of the phase transformations in Bi (I-II 2.55GPa, III-V 7.7 GPa) and/or ZnTe (at 9.6 GPa and 12.0 GPa) [*Lloyd, 1971; Kusaba et al., 1993*]. The reproducibility of the cell pressure for the same specimen is better than 3% for given oil pressure (or ram force).

The pressure gradient in the sample region was investigated by enclosing Bi in AgCl at either the center of the cell assembly (A) or at the surface in contact with the WC anvil (B) in two different runs (Figure 2a). As shown in Figure 2b, the pressures measured at the center of the cell are reproducible within 0.1 GPa from run to run, and are 0.6 GPa less than those at the edge of the cell assembly for a 3 mm sample (due to the well-known negative anvil effect). Consequently, we conclude that the pressure gradient in the sample region is less than 0.2 GPa/mm for our acoustic experiments.

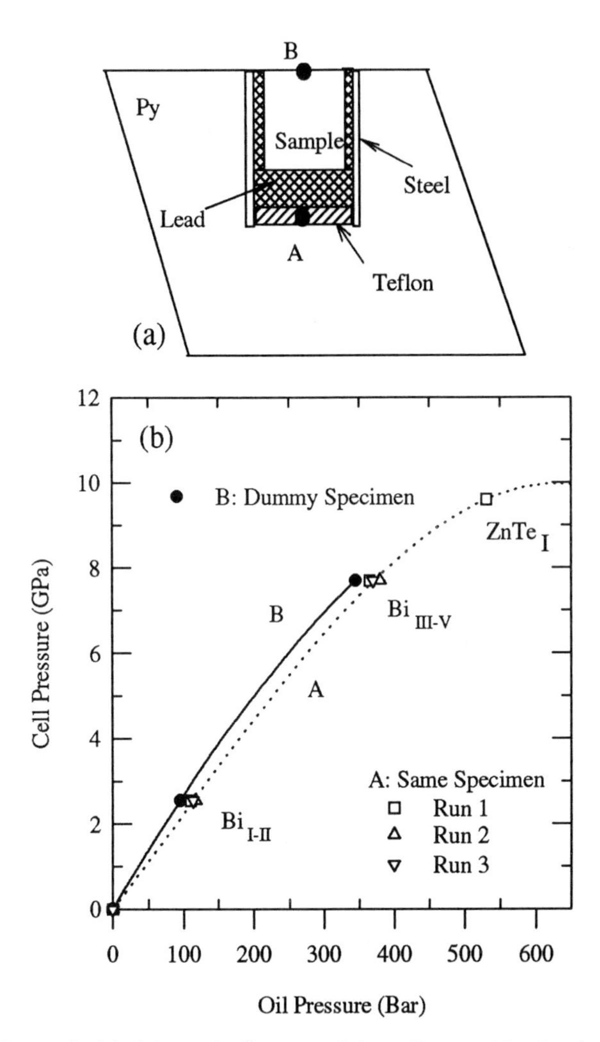

Figure 2. (a) Schematic diagram of the cell assembly showing the positions (A and B) of the pressure sensor in pressure calibration and acoustic experiments for polycrystalline alumina. (b) Comparison of the observed Bi phase transformation pressures vs. oil load from A and B in (a). The solid symbols and line are the results and fit for the top position of the cell (B); the open symbols are the observed Bi and ZnTe phase transformations at the bottom of the cell (A) in acoustic experiments for the same specimen . The dashed line is the fit to all data from acoustic experiments in run 1,2, and 3.

Ultrasonic Interferometry in Multianvil Apparatus

The acoustic signals are generated and received using disk-shaped (3.2 mm diameter), 40 MHz $LiNbO_3$ transducers (36° Y-cut for compressional wave and 41° X-cut for shear waves). The transducer is mounted onto the exposed corner of the buffer rod cube using Aremco Crystalbond. Truncations on the edges of the three bottom first-stage cylinder anvils, originally designed for thermocouple feedthroughs, provide a channel for the coaxial cables to connect the interferometer and the transducer. A spring-loaded sliding pin is placed in the vertical gap between the bottom three anvils to make contact with one electrode of the transducer while the buffer rod cube serves as the electrical ground. A 50 Ω resistor connected in parallel with the transducer provides appropriate electrical termination. At elevated pressure, the transducer remains stress-free since it is located in the gap between the first-stage anvils and the second-stage cubes, allowing precise travel-time measurements over a wide frequency range (20 to 70 MHz).

The ultrasonic phase-comparison method implemented on the Australian Scientific Instruments Ultrasonic Interferometer employed in our laboratories has been described in detail in previous studies [*Rigden et al., 1988; Niesler and Jackson, 1989, Rigden et al., 1992*]. Briefly, the output from a continuous wave source is gated to produce a pair of phase-coherent, high frequency pulses. These pulses are applied to the transducer which is bonded to the buffer rod. The elastic waves generated by each pulse are reflected and transmitted at the buffer rod/sample interface, and the transmitted portion reverberates inside the sample, resulting in a series of 'sample' echoes following the buffer rod echo. If the applied pulses are separated by the apparent two-way travel time through the sample, the first buffer echo from the second applied source pulse will superimpose with the first sample echo from the first source pulse. As the carrier frequency is varied, alternate constructive and destructive interferences between the superimposed signals will occur, resulting in a series of maxima and minima on the amplitude spectrum modulated by the transducer response envelope. Frequencies for pth and $(p+n)$th interference extrema, f_p and f_{p+n} can be used to estimate the apparent travel time by $t'_{est} = n / (f_{p+n} - f_p)$; then the p value is calculated from $p = f_p t'_{est}$ and the closest half or integral value is therefore assigned to frequency f_p and all remaining extrema can be assigned sequentially. In practice, the interference minima are normally used to reduce the travel-time data because they are sharper than the maxima. For the situation in which the amplitude ratio of the first buffer rod echo and the first sample echo (B1/S1) is very different from unity, the perturbation to travel time from the transducer response envelope has to be taken into consideration, especially for interference maxima [see *Jackson et al., 1981; Niesler and Jackson, 1989*].

The acoustic signals at different stages are compared in Figure 3 from an S wave measurement experiment on polycrystalline alumina (see details in following section). Initially, after the cell assembly is loaded into the press and before oil pressure is applied, one sees only a series of echoes from the bottom triangular face of the WC buffer rod (top photo), due to inadequate mechanical coupling between the buffer rod, the gold foil, and the specimen (Figure 1). When the ram load is increased to some 7 ~10 tons (estimated cell pressure 0.2 GPa), contact is made and multiple sample echoes appear following the buffer rod echoes (middle photo), and the amplitude ratio B1/S1 decreases rapidly with further pressurization. The B1/S1 reaches a stable level at pressures of about 1~2 GPa and varies only slightly with further increase of pressure. The bottom photo in Figure 3 shows the acoustic signals at ~10 GPa; it is evident that the quality of the acoustic signals is maintained as pressure is increased. The interference spectrum between B1 and S1 at ~10 GPa is shown in Figure 3b from 15 to 55 MHz, and the apparent travel times obtained from these interference extrema are shown in Figure 3c. The dispersion from 20 to 40 MHz (~ 1%) is much more pronounced than that at frequencies higher than 40 MHz in which range the travel times are insensitive to the frequency. This characteristic of the dispersion persists at all pressures in our experiments.

However, the observed dispersion in the frequency range lower than 40 MHz is not considered to be intrinsic to the sample (i.e. change of velocity with frequency). Instead, the dispersion is largely caused by departures from the assumption of wave propagation in an infinite medium, due to the small size of the sample as has been suggested by Rigden et al. [1992]. Small samples are likely to be effected by the sidewall reflections from the sample at certain low frequencies. Comparison of the results of this study with measurements for samples of the same material but larger diameters [I. Jackson, personal communication, 1996] show that the dispersion in the frequency range 20-50 MHz decreases from 1% to 0.3% as the sample diameter increases from 2.9 to 7 mm. Further measurements in the frequency range of 120-150 MHz on both polycrystalline alumina and forsterite samples (diameter 2.7-2.9 mm) used in this study show a very good agreement (better than 10^{-3}) with measurements at frequency range 40-70 MHz, confirming that the travel times at low-frequency range are influenced by the experimental configuration, and reliable travel times can be obtained at the high-frequency range [*Li et al., 1996b*, Figure 5]. The uncertainties in the measured travel times in this study are thus estimated to be about 0.3%.

As shown in Figure 1, a thin gold foil (2-μm thickness) is inserted between the buffer rod and the sample (both polished with 1-μm diamond paste finish) to enhance mechanical bonding by smoothing the interface at high pressures. However, the reverberation of the acoustic energy inside the bonding layer causes an unwanted phase shift in the buffer rod and sample echoes. The perturbation by these phase shifts have to be removed from the measured travel times. Quantitative studies on this subject have been performed both experimentally and theoretically

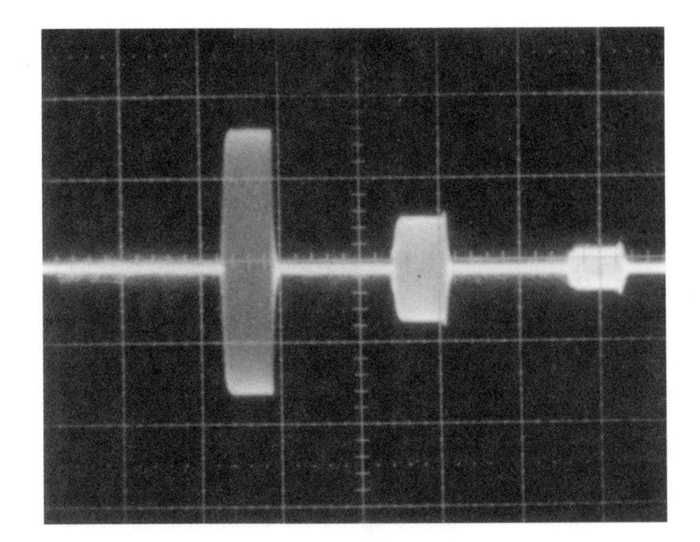

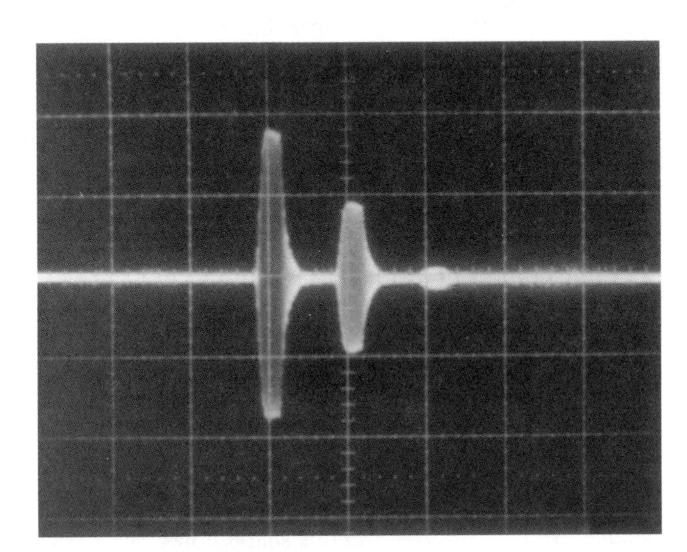

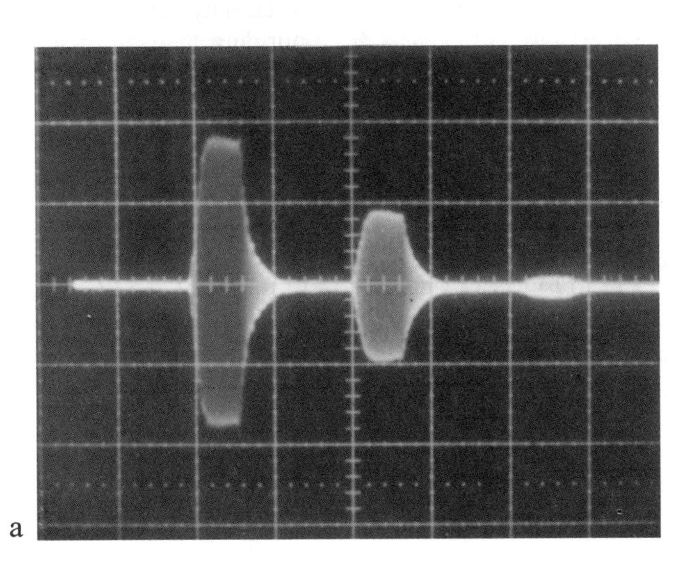

a

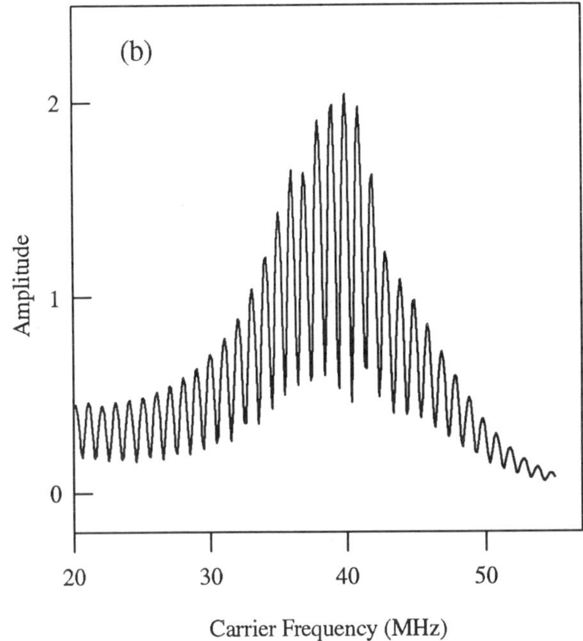

b

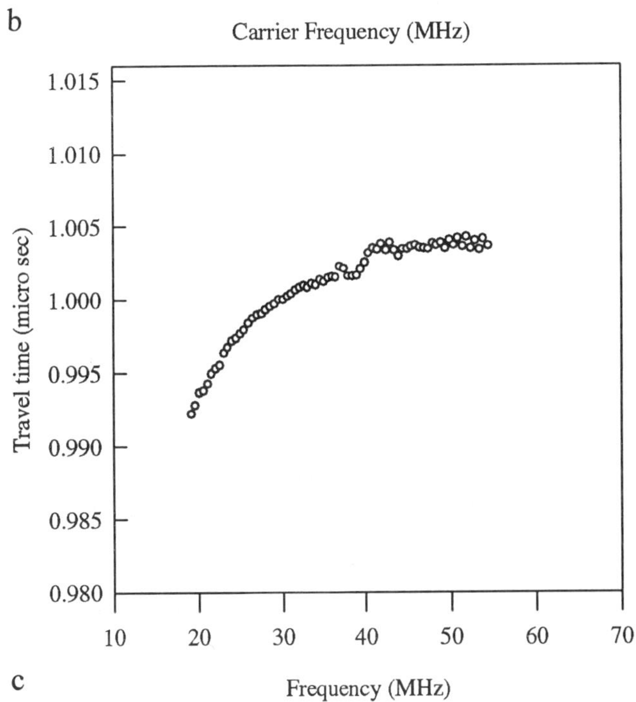

c

Figure 3. (a) Acoustic signals at different stages in an experiment for S wave velocity measurement on Lucalox alumina. Top: WC buffer rod echoes (B1, B2, B3) before pressurization (timescale = 5 μs/cm); Middle: WC buffer rod (B1) and sample echoes (S1, S2) at ~ 0.4 GPa (0.5 μs/cm); Bottom: WC buffer rod (B1) and sample (S1, S2) echoes at ~ 10 GPa (0.2 μs/cm).

(b) The interference spectrum between B1 and S1 shown in the bottom of (a) in the frequency range of 15 to 55 MHz.

(c) The travel times obtained from all the extrema in the interference spectrum of (b) from 20 to 55 MHz.

TABLE 1a. Elastic Properties of Lucalox Alumina

Pressure	Round-trip travel time		Density	Length	V_P	V_S	K_s	G
(GPa)	t_P (ms)	t_S (ms)	(g/cm^3)	(mm)	(km/s)	(km/s)	(GPa)	(GPa)
0.00	0.6120	1.0450	3.972	3.340	10.915	6.392	256.8	162.3
1.03	0.6081	1.0412	3.989	3.335	10.970	6.407	261.7	163.7
2.11	0.6035	1.0350	4.005	3.331	11.038	6.436	266.8	165.9
3.20	0.5992	1.0329	4.021	3.326	11.103	6.463	271.7	168.0
4.28	0.5949	1.0237	4.037	3.322	11.168	6.490	276.9	170.1
5.34	0.5913	1.0186	4.053	3.318	11.221	6.514	281.0	171.9
6.33	0.5880	1.0144	4.067	3.314	11.272	6.533	285.3	173.6
7.26	0.5848	1.0103	4.080	3.310	11.320	6.553	289.3	175.2
8.08	0.5822	1.0068	4.092	3.307	11.361	6.569	292.7	176.6
8.78	0.5798	1.0039	4.102	3.304	11.397	6.583	295.9	177.8
9.34	0.5776	1.0012	4.110	3.302	11.434	6.597	298.8	178.8
9.73	0.5769	0.9988	4.115	3.301	11.444	6.610	299.2	179.8

TABLE 1b. Comparison of Pressure Derivatives for Aluminum Oxide

Reference	Source of Specimen	Bench-top Velocity		P range	K_0'	G_0'
		V_P(km/s)	V_S (km/s)	(GPa)		
Schreiber and Anderson (1966)	Lucalox	10.845	6.373	0-0.4	3.98	1.74
Gieske and Barsch (1968)	Single Crystal	10.88*	6.40*	0-1.0	4.3	1.7
This work	Lucalox	10.92(3)	6.39(2)	0-9.7	4.44(2)	1.77(1)

*: Hashin-Shtrikman Averages.

[*Davies and O'Connell*, 1977, *Jackson et al.*, 1981]. Theoretically, the phase shift can be calculated by summing to infinity the reverberations within the bonding layer from the known thickness of the bond and the reflection/transmission coefficients at the interface determined by the acoustic impedances of buffer rod, bond material and sample. In this study, we have applied the same bond thickness (2 μm) corrections to the measured travel times at all pressures due to the negligible thickness change [estimated to be $l / l_0 \sim P / 3K$ (~ 2% at 10 GPa), where P is the pressure and K is the bulk modulus of gold (~ 166 GPa)].

VELOCITY MEASUREMENTS AT HIGH PRESSURE

Polycrystalline Lucalox Alumina

Alumina was chosen for our feasibility study [*Li et al., 1994, 1996b*] for the following reasons: (1) polycrystalline specimens of high acoustic quality were commercially available (GE Lucalox); (2) Lucalox has a high compressive strength (2.2 GPa); (3) elasticity data for both single crystals and Lucalox are available from hydrostatic experiments at low pressures (< 1 GPa) [see *Gieske and Barsch, 1968; Schreiber and Anderson, 1966*; respectively]; and (4) alumina is a very incompressible material (K_0 ~ 254 GPa) and does not undergo any phase transitions in the pressure range of this study. As such, it is a very suitable material for these ultrasonic experiments in a multianvil apparatus.

Lucalox is a translucent polycrystalline material with a bulk density of 3.972(2) g/cm^3 (or 99.6% of the X-ray density of 3.986 g/cm^3). Four specimens with diameters of 2.8-2.9 mm diameter were cored in Australian National University. Examination of the Lucalox ceramic by scanning electron microscopy (SEM) reveals a highly equilibrated texture with an average grain size of 20-30 μm, and a very small amount of spinel (approximately

$MgAl_2O_4$ composition) on the alumina grain boundaries [I. Jackson, personal communication, 1994]). The bench-top velocity measurements yield V_P = 10.92(3) km/s and V_S = 6.39(2) km /s (see Table 1) which are within 0.5% of the Hashin-Shtrikman (HS) averages of the single crystal data of Gieske and Barsch [1968] and earlier results on polycrystalline Lucalox alumina by Schreiber and Anderson [1966].

The most robust estimates of the pressure dependence of the travel times are obtained by averaging all data between 45 and 65 MHz where they are relatively frequency insensitive. The travel times (corrected for bond effect) and calculated elastic properties to 9.7 GPa for Lucalox are tabulated in Table 1. The overall travel-time decrease is about 8% for P wave and 5% for S wave for a pressure variation from ambient conditions to ~10 GPa. Using the internally consistent method developed by Cook [1957] based on the observed travel times for P and S waves versus pressure, as shown in Table 1a, the length change at ~10 GPa is calculated to be 1% (see Table 1a). Consequently, the actual change in wave speed for the Lucalox alumina is ~ 7% for P waves and ~ 4% for S waves over the pressure range investigated.

The new velocity results are also compared with extrapolations from low pressure measurements ($P \leq 1$ GPa) on single crystals [*Gieske and Barsch, 1968*] and polycrystals [*Schreiber and Anderson, 1966*]. The Voigt, Reuss, VRH, and HS bounds are calculated using the linear extrapolation of single-crystal elastic constants C_{ij} in conjunction with densities determined from the Murnaghan equation of state. Excellent agreement has been achieved between this study and the previous studies for pressure dependence of both compressional (P) and shear (S) wave velocity [see *Li et al., 1996b*].

The bulk (K_S) and shear (G) moduli for Lucalox alumina are calculated from the measured P and S wave velocities and compared with single crystal data of Gieske and Barsch [1968] in Figure 4. The bulk modulus from our study is about ~ 1% higher than the HS bounds calculated from single crystal data of Gieske and Barsch [1968]. The shear modulus is almost identical in the pressure range of this study. Both K_S and G show linear pressure dependences to 10 GPa and first order fits yield dK_S/dP = 4.4 and dG/dP = 1.8 which are in good accord with the data from single-crystal data of Gieske and Barsch [1969] and polycrystals data of Schreiber and Anderson [1966] (Table 1b).

The "sample friendly" cell assembly enables us to perform multiple measurements on a single sample. Upon recovery from each high-pressure experiment, these specimens were carefully examined, including

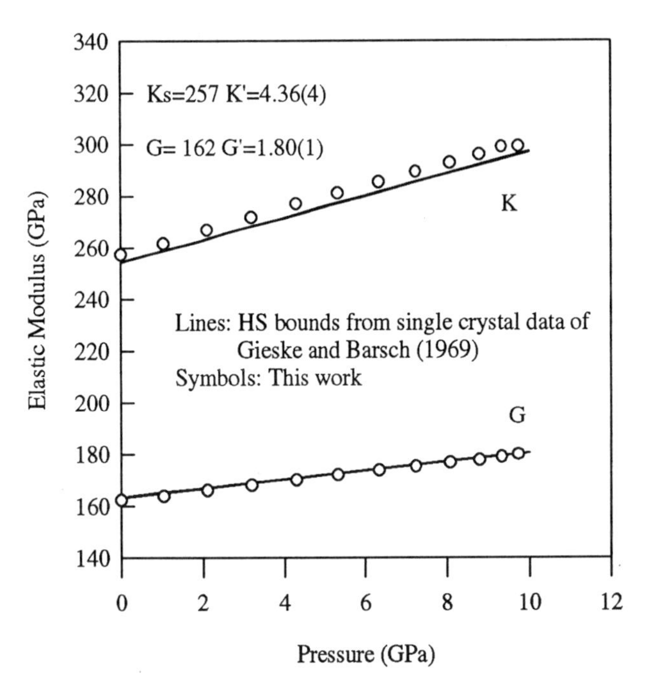

Figure 4. The bulk (K_S) and shear (G) moduli for Lucalox alumina calculated from the measured compressional and shear wave velocities. Symbols are results from this study and the lines are Hashin-Shtrikman bounds calculated from single crystal data of Gieske and Barsch [1968] and their extrapolations.

measurement of dimensions, flatness and wave speeds at room conditions. No noticeable changes in dimensions have been observed even after multiple measurements. Sometimes, minor repolishing of the sample surface is performed to remove residual gold foil. The results of bench-top velocity measurements are reversible. Slight indentations on the WC cube caused by the steel and the sample can be barely seen by the naked eye but are easily recognized from the interference pattern obtained with an optical flat. The overall reproducibility in travel time for duplicate high pressure runs is about 0.3% [see *Li et al., 1996b*, Figure 9].

The velocities at $P < 2$ GPa seem to be lower than the trend of the data at high-pressure and room conditions [see *Li et al., 1996b*, Figure 10]. This feature has also been noticed and reported in high-pressure velocity measurements for a variety of polycrystalline samples [*Gwanmesia et al., 1990b; Rigden et al., 1988, 1991, 1992*]. From the observation of the amplitude ratio between the buffer rod and the sample (B1/S1) and the travel time with increasing pressure, Rigden et al. [1992] have demonstrated that the pressures at which the amplitude ratio B1/S1 stabilizes correlate very well with the pressures above which the best fittings are obtained in travel times; i.e., when the amplitude ratio B1/S1 remains essentially

TABLE 2. Specimens Used for Acoustic Velocity Measurement

Sample	Hot-pressing P and T	Density (g/cm^3)	Vp (km/s)	Vs (km/s)
α-Mg2SiO4 (#2404)	12 GPa 1200 °C	3.203(6)	8.51(2)	4.99(1)
α-Mg2SiO4 (#1290)	12 GPa 1200 °C	3.210(6)	8.53(2)	5.01(1)
α-Mg2SiO4 (K211)	12 GPa 1200 °C	3.210(6)	8.51 (2)	4.99(1)
β-Mg2SiO4 (#1188)	14 GPa 1000 °C	3.440(6)	9.55(2)	5.60(1)

unchanged with increasing pressure, obvious deviation in travel times from the fitted mean disappears. Similar observation have been made in our current studies but at slightly higher threshold pressures. The threshold pressures are also sample-size dependent, being higher for smaller samples which are surrounded by thicker lead and teflon disks (e.g., beta phase #1188, later in this paper).

Olivine (α) and Beta (β) Polymorphs of Mg_2SiO_4

In the previous US-Japan High Pressure seminar [*Gwanmesia and Liebermann, 1992*], we reported major advances in hot-pressing polycrystalline aggregates of high-pressure phases of mantle minerals, which provided high-quality specimens for velocity measurements to 3 GPa for the beta and spinel phases of Mg_2SiO_4, pyrope-majorite garnets and stishovite [*Gwanmesia et al., 1990b, Rigden et al., 1991, 1992, 1994; Li et al., 1996a*]. The availability of these crack- and pore-free, fine-grained specimens has been critical to the success of the current generation of velocity studies in multianvil apparatus.

The polycrystalline samples used in this study were hot-pressed in a uniaxial split-sphere apparatus (USSA-2000): α-Mg_2SiO_4 (#2404) at P=12 GPa and T=1200°C and β-Mg_2SiO_4 (#1188) at P =15 GPa and T=1000 °C using the sintering techniques developed by Gwanmesia et al. [1990a, 1993]. A summary of the sample information of these specimens is given in Table 2. The diameters of the α-Mg_2SiO_4 and β-Mg_2SiO_4 samples are 2.7 and 2.5 mm, respectively. Transmission electron microscopy (TEM)

TABLE 3. Elastic Properties of Olivine Mg_2SiO_4

Pressure (GPa)	Round-trip Travel Time t_P (ms)	t_S (ms)	Density (g/cm^3)	Length (mm)	V_P (km/s)	V_S (km/s)	K_S (GPa)	G (GPa)
0.00	0.5710	0.9775	3.203	2.440	8.546	4.992	127.5	79.8
1.21	0.5636	0.9688	3.233	2.432	8.629	5.020	132.1	81.4
2.46	0.5526	0.9581	3.266	2.424	8.771	5.059	139.8	83.5
3.63	0.5446	0.9485	3.296	2.416	8.873	5.094	145.4	85.5
4.74	0.5376	0.9402	3.323	2.410	8.964	5.125	150.5	87.3
5.77	0.5316	0.9327	3.348	2.404	9.042	5.154	155.1	88.9
6.74	0.5266	0.9268	3.370	2.398	9.108	5.175	159.2	90.2
7.63	0.5216	0.9216	3.391	2.393	9.176	5.194	163.5	91.4
8.45	0.5176	0.9170	3.409	2.389	9.231	5.210	167.1	92.5
9.20	0.5136	0.9130	3.426	2.385	9.287	5.224	170.8	93.5
9.88	0.5106	0.9099	3.441	2.382	9.328	5.234	173.7	94.2
11.02	0.5056	0.9052	3.466	2.376	9.398	5.249	178.7	95.5
11.88	0.5021	0.9012	3.484	2.372	9.447	5.264	182.2	96.5
12.50	0.4991	0.8984	3.497	2.369	9.492	5.274	185.4	97.2

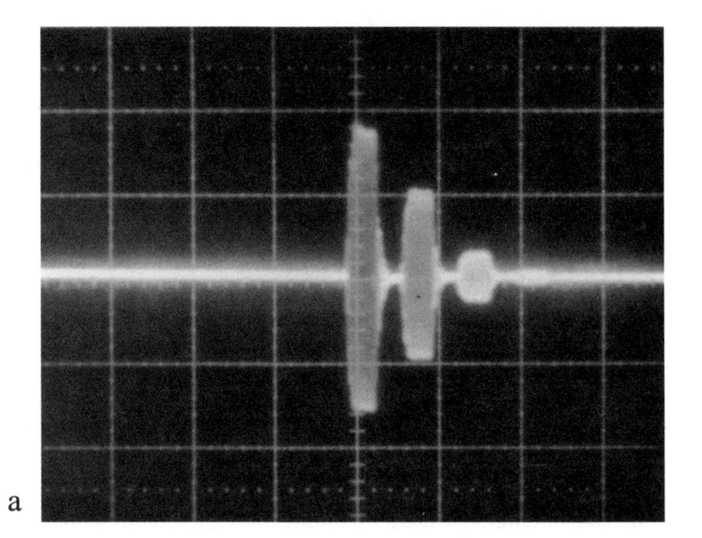

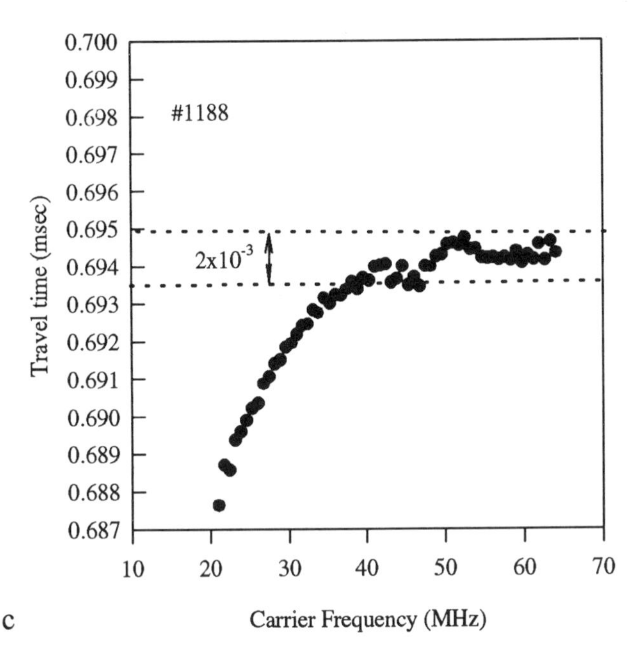

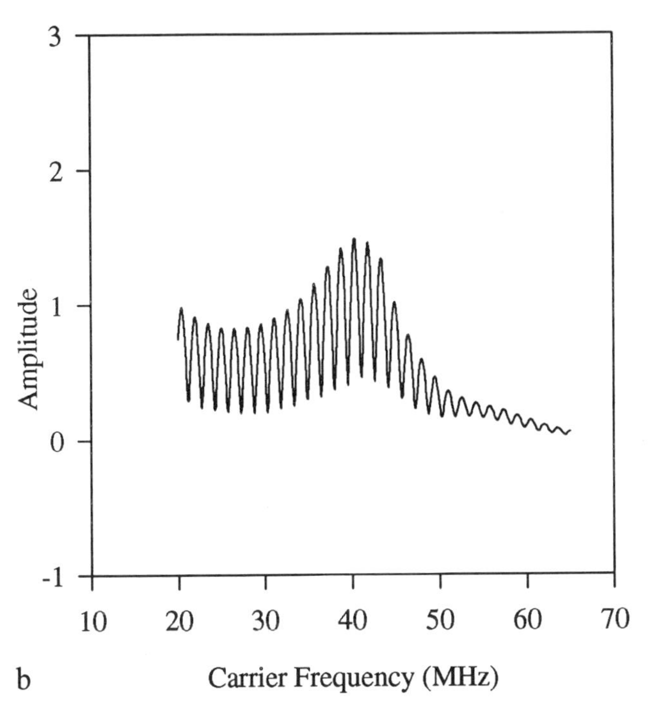

Figure 5. (a) The S wave acoustic signals at ~ 1.5 GPa for a polycrystalline specimen of the beta phase of Mg_2SiO_4 (#1188). Three echoes from left to right are the first WC buffer rod (B1), first sample (S1) and the second sample (S2) echoes. The sample length is 1.94 mm and the time scale is 1.0 μs/cm.
(b) The interference spectrum between B1 and S1 in (a) from 20 to 65 MHz.
(c) The calculated travel times from all the extrema in the interference spectrum in (b).

studies on similar specimens of the beta phase [see *Gwanmesia et al., 1990a*] show that the grain size is around 3-5 microns with an equilibrated texture and 120° triple junctions. The densities of these single-phased specimens are within 1% of the theoretical values and are acoustically isotropic with high quality acoustic signals at both room conditions and high pressures. The P and S wave velocities at room conditions are within 2% of the Hashin-Shtrikman average of single crystal data. An example of the acoustic signals at P ~ 1.5 GPa is illustrated in Figure 5a. Three echo trains are the first buffer-rod echo (B1), the first sample (S1), and the second sample echo (S2), respectively. There were no sidewall reflections or any other spurious reflections between buffer-rod and sample echoes, and the quality of the acoustic echoes is preserved at peak pressure (~ 12.5 GPa) of this study. The interference pattern between the B1 and S1 signals is shown in Figure 5b. The high signal-to-noise ratio and amplitude ratio S1/B1 assure a precise travel-time deduction from the interference spectrum of B1 and S1 (0.2%) with similar dispersion characteristics as in alumina data (Figure 5c).

α-Mg_2SiO_4. P and S wave experiments on specimen #2404 were carried out to P >12 GPa in separate runs and the results are shown in Table 3. The P and S wave velocities from our new measurements increase more slowly than the finite strain extrapolations of low-pressure data of Kumazawa and Anderson [1969] (0-0.2 GPa) and Graham and Barsch [1969] (0-1.0 GPa), differences reaching about 1% for P wave and ~ 2% for S wave at 12.5 GPa [see *Li et al., 1996c*, Figure 1]. However, they agree with the recent single-crystal data from ultrasonic (0-6 GPa) and Brillouin scattering (3.1-16.2 GPa) studies (Figure 6) [see also *Li et al., 1996c, Duffy et al., 1995; Zha*

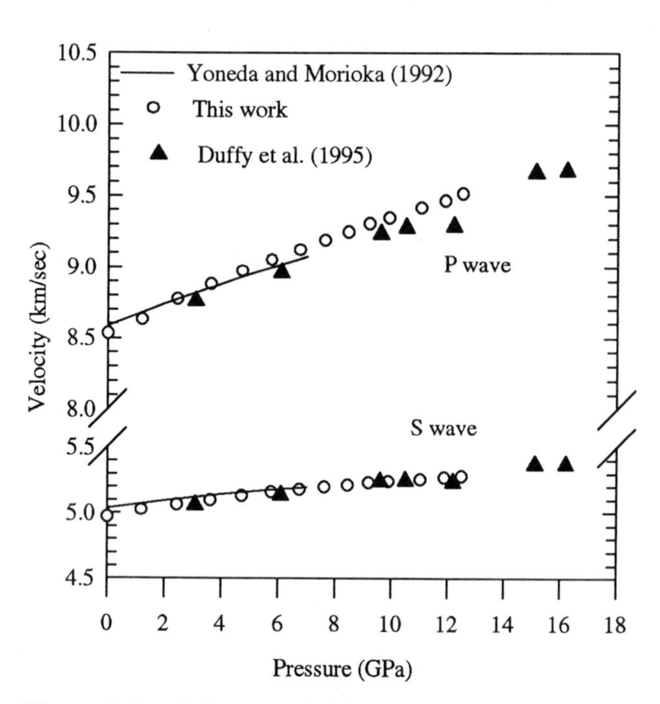

Figure 6. P and S wave velocities measured in this study and comparison with single crystal data from ultrasonic and Brillouin scattering techniques. The open circles are the experimental results from this study, the solid triangles are the Brillouin scattering data of Duffy et al. [1995], and the solid lines are calculated from ultrasonic measurements on single crystal forsterite of Yoneda and Morioka [1992].

et al., 1996; Yoneda and Morioka, 1992]. Considering the difference in acoustic techniques and source and form of specimens, there is remarkable consistency in these data, especially for P < 10 GPa. At higher pressures, the Brillouin scattering data exhibit larger scatter (especially in the compressional wave), but the Duffy et al. [1995] /Zha et al. [1996] data at 15-16 GPa agree with the extrapolation of our data and those of Yoneda and Morioka [1992]. The velocities at room conditions were measured and compared with the velocities before and after each high-pressure run; a reproducibility better than 0.2% is obtained.

The elastic bulk (K_S) and shear (G) moduli for α-Mg_2SiO_4 calculated from the measured P and S wave velocity are also tabulated in Table 3. Our ultrasonic moduli exhibit systematic increases of both K_S and G with pressure (44% and 21% for K_S and G at 12.5 GPa) and are characterized by smaller uncertainties than Brillouin scattering data. Linear fits of K_S and G versus pressure yield K_0'= 4.4 and G_0'= 1.3 for the current experiments. These values are distinctly lower than those from the earliest ultrasonic measurements of the pressure derivatives on forsterite olivine, i.e., K_0' = 5.1 and G_0'=1.8 at P < 1 GPa from Graham and Barsch [1969] and K_0' =4.8 and G_0' = 1.8 from Kumazawa and Anderson [1969] at P < 0.2 GPa. Compared to the other high pressure studies, our K_0' value is higher while our G_0' is lower than the others; clearly such behavior cannot be ascribed to the polycrystalline nature of our specimen.

β-Mg_2SiO_4. Duplicate measurements of both P and S wave velocities up to 9 GPa and 12.5 GPa were made with Bi and ZnTe as pressure markers, respectively. The results to the highest pressures are listed in Table 4. Although the data at P < 2 GPa may be affected by an inadequate bond between the buffer rod and sample as discussed earlier, these new velocity data agree very well with the data obtained on similar specimens up to 3 GPa by Gwanmesia et al. [1990b] using the same techniques but in a piston cylinder apparatus with hydrostatic pressure medium [see *Li et al., 1996c*, Figure 4]. Finite strain extrapolations of the Gwanmesia et al. [1990] data agree with our results within 1% for both P and S waves to the highest pressure of this study (12.5 GPa). Our data yield V_P = 10.30 km/s and V_S = 5.88 km/s at 12.5 GPa, which compare very well with V_P = 10.36 km/s and V_S = 5.95 km/s calculated from the extrapolation of the Gwanmesia et al. [1990b] data.

The elastic moduli for β-Mg_2SiO_4 have been calculated from the velocity data for the highest pressure run and are listed in Table 5 at each pressure. These new data are compared in Figure 7 with the extrapolations using the experimental results of K_0, K', G_0 and G' of Gwanmesia et al. [1990b]. The two experimental data sets agree with each other within ~2 % and ~ 4% to 12.5 GPa for K_S and G, respectively. As fitted with a linear function of pressure, our new data yield K_0' = 4.2 and G_0' = 1.5. These values are lower than those of the previous study to 3 GPa on similar polycrystalline specimens; this difference is reduced if we fit our new data by the finite strain method (yielding K_0' = 4.5 and G_0' = 1.6) to accommodate the slight curvature in our data at low pressures. Note the excellent agreement at zero pressure of the elastic moduli of our polycrystalline specimens (<1% X-ray density) with the Hashin-Shtrikman average of the single-crystal moduli of Sawamoto et al. [1984] (Table 5).

Single Crystals of Forsterite and San Carlos Olivine

These ultrasonic techniques in multianvil apparatus have also been successfully applied to study the elasticity of olivine single crystals to 13 GPa [see *Chen et al., 1996*]. The goals of these measurements are to demonstrate the feasibility of performing such experiments on single crystal samples and to re-examine the unusual behavior of the shear modulus C_{55} vs. pressure in San Carlos olivine [*Zaug*

TABLE 4. Elastic Properties of Beta Mg_2SiO_4

Pressure (GPa)	Round-trip Travel Time t_P (ms)	t_S (ms)	Density (g/cm^3)	Length (mm)	Vp (km/s)	Vs (km/s)	K_S (GPa)	G (GPa)
0.00	0.3780	0.6460	3.440	1.805	9.550	5.5960	170.1	107.7
1.20	0.3790	0.6414	3.465	1.801	9.502	5.6145	167.2	109.2
2.41	0.3727	0.6356	3.490	1.796	9.640	5.6530	175.6	111.5
3.55	0.3672	0.6303	3.512	1.793	9.762	5.6876	183.2	113.6
4.73	0.3633	0.6253	3.535	1.789	9.848	5.7214	188.5	115.7
5.74	0.3602	0.6212	3.554	1.786	9.914	5.7484	192.7	117.4
6.60	0.3573	0.6180	3.570	1.783	9.979	5.7698	197.1	118.8
8.31	0.3521	0.6121	3.601	1.778	10.097	5.8088	205.1	121.5
9.07	0.3501	0.6097	3.614	1.776	10.144	5.8244	208.4	122.6
9.76	0.3484	0.6076	3.626	1.774	10.181	5.8376	211.1	123.6
10.39	0.3467	0.6059	3.637	1.772	10.220	5.8488	214.0	124.4
10.96	0.3457	0.6044	3.647	1.770	10.242	5.8581	215.7	125.1
11.89	0.3441	0.6021	3.662	1.768	10.274	5.8721	218.2	126.3
12.57	0.3428	0.6005	3.674	1.766	10.302	5.8810	220.5	127.1

et al., 1993]. Cylindrical samples 2-3 mm in diameter and 1-2 mm in length were polished to optical quality. The San Carlos olivine samples were oriented by the X-ray diffraction precession method to better than 1°. The forsterite sample was cut from the [100] direction sample (oriented by Laue method to ~1°) used by Yoneda and Morioka [1992]. Examination of a recovered San Carlos olivine sample after it was subjected to ~10 GPa revealed that there were no internal cracks and no increase in the dislocation density in the single crystal sample [J. Ando, personal communication, 1996].

Figure 8a shows the acoustic echo pattern observed at 10 GPa for the S-waves (polarized in [001]) in forsterite using a 40 MHz $LiNbO_3$ transducer. The first echo is the

TABLE 5. The Elastic Moduli of Olivine and Beta Phase Polymorphs of Mg_2SiO_4

Mineral	Reference	P range (GPa)	K_0 (GPa)	K_0'	G_0 (GPa)	G_0'
α-Mg_2SiO_4	This work	12.5	128	4.44(2)	80	1.32(3)
	Duffy et al (1995) / Zha et al., (1996)	16.2	129	4.2	81	1.4
	Yoneda and Morioka (1992)	6.0	128.7	4.1[a]	81.7	1.6
	Kumazawa and Anderson (1969)	0.2	129.1	5.37	81.6	1.80
	Graham and Barsch (1969)	1.0	128.6	5.13	81.1	1.82
β-Mg_2SiO_4	This work	12.5	170	4.24(10)	108	1.49(3)
	Gwanmesia et al. (1990)	3.0	163	4.8	110	1.7
	Sawamoto et al.(1984)	Room	177[b]	--	112[b]	--

a: Refit with first order in pressure.

b: Hashin-Shtrikman averages.

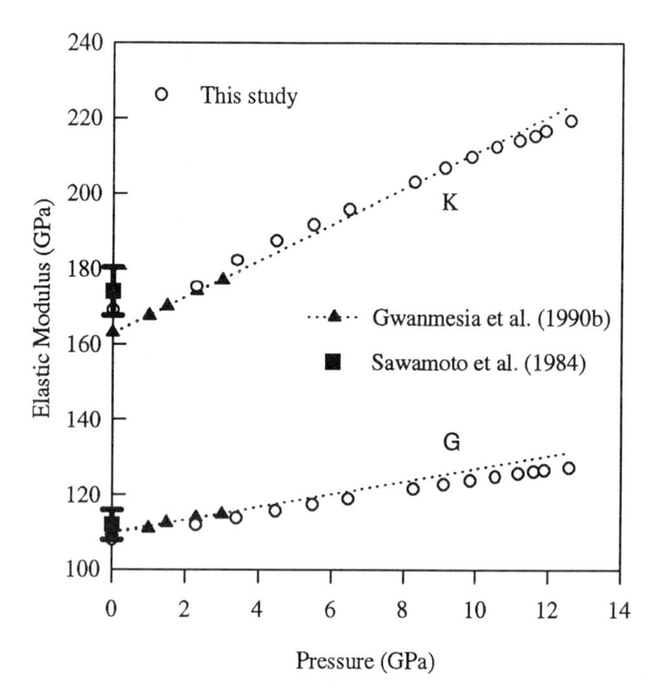

Figure 7. Comparison of the calculated bulk and shear modulus for beta phase of Mg_2SiO_4 as a function of pressure from this study with extrapolations from measurements on similar specimen to 3 GPa by Gwanmesia et al. [1990b] using ultrasonic interferometry. The open circles are experimental data from this study, the solid triangles are the experimental data of Gwanmesia et al. [1990b]; and the solid squares are the room condition measurements of Sawamoto et al. [1984] using Brillouin scattering techniques.

reflection from the end of the WC cube and the ensuing echoes are internal reflections within the sample; one can see three sample reflections. In Figure 8b, we illustrate the sharpness of the interference patterns between the WC buffer rod echo and the first sample echo for the S-waves in San Carlos olivine at 13 GPa. The travel time vs. frequency data from the amplitude data in Figure 8b are shown in Figure 8c; from 15 to 30 MHz the travel time increases and then from 30-60 MHz attains a steady value ($\pm$0.15%). A demodulation algorithm has been applied to the interference amplitude data to remove the amplitude trend effected by the electronic components and the transducer response (the resultant amplitude spectrum is shown in the inset in Figure 8b; details see Spetzler et al., [1993]).

Travel-time data and calculated elastic moduli versus pressure are given in Table 6 for the C_{55} mode in forsterite and for the C_{22} and C_{55} modes in San Carlos olivine. The conversion procedure from acoustic travel times to elastic moduli is outlined in Chen et al. [1996]. Our C_{55} data for forsterite agree well with the earlier ultrasonic study of Yoneda and Morioka [1992] to 6 GPa and the recent Brillouin scattering data of Duffy et al. [1995; see also Zha et al., 1996] to 16 GPa. Most significantly, C_{55} is a linear function of pressure in all three studies.

For San Carlos olivine, we measured first the longitudinal modulus C_{22} and obtained good agreement with the ultrasonic data of Webb [1989] to 3 GPa and the impulsive stimulated scattering (ISS) data of Zaug et al. (1993) to 12.5 GPa. Our new data for C_{55} versus pressure for San Carlos olivine are shown in Figure 9, together with those of Zaug et al. [1993] and Webb [1989]. To clarify the behavior of this mode, we carried out measurements on two different samples. One sample is oriented parallel to the [100] axis and the other sample is oriented parallel to the [001] axis. The first experiment was made using the [001] orientation sample (polarization of S-wave [100]) to 9.7 GPa, just above the first ZnTe phase transition and the second experiment was made using the [100] orientation (polarization of S-wave [001]) to ~13.5 GPa, above the second ZnTe phase transition. The two experiments yield consistent results for C_{55} of San Carlos olivine as a function of pressure; this redundancy provides a cross-check on the C_{55} behavior and demonstrates that neither measurement was contaminated by misorientation of the transducer and the sample. The new data are in excellent agreement with Webb's results to 3 GPa. In the low pressure range (to ~ 8 GPa), Zaug et al.'s measurements also agree with our new data; however, their data above 9 GPa deviate markedly from our results and imply a downward curvature of C_{55} versus pressure. Comparing the results at P = 13.5 GPa (equal to 410 km depth), our value of C_{55} is about 6% higher than that of Zaug et al. [1993].

The results from Webb, Zaug et al., and Duffy et al. suggest that the major disagreement for the pressure effect on the elasticity of olivine is in the C_{55} mode. Table 7 summarizes our new results on C_{55} of forsterite and San Carlos olivine and also the results from previous studies. Neither the Brillouin spectroscopy nor our new ultrasonic data exhibit the nonlinear dependence of C_{55} on pressure observed in the ISS data of Zaug et al. [1993].

ULTRASONICS AT SIMULTANEOUS HIGH P AND T

In the previous section, we have presented new data for the elastic wave velocity for olivine and the beta phase to P> 12 GPa at room temperature on millimeter-sized polycrystalline and single crystal specimens using ultrasonic interferometry in a multianvil apparatus. These measurements eliminate the necessity and the potential uncertainties in a few-fold extrapolation in pressure to reach conditions of the mantle transition zone.

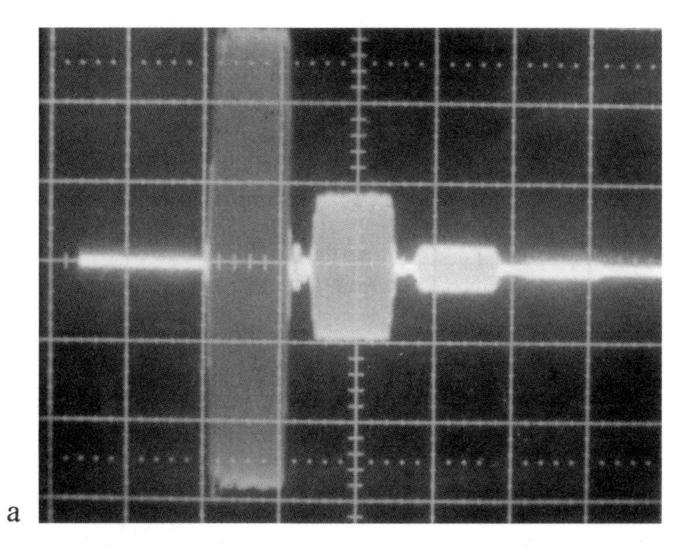

a

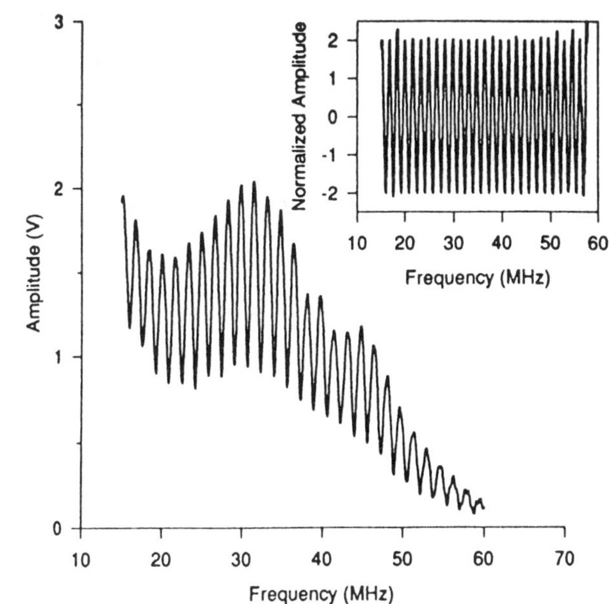

b

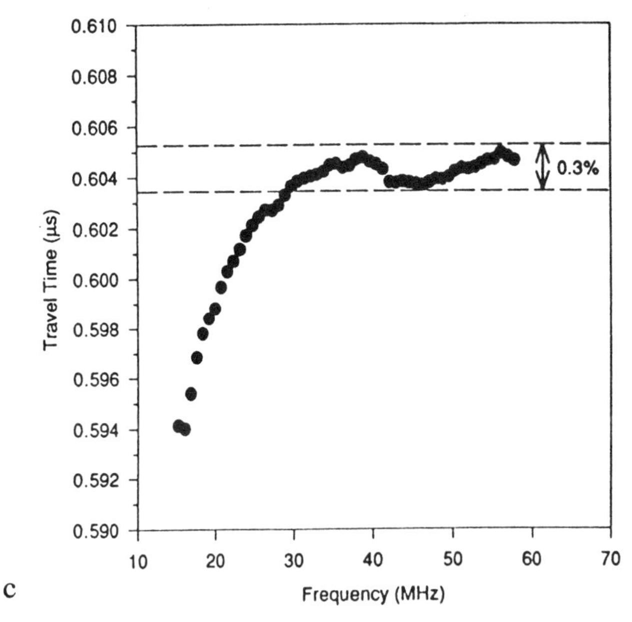

c

Figure 8. (a) S-wave acoustic signal for the single-crystal forsterite [100] sample (s-wave polarizing in [001]) at ~10 GPa. The four echoes from left to right are the WC cube buffer rod echo, first sample echo, second sample echo, and third sample echo.
(b) S-wave interference pattern from the single-crystal San Carlos olivine [100] sample (s-wave polarizing in [001]) at ~13.5 GPa. The inset shows the same interference data after applying a demodulation procedure to the raw data.
(c) Travel time data from the reduced interference data shown in the inset of (b).

However, without direct measurements of elastic wave velocities at simultaneous high pressure and temperature, comparisons with seismic models are still fraught with ambiguity. For instance, attempts to match the olivine-beta phase transformation with the observed seismic discontinuity at 410 km depth to infer the olivine content of the mantle have resulted in widely scattered results, ranging from 27-44% [*Duffy et al., 1995*] to 45-65% [*Gwanmesia et al., 1990b*], even when new measurements of dK_S/dT for the beta phase by Fei et al. [1992] and Meng et al. [1993] are incorporated. Obviously, the solution to this dilemma is to directly measure the velocities of the relevant mantle phases at the P and T conditions of the Earth's transition zone. Such experimental work is under development in our laboratory and others [*Li et al., 1995b; Knoche et al., 1995*]. In this section, we report the implementation of such experiments and present some preliminary results for polycrystalline forsterite.

Cell Assembly

The cell assembly for the USCA-1000 multianvil apparatus (Figure 1) has been modified to enable velocity measurements at simultaneous high pressures and temperatures to be made by inserting a cylindrical graphite furnace with caps at both ends into an octahedron of semi-sintered MgO (Figure 10). The furnace is surrounded by a sleeve and end plugs of zirconia which serve as thermal insulators to surround the sample; two MgO washers fill the remaining space inside the furnace. Holes in the zirconia end plugs accommodate an acoustic buffer rod and thermocouple feedthroughs. On one end, an alumina rod (Coors brand 998 - 3.2-mm diameter, 3.9-mm length) serves as another acoustic buffer rod; it is surrounded by metal foil (such as Pt or Cu) to provide electrical contact between the WC cube and the graphite

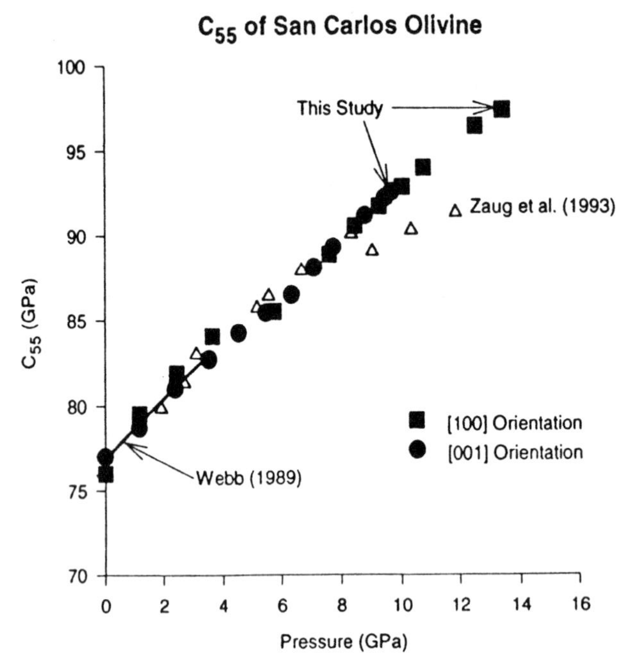

Figure 9. Comparison of C_{55} mode of San Carlos olivine from this study with data up to 12.5 GPa of Zaug et al., [1993] from the impulsive stimulated scattering (ISS) technique and ultrasonic data of Webb [1989] to 3 GPa.

furnace. On the other end of the cell, a piece of two-hole mullite tubing (1.7-mm diameter) contains the thermocouple wires (W3%Re/W25%Re); a Pt (or TZM) ring outside the mullite serves as the other furnace electrode. To protect the sample, the thermocouple junction does not touch the sample, but instead is located at the bottom of the NaCl cup in a position symmetrical to that of the sample/buffer rod interface. The high thermal conductivity of NaCl should reduce the temperature gradient across the 1.5-mm long sample. From measurements in a hot-pressing cell of similar design, we estimate the temperature gradient in the sample to be ~15°C/mm [*Gwanmesia and Liebermann, 1992*].

As in the room temperature experiments, the piezoelectric transducer is mounted onto the stress-free corner of the WC cube (with Aremco Crystalbond in the initial experiments, and later with high temperature epoxy 350ND from Epoxy-Tech). Both surfaces of the WC cube, the alumina rod, and the sample are polished with 1-μm diamond paste finish and the estimated parallelism is about 250 nm from previous studies [Rigden et al., 1988]. Gold foil (2.0 μm thick) is inserted between the WC cube and the alumina buffer rod and between the alumina and the sample. Alumina provides a good impedance contrast

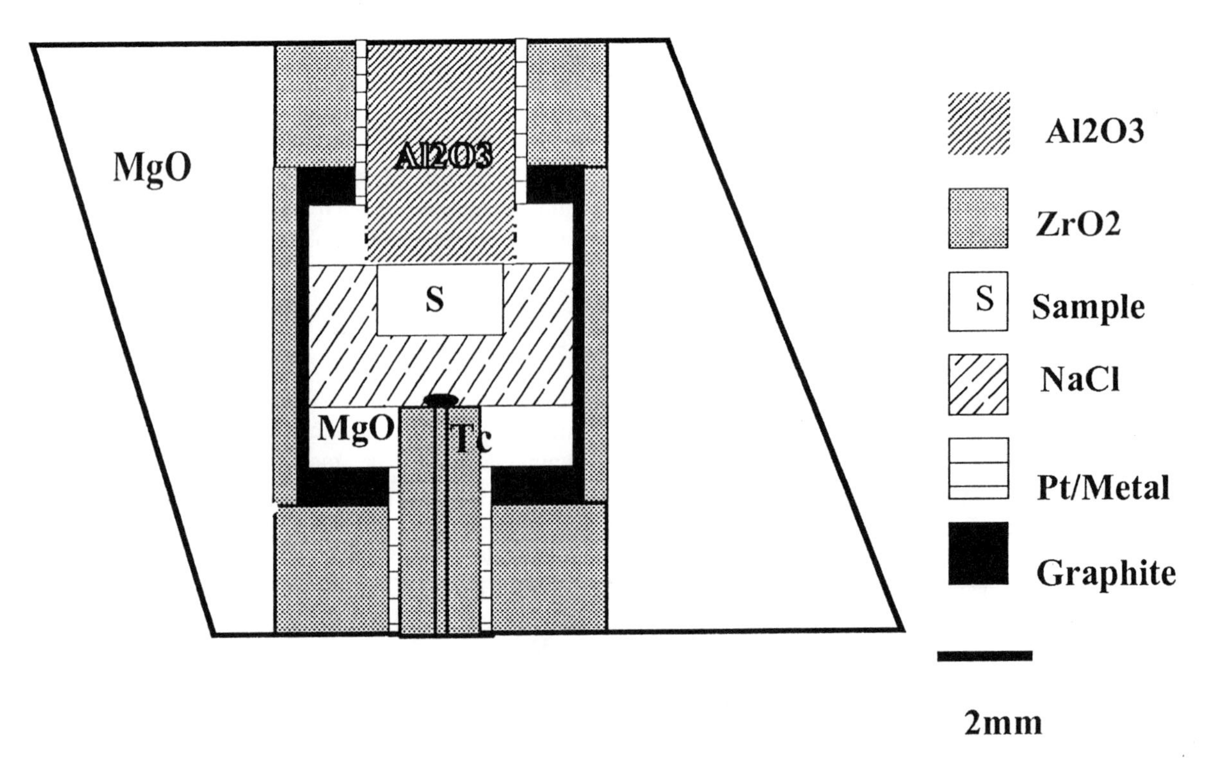

Figure 10. Modified cell assembly for simultaneous high pressure and temperature velocity measurements. W3%Re/W25%Re thermocouple wires are used.

TABLE 6. Travel Time and Modulus Data From Measurements on Single Crystal Olivines

(a) Olivine [010] P-wave

Pursuer (GPa)	Travel Time (ms)	C_{22} (GPa)
0.0	275	196
1.0	274	198
1.9	270	204
2.8	265	211
3.7	261	217
4.7	258	223
5.6	256	228
6.5	253	232
7.3	252	235
8.2	250	239
9.1	248	242
Uncertainty	2	0.5

(b) Forsterite [100] S-wave polarizing [001]

Pressure (GPa)	Travel Time (ns)	C_{55} (GPa)
0.0	686	81.9
1.3	682	83.1
2.5	675	85.3
3.7	670	87.2
4.8	664	89.1
5.8	660	90.4
6.7	657	91.6
7.5	656	92.2
8.3	653	93.4
9.0	650	94.3
9.7	648	95.2
Uncertainty	2	0.5

(c) Olivine [001] S-wave polarizing [100]

Pursuer (GPa)	Travel Time (ns)	C_{55} (GPa)
0.0	943	77.0
1.2	932	78.7
2.4	921	81.0
3.5	913	32.7
4.5	905	84.2
5.5	899	85.5
6.3	894	86.5
7.1	888	88.1
7.8	883	89.3
8.8	876	91.2
9.5	871	92.2
9.7	869	92.6
Uncertainty	2	0.5

(d) Olivine [100] S-wave polarizing [001]

Pursuer (GPa)	Travel Time (ns)	C_{55} (GPa)
0.0	668	76.0
1.2	655	79.5
2.4	647	81.9
3.6	640	84.1
5.7	637	85.5
7.6	627	88.9
8.5	622	90.6
9.3	619	91.7
10.1	616	92.9
10.8	613	94.0
12.5	607	96.5
13.5	605	97.4
Uncertainty	2	0.5

TABLE 7. Summary of C_{55} for Olivine and Forsterite as Function of Pressure

(a) Natural San Carlos Olivine -$(Mg_{0.9}Fe_{0.1})_2SiO_4$

Source	C_{55} ambient	$\partial C_{55}/\partial P$	Comment
This Study*	77.2 GPa	1.56(3)	to P>12 GPa
Webb (1989)	76.9	1.62	Refit with straight line (to 3 GPa)
Zaug et al. (1993)	77.0 GPa	2.18	$\partial^2 C_{55}/\partial P^2 =$ -0.16 /GPa
Kumazawa and Anderson (1969)	76.9 GPa	1.80	to 0.2 GPa

*: Linear fit to the two measurement data sets.

(b) Synthetic Pure Forsterite (Mg_2SiO_4)

Source	C_{55} (ambient)	$\partial C_{55}/\partial P$	Comment
This study	81.9 GPa	1.41(4)	to ~ 10 GPa
Duffy et al. (1995)/ Zha et al. (1996)	--	1.38	to 16 GPa
Yoneda and Morioka (1992)	81.2 GPa	1.40	Re-fit with straight line (to 6 GPa)
Bassett et al. (1982)	83.8 GPa	1.50	to 4 GPa
Graham and Barsch (1969)	81.4 GPa	1.65	to 1 GPa
Kumazawa and Anderson (1969)	78.1 GPa	1.64	to 0.2 GPa

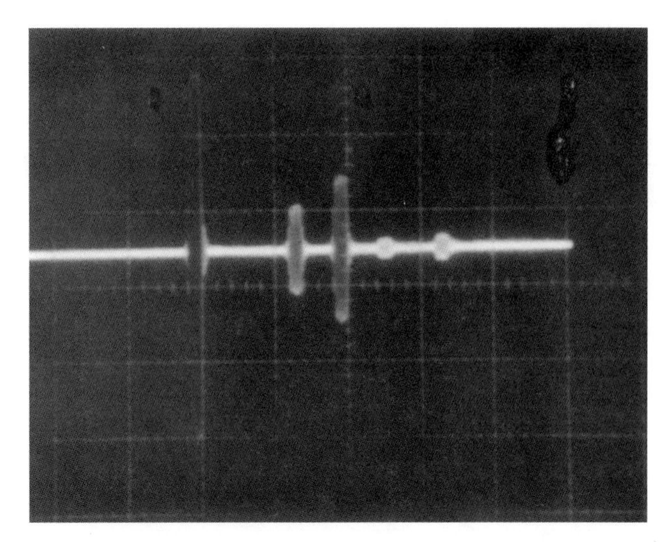

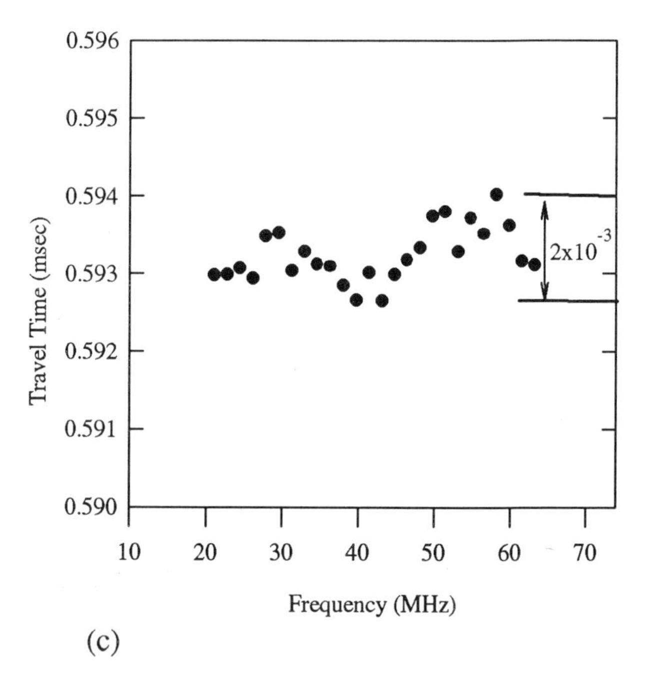

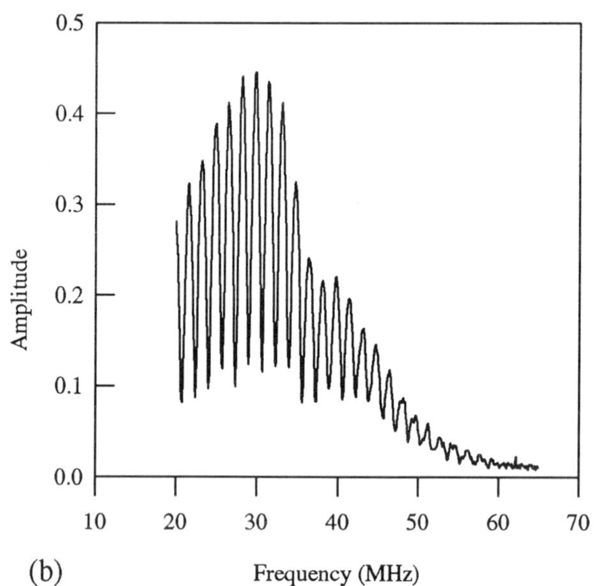

Figure 11. (a) The S wave acoustic signals at ~ 2 GPa 400 °C for polycrystalline forsterite sample #1290. Three strong echoes from left to right are the first WC buffer rod (B0), first alumina buffer rod (B1), first sample (S1) echoes . The weak signals are the second alumina buffer rod, sample and NaCl echoes.
(b) The interference spectrum between the alumina buffer rod (B1) and the first sample (S1) in (a) from 20 to 60 MHz.
(c) The travel times derived from the interference extrema in (b) as a function of frequency.

between the WC cube and the sample to ensure a high signal-to-noise ratio of the acoustic signals.

For our pilot studies at high P and T, we used two polycrystalline specimens of forsterite which were hot-pressed in either the USSA-2000 (specimen #1290) or USCA-1000 apparatus (specimen K211). These specimens are 2.7 mm in diameter and 1.54 and 1.14 mm in length, respectively, and have bulk densities and bench velocities within 1% of the theoretical densities and Hashin-Shtrikman averages of the isotropic velocities (Table 2).

Figure 11a shows an example of the S wave signals at 2 GPa and 400 °C for specimen #1290. The three strong echoes are the reflections from the ends of the WC cube, the alumina buffer rod, and the sample. The weak reflections following the sample echo are another cycle of the buffer rod and sample reflections plus possibly reflections from the bottom of the NaCl cup. The interference pattern between the alumina buffer rod echo and first sample echo is shown in Figure 11b and the calculated travel-times vs. frequency in Figure 11c. Note that there is much less dispersion for frequencies lower than 40 MHz than we typically observe in the acoustic experiments (see Figure 5c) in the room temperature cell (Figure 1). This observation (see also Figure 13 below) supports our interpretation that the dispersion at low frequency in the room temperature experiments is due to the relative surface area of the WC buffer rod and the sample.

Measurements at Simultaneous High P and T

Travel-time measurements have been performed along a specially designed P-T path to peak P and T conditions of 10 GPa and 1000 °C (Figure 12). First, oil pressure is raised slowly to ~ 80 bars (~2 GPa cell pressure) at room temperature (A to B); during this stage, the amplitude ratio B1/S1 and the resistance of the furnace reach stable values.

Considering that further compression at room temperature has the potential to damage the sample by the increasing shear stress imposed by solid pressure medium, a pre-heating to 400 °C is therefore performed (B to C). According to in situ X-ray studies by Weidner et al. [1992], NaCl reaches its yield strength at about 400 °C at P < 10 GPa. Therefore, the pressure environment of the sample after pre-heating should be close to hydrostatic. Experimentally, it might be possible to retain the acoustic signals by cold-pressing to higher pressures, but the potential plastic deformation would make the length estimation more difficult at high P and T conditions. Further pressurization is continued to the target pressure while maintaining temperature close to 400 °C (C to D). At the peak oil pressure, the temperature is then raised to the desired temperature (D to E).

The decompression and cooling path follows the recovery path reported in hot-pressing experiments [see *Gwanmesia and Liebermann, 1992*] in which the pressure and temperature are decreased simultaneously to about 600°C, then the temperature is held constant while slow decompression is continued. Simultaneous slow decompression and cooling resume for the last 50 bars on the pressure gauge to room pressure and about 250 °C. These specially designed P-T paths have proven to be effective to prevent permanent deformation of the sample and to anneal intergranular stresses as the samples are recovered from the high-pressures and temperatures.

No dimensional changes are observed in our specimens after the acoustic experiments, and the polished surfaces remain optically flat. However, specimen #1290 was broken into two half-cylinders when removed from the cell assembly. We believe that this was a consequence of additional uniaxial stress in the specimen caused by the use of an alumina ceramic for the thermocouple tubing and a TZM ring for the electrode; in our later experiments, these have been replaced by mullite tubing and a Pt ring, as described above.

The cell pressure at the peak oil pressure (400 bars) of the experiment was calibrated by the observation of the coesite/stishovite phase transformations according to the revised phase boundary by Zhang et al. [1996] and Liu et al. [1996] following the same P-T path as discussed above. Phase transformations were obtained at about 1250 °C which constrains the cell pressure to be 9.6 GPa for oil pressures of 400 bar (~ 300 tons).

Preliminary Results for Forsterite

The travel-times for S waves on specimen #1290 along the P-T path are shown in Figure 12. The variation of travel times during each cycle with either increasing pressure or temperature is clear and systematic. The room condition result (0.614 ms) is shown at the far right of the figure, at letter A. Initial pressurization to 80 bars (~ 2 GPa in cell pressure) from A to B shows a decrease of travel time from 0.614 ms to 0.592 ms. During the pre-heating to 400 °C from B to C, the travel time increases from 0.592 to 0.617 ms, yielding $dt/dT = 6.7 \times 10^{-5}$ ms/°C if we assume the pressure is constant during heating to 400°C. Under further pressurization from C to D, the travel times decreases from 0.617 ms to 0.575 ms. In the final heating cycle from D to E, the travel time increases linearly with temperature from 0.573 ms at 350 °C to 0.598 ms at 1029 °C, yielding $dt/dT = 3.7 \times 10^{-5}$ ms/°C at ~ 10 GPa with the assumption of constant pressure from 350 °C to 1029 °C.

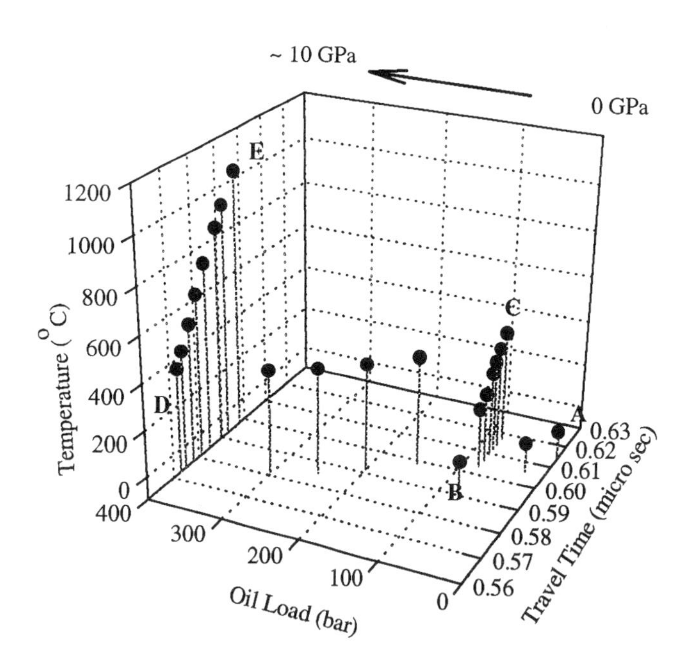

Figure 12. S wave travel times measured along specially designed P-T path. A: Room conditions; A-B: pressurizing to ~ 2 GPa at room T; B-C: preheating to ~ 400 °C at ~2 GPa; C-D: further pressuring to ~ 10 GPa at temperature close to 400 °C; D-E: heating at constant load to peak temperature of 1000 °C.

The dispersion curves for S waves in forsterite are plotted in Figure 13a as a function of pressure at T ~ 400°C (stage C-D) and in Figure 13b as a function of temperature at P ~10 GPa (stage D-E). The travel times are less scattered at temperatures below 900 °C with agreement of better than 0.5% in the frequency range 20 to 60 MHz. The results at 1029 °C show slightly larger scatter than at lower temperatures because of the very weak acoustic signals at this temperature. At this temperature, the WC cubes are heated up to the melting temperature of

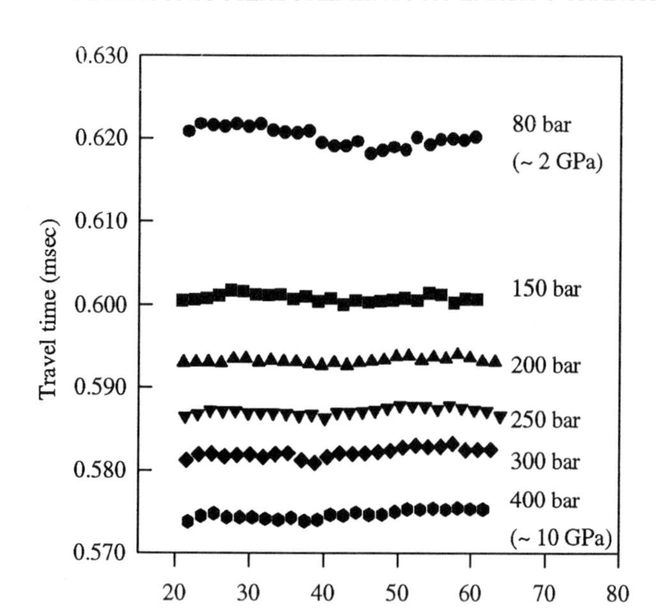

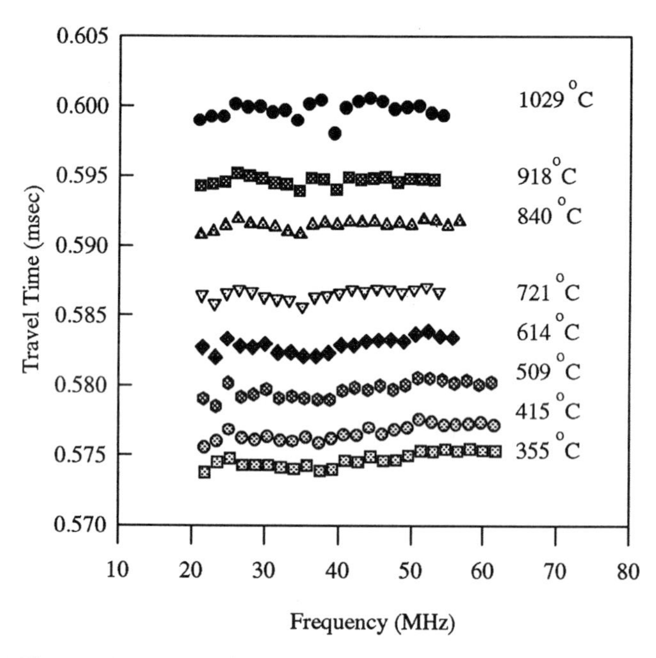

Figure 13. S wave travel times vs. frequency at various P and T conditions along the P-T path for specimen #1290 (a) At temperature close to 400 °C as a function of frequency at elevated pressures (stage C -D in Figure 12). (b). From 355 °C to 1029 °C at ~ 10 GPa (stage D-E in Figure 12).

the Crystalbond (~ 125 °C) used for mounting the transducer, resulting in substantial amount of energy loss. As a result, the interference extrema are very easily biased by the involvement of the stray energy from the background. However, even with the large scatter at 1029°C at 38-40 MHz, the average results at this frequency range are very consistent with the results at other frequencies. An overall uncertainty in the travel-time data is estimated to be less than 0.3%. Replacing the Crystalbond with high-T epoxy in later experiments has improved the signals at temperature above 1000 °C.

The cell assembly shown in Figure 10 has the capability to reach P ~ 14 GPa and T > 1400 °C, which will provide direct measurements of the sound velocity of mantle minerals at the P and T conditions of the transition zone. We are now utilizing these techniques to study a variety of polycrystalline and single crystal specimens of mantle minerals.

SUMMARY

We have demonstrated the feasibility of performing velocity measurements on millimeter-sized specimens to mantle transition zone pressures at room temperature in a multianvil apparatus using ultrasonic interferometry. The exploratory measurements on polycrystalline Lucalox alumina show a very good agreement with extrapolations of the previous measurements on single-crystal and polycrystalline samples at low pressures [*Li et al., 1996b*]. The "sample friendly" cell assembly preserves the sample at high pressures and the recovered samples are re-usable. Multiple measurements performed on a single specimen show a reproducibility of about 0.5% for the velocity results at high pressures.

We have applied these new techniques to the study of the high-pressure behavior of synthetic polycrystalline aggregates of both the olivine and beta phases of Mg_2SiO_4 [*Li et al., 1996c*] and to single crystals of synthetic forsterite and natural San Carlos olivine [*Chen et al., 1996*].

We have also developed cell assemblies for acoustic experiments at simultaneous high pressure and temperature condition in our USCA-1000 multianvil apparatus and reported the results of preliminary experiments on polycrystalline forsterite to 10 GPa and 1000°C. A specially-designed P-T path in pressurization and heating and decompression and cooling is used to preserve the sample, and the recovered the samples have no indications of dimensional changes due to plastic deformation. The pressure is calibrated using the coesite /stishovite phase boundary following the same P-T path. Systematic changes in travel times have been observed along the designated P-T path. The uncertainty of the measured travel time is better than 0.3%. Complete measurements on the olivine and beta phases are underway. These measurements will provide direct measurements of the elastic properties of these mantle minerals at the pressure and temperature conditions of the Earth's transition zone.

Acknowledgements. We would like to dedicate this paper to Orson L. Anderson, co-founder of the Mineral Physics Laboratory at the Lamont Geological Observatory in 1963, for his pioneering work in experimental physical acoustics and his continuing activity and leadership in this field. We are especially grateful for the key contributions made by Ian Jackson and Tibor Gasparik in the early stage of this project. We thank Carey Koleda, Herb Schay, and Ed Vorisek for fabricating tools and cell parts for this new class of high pressure experiments, and Ben Vitale for advice on electronic issues and computer software to operate the press. We also benefited from discussions with Don Weidner and Jianzhong Zhang. The manuscript has been improved by the constructive reviews of K. Kusaba and D.C. Rubie. These high-pressure experiments were conducted in the Stony Brook High Pressure Laboratory, which is jointly supported by the State University of New York at Stony Brook and the NSF Science and Technology Center for High Pressure Research (EAR-89-20239). This research was also supported by EAR-93-04502. MPI contribution No. 172.

REFERENCES

Bassett, W.A., H. Shimizu, and E.M. Brody, Pressure dependence of elastic moduli of forsterite by Brillouin scattering in the diamond cell, in *High-Pressure Research: Application to Earth and Planetary Sciences*, edited by M. Manghnani and Y. Syono, pp. 115-121, Center for Academic Publication, Tokyo, Japan, 1982.

Bina, C. R., and B. J. Wood, Olivine-spinel transitions: experimental and thermodynamic constraints for the nature of the 400 km seismic discontinuity, *J. Geophys. Res., 92*, 4853-4866, 1987.

Chen, G., B. Li, and R. C. Liebermann, Selected elastic moduli of single-crystal olivines from ultrasonic experiments to mantle pressures, *Science, 272*, 979-980, 1996.

Cook, R. K., Variation of elastic constants and static strains with hydrostatic pressure: A method for calculation from ultrasonic measurements, *J. Acoust. Soc. Am., 29*, 445-449, 1957.

Davis, G. F., and R. J. O'Connell, Transducer and bond phase shifts in ultrasonics, and their effects on measured pressure derivatives of elastic moduli, in *High Pressure Research: Application in Geophysics*, edited by M. Manghnani and S. Akimoto, pp. 536-562, Academic Press, New York, 1977.

Duffy, T. S., C. S. Zha, R. T. Downs, H. K. Mao, and R. J. Hemley, Elasticity of forsterite to 16 GPa and the composition of the upper mantle, *Nature, 378*, 170-173,1995.

Fei, Y., H. K. Mao, J. Shu, G. Parthasarathy, W. A. Bassett and J. Ko, Simultaneous high-P and high-T X-ray diffraction study of β–$(Mg,Fe)_2SiO_4$ to 26 GPa and 900 K. *J. Geophys. Res., 97*, 4489-4495, 1992.

Fujisawa, H. and E. Ito, Measurements of ultrasonic wave velocities in solid under high pressure, in *Proceedings of the 4th Symposium on Ultrasonic Electronics, Tokyo, 1983 Japan. J. Applied Physics, Supplement 23-1*, pp. 51-53, 1984.

Fujisawa, H. and E. Ito, Measurements of ultrasonic wave velocities of tungsten carbide as a standard material under high pressure up to 8 GPa, in *Proceedings of the 5th Symposium on Ultrasonic Electronics, Tokyo, 1984 Japan. J. Applied Physics, Supplement 24-1*, pp.103-105, 1985.

Gieske, J. H., and G. R. Barsch, Pressure dependence of the elastic constants of single crystalline aluminum oxide, *Phys. Stat. Sol., 29*, 121-131, 1968.

Gwanmesia, G. D., R. C. Liebermann, and F. Guyot, Hot-pressing and characterization of polycrystals of β–Mg_2SiO_4 for acoustic velocity measurements, *Geophys. Res. Lett., 17*, 1331-1334, 1990a.

Gwanmesia G. D., S. M. Rigden, I. Jackson, and R. C. Liebermann, Pressure dependence of elastic wave velocity for β–Mg_2SiO_4 and the composition of the Earth's mantle, *Science, 250*, 794-797, 1990b.

Gwanmesia, G. D., and R. C. Liebermann, Polycrystals of high-pressure phases of mantle minerals: Hot-pressing and characterization of physical properties, *High Pressure Research: Application to Earth and Planetary Sciences,* edited by Y. Syono and M. Manghnani, pp. 117-135, AGU, Washington, D.C., 1992.

Gwanmesia, G. D., B. Li, and R. C. Liebermann, Hot pressing of polycrystals of high pressure phases of mantle minerals in multianvil apparatus, in *Experimental Techniques in Mineral and Rock Physics,* edited by R. C. Liebermann, and C. H. Sondergeld, *PAGEOPH, 141*, pp. 467-484, 1993.

Jackson, I., and H. Niesler, The elasticity of periclase to 3 GPa and some geophysical implications, in *High Pressure Research: Application to Earth and Planetary Sciences,* edited by M. Manghnani and Y. Syono, pp. 93-113, Center for Academic Publication, Tokyo, Japan, 1982.

Jackson, I., H. Niesler, and D. J. Weidner, Explicit correction of ultrasonically determined elastic wave velocities for transducer-bond phase shift, *J. Geophys. Res., 86*, 3736-3748, 1981.

Kinoshita, H., N. Hamaya, and H. Fujisawa, Elastic properties of single-crystal NaCl under high pressures to 80 kbar, *J. Phys. Earth, 27*, 337-350, 1979.

Knoche, R., S. L. Webb, and D. C. Rubie, Ultrasonic velocity to 10 GPa and 1500 °C in the multianvil press: Measurements in polycrystalline olivine (abstract), EOS, Trans. AGU, *76*, Fall Meeting Suppl., p.563, 1995.

Kumazawa, M., and O. L. Anderson, Elastic moduli, pressure derivatives, and temperature derivatives of single-crystal olivine and single-crystal forsterite, *J. Geophys. Res., 74*, 5961-5972, 1969.

Kusaba, K., L. Galoisy, Y. Wang, M. T. Vaughan, and D. J.

Weidner, Determination of phase transition pressures of ZnTe under quasi-hydrostatic conditions, in *Experimental Techniques in Mineral and Rock Physics,* edited by R. C. Liebermann and C. H. Sondergeld, *PAGEOPH, 141,* 644-652, 1993.

Li, B., R. C. Liebermann, and I. Jackson, Measurements of elastic wave velocity of polycrystal Al_2O_3 to 10 GPa (abstract), *EOS, Trans. AGU, 75,* Fall Meeting Suppl., p596, 1994.

Li, B., R. C. Liebermann, G. D. Gwanmesia and I. Jackson, Elastic Wave velocities of mantle minerals to 10 GPa in multi-anvil apparatus by in situ ultrasonic techniques(abstract), *EOS, Trans. AGU, 76,* Spring Meeting Suppl., p277, 1995a.

Li, B., G. D. Gwanmesia, and R.C. Liebermann, Elastic wave velocity of olivine and beta polymorphs of Mg_2SiO_4 at transition zone pressures(abstract), *EOS, Trans. Amer. Geophys. Union, 76,* Fall Meeting Suppl., p619, 1995b.

Li, B., S. M. Rigden, and R. C. Liebermann, Elasticity of stishovite at high pressure, *Phys. Earth Planet. Inter.,* Ringwood Volume, edited by T. Irifune and S. Kesson, *96,* 113-127, 1996a.

Li, B., I. Jackson, T. Gasparik, and R. C. Liebermann, Elastic wave velocity measurement in multi-anvil apparatus to 10 GPa using ultrasonic interferometry, *Phys. Earth Planet Interi.,* IUGG XXI Special Volume, *98,* 79-91, 1996b.

Li, B., G. D. Gwanmesia, and R. C. Liebermann, Sound velocities of olivine and beta polymorphs of Mg_2SiO_4 at Earth's transition zone pressures, *Geophys. Res. Lett., 23,* 2259-2262, 1996c.

Liu, J., L. Topor, J. Zhang, A. Navrotsky, and R. C. Liebermann, Calorimetric study of coesite stishovite transformation and calculation of the phase boundary, *Phys. Chem. Min., 23,* 11-16, 1996.

Liebermann, R. C., A. E. Ringwood, D. J. Mayson, and A. Major, Hot-pressing of polycrystalline aggregate at very high pressure for ultrasonic measurements, in *Proceedings of the 4th Conference on High Pressure,* edited by J. Osugi, pp. 495-502, Physico-Chemical Society of Japan, Tokyo, 1975.

Lloyd, E. C., Accurate Characterization of the high pressure environment, *NBS Spec. Publ. No. 326, Washington D.C.,* 1-3, 1971.

McSkimin, H. J., Ultrasonic measurement techniques applicable to small solid specimens. *J. Acoust. Soc. Am., 22,* 413-418, 1950.

Meng, Y., D. J. Weidner, G. D. Gwanmesia, R. C. Liebermann, M. T. Vaughan, Y. Wang, K. Leinenweber, R. E. Pacalo, A. Yeganeh-Haeri, and Y. Zhao, In-situ high P-T X-ray diffraction studies on three polymorphs (α, β, γ) of Mg_2SiO_4, *J. Geophys. Res., 98,* 23199-23207, 1993.

Niesler, H., and I. Jackson, Pressure derivatives of elastic wave velocities from ultrasonic interferometric measurements on jacketed polycrystals, *J. Acoust. Soc. Am., 86,* 1573-1585, 1989.

Niesler H., and G. Fisher, Technique for ultrasonic measurement of samples at 2.0 GPa and 600 °C. *High Temperatures - High Pressures,* 24, 65-74, 1992.

Nolet G., S. P. Grand, and B. L. N. Kennett, Seismic heterogeneity in the upper mantle, *J. Geophys. Res., 99,* 23753-23766, 1994.

Rigden, S. M., G. D. Gwanmesia, I. Jackson, and R. C. Liebermann, Progress in high-pressure ultrasonic interferometry, the pressure dependence of elasticity of Mg_2SiO_4 polymorphs and constraints on the composition of the transition zone of the Earth's mantle, in *High Pressure Research: Application to Earth and Planetary Sciences,* edited by Y. Syono and M. Manghnani, pp. 167-182, Terra Scientific Publishing Co. and AGU, Tokyo and Washington, D.C., 1992.

Rigden, S. M., I. Jackson, H. Niesler, R. C. Liebermann, and A. E. Ringwood, Pressure dependence of the elastic wave velocities for Mg_2GeO_4 spinel up to 3 GPa, *Geophys. Res. Lett., 15,* 604-608, 1988.

Rigden, S. M., I. Jackson, R. C. Liebermann, and A. E. Ringwood, Elasticity of germanate and silicate spinels at high pressure, *J. Geophys. Res., 96,* 9999-10,006, 1991.

Rigden, S. M., G. D. Gwanmesia, and R. C. Liebermann, Elastic wave velocity of a pyrope-majorite garnet to 3 GPa, *Phys. Earth Planet. Inter., 86,* 35-44, 1994.

Sawamoto, H., D. J. Weidner, S. Sasaki, M. Kumazawa, Single-crystal elastic properties of the modified-spinel (beta) phase of magnesium orthosilicate, *Science, 224,* 749-751, 1984.

Schreiber, E., and O. L.. Anderson, Pressure derivatives of the sound velocities of polycrystalline alumina, *J. Am. Ceramic Soc., 49,* 184-190, 1966.

Spetzler, H., Equation of state of polycrystalline and single-crystal MgO to 8 kilobars and 800K, *J. Geophys. Res., 75,* 2073-2087, 1970.

Spetzler, H. A., G. Chen, S. Whitehead, and I. C. Getting, A new ultrasonic interferometer for the determination of equation of state parameters of sub-millimeter single crystals, in *Experimental Techniques in Mineral and Rock Physics,* edited by R. C. Liebermann, and C. H. Sondergeld, *PAGEOPH, 141,* 341-377, 1993.

Suito, K., M. Miyoshi, T. Sasakura, and H. Fujisawa, Elastic properties of obsidian, vitreous SiO_2, and vitreous GeO_2 under high pressure up to 6 GPa, in *High Pressure Research: Application to Earth and Planetary Sciences,* edited by Y. Syono and M. Manghnani, pp. 219-225, Terra Scientific Publishing Co., Tokyo and AGU, Washington, D. C., 1992.

Wang, H. and G. Simmons, Elasticity of some mantle crystal structures: 1. Pleonaste and Hercynite spinel, *J. Geophys. Res., 77,* 4379-4392, 1972.

Wang, Y., R. C. Liebermann, and J. N. Boland, Olivine as an in situ piezometer in high pressure apparatus, *Phys. Chem. Minerals, 15,* 493-497, 1988.

Webb, S. L., The elasticity of the upper mantle orthosilicates olivine and garnet to 3 GPa, *Phys. Chem. Miner., 16,* 684-692, 1989.

Weidner, D. J., and E. Ito, Mineral physics constraints on a uniform mantle composition in *High Pressure Research in Minerals Physics,* edited by M. H. Manghnani and Y. Syono, pp. 439-446, Terra Scientific Publishing Co., Tokyo and AGU, Whashington D. C., 1987.

Weidner, D. J., M. T. Vaughan, J. Ko, Y. Wang, X. Liu, A. Yeganeh-Haeri, R. E. Pacalo, and Y. Zhao, Characterization of stress, pressure, and temperature in SAM85, a DIA type high Pressure apparatus, in *High Pressure Research: Application to Earth and Planetary Sciences,* edited by Y. Syono and M. Manghnani, pp. 13-17, Terra Scientific Publishing Co., Tokyo and AGU, Washington, D.C., 1992.

Yoneda, A. and M. Morioka, Pressure derivatives of elastic constants of single-crystal forsterite, in *High Pressure Research: Application to Earth and Planetary Sciences,* edited by Y. Syono and M. Manghnani, pp. 157-166, Terra Scientific Publishing Co. and AGU, Tokyo and Washington, D.C., 1992.

Yoneda, A., Pressure derivatives of elastic constants of single crystal MgO and $MgAl_2O_4$, *J. Phys. Earth, 38*, 19-55, 1990.

Zaug, J. M., E. H. Abramson, J. M. Brown, and L. J. Slutsky, Elastic constants, equation of state and thermal diffusivity at high pressure, in *High Pressure research,: Applications to Earth and Planetary Sciences*, edited by Y. Syono and M. Manghnani, pp. 157-166, Terra Scientific Publishing Co., Tokyo and AGU, Washington, D.C., 1992.

Zaug, J. M., E. H. Abramson, J. M. Brown, and L. J. Slutsky, Sound velocity in olivine at Earth mantle pressures, *Science, 260*, 1487-1490, 1993.

Zha, C. S., T. S. Duffy, R. T. Downs, H. K. Mao, and R. J. Hemley, Sound velocity and elasticity of single-crystal forsterite to 16 GPa, *J. Geophys. Res.*, 101, 17535-17545, 1996.

Zhang, J., B. Li, W. Utsumi, and R.C. Liebermann, In situ X-ray observations of the coesite-stishovite transition: reversed phase boundary and kinetics, *Phys. Chem. Min.*, *23*, 1-10, 1996.

B. Li, G. Chen, and R. C. Liebermann. Department of Earth and Space Sciences, SUNY at Stony Brook, Stony Brook, NY 11794-2100

---

[1]Also at Department of Physics and Astronomy, Delaware State University, Dover, DE 19906.

[2]Center for High Pressure Research* and Department of Earth and Space Sciences, State University of New York at Stony Brook, Stony Brook, NY 11794.

* A NSF Science and Technology Center.

Can the Multianvil Apparatus Really Be Used for High-Pressure Deformation Experiments?

W. B. Durham* and D. C. Rubie

Bayerisches Geoinstitut, Universität Bayreuth, 95440 Bayreuth, Germany; +49 921 55-3700

Past claims of the suitability of the MA-8 multianvil press as a deformation apparatus may have been overstated. On the basis of measurements of final octahedron size and of guide block displacement as a function of time, using the 10/5, 14/8, and 18/11 assemblies (octahedron edge length in mm/truncation edge length in mm) with MgO octahedra and pyrophyllite gasketing, it appears that at run conditions of interest to most researchers there is no appreciable time-dependent creep of gaskets and octahedra. All inelastic deformation occurs at rather low pressures: below about 10 GPa for the 10/5, 7 GPa for the 14/8, and 6 GPa for the 18/11 assemblies, with substantial uncertainties in these pressures. Above these limits all deformation of the pressure medium is elastic. Pressure ramping as a means of increasing the inelastic deformation rate of a sample is probably ineffective. Displacement measured at the guide blocks, previously believed to indicate deformation of the gaskets and octahedron, appears now to be unrelated to creep of these components. The calibrations have not been exhaustive and there is considerable scatter in some of the size measurements, so the above conclusions are not unequivocal. The calibrations do not exclude the possibility of deformation of a few tens of microns after the attainment of high pressure. Efforts to impose permanent shape change to samples at high pressure and temperature simply by relying on long run durations must be viewed with skepticism. There may be possibilities for deformation in the multianvil apparatus if materials of contrasting elastic modulus are used to differentially load a sample during pressure ramping and/or by deforming a very thin sample in simple shear.

1. INTRODUCTION

Recent papers have suggested that the slightly nonhydrostatic stress environment of the large-volume high-pressure multianvil (MA-8) apparatus might be exploited to investigate the deformation of mantle materials in situ at high pressures and temperatures [e.g., *Liebermann and Wang*, 1992]. Important experimental results have also been published that purport to have imposed small amounts of permanent strain to samples at high pressures and temperatures to produce high-pressure

*W. B. Durham, University of California Lawrence Livermore National Laboratory, P. O. Box 808, Livermore, CA 94550, USA (email: durham1@llnl.gov)

Properties of Earth and Planetary Materials at High Pressure and Temperature
Geophysical Monograph 101

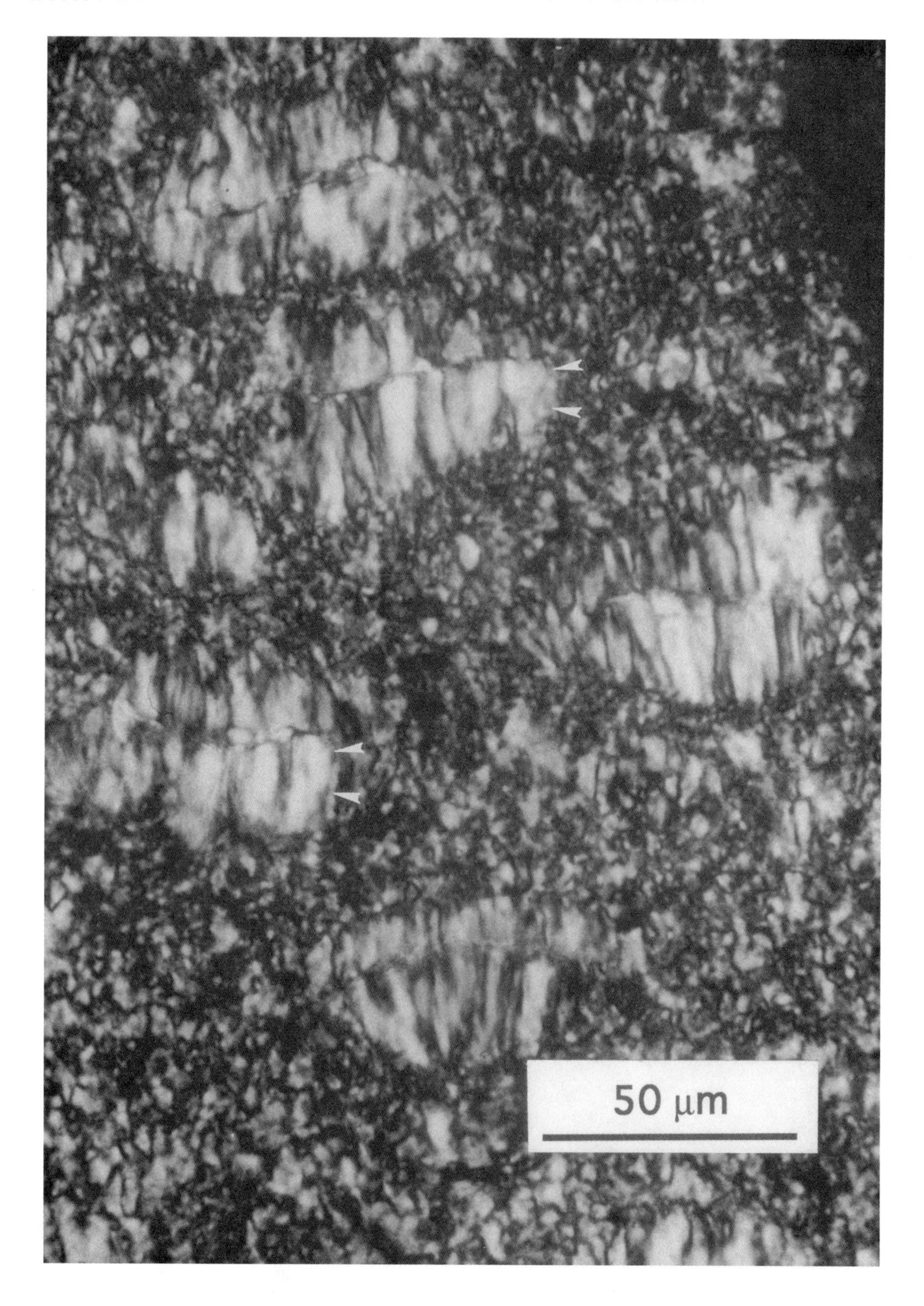

Figure 1. Optical photomicrograph of wadsleyite (two examples are indicated by white arrows) formed by the transformation of $Mg_{1.8}Fe_{0.2}SiO_4$ olivine. The individual wadsleyite grains are elongate because their growth rate is fastest in the direction of the maximum compressive stress, which runs vertically in the micrograph [c.f., *Vaughan et al.*, 1984; *Green*, 1986].

faulting (as related to deep-focus earthquakes) [*Green et al.*, 1990] and large amounts to measure the activation volume for creep in olivine [*Bussod et al.*, 1993]. The ability to deform materials at mantle conditions represents an important experimental advance that is of interest to many geophysicists.

There is convincing evidence that differential stresses can exist in the multianvil cell at high pressures and temperatures. We ourselves have produced in the multianvil cell the scalloped remnant olivine grains and elongated wadsleyite grains during the $(Mg,Fe)_2SiO_4$ olivine to wadsleyite phase transformation that *Vaughan et al.* [1984] and *Green* [1986], in studies of the Mg_2GeO_4 olivine to ringwoodite transformation, have shown identify a direction of maximum compressive stress when the transformation takes place under nonhydrostatic conditions (Figure 1). However, the evidence that such differential stresses endure over macroscopic strains is much more circumstantial. The sample assembly in the multianvil apparatus typically undergoes high inelastic strain during initial pressurization, and differentiating permanent strain that occurs at high pressure from that which occurs during initial pressurization can be problematic. We have been attempting to further develop multianvil techniques with the aim of performing deformation experiments under pressure-temperature conditions of the Earth's mantle, but in the course of trying to determine the true amount of imposed deformation at high pressure and temperature, a very different picture has emerged than previous studies have suggested. The purpose of this short paper is to bring expectations into line with reality as regards using the multianvil apparatus as a creep apparatus. We do not attempt to provide a rigorous calibration of the mechanical behavior of the apparatus.

2. MEASUREMENTS

Experiments described here were performed using the two MA-8 multianvil presses at the Bayerisches Geoinstitut in Bayreuth, Germany. A general description of the experimental configuration is given by *Ito et al.* [1984] and *Liebermann and Wang* [1992] and specific details of the pressure calibrations and the sample assemblies used can be found in *Rubie at al.* [1993a,b], *Bussod et al.* [1993], and *Canil* [1993]. We used three common pressure cell configurations: 18/11, 14/8, and 10/5, where the notation is octahedron edge length in mm/truncation edge length in mm. Machined octahedra consisting of sintered MgO (+5%CrO_2) provided the confining medium and gaskets were made of pyrophyllite. High-temperature sample assemblies used a $LaCrO_2$ furnace surrounded by an insulating sleeve of ZrO_2. In most cases standard sample assemblies were used and only in a few

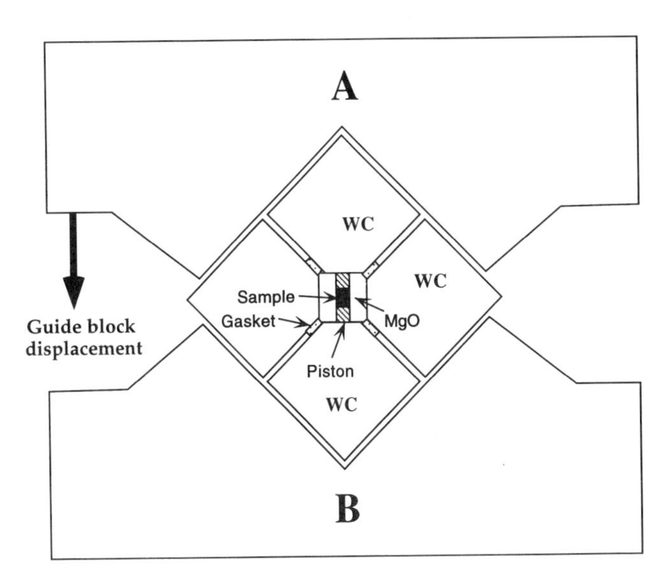

Figure 2. Schematic sketch of the experimental setup for MA-8 multianvil deformation experiments. The two guide blocks (A and B), which contain the outer set of six anvils, are compressed in a uniaxial press up to a load of 500-1000 tonnes. Also shown are four of the eight inner anvils, consisting of WC cubes, and the pressure assembly consisting of a MgO octahedron. In previous attempts to perform deformation experiments, the sample was located uniaxially between hard pistons (e.g., Al_2O_3) within the MgO octahedron. Measurements presented in this study consist of (1) measurements of the final octahedron size after experiments as a function of sample pressure, sample temperature, and experimental run duration; and (2) guide block displacement (arrow) as a function of experimental run time.

experiments were assemblies used that contained Al_2O_3 pistons designed specifically to stress the sample [e.g., *Green et al.*, 1990; *Bussod et al.*, 1993].

Bussod et al. [1993] have presented data suggesting that the sample strain, octahedron strain, and the relative displacement of the multianvil guide blocks are all related in multianvil experiments. If this were the case, the sample strain could be monitored in real time simply by measuring the guide block displacement. To test this possibility more rigorously, we have attempted to determine the amount of deformation that takes place under pressure by making two types of measurements: (1) final octahedron size after depressurization, as indicated by benchtop measurements of the distance between the four pairs of opposite parallel faces; and (2) relative displacement of the two guide blocks (Figure 2). Closure of the guide blocks is monitored by a single direct current differential transformer displacement transducer whose body is rigidly attached to one guide block and whose core moves rigidly with the other guide block. Both octahedron size and guide block displacement were measured at a range of pressures, temperatures, run times, and cell configurations. Many of the measurements were made during routine experiments

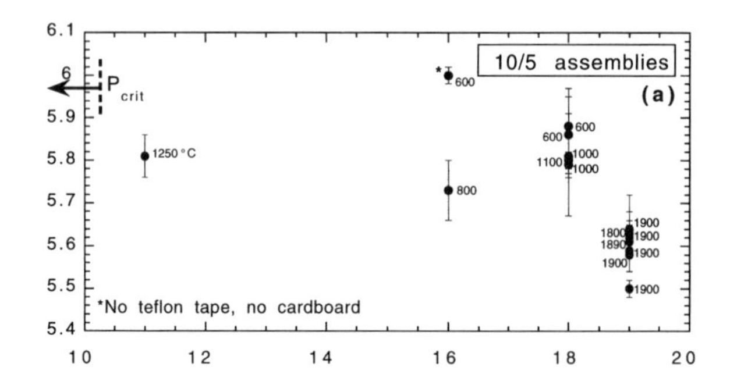

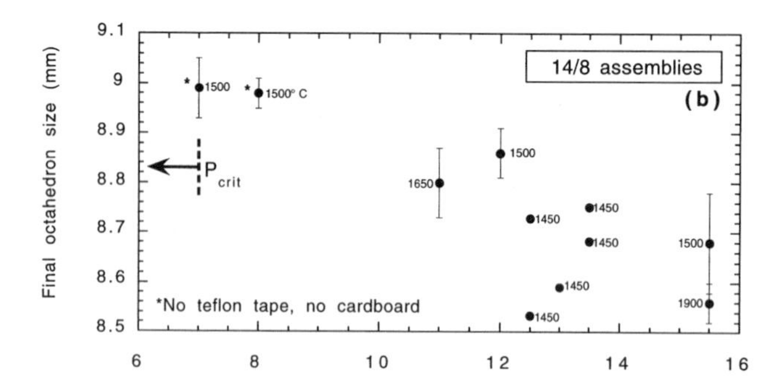

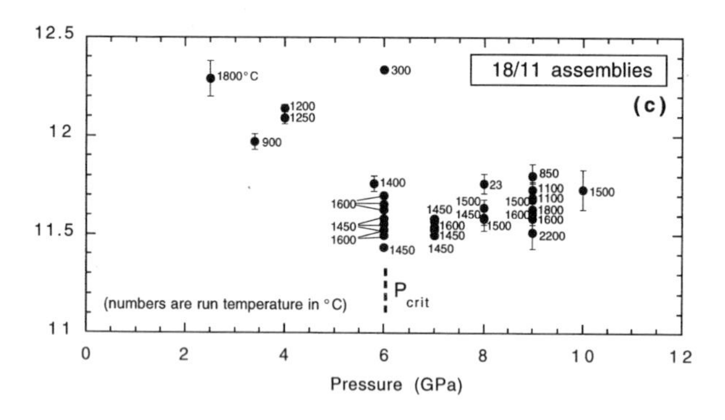

Figure 3. Final octahedron size, as indicated by the mean and total scatter (error bars) of distances between the 4 pairs of parallel faces, as a function of maximum run pressure for the (a) 10/5, (b) 14/8, (c) 18/11 multianvil assemblies (the notation is octahedron edge length in mm/ tungsten carbide cube truncation edge length in mm). Most of the points without error bars are taken from *Bussod et al.* [1993]. Points are labeled with maximum run temperature. One point in (a) and two points in (b) labeled with asterisks (*) were special calibration runs (see text). Their final size is anomalously large because they were constructed without the usual (see *Rubie at al.* [1993a, b], *Bussod et al.* [1993], and *Canil* [1993]) cardboard and teflon tape on the 12 facing pairs of internal faces of the WC cubes. For most of the points, run times (not given) varied from several seconds to 3 hours. Note that points are not very evenly distributed in (*P, T*); in particular, note in (a) that the runs at highest *P* are also the runs at highest *T*. The labels P_{crit} are our identification of the upper bound on the pressure at which inelastic compaction ceases (see text).

performed in the multianvil laboratory at the Bayerisches Geoinstitut in the course of other investigations.

3. OBSERVATIONS

3.1. *Octahedron Size vs Pressure*

The MgO octahedra have an initial porosity of about 20% and interfacial distances of 14.6 mm, 11.3 mm, and 8.2 mm, all ± 0.05 mm, respectively, for the three different sizes (edge length 18 mm, 14 mm and 10 mm). Final octahedron sizes for the three different assemblies are plotted as a function of pressure in Figure 3. The run conditions are not uniformly distributed in pressure-temperature space, but tend to cluster where researchers chose to perform their experiments, so attention must be paid to the temperature labels attached to each point. For example, for the 10/5 assemblies (Figure 3a), the highest pressure runs are also the highest temperature runs; the significant decrease in size from 18 to 19 GPa can be caused (based on Figure 3 alone) by either higher temperature, higher pressure, or both. Run times are not shown. Except for the few longer runs performed specifically for the purposes of creep calibration (see below), run times vary from a few seconds to 6 hours.

No trend is obvious in the 18/11 assembly above 6 GPa (Figure 3c), data are sparse and scattered for the 14/8 (Figure 3b), and the strongest trend for the 10/5 (Figure 3a) cannot distinguish the effect of pressure from that of temperature. Replotting the data against temperature, with pressure as an attached label (Figure 4), reveals a weak but consistent trend for all three configurations of decreasing octahedron size with increasing temperature. The trends might be less scattered were the 14/8 and 18/11 size measurements made by *Bussod et al.* [1993] not included (those in Figures 3 and 4 without error bars). Those sizes appear to be systematically lower than those measured by us, but without any a priori reason for excluding those data, we include them in the figures.

Taking the temperature trend in Figure 4c as a correction to the data in Figure 3c produces a clearer indication that the size of 18/11 assembly is independent of pressure (Figure 5) above 6 GPa. The same correction cannot be applied for the 10/5 data because of their biased locations in pressure-temperature space.

The perspective provided by two points at 1500°C and 7 and 8 GPa in Figures 3b and 4b deserves special note. These experiments were performed for the sole purpose of creep calibration, that is, specifically to determine the effectiveness of the pressure ramping technique [*Bussod et al.*, 1993], wherein the "run pressure" is not held constant but rather increased at a slow, steady rate, typically 1 GPa/h or slower, to produce a higher deformation rate in the sample. (A note in the Figure 3b caption explains why the two data points are displaced with respect to the other

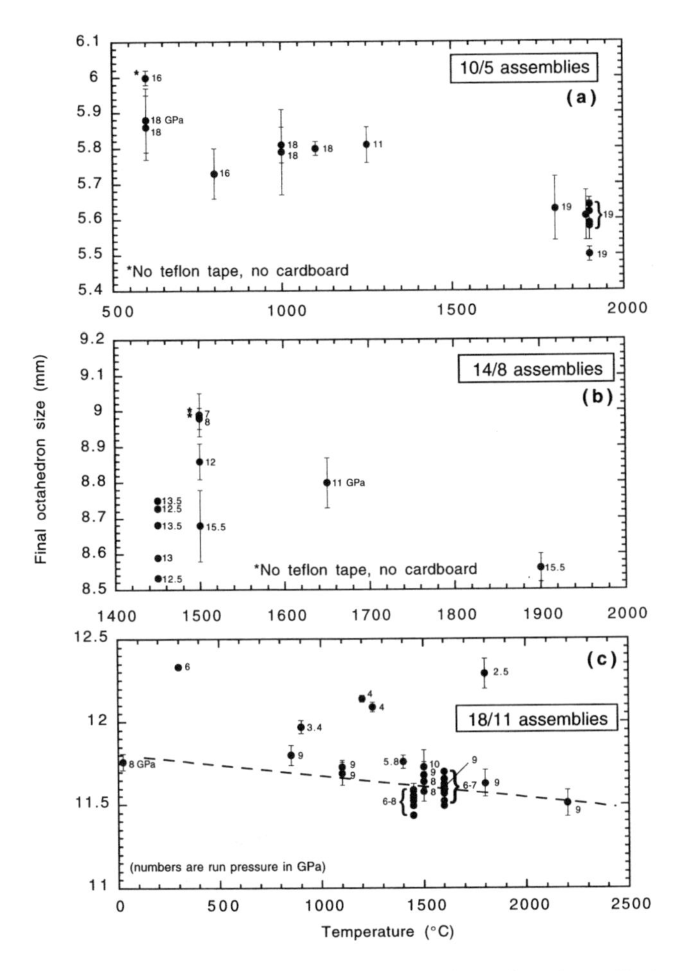

Figure 4. Same data as in Figure 3, but plotted as final octahedron size vs maximum run temperature for the (a) 10/5, (b) 14/8, and (c) 18/11 assemblies, with points labeled according to maximum run pressure. The straight dashed line in (c) is a least squares fit to all points for which $P > P_{crit}$ (6 GPa).

points in the plots.) The two assemblies were prepared simultaneously and run sequentially in the same multianvil press. Pressure-temperature-time histories differed only by the pressure ramp: both were taken to their run pressures (7 GPa in the one case; 8 GPa in the other) at the same rate and then heated to 1500°C. Both were held at highest (*P, T*) for 30 minutes, then quenched to room temperature, and finally depressurized at identical rates. Not only were the sizes of the two octahedra indistinguishable, but the fact that they experienced relatively low (for the 14/8 assembly) pressures where inelastic compaction might be expected to be higher than at more elevated pressures did not bode at all well for the use of the pressure ramping technique, or the multianvil apparatus in general, in deformation experiments.

3.2. *Guide Block Displacement vs Time*

Guide block displacement vs. time curves (Figure 6) give the distinct impression that creep occurs while the assembly is under pressure. For the duration of typical runs, usually a few minutes to a few hours at constant pressure and temperature, the displacement transducer reveals a steady, if not slightly decreasing, rate of advancement of the guide blocks of 10 mm or more per hour. Were this displacement rate imposed on a typical sample of length 1–2 mm, the strain rate would be 10^{-5} to 10^{-6} s^{-1} (e.g., as *Bussod et al.* [1993] have suggested), which is within the range of strain rates of conventional laboratory deformation experiments in geophysics. In no case, however, have we ever observed the guide block advancement to reach a steady, non-zero rate. Carried to longer and longer run times, the advancement rate can be seen to slow to insignificantly low values (for deformation experiments) and, finally, after about 24 hours duration, to stop completely.

It is important to point out that the fastest guide block advancement rates of >10 mm/h are not necessarily inconsistent with the observations of final octahedron size shown in Figures 3 and 4. This becomes apparent when the fine displacement scales of Figure 6 are compared with the much coarser vertical scales of Figures 3 and 4. It would not be difficult to hide, say, 0.05–0.1 mm of inelastic deformation in the scatter of Figures 3 and 4. The lack of a demonstrable decrease in octahedron size during pressure ramping, where guide block displacement rates are closer to 100 mm/h, can be seen in hindsight to indicate that inelastic strain was an illusion. This illusion probably persisted in earlier work because of the strength and consistency of the guide block displacement signal at constant pressure: the implications of the near constant octahedron size were simply overlooked given the steady, reproducible trace of guide block advancement vs time, and the concept was never seriously tested by experiment

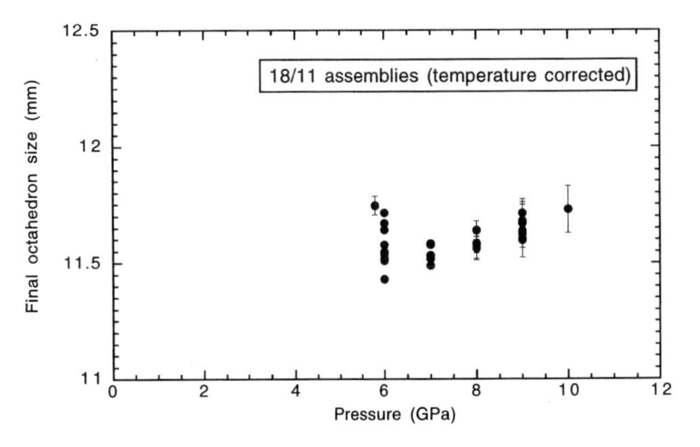

Figure 5. 18/11 data from Figure 3c corrected for temperature according to the linear trend in Figure 4c.

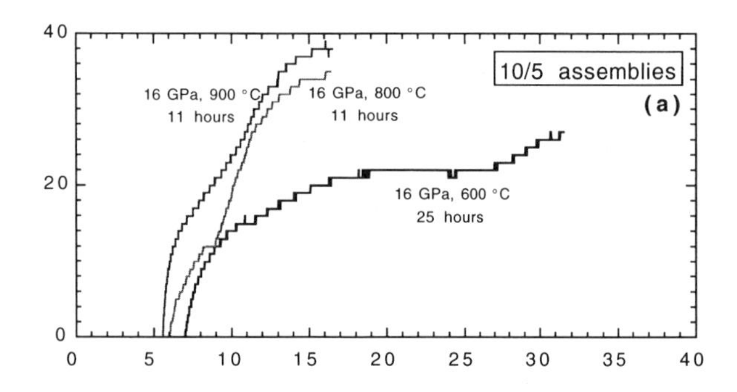

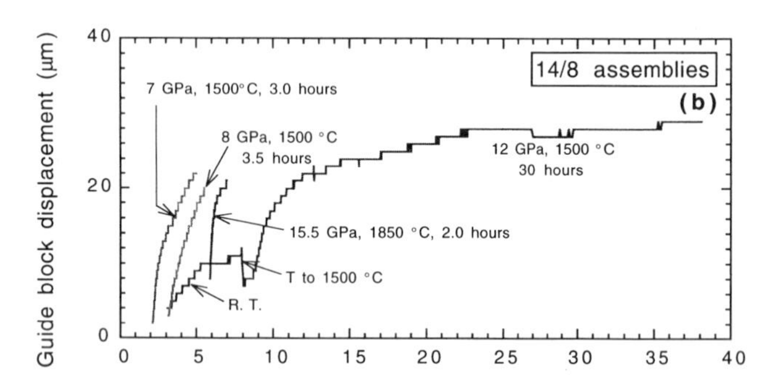

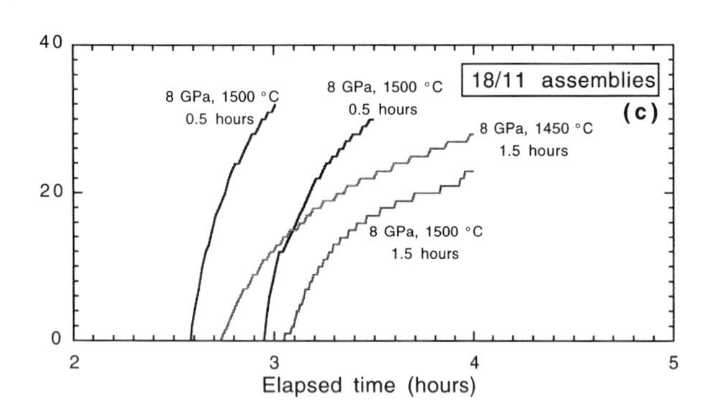

Figure 6. Guide block advancement as measured by a displacement transducer located between guide blocks as a function of elapsed time for the (a) 10/5, (b) 14/8, and (c) 18/11 assemblies. There is little significance to the absolute value of elapsed time because it is measured from the start of each experiment, and because there is considerable variation from run to run in the time required to reach run pressure and temperature. For run durations of a few hours, there appears to be a nearly steady rate of advance, suggesting that the gaskets and octahedron are creeping. However, over very long run times (a, b), the guide block advancement rate drops to insignificant levels. Note the behavior of displacement immediately after heating in the long duration 14/8 run in (b). The prompt response was an (apparent) widening of the inter-guide block distance by 4 mm, followed after 40 minutes by an (apparent) resumption of closure. Because temperatures in the neighborhood of the octahedron and gaskets reach steady-state in much less than 40 minutes, we believe that the most of the displacement in this curve (and by inference all the other curves) did not result from creep of the gaskets or octahedron.

until now. The minute decrease of the guide block advancement rate over the course of a short duration experiment could also be dismissed as gasket area increasing as octahedron and gasket extrude into the space between tungsten carbide anvils.

Another aspect of the behavior of the apparatus that can be misleading is the guide block response upon initial heating. In most runs, pressure is first raised to the desired level before the temperature is raised. Immediately upon heating, one almost invariably observes that the rate of guide block advancement and oil pumping to the ram immediately speed up, indisputable evidence that the thickness of the octahedron and gaskets are decreasing. Prompt thermal expansion should have exactly the opposite effect on displacement, so presumably the shrinking comes from weakness of the respective materials at high temperatures. We have difficulty reconciling this observation with the behavior shown in Figures 3–6. The weak negative dependence of final octahedron size upon temperature (Figure 4) is, at first glance, consistent with this idea because higher temperature usually causes materials to creep faster. However, if the sudden weakness in octahedron and gaskets occurs because of the higher temperature, one must question why the weakness is only short-lived even though high temperature persists. The spreading gasket hypothesis mentioned above does not work for large changes in displacement rate, since the change in displacement rate must be proportional to the change in gasket area, which would presumably be related in a complex, but distinct, way to final octahedron size. Work hardening of the MgO and pyrophyllite are difficult to prove and also do not provide a very satisfactory explanation because of the extreme magnitude of work hardening required. The very long duration experiment in Figure 6b is very revealing in this regard. In that experiment, pressure was held constant at 12 GPa for nearly 6 hours at room temperature before heating the sample chamber. In that experiment the prompt response to heating was a separation of the guide blocks. Acceleration in guide block closure did occur, but only after nearly one hour had elapsed. Thermal steady state at the octahedron and gaskets must have been reached in a few minutes, and certainly in less than an hour. It seems very unlikely, therefore, that the guide block closure indicated by the long-duration test in Figure 6b was related to displacements at the octahedron. The effect of the heat pulse as it encounters the mass of the press is too complex even for a qualitative discussion. One can speculate over various explanations for a divergence of the guide blocks (e.g., thermal expansion of the press frame) as well as for an apparent closure (e.g., warming up and thermal expansion of the displacement transducer).

Further evidence that the high displacement rates shortly after heating are illusory is provided by the measurements of final octahedron size as a function of run duration for seven nearly identical 10/5 experiments shown in Figure 7.

Those runs were carried out for relatively short durations of 35 minutes or less, the period during which apparent guide block displacement rates are most rapid (much of this early rapid displacement has been removed from the portions of the run records shown in Figure 6). Figure 7 shows no tendency for the octahedron size to decrease with time.

4. DISCUSSION AND CONCLUSIONS

Upon initial pressurization in the multianvil apparatus there is considerable inelastic compaction of the octahedral pressure medium, but on the basis of the data in Figures 3–7, significant inelastic compaction apparently ceases at pressures P_{crit} that are well below most common experimental run pressures. For the 18/11, 14/8, and 10/5 assemblies, we can identify approximate values or limits of P_{crit} of 6, ≤7, and ≤10 GPa, respectively (Figure 3). Excluding stress-relaxation tests, deformation experiments using the multianvil apparatus require that gaskets (and octahedron) creep at fixed sample pressure. The idea that gaskets creep in the multianvil apparatus [*Bussod et al.*, 1993] seems to have its origin not in direct measurements of gasket thickness vs time, but rather in consistent observations of guide block displacement vs. time. Creeping gaskets is not the only explanation for guide block displacement, and since the observations in the previous section are incompatible with gasket creep at $P > P_{crit}$, either at fixed pressure or during pressure ramping, it must now be presumed that the gaskets do not creep. Therefore, until such time that there is direct evidence of gasket creep, previously published results based on the concept of creeping gaskets [*Bussod et al.*, 1993] must be viewed with skepticism.

The calibration measurements in this work are not exhaustive. Measurement scatter in some of the figures is very high. However, more rigorous calibration will require a major effort, for which no plans currently exist. The point in publishing this work now is not to settle the issue of creeping gaskets, but rather to bring to the attention of the community the possibility that they may not creep.

It remains likely that a small amount of inelastic deformation will occur early during runs at high pressure and temperature in the MA-8 apparatus, mainly from locked-in elastic strains generated by first pressurizing the assembly at room temperature, where the parts of the cell assembly (including the sample) are sufficiently hard to support very high differential stresses without flowing. When temperature is raised at run pressure, the sample will soften and deform up to the amount of original locked-in differential elastic strain, perhaps as much as 1–2%. This amount of deformation is sufficient to produce a distinct dislocation microstructure [*Bussod et al.*, 1993; *Sharp et* al., 1994] and may be sufficient to produce high-pressure faulting [*Green et al.*, 1990].

The ability to perform deformation experiments at extreme pressures is still a noble goal of geophysical research, and in fact we have not fully abandoned the possibility of doing such experiments in the MA-8 apparatus. We are currently exploring the idea of using materials of strongly contrasting elastic bulk moduli inside the MgO octahedron. With proper choice and arrangement of parts, low-magnitude pressure ramping can be exploited to impose a small displacement to a sample. For example, we calculate that if tungsten pistons are used (Figure 2), pressure ramping of 1 GPa at some level above P_{crit} should force an inelastic shortening of several tens of micrometers to any material located between the pistons. Furthermore, special piston geometries can amplify this deformation to many tens of percent of strain in a sample [*Karato and Rubie*, 1997].

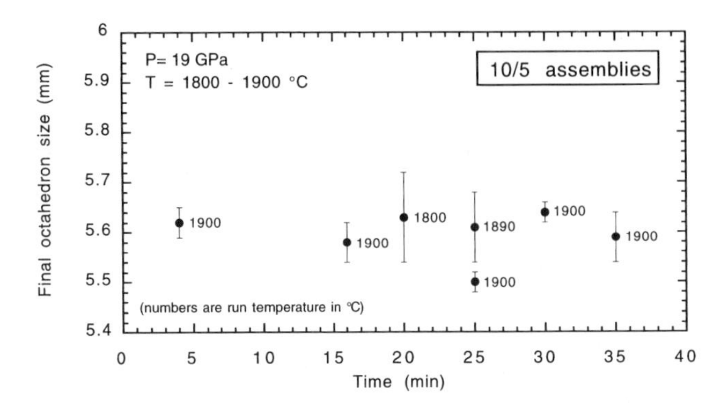

Figure 7. Final octahedron size as a function of run time for six 10/5 runs carried out with identical sample assemblies at identical pressures and temperature by the same researcher (N. Ross) on the same multianvil press. The lack of size change with run time even over a range of short duration runs is another argument in support of the displacements shown in Figure 6 not being related to octahedron size.

Acknowledgments. The paper was written while the first author was on sabbatical leave at the Bayerisches Geoinstitut, and he gratefully acknowledges the financial support of the Alexander von Humboldt Foundation during that leave. We want to thank the following staff members, students, and guest researchers at the Geoinstitut for cooperating with our intrusive measurements of their MgO octahedra following their tests: B. Poe, C. Geßmann, S. Karato, L. Kerschhofer, R. Knoche, M. Schmidt, N. Ross, and S. Webb. We are indebted to C. Dupas for contributing Figure 1. Discussions with E. Ohtani and T. Sharp contributed materially to this work and are greatly appreciated. The first author acknowledges that his work was performed under the auspices of the U. S. Department of Energy by the Lawrence Livermore National Laboratory under contract W-7405-ENG-48.

REFERENCES

Bussod, G. Y., T. Katsura, and D. C. Rubie, The large volume multi-anvil press as a high P-T deformation apparatus, *Pageoph, 141*, 579-599, 1993.

Canil, D., in *Short Course Handbook on Experiments at High Pressure and Applications to the Earth's Mantle*, edited by R. W. Luth, pp. 197-245, Mineralogical Society of Canada, Edmonton, Canada, 1993.

Green, H. W. II, T. E. Young, D. Walker, and C. H. Scholz, Anticrack-associated faulting at very high pressure in natural olivine, *Nature, 348*, 720-722, 1990.

Green, H. W. II, Phase transformation under stress and volume transfer creep, in *Mineral and Rock Deformation: Laboratory Studies (The Paterson Volume)*, *Geophys. Monogr. Ser.*, vol. 36, edited by B. E. Hobbs and H. C. Heard, pp. 201-211, AGU, Washington, D.C., 1986.

Ito, E., E. Takahashi, Y. Matsui, The mineralogy and chemistry of the lower mantle: an implication of the ultrahigh-pressure phase relations in the system MgO-FeO-SiO_2, *Earth Planet. Sci. Lett., 67*, 238-248, 1984.

Karato, S., D. C. Rubie, D. C., Towards an experimental study of deep mantle rheology: a new multi-anvil sample assembly for deformation studies under high pressures and temperatures, *J. Geophys. Res.*, submitted, 1997.

Liebermann, R. C. and Y. Wang, Characterization of sample environment in a uniaxial split-sphere apparatus, in *High-Pressure Research: Application to Earth and Planetary Sciences*, *Geophys. Monogr. Ser.*, vol. 67, edited by Y. Syono and M. H. Manghnani, pp. 19-31, Terra Scientific, Tokyo/ AGU, Washington D.C., 1992.

Rubie, D. C., C. R. Ross II, M. R. Carroll, and S. C. Elphick, Oxygen self diffusivity in $Na_2Si_4O_9$ liquid up to 10 GPa and estimation of high-pressure melt viscosities, *Am. Mineral., 78*, 574-582, 1993a.

Rubie, D. C., S. Karato, H. Yan, and H. St. C. O'Neill, Low differential stress and controlled chemical environment in multianvil high-pressure experiments, *Phys. Chem. Mineral., 20*, 315-322,1993b.

Sharp, T. G., G. Y. A. Bussod, and T. Katsura, Microstructures in β-$Mg_{1.8}Fe_{0.2}SiO_4$ experimentally deformed at transition-zone conditions, *Phys. Earth Planet. Int., 86*, 69-83, 1994.

Vaughan, P. J., H. W. Green, and R. S. Coe, Anisotropic growth in the olivine-spinel transformation of Mg_2GeO_4 under nonhydrostatic stress, *Tectonophysics, 108,* 299-322, 1984.

W. B. Durham, University of California Lawrence Livermore National Laboratory, P. O. Box 808, Livermore, CA 94550, USA (email: durham1@llnl.gov)

D. C. Rubie, Bayerisches Geoinstitut, Universität Bayreuth, D-95440 Bayreuth, Germany. (email: Dave.Rubie@uni-bayreuth.de)

GHz Ultrasonic Interferometry in a Diamond Anvil Cell: P-Wave Velocities in Periclase to 4.4 GPa and 207°C

A. H. Shen[1,*], H.-J. Reichmann[1], G. Chen[2], R. J. Angel[1], W. A. Bassett[3], and H. Spetzler[4]

P-wave velocities in periclase were determined up to 4.4 GPa and 207°C using a GHz ultrasonic interferometer and a resistance-heated diamond anvil cell. The samples were disks of synthetic periclase single crystal cut parallel to the (100) plane with dimensions of 300 μm in diameter and 46 μm in thickness. A 250-μm-thick tungsten-tantalum alloy as well as rhenium foils were used as gaskets. The signal was introduced through a buffer rod which was directly coupled to one of the diamond anvils by means of force applied to the buffer rod. The sample was coupled to one of the diamond anvils by means of normal force applied from the pressure medium, which was a fine-grain potassium bromide powder. The pressures in room temperature runs were determined by ruby fluorescence. Least squares fitting to the elastic constants derived from our experimental data yield c_{11} (GPa) = 296.8(±0.7) + 10.9(±0.9) × P (GPa). The results are in good agreement with previous determinations using ultrasonic interferometry on larger samples at lower frequencies and at lower pressures. The ultrasonic signals obtained in the high-temperature runs showed no deterioration in quality, and we were able to obtain travel-time data. By extrapolating the results from earlier work, we were able to estimate the pressure in the sample chamber while at high temperatures using the travel-time data.

INTRODUCTION

A knowledge of the elasticity of minerals is important in our quest to understand Earth's interior through the interpretation of seismic data. There are a number of techniques to measure elasticity at high pressures. X ray measurements yield dimensional changes directly and upon differentiation with pressure, elastic moduli [*Bassett et al.*, 1967]. No shear moduli can be obtained from X ray measurements, and thus neither compressional nor shear velocities can be calculated from X ray data. Other techniques such as Brillouin scattering [*Bassett and Brody*, 1977], impulsive stimulated scattering [*Zaug et al.*, 1993], transition metal ion fluorescence [*Chopelas*, 1992] and ultrasonic interferometry [*Jackson and Niesler*, 1982] yield the pressure derivatives of the dimensional changes directly, and both shear and compressional velocities can be determined. All these techniques except ultrasonic interferometry have been successfully used in the diamond anvil cells (DAC), and only recently [*Spetzler et al.*, 1996] has the ultrasonic interferometry joined these other techniques for measurements in the diamond anvil cell.

The major problem for performing ultrasonic interferometry in the diamond cell was the long wavelengths in the acoustic waves (a few hundreds μm for the traditional ultra-

[1]Bayerisches Geoinstitut, Universität Bayreuth, Germany

[2]CHiPR and Department of Earth and Space Sciences, SUNY at Stony Brook, New York

[3]Department of Geological Sciences, Cornell University, Ithaca, New York

[4]Department of Geological Sciences and Cooperative Institute for Research in Environmental Sciences (CIRES), University of Colorado, Boulder, Colorado

Properties of Earth and Planetary Materials
at High Pressure and Temperature
Geophysical Monograph 101

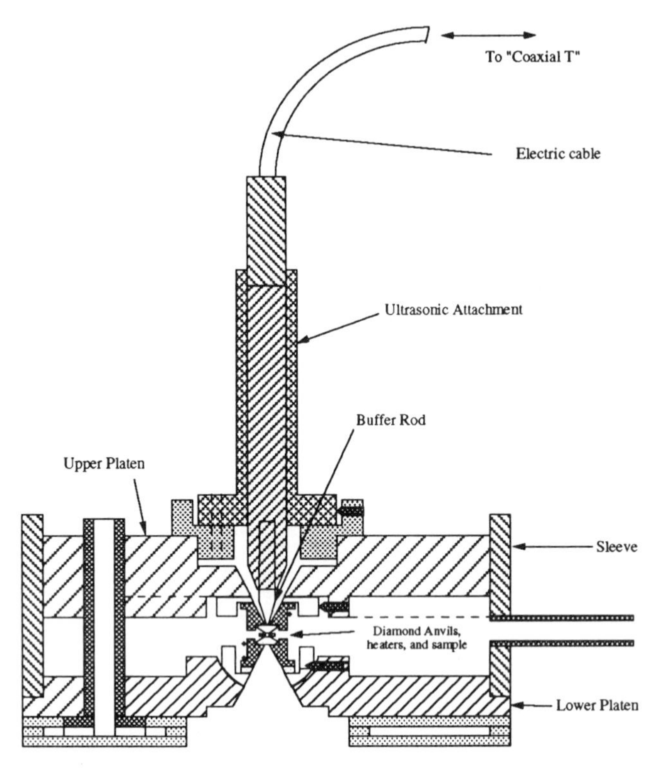

Figure 1. Cross section of the diamond anvil cell together with the ultrasonic attachment.

sonic interferometers operating in MHz range) in the sample. By increasing the frequency of the ultrasonic waves from tens of MHz to GHz, we were able to reduce the wavelength to a few micrometers. This short wavelength made it possible to perform ultrasonic interferometry on a very small sample loaded in the diamond cell. By adding ultrasonic GHz interferometry to the tools available in the DAC, the range of samples for which elasticity can be measured is significantly extended. This contribution reports p-velocity measurements on a 46-μm-thick sample of periclase to a pressure of 4.4 GPa and 207°C.

SAMPLES, APPARATUS, AND EXPERIMENTAL PROCEDURES

Sample

The sample used in this study was a synthetic periclase single crystal. The reasons for choosing periclase as our first sample are (1) periclase is an important mineral in Earth's interior and (2) the elasticity of periclase has been studied extensively by various methods [*Jackson and Niesler*, 1982; *Chopelas*, 1992], and this makes it a good candidate for comparing our results with others. The sample was oriented using a Weissenberg camera and carefully cut perpendicular to the [100] direction. It was then doubly polished to a thickness of 46 μm and cut into a 300-μm-diameter disc. The sample was checked again with the Weissenberg camera to ensure the orientation and was found to be within 0.5°. The thickness was measured with a micrometer under a microscope and the uncertainly is estimated to be about ± 0.5 μm.

Apparatus

The diamond anvil cell used in this study has been described by *Bassett et al.* [1993a, 1993b] and, thus, the details of operating this cell are referred to those publications. Figure 1 shows the diamond cell and the ultrasonic attachment. A diamond cell of this type is capable of achieving 10 GPa and 1000 °C. In this study, we used diamonds with 1-mm culet faces and thus limited the maximum pressure accessible to around 5 GPa. The gasket materials used were pure rhenium or 90% tantalum-10% tungsten alloy. We used foils with a thickness of 250 μm of either material and drilled 500-μm-diameter holes to serve as sample chambers. The gaskets were pre-indented to ensure the consistency of the size of the sample chamber.

The system setup is shown in Figure 2. It includes a pulse generator which is capable of generating two short pulses with fast rise and fall times (< 1 ns), a broad band RF synthesizer, and a high-frequency (20 GHz) sampling oscilloscope. Pulses were used to trigger the RF synthesizer to generate short packets of coherent sinusoidal waves. The transducer produces the acoustic waves from these packets. The reflections (echoes) from each interface encountered en route are received by the same transducer and converted into electrical wave forms to be detected by the oscilloscope. The whole system can be operated either

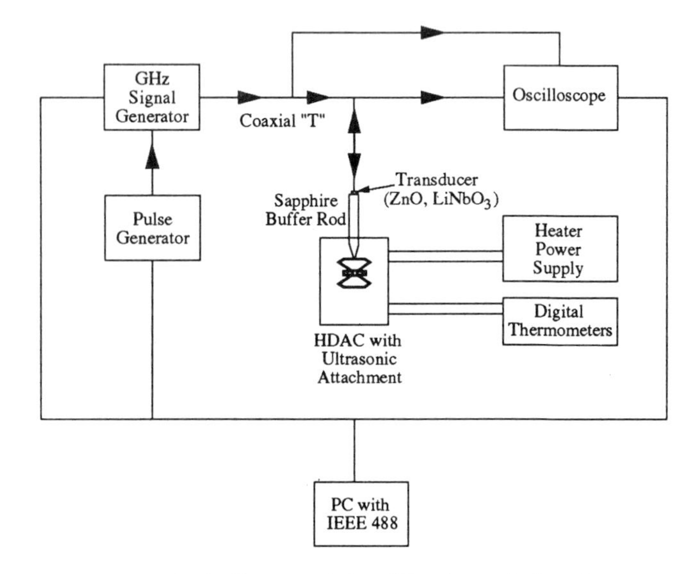

Figure 2. Block diagram of the GHz ultrasonic interferometer and the diamond anvil cell.

manually or controlled by a personal computer through an IEEE-488 bus. The actual data acquisition is coordinated through software which controls the functions of the synthesizer and the oscilloscope. The raw data consisting of sinusoidal traces of echoes at specified frequencies are stored on hard disks for later data reduction.

Experimental procedures

The sample loading procedure started with cementing the gasket to the lower diamond and filling the sample chamber with fine-grain potassium bromide powder with a few ruby grains situated close to the edge of the sample chamber. We then brought down the upper diamond to compact the KBr powder. After compacting the KBr powder, the periclase disk was placed upon the KBr wafer and was very gently compressed into the sample chamber. This procedure ensures good acoustic coupling between the diamond anvil and the sample without using any bonding material.

The ultrasonic attachment can be installed before or after assembling the diamond cell. In this experiment, we installed the ultrasonic attachment after the diamond cell had been assembled and loaded with the periclase sample. Because of the transparency of the sample, we were able to see through it and align the buffer rod. The attachment has three set screws for the translational adjustments and another three screws to provide rocking of the buffer rod assembly (Figure 1). Aligning the buffer rod assembly is a similar procedure to that used in aligning the diamond cell. First color fringes at the interface between the diamond and the buffer rod are used to guide the alignment. When good alignment is achieved, the whole interface appears dark or grey. The final adjustment of the alignment is done by maximizing the acoustic signal.

The principle of the ultrasonic interferometry in the diamond cell is depicted in Figure 3. More detailed description can be found in a previous publication [*Spetzler et al.*, 1996] and only the fundamentals are presented here. The top figure shows a sketch of the central parts of the diamond cell with the buffer rod attached. An acoustic signal consisting of one short packet of sinusoidal waves travels into the buffer rod and is reflected from each interface encountered en route. We identify three reflecting interfaces: the diamond-buffer rod interface, the sample-diamond interface, and the sample-pressure media interface. In the top figure, the reflections (echoes) were labeled 1, 2, and 3, corresponding to echo from each interface and will be called, hereafter, buffer rod echo, diamond echo, and sample echo, respectively. The amplitudes of the echoes depend on the relative impedances of the materials and on the quality of coupling. The bottom figure is a sketch showing the relative positions of the echoes as a function of time. The solid-outlined peaks represent echoes resulting from pulse P alone. The dashed-outlined and hashed peaks represent echoes resulting from a second pulse P' which is sent after P with time delay of one round trip travel-time in the sample. Because KBr has very low bulk modulus and therefore the acoustic impedance of the pressure medium in the pressure-temperature range covered in this study is lower than that of periclase, we believe that the diamond echo and the sample echo are in phase, i.e., the interference between the superimposed sample echo (3) and diamond echo (2') has integer interference order for constructive interference and half-integer for destructive interference. The following equation calculates the total phase shift in the nth echo during our experiments:

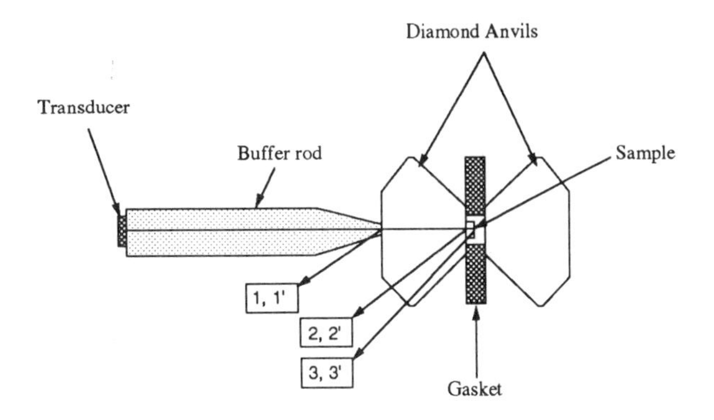

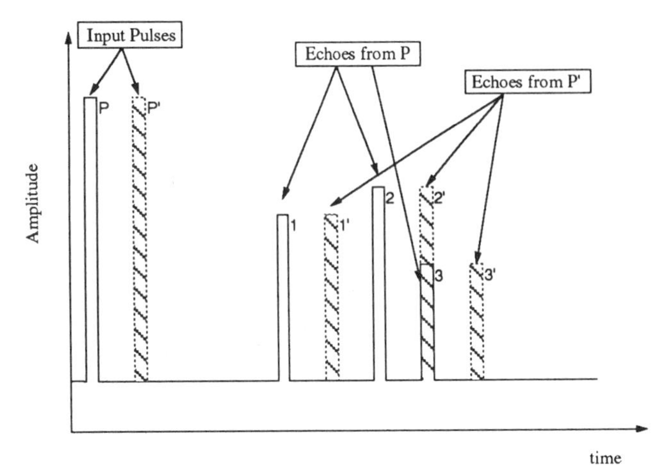

Figure 3. Schematic diagram depicts the principle of the ultrasonic interferometer when used to determine the elastic property of a sample loaded in a diamond anvil cell. P and P′ are the input pulses and the echoes 1, 2, 3 and 1′, 2′, 3′ are corresponding echoes from the diamond-buffer rod interface, from the diamond-sample interface and from the sample-pressure media interface, respectively.

$$\Phi = n\pi\left[1 + 2\left(\frac{2l}{\lambda} - m\right)\right] \quad (1)$$

where l the sample length (thickness), λ is the acoustic wavelength in the sample, and m is the integer number of whole wavelengths in $2l$. Maximum (constructive interfer-

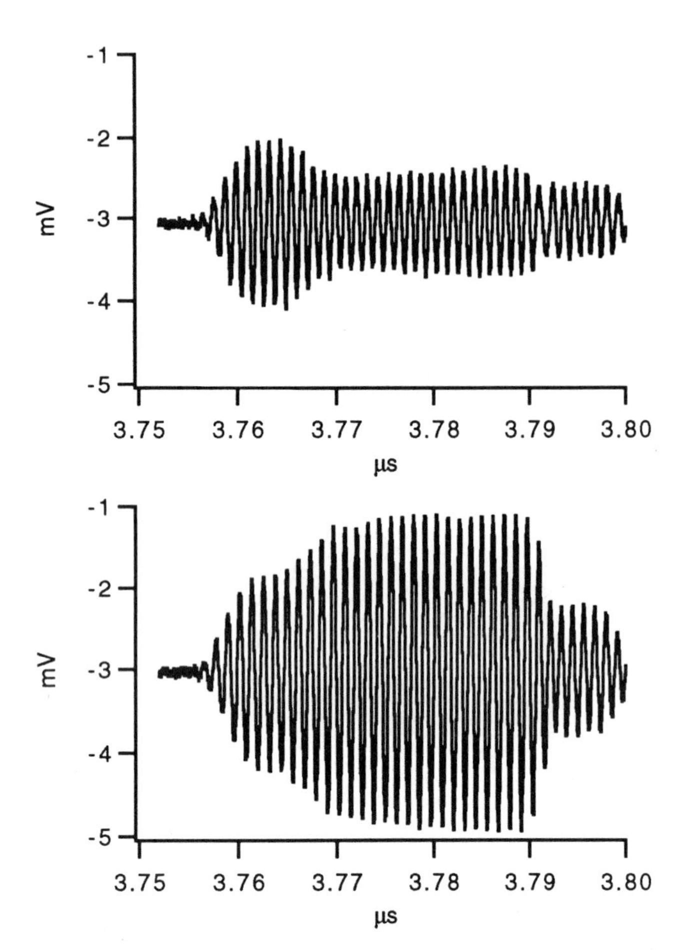

Figure 4. (top) Actual data at 850 MHz obtained using a 30 ns long pulse. The diamond echo arrived around 3.76 ms and the sample echo arrived around 3.77 ms. The destructive interference can be seen clearly around 3.78 ms. A few sine waves were taken between 3.76 and 3.77 ms and between 3.77 and 3.78 ms as the diamond echo and the interfered echo. These echoes were fitted with a sine function using a non-linear least squares procedure to find the amplitude. (bottom) Actual data from the same experimental run at 820 MHz; the constructive interference is clearly visible. The same data processing procedures were used.

ence) or minimum (destructive interference) amplitude between two interfering echoes occurs when $[(2l/\lambda)-m]$ is an integer or half integer.

In principle two sinusoidal wave packets must be superimposed to achieve the necessary conditions for making the interferometric measurement. However when the time between the onset of the two echoes is very short it may not be possible to work with two separate pulses. This is the case for the present sample. From previous studies [*Jackson and Niesler*, 1982; *Chopelas*, 1992], we know the p-velocity in the periclase sample to be about 10 µm/ns (km/s). Thus the round trip travel-time in our 49-µm-thick sample is about 10 ns. Rise- and fall-time limitations (and other electronic switching problems) control the minimum working pulse length to about 30 to 40 ns. We thus use a single pulse and stretch it to such time that the arrival time of the sample echo comes earlier than the end of the diamond echo and they overlap. In Figure 4 we show two oscilloscope traces depicting interference patterns for destructive (top) and constructive (bottom) interference. Note that only part of the signal shows clear interference. It is from within this part of the signal that five to ten sine waves are chosen and their amplitudes recorded over the entire frequency sweep. In all four experiments, we scanned the frequency from 300 MHz to 1.2 GHz and took amplitude data at 1.5 MHz intervals.

RESULTS

A total of four experimental runs was performed at room temperature and one each at 107°C and 207°C, respectively. The pressure ranged from 1.74 GPa to 4.36 GPa. The experimental results are tabulated in Table 1. Figure 5 shows echo amplitudes versus frequency for a typical run at 2.66 GPa. Complete data reduction involves demodulation of the interfered signal using the diamond echo and the sample echo as elaborated in *Spetzler et al.* [1993]. However, in this case because of the overlapping echoes, we could not single out the amplitude of the uninterfered sample echo, which is required for demodulation, and thus we used the amplitude of the uninterfered diamond echo to normalize the

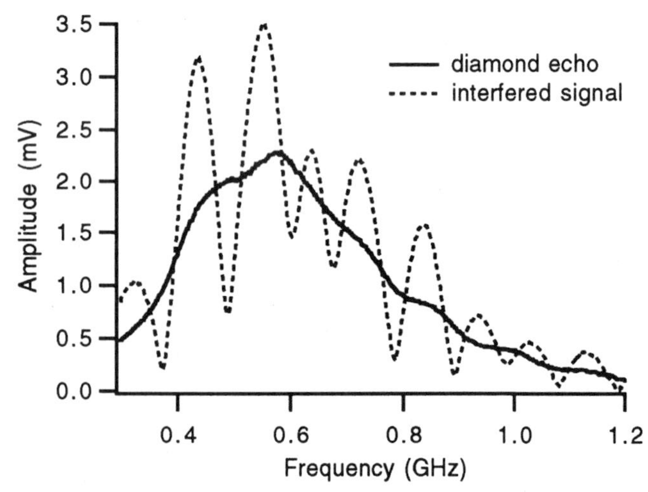

Figure 5. A plot of amplitude of the diamond echo and the interfered signal versus frequency. Notice only nine pairs of maxima-minima can be identified in the whole bandwidth of 900 MHz. These amplitudes were obtained by fitting parts of the signals shown in Figure 4 to sine functions. The low frequency modulation of the diamond echo is due to the impedance mismatch and other electronic parts (cables, etc.). This modulation has also influenced the interfered signal.

TABLE 1. Summary of the Experimental Runs

Run No.	Pressure (GPa)*	Travel time (ns)†	velocity (km/s)**	c_{11} (GPa)**
1105	1.74 (0.08)	9.87 (0.050)	9.29 (0.056)	312.1 (3.8)
1106	2.66 (0.08)	9.66 (0.051)	9.47 (0.057)	326.4 (4.0)
1016	3.43 (0.08)	9.61 (0.069)	9.51 (0.062)	330.3 (4.3)
1017	4.36 (0.08)	9.30 (0.080)	9.81 (0.068)	353.3 (4.9)
1203‡	1.94 (0.08)	9.28 (0.087)	9.34 (0.071)	315.9 (5.1)
1203a‡	0.84§	9.50 (0.046)		
1204‡	1.92§	9.45 (0.11)		

*Pressure determined by ruby fluorescence method for room temperature runs and uncertainty obtained from curve fitting.
†Travel time determined by using the extrema with frequencies > 700 MHz and method described in the text.
**Velocities were calculated with sample length corrected using a Birch-Murnaghan equation; the elastic constants were calculated using a room pressure density of 3.5798 g/cc. In addition, the error were calculated using the error propagation equation described in *Bevington and Robinson* [1992].
‡These runs used a sample of 43.5-μm thick. Run 1203a was made at 107 °C; run 1204 was made at 207 °C.
§These pressures were obtained by extrapolating the frequency-pressure-temperature relations in *Spetzler* [1970].

amplitude-frequencies plot shown in Figure 5. Figure 6 shows the result of such an exercise for the data collected at 2.66 GPa. Because the maxima or minima amplitude occurs when there is constructive or destructive interference according to equation (1), one can calculate the travel time according to the following relation:

$$t_{2l} = \frac{m_o + m_i}{f_{mi}} \tag{2}$$

where t_{2l} is the round trip travel time in the sample; f_{mi} is the frequency corresponding to the ith extrema; m_o is the integer number of whole wavelengths in $2l$, which is denoted as m(0) in the figure legend, and m_i is an integer when f_{mi} corresponds to a maximum amplitude, a half integer when f_{mi} corresponds to a minimum amplitude. The three curves (in different symbols) in Figure 6 show the correct (center) and the two adjacent interference order assignments. Least-squares fitting yields a travel time of 9.66 ns with a standard deviation of 0.051 ns, if we use only extrema with frequencies higher than 700 MHz. The same least squares procedure yields a travel time of 9.57 ns with a standard deviation of 0.15 ns, if all extrema are used. The travel times reported in this contribution used only the extrema with frequency higher than 700 MHz. *Spetzler et al.* [1993] recognized this behaviour in the GHz ultrasonic in-

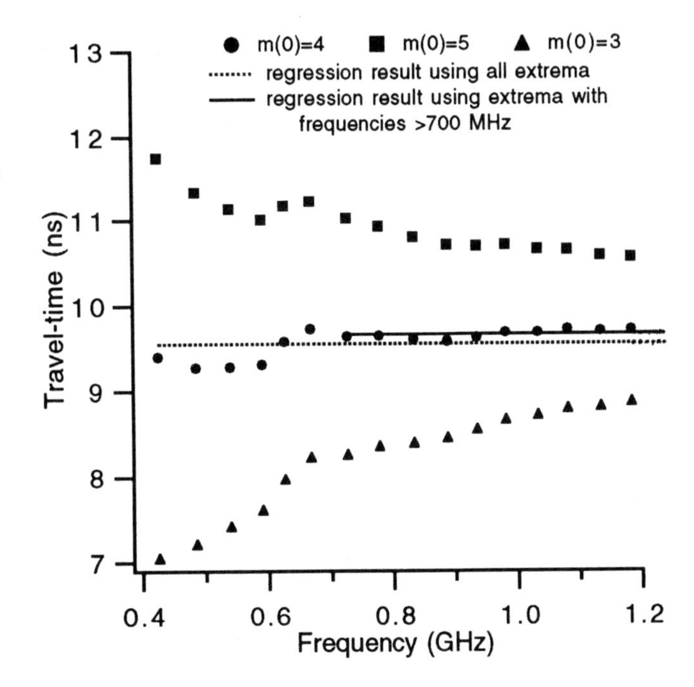

Figure 6. A plot showing the travel time determination. Each maximum and minimum in Figure 5 corresponds to a constructive and a destructive interference condition, respectively, and hence can be considered as an independent travel-time determination. The m(0) value given in the legend corresponds to the mo described in equation (2). The circles shows a nearly equivalent travel-time for each extremum with a m(0) value of 4. The squares and the triangles represent the neighboring fringes. The solid and dashed lines come from two different fitting schemes. The solid line resulted from regression using the extrema occurring at frequencies greater than 700 MHz (9.66 ± 0.051ns), and the dashed line resulted from using all extrema (9.57 ± 0.15 ns). These two procedures yielded different travel time and very different standard deviation. We used only the extrema with frequencies greater than 700 MHz to obtain the travel-time for each experimental run in this study.

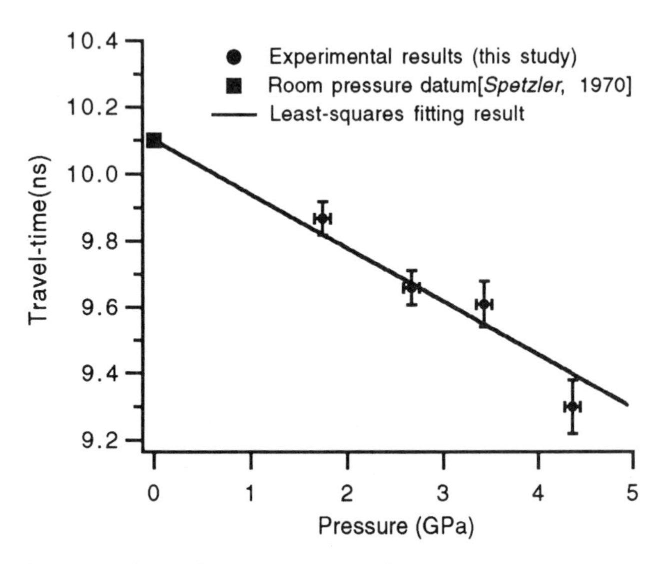

Figure 7. Plot of travel times from four room-temperature runs versus pressure. The room pressure datum was taken from *Spetzler* [1970]. The linear pressure dependence of the travel time can be seen from the regression line.

terferometer and suggested neglecting the low-frequency data during data reduction. The results from the four experimental runs are plotted as a function of pressure in Figure 7. The datum at room pressure was taken from *Spetzler et al.* [1970]. A weighted least-squares procedure was used to fit the experimental travel time data. The solid line represents the result of this procedure.

Figure 8 shows the derived c_{11} values from our travel-time data. The derivation process involves calculating the length of the sample while at high pressure; we have chosen the bulk modulus value from *Jackson and Niesler* [1982], because this facilitates comparing our results to theirs. The solid line in this figure was obtained by fitting the c_{11} elastic constants as a linear function of pressure using a weighted least-squares fitting procedure. We otained a linear pressure dependence of c_{11} (GPa) = 296.8 (±0.7) + 10.9 (±0.9) × P (GPa). The results from *Jackson and Niesler* [1982] are depicted as a dashed line. Considering that there exists uniaxial stress in our experiments, our c_{11} is in very good agreement with Jackson and Niesler's results.

The data we obtained at 107°C and 207°C showed no deterioration of the signal quality and thus demonstrate that there are, at least in principle, no serious problems in making travel-time measurements at elevated temperature. The starting pressure for these measurements was 1.94 GPa. We do not know the pressure at temperature since it was not possible with the experimental setup to transport the heated cell to the ruby fluorescence spectrometer. We used the known equation of state of MgO at simultaneous pressure and temperature [*Spetzler*, 1970] to estimate the pressure and found 0.84 GPa for the 107°C and 1.92 GPa for the 207°C.

DISCUSSION

In this contribution, we have shown that it is possible to perform ultrasonic interferometry on a sample loaded in a diamond cell and obtain meaningful elastic data on minerals at high pressure and moderate temperature conditions. However, there are a few points that still require further effort and improvement.

First, the sample dimension at ambient conditions was not very accurately measured, thus preventing accurate velocity determination. For minerals for which the velocities are accurately known at ambient conditions, the GHz interferometer can be used to measure the initial length. Various optical techniques are also available with which one can obtain higher precision in measurement than is available from a mechanical micrometer. For example, we modified a dial gauge by adding mirrors and used a Michelson interferometer to count fringes, achieving 1/2 wavelength (0.3 μm) precision.

Secondly, the sample loading method and the pressure media cannot create a hydrostatic environment for the sample. In some of our loading efforts we observed by polarized light microscopy some stress development within the

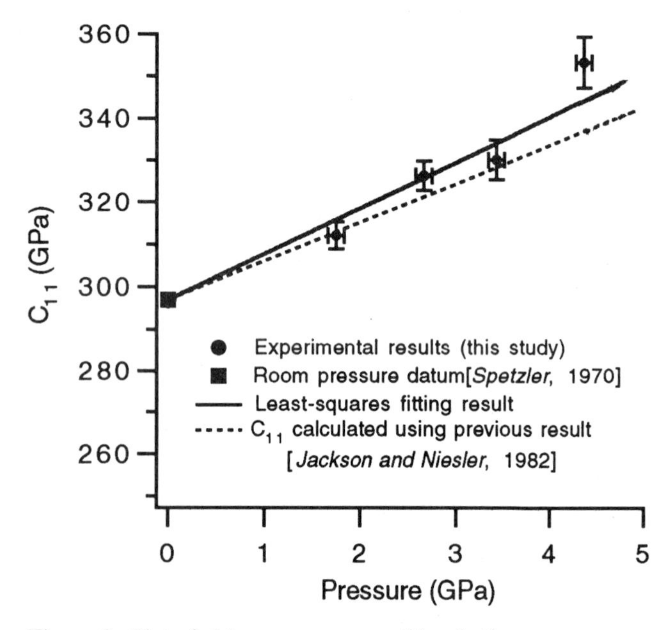

Figure 8. Plot of c11 versus pressure. The elastic constants were calculated using the travel-time determinations in this study and lengths corrected using the bulk modulus and the pressure derivative of the bulk modulus from *Jackson and Niesler* [1982]. The large error bars in the elastic constants come mainly from the uncertainty in our thickness determination (±0.5 mm). Our regression line yielded the same initial value, but somewhat different pressure dependence (slope). However, considering the possible high uniaxial stress in our runs, the agreement between these two studies is actually very good.

crystal and preliminary X ray diffraction experiments confirmed that we were working with a strained sample. However, using a fluid pressure medium such as a methanol:ethanol mixture can create severe problems in maintaining the sample-to-diamond bond and thus problems in transmitting ultrasound through this interface. Making good contact between the sample and the diamond in a hydrostatic environment without using a separate bonding material remains a challenge. We are therefore exploring various schemes for designing weak springs to hold samples in a hydrostatic environment.

Thirdly, coupling shear wave into the diamond cell is difficult. We have (once) successfully produced a 600-MHz shear transducer at the end of a buffer rod and made shear measurements on a garnet sample [*Chen et al,.* in press]. In the present effort, we were able to couple the acoustic waves through the diamond anvil, but we failed to obtain a sample echo in the diamond cell. We are still working on solving this problem and have begun a systematic study of the bonding and polishing involved in producing a 600-MHz transducer from a 25-MHz lithium niobate commercial transducer.

Acknowledgments. A.H.S. would like to acknowledge many helpful and fruitful discussions with S. Webb on various aspects of ultrasonic measurements. H.S. gratefully acknowledge the financial support of NSF through grants EAR-9304451 and EAR-8916327, the University of Colorado for the granting of a sabbatical leave, and the Alexander von Humboldt Foundation for awarding the Humboldt Award which made his extended stay at the Bavarian Geoinstitute possible. G.C. thanks CHiPR for a two week leave of absence which made this collaborative effort possible. Thorough reviews and suggestions from the reviewers greatly improved the readability and clarity of the original manuscript.

REFERENCES

Bassett, W. A., T. Takahashi, and P. W. Stook, X-ray diffraction and optical observations on crystalline solids up to 300 Kbar, *Rev. Sci. Instrum., 38,* 37-42, 1967.

Bassett, W. A., and E. M. Brody, Brillouin scattering: A new way to measure elastic moduli at high pressures, in *High-Pressure Research: Applications in Geophysics*, edited by M. H. Manghnani and S. Akimoto, pp. 519-532. Academic Press, New York, 1977.

Bassett W. A., A. H. Shen, M. J. Bucknum, and I-Ming Chou, A new diamond anvil cell for hydrothermal studies to 2.5 GPa and -190°C to 1200°C, *Rev. Sci. Instrum., 64,* 2340, 1993a.

Bassett W. A., A. H. Shen, M. Bucknum, and I-Ming Chou, Hydrothermal studies in a new diamond anvil cell up to 10 GPa and from -190 °C to 1200 °C, *PAGEOPH., 141,* 487-495, 1993b.

Bevington, P. R., and D. K. Robinson, *Data Reduction and Error Analysis for the Physical Sciences*, 328 pp., McGraw-Hill, New York, 1992.

Chen, G., R. Miletich, and H. Spetzler, Shear and compressional mode measurements in GHz ultrasonic interferometry and velocity-composition systematics for the pyrope-almandine solid solution series, *Phys. Earth Planet. Interior.*, in press.

Chopelas, A., Sound velocities of MgO to very high compression, *Earth Planet. Sci. Letters, 114,* 185-192, 1992.

Jackson, I., and H. Niesler, The elasticity of periclase to 3 GPa and some geophysical implications, in *High-Pressure Research in Geophysics*, edited by S. Akimoto and M. H. Manghnani, pp. 93-113, Center for Academic Publications/D. Reidel, Tokyo/Dordrecht, 1982.

Spetzler, H., Equation of state of polycrystalline and single-crystal MgO to 8 kilobars and 800°K, *J. Geophys. Res., 75,* 2073-2087, 1970.

Spetzler, H., G. Chen, S. Whitehead, and I. Getting, A new ultrasonic interferometer for the determination of equation of state parameters of sub-millimeter single crystals, *PAGEOPH, 141,* 341-377, 1993.

Spetzler, H., A. H. Shen, G. Chen, G. Herrmannsdoerfer, H. Schulze, and R. Weigel, Ultrasonic measurements in a diamond anvil cell, *Phys. Earth Planet. Interior, 98,* 93-99, 1996.

Zaug, J. M., E. H. Abramson, J. M. Brown, and L. J. Slutsky, Sound velocities in olivine at earth mantle pressure, *Science, 260,* 1487-1489, 1993.

R. J. Angel, H.-J. Reichmann, and A. H. Shen, Bayerisches Geoinstitut, Universität, Bayerishces, Bayreuth, D-95440, Germany

W. A. Bassett, Department of Geological Sciences, Cornell University, Snee Hall, Ithaca, NY 14853-1504, USA

G. Chen, CHiPR and Department of Earth and Space Sciences, SUNY at Stony Brook, Stony Brook, NY 11794, USA

H. Spetzler, Department of Geological Sciences and CIRES, University of Colorado, Boulder, CO 80309, USA

*Current address: Department of Earth Sciences, University of Cambridge, Downing Street, Cambridge CB2 3EQ, UK

A New Facility for High-Pressure Research at the Advanced Photon Source

Mark L. Rivers, Thomas S. Duffy,[1] Yanbin Wang, Peter J. Eng, Stephen R. Sutton, and Guoyin Shen

Consortium for Advanced Radiation Sources, The University of Chicago

The Advanced Photon Source (APS) is a third-generation synchrotron storage ring that became operational in 1996. A national user facility at the APS is being constructed for research in Earth, soil, and environmental sciences by the GeoSoilEnviroCARS (GSECARS) group of the Consortium for Advanced Radiation Sources (CARS). The GSECARS sector consists of an undulator and a bending magnet beamline, both of which have been designed to allow for a wide range of possible experiments on geological materials. An experimental station on the undulator beamline will be dedicated to high-pressure experiments using a multi-anvil press and diamond anvil cells. Energy-dispersive and monochromatic diffraction experiments will be performed in this and other stations using solid state and two-dimensional detectors. A high-pressure support laboratory is being developed concurrently. The first high-pressure experiments at the GSECARS sector were successfully conducted in December, 1996.

1. INTRODUCTION

Many advances in the study of condensed matter at high pressure have been made possible through the development of high-intensity synchrotron X ray sources. The drive to study ever smaller and more complex samples at increasingly high pressures and temperatures and with higher accuracy probes is now being aided by the development of a new generation of synchrotron facilities at locations in Europe, Japan, and the United States. Compared with existing second-generation synchrotron facilities, these third-generation synchrotron sources produce radiation with higher energy and higher brilliance, both of which are advantageous for high-pressure experiments. In this paper, we describe the high-pressure research facility that is being developed for a new third-generation synchrotron near Chicago, Illinois.

A synchrotron storage ring consists of a vacuum ring with alternating straight sections and arcs in which bunches of charged particles (e.g., positrons) travel at nearly the speed of light. Magnets are used to steer and focus the particle beam, and energy losses are replenished by radiofrequency cavities. Bending magnets are used to deflect the particles into a circular arc. The inward acceleration causes an intense, well-collimated beam of x-rays to be emitted along a tangential path from the arc. Insertion devices, consisting of arrays of pairs of alternate polarity magnets, are used to extract even higher intensity radiation from straight sections of the ring by causing the beam particles to oscillate as they travel.

[1]Now at Department of Geosciences, Princeton University, Princeton, NJ 08544.

Properties of Earth and Planetary Materials
at High Pressure and Temperature
Geophysical Monograph 101

TABLE 1. APS Parameters

Property	Value
Energy	7.0 GeV
Beam particle	Positron
Beam current	>100 mA
Lifetime	> 10 hrs
Fill time	3.7 min
Horizontal emittance	7 nm rad
Vertical emittance	0.7 nm rad
Undulator brilliance at 8 keV*	2 x 10^{18}
Bending magnet critical energy	19.6 keV
Positional stability	20 μm
Directional stability	2 μrad
Number of bunches	1-60
Bunch duration	73 ps
Circumference	1104 m
Number of insertion devices	35

*In units of photons/sec/$mrad^2$/mm^2/0.1% bandwidth.

The two types of insertion devices are called wigglers and undulators. For a given magnetic field strength, a wiggler generates radiation that adds to give a combined flux that is N times greater than a bending magnet, where N is the number of pairs of magnets comprising the wiggler. An undulator uses a weaker magnetic field to produce smaller particle displacements than a wiggler. In this case, the emitted radiation interferes to produce a spectrum of harmonic peaks. The wavelength of the harmonic peaks can be tuned by changing the magnetic gap, and hence field strength. If the undulator gap is tapered, rather than uniform, then the harmonics are broadened, and the energy distribution can approach the smooth spectrum of a wiggler.

The emitted radiation from a bending magnet or insertion device is directed down a beamline, which consists of a set of evacuated pipes transporting the beam through a series of enclosures. A shielding wall separates the storage ring from the experimental floor and beryllium windows are used to isolate separate evacuated portions of the beamline and storage ring. The radiation is first transported through one or more optics enclosures which can contain a variety of components including shutters, X ray mirrors, monochromators, and beam diagnostics equipment. The beam then reaches experimental stations where the user-controlled experimental apparatus is housed. Detailed description of the characteristics and operation of synchrotron radiation facilities can be found in *Koch* [1983], *Buras and Gerward* [1989], *Finger* [1989], *Bassett and Brown* [1990], *Brefeld and Gürtler* [1991], and *Smith and Rivers* [1995].

The Advanced Photon Source (APS) is a 7 GeV storage ring that is now in operation at Argonne National Laboratory in Argonne, Illinois. The main characteristics of the APS are summarized in Table 1 [*Moncton*, 1996]. The design goal for the facility (a current of 100 mA with a lifetime of 10 hrs) has been achieved. The eventual performance is expected to be much better than this (e.g., 300 mA current with 100 hr lifetime and fill time of less than 30 sec) [*Moncton*, 1996]. Compared to bending magnet beamlines at existing second-generation sources, the APS will deliver an X ray brilliance that is at least 10^2-10^3 times greater. The narrow, intense, short-wavelength beam at the APS is especially well-suited to high-pressure experiments which are limited by small sample volumes and absorption in the anvil materials.

The APS is a large facility situated on a 79 acre site. The storage ring has a circumference of 1100 m. The experimental floor is divided into 35 sectors, each of which consists of an insertion device and a bending magnet beamline. The sectors are being developed by independent collaborative access teams which are typically composed of members from universities, national laboratories, and industry. In the present phase of development, 20 sectors are being constructed. In addition to the beamlines on the experimental floor, each sector has 1200 ft^2 of laboratory space and an office area of 1500 ft^2. A user residence facility for housing visiting researchers is now open.

The Consortium for Advanced Radiation Sources (CARS) is a collaborative access team that is constructing three sectors at the APS. The GeoSoilEnviroCARS (GSECARS) group is designing a sector for research in Earth, soil, and environmental sciences. The other CARS sectors will be devoted to biological sciences (BioCARS) and chemistry and materials science (ChemMatCARS). The primary experimental techniques to be used at the GSECARS sector will be single-crystal and powder diffraction (at ambient and high pressure), X ray absorption spectroscopy, X ray fluorescence microprobe analysis, and microtomography.

High-pressure experiments will be conducted at the APS using two types of pressure-generating devices: the diamond anvil cell (DAC) and the large-volume press (LVP). The large-volume press is characterized by relatively large sample volumes (0.1-1 mm^3) and good pressure and temperature stability. The diamond cell, on the other hand, is compact, can easily achieve much higher pressures and temperatures and is transparent to a broad range of electromagnetic radiation.

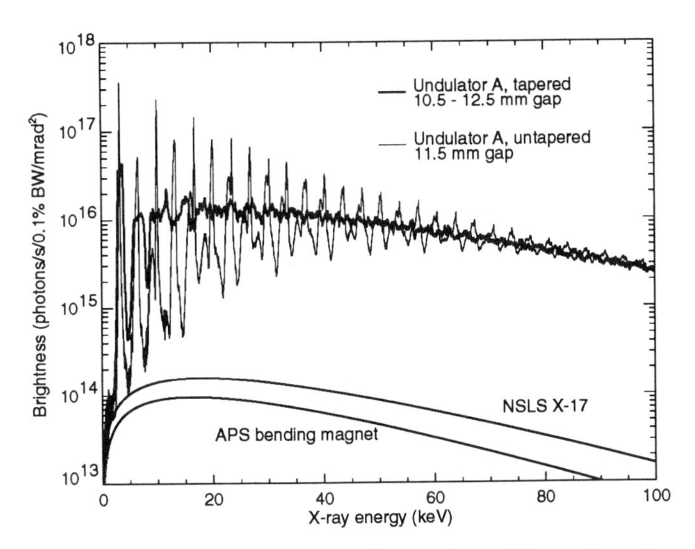

Figure 1. Brightness spectra for the APS undulator A and bending magnet. Spectra for the undulator are shown for both its normal configuration and using a tapered gap. For comparison, the spectrum of the superconducting wiggler (X17) at the NSLS is shown.

2. BEAMLINE COMPONENTS

The GSECARS sector consists of an undulator and a bending magnet beamline. The undulator is the standard APS undulator A, which is a 3.3-cm-period device. The first harmonic is tunable from 3.3 keV to 13 keV by opening the magnetic gap from 10.5 to 25 mm. In its standard configuration, the undulator produces a spectrum of harmonic peaks of approximately 1% FWHM. Harmonic peaks out to 100 keV are predicted and have been observed in the initial operation of the device. The undulator spectrum is ideal for monochromatic diffraction experiments. By changing to a tapered gap, the harmonic structure beyond 20 keV is nearly eliminated, and the spectrum resembles that from a wiggler but with higher on-axis brightness.

The spectrum from the bending magnet at the APS has a critical energy of 20 keV, which is the same at that of the superconducting wiggler at X17 of the National Synchrotron Light Source (NSLS). The bending magnet is an excellent source for energy-dispersive diffraction experiments when the brightness of the undulator is not required.

There are a number of possible figures of merit which can describe the characteristics of a synchrotron beam. The flux is the number of photons/sec/horizontal angle emitted in a relative energy bandwidth of $\delta\lambda/\lambda = 0.001$, integrated over the entire vertical angle. The brightness is flux/vertical angle which is essentially the number of photons/s per solid angle. This is relevant for experiments that use a pinhole or collimator to select a portion of the beam to interact with the sample. Finally, the brilliance is the brightness/source area. This is the appropriate measure when optics are used to focus the entire beam onto the sample. Figure 1 compares the brightness spectra for the APS undulator in tapered and untapered mode, the APS bending magnet, and the superconducting wiggler of the NSLS.

The layout of the GSECARS sector is shown in Figure 2. The two beamlines are referred to as 13-ID (insertion device) and 13-BM (bending magnet). Each beamline has 4 radiation enclosures: two optics enclosures (the A and B stations) and two experimental stations (the C and D stations). The two experimental stations of the bending magnet beamline are completely independent. The two stations on the insertion device beamline can operate independently for certain experiments and must share time for others.

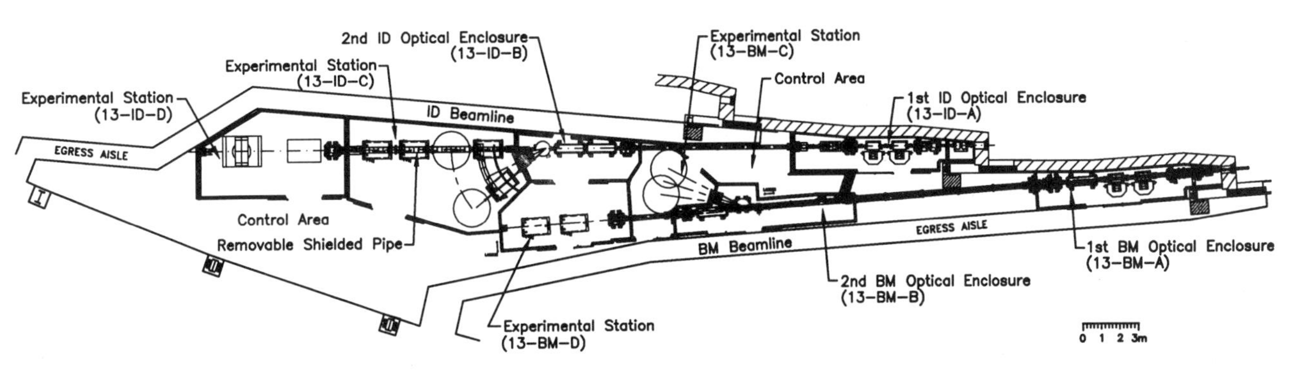

Figure 2. Layout of the GSECARS sector at the APS.

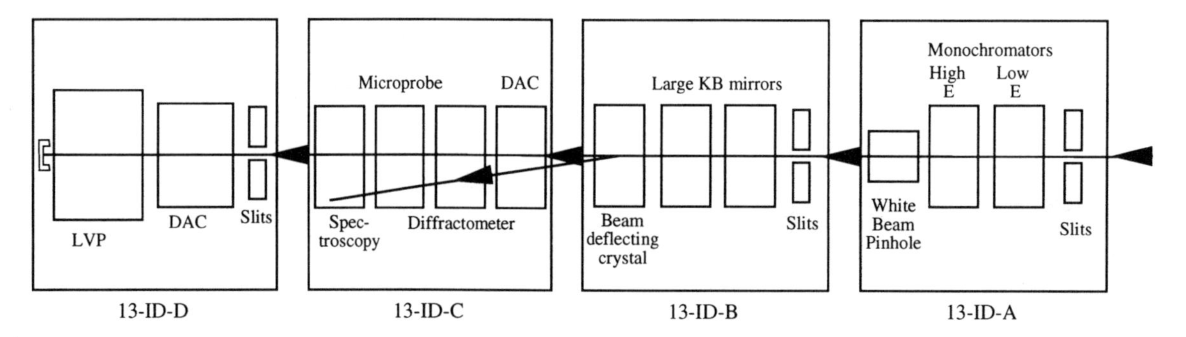

Figure 3. Schematic diagram of the main components of the insertion device beamline.

2.1 Insertion Device Layout

The first optics enclosure (FOE) of the undulator beamline (13-ID-A) will contain a set of primary slits, cryogenically cooled, high- and low-energy monochromators, a beam diagnostics tank, and a white beam pinhole. The low-energy monochromator will operate from 4.5 to 21 keV and the high-energy monochromator will be usable between 15 and 80 keV. The first crystal on the low-energy monochromator is thin, and the transmitted beam can be used directly or with the high-energy monochromator. Thus, the two monochromators can be used simultaneously. The low-energy monochromator produces a vertical +50 mm offset in the beam and the high-energy monochromator produces a -19 mm vertical offset. The pinhole is used to restrict the total white-beam power downstream of the FOE to less than 100 W (at 100 mA current) in order to simplify the power handling requirements of downstream optics. The pinhole dimensions are approximately 0.65 mm (vertical) x 1.1 mm (horizontal).

The second optics enclosure will contain additional slits, a pair of water-cooled ~1-m-long focusing mirrors, and a beam deflecting crystal. The large mirrors will be of the Kirkpatrick-Baez (KB) type: focusing is achieved by the grazing incidence reflection of a mirror bent to an elliptical shape. These mirrors will be capable of collecting the undulator beam (white or monochromatic) and focusing it to less than 100 μm. The actual demagnification ranges of the KB mirrors will be from 10:1 to 3:1 in the 13-ID-C and 13-ID-D stations. The Ge (111) beam deflecting crystal is capable of deflecting the low-energy monochromatic beam sideways into the 13-ID-C experimental station. The use of this crystal allows for simultaneous operation of the two experimental stations on the insertion device beamline. A schematic of the insertion device beamline components is shown in Figure 3.

Two experimental stations are planned for this beamline. The upstream station (13-ID-C) can be used in focused or unfocused white-beam mode for energy-dispersive experiments in the DAC. It can also be used for angle-dispersive DAC experiments. The end station (13-ID-D) will be dedicated to high-pressure experiments with a permanently installed large-volume press and an optical table for diamond cell experiments. Both monochromatic and white-beam experiments using either focused or unfocused beams will be performed here.

2.2 Bending Magnet Layout

The bending magnet beam is split into two independent beamlines (Figure 4). In 13-BM-A, the 6-mrad bending magnet fan is split by a fixed aperture plate into two components: a 2.5-mrad outboard fan for 13-BM-D and 1.5-mrad inboard fan for 13-BM-C. The first optics enclosure also houses slit tanks, high- and low-energy monochromators (for 13-BM-D), and a conical mirror for 13-BM-C. The second optics enclosure contains a bent flat vertically focusing mirror for 13-BM-D and a single bounce, horizontally focusing monochromator for 13-BM-C. The first station (13-BM-C) is a side station which is primarily dedicated to powder and single-crystal diffraction at 7.5-20 keV. The end station (13-BM-D) can use white beam, or radiation from either the high- or low-energy monochromators with sagittal focusing and a bent flat vertical focusing mirror. Both white and monochromatic diffraction experiments using either the diamond cell or the large-volume press can be carried out in the bending magnet end station.

3. HIGH-PRESSURE EXPERIMENTAL PROGRAM

High-pressure experiments generally require very small beam size, short-wavelength photons, and a high-brilliance source. Small X ray beams (~10 μm) are particularly needed for DAC experiments to minimize the effects of pressure and temperature gradients and to avoid signal from the gasket or pressure-transmitting medium. Absorption in the anvils requires the use of

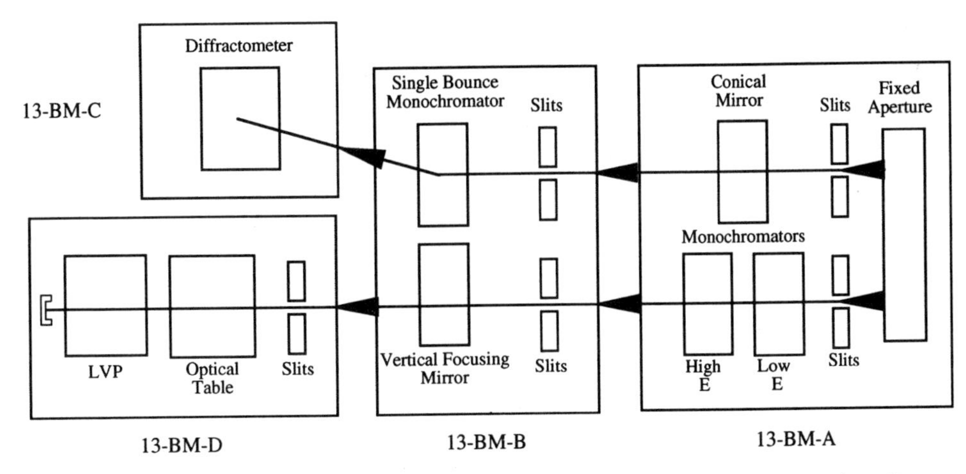

Figure 4. Schematic illustration of the primary components of the bending magnet beamline.

photons with energies above 15 keV for the diamond cell and above 20 keV for the large volume press. The extremely small volumes which can be held at high pressure (micro- to nanoliters for the large-volume press, nano- to picoliters for the diamond cell) requires the use of the highest brightness X ray source possible. The primary scientific goal of the high-pressure experimental program at the APS will be to significantly extend the range (e.g., pressure and temperature) of feasible high-pressure experiments. At the same time, there will be a major effort to improve the accuracy of the experimentally determined quantities.

3.1 Diamond Anvil Cell

Experimental stations for diamond cell research have been constructed at a number of second-generation synchrotrons [e.g., *Olsen*, 1992; *Brister*, 1992; *Hu et al.*, 1994; *Ancharov et al.*, 1995] and are currently being developed for third-generation sources at the European Synchrotron Radiation Facility [*Häusermann and Hanfland*, 1996], Spring-8 in Japan [*Shimomura et al.*, 1992*a*], and the APS. The diamond cell program at the APS will be divided into two phases. The first phase will use existing techniques (primarily energy-dispersive diffraction) to carry out frontier experiments utilizing the high brightness and coherence of the third-generation source. The second phase of experiments will use newer techniques (e.g., angle-dispersive diffraction, area detectors) for high precision studies.

A small Kirkpatrick-Baez mirror system has been developed for providing very small (<10 μm) X ray beams for diamond cell experiments [*Yang et al.*, 1995; *Eng et al.*, 1995]. These mirrors consist of 100-mm long Pt- or Rh-coated glass bent to an elliptical shape with a two-moment bender. In initial tests of this system, a 70 x 70 μm collection area was focused to a 3 x 9 μm focal spot with an efficiency of greater than 90% for wavelengths greater than 0.2 Å [*Eng et al.*, 1995].

An energy-dispersive diffractometer for diamond cell experiments has been constructed (Figure 5). This instrument consists of a two-circle horizontal diffractometer mounted on an optical table. The sample stage has both motorized and manual x-y-z translations and ω and χ rotations. A kinematic base is used for reproducibly positioning the diamond cell. An optical microscope with CCD camera is available for sample viewing. The incident X ray beam is controlled by a pair of adjustable slits made from WC cubes (from a multi-anvil device). Beam sizes down to 10 μm can be obtained in this way. The small KB mirrors are used to produce focused beams of less than 10 μm. The detector rotates

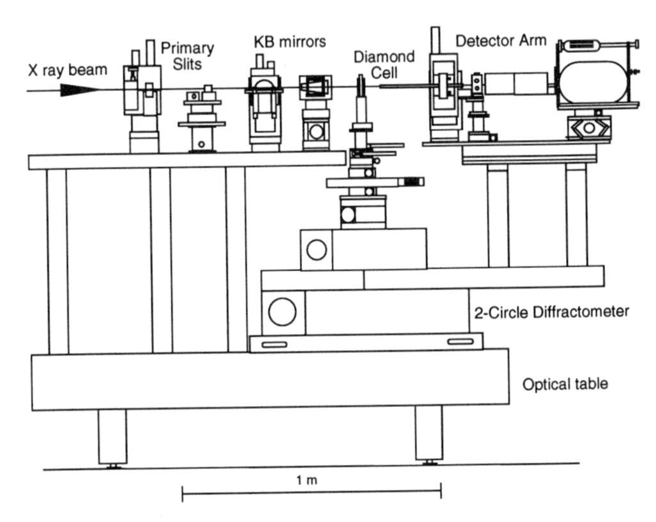

Figure 5. Energy-dispersive diffractometer for the high-pressure experimental station.

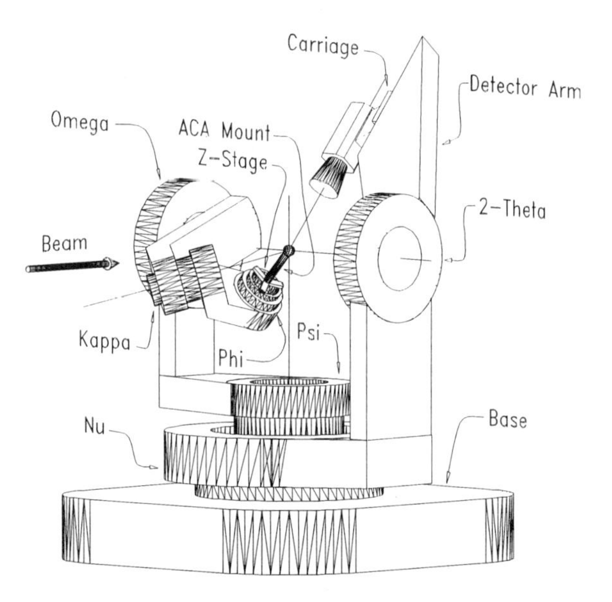

Figure 6. General layout for the 2+2+Kappa diffractometer for micro-crystal and diamond anvil cell experiments.

on a 2θ arm that has the same center of rotation as the sample stage. Two sets of slits define the diffraction angle and exclude scattered radiation from the diamonds and gasket. The diffracted intensity is recorded with an energy-sensitive solid state detector and multichannel analyzer. Overall, the system is similar to a previous one developed for beamline X17C of the NSLS [*Hu et al.*, 1994], but has been scaled up for increased stability and to handle larger and more complex sample assemblies.

The diffractometer was tested at X17B of the NSLS with a novel laser heating system [*Mao et al.*, this volume] consisting of a 100 W multimode YAG laser, optics to heat the sample from both sides, temperature stabilization, and a two-dimensional CCD for temperature measurement. This combined laser heating-energy dispersive diffraction system has been used to obtain *in situ* structural data on metals, alloys, and silicates to temperatures in excess of 1700 °C and pressures above 90 GPa (G. Shen et al., manuscript in preparation).

In addition to the two-circle diffractometer, a multi-axis diffractometer will be available for both energy dispersive and angle dispersive diamond cell diffraction. This diffractometer represents a new design, and is optimized for performing diffraction measurements on small samples contained in complex sample environments. The sample cradle consists of a fine motorized x-y-z stage capable of supporting large sample environments weighing up to 20 kg with negligible loss in sample centering over a wide angular range. The sample stage is also designed to accommodate a high-power liquid He and N_2 flow cryostat capable of rapidly cooling an object the size of a diamond anvil cell to 4.2 K. The stage is integrated into a Kappa goniometer with an inclination angle of 50°. To accommodate the restricted scattering geometry imposed by diamond anvil cells, the diffractometer has two additional degrees of freedom compared to a classic Kappa diffractometer. There is one additional horizontal ψ circle with a vertical rotation axis located below the sample cradle, and a second similar but independent ν circle located below a traditional horizontal axis 2θ detector circle. The geometry is referred to as 2+2+Kappa (Figure 6).

Having two degrees of freedom on the detector arm allows the diffraction plane to take on an arbitrary orientation and removes the need to rotate the diffraction vector of the sample onto a fixed diffraction plane. This reduces the range of sample rotations needed, resulting in better sample centering stability for monochromatic measurements, and can completely eliminate the need to rotate the sample for energy dispersive single crystal experiments. The detector arm can carry loads up to 40 kg without loss of centering and this allows for the possibility of mounting, at the same time, both an area detector and an energy dispersive detector for micro single crystal energy dispersive diffraction.

3.2 Multi-anvil Press

Multi-anvil high-pressure devices have been used extensively at second- generation synchrotron sources such as the Photon Factory, Tsukuba, Japan [*Shimomura et al.*, 1992*a,b*] and the National Synchrotron Light Source [*Weidner et al.*, 1992]. A wide range of experiments have been carried out, including solid-state and melt phase equilibria, P-V-T equation of state, strength and other rheological properties, as well as crystallography using monochromatic X rays. New multi-anvil press beam lines are being developed at Daresbury and Spring 8 [*Shimomura et al.*, 1992a].

The multi-anvil press at the APS will be capable of generating pressures beyond 30 GPa and temperatures up to 2800 °C. Our design philosophy for the large-volume presses is to decouple the hydraulic presses, which are used for applying the force for pressure generation, from the pressure tooling which can have various pressure generating mechanisms. Specifically, we will utilize two major types of pressure tooling, both having a long track record in large-volume high-pressure research. The first is the so-called DIA, which consists of the upper and lower pyramidal guide blocks (bolsters) installed on the heads of the hydraulic press rams, four trapezoid end blocks (thrust blocks), and six anvil holders, as indicated in Figure 7a. The inner surfaces of the guide blocks form a tetragonal pyramid. Two of the six anvils are along the center line of this pyramid and are fixed opposite to each other on each guide block. The other four anvils are horizontally located on the midpoints of the square edges of a bipyramid. This results in the formation of a cubic nest bounded by the flat faces of six anvils. A ram force applied along the vertical axis is thus decomposed into three pairs acting along three orthogonal directions, forcing the six anvils to advance synchronously toward the center of the cube.

The second is the so-called T-Cup, recently developed at the Center for High Pressure Research and Stony Brook [*Weidner et al.*, 1996]. This is a two-stage system. The first stage is a tool steel cylinder split into six parts, each with a corner truncated into a square, enclosing a cubic cavity (Figure 7b), which contains the second-stage anvil assembly. Conic slots are made on two of the first- stage pieces to allow X rays to enter and exit. The second stage is assembled outside the press and consists of eight cubes (WC or sintered diamond) separated by preformed gaskets and spacers. Each cube has one corner truncated into a triangular face; the eight truncations create an octahedral nest in which the pressure medium is compressed. Incident and diffracted X rays can either pass through the gaps between the anvils or through the anvils directly when sintered diamond anvils are used.

A 250-ton DIA-type cubic anvil apparatus will be installed at the end station of the bending magnet beam line. This apparatus will allow most kinds of experiments currently being conducted at second-generation synchrotron sources. A 1000-ton press will be installed at the end station of the insertion device beam line and will be capable of using various high pressure tooling for much wider range of experiments. For example, an eight-cube anvil assembly may be placed within a DIA; using sintered diamond cubes, this 6/8-in-DIA system has proven capable of generating over 35 GPa and 2000 °C [e.g., *Kato et al.*, 1992]. The 1000-ton capacity also makes it possible to develop a modified DIA tooling, where the top anvil is driven independently by a smaller hydraulic jack, installed within the top guide block. This configuration allows a controlled differential stress field to be applied to the sample; with X rays as a probe, deformation experiments can be conducted at conditions of pressure and temperature.

Both the bending magnet and insertion device beam lines are capable of switching between energy-dispersive and monochromatic modes. The two large-volume ap-

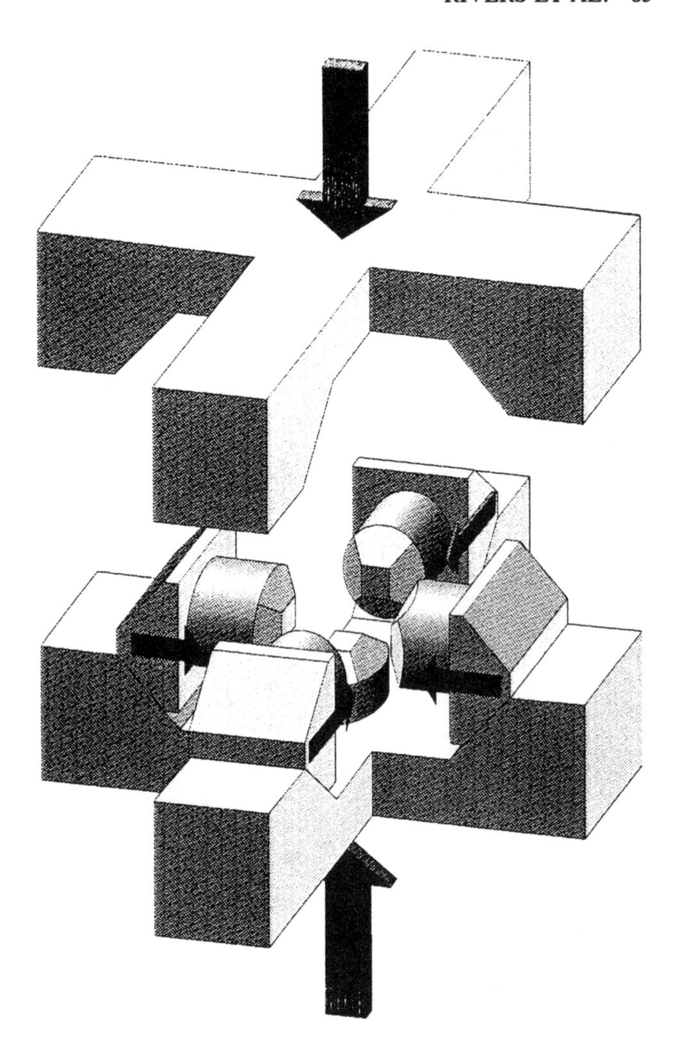

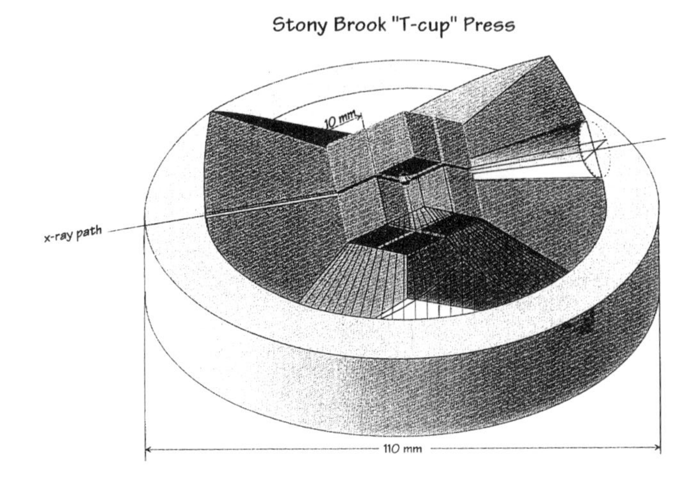

Figure 7. Pressure generating mechanisms of large volume press. (a) DIA arrangement, (b) T-cup assembly.

paratus will be able to operate with both modes. Energy-dispersive diffraction can be performed using a single or multi-element solid state detector, while angle-dispersive diffraction can be carried out using an image plate or a CCD detector.

3.3 Support Laboratory

In addition to the experimental stations, a high-pressure support laboratory is being constructed within the sector laboratory space. This laboratory will contain a variety of diamond cells including Mao-Bell, Merrill-Bassett, membrane, and inconel cells for resistive heating. Sample preparation and characterization facilities will include microscopes, a micromanipulator, glove box, mechanical microdrill and an electric discharge machine. A 3-kbar gas loading facility will allow for loading diamond cell samples which are gases at ambient conditions. A portable optical spectrometer will be used for calibrating pressure by the ruby fluorescence method. Micro-Raman spectroscopy will be carried out using a spectrograph equipped with CCD detector. Cryostats will be available for low-temperature diffraction studies.

There will also be on-site machine shop facilities for trained users and GSECARS staff. GSECARS will also provide office space and computer hardware and software for users to carry out complete on-site data analysis after their beamtime is over.

The existence of this laboratory will make it possible to carry out complete high-pressure experiments at the APS, rather than solely relying on samples prepared at the home institution. Furthermore, it makes it possible to carry out collaborative work between GSECARS staff and scientists who do not have high-pressure capabilities at their home laboratory.

4. RECENT PROGRESS AND FUTURE DIRECTIONS

First light at the CARS sector of the Advanced Photon Source occurred on September 17, 1996 in 13-BM-A. After completion of shielding verification, the first experiments within the CARS sector were carried out in this first optics enclosure of the bending magnet beamline in December, 1996. In these experiments, gold and rhenium were compressed in a diamond cell to 42 GPa. A beryllium gasket and a side-diffraction geometry were used to record the variation of d-spacing as a function of angle from the diamond cell stress axis.

The first optics enclosure of the insertion device beamline is expected to become operational in January, 1997. It is currently estimated that the first experimental stations (13-BM-D, 13-ID-C) will be in operation by summer, 1997. The endstation of the insertion device beamline (13-ID-D) is expected to be ready for white beam experiments at the end of 1997.

The GSECARS sector at the APS is a national user facility for research in Earth, soil, and environmental sciences. Beamtime will be assigned on the basis of competitive proposals. Those proposals that take advantage of the unique characteristics of third-generation sources will receive the highest priority.

Among the scientific goals of the high-pressure program at GSECARS is the study of crystal structures across the entire pressure-temperature spectrum of the terrestrial planets. Other major areas of focus will be on achieving ultra-high pressures with powders and single crystals, study of equations of state and phase transitions, high-pressure rheology, and high-accuracy pressure calibration at simultaneous high temperatures. Over the longer term, experiments will focus on complex problems such as accurate structural and electron density determinations at simultaneous high P and T, hydrothermal reactions, rheological properties, structural studies of melts and glasses, cryogenic studies, phase equilibria of multicomponent systems, and kinetics of phase transitions.

Acknowledgments. We thank the GSECARS design teams for their contributions to this project. J. Hu is thanked for providing experimental assistance. This work was supported by the National Science Foundation, U.S. Department of Energy, W. M. Keck Foundation, and the state of Illinois Board of Higher Education.

REFERENCES

Ancharov, A. I., B. P. Tolochko, R. Chidambaram, S. K. Sikka, S. N. Momin, V. Vijayakumar, and G. N. Kulipanov, An energy dispersive X ray diffraction station at a VEPP-4 synchrotron beam line for structural studies at high pressure, *Nucl. Instr. Meth. A, 359*, 206-209, 1995.

Bassett, W. A., and G. E. Brown, Synchrotron radiation: Applications in the Earth sciences, *Ann. Rev. Earth Planet. Sci., 18*, 387-447, 1990.

Brefeld, W., and P. Gürtler, Synchrotron radiation sources, in *Handbook on Synchrotron Radiation, vol. 4*, edited by S. Ebashi, M. Koch, and E Rubenstein, 269-296, North-Holland, Amsterdam, 1991.

Brister, K., High-pressure instrumentation at CHESS, *Rev. Sci. Instrum., 63*, 995-998, 1992.

Buras, B., and L. Gerward, Application of X ray energy dispersive diffraction for characterization of materials under high pressure, *Prog. Crystal Growth and Charact., 18*, 93-138, 1989.

Eng, P. J., M. Rivers, B. X. Yang, and W. Schildkamp, Micro-focusing 4 keV to 65 keV X rays with bent Kirkpartick-Baez mirrors, *Proc. SPIE, 2516*, 41-51, 1995.

Finger, L., Synchrotron powder diffraction, *Rev. in Mineral., 20*, 309-331, 1989.

Häusermann, D. and M. Hanfland, Optics and beamlines for high-pressure research at the European Synchrotron Radiation Facility, *High Pressure Research, 14*, 223-234, 1996.

Hu, J. Z., H. K. Mao, J. F. Shu, and R. J. Hemley, High pressure energy dispersive X ray diffraction technique with synchrotron radiation, in *High-Pressure Science and Technology - 1993*, edited by S. C. Schmidt, J. W. Shaner, G. A. Samara, and M. Ross, pp. 441-444, American Institute of Physics, New York, 1994.

Kato, T., E. Ohtani, N. Kamaya, O. Shimomura, and T. Kikegawa, Double-stage multi-anvil system with a sintered diamond anvil for X ray diffraction experiment at high pressures and temperatures. in *High Pressure Research: Applications to Earth and Planetary Sciences*, edited by Y. Syono and M. H. Manghnani, Terra Scientific, Tokyo, 33 - 36, 1992.

Koch, E. E., ed., *Handbook of Synchrotron Radiation, vol. 1*, 1165 pp., North-Holland, Amsterdam, 1983.

Mao, H. K., G. Shen, R. J. Hemley, and T. S. Duffy, X ray diffraction with a double hot-plate laser-heated diamond cell, this volume.

Moncton, D. E. Status of the Advanced Photon Source at Argonne National Laboratory, *Rev. Sci. Instrum. Suppl., 67*, 1-10, 1996.

Olsen, J. S., Instrumentation for high-pressure X ray diffraction research at HASYLAB, *Rev. Sci. Instrum., 63*, 1058-1061, 1992.

Shimomura, O., K. Tsuji, N. Hamaya, The Spring-8 project and the high pressure group in Japan, *High Pressure Research, 8*, 703-710, 1992a.

Shimomura, O., W. Utsumi, T. Taniguchi, T. Kikegawa, and T. Nagashima, A new high pressure and high temperature apparatus with sintered diamond anvils for synchrotron radiation use. in *High Pressure Research: Applications to Earth and Planetary Sciences, Geophy. Monogr. Ser.*, vol. 67, edited by Y. Syono and M. H. Manghnani, Terra Scientific, Tokyo, AGU, Washington, DC, 3-11, 1992b.

Smith, J. V. and M. L Rivers, Synchrotron X ray Microanalysis, in *Microprobe techniques in the Earth Sciences*, edited by P. J. Potts, J. F. W. Bowles, S. J. B. Reed, and M. R. Cave, pp. 163-233, Chapman and Hall, London, 1995.

Weidner, D. J., M. T. Vaughan, J. Ko, Y. Wang, X. Liu, A. Yeganeh-Haeri, R. E. Pacalo, and Y. Zhao, Characteristics of stress, pressure, and temperature in SAM85, a DIA type high pressure apparatus. in *High Pressure Research: Applications to Earth and Planetary Sciences, Geophys. Monogr. Ser.*, vol. 67, edited by Y. Syono and M. H. Manghnani, Terra Scientific, Tokyo, AGU, Washington, DC, 13-17, 1992.

Yang, B. X., M. Rivers, W. Schildkamp, and P. J. Eng, GeoCARS microfocusing Kirkpatrick-Baez mirror bender development, *Rev. Sci. Instrum., 66*, 2278-2280, 1995.

M. L. Rivers, T. S. Duffy, Y. Wang, P. J. Eng, S. R. Sutton, and G. Shen, Consortium for Advanced Radiation Sources, The University of Chicago, 5640 S. Ellis Avenue, Chicago, IL 60637

High Pressure Toroid Cell: Applications in Planetary and Material Sciences

L. G. Khvostantsev, V. A. Sidorov, and O. B. Tsiok

Institute for High Pressure Physics, Russian Academy of Sciences, Troitsk, Moscow region, Russia

The toroid-type high-pressure-high-temperature apparatus is described. It is rather simple in design and operation and allows to generate pressures in the range of 9–12 GPa with a working volume of 1–0.3 cm^3. The advantages of this apparatus are its convenience for introducing a fluid-filled capsule and numerous electrical leads in the high-pressure region and reliable operation of these leads. It makes possible measurements of electric, thermal, magnetic and volume properties of matter in a hydrostatic environment at room and elevated temperatures as well as material synthesis experiments. The volume change, phase transitions, and relaxation phenomena under pressure were studied for amorphous materials (GaSb, GeO_2) and porous media.

1. INTRODUCTION

The high pressure device capable of producing pressures greater than 5 GPa at high temperatures in a large working volume is an important part of high-pressure technology. Many experiments in physics, geoscience, and material science need the combination of large volume and high pressure conditions. Two main types of these devices are widely used: the belt and the multianvil press. The belt, capable of producing a few runs up to 8 GPa in a working volume of 1 cm^3 or more, is usually not used above 6 GPa to avoid rupture. The same is true for the multianvil press.

When the problem of synthesis of superhard materials (namely diamond) at high pressure arose in the Soviet Union, other kinds of high pressure devices were invented. First, around 1960, the so-called recess type anvil device was developed at the Institute for High Pressure Physics; it became widespread in the Soviet Union. The recess type anvil apparatus has working parameters similar to those of the belt: with the routine pressure of 6 GPa it is possible to make a few runs up to 8 GPa, but the electrical leads to the high pressure zone break often. However, it is cheaper than the belt and is more simple in operation. The powerful diamond industry was created in the Soviet Union based on the recess type anvil apparatus. The usable pressure of 6 GPa is sufficient to produce diamond powder and then to sinter this powder with the binder.

The next step in the progress of the Soviet diamond industry, high pressure physics, and geoscience was connected with the so-called toroid device, also invented at the Institute for High Pressure Physics [*Khvostantsev et.al.*, 1977; *Khvostantsev*, 1984] and protected by the patents of leading industrial states such as USA, Japan, Germany, France, Great Britain. The special features of the toroid type apparatus will be described below. In this section it should be noted that the higher usable pressure of the toroid device made it possible to produce large polycrystalline diamond samples, analogous to natural carbonado.

As a matter of fact, the toroidal and recess type anvil devices were nearly unknown to researchers outside the Soviet Union. The first in the West who evaluated the usefulness of the toroid device was Besson (France). For

Properties of Earth and Planetary Materials
at High Pressure and Temperature
Geophysical Monograph 101

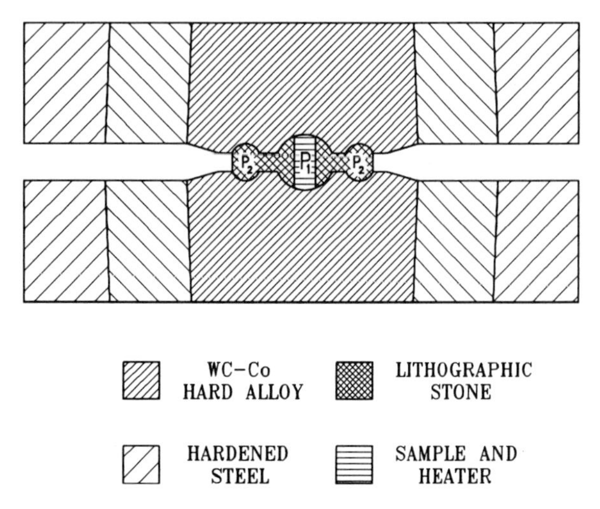

Figure 1. Toroid type high pressure device with recessed central part.

many years his group tried to develop a miniature toroid device for neutron diffraction experiments above 10 GPa. This attempt was successful [*Besson et al.*, 1992] and now the toroid device with 100 mm^3 usable volume operates in the nuclear centers in Grenoble, Edinburgh, and Los Alamos for structure determination of solids and liquids above 10 GPa.

Presented here is the original design of the toroid device manufactured at the Institute for High Pressure Physics. "Toroid" passed the test for more than twenty years of diamond and other materials synthesis and of other physical experiments in the pressure range of about 10 GPa.

2. BASICS OF THE TOROID DEVICE OPERATION

In the simple Bridgman anvils or recess type anvil device the compression of the sample takes place as a result of compression and extrusion of the gasket material. The deformation of anvils under pressure and excessive extrusion of gasket material hinder obtaining higher pressures in all high pressure anvil devices. In the toroid device, a new method of supporting the gasket material surrounding the sample and the central portion of the anvils is used, with the object of reducing the anvil deformation and gasket extrusion.

The toroid device with recessed central region is shown in Figure 1. The device consists of two cemented carbide parts (anvils) with a taper press-fitted into the steel supporting rings. The gasket, repeating the shape of the anvils, is compressed between them, the sample assembly being placed in the hole made in the central part of the gasket. The faces of the anvils are each provided with a circular groove concentric with their axis; the toroid-like space is filled by the gasket. When the anvils are moving towards each other under the force applied, the pressure P_2 is produced in this toroidal space, whilst the pressure in the central region near the sample is P_1 ($P_1 > P_2$). Thus the gasket material around the sample can be regarded as being in the closed space which is subjected to an external pressure P_2 directed from the circular grooves. As a result the extrusion of gasket material from the central part is diminished and even at maximum pressure the gap between anvils is large enough to provide the reliable operation of electrical leads to the sample region. For example, the thickness of gasket in the Toroid-15 device (15 mm diameter of the central recess) after the 10 GPa pressure run is about 1.5 millimeters (in the thinnest part). The gasket thickness for Toroid-25 device after a 9 GPa run is about 2.3 millimeters.

The pressure P_2 in the region of the toroidal groove causes the normal and tangential stresses in the anvil material near the groove. The tangential stresses in the groove region are smaller and in opposite sign than those acting in the central part of the anvil, which is subjected to maximum pressure P_1. So, the central part of the anvil is supported effectively, and this support increases with the increase of pressure. Also the deformation of the central part is smaller compared to the recess type anvils. It results in a higher usable pressure and longer lifetime of the anvils. The calibration curves for two basic cells, the Toroid-15 and Toroid-25, are shown in Figure 2.

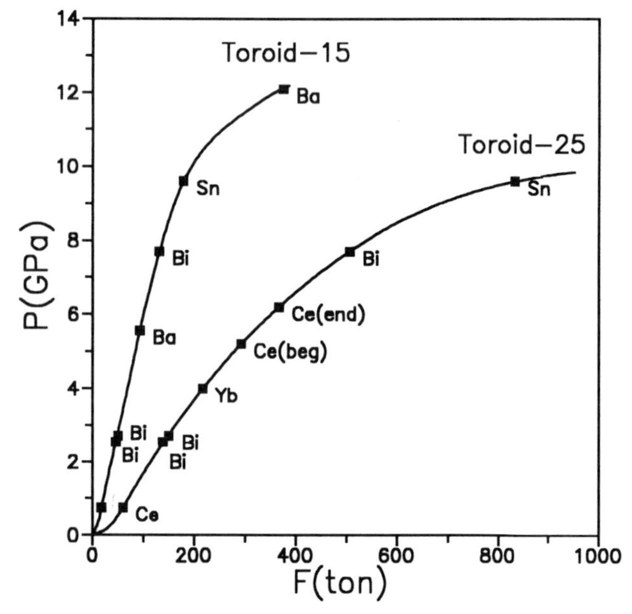

Figure 2. Pressure-load curves for toroid devices. The diam-eter of the central recess is 15 mm for Toroid-15 and 25 mm for Toroid-25.

The idea of gradual diminishing of tangential stresses in the body of strained anvils is very important. In the toroid device it is realized through the groove of special shape filled with a gasket material. In the general case for any high-pressure device some part of the loading force is applied directly to the sample region to generate high pressure. The remaining part of the loading force goes to support the strained parts of the high pressure device. In the case of the simple piston and cylinder type device, all the load acts on the piston to generate pressure and nothing is used to support the strained parts of this device. So, its pressure limit is usually not more than 3 GPa. The end-loaded piston and cylinder devices are capable of producing higher pressures (of about 5 GPa). In the toroid device a significant part of the load is used to support the anvil body and the usable pressure is above 9 GPa.

3. MAIN ADVANTAGES AND USEFUL FEATURES OF TOROID DEVICES

1. Simplicity in operation and use. Toroid is an anvil type device consisting of two parts of relatively small size and weight which can be installed in any hydraulic press. It is very easy to prepare the cell and sample assembly for experiment.
2. Reliable operation of electrical leads on loading and unloading (multiple loading-unloading cycles are also possible). The number of electrical leads is limited only by the diameter of the high pressure region and the thickness of the adjacent leads. Typically there are 12 leads in the experiments where the physical properties are measured.
3. The geometry of the toroid cell allows introduction of a liquid-filled ampoule in the high pressure zone. The combination of compression and flow of the gasket material permits the ampoule deformation under pressure to be uniform and the final shape of ampoule is nearly the same as before compression. The hydrostatic sample environment is favorable for work with single crystals.
4. The device can be easily adopted for material synthesis experiments. In this case the solid pressure transmitting medium, heater, and sample (or metal ampoule, containing liquid and sample) are placed in the central part of a gasket. The sample can be treated at temperature up to 2000 °C with the temperature control by a W-W(Re) thermocouple.

As a result, the toroid device is nearly ideal for studies of physical properties that can be realized on the basis of electrical measurements. The authors have succeeded in measurements of electrical resistivity, thermopower, thermal conductivity, differential thermal analysis (DTA), magnetic susceptibility, volume change under pressure via strain gauge technique and so on. It became possible to use most of the methods mentioned above at elevated temperatures up to 600°C (thermal stability limit of liquid) with precise control of pressure by a manganin gauge. In this way the positions of phase transition points can be located with high accuracy and P-T diagrams can be constructed. The recent development of the strain gauge measurements allowed us to study problems of disordered systems (amorphous solids and porous media compacted from nanocrystalline powders).

4. EXAMPLES OF EXPERIMENTAL METHODS

To illustrate the experimental possibilities of the toroid device, we consider briefly the design of a liquid-filled ampoule for high temperature measurements of magnetic susceptibility and the strain gauge technique for studies of the equation of state at room temperature.

The view of an ampoule is shown in Figure 3a. It consists of teflon parts and metal lids and is placed in a hole made in the central part of a gasket. The electrical leads through the gasket region are made from a wire (0.5 mm in diameter) enclosed in the tube made of the wire material. Thin wires connected to the sample or thermocouple are introduced into the same tube from the side of an ampoule. A very convenient method to prepare the enclosing tube is to wind it with appropriate wire and rod. These wound tubes help to prevent rupture of the leads. Pressure measurements are done with the use of a manganin gauge, its resistance change under pressure being linear up to 9 GPa in the hydrostatic environment [*Khvostantsev and Sidorov*, 1981]. The liquids used are methanol-ethanol (4:1) and petroleum ether. The pressure gauge is placed in the bottom part of an ampoule. The wire heater, sample, and thermocouple are placed in the top part, which is separated from the pressure gauge by porous thermal insulation material. This assembly allows us to make measurements of various physical properties or P-T treatment of the sample in well characterized pressure and temperature conditions. The sample assembly for measurement of a magnetic susceptibility is shown in Figure 3b. It was used to study the Curie point and spin reorientation transitions under pressure in some invars and Geusler alloys [*Sidorov and Khvostantsev*, 1994; *Gavriliuk et al.*, 1996]. A system of two measuring and one exciting coils is placed inside the heater. Measuring coils, connected in series, have equal number of turns and are wound in opposite directions. When excited by an alternating current via exciting coil, the measuring coil system exhibits a nearly zero voltage output signal without the sample or with the sample in nonmagnetic state. The

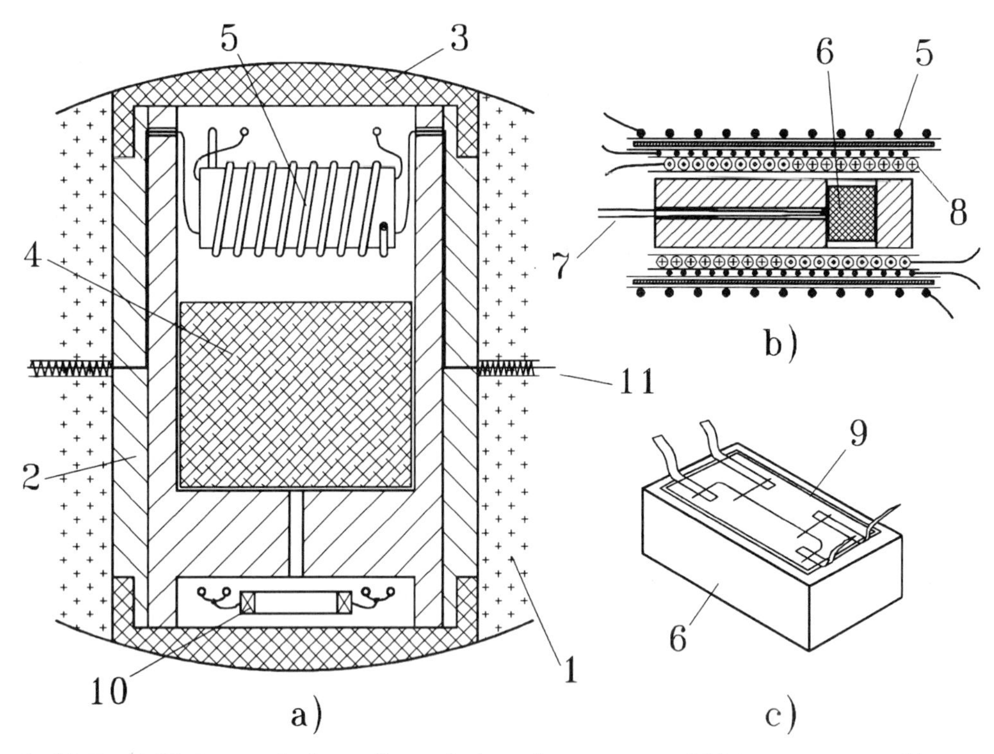

Figure 3. (a) liquid-filled ampoule for studies at hydrostatic pressures and high temperatures; (b) assembly for magnetic measurements placed in the top part of an ampoule; (c) sample with the strain gauge bonded; 1-gasket (lithographic stone), 2-teflon ampoule, 3-lid, 4-thermal insulation (asbestos), 5-heater, 6-sample, 7-thermocouple, 8-coil system for magnetic measure-ments, 9-strain gauge, 10-manganin pressure gauge, 11-electrical leads.

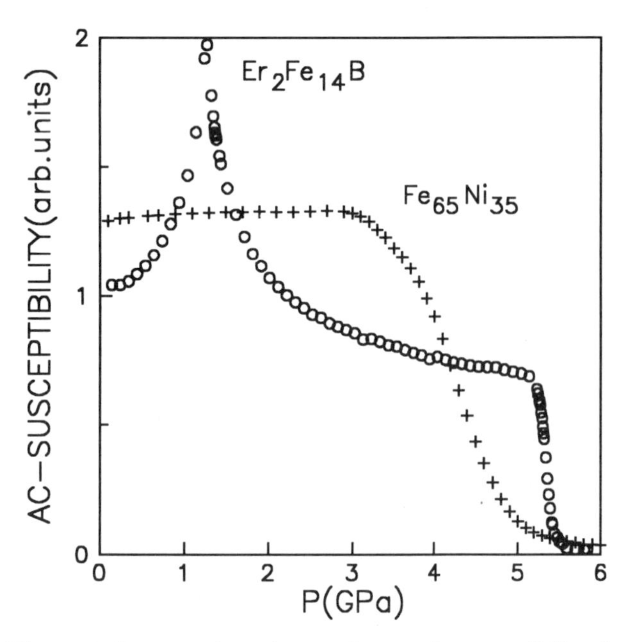

Figure 4. Pressure dependence of magnetic susceptibility for invar materials $Er_2Fe_{14}B$ and $Fe_{65}Ni_{35}$. Spin reorientation transition and disappearance of ferromagnetism are clearly seen.

sample, placed inside one of the coils, causes the large output signal which depends on its magnetic susceptibility. The example of pressure dependence of magnetic susceptibility of two invars can be seen in Figure 4. Some examples of differential thermal analysis, thermopower, and electrical resistance measurements for solid-solid and solid-liquid phase transitions at high temperature up to 600°C and pressure up to 9 GPa can be found elsewhere [*Khvostantsev and Sidorov*, 1984].

The strain gauge technique was developed for measurements of volume change at high pressure [*Tsiok et al.*, 1992]. A miniature single-wire strain gauge is bonded to the sample surface (Figure 3c). The gauge is made from 20 μm constantan wire. The sample may be single crystal, amorphous solid, or powder compact. In the latter case the sample is prepared by precompaction of a powder (confined in a latex envelope) by high-pressure liquid (at 0.2 GPa) and machining to a desired size. The sample with a strain gauge bonded is then covered by thin elastic envelope (latex) preventing the penetration of pressure transmitting liquid in the sample. The relation between length change of the sample and the resistance change of the gauge is derived and checked experimentally on the basis of NaCl and Al equations of state. This technique allows the precise quantitative measurement of volume change, the relative accuracy of measurements and resolution being much better than that attainable in x ray diffraction measurements. The resolution of a sample length change is 0.001%, which makes it possible to observe very weak peculiarities on the compression curve of a sample. The technique developed is very convenient for measurements in disordered systems (amorphous materials and compacts of ultrafine powder) where the use of diffraction methods is limited.

5. VOLUME CHANGE, PHASE TRANSITIONS, AND RELAXATION PHENOMENA IN DISORDERED SYSTEMS

5.1. *a-GaSb*

The electrical resistivity and compressibility of bulk amorphous semiconductors a-$(GaSb)_{1-X}(Ge_2)_X$ (x = 0 and 0.27) were investigated at high hydrostatic pressure [*Sidorov et al.*, 1994]. Bulk amorphous semiconductors were synthesized at 9 GPa and 700K with the use of the toroid device. Discontinuous crystallization of a-$(GaSb)_{0.73}(Ge_2)_{0.27}$ occurs at 4.7 GPa. In a-GaSb the continuous volume anomaly takes place at 3–7 GPa, whereas the semiconductor-to-metal transition occurs at 3–4 GPa (Figure 5). The metallization of a-GaSb seems to occur by a percolation mechanism involving the interaction of microregions with higher coordination. The conductive microregions with higher coordination (Z ~ 6) seem to be formed in a lower coordinated (Z = 4) amorphous matrix at increase of pressure. This process, depending on germanium content, nonhydrostatic stress, and temperature, can proceed as a very fast crystallization at certain pressure or continuously in a wide pressure range. At the final stage of the continuous process the microregions with higher coordination grow and become crystalline. The bulk moduli of amorphous compounds at normal pressure are less (35 and 40 GPa for x = 0 and x = 0.27 respectively) than that of crystalline GaSb (55 GPa) and exhibit strong softening long before the transitions.

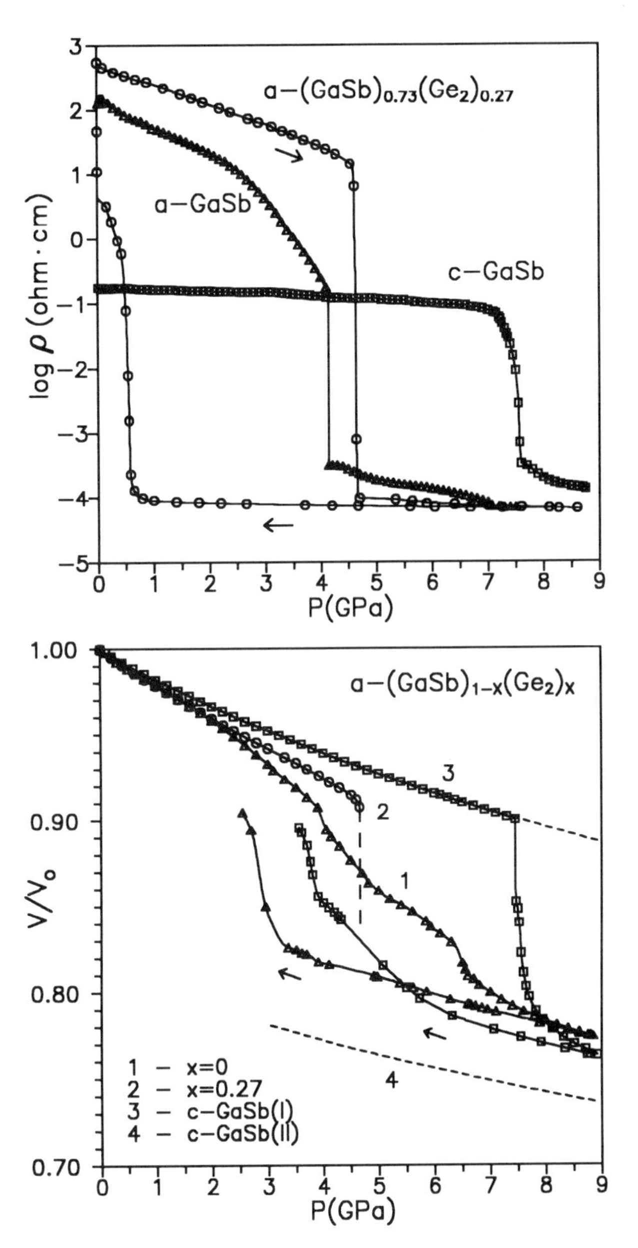

Figure 5. The electrical resistivity and volume change of amorphous $(GaSb)_{1-X}(Ge_2)_X$ and crystalline GaSb at high pressure. Dashed lines - the x ray diffraction data for crystalline GaSb(I) and GaSb(II) [*Weir et. al.*, 1987].

5.2. *Fine-powder Media.*

The densification of porous compacts prepared from nanocrystalline powders revealed the features typical for microscopic disordered systems (spin glasses, viscous fluids near freezing point). The relaxation law in these systems is logarithmic in time, which can be explained by the hierarchical structure. Namely, a number of processes with different scales and different relaxation times exist in hierarchically constructed systems deflected from equilibrium, resulting in log(t) response in some time interval. In the case of powder media, the relaxation of density at high pressure is due to sliding and repacking of particles. External pressure causes the particles to change their positions to arrive at a configuration with smaller

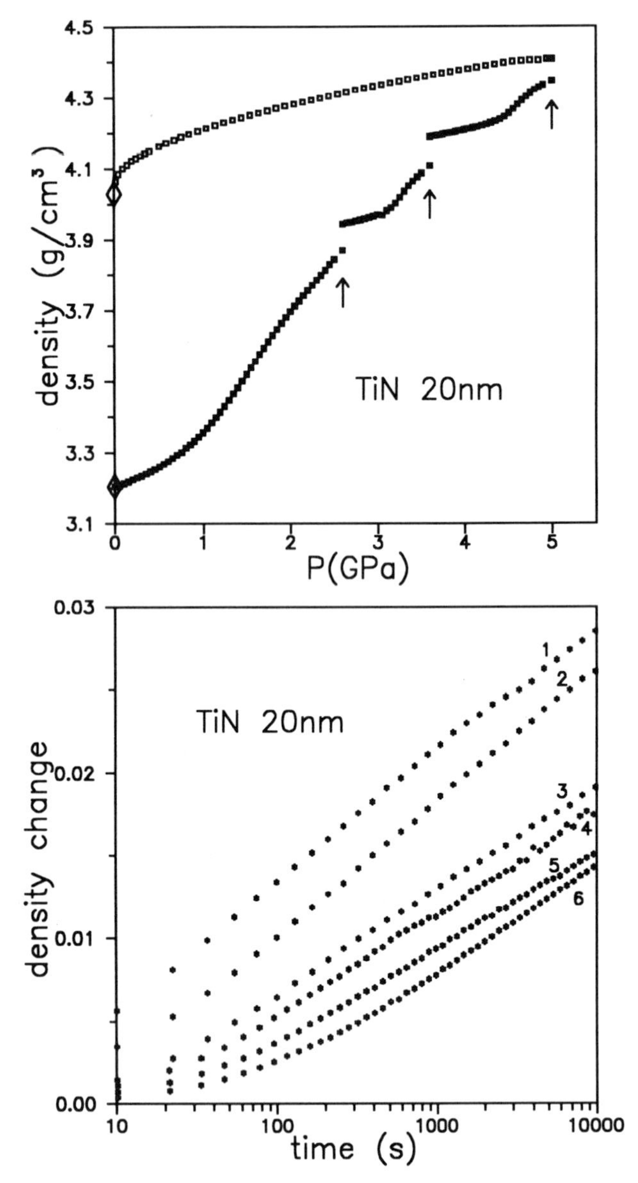

Figure 6. Density change under pressure and log(t) relaxation of density in TiN ultrafine powder compact at different pressures (1–0.48 GPa, 2–0.73 GPa, 3–1.23 GPa, 4–2.6 GPa, 5–3.6 GPa, 6–5.0 GPa). Arrows indicate points where the relaxation of density was recorded. Rhombii indicate the sample density, measured by Archimedes' method before and after experiment.

volume. Some selected particles can arrive at a more preferable position very quickly (the shortest relaxation time), but it will change the external conditions for its neighbors, causing movement. The larger volume element under consideration (the larger will be scale), the larger will be the time interval needed for it to come to an equilibrium state with the neighbor volume element of the same scale.

The experiment was done as follows. The powder sample was compressed at a given rate and then pressure was fixed and maintained constant (using a manganin gauge inside ampoule) during a long period of time. The density of the sample continued to rise at fixed pressure and this relaxation process was recorded. The results for 20 nm grain size powder TiN sample are shown in Figure 6. The main parameters of relaxation and densification processes are closely connected [*Tsiok et al.*, 1995].

5.3. *a- GeO_2*

The anomalous elastic properties of SiO_2 glass and irreversible effects of its densification at high pressure have been of interest for many years [*Zha et al.*, 1994]. Silica is an important geomaterial and exhibits interesting polymorphism at high pressure. The coordination change in silica glass were observed at pressures above 10 GPa [*Meade et al.*, 1992]. Amorphous GeO_2 exhibits similar phenomena, but at lower pressures. The change of the coordination of the germanium atom from 4 to 6 was observed in the XANES studies of a-GeO_2 above 6.5 GPa [*Itie et al.*, 1990]. Sound velocities in a-GeO_2 was studied by ultrasonic travel time technique and Brillouin scattering [*Suito et al.*, 1992; *Wolf et al.*, 1992]. These studies indicate that volume of a-GeO_2 should decrease drastically at pressures above 4 GPa when the coordination change starts. Nevertheless, there is no direct measurement of density and static bulk modulus of germania glass in this range of pressure.

We have done measurements of relative volume of a-GeO_2 with the use of the strain gauge technique. The coordination transition starts at ~3 GPa and is not complete at 9 GPa. In this pressure range irreversible changes of density and relaxation were found. Three runs are depicted in Figure 7. In one run the maximum pressure was 3.3 GPa, slightly above the beginning of volume anomaly. In this case the irreversible increase of sample density was ~1.5%. Two runs were done up to 9 GPa. One of them was with a continuous rise of pressure at a rate of 0.1 GPa/min. In the second run stops were made and the relaxation of density was recorded at the intermediate points. The relaxation phenomena at this transition are quite similar to those observed in porous disordered media (Figures 6 and 7). Decrease of pressure was continuous in all cases. The reverse transition starts at ~3 GPa in accordance with XANES studies [*Itie et al.*, 1990]. It is known that germanium in a-GeO_2 hysteretically reverts to tetrahedral coordination at decrease of pressure [*Itie et al.*, 1990; *Wolf et al.*, 1992]. The Ge-O distance in the recovered sample has the same value as before the pressure cycle [*Itie et al.*,

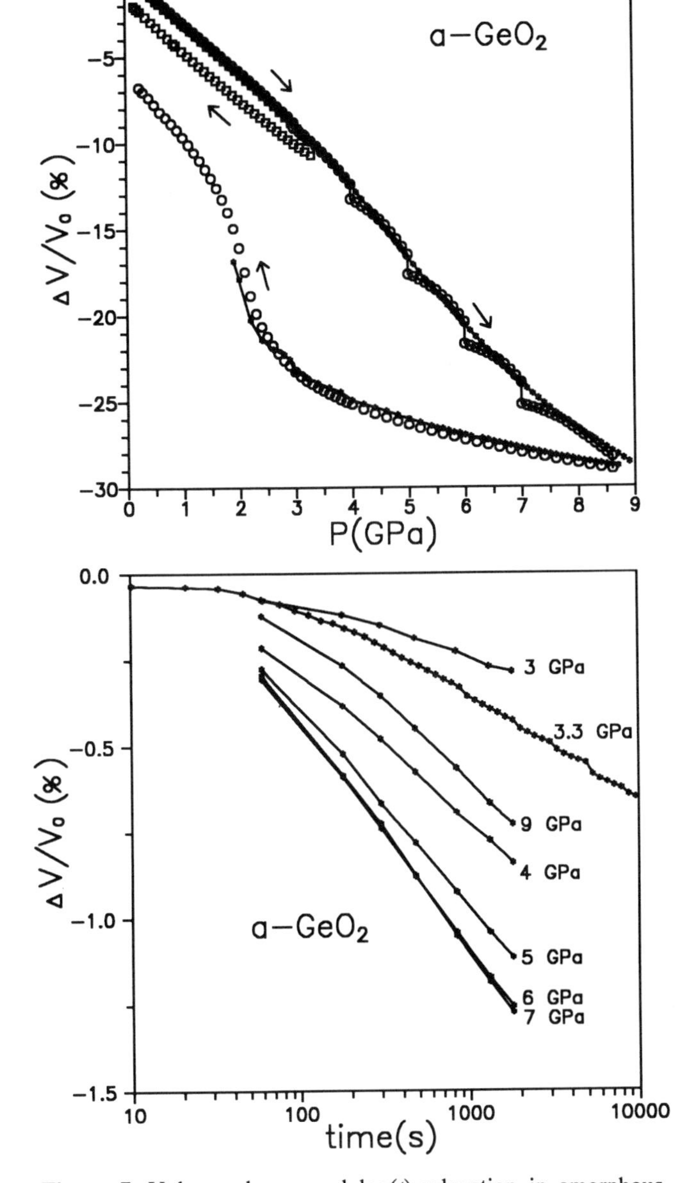

Figure 7. Volume change and log(t) relaxation in amorphous GeO_2 at the coordination change under high hydrostatic pressure. Three runs are shown in the top part. See text for details.

1990]. In our experiments the irreversible change of density takes place after reversible coordination change, which may indicate changes in the structure of a-GeO_2. It should be mentioned that at 9 GPa the transition is incomplete.

6. SUMMARY

The toroid high pressure device for producing high pressure at high temperature is described. It can be used for material science, geoscience, physics and other fields of high pressure research, where the combination of uniform high pressure up to ~10 GPa, high temperature and large sample volume are required.

REFERENCES

Besson J. M., G. Hamel, T. Grima, R. J. Nelmes, J. S. Loveday, S. Hull, and D. Hausermann, A large volume pressure cell for high temperatures, *High Pres. Res., 8*, 625-630, 1992.

Gavriliuk A. G., G. N. Stepanov, V. A. Sidorov, and S. M. Irkaev, Hyperfine magnetic fields and Curie temperature in the Heusler alloy Ni_2MnSn at high pressure, *J. Appl. Phys., 79*, 2609-2612, 1996.

Itie J. P., A. Polian, G. Calas, J. Petiau, A. Fontaine, and H. Tolentino, Coordination changes in crystalline and vitreous GeO_2, *High Pres. Res.,, 5*, 717-719, 1990.

Khvostantsev L. G., L. F. Vereshchagin, and A. P. Novikov, Device of toroid type for high pressure generation, *High Temp.- High Pres., 9*, 637-639, 1977.

Khvostantsev L. G., and Sidorov V. A., High pressure polymorphism of antimony. Thermoelectric properties and electrical resistance studies, *Phys. St. Sol. (a), 64*, 379-384, 1981.

Khvostantsev L. G., A verkh-niz (up-down) toroid device for generation of high pressure, *High Temp.- High Pres., 16*, 165-169, 1984.

Khvostantsev L. G., and Sidorov V. A., Phase transitions in antimony at hydrostatic pressure up to 9 GPa, *Phys. St. Sol. (a), 82*, 389-398, 1984.

Meade C., R. J. Hemley, and H .K. Mao, High-pressure x-ray diffraction of SiO_2 glass, *Phys. Rev. Lett., 69*, 1387-1390, 1992.

Sidorov V. A., and Khvostantsev L. G., Magnetovolume effects and magnetic transitions in the invar systems $Fe_{65}Ni_{35}$ and $Er_2Fe_{14}B$ at high hydrostatic pressure, *J. Magn. Magn. Mater., 129*, 356-360, 1994.

Sidorov V. A., V. V. Brazhkin, L. G. Khvostantsev, A. G. Lyapin, A. V. Sapelkin, and O. B. Tsiok, Nature of semiconductor-to-metal transition and volume properties of bulk tetrahedral amorphous GaSb and GaSb-Ge semiconductors under high pressure, *Phys. Rev. Lett., 73*, 3262-3265, 1994.

Suito K., M. Miyoshi, T. Sasakura, and H. Fujisawa, Elastic properties of obsidian, vitreous SiO_2 and vitreous GeO_2 under high pressure up to 6 GPa, in *High Pressure Research: Application to Earth and Planetary Sciences, Geophys. Monogr. Ser.*, vol. 67, edited by Y. Syono, and M. H. Manghnani, pp. 219-225, Terra Publishing Company, Tokyo / AGU, Washington, D.C., 1992.

Tsiok O. B., V. V. Bredikhin, V. A. Sidorov, and L. G. Khvostantsev, Measurements of compressibility of solids and powder compacts by a strain gauge technique at hydrostatic pressure up to 9 GPa, *High Pres. Res., 10*, 523-533, 1992.

Tsiok O. B., V. A. Sidorov, V. V. Bredikhin, L. G. Khvostantsev, V. N. Troitskiy, and L. I. Trusov, Relaxation effects during the densification of ultrafine powders at high hydrostatic pressure, *Phys. Rev.B, 51*, 12127-12132, 1995.

Weir S. T., Y. K. Vohra, and A. L. Ruoff, Phase transitions in GaSb to 110 GPa (1.1 Mbar), *Phys. Rev. B, 36,* 4543-4546, 1987.

Wolf G. H., S. Wang, C. A. Herbst, D .J. Durben, W. F. Oliver, Z. C. Kang, and K. Halvorson, Pressure induced collapse of tetrahedral framework in crystalline and amorphous GeO_2, in *High Pressure Research: Application to Earth and Planetary Sciences, Geophys. Monogr. Ser.*, vol. 67, edited by Y. Syono, and M. H. Manghnani, pp.503-517, Terra Publishing Company, Tokyo / AGU, Washington, D.C., 1992.

Zha C. S., R. J. Hemley, H. K. Mao, T. S. Duffy, and C. Meade, Brillouin scattering of silica glass to 57.5 GPa, in *High Pressure Science and Technology-1993*, edited by S. C. Schmidt, J. W. Shaner, G. A. Samara, and M. Ross, pp.93-96, AIP Conference Proceedings 309, Part I, New York, 1994.

L. G. Khvostantsev, V. A. Sidorov, and O. B. Tsiok, Institute for High Pressure Physics, Russian Academy of Sciences, 142092 Troitsk, Moscow Region, Russia.

Cryogenic Recovery of Unquenchable High-Pressure Samples Using a Multianvil Device

Kurt Leinenweber, Udo Schuelke, Shirley Ekbundit, and Paul F. McMillan

Materials Research Science and Engineering Center, Dept. of Chemistry and Biochemistry, Arizona State University

A simple modification has been made to the standard Walker-style 6-8 multianvil apparatus, to allow samples to be depressurized at low temperatures. This is done by introducing a coolant, in this case liquid nitrogen, into the sample area during decompression. Thus low-temperature recovery techniques, which have been used previously in a few instances in the diamond-anvil cell and piston-cylinder apparatus, have been extended to the multiple anvil, allowing large samples of some unquenchable phases to be recovered from high pressures. Two large (60 mg. each) samples of calcium hydroxide ($Ca(OH)_2$) and three samples of calcium deuteroxide ($Ca(OD)_2$), both in the unquenchable high pressure EuI_2 (baddeleyite-related) form, have been recovered from 9 GPa and 400 °C using this technique. The recovered samples are shown by Raman spectroscopy to retain the high pressure structure at 85 K. During heating at 10 K per minute, the material begins to revert to the low pressure portlandite phase at 223 K, and the reversion is complete by 275 K. This recovery technique can be used subject to certain kinetic conditions, but in general the outcome cannot be predicted and an experiment needs to be performed for each material of interest.

1. INTRODUCTION

Much of our knowledge of high pressure mineralogy is based on phases which have been synthesized within their stability field under high pressure conditions, and subsequently decompressed to ambient conditions for characterization of their structural and thermodynamic properties. On the other hand, a large number of unquenchable high pressure phases, or phases which are synthesized at high pressure but cannot be recovered to ambient pressure and temperature, are known. This occurs because of conversion of the high pressure phase to a lower pressure form during decompression at ambient temperature. The lower pressure form can be amorphous, as in the well-known case of $CaSiO_3$ perovskite, which converts to a glassy material during decompression [*Liu and Ringwood*, 1975]. In other cases the lower pressure form is crystalline. One example of this is $Ca(OH)_2$, which transforms to an EuI_2 structure (closely related to baddeleyite, ZrO_2) at high pressure, but reverts to the low-pressure portlandite form on decompression [*Kunz et al.*, 1996]; another is the high-pressure perovskite form of $FeTiO_3$, which transforms to a $LiNbO_3$-type structure on decompression, a phase which appears to be always metastable [*Mehta et al.*, 1994]. These and some other interesting examples drawn from mineral physics and solid state chemistry are summarized in Table 1.

For the most part, the structure and properties of these unquenchable phases are studied in situ at high pressure,

Properties of Earth and Planetary Materials
at High Pressure and Temperature
Geophysical Monograph 101

TABLE 1. Examples of normally unquenchable high pressure minerals and related phases of interest

High pressure phase	Quench product(s)	Reversion pressure	Hysteresis
$CaSiO_3$ perovskite[1]	Amorphous	0 - 1 GPa	(irreversible)
$Ca(OH)_2$, EuI_2-type[2]	$Ca(OH)_2$ portlandite	1 GPa	(irreversible)
$Ca(OH)_2$, $Sr(OH)_2$-type[3]	$Ca(OH)_2$ portlandite	5 GPa	1 GPa
SiO_2 cristobalite II[4]	α-cristobalite	0.5 GPa	1 GPa
SiO_2 cristobalite III[5]	α-cristobalite	(not given)	(not given)
$MnTiO_3$ perovskite[6]	$MnTiO_3$ lithium niobate	2.5 GPa	0.5 GPa
$FeTiO_3$ perovskite[7]	$FeTiO_3$ lithium niobate	16 GPa	2 GPa
$MgSiO_3$ low clinoenstatite	$MgSiO_3$ high clinoenstatite	5.3 GPa	2.5 GPa
$LiVO_3$ lithium niobate[9]	$LiVO_3$ clinoenstatite	4 GPa	(irreversible)
Mn_2SiSe_4 spinel[10]	Mn_2SiSe_4 olivine or spinelloids	2 GPa	(irreversible)

[1]Liu and Ringwood, 1975, Kudoh and Kanzaki , 1993, Wang and Weidner, 1994; [2]Kunz et al. (1996), Leinenweber et al. (1997); [3]Ekbundit et al. (1996); [4]Palmer and Finger (1994); [5]Tsuchida and Yagi (1990); [6]Ross et al. (1989); [7]Leinenweber et al. (1991); [8]Angel et al. (1992); [9]Grzechnik and McMillan (1995); [10]Grzechnik et al. (1997)

within their stability regime. A wide range of spectroscopic and structure determination methods have been developed for this purpose. However, there are inherent limitations and problems associated with such experiments, the seriousness of which depends on the technique employed and the information desired. In x-ray diffraction experiments, interference from the high pressure device and pressure medium, and reduction in data quality due to stresses in the sample, can hamper structure determinations. In spectroscopic studies, peak broadening due to non-hydrostatic stresses and high background due to anvil materials (usually diamond) and pressure transmitting media can reduct the quality of the spectra. Some in situ techniques, such as neutron diffraction and differential scanning calorimetry under pressure, are still in their early stages of development, while other techniques such as TEM are unlikely ever to be realized for samples at high pressure. For these reasons, a means of recovery of materials such as those listed in Table 1 to ambient conditions would be desirable in order to learn more about these materials.

In cases in which the back-transformation to low pressure materials involves a thermally activated step, decompression under cryogenic conditions can slow the reversion process sufficiently that the high pressure phase can be recovered to ambient pressure. Such a technique has been employed previously; for example to recover a phase of HgSe from a piston cylinder device for x-ray diffraction [*Kafalas et al.*, 1962], and more recently to recover high-pressure ice VIII [*Whalley et al.*, 1991] and a transparent high pressure form of carbon [*Miller et al.*, 1997]. The purpose of the present work is to develop and demonstrate a cryogenic recovery technique in the multianvil apparatus, for routine use in studies of high pressure phases such as those listed in Table 1. This technique has been used to recover the unquenchable high pressure phase of $Ca(OH)_2$.

Because this paper deals with the recovery of unquenchable phases, the meaning of the term "unquenchable" needs to be given a more specific definition. This term is in common usage for phases that cannot be brought to ambient pressure *at room temperature*, and it seems reasonable to retain this definition. To avoid the trivial paradox of "quenching" an unquenchable phase, the more general term "recovery" is used here to refer to the decompression and removal of an unquenchable high pressure phase *at cryogenic temperatures*.

2. EXPERIMENTAL

55-65 milligram samples of $Ca(OH)_2$ (Baker) and $Ca(OD)_2$ were enclosed in platinum, and placed in a multianvil assembly with 8 mm truncations, with a magnesia/alumina octahedral pressure medium (Aremco Ceramcast 584) and a graphite resistance furnace. In early runs, a welded Pt capsule was used to contain the sample. However, removal of the sample from the capsule under liquid nitrogen proved to be difficult, so that wrapped Pt foil was used in subsequent runs. The foil could be easily unwrapped under liquid nitrogen after a run.

The high pressure runs were performed in a Walker module [*Walker et al.*, 1990; *Walker*, 1991] at Arizona State University. The device was modified in order to permit liquid nitrogen to be pumped past the cell assembly. The source of nitrogen was a cryogenic Dewar (VWR 50 LD) with a commercial liquid withdrawal device (Taylor-Wharton), attached to a regulated 10 psi compressed air source in order to maintain an even flow. The nitrogen was pumped to the cell assembly through flexible corrugated Teflon FEP tubing, with 1/4" ID noncorrugated sections (Cole-Parmer). This tubing is still flexible at cryogenic temperatures, and can also be easily threaded through the ports in the multianvil device. Between the multianvil wedges, where the hoses might be torn during the compression, 1/4" stainless steel tubing was used instead; this was securely fitted to the noncorrugated section of the teflon tubing simply by joining them together. Two hoses were employed, one for entry and one for exit of liquid nitrogen from the sample area.

The thermocouple ports in the standard Walker module were widened in order to admit the teflon hoses. Also, slots were cut in 4 of the tungsten carbide anvils in order to admit a flow of nitrogen, using the geometry shown in Figure 1. In early runs extra gaskets (expanded polystyrene) were placed around the outer perimeter of the spaces between the carbide cubes, just inside the G-10 pads, in order to form an outer seal around the cell assembly; later, these were found to be unnecessary. The cell assembly can be adequately sealed simply by wrapping duct tape around the points at which the hoses enter the carbide cube assembly, and also placing a small piece of duct tape over each of the 10 openings between the carbide cubes which are not already covered by G-10 pads [see *Walker et al.*, 1990, for standard details about the multianvil apparatus].

In the present experiments, the $Ca(OH)_2$ and $Ca(OD)_2$ samples were heated to 400 °C for 10 minutes at 9 GPa, and temperature-quenched to 298 K in a few seconds while still under pressure. Following this step, the sample was decompressed partway, to about 3 GPa, which is still above the onset pressure of the reversion to portlandite. Liquid nitrogen was then admitted to the sample assembly area through the entry hose, and the remainder of the decompression was performed after liquid nitrogen could be seen coming out of the exit hose. The module was then opened and some of the carbide cubes removed without stopping the flow of liquid nitrogen onto the sample octahedron. Finally, the octahedron was quickly transferred to a container of liquid nitrogen, where the sample could be removed from it. The samples were stored in a long-term cryogenic storage refrigerator (VWR 10XT-11) This technique successfully allowed recovery of the high pressure phase, which is normally unquenchable.

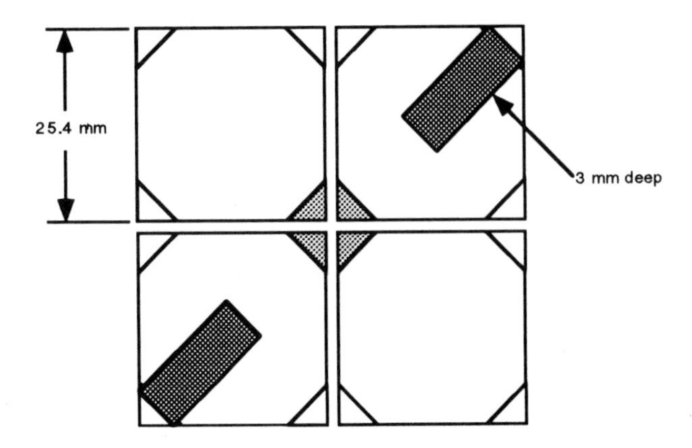

Figure 1. View of lower 4 cubes in multianvil experiment, showing position of sample octahedron (center) relative to notches cut for liquid nitrogen entry and exit.

Our experience after carrying out several runs of this type is that the tungsten carbide cubes do not appear to suffer more than usual breakage due to decompression at low temperatures. Also, no damage to the steel wedges of the first stage has occurred. It is, however, advisable to place the cold wedges in a dessicator after each run, to prevent rusting due to condensation of water on the wedge surfaces. Only a few liters of liquid nitrogen are needed per run, so the 50 liter storage Dewar can be used for several runs without refilling. Overall, it was found that with experience, cryogenic recovery in the multianvil is only marginally more difficult than normal sample recovery. The Walker geometry proved to be advantageous for this purpose, since the entire module could be easily removed and opened after decompression without stopping the flow of liquid nitrogen onto the sample.

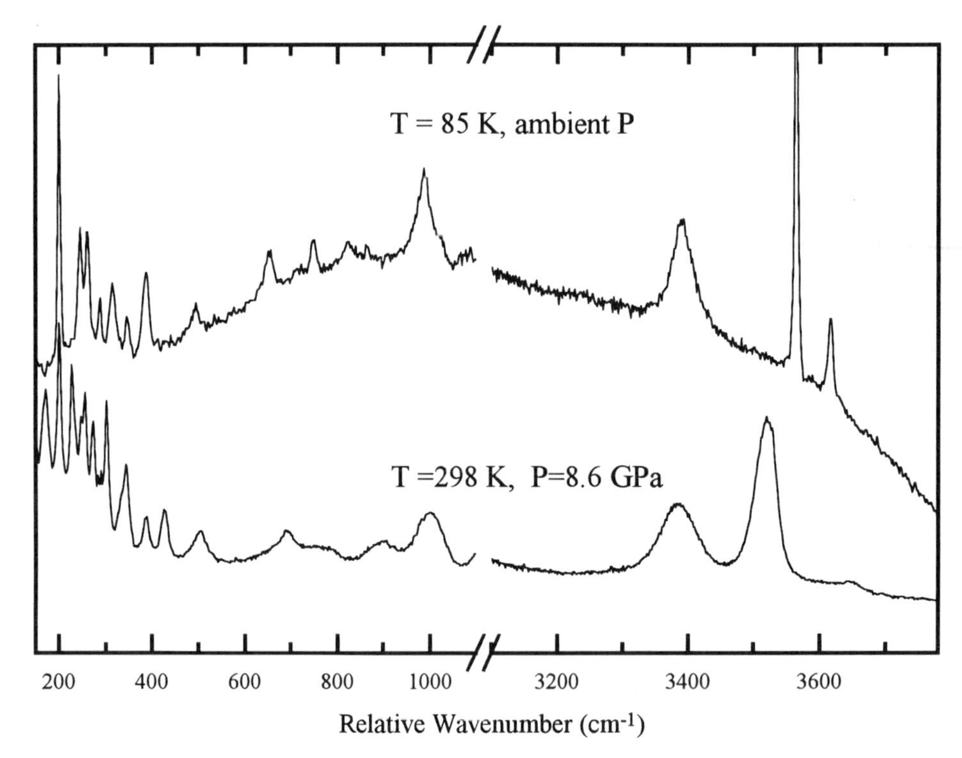

Figure 2. Comparison of the Raman spectra of the high pressure form of $Ca(OH)_2$ as cryogenically recovered (top spectrum) and in situ in the diamond cell (bottom spectrum).

3. RESULTS

The initial success of the recovery of the high pressure form of $Ca(OH)_2$ was judged simply by allowing small pieces of the samples to warm up to room temperature under visual observation. The recovered samples were coarsely crystalline at low temperature (77 K), but a few seconds after removal from liquid nitrogen, they suddenly turned white and powdery, and underwent a small but discernable volume expansion. Subsequent visual observation while heating at 10 °C per minute using a Linkam hot/cold stage showed that the transformation is complete by around 265 K.

Raman spectroscopy of the sample was performed using the cold stage in order to further study the back-transformation to portlandite. In Figure 2, the Raman spectrum of a recovered high-pressure sample is compared with a spectrum taken in situ at high pressure in a diamond-anvil cell; both correspond to the high pressure EuI_2 (baddeleyite-like) form of $Ca(OH)_2$, although the peaks are considerably sharper in the recovered sample, and some peaks which are marginal in the high pressure spectrum are clearly present in the ambient pressure spectrum, particularly in the OH-stretching region at 3300-3700 wavenumbers. The broadness of the peaks at high pressure may be due to the presence of nonhydrostatic stresses in the diamond cell, which was used without a pressure medium (because of the necessity of externally heating the diamond cell to make the EuI_2 form), but on the other hand could also be due to a change in the hydrogen bonding or a disordering of the hydrogen positions. The effect of heating the recovered sample from 85 K to 298 K is shown in Figure 3. The spectrum of the EuI_2-type phase has 3 peaks in the OH-stretching region. This persists as a pure spectrum to 173 K, but a small amount of portlandite is present at 223 K, as shown by the appearance of the portlandite OH-stretching peak (shown by an arrow in the Figure). By 268 K, only a small amount of the EuI_2 phase remains, and by 298 K, the entire sample has converted to portlandite. This transition is irreversible, and represents the same process that occurs during the decompression of the sample at room temperature.

Because the back-transformation of the high pressure phase is a thermally activated process which occurs under metastable conditions, the reversion kinetics depend in general upon many variables, including the sample grain size and morphology, the details of the decompression history and any subsequent sample treatment, and the temperature and pressure of storage of the metastable high pressure phase. In the present case, the EuI_2-structured phase was amenable to storage, grinding, and study at liquid nitrogen temperature (77 K). One sample has now been stored for over 6 months at 77 K, with no evidence yet for back-transformation. The three deuterated samples ($Ca(OD)_2$) have been used successfully for structure determination by powder neutron diffraction [*Leinenweber et al.*, 1997].

4. DISCUSSION

Whether cryogenic recovery will work for phases such as those listed in Table 1 is a question to be answered only by experimentation. However, some rough guidelines for establishing likely candidates can be suggested. First, and most obviously, high pressure phases which revert by a nearly second order mechanism with little or no activation barrier can not be recovered metastably by any technique. Typically, unquenchable phases with an activation barrier revert to lower pressure forms during the timescale of the decompression cycle, on the order of several minutes to hours (10^2 - 10^4 s). The key to the success of cryogenic recovery lies in slowing the rate of any back-transformation reaction to a "useful" range, preferably on the order of days to weeks or months. In Figure 4, we show back-transformation rates (expressed as the rate of disappearance of the metastable high pressure phase) as a function of temperature, for activation energies (E_a) ranging from 10 to 50 kJ/mol, based on a simple model and presented here for purpose of illustration. It is obvious that, for E_a < 30 kJ/mol, cryogenic temperatures would not sigificantly affect recovery of the high pressure phase. For E_a in the range 30 - 50 kJ/mol, cryogenic temperatures slow down the back-transformation significantly compared with ambient temperature, and hence would allow recovery of the unquenchable high pressure phase. For back-transformation processes with E_a greater than 100 kJ/mol, the high pressure phases are readily recovered and studied at room temperature.

The magnitude of the activation energy for a back-transformation process will depend greatly upon the nature of the reaction, and on details of the sample and the experimental procedure. The largest barriers will separate those phases which require the most energetically costly steps, such as breaking of strong bonds, so that materials which are always metastable can form the recovered products, rather than the thermodynamically stable phase. Recovery can also be affected by grain size and morphology of the high pressure phase [*Kawasaki et al.*, 1990], as well as any dislocations induced by the transition, because these may provide nucleation sites for back-transformation. In some cases, recovery of phases which are difficult to quench may be aided by annealing the samples at high pressure, prior to decompression.

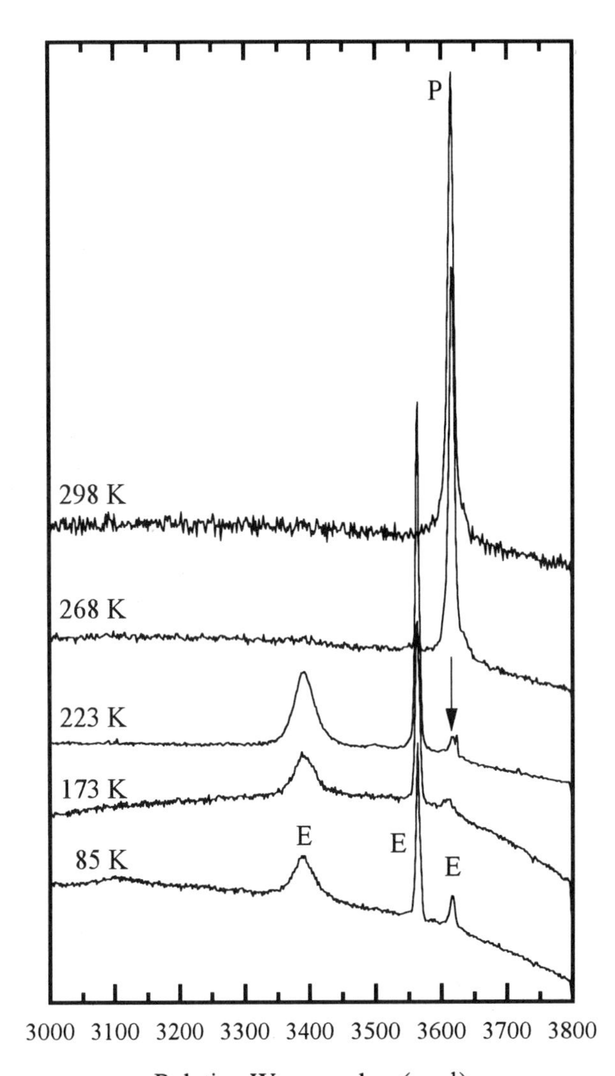

Figure 3. Raman bands of the cryogenically recovered $Ca(OH)_2$ during heating from 85 K to room temperature, in the OH-stretching region. Peaks marked E at 85 K are from the EuI_2 phase, while the band marked P at 298 K is the portlandite OH-stretching band. Arrow shows the first appearance of the portlandite peak, at 223 K.

A parameter which may help to predict the recoverability is the pressure hysteresis of the transition, combined with the absolute pressure of the transittion. A large hysteresis indicates a larger activation barrier, and a greater chance that lowering the temperature will suppress the transformation, while a lower transition pressure also results in a greater chance of recovery when the transition is suppressed. An attempt is made to list values of

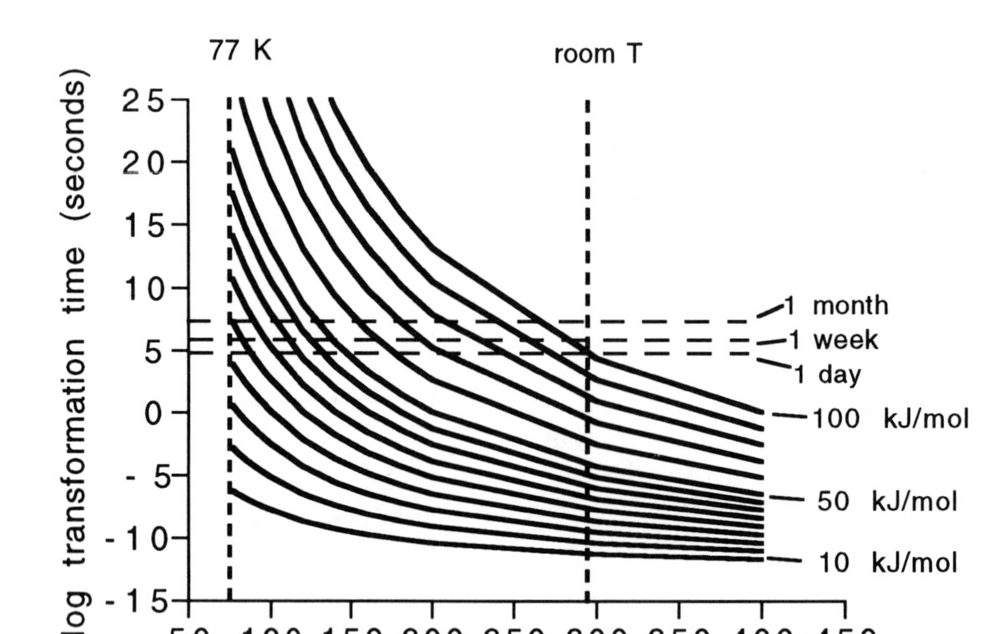

Figure 4. Kinetics of an activated back-transformation reaction B → A, where B is the high pressure phase. The rate of transformation is assumed to follow $-dB/dt = Z\ e^{E_a/RT}$, where E_a is the activation energy. The pre-exponential factor Z is expected to be close to a vibrational (attempt) frequency, $Z \sim 10^{-13}$ s. The logarithm of the transformation time is plotted as the inverse of the transformation rate for activation energies ranging (in 5 kJ/mol increments) from 10 kJ/mol to 50 kJ/mol. For $E_a < 30$ kJ/mol, the sample back-transforms within seconds to minutes, even at liquid nitrogen temperature (77 K). However, for transitions with E_a in the range 30-50 kJ/mol, cooling the sample to liquid nitrogen temperature during decompression significantly lengthens the back-transformation time, to times on the order of days to weeks. Materials with activation energy barriers significantly greater than 50 kJ/mol (e.g., 100 kJ/mol) can be readily recovered on cooling at ambient temperature.

pressure hysteresis in Table 2; however, this quantity is often not carefully reported in papers on high-pressure phase transitions, and also cannot be defined in the many cases where the reaction is irreversible, such as in the reversion of $CaSiO_3$ to glass.

The cryogenic techniques are presently being extended to include neutron and conventional x-ray diffraction to be performed on stress-free powder samples, for loading capillaries for synchrotron x-ray diffraction, and for cryogenic loading for TEM. Also, for back-transformations that occur rapidly (such as the EuI_2 to portlandite reversion in $Ca(OH)_2$), the measurement of transition enthalpy by differential scanning calorimetry (DSC) is being developed. In general, it is only necessary to develop cold-loading techniques for each application. Hopefully, with experience, such low temperature techniques can become routine tools for research on unquenchable high pressure phases.

Acknowledgements. This research was supported by NSF Materials Research Group grant DMR-9121570, and by NSF grant EAR-9628678 to K. Leinenweber. Thanks to Keiji Kusaba and Jeff Yarger for useful discussions.

REFERENCES

Angel, R.J.; A. Chopelas, and N.L. Ross, Stability of high-density climopyroxene at upper-mantle pressures, *Nature*, *358*, 322-324, 1992.

Ekbundit, S., K. Leinenweber, J.S. Robinson, M. Verhelst-Voorhees, and G.H. Wolf, New high pressure phase and pressure-induced amorphization of $Ca(OH)_2$: grain size effect, *Journal of Solid State Chemistry*, *126*, 300-307, 1997.

Grzechnik, A. and McMillan, P.F., High-temperature and high-pressure Raman study of $LiVO_3$, *Journal of Physics and Chemistry of Solids*, *56*, 159-164, 1995.

Grzechnik, A., P.F. McMillan, and G. Ouvrard, Olivine-spinel transitions in Mn_2SiSe_4, *Journal of Physics and Chemistry of Solids*, in press, 1997.

Kafalas J.A., H.C. Gatos, M.C. Lavine, M.D. Banus, High pressure phase transition in mercury selenide, *Journal of Physics and Chemistry of Solids*, *23*, 1541-1544, 1962.

Kawasaki, S., T. Yamanaka, S. Kume, and T. Ashida, Crystallite size effect on the pressure-induced phase transformation of ZrO_2, *Solid State Communications*, *76*, 527-530, 1990.

Kudoh, Y., and M. Kanzaki, On the symmetry of the $CaSiO_3$ perovskite quenched from 15 GPa and 1500 °C. Proceedings of the 14th AIRAPT Conference, Colorado Springs, Colorado, 1993.

Kunz, M., K. Leinenweber, J.B. Parise, T.-C. Wu, W.A. Bassett,K. Brister, D.J. Weidner,M.T. Vaughan, Y. Wang, The baddeleyite-type high pressure phase of portlandite, $Ca(OH)_2$, *High Pressure Research*, 14, 311-319, 1996.

Leinenweber, K., W. Utsumi,Y. Tsuchida, T. Yagi, K. Kurita, Unquenchable high-pressure perovskite polymorphs of $MnSnO_3$ and $FeTiO_3$, *Physics and Chemistry of Minerals*, *18*, 244-250, 1991.

Leinenweber, K., D. Partin, U. Schuelke, M. O'Keeffe, R.B. VonDreele, The structure of high pressure $Ca(OD)_2$ II from powder neutron diffraction: relationship to the ZrO_2 and EuI_2 structures, *Journal of Solid State Chemistry*, in press, 1997.

Liu, L., A.E. Ringwood, Synthesis of a perovskite-type polymorph of $CaSiO_3$. *Earth and Planetary Science Letters*, *28*, 209-211, 1975.

Mehta, A., K. Leinenweber, A. Navrotsky, M. Akaogi, M., Calorimetric study of high pressure polymorphism in $FeTiO_3$: Stability of the perovskite phase. *Physics and Chemistry of Minerals*, 21, 207-212, 1994.

Miller, E.D., D.C. Nesting, J.V. Badding, Quenchable transparent phase of carbon, *Chem. Mater.*, *9*, 18-22, 1997.

Palmer DC, L.W. Finger, Pressure-induced phase transition in cristobalite: An X-ray powder diffraction study to 4.4 GPa, *American Mineralogist*, *79*, 1-8, 1994.

Ross, N.L., J. Ko, C.T. Prewitt, A new phase transition in $MnTiO_3$: $LiNbO_3$ - perovskite structure, *Physics and Chemistry of Minerals*, *16*, 621-629, 1989.

Tsuchida, Y., and T. Yagi, New pressure-induced transformations of silica at room temperature, *Nature*, *347*, 267-269, 1990.

Walker, D., M.A. Carpenter, C.M. Hitch, Some simplifications to multianvil devices for high pressure experiments, *American Mineralogist*, *75*, 1020-1028, 1990.

Walker, D., Lubrication, gasketing, and precision in multianvil experiments, *American Mineralogist*, *76*, 1092-1100, 1991.

Wang Y, D.J. Weidner, Thermoelasticity of $CaSiO_3$ perovskite and implications for the lower mantle. *Geophysical Research Letters*, *21*, 895-898, 1994.

Whalley, E., D.D. Klug, Y.P. Handa, E.C. Svensson, J.H. Root,V.F. Sears, Some aspects of hydrogen bonding in the phases of water, *Journal of Molecular Chemistry* , *250*, 337-349, 1991.

Shirley Ekbundit, Kurt Leinenweber, Paul F. McMillan, Udo Schuelke, Department of Chemistry and Biochemistry, Arizona State University, Tempe, AZ, 85287-1604.

Compression of PON Cristobalite to 70 GPa

Kathleen J. Kingma, Rosemary E. Gerald Pacalo, and Paul F. McMillan

Materials Research Science and Engineering Center, Department of Chemistry and Biochemistry, Arizona State University, Tempe, AZ 85287

We have compressed phosphorus oxynitride (PON) in a diamond-anvil cell to 70 GPa. The PON has a distorted cristobalite structure and is isoelectronic with SiO_2. In situ micro-Raman spectroscopy and synchrotron X ray diffraction show that PON cristobalite undergoes a phase transformation in the range of 12 to 20 GPa. Based on the similarity of X ray and Raman spectra before and after the transition, the high-pressure phase also has a cristobalite-type structure. Remarkably, this open-framework structure remains stable to the maximum experimental pressures. Although there is evidence for a reversal transformation of the high-pressure phase near 17 GPa upon decompression, the crystalline phase quenched to ambient conditions is not identical to the starting material. No amorphization of crystalline PON is observed during the in situ compression and decompression measurements. However, X ray investigation of the decompressed sample at ambient pressure reveals a mixture of amorphous and crystalline PON components.

1. INTRODUCTION

The response of silica to pressure has been studied extensively, revealing a rich and complicated polymorphism at room temperature [for review, see *Hemley et al.* (1994)]. Pressurization of α-cristobalite, the high-temperature, low-pressure silica polymorph that exists metastably at ambient conditions, indicates several crystalline-crystalline phase transformations. Single-crystal X ray diffraction has demonstrated the high compressibility of this open-framework silica phase ($K_{0T} \approx 11.5$ GPa) [*Downs and Palmer*, 1994]. A first-order, displacive phase transformation of α-cristobalite has been identified near 1.5 GPa by the splitting of certain diffraction lines in high-resolution powder synchrotron X ray patterns [*Yeganeh-Haeri et al.*, 1990; *Palmer and Finger*, 1994; *Parise et al.*, 1994]; the new phase (cristobalite II) can be indexed on a monoclinic unit cell [*Palmer and Finger*, 1994]. Upon further examination of the cristobalite I→II transition by Raman spectroscopy [*Palmer et al.*, 1994], a second transition (cristobalite II→III) has been identified near 11 GPa. Although there have been reports of amorphization near 20 GPa under static compression [*Halvorson and Wolf*, 1990; *Halvorson*, 1992; *Palmer et al.*, 1994; *Hemley*, unpublished] and with shock compression to 30 GPa [*Gratz et al.*, 1993], earlier diffraction studies of α-cristobalite to higher pressures show no indication of amorphization [*Tsuchida and Yagi*, 1990]. Instead, Tsuchida and Yagi (1990) observed a reversible transformation of cristobalite to a new phase (silica XI) at 10 GPa. This observation is coincident with the cristobalite II→III transition more recently described by Palmer et al. (1994), thus silica XI and cristobalite III are likely identical. Tsuchida and Yagi (1990) report that silica XI (cristobalite III) undergoes a further transformation near 40 GPa to another high-pressure phase (silica XII). Upon quenching, silica XII reverts to a phase of unknown structure (silica XIII). Infrared examination of cristobalite II, silica XI

Properties of Earth and Planetary Materials
at High Pressure and Temperature
Geophysical Monograph 101

(cristobalite III), and a pressure-quenched sample of silica XIII suggests that each of these phases has silicon in tetrahedral coordination [*Yahagi et al.*, 1994].

There has also been considerable study of silica analogs in which M^{3+} and N^{5+} cations are substituted for Si^{4+} (e.g., $AlPO_4$, $GaPO_4$, $InPO_4$, $AlAsO_4$). For example, a recent study of cristobalite-structured $GaPO_4$ [*Robeson et al.*, 1994] indicates transformation to the *Cmcm* phase that was originally predicted for SiO_2 [*Tsuneyuki et al.*, 1989]. The transformation to a *Cmcm* phase has been observed and predicted for other $M^{3+}N^{5+}O_4$ cristobalite structures [*Sharma and Sikka*, 1995]. A Raman spectroscopic study of quartz-structured $AlPO_4$ indicates that a crystalline-crystalline transformation occurs at 14-15 GPa [*Gillet et al.*, 1995], rather than the amorphization transition previously reported [*Kruger and Jeanloz*, 1990].

Another scheme for creating potentially useful analogs of silica is to have two different anions instead of differing cations. An interesting series of silica analogs are generated when Si is replaced by P and charge balance is achieved by varying the remaining composition [*Marchand and Laurent*, 1991]. One such analog is crystalline phosphorus oxynitride ($P^{5+}O^{2-}N^{3-}$), which has an open-framework, distorted cristobalite structure at ambient conditions [*Bourkbir et al.*, 1989]. The chemical stability of PON has potential for use as an insulator as well as for fireproofing applications [*Marchand and Laurent*, 1991; and the references therein].

We have investigated the high-pressure vibrational and structural characteristics of PON cristobalite (which we will designate *c*-PON) using in situ Raman spectroscopy and synchrotron X ray diffraction. Here, we report results of our *c*-PON experiments and compare them with the previous high-pressure results for SiO_2 α-cristobalite.

2. EXPERIMENTAL

2.1. *Synthesis of Starting Material*

We have produced crystalline PON using a new synthesis technique [*Coffman*, 1996]. Under an inert helium atmosphere, P_3N_5 and P_2O_5 were ground together in a 1:1 stoichiometric mixture. The powder was transferred to a graphite crucible which was then placed in a silica ampoule. The ampoule was sealed under vacuum and heated at 780°C for 48 hours. As PON is reported to be stable in the atmosphere [*Bourkbir et al.*, 1989], the ampoule was opened in air. Most of the resultant PON poured freely from the graphite crucible; any crystals that remained on the crucible walls were carefully scraped and cleaned of graphite residue. Optical inspection of the product indicated an average crystallite size of less than 10 μm.

Characterization by conventional X ray powder diffraction (Table 1) confirms that our synthesized material has a pattern consistent with the $P\bar{4}$ cristobalite PON previously synthesized and described [*Bourkbir et al.*, 1989].

We tested PON cristobalite for relative hardness. Although *c*-PON did not appear to scratch corundum when placed between two discs of sapphire (a gem variety of corundum which has a Moh's hardness of 9, compared to diamond which has hardness of 10) or quartz (H=7), *c*-PON easily scratched a standard glass slide (H=5.5). Thus, *c*-PON has a Moh's hardness greater than 5.5, yet less than 7.

2.2. *Micro-Raman Spectroscopy*

The *c*-PON powder was compressed without a medium between two low-fluorescence type-I diamonds having 350-μm culets mounted in a Mao-Bell, megabar-type diamond-anvil cell. Chips of ruby were placed into the sample chamber (150-μm diameter, ~ 40-μm thick) to serve as a pressure calibrant using the ruby-fluorescence method [*Mao et al.*, 1986].

Raman measurements were made at room temperature (298 ± 3 K) to 65 GPa using a micro-Raman system designed for in situ high-pressure measurements of diamond-cell samples. This system uses an Instruments S. A. triple spectrometer (S3000) equipped with a Princeton Instruments liquid-nitrogen-cooled CCD detector (PI-1100); laser focusing and collection optics are built around a modified Olympus (BH-2) petrographic microscope. The 488.0-nm line of a Coherent 90-5 argon ion laser was used as an excitation source and experiments were performed using the backscattered 180° geometry. Laser power was approximately 150 mW (before attenuation by the diamonds) and the probe size was ~ 5 μm at the sample. Each spectrum was measured for 5 minutes.

2.3. *Energy-Dispersive Synchrotron X Ray Diffraction*

In order to investigate the structural changes of PON with increasing pressure, *c*-PON powder was compressed at room temperature without a medium in a diamond-anvil cell and diffraction was performed using energy-dispersive X ray synchrotron radiation at beam line X17-

C of the National Synchrotron Light Source at Brookhaven National Laboratory. The choice of synchrotron over conventional X ray sources allowed for rapid collection of high-quality diffraction data from the small sample. In order to reduce preferred orientation, the PON starting material was ground with a corundum mortar and pestle; optical examination showed the maximum grain size to be ~ 2 µm. The ground PON was loaded into the diamond cell without a medium and several small grains of gold were added to serve as a pressure calibrant [*Jamieson et al.*, 1982]. Although grain size was reduced by grinding of the sample, preferred orientation was apparent and was further enhanced by the small size of the X ray beam. Use of a small beam (~ 4 x 20 µm, controlled by a horizontal focusing mirror and a vertical slit) was helpful in minimizing the effects of pressure and stress gradients across the probed area. To bring as many reflections as possible into diffracting orientation, several locations were monitored in the sample chamber. Although some variability in relative peak intensities between diffraction measurements is observed, peak positions do not vary.

During grinding, Al_2O_3 corundum (from the mortar and pestle) was unknowingly introduced as a contaminant to the sample. Although diffraction lines from this contaminant were observed in the high-pressure diffraction patterns, the presence of the corundum in the load appears not to interfere with the PON experiment and subsequent interpretation of PON diffraction results (see below).

Diffraction was recorded at room temperature to ~ 70 GPa. Diffraction spectra were also collected during decompression and for the recovered, pressure-quenched samples from both the synchrotron diffraction and the Raman spectroscopy experiments. The corundum contaminant was not present in the Raman run.

3. RESULTS

3.1. *PON Micro-Raman Spectroscopy*

The ambient-condition *c*-PON Raman spectrum below 1200 cm^{-1} is characterized by a sharp peak at 223.5 cm^{-1}, a group of at least three strong bands between 380 and 580 cm^{-1}, and two weak and very broad features near 864 and 1080 cm^{-1} (Figure 1, Table 2). With nonhydrostatic compression, there are dramatic changes in the Raman spectrum, particularly the two dominant peaks in the 380-580 cm^{-1} region of the 1-bar spectrum. By 20 GPa, the strong 478 cm^{-1} peak becomes weak, and the 544 cm^{-1} peak, which was initially weaker than the 478 cm^{-1}, becomes the most intense feature in the Raman spectrum (Figure 2a). A crossover of the relative intensities of these peaks is observed near 3 GPa (Figure 2b). Between 12 and 14 GPa, there is a distinct change in the pressure dependencies of the 478 cm^{-1} and 544 cm^{-1} PON peaks, especially noted by the turnover of the 478 cm^{-1} band which begins to decrease in wavenumber with increasing pressure above 14 GPa (Figure 2c). There is no significant loss of peak intensities to 65 GPa, the highest experimental pressure.

TABLE 1. Interplanar Spacings (d_{hkl}) of *c*-PON Starting Sample Compared to Material Quenched from Nonhydrostatic, Room-Temperature Compression to 65 GPa.

c-PON starting material			pressurized *c*-PON
calculated d (Å)[a]	observed d (Å)[b]	hkl	observed d (Å)[c]
3.860	3.86	101	3.866
2.391	2.387	112	2.397
2.313	2.309	200	2.315
2.084	2.081	103	2.088
1.984	1.98	211	2.008
1.930	1.927	202	1.935
		?	1.790
1.751		004	1.769
1.636	1.632	220	
1.548	1.546	213	1.558
1.506	1.503	301	1.535
1.463	1.46	310	
1.396	1.394	204	1.399
1.35	1.347	312	1.350

[a] Calculated *c*-PON diffraction pattern using $P\bar{4}$ structural parameters of Bourkbir et al. (1989) with a = 4.6266(1) Å, c = 7.0037(3) Å, and c/a = 1.514.

[b] Powder diffraction results from our synthesized *c*-PON using a Siemens D5000 diffractometer. Peak assignments made after Bourkbir et al. (1989). Refinement of tetragonal unit-cell parameters using all observed diffraction lines yields a = 4.617(2) Å, c = 6.996(9) Å, and c/a = 1.515.

[c] The *c*-PON sample pressurized to 65 GPa during Raman experiment, then quenched to ambient conditions. Results are average of energy-dispersive synchrotron X ray diffraction patterns (see Figure 6) measured at different locations of the pressure-quenched sample after removal from the diamond cell. Refinement of tetragonal unit-cell parameters using diffraction lines indexed after Bourkbir et al. (1989) yields a = 4.66(2) Å, c = 7.03(3) Å, and c/a = 1.509.

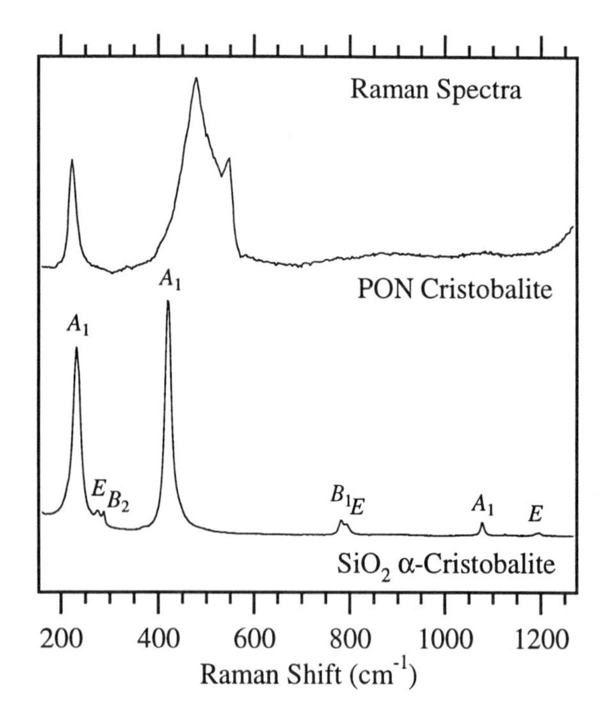

Figure 1. Comparison of PON cristobalite and SiO_2 α-cristobalite Raman spectra (1 bar, 298 K). The *c*-PON Raman spectrum was measured with the sample in the diamond cell before compression; the rising tail at the high-frequency limit of the spectrum is from a diamond Raman band.

The pressure evolution of the weaker features in the *c*-PON Raman spectrum is displayed in Figure 3. Similar to the features observed in the 380-580 cm^{-1} region of the 1-bar spectrum, other bands exhibit a kink in their pressure dependencies between 12 and 14 GPa, especially the 404 and 1080 cm^{-1} bands. With continued compression, a new feature appears above 20 GPa near 245 cm^{-1}, and above 30 GPa new peaks appear in the 480-540 cm^{-1} region. Above 20 GPa the weak 864 cm^{-1} feature can be fit by two broad peaks.

3.2. *PON Synchrotron X Ray Diffraction*

Compression to near 70 GPa shows two sets of X ray diffraction lines having unique pressure trends. The first group, corresponding to the *c*-PON lines, shows rapid decrease on compression relative to the second set of peaks (Figure 4). The second set includes weak peaks which compress much less rapidly along nearly linear trends; these peaks belong to the contaminant phase, corundum, which was introduced during grinding of the *c*-PON starting material. Using the observed corundum peaks for cell refinements, pressure-volume (*P-V*) data give $K = 234$ (±28) GPa and $K' = 7$ (±2). Our *P-V* data for the contaminant corundum are not consistent with a previous nonhydrostatic study of corundum [*Jephcoat et al.*, 1988], but instead with a quasihydrostatic study using argon as a pressure medium [*Richet et al.*, 1988]. This suggests that *c*-PON acted as a pressure-transmitting medium in our experiment for the small amount of corundum in the sample load. Comparison of the relative intensities of diffraction peaks from the two phases (e.g., see Figure 5) to relative intensities in diffraction patterns calculated for varying proportions of *c*-PON and corundum suggests that the Al_2O_3 constituted approximately 10 volume percent of the total sample in the diamond cell.

With compression to 20 GPa, there are subtle changes in the *c*-PON diffraction pattern. The PON 101 splits into two peaks (i.e., appears as an asymmetric feature) between 1.3 and 4.5 GPa (Figures 4, 5a). Between 14.3 and 21.1 GPa, the asymmetric feature becomes symmetric and loses overall intensity. This intensity loss can be accounted for by the disappearance of the second peak.

In addition to the changes observed in the PON 101, there are three sets of peaks which appear to merge into single peaks by 20 GPa (Figure 5). Specifically, these sets are (*1*) the PON 112 and 200, (*2*) the 211 and 202, and (*3*) the 220, 213, and 301. The characteristics of these mergings are similar in each case and are demonstrated in Figure 5b, which shows the behavior of the PON 112 and 200 reflections. Between 1.3 and 4.5 GPa, the 112 and 200 have merged into an asymmetric peak and by 21.1 GPa only a single symmetric peak is observed. Unlike the PON 101, there is no associated change in intensity observed between 14.3 and 21.1 GPa for the higher index peaks.

With compression above 20 GPa, only four PON peaks (labeled 101, 112, 211, and 220 on Figure 4) are observed; the PON 220 is lost near 40 GPa due to spectral cut-off of the diffraction pattern below ~ 1.4 Å (for the selected detector angle) by the synchrotron beam focusing mirror. Diffraction peaks from crystalline PON are observed to the highest experimental pressure (70 GPa). At this pressure, however, diffraction intensities of the observed peaks have been reduced by a factor of ~ 10 from intensities measured at ambient conditions in the diamond cell. Because the corundum contaminant does not display preferred orientation, the presence of corundum in the sample load serves as an internal calibrant for monitoring intensity variations of the PON with increasing pressure.

Upon decompression, only the stronger PON 101 and

TABLE 2. Observed Raman-Active Modes of PON Cristobalite (1 bar, 300 K).

c-PON			SiO$_2$ α-Cristobalite			
Frequency, ν_i (cm^{-1})	$\left(\frac{d\nu_i}{dP}\right)$ (cm^{-1}/GPa)	γ_i^a	Frequency^b (cm^{-1})	Frequency^c (cm^{-1})	γ_i^c	Symmetry Assignment^c
				32		B_1
			116.4	113.9	2.2	B_1
				145		
223.5	7.5(5)	2.68	232.1	230.7	1.1	A_1
			275.9	276.2	-0.01	E
			287.4	287.6	-0.2	B_2
			371.5	364.7	-0.7	A_1 or B_1
404	6.6(6)	1.31		380		E
478.0	7.6(6)	1.27	420.6	419.9	1.0	A_1 or B_2
525	3.2(2)	0.49				
544.5	7.8(4)	1.15		447		E
				485	0.9	E
864	4.6(2)	0.43	782.8	785		B_1
			794.9	794		E
				1035		
1080	2.3(4)	0.17	1078.1	1069		A_1
				1163		
			1194.0			

[a] Calculated values using $\gamma_i = \left(\frac{d\nu_i}{dP}\right)\frac{K}{\nu_i}$, where K = 80 GPa (see Figure 7).

[b] SiO$_2$ α-Cristobalite sample synthesized by B. Mysen; Raman spectrum that was previously reported [*Kingma and Hemley*, 1994] is shown in Figure 1.

[c] SiO$_2$ α-Cristobalite data from Palmer et al. (1994) and the references therein.

112 peaks can be followed. A plot of these d values versus pressure shows hysteresis between 70 and 20 GPa with respect to their compressional curves, in that they follow a shallower d(P) trend (open symbols, Figure 4). Between 20 and 17 GPa, the peak positions jump back to their compressional trends, indicating release of stored strain energy in the high-pressure sample. During this decompression pressure interval, the change in the decompressional trends is accompanied by a marked increase in intensity of the PON 101. There is no change in relative intensity of the 112 peak compared to its intensity observed at the maximum pressure.

Diffraction measurements were performed on the pressure-quenched samples recovered from the Raman (Figure 6) and diffraction experiments; measurements at different locations across each sample confirmed the presence of preferred orientation. There were no differences between the PON d values from the quenched Raman and diffraction samples, thus demonstrating that the inclusion of the corundum contaminant in the diffraction experiment did not affect the high-pressure behavior of c-PON. Underlying the diffraction peaks from crystalline PON, all patterns show a broad feature centered near 16 keV that originates from amorphous PON (Figure 6). There was no indication of the growth of this feature in the spectra taken in situ at high pressure or during the decompression pathway; the broad peak only appeared once the samples were removed from the diamond-anvil cell (e.g., on decompression below 1-2 GPa).

4. DISCUSSION

4.1. *Starting Material*

Crystalline PON, initially thought to crystallize in the $I4_122$ cristobalite-type structure which has one unique

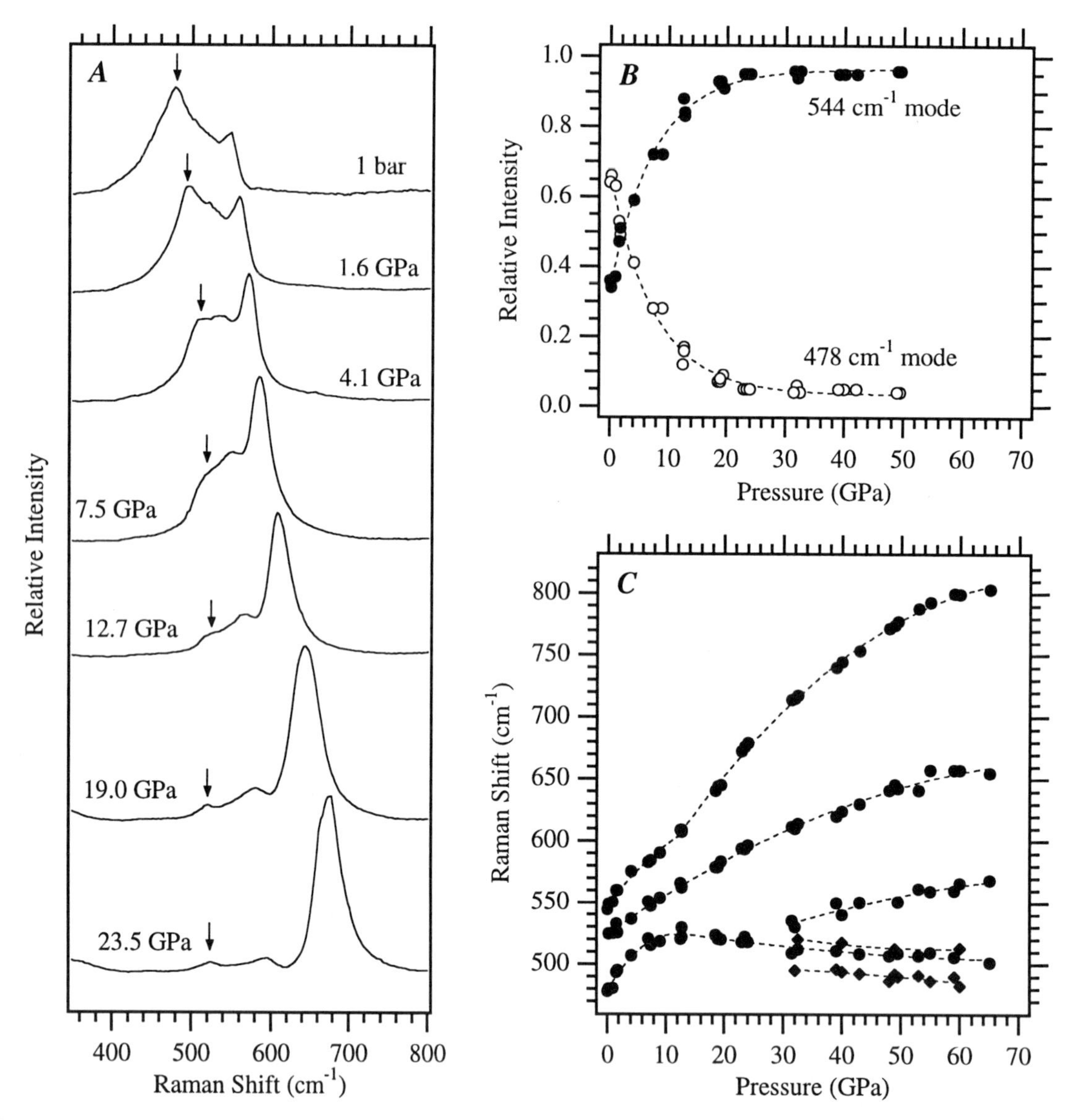

Figure 2. Pressure behavior of strongest *c*-PON Raman features. (*a*) Representative Raman spectra to 23.5 GPa. Arrows indicate the pressure trend of the 478 cm^{-1} band. The relative intensities of the Raman features in this region do not vary significantly from those shown in the 23.5 GPa spectrum to the highest experimental pressure, 65 GPa. (*b*) Relative normalized intensities of the 478 and 544 cm^{-1} modes at each pressure. (*c*) Pressure dependence of Raman band frequencies showing discontinuity near 12 to 14 GPa.

cation and one unique anion position [*Waerstad and Sullivan*, 1976], has more recently been characterized by time-of-flight neutron diffraction [*Bourkbir et al.*, 1989]. Selection of the less symmetric space group $P\bar{4}$, which yields two anion and three cation positions in a cristobalite-type structure, gave much better agreement than refinement of the neutron data using $I\bar{4}2d$. Refinement showed that the oxygen and nitrogen atoms are randomly distributed over the two unique $P\bar{4}$ anion positions. Although we expect each phosphorus to be bonded to two oxygens and two nitrogens in order to maintain local charge balance, there is complete disorder in the connectivity of P-N-P and P-O-P linkages.

Assuming $P\bar{4}$ (S_4^1) space-group symmetry for *c*-PON and four PON formula units per unit cell [*Bourkbir et al.*, 1989], the irreducible representation for the optical and acoustic modes of PON cristobalite can be determined

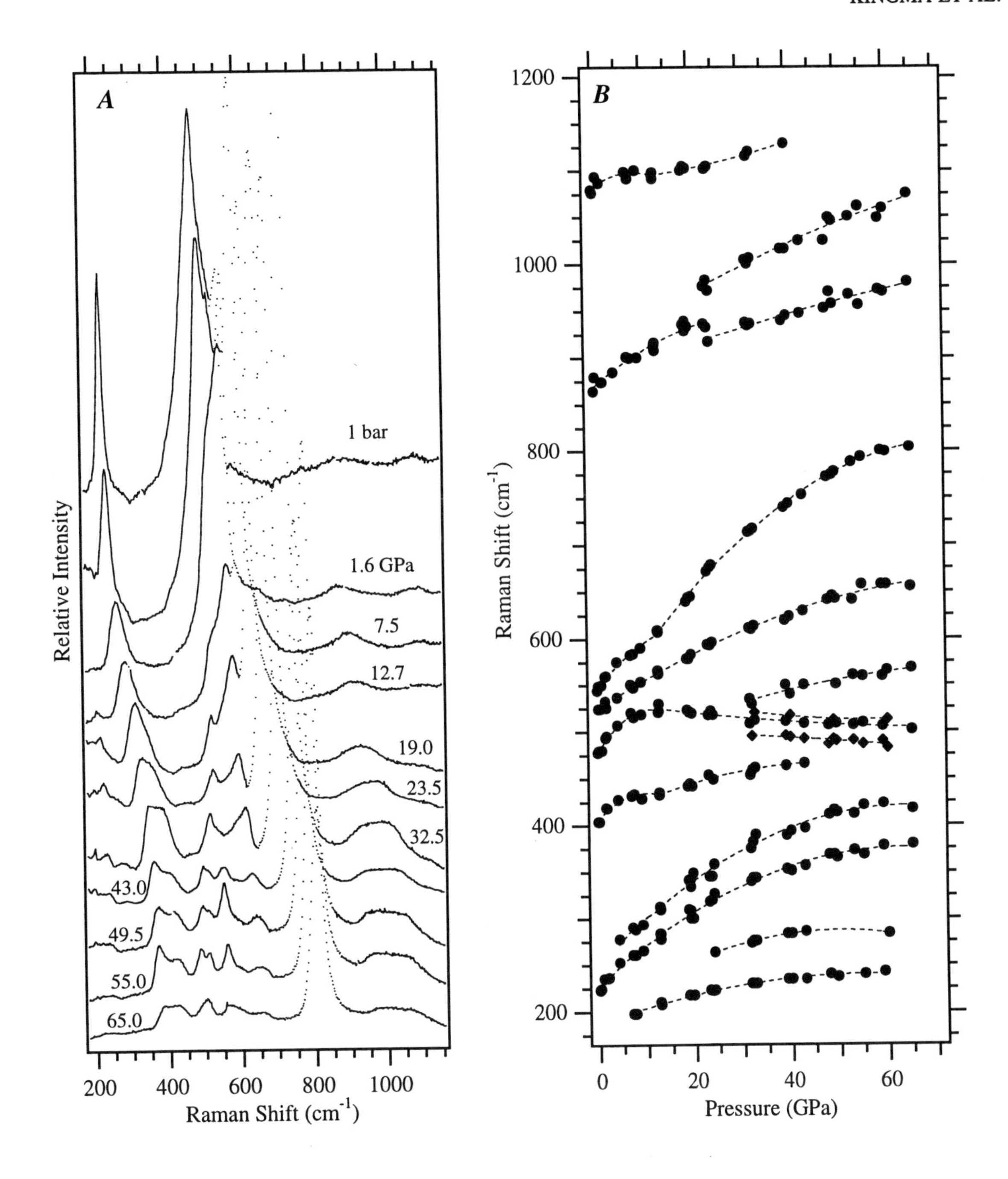

Figure 3. Pressure behavior of lower intensity *c*-PON Raman features. (*a*) Representative Raman spectra to 65 GPa. Intensity (vertical axis) has been expanded as compared to Figure 2. Because the 544 cm^{-1} band (which is featured in Figure 2) is so strong, it has been plotted as dots to guide the eye to the weaker peaks of the PON spectrum. (*b*) Pressure dependence of all observed *c*-PON Raman bands. Plotted peak positions report only those features that are recognizable as individual peaks. Because *c*-PON has a very complicated Raman spectrum, the sudden appearance (or disappearance) of features may result from overlap of Raman bands in a different pressure range. For instance, the appearance of new features in the 480-540 cm-1 region above 30 GPa may result from the simultaneous weakening of the strong 478 cm^{-1} peak which possibly obscured the weaker bands at lower pressure.

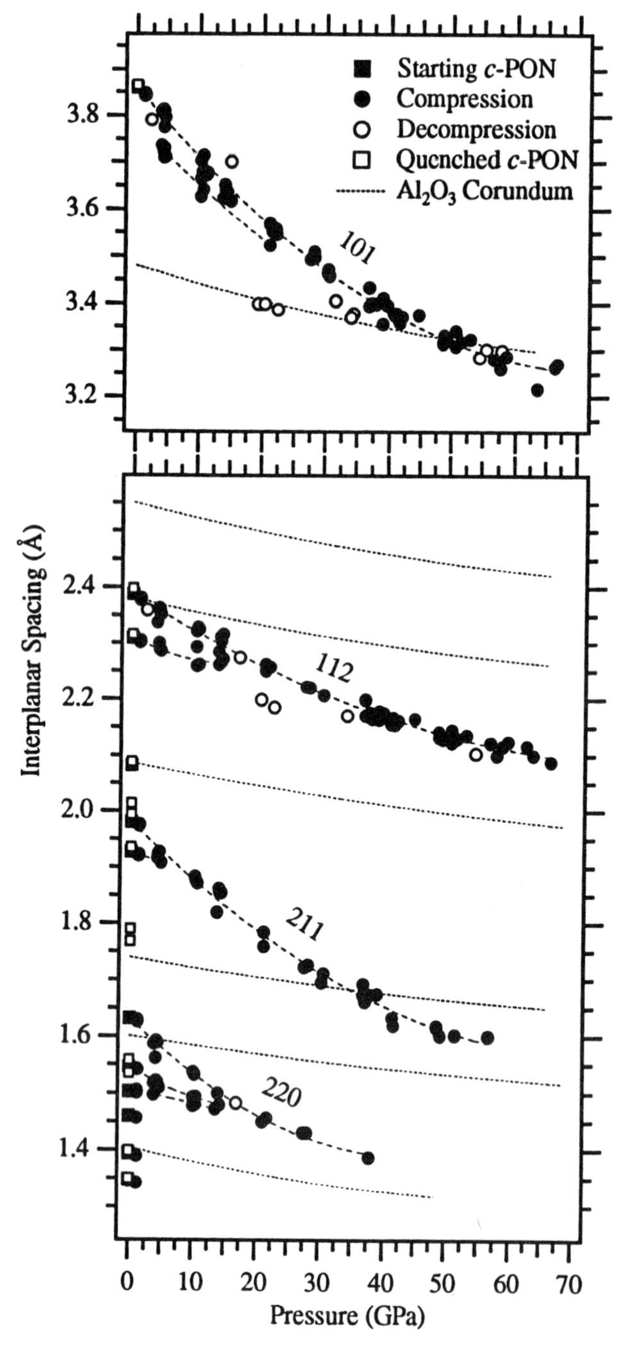

Figure 4. Variation in *c*-PON interplanar spacings (Å) with compression to nearly 70 GPa (filled symbols) and with decompression to ambient conditions (open symbols). Fine dashed lines summarize interplanar spacings of the corundum contaminant which was introduced during grinding of the *c*-PON starting sample. Although the decompressional curve of the PON 101 follows the trend of the contaminant corundum 012 peak, the nearly coincident peaks (between 70 and 20 GPa) were distinguished from each other based upon their differences in peak intensities (i.e., the PON 101 was stronger).

by factor group analysis [*Fateley et al.*, 1972]:

$$\Gamma_{op} = 7A + 8B + 9E$$
$$\Gamma_{ac} = B + E,$$

where each A, B, and E optical mode is both Raman- and infrared-active. Of the 24 Raman-active species predicted for $P\bar{4}$ PON cristobalite, we observe seven bands in the 1-bar spectrum (Figure 1, Table 2).

Comparison of the *c*-PON Raman spectrum with the well-understood spectrum of SiO_2 α-cristobalite [$P4_12_12$ (D_4^4), $\Gamma_{op} = 4\ A_1^R + 4\ A_2^R + 5\ B_1^R + 4\ B_2^{IR} + 8\ E^{IR}$] provides suggestions for the symmetry assignments of the *c*-PON Raman bands. In particular, the *c*-PON band near 233 cm^{-1} is likely of A symmetry, corresponding to the soft A_1 mode (232 cm^{-1}) which drives the temperature-induced α- to β-cristobalite transformation in SiO_2. The broad feature in the *c*-PON spectrum between 380 and 580 cm^{-1} contains several bands, as compared to a single peak in this region for SiO_2-cristobalite. The strong 420 cm^{-1} A_1 mode in the SiO_2 spectrum is associated with Si-O-Si/O-Si-O bending; the PON features in this region likely result from P-(O,N)-P/(O,N)-P-(O,N) bending. In the SiO_2 α-cristobalite Raman spectrum, features above 700 cm^{-1} originate from the Si-O stretching modes. By correlation, the weak, broad features in the PON spectrum at 864 and 1080 cm^{-1} are likely from P-(O,N) stretching.

We propose that the broadness of the *c*-PON Raman bands as compared to its SiO_2 analog (Figure 1) results from disorder in the (O,N) sites of the *c*-PON structure. This is consistent with a Raman study of nitrided $NaPO_3$ glasses which showed significant broadening of sharp stretching and bending modes and the appearance of new bands when N was incorporated in the parent oxide glass [*Bunker et al.*, 1987].

4.2. *Pressure-Induced Amorphization?*

Many of the silica polymorphs have been shown to undergo pressure-induced amorphization at 25 to 30 GPa [for review, see *Hemley et al.* (1994)]. Although intensities of the *c*-PON diffraction peaks have been reduced by a factor of ~ 10 by the highest pressures, neither the diffraction nor the spectroscopic spectra show growth of background intensity which might indicate heterogeneous transformation of crystalline PON to an amorphous phase during compression or decompression. Such background intensity in diffraction patterns has been previously used as in situ evidence of pressure-induced amorphization of SiO_2 quartz [*Kingma*

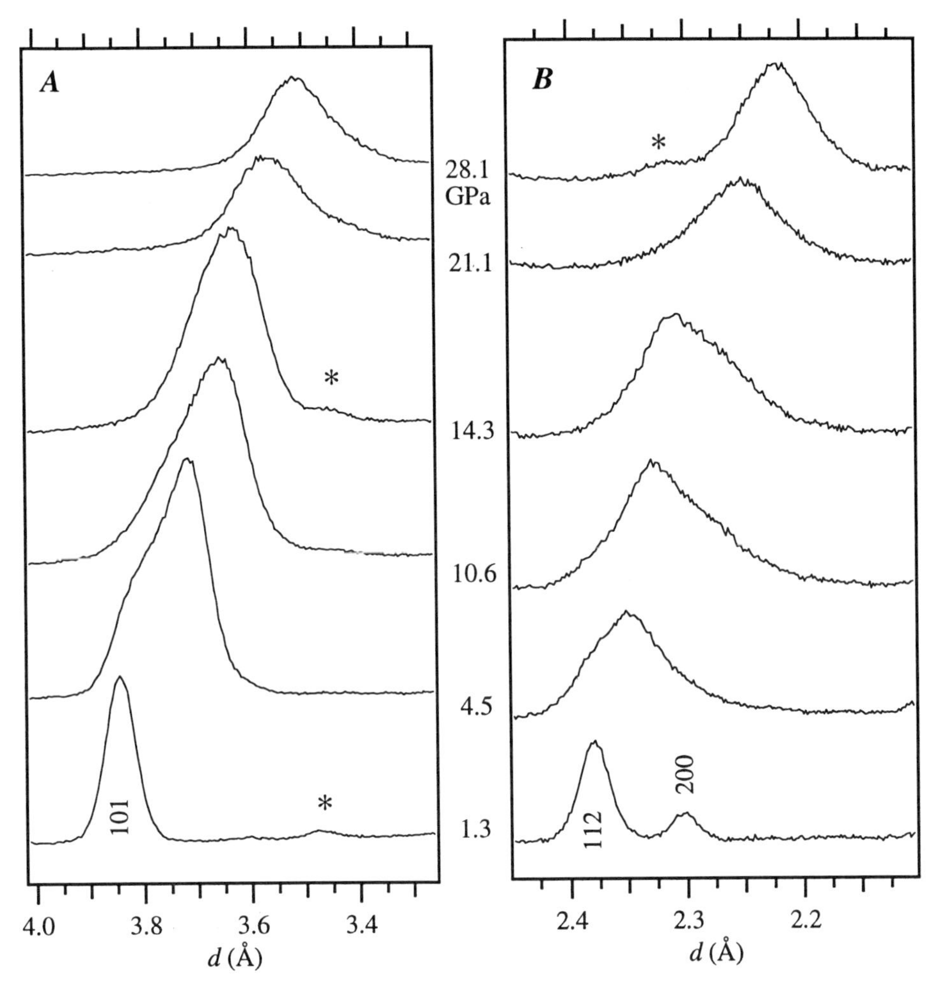

Figure 5. Representative energy-dispersive synchrotron X ray diffraction patterns for *c*-PON to 28 GPa. Star symbols (*) indicate corundum contaminant peaks. (*a*) A new peak of lower *d* value is observed above 1.3 GPa near the PON 101. Between 14.3 and 21.1 GPa, the asymmetric peak formed by the PON 101 and the new line becomes symmetric and looses intensity. (*b*) Merging of the *c*-PON 112 and 200 peaks above 1.3 GPa. Between 14.3 and 21.1 GPa, the asymmetric peak formed by the PON 112 and 200 lines becomes symmetric, but without an associated loss of intensity.

et al., 1993; *Kingma et al.*, 1996]. In the case of our PON experiments, the loss of diffraction peak intensity may be due to both thinning of the sample with increasing pressure (which was not corrected for in the plots of diffraction patterns) and loss of crystallinity (e.g., crystallite size reduction, which may be associated with formation of defects). After the samples were removed from the diamond cells, however, a heterogeneous mixture of amorphous and crystalline PON components is observed by diffraction (Figure 6, see below).

4.3. *High-Pressure Phase Transformation*

The diffraction and spectroscopic data suggest that *c*-PON undergoes a crystalline-crystalline phase transition in the 12-20 GPa range. The change in intensity of the PON 101 diffraction peak (Figure 5a) and the modification of the asymmetric diffraction peaks (multiple peaks) to become symmetric (a single peak) indicate that a transition occurs in *c*-PON in the pressure range between 14.3 and 21.1 GPa. The sharp change in pressure dependencies of several *c*-PON Raman bands supports occurrence of a transition beginning between 12 and 14 GPa (Figure 3b).

More subtle structural changes take place at lower pressure, indicated particularly by the continuous and gradual change in relative intensities of the 478 and 544 cm^{-1} Raman bands (Figure 2), which may be premonitory to the phase transition. It is interesting to note that these observations are reminiscent of the

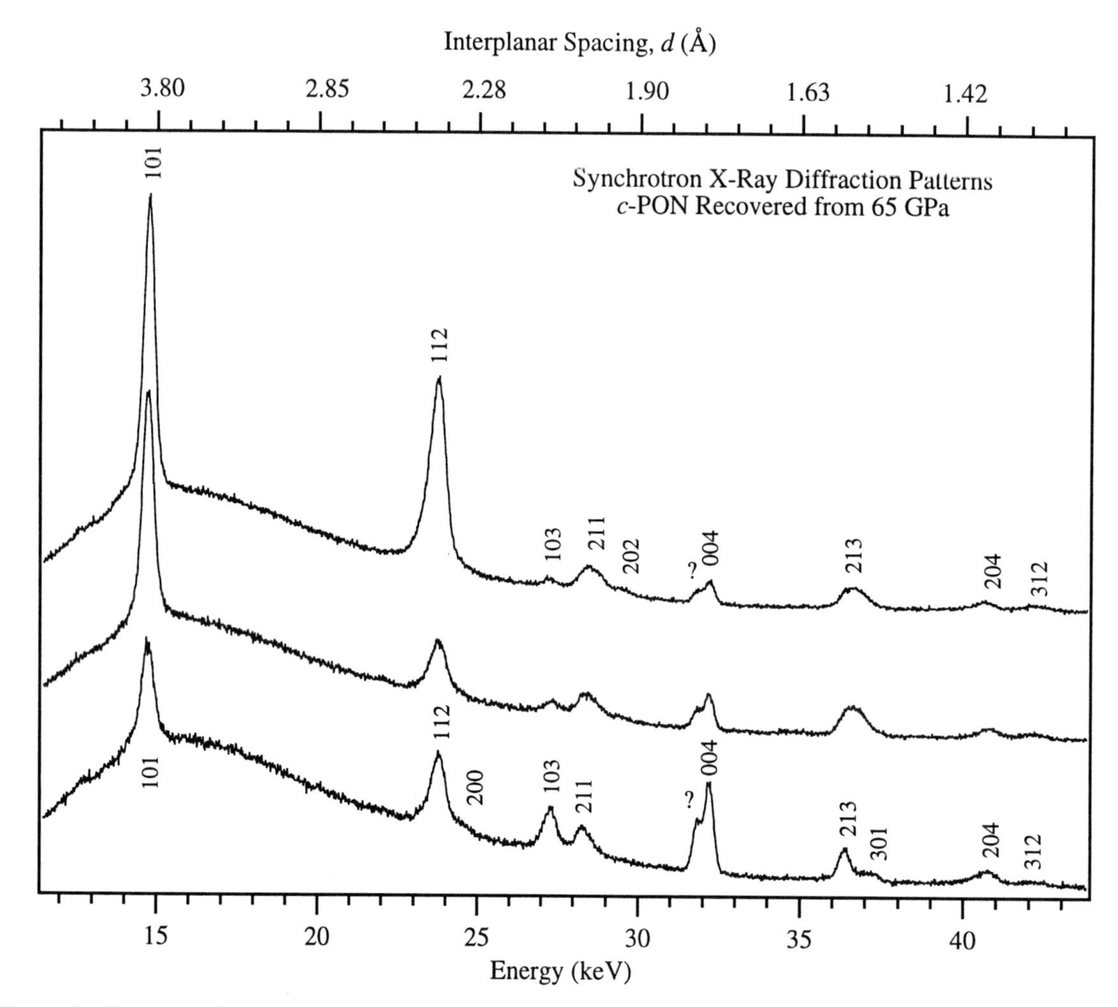

Figure 6. Representative energy-dispersive synchrotron X ray diffraction patterns of pressurized *c*-PON recovered from nonhydrostatic compression to 65 GPa. The three diffraction patterns were measured at different locations within the same sample, which was recovered from the Raman experiment and did not contain the corundum contamination. Intensity variations of diffraction lines between the three spectra demonstrate the existence of preferred orientation in the pressure-quenched sample. Peak positions are listed in Table 1. The broad peak in the background intensity centered near 16 keV originates from an amorphous PON component present in the pressure-quenched sample.

changes observed in the Raman spectrum of SiO_2 α-cristobalite above 1 GPa [*Palmer et al.*, 1994]. Also, the intertetrahedral angle in SiO_2-cristobalite has collapsed from its ambient-condition angle of 146.5° to 140.5° at 1.05 GPa, approaching the average of P-(O,N)-P angles in *c*-PON at ambient pressure (131.6-132.9°) [*Bourkbir et al.*, 1989]. These observations may suggest an analogy between the structure of *c*-PON at ambient pressure, and the structure of SiO_2 cristobalite II above 1.5 GPa.

From the X ray and Raman data, it is likely that the high-pressure PON phase also has a distorted cristobalite structure and that the phase transition is displacive. Compression shows a distinct merging (by 4.5 GPa) and eventual loss (above 14 GPa) of several sets of *c*-PON diffraction peaks. The pressure curves of the remaining four diffraction lines (101, 112, 211, and 220) continue unbroken and without any kinks above 20 GPa (Figure 4). Similarly, all Raman bands continue beyond the transition pressure without any dramatic or sudden changes in intensities or spectral characteristics (Figures 2, 3). The similarities of the Raman and the diffraction spectra before and after the transition pressure suggest that the high-pressure phase has a structure very similar to the low-pressure phase. The subtle changes in symmetry are more apparent in the Raman data compared with the diffraction data, as evidenced by the sharp kink in the frequency versus pressure curves of some Raman bands near 12 to 14 GPa (Figure 3b).

Using the strongest *c*-PON diffraction peaks, cell parameters were refined, first using a tetragonal cell and then using orthorhombic constraints. A plot of cell parameters from the orthorhombic refinements shows no deviation within error from tetragonal (Figure 7). Because of the likely structural similarities of the low- and high-pressure phases of PON, we have fit all *P-V* data to the Birch-Murnaghan equation-of-state, producing a rough estimate of *c*-PON's bulk modulus, K = 80 (±5) GPa, and its first derivative, K' = 2.6 (±0.2). Figure 7 demonstrates the stiffness of the *c*-PON structure compared to its silica cristobalite analog, and other SiO_2 polymorphs. *c*-PON is much less compressible than *c*-SiO_2, and has a compressibility intermediate between those of SiO_2 quartz and coesite.

In the SiO_2 polymorphs, the main compression mechanism at low pressure involves narrowing the intertetrahedral Si-O-Si angle [*Hazen and Finger*, 1979]. The observed stiffness of *c*-PON suggests that there is greater resistance to further bending of the P-(O,N)-P linkages between the $P(O,N)_4$ tetrahedra. At ambient pressure, the Si-O-Si angle in α-cristobalite is 146.5° [*Downs and Palmer*, 1994] (in the β-cristobalite *Fd*3*m* structure, it is 147.7° at 221° C [*Peacor*, 1973]), whereas the P-(O,N)-P angles in *c*-PON range between 131.6 and 132.9° [*Bourkbir et al.*, 1989]. The narrower intertetrahedral angle in PON requires that the metal (P) atoms are in closer contact at ambient pressure than are those (Si) in SiO_2 cristobalite. This will result in increased non-bonded repulsion as *c*-PON is compressed [*O'Keeffe and Hyde*, 1976], which is consistent with our observations.

Finally, it is interesting to compare the behavior of the SiO_2 polymorphs with *c*-PON at very high pressure when compressed at room temperature. The tetrahedral structures of the SiO_2 polymorphs show instability above

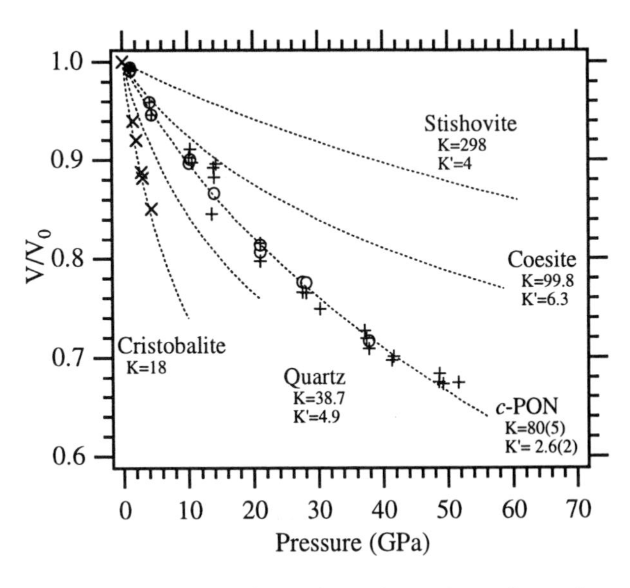

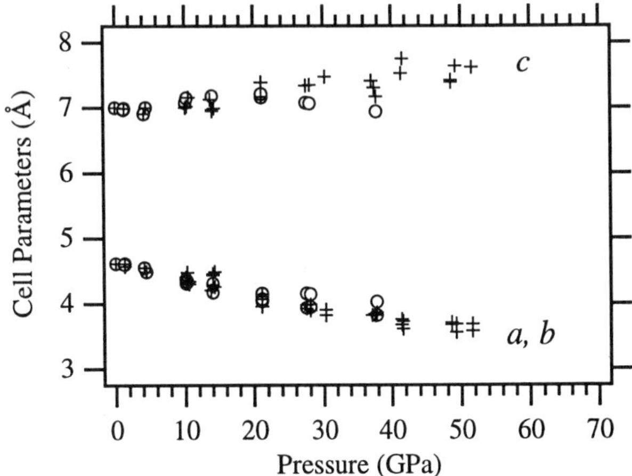

Figure 7. (*Bottom*) Pressure dependence of the *c*-PON cell parameters refined using orthorhombic constraints. Open circles designate refinements from patterns which had four or more diffraction lines. Above 40 GPa, only three strong peaks (the *c*-PON 101, 112, and 211) were observed in the diffraction spectra; for comparison, cell refinements were also performed for each measurement using only these three lines (crosses). (*Top*) *c*-PON equation-of-state compared to SiO_2 polymorphs. A fit of the PON volume data that was calculated using four or more diffraction lines (open circles) to the Birch-Murnaghan equation-of-state,

$$P=\frac{3}{2}K_0\left(y^{-7/3}-y^{-5/3}\right)\left[1-\frac{3}{4}\left(4-K_0'\right)\left(y^{-2/3}-1\right)\right],$$ where

$y=\frac{V}{V_0}$, gives a bulk modulus, K, of 83 (± 2) GPa, and its first derivative, K' equal to 2.4 (± 0.1). The dashed line through the *c*-PON *P-V* data [K = 80 (± 5) GPa and K' = 2.6 (± 0.2)] is from fit of the volume data that was refined at each pressure from only three diffraction lines (crosses). Data shown for SiO_2 cristobalite (cristobalite I at ambient conditions and cristobalite II above 1.5 GPa) is from Palmer and Finger (1994). Downs and Palmer (1994) report K = 11.5 (± 0.7) GPa and K' = 9 (± 2) for cristobalite I. The plotted equation-of-state [K = 18 (± 1) GPa] is from X ray diffraction data collected to ~ 10 GPa [*Tsuchida and Yagi*, 1990]; this fit does not take into account the cristobalite I→II transition. The plotted equations-of-state and listed K and K' values for the other SiO_2 polymorphs are from data previously reviewed by Hemley et al. (1994).

the 25 to 30 GPa range associated with a [4]Si to [6]Si coordination change, resulting in pressure-induced amorphization or crystalline-crystalline transitions. This is evident in the characteristic behavior of the Raman spectra of the crystalline and amorphous silica phases, in which the pressure dependence of the dominant Si-O-Si bending mode saturates at a limiting value near 620 cm^{-1} and is associated with a marked decrease in intensity [*Halvorson*, 1992; *Williams et al.*, 1993]. This behavior, which has also been observed for the silica analog GeO_2 [*Wolf et al.*, 1992], may be associated with addition of a third bond to the bridging oxygen in the Si-O-Si linkage (producing an OSi_3 unit) resulting from the [4]Si to [6]Si coordination change. No analogous saturation of the pressure dependence of the P-(O,N)-P bending mode frequency and loss of intensity is observed for *c*-PON; the frequency of the P-(O,N)-P bending vibration continues to increase with increasing pressure and the band remains strong to the highest pressure examined by Raman spectroscopy, 65 GPa (Figure 2). We conclude that the bridging O and N atoms are not involved in a coordination change in this pressure regime and that no coordination increase occurs around the P atoms. This may be due to the inability to form appropriate valence sums around octahedrally coordinated phosphorous (e.g., in NP_3 or OP_3 units) within candidate high-pressure PON structures.

5. SUMMARY

The present Raman spectroscopic and synchrotron X ray diffraction data indicate that cristobalite-structured phosphorus oxynitride (*c*-PON) begins a phase transformation at 12 to 14 GPa when nonhydrostatically compressed at room temperature. The transition, which is complete by 20 GPa, is likely displacive, with the high-pressure phase also having a cristobalite-type structure. The high-pressure PON phase remains stable to the maximum experimental pressure, 70 GPa, with no in situ evidence for pressure-induced amorphization or further crystalline-crystalline transformation. This is in sharp contrast to the high-pressure behavior of the tetrahedral silica polymorphs, demonstrating the reluctance of P to take high (VI-) coordination to O or N under the present experimental conditions. It will be of interest to see if a high-pressure PON phase analogous to SiO_2 stishovite can be synthesized at higher pressure and/or temperature.

Acknowledgments. We thank P. Coffman for assistance with the synthesis of the starting material. The work was supported by the ASU Materials Research Group grant NSF DMR-9121570 and by the ASU MRSEC.

REFERENCES

Bourkbir, L., R. Marchand, Y. Laurent, P. Bacher, and G. Roult, Preparation and time-of-flight neutron diffraction study of the cristobalite-type PON phosphorus oxynitride, *Ann. Chim. Fr.*, 14, 475-481, 1989.

Bunker, B. C., D. R. Tallant, C. A. Balfe, R. J. Kirkpatrick, G. L. Turner, and M. R. Reidmeyer, Structure of phosphorus oxynitride glasses, *J. Am. Ceram. Soc.*, 70, 675-681, 1987.

Coffman, P. R., Synthesis, Processing, and Characterization of Several Group IV, V, and VI Nitrides and Related Compounds, Ph.D. Thesis, Arizona State University, 1996.

Downs, R. T., and D. C. Palmer, The pressure behavior of α cristobalite, *Am. Mineral.*, 79, 9-14, 1994.

Fateley, W. G., F. R. Dollish, N. T. McDevitt, and F. F. Bentley, *Infrared and Raman Selection Rules for Molecular and Lattice Vibrations: The Correlation Method*, 222 pp., Wiley-Interscience, New York, 1972.

Gillet, P., J. Badro, B. Barrel, and P. F. McMillan, High-pressure behavior in α-$AlPO_4$: Amorphization and the memory-glass effect, *Phys. Rev. B*, 51, 11262-11269, 1995.

Gratz, A. J., L. D. DeLoach, T. M. Clough, and W. J. Nellis, Shock amorphization of cristobalite, *Science*, 259, 663-666, 1993.

Halvorson, K. E., Pressure Induced Glass Formation Studied by Raman Spectroscopy and Nuclear Magnetic Resonance Spectroscopy, Ph.D. Thesis, Arizona State University, 1992.

Halvorson, K., and G. Wolf, Pressure-induced amorphization of cristobalite: Structural and dynamical relationships of crystal-amorphous transitions and polymorphic glass transitions in silica polymorphs (abstract), *EOS Trans. AGU*, 71, 1671, 1990.

Hazen, R. M., and L. W. Finger, Polyhedral tilting: A common type of pure displacive phase transition and its relationship to analcite at high pressure, *Phase Transitions*, 1, 1-22, 1979.

Hemley, R. J., C. T. Prewitt, and K. J. Kingma, High-pressure behavior of silica, in *Silica: Physical Behavior, Geochemistry, and Materials Applications*, edited by P. J. Heaney, C. T. Prewitt, and G. V. Gibbs, 29, 41-81, 1994.

Jamieson, J. C., J. N. Fritz, and M. H. Manghnani, Pressure measurement at high temperature in x-ray diffraction studies: Gold as a primary standard, in *High-Pressure Research in Geophysics*, edited by S. Akimoto and M. H. Manghnani, 12, 27-48, 1982.

Jephcoat, A. P., R. J. Hemley, and H.-K. Mao, X-ray diffraction of ruby (Al_2O_3:Cr^{3+}) to 175 GPa, *Physica B*, 150, 115-121, 1988.

Kingma, K. J., and R. J. Hemley, Raman spectroscopic study of microcrystalline silica, *Am. Mineral.*, 79, 269-273, 1994.

Kingma, K. J., R. J. Hemley, H.-K. Mao, and D. R. Veblen, New high-pressure transformation in α-quartz, *Phys. Rev. Lett.*, 70, 3927-3930, 1993.

Kingma, K. J., H.-K. Mao, and R. J. Hemley, Synchrotron x-ray diffraction of SiO_2 to multimegabar pressures, *High Pressure Research*, 14, 363-374, 1996.

Kruger, M., and R. Jeanloz, Memory glass: an amorphous material formed from $AlPO_4$, *Science*, 249, 647-649, 1990.

Mao, H. K., J. Xu, and P. M. Bell, Calibration of the ruby pressure gauge to 800 kbar under quasi-hydrostatic conditions, *J. Geophys. Res.*, 91, 4673-4676, 1986.

Marchand, R., and Y. Laurent, Nitrogen-oxygen substitution in the PO_4 tetrahedron, *Eur. J. Solid State Inorg. Chem.* , 28, 57-76, 1991.

O'Keeffe, M., and B. G. Hyde, Cristobalites and topologically related structures, *Acta. Cryst.*, B32, 2923-2936, 1976.

Palmer, D. C., and L. W. Finger, Pressure-induced phase transition in cristobalite: An x-ray powder diffraction study to 4.4 GPa, *Am. Mineral.*, 79, 1-8, 1994.

Palmer, D. C., R. J. Hemley, and C. T. Prewitt, Raman spectroscopic study of high-pressure phase transitions in cristobalite, *Phys. Chem. Miner.*, 21, 481-488, 1994.

Parise, J. B., A. Yeganeh-Haeri, D. J. Weidner, and M. A. Saltzberg, Pressure-induced phase transition and pressure dependence of crystal structure in low (α) and Ca/Al-doped cristobalite, *J. Appl. Phys.*, 75, 1361-1367, 1994.

Peacor, D. R., High-temperature single-crystal study of the cristobalite inversion, *Z. Kristallogr.*, 138, 274-298, 1973.

Richet, P., J.-A. Xu, and H.-K. Mao, Quasi-hydrostatic compression of ruby to 500 kbar, *Phys. Chem. Minerals*, 16, 1988.

Robeson, J. L., R. R. Winters, and W. S. Hammack, Pressure-induced transformations of the low-cristobalite phase of $GaPO_4$, *Phys. Rev. Lett.*, 73, 1644-1647, 1994.

Sharma, S. M., and S. K. Sikka, Comment on "Pressure-induced transformations of the low-cristobalite phase of $GaPO_4$", *Phys. Rev. Lett.*, 74, 3301, 1995.

Tsuchida, Y., and T. Yagi, New pressure-induced transformations of silica at room temperature, *Nature*, 347, 267-269, 1990.

Tsuneyuki, S., Y. Matsui, H. Aoki, and M. Tsukada, New pressure-induced structural transformations in silica obtained by computer simulation, *Nature*, 339, 209-211, 1989.

Waerstad, K. R., and J. M. Sullivan, Crystal data for phosphorus oxynitride, *J. Appl. Cryst.*, 9, 411, 1976.

Williams, Q., R. J. Hemley, M. B. Kruger, and R. Jeanloz, High pressure infrared spectra of α-quartz, coesite, stishovite, and silica glass, *J. Geophys. Res.*, 98, 22157-22170, 1993.

Wolf, G. H., S. Wang, C. A. Herbst, D. J. Durben, W. F. Oliver, Z. C. Kang, and K. Halvorson, Pressure induced collapse of the tetrahedral framework in crystalline and amorphous GeO_2, in *High-Pressure Research: Application to Earth and Planetary Sciences*, edited by Y. Syono and M. H. Manghnani, 503-517, 1992.

Yahagi, Y., T. Yagi, H. Yamawaki, and K. Aoki, Infrared absorption spectra of the high-pressure phases of cristobalite and their coordination numbers of silicon atoms, *Solid State Commun.*, 89, 945-948, 1994.

Yeganeh-Haeri, A., D. J. Weidner, J. Parise, J. Ko, M. T. Vaughan, X. Liu, Y. Zhao, Y. Wang, and R. Pacalo, A new polymorph of SiO_2 (abstract), *EOS Trans. AGU*, 71, 1671, 1990.

K. J. Kingma, P. F. McMillan, and R. E. Gerald Pacalo, Arizona State University, Department of Chemistry and Biochemistry, Box 871604, Tempe, AZ 85287

Measurements of Acoustic Wave Velocities at P-T Conditions of the Earth's Mantle

Ruth Knoche, Sharon L. Webb[1], and David C. Rubie

Bayerisches Geoinstitut, Universität Bayreuth, Bayreuth, Germany

A method is being developed to determine acoustic wave velocities in polycrystalline samples at pressures and temperatures of the Earth's mantle using an MA-8 multianvil apparatus. Ultrasonic sound waves, generated by a $LiNbO_3$ transducer, are transmitted through one of the eight tungsten carbide anvils and a platinum buffer rod to the sample which is situated at the center of the multianvil high P-T assembly. Travel times are determined by interferometry using the phase comparison method. Currently, the main uncertainty in determining velocities arises from estimating the changes in sample length at high pressure and temperature. Potentially, however, both compressional and shear wave velocities can be measured simultaneously using this technique. This leads to the possibility of determining the compressibility and resulting changes in sample length with pressure in a self consistent manner on the same sample, thus greatly improving the accuracy of the determined velocities. The technique has the potential for improving our knowledge of acoustic velocities in single phase and polyphase aggregates relevant to the Earth's mantle without the need for large extrapolations of experimental data.

1. INTRODUCTION

The comparison of the velocities of compressional (P) and shear (S) waves measured on samples in the laboratory with observed seismic wave velocities in the Earth is the main constraint on the composition and mineralogy of the Earth's interior [e.g., *Anderson*, 1989; *Duffy and Anderson*, 1989]. Currently, such comparisons leave many important questions open. These include how much olivine is present in the upper mantle, whether or not the 410-km discontinuity can be explained entirely by the $(Mg,Fe)_2SiO_4$ $\alpha-\beta$ phase transformation, what the origin is of the steep seismic velocity gradients in the mantle transition zone, and whether or not the 660-km discontinuity is caused by a change in bulk composition in addition to the phase transformation of $(Mg,Fe)_2SiO_4$ ringwoodite to perovskite + magnesiowüstite.

Further refinements of models of mantle mineralogy depend on being able to better determine acoustic velocities in minerals such as olivine, pyroxene, garnet, and their high-pressure polymorphs at the pressure-temperature (P-T) conditions of the Earth's interior. Previous measurements of acoustic velocities and elastic properties of mantle minerals have been obtained using a variety of techniques, including ultrasonic interferometry [e.g., *Graham and Barsch*, 1969; *Webb*, 1989; *Niesler and*

[1] Present address: Research School of Earth Sciences, Australian National University, Canberra ACT 0200, Australia

Properties of Earth and Planetary Materials at High Pressure and Temperature
Geophysical Monograph 101

Jackson, 1989; *Rigden et al.*, 1992], Brillouin scattering [*Duffy et al.*, 1995], and impulsive stimulated scattering [*Zaug et al.*, 1993]. Until recently, most measurements had been made either at ambient temperature and modest pressures (e.g., ≤3 GPa; *Rigden et al.*, 1994) or, in a few studies, at high temperatures (≤1400°C) and ambient pressure [e.g., *Isaak et al.*, 1989; *Isaak*, 1992]. Large extrapolations in both temperature and pressure are therefore necessary to use such data to estimate P- and S-wave velocities under mantle conditions. The recent study of *Duffy et al.* [1995] suggests that the error associated in extrapolating velocity data for olivine from 1 to 14 GPa is typically of the order of 3%. In terms of comparing seismic and experimentally determined acoustic velocity data, this is a large uncertainty. For some minerals, such as orthopyroxene, the uncertainty in extrapolation is likely to be even larger because of changes in compression mechanisms (and therefore in elastic properties) which cannot be predicted from low pressure data [e.g., *Hugh-Jones and Angel*, 1994; *Angel and Ross*, 1996].

In recent years, techniques have been developed to determine acoustic velocities at much higher pressures than 3 GPa, thus eliminating some of the uncertainties of extrapolation. Using a liquid-solid hybrid sample cell in a multianvil apparatus, *Yoneda* [1990] and *Yoneda and Morioka* [1992] determined the pressure derivatives of single crystal MgO, $MgAl_2O_4$ and Mg_2SiO_4 up to 8 GPa. The elastic moduli of olivine single crystals have been measured up to 16 GPa using Brillouin scattering and impulsive stimulated scattering techniques combined with the diamond anvil cell [*Zaug et al.*, 1993; *Duffy et al.*, 1995]. In addition, important developments have enabled in situ ultrasonic interferometry measurements to be made at pressures up to 10 GPa on both single crystals and polycrystals using the multi-anvil apparatus [*Chen et al.*, 1996; *Li et al.*, 1996]. However, because such measurements have been made at ambient temperature, uncertainties in temperature extrapolations still remain when estimating velocities under mantle conditions. The cross derivative of velocity $\partial^2 v/\partial P \partial T$, for example, has not been measured at high P and T and its value is therefore uncertain.

In situ measurements of shear wave velocities in olivine were made by *Fukizawa and Kinoshita* [1982] at elevated temperatures and pressures more than fourteen years ago. These experiments were performed on Fe_2SiO_4 up to 5.2 GPa and 700°C, and from time-of-flight measurements they were able to measure the velocity jump at the olivine-spinel phase transition. The purpose of the present study is to further develop in situ acoustic velocity measurements and to extend the P-T range to that of the mantle transition zone with the aim of eliminating the need for extrapolation. Here we present preliminary results of P- and S-wave velocity measurements on polycrystalline samples of natural San Carlos olivine at upper mantle conditions (up to 10.5 GPa and 1500°C) and discuss the uncertainties and experimental problems which are currently associated with such in situ measurements.

2. EXPERIMENTAL METHODS

Experiments at high P and T are performed using a 1000 tonne split-cylinder (6-8 type) multianvil apparatus [e.g., *Ito et al.*, 1984; *Liebermann and Wang*, 1992] at the Bayerisches Geoinstitut. The sample assembly consists of a MgO (+5 wt% Cr_2O_3) octahedron with an 18 mm edge length (Figure 1). High temperatures are generated using a $LaCrO_3$ cylindrical heater with a stepped wall thickness in order to minimize thermal gradients within the sample [*Rubie et al.*, 1993a]. The octahedron is compressed using eight tungsten carbide cubes (WC cubes, Toshiba F-grade) with corners truncated to an edge length of 11 mm (Figure 2). For the ultrasonic measurements, we use the experimental setup shown in Figure 2 in which two parallel gold-plated 20 MHz $LiNbO_3$ transducers (P- and S-wave respectively) with diameters of 6.35 mm are glued to the polished, truncated corners of the WC cubes which are located opposite the two ends of the cylindrical sample. The ultrasonic waves generated by each of these transducers travel through the WC cube and are transmitted to the sample through a 2-mm-diameter Pt buffer rod (Figures 1 and 2). Using this arrangement, the transducers are located in a stress-free and low-temperature environment [see also *Li et al.*, 1996; *Chen et al.*, 1996]. In contrast, in the earlier experiments of *Fukizawa and Kinoshita* [1982], the transducer was contained within the pressure cell and was thus subjected to both high stresses and high temperatures. The advantage of using two transducers and two buffer rods is that S- and P-wave velocities can be measured during the same experiment at identical P-T conditions, and the length changes of the sample (currently the greatest source of uncertainty in determining acoustic velocities) at high P and T can be calculated from the compressibility determined at pressure and temperature without needing to combine data from two different experiments on potentially different samples.

A 25 μm thick gold foil was placed between the WC cubes and the Pt buffer rods. No bonding agents were used between the Pt buffer rods and the sample because these interfaces were close to ideal at the pressures studied (>1.8 GPa). The absence of a bond at the sample-buffer rod interfaces considerably eases calculations of the sample

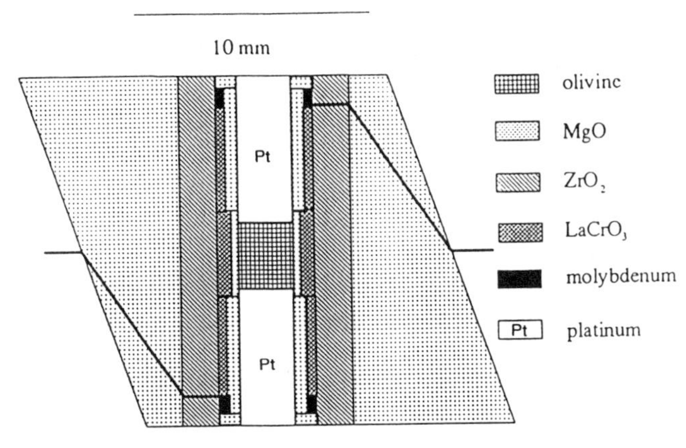

Figure 1. Schematic cross section showing details of the octahedral pressure cell. The stepped geometry of the cylindrical $LaCrO_3$ heater ensures low temperature gradients in the sample region and the ZrO_2 sleeve improves the thermal insulation. The Pt buffer rods, located at both ends of the olivine sample, transmit the ultrasonic waves (which are generated at the transducers) from the WC cubes to the sample. Copper wires are used to conduct electric current to the $LaCrO_3$ heater (the standard technique of passing current through the WC cubes cannot be used because of the presence of the Pt buffer rods). The sample and Pt buffer rods are surrounded by 30 μm thick Pt foil (not shown here) to provide a uniform chemical environment around the sample and to separate it from the adjacent MgO components.

travel times because they can be determined directly without application of a correction term for the thickness and properties of the bond. The Pt buffer rods and cylindrical sample were surrounded by Pt foil (30 μm thick) to provide a uniform chemical environment around the sample and to avoid contact (and possible contamination) between the sample and the MgO of the cell assembly. Due to the limited time spent at high temperature, no detectable reaction occurred between the olivine and the Pt (see also *Rubie et al.*, 1993a, their Figure 7).

The sample pressure was calibrated as a function of the hydraulic oil pressure using Bi phase transitions at room temperature and using the quartz-coesite, coesite-stishovite, and Fe_2SiO_4 olivine-spinel phase transitions at 1000°C and 1450°C [c.f. *Rubie et al.*, 1993b]. As a result of heating to 1000°C, the sample pressure increases by about 20% compared with the pressure at room temperature (at a constant hydraulic oil pressure). The results of pressure calibration experiments at 1450°C are identical to those obtained at 1000°C, indicating that the sample pressure is independent of temperature in the range 1000-1450°C. The pressure uncertainty is estimated to be 0.5 GPa, based on uncertainties in the respective phase equilibria as well as on the reproducibility of the calibration experiments.

Because of the presence of the Pt buffer rods, it was not possible to measure temperature using a thermocouple inserted along the axis of the cylindrical heater [c.f. *Rubie et al.*, 1993a, their Figure 1). The alternative is to insert a thermocouple through the wall of the heater, which is likely to generate localized temperature gradients. Therefore (and also for the sake of simplicity), we have estimated temperature from a temperature-power calibration in these preliminary experiments. Based on this calibration, the temperature uncertainty is ±50°C. To avoid interferences of the electric current with the transducers during heating, the heater was electrically insulated from the WC cubes and electric current was supplied to the heater through Cu leads (Figure 1).

The ultrasonic waves generated by the transducers are reflected at the interfaces between (1) the WC cube and the Pt buffer rod, (2) the first buffer rod and the sample, and (3) the sample and the second buffer rod. In order to determine travel times in the sample, we have used the phase comparison technique in which interference between

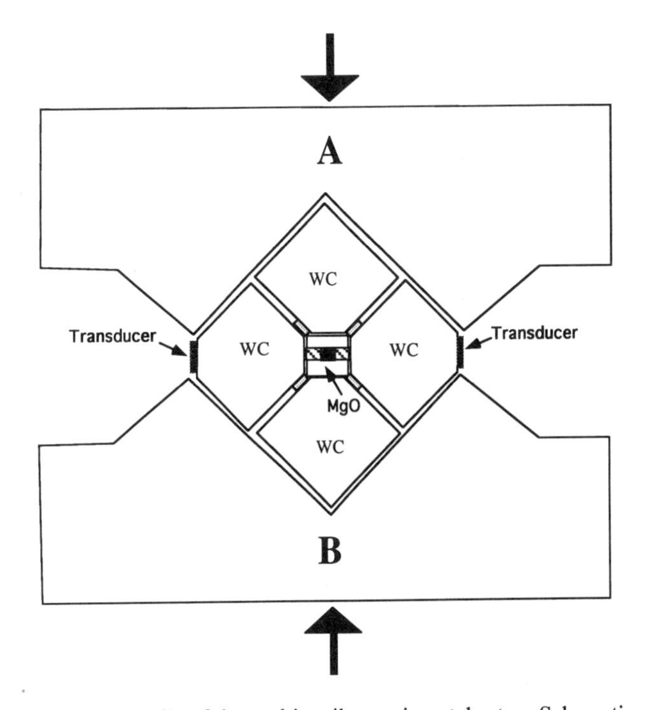

Figure 2. Details of the multianvil experimental setup. Schematic sketch showing four of the eight WC cubes (WC) which are compressed in a 1000 tonne uniaxial press by a system of hardened steel anvils (A and B). The pressure cell, consisting of a MgO octahedron, is compressed by the inner truncated corners of the WC cubes. The 20-MHz transducers (P- and S-wave respectively) are glued to outer truncated corners of two of the WC cubes and are located diagonally opposite to the buffer rods and sample (black) contained in the pressure cell (see Figure 1). The WC cubes are separated from each other by thin pyrophyllite gaskets (stippled).

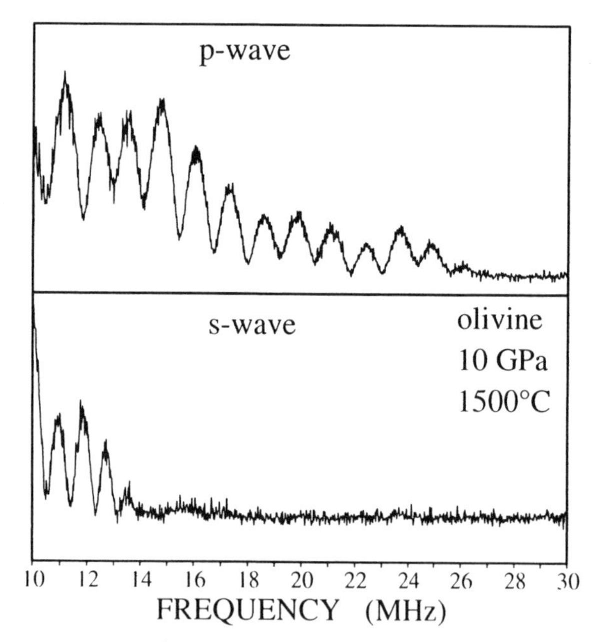

Figure 3. P- and S-wave interference patterns as a function of frequency for olivine at 10.5 GPa and 1500°C.

the waves reflected from each end of the sample is measured as a function of frequency [*Niesler and Jackson*, 1989; *Rigden et al.*, 1992]. This velocity determination is dependent on the sample length and on the acoustic quality of the two interfaces but not on the remainder of the path. Thus, effects arising from travel times in the WC cubes and buffer rods do not influence the accuracy of the results.

In order to maintain the validity of the pressure calibration during the ultrasonic measurements, the same P-T paths were followed as in the calibration experiments. First the pressure was increased at room temperature, with travel time measurements being made at intervals of ~0.5 GPa. After reaching a desired pressure, the sample was heated to the final desired temperature, keeping the hydraulic oil pressure at a constant value. During this stage, travel time measurements were made at intervals of ~250°C.

3. SAMPLE PREPARATION AND CHARACTERISTICS

Samples of polycrystalline olivine of composition $Mg_{1.8}Fe_{0.2}SiO_4$ were prepared by hot pressing San Carlos olivine powder in a molybdenum capsule at 4 GPa and 1200°C for 3.5 hours in a multianvil press. A small amount (5 wt%) of Bamble enstatite of composition $Mg_{0.85}Fe_{0.15}SiO_3$ was also added to the powdered sample prior to hot pressing. Although this second phase complicates the interpretation of the velocity measurements, its presence inhibits grain growth in the sample. As discussed below, grain growth enhances microfracturing during decompression which creates a significant problem when estimating the sample length. The grain size of the hot-pressed samples was ~ 25 μm. After hot pressing, the ends of the recovered samples were ground and polished flat and parallel to a precision of ±1 μm. The dimensions of the samples prepared for the ultrasonic experiments were ~1.8 mm diameter and 3.2-3.5 mm long. The length was measured to an accuracy of ±1 μm using digital calipers.

Several types of porosity were present in the hot-pressed samples:

1. A small number of decompression cracks, 0.5-2 μm wide, were oriented perpendicular to the cylindrical axis of the samples. Although the occurrence of such fractures can be reduced by hot pressing the samples in a hydrostatic stress environment [*Gwanmesia and Liebermann*, 1992], this would have been only possible at the expense of the sample size in the assemblies used at the moment.

2. Microcracks, 0.1 to 0.2 μm wide, were present along many of the grain boundaries. These microcracks form in the polycrystalline samples after hot pressing because of stresses which develop during quenching and decompression from high pressure as a result of the anisotropic compressibility and thermal expansivity of olivine combined with the relatively large grain size [*Cleveland and Bradt*, 1978].

3. Sub-equant pores were present at some grain boundary triple junctions and, unlike the other two types of porosity described above, were certainly present at high temperature and pressure. This type of porosity amounts to <0.5% and is much less abundant than the porosity due to microcracks.

The effects of these three types of sample porosity on both acoustic velocities and estimates of the sample length at high pressure are discussed below.

4. RESULTS

We have measured travel times from interference patterns obtained over the frequency range 10-30 MHz for P- and S-waves at pressures of 2.0 to 8.7 GPa at room temperature and at temperatures of 1000-1500°C at 8.0 and 10.5 GPa. For this range of frequencies, the wavelength of the P-waves is 0.8-0.3 mm. Thus the wavelength of the propagating P-wave is significantly less than the diameter of the sample, and the compressional wave velocity is determined [*Kolsky*, 1963]. Typical interference patterns are shown in Figure 3. The P-wave travel time measurements obtained at pressures up to 8.7 GPa at room

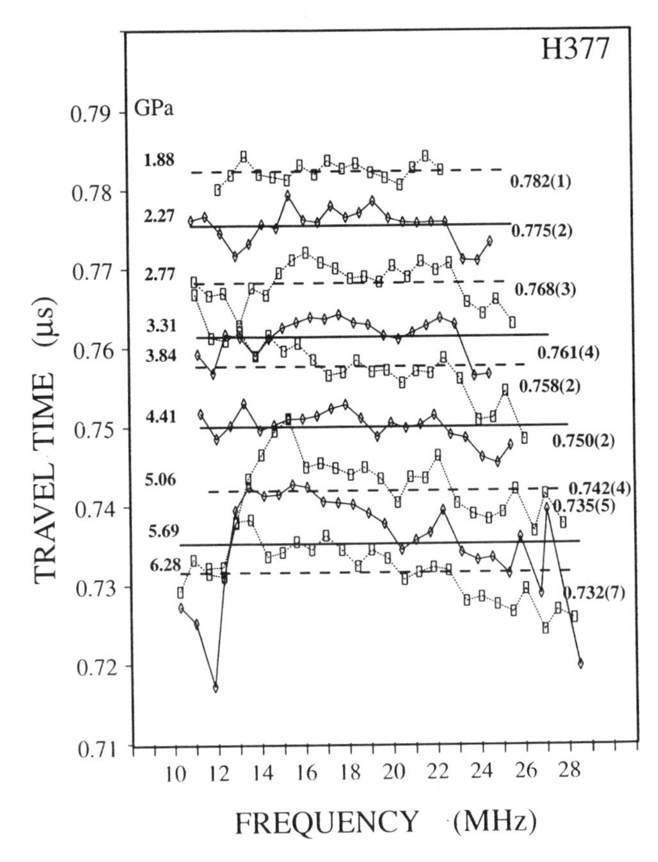

Figure 4. Calculated travel times for sample H-377, as a function of frequency, for increasing pressure at room temperature.

temperature are independent of frequency and therefore of wavelength (see Figure 4). Only the P-wave data are presented here because further technical developments are required in order to improve the quality of the S-wave signals (see Figure 3).

Following the ultrasonic experiments, the samples were decompressed at room temperature. Each sample, still enclosed in the pressure cell, was then mounted in epoxy, sectioned, and polished for optical microscope and scanning electron microscope (SEM) observations (Figure 5). The flatness and parallelism of the interfaces between the sample and buffer rods were found to be unchanged after the ultrasonic experiments. The sample length was measured using an optical microscope fitted with a vernier X-Y stage and was found to be identical to the original length within the precision (±3 μm) of this measuring instrument. Thus, there was no detectable plastic deformation of the samples during the experiments. The octahedral sample assembly undergoes a large amount of shortening during pressurization but this was accommodated in the direction of the cylindrical axis of the sample by deformation of the Pt buffer rods which shortened by ~ 25%. The high ductility of the Pt ensures that the differential stress on the sample is low (see below) and also ensures that the interfaces between the sample and the buffer rods are of high acoustic quality at high pressures. Microfractures on grain boundaries are comparable to those in the hot-pressed samples but decompression cracks (as defined above) are absent – this also suggests that the differential stress was low because of the presence of the Pt buffer rods. The porosity due to sub-equant pores at grain boundary triple junctions was greatly reduced following the high P-T ultrasonic experiments and is almost undetectable (<<0.5%).

The differential stress acting on the sample at high pressure and temperature can be estimated from the magnitude of the dislocation density [e.g., *Kohlstedt et al.*, 1976]. Dislocation microstructures in the sample from an experiment at 10.5 GPa and 1500°C were observed using SEM after dislocation decoration by oxidation [*Karato*, 1987]. The spacing of dislocations indicates a dislocation density of about 10^{12} m^{-2} (Figure 6). Using the dislocation density/stress calibration of *Kohlstedt et al.* [1976], a differential stress of 100-300 MPa can be estimated. It is possible that this differential stress arises at least partly from the porosity reduction which is observed to occur at high P and T. It must also be noted that dislocation recovery is very slow at 10 GPa and 1500°C [*Karato et al.*,

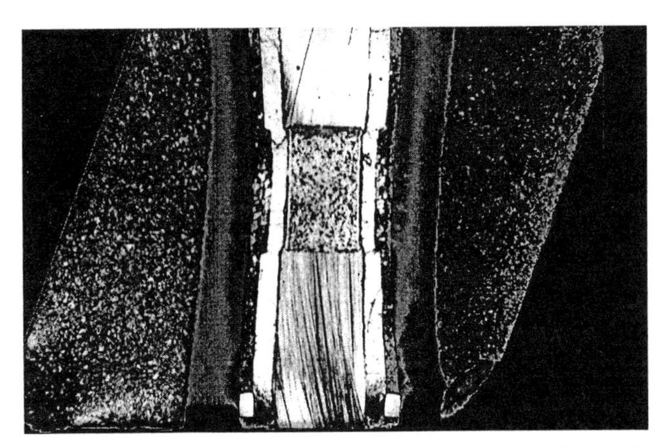

Figure 5. Cross section of the pressure cell after experiment H-377 up to 8 GPa and 1250°C. For a detailed description of the components of the pressure cell see Figure 1. For better reproduction false colors are used, which cause, for example, the black, stepped $LaCrO_3$ heater to appear white. The sample length was found to be the same as before the experiment (3.22 mm) indicating that only elastic deformation took place during the experiment. The Pt rods were shortened by ~ 25% after the experiment (and were scratched considerably during polishing of the cross section). The flatness and parallelism of the interfaces between the sample and the Pt buffer rods were essentially unchanged after the experiment. (Note that some plucking of the sample has occurred at the upper interface during polishing.)

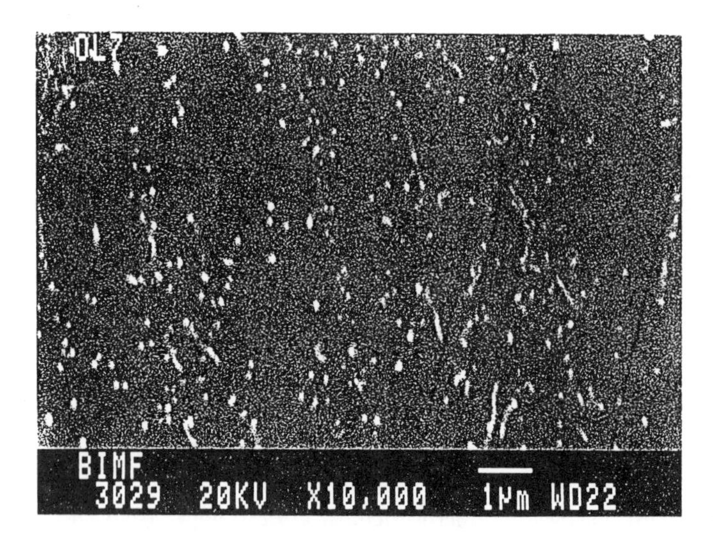

Figure 6. SEM backscattered electron image of dislocation microstructures in sample H-391 after ultrasonic measurements to 10.5 GPa and 1500°C. Dislocations have been decorated by oxidation in air for 1 hour at 900°C.

1993] so that if the observed dislocation density had developed during an early stage of the experiment at low temperature (during compression, for example) the dislocations would not have annealed out significantly at 1500°C. The stress estimate of 100-300 MPa can therefore be regarded as an upper limit at high temperature.

5. DISCUSSION

P-wave velocities calculated from the measured travel times allowing for elastic compression of the sample length, are shown in Figure 7a for two olivine samples. The sample length was corrected for elastic deformation using a pressure-dependent bulk modulus from previous ultrasonic measurements to 3 GPa [*Webb*, 1989]. The thermal expansion data of *Suzuki* [1975] were used to calculate the length of the sample after heating to high temperature. A comparison of the results obtained for these two samples at ambient temperature demonstrates that the data are reproducible within the uncertainties.

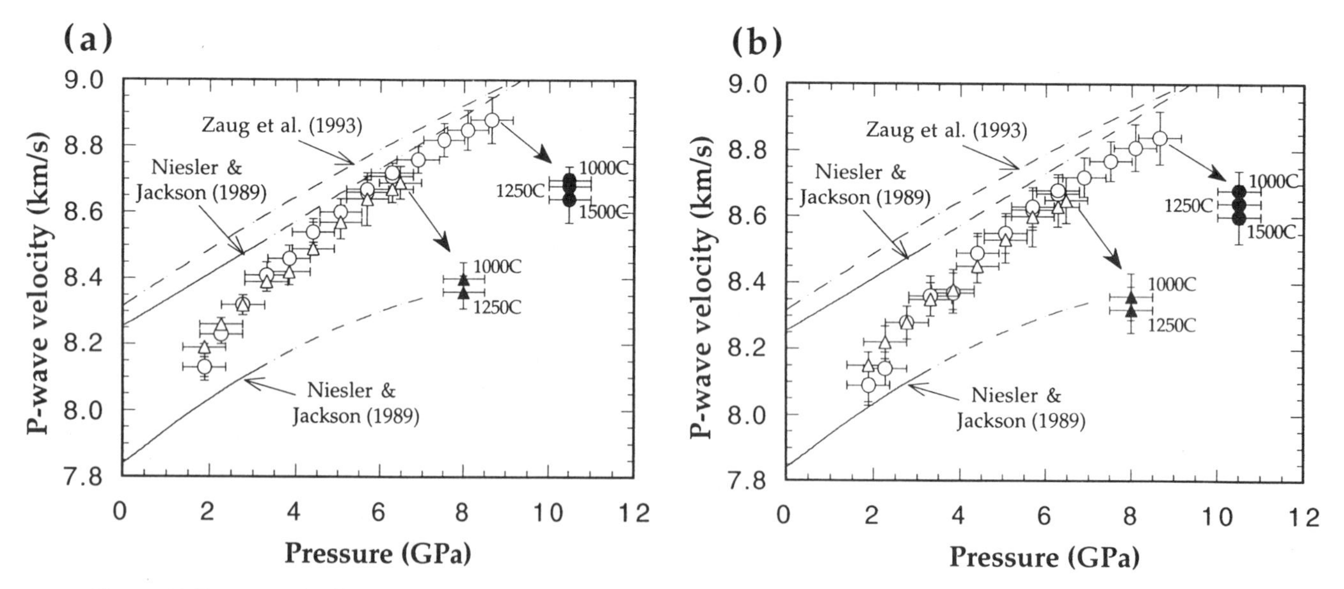

Figure 7. P-wave velocities determined as a function of pressure in two experiments, plotted as circles (H-391) and triangles (H-377) respectively. Room temperature data are indicated by open symbols and high temperature data by filled symbols. The arrows show the changes in velocity and pressure which resulted upon heating to 1000°C during the two experiments. An extrapolation of P-wave velocities in San Carlos olivine, measured up to 3 GPa by *Niesler and Jackson* [1989] in two polycrystalline samples, is shown for comparison; these samples had porosities of 0.3% (high velocity data) and 4.8% (low velocity data). Also shown are calculated velocities for polycrystalline olivine, based on the single crystal data of *Zaug et al.* [1993].
(a) Our velocities have been calculated from measured travel times using sample lengths corrected for pressure assuming elastic deformation and hydrostatic stress.
(b) Velocities determined for pure polycrystalline olivine from our travel times using sample lengths corrected for crack closure, in addition to elastic deformation as in (a). A correction of +0.5% has been applied to subtract the effect of enstatite (see text).

TABLE 1. Measured Travel Times and Calculated P- and S-Wave Velocities for two Olivine Samples.

Pressure GPa	run H-391 travel time μsec	run H-391 velocity $m\ s^{-1}$	run H-377 travel time μsec	run H-377 velocity $m\ s^{-1}$	Temperature °C
1.88	0.861(3)	8130(30)	0.782(1)	8190(10)	20
1.88	0.861(4)	8130(40)	-	-	20
2.27	0.850(3)	8230(30)	0.775(2)	8260(20)	20
2.27	0.850(3)	8230(30)	-	-	20
2.77	0.840(3)	8320(30)	0.768(3)	8320(30)	20
3.31	0.830(4)	8410(40)	0.761(2)	8390(30)	20
3.84	0.824(4)	8460(40)	0.758(4)	8420(40)	20
4.41	0.815(3)	8540(30)	0.750(2)	8490(20)	20
4.41	0.815(4)	8540(40)	-	-	20
5.06	0.808(4)	8600(40)	0.742(5)	8570(50)	20
5.69	0.801(4)	8660(40)	0.735(7)	8640(80)	20
5.69	0.800(4)	8670(40)	-	-	20
6.28	0.795(3)	8710(30)	0.732(4)	8670(40)	20
6.28	0.795(3)	8720(30)	-	-	20
6.28	0.795(3)	8720(30)	-	-	20
6.47	-	-	0.730(4)	8690(50)	20
6.89	0.790(4)	8760(40)	-	-	20
7.51	0.784(4)	8820(50)	-	-	20
8.00	-		0.760(5)	8400(50)	1000(50)
8.00	-		0.764(1)	8360(50)	1250(50)
8.09	0.780(5)	8850(60)	-	-	20
8.66	0.776(6)	8880(70)	-	-	20
10.5	0.795(4)	8720(40)	-	-	1000(50)
10.5	0.800(4)	8680(40)	-	-	1250(50)
10.5	0.806(7)	8640(70)	-	-	1500(50)
±0.5 GPa					

Sample lengths at 1 atmosphere pressure: H-391 - 3.516±0.001 mm; H-377 - 3.219±0.001 mm. Velocities have been calculated allowing for elastic compression and thermal expansion.

The velocities shown in Figure 7a and Table 1 must be corrected for changes in sample length resulting from crack closure during pressurization. In order to calculate the P-wave velocities for pure olivine, we also make a correction for the presence of 5% enstatite in the sample. The length correction due to crack closure is estimated to be -1.0% (±0.5%) based on SEM observations of crack densities and widths. To correct for the presence of 5% enstatite, we used the high pressure, room temperature data of *Hugh-Jones and Angel* [1994] to calculate the Reuss and Voigt bounds for composite materials [*Watt et al.*, 1976] and estimate that our measured velocities are reduced by 0.5% compared with velocities in pure polycrystalline olivine. The room temperature data for enstatite were also used to calculate the high-temperature olivine velocities as there are no elasticity data for this phase at combined high P and T. The combined effect of 1% shortening of the sample caused by crack closure together with the mathematical removal of the enstatite component reduces the velocities presented in Table 1 by 0.5%.

The calculated velocities for crack-free polycrystalline San Carlos olivine are shown in Figure 7b together with extrapolations of low pressure data (up to 3 GPa) from *Niesler and Jackson* [1989] for two samples of hot-pressed San Carlos olivine. Also shown are calculated velocities for polycrystalline olivine, based on the single crystal data of *Zaug et al.* [1993]. Several aspects of the results of Figure 7b require discussion. First, below ~ 5 GPa the pressure-dependence of our velocity data (dv_P/dP) is unusually high compared with the other data shown in Figure 7b. A high-pressure dependence of velocity is commonly observed in ultrasonic studies of polycrystalline materials at low pressure but normally only below pressures of about 1 GPa and is generally explained by the closure of porosity [e.g., *Niesler and Jackson*, 1989]. We believe that this is also the cause of the trend of the data shown in Figure 7b, which would indicate that porosity in our samples only becomes closed above 4-5 GPa. As discussed above, most porosity in our samples consists of microfractures on grain boundaries. These developed during decompression as a consequence of stresses caused by the anisotropic compressibility and thermal expansivity of olivine. The magnitude of the stresses is dependent on grain size [*Cleveland and Bradt*, 1978]. Because our samples are coarse-grained relative to those of most previous ultrasonic studies, relatively high stresses, and therefore high pressures, are required to close the open grain boundaries. Thus, the fact that our data indicate that porosity is being squeezed out continuously to 5 GPa, whereas in previous studies this critical pressure is much lower (<1.0 GPa), is probably a consequence of the large grain size of our samples. It is also notable that the pressure at which the porosity apparently becomes closed is near to the pressure of 4 GPa at which the samples were hot-pressed and therefore microstructurally equilibrated.

At pressures above 5 GPa, dv_P/dP for our measurements is comparable to that of *Niesler and Jackson* [1989] and *Zaug et al.* [1993] (Figure 7b). The absolute values of our velocities are slightly lower (by <2%). This difference could be the result of residual porosity, errors in estimating the sample length at high pressure, differential stress, the presence of dislocations, or preferred orientation. Much of the discrepancy between our high-pressure data and velocities calculated from elastic moduli is likely to be due to porosity. Sub-equant pores are present in our samples and constitute up to 0.5% of the sample prior to the high-temperature stage of the ultrasonic experiments. Above 5 GPa, our velocities are slightly lower than those of *Niesler and Jackson*'s [1989] low-pressure measurements on a sample of San Carlos olivine with 0.3% porosity (Figure 7b). In contrast, velocities for a sample with 4.8% porosity [*Niesler and Jackson*, 1989] are much lower than our data (Figure 7b). If there is indeed an error in our room temperature measurements due to the presence of up to 0.5% porosity, the error on the velocity measurements at high temperatures should be reduced because the porosity is observed to decrease significantly during annealing at high temperatures.

Our calculated velocities will be too low if our estimates of sample length (based on the corrections detailed above) are too small. One of the largest uncertainties in sample length currently results from the presence of numerous microcracks. The uncertainties in sample length could be reduced by eliminating microcracks in hot-pressed samples, firstly by reducing the grain size to <5 μm [*Farver et al.*, 1994] and secondly by hot-pressing the sample in a hydrostatic pressure medium [*Gwanmesia and Liebermann*, 1992]. Uncertainties resulting from elastic deformation and thermal expansion at conditions of simultaneous pressure and temperature will still remain. Measuring the shear and longitudinal velocities simultaneously using the two buffer rods in this assembly will allow us to calculatc dK/dP and thus to calculate the change in length of the sample due to elastic compression for each experiment instead of relying upon literature data. The calculation of the sample length from the measured velocities will also allow a better estimate of the effects of crack and porosity closure on the sample length.

Differential stress in our experiments is estimated to be 100-300 MPa or less, as discussed above. The measurements of *Bateman et al.* [1961] show that 100 MPa of differential stress has roughly the same effect on elastic moduli as 100 MPa of confining pressure. The stresses in our experiments should therefore not significantly affect the measurements at high P (and T) because they are much lower than the maximum confining pressures of 8-10 GPa. Under these latter conditions we estimate that a differential stress of 100 MPa will affect the measured velocities by about 0.1%. However, the effect of stress on the low pressure measurements (i.e., <5 GPa at room temperature) may be larger.

The effect of dislocations on acoustic velocities has been studied by Spetzler and Karato (Karato, pers. communication, 1996) and shown to be negligible. The effect of dislocation densities comparable to those observed in our samples could not be detected in their high-resolution velocity measurements on garnet single crystals.

6. CONCLUSIONS

We have shown that ultrasonic determinations of the velocity of compressional and shear waves at simultaneous pressures and temperatures approaching mantle conditions are possible. Uncertainties in our preliminary results, which originate mainly from the presence of cracks and porosity in the hot-pressed polycrystals, can be largely eliminated by the use of samples with smaller grain sizes, synthesized in a hydrostatic pressure environment. We have observed that the sample in the present assembly is not plastically deformed. This means that once the current uncertainties in sample length, associated with crack closure, are eliminated, it will be possible to determine velocities with the accuracy required to compare the laboratory data with seismic data. Further miniaturization of the cell assembly should allow these measurements to be performed in smaller MgO octahedra with which pressures of 16 GPa can be achieved. Thus, the pressure and temperature conditions of the mantle transition zone will be accessible.

Acknowledgments. Technical support from H. Schulze and H. Küffner is greatly acknowledged. We thank S. Karato and I. Jackson for helpful discussions and R. Liebermann and an anonymous reviewer for reviews which have improved the manuscript. This work was partially funded by the EU "Human Capital and Mobility - Access to Large Scale Facilities" programme (Contract No. ERBCHGECT940053 to D.C. Rubie).

REFERENCES

Anderson, D. L., *Theory of the Earth*, 366 pp., Blackwell, Oxford, 1989.

Angel, R. J., and N. L. Ross, Compression mechanisms and equations of state, *Phil. Trans. Roy. Soc. A.*, *354*, 1-11, 1996.

Bateman, T., W. P. Mason, and H. J. McSkimin, Third-order elastic moduli of germanium, *J. Appl. Phys.*, *32*, 928-936, 1961.

Chen, G., B. Li, and R. C. Liebermann, Selected elastic moduli of single-crystal olivines from ultrasonic experiments to mantle pressures, *Science, 272*, 979-980, 1996.

Cleveland, J. J., and R. C. Bradt, Grain size/microcracking relations for pseudobrookite oxides, *J. Am. Ceramic Soc.*, *61*, 478-481, 1978.

Duffy, T. S., and D. L. Anderson, Seismic velocities in mantle minerals and the mineralogy of the upper mantle, *J. Geophys. Res.*, *94*, 1895-1912, 1989.

Duffy, T. S., C. Zha, R. T. Downs, H. Mao, and R. J. Hemley, Elasticity of forsterite to 16 GPa and the composition of the upper mantle, *Nature*, *378*, 170-173, 1995.

Farver, J. R., R. A. Yund, and D. C. Rubie, Magnesium grain boundary diffusion in forsterite aggregates at 1000°-1300°C and 0.1 MPa to 10 GPa, *J. Geophys. Res.*, *99*, 19809-19819, 1994.

Fukizawa, A., and H. Kinoshita, Shear wave velocity jump at the olivine-spinel transformation in Fe_2SiO_4 by ultrasonic measurements in situ, *J. Phys. Earth*, *30*, 245-253, 1982.

Graham, E. K., and G. R. Barsch, Elastic constants of single crystal forsterite as a function of temperature and pressure, *J. Geophys. Res.*, *74*, 5949-5960, 1969.

Gwanmesia, G. D., and R. C. Liebermann, Polycrystals of high-pressure phases of mantle minerals: hot-pressing and characterization of physical properties, in *High-Pressure Research: Application to Earth and Planetary Sciences*, edited by Y. Syono and M. H. Manghnani, pp. 117-136, AGU, Washington, D.C., 1992.

Hugh-Jones, D. A., and R. J. Angel, A compressional study of $MgSiO_3$ orthoenstatite up to 8.5 GPa, *Am. Mineral.*, *79*, 405-410, 1994.

Isaak, D. G., High temperature elasticity of iron-bearing olivines, *J. Geophys. Res.*, *97*, 1871-1885, 1992.

Isaak, D. G., O. L. Anderson, and T. Goto, Elasticity of single-crystal forsterite measured to 1700 K, *J. Geophys. Res.*, *94*, 5895-5906, 1989.

Ito, E., E. Takahashi, and Y. Matsui, The mineralogy and chemistry of the lower mantle: an implication of the ultrahigh-pressure phase relations in the system MgO-FeO-SiO_2, *Earth Planet. Sci. Lett.*, *67*, 238-248, 1984.

Karato, S., Scanning electron microscope observations of dislocations in olivine, *Phys. Chem. Mineral.*, *14*, 245-248, 1987.

Karato, S., D. C. Rubie, and H. Yan, Dislocation recovery in olivine under deep upper mantle conditions: implications for creep and diffusion, *J. Geophys. Res.*, *98*, 9761-9768, 1993.

Kohlstedt, D. L., C. Goetze, and W. B. Durham, Experimental deformation of single crystal olivine with application to flow in the mantle, in *The Physics and Chemistry of Minerals and Rocks*, edited by R.G. Strens, pp. 35-49, Wiley, New York, 1976.

Kolsky, H., *Stress Waves in Solids*, 213 pp., Dover Publications, New York, 1963.

Li, B., I. Jackson, T. Gasparik, and R. C. Liebermann, Elastic wave velocity measurement in multi-anvil apparatus to 10 GPa using ultrasonic interferometry, *Phys. Earth Planet. Inter., 98*, 79-91, 1996.

Liebermann, R. C., and Y. Wang, Characterization of sample environment in a uniaxial split-sphere apparatus, in *High Pressure Research: Application to Earth and Planetary Sciences*, edited by Y. Syono and M.H. Manghnani, pp. 19-31, AGU, Washington, D.C., 1992.

Niesler, H., and I. Jackson, Pressure derivatives of elastic wave velocities from ultrasonic interferometric measurements on jacketed polycrystals, *J. Acoust. Soc. Am.*, *86*, 1573-1585, 1989.

Rigden, S. M., G. D. Gwanmesia, I. Jackson, and R. C. Liebermann, Progress in high-pressure ultrasonic interferometry, the pressure dependence of elasticity of Mg_2SiO_4 polymorphs and constraints on the composition of the transition zone of the Earth's mantle, in *High-Pressure Research: Application to Earth and Planetary Sciences*, *Geophys. Monogr. Ser.*, vol. 67, edited by Y. Syono and M.H. Manghnani, pp. 167-182, AGU, Washington, D.C., 1992.

Rigden, S. M., G. D. Gwanmesia, and R. C. Liebermann, Elastic wave velocities of a pyrope majorite garnet to 3 GPa, *Phys. Earth Planet. Int.*, *86*, 35-44, 1994.

Rubie, D. C., S. Karato, H. Yan, and H. St. C. O'Neill, Low differential stress and controlled chemical environment in multianvil high-pressure experiments, *Phys. Chem. Mineral.*, *20*, 315-322, 1993a.

Rubie, D. C., C. R. Ross II, M. R. Carroll, and S. C. Elphick, Oxygen self-diffusion in $Na_2Si_4O_9$ liquid up to 10 GPa and estimation of high-pressure melt viscosities, *Am. Mineral.*, *78*, 574-582, 1993b.

Suzuki, I., Thermal expansion of periclase and olivine, and their anharmonic properties, *J. Phys. Earth*, *23*, 145-159, 1975.

Watt, J. P., G. Davies, and R. J. O'Connell, The elastic properties of composite materials, *Rev. Geophys. Space Phys.*, *14*, 541-563, 1976.

Webb, S. L., The elasticity of the upper mantle orthosilicates olivine and garnet to 3 GPa, *Phys. Chem. Minerals*, *16*, 684-692, 1989.

Yoneda, A., Pressure derivatives of elastic constants of single crystal MgO and $MgAl_2O_4$, *J. Phys. Earth*, *38*, 19-55, 1990.

Yoneda, A., and M. Morioka, Pressure derivatives of elastic constants of single crystal forsterite, in *High-Pressure Research: Application to Earth and Planetary Sciences*, *Geophys. Monogr. Ser.*, vol. 67, edited by Y. Syono and M.H. Manghnani, pp. 207-214, AGU, Washington, D.C., 1992.

Zaug, J. M., E. H. Abramson, J. M. Brown, and L. J. Slutski, Sound velocities in olivine at Earth mantle pressures, *Science*, *260*, 1487-1489, 1993.

Ruth Knoche, Sharon L. Webb, and David C. Rubie, Bayerisches Geoinstitut, Universität Bayreuth, D-95440 Bayreuth, Germany

High-Pressure Raman Scattering Study of Majorite-Garnet Solid Solutions in the System $Mg_4Si_4O_{12}$–$Mg_3Al_2Si_3O_{12}$

M. H. Manghnani and V. Vijayakumar[1]

Mineral Physics Group, Hawaii Institute of Geophysics and Planetology, University of Hawaii, Honolulu, Hawaii

J. D. Bass

Department of Geology, University of Illinois, Urbana, Illinois

We report here the results of Raman scattering measurements to 30 GPa on polycrystalline end-member Mj_{100} tetragonal majorite and two majorite-garnet solid solutions, $Mj_{50}Py_{50}$ and $Mj_{80}Py_{20}$ (where Mj = $Mg_4Si_4O_{12}$ and Py = $Mg_3Al_2Si_3O_{12}$), synthesized at 17.7 GPa and 2000°C. The results are discussed in light of previous Raman, X ray, and elasticity studies. At ambient pressure, variations in the Raman spectra indicate changes in the structure and disorder with increasing Py content. For example, the spectra of Mj_{100} confirm its tetragonal structure, and the two bands in Mj_{100} at ~885 cm^{-1} and ~928 cm^{-1} merge in the case of $Mj_{80}Py_{20}$ and $Mj_{50}Py_{50}$. The spectra of $Mj_{50}Py_{50}$ and $Mj_{80}Py_{20}$ show similarity to those of Py_{100} (cubic garnet structure) and $(Mg_{.79}Fe_{.21})_4Si_4O_{12}$ majorite from the Catherwood meteorite, which also has cubic structure. The pressure dependencies of the Raman frequencies for all the modes are found to be linear, without discontinuities, throughout the pressure range of this study, implying no major structural changes. Using the pressure shift of the Raman bands, $(\partial \nu_i/\partial P)$, the mode Grüneisen parameters, γ_i, are calculated using the bulk modulus values measured by Brillouin spectroscopy. The averaged values of mode γ_i for the three Mj–Py samples lie in the range 0.81–1.13, lower than those of pyrope garnet (1.5) but close to γ_{th} ~1.1 calculated for $Mj_{58}Py_{42}$.

INTRODUCTION

Majorite ($Mg_4Si_4O_{12}$) is a garnet-structured high-pressure phase of pyroxene ($MgSiO_3$) and contains octahedrally coordinated silicon. It has been synthesized at pressures of 16–23 GPa and temperatures above 1600°C [*Kato and Kumazawa*, 1985; *Gasparik*, 1989]. Under high pressure, majorite (Mj) forms extensive solid solution with aluminous garnet (pyrope = Py) over a wide compositional range in the system $Mg_4Si_4O_{12}$ (Mj)–$Mg_3Al_2Si_3O_{12}$ (Py) [*Akaogi and Akimoto*, 1977; *Liu*, 1977; *Akaogi et al.*, 1987; *Kanzaki*, 1987; *Gasparik*, 1989]. Phase equilibria studies indicate that majorite-garnet is a stable phase under the pressure-temperature conditions in the transition zone (depths 300–700 km). Thus, β- and γ-spinel polymorphs of olivine and majorite-garnets are considered to be among the most abundant phases in the transition zone. Crystal-chemical, elastic, and thermodynamic properties of majorite-garnets have an important bearing on models of

[1] High Pressure Physics Division, Bhabha Atomic Research Center, Trombay, Mumbai, India

Properties of Earth and Planetary Materials
at High Pressure and Temperature
Geophysical Monograph 101

mineralogy and thermal state of the Earth's mantle inferred from seismological data. From that point of view, the elastic properties of Mj–Py solid solution series have been studied by several investigators: *Bass and Kanzaki* [1990]; *Yeganeh-Haeri and Weidner* [1990]; *Rigden et al.* [1994]; *Sinogeikin et al.*, [1997]; *Pacalo and Weidner* [1997], *Jeanloz* [1981] and *Sinogeikin et al.* [1997a]. Further, although discrepancies in the elasticity data base for majorite-garnets have been conclusively resolved by Brillouin scattering experiments [*Sinogeikin et al.*, 1997b], the possibility of more subtle variations in elastic properties due to crystal-structural differences remains [*Sinogeikin et al.*, 1997a]. It is therefore of interest to determine the composition range in which the tetragonal-cubic transition would take place and whether pressure would induce such a transition.

Thermodynamic and vibrational properties have been the subject of only a limited number of studies. For example, *Yagi et al.* [1987] have studied the thermal expansion and compression of $Mj_{58}Py_{42}$ and *McMillan et al.* [1989] have studied the IR and Raman spectra of Mj_{100}, $Mj_{58}Py_{42}$, and Py_{100}. Vibrational spectroscopy is particularly sensitive to the structural state of materials. In this respect, the Raman spectra of pyrope and other garnets have been studied at ambient and high pressure, for example, *Hofmeister and Chopelas* [1991], *Gillet et al.* [1992], *McMillan et al.* [1989], and *Liu et al.* [1994]. To our knowledge the Mj–Py series has not been systematically investigated under high pressure.

We present here the results of high-pressure Raman scattering measurements on three synthetic majorite-pyrope (Mj–Py) samples (Mj_{100}, $Mj_{80}Py_{20}$, and $Mj_{50}Py_{50}$) from 1-bar to 30 GPa. One of the purposes of this study is to compare the 1-bar results with similar measurements on Mj_{100}, Py_{100}, and $(Mg_{.79}Fe_{.21})_4Si_4O_{12}$ majorite (cubic structure) recovered from the Catherwood meteorite [*McMillan et al.*, 1989], in light of the tetragonal-cubic transition. The second purpose is to investigate the pressure dependencies of Raman shifts and to determine the mode Grüneisen and their averaged values, and to compare these with γ_{th} and other related data.

PREVIOUS STUDIES ON Mj–Py SOLID SOLUTIONS

1. *Elastic and Vibrational Properties*

Sinogeikin et al. [1997a] measured the adiabatic bulk (K_S) and shear (μ) elastic moduli of polycrystalline samples of $Mj_{50}Py_{50}$, $Mj_{80}Py_{20}$, and Mj_{100} compositions by Brillouin scattering, using the same samples employed in the present experiments. The *Sinogeikin et al.* [1997a] reported K_S values for the three samples of 173.1, 162.6, and 166(5) GPa, respectively, and they compared their values with other elastic moduli data for the Mj-Py system. An important conclusions of their work is that the elastic moduli for more-majorite-rich composition are lower and that the K_S and μ values are constant over the compositional range from Py_{100} to a majorite content of 70-80%, whereupon there is a decrease in the moduli with increase in majorite content.

Jeanloz [1981] reported vibrational (1R) and compressional properties (bulk modulus and its pressure derivative) of majorite $(Mg_{.79}Fe_{.21})SiO_3$ recovered from the Catherwood meteorite. *McMillan et al.* [1989] have studied the Raman and IR spectra of pure Mj_{100} and 3 Mj-Py solid solutions, the most majorite-rich solution being $Mj_{58}Py_{42}$. The broadness of the peaks led these authors to conclude that the intermediate samples are disordered. If the solid solutions are disordered, then they are most likely cubic since it is cation ordering on octahedral sites in the garnet structure which makes $MgSiO_3$ majorite tetragonal. Their paper seems to indicate that the cubic→tetragonal transition occurs between Mj_{58} and Mj_{100}.

2. *X Ray Studies*

Most of the X ray diffraction (XRD) studies [e.g., *Angel et al.*, 1989] have shown the structure of Mj_{100} to be tetragonal, in contrast to the cubic structure of pyrope garnet. *Liu* [1977] studied $Mj_{80}Py_{20}$ by XRD under high pressure to 20 GPa and did not observe splitting of any diffraction lines. *Yagi et al.* [1987] determined the high P-T equation of state of two majorites, (Mj_{58}-Py_{42} and Fs_{18}-Alm_{82}) by XRD and also did not observe splitting. *Angel et al.* [1989] found the structure of single-crystal $MgSiO_3$ garnet (Mj_{100}), synthesized at 17 GPa and 1800°C, to be tetragonal (a = 11.501(1)Å, c = 11.480(2)Å). The two Mg positions are coordinated by eight anions and Mg and Si are at least partially ordered over the octahedral sites.

Hatch and Ghose [1989] suggested that Mj_{100} is actually cubic at elevated temperature of about 1800°C and that the tetragonal structure forms by rapid ordering of Mg and Si on quenching to ambient conditions. They concluded this on the basis of the twinning observed in the sample by XRD [*Angel et al.*, 1989]. In a recent study, *Parise et al.* [1996] have shown that the cubic-tetragonal structural phase transition in garnet along the Mj–Py join occurs at composition less majorite rich than Mj_{75}.

3. *NMR Studies and Order/Disorder*

The order/disorder in Mj–Py solid solutions involves Mg-Si-Al cations on the octahedral sites. The NMR study of *Phillips et al.* [1992] shows that Mj_{100} has a high degree (88%) of Mg, Si order on the octahedral sites. Most peaks of their samples are sharp and well resolved. They modeled the spectrum of ^{29}Si well with 88% of order on the octahedral sites (two octahedral sites, one with 88% Si and 12% Mg, the other with 12% Si and 88% Mg), consistent with the X ray results of *Angel et al.* [1989], who got 80% order. *Phillips et al.* [1992] also studied the Mj_{80} sample (the same sample used in this study) and were able to model the spectra assuming a totally disordered cation distribution on the octahedral sites. The peaks are certainly broader, indicating disorder.

EXPERIMENTAL METHODS

A modified 4-pin Merrill-Bassett cell with low-fluorescence type diamond-anvils of 400 μm culet size was employed for high-pressure Raman scattering measurements. The sample (~75 μm across) and a few tiny chips of ruby for pressure determination were loaded in ~100 μm hole, made by spark erosion, in stainless steel gasket preindented to 15 GPa. A 4:1 mixture of methanol:ethanol was used as a pressure transmitting medium. Pressure was determined by a conventional ruby fluorescence measurement [*Piermarini et al.*, 1975]. Raman spectra were collected in the 180° (backscattering) geometry in the wavenumber range of 200–1400 cm^{-1} by means of a DILOR X-Y micro-Raman system equipped with a confocal microscope. The backscattered signal was directed into a liquid-nitrogen cooled multichannel charge-coupled detector (CCD) locked at -110°C. The excitation light (λ=514.5 nm) was emitted from a Spectra Physics model 2020 Ar^+ laser. The laser power of 0.5W was focused on the sample for recording the spectra. The accumulation time was 30–60 minutes.

Samples. The three majorite-garnet samples (Mj_{100}, $Mj_{80}Py_{20}$, $Mj_{50}Py_{50}$) employed in this study are the same samples used in a recent Brillouin scattering study [*Sinogeikin et al.*, 1997a] for determination of the elastic properties. The samples were synthesized at 17.7 GPa and 2000°C in a multianvil apparatus by T. Gasparik at SUNY. Powder x-ray diffraction (XRD) spectra of $Mj_{50}Py_{50}$ and $Mj_{80}Py_{20}$ samples did not detect splitting of diffraction lines [*Sinogeikin et al.*, 1997a], although some peaks were broad and our XRD spectra do not definitively distinguish cubic from tetragonal symmetry.

RESULTS AND DISCUSSION

In a pioneering study, *McMillan et al.* [1989] have assigned the Raman and IR vibrational modes in the Mj-Py series by comparing their spectra with those of $MgSiO_3$ garnet (Mj), pyrope (Py), and $(Mg_{.79}Fe_{.21})_4Si_4O_{12}$ majorite phase present in the Catherwood meteorite. Here, we have closely followed their work in the discussion of our results. In general, the peaks above 800 cm^{-1} in the vibrational spectra were assigned to stretching vibrations of the SiO_4 tetrahedral groups, as in the case of orthosilicates. For a cubic (0_h) garnet structure, the ν_1 and ν_3 symmetric and asymmetric stretching vibrations of the SiO_4 groups give rise to following Raman active modes [*Moore et al.*, 1971; *McMillan et al.*, 1989]:

$$\nu_1\text{: } A_{1g}\,(R) + E_g\,(R)$$
$$\nu_3\text{: } E_g\,(R) + 3T_{2g}\,(R)$$

The ν_1 (A_{1g}) mode (the strongest mode) and ν_3 (E_g) modes are observed in all samples at ~928 cm^{-1} and ~1060 cm^{-1}, respectively. *McMillan et al.* [1989] have shown that, on distortion to a tetragonal D_{4h} structure, the SiO_4 stretching vibrations would give rise to fifteen modes, of which they observed eight:

$$\nu_1\text{: } 2A_{1g}(R)B_{1g}(R) + E_g(R)$$
$$\nu_3\text{: } A_{1g}(R) + 2B_{1g}(R) + 3B_{2g}(R) + 5E_g(R)$$

Figure 1 compares the 1-bar Raman spectra for our three synthesized majorite-garnet samples (Mj_{100}, Mj_{80}, and Mj_{50}) with those of Mj_{100}, Catherwood majorite-garnet, and Py_{100} reported by *McMillan et al.* [1989]. In general, the number of vibrational modes recorded for Mj_{100} is higher than for $Mj_{80}Py_{20}$ and $Mj_{50}Py_{50}$ (11 vs. 8), and the peaks are sharper. There is a close resemblance between the spectra of our Mj_{100} and their Mj spectra, marked by Mc. The strongest A_{1g} mode is observed at 931 cm^{-1} in their sample and at 928 cm^{-1} in our sample. A splitting of this A_{1g} mode and the ~1060 cm^{-1} mode is a reasonable indication for tetragonal structure in our Mj_{100} sample. Furthermore, a correlation between the strong peaks at 597 cm^{-1} (in Mj_{100}) and 602 cm^{-1} (in Mj), assigned as SiOSi linkage vibrations [*McMillan et al.*, 1989], confirms the tetragonal structure in our Mj_{100} sample. *McMillan et al.* [1989] have shown that the Mj_{100} is highly ordered, and that if there is order in Mj_{80} (or Mj_{50}), it is much less than in Mj_{100}.

The spectra of $Mj_{50}Py_{50}$ and $Mj_{80}Py_{20}$ are similar but clearly different from Mj_{100}. Based on the resemblance between the spectra of pyrope reported by *McMillan et al.*

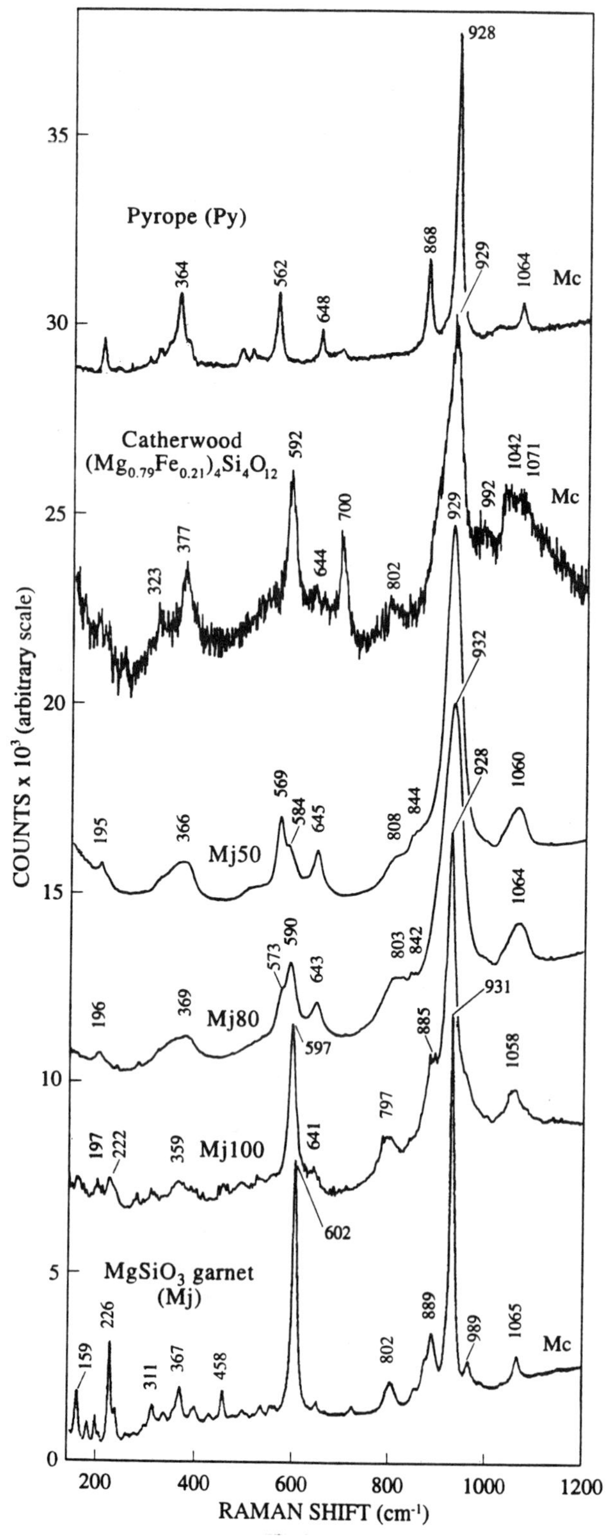

Figure 1. Raman spectra of the Mj–Py solid solution series. The spectra of our three samples (Mj_{50}, Mj_{80}, and Mj_{100}) are compared with those for Py, Catherwood majorite and Mj (marked by Mc) reported by *McMillan et al.* [1989].

[1989] and our $Mj_{80}Py_{20}$ and $Mj_{50}Py_{50}$ sample, the structure of $Mj_{80}Py_{20}$ and $Mj_{50}Py_{50}$ appears as cubic. Similarity between the spectra of Catherwood and Py and between $Mj_{50}Py_{50}$ and $Mj_{80}Py_{20}$, in terms of number of modes and band features, also lends support for the cubic structure in Mj_{50} and Mj_{80}. However, we observe very small changes in wavenumber and relative intensities but slightly increasing broadness of the Raman peaks in going from Mj_{80} to Mj_{50}, signifying increasing structural disorder over lattice sites as pointed out by *McMillan et al.* [1989].

The results of the NMR work on the same Mj_{100} sample [*Phillips et al.*, 1992] show that it has a high degree of (Mg, Si) order on the octahedral sites. Their ^{29}Si spectrum fits well with 88% order.

There is uncertainty about the boundary of the cubic→tetragonal transformation. Perhaps the best available evidence is the NMR work of *Phillips* [1992], which suggests that the boundary is between 80-100% majorite. This study does not, however, rule out a boundary at a more pyrope-rich composition because the NMR experiments cannot absolutely exclude tetragonal symmetry in the $Py_{20}Mj_{80}$ sample. In contrast to the conclusions of *Phllips* [1992], *Parise et al.* [1996] found XRD peaks violating the Ia3d spacegroup of cubic garnet for their $Py_{25}Mj_{75}$ sample. However, such features were not evident in the XRD spectra for their Mj_{79}, contradicting their conclusion that the tetragonal-cubic transition occurs at a more pyrope-rich composition than Mj_{75}. Therefore, the composition at which the cubic-tetragonal boundary is located remains ambiguous. The apparent decrease in elastic moduli for Mj_{80} is suggestive of a different structure from that of more pyrope rich compositions [*Sinogeikin et al.*, 1997a].

1. *Pressure Dependencies of Raman Spectra and* γ_i

Figures 2, 3, and 4 show the Raman spectra obtained for the three Mj–Py samples to ~30 GPa. The Raman peak shift with pressure for the three samples is shown in Figures 5, 6, and 7 respectively.

The pressure dependencies of the respective Raman shifts can be used to calculate the mode Grüneisen parameters γ_i from the relation

$$\gamma_i = -\frac{V}{\nu_i}\left(\frac{\partial \nu_i}{\partial V}\right) = \frac{K_T}{\nu_i}\left(\frac{\partial \nu_i}{\partial P}\right) \qquad (1)$$

where V is volume; K_T is isothermal bulk modulus; and ν_i mode frequency. Table 1 summarizes the observed Raman frequencies at 1 bar, their pressure dependencies, and calculated mode γ_i for the three Mj–Py samples, using

TABLE 1. Observed 1-bar Mode Raman Frequencies ν_i, Their Pressure Dependencies ($\partial\nu_i/\partial P$), and Calculated γ_i for the Three Majorite Samples. Also Shown, for Comparison, are 1-bar Mode Frequencies Reported for Catherwood Majorite-Garnet [*McMillan et al.*, 1989]. The Values for the Major Raman Bands are Shown in Bold. MFS Denotes Mean Frequency Shift and n is the Number of Modes for Calculating MFS.

Mj_{100}			$Mj_{80}Py_{20}$			$Mj_{50}Py_{50}$			$(Mg_{.79}Fe_{.21})_4Si_4O_{12}$ Catherwood
ν_i (cm^{-1})	($\partial\nu_i/\partial P$) (cm^{-1}/GPa)	γ_i	ν_i (cm^{-1})	($\partial\nu_i/\partial P$) (cm^{-1}/GPa)	γ_i	ν_i (cm^{-1})	($\partial\nu_i/\partial P$) (cm^{-1}/GPa)	γ_i	ν_i (cm^{-1})
197	0.99	0.84	196			195			
222	2.11	1.56							
			337*	4.39	2.12	332	3.67	1.92	323
359	1.65	0.74	369			366			
			388	4.89	2.05	377	4.42	2.03	377
456	1.91	0.69							
			573			**569**	**2.83**	**0.86**	
597	**2.54**	**0.70**	**590**	**2.05**	**0.56**	584			**592**
641	**2.43**	**0.63**	**643**	**3.29**	**0.82**	**645**	**2.52**	**0.77**	**644**
						721	3.10	0.74	700
743*	3.85	0.86							
797	4.29	0.89	803	5.84	1.76	808			802
			842						
885	**3.91**	**0.72**	917*	2.92	0.52	844*	4.55	0.89	
928	**3.67**	**0.66**	**932**	**3.52**	**0.61**	**929**	**3.88**	**0.72**	**929**
									992
1058*	3.72	0.59	1064*	3.94	0.60	1060	4.82	0.78	1042
									1071
Averaged γ_i (all modes)		0.81			1.13			1.09	
Averaged γ_i (principal modes)		0.68			0.66			0.75	
n	11			8			6		
MFS†	4.88			4.02			4.55		

* denotes that the line was resolved/appeared at high pressure and extrapolated to zero-pressure frequency value

†Mean frequency shift (MFS) in units of 10^{-3} GPa^{-1}

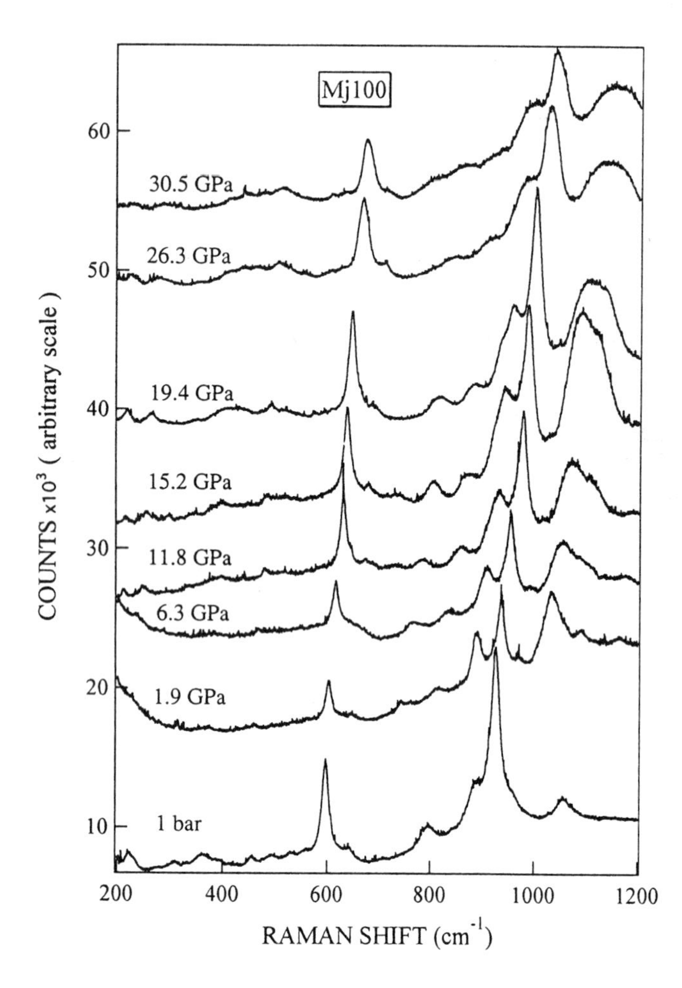

Figure 2. Raman spectra of Mj_{100} as a function of pressure to 30.5 GPa.

equation (1). The values for the principal vibrational modes (most prominent peaks) are shown in bold. It is interesting to note that, except in the case of low frequency modes (200–250 cm^{-1}), the $\partial\nu_i/\partial P$ slopes for the principal modes (shown in bold) are quite similar (2.05–3.91) for the three samples, and the averaged γ_i values for all the modes considered are in the range of 0.81 to 1.13.

Gillet et al. [1992] have correlated the mean relative pressure-induced frequency shifts MFS = $\Sigma(\partial \ln \nu_i/\partial P)/n$ versus bulk modulus for different minerals in discussing the individual polyhedral compressibility. Their reported MFS values (e.g., 4.8–6.6 for garnets, ~8.1 for forsterite, 16.2 for α-quartz, and 3.27 for stishovite) show inverse relationship with K_T. The MFS values for our three Mj-Py samples range from 4.02 to 4.88 (Table 1). Using the inverse relationship between MFS and K_T [figure 8, *Gillet et al.*, 1992] we obtain K_T values of 172–167 GPa, which are close to the reported Brillouin and X ray values for the Mj–Py series of samples [*Yagi et al.*, 1987; *Sinogeikin*, 1997a].

2. *Comparison of γ_i with $\gamma_{acoustic}$ and γ_{th}*

One could calculate $\gamma_{acoustic}$ ($\approx -1/6 + (1/2)K_T'$) and compare it with averaged mode γ_i. However, values of K_T' for the Mj–Py series are not as well known. *Yagi et al.* [1987] have reported the values of thermal expansion $\alpha = 1.7\times10^{-5}/°C$ and a value of $K_T' \sim 2.0$ for $Mj_{58}Py_{42}$. This K_T' value is significantly lower than the reported values for other garnets (4.25 for grossular, 4.5 for pyrope, and 5.45 for almandine-pyrope solid solution) [see references in *Yagi et al.*, 1987], and yields $\gamma_{acoustic}$ value of slightly less than 1, which is consistent with low averaged γ_i value (1.09) calculated from Raman frequency mode shifts for $Mj_{50}Py_{50}$. Next we consider correlation between γ_i calculated from $(\partial\nu_i/\partial P)$ and the thermal Grüneisen parameter, defined as

$$\gamma_{th} = \alpha K_s / \rho C_p$$

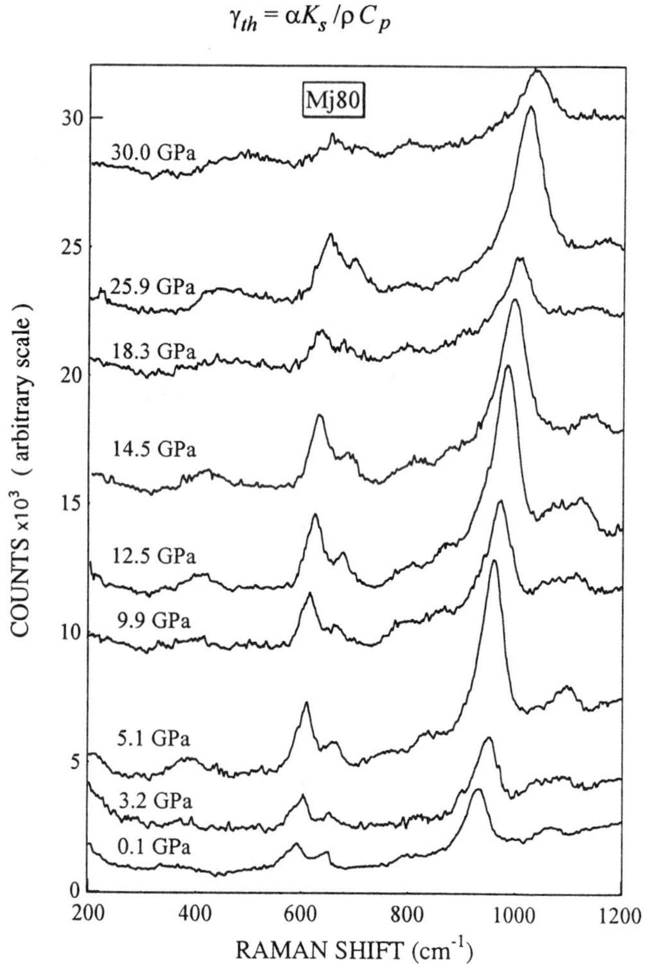

Figure 3. Raman spectra of Mj_{80} as a function of pressure to 30 GPa.

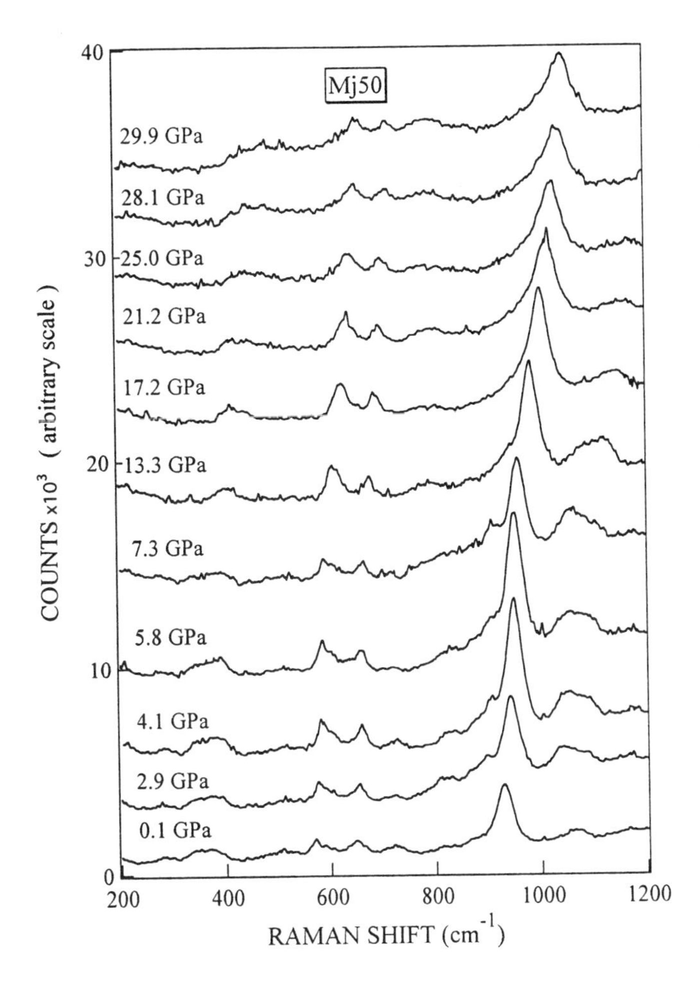

Figure 4. Raman spectra of Mj_{50} as a function of pressure to 29.9 GPa.

where α is the coefficient of volumetric thermal expansion, K_s is the adiabatic bulk modulus = K_T (1+αγT), ρ is density, and C_p is specific heat at constant volume. Using the *Yagi et al.* values of K_T = 164.2 GPa and α, and assuming C_p = 0.726 J/g K (same as for pyrope garnet) [*Anderson and Isaak*, 1993] and ρ = 3.538 g/cm^3 for $Mj_{58}Py_{42}$ (estimated from *Sinogeikin et al.*, 1997a), we estimate γ_{th} = 1.08, which again is in good agreement with the averaged value of mode γ_i (1.09). This γ_{th} value (1.08) value is lower than that reported for pyrope (1.50), which has a higher α than $Mj_{58}Py_{42}$ (2.36 × 10^{-5} deg^{-1} vs 1.6 × 10^{-5} deg^{-1}) [*Yagi et al.*, 1987; *Anderson and Isaac*, 1995]. In a nutshell, there is consistency between the averaged γ_i and calculated γ_{th} values for at least one of the majorite compositions examined. The relatively low calculated γ_{th} value can be attributed to the low α value reported for $Mj_{58}Py_{42}$. It would be valuable to obtain thermodynamic data (α and C_p) for other Mj–Py samples to confirm this.

CONCLUSIONS

1. The Raman spectra of Mj_{100} at ambient pressure are clearly different from those of $Mj_{80}Py_{20}$ and $Mj_{50}Py_{50}$ in terms of wavenumber and band features, and closely resemble that for Mj_{100} reported by *McMillan et al.* [1989]. Our results support the Mj_{100} to be tetragonal in structure.

2. Only minor differences exist between the spectra of $Mj_{80}Py_{20}$ and $Mj_{50}Py_{50}$, indicating that these samples have the same symmetry, and both have similar Raman spectra to those of $(Mg_{.79}Fe_{.21})_4Si_4O_{12}$ and Py_{100}, suggesting that all four samples have a cubic structure.

3. The pressure dependencies of all the vibrational frequencies are linear and positive throughout the range of this study (30 GPa), indicating that there is no phase transformation.

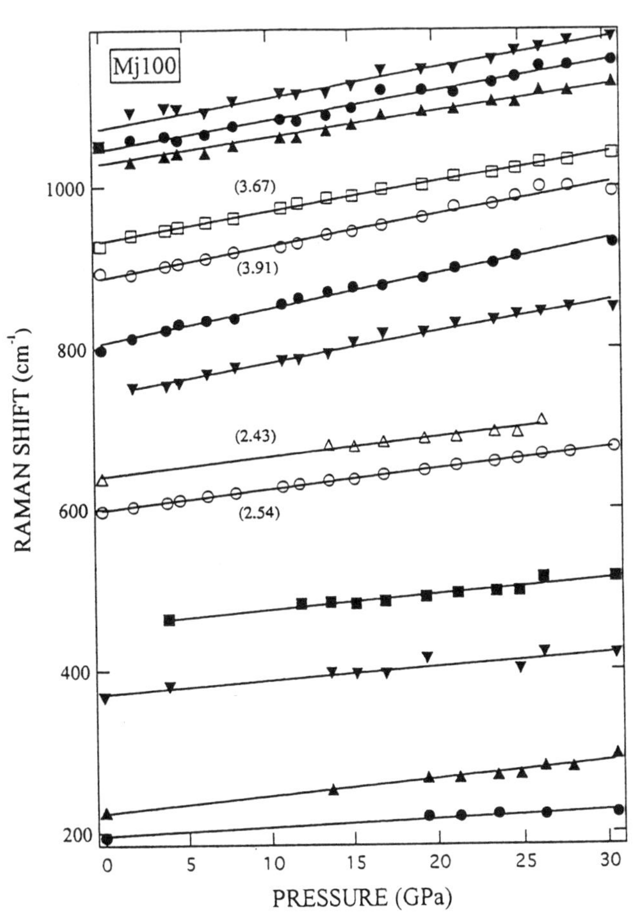

Figure 5. Pressure dependencies of the Raman shift for various modes observed in Mj_{100}. Values in parentheses are the linear slopes of the four major modes.

4. The averaged mode Grüneisen parameters for the three samples, calculated from $(\partial \nu_i/\partial P)$, lie between 0.81 and 1.13 and increase slightly with Py content. However, all the γ_i values for the three samples are lower than those for pyrope (1.5). The bulk modulus values calculated from MFS values (172–167 GPa) are in close agreement with previously reported values, based on Brillouin scattering and X ray measurements.

Acknowledgments. The authors thank T. Gasparik, Center for High Pressure Research (CHiPR), SUNY, Stonybrook for synthesizing the samples used in this study. They also thank Li Chung Ming for guidance in experimental work and for constructive comments on the manuscript and John Balogh for maintaining the equipment. School of Ocean and Earth Science and Technology contribution no. 4553.

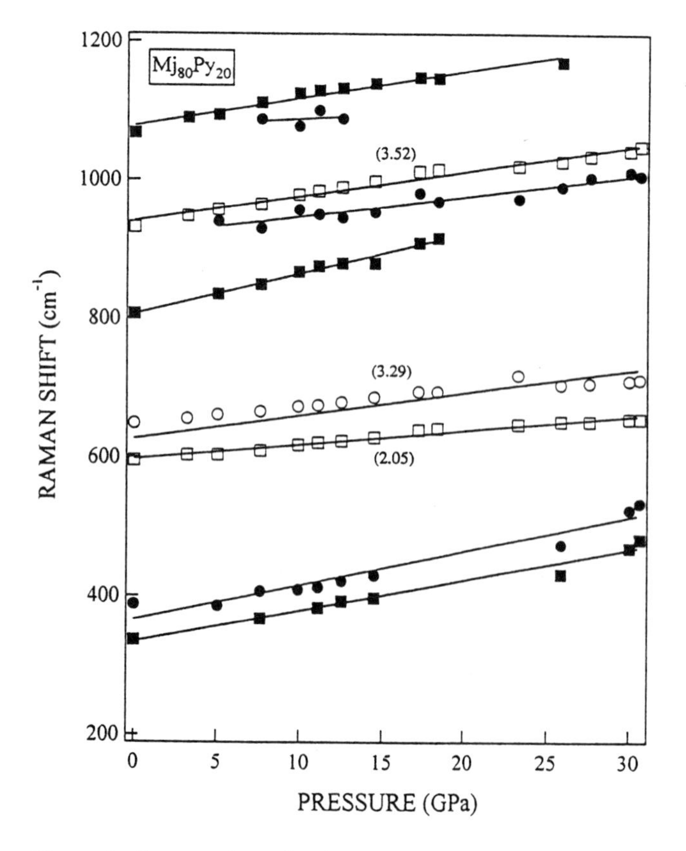

Figure 6. Pressure dependencies of the Raman shift for various modes observed in Mj_{80}. Values in parentheses are the linear slopes of the three major modes.

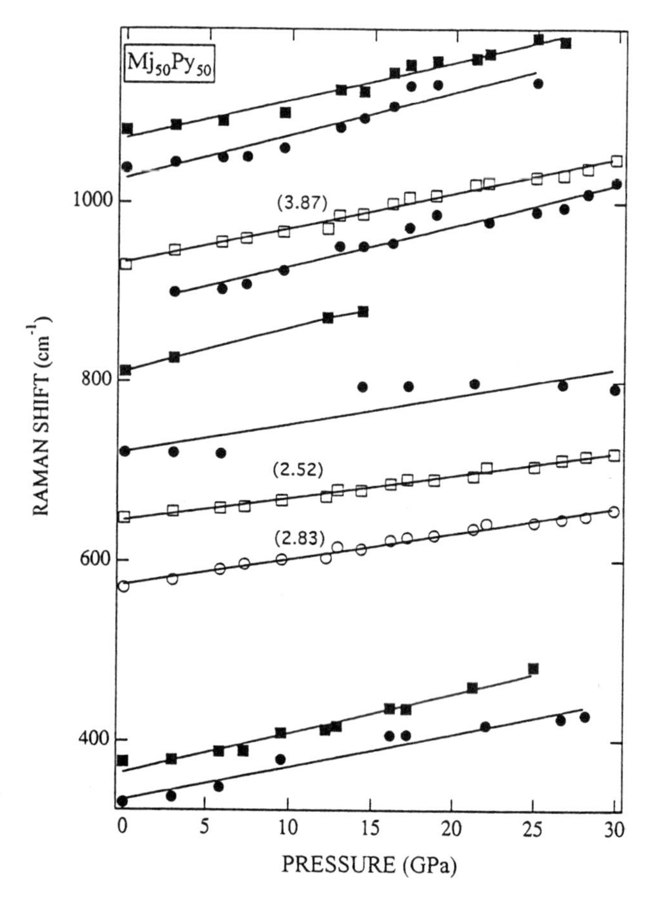

Figure 7. Pressure dependencies of the Raman shift for various modes observed in Mj_{50}. Values in parentheses are the linear slopes of the three major modes.

REFERENCES

Akaogi, M., and S. Akimoto, Pyroxene-garnet solid solution equilibria in the system $Mg_4Si_4O_{12}$-$Mg_3Al_2Si_3O_{12}$ and $Fe_4Si_4O_{12}$-$Fe_3Al_2Si_3O_{12}$ at high pressures and temperatures, *Phys. Earth Planet Int.*, *15*, 90-106, 1977.

Akaogi, M., A. Navrotsky, T. Yagi and S. Akimoto, Pyroxene-garnet transformations: thermochemistry and elasticity of garnet solid solutions, and application to a pyrolite mantle, *High-Pressure Research in Mineral Physics, Geophys. Monogr. Ser.*, vol. 39, edited by M. H. Manghnani and Y. Syono, pp. 251-260, Terra Scientific Publishing, Tokyo/AGU, Washington, D.C., 1987.

Anderson, O. L., and D. G. Isaak, Elastic constants of mantle minerals at high temperature, in *Mineral Physics & Crystallography: A Handbook of Physical Constants*, edited by T. J. Ahrens, AGU Reference Shelf 2, 64-97, AGU, Washington, D.C., 1995.

Angel, R. J., L. W. Ringer, R. M. Hazen, M. Kanzaki, D. J. Weidner, R. C. Liebermann, and D. R. Veblen, Structure and twinning of single-crystal MgSiO, garnet synthesized at 17 GPa and 1800°C, *Am. Mineral.*, *74*, 509-512, 1989.

Bass, J. D. and M. Kanzaki, Elasticity of majorite-pyrope solid solution, *Geophys. Res. Lett.*, *17*, 1989-1992, 1990.

Chen, M., T. G. Sharp, A. E. Goresy, B. Wopenka, and X. Xie, The majorite-pyrope + magnesiowustite assemblage: Constraints on the history of shock veins in chondrites, *Science*, *271*, 1570-1573, 1996.

Faust, J., and E. Knittle, The stability and equation of state of majorite garnet synthesized from natural basalt at mantle conditions, *Geophys. Res. Lett.*, *23*, 3377-3380, 1996.

Gasparik, T., Transformation of enstatite-diopside-jadeite pyroxenes to garnet, *Contrib. Mineral. Petrol.*, *102*, 389-405, 1989.

Gillet P., G. Fiquet, J.-M. Malezieux, and C. C. Geiger, High-pressure and high temperature Raman spectroscopy of end-member garnets: pyrope, grossular and andradite, *Eur. J. Mineral.*, *4*, 651-664, 1992.

Hatch, D. M., and S. Ghose, Symmetry analysis of the phase transition and twinning in MgSiO garnet: Implications to mantle mineralogy, *Am. Mineral.*, *74*, 1221-1224, 1989.

Heinemann, S., T. G. Sharp, F. Seifert, and D. C. Rubie, The cubic-tetragonal phase transition in the system majorite ($Mg_4Si_4O_{12}$)–pyrope ($Mg_3Al_2Si_3O_{12}$), and garnet symmetry in the Earth's transition zone, *Phys. Chem. Minerals*, *24*, 206-221, 1997.

Hofmeister A., and A. Chopelas, Vibrational spectroscopy of end-member silicate garnets, *Phys. Chem. Minerals*, *17*, 503-526, 1991.

Jeanloz, R., Majorite: Vibrational and compressional properties of a high-pressure phase, *J. Geophys. Res.*, *86*, 6171-6179, 1981.

Kanzaki, M., Ultrahigh-pressure phase relations in the system $Mg_4Si_4O_{12}$–$Mg_3Al_2Si_3O_{12}$, *Phys. Earth Planet. Int.*, *49*, 168-175, 1987.

Kato, T., and M. Kumazawa, Garnet phase of $MgSiO_3$ filling the pyroxene–ilmenite gap at very high temperature, *Nature*, *316*, 803-805, 1985.

Liu, L., The system enstatite-pyrope at high pressures and temperatures and the mineralogy of the Earth's mantle, *Earth Planet. Sci. Lett.*, *36*, 237-245, 1977.

Liu, L.-g., T. P. Mernagh, and T. Irifune, Raman spectra of pyrope and $MgSiO_3$-$10Al_2O_3$ garnet at various pressures and temperatures, *High Temperatures-High Pressures, 26*, 363-374, 1994.

Matsubara, R., H. Toraya, S. Tanaka, and H. Sawamoto, Precision lattice-parameter determination of (Mg,Fe)SiO_3 tetragonal garnets, *Science*, *24*, 697-699, 1990.

McMillan, P., M. Akaogi, E. Ohtani, Q. Williams, R. Nieman, and R. Sato, Cation disorder in garnets along the $Mg_3Al_2Si_3O_{12}$-$Mg_4Si_4O_{12}$ join: an infrared, Raman and NMR study, *Phys. Chem. Minerals*, *16*, 428-435, 1989.

McMillan, P. F., R. J. Hemley, and P. Gillet, Vibrational spectroscopy of mantle minerals, *Mineral Spectroscopy: A Tribute to Roger G. Burns*, The Geochemical Society, Special Publication No. 5, 175-213, 1996.

Mernagh, T. P., and L.-g. Liu, Pressure dependence of Raman spectra from the garnet end-members pyrope, grossularite and almandite, *J. Raman Spect.*, *21*, 305-309, 1990.

Moore, R. K., W. B. White, and T. V. Long, Vibrational Spectra of the common silicates: I. The garnets., *Am. Mineral.*, *56*, 54-71, 1971.

Ono, S., and A. Yasuda, Compositional change of majorite garnet in a MORB composition from 7 to 17 GPa and 1400 to 1600°C, *Phys. Earth Planet. Int.*, *96*, 171-179, 1996.

Pacalo, R. E. G., and D. J. Weidner, Elasticity of majorite, $MgSiO_3$ tetragonal garnet, *Phys. Earth Planet. Int.*, *99*, 145-154, 1997.

Pacalo, R. E. G., D. J. Weidner, and T. Gasparik, Elastic properties of sodium-rich majorite garnet, *Geophys. Res. Lett.*, *19*, 18, 1895-1898, 1992.

Parise, J. B., Y. Wang, G. D. Gwanmesia, J. Zhang, Y. Sinelnikov, J. Chmielowski, D. J. Weidner, and R. C. Liebermann, The symmetry of garnets on the pyrope ($Mg_3Al_2Si_3O_{12}$)-majorite ($MgSiO_3$) join, *Geophys. Res. Lett.*, *23*, 3799-3802, 1996.

Phillips, B. L., D. A. Howell, R. J. Kirkpatrick and T. Gasparik, Investigation of cation order in $MgSiO_3$-rich garnet using ^{29}Si and ^{27}Al MAS NMR spectroscopy, *Amer. Mineral.*, 77, 704-712, 1992.

Piermarini, G. J., S. Block, J. D. Barnett, and R. A. Forman, Calibration of the pressure dependence of the R1 ruby fluorescence line to 195 kbar, *J. Appl. Phys.*, *46*, 2774-2780, 1975.

Rigden, S. M., G. D. Gwanmesia, and R. C. Liebermann, Elastic wave velocities of a pyrope-majorite garnet to 3 GPa, *Phys. Earth Planet . Int.*, *86*, 35-44, 1994.

Sinogeikin, S. V., J. D. Bass, B. O'Neill, and T. Gasparik, Elasticity of tetragonal end-member majorite and solid solutions in the system $Mg_4Si_4O_{12}$–$Mg_3Al_2Si_3O_{12}$, *Phys. Chem. Minerals*, *24*, 115-121, 1997a.

Sinogeikin, S. V., J. D. Bass, A. Kavner, and R. Jeanloz, Elasticity of natural majorite and ringwoodite from the Catherwood meteorite, *Geophys. Res. Lett.*, in press, 1997b.

Yagi, T., M. Akaogi, O. Shimomura, H. Tamai, and S. Akimoto, High pressure and high temperature equations of state of majorite, in *High-Pressure Research in Mineral Physics*, *Geophys. Monogr. Ser.*, vol. 39, edited by M. H. Manghnani and Y. Syono, pp. 141-147, Terra Scientific Publishing, Tokyo/AGU, Washington, D.C., 1987.

Yagi, T., Y. Uchiyama, M. Akaogi, and E. Ito, Isothermal compression curve of $MgSiO_3$ tetragonal garnet, *Phys. Earth Planet. Int.*, *74*, 1-7, 1992.

Yeganeh-Haeri, A., and D. J. Weidner, Elastic properties of the pyrope–majorite solid solution series, *Geophys. Res. Lett.*, *17*, 2453-2456, 1990.

J. D. Bass, Cornell University, Department of Geological Sciences, Ithaca, NY 14853-1504 Bassett@geology.geo.cornell.edu

M. H. Manghnani, University of Hawaii, Hawaii Institute of Geophysics and Planetology, 2525 Correa Rd, Honolulu, HI 96822, murli@soest.hawaii.edu

V. Vijayakumar, High Pressure Physics Division, Bhabha Atomic Research Center, Trombay, Mumbai, India

The Imaging Plate System Interfaced to the Large-Volume Press at Beamline X17B1 of the National Synchrotron Light Source

J. Chen, J. B. Parise, R. Li, D. J. Weidner, and M. Vaughan

Center for High Pressure Research (CHiPR) and
Department of Earth and Space Sciences, State University of New York, StonyBrook, New York

A double imaging plate is interfaced to the Large Volume High Pressure Device (LVHPD) SAM-85 for the collection of in situ X ray powder diffraction data suitable for Rietveld analysis. The two-dimensional detector, combined with the uniform pressure and temperature environments of the LVHPD and the wide energy range available from a new Laue-Bragg monochromator at X17B1, results in high-quality diffraction data suitable for Rietveld refinements. A disk-type heater is used to minimize extrinsic diffraction peaks from the surrounding materials. Diffraction effects resulting from the boron-epoxy pressure transmitting medium are removed by subtraction. To demonstrate the quality of data obtained from the imaging-plate system, the cation distributions over the available sites in the crystal structures of $NiAl_2O_4$-spinel and $(Ni,Mg)_2SiO_4$-olivine have been refined from data collected at high pressures and temperatures. In both cases cation ordering is observed to increase with pressure.

1. INTRODUCTION

Knowledge of the crystal structures adopted by materials at high pressures and temperatures is fundamental to an understanding of their properties under these conditions. Traditionally, changes as a function of temperature and pressure have been interpreted using quench techniques; the presumption being the quench products represent a facsimile of the state existing under the conditions of synthesis. In situ studies, in which both temperature and pressure are maintained as data suitable of full structural refinement are collected, are more rare [*Hazen and Finger*, 1982]. However, recent technological developments vaticinate a period where such studies will become more routine. These developments include the availability of reliable large volume high pressure apparatus (LVHPA) interfaced to bright, high-energy X ray sources at synchrotrons and coupled with area detectors, such as imaging plates [*Shimomura et al.*, 1992]. The high-energy (>35 keV) monochromatic beams now available at second and third generation synchrotron sources allow collection of reliable step scan data, which can reveal subtle structural details such as octahedral tilting as a function of pressure in perovskites [*Zhao et al.*, 1994]. When coupled with imaging plate (IP) detectors, this powder X ray diffraction data can be collected in a matter of minutes rather than hours, raising the possibility of real-time studies [*Chen et al.*, 1997a,b].

In order to obtain reliable crystallographic information, accurate diffraction intensities are required. In a high-pressure experiment, the quality of this powder diffraction data can be compromised by a number of systematic errors, such as sample quality, deviatoric stress and absorption. The large-volume press offers a quasi-hydrostatic pressure environment, especially at

Properties of Earth and Planetary Materials
at High Pressure and Temperature
Geophysical Monograph 101

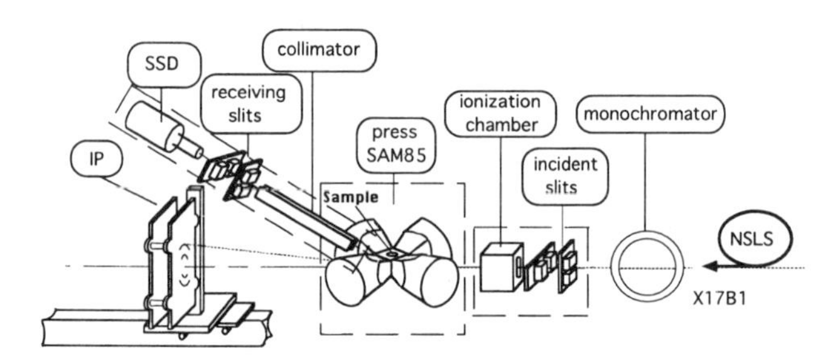

Figure 1. Schematic of the imaging plate diffraction system for the large volume press at NSLS beamline X17B1. IP: an imaging plate, SSD: a solid state detector, SAM85: a cubic-type multianvil high pressure device.

temperatures above the yield strength of the material, and allows a large amount of diffracting sample, which ensures good powder statistics. Further, cell assemblies to allow easy and prolonged heating with low temperature gradients are available. The wiggler or undulator synchrotron radiation to which the LVHPA can be coupled are the best X ray sources for high-pressure diffraction, since they provide tunable high-energy radiation suitable for penetrating cell assemblies and minimizing the effects of absorption. Improvements in the detecting system for the LVHPA have recently been made by coupling it to IP detectors at the Photon Factory in Japan [*Iwasaki et al.*, 1995; *Kikegawa et al.*, 1995; *Chen et al.*, 1997a]. In this paper, we describe the set-up and application of an IP detector to the LVHPA installed at the X17B1 synchrotron wiggler beamline of the NSLS. Several features make this installation unique for the determination of crystal structure under pressure using monochromatic radiation. The quality of the diffraction data that can be collected using this new system is highlighted in two studies where the pressure dependence of cation ordering in $NiAl_2O_4$-spinel and $NiMgSiO_4$-olivine are measured.

2. THE IMAGING PLATE SYSTEM AT X17B1

2.1. *Experimental Design*

A schematic of the imaging plate system, installed at the high-energy X ray synchrotron wiggler beamline X17B1 of the NSLS is shown in Figure 1. Details of this beamline, its layout, the beam characteristics and the LVHPA are published [*Chapman*, 1989; *Chapman and Tomlinson*, 1984; *Weidner et al.*, 1992; *Weidner and Mao*, 1993; *Schulze et al.*, 1994]. Briefly, the NSLS typically operates at 2.5 GeV and 250mA. The 13.3 milliradian beam for X17, generated in a 5-pole wiggler operating at 4.7 Tesla, is split into three beams of 2.15, 5.00, and 2.15 milliradians by a front end aperture-splitter-shutter assembly. These are the X17A, B and C beams, respectively. A graphite filter assembly, designed to remove the low X ray energies from the wiggler spectrum, removes about 40% of the wiggler power.

Another aperture-splitter-shutter assembly is used to define the beam falling onto a Laue-Bragg Si(220) monochromator placed approximately 10 meters from the sample (Figure 1). This monochromator (Figures 1 and 2) has a number of desirable features for high pressure work including the ability to quickly switch between monochromatic and polychromatic radiation mode by simply translating a beam-stop and the LVHPD vertically (Figure 2). The beam transmitted through the bottom crystal, set in Laue mode, is used for polychromatic experiments (Figure 2) where a solid state detector (SSD) is used for real-time energy dispersive X ray diffraction (EDXD) measurement to map the cell geometry, determine sample pressure, instrument alignment, mapping of phase boundaries and general test shots. The SSD is also used for calibrating the energy of the monochromatic X ray. It is mounted on a 2θ goniometer arm which can move from +35° to -35°. A set of receiving slits is installed in front of the SSD to collimate X rays diffracted from the sample. In monochromatic mode an imaging plate records accurate diffraction data.

The imaging plates are mounted in a specially designed holder held on an optical rail (Figure 1). This can be tilted vertically and horizontally to the incident X ray direction. The stage is movable along the incident X ray direction to change the sample-to-IP distance (400–870 mm), and can be easily taken off from the guide block to allow the goniometer arm to move down for EDXD measurements.

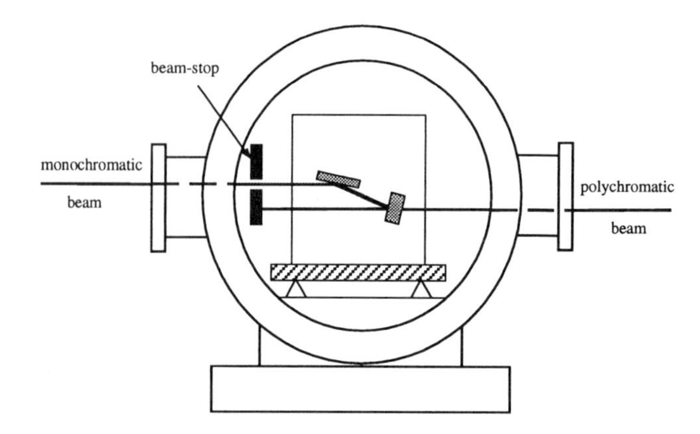

Figure 2. Laue-Bragg monochromator at NSLS beamline X17B1. The first crystal is set in transmission mode; translating the beam-stop can easily switch the X rays between monochromatic and polychromatic radiation.

A vacuum ensure the FUJI ST-V type plates, either 200 mm ×250 mm or 200 × 450 mm size, are kept flat to the holder during exposure. Exposed plates are read using a FUJI BAS2000 reader at a Pixel size resolution of 0.1 mm × 0.1 mm. When the imaging plate is used to record the diffraction pattern, the goniometer arm is rotated up to +35°. The direct beam is allowed to fall on the top of the imaging plate in order to allow the largest number of diffraction lines to be detected.

In the cubic-type multianvil LVHPD [*Shimomura et al.*, 1992; *Weidner and Mao*, 1993] shown schematically in Figure 1, X rays can only go through the gaps in the [110] cubic-directions of the four side anvils in the horizontal plane of the six anvil assembly; the top and bottom anvils are not shown in Figure 1 for the sake of clarity. Under pressure, the anvil gap ensures most of the diffracted X rays are masked by the side anvils. The diffraction pattern recorded on the IP is a set of arcs from the full Debye-Scherrer rings; an example diffraction pattern is shown in Figure 3.

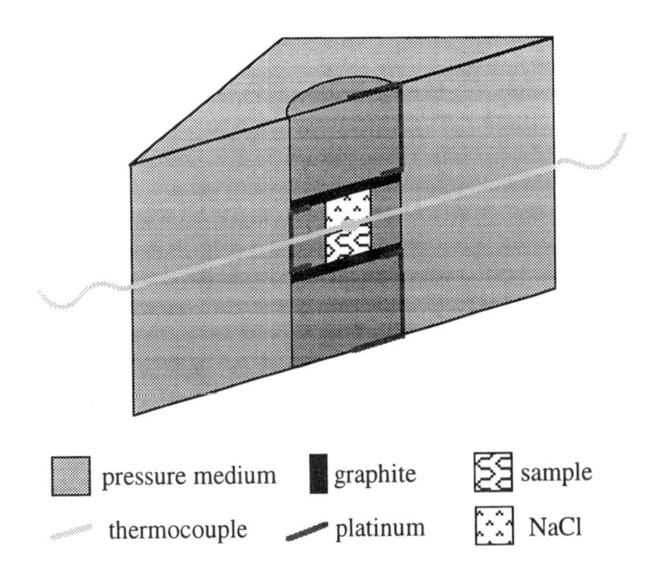

Figure 4. High-pressure cell assembly. The pressure medium is made of a mixture of amorphous boron and epoxy resin with the weight ratio 4:1, and the thermocouple is W5%Re/W26%Re.

2.2. *Calibration and Alignment*

The relationship of energy to channel number for the SSD is calibrated using fluorescence lines from various elements as well as a cocktail of three standard materials. The lines from the energy dispersive diffraction pattern are fit to a third order polynomial equation. The energy of the monochromatic X ray beam is then set close to the Ba edge by observing the disappearance of the transmitted polychromatic beam through a single crystal of BaF_2. The wavelength is determined with an uncertainty of 0.01 keV using the SSD. Alignment of the imaging plate is accomplished by adapting a technique proposed by *Shimomura et al.* [1992]. Because most of the Debye-Scherrer rings are masked by the anvils (Figure 3), the alignment is best done before the sample is loaded into the press in order to obtain the entire rings. The inclination angle and tilt of the rail are adjusted to give Debye-Sheerer rings concentric about the incident beam. Although the sample-to-IP distance (L) could not be measured directly, the double plate design allows this to be determined with one exposure since the distance (D) between the two plates is known accurately and L=rD/(R-r); here R and r the radii of the same Debye-Scherrer ring on the back and front plates, respectively (Figure 1).

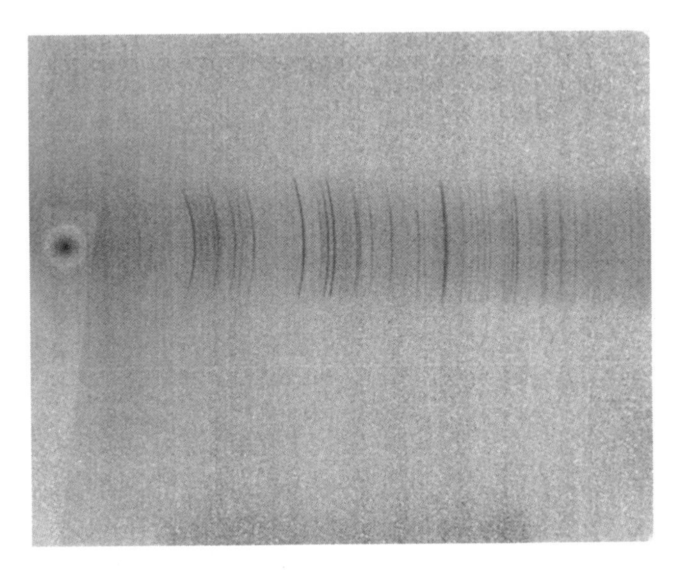

Figure 3. A diffraction pattern for NiMg-olivine at 4 GPa and 800 °C recorded on an imaging plate for 3 minutes at an X ray energy of 40.64 keV, incident beam size: 0.2 mm × 0.5 mm, IP-to-sample distance: 862 mm.

2.3. *Sample Assembly and Extrinsic Diffraction Peaks*

Because there is no slit between the imaging plate and sample, the imaging plate records all the diffraction from materials which intercept the incident X rays. To minimize the extrinsic diffraction from the materials other than sample, a cell assembly with disk heaters is adopted (Figure 4). The pressure medium is made of a mixture of four parts amorphous boron and one part epoxy resin.

The pressure medium has very little absorption at the X ray energies used (>35 keV) however it gives several weak diffraction peaks at low angles in addition to expected

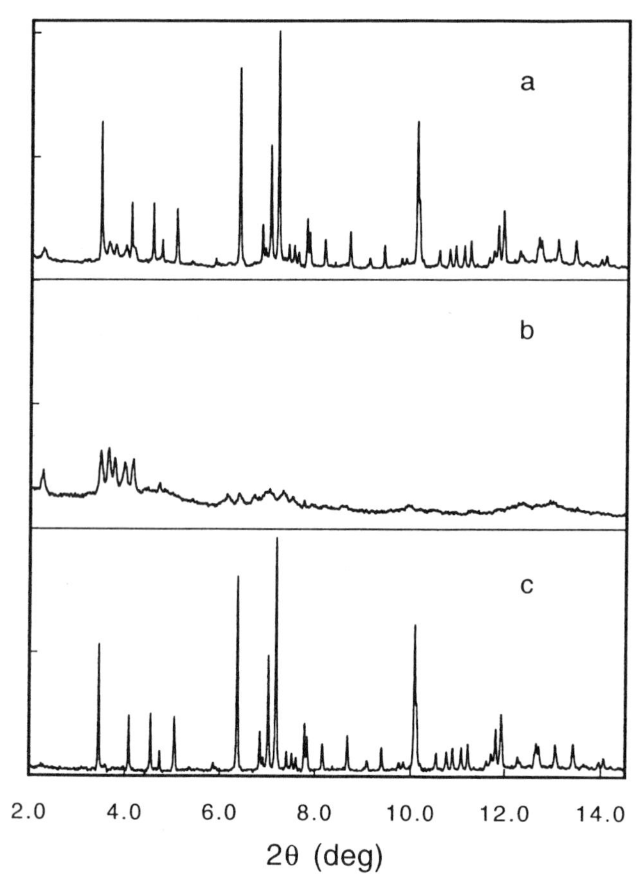

Figure 5. Subtraction of the diffraction patterns, pattern a: raw data from the sample and the pressure medium (by integrating the diffraction pattern shown in Figure 3), pattern b: diffraction data from the pressure medium close to the sample, pattern c: final data for the sample by subtracting the scaled pattern b from pattern a.

diffuse maxima from the amorphous material. These diffraction peaks are broadened making them difficult to fit during data processing. However by recording a pattern of the cell assembly material close to the sample, by translating the press vertically and exposing a fresh IP, subtraction of this scaled pattern from that containing both sample and assembly peaks is possible (Figure 5). In this way patterns free from the contamination from the cell assembly material (Figure 4) are obtained without the need to modify the cell assembly further [*Chen et al.*, 1997a].

2.4. *Experimental Procedures*

In order to reach a desired pressure and temperature, the sample is first compressed stepwise at room temperature. Diffraction patterns of an NaCl internal pressure marker and the sample are taken at each step using the SSD to determine the pressure by referring to Decker's scale [*Decker*, 1971], and to observe structure changes in the sample. When the desired pressure is reached or the diffraction pattern characteristic of a desired phase is observed, the sample is heated. A W5%Re/W26%Re thermocouple is used to determine the sample temperature. Once the temperature has reached a desired value or the diffraction pattern shows an expected characteristic, a diffraction pattern of the pressure marker is taken to check the pressure and the location of the sample is mapped using the SSD. The radiation is then switched to monochromatic (Figure 2), and the press is moved to match the position of the incident beam. An imaging plate is mounted to its preset position while the goniometer arm is rotated up to +35°. Either a direct beam is exposed on the imaging plate for a second, or a specially machined beam-stop allows enough photons to impinge on the plate and so identify the direct beam position to determine the center of the diffraction pattern.

2.5. *Data Processing*

The imaging plate is read off line and digitized. The diffraction intensities are derived by integrating portions of the Debye-Scherrer rings recorded on the imaging plate (Figure 3) using the program IPH developed for the DAC-IP system at Photon Factory, KEK, [*Shimomura et al.*, 1992]. Rietveld structure refinement [*Rietveld*, 1969] was carried out using the program GSAS [*Larson and Von Dreele*, 1986].

3. CATION DISTRIBUTIONS IN $NiAl_2O_4$-SPINEL AND $NiMgSiO_4$-OLIVINE AT HIGH PRESSURE

The effects of pressure on the state of cation order-disorder has recently been reviewed by *Hazen and Navrotsky* [1997]. They introduce a definition of volume of disordering

$$\Delta V_{dis} = V_{disordered} - V_{ordered}$$

and predict that certain structure types are likely to display this phenomenon [*Hazen and Navrotsky*, 1997]. The $NiAl_2O_4$-spinel and $NiMgSiO_4$-olivine are two such structure types .

3.1. *$NiAl_2O_4$ Spinel*

In $NiAl_2O_4$-spinel ($Fd\bar{3}m$, , Z=8) oxygen is located at positions 32*e* and the cations occupy sites 8*a* and 16*d*. Cations are in tetrahedral (8*a*) or octahedral (16*d*) coordination with nearest neighbor oxygen ions. While in normal AB_2O_4-oxide spinels cations occupy tetrahedral sites and B cations the octahedral sites, inverse spinels have the tetrahedral sites fully occupied by B cations and the octahedral sites equally occupied by A and B cations. As synthesized $NiAl_2O_4$ is a partly inverse spinel with a cation distribution given by the general formula $(Ni_{1-X}Al_X)^{[IV]}(Ni_XAl_{2-X})^{[VI]}O_4$, where [IV] and [VI] represent the tetrahedral and octahedral sites, and X represents the degree of inversion. The effect of temperature on the value of X has been studied by quench experiments at room pressure [*Mocala and Novrotsky*, 1989; *Roelofsen et al.*, 1992] and these show a tendency toward disorder over the A and B sites at high temperature. In our experiments, we observe a tendency toward increased ordering with increasing pressure.

The $NiAl_2O_4$ spinel was synthesized at 1300 °C by solid-state reaction from the oxides. Multiple cycles of grinding and firing lead to complete reaction as judged from X ray powder diffractometry. Two diffraction patterns were taken, one at ambient and one at high pressure (6 GPa) after annealing the sample at 850 °C, using the experimental procedures described above. The wavelength of monochromatic X rays was 0.3312 Å. The structural parameters derived from the Rietveld refinements [*Rietveld*, 1969; *Larson and Von Dreele*, 1986] are listed in Table 1. The X value increases from 0.857 at room pressure to 0.915 at 6 GPa indicating that the cation distribution tends toward a more inverse distribution under pressure.

3.2 *Order-Disorder Phenomena over the M1/M2 Octahedral Sites in Ni-Mg Olivine*

The olivine structure (***Pbnm***), M_2SiO_4 (for example M = Fe^{2+}, Mn^{2+}, Mg^{2+}, Ni^{2+}), consists of an hcp array of oxygen with Si^{4+} in one-eighth of the available tetrahedral sites and the 2+-cations distributed over half the octahedral interstices in two symmetry-independent sites. For this experiment an olivine was synthesized from a mixture of dried oxides, mixed to give the stoichiometry $(Ni_{0.5}Mg_{0.5})SiO_4$. After three cycles of regrinding and firing at 1500 °C, X ray diffraction data indicated a single olivine phase. Data suitable for structure analysis were collected under the following conditions in the apparatus

TABLE 1.. Final Structural Parameters for $NiAl_2O_4$* Spinel

Pressure	a(Å)	u*	X*
ambient	8.0484(1)	0.2554(1)	0.8571(1)
6 GPa, 25 °C	7.9590(2)	0.2583(3)	0.914(6)

* space group ***Fd3m***, Z=8, origin at center; $(Ni_{1-X}Al_X)^{[IV]}(Ni_XAl_{2-X})^{[VI]}O_4$, where [IV] and [VI] represent the tetrahedral and octahedral sites at 8*a* and 16*d*, respectively; O at site 32*e* (*u,u,u*); *X* represents the degree of inversion.

TABLE 2. Final Structural Parameters for NiMg-Olivine ($Ni_{0.5}Mg_{0.5}SiO_4$) at Ambient and after 69 Minutes at 4 GPa and 800 °C

		Room Pressure* (800 °C quenched)	4 Gpa, 800 °C§ (69 min)
Kd		8.27	10.25
M1†	U	0.015(3)	0.016(1)
	X(Ni)	0.742(3)	0.762(2)
M2	x	0.9898(9)	0.9892(12)
	y	0.2752(2)	0.2762(3)
	U	0.015(3)	0.016.(1)
	X(Ni)	0.258(3)	0.238(2)
Si	x	0.4256(8)	0.4260(10)
	y	0.0932(4)	0.0939(4)
	U	0.017(3)	0.017(2)
O1	x	0.7719(14)	0.7687(14)
	y	0.0916(8)	0.0890(9)
	U	0.009(3)	0.010(1)
O2	x	0.2208(12)	0.2259(13)
	y	0.4447(8)	0.4450(9)
	U	0.009(3)	0.010(1)
O3	x	0.2768(8)	0.2804(9)
	y	0.1640(5)	0.1626(5)
	z	0.0319(10)	0.0326(13)
	U	0.009(3)	0.100(1)

*a = 4.7362(2), b = 10.1682(3), c = 5.9401(2)Å, V = 286.07(2)Å^3, wR_p=0.08
§a = 4.7322(2), b = 10.1453(4), c = 5.9309(2)Å V = 284.75(2)Å^3, wR_p=0.08
† M1 at (0,0,0), M2, Si, O1 and O2 at (*x*, *y*, ¼) ***Pbnm***, Z = 4

described above: (1) at room temperature before pressurizing, (2) at 4 GPa and 800 °C; several data sets were collected at 20–30 minute intervals at these later conditions. Rietveld refinements show the distribution coefficient K_d ($K_d = [X_{Ni} / X_{Mg}]^{M1} / [X_{Ni} / X_{Mg}]^{M2}$) of the starting material is 8.27 at ambient. At 4 GPa/800 °C, K_d is larger (Table 2) and increases as a function of time. The wavelength of monochromatic X ray was 0.3051 Å and diffraction patterns were recorded up to 15° in 2θ on the 20 mm × 25 mm imaging plate. Figure 6 shows the results of Rietveld refinement of the high-pressure and temperature data. During refinement (Table 2) all sites were considered fully occupied, and the side-occupation fractions for M1 and M2 site were constrained by the composition of starting material. Although equilibrium was not achieved after three hours at 800 °C, the data clearly indicate (Table 2) an increase in ordering compared with the sample at ambient. Details of the kinetics of ordering, along with complete details of refinements at higher temperatures, will be published separately [*Chen et al.*, 1997b].

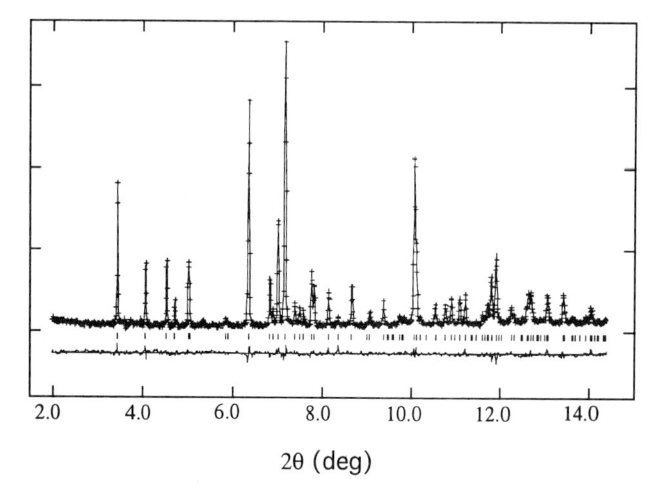

Figure 6. Final observed (+) and calculated (solid line) monochromatic X ray powder diffraction pattern for $NiMgSiO_4$ olivine at 4 GPa and 800 °C. Tick marks below the profile indicate the positions of allowed reflections. Profiles for the other data discussed in the text can be obtained from the authors.

4. CONCLUSION

Application of the imaging plate to the large-volume high-pressure apparatus at the X17B1 beamline of the NSLS allows the acquisition of quantitative monochromatic diffraction data at high pressures and temperatures with good resolution and peak-to-background discrimination. The combination of the imaging-plate system with high-energy synchrotron X radiation enables us to obtain high quality diffraction data, free of extrinsic scattering effects from the cell assembly for most samples. These data are suitable for the Rietveld refinement of crystal structure. In the two systems studied, $NiAl_2O_4$-spinel and NiMg-olivine, pressure increases the tendency toward ordering over the available cation sites in accord with predictions based on thermodynamic considerations.

Acknowledgments. We would like to thank Drs. Wang and Shields who helped with the collection of data. The imaging plate system was expertly constructed by Carey Kolida. This work was performed at the X-17B1 beamline of the NSLS, a facility funded by the DOE. This work was supported in part by the NSF EAR program with grant 95-06483 to JBP and DJW; the LVHPA and certain facilities at X17B1 were supported by the NSF Science and Technology Center grant to CHiPR.

REFERENCES

Chapman, D., X17B1 operations training manual, National Synchrotron Light Source, 1989.

Chapman, D., and W. Tomlinson, NSLS superconducting wiggler beam line (X17) conceptual design report, Brookhaven National Laboratory, 1984.

Chen, J., T. Kikegawa, O. Shimomura, and H. Iwasaki, Application of an Imaging Plate to the Large-Volume Press MAX80 at the Photon Factory, *J. Synchrotron Rad.*, 1997a (in press).

Chen, J., J. B. Parise, R. Li, and D. J. Weidner, Pressure induced ordering in Ni-Mg olivine, *Am. Mineral.*, 1997b (in press).

Decker, D. L., High-pressure equation of state for NaCl, KCl and CsCl, *J. Appl. Phys.*, *42*, 3239-3244, 1971.

Hazen, R. M., and L. W. Finger, *Comparative Crystal Chemistry*, John Wiley & Sons, New York, 1982.

Hazen, R. M., and A. Navrotsky, Effects of pressure on ordering-disordering reactions. *Am. Mineral.*, 1997 (in press).

Iwasaki, H., J. Chen, and T. Kikegawa, Structural study of the high pressure phases of bismuth using high-energy synchrotron radiation, *Rev. Sci. Instrum.*, *66*, 1388-1390, 1995.

Kikegawa, T., J. Chen, K. Yaoita, and O. Shimomura, DDX diffraction system: A combined diffraction system with EDX and ADX for high pressure structure studies, *Rev. Sci. Instrum.*, *66*, 1335-1337, 1995.

Larson, A. C., and R. B. Von Dreele, GSAS, General Structure Analysis System, LAUR Report 86-748, Los Alamos National Laboratory, 1986.

Mocala, K., and A. Novrotsky, Structural and thermodynamic variation in nickel aluminate spinel *J. Am. Ceram. Soc.*, *72* [5], 826-832, 1989.

Rietveld, H. M., A profile refinement method for nuclear and magnetic structures, *J. Appl. Crystallog.*, *2*, 65-71, 1969.

Roelofsen, J. N., R. C. Peterson, and M. Raudsepp, Structural variation in nickel aluminate spinel ($NiAl_2O_4$), *Am. Mineral.*, *77*, 522-528, 1992.

Schulze, C., P. Suortti, and D. Chapman, Test of a bent Laue double crystal fixed exit monochromator, *Syn. Rad. News*, *7*, 8-11, 1994.

Shimomura, O., K. Takemura, H. Fujihisa, Y. Fujii, Y. Ohishi, T. Kikegawa, Y. Amemiya, and T. Matsushita, Application of an imaging plate to high pressure x-ray study with a diamond anvil cell, *Rev. Sci. Instrum.*, *63*, 967-703, 1992.

Weidner, D. J., and H. Mao, Photons at High Pressure, *NSLS Newletter*, pp. 1-3, 1993.

Weidner, D. J., M. T. Vaughan, J. Ko, Y. Wang, X. Liu, A. Yeganeh-Haeri, R. E. Pacalo, and Y. Zhao, Characterization of stress, pressure, and temperature in SAM85, a DIA type high pressure apparatus, in *High-Pressure Research: Application to Earth and Planetary Sciences, Geophys. Monogr. Ser.*, vol. 67, edited by Y. Syono and M. H. Manghnani, pp. 13-17. TERRAPUB, Tokyo / AGU, Washington, D.C.

Zhao, Y., J. B. Parise, Y. Wang, K. Kusaba, M. T. Vaughan, D. J. Weidner, T. Kikegawa, J. Chen, and O. Shimomura, High-pressure crystal chemistry of $NaMgF_3$ perovskite: An angle dispersive diffraction study using monochromatic synchrotron x-radiation, *Am. Mineral.*, *79*, 615-621, 1994.

J. Chen, J. B. Parise, R. Li, D. J. Weidner, and M. Vaughan, Center for High Pressure Research (CHiPR) and Department of Earth and Space Sciences, State University of New York, StonyBrook, NY 11794-2100

Computational Modeling of Crystals and Liquids in the System Na_2O-CaO-MgO-Al_2O_3-SiO_2

Masanori Matsui

Department of Earth and Planetary Sciences, Faculty of Science, Kyushu University, Fukuoka, Japan

A transferable interatomic potential model has been developed for use in computer simulation of crystals and liquids in the system Na_2O-CaO-MgO-Al_2O_3-SiO_2 (NCMAS). The potential energy of the system is taken as the sum of pairwise additive Coulomb, van der Waals, and repulsive interactions. The net charges on the Na, Ca, Mg, Al, Si, and O ions are constrained to be $2q(\mathrm{Na}) = q(\mathrm{Ca}) = q(\mathrm{Mg}) = {}^{2}/_{3}\, q(\mathrm{Al}) = {}^{1}/_{2}\, q(\mathrm{Si}) = -q(\mathrm{O})$ in order to preserve the requirement of transferability between phases with different composition in the NCMAS system. To assess the potential model, the molecular dynamics (MD) method with the potential is applied to a wide structural variety of 29 crystals in the NCMAS system and the five silicate liquids with the compositions enstatite ($MgSiO_3$), wollastonite ($CaSiO_3$), diopside ($CaMgSi_2O_6$), anorthite ($CaAl_2Si_2O_8$), and albite ($NaAlSi_3O_8$). The MD simulated structures and bulk moduli of the 29 crystals and the MD values of the temperature-pressure-volume equation-of-state parameters for the five silicate liquids are found to compare well with the available experimental data. The MD technique is then used to simulate the pressure dependence of the molar volume of both $MgSiO_3$ perovskite and liquid at pressures up to 150 GPa and at the temperatures of 2500 and 3500 K, with the results that no volume reversal is found between $MgSiO_3$ perovskite and liquid at these temperature and pressure conditions. This indicates the perovskite melting curve has no maximum over the pressure range in the lower mantle.

1. INTRODUCTION

Molecular dynamics (MD) simulation has been used extensively to investigate and predict the structural, elastic, thermodynamic, and dynamic properties of molten and crystalline materials. The MD simulation is particularly useful and powerful for the study of properties under extreme conditions of pressure and temperature that may be inaccessible in the laboratory. Work on silicate crystals includes *Y. Matsui and Kawamura* [1987], *Tsuneyuki et al.* [1988], *Kapsta and Guillop* [1988], *Wall and Price* [1989], *Winkler and Dove* [1992], *Belonoshko* [1994], *and Matsui et al.* [1994]. Application of the MD technique to silicate liquids includes publications by *Woodcock et al.* [1976], *Y. Matsui and Kawamura* [1980, 1984], *Angell et al.* [1982], *Kubicki and Lasaga* [1991], *Wasserman et al.* [1993], *Stein and Spera* [1995], and *Matsui* [1996a]. *Wentzcovitch et al.* [1993, 1995] have recently developed the first-principle variable-cell-shape MD method to investigate the structural and elastic properties of magnesium silicates.

Properties of Earth and Planetary Materials
at High Pressure and Temperature
Geophysical Monograph 101

TABLE 1. Optimized Energy Parameters Used for Simulation

	$q/\|e\|$	A/Å	B/Å	C/[Å^3(kJ/mol)$^{1/2}$]
Na	0.4725	1.1600	0.040	40.91
Mg	0.945	0.8940	0.040	29.05
Ca	0.945	1.1720	0.040	45.00
Al	1.4175	0.7852	0.034	36.82
Si	1.890	0.7204	0.023	49.30
O	-0.945	1.8215	0.138	90.61

Crystals and liquids in the Na_2O-CaO-MgO-Al_2O_3-SiO_2 system (called NCMAS) are most important constituents in the Earth's crust and mantle. Recently we have developed a transferable potential model (CMAS94 potential) applicable to both crystals and silicate liquids in the CaO-MgO-Al_2O_3-SiO_2 system. The reliability and applicability of the CMAS94 potential has been fully tested using MD simulation, with the results that the simulation has succeeded in reproducing accurately (1) the observed structures and measured bulk moduli of a wide variety of 27 crystals in the CaO-MgO-Al_2O_3-SiO_2 system [*Matsui*, 1996b], and (2) the observed temperature-pressure-volume equation-of-state parameters of the four silicate liquids with the compositions enstatite ($MgSiO_3$), wollastonite ($CaSiO_3$), diopside ($CaMgSi_2O_6$), and anorthite ($CaAl_2Si_2O_8$) [*Matsui*, 1996a].

In the present paper we attempt to extend the CMAS94 potential to the NCMAS system. Here we apply MD to albite ($NaAlSi_3O_8$) liquid, as well as to the two Na-bearing silicates, jadeite ($NaAlSi_2O_6$) and low albite. The MD technique is further used to investigate the possible occurrence of the density inversion between $MgSiO_3$ perovskite and liquid at high temperatures and high pressures, in order to give some insight into the melting slope of $MgSiO_3$ perovskite at lower mantle conditions.

2. SIMULATION METHODS

Following our previous studies [*Matsui*, 1996a,b], the potential energy of the system was taken to be the sum of pairwise interactions between atoms of this form:

$$V(r_{ij}) = q_i q_j r_{ij}^{-1} - C_i C_j r_{ij}^{-6} + f\left(B_i + B_j\right) \times \exp\left[\left(A_i + A_j - r_{ij}\right) / \left(B_i + B_j\right)\right]$$

where r_{ij} is the interatomic distance between atoms i and j; q_i, A_i, B_i, and C_i are the energy parameters peculiar to the kind of atom i; and f is a standard force of 4.184 kJ Å^{-1}mol^{-1}. The net charges, q_i's, were constrained as $2q(\mathrm{Na}) = q(\mathrm{Ca}) = q(\mathrm{Mg}) = 2/3\, q(\mathrm{Al}) = 1/2\, q(\mathrm{Si}) = -q(\mathrm{O})$ in order to preserve the requirement of transferability between phases with different composition in the NCMAS system. The necessary energy parameters of the ions except Na were from our previous works [*Matsui*, 1996b]. The energy parameters for Na were obtained empirically by fitting them to the observed structures of both jadeite [*Cameron et al.*, 1973] and low albite [*Harlow and Brown*, 1980] as well as the measured bulk modulus of jadeite [*Kandelin and Weidner*, 1988]. The optimized energy parameters are listed in Table 1.

The MD simulations were carried out in the isothermal-isobaric (constant temperature T, constant pressure P, and constant number of particles N in the system) or the canonical (constant T, V, and N) ensemble. The usual periodic boundary conditions were imposed, and the equations of motion were solved numerically with the time increment of 1.0 or 2.0 fs, depending on the temperature conditions used for simulation. All the six MD basic-cell parameters (the three cell lengths and the three cell angles) were relaxed for crystal simulations, whereas they were constrained as cubic boxes with variable cubic-cell lengths for liquid simulations. For the case of crystal simulations, quantum corrections to MD results were made based on the Wigner-Kirkwood expansion of the free energy using the technique proposed by *Matsui* [1989]. Equilibrium structures were calculated from time-averages taken over sufficiently long time intervals (5,000~10,000 steps for crystals and 20,000~30,000 steps for liquid, usually). For crystals, the isothermal bulk moduli at zero pressure, K_0, were estimated from a numerical linear interpolation using MD simulated molar volumes at three different pressures, zero, $+P$, and $-P$ (P = 2~40 kb, depending on the magnitude of K_0). On the other hand, for liquids, V, K_0, and dK_0/dP were estimated by fitting the simulated P-V relations at P between 0 and 5 GPa to the third-order Birch-Murnaghan equation, and the volume thermal expansivities, α, were obtained by a linear fitting to V as a function of T between 1900 and 2700 K.

3. RESULTS AND DISCUSSION

3.1. *Crystal Simulations*

We studied 29 crystals in the NCMAS system, including simple oxides, a wide structural variety of silicates such as neso-, soro-, chain-, and tecto-silicates, and their high temperature and high pressure phases. The coordination numbers of the cations in the 29 crystals range from 4 to 6 for Al, 4 to 12 for Mg, 6 to 12 for Ca, 7 and 8 for Na, and 4 and 6 for Si. We have already shown [*Matsui*, 1996b] detailed MD results on the 27 crystals, except for jadeite and low albite, both of which are described below.

Jadeite crystals are monoclinic, space group $C2/c$, with

Si, Al, and Na ions in tetrahedral, octahedral, and 8-fold coordinations, respectively. Low albite crystals are triclinic, space group $C\bar{1}$, with both Si and Al ions in tetrahedral coordinations, and Na ions being surrounded by seven O ions at distances less than 3.0 Å [*Harlow and Brown*, 1980]. The MD basic cells are taken to be composed of 16 unit cells (2**a** × 2**b** × 4**c**, containing 640 atoms) for jadeite and 18 unit cells (3**a** × 2**b** × 3**c**, containing 936 atoms) for low albite. Table 2 gives the MD values for the structures and bulk moduli of jadeite and low albite at 300 K and 0 GPa, together with the observed values for comparison. The errors in the MD simulated cell lengths and cell angles are less than 1.3% and 1.5% respectively for jadeite, and 5.7% and 0.9% respectively for low albite. The MD values of the average nearest-neighbor cation-oxygen distances in jadeite and low albite also compare well with experiment, as shown in Table 2. There are four, three, and four non-equivalent nearest-neighbor Si-O, Al-O, and Na-O distances in jadeite, and twelve, four, and seven non-equivalent Si-O, Al-O, and Na-O distances in low albite. The maximum errors in the MD values in these individual nearest-neighbor Si-O, Al-O, and Na-O distances are 0.03Å (2% error), 0.09Å (4%), and 0.10Å (4%), respectively, in jadeite, and 0.04Å (2%), 0.02Å (1%), and 0.20Å (7%), respectively, in low albite. The calculated V differ by 4.0% for jadeite and 2.4% for low albite. The MD K_0 value of jadeite, 147 GPa, agrees with experiment within 2σ; however, that of low albite, 44 GPa, is much smaller, compared to the measured value of 70 GPa.

Table 3 gives a comparison of the observed and simulated V and K_0 values of the 29 crystals at 300 K and 0 GPa. The MD simulation reproduces well the available experimental data for the 29 crystals studied. The average error of the MD values in Table 3, which is defined as Σ|MD-Obs|/ ΣObs, is 2.2 % for V, and 7.3 % for K_0.

TABLE 2. Observed and Simulated Structures: Jadeite and Low Albite at 300 K, 0 GPa

	jadeite ($C2/c$) $NaAlSi_2O_6$		low albite ($C\bar{1}$) $NaAlSi_3O_8$	
	Obs[a]	MD	Obs[a]	MD
a/Å	9.42	9.47	8.14	7.70
b	8.56	8.67	12.79	13.06
c	5.22	5.29	7.16	7.31
α	90.0	90.0	94.2	94.4
β	107.6	106.0	116.6	117.6
γ	90.0	90.0	87.7	87.3
<Si(IV)-O>/ Å	1.62	1.63	1.61	1.63
<Al(IV)-O>			1.74	1.74
<Al(VI)-O>	1.93	1.97		
<Na(VII)-O>			2.63	2.62
<Na(VIII)-O>	2.47	2.52		
V/(cm^3/mol)	60.5	62.9	100.0	97.7
K/(10 GPa)	14.3(2)	14.7	7	4.4

[a] Structural data from *Cameron et al.* [1973] for jadeite, and from *Harlow and Brown* [1980] for low albite; bulk modulus from *Kandelin and Weidner* [1988] for jadeite with e.s.d. in parenthesis, and from *Angel et al.* [1988] for low albite.

3.2. *Silicate Liquid Simulations*

The five silicate liquids with the compositions enstatite, wollastonite, diopside, anorthite, and albite were studied. Of these, the four liquids except albite have been described in detail in a recent publication [*Matsui*, 1996a]. The MD basic cells are composed of 1080 (containing 216 formula units), 1080 (216), 1800 (180), 1625 (125), and 1664 (128) atoms for enstatite, wollastonite, diopside, anorthite, and albite liquids, respectively. We selected a reference temperature of 1900 K, which is above the liquidus temperature, T_l, of the five liquids; T_l = 1830, 1820, 1665, 1830, and 1373 K for enstatite, wollastonite, diopside, anorthite, and albite, respectively. Table 4 shows a comparison of the observed and MD values of V, K_0, dK_0/dP, and α for the five liquids at 1900 K and 0 GPa. In order to test the applicability of the third-order Birch-Murnaghan equation to silicate liquids at high temperature, we also calculated V, K_0, and dK_0/dP using the equation of state proposed by *Vinet et al.* [1987], who derived the equation by differentiating the cohesive energy of solids. For each liquid, the V, K_0, and dK_0/dP values using the *Vinet et al.* equation agree within their e.s.d.'s with the values from the third-order Birch-Murnaghan equation listed in Table 4.

As can be seen in Table 4, the MD calculated V compares well with experiment for each silicate liquid. The MD values for K_0 also agree within the scatter of the measured values [*Lange and Carmichael*, 1987, 1990; *Rivers and Carmichael*, 1987; *Rigden et al.*, 1989] for the five liquids. The MD simulated dK_0/dP values for both diopside and anorthite liquids again compare well with the observed values by *Rigden et al.* [1989]. The computed values for α for the five liquids are found to be all larger than the experimental data; however, it is to be noted that the experimental values are obtained indirectly from extrapolation using the partial molar volume data for component oxides at T between 1573 and 1873 K (see *Lange and Carmichael* [1987, 1990]). The bulk moduli are simulated to decrease substantially with T, with the predicted dK_0/dT values between 1900 K and 2700 K to be -0.012, -0.011, -0.011, and -0.009 GPa/K for enstatite, wollastonite, diopside, and anorthite liquids, respectively.

TABLE 3. Comparison[a] of the Observed and Simulated Molar Volumes and Bulk Moduli at 300 K and 0 GPa

	molar volume/(cm^3/mol)			bulk modulus/(10 GPa)		
Structure	Obs	MD	error	Obs	MD	error
Oxides						
MgO	11.2	10.8	-0.4	16.2(1)	18.0	1.8
CaO	16.7	16.6	-0.1	11.1(0)	11.1	0.0
Al_2O_3	25.5	25.3	-0.2	25.2(2)	25.8	0.6
$MgAl_2O_4$	39.7	39.5	-0.2	19.4(6)	20.2	0.8
SiO_2 system						
quartz	22.7	23.9	1.2	3.8(3)	3.8	0.0
cristobalite	25.9	25.6	-0.3	1.6(1)	2.2	0.6
coesite	20.6	21.8	1.2	9.6(3)	9.7	0.1
stishovite	14.0	14.9	0.9	30.6(4)	29.6	-1.0
$MgSiO_3$ system						
enstatite	31.3	31.7	0.4	10.8(3)	10.3	-0.5
clinoenstatite	31.2	31.4	0.2	11.1(3)	11.1	0.0
garnet	28.6	29.9	1.3	16.1(4)	14.6	-1.5
ilmenite	26.4	26.6	0.2	21.2(4)	22.4	1.2
perovskite	24.4	25.3	0.9	26.6(6)	25.3	-1.3
Mg_2SiO_4 system						
olivine	43.6	43.2	-0.4	12.7(1)	12.7	0.0
-spinel	40.5	40.6	0.1	17.4(3)	17.3	-0.1
spinel	39.5	40.0	0.5	18.4(2)	19.5	1.1
Ca(,Mg)-silicates						
wollastonite	39.8	40.9	1.1	-	4.1	-
perovskite	27.3	28.4	1.1	28.1(4)	25.8	-2.3
olivine	58.0	59.9	1.9	-	8.0	-
diopside	66.1	67.4	1.3	11.4(4)	10.2	-1.2
monticellite	51.5	52.2	0.7	11.3(3)	9.7	-1.6
silicates						
andalusite	51.6	52.7	1.1	16.2(2)	10.4	-5.8
sillimanite	50.0	50.1	0.1	17.1(2)	17.5	0.4
kyanite	44.2	44.3	0.1	18.4(6)	19.7	1.3
jadeite	60.5	62.9	2.4	14.3(2)	14.7	0.4
garnets and feldspars						
pyrope	113.2	117.4	4.2	16.9(1)	15.7	-1.2
grossular	125.1	128.5	3.4	16.7(1)	16.5	-0.2
anorthite	100.7	101.2	0.5	9.2	6.0	-2.8
low albite	100.0	97.7	-2.3	7	4.4	-2.6

[a] From *Matsui* [1996b] for the crystals, except jadeite and low albite. The observed bulk modulus of kyanite is from Inutsuka et al. [1995].
Estimated standard errors of the observed bulk moduli are given in parentheses when they are reported.

3.3. *Melting Curve of $MgSiO_3$ Perovskite at High Pressures*

The lower mantle is considered to be mainly composed of (Mg,Fe)SiO_3 perovskite, with up to ~20% (Mg,Fe)O magnesiowüstite and minor proportions of Ca and Al bearing minerals. Accurate knowledge of the melting curve of $MgSiO_3$ perovskite at lower mantle pressures is thus fundamental to investigating the differentiation processes in the early earth, the temperature distribution of the lower mantle, and the rheological and dynamical properties of the deep earth. The melting temperature of $MgSiO_3$ perovskite at lower pressures is relatively well constrained experimentally to be about 3000 K at 25 GPa [*Heinz and Jeanloz*, 1987; *Knittle and Jeanloz*, 1989; *Ito and Katsura*, 1992; *Zerr and Boehler*, 1993; *Sweeney and Heinz*, 1993]. However, there is much controversy about the dT_m/dP melting slope of this phase. Based on laser heated diamond anvil cell experiments, dT_m/dP of (Mg,Fe)SiO_3 perovskite is reported to be nearly zero over the P range between 22 and 65 GPa [*Heinz and Jeanloz*, 1987] or even slightly negative, -2.5 ± 0.6 K/GPa between 30 and 94 GPa [*Sweeney and Heinz*, 1993]. *Stixrude and Bukowinski* [1990] calculated the melting curve of $MgSiO_3$ perovskite, based on the estimated difference in Helmholtz free energy between the solid and liquid phases. They obtain that the melting curve possibly has a maximum under a lower mantle pressure, which is compatible with the experimental data *by Heinz and Jeanloz* [1987], and *Sweeney and Heinz* [1993]. The zero or negative dT_m/dP melting slope indicates that $MgSiO_3$ perovskite and liquid volumes converge or cross over under lower mantle conditions.

On the contrary, *Zerr and Boehler* [1993] measured the melting of (Mg,Fe)SiO_3 perovskite using a CO_2 laser heated diamond anvil cell with an argon pressure medium and reported that the melting temperature increases substantially with P, having the dT_m/dP slope of about 60 K/GPa at 22 GPa and of about 40 K/GPa at 60 GPa. Using a uniaxial split-sphere apparatus, *Ito and Katsura* [1992] found the melting slope of $MgSiO_3$ perovskite to be 30 ± 5 K/GPa in the P range 21-25 GPa. Based on the Clausius-Clapeyron equation, *Ohtani* [1983] estimated the dT_m/dP slope of 77 K/GPa at 23 GPa. He then used the Kraut-Kennedy or Simon equation to extrapolate the melting temperature at the core-mantle boundary (CMB) of 7500 ± 500 K. Using the Lindemann melting law, *Poirier* [1989] obtained the melting temperature at CMB of about 5000 K.

In order to investigate whether the volume inversion between $MgSiO_3$ perovskite and liquid occurs under lower mantle conditions, we applied the MD method to this system at P up to 150 GPa and T of 2500 K and 3500 K. We also used MD to calculate the volume compression at

TABLE 4. The Observed and Simulated Molar Volumes *V*, Bulk Moduli K_0, Pressure Derivatives of Bulk Moduli dK_0/dP, and Volume Thermal Expansivities α, of Enstatite, Wollastonite, Diopside, Anorthite, and Albite Liquids at 1900 K, 0 Gpa.

	$V/(cm^3/mol)$		K_0/GPa		dK_0/dP		$\alpha/(10^{-5}K^{-1})$	
liquids	Obs[a]	MD	Obs	MD	Obs[b]	MD	Obs[a]	MD
enstatite	38.9(2)	38.1(1)[c]	24(2)[a], 20[d]	24(2)	6.3(8)	7(2)	11.3(3)	
wollastonite	44.1(2)	45.4(1)	23(1)[a], 25[d]	18(1)	6.1(6)	7(2)	15.4(5)	
diopside	83.1(4)	83.6(2)	24(1)[a], 22[d], 23[b]	20(2)	6.9	6.1(9)	7(2)	13.5(3)
anorthite	108.7(4)	103.9(3)	21(1)[a], 18[d], 20[b]	22(2)	5.3	6.2(11)	5(1)	9.2(2)
albite	114.7(7)	108.0(3)	23(1)[a], 16[d]	15(1)		8.6(12)	4(1)	6.5(3)

[a]Extrapolated to 1900 K and 0 GPa, using partial molar data for component oxides at 1673 K, 0 GPa reported *by Lange and Carmichael* [1987; 1990].
[b]From *Rigden et al.* [1989]; values at 1773 K for diopside, and at 1923 K for anorthite.
[c]Estimated standard errors in parentheses refer to the last digits.
[d]From *Rivers and Carmichael* [1987]; K_0 obtained at 1913 K for enstatite, 1836 K for wollastonite, 1758 K for diopside, 1833 K for anorthite, and 1673 K for albite.

2500 K of high-clinoenstatite (space group *C2/c* recently found by *Angel et al.* [1992]) and $MgSiO_3$ garnet, for comparison. These two phases crystallize from liquid at *P* between 12 and 16 GPa (high-clinoenstatite) and between 16 and 22 GPa (garnet) [*Kato and Kumazawa*, 1985; *Presnall and Gasparik*, 1990].

The MD basic cells contained 1080 atoms (216 formula units) for liquid, 640 atoms (2**a** × 2**b** × 4**c**) for high-clinoenstatite, 1280 atoms ($2\mathbf{a}_1 \times 2\mathbf{a}_2 \times 2\mathbf{c}$) for garnet, and 960 atoms (4**a** × 4**b** × 3**c**) for perovskite. Figure 1(a) shows the MD simulated pressure dependence of *V* of high-clinoenstatite, $MgSiO_3$ garnet, perovskite, and liquid at 2500 K, and Figure 1(b) of $MgSiO_3$ perovskite and liquid at 3500 K. Each phase is simulated to preserve its structure metastably at the *T* and *P* conditions shown in Figures 1(a) and (b), ignoring surface effects [*Matsui and Price*, 1991] and the time needed for crystallization in the MD calculation. As Figures 1(a) and (b) show, at either *T* of 2500 or 3500 K, the volume difference between $MgSiO_3$ perovskite and liquid decreases substantially with *P*, but a volume inversion between the two phases never occurs at pressures up to 150 GPa. The possible occurrence of the

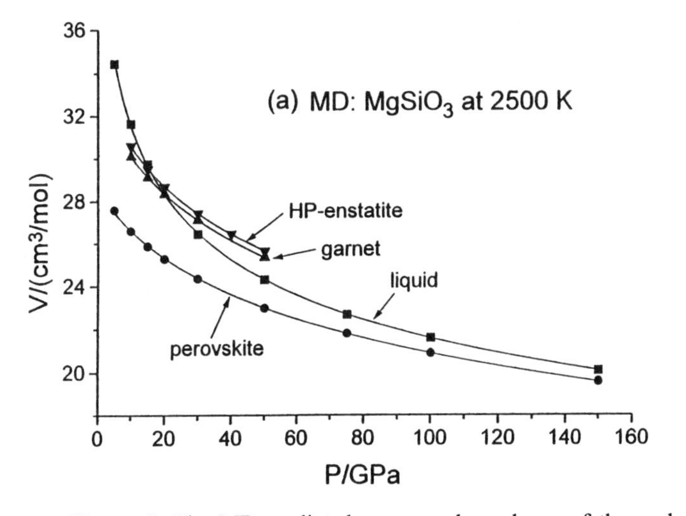

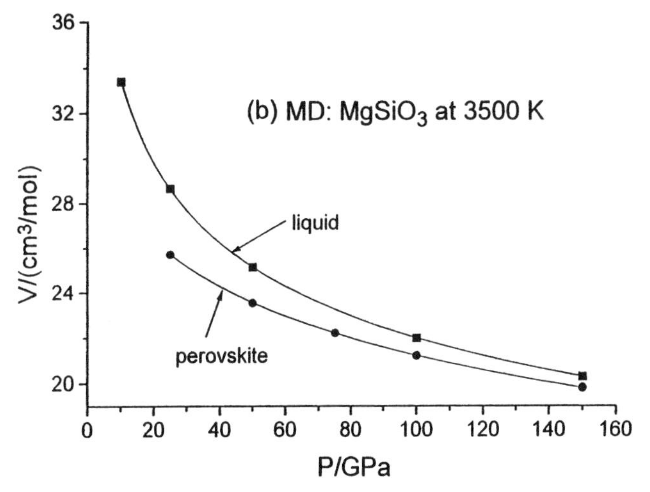

Figure 1. The MD-predicted pressure dependence of the molar volume of the crystals and liquid in the $MgSiO_3$ system at pressures up to 150 GPa. (a) High-clinoenstatite (HP-enstatite), garnet, perovskite and liquid at 2500 K. (b) perovskite and liquid at 3500 K.

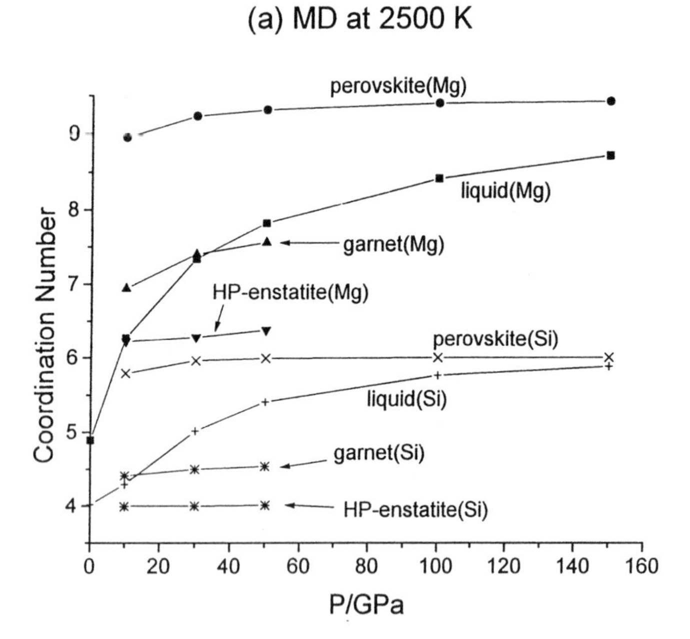

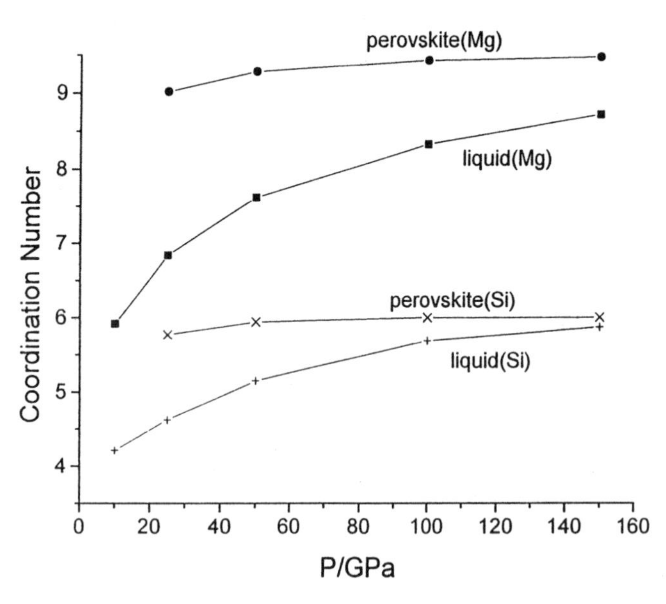

Figure 2. The MD-simulated coordination number of Mg and Si in the crystals and liquid in the $MgSiO_3$ system at pressures up to 150 GPa. The cutoff distances in the MD radial distribution functions are 2.90Å for Mg-O and 2.25Å for Si-O, following our previous work [*Matsui, 1996a*]. (a) High-clinoenstatite (HP-enstatite), garnet, perovskite and liquid at 2500 K. (b) perovskite and liquid at 3500 K. Ionic species are shown in parentheses.

volume inversion might be simply described from the difference in the cation coordination between solid and liquid. Thus we calculated the coordination numbers of Mg and Si in the crystal and liquid phases from the MD radial distribution functions. As can be seen in Figures 2(a) and (b), for either T of 2500 or 3500 K, the MD-calculated coordination number of either Mg or Si in the liquid is found never to become higher than that in perovskite over pressures studied. *Matsui and Price* [1991] simulated the melting behavior of $MgSiO_3$ perovskite using the MD technique, but with different pair potentials with those listed in Table 1. They found that, along the melting curve, V of the liquid is always greater than that of perovskite over the whole P range in the lower mantle, which is quite consistent with our present results. Thus we predict no volume cross-over or, in other words, no maximum in the perovskite melting curve under lower mantle conditions, unlike the experimental results of *Heinz and Jeanloz* [1987] and *Sweeney and Heinz* [1993] or the thermodynamic calculations of *Stixrude and Bukowinski* [1990]. Finally we note that at 2500 K, the MD values of V of high-clino-enstatite and garnet are both found to become larger than that of the liquid at P about 16 and 20 GPa, respectively, as shown in Figure 1(a). In each of high-clinoenstatite and garnet, as Figure 2(a) shows, the Mg and Si coordination numbers in the solid are both calculated to become smaller than those in the liquid at near the pressure where the volume inversion is predicted.

Acknowledgments. The author appreciate the comments of T. Suzuki, and an anonymous reviewer. This research is supported by a Grant-in-Aid for Scientific Research from the Ministry of Education, Science, Sports and Culture.

REFERENCES

Angel, R. J., A. Chopelas, and N. L. Ross, Stability of high-density clinoenstatite at upper-mantle pressures, *Nature*, *358*, 322-324, 1992.

Angel, R. J., R. M. Hazen, T. C. McCormick, C. T. Prewitt, and J. R. Smyth, Comparative compressibility of end-member feldspars, *Phys. Chem. Minerals*, *15*, 313-318, 1988.

Angell, C. A., P. A. Cheeseman, and S. Tamaddon, Pressure enhancement of ion mobilities in liquid silicates from computer simulation studies to 800 kbars, *Science*, *218*, 885-887, 1982.

Belonoshko, A. B., Molecular dynamics of silica at high pressures: equation of state, structure, and phase transitions, *Geochim. Cosmochim. Acta*, *58*, 1557-1566, 1994.

Cameron, M., S. Sueno, C. T. Prewitt, and J. J. Papike, High-temperature crystal chemistry of acmite, diopside, hedenbergite, jadeite, spodumene, and ureyite, *Am. Mineral.*, *58*, 594-618, 1973.

Harlow, G. E., and G. E. Brown, Jr, Low albite: an X-ray and neutron diffraction study, *Am. Mineral.*, *65*, 986-995, 1980.

Heinz, D. L., and R. Jeanloz, Measurement of the melting curve of $Mg_{0.9}Fe_{0.1}SiO_3$ at lower mantle conditions and its geophysical implications, *J. Geophys. Res.*, *92*, 11,437-11,444, 1987.

Inutsuka, S., M. Yamakata, T. Kondo, and T. Yagi, Isothermal compression curve of kyanite, Program and Abstracts, 36th High Pressure Conference of Japan, Tsukuba, 239, 1995.

Ito, E., and T. Katsura, Melting of ferromagnesian silicates under the lower mantle conditions, in *High-Pressure Research: Application to Earth and Planetary Sciences*, Geophysical Monograph 67, edited by Y. Syono and M. H. Manghnani, Terra Sci. Pub. Co., Tokyo/ AGU, Washington, D.C., pp. 315-322, 1992.

Kandelin, J., and D. J. Weidner, The single-crystal elastic properties of jadeite, *Phys. Earth Planet. Inter.*, *50*, 251-260, 1988.

Kapsta, B., and M. Guillop , High ionic diffusivity in the perovskite $MgSiO_3$: a molecular dynamics study, *Philos. Mag.*, *58*, 809-816, 1988.

Kato, T., and M. Kumazawa, Garnet phase of $MgSiO_3$ filling the pyroxene-ilmenite gap at very high temperature, *Nature*, *316*, 803-805, 1985.

Knittle, E., and R. Jeanloz, Melting curve of $(Mg,Fe)SiO_3$ perovskite to 96 GPa: evidence for a structural transition in lower mantle melts, *Geophys. Res. Lett.*, *16*, 421-424, 1989.

Kubicki, J. D. and A. C. Lasaga, Molecular dynamics simulations of pressure and temperature effects on $MgSiO_3$ and Mg_2SiO_4 melts and glasses, *Phys. Chem. Minerals*, 17, 661-673, 1991.

Lange, R. A. and I. S. E. Carmichael, Densities of Na_2O-K_2O-CaO-MgO-FeO-Fe_2O_3-$Al_2O_3TiO_2$-SiO_2 liquids: new measurements and derived partial molar properties, *Geochim. Cosmochim. Acta*, *51*, 2931-2946, 1987.

Lange, R. L. and I. S. E. Carmichael, Thermodynamic properties of silicate liquids with emphasis on density, thermal expansion and compressibility, in *Modern Methods of Igneous Petrology: Understanding Magmatic Processes*, edited by J. Nicholls and J. K. Russell., pp. 25-64, Mineralogical Soc. America, Washington, D.C., 1990.

Matsui, M., Molecular dynamics study of the structural and thermodynamic properties of MgO crystal with quantum correction, *J. Chem. Phys.*, *91*, 489-494, 1989.

Matsui, M., Molecular dynamics simulation of structures, bulk moduli, and volume thermal expansivities of silicate liquids in the system CaO-MgO-Al_2O_3-SiO_2, *Geophys. Res Lett.*, *23*, 395-398, 1996a.

Matsui, M., Molecular dynamics study of the structures and bulk moduli of crystals in the system CaO-MgO-Al_2O_3-SiO_2, *Phys. Chem. Minerals*, *23*, 345-353, 1996b.

Matsui, M., and G. D. Price, Simulation of the pre-melting behaviour of $MgSiO_3$ perovskite at high pressures and temperatures, *Nature*, *351*, 735-737, 1991.

Matsui, M., G. D. Price, and A. Patel, Comparison between the lattice dynamics and molecular dynamics methods: calculation results for $MgSiO_3$ perovskite, *Geophys. Res. Lett.*, *21*, 1659-1662, 1994.

Matsui, Y., and K. Kawamura, Instantaneous structure of an $MgSiO_3$ melt simulated by molecular dynamics, *Nature*, *285*, 648-649, 1980.

Matsui, Y., and K. Kawamura, Computer simulation of structures of silicate melts and glasses, in *Materials Science of Earth's Interior*, edited by I. Sunagawa, Terra Sci. Pub. Co., Tokyo, pp. 3-23, 1984.

Matsui, Y., and K. Kawamura, Computer-experimental synthesis of silica with the α-PbO_2 structure, *in High Pressure Research in Mineral Physics*, Geophysical Monograph 39, edited by M. H. Manghnani and Y. Syono, Terra Sci. Pub. Co., Tokyo/ AGU, Washington D.C., pp. 305-311, 1987.

Ohtani, E., Melting temperature distribution and fractionation in the lower mantle, *Phys. Earth Planet. Inter.*, 33, 12-25, 1983.

Poirier, J. P., Lindemann law and the melting temperature of perovskites, *Phys. Earth Planet. Inter.*, *54*, 364-369, 1989.

Presnall, D. C., and T. Gasparik, Melting of enstatite($MgSiO_3$) from 10 to 16.5 GPa and the forsterite (Mg_2SiO_4)-majorite ($MgSiO_3$) eutectic at 16.5 GPa: implications for the origin of the mantle, *J. Geophys. Res.*, *95*, 15771-15777, 1990.

Rigden, S. M., T. J. Ahrens, and E. M. Stolper, High-pressure equation of state of molten anorthite and diopside, *J. Geophys. Res.*, *94*, 9508-9522, 1989.

Rivers, M. L. and I. S. E. Carmichael, Ultrasonic studies of silicate melts, *J. Geophys. Res.*, *92*, 9247-9270, 1987.

Stein, D. J., and F. J. Spera, Molecular dynamics simulations of liquids and glasses in the system $NaAlSiO_4$-SiO_2: methodology and melt structures, *Am. Mineral.*, *80*, 417-431, 1995.

Stixrude, L., and M. S. T. Bukowinski, Fundamental thermodynamic relations and silicate melting with implications for the constitution of D", *J. Geophys. Res.*, *95*, 19311-19325, 1990.

Sweeney, J. S., and D. L. Heinz, Melting of iron-magnesium-silicate perovskite, *Geophys. Res. Lett.*, *20*, 855-858, 1993.

Tsuneyuki, S., M. Tsukada, H. Aoki, and Y. Matsui, First-principles interatomic potential of silica applied to molecular dynamics, *Phys. Rev. Lett.*, *61*, 869-872, 1988.

Vinet, P., J. Ferrante, J. H. Rose, and J. R. Smith, Compressibility of solids, *J. Geophys. Res.*, *92*, 9319-9325, 1987.

Wall, A., and G. D. Price, Electrical conductivity of the lower mantle: a molecular dynamics simulation of $MgSiO_3$ perovskite, *Phys. Earth Planet. Inter.*, *58*, 192-204, 1989.

Wasserman, E. A., D. A. Yuen, and J. R. Rustad, Molecular dynamics study of the transport properties of perovskite melts under high temperature and pressure conditions, *Earth Planet. Sci. Lett*, *114*, 373-384, 1993.

Wentzcovitch, R. M., J. L. Martins, and G. D. Price, Ab initio molecular dynamics with variable cell shape: application to $MgSiO_3$-perovskite, *Phys. Rev. Lett.*, *44*, 2-5, 1993.

Wentzcovitch, R. M., D. A. Hugh-Jones, R. J. Angel, and G. D. Price, Ab initio study of $MgSiO_3$ *C2/c* enstatite, *Phys. Chem. Minerals*, *22*, 453-460, 1995.

Winkler, B., and M. T. Dove, Thermodynamic properties of $MgSiO_3$ perovskite derived from large scale molecular dynamics simulations, *Phys. Chem. Minerals*, *18*, 407-415, 1992.

Woodcock, L. V., C. A. Angell, and P. Cheeseman, Molecular dynamics studies of the vitreous state: simple ionic systems and silica, *J. Chem. Phys.*, *65*, 1565-1577, 1976.

Zerr, A., and R. Boehler, Melting of $(Mg,Fe)SiO_3$-perovskite to 625 kilobars: indication of a high melting temperature in the lower mantle, *Science*, *262*, 553-555, 1993.

Masanori Matsui, Department of Earth and Planetary Sciences, Faculty of Science, Kyushu University, Hakozaki, Fukuoka 812-81, Japan.

First-Principles Studies on the Transformations of Graphite to Diamond

Y. Tateyama, T. Ogitsu, K. Kusakabe, and S. Tsuneyuki

Institute for Solid State Physics, University of Tokyo, Roppongi, Minato-ku, Tokyo, Japan

Constant-pressure first-principles molecular dynamics method is applied to study of the graphite-to-diamond transformations under pressure and to theoretical prediction of structural properties of BC_2N with diamond-like structures. The transition states of the transformations to cubic and hexagonal diamonds are investigated to clarify the difference of the mechanisms between them. It is shown that cubic diamond is formed with less activation energy than that for hexagonal diamond, provided that collective slide of graphite planes is not prohibited. As to BC_2N, equilibrium lattice parameters and bulk moduli are calculated for some hypothetical arrangements of atoms. Among them, an arrangement which contains more C-C and B-N bonds turns out to be the most stable and the hardest, in spite of breaking the Grimm-Sommerfeld rule.

1. INTRODUCTION

A constant-pressure first-principles molecular dynamics (FPMD) method [*Wentzcovitch et al.,* 1993] has recently attracted much attention for its wide applicability to phenomena under high pressure. In this method, electronic states, forces on atoms, and stress on a unit cell of the system are calculated quantum-mechanically and nonempirically. Moreover, a molecular dynamics simulation or a structure optimization under arbitrary pressure is realized using these forces and stress. Investigations with this method are expected to play a complementary role to experiments because it can describe the real system with sufficient accuracy. It is also possible, in principle, to predict structures and phenomena which have not been found in experiments. In addition, since external pressure can be arbitrarily set in the method, it enables us to examine phenomena in the ultra-high pressure region which is difficult to realize in experiments.

In studies of phenomena where both electronic states and structural properties are essential, this method is more efficient than other theoretical approaches such as the classical MD or the constant-volume FPMD. The classical MD enables us to perform a long time simulation for a large system, while electronic states are not treated directly in it. In the constant-volume FPMD, it is difficult to investigate phenomena with symmetry change of the unit cell under pressure. On the other hand, the constant-pressure FPMD can treat electronic states and structural properties, including cell shape change, together. Thus one can easily investigate structural transformations with change of electronic states and stability of unknown structures under arbitrary pressure with this method.

In this paper, we report two applications of this technique in high-pressure physics. One is the study on the pressure-induced structural transformations of graphite into diamonds. Here we discuss the difference of the

Properties of Earth and Planetary Materials
at High Pressure and Temperature
Geophysical Monograph 101

microscopic mechanisms between the transformation to cubic diamond and that to hexagonal diamond by comparing the intermediate paths and the activation barriers for the transformations. The other is the study of the stability and hardness of BC_2N with diamond-like structures. This material is expected to have hardness comparable to cubic diamond and zinc-blende boron nitride, although there are only a few reports of it being synthesized and the atomic arrangement has not been clarified yet. Here we investigate the most stable and the hardest polymorph of BC_2N among some hypothetical atomic arrangements. We also discuss the relation of the structural stability with atomic arrangements. The power of the constant-pressure FPMD method would be clearly demonstrated in these two topics.

2. METHOD

In our calculations, electronic states are optimized at each MD step by means of the conjugate-gradient method. The forces and stress are calculated with the optimized wave functions. Using these forces and stress, dynamical simulation of the atoms and the unit cell are performed according to a constant-pressure-type Lagrangian [*Wentzcovitch et al.,* 1993]. Calculations of electronic states are based on the density functional theory with local density approximation [*Kohn and Sham,* 1965; *Wigner,* 1938]. We use plane wave bases for expansion of the wave functions and norm-conserving soft pseudopotential for electron-ion interactions [*Troullier and Martins,* 1991]. In order to retain accuracy at different volumes, the constant energy-cutoff technique is used [*Bernasconi et al.,* 1995].

For investigation of transition states in structural transformations under arbitrary pressure, we use a new efficient method which we have developed on the basis of the constant-pressure FPMD. This method enables us to find a saddle point of the potential energy surface in the multidimensional configuration space automatically, even if the whole potential energy surface is unknown. Indeed, using this method, we perform a structure optimization with a fictitious force which is made by inverting the projection of the real force to a certain reference line in the configuration space. Thus the system always climbs up the potential surface in the direction parallel to the reference line, while it goes down the surface in the direction perpendicular to the line. When the system arrives at the state where all the forces vanish, the point is a saddle point of the potential surface at the pressure. The conditions for the reference line and the starting point for the saddle-point search is discussed elsewhere [*Tateyama et al.,* 1996].

3. TRANSFORMATIONS OF GRAPHITE TO DIAMOND UNDER PRESSURE

It is well known that graphite (Gr) transforms into cubic diamond (CD) by compression at high temperature. At room temperature, on the other hand, observations of the two kinds of graphite-to-diamond transformations are reported. One is the transformation to lonsdalite, namely, hexagonal diamond (HD) from kish graphite at 14 – 18 GPa under quasihydrostatic pressure [*Yagi et al.,* 1992]. In the experiment, the c-axis of HD was perpendicular to that of graphite. The other is the transformation to CD from polycrystalline graphite, which was observed by heating the sample under $\sim$ 20 GPa after static compression at room temperature [*Endo et al.,* 1994]. Although the differences in these two experiments might be attributed to the quality of the graphite sample, conditions and microscopic mechanisms for the diamond formation are still unclear. In this context, we investigate the intermediate paths and the activation barriers for these transformations.

There are calculations on the activation barriers of the transformations from rhombohedral Gr to CD [*Fahy et al.,* 1986] and from AA-stacking Gr to HD [*Fahy et al.,* 1987] using the constant-volume scheme. In those works, the path is fixed so that the c-axis is always parallel to that of the initial graphite during the transformations. This relation of the c-axis is different from that in the experiment for the Gr-HD transformation. In the present work, we investigate the transformations from hexagonal Gr to HD and CD allowing change of layer-stacking without any constraint on paths, where we consider the Gr-HD transformation with the orientation relationship corresponding to the experiment.

Here we consider the transformations without diffusion of atoms because those with breaking bonds are less probable at low temperature due to requiring high energy. To investigate such martensitic transformations, we adopt the monoclinic unit cell (common space group $P112/m$) containing 8 carbon atoms with periodic boundary condition. The unit cell in each phase is schematically displayed in Figure 1. The parameters to characterize the atomic geometry are also shown in Figure 1. The parameter u indicates the displacement of atoms perpendicular to the graphite planes in the internal coordinates of the unit cell, while R shown in Å is the distance between atoms in adjacent layers which are connected by formation of sp^3 bonding. For the electronic-structure calculations, 16 and 81 k points are used for structure optimizations and estimations of the barrier height, respectively. We employ 60 Ry for the cutoff energy. The external pressures we examined are 0 and 20 GPa which correspond to the pressures be-

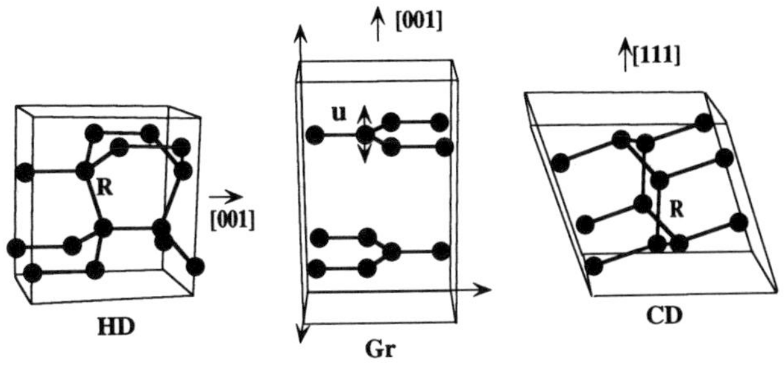

Figure 1. Hexagonal diamond, graphite, cubic diamond in the monoclinic unit cell used in this work are shown with the orientations in the conventional cells. The structural parameters u and R are also displayed.

low and above the critical ones for the transformations, respectively.

The structural parameters at the equilibrium structures of graphite and diamonds, and at the transition states obtained in this work are listed in Table 1. At the transition states, namely, saddle points of the potential energy surface, R has almost similar values (2.07 – 2.09 Å) regardless of the transformations or external pressures. This suggests that this critical R seems to be universal in the transformations between sp^2 and sp^3 bondings of carbon. Thus the saddle points might be determined by the distance between atoms combining the adjacent layers.

As to the transformation paths to the saddle points, there is little difference in the reduction of the interlayer distance. The c-axis is reduced from that of graphite similarly in both transformations (for example, 60% at 20 GPa). However, there exists difference in the lateral displacement of atoms. Indeed, we can clearly see a difference in the deformation angle of the unit cell, α. In the Gr-HD transformation, α remains 90°, while α in the Gr-CD transformation has already changed up to 106.5°, which is almost the same as that of CD. As to the out-of-layer displacement of atoms, u in the Gr-HD transformations changes over 50% from Gr to HD, while that in the Gr-CD transformations has values closer to graphite. These results indicate that the collective sliding of the layers, which affects the change of stacking, is significant in the Gr-CD transformation, while local change of atomic configurations, such as the buckling of planes, is more required in the Gr-HD because there is no need to change the stacking.

Next we describe the activation barriers in the transformations at the resulting saddle points. A schematic view of the potential surface at 0 and 20 GPa is shown in Figure 2 with the calculated values for the relative enthalpy to graphite, ΔH. At 0 GPa, the activation barriers from graphite are 0.347 (0.414) eV/atom for the transformations to CD (HD), while those at 20 GPa fall to 0.148 (0.222) eV/atom. Comparing two barriers, the activation barrier to CD is lower than that to HD by 0.067 eV/atom at 0 GPa and 0.074 eV/atom at 20 GPa.

From the potential surface shown in Figure 2, all the three phases can survive as metastable phases at ambient pressure owing to the existence of the high barriers between them at 0 GPa. Under high pressure, on the

TABLE 1. Structural Parameters for the Equilibrium Structures and at the Activation Barriers Between Them Under External Pressure P_{ext} = 0 and 20 GPa

	HD	HD-Gr	Gr	Gr-CD	CD
P_{ext} = 0 (GPa)					
c(Å)	4.348 (4.36)	4.895	6.759 (6.70)	4.966	4.369 (4.36)
α(∘)	90 (90)	90	90 (90)	107.83	109.42 (109.47)
u	0.0833 (1/12)	0.0482	0.0 (0)	-0.0295	-0.0624 (-1/16)
R(Å)	1.5422	2.073	3.453	2.085	1.545
P_{ext} = 20 (GPa)					
c(Å)	4.290	4.795	5.614	4.904	4.301
α(∘)	90	90	90	106.52	109.21
u	0.0834	0.0438	0.0	-0.0277	-0.0625
R(Å)	1.522	2.077	2.895	2.091	1.523

The experimental values are shown in brackets. Cell parameters c and α in this table are those in the monoclinic cell and the definitions of other structural parameters are explained in the text.

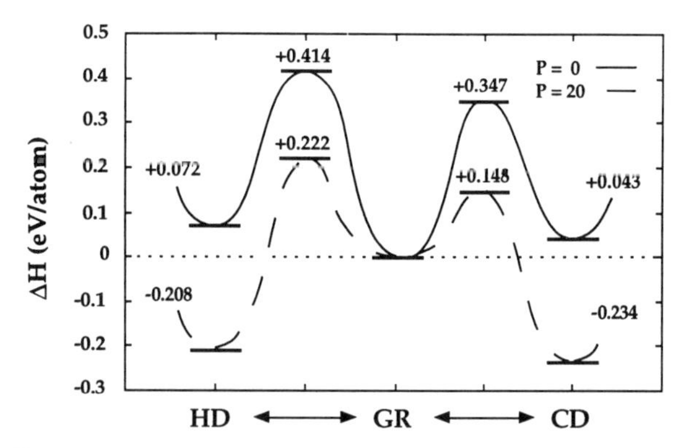

Figure 2. Schematic view of the potential energy surface at 0 GPa with solid line and at 20 GPa with broken line. The lines are guide for the eye. The relative enthalpy $H = E + P_{ext}\Omega$ (eV/atom) to that of graphite is shown for each pressure.

other hand, both barriers become sufficiently small for the system to transform at room temperature. By comparing these two transformations, it is confirmed that, at high pressure, the transformation to cubic diamond from graphite takes place under thermal equilibrium conditions. The present result also indicates that the transition probability to cubic diamond is higher than that to hexagonal diamond even under nonequilibrium condition from a view point of chemical kinetics because of the lower barrier for CD. Indeed, the transition probability to hexagonal diamond from graphite is roughly estimated to be 10% of that to cubic diamond at room temperature under 20 GPa.

Now we turn to the first question about the condition and the microscopic mechanisms for the Gr-HD and Gr-CD transformations. A possible explanation based on the present result is as follows. If a well-crystallized graphite sample of macroscopic size is compressed, the collective slide of graphite layers definitely required for the Gr-CD transformation would be prohibited, so that the Gr-HD transformation is expected to prevail. On the other hand, if the collective slide is allowed due to small size of the crystal, the Gr-CD transformation with smaller activation energy would take place. This picture will still hold even if we consider transformations initiated by local corrugation of the layers, which is likely to occur in real crystals but is neglected in the present calculations due to a small unit cell [*Tateyama et al.,* 1996].

4. STRUCTURAL STABILITY AND HARDNESS OF BC_2N

Compounds of boron, carbon, and nitrogen with tetrahedral coordinations are expected to be very hard because of the similarity of structures to cubic diamond and zinc-blende boron nitride (z-BN). However, there are yet only a few reports on the synthesis of them in experiments. In this work, we have focused on BC_2N with diamond-like structures among them and investigated the structural stability and the hardness for three different atomic arrangements of BC_2N. Furthermore, we discuss the relation between the stability and the arrangements.

We have examined three atomic arrangements, *i.e.* layered (space group $P\bar{4}m2$), chalcopyrite (space group $I\bar{4}2m$), and orthorhombic (space group Pmm2) structures, which are shown in Figure 3. The last structure is chosen because it can be made martensitically from a graphitic structure of BC_2N whose layers have the atomic arrangement previously suggested by the theoretical works on BC_2N monolayer [*Liu et al.,* 1989; *Nozaki and Itoh,* 1996]. We have performed structure optimizations for these structures under ambient pressure, where the ideal diamond configuration in a tetragonal unit cell is used as the starting one. In these calculations, we used 48 Ry for cutoff energy and 16 points for k-point sampling. The unit cell in this work contains 16 atoms.

Total energies, bulk moduli, and some structural parameters obtained in this work are listed in Table 2. Among the three arrangements, we have found that the orthorhombic one is the most stable and the hardest structure. The optimized structure of this arrangement has large deformation of tetrahedra as is known from the bond-length and the bond-angle. Furthermore, the angle γ in the present unit cell changes from 90° . The bulk modulus of it turns out to be 397 ± 10 GPa, which is similar to that of z-BN (414 ± 15 GPa) calculated using the same condition. This hard structure has the possibility of being synthesized in practice, because of its simple structural relation with the graphitic BC_2N.

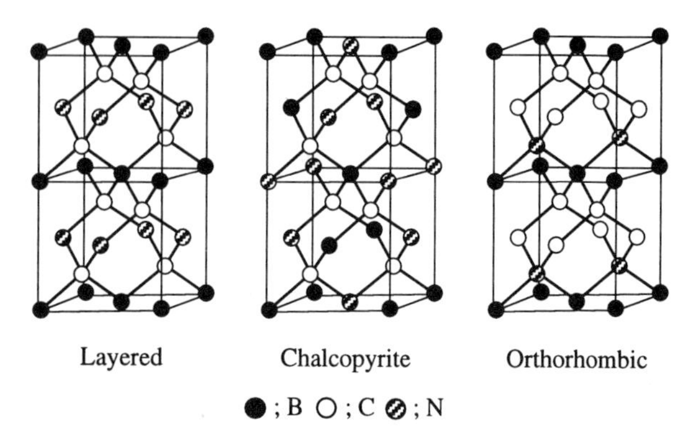

Figure 3. Layered, chalcopyrite, and orthorhombic structures of BC_2N in the tetragonal unit cell with 16 atoms used in this work.

TABLE 2. Calculated Total Energies, Bulk Moduli, and Structural Parameters of Layered, Chalcopyrite, and Orthorhombic Structures of BC_2N

	Layered	Chalcopyrite	Orthorhombic
E_{tot} (eV/atom)	-163.97	-164.05	-164.42
B_0 (GPa)	323 ± 40	317 ± 25	397 ± 10
γ(∘)	90	90	90.35
bond length (Å)	1.574 (B-C) 1.568 (C-N)	1.599 (B-C) 1.539 (C-N)	1.580 (B-C) 1.570 (C-N) 1.571 (B-N) 1.527 (C-C)
bond angle (°)	109.1~109.7	106.0 ~ 112.2	108.2 ~ 110.1

The angle γ denotes the value in the present unit cell.

From these results, we have found that the stability and the hardness of BC_2N with diamond-like structures have a relation with the sum of bonding energy of all bonds. Since the layered and chalcopyrite structures have the same number of C-B and C-N bonds, the energy and bulk modulus are similar in both structures. On the other hand, the orthorhombic structure is more stable than the others due to existence of C-C and B-N bonds. This is because C-C and B-N bonds have greater bonding energy than C-B and C-N bonds. These results indicate that the arrangements which have more C-C and B-N bonds would be more stable. Thus, atomic arrangements with separate domains or phases of diamond and z-BN would be more stable and favored. This trend on the structural stability is also seen in graphitic structures of BC_2N.

For ternary compounds with tetrahedral coordination, in general, there is the Grimm-Sommerfeld rule stating that only atomic arrangements ensuring local charge neutrality are allowed. However, the orthorhombic structure which is the most stable here violates this rule. This suggests that the energy loss from the violation of the rule can be compensated by the large lattice distortion with charge transfer.

5. SUMMARY

We have demonstrated the efficiency of the constant-pressure first-principles molecular dynamics method for investigations of structural transformation under pressure and for explorations of new materials with unknown structures. In particular, this method is suitable for phenomena under pressure where electronic states as well as structural properties are variable, because calculations with the method are performed quantum-mechanically and nonempirically at arbitrary pressure. Although it is still difficult to treat a large system due to huge computational requirements, with progress of algorithms and performance of computers, this method might play a more important role for high pressure physics.

Acknowledgments. We are grateful to Wataru Utsumi for useful discussions and to Satoshi Itoh for stimulative suggestion. We also thank Kazuaki Kobayashi for providing us with pseudopotential data. The computation in this work has been done using the facilities of the Supercomputer Center, Institute for Solid State Physics, University of Tokyo.

REFERENCES

Bernasconi, M., G. L. Chiarotti, P. Focher, S. Scandolo, E. Tosatti, and M. Parrinello, First-principle constant-pressure molecular dynamics, *J. Phys. Chem. Solids, 56,* 501, 1995.

Endo, S., N. Idani, R. Oshima, K. Takano, and M. Wakatsuki, X-ray diffraction and transmission-electron microscopy of natural polycrystalline graphite recovered from high pressure, *Phys. Rev. B, 49,* 22, 1994.

Fahy, S., S.G. Louie, and M. L. Cohen, Pseudopotential total-energy study of the transition from rhombohedral graphite to diamond, *Phys. Rev. B, 34,* 1191, 1986.

Fahy, S., S.G. Louie, and M. L. Cohen, Theoretical total-energy study of the transformation of graphite into hexagonal diamond, *Phys. Rev. B, 35,* 7623, 1987.

Kohn, W., and L. J. Sham, Self-consistent equations including exchange and correlation effects, *Phys. Rev., 140,* A1133, 1965.

Liu, A. Y., R. M. Wentzcovitch, and M. L. Cohen, Atomic arrangement and electronic structure of BC_2N, *Phys. Rev. B, 39,* 1760, 1989.

Nozaki H., and S. Itoh, Structural stability of BC_2N, *J. Phys. Chem. Solids, 57,* 41, 1996.

Tateyama,Y., T. Ogitsu, K. Kusakabe, and S. Tsuneyuki, Constant-pressure first-principle studies on the transition states of the graphite-diamond transformations under pressure, *Phys. Rev. B, 54,* 14994, 1996.

Troullier, N., and J. L. Martins, Efficient pseudopotentials for plane-wave calculations, *Phys. Rev. B, 42,* 1993, 1991.

Wentzcovitch, R. M., J. L. Martins, and G. D. Price, Ab initio molecular dynamics with variable cell shape: Application to $MgSiO_3$, *Phys. Rev. Lett, 70,* 3947, 1993.

Wigner, E. P., Effects of the electron interaction on the energy levels of electron in metals, *Trans. Faraday. Soc., 34,* 678, 1938.

Yagi, T., W. Utsumi, M. Yamakata, T. Kikegawa, and O. Shimomura, High-pressure *in situ* x-ray-diffraction study of the phase transformation from graphite to hexagonal diamond at room temperature, *Phys. Rev. B, 46,* 6031, 1992.

Y. Tateyama, T. Ogitsu, K. Kusakabe, and S. Tsuneyuki, Institute for Solid State Physics, University of Tokyo, 7-22-1 Roppongi, Minato-ku, Tokyo 106, Japan.

First-Principles Investigations of Solid Iron at High Pressure and Implications for the Earth's Inner Core

Lars Stixrude and Evgeny Wasserman[1]

School of Earth and Atmospheric Sciences, Georgia Institute of Technology

Ronald E. Cohen

Geophysical Laboratory and Center for High Pressure Research, Carnegie Institution of Washington

Density functional theory is used to investigate the equation of state, phase stability, magnetism, elasticity, and high temperature properties of iron. The equations of density functional theory are solved with the state-of-the-art linearized augmented plane wave (LAPW) method with gradient-corrected exchange-correlation functionals. A parametric tight-binding Hamiltonian, which is fit to the results of the elaborate LAPW calculations, is used to investigate elasticity and high temperature properties, the latter in the cell model approximation. The results of these calculations show that (1) the bcc phase of iron is mechanically unstable above 150 GPa and is unlikely to exist in the inner core (2) the anisotropy of hcp iron is similar in its magnitude and symmetry to that of the inner core, indicating that this is the stable phase of iron in this region and that the inner core is either very strongly textured, or that it consists of a single crystal, and (3) iron is significantly denser than the inner core, indicating the presence of a few weight percent light alloying elements in the solid inner core.

INTRODUCTION

The inner core is a relatively small portion of the Earth's interior, with a mass comparable to the Earth's moon. Despite its small size, it has been the subject of increasing attention over the past decade. The inner core is now known to be anisotropic [*Morelli et al.*, 1986; *Woodhouse et al.*, 1986; *Tromp*, 1993], a surprising and unpredicted observation which reveals the structure of this most remote region of the Earth's interior in unprecedented detail and promises to shed new light on the origin and evolution of the core. With this discovery come new questions relating to the origin of the anisotropy, which is still unknown, the dynamics of the inner core [*Jeanloz and Wenk*, 1986; *Song and Richards*, 1996], its growth, and its interaction with the magnetic field [*Karato*, 1993].

The effect of the inner core on the magnetic field has also received new attention. Recent numerical models have shown that the inner core can have a major effect on the typical length scale of lateral variability in the field, its time variability and on its reversal frequency [*Hollerbach and Jones*, 1993]. Conversely, the magnetic field may exert torques on the inner core which cause it to rotate faster than the mantle [*Glatzmaier and Roberts*, 1995].

These new results in seismology and geomagnetism,

---

[1]Now at Battelle, Pacific Northwest National Laboratory, Richland, Washington

Properties of Earth and Planetary Materials
at High Pressure and Temperature
Geophysical Monograph 101

as well as long-standing questions of the composition and temperature of the inner core, focus our attention on the physics of its most abundant constituent at the extreme pressures and temperature that prevail. The fact that the pressures and temperature exceed those that are readily accessed in the laboratory means that our understanding of the physics of iron in the core is still limited. For example, the relative stability of the different phases of iron at inner core conditions, and therefore the crystalline structure of the inner core, is unknown. The elastic properties of iron are unknown at high pressure, limiting our ability to interpret seismological observations of the anisotropic, or even isotropic, structure of the inner core. The electromagnetic properties of iron such as the conductivity, or magnetic susceptibility, that govern the influence of the inner core on the field, are unknown beyond zero pressure.

First principles theory is an ideal complement to experimental approaches in this context. First principles approaches, such as density functional theory (DFT) solve the fundamental quantum mechanical equations that govern the behavior of matter with a minimum of approximations. They are completely independent of experiment, and yet are now capable of reproducing many experimental measurements. There is no difficulty in applying these methods at high pressure - calculations at high pressure are in fact marginally faster, in terms of computational load, primarily because the charge density tends to become more uniform under compression. Finally, these methods permit access to the fundamental physical mechanisms that underly bulk behavior.

We review here the contributions of the density functional theory to our understanding of iron at high pressure and of the inner core. We begin with an overview of computational methods, including the state-of-the-art linearized augmented plane wave (LAPW) method and the parametric tight binding model that is derived from the LAPW calculations [*Cohen et al.*, 1994a]. We also discuss methods that have been developed to treat the effect of temperature on the properties of iron. We then discuss applications of the LAPW method to perfect lattices of the observed phases of iron: body-centered cubic (bcc), face-centered cubic (fcc), and hexagonal close-packed (hcp), including their relative stability and their equations of state. The LAPW calculations of the bcc phase also illustrate the effect of pressure on the magnetism of iron. Predictions of the elastic constants of fcc and hcp phases allow us to determine the anisotropy in elastic wave velocities in iron and to compare with the anisotropy of the inner core. Finally, the cell model is used to determine the equation of state of iron up to core temperatures. The implications of these results for the crystalline structure of the inner core, the origin and nature of its anisotropy, its magnetic state, its composition, and temperature are discussed.

COMPUTATIONAL METHODS

Band Structure and Total Energy

All the computational methods discussed here are ultimately based on density functional theory [*Hohenberg and Kohn*, 1964; *Kohn and Sham*, 1965]. The essence of this theory is the proof that the ground state properties of a material, including its ground state total energy

$$E = T + U\left[\rho(\vec{r})\right] + E_{xc}\left[\rho(\vec{r})\right] \tag{1}$$

are a unique functional of the charge density $\rho(\vec{r})$. T is the kinetic energy of a system of non-interacting electrons with the same charge density as the interacting system, U is the electrostatic (Coulomb) energy, including the electrostatic interaction between the nuclei, and E_{xc} is the exchange-correlation energy. A variational principle leads to a set of single-particle, Schrödinger-like, Kohn-Sham equations, with an effective potential given by

$$V_{KS} = V_{e-n}\left[\rho(\vec{r})\right] + V_{e-e}\left[\rho(\vec{r})\right] + V_{xc}\left[\rho(\vec{r})\right] \tag{2}$$

where the first two terms are Coulomb potentials due to the nuclei and the other electrons, and the last is the exchange-correlation potential. The power of density functional theory is that it allows one to calculate, in principle, the exact many-body total energy of a system from a set of single-particle equations. In practice, this is not possible because the exact exchange-correlation functional E_{xc} is unknown. Fortunately, simple approximations to this potential have been very successful. The success of the local density approximation (LDA), which replaces V_{xc} at every point in the crystal by the accurately known exchange-correlation potential for a homogeneous electron gas of the same local charge density, can be understood in terms of the satisfaction of exact sum rules for the exchange-correlation hole [*Gunnarson and Lundqvist*, 1976]. The LDA has been shown to yield excellent agreement with experiment for a wide variety of insulators, metals, and semiconductors, but fails to predict the correct ground state for iron. More recently proposed functionals, such as the Generalized Gradient Approximation (GGA) [*Perdew and Wang*, 1992], which uses information about local charge density gradients as well as the local density, yield much improved results for 3d transition metals [*Bagno et al.*, 1989; *Leung et al.*, 1992].

The solution to the Kohn-Sham equations for the single-particle (quasi-electronic) eigenvalues, $\epsilon(\vec{k})$, and eigenvectors, $\Psi(\vec{k})$, where $\vec{k}$ is a vector in reciprocal space, is that of the set of coupled generalized eigenvalue equations (units such that $\hbar^2/2m$=1)

$$H_{ij}\Psi_j = \epsilon O_{ij}\Psi_j \quad (3)$$

$$H_{ij}(\vec{k}) = \int \Psi_i^* \left(-\nabla^2 + V_{KS}\right) \Psi_j d\vec{r} \quad (4)$$

$$O_{ij}(\vec{k}) = \int \Psi_i^* \Psi_j d\vec{r} \quad (5)$$

where **H** and **O** are the Hamiltonian and overlap matrices, respectively. Because the Kohn-Sham potential is a functional of ρ, the eigenvalue equations must be solved self-consistently with the definition of the charge density in terms of the wavefunctions. First principles methods solve Equations (3-5) by expanding the wavefunctions and the potential in a complete and computationally convenient basis, $\Psi_i = c_{ij}\phi_j$.

The choice of basis functions in the LAPW method [*Wei and Krakauer*, 1985] explicitly treats the first-order partitioning of space into near-nucleus regions, where the charge density and its spatial variability are large, and interstitial regions, where the charge density varies more slowly. A dual-basis set is constructed, consisting of plane-waves in the interstitial regions that are matched continuously to more rapidly varying functions inside spheres centered about each nucleus. The advantages inherent in the LAPW method - no approximations to the shape of the charge density, potential, or to the nature of bonding - are expected to be particularly important in high pressure studies where qualitative changes in the nature of the electronic structure, such as insulator-metal transitions, changes in valence state, and coordination, may be induced by large compressions. The method is equally applicable to essentially all the elements of the periodic table, to metals, insulators, semi-conductors, and magnetic materials. The only essential limitations of the method are those inherent in the LDA or GGA approximations.

There are two major differences between tight binding methods, originally formulated by *Slater and Koster* [1954], and the first principles approaches discussed so far. First, the basis functions are chosen to be centered on the nuclei. For basis functions $\phi_{i\alpha}(\vec{r} - \vec{R}_i)$, where α labels the type of orbital (e.g., s, p, d, ...), and i labels the atom, the Hamiltonian matrix then consists of elements

$$H_{i\alpha j\beta}(\vec{k}) = \sum_{l=0}^{\infty} \exp(i\vec{k} \cdot \vec{R}_{ij}(l)) S_{\alpha\beta}(\hat{R}_{ij}(l)) h_{\alpha\beta}(R_{ij}i(l)) \quad (6)$$

$$O_{i\alpha j\beta}(\vec{k}) = \sum_{l=0}^{\infty} \exp(i\vec{k} \cdot \vec{R}_{ij}(l)) S_{\alpha\beta}(\hat{R}_{ij}(l)) o_{\alpha\beta}(R_{ij}i(l)) \quad (7)$$

where $R_{ij}(l)$ is the distance between the i-th atom in the reference unit cell (labeled $l = 0$) and the j-th atom in the l-th unit cell, the $S_{\alpha\beta}$ are functions of direction only and, in the two-center approximation, the $h_{\alpha\beta}$ and $o_{\alpha\beta}$ are functions only of internuclear distance. Indices i and j run over all atoms in the unit cell, and l runs over all unit cells. Under the assumption that the basis set consists of functions with the symmetry of s, p, d, ... atomic orbitals, the functions $S_{\alpha\beta}$ can be written in terms of spherical harmonics. The distance dependent functions, $h_{\alpha\beta}$ and $o_{\alpha\beta}$, are taken to be parametric functions of distance, with parameters chosen such that first principles results are reproduced. In this way, all explicit reference to the wavefunctions or charge density is eliminated. This simplifies the calculations tremendously, but renders the calculation non-self-consistent.

The non self-consistency of the tight binding approach has a very important consequence which has not been widely recognized [*Cohen et al.*, 1994a]. In general, the total energy can be written

$$E = \sum_i \int \epsilon_i(\vec{k}) d\vec{k} + F\left[\rho(\vec{r})\right] \quad (8)$$

where the first term is a sum over the self-consistent eigenvalues, and the second term, a functional of the charge density, contains all non-band structure contributions to the energy. The band structure now contains an arbitrary zero which must be fixed in order to calculate the total energy. The arbitrariness of the energy zero in the tight binding method can be exploited to recast the total energy as

$$E = \sum_i \int \epsilon_i'(\vec{k}) d\vec{k} \quad (9)$$

where the new eigenvalues are shifted in energy such that $\epsilon_i' = \epsilon_i - F\left[\rho(\vec{r})\right]$. With this formulation, the total energy is given simply as a sum over the bands, eliminating the need for pair potential repulsive terms which are often included in other treatments. The parameters of the tight binding model are determined by fitting to accurate LAPW band structures and total energies. This approach has been very successful in describing the properties of monatomic systems, such as iron [*Cohen et al.*, 1994b].

For a given arrangement of nuclei (crystal structure) the LAPW or tight binding total energy methods allow one to determine the total energy and charge density, and the quasi-particle eigenvalue spectrum (electronic band structure). By examining the dependence of the total energy on perturbations to the volume V or shape of the crystal (described by the deviatoric strain tensor, ϵ'_{ij}), or to the positions of the atoms, the Helmholtz free energy as a function of V, ϵ'_{ij}, and T can in principle be deduced. Assuming that the deviatoric strain is small and that the stress is hydrostatic [*Wallace*, 1972]

$$F(V, \epsilon'_{ij}, T) =$$

$$F_0(V) + F_{TH}(V,T) + \frac{1}{2} C_{ijkl}(V,T) \epsilon'_{ij} \epsilon'_{kl} \qquad (10)$$

where F_0 is the static (zero temperature) contribution, F_{TH} is due to the thermal excitation of electrons and phonons, and C_{ijkl} is the elastic constant tensor. This equation shows that the difference in Helmholtz free energy between a strained and unstrained lattice is in general related to a combination of elastic constants. We have made use of this to determine the full elastic constant tensor by applying a minimal set of high symmetry, volume conserving strains [*Mehl et al.*, 1990; *Stixrude and Cohen*, 1995a].

Thermal Properties

The temperature in the core far exceeds its Debye temperature so that the phonon spectrum is fully populated. In this limit, we can write F_{TH} in (10) as

$$F_{TH}(V,T) = F_{el}(V,T) - kT \ln Z \qquad (11)$$

where F_{el} is due to the thermal excitation of electrons and Z is the (classical) statistical mechanical partition function associated with atomic vibrations in the canonical ensemble

$$Z = \lambda^{-3N} \int \exp\left(-\beta(U(\vec{r}^{\,N}))\right) d\vec{r}^{\,N} \qquad (12)$$

where $\lambda = h/(2\pi m k_B T)^{1/2}$ is the de Broglie wavelength of the atoms, k_B is Boltzmann's constant, h is Planck's constant, $\beta = (k_B T)^{-1}$, U is the total energy, and $\vec{r}^{\,N}$ indicates integration over the Cartesian coordinates of N atoms ($N \approx 10^{23}$). Evaluation of this quantity requires an integration of a functional of the total energy over all vibrational degrees of freedom, an impossible task using a purely first principles approach.

In order to treat the vibrational degrees of freedom, we make use of a mean-field approximation known as the cell model [*Holt et al.*, 1970; *Ree and Holt*, 1973; *Cowley et al.*, 1990]. The motivation for using the cell model is that it accounts for anharmonicity and is computationally efficient, permitting precise determinations of all thermodynamic properties including free energies. Alternative methods for evaluating the vibrational partition function either ignore anharmonicity (e.g., quasi-harmonic lattice dynamics), or, as in the case of molecular dynamics, make fewer approximations, but are tens of thousands of times less efficient computationally.

The essential feature of the cell model is that the vibration of an atom is assumed to be uncorrelated with that of its neighbors, an approximation which is expected to be good at high temperature. In this limit, the partition function factorizes

$$Z_{cell} = \lambda^{-3N} \left[\int_{\Delta} \exp\left(-\beta(U(\vec{r}) - U_0)\right) d\vec{r} \right]^N , \qquad (13)$$

where the integral is now over the coordinates of a single atom within its Wigner Seitz cell Δ. Here U_0 is the potential energy of the system with all atoms on ideal lattice sites, $U(\vec{r})$ is the potential energy of the system with the wanderer atom displaced by the radius-vector $\vec{r}$ from its equilibrium position, and N is the total number of atoms in the system.

The pressure due to the thermal excitation of electrons is readily evaluated and is calculated with the same accuracy as the static pressure. This contribution is determined by performing self-consistent high temperature density functional calculations [*Mermin*, 1965; *McMahan and Ross*, 1977]. We have shown [*Wasserman et al.*, 1996] that this term is well approximated by a rigid band picture in which the band structure is assumed to be independent of temperature. We have assessed the accuracy of this assumption by comparing with fully self-consistent high temperature LAPW calculations. The temperature dependence of the band structure - which arises because the charge density, and therefore the potential change as higher lying states are populated - has a negligible effect on bulk thermodynamic properties.

RESULTS

Total Energy of Perfect Lattices

Calculations of the total energy of bcc, fcc, and hcp phases of iron over a range of volumes that span the pressure regime of the Earth's interior, show that density functional calculations are capable of describing the compression, phase stability, and magnetism of iron. They also clearly show the advantages of the GGA approximation to the exchange-correlation functional over the older LDA approximation [*Stixrude et al.*, 1994].

Both GGA and LDA correctly predict that hcp iron is the equilibrium low-temperature structure at high pressure (Figure 1). Moreover, GGA correctly predicts that

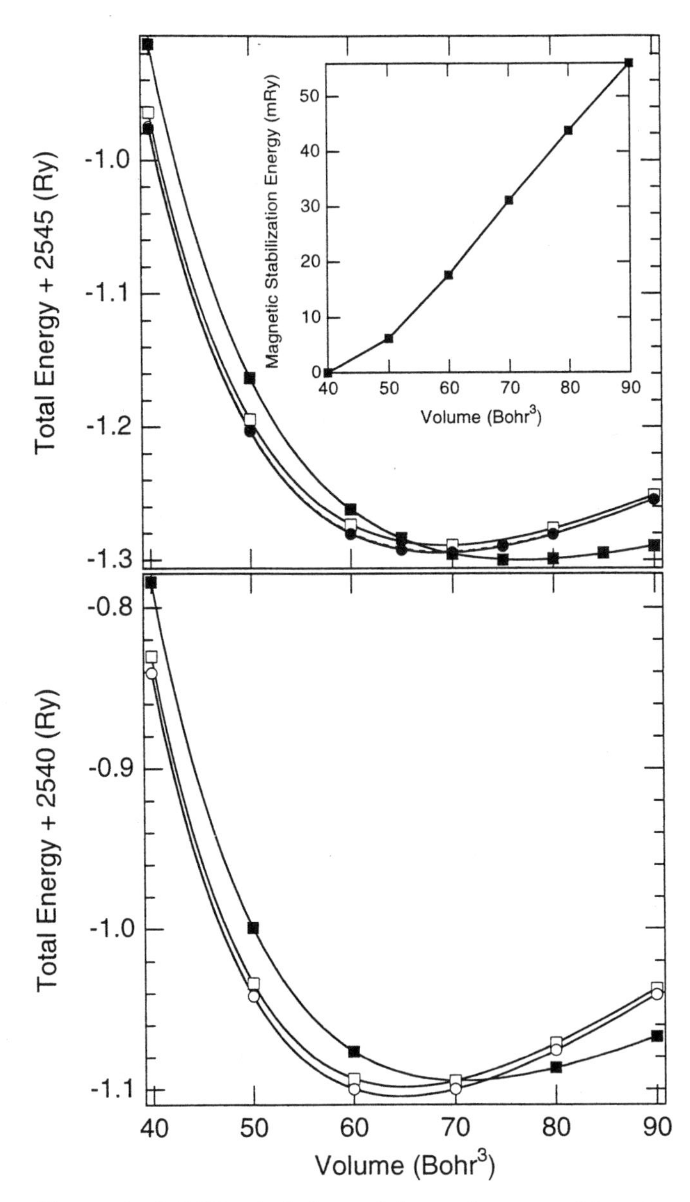

Figure 1. LAPW calculations of the total energy of iron in GGA (top) and LDA (bottom) approximations. Results for ferromagnetic bcc (filled squares), non-magnetic fcc (open squares), and hcp (circles) are shown. Results for both ideal (open circles) and minimum energy (filled circles) *c/a* ratios are shown for the hcp phase. The inset shows the magnetic stabilization energy of bcc: the difference in total energy between non-magnetic and ferromagnetic states. From *Stixrude et al.* [1994].

ferromagnetic bcc is the ground state structure, stable at zero pressure, as first shown by *Bagno et al.* [1989]. The incorporation of gradient corrections (the GGA functional) thus corrects a well known deficiency of the LDA, that it does not recover the correct ground state. We used the common tangent construction to determine the phase transition pressure from bcc to hcp predicted by the GGA calculations (Figure 2). The result, 11 GPa, is in excellent agreement with experimental observations and other density functional calculations [*Asada and Terakura*, 1992; *Söderlind et al.*, 1996]. This is an important illustration of the accuracy of density functional theory and the GGA - phase transition pressures are among the most challenging of experimental observations to reproduce theoretically because they depend on small differences between large numbers. For example, to accurately reproduce the bcc to hcp total energy difference at zero pressure requires one to calculate the total energy correctly to 8 significant figures.

In order to determine the equation of state of the three known phases of iron, we fit our total energy results to a *Birch* [1952] Eulerian finite strain expansion (Figure 3). Expansions to third order in finite strain were found to yield excellent fits to the total energies over the more than twofold compression range of our study (rms misfits were slightly less than the precision of the calculations, i.e., 0.1 mRy). The equation of state parameters determined from our GGA calculations [*Stixrude et al.*, 1994] show good agreement with the experimentally measured equation of state of bcc iron. The zero pressure volume and bulk modulus of

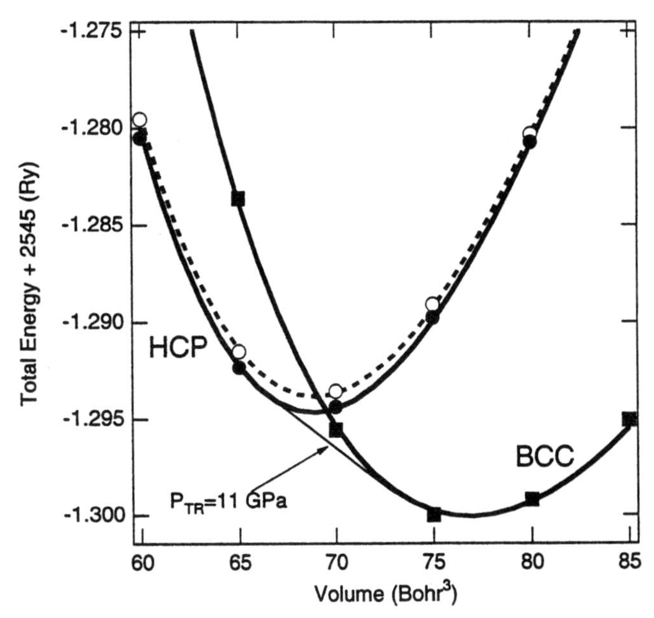

Figure 2. Detail of Figure 1 (top) showing the region of the bcc to hcp transition in the GGA approximation. Total energies of ferromagnetic bcc (closed squares), hcp with ideal *c/a* ratio (open circles and dashed line), and hcp with minimum energy *c/a* ratio (closed circles with solid line) are shown. The short thin line is the common tangent to bcc and hcp curves and represents the phase transition pressure (P_{TR}) between the two phases.

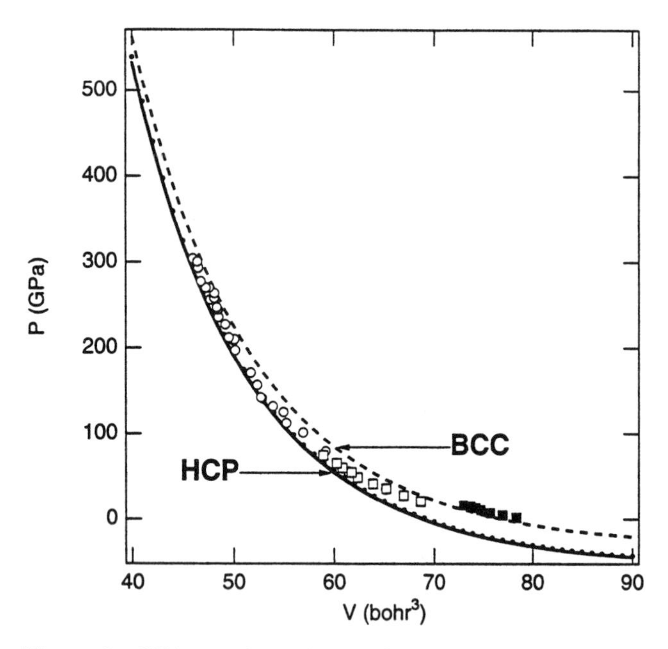

Figure 3. GGA equations of state of bcc (dashed), fcc (dotted), and hcp (solid) phases of iron compared with the experimental data of *Jephcoat et al.* [1986] (bcc, filled squares; hcp, open squares) and *Mao et al.* [1990] (hcp, open circles). From *Stixrude et al.* [1994].

the bcc phase predicted by GGA are within 3 and 10% of experiment. The agreement between GGA and the experimentally determined zero pressure properties of hcp iron is not as good. The theoretical zero pressure volume is 9% smaller than the experimental value extrapolated to zero pressure. This level of disagreement is surprising in light of the much better agreement with the zero pressure properties of the bcc phase and the performance of density functional theory in other systems, including nonmagnetic fcc 3d transition metals [*Leung et al.*, 1992]. We speculate that the discrepancy may be caused by the presence of magnetism in hcp iron at very low pressures. We have investigated only the non-magnetic state of the hcp phase as previous calculations have indicated that the ferromagnetic moment is zero for non-negative pressures [*Asada and Terakura*, 1992]. In all cases GGA agrees better with experiment than does LDA. The gradient corrections partially correct the well known tendency of local functionals to underestimate the volume.

The agreement between experiment and theory is significantly better at high pressure. At the lowest pressure at which the volume of hcp iron is actually measured, GGA and experiment differ by 6% in volume. Significantly, at the pressures of the Earth's core (136-363 GPa) GGA reproduces the experimental equation of state to within 2 %, and to within 1 % at inner core pressures (328-363 GPa). These results have been confirmed by *Sherman* [1995] also using the LAPW method and by *Söderlind et al.* [1996] using the full potential linear muffin tin orbital (LMTO) method. The agreement between different theoretical results is significant because it demonstrates the reproducibility of first principles calculations even, as in the case of *Söderlind et al.*, when different computational methods are used to solve the Kohn-Sham equations.

We use our results to address the relative stability of bcc, fcc, and hcp phases at high temperature [*Stixrude and Cohen*, 1995b]. The relative stability of these phases at core temperatures will depend on thermal contributions to their free energy. Without evaluating these thermal contributions explicitly, we find that the static total energy of the bcc structure at high pressure is so much higher than that of hcp that its stability at any temperature is highly unlikely. The total enthalpy, $H = E + PV$ of the bcc structure becomes much larger than that of hcp at high pressure. This is primarily a result of the larger volume of the bcc structure (Figure 4). The relative stability of bcc and hcp structures is governed by the difference in Gibbs free energies, G

$$\Delta G = \Delta H - T\Delta S \tag{14}$$

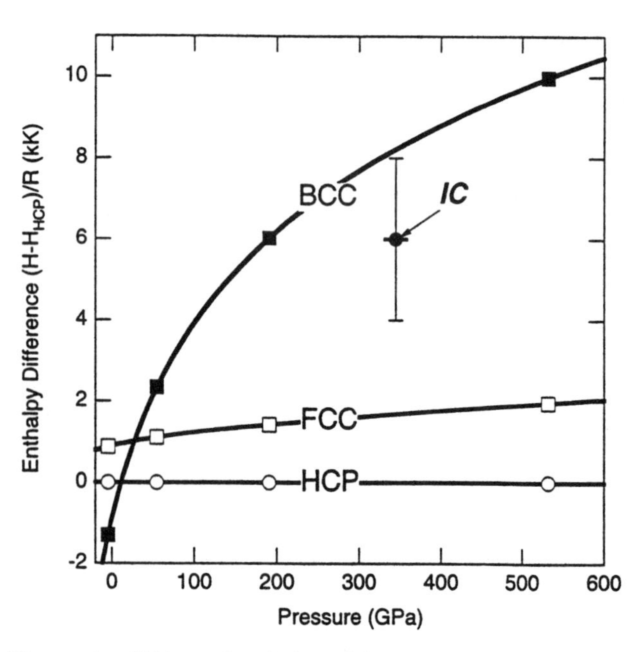

Figure 4. GGA total enthalpy of bcc (solid squares) and fcc (open squares)relative to that of hcp (open circles) as a function of pressure. The enthalpy difference is divided by the universal gas constant (R) resulting in units of temperature. Enthalpy differences are compared with the pressure range (horizontal bar) and range of estimates of the temperature (vertical error bar) of the inner core (IC). From *Stixrude and Cohen* [1995b].

where $\Delta G = G_{bcc} - G_{hcp}$. For the bcc structure to be stable with respect to the hcp structure at temperature T, we require

$$T > \Delta H / \Delta S \quad (15)$$

If we take typical entropies of melting as an upper bound on the entropy difference, $\Delta S < R$, where R is the gas constant, then the bcc structure is stable for temperatures $T > 8000$ K at inner core pressures. This temperature exceeds even the highest estimates of the temperature in the Earth's inner core. Moreover, it exceeds the highest estimates of the melting temperature of iron at inner core pressures. The stability of the bcc structure at inner core conditions is thus highly unlikely. The energetic unfavorability of bcc at high pressure was recently confirmed by *Moroni et al.* [1996], who combined LMTO calculations with an approximate treatment of the vibrational contribution to the free energy.

Magnetism

In the crustal environment, temperature plays a crucial role in determining the magnetic state of metals and transition metal-bearing silicates and oxides. The entropy of orientational disorder of magnetic moments becomes favorable at high temperatures and leads to a vanishing net magnetic moment above the Curie temperature. However, local magnetic moments survive and are often essentially unchanged in magnitude from their values in the magnetic low temperature structure. At core pressures, a different phenomenon occurs; the local magnetic moments themselves vanish under compression.

At zero pressure our GGA calculations of the ferromagnetic state of bcc iron [*Stixrude et al.*, 1994] show that the theoretical magnetic moment is in excellent agreement with experiment (2.174 vs. 2.12 μ_B) (Figure 5). Interestingly, and despite the failure of the LDA to recover the correct ground state for iron, the magnetic moment that it predicts for (metastable) ferromagnetic bcc iron is essentially identical to that predicted by GGA at the same volume.

Our GGA and LDA calculations show that the magnetic moment of ferromagnetic bcc iron vanishes at high pressure (Figure 5). The density at which the moment approaches zero is similar to that of the inner core ($V \approx 48$ Bohr^3). The dependence of the magnetic moment on volume is quasi-linear between V=60 and 90 Bohr^3, but then begins to decrease much more rapidly at smaller volumes. These results have been confirmed by *Söderlind et al.* [1996], whose LMTO calculations show an essentially identical dependence of magnetic moment on compression.

The collapse of the magnetic moment in iron can be understood in terms of band broadening. Many

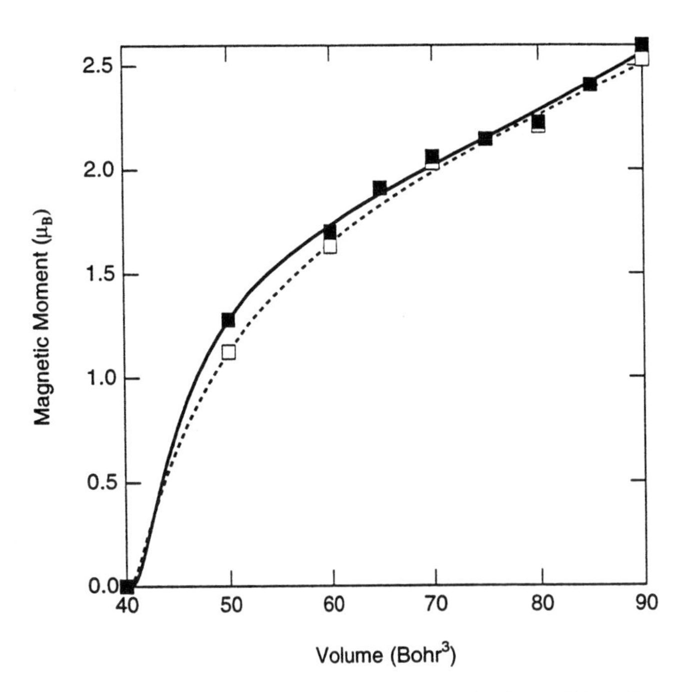

Figure 5. Magnetic moment of bcc iron in the GGA (solid squares, solid line) and LDA (open squares, dashed line) approximations in units of Bohr magnetons. From *Stixrude et al.* [1994].

transition metals are metallic because of the strong localization characteristic of d-states. These localized states produce a very narrow band with little dispersion through the Brillouin zone. However, under compression, the dispersion of these states increases because of enhanced d-d hopping, broadening the d-band and destroying their unique character. These arguments can be formalized in terms of a Stoner model [*Marcus and Moruzzi*, 1988].

Iron at densities comparable to, or slightly higher than that of the inner core, is expected to be nonmagnetic. It is worth re-emphasizing that this is a consequence of pressure, not of temperature. Pressure causes local magnetic moments to vanish. Without local moments, the concept of a Curie temperature, and the ability of high temperature to disorder moments has no significance.

Elasticity

The simplest elastic distortion of the cubic phases of iron - a stretching along one crystallographic axis - reveals a structural relationship between the bcc and fcc lattices, and the mechanical instability of the bcc phase [*Stixrude et al.*, 1994; *Stixrude and Cohen*, 1995b]. The bcc structure is recovered for c/a=1, while the fcc structure is obtained when $c/a = \sqrt{2}$. We investigated the total energy of iron as a function of the c/a ratio of this

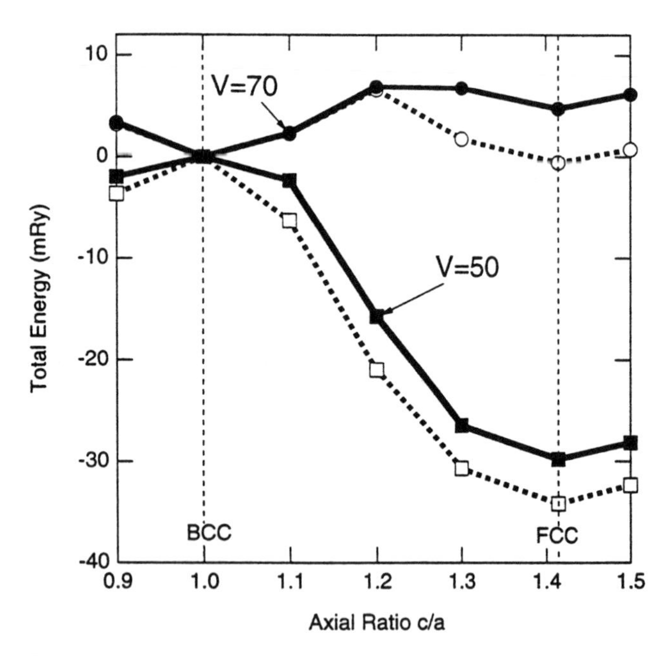

Figure 6. Total energy of iron as a function of *c/a* ratio. The values of *c/a* which correspond to the bcc (1.0) and fcc (√2) are indicated. Results at a volume of 70 Bohr^3 (near zero pressure, circles) are compared with those at 50 Bohr^3 (near 200 GPa, squares). Ferromagnetic GGA results are indicated by the filled symbols, ferromagnetic LDA results by the open symbols. From *Stixrude et al.* [1994].

tetragonal lattice with the LAPW method and using both LDA and GGA exchange-correlation functionals.

At low pressure, the total energy as a function of c/a ratio displays two local minima, one corresponding to the bcc structure and one to the fcc structure (Figure 6). This is in agreement with experimental observation. Both of these phases are known to be elastically stable at low pressure. In both theory and experiment, as the lattices are strained, their total energy rises, corresponding to an elastic restoring force. At high pressure, a completely different picture is revealed by our theoretical results. While the fcc phase is still elastically stable, the bcc phase is elastically unstable. As the bcc lattice is strained by small amounts, its energy is lowered. This means that there is no restoring force that preserves the bcc structure at high pressure. In the presence of infinitesimal thermal fluctuations, the bcc lattice will undergo a spontaneous distortion to the fcc structure. The mechanical instability of the bcc lattice was recently confirmed by *Söderlind et al.* [1996], whose LMTO calculations show mechanical instability for volumes less than approximately 55 Bohr^3, consistent with our results.

The mechanical instability of the bcc lattice can also be understood in terms of the Born stability criterion. The second derivative of the total energy with respect to c/a is proportional to the combination of elastic constants $C_S = C_{11} - C_{12}$ which must be greater than zero for a mechanically stable structure. The negative curvature exhibited by bcc above 150 GPa corresponds to a negative value of C_S and violation of the Born criterion.

There is no evidence for a stable tetragonal structure in iron at any pressure. Such a structure was proposed by *Söderlind et al.* [1996] on the basis of the apparent local minimum in total energy for $c/a < 1$ (Figure 7). It is important to recognize that this minimum is merely a saddle point and does not correspond to a mechanically stable structure. This becomes clear when one considers the orthorhombic strain energy surface. The fcc structure can be derived from the bcc structure in one of two ways: (1) by increasing the c/a ratio or (2) by decreasing c/a and b/a. By decreasing c/a below unity, a structure is generated which is intermediate between bcc and fcc. Tight-binding calculations at a series of volumes that span inner core conditions ($V = 40 - 50$ Bohr^3) confirm that the apparent local minimum is a saddle point and does not represent a mechanically stable structure.

Because the bcc phase is not only energetically unfavorable, but also mechanically unstable, it is highly unlikely to exist in the Earth's core. We investigated the elasticity of the remaining observed phases of iron, fcc

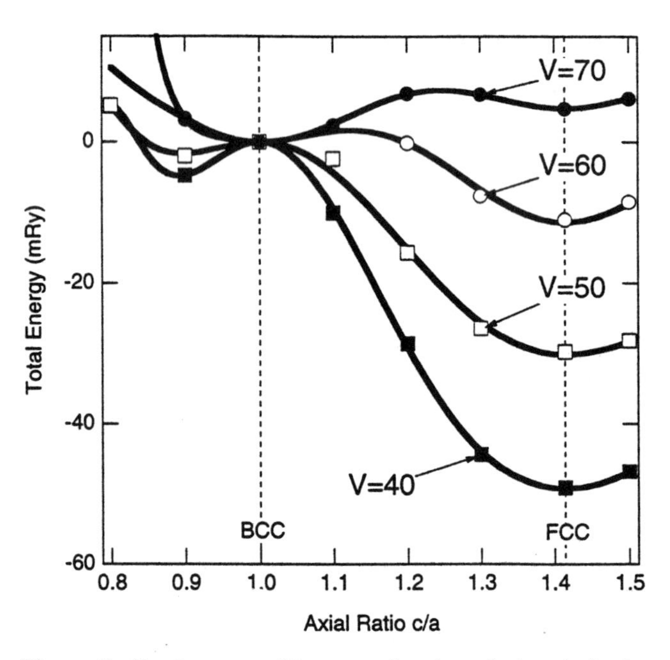

Figure 7. Total energy of iron as a function of *c/a* ratio in the GGA approximation. Results at four volumes are shown as indicated. The horizontal axis is extended towards smaller *c/a* ratios to illustrate the apparent minimum which develops near *c/a*=0.9 for *V*=50 and 60 Bohr^3.

and hcp in more detail with the tight binding method. We used this model to determine the full elastic constant tensor of both phases.

The elastic constants as a function of pressure of fcc and hcp iron are shown in Figure 8. We find that the elastic constants depend sublinearly on pressure. In the case of hcp, the diagonal elastic constants are similar in magnitude with C_{33} slightly larger than C_{11} at all pressures (by 6% at inner core densities). Similarly, $C_{66} = (C_{11} - C_{12})/2$ is slightly larger than C_{44} (by 5% at inner core densities). The differences between these pairs of elastic constants correspond to anisotropies in P- and S-wave velocities.

Our predicted fcc and hcp elastic constants are in generally good agreement with LMTO results [*Söderlind et al.*, 1996]; the RMS difference between the two sets of predictions is 70 GPa, or 6% at inner core densities. The largest difference occurs in C_{12} of hcp: while we find that C_{12} and C_{13} are similar, differing by no more than 6%, the LMTO study finds that C_{12} is more than 50% smaller than C_{13} at inner core densities. While the cause of this discrepancy is unclear, we note that experimental data on other hcp transition metals [*Brandes*, 1983] show $C_{12} \approx C_{13}$, consistent with our results. Moreover, *Söderlind et al.* performed their elastic constant calculations using the ideal, rather than the equilibrium value of c/a. We speculate that this may have biased their results.

The elastic constants completely specify elastic wave propagation in a single crystal as a function of propagation direction $\vec{n}$ and polarization direction $\vec{w}$. The wave velocities V are given by the eigenvalues of the Cristoffel equation

$$C_{ijkl} n_j w_k n_l = \rho V^2 w_i \qquad (16)$$

where ρ is the density. We find that the magnitude of the anisotropy is threefold greater in the cubic phase and threefold greater for S-waves than for P-waves: at the density of the inner core (V=48 Bohr^3) P-wave velocities vary with propagation direction by 10 and 3%, respectively in fcc and hcp, while S-wave velocities are 30 and 10% anisotropic (Figure 9). The greater anisotropy of the cubic structure, the directions of maximum and minimum velocities, and the greater anisotropy of S-waves can be understood in terms of a parameter free nearest neighbor central force model [*Born and Huang*, 1954]. The magnitude of the P-wave anisotropy in hcp is very similar to that observed seismologically in the inner core.

We may also compare the absolute values of elastic wave velocities determined theoretically with those of radial seismological models. The theoretical P- and S-wave velocities of isotropic aggregates of iron are higher

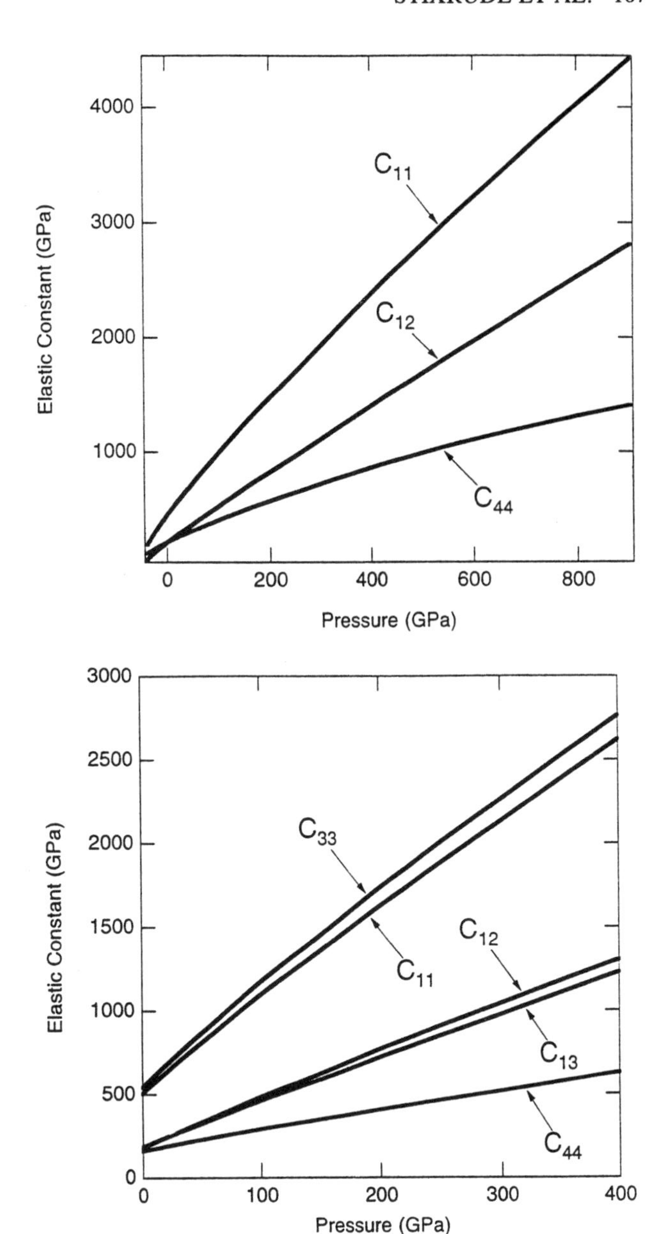

Figure 8. Elastic constants of fcc (top) and hcp (bottom) iron as a function of pressure.

than those observed for the inner core by 5 and 50%, respectively [*Stixrude and Cohen*, 1995a]. This difference is due to the difference in temperature between the static calculations and the inner core, and to the relatively low frequencies of seismic waves; our theoretical elastic constants correspond to infinite frequency values. One way to illustrate this is to compare the Poisson's ratio σ from our calculations with that of the inner core. We find that $\sigma \approx 0.31$ at inner core densities, a typical value for solids and much lower than that of the inner core (0.44). This comparison suggests

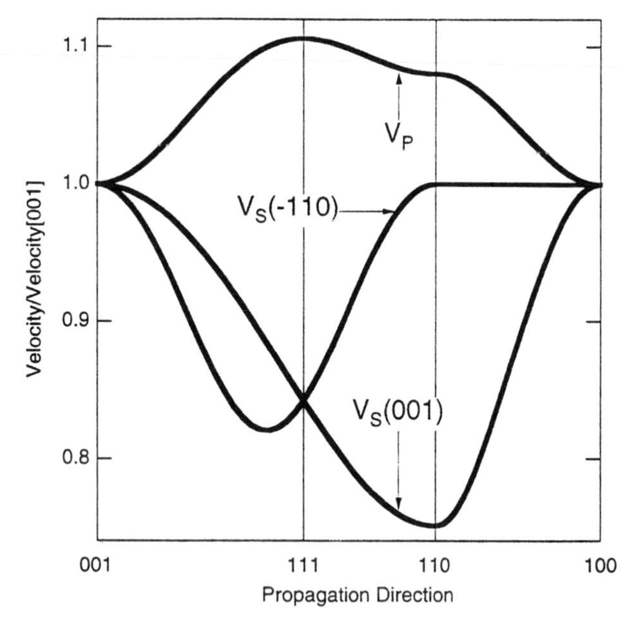

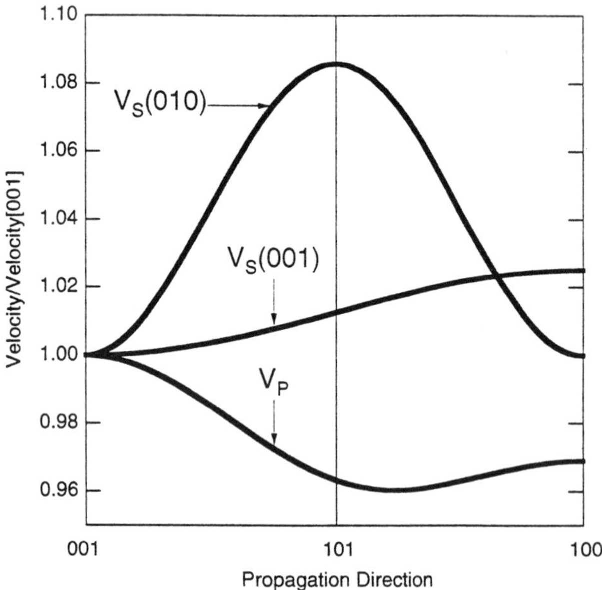

Figure 9. Velocity anisotropy in fcc (top) and hcp (bottom) iron at the density of the inner core (V=48 Bohr^3). Velocities are normalized to those for propagation in the [001] direction. Planes of polarization of S-waves are indicated. From *Stixrude and Cohen* [1995a].

that the anomalously high σ of the inner core may be due to anelasticity and dispersion at high temperature [*Jackson*, 1994].

High Temperature Properties

Our calculations of the static equation of state of iron show that the largest part of the pressure in the core is athermal - the static pressure of hcp iron at the density of the inner core is 240 GPa, more than 65% of the pressure at Earth's center. However, the thermal contribution to the pressure is non-negligible. In a metal, such as iron, there are two important contributions: (1) The pressure due to the thermal excitation of electrons. This term is unique to metals and arises from the population of high energy electronic states according to the Fermi-Dirac distribution. (2) The thermal pressure due to atomic vibrations. This contribution is also present in insulators.

At inner core densities, we find that the pressure due to the thermal excitation of electrons is approximately 10 GPa. The pressure due to atomic vibrations contributes approximately 50 GPa to the total pressure at inner core densities or 15% of the total pressure. The three contributions to the total pressure are shown separately along the T=6000 K isotherm in Figure 10. By combining LAPW calculations of the static pressure and that due to the thermal excitation of electrons, with cell model determinations of the thermal pressure, we obtain a complete description of the equation of state of iron.

Our theoretical high temperature equation of state of iron is in good agreement with available experimental data [*Wasserman et al.*, 1996]. We have shown that the Hugoniot calculated with the cell model and our tight binding Hamiltonian is within a few GPa of the experimental Hugoniot at core pressure. Our predicted Hugoniot temperatures are in excellent agreement with the estimates of *Brown and McQueen* [1986], but fall approximately 800 K lower than the values measured by *Yoo et al.* [1993]. We have also compared our equation

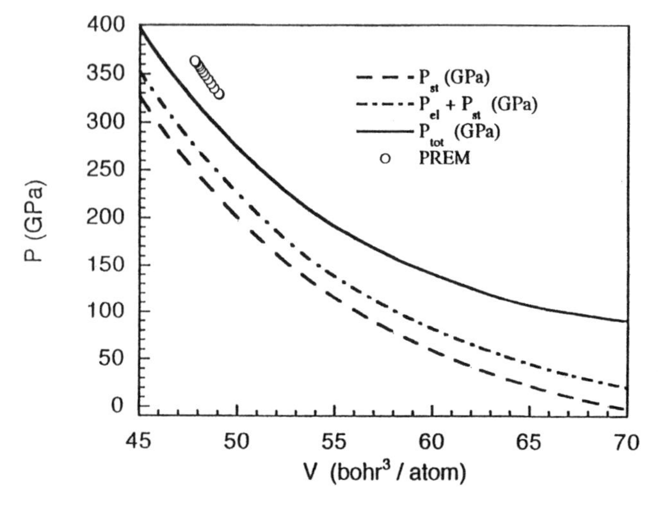

Figure 10. Different contributions to pressure at T=6000 K: static pressure (dashed line), static +electronic pressure (dot-dashed line), total pressure (solid line). Also shown is the pressure in the inner core according to PREM model (circles).

of state to measurements of the thermal expansivity of hcp and fcc iron and found good agreement.

We compare our equation of state of iron to the properties of the inner core as determined seismologically (Fig. 11). For temperatures in the middle of the range typically estimated for the inner core (6000 K) the density of iron is 3.5% greater than that of the inner core. This difference is larger than the estimated uncertainty in our theoretical equation of state - the difference between our results and comparable experimental data (e.g., the Hugoniot) is substantially less (1%). The difference between the density of iron and of the inner core also lies outside uncertainties in the seismological estimates of the inner core density. [*Masters and Shearer*, 1990]. We note that even for temperatures as high as 8000 K, somewhat higher than the highest estimates of inner core temperatures, iron is still 1% denser than the inner core.

We find that temperatures in excess of 8000 K are required for the density of iron to coincide with that of the inner core. The conclusion that very high temperatures are necessary is consistent with previous results, although the requisite temperatures found here are somewhat higher than those found by *Jephcoat and Olson* [1987]. These authors found that a temperature of 7000 K was adequate to yield agreement between the equation of state of iron and the inner core. The differences between our result and theirs is due to the different equations of state used. Because of the limited experimental data then available, Jephcoat and Olson were forced to extrapolate their semiempirical equation of state well outside the range of measurements. They also did not include an explicit division of vibrational and electronic thermal pressures.

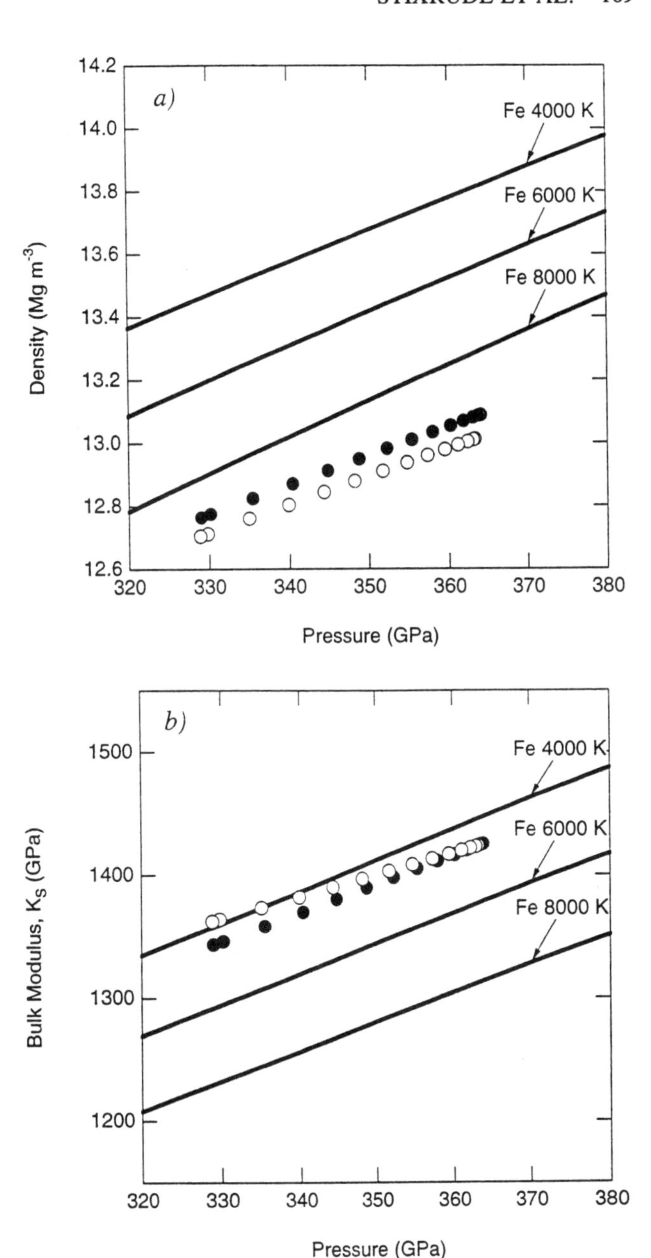

Figure 11. Density of pure iron along three isotherms (solid lines) compared with that of the inner core. The properties of the inner core are represented by the seismological models of *Dziewonski and Anderson* [1981] (solid circles) and *Dziewonski et al.* [1975] (open circles).

DISCUSSION AND CONCLUSIONS

It is now possible to understand the physics of iron at the pressures and temperatures of the Earth's core on the basis of density functional theory. LAPW calculations, and parametric extensions of first principles methods, such as our tight binding method, allow one to predict, independently of experimental data, the physical behavior of iron at extreme conditions. Together with experimental measurement, these results place new constraints on the composition of the core, its thermal state, the crystalline structure of the inner core, and the origin of its anisotropy.

Density functional theory places some of the first constraints on the crystalline structure of the inner core. All of the observed structures of iron have been considered as the stable phase at inner core conditions [*Anderson*, 1986; *Ross et al.*, 1990; *Jeanloz* 1990]. Density functional theory effectively eliminates one of these phases, bcc, as a likely constituent of the inner core. We find that it is highly unfavorable energetically with respect to hcp and fcc phases at high pressure, primarily because of its larger volume. More importantly, we find that it is mechanically unstable - it violates the Born stability criterion above 150 GPa.

Our calculations favor a hexagonal phase as the crystalline structure of the inner core. We find that hcp is the stable low temperature phase from 11 GPa, to pressures well beyond those of the inner core. Moreover, we find that the elastic anisotropy of the hcp phase provides a natural explanation of the anisotropy of the inner core. We find that the P-wave anisotropy of hcp iron is very similar to that of the bulk inner core. Assuming that the inner core is a polycrystalline aggregate, we have shown that very simple textural models, combined with our predicted elastic constants, can account for seismic travel time observations. An aggregate in which the c-axes of the constituent hcp crystals are nearly aligned with the spin axis can account for 60% of the variance in BC-DF travel time anomalies. An alternative model, one which accounts for the seismic data equally well and is motivated by the strong alingment that is required, is that the inner core consists of a single crystal of hcp iron [*Stixrude and Cohen*, 1995a]. Based on these results, we cannot rule out the possibility that the inner core is composed of a different hexagonal phase (e.g., dhcp) which may be similar to hcp elastically and energetically.

First principles calculations show that the magnetic state of iron at high pressures is likely to be very different from its familiar low pressure state. The hcp and fcc phases are expected to be nonmagnetic at high pressure. This means that not only is the net magnetic moment zero, but that magnetic moments do not exist, even locally: spin pairing is complete. This is to be contrasted with the state of iron at zero pressure, where, above its Curie temperature, the net magnetic moment may be zero, but local moments, while disordered, remain virtually undiminished in magnitude from their values in the ferromagnetic, low-temperature state.

Our theoretical equation of state of iron at high pressures and temperatures shows that the Earth's inner core is not likely to be composed of pure iron. The alloying elements must be lighter on average than iron. The amount of the light element in the inner core remains uncertain, but is likely to be several weight %, depending on the identity of the light element (S, O, ...), and the as yet poorly constrained properties of the relevant alloys.

Acknowledgments. This work was supported by the National Science Foundation under grants EAR-9305060 (LPS) and EAR-9304624 (REC). Calculations were performed on the Cray C90 at the Pittsburgh Supercomputer Center, the IBM SP2 at the Cornell Theory Center, and the Cray J90 at the Geophysical Laboratory.

REFERENCES

Anderson, O. L., Properties of iron at the earth's core conditions, Geophys. *J. R. Astr. Soc., 84*, 561-579, 1986.

Asada, T., and K. Terakura, Cohesive properties of iron obtained by use of the generalized gradient approximation, *Phys. Rev. B, 46*, 13,599-13,602, 1992.

Bagno, P., O. Jepsen, and O. Gunnarson, Ground-state properties of third-row elements with nonlocal density functionals, *Phys. Rev. B, 40*, 1997-2000, 1989.

Birch, F., Elasticity and composition of the Earth's interior, *J. Geophys. Res.*, 57, 227-286, 1952.

Born, M., and K. Huang, *Dynamical Theory of Crystal Lattices*, pp. 140-149, Oxford University Press, Oxford, 1954.

Brandes, E. A. (ed.), *Smithells Metals Reference Book*, Butterworths, London, 1983.

Brown, J. M., and R. G. McQueen, Phase transitions, Grüneisen parameter, and elasticity for shocked iron between 77 GPa and 400 GPa, *J. Geophys. Res., 91*, 7485-7494, 1986.

Cohen, R. E., M. J. Mehl, and D. A. Papaconstantopoulos, Tight-binding total-energy method for transition and noble metals, *Phys. Rev. B, 50*, 14,694-14,697, 1994a.

Cohen, R. E., L. Stixrude, and D. A. Papaconstantopoulos, A new tight-binding model of iron, towards high temperature simulations of the earth's core, in *High Pressure Science and Technology-1993*, edited by S. C. Schmidt, J. W. Shaner, G. A. Samara, and M. Ross, pp. 891-894, American Institute of Physics, 1994b.

Cowley, E. R., J. Gross, Z. X. Gong, and G. K. Horton, Cell-cluster and self-consistent calculations for a model sodium chloride crystal, *Phys. Rev. B, 42*, 3135-3141, 1990.

Dziewonski, A. M., A. L. Hales, and E. R. Lapwood, Parametrically simple earth models consistent with geophysical data, *Phys. Earth Planet. Int., 10*, 12-48, 1975.

Dziewonski, A. M., and D. L. Anderson, Preliminary reference Earth model, *Phys. Earth Planet. Int., 25*, 297-356, 1981.

Glatzmaier, G. A., and P. H. Roberts, A three-dimensional convective dynamo solution with rotating and finitely conducting inner core and mantle, *Phys. Earth Planet. Int., 91*, 63-75, 1995.

Gunnarson, O., and B. I. Lundqvist, Exchange and correlation in atoms, molecules, and solids by the spin-density-functional formalism, *Phys. Rev. B, 13*, 4274-4298, 1976.

Hohenberg, P., and W. Kohn, Inhomogeneous electron gas, *Phys. Rev., 136*, B864-B871, 1964.

Hollerbach, R., and C. A. Jones, Influence of the earth's inner core on geomagnetic fluctuations and reversals, *Nature, 365*, 541-543, 1993.

Holt, A. C., W. G. Hoover, S. G. Gray, and D. R. Shortle, Comparison of the lattice-dynamics and cell-model approximations with Monte-Carlo thermodynamic properties, *Physica, 49*, 61-76, 1970.

Jackson, I., Viscoelastic relaxation in iron and the shear modulus of the inner core, in *High Pressure Science and Technology-1993*, edited by S. C. Schmidt, J. W. Shaner, G. A. Samara, and M. Ross, pp. 939-942, American Institute of Physics, 1994.

Jeanloz, R., The nature of the earth's core, *Ann. Rev. Earth Planet. Sci, 18*, 357-386, 1990.

Jeanloz, R. and H. R. Wenk, Convection and anisotropy of the inner core, *Geophys. Res. Lett., 15*, 72-75, 1988.

Jephcoat, A. P., H. K. Mao, and P. M. Bell, The static compression of iron to 78 GPa with rare gas solids as pressure transmitting media, *J. Geophys. Res., 91*, 4677-4684, 1986.

Jephcoat, A. P., and P. Olson, Is the inner core of the Earth pure iron, *Nature, 325*, 332-335, 1987.

Karato, S., Inner core anisotropy due to the magnetic field induced preferred orientation of iron, *Science, 262*, 1708-1711, 1993.

Kohn, W., and L. J. Sham, Self-consistent equations including exchange and correlation effects, *Phys. Rev., 140*, A1133-A1138, 1965.

Leung, T. C., C. T. Chen, and B. N. Harmon, Ground-state properties of Fe, Co, Ni, and their monoxides: results of the generalized gradient approximation, *Phys. Rev. B, 44*, 2923-2927, 1991.

Mao, H. K., Y. Wu, L. C. Chen, J. F. Shu, and A. P. Jephcoat, Static compression of iron to 300 GPa and $Fe_{0.8}Ni_{0.2}$ alloy to 260 GPa - implications for composition of the core, *J. Geophys. Res., 95*, 21,737-21,742, 1990.

Marcus, P. M., and V. L. Moruzzi, Stoner model of ferromagnetism and total-energy band theory, *Phys. Rev. B, 38*, 6949-6953, 1988.

Masters, T. G., and P. M. Shearer, Summary of seismological constraints on the structure of the Earth's core, *J. Geophys. Res., 95*, 21,691-21,695, 1990.

Mehl, M. J., J. E. Osburn, D. A. Papaconstantopoulos, and B. M. Klein, *Phys. Rev. B, 41*, 10,311, 1990.

McMahan, A. K. and M. Ross, High-temperature electron-band calculations, *Phys. Rev. B, 15*, 718-725, 1977.

Mermin, N. D., Thermal properties of the inhomogeneous electron gas, *Phys. Rev., 137*, A1441-A1444, 1965.

Morelli, A., A. M. Dziewonski, and J. H. Woodhouse, Anisotropy of the inner core inferred from PKIKP travel times, *Geophys. Res. Lett., 13*, 1545-1548, 1986.

Moroni, E. G., G. Grimvall, and T. Jarlborg, Free energy contributions to the hcp-bcc transformation in transition metals, *Phys. Rev. Lett., 76*, 2758-2761, 1996.

Perdew, J. P. and Y. Wang, Accurate and simple analytic representation of the electron-gas correlation energy, *Phys. Rev. B, 45*, 13,244-13,249, 1992.

Ree, F. H. and A. C. Holt, Thermodynamic properties of the alkali-halide crystals, *Phys. Rev. B, 8*, 826-842, 1973.

Ross, M., D. A. Young, and R. Grover, Theory of the iron phase diagram at Earth core conditions, *J. Geophys. Res., 95*, 21,713-21,716, 1990.

Sherman, D. M., Stability of possible Fe-FeS and Fe-FeO alloy phases at high pressure and the composition of the earth's core, *Earth Planet. Sci. Lett., 132*, 87-98, 1995.

Slater, J. C. and G. F. Koster, Simplified LCAO method for the periodic potential problem, *Phys. Rev., 94*, 1498-1524, 1954.

Söderlind, P., J. A. Moriarty, and J. M. Willis, First-principles theory of iron up to earth-core pressures: structural, vibrational, and elastic properties, *Phys. Rev. B, 53*, 14,063-14,072, 1996.

Song, X. D. and P. G. Richards, Seismological evidence for differential rotation of the earth's inner core, *Nature, 382*, 221-224, 1996.

Stixrude, L., R. E. Cohen, and D. J. Singh, Iron at high pressure: Linearized-augmented-plane-wave computations in the generalized-gradient approximation, *Phys. Rev. B*, 50, 6442-6445, 1994.

Stixrude, L., and R. E. Cohen, High pressure elasticity of iron and anisotropy of earth's inner core, *Science, 267*, 1972-1975, 1995a.

Stixrude, L., and R. E. Cohen, Constraints on the crystalline structure of the inner core - mechanical instability of bcc iron at high pressure, *Geophys. Res. Lett., 22*, 125-128, 1995b.

Tromp, J., Support for anisotropy of the earth's core from free oscillations, *Nature, 366*, 678-681, 1993.

Wallace, D. C., Thermodynamics of Crystals, John Wiley \& Sons, New York, 1972.

Wasserman, E., L. Stixrude, and R. E. Cohen, Thermal properties of iron at high pressures and temperatures, *Phys. Rev. B*, 53, 8296, 1996.

Wei, S., and H. Krakauer, Local-density-functional calculation of the pressure-induced metallization of BaSe and BaTe, *Phys. Rev. Lett., 55*, 1200, 1985.

Woodhouse, J. H., D. Giardini, and X. D. Li, Evidence for inner core anisotropy from free oscillations, *Geophys. Res. Lett., 13*, 1549-1552, 1986.

Yoo, C. S., N. C. Holmes, M. Ross, D. J. Webb, et al., Shock temperatures and melting of iron at earth core conditions, *Phys. Rev. Lett., 70*, 3931-3934, 1993.

R. E. Cohen, Geophysical Laboratory, 5251 Broad Branch Rd. NW, Washington, DC 20015-1305.

L. Stixrude, School of Earth and Atmospheric Sciences, Georgia Institute of Technology, Atlanta, GA 30332-0340.

E. Wasserman, 3200 Q Av. ETB-K9-77 Battelle, Pacific Northwest National Laboratory, Richland, WA 99352.

Static Compression Experiments on Low-Z Planetary Materials

Russell J. Hemley and Ho-kwang Mao

Geophysical Laboratory and Center for High Pressure Research, Carnegie Institution of Washington, Washington, D. C.

Recent static high-pressure studies carried out with diamond anvil cells provide key information on properties of low-Z materials believed to be major constituents of the outer planets and their satellites. Profound pressure-induced changes in physical and chemical properties occur in such materials (i.e., gases and ices) as a result of the high compressions reached at planetary interior conditions. The equation of state of hydrogen and deuterium has been measured by single-crystal X ray diffraction to pressures above 100 GPa. Measurements of the optical properties of the hydrogen isotopes at megabar pressures indicate that the material transforms to a semiconducting charge transfer state with a complex phase diagram rather than to a semimetal at 150 GPa and low temperatures. The results may be compared with the reported metallization of the fluid at comparable pressures and high temperatures, as well as to theoretically predicted plasma phase transition. Infrared reflectivity measurements demonstrate that H_2O-ice transforms at 60 GPa to a nonmolecular, ionic phase, which persists to at least 210 GPa. Vibrational spectra strongly suggest a phase having symmetric hydrogen bonds. Mixtures of simple molecular materials at moderate pressures (e.g., beginning at <1 GPa) are found to exhibit a new high-pressure chemistry associated with the formation of stoichiometric compounds (or order alloys). For example, four new compounds are found in the CH_4-H_2 binary system at pressures below 8 GPa. The stability as a function of pressure has been studied to higher pressure, where unusual infrared properties are observed. Implications of these experiments for the outer solar system are discussed.

1. INTRODUCTION

Recent progress in planetary astronomy has opened a new vista on the outer solar system. Geophysical observations provide increasingly accurate determinations of gravitational moments, and spectroscopic measurements have constrained the abundance of compounds on planetary atmospheres and surfaces [*Lewis*, 1995]. Moreover, such studies have been supplemented by direct measurements of the Jovian atmosphere by the recent Galileo probe. Although the distribution of dominant molecular species (H_2, He, NH_3, CH_4, H_2O, CO_2, N_2, heavier rare gases) has been assumed for years, the direct measurements by the Galileo probe suggest major revisions of conventional views (e.g., *Neimann et al.*, [1996]). The myriad observations of the Shoemaker-Levy impact revealed chemical features in the dense atmosphere that are not yet fully understood. In addition, new generations of the evolutionary models provide increasingly realistic scenarios for planetary formation [*Wetherill*, 1994]. Moreover, the recent observations of planetary bodies outside the solar system, including the evidence for molecular species in their atmospheres [*Oppenheimer et al.*, 1995], suggest an evolutionary

Properties of Earth and Planetary Materials
at High Pressure and Temperature
Geophysical Monograph 101

continuum between large planets and brown dwarfs [*Boss*, 1995].

Despite these advances, the nature of the materials that form the bulk of the interiors of the large planets still poses a major problem. For many years, studies of candidate planetary materials were the exclusive domain of shock-wave (or dynamic) compression techniques and theoretical calculations [*Nellis et al.*, 1992]. At these pressures and temperatures, most of this material exists as high density fluids. Under the most extreme conditions (e.g., reaching 5-7 TPa in Jupiter), chemical interactions characteristic of atoms and molecules under ambient conditions are lost as the materials form dense plasmas [*Ichimaru and Ogata*, 1995]. Although beyond the range of most experimental techniques, theoretical calculations are increasingly accurate in this regime. Hydrogen is the dominant element in the solar system, and its behavior at high pressure is a problem of first rank. Dense forms of planetary ices are proposed constituents of the deeper levels of the large planets, particularly for Uranus and Neptune [*Hubbard et al.*, 1995]. Under such conditions, the bonding presumably differs from that observed in ambient pressure ices but information on the breakdown of the molecules and the form of speciation is lacking. Observational data also require the existence of deep cores consisting of solid phases [*Zharkov*, 1986; *Zharkov and Gudkova*, 1992], but their properties under these conditions are poorly known.

Recent developments in static high-pressure techniques are beginning to provide information on fundamental properties of low-Z materials important for understanding the interiors of the large planets and their satellites. This information is crucial for regimes of compression in which chemical interactions of the elements are strongly altered but not destroyed, as in full plasma states produced by the combined effects of ultrahigh temperatures and pressures resulting from adiabatic compression. The potential importance of such static pressure studies for planetary science was articulated by *Stevenson* [1985]. Here we discuss this emerging area of static high-pressure research. Specifically, we review four examples of recent experiments: the equation of state of hydrogen above 100 GPa, the current results on hydrogen metallization, compression of H_2O ice, and high-pressure chemistry in molecular mixtures. For such materials, there are major pressure-induced changes in physical and chemical properties, in some cases occurring over relatively modest pressures (to several GPa). We focus on the effects of pressure in order to establish low-temperature properties necessary for understanding higher *P-T* phenomena. We also briefly discuss the implications of these experiments for current problems for the outer planets.

2. EQUATION OF STATE OF HYDROGEN

Historically, the equation of state (EOS) of the molecular phase of hydrogen at megabar pressure has been a major experimental and theoretical challenge. Theoretically, the problem has been considerably more difficult than the atomic, metallic phase because of the complexity of the interactions in the dense molecular state [*Ashcroft*, 1995]. Since the first calculation by *Abrikosov* [1954], a large number of calculations have been performed to date [*Hemley and Mao*, 1995]. Pioneering and highly accurate measurements of the compressibility of hydrogen were performed to pressures up to 2.5 GPa by the use of direct observations of volume displacement [*Anderson and Swensen*, 1973; *Stewart*, 1956]. Accurate higher pressure measurements required in an in situ measurement of the unit cell volume. In fact, X ray diffraction of solid hydrogen was first performed at atmospheric pressure and 4 K by *Keesom et al.* [1930], who correctly determined the crystal structure to be hexagonal close packed (para-hydrogen) under these conditions. The first high-pressure X ray diffraction measurements in the gigapascal range were carried out to pressure of 26.5 GPa by X rays [*Mao et al.*, 1988] and to 30 GPa by neutrons [*Glazkov et al.*, 1988]. The measurements provided the first accurate equation of state for the molecular phase valid up to megabar pressures [*Hemley et al.*, 1990]. Subsequent measurements were performed to 42 GPa [*Mao and Hemley*, 1994]. Similar measurements were also carried out on solid helium to 60 GPa (also as a function of temperature) [*Loubeyre et al.*, 1993].

Recent developments of several new techniques have allowed the single-crystal X ray measurements of hydrogen to be extended to significantly higher pressures. The X ray technique required the preservation of single crystal over the range of the measurements. In fact, the limit in the previous work was the breakup of the single crystal. This was solved by loading hydrogen together with helium and growing a single crystal within the helium medium [*Loubeyre et al.*, 1996]. With the use of the helium medium, the loss in diffraction intensity from the single-crystal hydrogen that begins at 30-40 GPa when the material is compressed alone does not occur, and single crystal reflections can be measured to the highest pressures possible with a new megabar single-crystal diamond-cell. In the highest pressure experiment, three classes of reflections were followed to 105 GPa for H_2 and to 119 GPa for D_2. The experiments were carried out at the European Synchrotron Radiation Facility, Grenoble. The results confirm that the structure remains hcp over this pressure range (phase I at 300 K; [*Mao and Hemley*, 1994]), with no measurable isotope effect on the EOS under these conditions. The *P-V* data for H_2 and D_2, together with a Vinet EOS fit [*Vinet et al.*, 1987], are shown in Figure 1. The EOS is essentially indistinguishable from that determined previously by X ray diffraction to 26.5 GPa [*Hemley et al.*, 1990] but shows more compressibility at the highest pressures; this effect was also evident in the measurements to 42 GPa and motivated the construction of a new pair potential [*Duffy et al.*, 1994].

The calculated pressure-volume point corresponding to the reported metallization of the fluid is also shown [*Weir et al.*, 1996] (Figure 1). This point was calculated with the EOS of *Kerley* [1983], which was determined before the accurate measurements for the solid were available. It is well known that these early equations of state overestimated the volume (underestimated the compression) at high pressures. Even taking into account thermal effects and the fluid state reached in shock experiment, the point lies considerably above the accurate EOS for the (lower temperature) solid. The observation suggests the need to revise the *P-V* conditions that

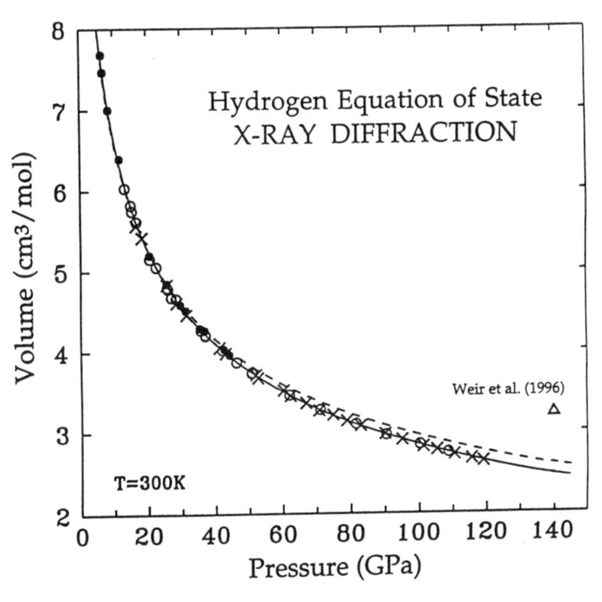

Figure 1. Equation of state of hydrogen from single-crystal X ray diffraction [*Loubeyre et al.*, 1996]. Solid circles, H_2 (no medium); open circles, H_2 (He medium); crosses, D_2 (He medium). The solid line is a Vinet EOS fit [*Vinet et al.*, 1987] to all of the data, with parameters V_0 = 25.43 cm³/mol, K_0 = 0.162 GPa, and K_0' = 6.813. The dashed is the fit to the earlier data to 26.5 GPa [*Hemley et al.*, 1990]. The triangle is the *P-V* point of the reported fluid metallization [*Weir et al.*, 1996] calculated using the EOS of *Kerley* [1983].

correspond to the reported fluid metallization. *Holmes et al.* [1996] report measurements of shock temperature and find lower values than that expected from previous EOS calculated from effective pair potentials [*Ross et al.*, 1983]. They therefore propose a model that includes dissociation and thus lower temperatures. *Duffy et al.* [1994] obtained a pair potential that fit the static compression and sound velocity data for the solid as well as the Hugoniot data (double shock points). Calculations of the temperature along the Hugoniot with this potential are lower than previous results that do not invoke dissociation; the temperatures calculated with the pair potential are close to the shock measurements (i.e., V > 4 cm³/mol). The non-uniqueness of the fit suggests there is some uncertainty in the degree of dissociation in the high-temperature shock.

The higher compressibility of hydrogen obtained from the new equations of state indicates enhanced stability of the molecular phase at high pressure. Integrating the equation of state, the total energy can be determined as a function of volume and compared with calculations of the total energy of theoretically predicted atomic phases. We use quantum Monte Carlo calculations for the lowest energy monatomic form which has the diamond structure [*Natoli et al.*, 1993]. Using the extrapolation of the previous X ray equation of state (Vinet formulation) [*Hemley et al.*, 1990], we find the transition pressure to be about 500 GPa. The new equation of state shifts the transition to ~620 GPa. Although these represent large extrapolations, the calculations suggest an enhanced pressure field of stability for the molecular phase, a result that further points to the notion that band-overlap metallization (i.e., metallization within the molecular phase) may precede pressure-induced dissociation [*Mao and Hemley*, 1989]. Finally, the new X ray measurements provide an improved baseline for the determination of an accurate *P-V-T* EOS for modeling planetary interiors and shock-wave data.

3. ELECTRON PROPERTIES AND METALLIZATION OF HYDROGEN

The effect of pressure on the electronic properties of molecular hydrogen, specifically testing the prediction of pressure-induced band overlap metallization, has been a major problem in the physics of dense hydrogen [*Mao and Hemley*, 1994]. This has driven the development of new optical techniques for probing the electronic structure of hydrogen at megabar pressures. Recently, we have extended our synchrotron infrared spectroscopic technique to mid infrared wavelengths in order to examine possible intraband and interband electronic excitations associated with the predicted transformation of solid molecular hydrogen to a semimetallic state at 150 GPa (phase III) [*Hemley et al.*, 1996]. This has been of particular interest because previous experiments revealed a number of new phenomena in the low-temperature solid at these pressures, including a complex phase diagram, unusual infrared properties, evidence for bond weakening, and a decreasing band gap [*Mao and Hemley*, 1994]. The recent infrared absorption and reflectivity measurements have established that hydrogen is transparent in the mid-infrared down to approximately 1200 cm^{-1} (0.15 eV) to at least 200 GPa [*Hemley et al.*, 1996]. The measurements provide evidence that the solid remains an insulator in phase III, at least just above the 150-GPa transition. Alternatively, if the phase is metallic at these pressures, the plasma frequency w_p is less than 0.3 eV at 77 K. Theory predicts closure of the band gap at these pressures, but this is strongly dependent on the orientational state of the molecules [*Chacham and Louie*, 1991; *Kaxiras et al.*, 1991]. Most recent calculations (that go beyond the standard local density approximation) are consistent with this interpretation [*Ashcroft*, 1995].

These results may be compared with recent dynamic compression studies. Conductivity measurements have been used to estimate the pressure dependence of the band gap of the shock-compressed fluid at lower pressures. The results were close to theoretical calculations appropriate for the fluid [*Chacham and Louie,* 1991] and to optical data for the low-temperature solid (Figure 2) [*Mao and Hemley,* 1994]. Recently, *Weir et al.* [1996] reported metallization of fluid hydrogen under dynamic compression using a reverberation technique to increase the pressure range without the problematic temperature increase associated with Hugoniot studies of hydrogen. *Weir et al.* [1996] report a plateau in the conductivity when shock pressure reached 140 GPa and 3000 K, evidence for a fully conducting state. Moreover, the authors interpret their results in terms of closure of the band gap, which implies the continued stability of the molecular species in the conducting fluid (Figure 2). This interpretation requires there to be large temperature effects on the band gap. Alternatively, the carriers may be ions produced by molecular dissociation. Although the magnitude of the conductivity is reportedly consistent with partial dissociation (5%), the degree of dissociation of hydrogen at high temperature and pressure is not yet well constrained [*Weir et al., 1996; Hemley and Ashcroft*, 1996]. In addition, the pressure-volume state calculated from the model differs from that expected from accurate lower temperature data [*Loubeyre et al.*, 1996], even with

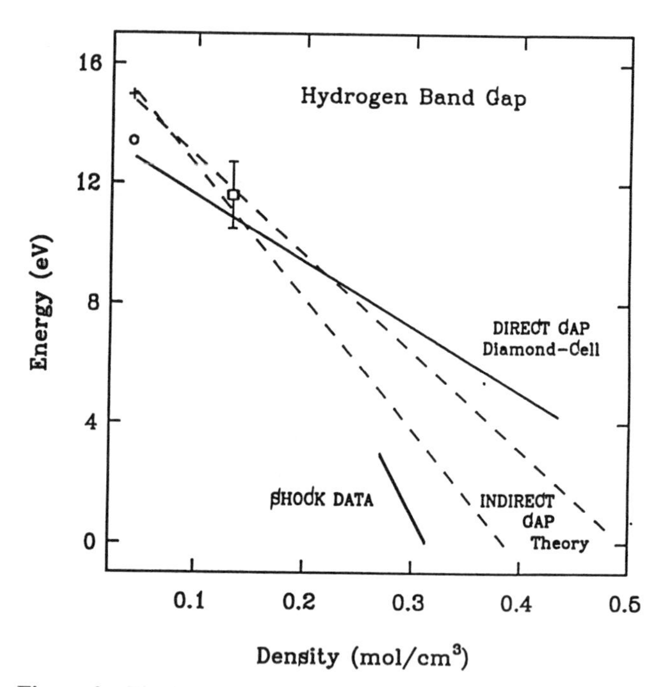

Figure 2. The band gap for hydrogen as a function of density. The bold line is the fit to the shock-wave data of *Weir et al.* [1996]. The square is the earlier shock-wave result [*Nellis et al.*, 1992]. The circle and thin line are dielectric model fits to diamond-anvil cell optical data. The dashed lines are the results of quasiparticle calculations for solid hydrogen (T = 0 K) by *Chacham and Louie* [1991]. The lower of the two is for the *c*-axis structure originally proposed by *Abrikosov* [1954]; the higher curve is the calculation for the hcp structure with spherically averaged molecules.

the inclusion of the expected thermal pressure (e.g., *Hemley et al.* [1990]). Although this difference could also be associated with an unusually large volume change on melting, it more likely indicates the need for softening the model EOS of the fluid. Such a correction will shift the onset of reported conductivity to higher densities.

A major question is the connection of these results to both the theoretically predicted plasma phase transition and the lower temperature data for the solid. *Saumon and Chabrier* [1989] proposed on the basis of semi-empirical calculations that hydrogen undergoes a plasma phase transition at high temperatures and pressures (Figure 3). This has gained theoretical support with the recent quantum Monte Carlo calculations of *Magro et al.* [1996], which are among the most reliable calculations available for hydrogen. Interestingly, the extrapolated boundary for the plasma phase transition from the latter study to lower temperatures and pressures is close to the metallization point reported by *Weir et al.* [1996] (Figure 3). Moreover, this also occurs at pressures at which the new phenomena

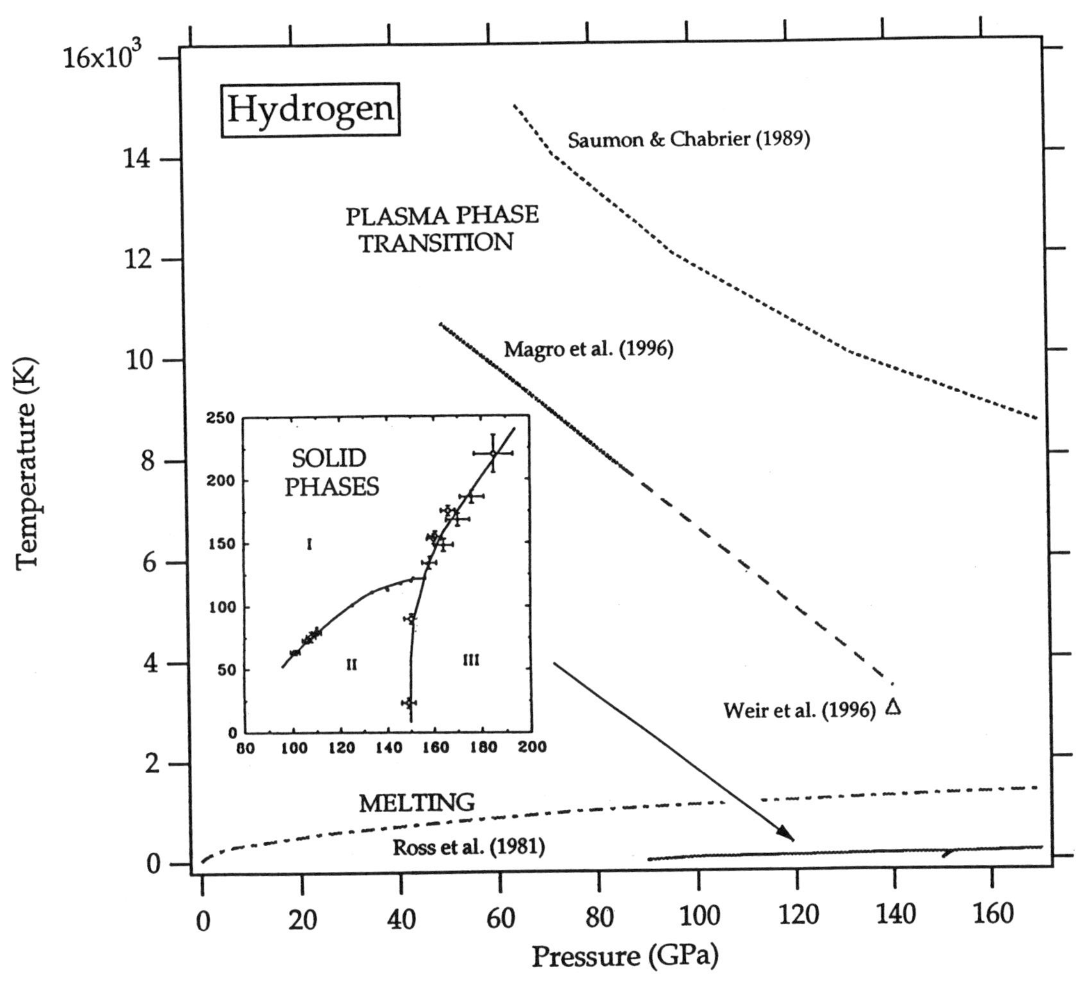

Figure 3. Phase diagram of hydrogen to megabar pressures and at high temperatures. The melting curve has been determined experimentally to 10 GPa; the results show the calculation of Ross and co-workers [*Ross et al.*, 1981]. The coexistence lines for the theoretically predicted plasma phase transition by *Saumon and Chabrier* [1989] and *Magro et al.* [1996] are also shown. Both curves terminate in a predicted critical point at lower pressures. The dashed line is extrapolated phase line given by *Magro et al.* [1996]. The inset shows the low temperature phase diagram for the solid at 80-200 GPa.

noted above are documented in the static experiments. Rather than become metallic, the low temperature solid appears to transform to a state characterized by (localized) intramolecular or intermolecular charge transfer. Moreover, it has also suggested that there is localization of electrons due to persistent rotational disorder [*Ashcroft*, 1995]. This proposal suggests an intriguing coupling between localized and delocalized electronic state; i.e., between the rotational disorder (which localizes the electrons) and electronic excitations across the band gap (into itinerant states).

The charge transfer effect in phase III may be quantified by the effective charge Q^* associated with the vibron [*Hemley et al.*, 1997], defined as $Q^*=Q_i+Q_d= dD/du$, where D is induced dipole moment, Q_i is ionic charge, and u is ionic displacement. For a vibron, $Q_d= R[dQ_i(R)/dR]$, where R is the H-H bond length. The charge is related to the oscillator strength f as $Q^* =e\ (Mf/2pm)^{1/2}$, where e is the electron charge, and M is the reduced mass of the vibrational mode. At the highest pressures of the measurements in phase III (230 GPa), Q^* reaches a value of 0.04e (Figure 4). The results indicate strong interaction between molecules at these pressures, but the magnitude of Q^* puts strong constraints on the hypothesis [*Baranowski*, 1992] that phase III is an ionic state consisting of H^+H^- ions. The magnitude of the absorption represents a

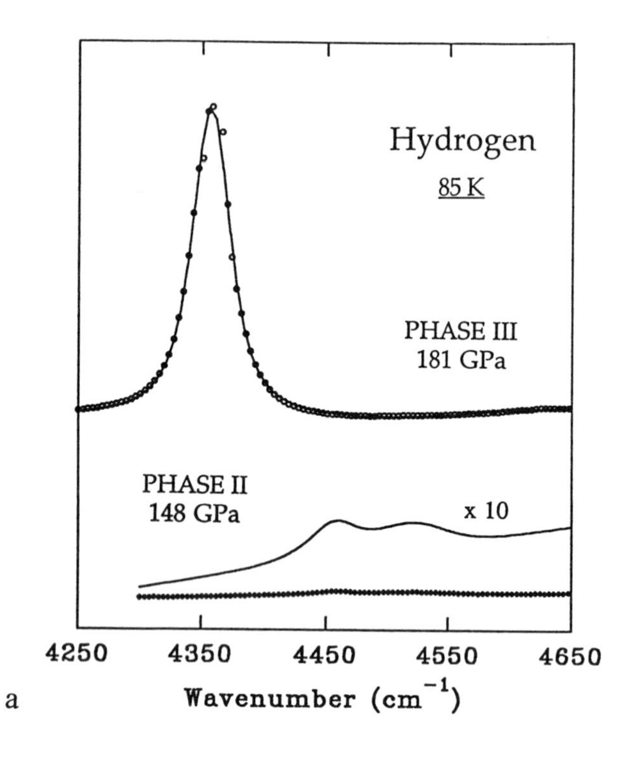

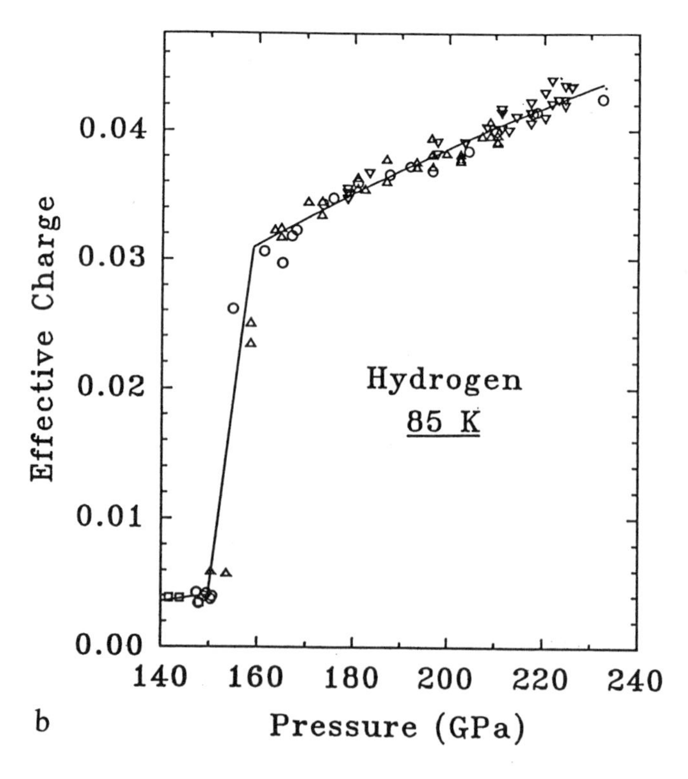

Figure 4. (a) Representative infrared absorption corresponding to the intramolecular H-H stretching mode (vibron) of hydrogen. (b) Effective charge as a function of pressure in phases II and III [*Hemley et al.*, 1997].

significant increase over that found at low pressures in phase I and II. Notably, in phase I, the absorption increases with the square of the density (r^2) [*Hanfland et al.*, 1992], which is the same dependence documented for collision induced absorption in low-density gases and used in opacity calculations for planetary atmospheres [*Guillot et al.*, 1994].

4. COMPRESSION OF ICE

The high-pressure behavior of end-member planetary ices (e.g., H_2O, NH_3, CH_4) is important for understanding fundamental properties of the outer planets. Gravitational data indicated that material in the mantles of Uranus and Neptune is intermediate in density between that of H_2-He mixtures and silicate-iron of the terrestrial planets [*Hubbard et al.*, 1995]. Hence, it has long been thought that dense planetary ice components comprise the deep layers of these planets (maximum pressures estimated to be 580 GPa and 800 GPa for Uranus and Neptune, respectively [*Zharkov and Gudkova,* 1992]). Recently, there has been progress on understanding the behavior of H_2O at megabar pressures. Under sufficient compression, ice has been predicted to transform from a molecular to a non-molecular phase having symmetric hydrogen bonds. Moreover, theoretical calculations predict transitions to denser forms at still higher pressures [*Benoit et al.*, 1996; *Hama et al.*, 1992]. Identification of higher pressure phases has been inconclusive because of lack of information on the behavior of the hydrogen (or deuterium) sublattice at the requisite pressures. X ray diffraction obtained to 128 GPa [*Hemley et al.*, 1987] is consistent with a structure based on the bcc oxygen sublattice. Raman measurements reveal softening of O-H stretching mode fundamentals but have been limited to below 60 GPa [*Pruzan*, 1994].

Recently, we used infrared reflectivity and absorption spectroscopy to study the high-pressure behavior of H_2O-ice to 210 GPa (Figure 5) [*Goncharov et al.*, 1996; *Struzhkin et al.*, 1997]. At 25–45 GPa, the O-H stretching vibration undergoes a cascading series of resonances with other infrared active vibrational modes [*Struzhkin et al.*, 1997]. With further increase of pressure, the sign of the pressure shift of the mode changes, and small discontinuities in other modes are observed. Reflectance spectra measured at 85 K also change abruptly: the n_3' band becomes much weaker and a new very strong low-frequency IR band is observed. Reflectance spectra at pressures higher than 60 GPa show systematic increase with pressure of the reflectivity in the 600–2600 cm^{-1} spectral range. These changes are indicative of a transition

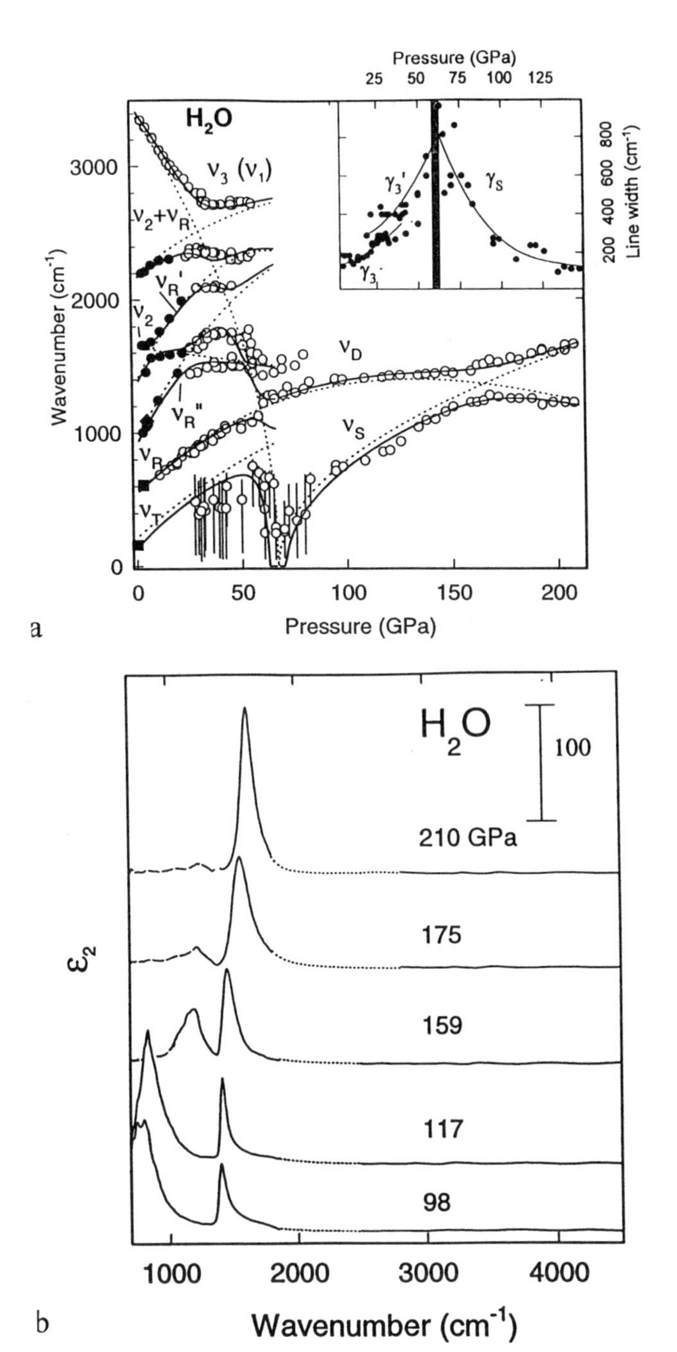

Figure 5. (a) Vibrational frequencies of H_2O to 210 GPa showing the phase transition to the symmetric ice structure at 60 GPa (300 K) [*Struzhkin et al.*, 1997]. The inset shows the pressure dependence of the line width. (b) Representative ϵ_2 spectra to 210 GPa obtained from the infrared reflectivity measurements [*Goncharov et al.*, 1996].

from a structure described by the ice rules (i.e., characterized by intramolecular covalent bonds and conventional hydrogen bonds between the molecules) to a non-molecular or ionic state. Moreover, the vibrational spectra are consistent with the predicted symmetric hydrogen bonded phase. The spectroscopic features of the phase persist to 210 GPa, although a weak structural (or order-disorder) transition along with a Fermi resonance, cannot be ruled out at 140–150 GPa.

The observed transition pressure may be compared to several previous predictions. *Pruzan* [1994] predicted the symmetrization transition should occur at a critical O-O distance ($R_{O\text{-}O}$) of 2.37 Å. According to the quasihydrostatic equation of state at 60 GPa, $R_{O\text{-}O}$ = 2.40 Å, which is slightly larger than the critical value [*Pruzan*, 1994]. Evidence for small change in compressibility at 60–70 GPa was observed in the X ray diffraction study to 128 GPa [*Hemley et al.*, 1987]. Hama and co-workers [*Hama and Suito*, 1992; *Hama et al.*, 1992] carried out an analysis of these data and found evidence for transitions at 40 and 70 GPa. As discussed by *Hemley et al.* [1987], systematic errors in the pressure determination arising from the large pressure gradients in the sample precluded the identification of a transition on the basis of the limited X ray diffraction data of the earlier study. It is likely that the shifts observed previously arose from changes in strength of the sample at the transition (i.e., during phase coexistence) rather than a volume discontinuity, and this is consistent with more recent measurements [*Hemley et al.*, in preparation]. The observed symmetrization pressure of 60 GPa is in very good agreement with ab initio simulations, which predicted the transition of 49 GPa [*Lee et al.*, 1992, 1993]. The agreement is especially close considering the fact that the proton dynamics were treated classically in the calculations. It may be of interest to note that *Bernal and Fowler* [1933] considered that H_2O at high pressure might adopt a quartz-type structure. Recently, the stability of a quartz-type polymorph of H_2O was predicted by molecular dynamics simulations. Experimentally, we find no evidence for such a structure to at least 210 GPa.

5. HIGH--PRESSURE MOLECULAR CHEMISTRY

The gases and ices described above interact only very weakly under normal low-pressure conditions (through van der Waals forces or hydrogen bonding). There is little information on the chemical interactions among these components and their physical behavior under the extreme *P-T* conditions of these planetary interiors. Experiments carried out during the past five years have uncovered a rich, new chemistry that occurs when these simple gases and liquids are mixed under pressure. This was first observed in mixtures of helium and nitrogen, which forms the compound $He(N_2)_{11}$ at 8 GPa [*Vos et al.*, 1992] and later observed in the He-Ne system [*Loubeyre et al.*, 1993]. The compounds formed in simple molecular mixtures with hydrogen are of potential planetary importance. The first example of such pressure-induced chemistry in hydrogen was its interaction with water under pressure [*Vos et al.*, 1993, 1994]. Under low-pressure conditions, water molecules form clathrate hydrates, which consist of networks of cages containing guest molecules. They are unstable at moderate pressures as the open networks break down under compression. Nevertheless, *Stevenson* [1985] suggested that H_2O at high pressure may accommodate a number of different components, although no structural basis for this property was proposed. In a high-pressure study of H_2-H_2O binary system, a novel type of clathrate with 1:1 ratio was discovered. In this high-pressure clathrate, H_2O and H_2 form two interlocking diamond networks. With the efficient molecular packing afforded by this structure, the compound is stable to at least 30 GPa. Study of the vibrational spectrum [*Vos et al.,* 1996] showed that the compound approaches a symmetrically hydrogen bonded state at < 40 GPa, that is, at lower pressures than in pure H_2O, as described above.

Recent studies of the H_2-CH_4 system shows an even richer high-pressure chemistry [*Somayazulu et al.*, 1996; to be published]. Four new solid compounds having H_2:CH_4 molar ratios 1:2, 1:1, 2:1, and 4:1 were discovered and characterized (Figure 6a). The crystal structures were determined by single-crystal X ray diffraction. Interestingly, the 1:1 compound is stable to at least 30 GPa, the maximum pressure studied. Compositions were verified from the bulk compositions of the mixtures, vibrational spectroscopy, and single-crystal and powder X ray diffraction. In addition, infrared measurements provide evidence for unusual charge transfer processes in the H_2-CH_4 compound at pressures above 30 GPa (Figure 6b) [*Somayazulu et al.*, to be published]. Similar behavior has also been observed in a high-pressure infrared study of the $Ar(H_2)_2$ compound to pressures of 220 GPa [*Datchi et al.*, 1996]. In contrast to the behavior of these binaries, high-pressure studies of the H_2-He [*Loubeyre et al.*, 1993] and H_2-NH_3 [*Lazor et al.*, 1996] system show no evidence for compound formation to the maximum pressures studied (at room temperature). The H_2-He system has been investigated to 200 GPa, in part to explore the use of He as a pressure medium in the megabar single-crystal X ray diffraction studies of H_2 [*Loubeyre et al.*, 1996], but these studies have been limited to room temperature and below. The high-temperature miscibility problem is thus still a theoretical issue. Recent molecular dynamics calculations of He-H_2 predict a lower temperature of miscibility [*Pfaffenzeller et al.*, 1995] than earlier static lattice computations [*Klepeis et al.*, 1991]..

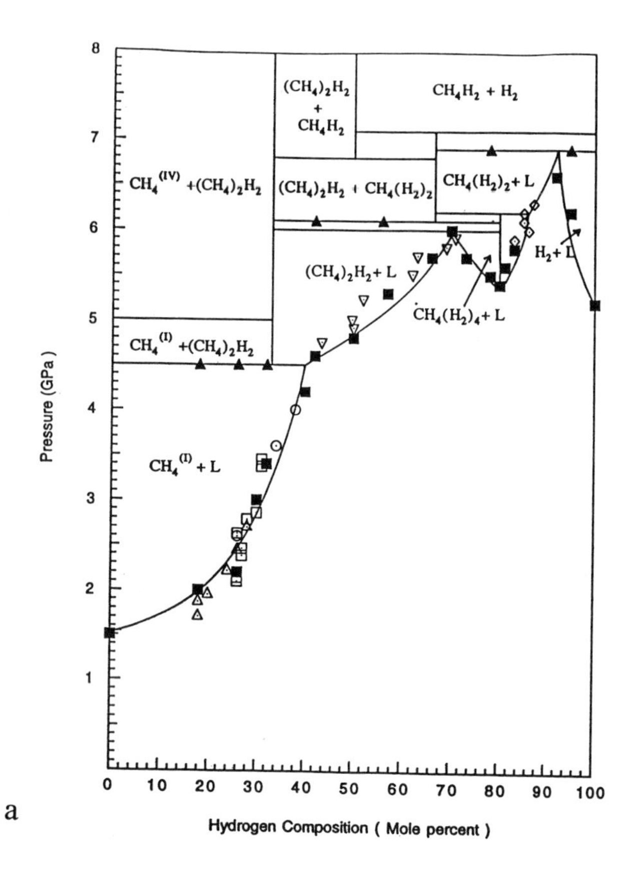

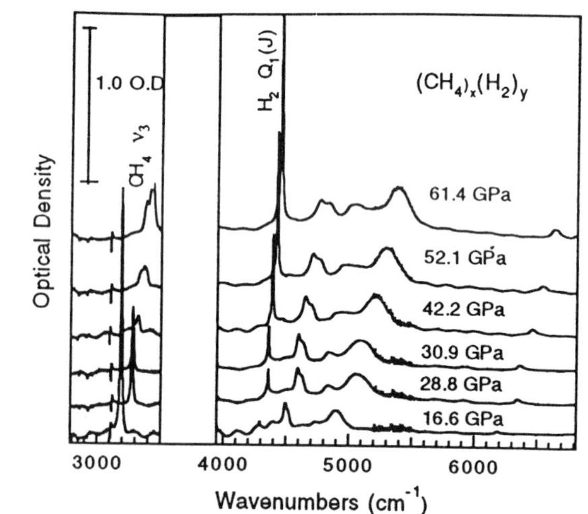

Figure 6. (a) Methane-hydrogen phase diagram at to 8 GPa at 295 K [*Somayazulu et al.*, 1996]. (b) Infrared absorption of the CH_4H_2 compound as a function of pressure [*Somayazulu et al.*, to be published].

6. DISCUSSION

We now briefly discuss a variety of implications of these results for the outer solar system. The origin and mechanism of the magnetic fields generated within the giant planets, and most recently observed within a Jovian satellite, are not understood. Indeed, one of the most remarkable features of Jupiter is the intensity of the field, presumably produced by convection of metallic hydrogen deep in the Jovian interior [*Zharkov*, 1986]. The possibility of forming a conducting fluid at the lower pressures indicates that the magnetic field is generated at shallower depths than previously thought. Moreover, the new results suggest that the boundary between the insulating molecular and the metallic, non-molecular layers is continuous, rather than forming a sharp, discontinuous boundary (e.g., at ~300 GPa [*Zharkov*, 1986]). It is this unclear whether the new shock-wave results rule out the existence of the plasma phase transition, as suggested by *Weir et al.* [1996]. It may be useful to consider the contribution to the magnetic field in the giants planets from ionic conductivity (e.g., from H_2O and NH_3). This mechanism has been proposed for the origin of the field in the icy satellites. Theoretical calculations predict diffusive motion of the protons in symmetric hydrogen-bonded state [*Benoit et al.*, 1996]; the contribution of such motion to field magnetic field generation in mixtures at high pressures and temperatures should be examined.

Another key question is understanding the non-solar abundances observed at the surface of the planets or inferred as existing at depth. Like studies of the solid Earth, this is related to style of convection within the large planets, the existence of stratified layers, and compositional gradients [*Zhang and Schubert,* 1996]. For example, H_2-He ratio in Jupiter determined by the Galileo probe is close to solar, but the amount of H_2O was much lower than expected, rare gas abundances varied markedly relative to solar ratios, and an excess of carbon was observed [*Neimann et al.,* 1996]. It is important to note that changes in bulk chemical compositions can be coupled with first-order phase transitions at depth, as discussed in regard to the 670 km discontinuity in the Earth [*Mao*, 1988; *Bina and Kumazawa*, 1993]. It has been suggested that this coupling occurs at the predicted plasma phase transition [*Chabrier et al.*, 1992]. *Guillot* [1995] predicts that condensation of relatively heavy molecules such as CH_4 and H_2O could inhibit convection in the dense atmosphere resulting in superadiabatic temperature gradients. It is not known whether the predicted dense clouds of condensed compounds (not observed by the Galileo probe) are insufficient to prevent layered convection even with the aid of a radiative window on the interior [*Guillot et al.*, 1995]. The radiative window arises from off-setting opacities; a radiative window near the surface leads to lower internal temperatures. We suggest that the newly discovered high-pressure molecular compounds may also exist in at depth; i.e., form dense cloud layers. However, the stability of these compounds, and more generally the miscibility of the molecular components, needs to be explored over a wide range of low and high temperatures appropriate for planetary and satellite interiors.

Finally, we comment on the materials behavior at pressures of the Jovian-type cores. The structures of phases under such conditions are not known experimentally, but density considerations indicate that these should be non-molecular and probably close-packed [*Ross,* 1981]. Both pressure and temperature favor a breakdown of molecular species, and it is useful to separate the effects of each. The observation of an ionic form of H_2O at high pressure (and low temperature) may be contrasted with the inferred ionic state at high temperatures. Thermodynamic calculations indicate that H_2O is more stable than phase separation to form the elements up to 200 GPa. Calculations for NH_3 indicate a similar result stability of the compound relative to elemental decomposition. Shock-wave studies showed evidence for breakdown of CH_4 at 100 GPa, suggesting that this material may be pyrolyzed to diamond or a carbon-rich phase within planetary bodies. Thermodynamic calculations carried out using the new EOS data indicate that solid CH_4 is unstable relative to the elements at pressures as low as 15 GPa. Theoretical calculations have provided a variety of increasingly accurate predictions. For example, *Bernasconi et al.* [1995] predicted the existence of new dense hydrocarbon phases produced from polyacetylene below 100 GPa. More recently, *Ancilotto et al.* [1997] predict condensation reactions involving methane under similar pressure conditions. A number of theoretical studies of elements at higher pressure have been performed. The lowest energy post-diamond structure predicted for carbon is BC8 and is calculated to be stable at 12 TPa; experimental studies to date indicate that diamond is stable to at least 400 GPa (see *Bundy et al.* [1996]).

Acknowledgments. This work was supported by NASA, NSF, and the DOE (NSLS). The work described here was performed with the excellent help of A. F. Goncharov, V. V. Struzhkin, M. Somayazulu, P. Loubeyre, R. LeToullec, D. Häusermann, M. Hanfland, I. I. Mazin, and L. W. Finger.

REFERENCES

Abrikosov, A. A., Equation of state of hydrogen at high pressures, *Astron. Zh.*, *31*, 112, 1954.

Ancilotto, F., G. L. Chiarotti, S. Scandolo, and E. Tosatti, Dissociation of methane into hydrocarbons at extreme (planetary) pressures and temperatures, *Science, 275*, 1288-1290, 1997.

Anderson, M. S., and C. A. Swensen, Experimental compressions for normal hydrogen and deuterium to 25 kbar at 4.2 K, *Phys. Rev. B, 10*, 5184-5191, 1973.

Ashcroft, N. W., The dense hydrogen plasma: translational, orientational, and electronic structure, in *Elementary Processes in Dense Plasmas*, edited by S. Ichimaru and S. Ogata, pp. 251-270, Addison-Wesley, Reading, Mass., 1995.

Baranowski, B., A hypothesis concerning the low temperature phase transition in solid hydrogen and deuterium at about 150 GPa, *Polish J. Chem., 66*, 1637-1640, 1992.

Benoit, M., M. Bernasconi, P. Focher, and M. Parrinello, New high-pressure phase of ice, *Phys. Rev. Lett., 76*, 2934-2936, 1996.

Bernal, J. D., and R. H. Fowler, A theory of water and ionic solution, with particular reference to hydrogen and hydroxyl ions, *J. Chem. Phys., 1*, 515, 1933.

Bernasconi, M., M. Parrinello, G. L. Chiarotti, P. Focher, and E. Tosatti, Anisotropic a-C:H form compression of polyacetylene, *Phys. Rev. Lett., 76*, 2081-2084, 1995.

Bina, C. R., and M. Kumazawa, Thermodynamic coupling of phase and chemical boundaries in planetary interiors, *Phys. Earth Planet. Inter., 76*, 329-341, 1993.

Boss, A., Proximity of Jupiter-like planets to low-mass stars, *Science*, 267, 360-362, 1995.

Bundy, F. P., M. A. Weathers, W. A. Bassett, R. J. Hemley, H. K. Mao, and A. F. Goncharov, Phase and transformation diagram of carbon; updated through 1994, *Carbon*, 1996.

Chabrier, G., D. Saumon, W. B. Hubbard, and J. I. Lunine, The molecular-metallic transition of hydrogen and the structure of Jupiter and Saturn, *Astrophys. J., 391*, 817-826, 1992.

Chacham, H., and S. G. Louie, Metallization of solid hydrogen at megabar pressures: A first-principles quasiparticle study, *Phys. Rev. Lett., 66*, 64-67, 1991.

Datchi, F., R. LeToullec, P. Loubeyre, A. F. Goncharov, M. Somayazulu, R. J. Hemley, and H. K. MaoSynchrotron infrared spectroscopy of $Ar(H_2)_2$ to 220 GPa, *Bull. Am. Phys. Soc., 41*, 564, 1996.

Duffy, T. S., W. Vos, C. S. Zha, R. J. Hemley, and H. K. Mao, Sound velocity in dense hydrogen and the interior of Jupiter, *Science, 263*, 1590-1593, 1994.

Glazkov, V. P., S. P. Besedin, I. N. Goncharenko, A. V. Irodova, I. N. Makarenko, V. A. Somenkov, S. M. Stishov, and S. Sh. Shilsteyn, Equation of state of deuterium to high pressures, *JETP Lett.. 47*, 661, 1988.

Goncharov, A. F., V. V. Struzhkin, M. Somayazulu, R. J. Hemley, and H. K. Mao, Compression of H_2O to 210 GPa: Infrared evidence for a symmetric hydrogen bonded phase, *Science, 273*, 218-220, 1996.

Guillot, T., Condensation of methane, ammonia, and water and the inhibition of convection in giant planets, *Science, 269*, 1697-1699, 1995.

Guillot, T., G. Chabrier, D. Gautier, and P. Morel, Effect of radiative transport on the evolution of Jupiter and Saturn, *Astrophys. J.*, 450, 463-472, 1995.

Guillot, T., G. Chabrier, P. Morel, and D. Gautier, Nonadiabatic models of Jupiter and Saturn, *Icarus, 112*, 354-367, 1994.

Hama, J., and K. Suito, Evidence for a new phase of ice above 70 GPa from the analysis of experimental data using a universal equation of state, *Phys. Lett. A, 187*, 346-350, 1992.

Hama, J., K. Suito, and M. Watanabe, Equation of state and insulator-metal transition of ice under ultra-high pressures, in *High Pressure Research in Mineral Physics: Application to Earth and Planetary Sciences*, edited by Y. Syono and M. Manghnani, pp. 403-408, 1992.

Hanfland, M., R. J. Hemley, H. K. Mao, and G. P. Williams, Synchrotron infrared spectroscopy at megabar pressures: vibrational dynamics of hydrogen to 180 GPa, *Phys. Rev. Lett., 69*, 1129-1132, 1992.

Hemley, R. J., and N. W. Ashcroft, Shocking states of matter, *Nature, 380*, 671-672, 1996.

Hemley, R. J., A. P. Jephcoat, H. K. Mao, L. W. Finger, C. S. Zha, and D. E. Cox, Compression of H_2O-ice to 128 GPa (1.28 Mbar), *Nature, 330*, 737-740, 1987.

Hemley, R. J., and H. K. Mao, Progress on hydrogen at ultrahigh pressures, in *Elementary Processes in Dense Plasmas*, edited by S. Ichimaru and S. Ogata, pp. 269-280, Addison-Wesley, Reading, Mass., 1995.

Hemley, R. J., H. K. Mao, L. W. Finger, A. P. Jephcoat, R. M. Hazen, and C. S. Zha, Equation of state of solid hydrogen and deuterium from single-crystal X ray diffraction to 26.5 GPa., *Phys. Rev. B, 42*, 6458-6470, 1990.

Hemley, R. J., H. K. Mao, A. F. Goncharov, M. Hanfland, and V. V. Struzhkin, Synchrotron infrared measurements of H_2 and D_2 to 0.15 eV at megabar pressures, *Phys. Rev. Lett., 76*, 1667-1670, 1996.

Hemley, R. J., I. I. Mazin, A. F. Goncharov, and H. K. Mao, Vibron effective charges in dense hydrogen, *Europhys. Lett.*, 37, 403-407, 1997.

Holmes, N. C., W. J. Nellis, and M. Ross, Temperature measurements and dissociation of shock-compressed liquid deuterium and hydrogen, *Phys. Rev. B, 52*, 15835-15845, 1996.

Hubbard, W. B., M. Podolak, and D. J. Stevenson, The interior of Neptune, in *Neptune and Triton*, edited by D. P. Cruikshank, pp. 109-138, University of Arizona Press, Tucson, 1995.

Ichimaru, S., and S. Ogata (eds.), *Elementary Processes in Dense Plasmas*, Addison-Wesley, Reading, Mass., 1995.

Kaxiras, E., J. Broughton, and R. J. Hemley, Onset of metallization and related transitions in solid hydrogen, *Phys. Rev. Lett., 67*, 1138-1141, 1991.

Keesom, W. H., J. de Smedt, and H. H. Mooy, On the crystal structure of para-hydrogen at liquid helium temperatures, *Proc. Royal Acad. Amsterdam, 33*, 814-819, 1930.

Kerley, G. I., A model for the calculation of thermodynamic properties of a fluid, in *Molecular-Based Study of Fluids*, edited by J. M. Haile and G. A. Mansoori, pp. 107-138, American Chemical Society, Washington, D.C., 1983.

Klepeis, J. K., K. J. Schafer, T. W. Barbee, and M. Ross, *Science, 254*, 986-988, 1991.

Lazor, P., R. J. Hemley, and H. K. Mao, High-pressure study of the NH_3-H_2 system, *Bull. Am. Phys. Soc., 41*, 564, 1996.

Lee, C., D. Vanderbilt, K. Laasonen, R. Car, and M. Parrinello, Ab initio studies on high pressure phases of ice, *Phys. Rev. Lett., 69*, 462-465, 1992.

Lee, C., D. Vanderbilt, K. Laasonen, R. Car, and M. Parrinello, Ab initio studies on the strutural and dynamical properties of ice, *Phys. Rev. B, 47*, 4863-4872, 1993.

Lewis, J. S., *Physics and Chemistry of the Solar System*, Academic, New York, 1995.

Loubeyre, P., M. Jean-Louis, R. LeToullec, and L. Charon-Gerard, High pressure measurements of the He-Ne binary phase diagram at 296 K: Evidence for the stability of a stoichiometric $Ne(He)_2$ solid, *Phys. Rev. Lett., 70*, 178-181, 1993.

Loubeyre, P., R. LeToullec, D. Häusermann, M. Hanfland, R. J. Hemley, H. K. Mao, and L. W. Finger, X-ray diffraction and equation of state of hydrogen above one megabar, *Nature, 383*, 702-704, 1996.

Loubeyre, P., R. LeToullec, J. P. Pinceaux, J. Hu, H. K. Mao and R. J. Hemley, Phase diagram and equation of state of solid ^{4}He from single-crystal x-ray diffraction over a large P-T domain, *Phys. Rev. Lett., 71*, 2272-2275, 1993.

Magro, W. R., D. M. Ceperley, C. Pierleoni, and B. Bernu, Molecular dissociation in hot, dense hydrogen, *Phys. Rev. Lett., 76*, 1240-1243, 1996.

Mao, H. K., The 670-km discontinuity in the mantle: A bulk chemical composition boundary driven by phase transformation, *Eos Trans. Am. Geophys. Union*, 69, 1420, 1988.

Mao, H. K., and R. J. Hemley, Optical observations of hydrogen above 200 gigapascals: evidence for metallization by band overlap, *Science, 244*, 1462-1465, 1989.

Mao, H. K. and R. J. Hemley, Ultrahigh pressure transitions in solid hydrogen, *Rev. Mod. Phys.*, 66, 671-692, 1994.

Mao, H. K., A. P. Jephcoat, R. J. Hemley, L. W. Finger, C. S. Zha, R. M. Hazen, and D. E. Cox, Synchrotron X ray diffraction measurements of single-crystal hydrogen to 26.5 GPa, *Science, 239*, 1131-1134, 1988.

Natoli, V., R. M. Martin and D. M. Ceperley, The crystal structure of atomic hydrogen, *Phys. Rev. Lett., 70*, 1952, 1993.

Neimann, H. B., S. K. Atreya, G. R. Carignan, T. M. Donahue, J. A. Haberman, D. N. Harpold, R. E. Hartle, D. M. Hunten, W. T. Kasprzak, P. R. Mahaffy, T. C. Owen, N. W. Spencer, and S. H. Way, The Galileo probe mass spectrometer: Composition of Jupiter's atmosphere, *Science, 272*, 846-849, 1996.

Nellis, W. J., A. C. Mitchell, N. C. Holmes, and P. C. McCandless, Properties of planetary fluids at high shock pressures and temperatures, in *High Pressure Research in Mineral Physics: Application to Earth and Planetary Sciences*, edited by Y. Syono and M. Manghnani, pp. 387-391, Terra Scientific Publishing, Tokyo/AGU, Washington, D.C., 1992.

Nellis, W. J., A. C. Mitchell, P. C. McCandless, D. J. Erskine, and S. T. Weir, Electronic energy gap of molecular hydrogen from electrical conductivity measurements at high shock pressures, *Phys. Rev. Lett., 68*, 2937-2940, 1992.

Oppenheimer, B. R., S. R. Kulkarni, K. Matthews, and T. Nakajima, Infrared spectrum of the cool brown dwarf GI 229B, *Science, 270*, 1478-1479, 1995.

Pfaffenzeller, O., D. Hohl, and P. Ballone, Miscibility of hydrogen and helium under astrophysical conditions, *Phys. Rev. Lett.*, 74, 2599-2602, 1995.

Pruzan, P., Pressure effects on the hydrogen bond in ice up to 80 GPa, *J. Mol. Struct., 322*, 279-286, 1994.

Ross, M., The ice layer in Uranus and Neptune - diamonds in the sky, *Nature*, 292, 435-436, 1981.

Ross, M., H. C. Graboske, and W. J. Nellis, Equation of state experiments and theory relevant to planetary modelling, *Phil. Trans. R. Soc. Lond. A, 303*, 303-313, 1981.

Ross, M., F. R. Ree, and D. A. Young, *J. Chem. Phys., 79*, 1487, 1983.

Saumon, D., and G. Chabrier, Fluid hydrogen at high density: The plasma phase transition, *Phys. Rev. Lett., 62*, 2397-2400, 1989.

Somayazulu, M., L. W. Finger, R. J. Hemley, and H. K. Mao, High-pressure compounds in methane-hydrogen mixtures, *Science, 271*, 1400-1402, 1996.

Somayazulu, M., R. J. Hemley, A. F. Goncharov, H. K. Mao, and L. W. Finger, High-pressure compounds in the methane-hydrogen system: origin of stability and intermolecular interactions, *Eur. J. Solid State Inorg. Chem.*, to be published.

Stevenson, D. J., Cosmochemistry and structure of the giant planets and their satellites, *Icarus, 62*, 4-15, 1985.

Stewart, J. W., Compressibility of hydrogen and deuterium, *J. Phys. Chem. Solids, 1*, 146, 1956.

Struzhkin, V. V., A. F. Goncharov, R. J. Hemley, and H. K. Mao, Cascading Fermi resonances and the soft mode in dense ice, *Phys. Rev. Lett., 78*, 4446-4449, 1997.

Vinet, P., J. Ferrante, J. H. Rose, and J. R. Smith, Compressibility of solids, *J. Geophys. Res., 92*, 9319-9325, 1987.

Vos, W., L. W. Finger, R. J. Hemley, J. Z. Hu, H. K. Mao, and J. A. Schouten, A high-pressure van der Waals compound in solid nitrogen-helium mixtures, *Nature, 358*, 46-48, 1992.

Vos, W., L. W. Finger, R. J. Hemley, and H. K. Mao, Pressure dependence of hydrogen bonding in a novel H_2O-H_2 clathrate, *Chem. Phys. Lett., 257*, 524-530, 1996.

Vos, W. L., L. W. Finger, R. J. Hemley, and H. K. Mao, Novel H_2-H_2O clathrates at high pressures, *Phys. Rev. Lett., 71*, 3150-3153, 1993.

Vos, W. L., L. W. Finger, H. K. Mao, R. J. Hemley, and H. S. Yoder, Phase behavior of H_2-H_2O at high pressure, in *High-Pressure Science and Technology - 1993*, edited by S. C. Schmidt, J. W. Shaner, G. A. Samara, and M. Ross, pp. 441-444, AIP Press, New York, 1994.

Weir, S. T., A. C. Mitchell and W. J. Nellis, Metallization of fluid molecular hydrogen at 140 GPa (1.4 Mbar), *Phys. Rev. Lett., 76*, 1860-1863, 1996.

Wetherill, W., Provenance of the terrestrial planets, *Icarus, 58*, 4513-4520, 1994.

Zhang, K. and G. Schubert, Penetrative convection and zonal flow on Jupiter, *Science, 273*, 941-943, 1996.

Zharkov, V. N., *Interior Structure of the Earth and Planets*, Harwood Academic Publishers, New York, 1986.

Zharkov, V. N. and T. V. Gudkova, Modern models of giant planets, in *High Pressure Research in Mineral Physics: Application to Earth and Planetary Sciences*, edited by Y. Syono and M. Manghnani, pp. 393-401, Terra Scientific Publishing, Tokyo/AGU, Washington, D.C., 1992.

R. J. Hemley and H. K. Mao, Geophysical Laboratory and Center for High Pressure Research, Carnegie Institution of Washington, 5251 Broad Branch Rd. NW, Washington, DC 20015 USA.

The Melting Curve and Premelting of MgO

Ronald E. Cohen and J. S. Weitz

Geophysical Laboratory and Center for High Pressure Research,
Carnegie Institution of Washington, Washington D.C. 20015

The melting curve for MgO was obtained using molecular dynamics and a non-empirical, many-body potential. We also studied premelting effects by computing the dynamical structure factor in the crystal on approach to melting. The melting curve simulations were performed with periodic boundary conditions with cells up to 512 atoms using the ab-initio Variational Induced Breathing (VIB) model. The melting curve was obtained by computing ΔH_m and ΔV_m and integrating the Clapeyron equation so that we avoid all problems with nucleation and hysteresis of melting and crystallization. System size dependencies were also carefully checked and accuracy of our melting curve is not limited by small periodic cell sizes. Our ΔH_m is in agreement with previous estimates and we obtain a reasonable ΔV_m, but our melting slope dT/dP (114 K/GPa) is three times greater than the experimental determination of Zerr and Boehler (35 K/GPa), suggesting a problem with the experimental melting curve or an indication of exotic, non-ionic behavior of MgO liquid. We computed $S(q,\omega)$ from simulations of 1000 atom clusters using the Potential Induced Breathing (PIB) model. A low frequency peak in the dynamical structure factor $S(q,\omega)$ arises below the melting point which are related to the onset of bulk diffusion below the melting point. Diffusion rapidly increases to near liquid values below the melting point with a large activation energy, with a mechanism of many-atom exchanges and dynamical defect clusters. This destabilizes the crystalline state and leads to the melting transition.

INTRODUCTION

Understanding melting is crucial for understanding the evolution and dynamics of the Earth, and in order to trace the development of the Earth from its origin until now, it is important to know the melting temperatures, enthalpy of melting, and density of melts and solids as functions of composition, spanning the major element chemistry of the Earth. There is also fundamental interest in understanding the melting process. Why and how do crystals melt, and are there any precursors to the melting transition evident in the crystalline phase? Here we are particularly interested in how pressure might affect melting and premelting behavior, and since we are interested in eventually understanding melting in the Earth, we start with the simplest oxide, MgO, and study its melting and premelting in the crystalline phase as a function of compression.

Measuring melting curves to extreme pressures is

Properties of Earth and Planetary Materials
at High Pressure and Temperature
Geophysical Monograph 101

very difficult, and there have been significant discrepancies among laboratories on melting curves of the important materials Fe [*Anderson and Ahrens, 1996; Boehler, 1996*] and $MgSiO_3$ perovskite [*Sweeney and Heinz, 1993; Zerr and Boehler, 1993; Heinz et al., 1994; Sweeney and Heinz, 1995*]. It is also very difficult to calculate melting curves theoretically in spite of many attempts to develop predictive models for melting. Calculation of melting curves from fundamental physics is a difficult undertaking as well, because accurate potentials or electronic structure methods are needed to obtain the forces among atoms, long simulations are needed to equilibrate and obtain thermodynamic properties of the liquid, and since free-energies cannot be directly calculated, one must perform thermodynamic integrations or reversals [*Cohen and Gong, 1994*] in order to obtain the melting point.

Cohen and Gong [*1994*] predicted the melting curve of MgO to 300 GPa using molecular dynamics simulations for finite clusters. The interactions among atoms were obtained using the non-empirical potential induced breathing (PIB) model which had been shown previously to give excellent agreement with experiment for thermoelastic properties of MgO to high pressures [*Isaak et al., 1990*]. Molecular dynamics (MD) simulations using the closely related VIB (variationally induced breathing) model (described below) show exceptional accuracy for the equation of state of MgO, including high order properties such as the change in thermal expansivity with pressure [*Inbar and Cohen, 1995*]. Given the accuracy of the model for properties of the solid, it was surprising when the first experimental measurements for the melting curve of MgO showed a discrepancy of over a factor of three in the dT/dP slope, with the experiments showing a much shallower slope. Since the MD and lattice dynamics calculations showed that crystalline properties of MgO were well predicted by the models, such a large discrepancy could indicate a problem with the liquid simulations. Since the potentials do not use any information that is particular to the crystalline state and are based on fundamental physics, only one possibility seemed open – that there was a problem with the liquid simulations due to the use of finite clusters. *Cohen and Gong* [*1994*] used finite clusters of 64 to 1000 atoms and then extrapolated to the bulk by fitting the finite cluster results as functions of $1/L$ and then extrapolated as $1/L \rightarrow 0$, where L is the length (i.e. dimension) of the clusters. The extrapolation to bulk is effectively over about 20 orders of magnitude of system size, so there is cause for concern that this could introduce errors in the predicted melting curve. Thus here we have effectively eliminated size effects by using another technique based on similar potentials, but with periodic boundary conditions and no surfaces, to obtain the melting slope.

We also further continue the study of melting in clusters using PIB, and look for dynamical premelting effects by computing and studying the power spectrum $S(q,\omega)$ in the crystalline phase. *Cohen and Gong* [*1994*] found evidence of an intrinsic instability in the crystal near melting in the Lindemann ratio u_{rms}/a where u_{rms} is the r.m.s. displacement and a is the mean near neighbor distance. They found this ratio to be constant along the melting curve spanning 300 GPa and 15,000 K. Whereas such scaling is expected in power law potentials [*Ross, 1969*] where liquid and solid structures are constant along the melting curve, it is not constrained to behave so with realistic potentials. The large changes in liquid structure along the melting curve indicate that the constancy of the Lindemann ratio must have a deeper origin. *Cohen and Gong* [*1994*] hypothesized that in pure systems such as MgO, there might be an underlying instability in the crystal leading to melting. The fact that melting is a first-order transition in no way negates this possibility; for instance in $BaTiO_3$ there is universal agreement that there is an underlying, observable overdamped soft mode, in spite of the fact that the transition itself appears to be order-disorder and is first-order. *Cohen and Gong* proposed a similar scenario for MgO, but there was no evidence of what the underlying instability might be. The best candidate is the shear instability c_{11}-c_{12}, which vanishes at the melting point at zero pressure, but the instability occurs at higher temperatures than the melting point with increasing pressure; the meaning of this behavior is still a mystery. Here we try to understand better whether there is an underlying dynamical instability in the lattice and what it is by studying the dynamical structure factor or power spectrum $S(q,\omega)$ in crystalline clusters of 1000 atoms as melting is approached.

Premelting phenomenon has been seen in many types of crystals [*Ubbelohde, 1978*]. Experimental observations include anomalous increases in diffusion below melting [*Dimanov and Ingrin, 1995*], increases in heat capacity [*Richet and Fiquet, 1991*], and spectroscopic changes [*Richet et al., 1996*]. Understanding premelting has many practical applications, in that it is responsible for the phenomenon of frost heave in the ice-water system, and can be understood in terms of surface energies [*Ferrando et al., 1993; Wettlaufer et al., 1996*]. Much work has been done on premelting in clusters and finite systems concentrating on surface melting and roughening transitions [*Bastiannsen and Knops, 1996; Nagaev and Zil'bervarg, 1996*]. There is also considerable evidence that small clusters do not undergo discontinuous phase transitions, but rather go through a region of fluctuations between solid-like and liquid-like configurations [*Wells and Berry, 1994; Nayak et al., 1995; Bhattacharya et al., 1996*]. The clusters we are studying here of 1000 atoms are significantly larger than those discussed in the latter studies which range up to 55 atoms. Evidence was seen in *Cohen and Gong* [*1994*] for coexistence and slow fluctuations between solid and liquid in the van der Waals loop region of the transition, which is probably closely related to what is observed in the smaller clusters. Our interest in studying S(q,ω) is not however to understand better the dynamics of this fluctuation/coexistence regime, but rather to look for evidence of an approaching dynamical instability in the crystalline field as the melting point is approached.

METHODS

We performed classical molecular dynamics simulations for periodic and cluster systems, using the nonempirical VIB and PIB models, respectively. VIB [*Wolf and Bukowinski, 1988*] and PIB [*Cohen et al., 1987a,b*] are very similar ionic Gordon-Kim [*Gordon and Kim, 1972*] type models, in which the total charge density is modeled by overlapping ionic charge densities, which are computed from quantum mechanical atomic calculations with no adjustable parameters. Only the ionic charge and nuclear charge are input, and the ionic charges used are the nominal 2+ and 2- for Mg and O, respectively. In Gordon-Kim models, the total energy is a sum of three terms, the long-range electrostatic or Madelung energy, the self-energy of each atom or ion, and the short-range interaction energy which is a sum of the kinetic energy, short-range electrostatic, and exchange-correlation energy, all of which are functions of the model charge density. We use the Hedin-Lundqvist [*Hedin and Lundqvist, 1971*] parametrization of the exchange-correlation energy and the Thomas-Fermi kinetic energy in the interaction energy. For the self- energies we use the Kohn-Sham total energy [*Kohn and Sham, 1965*].

Since O^{2-} is not stable in the free state, it is stabilized with a sphere of +2 charge (called a "Watson sphere") in the atomic calculations, and the radius of this sphere is chosen to be different in the PIB and VIB models; this is the only difference between PIB and VIB. In the PIB model, the radius of the Watson sphere, r_{Wat} is chosen so that the electrostatic potential inside the sphere is the same as the Madelung (i.e. electrostatic) potential at the site the crystal in order to model the electrostatic stabilization of the O^{2-} ion by the crystal field. In the VIB model, r_{Wat} is chosen to minimize the total energy in the crystal for a given configuration of atoms. Both VIB and PIB give very similar results except at very high pressures where the VIB potential is softer and more accurate due to the fact that it includes short-range contributions to the O^{2-} size, as well as the long-range Madelung contributions [*Inbar and Cohen, 1995*]. They also give different LO-TO splittings; VIB gives the rigid ion LO-TO splitting whereas PIB gives reduced values that are closer to experiment (although for the wrong reasons) when a simple correction to reference the Madelung potential to the local average potential is included. This is due to the dependence in PIB of r_{Wat} on the magnitude of the potential, as opposed to potential differences. [*Cohen et al., 1987a, b*]. Since the VIB model is better behaved we used the VIB choice of r_{Wat} here for the periodic boundary condition computations. We employ the pair approximation, calculate pairwise interactions as functions of the distance between atoms and r_{Wat} on the anions, and fit an analytic function to the calculated energies as functions of distance and r_{Wat}. This function, which has up to 21 parameters for O-O interactions, can be evaluated much more rapidly than doing the full quantum calculations at each time step. The resulting potential has been severely tested for thermal properties of MgO, and we have great confidence in the potential.

In the molecular dynamics simulations, Newton's equation $\mathbf{F} = m\mathbf{a}$ is integrated forward in time. In VIB, the energy is minimized with respect to all of the $\mathbf{r}_{Wat}$ for each time step. In PIB, the Madelung potentials are computed at each time step. In both cases, the forces are obtained analytically for each configuration of atoms and Watson sphere radii r_{Wat}. The pair interactions and the self-energies of the O^{2-} anions are functions of r_{Wat} for each anion. The process of optimizing the Watson sphere radii at each time step follows the well known adiabatic, or Born-Oppenheimer approximation, which is appropriate for the VIB model in an insulator. Non-adiabatic processes which are important in hydrogen and possibly some metals are not important in a wide-gap insulator such as MgO. Thus this is a rigorous procedure for studying the dynamics of MgO. The many-body interactions in VIB are important to give proper elastic behavior (such as the Cauchy violations) in MgO [*Mehl et al., 1986; Isaak et al., 1990*], and to give agreement with LAPW charge densities and structures [*Mehl, 1988*], and they may be important for melting as well.

Melting Curve

We have simulated crystalline and liquid MgO both with periodic boundary conditions, i.e., with no surfaces, using the VIB model to obtain the melting curve. Periodic boundary conditions introduce long-range lattice structure onto a liquid, which should not be present, and this can cause systematic errors especially in ionic crystals with long-range forces. However, by studying two very different periodic cell sizes, 64 atoms and 512 atoms, we can test whether the quantities we calculate, V and T for given P and E, are affected. We obtain the change in enthalpy and volume, $\Delta H = \Delta E + P\Delta V$ and ΔV as functions of T and P between the solid and liquid, which at the melting point T_m gives us the melting slope through the Clapeyron equation,

$$\frac{dT}{dP} = \frac{T_m \Delta V_m}{\Delta H_m}. \tag{1}$$

Since the primary discrepancy with the experiment is the melting slope, and both theory and experiment agree on the melting point at zero pressure, we fix the zero pressure melting point at 3200K and then integrate the Clapeyron equation to give the melting curve.

This procedure circumvents the problems of hysteresis and problems associated with MD simulations for small periodic cells. Only the volume and enthalpy of the liquid and solid are needed at the melting point, and as shown below these are very stable quantities. The same procedure was recently used by *Boehler et al. [1996]* for CsI.

The systems consisted of a sample of 64 atoms, initially arranged in a cubic lattice. In order to check for system-size effects, some simulations were also carried out on 512 atom systems. Periodic boundary conditions were employed to eliminate surface effects, and timesteps of 1 fs were used. The equations of motion were numerically integrated using a fifth-order Gear predictor-corrector method [*Gear, 1966*]. Throughout our simulations, enthalpy was conserved to approximately 1 part in 10^6 per iteration.

In the periodic boundary condition simulations, we employed the variable-cell-shape technique of *Parrinello and Rahman* [*1980*]. In this technique, extra fictitious dynamical degrees of freedom associated with the shape of the computational cell are introduced to allow the computational cell volume and shape to vary. This technique conserves enthalpy, rather than energy, and the external pressure is kept constant, rather than the volume. The off-diagonal elements of the strain matrix were not allowed to vary in either crystal or liquid simulations to avoid problems with large fluctuations in the shape of the periodic cell in the liquid state. Diagonal elements were allowed to vary independently. Tests showed that this approximation did not affect the resulting enthalpies or volumes, but greatly increased the stability of the simulations. The fact that 64 atom and 512 atom cells gave identical results is also evidence that this approximation introduced negligible error.

Simulations were performed at P= 0, 12.5, 25, 50 and 100 GPa. At each pressure, MD runs were performed at various temperatures near the expected melting point in both solid and liquid. Initially, the kinetic energies of the atoms were scaled to obtain approximately the desired temperature. After equilibration, which lasted for 2 ps, we ran each simulation for an additional 6-15 ps during which the system volume, enthalpy, and kinetic energy were monitored each iteration. The length of each run was determined by the convergence of the average volume of the system. The enthalpy was a constant of the motion. The temperature was calculated from the average kinetic energy, and the volume was averaged over the length of the equilibrated run.

After performing these simulations at several temperatures in the solid, the limit of superheating was reached, and the solid melted. This was determined both from the presence of diffusion and from the dramatic reduction in intensity of a simulated Bragg reflection intensity. After melting, the temperature dropped because of the conversion of the latent heat of melting to potential energy. Simulations were then performed at various temperatures in the liquid state by cooling or heating the liquid via velocity scaling, followed by reequilibration, and volumes and enthalpies of the liquid were obtained. Only the volume and enthalpies of the liquid and crystal were used, via equation 1, to obtain the melting curve. The point at which melting occurred during superhearing in the simulations was always much higher than the equilibrium melting curve.

Power Spectrum

Since the previous cluster calculations [*Cohen and Gong, 1994*] used PIB as opposed to VIB, we use PIB here for the study of the dynamic structure factor S(q,ω) in clusters. The power spectrum S(q,ω) was obtained as follows. Simulations for clusters of 1000 atoms were performed for MgO at P=0, 100 and 150 GPa at a series of temperatures. At zero pressure the cluster had free boundary conditions, and for the high pressure runs pressure was imposed by enclosing the cluster in a cubic elastic box as in *Cohen and Gong.* Atoms that hit the box walls reflect specularly; the momentum component perpendicular to the wall is reversed. The Verlet algorithm was used to integrate the classical Newton's equations with a time step of 1-3 fs. Simulations were started with equilibrated system prepared by *Cohen and Gong* [*1994*] and were run for 20,000-40,000 time steps. Frames of the atomic positions were saved every 4-8 time steps for computation of S(q,ω). The density function ρ is defined as

$$\rho(\mathbf{r},t)=\sum_{i=1}^{N}\delta(\mathbf{r}-\mathbf{r}_i(t)) \quad (2)$$

and the transform is

$$p(\mathbf{k},t)=\sum_{i=1}^{N}e^{i\mathbf{k}\cdot\mathbf{r}_i(t)}. \quad (3)$$

The dynamical structure factor is defined as

$$S(\mathbf{q},\omega)=\int dtF(\mathbf{q},t)e^{i\omega t}W(t) \quad (4)$$

where $\mathbf{q}=2\pi\mathbf{k}/\mathbf{l}$ for mean cubic cell lattice constant l and $F(\mathbf{q},t)$ is the intermediate structure function defined as the correlation function

$$F(\mathbf{q},t)=\langle p(\mathbf{q},t)p^{*}(\mathbf{q},0)\rangle \quad (5)$$

where $\langle\rangle$ indicates an average over all time origins. We used the Blackman-Harris exact three parameter window function [*Harris, 1978*]

$$\begin{aligned}W(t)= &\ 0.42659071+0.49656062\cos(2\pi t/T)\\ &+0.07684867\cos(4\pi t/T) \quad (6)\end{aligned}$$

where $T=2\tau$, and τ is the total time), in order to minimize artifacts from the finite time series. Equation 4 is solved by a discrete fast Fourier transform over the time slices. In order to improve statistics [*Press et al., 1992*] runs were divided into eight segments and seven overlapping transforms were performed and averaged. We also averaged over equivalent q's and the final $S(\mathbf{q},\omega)$ was smoothed with a running average over ±1 cm^{-1}.

There are non-trivial issues regarding **q** dependent quantities such as $S(\mathbf{q},\omega)$ for clusters. In periodic boundary conditions, the meaningful **q**-space is quantized according to the size of the periodic cell. However in a cluster, results can be obtained for any q. This is because we have free boundary conditions at the surfaces so that the waves do not need to have nodes at the surfaces. However for almost all q's, there is a large peak at $S(q,\omega=0)$ due to the fact that there are not equal numbers of positive and negative displacements for most q's. At these q's it is difficult to obtain a clear spectrum of $S(q,\omega)$ because aliasing and spillout of the $\omega=0$ peak results from the finite time sampling and windowing, thus masking the physically important behavior at small omega. Thus we pick q's that are minima in $S(q,0)$, and these turn out to be close to the commensurate q's at $(\frac{q}{2}\frac{q}{2}\frac{q}{2})$, $(qq0)$, and $(q00)$ where q=0.1,0.2,0.3,0.4, and 0.5 $4\pi/a$ where a is the cubic lattice constant (note that for an fcc lattice the X point is at $(2\pi/a,0,0)$ and the L point is at $(\pi/a,\pi/a,\pi/a)$ where a is the cubic lattice constant). A further complication is that at finite temperatures the minima are sharp single peaks along (q00), sharp double peaks along (qq0), and sharp triple peaks along (qqq) due to the thermal motions. Thus we displaced our choice of q slightly from the commensurate q's to obtain the best spectra. When q is in the first Brillouin zone $S(\mathbf{q},\omega)$ gives to first order only the longitudinal excitations since the particle positions enter only as $q\cdot r$ [*Kaneko and Ueda, 1989*]. We have examined tranverse oscillations as well by studying q's such as $(q,0,8\pi/a)$ which approximately show the transverse behavior at $(q,0,0)$.

RESULTS AND DISCUSSION

Melting Curve

Results of our simulations showing the enthalpies and volumes of solid and liquid at P = 0 are shown in Figure 1. As a check on system-size effects, the calculations were repeated using a 512 atom system. No system-size effects were seen in volume or enthalpy. For any temperature, one may obtain the difference in volume and enthalpy between solid and liquid, either by interpolating or extrapolating. From the figure one can see that there is some temperature variation in both of these quantities, but the differences, ΔV and ΔH, are quite stable.

To obtain the melting curve, the melting temperature at each pressure was estimated to be the temperature of the liquid just beyond the limit of superheating of the solid. The ΔV and ΔH of melting were then calculated for those temperatures, and equation 1 was numerically integrated to obtain new estimates of the melting temperatures. This process was repeated until the melting temperature converged. As mentioned above, the melting point at 0 GPa was anchored to 3200 K, which is consistent with the melting reversals in *Cohen and Gong* [*1994*]. The resulting melting curve is shown in Figure 2.

Now we compare our melting curve with other recent molecular dynamics results. Our melting curve is in excellent agreement with that of Vocadlo and Price [*1996*], based on empirical pair potentials. Agreement is less good with Belonoshko and Dubrovinsky [*1996*] (BD

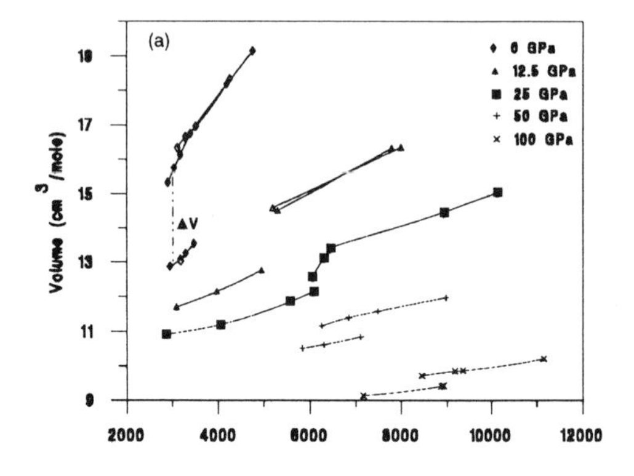

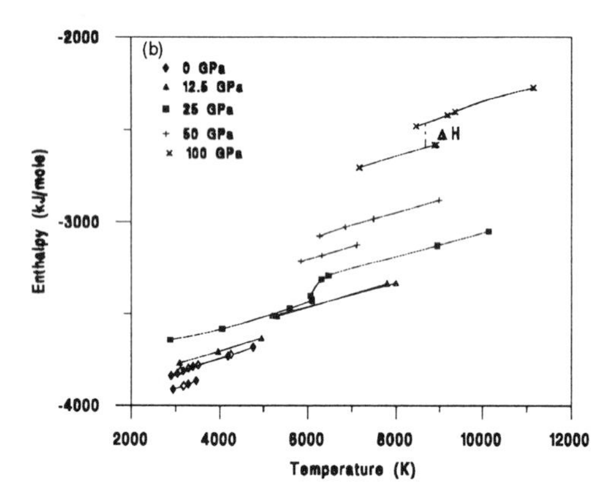

Figure 1. Volumes and enthalpies of crystalline and liquid MgO as functions of temperature at different pressures. The open diamonds and triangles are for 512 atom supercells; other symbols are for 64 atom supercells. The agreement for the different size periodic systems indicates that we are converged for the volume and enthalpy. ΔV and ΔH are indicated at 0 GPa and 100 GPa respectively. (a) volume, (b) enthalpy.

below), and their results deviate sharply both from our melting curve and Vocadlo and Price's with increasing pressure, so that BD obtain T_m = 7000 K at above 140 GPa, compared with 70 GPa for Vocadlo and Price and this study. Two possible sources for this discrepancy are differences in potentials and the method of simulation. Empirical potentials by design reproduce the data to which they are fit, but may not extrapolate well or predict properties outside the fitted range. The fact that BD reproduced the NaCl melting curve is not evidence that their MgO potential is accurate; many-body forces are very important in oxides [*Cohen et al., 1987a,b*], but negligible in NaCl. Secondly, DB used visual inspection of snapshots to distinguish melt from solid, yet *Cohen and Gong* showed that melt and crystal structures become more similar with increasing pressure, and more difficult to distinguish from snapshots of atomic positions; perhaps BD are misidentifying the melting transition at high pressure. Unlike BD, the method we use here to determine the melting curve has no problems with hysteresis, does not require bracketing, and does not require subjective evaluations of liquid versus crystalline structure.

The melting curve we obtain here is significantly lower than the cluster melting of *Cohen and Gong* [*1994*] in the intermediate pressure range, though they seem to converge to similar values both at high pressures and at zero pressure. BD speculate at length about the *Cohen and Gong* simulations and we respond to their assertions here. First BD claimed that the hysteresis at zero pressure was too large to judge how well the experimental value is reproduced. The hysteresis was discussed in detail in *Cohen and Gong*, and disappears by 5 GPa. Since the goal of that study was to understand the physics of melting at pressure up to 300 GPa, the uncertainty in the 0-5 GPa region is immaterial. Secondly, BD claim that the use of a box to enclose the cluster to simulate pressure introduces artifacts from the wall effect. This is true, and is why *Cohen and Gong* did simulations as a function of sample size and extrapolated to infinite sample size to remove size dependent effects. BD suggest that since the equation of state of their potential and Cohen and Gong's was similar, melting differences cannot be due to the potential. However, melting is not a function of the equation of state alone. Most important are the shear properties. The PIB and VIB models are many-body potentials, unlike BD's, and give proper Cauchy violations and shear behavior, thus should be more reliable for melting simulations. Finally, it should be emphasized that PIB and VIB are not identical potentials, so that the differences in the *Cohen and Gong* melting curve and the present work are not due entirely to the differences between the simulation methods. The *ab initio* PIB and VIB potentials have computational burdens orders of magnitude greater than simple pair potentials such as those used by BD, but have the advantage of being based on fundamental physics.

Nevertheless, internal evidence in *Cohen and Gong* [*1994*] does suggest that the low pressure part of the melting curve was too high compared with bulk melting, because the clusters melted at higher temperatures than the bulk elastic instability c_{11}-c_{12}, which must be an upper bound for the bulk melting curve. The reason for this discrepancy is not clear, since the size dependence found at pressures below 20 GPa was fairly small. An unsolved mystery observed in *Cohen and Gong*, was the fact that melting coincided with the elastic instability at low pressures, but occurred well below the elastic instability at high pressures. This will be addressed fur-

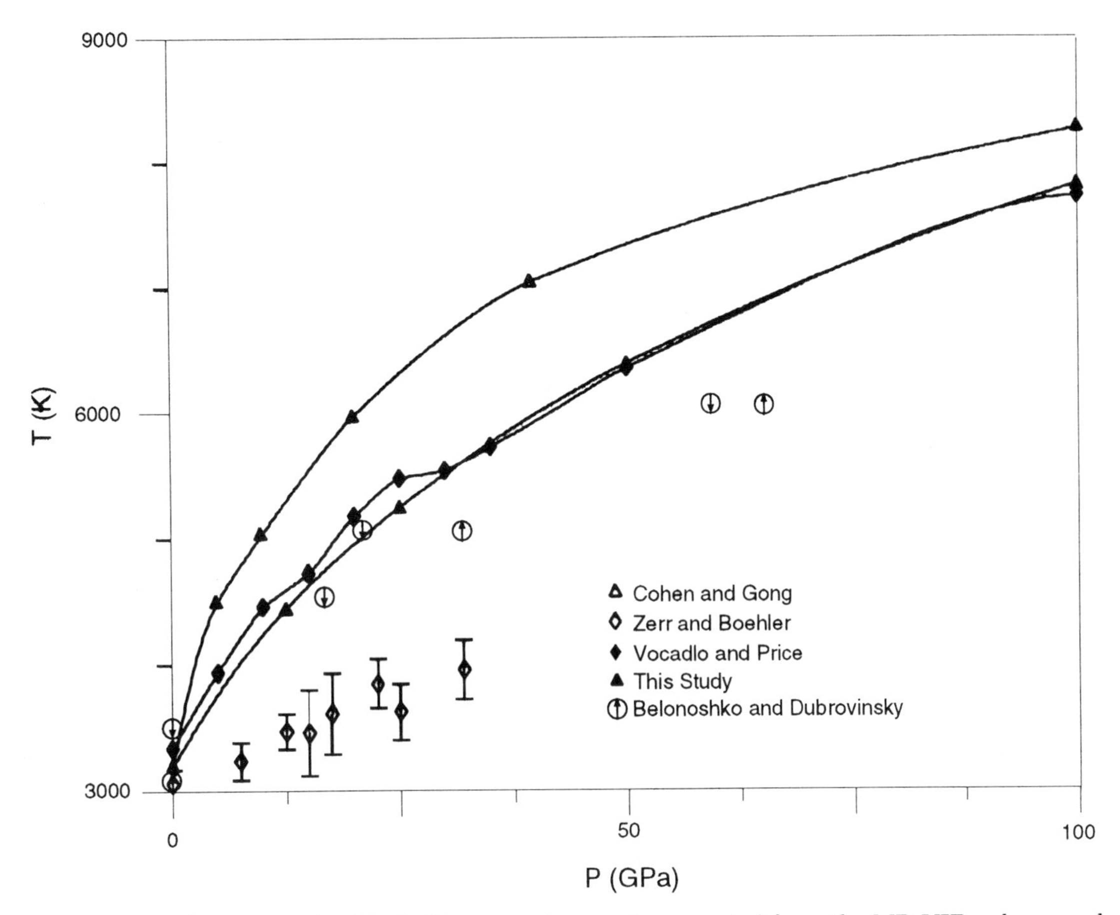

Figure 2. Melting curve for MgO. The present curve is computed from the MD VIB volume and enthalpy data by integrating the Clapeyron equation (eq. 1) starting at 3200 K at 0 GPa. The *Cohen and Gong [1994]* curve was obtained by reversing the melting transition for clusters and extrapolating to infinite system size using the PIB model. The *Vocadlo and Price* [*1996*] curve was obtained by melting periodic systems with an empirical potential. There is almost perfect agreement between the current results and the Vocadlo and Price results. There is a factor of three discrepancy in the slope with the experimental results of *Zerr and Boehler [1994]*. The differences between the Gong and Cohen curves and the present results are primarily due to the use of clusters in the earlier study, with extrapolation to bulk melting. The present computation is much more straightforward, and is more appropriate to the bulk. Agreement with the earlier Gong and Cohen computations is quite good considering the differences in potentials and methods, and give further credence to the present results. The melting brackets from BD are significantly lower than those of this study and Vocadlo and Price [*1996*] at high pressures.

ther below, but here we observe that the superheating of clusters relative to the bulk melting at low pressures may be due to the lack of long-range elastic modes in small clusters. This may not be only a matter of hysteresis, but also the melting temperature itself since the elastic modes contribute to the free energy. If the elastic instability onsets at small q, it may not be apparent from size dependence studies with small clusters.

There is a large discrepancy between our predicted melting curve and the experimental melting results of Zerr and Boehler, amounting to a factor of three in the slope dT/dP (114 K/GPa versus 35, respectively). This implies a discrepancy in ΔV_m and/or ΔH_m through the Clapeyron equation (equation 1). Unfortunately, neither ΔH_m nor ΔV_m have been measured directly for any alkaline earth oxide. Nevertheless, our value for ΔH_m is consistent with literature estimates [*Chase, Jr. et al., 1985*]. With our methods it would be very unexpected to obtain a reasonable ΔH_m and an anomalous ΔV_m because the potentials vary exponentially with distance and a large error in ΔV_m would give a huge error in ΔH_m. Also ΔV_m is in the range expected for

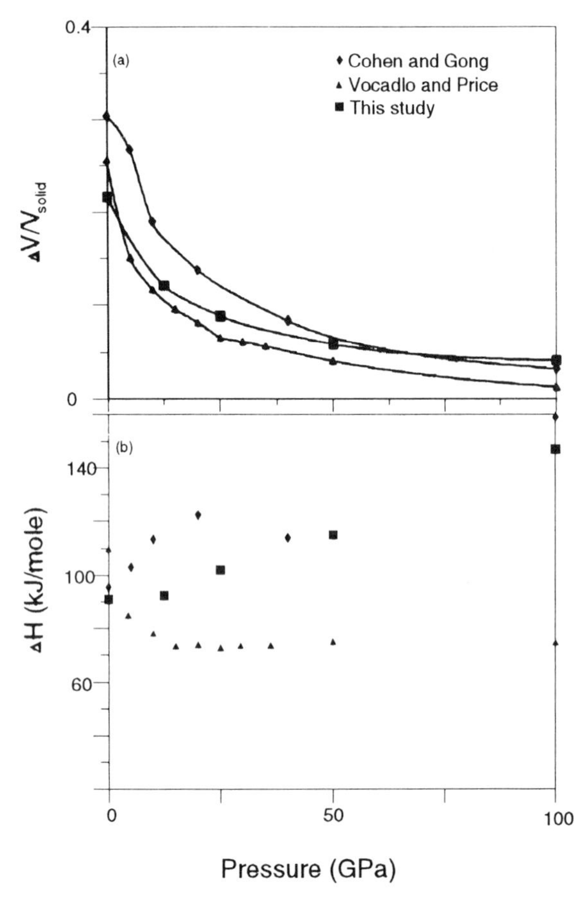

Figure 3. Fractional (a) volume of melting and (b) enthalpy of melting versus pressure.

ionic crystals [*Ubbelohde, 1978*]. Diamond anvil measurements on melting of alkali halides, which agree with other measurements, are not evidence for the correctness of the MgO melting curve in the same lab; not only is MgO a different material, but the experimental design is different. For MgO the sample was heated with a CO_2 laser in an Ar pressure medium [*Zerr and Boehler, 1994*], whereas the alkali halides were heated via a YLF laser on a tungsten foil and there was no pressure medium [*Boehler et al., 1996*]. One possible explanation of the origin of the discrepancy would be Ar solubility in the MgO melt, thus depressing the melting curve at high pressures. Another possibility is the melting criterion, which perhaps becomes more difficult with increasing pressure. Finding the origin of the discrepancy is important since similar methods are being used to obtain melting curves for other materials as well.

We show the fractional change in volume on melting and ΔH_m versus pressure in Figure 3. Agreement is quite good for the volume of melting with *Cohen and Gong* [*1994*] and with *Vocadlo and Price* [*1996*]. The enthalpy of melting agrees within the precision of the earlier cluster results of *Cohen and Gong*, but there is an explained discrepancy with the results of Vocadlo and Price which do not seem consistent with their melting curve.

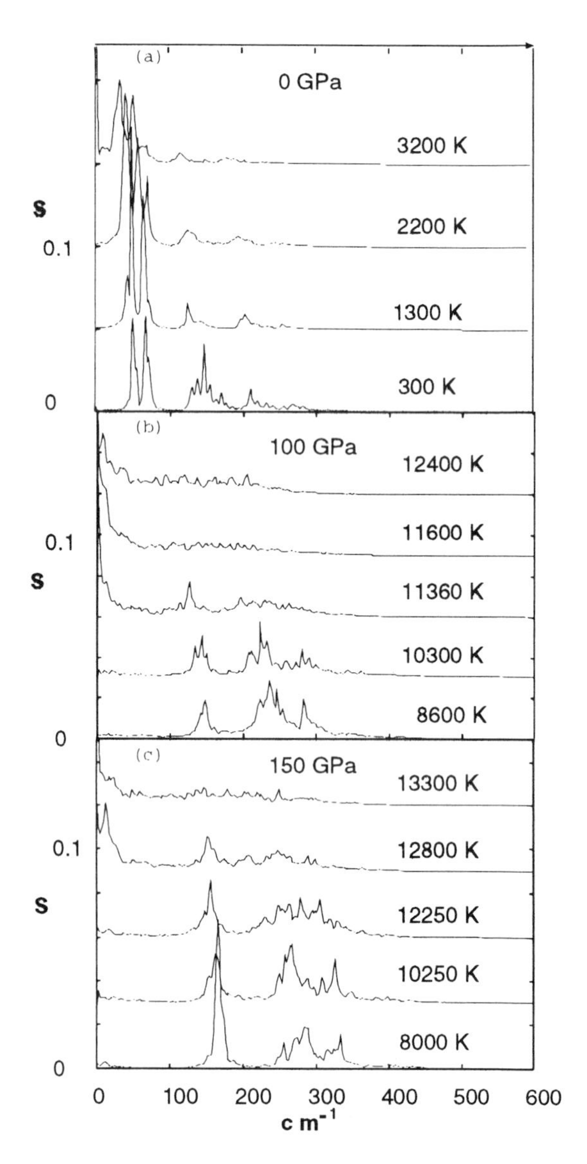

Figure 4. $S(q,\omega)$ for $q = (2\pi/5a, 2\pi/5a, 0)$ at (a) 0 GPa, (b) 100 GPa, and (c) 150 GPa as functions of temperature. The curves are offset for each temperature shown. At 0 GPa, the crystal has not yet melted at 3200 K, but the low frequency peak is evident. At 100 GPa, melting occurs between 11360 and 11600 for this cluster size, and a low frequency peak arises before melting. At 150 GPa melting occurs between 12800 and 13300K, and there is a low frequency peak at 12800 K.

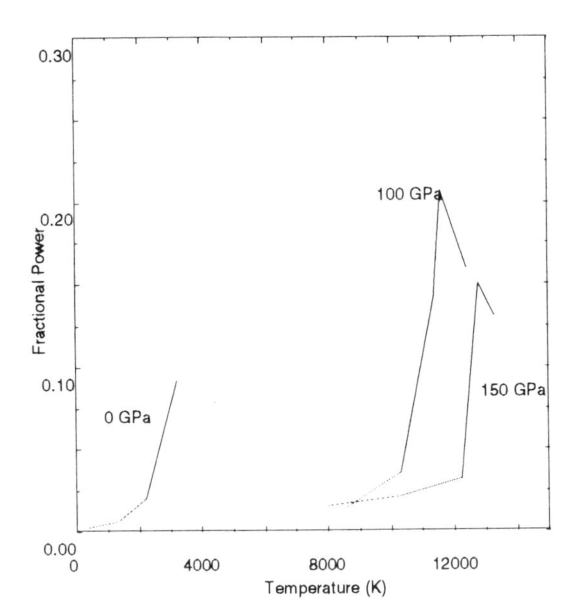

Figure 5. Low frequency power (integrated from 3-19 cm^{-1}) as a function of temperature at different pressures. At the melting transition the low frequency power seems to drop. Note that melting requires superheating at zero pressure, but little hysteresis is observed at high pressures. Also, note that melting occurs at higher temperatures than the bulk at high pressures.

Power Spectrum

The power spectrum $S(q,\omega)$ is quite complicated for our clusters. The complication over periodic boundary conditions is that we observe not only phonon-like excitations, but also free oscillations of the cluster [*Ozaki et al., 1991*], which are particularly evident at zero pressure due to the free surfaces. We find both phonon-like excitations which are dispersive in q, and free oscillation modes for the cluster which are not dispersive. Similar behavior was observed in tiny Ar_{13} clusters [*Bhattacharya et al., 1996*]. The phonon peaks are also split due to the shape of the cluster and its small finite size.

We observe interesting behavior in the cluster $S(q,\omega)$ on increasing temperature towards melting. These results are shown in Figure 4. We found growth in the low frequency response significantly before the melting transition. Figure 5 shows the low frequency part of the response (3-19 cm^{-1}) as a function of temperature. The rapid rise below the melting transition is evidence of premelting behavior which may be related to increases in heat capacity on approach to melting observed in many systems [*Richet and Fiquet, 1991*]. It may also be related to instabilities driven by dislocations in larger systems [*Lund, 1992*].

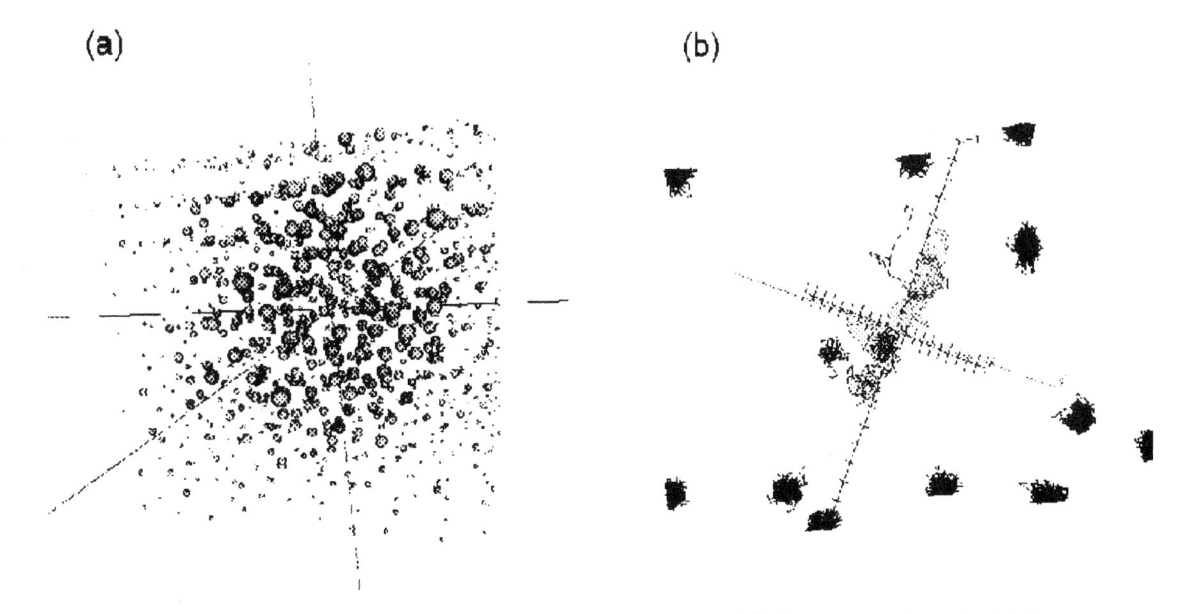

Figure 6. Representations of origin of low frequency power. (a) Representation of which atoms contribute to the low frequency power at 100 GPa and T=11360 in the crystal before melting. Note that even though many atom exchanges are common, a snapshot looks perfectly crystalline. (b) Representation of trajectories of selected atoms, with the intensity of color chosen to show the contribution to the low frequency power at P=100 GPa, and T= 1136 K. The black trajectories show atoms that do not contribute to the low frequency response–they show only oscillating, non-diffusive motions. Atoms that contribute to the low frequencies response do diffuse and seem to exchange with other atoms in complicated motions.

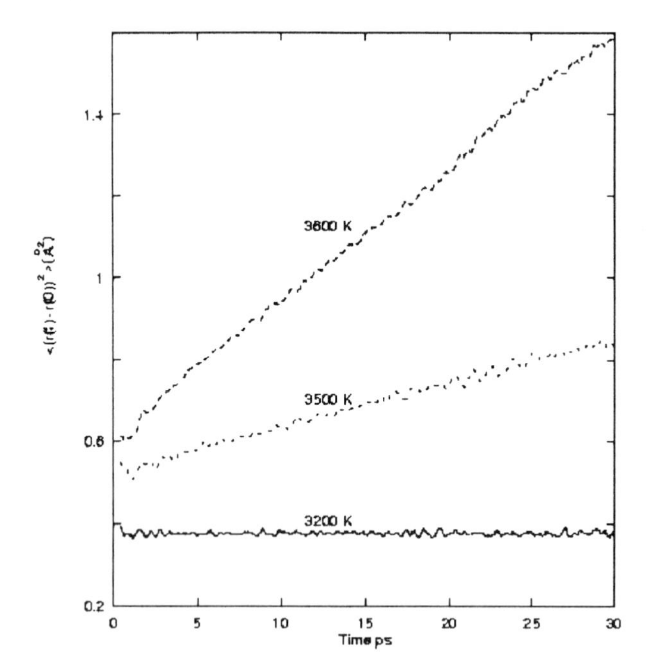

Figure 7. Mean squared displacement as a function of time at 0 GPa for a thousand atom cluster. The slope at moderate times is given by $6D$ where D is the diffusion constant. No diffusion is statistically detectable at 3200 K, but below melting in superheated crystal diffusion becomes very large approaching liquid values. At 3500 K we obtain $D = 1.7\times10^{-7}$cm^2/s and 5.5×10^{-7}cm^2/s at 3500K in the liquid.

Since our clusters have surfaces, the first obvious question is whether the low frequency power we observed on approach to melting is related to surface melting or other surface changes, or is in the bulk of the cluster. In order to determine where the low frequency power is localized we filtered the Fourier representation of the density function (equation 3) and then transformed back to real space. Figure 6 shows a representative snapshot of resulting low frequency, low q weight contribution on each atom, with the radii of the atoms proportional to their contribution to the low frequency response at 100 GPa and T= 11360 K. It is clear that the low frequency power is a bulk effect and not localized at the surface. The low frequency response is stronger in the longitudinal than transverse response, but it appears in both. The surface layer probably contributed low frequency power as well, at lower frequencies than our window.

Next we consider what dynamics are involved in the low frequency part of $S(q, \omega)$ by tracing out selected trajectories of atoms and coloring the trajectories by their contribution to the low frequency power (figure 6b). If all trajectories were depicted it would be difficult to make anything out, but it is clear from examining many trajectories and snapshots that at high pressures the low frequency power is due to complicated many atom diffusive exchanges.

Diffusion gives rise to a central peak in $S(q, \omega)$ with peak width $2Dk^2$ [*March and Tosi, 1976*] and the low frequency response and the trajectories we observe are consistent with the onset of large scale bulk diffusion below the melting temperature. To further test this we have computed $< (r(t) - r(0))^2 >$ (Figure 7) which equals $A + 6Dt$ at medium times (at small times the mean squared displacement vanishes and it long-times it saturates due to the finite cluster size). Below the premelting region the diffusion is so slow as to be unresolvable in these simulations, but in the premelting region large diffusion constants are obtained; at zero pressure 3500K in the superheated solid we obtain $D = 2\times10^{-7}$ cm^2/s, only a factor of 6 smaller than the value we obtain in the liquid at 3600 K of 6×10^{-7}. At 100 GPa, we get 2×10^{-5} cm^2/s at 11450 K in the solid, and 2×10^{-4} cm^2/s in the liquid at 11640 K, whereas $D = 6\times10^{-10}$ at 8570 K, below the premelting region. Figure 8 shows $\log(D)$ versus $1/T$; diffusion in the premelting region does not appear Arrhenius-like. The effective activation energy is very large, reaching 25 MJ/mole right before melting, and averaging 6500 kJ/mole from 8570 K to 11450 K. It is possible that the increase of diffusivity to liquid values in the crystalline phase may be due to the continuous melting transition found in clusters, as opposed to first-order melting transitions appropriate to thermodynamic bulk systems. Studies in periodic boundary conditions would show whether this is the case. Experimental observation of premelting effects with large increases in diffusivity before melting suggest that the phenomena we observe are applicable to the bulk.

Large increases diffusion and in effective activation energy have been observed in the premelting region experimentally, and an activation energy of 1524 kJ/mole is found for Ca diffusion in diopside ($CaMgSi_2O_6$) [*Dimanov and Ingrin, 1995*] for example. The authors ascribe this activation energy to the activation energy for formation of Frenkel (vacancy plus interstitial) defects, but this value (and our larger values) are too large to be ascribed to simple defect pairs. An MD simulation for premelting in Si found increased diffusion and activation energy approaching melting and showed that it was due to the formation of large defect clusters during multiatom exchanges [*Smargiassi and Car, 1996*]. This is consistent with trajectories observed in animations of our simulations in the premelting region (Figure 6). It is intriguing to speculate that it is the onset of mul-

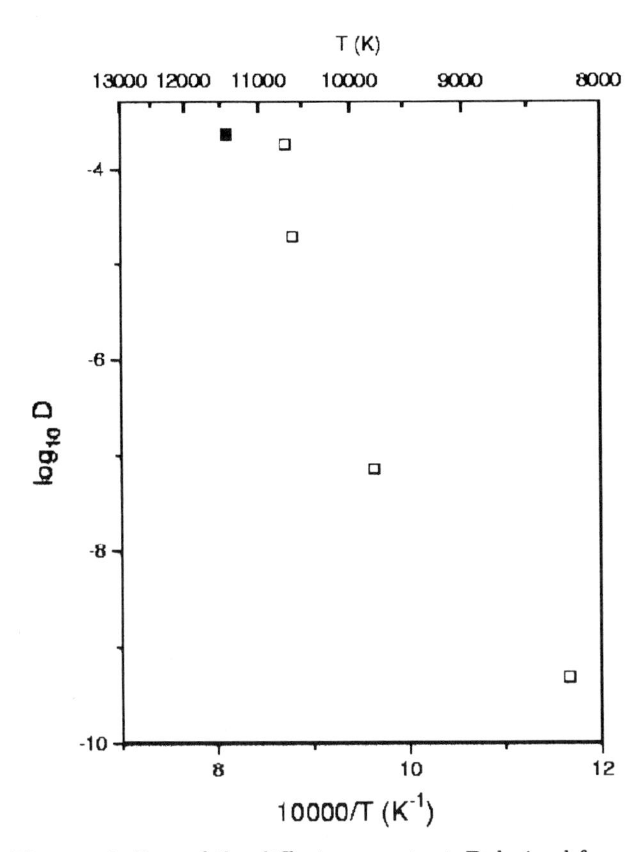

Figure 8. Log of the diffusion constant D derived from the mean-squared displacement at 100 GPa for a thousand atom cluster. Non-Arrhenius behavior is observed approaching melting (between 11453 and 12330 K) and the effective activation energy becomes very large, perhaps even diverging.

tiatom diffusion and defect clusters that leads to the instability of the crystalline phase and to melting.

Finally, we consider again the observation in *Cohen and Gong* that melting coincided with the c_{11}-c_{12} elastic instability at low pressures, but occurred well below the elastic instability at high pressures. One interpretation is a change in the driving force for melting with pressure, so that destabilization of the crystal is due primarily to the elastic instability at low pressures, but the onset of large scale diffusion drives melting at higher pressures. The relative importance of diffusional instabilities and elastic instabilities thus may be system and pressure dependent. There in fact can be a continuous change betyween these two methods, characterized by the dispersion of the instabilities.

CONCLUSIONS

We have performed MD simulations of periodic bulk and finite 1000 atom clusters of the melting of MgO. We obtain a melting curve that has a slope three times greater than that obtained experimentally, but obtain reasonable volumes and enthalpies of melting. The experiments should be repeated using different techniques. If indeed the ionic model is as far off from experiment as the current results show, something exotic must occur in the electronic structure and bonding of MgO liquid. The alternative is that there is some systematic error in the experiment. We have also studied the power spectrum on heating in MgO clusters and find evidence for premelting phenomena involving many atom exchanges. It is possible that the onset of these many atom exchanges and dynamical defect clusters leads to the melting instability, and to the thermodynamic destabilization of the crystal.

Acknowledgmens. Mark Kluge performed technical assistance and performed the periodic bulk simulations. We thank Iris Inbar and Joe Feldman for helpful discussions. This work was supported by NSF EAR-9418934, and computations were performed on the Cray J90 at the Carnegie Institution of Washington.

REFERENCES

Anderson, W. W. and T. J. Ahrens, Shock temperature and melting in iron sulfides at core pressures, *J. Geophys. Res., 101*, 5627-5642, 1996.

Bastiannsen, P. J. M. and H. J. F. Knops, Is surface melting a surface phase transition? *J. Chem. Phys., 104*, 3822-3831, 1996.

Belonoshko, A. B. and Dubrovinsky, L. S., Molecular dynamics of NaCl (B1 and B2) and MgO (B1) melting: Two-phase simulation, *Amer. Mineral., 81*, 303-316, 1996.

Bhattacharya, A., B. Chen, and S. D. Mahanti, Structural dynamics of clusters near melting, *Phys. Rev. E, 53*, R33-R36, 1996.

Boehler, R., Experimental constraints on melting conditions relevant to core formation, *Geochim. Cosmochim. Acta, 60*, 1109-1112, 1996.

Boehler, R., M. Ross, and D. B. Boercker, High-pressure melting curves of alkali halides, *Phys. Rev. B, 53, 556-563*, 1996.

Chase, M. W., Jr., C. A. Davies, J. R. Downey, Jr., D. J. Frurip, R. A. McDonald, and A. N. Syverud, JANAF Thermochemical Tables, Third, *J. Phys. Chem. Ref. Data, 14*, Suppl., 1985, p. 1470.

Cohen, R. E., L. L. Boyer, and M. J. Mehl, Lattice dynamics of the potential induced breathing model: First principles phonon dispersion in the alkaline earth oxides, *Phys. Rev. B, 35*, 5749-5760, 1987a.

Cohen, R. E., L. L. Boyer, and M. J. Mehl, Theoretical studies of charge relaxation effects on the statics and dynamics of oxides, *Phys. Chem. Minerals, 14*, 294-302, 1987b.

Cohen, R. E. and Z. Gong, Melting and melt structure of MgO at high pressures, *Phys. Rev. B, 50*, 12301-12311, 1994.

Dimanov, A. and J. Ingrin, Premelting and high-temperature diffusion of Ca in synthetic diopside: an increase of the cation mobility, *Phys. Chem. Minerals*, *22*, 437-442, 1995.

Ferrando, R., R. Spadacini, and G. E. Tommei, Theory of diffusion in premelting systems, *Nuovo Cimento D (Italy)*, *15D*, 557-563, 1993.

Gear, C. W., The numerical integration of ordinary differential equations of various orders, Argonne National Laboratory, ANL 7126, 1966.

Gordon, R. G. and Y. S. Kim, Theory for the forces between closed-shell atoms and molecules, *J. Chem. Phys.*, *56*, 3122-3133, 1972.

Harris, F. J. On the use of windows for harmonic analysis w/discrete Fourier transforms, *Proc. IEEE, 66*, 51-83, 1978.

Hedin, L. and B. I. Lundqvist, Explicit local exchange-correlation potentials, *J. Physics C, 4*, 2064-2083, 1971.

Heinz, D. L., E. Knittle, J. S. Sweeney, Q. Williams, and R. Jeanloz, High-Pressure Melting of (Mg,Fe)SiO_3 Perovskite, *Science, 264*, 279-280, 1994.

Inbar, I. and R. E. Cohen, Origin of ferroelectricity in $LiTaO_3$ and $LiNbO_3$: LAPW total energy calculations, *Ferroelectrics, 164*, 45-56, 1995.

Isaak, D. G., R. E. Cohen, and M. J. Mehl, Calculated elastic and thermal properties of MgO at high pressure and temperatures, *J. Geophys. Res.*, *95*, 7055-7067, 1990.

Kaneko, Y. and A. Ueda, Dynamical structure factor of α-AgI, *Phys. Rev. B, 39*, 10281-10287, 1989.

Kohn, W. and L. J. Sham, Self-consistent equations including exchange and correlation effects, *Phys. Rev. A, 140*, 1133-1140, 1965.

Lund, F., Instability driven by dislocation loops in bulk elastic solids: Melting and superheating, *Phys. Rev. Lett.*, *69*, 3084-3087, 1992.

March, N. H. and M. P. Tosi, Atomic Dynamics in Liquids, Dover, New York, 330 pp., 1976.

Mehl, M. J., LAPW electronic structure calculations for MgO and CaO, *J. Geophys. Res.*, *93*, 8009,1988.

Mehl, M. J., R. J. Hemley, and L. L. Boyer, Potential induced breathing model for the elastic moduli and high-pressure behavior of the cubic alkaline-earth oxides, *Phys. Rev. B, 33*, 8685-8696, 1986.

Nagaev, E. L. and V. E. Zil'bervarg, Double surface-bulk melting and suppression of overheating at first-order phase transitions, *Phys. Rev. B, 53*, 5011-5014, 1996.

Nayak, S. K., R. Ramaswamy, and C. Chakravarty, 1/f spectra in finite atomic clusters, *Phys. Rev. Lett.*, *74*, 4181-4184, 1995.

Ozaki, Y., M. Ichihashi, and T. Kondow, Analysis of breathing vibration of nearly spherical Ar clusters based on a dense sphere model, *Chem. Phys. Lett.*, 182, 57-62, 1991.

Parrinello, M. and A. Rahman, Crystal structure and pair potentials: A molecular dynamics study, *Phys. Rev. Lett.*, *45*, 1196-1199, 1980.

Press, W. H., S. A. Teukolsky, W. T. Vetterling, and B. P. Flannery, *Numerical Recipes in FORTRAN: The Art of Scientific Computing*, Second, Cambridge University Press, New York, 1992.

Richet, P., D. Andrault, and B. O. Mysen, Melting and premelting of silicates: Raman spectroscopy and X-ray diffraction of Li_2SiO_3 and Na_2SiO_3, *Phys. Chem. Miner.*, *23*, 157-172, 1996.

Richet, P. and G. Fiquet, High-temperature heat capacity and premelting of minerals in the system MgO-CaO-Al_2O_3-SiO_2, *J. Geophys. Res.*, *96*, 445-456, 1991.

Ross, M.,Generalized Lindemann melting law, *Phys. Rev.*, *184*, 233-242, 1969.

Smargiassi, E. and R. Car, Dynamical effects and vacancy motion in silicon at high temperature, *Intl. J. Mod. Phys. C (Physics and Computers)*, *7*, 57-64, 1996.

Sweeney, J. S. and D. L. Heinz, Melting of iron-magnesium-silicate perovskite, *Geophys. Res. Lett.*, *20*, 855-858, 1993.

Sweeney, J. S. and D. L. Heinz, Irreversible melting of a magnesium-iron-silicate perovskite at lower mantle pressures [abstract], *Eos Trans. AGU*, *76*, F553,1995.

Ubbelohde, A. R., *The Molten State of Matter*, Wiley, New York, 1978, 454 pp.

Vocadlo, L. and G. D. Price, The melting of MgO—computer calculations via molecular dynamics, *Phys. Chem. Minerals*, *23*, 42-49,1996.

Wells, D. J. and R. S. Berry, Coexistence in finite systems, *Phys. Rev. Lett.*, *73*, 2875-2878, 1994.

Wettlaufer, J. S., M. G. Worster, L. A. Wilen, and J. G. Dash, A theory of premelting dynamics for all power law forces, *Phys. Rev. Lett.*, *76*, 3602-3609, 1996.

Wolf, G. H. and M. S. T. Bukowinski, Variational stabilization of the ionic charge densities in the electron-gas theory of crystals: Applications to MgO and CaO, *Phys. Chem. Minerals*, *15*, 209-220, 1988.

Zerr, A. and R. Boehler, Melting of (Mg,Fe)SiO_3 perovskite to 625 kilobars: indication of a high melting temperature in the lower mantle, *Science*, *262*, 553-555, 1993.

Zerr, A. and R. Boehler, Constraints on the melting temperature of the lower mantle from high-pressure experiments on MgO and magnesiowustite, *Nature*, *371*, 506-508, 1994.

Ronald E. Cohen, Geophysical Laboratory and Center for High Pressure Research, Carnegie Institution of Washington, Washington D.C. 20015

J. S. Weitz, Geophysical Laboratory and Center for High Pressure Research, Carnegie Institution of Washington, Washington D.C. 20015

Laser-Heating Through a Diamond-Anvil Cell: Melting at High Pressures

J. S. Sweeney[1] and D. L. Heinz[2]

Department of the Geophysical Sciences, University of Chicago, Illinois.

A laser-heating system was constructed that can stably heat through a diamond-anvil cell, probe regions as small as 6.7 μm in diameter, and accurately measure visible spectral radiation at frequencies up to 30 Hz. Spectra can be analyzed to measure temperature and to identify phenomena of interest such as melting. Reflecting optics were used where dispersion is critical. Temperature is stabilized by feedback from the thermal emissions of laser-heated samples and by attenuating the laser beam using a liquid-crystal variable waveplate and a fixed polarizer. The response time of the stabilizer is estimated to be 1.25 ms. The laser-heating system is suitable for experiments at the pressures and temperatures of the Earth's lower mantle and core.

With this apparatus, a magnesium–iron–silicate perovskite, $(Mg_{.88}Fe_{.12})\,SiO_3$, was melted in a diamond-anvil cell at pressures between 25 GPa and 85 GPa. Natural bronzite and synthetic enstatite were the starting materials. Polycrystalline samples with no pressure medium and single crystals in a NaCl pressure medium were melted. Melting was determined in situ by thermal analysis and corroborated by the appearance of glass in temperature-quenched samples. Corrections for radial temperature gradients in the samples were obviated by aperturing the collecting optics of the spectrometer. Corrections for axial temperature gradients were estimated from a simple model. The slope of the melting curve between 35 GPa and 85 GPa is 5.2 ± 0.8 K/GPa and sub-adiabatic. The melting temperature extrapolated to the core-mantle boundary is 4502 ± 176 K. Both are significantly less than recent estimates.

1. INTRODUCTION

Laser heating through a diamond-anvil cell was invented specifically to study the mineralogy of the Earth's lower mantle [*Bassett and Ming*, 1972; *Ming and Bassett*, 1975]. Silicate perovskite, for example, was first synthesized this way [*Liu*, 1974, 1975, 1976]. Techniques for laser heating have advanced incrementally since the first experiments of *Bassett and Ming* [1972] with a pulsed ruby laser. Common now are continuous-wave near-infrared Nd:YAG lasers [*e.g.*, *Ming and Bassett*, 1974], and mid-infrared CO_2 lasers [*e.g.*, *Boehler and Chopelas*, 1992]. Temperatures during laser heating have been stabilized by various feedback systems [*Heinz and Jeanloz*, 1987b; *Heinz et al.*, 1991]. Radial temperature gradients have been measured [*Jeanloz and Heinz*, 1984; *Boehler et al.*, 1990]. Spectroscopic measurements are now more accurate [*Jeanloz and Heinz*, 1984] and faster [*Mao et al.*, 1987; *Heinz et al.*, 1991]. And recently, laser-heating experiments have been automated for thermal analysis [*Sweeney and Heinz*, 1993b].

Here, a new laser-heating system built at the University of

[1]Now at The Center for High Pressure Research, State University of New York at Stony Brook.

[2]Also The James Franck Institute, University of Chicago, Illinois.

Properties of Earth and Planetary Materials
at High Pressure and Temperature
Geophysical Monograph 101

Table 1. Optical components for the laser-heating system described in Figure 1.

label	*description*
lenses	
L1	10× beam expander
L2	65 -mm focal length plano-convex lens
L3	15× Burch reflecting objective
L4	achromatic triplet relay lens
L5	viewing eyepiece, 10×
mirrors	
M1	laser steering mirror
M2	laser steering mirror
M3	gold-coated laser mirror
M4	80-20% neutral-density beam splitter
M5	neutral-density beam splitter
M6	neutral-density beam splitter
filters	
F1	1064 -nm laser-line filter
F2	short-pass filter, laser blocking
F3	neutral density filters
F4	filter, Schott glass *OG*570
F5	short-pass filter, laser blocking
detectors	
D1	silicon photodiode detector
D2	photomultiplier tube (PMT)
D3	optical multichannel analyzer (OMA)
D4	CCD monochrome camera
other components	
W1	liquid-crystal variable waveplate
P1	laser-glan polarizer
S1	spectrometer
DAC	Mao-Bell type diamond anvil cell, or neon lamp for wavelength calibration, or tungsten lamp for spectral calibration
Y1	Nd:YAG laser, 18 W_{CW}, TEM_{00}-mode
A1	pinhole aperture, 100 -μm diameter

Chicago is described. To illustrate its utility, results are reported for melting of a magnesium–iron–silicate perovskite at lower mantle pressures. This is an important experimental problem because magnesium–iron–silicate perovskite is the principal constituent of the Earth's lower mantle and the most abundant silicate mineral in the Earth [*Liu*, 1975, 1976; *Knittle and Jeanloz*, 1987]. Its melting curve is an upper bound for the temperature of the lower mantle. The melting curve may also constrain the viscosity, and thereby the geodynamics, of the lower mantle.

2. LASER HEATING SYSTEM

The laser-heating system was designed to heat samples at high pressures, control experiments, and measure thermal emissions (Figure 1 and Table 1). It is similar in some ways to previous systems built at the University of Chicago [*Heinz et al.*, 1991; *Sweeney and Heinz*, 1993b]. Perhaps the most obvious difference is that this system is constructed on an optical table rather than around a microscope. Components are more securely mounted so that alignment is more stable. Moreover, rebuilding was an opportunity to reconsider and improve other aspects of the apparatus.

Laser-heating experiments are monitored and controlled by the signals diagramed in Figure 2 and described below. Signals to the computer are captured by a multiplexed analog-to-digital converter and stored for analysis. Outputs for controlling the experiments are from a waveform generator in the computer. Simple computer programs coordinate the input and output and provide a rudimentary user interface to the laser-heating system. Spectra of thermal emissions from *D3* are stored by the OMA controller and transferred to the computer at the end of each experiment. Thermal radiation emitted during laser heating is stabilized by a feedback loop composed of the preamplifier, the attenuator controller, the attenuator, and detector *D2*.

2.1. Spectroscopic Optics

The aperture *A1* in Figure 1 is located at the image plane of objective lens *L3*. It is a spatial filter that passes thermal emissions from the hottest region of a sample to the spectrometer and obscures cooler regions. The aperture diminishes errors in temperature that result from the radial temperature gradient discussed below. A 100 -μm pinhole at *A1*, with the 15× objective *L3*, effectively selects thermal emissions from a region 6.7 μm in diameter. The probe diameter, chosen by trial and error, is a reasonable compromise between signal strength and the size of the laser-heated spot. With a much smaller 25 -μm pinhole, the signal-to-noise ratio is poor and diffraction through the aperture becomes significant.

The probe is centered on the laser-heated spot where the radial temperature gradient is minimal. Final alignment is checked prior to each experiment by gently heating the sample to 1500 – 2000 K with the stabilizer on. Then the aperture is translated normal to the optical axis until the laser power monitored by detector *D1* is minimized.

While conceptually simple, the optical requirements for an apertured spectroscopic system are exacting. For example, the same pinhole was tried with another laser-heating system [*i.e., Heinz et al.*, 1991] with a Leitz L25/0.22 refracting objective. The L25 objective has a long working distance, 14.8 mm, that is ideal for our diamond-anvil cell. It is a Leitz UT40/0.34 universal stage lens that can be used without its semispherical segment. Chromatic aberration with this lens was so severe and sensitive to the focus that reproducible temperature measurements were impossible. Subsequently, a number of objective lenses were tested for axial chromatic aberration including the Leitz L25, the Ealing 15× reflecting objective *L3*, a Nikon M Plan 20/0.35 SLWD, a Bosch & Lomb 6/0.17, a Leitz UT16, and a pair of Beck 15× reflecting objectives. For these tests, all the optical components between an objective lens and spectrometer *S1* were removed. A slit at the object plane was back-lit with a neon calibration lamp. Peak shape from the neon spectrum was used to determine focus at the entrance to *S1*. Distance to the spectrometer was nominally equivalent to the standard 160-mm microscope tube length and it was adjusted to focus at each color. The tube length at two wavelengths, 585.25 nm and 743.89 nm, was measured and the difference was recorded as a measure of axial chromatic aberration. The difference for the Leitz L25 refracting objective was 27 μm while for the reflecting objective *L3* it was 3 μm.

Temperatures were also measured to estimate the effects of chromatic aberration. Focus was determined as above with a neon lamp, then this lamp was replaced with a standard tungsten ribbon-filament lamp and its temperature was measured radiometrically. A 100-μm pinhole was placed at the entrance slit of the spectrometer to simulate the aperture *A1* in Figure 1. With refracting objectives, temperatures varied up to 840 K when the instrument was focused first at 585.25 nm and then at 743.89 nm. Temperature differences with the reflecting objectives were negligible.

These tests were a major impetus for constructing a laser heating system using reflecting optics. They confirmed our suspicions and agreed with the observations of *Boehler et al.* [1990] who were early proponents of reflecting optics for laser-heating experiments. Our current system is based on a broadband visible-light spectroscopic system developed by

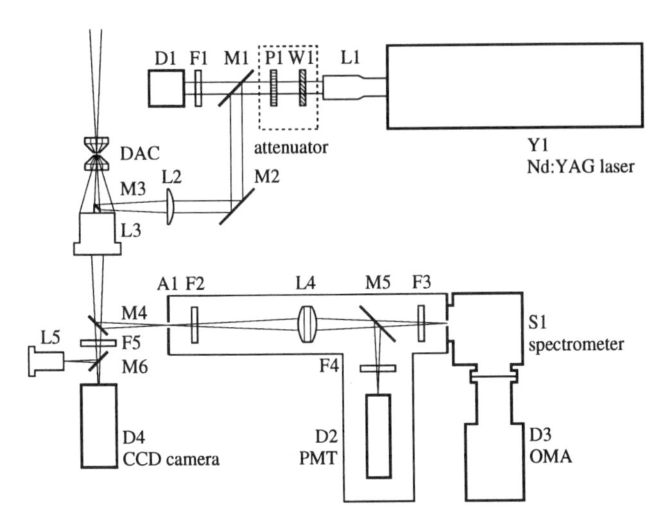

Figure 1. Optical diagram of the laser-heating and spectroscopic systems. Laser *Y1* heats samples at high-pressure through diamond anvil cell *DAC*. The variable waveplate *W1* and polarizer *P1* control the heating with feedback from photomultiplier *D2*. Photodiode *D1* measures laser power, and optical multichannel analyzer *D3* (OMA) measures the intensity, temperature, and emissivity of heated samples. Experiments are observed through eyepiece *L5* and camera *D4*.

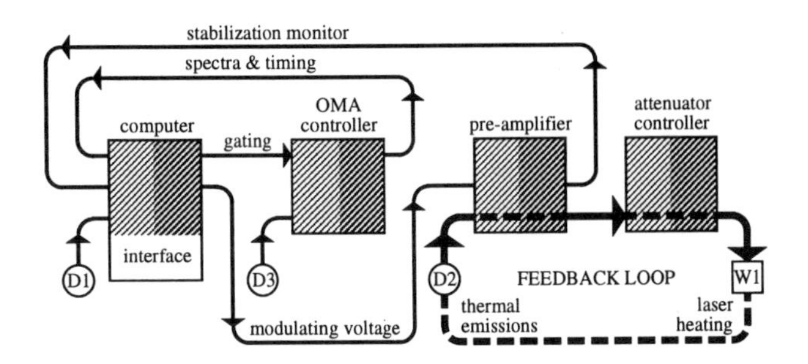

Figure 2. Signals that monitor and control the laser-heating system. Input is on the left side of each diagramed component and output is on the right. Signals originate from detectors *D1*, *D2*, and *D3*. The feedback loop for stabilizing thermal emissions is the heavy dashed line.

Syassen and Sonnenschein [1982]. All the optics, except one diamond anvil, are dispersionless in the critical region between the sample and aperture *A1*. A dispersionless optic is typically mirrored on its first surface and dispersionless because the index of refraction is unity everywhere. Thus for example, beamsplitter *M4* has its first surface turned toward aperture *A1*. Without dispersion and with negligible diffraction, there can be no chromatic aberration from such components.

Objective *L3* is a Schwarzschild objective constructed from two first-surface mirrors. The Schwarzschild objective is similar to a Cassegrain telescope but with the object plane near the lens instead of at infinity. A special class of Schwarzschild objectives was described by *Burch* and these lenses often carry his name [for a survey see *Burch*, 1947].

One diamond anvil is necessarily included in the optical path of the spectroscopic system and no compensation was made for dispersion through it. However, the diamond has plane-parallel surfaces and it is thin compared to a typical lens. Moreover, objective *L3* has a large f-number compared to normal microscope objectives because of its long working distance. For these reasons, dispersion through the diamond anvil is minimal and no compensating lens was included. The focus of the system at *A1* was checked with a neon lamp similar to the optical tests described above. No axial chromatic aberration was measurable even when a diamond anvil was included in the optical path.

Between aperture *A1* and spectrometer *S1*, the spectroscopic system is treated as a non-imaging optical system. Here, chromatic aberrations are not a concern since the apertures defined by the edges of the optical components are large, and the entrance slit at *S1* is underfilled.

3. STABILITY AND CONTROL OF THERMAL EMISSIONS

Early workers understood the effects of laser instability on sample temperatures [*Jeanloz and Heinz*, 1984]. Com-

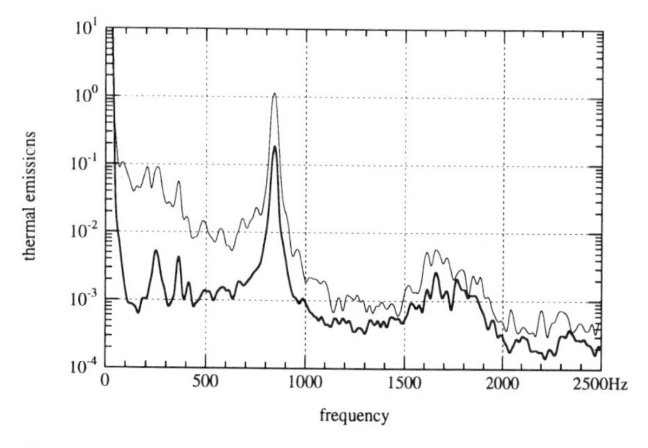

Figure 3. Thermal emissions measured with PMT *D2* at a sampling rate of 15.6 kHz and Fourier-transformed to the frequency domain. The lighter line is thermal emissions with the stabilizer off. The heavier line is thermal emissions measured with the stabilizer on. Both were measured while laser heating a silicate perovskite sample. Thermal emissions above 2.5 kHz were mimimal. The prominent peak is at 835 Hz.

monly, Nd:YAG solid-state lasers have rated root-mean-square (RMS) power fluctuations of 5%, and fluctuations up to 7% have been measured at high power settings. Power fluctuations of this magnitude lead to temperature fluctuations up to 1000 K in laser-heating experiments [*Jeanloz and Heinz*, 1985]. To minimize laser power fluctuations, *Heinz and Jeanloz* [1987a] isolated their laser from vibrations and monitored its beam for a feedback loop that controlled current to the laser's pump lamp. With these techniques, RMS power fluctuations were reduced to 1-3% but temperature fluctuations were still considerable, about 500 K.

Lately, thermal emissions have been monitored to stabilize and control temperatures during laser heating [*Heinz et al.*, 1991]. We consider the method superior to those that monitor only laser power for reasons discussed below. Furthermore, the laser beam is attenuated without adjusting current to the laser's pump lamp. When the pump lamp is operated at a constant current, the laser can be tuned so that its pointing, mode structure, and power are more stable [*Heinz and Jeanloz*, 1987b; *Heinz et al.*, 1991].

3.1. The Feedback Loop

With several refinements, a stabilization and control system was implemented similar to that of *Heinz et al.* [1991]. A region of the laser-heated spot is selected by aperture *A1*. Thermal emissions from the region are monitored by photomultiplier *D2*; the photomultiplier current is conditioned by a preamplifier; output from the preamplifier is fed to a controller for the laser attenuator; and laser power is attenuated to stabilize the thermal emissions.

Temperatures are also stabilized by this feedback loop since temperature and thermal emissions, $I = \int I_\lambda \, d\lambda$, are related by the Stefan-Boltzmann relation,

$$I = \varepsilon \sigma T^4, \tag{1}$$

where ε is emissivity, and σ is the Stefan-Boltzmann constant. Equation 1 must be integrated over the region from which the thermal emissions are measured. Here, the area of the region is determined by the aperture *A1* and the size of the laser-heated spot can be neglected because it overfills the aperture. Temperature gradients, discussed below, are steady because the region is centered on the laser-heated spot and the laser has good pointing stability. The emissivity in Equation 1 is a property of state thermodynamic variables and the configuration of the sample. Because the pressure, temperature, volume, and configuration vary smoothly over time, the emissivity does also. For these reasons, the thermal emissions are well approximated by the Stefan-Boltzmann relation and they are a sensitive metric of temperature stability because they increase proportional to the fourth power of temperature. In this instance, the thermal emissions are an excellent proxy for temperature because they can be sampled faster than the laser power fluctuates.

The operation of the stabilizer is illustrated by a simple thought experiment. Suppose we move the region selected by aperture *A1* away from the center of the laser-heated spot. Then, the attenuator simultaneously adjusts the laser power upward. The temperature at the center of the laser-heated spot increases but it is no longer measured since it is outside the field of view. In the field of view, the observed temperature does not change. All that changes are the temperature gradients. We can move further from the center until the the attenuator is passing all the laser power. Beyond that distance, which depends on the temperature of the selected region, stability cannot be maintained. Experiments in the time domain are similar but the radial temperature gradient does not change since the region selected by the aperture remains centered on the laser-heated spot.

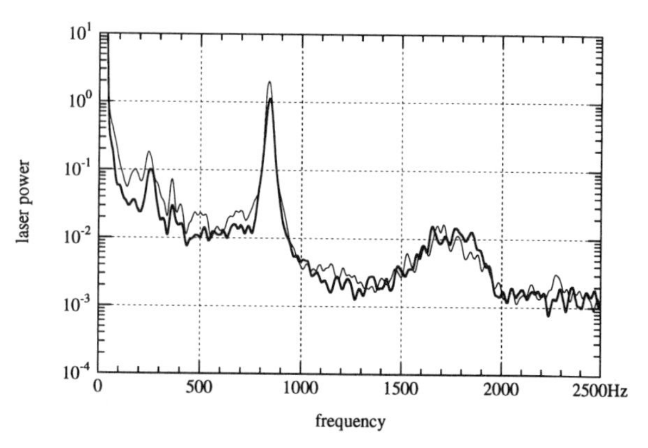

Figure 4. Similar to Figure 3 but laser power instead of thermal emissions. The lighter line is with the stabilizer off and the heavier line is with the stabilizer on. Laser power was measured at *D1*, after the liquid-crystal attenuator.

The laser beam is attenuated by passing it through a birefringent liquid-crystal waveplate, *W1*, followed by a fixed polarizer, *P1*. The polarization of the beam is rotated by the liquid crystal and a portion of the beam is rejected at the fixed polarizer. The birefringence of the liquid crystal and the attenuation are proportional to a voltage applied across the liquid crystal by the attenuator controller [*Jacobs et al.*, 1988; *Heinz et al.*, 1991].

The birefringence of the liquid crystal *W1* is also strongly temperature dependent. To minimize heating at the liquid crystal, laser-power density was reduced by means of the 10× beam expander *L1*. The liquid-crystal cell was also epoxied to a water-cooled aluminum heat sink maintained at 16 °C by a recirculating chiller. These modifications substantially improved the long-term stability of the laser-beam attenuator. The efficiency of the attenuator was also improved over previous versions by antireflection coating *W1* and *P1* for the laser wavelength, 1.064 μm.

The attenuator controller outputs an amplitude modulated square wave at a frequency of 2 kHz to indium-tin-oxide electrodes that coat the surfaces of the liquid-crystal cell. Alternating current is used since the cell is a large capacitive load and it would require a much larger power supply if driven with direct current. By design, the period of the square wave is shorter than the response time of the liquid-crystal cell so the square wave is not apparent in power frequency distributions of thermal emissions or laser power (Figures 3 and 4).

The detector for the feedback loop is a Burle Model 8645 photomultiplier, PMT *D2* in Figure 1. The visible thermal radiation emitted by laser-heated samples is collected by objective lens *L3*, and a portion of this light is reflected by beam splitter *M5* to the PMT. The S20 spectral sensitivity of *D2* is tuned by filter *F4* to more closely match the sensitivity of the optical multichannel analyzer *D3*. With filter *F4*, the signal from *D2* is comparable to the spectrally integrated signal from *D3* (*e.g.*, Figure 5).

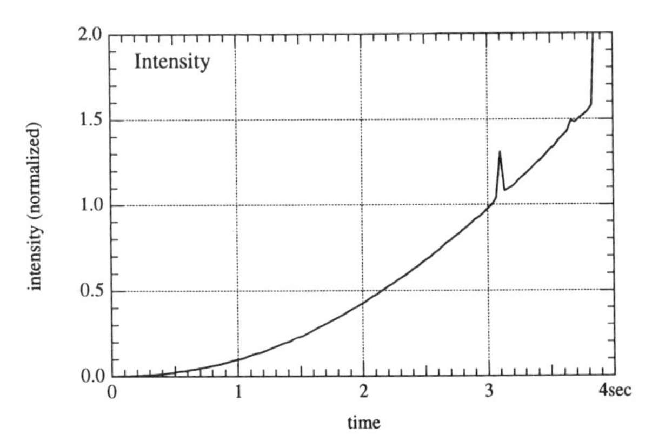

Figure 5. Integrated intensity during experiment *p398*. Similar to Figure 13.

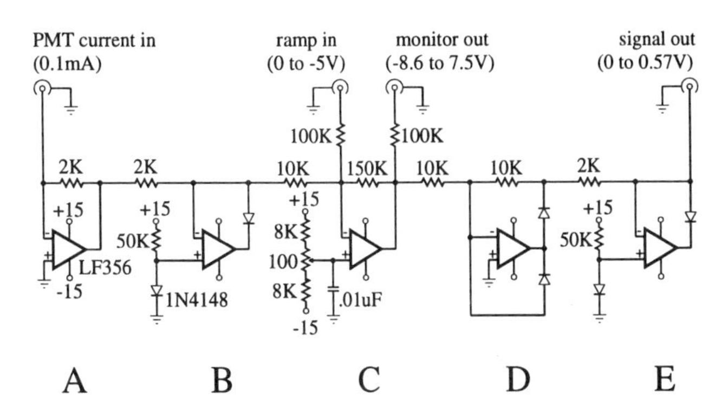

Figure 6. The pre-amplifier circuit conditions and amplifies the signal from PMT *D2* and outputs a voltage to the attenuator controller. Stage *A* is a trans-resistance stage that converts current to voltage. Stage *B* provides over-voltage protection for subsequent pre-amplifier stages. Stage *C* inverts and amplifies the signal with a gain of 15×, and adjusts the DC-offset. This stage also has an input for a modulating voltage ("ramp in") and an output for monitoring the signal. Stage *D* inverts and rectifies the signal. Stage *E* provides over-voltage protection for the attenuator controller.

3.2. Preamplifier

Fluctuations in the thermal radiation emitted by a laser-heated sample are the feedback for stabilization. But the fluctuations sit on top of large mean thermal emissions that must be subtracted by appropriate circuitry. Moreover, the fluctuations are larger at higher temperatures. Ideally, preamplifier gain should scale inversely with temperature so that thermal emissions could be stabilized equally well at low and high temperatures. Practically, a fixed-gain multistage preamplifier was used and protection was built into the preamplifier to clip fluctuations that were too large (Figure 6).

A principal criteria for the preamplifier was that all its stages and the attenuator controller remain within their current and voltage limits. If a feedback system is overloaded, it may recover but only after tens of milliseconds. During that time, control and stability are lost, perhaps unknowingly. We believe that such overloading may have been misinterpreted as melting by *Sweeney and Heinz* [1993a].

3.3. Response Time

The response time of the stabilizer is limited by the variable waveplate *W1*, the slowest component of the system. It is many times slower than the PMT *D2*, the preamplifier, and the attenuator controller. The response time of *W1* is a function of its thickness, the amplitude of the square wave from the controller (0 V to ±5.5 V), and the viscosity and dielectric susceptibility of the liquid crystal.

The response can be estimated from frequency distributions for stabilized and unstabilized thermal emissions. These are compared in Figure 3. Thermal emissions below 800 Hz are attenuated by one to two orders of magnitude by

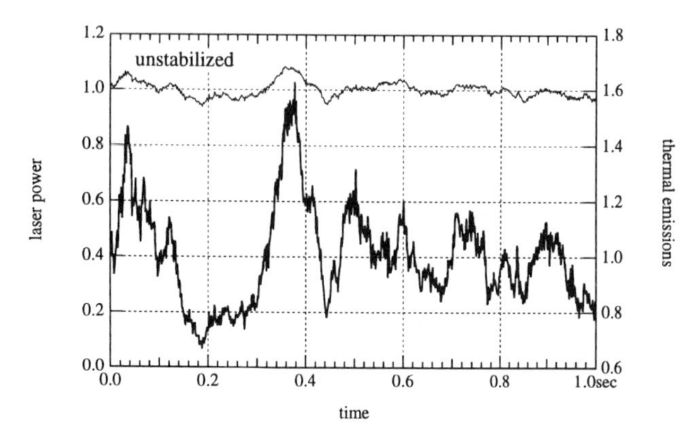

Figure 7. Laser power and unstabilized thermal emissions during heating of a silicate perovskite sample. The light line is laser power and the heavy line is thermal emissions from PMT *D2*, both normalized to unity and offset for clarity. The data were collected at 15.6 kHz and smoothed to highlight lower frequency fluctuations.

stabilization. The peak at 835 Hz is partially attenuated, and above 1000 Hz the thermal emissions are unchanged but they are small. The response time of the liquid-crystal attenuator estimated from Figure 3 is about 1.25 ms.

3.4. Signal-to-Noise

A signal-to-noise ratio for stabilization can be estimated by comparing the amplitudes of fluctuations to their mean since fluctuations are the signal for the feedback loop and their mean can be considered noise in this case. Figure 7 shows unstabilized laser power and thermal emissions in the time domain. Their RMS fluctuations are 2.89% and 18.12%, respectively. From these fluctuations, the signal-to-noise ratio for stabilizing by thermal emissions is about six times better than that for stabilizing by laser power. This is one reason why we consider stabilizing by feedback from thermal emissions superior to the alternate method of stabilizing by feedback from laser power. If both laser power and thermal emissions were stabilized then the effective dynamic range of the system would be greater and the thermal analysis described below could be augmented because stabilized laser power is more appropriate for those comparisons.

3.5. Coupling of Laser Power and Thermal Emissions

At frequencies of several hundred hertz, some attenuation of fluctuations in thermal emissions has been attributed to thermal damping within laser-heated samples [*Jeanloz and Heinz*, 1984; *Heinz and Jeanloz*, 1987b]. Presumably, thermal damping results from the heat capacities and the conductive and radiative properties of a sample and its assembly. Conversely, *Jeanloz and Kavner* [1995] have estimated a quench time for laser-heated silicate samples of about 0.1 ms, implying no attenuation from thermal damping below frequencies of about 10 kHz. The present study is consistent with this estimate. Comparing Figures 3 and 4, unstabilized thermal emissions mimic laser fluctuations with no thermal damping to at least 7808 Hz, the Nyquist frequency of these measurements. *Jeanloz and Heinz* [1984] also noted a phase lag between fluctuations in laser power and those in thermal emissions. Their sample was Al_2O_3 with a few percent platinum added to absorb the laser light. With the silicate perovskite samples used for our study, no phase lag was found to 7808 Hz. None is apparent in Figure 7 comparing laser power and unstabilized thermal emissions, or in Figure 8 comparing laser power and stabilized thermal emissions.

The attenuations of laser power that are required for stabilization are less than intrinsic laser power fluctuations. So while the RMS fluctuations of thermal emissions are reduced to 1.01% by stabilization, the RMS laser power fluctuations of 2.89% only diminish to 2.12% [*e.g.* Figure 4]. We infer from this that thermal emissions and laser power are coupled in a manner that is not well understood but that may depend on the time variance of laser power as well as absolute laser power. This is the principal reason why we prefer stabilizing by feedback from thermal emissions rather than from laser power.

4. TEMPERATURE MEASUREMENTS

The first temperature measurements during laser heating through a diamond-anvil cell were by optical pyrometry [*Ming and Bassett*, 1974; *Yagi et al.*, 1979] and spectral radiometry [*Jeanloz and Heinz*, 1984]. With the laser-heating system described here, temperatures are also measured by spectral radiometry. However, speed and accuracy were improved by dispersing thermal emissions with a grating spectrometer and collecting the resulting spectra with an optical

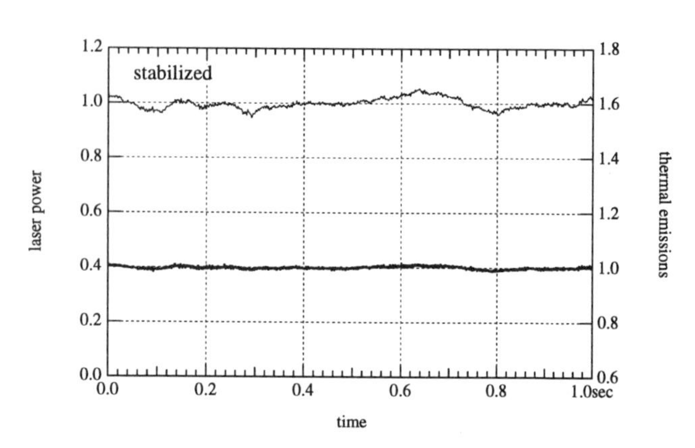

Figure 8. Similar to Figure 7 but with the stabilizer turned on. Thermal emissions are proportional to the fourth power of temperature (see text and Equation 1). Here and in Figures 3 and 7, thermal emissions are a useful proxy for temperature because temperature is not easily measured at frequencies above 30 Hz.

multichannel analyzer [after *Mao et al.*, 1987; *Heinz et al.*, 1991].

The optical multichannel analyzer, *D3*, is calibrated for wavelength and spectral intensity, and temperature is determined by means of Planck's blackbody equation in the usual way:

$$I_\lambda = \frac{\varepsilon c_1}{\lambda^5 \left[\exp\left(c_2/\lambda T\right) - 1\right]}, \tag{2}$$

where I_λ is spectral intensity, ε is emissivity, λ is wavelength, T is temperature, and c_1 and c_2 are physical constants. Emissivity is assumed constant with wavelength, *i.e.*, the greybody approximation. Spectral intensity, typically from 550 nm through 750 nm, and wavelength are fit to determine parameters T and ε. One spectrum from a melting experiment on magnesium–iron–silicate perovskite and its fit to Equation 2 are shown in Figure 9. The fit can be made linear with the Wien approximation to Planck's blackbody equation:

$$J = \ln(\varepsilon) - \frac{\omega}{T} \tag{3}$$

where $J = \ln(I\lambda^5/c_1)$ and $\omega = c_2/\lambda$ are observables [*Jeanloz and Heinz*, 1984; *Heinz and Jeanloz*, 1987b; *Heinz et al.*, 1991]. The spectrum, in coordinates of J and ω, and its fit to Equation 3 are shown in Figure 10. Temperature from the fit is 3618 K. For the results reported here, the 1σ errors in temperature from fits to the Wien approximation are about 0.1%. Temperature measurements by spectral radiometry have been validated by measuring the melting temperatures of metals at ambient pressure [*Jeanloz and Heinz*, 1984; *Heinz et al.*, 1991]. Using literature values for wavelength dependent emissivity, these measured metal melting temperatures are within 3.3% of their accepted published values [*Heinz et al.*, 1991].

Most materials are only approximately greybodies since they have wavelength dependent emissivity. Moreover, temperature estimates from the greybody approximation are worse at higher temperatures. Ideally, wavelength dependent emissivity should be included in fits to Equation 2 or 3 [*Heinz et al.*, 1991]. Unfortunately, temperature and emissivity are somewhat anticorrelated in these fits and with current instrumental error it is not possible to determine parameters for wavelength dependent emissivity.

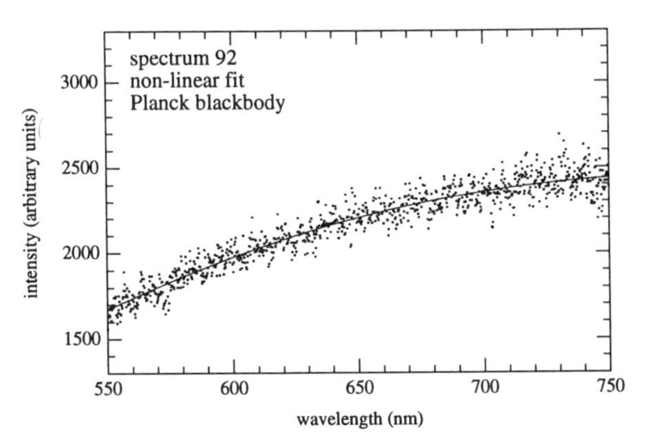

Figure 9. Typical spectrum of thermal emissions after calibration, spectrum 92 of experiment *p398*, and its fit to Planck's blackbody equation.

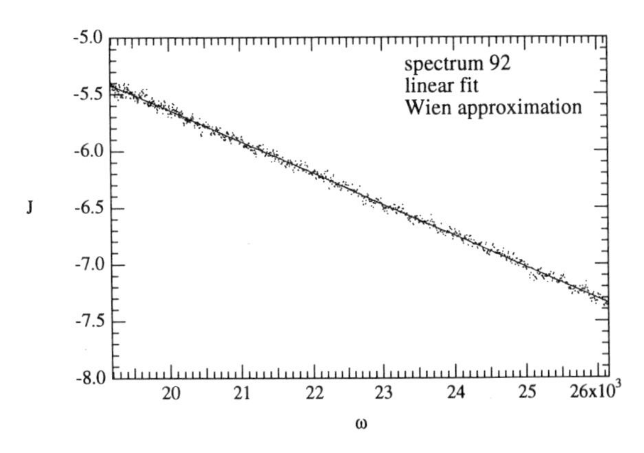

Figure 10. Spectrum 92 of experiment *p398*, and its fit to the Wien approximation. The coordinates of the graph are $J = \ln(I\lambda^5/c_1)$ and $\omega = c_2/\lambda$. Temperature from the linear fit is 3618 K.

5. MELTING CRITERIA

Several criteria have been used to identify melting of laser-heated samples. Many laser-heated silicate and oxide liquids can be temperature-quenched to glasses at high pressures simply by closing the laser's shutter. Typically, the quenched glass blobs are about 20-μm in diameter and they can be viewed with a microscope at high pressures through the diamond anvils. Melting temperatures can be determined using only this technique by repeatedly heating and quenching a sample. Melting is bracketed below by the highest temperature where no glass is quenched and above by the lowest temperature where glass is quenched [*Jeanloz and Heinz*, 1984; *Heinz*, 1986; *Heinz and Jeanloz*, 1987a].

For the results reported below, melting was principally determined by thermal analysis [below and *Sweeney and Heinz*, 1992, 1993b], since it is unambiguous and less subjective than other criteria. Melting determined this way was exactly corroborated by observations of temperature-quenched glass blobs at the melted spots. If a laser-heated spot melted as determined by thermal analysis then glass blobs were visible, and vice versa. Here, each experiment was heated and quenched only once to minimize the subsolidus diffusion of iron.

Below about 3000 K, other workers have observed the onset of "fluid-like motion" during laser heating and they have interpreted this as melting. But at higher temperatures, these observations are impaired by the bright thermal radiation

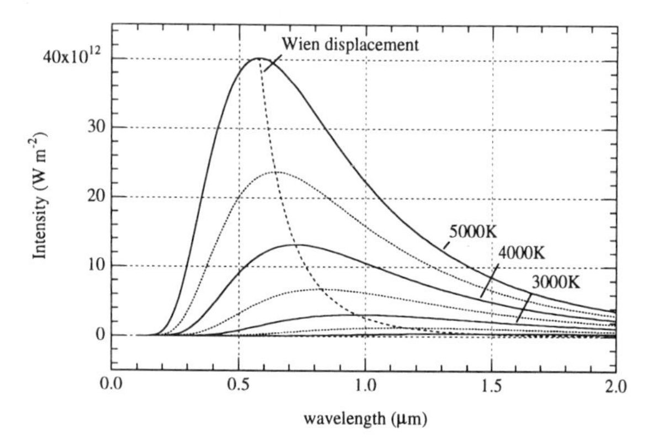

Figure 11. Spectral distribution for blackbodies at temperatures from 2000 – 5000 K. The dotted line is the Wien displacement of λ_{max} toward higher energy at higher temperatures.

emitted from laser-heated spots [*Williams et al.*, 1987, 1991; *Williams and Jeanloz*, 1990].

To enhance contrast during such in situ observations, laser-heated spots have been illuminated with an argon-ion laser and viewed through a matching laser-line filter at 488 nm [*Lazor et al.*, 1993; *Shen et al.*, 1993b; *Zerr and Boehler*, 1993]. For a blackbody at high temperatures, thermal radiation is considerable at 488 nm (Figure 11). Illumination from an argon-ion laser must be much brighter than this to provide good visual confirmation of fluid-like motion, particularly when viewed through the speckle that is a characteristic of coherent laser light sources. Moreover, from such observations, *Shen et al.* [1993b] suggested that melting may occur at temperatures below the lower bound determined by temperature quenching. However, our results show that melting temperatures determined by thermal analysis are exactly corroborated by temperature quenching. For these reasons, we use thermal analysis and not observations of fluid-like motion to determine melting.

Boehler et al. [1990], *Lazor et al.* [1993], and *Shen et al.* [1993b] measured temperature as a function of laser power and they associated discontinuities in their data with melting. Such measurements are conceptually similar to the thermal analysis of *Sweeney and Heinz* [1992, 1993b], but laser power is not a sensitive indicator of melting since small changes in laser power can yield large changes in the temperature of a sample [above and *Jeanloz and Heinz*, 1985].

5.1. Thermal Analysis

Melting determined by thermal analysis was described by *Sweeney and Heinz* [1993b]. For the experiments reported below, the thermal emissions from a sample, and therefore its temperature, were increased at a prescribed rate by summing a modulating voltage to the feedback from photomultiplier *D2*. The modulating voltage is a computer-generated waveform that is triggered at the beginning of each experiment. It was typically a power-law function of time, t:

$$V = V_o + at^c, \tag{4}$$

with V between zero and five volts. V_o was adjusted so that thermal emissions were barely visible at the beginning of an experiment, with the temperature less than about 1500 K. Parameter a was chosen by trial and error so that thermal emissions and temperature increased through the melting transition. Parameter c was usually 1 or 2.

Some data describing the state of the system were collected at a frequency of 1024 Hz. Those included the modulating voltage, laser power at photodiode *D1*, the stabilizer signal from PMT *D2* at the preamplifier, and a 30-Hz timing signal from the optical multichannel analyzer. Simultaneously, the optical multichannel analyzer *D3* collected spectra of the thermal emissions from 550 nm to 750 nm. Usually 128 spectra were collected for each experiment at 30 Hz. Temperature, emissivity, and χ^2 were determined for each spectrum from a fit to the Wien approximation. The spectra were also integrated over their bandwidth for comparison to the modulating voltage.

Melting for a typical experiment is apparent near the end of spectrum 93 when compared to spectrum 92 (Figures 12 and 9). Typical for $(Mg_{.88}Fe_{.12})\,SiO_3$, melting was accompanied by a sudden broadband increase in thermal emissions lasting about 0.2 ms and faster than the response time of the liquid-crystal attenuator. The anomalous spectrum 93 is evident at 3.07 s in Figures 5, 13, and 14.

6. MELTING TEMPERATURES

The melting experiments were not reversed, and for that reason, the hottest temperatures measured before melting could be considered a lower bound for the melting curve. Here, more robust estimates of melting temperatures, their probability distributions, and confidence limits are calculated. These calculations take advantage of the discrete and

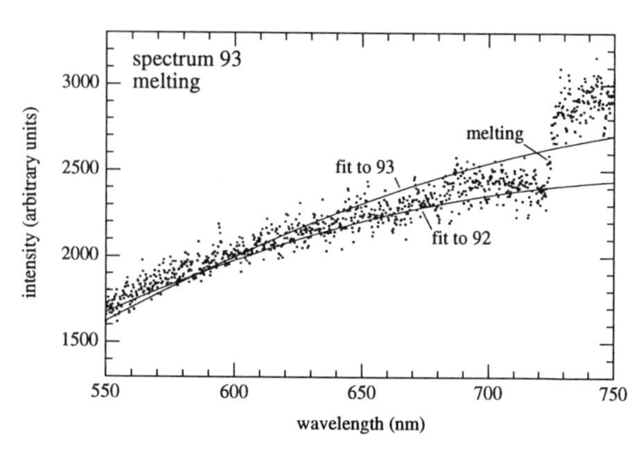

Figure 12. Spectrum 93 from experiment *p398* and its fit to Planck's blackbody equation. Melting occurred while this spectrum was accumulating. Note the step in intensity at the right end of the spectrum. Compare this spectrum with spectrum 92 shown in Figure 9.

regular sampling used for thermal analysis by interpolating or extrapolating the data where they vary smoothly to determine temperatures between or beyond the discrete measurements. We demonstrate for each experiment that the melting temperature is on average only 7.8 K above the hottest measured subsolidus temperature. We conclude that the significant scatter among the data cannot simply result from our approach to the melting curve from below and the data do not represent only a lower bound for the melting curve.

In detail, an observed temperature, T_n determined from spectrum n (*e.g.*, spectrum 92, Figure 9), is the hottest and last temperature measured before melting and melting occurs before the next spectrum, $n+1$, is completely collected. Therefore, the melting temperature, T_m, is hotter than T_n and cooler than T_{n+1}. The time between n and $n+1$ is 33 ms, the collection time of the optical multichannel analyzer. Uncertainty arises since melting occurs at some unknown time during that interval. However, if the temperatures, T_n and T_{n+1}, and their instrumental errors, σ_n and σ_{n+1}, are known then the melting temperature and its error can be estimated [after *Demarest and Haselton*, 1981].

The cooler temperature, T_n, and its error, σ_n, are determined by fitting spectrum n to the Wien approximation in the usual way. However, the laser-heated spot melted as the optical multichannel analyzer was collecting spectrum $n+1$. For that reason, spectrum $n+1$ poorly approximates a greybody spectrum and the Wien approximation is not a good model function for determining the temperature T_{n+1} (*e.g.*, spectrum 93, Figure 12).

Instead, T_{n+1} can be extrapolated from the subsolidus temperatures by fitting T_{n-i} through T_n to an appropriate function. For the experiments described below, a second-order polynomial was sufficient for this extrapolation with $i = 9$. In the same way, the error at $n+1$ was estimated by fitting

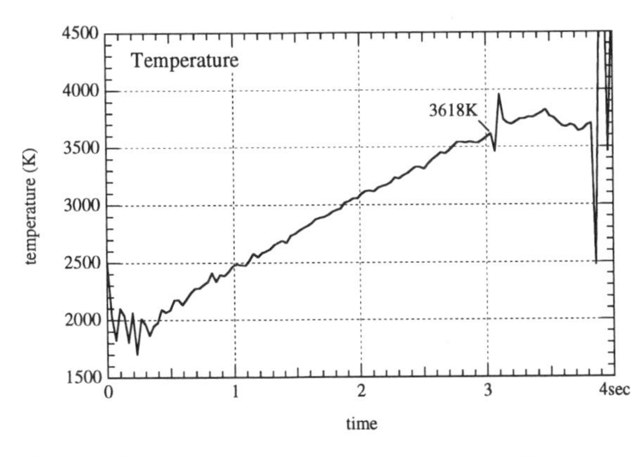

Figure 13. Temperature during experiment *p398* from spectra measured by the optical multichannel analyzer *D3*. Each spectrum is fit to the Wien approximation to determine temperature and emissivity. The last temperature measured before melting, T_{92}, is 3618 K at 3.04 s. The spectra were collected at 30 Hz. Melting occurred near the end of spectrum 93 at 3.07 s.

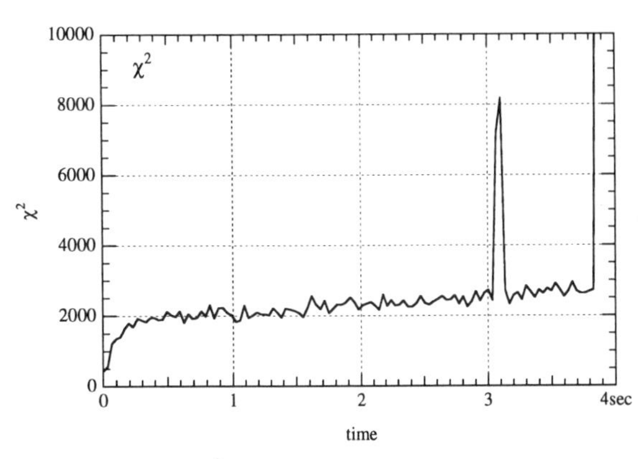

Figure 14. The χ^2 for the fit to the Wien approximation during experiment *p398*. Similar to Figure 13.

from σ_{n-i} through σ_n and extrapolating to σ_{n+1}. For most experiments, uncertainty from the fit of T_{n-i} through T_n was smaller than the extrapolated instrumental error, σ_{n+1}. When it was larger, it replaced the instrumental errors at both T_n and T_{n+1}.

With these temperatures, the probability distribution for a melting temperature can be calculated. The probability for melting near T_n is the integral of the normal (Gaussian) probability distribution parameterized by σ_n [*Demarest and Haselton*, 1981]. This integral is an error function [*Abramowitz and Stegun*, 1964]:

$$p_n = \frac{1}{2}\left[\operatorname{erf}\left(\frac{T - T_n}{\sqrt{2}\sigma_n}\right) + 1\right]. \tag{5}$$

The probability for melting asymptotically approaches zero below T_n and a maximum above T_n. Symmetrically for T_{n+1}, the probability is

$$p_{n+1} = \frac{1}{2}\left[\operatorname{erf}\left(\frac{T_{n+1} - T}{\sqrt{2}\sigma_{n+1}}\right) + 1\right]. \tag{6}$$

The probability distribution for the melting temperature, p_m, is just the product of Equations 5 and 6 illustrated in Figure 15. A normalization function, a, is defined:

$$a = \int_{-\infty}^{\infty} p_m \, dT, \tag{7}$$

so that the integral of the probability over all temperatures equals one. Then, the mean, T_μ, is

$$T_\mu = \frac{1}{a}\int_{-\infty}^{\infty} T \, p_m \, dT; \tag{8}$$

and the variance, v_m, is

$$v_m = \frac{1}{a}\int_{-\infty}^{\infty} T^2 \, p_m \, dT - T_\mu^2. \tag{9}$$

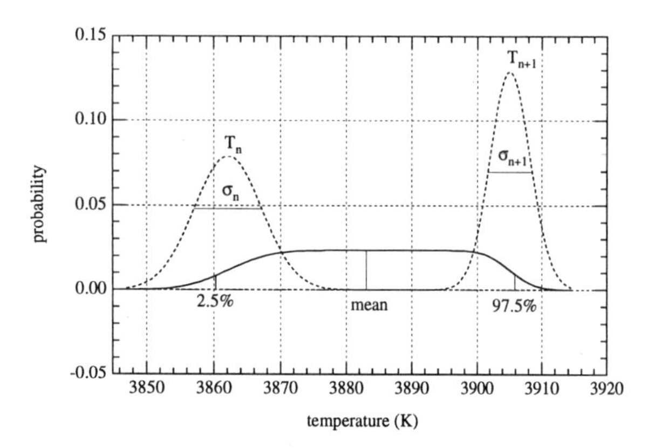

Figure 15. Probability distribution for an estimated melting temperature. Melting occurred between temperatures T_n and T_{n+1}. Instrumental errors are shown dotted. The solid line is the probability distribution for the estimated melting temperature. The 95% confidence limits and the mean, T_μ, are indicated by vertical lines. The data are representative except that σ_n was exaggerated to emphasize the possible asymmetry of the distribution.

The variance bears little relation to the confidence limits since the probability distribution is not Gaussian. A confidence limit, δ, is determined from the roots of the integral equation [*Demarest and Haselton*, 1981; *Bevington and Robinson*, 1992]:

$$\frac{1 \pm \delta}{2} = \frac{1}{a} \int_{-\infty}^{T} p_m \, dT. \qquad (10)$$

If the spread between the confidence limits is large compared to the instrumental errors, σ_n and σ_{n+1}, then the mean melting temperature, T_μ, is significant. Consequently, mean melting temperatures should be fit to determine a melting curve and not T_n, the hottest temperatures measured before melting. For the results reported below, the average of $T_\mu - T_n$ is 7.8 K and independent of temperature.

A simpler and more accurate treatment is possible if the internal timing of the optical multichannel analyzer is known. In that case, the time of melting, t_m, can be determined from spectra like that in Figure 12 by noting the channel where melting occurred and calculating the time when that channel was read. Then, T_m and σ_m can be extrapolated directly from the subsolidus temperatures T_{n-i} through T_n and their errors σ_{n-i} through σ_n.

7. TEMPERATURE GRADIENTS

Temperature gradients are a significant concern for laser-heated diamond-anvil cell experiments. They depend on the power distribution of the heating laser (Gaussian in TEM_{oo} mode) and the spectroscopic and thermal properties of the sample, the gasket, and the diamond anvils. While both radiative and conductive heat transfer are involved, the gradients are at least symmetric from the symmetries of the sample assembly and the laser power distribution.

Two temperature gradients arise from this symmetry: one axial, along the optical axis, and another radial, normal to the optical axis. For the radial temperature gradient, temperature diminishes from a maximum at the center of the heated spot and approaches the ambient temperature at tens of micrometers from the center. At the center and at the edge of the heated spot, the radial temperature gradient approaches zero [*Jeanloz and Heinz*, 1984; *Heinz and Jeanloz*, 1987b; *Bodea and Jeanloz*, 1989]. For the axial temperature gradient, temperature is maximum near the middle of the sample and it approaches the ambient temperature at the diamond anvils. The axial temperature gradient is zero at the temperature maximum but it is large at the diamond anvils since almost all the heat loss occurs there. Conductive heat is lost since the thermal conductivity of diamonds is high. Diamond is also transparent to most radiated energy at the temperatures of typical experiments so radiative heat is also lost through the diamond anvils.

7.1. Weighting of Observed Temperatures

One consequence of temperature gradients is that a measured spectrum includes thermal emissions from portions of the sample that are at different temperatures. Therefore, the fit of the spectrum to Equation 3 gives an observed temperature that is lower than the maximum temperature, and a correction is required to compensate for this effect.

If the observed temperature was simply a spatial average of the sample's temperature then it would be near the ambient temperature since most of the sample is unheated. But temperature is measured radiometrically and the observed temperature is weighted toward the hotter portions of the sample since thermal emissions are larger there. This weighting is predominantly from the Stefan-Boltzmann relation, Equation 1 (Figure 11).

To a lesser degree, the spectral distribution of thermal emissions also weights temperature measurements toward the hottest part of a sample since thermal emissions, I_λ, shift toward the bandwidth and sensitivity of the spectometic system at higher temperatures. The shift is described by the Wien displacement law, $\lambda_{max} T = k_w$, where k_w is the Wien constant (Figure 11). The weighting depends on the spectral response of the system shown in Figure 16.

7.2. Radial Temperature Gradient

For the experiments reported below, the temperatures of interest were those at the center of the laser-heated spots where melting occurred. Thermal emissions from lower temperature regions were effectively masked out by the aperture *A1*. In this way, systematic errors from the radial temperature gradient are practically obviated. The slight remaining correction from an observed temperature to the maximum temperature is about 1% with a 6.7-μm aperture. The correction was calculated with the greybody approximation, a Gaussian temperature distribution, and a flat spectral response for the optics and detectors. The calculation assumes a diameter of about 36 μm for the laser-heated spots at half their maximum temperature (FWHM). The FWHM was estimated by comparing silicate perovskite experiments with unpublished experiments on iron foil where the radial temperature gradient was measured using the inverse technique of *Jeanloz and Heinz* [1984].

Two methods have been used to measure radial temperature gradients. *Jeanloz and Heinz* [1984] first measured such a gradient by stepping a slit across an image of a laser-heated spot. The data from this line probe were inverted to determine a radial temperature profile using a tomographic technique called the Abel transform [see also *Heinz and Jeanloz*, 1987b]. *Boehler et al.* [1990] measured radial temperature profiles but by stepping a small pinhole across an image of the spot. No inverse method is required for such a point probe but its signal-to-noise ratio is much worse because, unlike the Abel transform, it does not take advantage of the symmetry of the laser-heated spot.

Boehler and Chopelas [1992] moderated radial temperature gradients by enlarging their laser-heated spots. A 120-W CO_2 laser was used to maintain adequate laser power density. The CO_2 laser-heated spots were two or three times larger in diameter than Nd:YAG laser-heated spots and radial temperature gradients were proportionally decreased. An aperture was used to measure temperature at the center of their spots. Without such an aperture, the radial gradient should be measured during each experiment [*e.g.*, *Jeanloz and Kavner*, 1995], or temperatures should be corrected based on previous measurements [*e.g.*, *Heinz and Jeanloz*, 1987a]. Such corrections are practically independent of the spot size because temperatures determined by a fit to the Wien approximation are not proportional to the absolute magnitude of the thermal emissions. Rather, they depend on the spectral distribution of thermal emissions and that is not a function of the spot's diameter.

7.3. Axial Temperature Gradient

For opaque samples such as metallic iron it is sufficient to consider only the radial temperature gradient since temperature at a surface is measured. However, many dielectric materials like silicate perovskite are nearly transparent so they absorb and radiate within their bulk. For that reason, an axial temperature gradient must also be considered. Furthermore, the axial gradient is larger than the radial gradient since the samples are thin and heat is lost primarily through the diamond anvils.

An axial temperature profile has not been measured for a laser-heated sample because nobody has devised a method for making this measurement. *Zerr and Boehler* [1993] have tried to minimize axial temperature gradients by constructing special sample assemblies, but there have been objections to their approach [*Heinz et al.*, 1994].

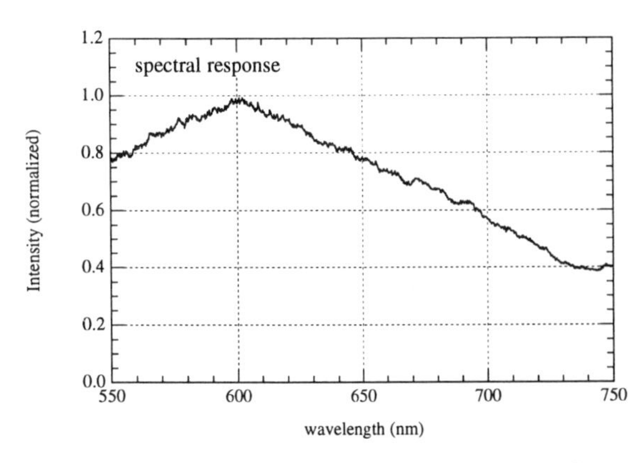

Figure 16. Normalized response of the spectrometric system to an input of constant intensity from 550 – 750 nm. The response is due to wavelength dependent transmission or reflection of the optical components and the spectral sensitivity of the optical multichannel analyzer.

Instead of measuring or eliminating the axial gradient, it can be estimated by modeling. *Bodea and Jeanloz* [1989] calculated axial and radial temperature profiles by numerically solving for conductive heat flow in two dimensions assuming temperature independent thermal conductivity. An analogous heat flow problem is solved here [after *Sweeney and Heinz*, 1995; *Sweeney*, 1996], but the radial temperature gradient is safely ignored because it is several times smaller than the axial gradient and its effect on measured temperatures has been ameliorated by the aperture *A1*. Results from the *1d* model are similar to those from the *2d* model and from more sophisticated *1d* analyses that include temperature dependent thermal conductivity [*Jeanloz and Manga*, 1995; *Manga and Jeanloz*, 1996]. However, the *1d* case with temperature independent thermal conductivity is easily solved analytically and sufficient for our correction given the uncertainty of the thermal and radiative properties involved.

All the models describe worse-case axial temperature profiles since they do not include radiative transfer. While radiative transfer may be significant for some experiments, it is often difficult to quantify since absorptivity at high temperatures is usually not well known. For this reason, the complexity added by radiative components was deemed unwarranted and only the qualitative effects of radiative transfer are considered.

For the *1d* model, the sample is assumed homogeneous and isotropic, and thermal conductivity and optical absorption are assumed to be independent of temperature. The diamond culets are assumed to be near the ambient temperature because of the high thermal conductivity of diamond. Heating is assumed uniform through the sample since absorption is weak for this silicate at the laser wavelength, 1064 nm. The heat flow is assumed to be in a steady state since the sample is small and quickly equilibrates.

With these assumptions, we have *1d* heat-flow in a unitop, $z = 0, T = T_o, q = q_o$

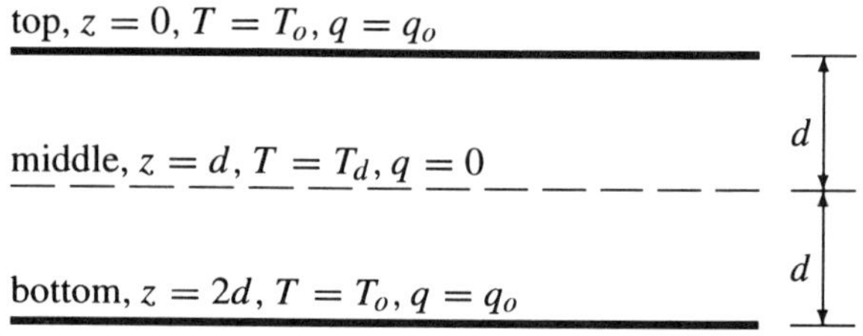

middle, $z = d, T = T_d, q = 0$

bottom, $z = 2d, T = T_o, q = q_o$

d

d

Figure 17. A simple one-dimensional axial heating model. z is depth, T is temperature, and q is heat flux.

form medium with constant internal heating. The model is diagramed in Figure 17. Temperature is symmetric about the middle, where it has a maximum, and where heat flux, q, is zero.

To determine temperature, begin with the steady state heat flow equation with an internal heating term, H:

$$0 = \kappa \frac{d^2T}{dz^2} + H, \tag{11}$$

where κ is the thermal conductivity of the sample. Integrating twice with the given boundary conditions gives

$$T - T_o = \frac{Hd}{\kappa} z - \frac{H}{2\kappa} z^2. \tag{12}$$

Equation 12 can be expressed in terms of the temperature difference and the thickness by evaluating at $z = d$:

$$\Delta = T_d - T_o = \frac{Hd^2}{2\kappa}. \tag{13}$$

Substituting Equation 13 in Equation 12 gives

$$T = T_o + \frac{2\Delta}{d} z - \frac{\Delta}{d^2} z^2. \tag{14}$$

Equation 14 can be represented in nondimensional form with $(T - T_o)/\Delta = \theta$ and $z' = z/d$:

$$\theta = 2z' - z'^2. \tag{15}$$

Temperature is everywhere determined when the temperature at the anvil and the temperature at the middle of the sample are known. Heat flux is found by differentiating Equation 14.

Typical temperature and intensity profiles are illustrated in Figure 18 for a source with emissivity independent of wavelength (*i.e.*, a greybody). The axial temperature profile compares favorably with that of *Bodea and Jeanloz* [1989] though the *1d* model is much more tractable. The observed temperature for the illustrated model is 4144 K for a maximum temperature, T_d, of 4500 K. As with radial temperature profiles, the observed temperature is weighted toward the maximum temperature. Due to this weighting, observed temperatures are insensitive to T_o when T_o is below about 1000 K. The high thermal conductivity of diamond suggests that this is always the case.

A model function can be formed from Equation 14 using Planck's blackbody law, Equation 2:

$$I_\lambda = 2d \frac{\varepsilon c_1}{\lambda^5} \int_0^1 \left[\exp\left(\frac{c_2 \lambda^{-1}}{T_o + 2\Delta z' - \Delta z'^2} \right) - 1 \right]^{-1} dz. \tag{16}$$

By forward modeling, we can estimate an observed temperature, T_{obs}, for the given temperatures, $\Delta = T_d - T_o$. Equation 16 was numerically integrated at many wavelengths for the selected parameters T_d, T_o, d, and ε. An observed temperature was determined by fitting the resulting spectrum to the Wien approximation.

This procedure was repeated for relevant ranges of the parameters. From Equation 16, intensity, I_λ, scales with the thickness of the sample, $2d$, and with emissivity, ε. These parameters can be combined to form a normalized intensity so that the observed temperature, T_{obs}, depends only on T_d, T_o, and the spectral distribution of the intensity. The model is insensitive to T_o due to the weighting described above, so this parameter was fixed at 350 K. T_d is shown as a function of T_{obs} from 2500 K to 5500 K in Figure 19. At T_{obs} = 2500 K, the temperature difference, $T_d - T_{obs}$, is 127 K. And at T_{obs} = 5500 K, $T_d - T_{obs}$ is 689 K. Practically, the temperature difference as a function of observed temperature is well approximated by the polynomial:

$$\begin{aligned} T_d - T_{obs} = & \\ & -0.8187 \times 10^2 + (0.9227 \times 10^{-1})\, T_{obs} - \\ & (0.1358 \times 10^{-4})\, T_{obs}^2 + (0.4052 \times 10^{-8})\, T_{obs}^3. \end{aligned} \tag{17}$$

Spectra can also be fit to Equation 16 by numerical integration and non-linear regression [*Bevington and Robinson*, 1992; *Press et al.*, 1992]. But the procedure is slow and unwarranted for such an approximate model.

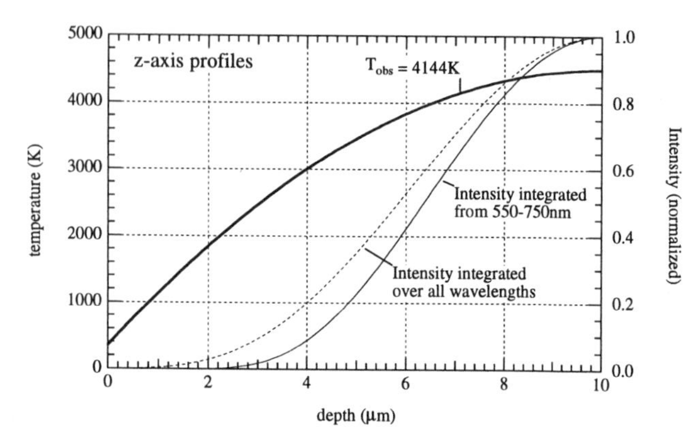

Figure 18. Temperature and intensity profiles for a model with T_d =4500 K and T_o =350 K, and a sample thickness of 20 μm.

Compared to purely conductive heat flow models, observed temperatures more closely approximate peak temperatures when radiative transfer is considered. Radiative energy is emitted approximately proportional to the fourth power of temperature and absorbed by nearby regions. When a steady state is achieved, the net result is that hotter regions near the middle of the sample are cooled, nearby cooler regions are heated, and regions near the anvil interface lose some of their energy by radiating through the diamond anvils. Thus, radiative transfer flattens the temperature curve of Figure 18 near the middle of the sample and steepens the curve near the anvil interface [see Figure 2 in *Manga and Jeanloz*, 1996]. Though the gradient is steeper near the anvil, the sample's temperature is more uniform and the difference between T_{obs} and T_d is smaller.

While the sign of the radiative transfer effect is known, the size of the effect is difficult to estimate. Therefore, a correction for estimating maximum temperatures may range from no correction to the purely conductive model of Figure 19.

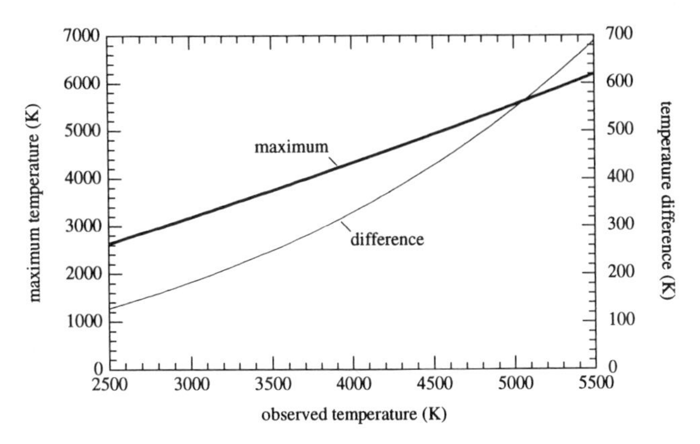

Figure 19. Maximum temperature, T_d, as a function of observed temperature, T_{obs}, for the *1d* conductive model. The heavier line is T_d whose scale is given on the left axis. The lighter line is the difference $T_d - T_{obs}$ whose scale is given on the right axis.

8. MELTING OF MAGNESIUM–IRON–SILICATE PEROVSKITE

With the techniques described above, magnesium–iron–silicate perovskite was melted by laser heating through a Mao-Bell diamond-anvil cell. The anvils were Type IA, brilliant cut, 0.25 carat, with 350-μm culets. Pressure was measured by strain gauges mounted on the cell and calibrated against the ruby pressure scale [*Mao et al.*, 1978; *Bell et al.*, 1986; *Mao et al.*, 1986]. The 1σ error was 1.3 GPa for repeated calibrations.

The 200-μm-thick stainless-steel gaskets were work hardened in the diamond-anvil cell by pressing to about 25 GPa before drilling. The sample chambers were typically 150 μm in diameter and 30 μm deep before pressure was applied.

Typically, polycrystalline enstatite starting materials were loaded without a pressure medium and converted to silicate perovskite by rastering the stabilized heating laser across the samples. The pressure dropped several gigapascals due to the volume change of the phase transformation. This was compensated to some extent by thermal pressure during heating so no pressure correction was applied [*Heinz*, 1990].

8.1. Starting Materials

For most experiments, the starting material was Webster bronzite, a bronzite type enstatite from the Webster-Addie ultramafic ring in Jackson County, North Carolina. This bronzite has been extensively studied and it is exceptionally clean [*Hess and Phillips*, 1940; *Miller*, 1953; *Akella and Boyd*, 1974]. Webster bronzite has an Fe:Mg+Fe mole-ratio of 0.1169 ± 0.0023. A synthetic enstatite, $(Mg_{.9}Fe_{.1})SiO_3$, was also used for some experiments. Of high purity, it was obtained from A. Navrotsky, Princeton University, from an aliquot used for calorimetric studies. It was synthesized in a multianvil press by Y. Fei at The Carnegie Institution of Washington.

For some experiments, single crystals of Webster bronzite were loaded in a NaCl pressure medium. Results from synthetic enstatite and from single crystals of Webster bronzite were congruent with results from the typical experiments with Webster bronzite with no pressure medium.

Some earlier studies of silicate-perovskite melting used a bronzite-type enstatite from Bamble, Norway as a starting material [*Heinz*, 1986; *Heinz and Jeanloz*, 1987a; *Knittle and Jeanloz*, 1989; *Sweeney and Heinz*, 1993a]. For comparison, some experiments were tried with the same aliquot of Bamble enstatite. Melting temperatures for the Bamble enstatite experiments were scattered between 2462 K and 3495 K and they were as much as 1500 K below melting temperatures for experiments using Webster bronzite and synthetic enstatite. These disparate melting temperatures are tentatively attributed to alteration of the Bamble enstatite and contamination by accessory phases. For this reason, the Bamble enstatite experiments were excluded from the data analysis.

8.2. Melting Temperatures

Melting temperatures were estimated as described as above. A mean melting temperature, T_μ, and variance, ν, were calculated for each experiment and the variance was used as a weighting term for fits to model melting curves. The probability distributions for the estimated temperatures appear significant since their confidence intervals, $\delta_{95} = 30 \pm 17$ K, are large compared to the instrumental errors in the measurements, $2\sigma_n = 9.6 \pm 5.0$ K. Yet, even these comparatively broad probability distributions do not encompass the scatter among the data.

Melting temperatures were also corrected for an axial temperature gradient estimated from the conductive model de-

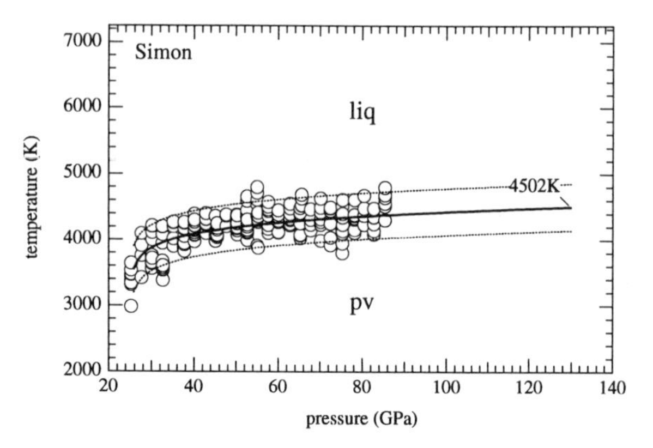

Figure 20. Fit of melting temperatures to the Simon equation. Circles indicate mean melting temperatures after correcting for the axial temperature gradient. The solid line is the Simon equation extrapolated to the core-mantle boundary, about 130 GPa. The 1σ error in temperature is 176 K. The dotted lines are 2σ errors.

scribed above. The correction depends on radiative transfer so it should be less than the correction given by Equation 17. Since the spectroscopic properties of silicate perovskite at high temperatures and pressures are not well known [*e.g.*, *Heinz and Jeanloz*, 1987a; *Shen et al.*, 1993a], a somewhat arbitrary correction of half the conductive model was applied to the data shown in Figure 20. It is systematic and accounts for none of the scatter.

8.3. *Thermodynamic and Kinetic Considerations*

Silicate perovskites synthesized from Webster bronzite and synthetic enstatite were examined by X ray diffraction and electron-microprobe analyses. No evidence was found for decomposition or for iron partitioning into an auxiliary phase [*e.g.*, *Fei et al.*, 1992; *Fei and Wang*, 1992]. If trace amounts of (Mg, Fe) O are present, they exist in domains that are too small for detection. It is unlikely that a heterogeneous distribution of these microdomains could account for all the scatter among the melting data.

In the absence of kinetic effects, thermodynamic equilibrium should be approached near the center of a laser-heated spot since temperature and pressure gradients are mimimal there. One kinetic effect, the Soret diffusion of iron [*e.g.*, *Heinz*, 1986], was diminished by heating rapidly from about 1500 K up to melting in 0.4 – 12.7 s. Rapid heating suggests another kinetic effect *i.e.*, superheating of a solid above its thermodynamic melting temperature. Superheating is likely when heating is rapid, the melt is viscous, and its thermal diffusivity is low. For example, quartz has bee superheated 450 K above its melting point [*Uhlmann*, 1980]. For our experiments, melting temperatures are uncorrelated with heating rates. From that evidence and from rough estimates of the viscosity and thermal diffusivity of the melt at high temperatures and pressures, superheating is deemed unlikely. We conclude that the measured melting temperatures should approximate equilibrium melting temperatures for these experiments.

Given its composition, $(Mg_{.88}Fe_{.12})\,SiO_3$–perovskite most likely melts incongruently so that its phase diagram has a region of partial melting. However, experiments that include the end-member $MgSiO_3$–perovskite [*Zerr and Boehler*, 1993] suggest that partial melting is negligible. This is also consistent with our observations of a single abrupt melting event for each experiment (*e.g.*, Figure 12).

For the following anaylsis we assume that melting temperatures are normally distributed about a univariant melting curve since kinetic effects and incongruent melting do not adequately explain the envelope of melting temperatures measured for $(Mg_{.88}Fe_{.12})\,SiO_3$–perovskite.

8.4. *Melting Equation*

An equation is required for extrapolating the data to the pressure of the core-mantle boundary, about 130 GPa. One simple melting equation is that of *Simon* [1937]. A convenient form of the Simon equation is

$$P - P_o = \frac{a}{B}\left[\left(\frac{T}{T_o}\right)^B - 1\right], \tag{18}$$

where P and T are pressure and melting temperature, P_o and T_o are at the garnet–perovskite–liquid triple point for $(Mg_{.88}Fe_{.12})\,SiO_3$, and a and B are empirical parameters [*Gilvarry*, 1956]. The data were fit by nonlinear regression using the triple-point temperature of *Ito and Katsura* [1992]. For consistency with the pressure scale of the present study, the triple-point pressure was taken to be the lowest pressure where silicate perovskite could be synthesized in the diamond-anvil cell. From the fit, parameter a is 0.084 GPa,

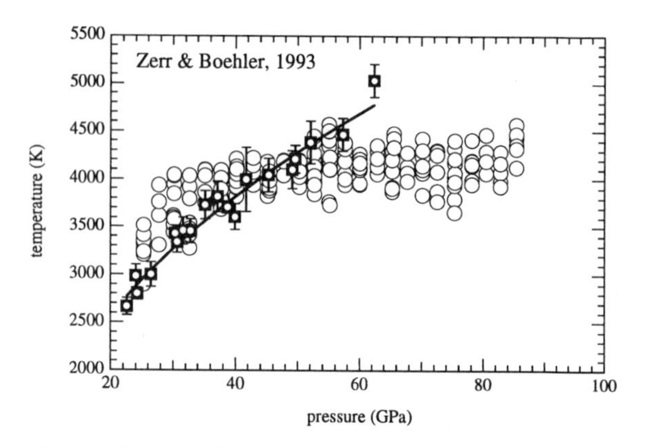

Figure 21. The silicate perovskite melting temperatures of *Zerr and Boehler* [1993] are the squares with circular holes. For comparison, the open circles are data from Figure 20 without a correction for the axial temperature gradient.

the dimensionless parameter B is 20.1, and the 1σ error in temperature is 176 K. At 130 GPa the melting temperature is 4502 K (Figure 20). The high-pressure Clapeyron slope is $5.2 \pm 0.8\,\mathrm{K\,GPa^{-1}}$, estimated by fitting the data above 35 GPa to a straight line.

9. THE EARTH'S LOWER MANTLE

While the melting curve was measured for only one composition of a single phase, $(Mg_{.88}Fe_{.12})\,SiO_3$ is a plausible composition and silicate perovskite is the dominant phase in the Earth's lower mantle [*Liu*, 1975, 1976; *Ito and Yamada*, 1982; *Knittle and Jeanloz*, 1987; *Ito and Takahashi*, 1989]. The melting curve is an upper bound for the temperature of the lower mantle since from seismic data we know that the lower mantle is substantially solid.

Numerous workers have measured melting temperatures for $(Fe_xMg_{1-x})\,SiO_3$ at high pressures by laser heating through a diamond-anvil cell [*Heinz*, 1986; *Heinz and Jeanloz*, 1987a; *Knittle and Jeanloz*, 1989; *Sweeney and Heinz*, 1993a; *Zerr and Boehler*, 1993; *Shen and Lazor*, 1995], or with a uniaxial split-sphere multianvil press [*Ito and Katsura*, 1992]. For example, our data and data from *Zerr and Boehler* are plotted together in Figure 21. Considering the scatter, only their highest pressure point is an outlier when the two data sets are taken together. However, systematic differences are apparent and they can be emphasized by fitting both to melting curves. In Figure 21 the *Zerr and Boehler* data are fit to the Simon equation, Equation 18. Extrapolating to the core-mantle boundary yields a melting temperature of 6770 K, somewhat below their estimate of 7000 K to 8500 K but 2270 K hotter than the extrapolation of our data. Such a hot melting curve would suggest a more viscous lower mantle [*Weertman and Weertman*, 1975] with implications for the style and vigor of mantle convection.

Finally, others have suggested that the Earth's mantle was at least partially if not substantially molten during its early history, whether from adiabatic and radiogenic heating during accretion [*Safronov*, 1978]; or from the gravitational energy of core formation [*Flasar and Birch*, 1973; *Verhoogen*, 1980]; or by giant impacts during the late stages of accretion [*Benz et al.*, 1986, 1987; *Stevenson*, 1987; *Newsom and Taylor*, 1989; *Benz and Cameron*, 1990; *Melosh*, 1990; *Cameron and Benz*, 1991]; and perhaps aided by the insulating effects of a protoatmosphere [*Abe and Matsui*, 1986; *Matsui and Abe*, 1986; *Zahnle et al.*, 1988; *Sasaki*, 1990]. Such a putative magma ocean may have been deep enough to involve the entire mantle [*Tonks and Melosh*, 1993]. Here, we only mention that the cooling history of a deep magma ocean would be substantially affected by the melting curve of magnesium-iron-silicate perovskite [*e.g.*, *Miller et al.*, 1991].

More detailed discussions of the geophysical implications and comparisons to other melting curves are in preparation.

REFERENCES

Abe, Y., and T. Matsui, Early evolution of the Earth: accretion, atmosphere formation, and thermal history, *J. Geophys. Res.*, *91*, E291–E302, 1986.

Abramowitz, M., and I. Stegun, eds., *Handbook of Mathematical Functions with Formulas, Graphs, and Mathematical Tables*, p. 1046, United States. National Bureau of Standards. Applied Mathematics Series, 55, U.S. Government Printing Office, Washington D.C., 1964.

Akella, J., and F. R. Boyd, Petrogenetic grid for garnet peridotites, *Carnegie Inst. Washington, Yearb.*, *73*, 269–273, 1974.

Bassett, W. A., and L.-C. Ming, Disproportionation of Fe_2SiO_4 to $2 \cdot FeO + SiO_2$ at pressures up to 250kbar and temperatures up to 3000°C, *Phys. Earth Planet. Inter.*, *6*, 154–160, 1972.

Bell, P. M., J. Xu, and H.-K. Mao, Static compression of gold and copper and calibration of the ruby pressure scale to pressures to 1.8 megabars, in *Proceedings of the Fourth American Physical Society Topical Conference on Shock Waves in Condensed Matter*, edited by Y. Gupta, pp. 125–130, Plenum Publishing, New York, 1986.

Benz, W., and A. G. W. Cameron, Terrestrial effects of the giant impact, in *Origin of the Earth*, edited by H. E. Newsom, and J. H. Jones, pp. 61–67, Oxford University Press, Oxford, 1990.

Benz, W., W. L. Slattery, and A. G. W. Cameron, The origin of the moon and the single impact hypothesis, I, *Icarus*, *66*, 515–535, 1986.

Benz, W., W. L. Slattery, and A. G. W. Cameron, The origin of the moon and the single impact hypothesis, II, *Icarus*, *71*, 30–45, 1987.

Bevington, P. R., and D. K. Robinson, *Data Reduction and Error Analysis for the Physical Sciences*, McGraw Hill, Inc., New York, 2nd edn., 1992.

Bodea, S., and R. Jeanloz, Model calculations of the temperature distribution in the laser-heated diamond cell, *J. Appl. Phys.*, *65*(12), 4688–4692, 1989.

Boehler, R., and A. Chopelas, Phase transitions in a 500 kbar–3000 K gas apparatus, in *High-Pressure Research: Application to Earth and Planetary Sciences*, edited by Y. Syono, and M. H. Manghnani, pp. 55–60, American Geophysical Union, Washington, DC, 1992.

Boehler, R., N. von Bargen, and A. Chopelas, Melting, thermal expansion, and phase transitions of iron at high pressures, *J. Geophys. Res.*, *95*(B13), 21731–21736, 1990.

Burch, C. R., Semi-aplanat reflecting microscopes, *Proc. Phys. Soc.*, *57*, 47–49, 1947.

Cameron, A. G. W., and W. Benz, The origin of the Moon and the single impact hypothesis IV, *Icarus*, *92*, 204–216, 1991.

Demarest, H. H., Jr., and H. T. Haselton, Jr., Error analysis for bracketed phase equilibrium data, *Geochim. Cosmochim. Acta*, *45*, 217–224, 1981.

Fei, Y., and Y. Wang, The maximum solubility of iron in $(Mg,Fe)SiO_3$-perovskite as a function of temperature, *Trans., Am. Geophys. Union, suppl.*, *73*(43), 596, 1992.

Fei, Y., Y. Wang, D. Virgo, B. O. Mysen, and H.-K. Mao, Ferric iron in $(Mg,Fe)SiO_3$-perovskite: A Mössbauer spectroscopic study, *Trans., Am. Geophys. Union, suppl.*, *73*(14), 300, 1992.

Flasar, F. M., and F. Birch, Energetics of core formation: a correction, *J. Geophys. Res.*, *78*, 6101–6103, 1973.

Gilvarry, J. J., Equation of the fusion curve, *Phys. Rev.*, *102*(2), 325–331, 1956.

Heinz, D. L., Melting curve of magnesian silicate perovskite, Ph.D., University of California, Berkeley, 1986.

Heinz, D. L., Thermal pressure in the laser-heated diamond anvil cell, *Geophys. Res. Lett.*, *17*(8), 1161–1164, 1990.

Heinz, D. L., and R. Jeanloz, Measurement of the melting curve of $Mg_{0.9}Fe_{0.1}SiO_3$ at lower mantle conditions and its geophysical implications, *J. Geophys. Res.*, *92*(B11), 11437–11444, 1987a.

Heinz, D. L., and R. Jeanloz, Temperature measurements in the laser-heated diamond cell, in *High-Pressure Research in Mineral Physics*, edited by M. H. Manghnani, and Y. Syono, pp. 113–127, American Geophysical Union, Washington D.C., 1987b.

Heinz, D. L., J. S. Sweeney, and P. Miller, A laser heating system that stabilizes and controls the temperature: diamond anvil cell applications, *Rev. Sci. Instrum.*, *62*(2), 1568–1575, 1991.

Heinz, D. L., E. Knittle, J. S. Sweeney, Q. Williams, and R. Jeanloz, High-pressure melting of (Mg,Fe)SiO_3-perovskite, *Science*, *264*, 279–281, 1994.

Hess, H. H., and A. H. Phillips, Optical properties and chemical composition of magnesian orthopyroxenes, *Am. Mineral.*, *25*, 271–285, 1940.

Ito, E., and T. Katsura, Melting of ferromagnesian silicates under the lower mantle conditions, in *High-Pressure Research: Applications to Earth and Planetary Sciences*, edited by Y. Syono, and M. H. Manghnani, pp. 315–322, American Geophysical Union, Washington, D. C., 1992.

Ito, E., and E. Takahashi, Postspinel transformations in the system Mg_2SiO_4-Fe_2SiO_4 and some geophysical implications, *J. Geophys. Res.*, *94*(B8), 10637–10646, 1989.

Ito, E., and H. Yamada, Stability relations of silicate spinels, ilmenites, and perovskites, in *High-Pressure Research in Geophysics*, edited by S. Akimoto, and M. H. Manghnani, pp. 405–419, Center for Academic Publications Japan, Tokyo, 1982.

Jacobs, S. D., K. A. Cerqua, K. L. Marshall, A. Schmid, M. J. Guardalben, and K. J. Skerrett, Liquid-crystal laser optics: design, fabrication, and performance, *J. Opt. Soc. Am., B*, *5*(9), 1962–1979, 1988.

Jeanloz, R., and D. L. Heinz, Experiments at high temperature and pressure: laser heating through the diamond cell, *J. Physique*, *45*, 83–92, 1984.

Jeanloz, R., and D. L. Heinz, Measurement of the temperature distribution in CW-laser heated materials, in *Laser Welding, Machining and Materials Processing: International Conference on Applications of Lasers and Electro-optics*, edited by C. Albright, pp. 239–243, Springer-Verlag, San Francisco, California, USA, 1985.

Jeanloz, R., and A. Kavner, Melting criteria and imaging spectoradiometry in laser-heated diamond-cell experiments, *Philos. Trans. R. Soc. London, Ser. A*, *submitted*, 1–30, 1995.

Jeanloz, R., and M. Manga, Modelling the temperature distribution in the laser-heated diamond cell: Effect of temperature gradients and the measurement of thermophysical properties, *Trans., Am. Geophys. Union, suppl.*, *76*(46), F586, 1995.

Knittle, E., and R. Jeanloz, Synthesis and equation of state of (Mg,Fe)SiO_3 perovskite to over 100 gigapascals, *Science*, *235*, 668–670, 1987.

Knittle, E., and R. Jeanloz, Melting curve of (Mg,Fe)SiO_3 perovskite to 96 GPa: Evidence for a structural transition in lower mantle melts, *Geophys. Res. Lett.*, *16*(5), 421–424, 1989.

Lazor, P., G. Shen, and S. K. Saxena, Laser-heated diamond anvil cell experiments at high pressure: melting curve of nickel up to 700 kbar, *Phys. Chem. Minerals*, *20*, 86–90, 1993.

Liu, L.-G., Silicate perovskite from phase transformations of pyrope-garnet at high pressure and temperature, *Geophys. Res. Lett.*, *1*(6), 277–280, 1974.

Liu, L.-G., Post-oxide phases of olivine and pyroxene and mineralogy of the mantle, *Nature*, *258*, 510–513, 1975.

Liu, L.-G., The high-pressure phases of $MgSiO_3$, *Earth Planet. Sci. Lett.*, *31*, 200–208, 1976.

Manga, M., and R. Jeanloz, Axial temperature gradients in dielectric samples in the laser-heated diamond cell, *Geophys. Res. Lett.*, *23*(14), 1845–1848, 1996.

Mao, H.-K., P. M. Bell, J. W. Shaner, and D. J. Steinberg, Specific volume measurements of Cu, Mo, Pd, and Ag and calibration of the ruby R1 fluorescence pressure gauge from 0.06 to 1 Mbar, *J. Appl. Phys.*, *49*(6), 3276–3283, 1978.

Mao, H.-K., J. Xu, and P. M. Bell, Calibration of the ruby pressure gauge to 800 kbar under quasi-hydrostatic conditions, *J. Geophys. Res.*, *91*(B5), 4673–4676, 1986.

Mao, H.-K., P. M. Bell, and C. Hadidiacos, Experimental phase relations in iron to 360 kbar, 1400°C, determined in an internally heated diamond-anvil apparatus, in *High-Pressure Research in Mineral Physics*, edited by M. H. Manghnani, and Y. Syono, pp. 135–138, American Geophysical Union, Washington, D.C., 1987.

Matsui, T., and Y. Abe, Formation of a "magma ocean" on the terrestrial planets due to the blanketing effect of an impact-induced atmosphere, *Earth Moon Planets*, *34*, 223–230, 1986.

Melosh, H. J., Giant impacts and the thermal state of the early Earth, in *Origin of the Earth*, edited by H. E. Newsom, and J. H. Jones, pp. 69–83, Oxford University Press, Oxford, 1990.

Miller, G. H., E. M. Stolper, and T. J. Ahrens, The equation of state of a molten komatiite 2. application to komatiite petrogenesis and the hadean mantle, *J. Geophys. Res.*, *96*(B7), 11849–11864, 1991.

Miller, R., III, The Webster-Addie ultramafic ring, Jackson County, North Carolina, and secondary alteration of its chromite, *Am. Mineral.*, *38*(11-12), 1134–1147, 1953.

Ming, L.-C., and W. A. Bassett, Laser heating in the diamond anvil press up to 2000°C sustained and 3000°C pulsed at pressures up to 260 kilobars, *Rev. Sci. Instrum.*, *45*(9), 1115–1118, 1974.

Ming, L.-C., and W. A. Bassett, High-pressure phase transformations in the system of $MgSiO_3$-$FeSiO_3$, *Earth Planet. Sci. Lett.*, *27*, 85–89, 1975.

Newsom, H. E., and S. R. Taylor, Geochemical implications of the formation of the Moon by a single great impact, *Nature*, *338*, 29–34, 1989.

Press, W. H., S. A. Teukolsky, W. T. Vetterling, and B. P. Flannery, *Numerical Recipes in FORTRAN: The Art of Scientific Computing*, Cambridge University Press, New York, 2nd edn., 1992.

Safronov, V. S., The heating of the Earth during its formation, *Icarus*, *33*, 3–12, 1978.

Sasaki, S., The primary solar-type atmosphere surrounding the accreting Earth: H_2O–induced high surface temperature, in *Origins of the Earth*, edited by H. E. Newsom, and J. H. Jones, pp. 195–209, Oxford University Press, Oxford, 1990.

Shen, G., and P. Lazor, Measurement of melting temperatures of some minerals under lower mantle pressures, *J. Geophys. Res.*, *100*(B9), 17699–17713, 1995.

Shen, G., Y. Fei, U. Halenius, and Y. Wang, Optical absorption spectra of (Mg,Fe)SiO_3 silicate perovskites, *Phys. Chem. Minerals*, *20*, 478–482, 1993a.

Shen, G., P. Lazor, and S. K. Saxena, Melting of wüstite and iron up to pressures of 600 kbar, *Phys. Chem. Minerals*, *20*, 91–96, 1993b.

Simon, F., On the range of stability of the fluid state, *Trans. Faraday Soc.*, *33*, 65–73, 1937.

Stevenson, D. J., Origin of the Moon – the collision hypothesis, *Annu. Rev. Earth Planet. Sci.*, *15*, 271–315, 1987.

Sweeney, J. S., Irreversible melting of a magnesiun-iron-silicate perovskite at lower mantle pressures, Ph.D., University of Chicago, 1996.

Sweeney, J. S., and D. L. Heinz, Melting of iron-magnesium-silicate perovskite, *Trans., Am. Geophys. Union, suppl.*, *73*(14), 368, 1992.

Sweeney, J. S., and D. L. Heinz, Melting of iron-magnesium-silicate perovskite, *Geophys. Res. Lett.*, *20*(9), 855–858, 1993a.

Sweeney, J. S., and D. L. Heinz, Thermal analysis in the laser-heated diamond anvil cell, *Pure Appl. Geophys.*, *141*(2/3/4), 497–507, 1993b.

Sweeney, J. S., and D. L. Heinz, Irreversible melting of magnesium-iron-silicate perovskite at lower mantle pressures, *Trans., Am. Geophys. Union, suppl.*, *76*(46), F553, 1995.

Syassen, K., and R. Sonnenschein, Microoptic double beam system for reflectance and absorption measurements at high pressure, *Rev. Sci. Instrum.*, *53*(5), 644–650, 1982.

Tonks, W. B., and H. J. Melosh, Magma ocean formation due to giant impacts, *J. Geophys. Res.*, *98*(E3), 5319–5333, 1993.

Uhlmann, D. R., On the internal nucleation of melting, *J. Non-Cryst. Solids*, *41*, 347–357, 1980.

Verhoogen, J., *Energetics of the Earth*, National Academy of Sciences, Washington, D.C., 1980.

Weertman, J., and J. R. Weertman, High temperature creep of rock and mantle viscosity, *Annu. Rev. Earth Planet. Sci.*, *3*, 293–315, 1975.

Williams, Q., and R. Jeanloz, Melting relations in the iron-sulfur system at ultra-high pressures: implications for thermal state of the Earth, *J. Geophys. Res.*, *95*(B12), 19299–19310, 1990.

Williams, Q., R. Jeanloz, J. Bass, B. Svendsen, and T. J. Ahrens, The melting curve of iron to 250 gigapascals: A constraint on the temperature at Earth's center, *Science*, *236*, 181–182, 1987.

Williams, Q., E. Knittle, and R. Jeanloz, The high-pressure melting curve of iron: a technical discussion, *J. Geophys. Res.*, *96*(B2), 2171–2184, 1991.

Yagi, T., P. M. Bell, and H.-K. Mao, Phase relations in the system MgO-FeO-SiO_2 between 150 and 700 kbar at 1000° C, *Carnegie Inst. Washington, Yearb.*, *78*, 614–618, 1979.

Zahnle, K. J., J. F. Kasting, and J. B. Pollack, Evolution of a steam atmosphere during Earth's accretion, *Icarus*, *74*, 62–97, 1988.

Zerr, A., and R. Boehler, Melting of (Mg,Fe)SiO_3-perovskite to 625 kilobars: indication of a high melting temperature in the lower mantle, *Science*, *262*, 553–555, 1993.

J. S. Sweeney, Center for High Pressure Research, SUNY at Stony Brook, Stony Brook, NY 11794-2100.

D. L. Heinz, Department of The Geophysical Sciences, University of Chicago, 5734 S. Ellis Ave., Chicago, IL 60637.

Metal/Silicate Partitioning of Mn, Co, and Ni at High-Pressures and High Temperatures and Implications for Core Formation in a Deep Magma Ocean

E. Ito* and T. Katsura

Institute for Study of the Earth's Interior, Okayama University, Misasa, Tottori-ken, 682-01 Japan

T. Suzuki

Department of Chemistry, Gakushuin University, Tokyo, 171 Japan

On melting of natural peridotite at about 30 GPa, magnesiowüstite is the first liquidus phase and perovskite appears at slightly lower temperature than the liquidus, implying that these phases would be potential solid phases in a deep magma ocean. Thus, in order to assess the chemical equilibrium of core formation, partitionings of Mn, Ni and Co between molten iron (MI), silicate liquid (SL), and magnesiowüstite (Mw) were investigated in the olivine-Fe and peridotite-Fe alloy systems within a MgO capsule at 20, 24, and 26 GPa at about 2600°C. With increasing pressure, Ni and Co become less siderophile and Mn becomes less lithophile. The exchange partition coefficient between molten iron and silicate liquid, K'(MI/SL), is definitely larger than that between molten iron and magnesiowüstite, K'(MI/Mw), for both Ni and Co. Pressure decreases the latter more conspicuously than the former. Convergence of partition coefficents of Ni and Co between MI and SL suggested by Li and Agee [1996] was not observed in the present study. Comparing the K' values extrapolated to higher pressure with the partition coefficients of Ni and Co between the core and the mantle derived from mass balance calculation, it is implied that the core material would have been equilibrated with magnesiowüstite and silicate liquid at the bottom of a magma ocean with a depth greater than 900 km. It is also suggested that the fraction of silicate liquid increases for the core segregation in the deeper magma ocean. As K's of Ni and Co between molten iron and perovskite have remarkably larger values than those of K'(MI/Mw) and K'(MI/SL), possible coexistence of this phase requires a significantly greater depth for the core seggregation.

INTRODUCTION

It has been realized that abundances of siderophile elements in the primitive Earth (silicate part of the earth) are considerably lower than those of CI chondrites, the class of meteorite closest in composition to the sun [*Wänke*

Properties of Earth and Planetary Materials
at High Pressure and Temperature
Geophysical Monograph 101

et al., 1984]. Although some volatile elements would have been lost before or during accretion, the depletion of siderophile elements is generally accepted as a result of partitioning of these elements in the core. However, experimental metal/silicate partitioning (acquired at low pressure and temperatures up to 1500°C) cannot account for the depletion: the calculated depletion for most of the refractory siderophile elements is larger than the observed depletion by some orders of magnitude [e.g., *Newsom*, 1990]. This discrepancy led to the proposal of special processes, such as heterogeneous accretion [*Ringwood*, 1984; *Newsom and Sims*, 1991] and inefficient core formation [*Jones and Drake*, 1986].

Recent planet accretion theories [e.g., *Sasaki and Nakazawa*, 1986] suggest that the terrestrial core would have been formed by sinking of molten iron through a deep magma ocean. Moreover, if a giant impact [*Wetherill*, 1985], which might have triggered formation of the moon, would have occurred later, considerable remixing of the protomantle and protocore should have taken place together with an extensive melting of the Earth [*Melosh*, 1990]. *Murthy* [1991] suggested that the metal-silicate partition coefficients D(M/S)'s of all siderophile elements will converge on unity with increasing temperature and, consequently, the mantle abundances of siderophile elements can be explained simply by a high temperature (>3000°C) equilibrium between metal and silicate liquid in a magma ocean. Murthy's extrapolation method, however, was criticized by several authors with more rigorous thermodynamic analyses [*Jones et al.*, 1992; *O'Neill*, 1992; *Capabianco* et al., 1993]. Experimental results by *Walker et al.* [1993] and *Thibault and Walter* [1995] showed that, for some siderophile elements, D(M/S) decreases with temperature, although not at a rate as large as predicted by Murthy [1991]. In order to assess core formation models, therefore, it is indispensable to determine the metal/silicate partitioning of siderophile elements under the high-pressure conditions corresponding to a deep magma ocean.

In this context, the behavior of Co and Ni has been considered to provide an important clue, both being refractory elements and present in the mantle in a near-chondritic Ni to Co ratio [e.g., *Newsom*, 1990]. *Hillgren et al.* [1994] studied partitioning of Fe, Ni, Co, W, and Mo in the iron-basaltic silicate system at 10.0 GPa and 2000°C. They demonstrated an important decrease in D(M/S) for Co and Ni, compared to values at 0 GPa and 1260°C. *Thibault and Walter* [1995] investigated the molten iron/silicate liquid partitioning of Ni and Co in a model cl chondrite at pressures of 1.2–12.0 GPa and temperatures of 2123–2750 K. They showed that both increasing pressure and temperature result in lowering the exchange partition coefficients of Ni and Co and that the pressure effect is significantly more important than the temperature effect for reasonable geotherms. Recently *Li and Agee* [1996] have determined the partitioning of Ni and Co between S-bearing iron and silicate liquid in the Allende meteorite up to 20 GPa. By extrapolating the partition coefficients to higher pressures, they have concluded that the observed abundances of Ni and Co and the Ni to Co ratio in the mantle can be explained if the core segregation would have taken place in the magma ocean with a depth of 750–1100 km.

We have also examined iron/silicate partitioning of Mn, Ni, and Co at high pressures. It is highly likely that some refractory solid phases would have been present in the magma ocean and made significant contribution to core formation. In the present study, therefore, we first examined the melting relations of natural peridotites at about 30 GPa to specify the refractory solid phases. The results indicate that both magnesiowüstite and perovskite appear in the liquidus. Then, we conducted a series of experiments to determine the partitioning between molten iron, silicate liquid, magnesiowüstite, and perovskite. The experimental pressure was varied from 20 to 26 GPa at about 2600°C. In this paper we will first show the result of melting of natural peridotite and then report the partitioning of Mn, Ni, and Co with some implications for core formation.

EXPERIMENTAL

Two natural peridotites, KLB-1 [*Takahashi*, 1986] and KR-4003 [*Xue et al.*, 1990], were employed as the starting materials of the melting experiment. Fine powder of the peridotite was put into a small cylindrical Re heater embedded in a magnesia octahedron as described by *Ito and Katsura* [1992]. However, pressurization was carried out using sintered diamond cubes with 10-mm edge length and 2-mm truncation. The cubic assembly was squeezed in the original double-staged, split-sphere vessel [*Kawai and Endo*, 1970]. Although the experimental pressure was not determined precisely, it would have been higher than 30 GPa as judged from the calibration based on the ilmenite-perovskite transformation in $MgSiO_3$ at 1600°C [*Ito and Takahashi*, 1989]. After pressurization, the sample was heated to 2600°C for 2 minutes and then quenched by shutting off the electric power.

In the partitioning experiments, two types of starting materials were prepared. One was a mixture of olivine and sponge iron (purity 99.99%) with a 3:1 volume ratio, in which the olivine was $(Mg_{0.9}Fe_{0.1})_2SiO_4$ doped with 5%

TABLE 1. Experimental Conditions and Starting Materials

Run No.	Starting Material	P(GPa)	T(°C)	t(min)
Tr01	Ol+Fe[a]	20	2600	1.0
Tr02	Ol+Fe	24	2600	2.0
Tr03	Ol+Fe	26	2600	1.5
Tr04	Per+Fe-alloy[b]	20	2600	3.0
Tr05	Per+Fe-alloy	24	2600	1.5
Tr06	Per+Fe-alloy	26	2600	2.0

[a]3{$(Mg_{0.9}Fe_{0.1})_2SiO_4$ olivine + 5%Mn_2SiO_4 + 5%Ni_2SiO_4 + 5Co_2SiO_4}
+1Fe (in volume)
[b]6KR-4003 + 1 $Fe_{100}Ni_{16}Co_5$ (in volume)

each of Mn_2SiO_4, $Ni_2SiO_4$and Co_2SiO_4. Another was a mixture of peridotite KR-4003 and an alloy $Fe_{100}Ni_{16}Co_5$ (in atomic) with a 6:1 volume ratio. Therefore the mixtures reverse each other on partitioning of the transition elements between metal and silicate. The experimental procedures were similar to those described elsewhere [*Ito and Katsura*, 1992; *Ito et al*., 1995]. The starting materials were put into a MgO capsule inserted in a small cylindrical rhenium heater. Run duration was limited to a few minutes because of the risk of reaction between molten iron in the sample and the rhenium heater. The experiments were carried out at 20, 24, and 26 GPa at about 2600°C, using a double-staged split-sphere apparatus installed at our institute (USSA5000). Pressure values were determined based on the calibration curve constructed at 1600°C [*Ito and Takahashi*, 1989]. The uncertainty in temperature determination might be within ±100°C. The experimental conditions and the starting materials are summarized in Table 1.

As the coexistence of molten iron and perovskite was not observed in the run products of the above starting materials (see later), the partitioning of Mn, Ni, and Co between molten iron and perovskite was determined indirectly by combining partitioning results for perovskite and magnesiowüstite and those for molten iron and magnesiowüstite. In order to determine the partitioning between perovskite and magnesiowüstite, an experiment at 26 GPa and about 2500°C was conducted using a mixture of KR-4003 plus 5% of Mn_2SiO_4, Ni_2SiO_4 and Co_2SiO_4 olivines as the starting material.

Polished sections of the quenched samples were examined by electron microscopy and then analyzed by the wavelength dispersive electron microprobe analyzer JEOL 8800 at our institute. Identification of phases present was made from textural observation and the chemical compositions.

RESULTS AND DISCUSSION

Melting of natural peridotite

In the run product quenched at about 30 GPa 2600°C, melting was observed along the wall of the rhenium heater in the central portion as shown in Figure 1. The upper portion (close to the rhenium heater) is composed of quenched liquid and magnesiowüstite of a rounded shape, whereas the lower portion is mostly an aggregate of magnesiowüstite and perovskite with small amount of liquid between the grain boundaries. Therefore it is seen that magnesiowüstite is the liquidus phase and both perovskite and magnesiowüstite coexist with liquid slightly below the liquidus. Almost the same results were obtained in the experiment on KR-4003 peridotite. The observation is consistent with the results of melting experiments by

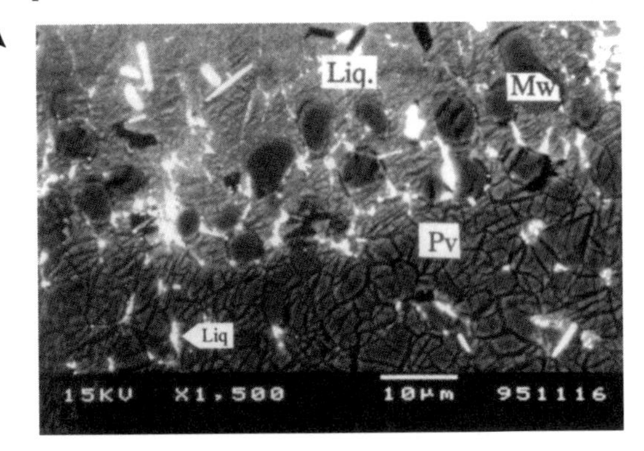

Figure 1. Back-scattered electron image of the polished section of peridotite KLB-a quenched at about 30 GPa and 2600°C. Liq, quenched liquid; Mw, magnesiowüstite; Pv, perovskite. Phases present are Liq, Liq + Mw, Liq + Mw + Pv, and Pv + Mw + trace liq with decreasing temperature.

High Temperature ←

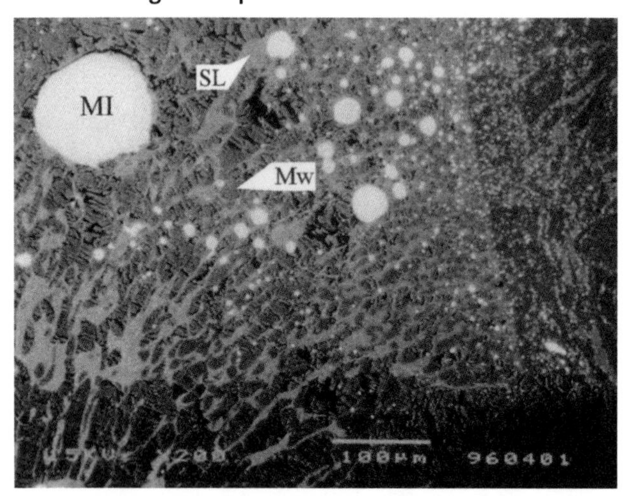

Figure 2. Back-scattered electron image of the polished section of Per + Fe-all quenched at 24 GPa and about 2600°C (run Tra05). MI, molten iron; Mw, magnesiowüstite; silicate liquid. Upper left portion was molten and right had portion was mostly solid (see text).

Zhang and Herzberg [1994] and *Herzberg and Zhang* [1996] on KLB-1 peridotite and by *Agee et al.* [1995] on the Allende meteorite. Therefore, magnesiowüstite would be the primary liquidus phase in a deep magma ocean, but perovskite as a secondary liquidus phase cannot be ruled out.

Results of Partitioning of Mn, Ni, and Co

Molten iron/magnesiowüstite and molten iron/silicate liquid. The general features observed in the experimental charges of a silicate-iron system were similar to those described by *Ito et al.* [1995]. All the run products listed in Table 1 showed evidence for melting of both iron and silicate in the central portions.

A typical view of the samples is shown in Figure 2, a back-scattered electron image of the run product of KR-4003 + Fe-alloy quenched at 24 GPa and 2600°C (run Tr05). In Figure 2, the upper left is the molten portion, which is composed of quenched silicate liquid, magnesiowüstite, and molten iron with a rounded shape. Perovskite was not observed in this region. Large molten iron grains (larger than 40 microns) contain many small blobs, indicating the dissolution of Si and O in the molten iron [*Ito et al.*, 1995]. The right-hand portion, separated from the molten portion by a sharp boundary, is a fine-grained aggregate of mostly perovskite and magnesiowüstite, and the degree of melting in this portion would have been very small. The rather dark portion in the bottom was originally the MgO capsule, into which a small amount of silicate liquid intruded. The constituents of the other run products were similar to those described above except that in the run products quenched at 20 GPa, both sides of the central molten portion were an assemblage of magnesiowüstite and spinel (run Tr01) or of magnesiowüstite and majorite (run Tr04). These assemblages would have coexisted with a small amount of liquid [*Zhang and Herzberg*, 1994].

TABLE 2. Chemical Compositions of Molten Iron, Magnesiowüstite, and Silicate Melt in the Run Product Per+Fe-all quenched at 24 GPa and about 2600°C

Molten iron(n=5)		Magnesiowüstite(n=4)		Silicate liquid(n=5)
Fe	72.07(1.89)	Na_2O	0.06(0.01)	0.06(0.02)
Ni	13.66(1.10)	MgO	85.10(0.70)	31.15(1.04)
Co	4.93(0.17)	SiO_2	0.13(0.01)	48.14(1.98)
Mn	0.02(0.01)	Al_2O_3	1.83(0.04)	5.45(0.38)
Cr	0.57(0.10)	FeO	11.63(0.01)	5.59(0.72)
Re	0.01(0.01)	NiO	0.54(0.09)	0.14(0.03)
O	1.29(0.80)	CoO	0.34(0.03)	0.10(0.02)
Si	0.91(0.04)	MnO	0.12(0.02)	0.04(0.02)
Total	93.46	CaO	0.06(0.02)	5.12(0.05)
		Cr_2O_3	0.66(0.12)	0.85(0.02)
		TiO_2	0.02(0.01)	0.16(0.07)
		La_2O_3	0.01(0.01)	4.15(1.01)
		ReO_2	0.1(0.1)	0.07(0.06)
		Total	100.60	101.71

TABLE 3. Analytical Results and Partition Coefficients

Run No	Tr01	Tr04	Tr02	Tr05	Tr03	Tr06
P/GPa	20	20	24	24	26	26
Starting material	Ol+Fe	Per+Fe-al	Ol+Fe	Per+Fe-al	Ol+Fe	Per+Fe-al
Molten iron(at.%)						
n	7	5	4	5	4	7
Fe	91.87	69.77	90.84	74.57	91.06	70.03
Ni	3.44	22.06	3.19	13.46	3.82	22.75
Co	4.13	7.56	3.71	4.83	4.78	5.83
Mn	0.18	-	0.23	-	0.31	-
Magnesiowüstite(mol %)						
n	6	5	3	4	2	6
MgO	92.06	89.21	92.68	90.3	91.65	86.51
FeO	6.80	7.72	6.34	6.93	6.96	9.00
NiO	0.05	0.61	0.06	0.31	0.10	0.86
CoO	0.10	0.33	0.11	0.19	0.18	0.38
MnO	0.57	-	0.35	-	0.47	-
Silcate Liquid(mol %)						
n	6	5	4	5	5	8
MgO	37.19	44.10	35.25	40.80	34.45	42.99
SiO_2	26.95	33.78	28.53	42.26	28.71	38.91
FeO	3.75	6.73	3.38	4.11	3.83	6.22
NiO	0.017	0.25	0.017	0.10	0.25	0.32
CoO	0.042	0.17	0.042	0.07	0.06	0.14
MnO	0.53	-	0.35	-	0.56	-
K'Ni/Fe(MI/Mw)[a]	5.05(0.50)	4.00(0.37)	3.34(0.71)	3.60(0.53)	2.92(0.15)	3.40(0.16)
K'Co/Fe(MI/Mw)	3.06(0.25)	2.53(0.32)	2.23(0.25)	2.32(0.26)	2.09(0.21)	1.98(0.10)
K'Mn/Fe(MI/Mw)	0.024(0.001)	-	0.046(0.004)	-	0.051(0.020)	-
K'Ni/Fe(MI/SL)	8.41(2.30)	8.47(2.78)	7.02(2.50)	7.20(0.60)	6.31(1.55)	6.37(1.28)
K'Co/Fe(MI/SL)	4.05(1.25)	4.22(1.01)	3.32(0.39)	3.82(0.46)	3.16(1.00)	3.61(0.65)
K'Mn/Fe(MI/SL)	0.014(0.03)	-	0.024(0.006)	-	0.023(0.008)	-
fo_2 (in log unit relative to IW)	-2.26	-1.91	-2.31	-2.06	-2.23	-1.78

[a] MI: Molten iron, Mw:Magnesiowüstite; SL: Silicate liquid

Coexisting sets of molten iron, silicate liquid, and magnesiowüstite were chosen for chemical analysis. The chemical compositions of the coexisting phases obtained for the run Tr05 conducted at 24 GPa and 2600°C are shown in Table 2 as typical results. All the analytical results are summarized in Table 3 with experimental conditions and the starting materials employed. It is shown that Ni and Co are preferentially partitioned to molten iron over magnesiowüstite and silicate liquid, whereas Mn is partitioned more uniformly in these phases. The oxygen fugacities of the experimental charges were estimated to be about two logarithmic units lower than that of the iron-wüstite buffer from the Fe content of molten iron and the FeO contents of magnesiowüstite.

The partitioning of Mn, Ni, and Co between molten iron and magnesiowüstite or silicate liquid is recognized as an exchange reaction expressed as follows:

$$\text{Fe (MI)} + \text{MO (Mw or SL)} = \text{M (MI)} + \text{FeO (Mw or SL)} \quad (1)$$

and the exchange partition coefficient, K', is written:

$$K'_{M/Fe}\,(\text{MI/Mw or SL}) = (X_M/X_{Fe})^{MI}/(X_M/X_{Fe})^{\text{Mw or SL}} \quad (2)$$

where MI, Mw, and SL are molten iron, magnesiowüstite, and silicate liquid, respectively; M represents Mn, Co, or Ni; and X_i is the molar fraction of element i of each phase. An advantage of treating the elemental partitioning as an exchange reaction is that, at constant pressure and temperature, the partition coefficient K' is constant at varying oxygen fugacity [e.g., *O'Neill*, 1992]. The K' values were calculated from the analytical results and are shown in Table 3. As the Mn contents of molten iron in

TABLE 4. Chemical Compositions of Perovskite and Magnesiowüstite in the Run Product of KR-4003+5%Mn_2SiO_4 + 5%Ni_2SiO_4 quenched at 26 GPa and 2500°C

	Perovskite(n=5)	Magnesiowustite(n=5)
Na_2O	0.08(0.03)	0.74(0.03)
MgO	32.07(0.15)	61.37(0.11)
Al_2O_3	5.90(0.14)	1.71(0.05)
SiO_2	51.54(0.16)	0.22(0.04)
NiO	0.66(0.02)	12.63(0.08)
CoO	1.15(0.02)	10.86(0.02)
MnO	1.76(0.05)	3.02(0.03)
FeO	4.27(0.04)	9.95(0.05)
TiO_2	0.17(0.05	0.04(0.01)
CaO	0.94(0.03)	0.03(0.01)
Cr_2O_3	0.28(0.05)	0.44(0.01)
total	98.82	101.01

Exchange partition coefficients:

K'Fe/Mg(Pv/Mw)=0.82(0.02)

K'Ni/Fe(Pv/Mw)=0.12(0.02) K'Ni/Fe(MI/Pv)=26.64(4.50)*

K'Co/Fe(Pv/Mw)=0.26(0.05) K'Co/Fe(MI/Pv)=8.57(1.91)

K'Mn/Fe(Pv/Mw)=1.36(0.19) K'Mn/Fe(MI/Pv)=0.046(0.007)

* K'(MI/Pv) is calculated from K'(Pv/Mw) and K'(MI/Mw) in Table 3.

runs Tra04, Tra 05 and Tra06 were less than 200 ppm with large errors, K's of Mn were not calculated from these runs.

The K' values for Ni and Co obtained at the same pressure but from the different starting materials are very close each other and almost identical within uncertainties (see Table 3), indicating that our experimental charges were very close to equilibrium with respect to elemental partitioning.

Molten iron/perovskite. The back-scattered electron image of the run product of KR-4003 peridotite plus 5% of Mn_2SiO_4, Ni_2SiO_4, and Co_2SiO_4 olivines quenched at 26 GPa and 2500°C is reproduced in Figure 3. In this run, however, temperature was first raised to 2600°C and then was lowered to 2500°C and kept for 2 minutes. Therefore the perovskite and magnesiowüstite with large grain sizes in the upper right portion would have been crystallized from melt under hypersolidus conditions. Usually the grain sizes of these phases formed by the dissociation of spinel under subsolidus conditions are small and inadequate for EPMA analysis, as seen from the lower left portion of Figure 3 (see also *Katsura and Ito* [1996]).

The chemical compositions of perovskite and magnesiowüstite coexisting in the coarse-grained region are shown in Table 4, from which K's between perovskite and magnesiowüstite are calculated. Combining these values with those between molten iron and magnesiowüstite, we have partition coefficients of Mn, Ni, and Co between molten iron and perovskite, neglecting the effect of a temperature difference of 100°C and that of difference in chemical composition of magnesiowüstite (Table 4).

Figure 3. Back-scattered electron image of the polished section of peridotite KR-4003 + 5%Mn_2SiO_4 + 5%Ni_2SiO_4 + 5%Co_2SiO_4 sample quenched at 26 GPa and 2500°C. Pv, perovskite; Mw, magnesiowüstite. Large grained Pv and Mw in the upper right portion would have crystallized from liquid.

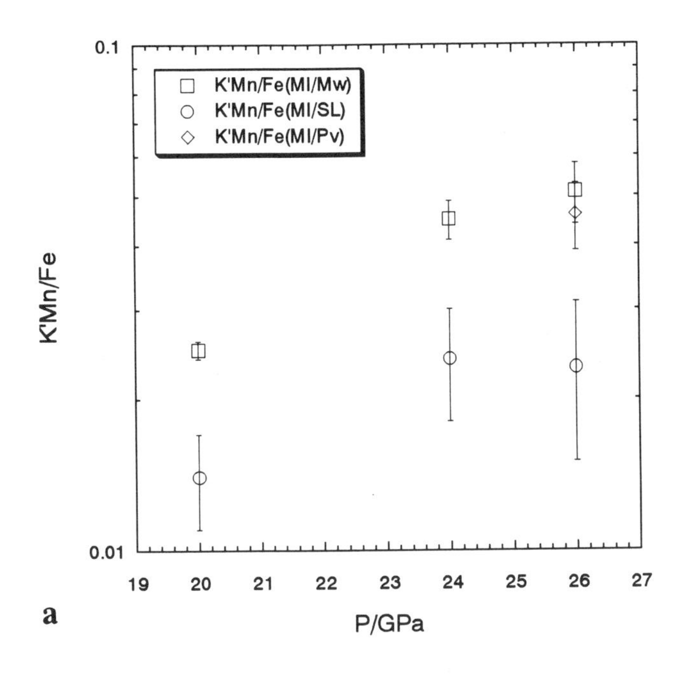

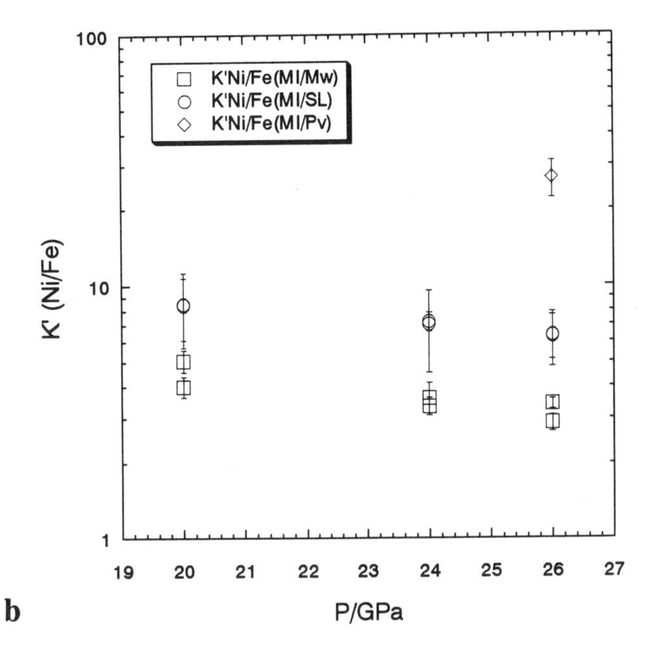

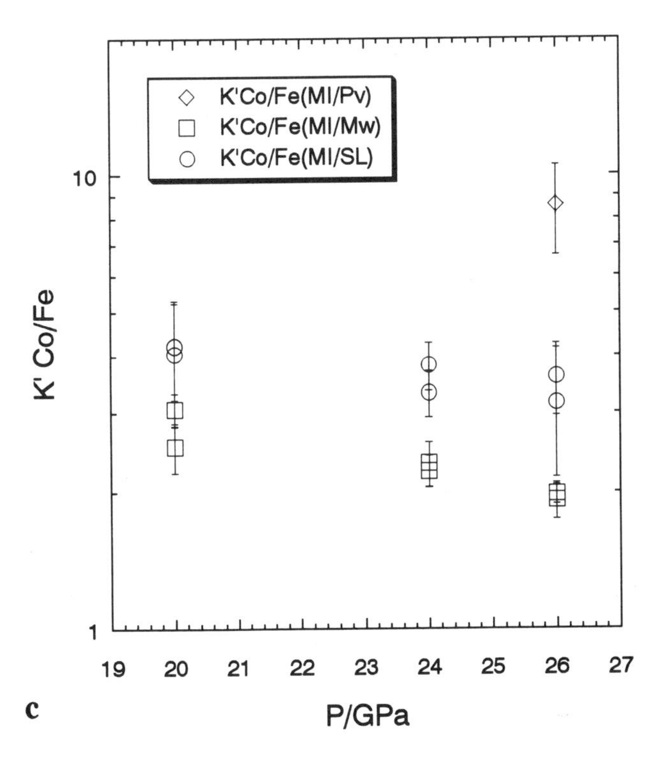

Figure 4. Variatiohn of the exchange partition coefficients K'M/Fe(MI/Pv), K'M/Fe(MI/Mw), and K'M/Fe(MI/SL) as a function of pressure for sets of experiments at 2600°C. M denotes Mn, Ni, and Co. MI, molten iron; Mw, magnesiowüstite; SL, silicate liquid. 4a: Mn, 4b: Ni, 4c: Co. The error bars represent the uncertainties displayed in Tables 3 and 4.

Pressure Dependence of the Partition Coefficients and Comparison with Previous Studies

Partition coefficients of Mn, Ni, and Co between molten iron and magnesiowüstite, molten iron and silicate liquid, and molten iron and perovskite, obtained in the present study, are plotted against pressure in Figures 4a, b, and c, respectively. Change in partition coefficients of Ni and Co with pressure is compared in Figure 5. The K's depend not only on pressure and temperature but on chemical compositions of the phases concerned; it is shown that K'(MI/Mw), especially for Ni, decreases considerably with increasing MgO content of magnesiowüstite [*Urakawa*, 1991; *Suzuki et al.*, 1994]. In the present study, Mg-numbers of magnesiowüstite are in a range of 86 to 93 and the variation in K'(MI/SL) due to change in composition of silicate liquid is not recognized (see Table 3). Therefore, changes in K's plotted in Figures 4a-4c display predominantly the pressure effects on the partition coefficients.

In Figure 4a, although both K'Mn/Fe(MI/Mw) and K'Mn/Fe(MI/SL) are smaller than unity, the former is larger than the latter: i.e., Mn prefers silicate liquid to magnesiowüstite. However both increase with increasing pressure, indicating a less lithophile nature of Mn at high pressures. It is also suggested that K'Mn/Fe(MI/Mw) increases more rapidly than K'Mn/Fe(MI/SL) in a pressure range of 20–26 GPa.

Ohtani et al. [1991a,b] determined partitioning of transition elements between solid iron and

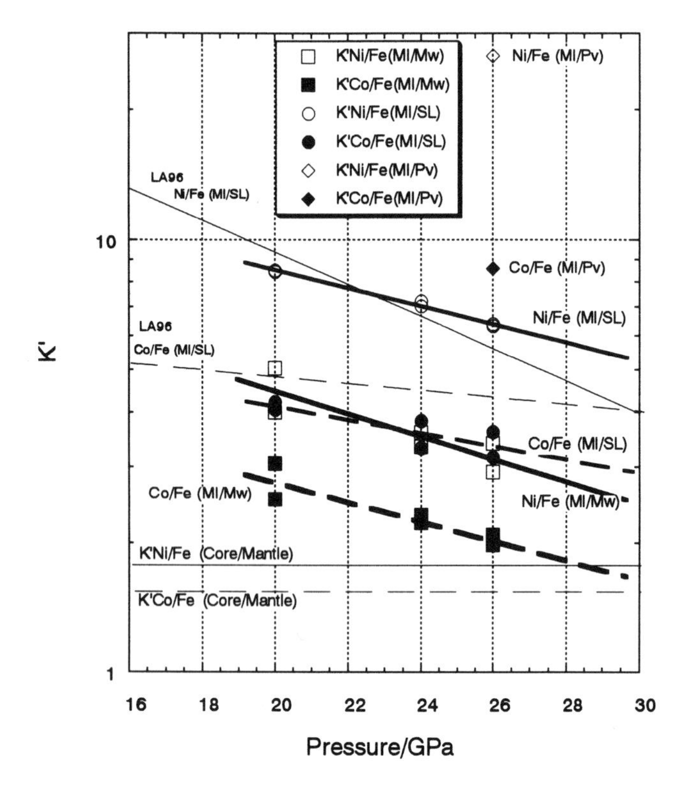

Figure 5. Variation of the exchange partition coefficients for Ni and Co as a function of pressure. Open and filled symbols are for Ni and Co, respectively. Squares, circles, and diamonds denote partitioning between molten iron/magnesiowüstite, molten iron/silicate liquid, and molten iron/perovskite, respectively. Solid and dashed lines indicate the slopes of the relations between partition coefficients and pressure for Ni and Co, respectively. Two horizontal lines marked as K'Ni/Fe(Core/Mantle) and K'Co/Fe(Core/Mantle) indicate the partition coefficients between the core and the mantle for Ni and Co (see text). LA96Ni/Fe(MI/SL) and LA96Co/Fe(MI/SL) are slopes for Ni and Co indicated by *Li and Agee* [1996].

magnesiowüstite or perovskite at 1700°C and 27 GPa. The calculated values of K'Mn/Fe(Fe/Mw) and K'Mn/Fe(Fe/Pv) from their analytical data are 0.03 and 0.024, respectively, which are generally consistent with the present results, in spite of the difference in the experimental temperature and the phase of iron, solid and molten, in both studies. *Suzuki et al.* [1994] studied partitioning of Mn, Ni, and Co between molten iron and magnesiowüstite ore olivine in the hydrous iron-silicate system at pressures of 3.0–11.2 GPa and at 1400°C. They obtained values around 0.01 for K'Mn/Fe(MI/Mw) at pressures of 3–11 GPa. *Walker et al.* [1993] and *Hillgren et al.* [1994] reported results of partitioning of Mn between molten iron and silicate liquid at 10 GPa and temperatures of 2000°C or higher. These works give the K'Mn/Fe(MI/SL) values of 0.0067 at 2180°C and 0.085 at 2800°C for the S-bearing molten iron/ultrabasic silicate liquid pair [*Walker et al.*, 1993] and 0.0025 at 2000°C for the molten iron/basaltic liquid pair [*Hillgren et al.*, 1994]. Although there are significant differences in the experimental temperature and notable variation in the chemical compositional of the phases concerned among the present and the previous studies, it can be concluded that both K'Mn/Fe(MI/Mw) and K'Mn/Fe(MI/SL) considerably increase from low pressure to 26 GPa.

As seen from Figures 4b and 4c, both K'(MI/Mw) and K'(MI/SL) of Ni and Co show a clear tendency to decrease with increasing pressure; i.e., both Ni and Co become less siderophile. The pressure effect on K'(MI/Mw) is more notable than that on K'(MI/SL) (see also Figure 5). It is also evident that K'(MI/SL) is larger than K'(MI/Mw) for both Ni and Co, and the values of K'(MI/Pv) at 26 GPa are much larger than those of K'(MI/SL) and K'(MI/Mw) at 26 GPa. Therefore, phase preference of both Ni and Co is in the order of magnesiowüstite, silicate liquid, and perovskite at pressures of 20–26 GPa and at 2600°C.

The K's for Ni and Co between solid iron and magnesiowüstite calculated from the results of *Ohtani et al.* [1991a,b] at 27 GPa and 1700°C are 6.0 and 4.5 respectively, which may be again consistent with the present results considering the difference in temperature and in the iron phase. However, considerably smaller values, 4.5 for Ni and 2.9 for Co, are calculated for the solid iron/perovskite partitioning from their results. The reason for the discrepancy between the studies is not clear. *Suzuki et al.* [1994] showed that K'(MI/Mw) values of Ni and Co decrease almost linearly with pressure up to 11.2 GPa and K' values of Ni are definitely larger than those of Co. These features are in harmony with the present results.

Pressure dependence of K'(MI/SL) for Ni and Co has been investigated by several authors [e.g., *Peach and Mathez*, 1993; *Walker et al.*, 1993; *Hillgren et al.*, 1994; *Thibault and Walter*, 1995; *Li and Agee*; 1996]. All of these works have demonstrated that the K'(MI/SL) values decrease remarkably with increasing pressure and that, at constant pressure, K'Ni/Fe(MI/SL) is distinctly larger than K'Co/Fe(MI/SL) in the pressure range studied, up to 20 GPa. The observation is quite consistent with the present results. Comparing their own results and those of *Thibault and Walter* [1995], *Li and Agee* [1996] pointed out that the effect of S in molten iron on partition coefficients becomes less important at high pressures.

The pressure dependence of K'Ni/Fe(MI/SL)' in the range of 20–26 GPa obtained in the present study is slightly less than that in the pressures up to 20 GPa by *Li and Agee* [1996] as shown in Figure 5. In regard to

K'Co/Fe(MI/SL), linear interpolation of our results to lower pressure suggests slightly lower values than those obtained by *Thibault and Walter* [1995] and *Li and Agee* [1996], which would be due to the difference of the experimental temperature. The changes in pressure dependence of K's would be due to changes in coordination states of Ni and Co in silicate liquid with increasing pressure. In a spectroscopic study on Ni- or Co-doped albite glasses quenched at up to 2900°C and 10 GPa, *Keppler and Rubie* [1993] showed that about 50% of originally tetrahedrally coordinated shifted to octahedral coordination at 10 GPa. For Ni, they suggested a pressure-induced coordination change of the small fraction of originally tetrahedral Ni^{2+} or a pressure-enhanced distortion of the octahedral site of Ni^{2+}. Both the coordination change of Co^{2+}(and possibly of Ni^{2+}) and the distortion of the octahedral site of Ni^{2+} result in large increases in stabilities of both irons in silicate liquid. The apparent difference in pressure effect on K's below and above 20 GPa may be responsible for the pressure-induced structural changes of silicate liquid. *Keppler and Rubie* [1993] suggested that the reduction of the K' of Co by the coordination change would hold up to about 20 GPa. Nevertheless, it is noted that the K' values of Ni and Co at 20 GPa obtained in the present study and those *by Li and Agee* [1996] are very close; 8.41 and 8.47 versus 8.57 for Ni and 4.04 and 4.22 versus 3.82 for Co, respectively.

Implication to the Core Formation

Extrapolation of partition coefficients D(MI/SL)'s of Ni and Co by *Li and Agee* [1996] suggests that, at about 28 GPa, both will converge and reach 29 and 26, respectively, the values required for the observed core-mantle partitioning for Ni and Co. Therefore *Li and Agee* [1996] concluded that the core material of molten iron alloy would have been equilibrated with silicate liquid during core extraction in a magma ocean with a depth of 570–1100 km. In our partitioning results at 20–26 GPa, however, the pressure effect on the partitioning of Ni is slightly less than that expected by Li and Agee from their low pressure trend and is comparable to the pressure effect on the partitioning of Co. Therefore our experimental results may not support their conclusion..

In Figure 5, two horizontal lines marked as Ni/Fe(Core/Mantle) and Co/Fe(Core/Mantle) indicate the partition coefficients for Co and Ni between the core and the mantle which are calculated from the abundances of these elements in the mantle [*Wänke et al.*, 1984] and the cosmic abundance [*Anders and Grevesse*, 1989], assuming that no Mg enters in the core. These are 1.77 for Ni and 1.53 for Co. It is seen that the trends of decrease in K'(MI/SL) of both Co and Ni with pressure require very high pressure, presumably larger than 50 GPa, for the K's to reach these values. In other words, core segregation would have occurred at depths greater than 1300 km, if the molten iron is equilibrated with only silicate liquid in the magma ocean and receives no significant change in composition through later movement to the center. The K'(MI/Mw)'s which are definitely smaller than K'(MI/SL)'s, on the other hand, decrease more rapidly than K'(MI/SL) and will become less than K'(Core/Mantle)'s at pressures around 30 GPa. Considering that magnesiowüstite is the liquidus phase in melting in peridotite at high pressures, the present experimental results imply that the partitioning of Co and Ni between the core and the mantle can be accounted for if the core material would have been equilibrated with both silicate liquid and magnesiowüstite at the bottom of a magma ocean extended to a depth greater than 900 km. It should also be noted that the fraction of silicate liquid increases in a deeper magma ocean. However, coexistence of perovskite with molten iron in the magma ocean is also likely because this phase is present at temperatures slightly lower than the liquidus (see Figure 1). As K'(MI/Pv)'s have much higher values than K'(MI/Mw) and K'(MI/SL), perovskite makes the mantle depleted in Ni and Co, and a greater depth would be required for the core segregation. The pressures at which K'(MI/SL)'s meet K'(Core/Mantle)'s may provide important constraints for depth of magma ocean of the core segregation. Nevertheless it is likely that possible structural change in silicate liquid [*Knittle and Jeanloz*, 1989] at higher pressures will significantly affect the partitioning between molten iron and silicate liquid. Therefore, in order to make critical evaluation of the core formation models, experiments on melting of the mantle material and elemental partitioniong at higher pressures at higher pressures are needed.

SUMMARY

1 . Melting experiments of natural peridotite at about 30 GPa have shown that magnesiowüstite is the first liquidus phase and perovskite appears at a slightly below the liquidus. Therefore both the phases would be the solid phases in the deep magma ocean.

2. Exchange partition coefficients of Ni, Co, and Mn between molten iron and magnesiowüstite or silicate liquid were determined at 20, 24, and 26 GPa and about 2600°C. Those between molten iron and perovskite were obtained indirectly using the partition coefficients between perovskite and magnesiowüstite at 26 GPa and 2500°C.

with increasing pressure Ni and Co become less siderophile and Mn becomes less lithophile. In the pressure range studied, phases preference of Ni and Co is in the order of magnesiowüstite, silicate liquid, and perovskite whereas that of Mn is in silicate liquid, perovskite, and magnesiowüstite. The partition coefficients of both Ni and Co between molten iron and magnesiowüstite decrease more rapidly than those between molten iron and silicate liquid with increasing pressure.

3. Present experimental results suggest that the terrestrial core material would have been equilibrated with magnesiowüstite and silicate liquid at a depth greater than 900 km. The fraction of silicate liquid should become larger for deeper equilibration depths. Possible coexistence of perovskite renders the depth of core segregation still deeper.

Acknowledgments. The authors thank M. J. Walter and S. Urakawa for discussion. They are grateful to E. Takahashi and M. J. Walter for their kind supply of natural peridotites. Critical reviews by C. Herzberg, E. Ohtani, and K. T. M. Johnson were helpfull in improving the final version of the paper. This research was made possible by grant 05102003 from the Ministry of Education, Science and Culture, Japan.

REFERENCES

Agee, C. B., J., Li, M. C. Shannon, and S. Circone, Pressure-temperature phase diagram for the Allende meteorite, *J. Geophys. Res.*, *100*, 17,725-17,740, 1995.

Anders, E., and N. Grevesse, Abundances of the elements: Meteoritic and solar, *Geochim. Cosmochim. Acta*, *53*, 197-214, 1989.

Capobianco, C. J., J. H. Jones, and M. J. Drake, Metal-silicate thermochemistry at high temperature: Magma oceans and the "excess siderophile element problem" of the Earth Upper Mantle, *J. Geophys. Res.*, *98*, 5,433-5,443, 1993.

Herzberg, C., and J. Zhang, Melting experiments on anhydrous peridotite KIB-1: Compositions of magmas in the upper mantle and transition zone, *J. Geophys. Res.*, *101*, 8271-8295, 1996.

Hillgren, V. J., M. J. Drake, and D. C. Rubie, High-pressure and high-temperature experiments on core-mantle segregation in the accreting earth, *Science*, *264*, 1442-1445, 1994.

Ito, E., and T. Katsura, Melting of ferromagnesian silicate under the lower mantle conditions, in *High-Pressure Research; Application to Earth and Planetary Science, Geophys. Monogr. Ser.*, vol. 67, edited by Y. Syono and M. H. Manghnani, pp. 315-322, Terra Scientific Publishing, Tokyo/ AGU, Washington, D. C., 1992.

Ito, E., K. Morooa, O. Ujike, and T. Katsura, Reactions between molten iron and silicate melts at high pressure: Implication for the chemical evolution of the Earth's core, *J. Geophys. Res.*, *100*, 5901-5910, 1995.

Ito, E. and E. Takahashi, Postspinel transformation in the system Mg_2SiO_4-Fe_2SiO_4 and some implications, *J. Geophys. Res.*, *94*, 10,637-10,646, 1989.

Jagoutz, E., H. Palme, H. Baddenhausen, K. Blum, M. Cenales, G. Dreibus, B. Spettel, V. Lorenz, and H. Wänke, The abundances of major, minor and trace elements in the earth's mantle as derived from primitive ultramafic rocks, *Proc. 10th Lunar Planet. Sci. Conf.*, 2031-2050, 1979.

Jones, J. H., C. J. Capobianco, and M. J. Drake, Siderophile elements and the Earth's formation, *Science*, *257*, 1281-1283, 1992.

Jones, J. H., and M. J. Drake, Geochemical constraints oncore formation in the earth, *Nature*, *322*, 221-228, 1986.

Katsura, T., and E. Ito, Determination of Fe-Mg partitioning between perovskite and magnesiowüstite, *Geophys. Res. Lett.*, 1996, in press.

Kawai, N. and S. Endo, The generation of ultrahigh hydrostatic pressure by a split sphere apparatus, *Rev. Sci. Instr.*, *4*, 425-428, 1970.

Keppler, H. and D. C. Rubie, Pressure-induced coordination changes of transition metal ions in silicate melts, *Nature*, *364*, 54-56, 1993.

Knittle, E. and R. Jeanloz, Melting curve of $(Mg,Fe)SiO_3$ perovskite to 96 GPa: Evidence for structural transition in lower mantle melts, *Geophys. Res. Lett.*, *16*, 421-424, 1989.

Li, J. and C.B. Agee, Geochemistry of mantle-core diferentiation at high pressure, *Nature*, *381*, 686-689.

Melosh, H. J., Giant impacts and the thermal state of the early earth, in *Origin of the Earth*, edited by H. E. Newton and J. H. Jones, pp.69-83, Oxford University Press, New York, 1990.

Newsom, H. E., Accretion and core formation in the earth: evidence from siderophile elements, in *Origin of the Earth*, edited by H. E. Newson and J. H. Jones, pp. 273-288, Oxford University Press, New York, 1990.

Newson, H. E., and K. W. W. Sims, Core formation during early accretion of the earth, *Science*, *252*, 926-933, 1991.

Ohtani, E., T. Kato, and E. Ito, Transition metal partitioning between lower mantle and core materials at 27 GPa, *Geophys. Res. Lett.*, *18*, 85-88, 1991a.

Ohtani, E., T. Kato, and E. Ito, Correction to "Transition metal partitioning between lower mantle and core materials at 27 GPa", *Geophys. Res. Lett.*, *18*, 796, 1991b.

O'Neill, H., Siderophile elements and the Earth's formation, *Science*, *257*, 1281-1285, 1992.

Peach, C. L., and E. A. Mathez, Sulfide melt-silicate melt distribution coefficients for nickel and iron and implications for the distribution of other chalcophile elements, *Geochim. Cosmochim. Acta*, *57*, 3013-3021, 1993.

Ringwood, A. E., The earth's core: its composition, formation and bearing upon the origin of the earth, *Proc. R. Soc. London*, Ser. *A395*, 1-46, 1984.

Sasaki, S. and K. Nakazawa, Metal-silicate fractionation in the growing earth: Energy source of the terrestrial magma ocean, *J. Geophys. Res.*, *91*, 9231-9238, 1986.

Suzuki, T., M. Akaogi, and T. Yagi, Pressure effect on Mn, Co and Ni partitioning between iron hydride and mantle minerals, *Proc. Jpn. Acad.*, *70*, Ser. B, 31-36, 1994.

Takahashi, E., Melting of a dery perodotite KLB-1 up to 14 GPa; Implications on the origin of peridotitic upper mantle, *J. Geophys. Res.*, *91*, 9367-9382, 1986.

Thibault, Y., and M. J. Walter, The influence of pressure and temperature on the metal-silicate partition coefficients of nickel and cobalt in a model Cl chondrite and implications for metal segregation in a deep magma ocean, *Geochim. Cosmochim. Acta*, *59*, 991-1002, 1995.

Urakawa, S., Partitioning of Ni between magnesiowüstite and metal at high pressure: implications for core-mantle equilibrium, *Earth Planet. Sci. Lett.*, *105*, 293-313, 1991.

Walker, D., L. Norby, and J. H. Jones, Super heating effects on metal-silicate partitioning of siderophile elements, *Science*, *262*, 1858-1861, 1993.

Wänke, H., G. Dreibus, and E. Jagoutz, Mantle chemistry and accretion history of the Earth, in *Archean Geochemistry*, edited by A. Kröner, G. N. Hanson, and A M. Goodwin, pp. 1-24, Springer-Verlag, 1984.

Wetherill, G. W., Occurrence of gaint impacts during the growth of the terrestrial planets, *Science, 228*, 877-879, 19??.

Xue, X., H. Baadsgaard, A. J. Irving, and C. M. Scarfe, Geochemical and isotopic characteristics of lithospheric mantle beneath West Kettle River, British Columbia: Evidence from ultramafic xenoliths, *J. Geophys. Res.*, *95*, 15,879-15,891, 1990.

Zhang, J., and C. Herzberg, Melting experiments on an hydrous peridotite KLB-1 from 5.0 to 22.5 GPa, *J. Geophys. Res.*, *99*, 17,729-17,742, 1994.

E. Ito and T. Katsura, Institute for Study of the Earth's Interior, Okayama University, Misasa, Tottori-ken 682-01, Japan.

T. Suzuki, Department of Chemistry, Gakushuin University, Tokyo 171, Japan.

Flotation of Olivine and Diamond in Mantle Melt at High Pressure: Implications for Fractionation in the Deep Mantle and Ultradeep Origin of Diamond

Eiji Ohtani, Akio Suzuki, and Takumi Kato[1]

Institute of Mineralogy, Petrology, and Economic Geology, Tohoku University, Sendai, Japan

The density of the three ultramafic melts, PHN1611, IT8720, and MA at high pressure was determined by the sink-float method using olivine and diamond as density markers. Using these data, we obtain K and K' of the Birch-Murnaghan equation of state for the compression curves of the silicate melts. The density relation between the mantle minerals and the partial melt provides a possible fractionation mode in the deep mantle; accumulation of olivine and pyroxene could occur in the region just above the transition zone, and separation of the partial melt and garnet could provide a potential mechanism for enrichment of the basaltic component in the transition zone. The density relation between diamond, the mantle minerals, and the melt suggests a possibility for accumulation of diamond in the transition zone. Recent discovery of the high-pressure minerals as inclusions in diamond may be consistent with the present model of accumulation of diamond in the deep mantle. The compositional features of the high-pressure mineral inclusions in diamond are discussed on the basis of the mineral-melt partition coefficients determined in the laboratory. The present analysis implies that the compositions of the high-pressure phases in diamond, especially a high concentration of REE in Ca-perovskite cannot be explained by a simple chemical equilibrium at high temperature above 2000°C, but can be explained by (1) chemical equilibrium at low temperature around 1000-1200°C perhaps in the hydrous slab conditions, or (2) multiple events of partial melting and fractional crystallization, that preclude chemical equilibrium among mineral inclusions.

1. INTRODUCTION

Several factors controlled the fractionation process in the mantle that occurred in the magma ocean in the earliest stage of the terrestrial evolution and during succeeding igneous activities. One of the important factors is the partitioning of elements between the mantle minerals and magmas. The density relation between the mantle minerals and magma is also an important factor [e.g., *Ohtani*, 1984; *Agee and Walker*, 1993].

In this study, we report our recent results on the measurement of the silicate melt density by using the sink-float experiments of olivine and diamond in the pressure range up to 23 GPa. We also summarize our recent data on the partition coefficients of some lithophile elements between Mg-perovskite, Ca-perovskite, majorite, and

[1]Also at Institute of Geosciences, University of Tsukuba, Tsukuba 305, Japan.

Properties of Earth and Planetary Materials
at High Pressure and Temperature
Geophysical Monograph 101

Figure 1a. The DIA type guide block system for operation of the multianvil apparatus installed at Tohoku University.

liquid, which were determined in our laboratory.

Using this data set, we discuss the possibility of chemical fractionation and accumulation of diamond in the deep mantle. Recent studies of mineral inclusions in diamond suggested that some diamonds have originated from the mantle transition zone or the lower mantle [e.g., *Moore et al.*, 1991]. The chemical compositions of such mineral inclusions might provide information on the chemical environments in the transition zone and the lower mantle [e.g., *Harte et al.*, 1994]. We also discuss the origin of chemical features of mineral inclusions in diamond on the basis of the partition coefficients between the high-pressure minerals and the silicate melt.

2. HIGH PRESSURE AND HIGH TEMPERATURE EXPERIMENT

We have conducted density measurements and partitioning experiments at high pressure using a multianvil apparatus. We used a large DIA type cubic guide block system driven by a 3000-ton uniaxial press together with a split-cylinder guide block system at Tohoku University. The DIA type cubic guide block system (Figure 1a and b) was designed and manufactured by TRY Engineering Co., Tokyo, and it enables us to drive the multianvil system with 26-mm edge cubes routinely up to 1300 ton. The truncated edge length of the anvils of the DIA type guide block is usually 50 mm, whereas the anvils with a smaller edge length of 20mm is used to operate a sintered diamond multianvil system used by *Kondo et al.* [1993]. The details of the performance for operating the sintered diamond multianvil will be given elsewhere. In the present experiments, we used the usual tungsten carbide cube anvils with 26 mm edge. The experiments for determining the partition coefficients were conducted by using the multianvil apparatus with a 3.5-mm or 2.0-mm anvil truncation, whereas the anvils with 12-mm, 7-mm, and 3.5-mm truncation were used for the density measurement at high pressure.

The pressure was calibrated on the basis of the phase boundary of garnet-perovskite transition in $CaGeO_3$, P(GPa)=6.9 - 0.0008T(°C) [*Susaki et al.*, 1985]; coesite-stishovite transition in SiO_2, P(GPa)=6.1 + 0.0026T(°C) [*Zhang et al.*, 1996]; α-β transition of Mg_2SiO_4, P(GPa)=9.3 + 0.0036T(°C) [*Morishima et al.*, 1994]; and ilmenite-perovskite transition in $MgSiO_3$. The phase boundary curve of ilmenite-perovskite transition adopted here was obtained by combining the phase transition pressure of 24 GPa at 760°C [*Kato et al.*, 1995] and the slope of the phase boundary deduced from the thermodynamic data [*Yusa et al.*, 1993]. The calibration curve used for the present study is given elsewhere [*Suzuki*

Figure 1b. 3000-ton uniaxial press with a large DIA type guide block (MAP-3000).

et al., 1997]. The uncertainty of the pressure at high temperature is estimated to be about ±0.5 GPa at around 10 GPa, and about ±1 GPa at around 20 GPa.

The typical cell assemblies used for the experiments are composed of a sintered zirconia pressure medium, a $LaCrO_3$ heater and Ta or TiC electrodes for the density measurements made above 8 GPa and for the experiments on partitioning between Ca-perovskite and liquid. We also used a TiC+diamond composite material for the heating material and TiC for electrodes in the Mg-perovskite/liquid partitioning run. Heater and electrode made of graphite were employed for the density measurement below 8 GPa. A W25%Re/W3%Re thermocouple was used for monitoring the temperature of the charge. The effect of pressure on the emf of the thermocouple was ignored in the present experiments.

We employed the sink-float method for determination of the density of the silicate melts at high pressure. A diamond single crystal with a diameter of around 200 μm or olivine crystals with compositions of Fo_{90}, Fo_{95}, and Fo_{100} with a diameter of 500 μm were used as the buoy to constrain the density of the melt at high pressure. We used a graphite capsule for most buoyancy test experiments except for those of the MA melt using olivine density marker; a molybdenum capsule was used in these runs. We placed a density marker of olivine or diamond in the center of the capsule. After heating at high pressure and temperature, we determine the flotation or sinking of the density marker by observing the position of the marker in the polished section of the recovered run products.

The parameters of the equation of state for olivine and diamond buoys are summarized in the appendix. The densities of the olivine and diamond markers were calculated on the basis of the third order Birch-Murnaghan equation of state,

TABLE 1. Chemical Compositions of the Silicate Melts

	PHN1611[a]	Pyrolite[b]	IT8720[c]	MA[d]
SiO_2	45.1	46.2	41.2	42.1
Al_2O_3	2.8	3.6	3.7	6.5
FeO	10.4	8.7	15.1	16.0
MgO	38.4	38.3	33.0	30.2
CaO	3.4	3.2	7.0	5.3
Total	100.0	100.0	100.0	100.0
Mg#	86.9	88.8	79.6	77.1

Mg#, MgO/(MgO+FeO)x100 in molar fraction
[a] Simplified composition of garnet Iherzolite, PHN1611 [*Nixon and Boyd,* 1973]. [b] Pyrolite [*Ringwood,* 1975]. [c] Melt formed by partial melting of PHN1611 at 20 Gpa [*Ito and Takahashi,* 1987]. [d] Model composition of the Martian mantle [*Morgan and Anders,* 1979].

$$P=(3/2)K_T\{(\rho/\rho_0)^{7/3}-(\rho/\rho_0)^{5/3}\}[1-(3/4)(4-K')\{(\rho/\rho_0)^{2/3}-1\}],$$

where ρ is the density at high pressure, ρ_0 is the zero pressure density, K_T is the isothermal bulk modulus, and K' is the pressure derivative of K_T. The physical parameters of the olivine solid solutions were calculated from linear combination of the forsterite and fayalite data.

The starting materials used as the peridotite melts for the density measurement were simplified composition of a primitive peridotite PHN1611 and pyrolite [*Nixon and Boyd,* 1973; *Ringwood,* 1975] and the melt formed by partial melting of the primitive peridotite at 20 GPa (IT8720; *Ito and Takahashi,* [1987]). The Martian mantle composition, MA [*Morgan and Anders,* 1978], was also used as the starting composition. The compositions of the melts used for the density measurements are summarized in Table 1. The starting materials were synthesized by heating the stoichiometric mixture of reagents, MgO, $CaCO_3$, Al_2O_3, SiO_2, and Fe_2O_3 at about 1000°C with the

APPENDIX. Parameters for Calculation of the Density at High Pressure and Temperature

	Molar volume at 25°C	Thermal expansion coefficients				Bulk modulus at 25°C			Pressure derivative of K
		$\alpha(K^{-1})=\alpha_0+\alpha_1T+\alpha_2T^{-1}+\alpha_3T^{-2}$				K	$dK/dT\times10^2$	$d^2K/dT^2\times10^5$	K'
	cc/mol	$\alpha_0\times10^6$	$\alpha_1\times10^8$	$\alpha_2\times10^3$	α_3	GPa	GPa/K	GPa/K^2	
Diamond	3.417[a]	2.43[b]	1[b]	0[b]	0[b]	442[c]	-1.54[c]	-2.01[c]	4.03[d]
Forsterite	43.79[e]	2.01[f]	1.39[f]	1.627[f]	-0.338[f]	129[g]	-2.3[g]		5.39[h]
Fayalite	46.39[e]	5.673[b]	16.315[b]	-2.5186[b]	-1.61331[b]	135.1[g]	-2.7[g]		5.2[i]

[a] *Gustafson* [1986]; [b] *Saxena et al.* [1993]; [c] calculated from the elastic constant and its pressure derivative [*Zouboulis and Grimsditch, 1992; McSkimin and Andreatch, 1972]*; [d] *McSkimin and Andreatch [1972]*; [e] *Robie et al. [1978]*; [f] *Saxena and Shen [1992]*; [g] *Anderson et al.* [1992]; [h] *Kumazawa and Anderson* [1969]; [i] *Graham et al.* [1988]

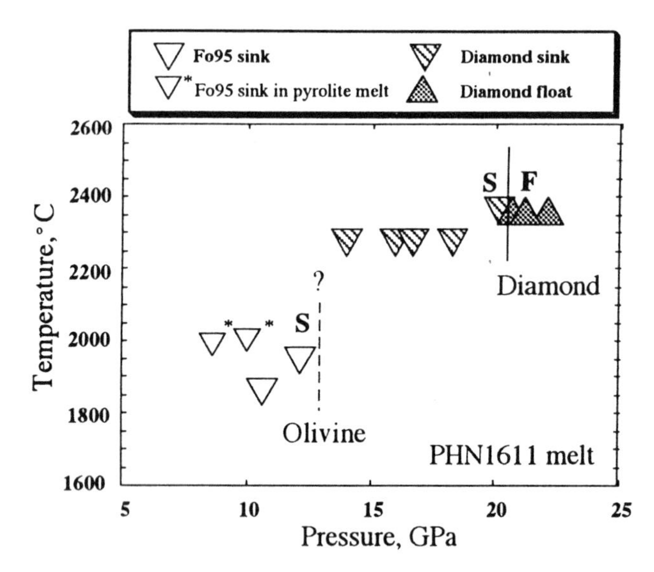

Figure 2a. Results of experiments on the sink-float test of olivine and diamond in the PHN1611 peridotite melt.

oxygen fugacity around the quartz-fayalite-magnetite (QFM) buffer for more than 20 hours. The starting materials thus synthesized were composed of olivine, orthopyroxene, clinopyroxene, and anorthite.

The starting material used for the Mg-perovskite/magma partitioning run was a mixture of olivine, pyroxenes, and small amount of anorthite. The bulk composition was as follows; 38.2 wt.% MgO, 53% SiO_2, 7.6% FeO, 0.5% CaO, 0.6% Al_2O_3 with 0.15 % each of La, Ce, Sm, Gd, Yb, Sc, Zr, and Hf as minor elements. It was synthesized by heating the mixture of the reagents at about 1200°C for 10 hours under the atmosphere with controlling the oxygen fugacity around the QFM buffer. The starting material for the Ca-perovskite/magma partitioning run was a mixture of 50 wt.% synthetic orthoenstatite, 50% natural kimberlite (aphanitic Group I kimberlite from the Wesselton mine, South Africa), and oxide dopants of La, Ce, Sm, Gd, Yb, Sc, Zr, and Hf in 0.1wt.% level. The mixture was fired at 1500°C for an hour in a controlled oxygen fugacity around the QFM buffer. The chemical composition of the starting material thus prepared is 35.8wt.% MgO, 45.2% SiO_2, 1.84% TiO_2, 1.80% Al_2O_3, 0.13% Cr_2O_3, 6.23% FeO, 0.11% MnO, 8.37% CaO, 0.15% Na_2O, 0.34% K_2O, and 0.1% La_2O_3, Ce_2O_3, Sm_2O_3, Gd_2O_3, Sc_2O_3, ZrO_2, and HfO_2.

All of the furnace components were fired at around 1000°C and stored in an oven for a few hours just before the experiment in order to avoid the effect of the absorbed water for the experiment. The charge was compressed to the desired pressure and then the temperature was increased quickly and held constant for several minutes. The charge was quenched by shutting off the electric power supply to the furnace before releasing pressure. The position of the density marker was determined by observing the polished section of the recovered sample with optical and scanning electron microscopes. Compositions of the quenched melts were analyzed using an electron probe microanalyser.

3. DENSITY MEASUREMENT OF SILICATE MELTS AT HIGH PRESSURE

For sink-float test experiments of olivine crystal at high pressure, the duration of the experiments was reduced to one minute in order to avoid dissolution of olivine in the peridotite melts. The run duration of the buoyancy test experiments using a diamond marker was five minutes. We observed the density marker of olivine or diamond in the top or bottom of the capsule, which implies that flotation or sinking of the marker respectively. All of the experiments were conducted above the liquidus temperature. We observed very fine dendrites of the quench crystals in the run products, which implies that the peridotites were completely molten during the experiment. Figures 2 a, b, and c show the experimental conditions and the results of sink-float experiments on olivine and diamond for three melt compositions, PHN1611 (and pyrolite), IT8920 and MA.

We observed sinking of olivine crystal Fo_{95} in the pyrolite melt at 8 GPa and 10 GPa at around 2000°C. Olivine crystal with Fo_{95}, which is in equilibrium with the peridotite melt PHN1611, sank at 10 GPa and 12 GPa at around 2000°C. The density crossover between the

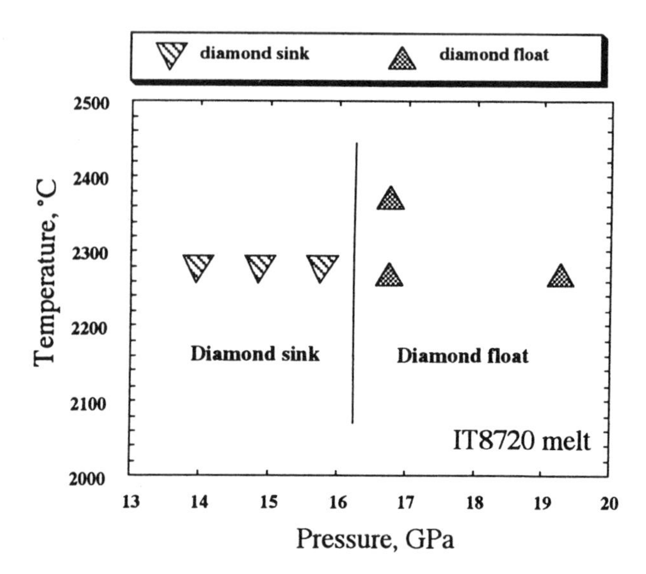

Figure 2b. Results of experiments on the sink-float test of diamond in the IT8720 peridotite melt.

peridotite melt and the equilibrium olivine is at a pressure above 12 GPa at about 2000°C. We observed sinking of diamond at 20.2 GPa and 2360°C, whereas we observed flotation of diamond at 20.7 GPa and 2360°C. Thus the density crossover between the peridotite melt and diamond is at a pressure around 20.5 GPa at this temperature, and the density of the melt is calculated to be 3.59±0.02 g/cm^3 in this condition on the basis of the third-order Birch-Murnaghan equation of state of diamond with the parameters summarized in the appendix. The density of the silicate melts may be expressed by the Birch-Murnaghan equation of state [e.g., *Stolper et al.*, 1981]. The isothermal density curve of the peridotite melt PHN1611 is expressed by this equation of state with the zero pressure density ρ_0= 2.62±0.12g/cm^3 and K=36±8GPa assuming K'=4 at 2360°C. In this calculation of the density curve of the melt, we estimated the density of the melt at one atmosphere and high temperature on the basis of the partial molar volume of oxides in the melts obtained by *Lange and Carmichael* [1990].

It is important to study the density of the melt formed by partial melting in order to discuss the fractionation of magmas formed in the mantle. Therefore, we made density measurement of the melt IT8720, which was a reported partial melt of PHN1611 at 20 GPa [*Ito and Takahashi,* 1987]. Sinking of diamond up to 15.8 GPa at 2270°C and flotation of diamond at 16.7 GPa and 2270°C in the IT8720 melt was confirmed in the present experiments. The density crossover between diamond and the melt is at 16.3±0.5GPa and 2270°C. The density of the melt was calculated to be 3.56±0.03g/cm^3 in this condition. Diamond floats in the conditions of the formation of the partial melt IT8720 at 20 GPa. The isothermal density curve of the melt can be written by the third order Birch-Murnaghan equation of state with the zero pressure density ρ_0= 2.71±0.12g/cm^3 [*Lange and Carmichael,* 1990] and K=36±9 GPa assuming K'=4 at 2270°C. The experimental conditions for the density measurement of the PHN1611 and IT8720 melts are briefly given in *Suzuki et al.* [1995].

Figure 2c summarizes the experiments for a model Martian mantle composition, MA [Morgan and Anders, 1979]. The Martian mantle melt with Mg$^\#$=(100 MgO/(MgO+FeO))= 77.1 is in equilibrium with San Carlos olivine with Fo_{90}. Our result implies that the San Carlos olivine sinks in the MA melt at 6.3 GPa and 1860°C, whereas it floats at 8.1 GPa and 1980°C, and the density crossover between the melt and the equilibrium olivine is at 7.4±1.1 GPa and 1890±90°C. Diamond sinks at 15.1 GPa, whereas it floats at 16.9 GPa in the MA melt at 2330°C. The expression of the Birch-Murnaghan equation of state of the melt can be given as follows: ρ_0=

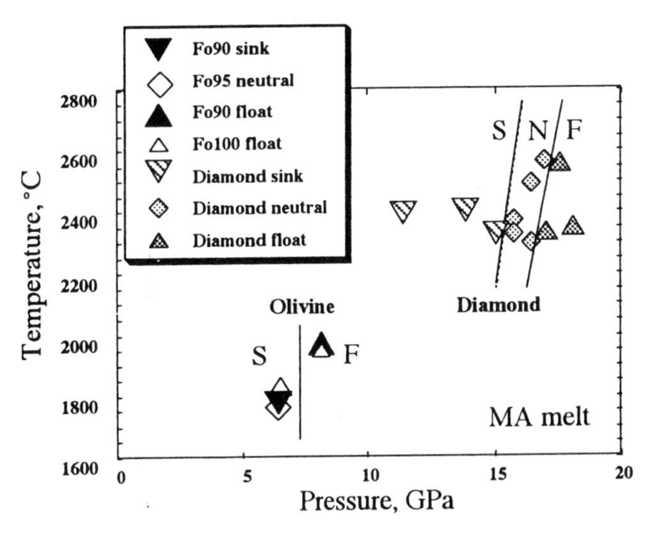

Figure 2c. Results of experiments on the sink-float test of olivine and diamond in the MA melt.

2.7±0.12 g/cm^3 at 2330°C and K=34±7 GPa assuming K'=4. The physical properties of the MA melt such as the density, bulk modulus, and the thermal expansion are also discussed in details in *Suzuki et al.* [1997].

Figure 3a shows the density curves of the melts summarized in this study. It is not possible to determine the density curve of the peridotite melts uniquely on the basis of the present diamond and olivine flotation experiments, but we can estimate the possible relationship between the bulk modulus K and its pressure derivative K' for the Birch-Murnaghan equation of state as shown in Figure 3b. In this calculation of the density curves of the peridotite melts, we estimated the densities of the melts at one atmosphere and high temperature on the basis of the partial molar volume of oxide in the melts obtained by *Lange and Carmichael* [1990]. The difference is not detected in the K and K' relationship for all peridotite melt compositions, PHN1611, IT8720, and MA.

Recently *Courtial et al.* [1997] determined the one bar density of the iron bearing silicate melts such as the MA melt, model PHN1611 melt, the komatiite melt, and the fayalite melt up to 2000°C. Lange and Carmichael's partial molar volume data was obtained from the density of basaltic and andesitic melts enriched in SiO_2 compared to the present ultramafic melt. The new data of the ultramafic melt by *Courtial et al.* (1997) give smaller densities than those calculated from the *Lange and Carmichael's* model. When combining the new zero pressure density data with the present results at high pressure, we obtain about 10-30% smaller bulk modulus values for the same K' value: ρ_0= 2.53g/cm^3 and K=22.9 GPa assuming K'=4 at 2330°C

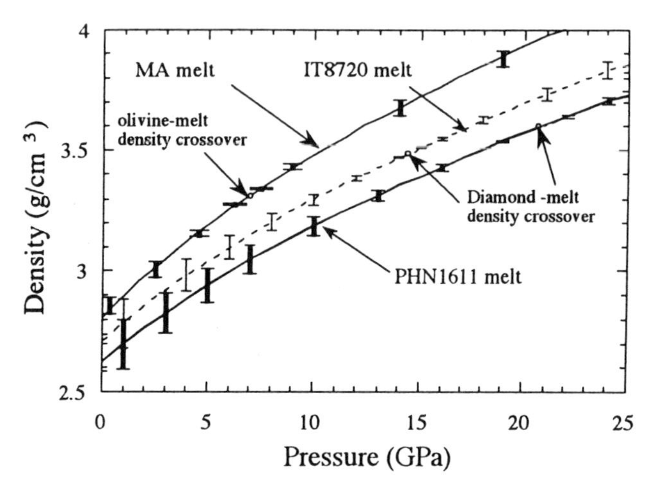

Figure 3a. Isothermal compression curves of the silicate melts PHN1611 at 2360°C, IT8720 at 2270°C, and MA at 1980°C. Error bars of the compression curves originate from uncertainty of the melt density by extrapolation to higher temperatures at one atmosphere [*Lange and Carmichael*, 1990].

for the MA melt, and ρ_0 = 2.6g/cm^3 and K= 32.1 GPa assuming K'=4 at 2360°C for the PHN1611 melt.

Agee and Walker [1993] estimated that the density crossover between peridotite melt and the equilibrium olivine is around 11 GPa and 2000°C. The present density constraints imply that we will not observe the density crossover between olivine and the primitive peridotite PHN1611 (and pyrolite) melt at least up to 12 GPa at 2360°C. The density crossover between the primitive melt PHN1611 and the equilibrium olivine at 2360°C and that between the IT8720 melt and olivine at 2270°C are expected to be located at around 17 GPa and 13 GPa respectively by assuming K'=4, although a slightly lower pressure of the density crossover may be expected if we adopt a larger K' value. The present experiment implies that olivine flotation in the partial melt could occur in a relatively narrow pressure range between 13 and 17 GPa above which olivine transforms into β-phase in the transition zone.

4. PARTITIONING OF ELEMENTS BETWEEN Mg-PEROVSKITE, Ca-PEROVSKITE, AND LIQUID AT HIGH PRESSURE

We have measured the partition coefficients between Mg-perovskite and melt and Ca-perovskite and melt. These experiments were conducted at 25 GPa and 2200°C. The run duration for the Mg-perovskite/liquid partitioning run was 5 minutes, whereas that for the Ca-perovskite/liquid partitioning run was 3 minutes. The point focused beam of EPMA was used for analysis of the crystals, whereas the composition of the melt was determined with a broad beam of 30 μm in diameter.

The partition coefficients of elements including REE between Mg-perovskite, majorite, and the silicate liquid are given in Table 2 together with the results determined by various authors. The compositions of Mg-perovskite and the coexisting melt at 25 GPa and 2200°C are also given in the same table. This table implies that LREE is depleted in Mg-perovskite and majorite, whereas HREE is enriched in these phases. The partition coefficients of REE between Ca-perovskite and the silicate liquid are also given in Table 2. The Ca-perovskite/liquid partition coefficients of La and Sm are greater than those of HREE such as Yb. Zr and Hf are relatively more favored in Mg- and Ca-perovskites compared to majorite as is shown in this table.

The temperature dependence of the partition coefficient D for a given element can be written generally as $\ln D = C_1/T + C_2$ [*McIntire,* 1963], where C_1 and C_2 are constant. However, *Murthy* [1992] made rather simple thermodynamic analysis of the temperature effect of the partition coefficient. The partition coefficient, Di, can be written as follows; $D_i = \exp(\Delta\mu_0/RT)$, where $\Delta\mu_{0i} = \mu_{0ia} - \mu_{0ib}$. Since $\Delta\mu_{0i}$ is a chemical potential difference between the hypothetical pure phases at the temperature T, $\Delta\mu_{0i}$ at T=T* can be expressed as follows by using the measured partition coefficients at temperature T* as $\Delta\mu_{0i} = RT^* \ln D^*$. Although $\Delta\mu_{0i}$ is generally a function of pressure and temperature [*Jones et al.*, 1992; *O'Neill*, 1992], we may be able to ignore its temperature dependency for simplicity. In such a case, the partition coefficient at a temperature T can be expressed by $D = \exp(T^* \ln D^*/T)$.

The above analysis is clearly oversimplified by ignoring

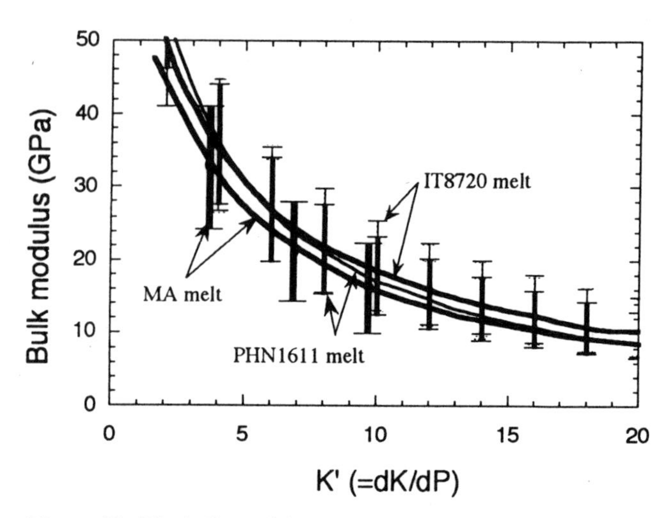

Figure 3b. The bulk modulus K and its pressure derivative K' for the silicate melts, PHN1611, IT8720, and MA.

TABLE 2. Partition Coefficient between Minerals and Melt

Element	25GPa* 2200°C perovskite	liquid	D(MgPv/L) 25GPa 2200°C	25GPa 2200°C (1)	24.5GPa 2100-2300°C (2)	25GPa 2500°C (3)	23GPa 1500-1600°C (4)	D(CaPv/L) 25GPa 2200°C	22-24GPa >2100°C (2)	23GPa 1500-1600°C (4)	D(Mj/L) 20GPa 2000°C (5)
Mg	21.9wt.%	26.8wt.%	0.82(0.06)**	6.1		1.0, 0.8	0.9	0.3	0.08	<0.1	
Al	3.64	2.22	1.6(0.1)	2.1	1.3, 0.5	1.2, 2.3	3.1	1.4	0.6	3.0, 9.7	3.06
Si	26.2	20.01	1.3(0.1)	1.2		1.1, 1.9	2.2	1	0.75	1.6, 3.1	
Ca	3.25	4.11	0.79(0.06)	0.02	0.1, 0.2	0.2, 0.7	<0.1	1	2.4	4.0, 3.7	0.47
Sc	1.52	1.31	1.2(0.1)	15	2.8, 5	1.6	8.9	2(2)	0.2	0.6, 1.0	1.64
Ti	-	-		0.74(0.1)	1.3, 3	0.6	1.4	1.7(0.1)	4	5.1, 26.2	0.63
Fe	4.13	7.26	0.57(0.04)	3.5				0.6			0.42
Zr	-	-		1.6	6, 9			1.3(1.0)	5		0.46
La	280ppm	1620ppm	0.17(0.01)	0.02(0.02)				3.1(1.5)	3.8, 5		0.09
Ce	470	1730	0.27(0.02)	0.1(0.1)				3.8(2.0)			
Sm	1020	2890	0.35(0.02)	0.2(0.2)	<0.17, 0.3	0.2	<0.2	2.6(1.5)	5.5	1.0, 60.6	0.2
Gd	410	1260	0.33(0.02)	1.0(0.8)				3.3(2.8)			0.26
Yb	920	830	1.1(0.08)	0.5(0.3)	0.6, 2			2.6(1.5)	1.2		1.05
Hf	2290	890	2.6(0.2)	2.3(1.7)	2.4, 14			2.0(1.5)	6		0.69

*Run duration is five minutes; **The values in parentheses are uncerataintes of the partition coefficients based on the error in analyses. (1) Compositions of the coexisting Mg-perovskite, Ca-perovskite, and liquid are given in *Kato et al.* [1996]; the silicate liquid has a calcic ultramafic composition; SiO_2 44.5wt.%; TiO_2, 2.32%; Al_2O_3, 1.22; Cr_2O_3, 0.05; MgO, 5.92; FeO, 0.88; MnO, 0.03; NiO, 0.03; CaO, 40.6; Na_2O, 0.02; K_2O,0.12; Total, 95.7. The calcic ultramafic liquid contains La, 2100ppm; Ce, 1000ppm; Sm, 1100ppm; Gd, 300ppm; Yb, 900ppm; Sc. 100ppm; Ze, 900ppm; Hf, 400ppm. The run duration of the experiment was 3 minutes. (2) *Kato et al.* [1988]. (3) *Drake et al.* [1992]. (4) *Gasparik and Drake* [1995]. (5) *Ohtani et al.* [1989].

the temperature dependency for $\Delta\mu_{0i}$ and the effect of change of the melt and solid compositions with temperature. Nevertheless, the present analysis gives qualitatively a reasonable temperature dependency of the partition coefficients which approaches to unity at very high temperature and deviates from unity at low temperature. A large value of the partition coefficients of La and Sm between Ca-perovskite and liquid and a small value between Mg-perovskite and the liquid determined by *Gasparik and Drake* [1995] at around 1500-1600°C are consistent with the temperature effect estimated by the present analysis, although their partition coefficients may be affected not only by the temperature but also by water added in their experiments, i.e. by the compositional dependency of the silicate melt.

On the basis of the above analysis of temperature dependency of the partition coefficients, we estimated the partition coefficients of La and Sm between Ca-perovskite and the melt, for example, to be greater than 10, and that of Yb to be about 5 at around 1000°C. The estimated partition coefficients between majorite, Mg-perovskite, and the melt at the lower temperatures would be much less than those determined experimentally; less than 0.1 for La and Sm, and less than 0.5 for Yb at 1000°C.

5. DISCUSSIONS

5.1. *Fractionation of the deep mantle and accumulation of diamond in the transition zone*

The density relations among mantle minerals, diamond, and the mantle partial melts IT8720 are summarized in Figure 4. The parameters used for calculation of the density curves of minerals are given in the appendix. According to the present experiments, the density crossover between olivine and the partial melt IT8720 is estimated to be at around 13GPa at 2270°C. Since the density of pyroxene is close to that of olivine, the density crossover between pyroxene and the melt will also locate at the same pressure range. Pyroxene gradually dissolves into the garnet structure to form majorite in a wide pressure interval up to about 15-17 GPa [*Zhang and Herzberg,* 1994]. In the alumina-poor system, pyroxene transforms to majorite at around 17 GPa at temperatures

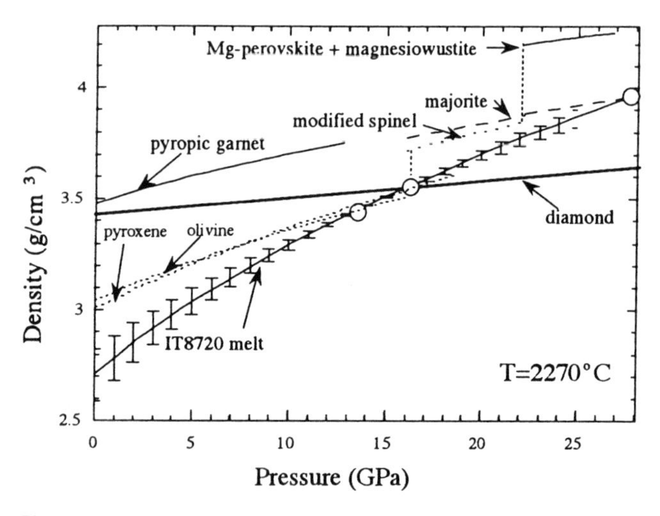

Figure 4. Density relations among mantle minerals, diamond, and the mantle melt IT8720 at high pressure.

above 1600°C, whereas it decomposes into β-phase and stishovite at the same pressure at temperatures below 1600°C [*Presnall and Gasparik,* 1990; *Yusa et al.,* 1993]. Therefore, the pressure limit of the stability field of pyroxene is greater than that of olivine. Pyropic garnet and majorite are denser than the melt in the pressure range at least up to 20 GPa. Extrapolation of the density curve of the partial melt to higher pressure suggests existence of the density crossover between majorite and the melt in the uppermost part of the lower mantle as suggested in Figure 4.

The present density relation between the silicate melt and mantle minerals indicates that olivine-liquid and pyroxene-liquid density crossover are in the similar pressure range. The depth interval of the olivine flotation is limited in a narrow depth range just above the transition zone, whereas pyroxene flotation could occur even in the upper part of the transition zone. This density relation suggests that accumulation of olivine and pyroxene, i.e., harzburgite assemblage, could occur in the region just above the transition zone. Further, the absence of a density crossover between garnet and partial melt up to 25 GPa implies that separation and descent of partial melt and garnet from the harzburgite assemblage of olivine and pyroxene occur due to partial melting. This could provide a potential mechanism for enrichment of the basaltic component in the transition zone. We propose that sinking and floating of crystals could produce chemical fractionation in the deep mantle. On the other hand, some authors argue that fractionation by this process does not work effectively in a large scale magma ocean in the early stage of the terrestrial evolution due to convective mixing of the system (e.g. *Tonks and Melosh,* [1990]; *Solomatov and Stevenson,* [1993]).

Density of diamond is also shown in Figure 4. According to this figure, diamond is denser than the partial melt IT8720, olivine, and pyroxene in the upper mantle conditions, and the density crossover between diamond and the melt is at around 16 GPa. Diamond is less dense than the melt and the minerals composing the transition zone and the lower mantle, such as β- and γ-phases, majorite, perovskite, and magnesiowüstite. Thus diamond is denser than the mantle peridotite and the ultramafic melts in the upper mantle, whereas it is less dense than the mantle melt and high pressure minerals in the transition zone and the lower mantle. These observations imply that the diamond crystals formed in the lower mantle and the bottom of the transition zone tend to float upwards, whereas diamond crystallized in the upper mantle tends to sink into the transition zone, resulting in accumulation of diamond in the transition zone. Thus, the transition zone could be a diamond enrichment zone, a potential reservoir of diamond. The schematic figure for the mechanism for accumulation of diamond in the transition zone is given in Figure 5. Recent discovery of the high-pressure minerals such as majorite as inclusions in diamond [e.g., *Moore et al.,* 1991] is consistent with the present model of accumulation of diamond in the transition zone.

5.2. *Chemistry of the silicate inclusions in diamond and nature of the lower mantle*

Existence of the high pressure minerals such as majorite as inclusions in diamond [e.g., *Moore et al.,* 1991; *Harte et*

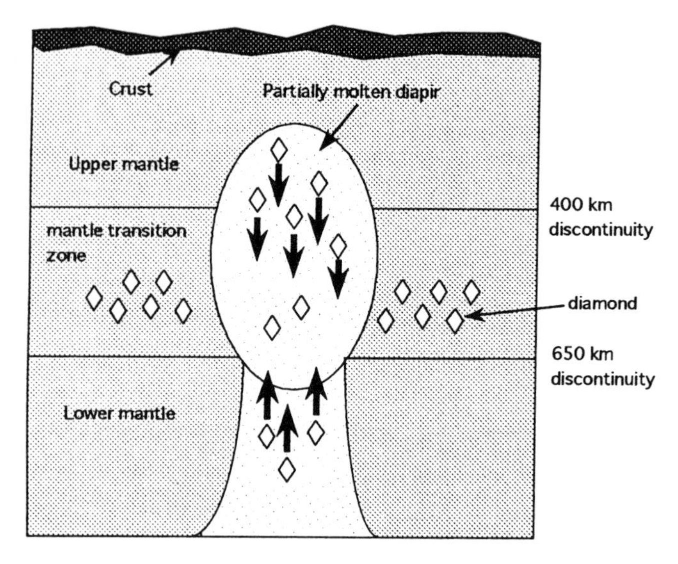

Figure 5. Accumulation of diamond in the transition zone of the Earth's mantle.

TABLE 3. Relative Concentration of Elements in Mineral Inclusions in Diamond [*Harte et al.*, 1994] and Partition Coefficients Between the Minerals

	Inclusion in diamond [*Harte et al.*, 1994]				D(Capv/Mgpv)*		D(Capv/Mj)**		D(Mgpv/Mj)***	
	Capv/Mgpv	Capv/Gt	Mgpv/Gt	Mgpv/Mw	2200°C	1000°C	2200°C	1000°C	2200°C	1000°C
La	6100	1500	0.25	0.45	18-160	280-18000	26-42	550-1400	0.17-1.4	0.03-2.0
Ce	11000	3900	0.34	1.1	14-38	170-1200				
Pr			0.65							
Nd	1800	390	0.22	0.42						
Sm	120	180	1.5		7.4-13	50-150	11-23	100-440	0.8-1.4	0.7-2.0
Gd		190			3.3-10	10-88	11	100	1.1-3.3	1.2-10
Tb	650	180	0.27	2.4						
Ho	2000	140	0.07							
Er	390	93	0.24							
Yb					1.1-4.3	1.2-17	1.2-2.5	1.3-5.9	0.43-1.0	0.2-1.0
Lu	37									
Y	4200	3900	0.95	0.82						
Zr	49	66	1.3	17	0.22-0.8	0.05-0.59	2.6-10	6-86	2.8-12	7.6-125
Nb			32	34						

* D(Capv/L) and D(Mgpv/L) determined at 25 GPa and 2200°C in Table 2 are used for calculation. ** D(Capv/L) by *Kato et al.* [1988, 1996] and D(Mj/L) by *Ohtani et al.* [1989] are used for calculation; the measured value of D(Mj/L) at 2000°C was extrapolated to 2200°C and used for calculation. *** D(Mgpv/L) determined at 25 GPa and 2200°C and D(Mj/L) by *Ohtani et al.* [1989] are used for calculation.

al.,1994] has been reported recently. Majorite crystals in diamond imply that the diamond originated from the depths greater than the transition zone in the mantle. They also observed magnesiowüstite, $MgSiO_3$, and $CaSiO_3$ phases. The latter two phases are believed to be crystallized as the perovskite phases in the lower mantle [e.g., *Kesson and FitzGerald.,* 1991]. *Harte et al.* [1994] determined the major and minor element compositions of the mineral inclusions such as $MgSiO_3$, $CaSiO_3$, garnet phases, and magnesiowüstite. If these inclusions are lower mantle fragments, they might provide information on the lower mantle composition. On the basis of the compositions of $MgSiO_3$, $CaSiO_3$ phases, garnet, and magnesiowüstite included in diamond, and assuming that the major element composition of the lower mantle is close to pyrolite, *Harte et al.* [1994] estimated REE and HFSE abundances in the lower mantle. According to their estimation, the lower mantle seems to be relatively depleted in heavy REE. The concentration of light REE such as La and Ce is about 17-20 times greater than the chondritic abundances, and heavy REE such as Er and Lu are about 3-6 times greater than chondritic abundances; the C1 chondrite and Si normalized abundances of the REE in the primitive (upper) mantle are about 1.45 [e.g., *Kargel and Lewis,* 1993], whereas those of light REE and heavy REE in the lower mantle [*Harte et al.*,1994] are about 8-10 and 1.5-3 respectively. If we assume a homogeneous lower mantle, the estimated lower mantle composition gives significantly high concentration of REE in the bulk silicate Earth compared to the chondritic abundance. Therefore, there is a possibility that the mineral inclusions in diamond do not represent the lower mantle mineral assemblage.

The test of chemical equilibrium among the phases in diamond may be made by comparing the compositional ratios obtained by *Harte et al.* [1994] and the partition coefficients of the elements between the phases determined experimentally. The observed compositional ratio and the partition coefficients between the minerals calculated from the experimental data are summarized in Table 3. The partition coefficients between the mantle minerals are obtained from the ratio of the crystal-melt partition coefficients determined experimentally at the temperature above 2000°C: e.g. D(Mg-pv/Ca-pv)=X_{Mgpv}/X_{Capv}= D(Mg-pv/L)/D(Ca-pv/L). According to this table, there is a large discrepancy between the partitioning behavior observed in inclusions in diamond and that determined experimentally at around 2000°C. Ca-perovskite contains a large amount of REE, which can never be produced by a simple chemical equilibrium among Ca-perovskite, Mg-perovskite, majorite, and magma when we adopt the experimentally observed partition coefficients. Thus, Ca-perovskite, Mg-perovskite, and majorite observed as inclusions in diamond may not be in equilibrium with one another.

5.3. *Origin of high concentration of REE in Ca-perovskite in the diamond inclusion*

It is difficult to account for the high concentration of REE in Ca-perovskite observed as inclusions in diamond, when we use the reported partition coefficients of REE. Since the partition coefficient has a temperature dependency, the high REE concentration in Ca-perovskite may be accounted for by the equilibrium partitioning among the phases at lower temperature.

In the previous section 4, the temperature dependence of the partition coefficients was estimated. The partition coefficients between the high pressure minerals at 1000°C are calculated and shown in Table 3. According to Table 3, the relatively high concentration of light REE and Sm may be accounted for by the equilibrium partitioning at low temperature around 1000-1200°C. However, a high concentration of HREE such as Yb may not be explained by the temperature effect as shown in Table 3, since the temperature dependency of the partition coefficient with a value close to unity has a very weak temperature dependency.

The above suggestion of equilibrium among the mineral inclusions in diamond at low temperature may provide important information on growth conditions of diamond. Spiral growth textures recorded on surfaces of some diamonds indicate growth from melt or fluid. The possibility of equilibrium at low temperature may suggest that some diamonds were grown in hydrothermal fluid or in hydrous magma. The high concentration of REE in Ca-perovskite inclusions in diamond may be also caused by the mineral/fluid partitioning at relatively low temperatures as shown by *Gasparik and Drake* [1995].

It is well known that cold slabs have been descending into the transition zone or even into the lower mantle [e.g., *Fukao,* 1992]. Such cold slabs may bring some amount of water into the lower mantle with hydrous minerals stabilized at high pressure [e.g., *Ohtani et al.,* 1995]. The hydrous low temperature conditions of the diamond formation may imply that some diamonds might have been grown in these slab materials descending into the transition zone and the lower mantle. The present model of diamond formation might support the model of the slab related origin of diamond by *Kesson and Ringwood* [1989], although the depth of diamond formation by the slab-mantle reaction is much shallower, i.e., the pressure less than 10 GPa in their model. If the present model is correct, the mineral inclusions in diamond do not represent the normal lower mantle but represent the cold slab materials.

The other way for explaining the high concentration of REE in Ca-perovskite may be a combination of melting and fractional crystallization processes: partial melting and succeeding separation of the melt, and fractional crystallization in the separated melt. The most simple way to evaluate this process is to apply the equation of batch melting for partial melting, $C_L/C_0=1/\{D(1-F)+F\}$ assuming modal melting [*Shaw,* 1970], and the equation of the Rayleigh fractionation for fractional crystallization, $C_L/C_{L0}=f^{(D-1)}$ [*Neuman et al.,*1954]. Here, C_L, C_0, and C_{L0} are the weight concentrations in the partial melt, parent material, and the parent melt; F is weight fraction of the melt relative to original parent; f is weight fraction of the melt remaining during fractional crystallization.

Observed crystal-melt partition coefficients for light REE such as La are small for Mg-perovskite, majorite, and magnesiowüstite and may be considered to be in the range about 0.1. During partial melting and most stages of the succeeding fractional crystallization, we may be able to assume that Ca-perovskite is absent in the solid coexisting with the melt, since the phase relations of peridotite indicate that Ca-perovskite appears only along the solidus curve at lower temperature [e.g., *Zhang and Herzberg,* 1994]. About 10% partial melting can concentrate light REE with D=0.1 up to about 5 times that of the parent (which may be assumed to be the C1-chondritic abundance). Succeeding 90% fractional crystallization of the separated melt in the absence of Ca-perovskite may concentrate the elements in the evolved melt to 8 times that of the initial magma. Crystallization of Ca-perovskite from the evolved melt in the final stage can produce high concentration of LREE in Ca-perovskite; if we adopt the partition coefficients between Ca-perovskite and melt of about 5 (see Table 2), we can finally obtain a very high concentration of LREE in Ca-perovskite, around 200 times of the chondritic value, which can explain the observed high concentration of these elements in inclusions of diamond, about 240-270 times of chondrite. Even if we adopt a different degree of partial melting, we can produce such a high concentration of the light REE in Ca-perovskite; 100% melting and 98-99% fractional crystallization of crystals in absence of Ca-perovskite, and Ca-perovskite crystallization in the final stage can produce a similar high concentration of light REE in Ca-perovskite. The present analysis implies that the mineral inclusions in diamond do not represent normal lower mantle, but they may represent some complicated melting and fractionation processes.

In the above analysis of Ca-perovskite crystallization, we assumed that Ca-perovskite crystallizes only in the latest stage along the solidus, and a very small amount of the partial melt can coexist with this phase. However, we have no reliable data of the initial melt fraction coexisting with Ca-perovskite, Mg-perovskite, magnesiowüstite, and majorite along the solidus of the mantle peridotite. If Ca-perovskite can coexist with the initial melt of relatively a large melt fraction, it is not possible to produce a high concentration of REE in Ca-perovskite as was observed in $CaSiO_3$ phase in the diamond.

The high concentration of heavy REE such as Yb may be difficult to account for when we adopt a relatively large partition coefficient for heavy REE for the mean value of Mg-perovskite, majorite, and magnesiowüstite, around 0.75 (see Table 2). The similar process can produce heavy REE of 6-12 times that of the C1-chondrite abundance in Ca-perovskite, which is significantly lower than the observed concentration in heavy REE, about 30-65 times that of chondrite, although there are no data on the concentration of Yb in Ca-perovskite in diamond. The relatively high concentration of heavy REE in Ca-perovskite might represent the nature of the parent material, i.e., the source material in the upper part of the lower mantle may be enriched in heavy REE, which is a signature of majorite accumulation. Possible existence of the density crossover between majorite and the melt in the uppermost part of the lower mantle might provide a mechanism for concentration of garnet phase in the early global melting stage or during geological history.

6. CONCLUSIONS

We made sink-float experiments of olivine and diamond and partitioning experiments of some lithophile elements between mantle minerals and melt by using multianvil apparatus.

(1) We measured the density of the three ultramafic melts, PHN1611, IT8720, and MA at high pressure by the sink-float method using olivine and diamond as the density marker. We observed flotation of the equilibrium olivine in the MA melt at 8.1 GPa and 1890°C. We also observed flotation of diamond in the PHN1611 melt at 20.2 GPa and 2360°C, in the IT8720 melt at 16.7 GPa and 2270°C, and in the MA melt at 16.9 GPa and 2330°C. Using these data we obtained a set of the parameters K and K' of the Birch-Murnaghan equation of state for the compression curve of silicate melt.

(2) The density relation between the mantle minerals and the partial melt IT8720 provides a possible fractionation mode in the deep mantle; accumulation of olivine and pyroxene could occur in the region just above the transition zone, and separation of the partial melt and garnet could provide a potential mechanism for enrichment of the basaltic component in the transition zone. The density relation between diamond, the mantle minerals, and the melt suggests a possibility for accumulation of diamond in the transition zone.

(3) The partition coefficients determined in our laboratory together with those determined previously were summarized. The partition coefficients between the mantle minerals and the temperature dependency of the partition coefficients were estimated based on these data. The present analysis implies that the compositions of the high-pressure phases in diamond, especially a high concentration of REE in Ca-perovskite cannot be explained by simple chemical equilibrium at high temperature above 2000°C. The compositional feature of the mineral inclusions in diamond can be explained by the following mechanisms; (a) the compositions of the minerals imply the chemical equilibrium at low temperature around 1000-1200°C perhaps under hydrous slab conditions, alternatively (b) The mineral inclusions in the diamond were formed by the multiple events such as partial melting and fractional crystallization.

Acknowledgments. The authors appreciate K. Onuma for useful discussions and comments. We also thank D. Yamazaki, T. Kubo, and K. Takanashi for their help during the course of the high pressure experiments using the DIA type guide block/uniaxial press system of Tohoku University, and Y. Ito for EPMA analysis. We thank D. Walker and an anonymous referee for thorough review to improve the manuscript. This work was supported by the grant-in-aid of Ministry of Education, Science, Sports, and Culture of the Japanese government.

REFERENCES

Agee, C. B., and D. Walker, Olivine flotation in mantle melt. *Earth Planet. Sci. Lett.*, *90*, 144-156, 1993.

Courtial, F., E. Ohtani, and D. B. Dingwell, High-temperature densities of some mantle melts, *Geochim. Cosmochim. Acta,* in press, 1997.

Drake, M.J., E.A. McFarlane, T. Gasparik, and D.C. Rubie, Mg-perovskite/silicate melt partition coefficients in the system $CaO-MgO-SiO_2$ at high temperature and pressure. *J. Geophys. Res.*, *98*, 5427-5431, 1992.

Fukao, Y., M. Obayashi, H. Inoue, and M. Nenbai, Subducting slabs stagnant in the mantle transition zone. *J. Geophys. Res.*, *97*, 4809-4822, 1992.

Gasparik, T., and M. J. Drake, Partitioning of elements among two silicate perovskites, superphase B, and volatile-bearing melt at 23 GPa and 1500-1600°C. *Earth Planet. Sci. Lett.*, 307-318, 1995.

Graham, E. K., J. A. Schwab, S. M. Sopkin, and H. Takei, The pressure and temperature dependence of the elastic properties of single-crystal fayalite Fe_2SiO_4. *Phys. Chem. Mineral.*, 16, 186-198, 1986.

Gustafson, P., An evaluation of the thermodynamic properties and the P, T phase diagram of carbon. *Carbon*, *24*, 169-176, 1986.

Harte, B., M. T. Hutchson, and J. W. Harris, 1994. Trace element characteristics of the lower mantle: an ion probe study of inclusions in diamonds from Sao Luiz, Brazil, *Mineral. Magazine*, *58A*, 386-387, 1994.

Ito, E., and E. Takahashi, Melting of peridotite at upper most lower-mantle. *Nature*, *328*, 514-517, 1987.

Jones, J. H., C. J. Capobianco, M. J. Drake, Siderophile elements and the Earth's formation, *Science, 257,* 1281-1282, 1992.

Kargel. J. S., and J. S. Lewis, The composition and early evolution of the Earth, *Icarus*, *105*, 1-25, 1993.

Kato, T., A.E. Ringwood, and T. Irifune, Experimental determination of element partitioning between silicate perovskites, garnets, and liquids: constraints on early differentiation of the mantle, *Earth Planet. Sci., Lett.*, *89*, 123-145, 1988.

Kato, T., E. Ohtani, H. Morishima, D. Yamazaki, A. Suzuki, M. Suto, T. Kubo, T. Kikegawa, and O. Shimomura, In situ X-ray observation of high pressure phase transitions of $MgSiO_3$ and thermal expansion of $MgSiO_3$ perovskite. *J. Geophys. Res.*, *100*, 20,475-20,481, 1995.

Kato, T., E. Ohtani, Y. Ito, and K. Onuma, Element partitioning between silicate perovskites and calcic ultrabasic melt, *Phys. Earth Planet. Inter.*, *96*, 201-207, 1996.

Kesson, S. E., and J. D. Fitz Gerald, Partitioning of MgO, FeO, FeO, NiO, MnO, and Cr_2O_3 between magnesian silicate perovskite and magnesiowüstite: implications for the origin of inclusions in diamond and the composition of the lower mantle, *Earth Planet. Sci. Lett.*, *111*, 229-240, 1991.

Kesson, S. E., and A. E. Ringwood, Slab-mantle interactions, 2. The formation of diamonds. *Chem. Geol.*, 78, 97-118, 1989.

Kondo, H., H. Sawamoto, A. Yoneda, M. Kato, A. Mitsumuro, and T. Yagi, Ultrahigh-pressure and high-temperature generation by use of the MA8 system with sintered-diamond anvils., *High Temp.-High Press.*, *25*, 105-112, 1993.

Kumazawa, M. and O. L. Anderson, Elastic moduli, pressure derivatives, and temperature derivatives of single-crystal olivine and single-crystal forsterite. *J. Geophys. Res.*, *74*, 5961-5972, 1969.

Lange, R. L. and I. S. E. Carmichael, Thermodynamic properties of silicate liquids with emphasis on density, thermal expansion, and compressibility. *Rev. Mineral.*, *24*, 25-64, 1990.

McIntire, W. L., Trace element partition coefficients - a review of theory and application to geology, *Geochim. Cosmochim. Acta* , *27*, 1209, 1963.

McSkimin, H. J., and P. Jr. Andreatch, Elastic moduli of diamond as function of pressure and temperature. *J. Appl. Phys.*, *43*, 2944-2948. 1972.

Moore, R. O., J. J. Gurney, W. L., Griffin, and N. Shimizu, Ultrahigh pressure garnet inclusions in Monastery diamonds: trace element abundance patterns and conditions of origin, *Eur. J. Mineral.*, *3*, 213,-230, 1991.

Morgan, J. W., and E. Anders, Chemical composition of Mars. *Geochim. Cosmochim. Acta*, *43*, 1601-1610, 1979.

Morishima, H., T.Kato, M.Suto, E. Ohtani, S. Urakawa, W. Utsumi, O. Shimomura, and T. Kikegawa, The phase boundaries between α- and β-Mg_2SiO_4 determined by in situ x-ray observation, *Science*, *265*, 1202-1203, 1994.

Murthy, V. R., Early differentiation of the Earth and the problem of mantle siderophile elements: a new approach. *Science*, *253*, 303-306, 1991.

Murthy, V. R., Geochemical evidence for an initially molten Earth. *Phys. Earth Planet. Inter.*, *71*, 46-51, 1992.

Neuman, H., J. Mead, and C. J. Vitaliano, Trace element variations during fractional crystallization as calculated from the distribution law, *Geochim. Cosmochim. Acta.*, *6*, 90, 1954.

Nixon, P. H., and F. R. Boyd, Petrogenesis of the granular and sheared ultramafic nodule suite in *Kimberlites*, edited by P.H. Nixon, Lesotho National Development, Maseru, 48-56, 1973.

Ohtani, E., Generation of komatiite magma and gravitational differentiation in the deep upper mantle, *Earth Planet. Sci. Lett.*, *67*, 261-272, 1984.

Ohtani, E., and H. Yurimoto, Element partitioning between metallic liquid, magnesiowüstite, and silicate liquid at 20 GPa and 2500°C: A secondary ion mass spectrometric study. *Geophys. Res. Lett.*, *23*, 1933-1996, 1996.

Ohtani, E., I. Kawabe, J. Morishima, and Y. Nagata, Partitioning of elements between majorite garnet and melt and implications for petrogenesis of komatiite, *Contrib. Mineral, Petrol.*, *103*, 263-269, 1989.

Ohtani, E., H. Yurimoto, T. Segawa, and T. Kato., Element partitioning between $MgSiO_3$ perovskite, magma, and molten iron: Constraints for earliest processes of the Earth-Moon system, the Earth's Central Part: Its Structure and Dynamics, edited by T. Yukutake, 287-300, TERRAPUB., Tokyo, 1995.

Ohtani, E., T. Shibata, T. Kubo, T. Kato, Stability of hydrous phases in the transition zone and the upper most part of the lower mantle., *Geophys. Res, Lett.*, *22*, 2553-2556, 1995.

O'Neill, H., Siderophile elements and the Earth's formation, *Science, 257*, 1282-1284, 1992.

Presnall, D. C., and T. Gasparik, Melting of enstatite ($MgSiO_3$) from 10 to 16.5 GPa: Implication for origin of the mantle. *J. Geophys. Res.*, *95*, 15,853-15,858, 1990.

Ringwood, A. E., 1975. Composition and petrology of the Earth's mantle. McGraw-Hill, New York, N.Y., 618 pp.

Ringwood, A. E., and T. Irifune, Constraints on element partition coefficients between $MgSiO_3$ perovskite and liquid determined by direct measurements. *Earth Planet. Sci. Lett.*, *90*, 65-68, 1988.

Robie, R. A., B. S. Hemingway, and J. R. Fisher, Thermodynamic properties of minerals and related substances at 298.15K and 1 bar (10^5 pascals) pressure and at higher temperatures. *U.S. Geol. Survey Bull.*, 1452, p. 456, 1978.

Saxena, S, K., and G. Shen, 1992. Assessed data on heat capacity, thermal expansion, and compressibility for some oxides and silicates. *J. Geophys. Res.*, *97*, 19,813-19,825.

Saxena, S. K., N. Chatterjee, Y. Fei, and G. Shen, *Thermodynamic data on oxides and silicates.* Springer-Verlag, Berlin, 428 pp., 1993.

Shaw, D. M., Trace element fractionation during anatexis, *Geochim. Cosmochim. Acta.*, *34*, 237, 1970.

Solomatov, V. S., and D. J. Stevenson, Nonfractional crystallization of a terrestrial magma ocean. *J. Geophys. Res.*, 98, 5391-5406, 1993.

Stolper, E., D. Walker, B. H. Hager, and J. F. Hays, Melt

segregation from partially molten source regions: The importance of melt density and source region size. *J. Geophys. Res.*, *86*, 6261-6271, 1981.

Susaki, J., M. Akaogi, S. Akimoto, and O. Shimomura, Garnet-perovskite transition in $CaGeO_3$: In-situ X-ray measurements using synchrotron radiation. *Geophys. Res. Lett.*, *12*, 729-7332, 1985.

Suzuki, A., E. Ohtani, and T. Kato, Flotation of diamond in mantle melt at high pressure. *Science*, *269*, 216-218, 1995.

Suzuki, A., E. Ohtani, and T. Kato, Density and thermal expansion of a peridotite melt at high pressure. *Phys. Earth Planet. Inter.*, in press, 1997.

Tonks, W. B., and H. J. Melosh, In H. E. Newsom and J. H. Jones (eds.) The physics of crystal settling and suspension in a turbulent magma ocean, *Origin of the Earth*, Oxford Univ. Press, 151-174, 1990.

Yusa, H., M. Akaogi, and E. Ito, Calorimetric study of $MgSiO_3$ garnet and pyroxene: Heat capacities, transition enthalpies, and equilibrium phase relations in $MgSiO_3$ at high pressures and temperatures. *J. Geophys. Res.*, *98*, 6453-6460, 1993.

Zhang, J., and C. Herzberg, Melting experiments on anhydrous peridotite KLB-1 from 5.0 to 22.5 GPa. *J. Geophys. Res.*, *99*, 17729-17742, 1994.

Zhang, J., B. Li, W. Utsumi, R. C. Liebermann, In situ X-ray observations of the coesite-stishovite transition: reversed phase boundary and kinetics. *Phys. Chem. Miner.*, *23*, 1-10, 1997.

Zouboulis, E. S., and M. Grimsditch, Elastic constants of diamond up to 1600K. *High Press. Res.*, *9*, 218-224, 1992.

T. Kato , E. Ohtani, and A. Suzuki, Institute of Mineralogy, Petrology, and Economic Geology, Tohoku University, Sendai, 980-77 JAPAN.

X Ray Diffraction Analysis of Molten KCl and KBr under Pressure: Pressure-Induced Structural Transition in Melt

Satoru Urakawa[1], Naoki Igawa[2], Osamu Shimomura[3], and Hideo Ohno[4]

The structures of molten KCl and KBr were studied by energy dispersive X ray diffraction experiments under high temperature and high pressure using synchrotron radiation. The radial distribution function $g(r)$ for molten KCl and KBr was obtained at several points just above the melting temperatures up to 4 GPa. The second neighbor ionic distance decreases with pressure, although the nearest neighbor distance is almost constant. The coordination number of the nearest neighbor ions increases with pressure. Both melts transform from an open, simple-cubic-like structure into a more highly coordinated structure, probably a body-center-cubic-like structure. The structure of molten KCl changes before the crystalline phase but the molten KBr transforms into a dense phase about the same time as the solid. In both KCl and KBr melts the structural transition occurs over a narrow pressure range.

INTRODUCTION

The structure of molten alkali halides at atmospheric pressure has been investigated by X ray and neutron diffraction experiments coupled with computer simulation such as molecular dynamics (MD), Monte Carlo, and hypernetted-chain (HNC) equation calculations and has been established to be an open structure with 3-5 nearest neighbor ions, in which the nearest ionic distance is smaller than that of solids [*Ohno et al.*, 1994]. In molten alkali halides, the short-range repulsive force is balanced by the long-range Coulombic force, so that the packing of ions shows the charge ordering. In other words, the alkali halide melts are characterized by a shell structure with the nearest neighbor shell of mainly oppositely charged ions, the second neighbor shell of like-charged ions and so on. It is of interest to know what the effect of pressure is on such a structure.

The application of pressure has been generally believed to result in the increase of the coordination number of molten alkali halides. In crystalline state, many alkali halides have a sixfold-coordinated NaCl-structure at low pressure and transform into an eightfold-coordinated CsCl-structure at high pressures. With increasing pressure, the slope of the melting curve of an alkali halide with NaCl-structure becomes moderate and seems to be approaching a maximum [*Kawai and Inokuchi*, 1970]. According to the Clausius-Clapeyron relation, this suggests that the density of alkali halide melt increases to be close to that of solid. *Kawai and Inokuchi* [1970] and *Tallon* [1979] suggested that this behavior could be explained by the continuous transformation of molten alkali halide from a sixfold-coordinated state into an eightfold-coordinated state. The HNC calculations showed the structure changes gradually from the simple-cubic-like structure with six opposed

[1]Department of Earth Sciences, Okayama University, Okayama, Japan
[2]Depertment of Materials Science and Engineering, Japan Atomic Energy Research Institute, Tokai, Japan
[3]Photon Factory, National Laboratory for High Energy Physics, Tsukuba, Japan
[4]Department of Synchrotron Radiation Facilities Project, Japan Atomic Energy Research Institute, Kamigori, Japan

Properties of Earth and Planetary Materials
at High Pressure and Temperature
Geophysical Monograph 101

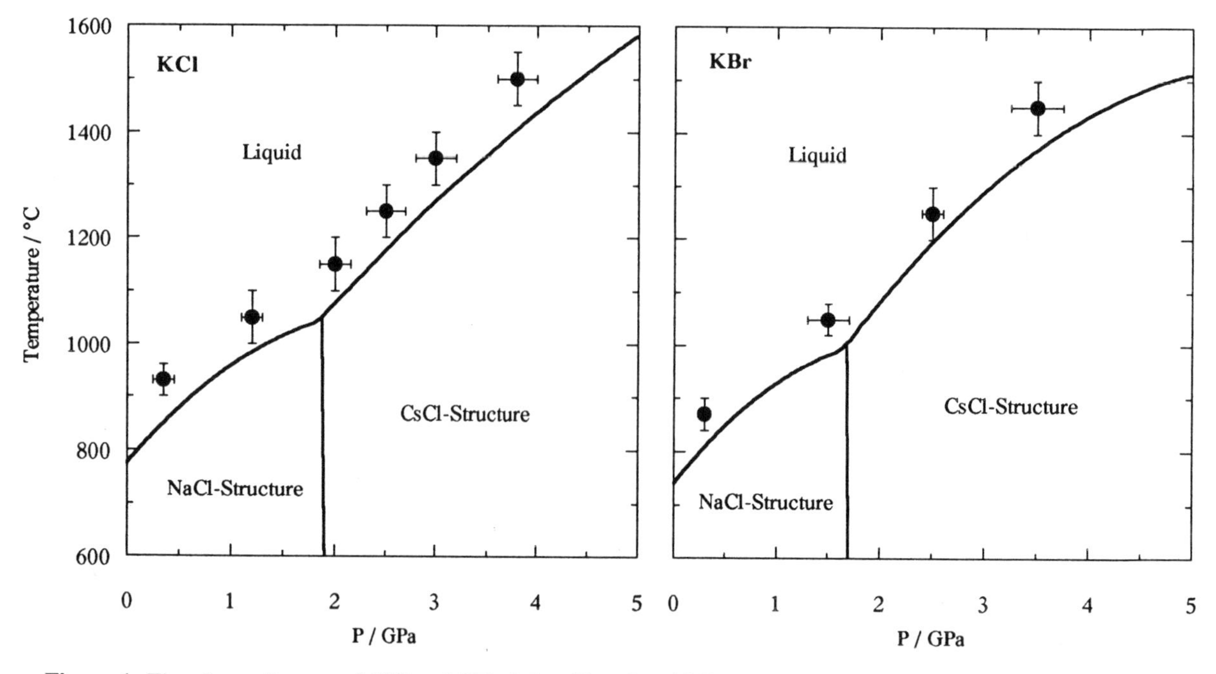

Figure 1. The phase diagram of KCl and KBr [after *Pistorius*, 1964 and 1965]. The experimental data points are shown as solid circles.

charged ions of the nearest neighbor shell to the rare gas-like closely packed structure with twelve nearest neighbor mixed charged ions up to several tens GPa [*Adams*, 1976; *Ross and Rogers*, 1985; *Boehler et al.*, 1996]. The same authors also concluded that the large curvature of the melting curve results from the density increase accompanied by this structure change. However, the detail of structural transition of molten alkali halides along the melting curves has never been experimentally clarified. The transition pressures of KCl and KBr from the NaCl-structure to the CsCl-one are about 2 GPa (see Figure 1), so that it is expected that such a structural transition in melt occurs at low pressure. Indeed, our preliminary study showed the molten KCl transforms into a more dense state with the increase of the nearest neighbor coordination number [*Urakawa et al.*, 1996]. From the Monte Carlo simulation, *Ross and Wolf* [1986] also showed that the coordination number of molten KCl gradually increases from 4 at atmospheric pressure to about 5 at 4 GPa.

Recently we have been able to investigate the structure of liquid under pressure by in situ X ray observation using a large volume multianvil apparatus and synchrotron radiation [*Tsuji et al.*, 1989; *Urakawa et al.*, 1996]. In this study we conducted X ray diffraction analysis on molten KCl and KBr at high pressures and obtained the radial distribution functions of these melts.

EXPERIMENTATION

The high-pressure and high-temperature experiments were carried out by using the cubic type high-pressure apparatus MAX90 [*Shimomura et al.*, 1992], which was installed at Photon Factory (PF) in Tsukuba, Japan. The tungsten carbide anvils had an 8-mm-edge length.

The sample cell assembly for high-pressure and high-temperature experiments is illustrated in Figure 2. The pressure transmitting medium was a cube with 12-mm-edge-length, which was made of the mixture of the amorphous boron and the epoxy resin with a weight ratio of 4 to 1. Samples were a reagent powder of KCl and KBr (99.99% pure) and were formed into a pellet with a diameter of 3 mm. The sample was capsuled in the boron nitride container, the graphite tube heater, and the boron-sodium silicate glass cylindrical thermal insulator. These were embedded in a cubic pressure transmitting medium. Temperature was measured at the outside of the heater by a Pt6%Rh-Pt30%Rh thermocouple. In this study, the pressure was determined based on the equation of state of h-BN that was used as a sample capsule. The X ray diffraction profiles were collected at 50-100°C above the melting point up to about 4 GPa. The P-T conditions of experimental data points are shown in Figure 1.

The in situ X ray observation was conducted at BL-14C

in PF, the vertical wiggler beamline, in which the brilliant white X rays up to 120 keV are available. The energy dispersive X ray diffraction method with a transmitting geometry was applied (Figure 2). The diffracted X rays were detected by the Ge-SSD fixed at several Bragg angles between 3 and 25°, in order to cover the wide Q-range (1-15 $Å^{-1}$). A typical exposure time was 1000 s at each angle. The X ray scattering power of melt is some orders of magnitude less than that of solid. Therefore, in order to eliminate the diffraction peak from the sample capsule, the heater and the pressure medium, we used the large sample and the sharp slit system (see Figure 2). An example of the measured X ray diffraction profile is shown in Figure 3.

REDUCTION TO THE STRUCTURE FACTOR

We obtained the structure factor $S(Q)$ of molten KCl and KBr from the measured energy spectrum of X rays by the following process.

Ignoring the multiple scattering and the background intensities, we can express the observed spectrum $I_{ob}(E)$ as

$$I_{ob}(E) = k\Big(I_0(E)\ f(Q)^2 S(Q) A(E,\theta) PL(E,\theta) + I_0(E') I_{comp}(Q) A(E,E',\theta) PL(E',\theta) K(dE'/dE)\Big) \quad (1)$$

$$Q = 4\pi E \sin\theta / 12.398$$

where E is the energy of X ray in keV, k is a scale factor, $I_0(E)$ is an energy profile of incident X rays, $f(Q)$ is an atomic scattering factor, $S(Q)$ is a structure factor, $A(E,q)$ is an absorption factor, $PL(E,q)$ is a polarization-Lorenz factor, $I_{comp}(Q)$ is a Compton scattering intensity, and $K(dE'/dE)$ is a Breit-Dirac factor, respectively. In equation (1), the first term and the second term correspond to the coherent and the incoherent scattering of the sample, respectively.

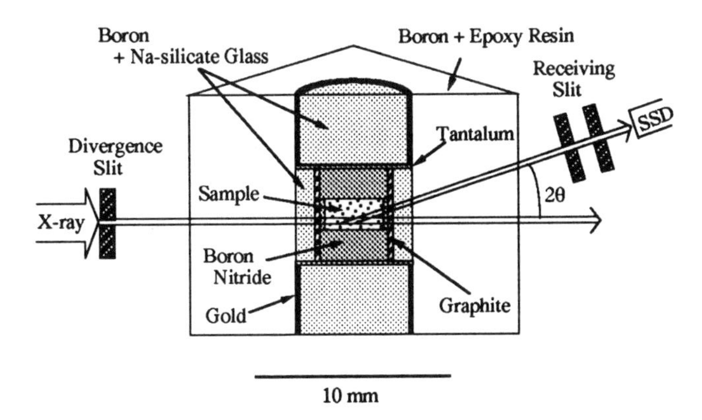

Figure 2. The schematics of the high pressure cell assembly, the slit system and X ray path.

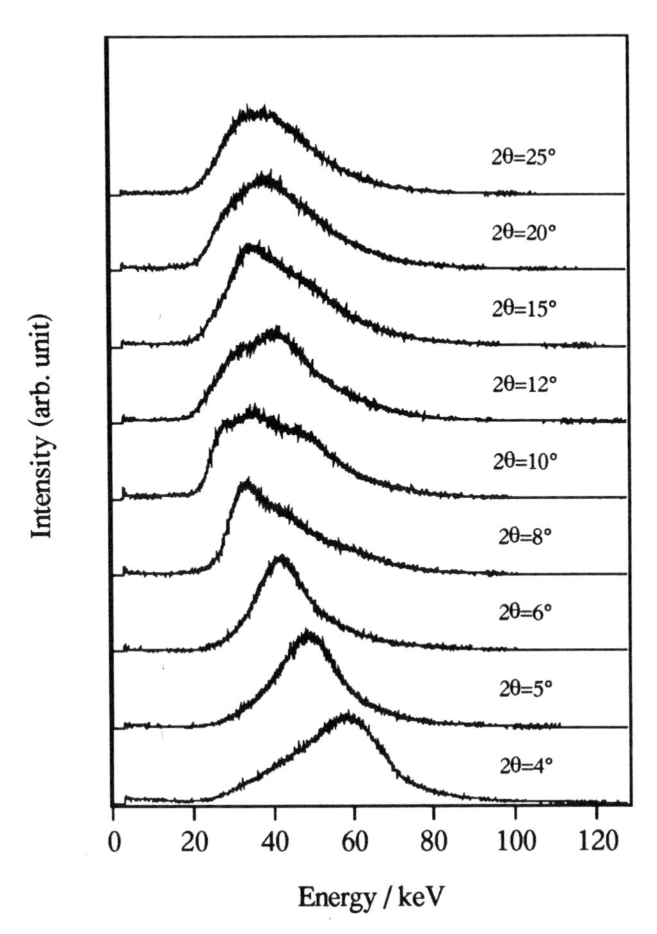

Figure 3. An example of the energy spectrum of the diffracted X ray from molten alkali halides. These profiles for molten KCl were measured at 3.0 GPa and 1350°C.

There are some problems in the reduction of $I_{ob}(E)$ to $S(Q)$. In the energy-dispersive method with transmitting geometry, the absorption term depends on not only photon energy but also diffraction angle, because the X ray path length varies in the case of a parallel piped sample. As the length and the density of the material in the X ray path change with compression, it is difficult to estimate the absorption term accurately. This correction is, especially, almost impossible in low photon energy. Although we used the energy spectrum above 35 keV, the uncertainty of several percent was included in the absorption correction. Further, the energy spectrum of incident X rays from the synchrotron radiation source remains as an unknown factor. This also makes the data analysis difficult. However, we can deduce a structure factor by the data analysis method proposed by *Tsuji et al.* [1989] without a priori information of the incident X rays profile.

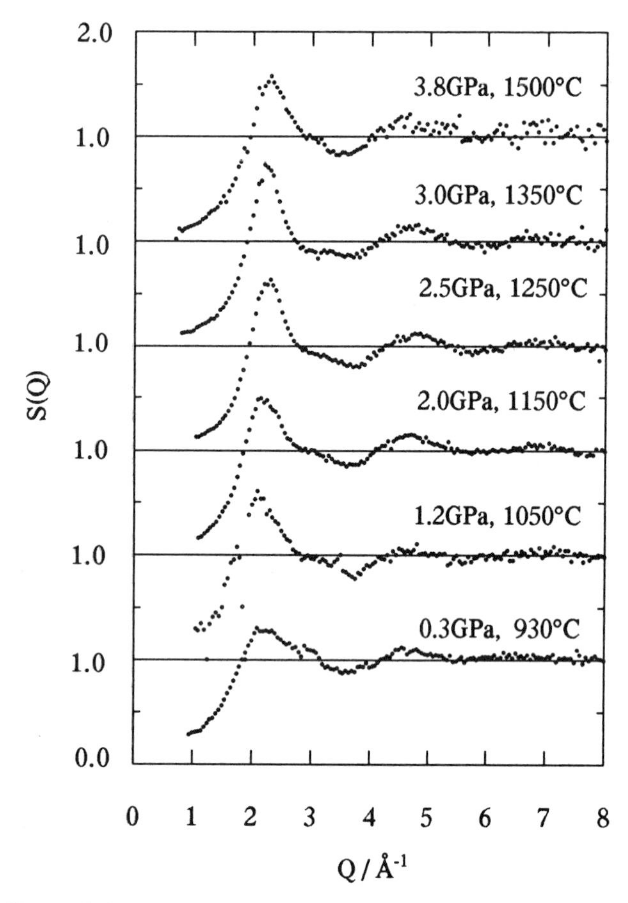

Figure 4. Structure factor $S(Q)$ for molten KCl as a function of pressure.

In order to apply Tsuji method, we used the following approximation. The energy shift of Compton scattering is dependent on the Bragg angle as well as the photon energy. The energy shift reaches one percent at $2\theta = 25°$ and 100 keV, but it is less than 0.1 percent at $2\theta < 20°$. Here we ignored the Compton shift and assumed $E = E'$, so that $K(dE'/dE)$ becomes unity [*Nishikawa and Iijima*, 1984]. In this approximation, equation (1) can be rewritten in the simple form as

$$I_{ob}(E) = k\, I_0(E)\, A(E,\theta) PL(E,\theta)\left(f(Q)^2 S(Q) + I_{comp}(Q)\right) \quad (2)$$

According to the Tsuji method, we defined an energy dependent function $B(E)$ as follows,

$$\begin{aligned} B(E) &= I_{ob}(E) \,/\, k\, A(E,\theta) PL(E,\theta)\left(f(Q)^2 + I_{comp}(Q)\right) \\ &= I_0(E)\left[\alpha(Q) S(Q) + \beta(Q)\right] \\ &= S'(Q)\, I_0(E) \end{aligned} \quad (3)$$

where $\alpha(Q)$ is $f(Q)^2 / \left(f(Q)^2 + I_{comp}(Q)\right)$, $\beta(Q)$ is $I_{comp}(Q)$ $\left(f(Q)^2 + I_{comp}(Q)\right)$ and $S'(Q)$ is $(a(Q)S(Q)+b(Q))$. Several $B(E)$'s were calculated for various 2θ. The scale factor k's were determined for all $B(E)$ curves to overlap. As $S'(Q)$ oscillates around unity like as $S(Q)$, $B(E)$ also oscillates around $I_0(E)$. At Q higher than 10 Å^{-1}, an amplitude of oscillation of $S'(Q)$ is so small that $B(E)$ becomes close to $I_0(E)$. Therefore, we estimated $I_0(E)$ from $B(E)$'s obtained at high Bragg angles. Then $S'(Q)$'s were calculated at the measured 2θ, which had the different Q-range correspond to 2θ. These $S'(Q)$ curves were connected to a single $S'(Q)$ curve by a trial-and-error method. Then structure factor $S(Q)$ was derived from $S'(Q)$.

The obtained $S(Q)$'s for molten KCl and KBr are shown in Figures 4 and 5, respectively. The oscillation of $S(Q)$ is rapidly damping with Q. Overall features of $S(Q)$'s for KCl and KBr obtained at 0.3 GPa are similar to those at atmospheric pressure reported by *Takagi et al.* [1979] and *Ohno et al.* [1983]. In these $S(Q)$'s, we can observe two peaks: the first peak around $Q = 2.2$–2.4 Å^{-1} and the second peak around $Q = 4$–5 Å^{-1}. The third peak should exist around $Q = 7$ Å^{-1}, but is not clear. A subpeak at high Q side of the

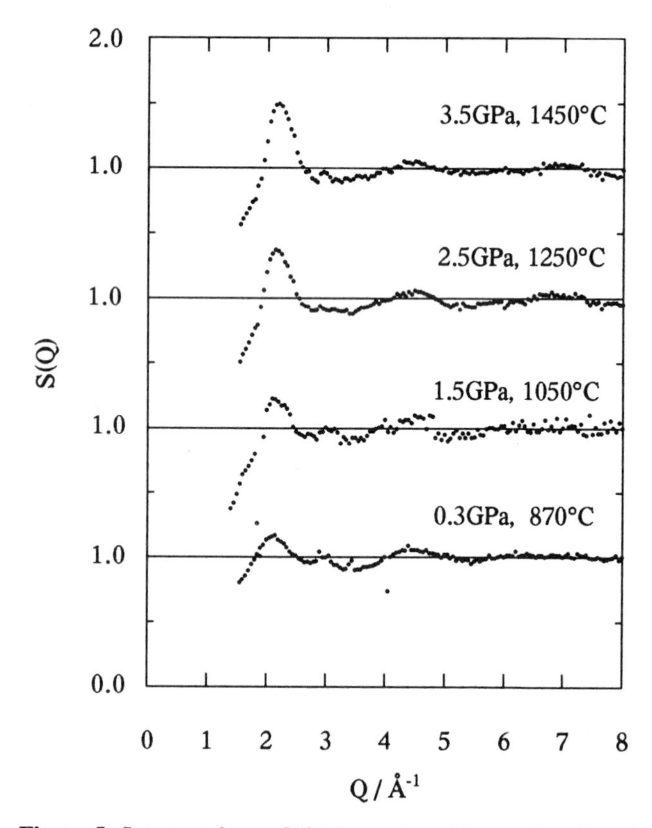

Figure 5. Structure factor $S(Q)$ for molten KBr as a function of pressure.

first peak is recognized for both melts at low pressure. These peaks' positions do not seem to shift much with pressure. However, the height of the first peak clearly increases with pressure. These pronounced features of $S(Q)$ suggest that local structure in molten KCl an KBr may change under pressure.

RADIAL DISTRIBUTION FUNCTION

The radial distribution functions for molten KCl and KBr were calculated by means of a Fourier transformation according to

$$g(r) = 1 + \sum_{m} (\bar{K}_m^2) / \left(2\pi^2 g_0 r\right) \int_0^{Q_{\max}} Q\left(S(Q) - 1\right) \sin\left(rQ\right) d \tag{4}$$

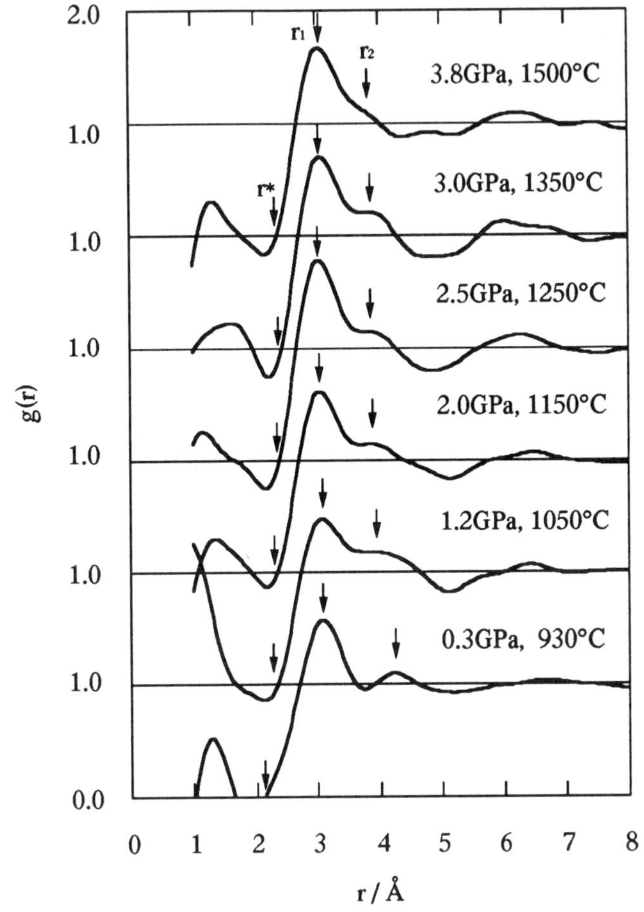

Figure 6. The radial distribution function $g(r)$ for molten KCl were calculated by the Fourier transform of equation (4) from $S(Q)$ shown in Figure 4. The arrows denote the closest approach distance (r^*), the first peak position (r_1) and the second peak position (r_2).

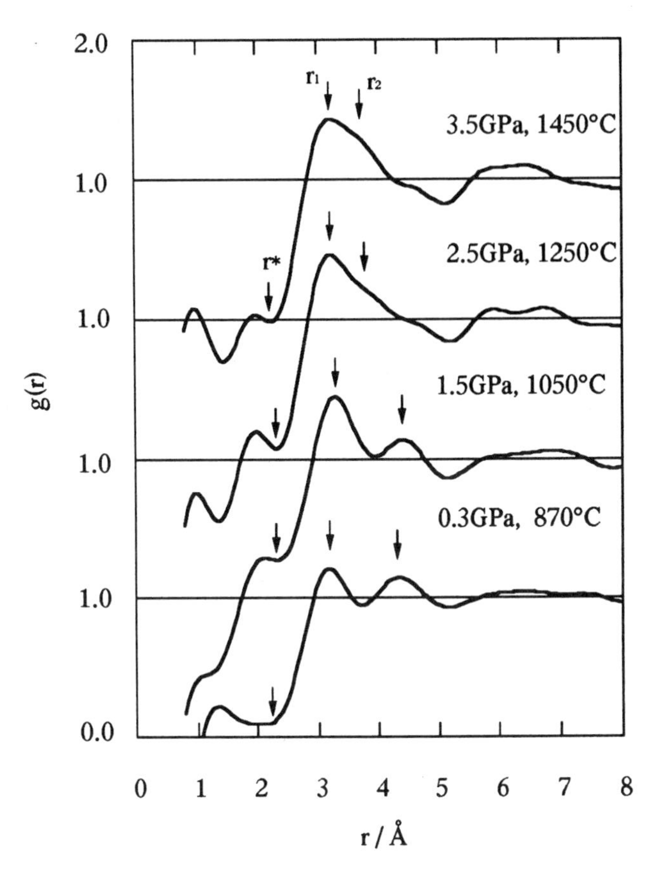

Figure 7. The radial distribution function $g(r)$ for molten KBr were calculated from $S(Q)$ shown in Figure 5.

where g_0 is the average density of atoms and K_m is the average effective electron number for atoms of type m. Here the densities of melts were estimated by the equation of states for KCl and KBr and Clausius-Clapeyron relation. The calculated radial distribution functions $g(r)$ for molten KCl and KBr are shown in Figures 6 and 7, respectively. The ripples that appear before the first peak originate from terminating the Fourier transformation at Q_{max}. These ghost peaks did not critically affect the estimation of the inter-ionic distance and the coordination number. We carefully checked the positions of both real and ghost peaks by changing Q_{max} from 7 to 10 $Å^{-1}$. The arrows in these figures denote the positions of the first and second peaks of $g(r)$. The first peak of $g(r)$ generally corresponds to the nearest neighbor cation-anion distance (r_1), and the second one represents the second neighbor cation-cation or anion-anion distances (r_2). The rapid damping in oscillation of $g(r)$ implies the long range ordering is insignificant in these melts under pressure.

Figures 6 and 7 clearly show the second neighbor distance decreases with pressure in both KCl and KBr melts, although the nearest neighbor distance is almost constant.

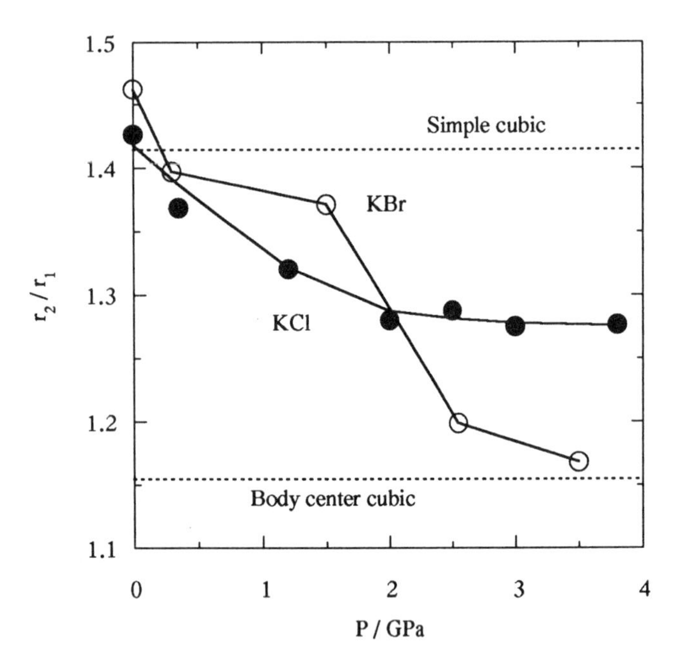

Figure 8. The ratio of the first peak and the second peak positions of $g(r)$ for molten KCl and KBr are shown as a function of pressure. Solid circles denote CN for molten KCl and open circles indicate that for molten KBr. The values at atmospheric pressure were reported by *Takagi et al.* [1979] and *Ohno et al.* [1983]. Dotted lines are values of r_2/r_1 for the simple cubic and the body centered cubic structures.

Figure 8 shows the ratio of the second peak and the first peak positions (r_2/r_1) as a function of pressure. At low pressure, the estimated second peak positions (r_2) are near $1.4r_1$, which is comparable to the value of simple cubic structure. In the case of KCl, the second peak shifts toward the first peak with increasing pressure and r_2/r_1 decreases with pressure and becomes nearly constant at pressure higher than 2 GPa. On the other hand, r_2/r_1 for molten KBr stays around 1.4 up to 1.5 GPa, but it abruptly decreases to less than 1.2 above 2.5 GPa

The coordination number of nearest neighbor ions (CN) is generally calculated by the integration of $4pr^2g_0g(r)$ curves around the first peak. However, it is difficult to distinguish the first peak from the second one at high pressures as shown in Figures 6 and 7. In this study we calculated CN by terminating the integration of $4pr^2g_0g(r)$ at $2r_1 - r^*$, where r^* is the closest approach distance. The estimation of CN includes the error originated from the uncertainty of melt density at high pressure. Figure 9 shows the CN for molten KCl and KBr as a function of pressure. In both cases, CN rapidly increases with pressure and becomes nearly constant above 2 GPa. This CN does not mean the coordination number of the oppositely charged ions, because the nearest neighbor shell becomes a mixed charged state with compression [*Ross and Wolf*, 1986; *Urakawa et al.*, 1996].

DISCUSSION

We can deduce the local structures of molten KCl and KBr under pressure from $g(r)$ and CN data. Data obtained at 0.3 GPa are similar to those at atmospheric pressure reported by *Takagi et al.* [1979] and *Ohno et al.* [1983], so that, at 0.3 GPa, both alkali halide melts have an open structure resembling a simple-cubic configuration.

While molten KBr keeps the simple-cubic-like structure up to 1.5 GPa, its coordination number increases to become close to crystalline phase. The rapid increase of CN implies the vacancy sites in the simple-cubic-like KBr melt diminish by compression. Thus the density of the KBr melt increases with pressure to be comparable to that of a solid at 1.5 GPa, which results in the large curvature of the melting curve. Between 1.5 and 2.5 GPa, the structure of the molten KBr changes to a body-center-cubic-like structure, but its CN change is small.

The rapid increase of CN with pressure also occurs before the structural change in molten KCl. The second peak of $g(r)$ at 1.2 GPa is broad and seems to be a transient state (see Figure 6). Thus the structural change in molten KCl goes on with a transition region up to 2 GPa, which is

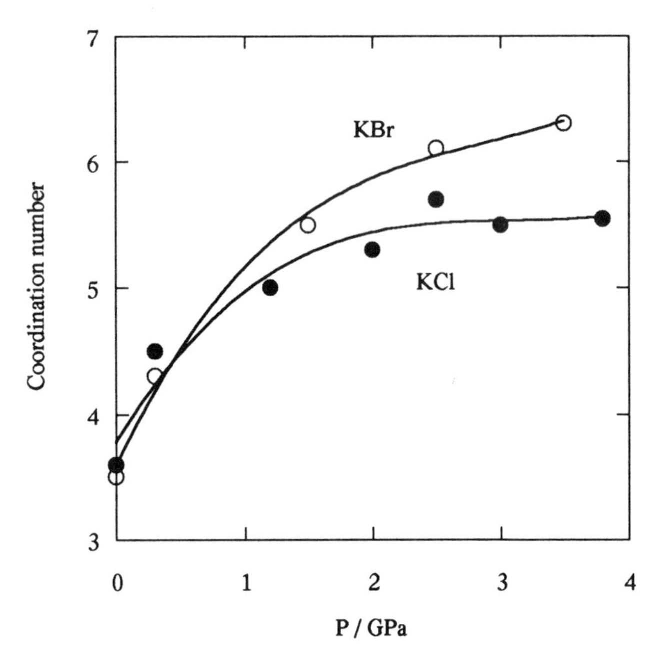

Figure 9. The coordination number (CN) of the nearest neighbor ions. Solid circles denote CN for molten KCl and open circles indicate that for molten KBr. CN's at atmospheric pressure were recalculated using $g(r)$ reported by *Takagi et al.* [1979] and *Ohno et al.* [1983].

accompanied by the increase of CN. It is difficult to speculate on the structure of high-pressure dense KCl melt from $g(r)$ and CN. It is generally seen, in many alkali halides, that the nearest ionic distance decreases but the second ionic distance slightly increases during melting [*Ohno et al.*, 1994]. Therefore, although r_2/r_1 is an intermediate value between simple-cubic and a body-center-cubic structures, we thought the structure of dense KCl melt was closely related to a body-centered-cubic one. The model simulation, however, is necessary to reveal how the ions pack in a high-pressure KCl melt.

As mentioned above, the structures of high-pressure KCl and KBr melts seem to be analogous with the high-pressure crystalline phases. As CN of high-pressure body-center-cubic-like melts is smaller than that of crystals, these melts are thought to have as much vacancy in their structures as the low-pressure molten alkali halides.

At pressure higher than 2 GPa, the density difference between crystal and melt at melting point may be smaller in the case of KBr rather than with KCl, because of the high CN and small r_2/r_1 ratio. This leads to the smaller pressure derivative of the melting slope in CsCl structured KBr, which is consistent with the reported curvature of the melting curve (see Figure 1).

It is of interest that the position of the first peak in $g(r)$ does not change with pressure in both melts. The MD simulation also showed the constant nearest neighbor ionic distance with pressure [*Urakawa et al.*, 1996]. The nearest neighbor ionic distance must contract with compression but it must expand with the increase of CN and the thermal expansion. In P-T conditions along the melting curve, these effects may be balanced to result the nearly constant r_1.

The structural transition of molten KCl starts instantaneously with compression and precedes the crystalline phase. On the other hand, the structure change in molten KBr occurs about the same time as the phase transition of solid KBr. This difference can be explained by the ionic radius and the polarizability of ions. At atmospheric pressure, KCl melt has a large cation-anion radius ratio (r_c/r_a) of 0.79, while the molten KBr has a low r_c/r_a of 0.68. The molten KCl with larger r_c/r_a and lesser polarizability prefers the highly coordinated configuration to the octahedral coordinated prefered by molten KBr, as a general rule from crystal chemistry. In crystalline state, the compressibility of KBr is known to be larger than KCl. This implies the Br^- ion is softer than the Cl^- ion. Thus r_c/r_a in KBr melt increases with pressure to become compatible with an eightfold-configuration above 2 GPa.

Adams [1976] and *Ross and Rogers* [1985] showed the continuous structural change in molten alkali halide with increasing pressure from the simple-cubic-like structure to the rare-gas-like structure from the HNC equation calculations. However, the present study shows the structural transformation in molten KCl and KBr occurs over a narrow pressure range like as a first order phase transition in solid. *Kawai and Inokuchi* [1970] and *Tallon* [1979] interpreted the shape of the melting curve of alkali halides as a result of the continuous transition in the melt from a sixfold-coordinated state to an eightfold-coordinated state. Their conclusions are confirmed by the present results, except that the melt undergoes the structural change with a narrow transition region.

From a geophysical standpoint, it has been believed that the structural changes occur in silicate melts by compression. The changes in polymerization of SiO_4 tetrahedrons and the coordination number of Si are of interest for the effect of pressure on the physical properties of silicate melts such as density and viscosity. Here we, however, turn our attention to the transition-metal ions in silicate melts, which are highly ionic. At atmospheric pressure, the coordination number of the nearest neighbor oxygen of transition-metal ion in melt is smaller than six in crystal [*Waseda and Toguri*, 1978]. It is easy to predict that the application of pressure leads to the increase of CN as in molten alkali halides. From their spectroscopic study on silicate glass quenched at high pressure, *Keppler and Rubie* [1993] concluded that the sixfold-coordinated Co and Ni dominate in silicate melts at high pressures and the crystal field stabilization energies (CFSE) of Co and Ni in melt increase with pressure. It is, however, difficult to speculate the melt structure based on the quenched glass, because, as mentioned above, the coordination number does not correspond to the local structure in liquid state. Nevertheless, as the melt is deformable, it is possible the transition elements with six nearest neighbor oxygens have a high CFSE in silicate melt. Thus the partitioning of transition elements into silicate melts from silicate crystal and the metal under pressure would be enhanced by the structural change in silicate melts in terms of CFSE increase as well as the volume decrease of melt.

CONCLUSIONS

In-situ X ray diffraction experiments were carried out on the molten KCl and KBr under pressure and temperature by using synchrotron radiation. From the calculated radial distribution function, the inter-ionic distance and the coordination number were evaluated. From these data, we can derive the following conclusions: (1) Both the molten KCl and KBr have an open simple-cubic-like structure at low pressure and transform into a more dense structure above 2 GPa, probably an open body-center-cubic-like structure. (2) The coordination number of nearest neighbor ions (CN) rapidly increases with pressure to 2 GPa, which implies the decrease of vacancy in the melt. However, the increase of CN accompanied by structural change is relatively small. (3) The structural change of molten KCl precedes that of crystalline phase, but the molten KBr transforms into a dense phase about the same time as the solid. (4) The molten KCl and KBr undergo a structural transformation with a narrow transition region.

Acknowledgments. The authors are grateful to K. Igarashi, N. Umesaki, and T. Kondo for their help in experiments. They are indebted to K. Tsuji for instruction of data analysis method. They also thank two anonymous reviewers for their helpful comments on this manuscript. This study has been performed under the approval of the Photon Factory Program Advisory Committee (Proposal Nos. 92-125 and 94G141).

REFERENCES

Adams, D.J., Monte Carlo and HNC calculations for molten potassium chloride, *J. Chem. Soc. Faraday Trans. II*, *72*, 1372-1384, 1976.

Boehler, R., M. Ross, and D. B. Boercker, High-pressure melting curves of alkali halides, *Phys. Rev. B*, *53*, 556-563, 1996.

Kawai, N., and Y. Inokuchi, High pressure melting of general compounds, and with some physical models, *Jpn. J. Appl. Phys.*, *9*, 31-48, 1970.

Keppler, H., and D. C. Rubie, Pressure-induced coordination changes of transition-metal ions in silicate melts, *Nature*, *363*, 54-56, 1993.

Nishikawa, K., and T. Iijima, Correction for intensity data in energy-dispersive X ray diffractometry of liquids. Application to carbon tetrachloride, *Bull. Chem. Soc. Jpn.*, *57*, 1750-1759, 1984.

Ohno, H., K. Furukawa, and R. Takagi, X ray diffraction analysis of molten Potassium Bromide, *J. Chem. Soc. Faraday Trans. II*, *79*, 463-471, 1983.

Ohno, H., K. Igarashi, N. Umesaki, and K. Furukawa, *X ray diffraction analysis of ionic liquids*, Trans Tech Publications, Switzerland, 1994.

Pistorius, C. W. F. T., Polymorphic transitions of the alkali chlorides at high pressures to 200°C, *J. Phys. Chem. Solids*, *25*, 1477-1481, 1964.

Pistorius, C. W. F. T., Melting curves of the potassium halides at high pressures, *J. Phys. Chem. Solids*, *26*, 1543-1548, 1965.

Ross, M., and F. J. Rogers, Structure of dense shock-melted alkali halides: Evidence for a continuous pressure-induced structural transition in the melt, *Phys. Rev. B*, *31*, 1463-1468, 1985.

Ross, M., and G. Wolf, High-pressure melting curve of KCl: Evidence against lattice-instability theories of melting, *Phys. Rev. Lett.*, *57*, 214-217, 1986.

Shimomura, O., W. Utsumi, T. Taniguchi, T. Kikegawa, and T. Nagashima, A new high pressure and high temperature apparatus with sintered diamond anvils for synchrotron radiation use, in *High Pressure Research: Application to Earth and Planetary Sciences*, *Geophys. Monogr. Ser.*, Vol. 67, edited by Y. Syono and M. H. Manghnani, pp. 3-12, Terra Scientific Publishing Co., Tokyo / AGU, Washington, D. C., 1992.

Takagi, R., H. Ohno, and K. Furukawa, Structure of molten KCl, *J. Chem. Soc. Faraday Trans. I*, *75*, 1477-1486, 1979.

Tallon, J. L., The pressure dependence of melting temperature for alkali halides, *Phys. Lett.*, *72A*, 150-152,1979.

Tsuji, K., K. Yaoita, M. Imai, O. Shimomura, and T. Kikegawa, Measurements of X ray diffraction for liquid metals under pressure, *Rev. Sci. Instrum.* *60*, 2425-2428, 1989.

Urakawa, S., N. Igawa, N. Umesaki, K. Igarashi, O. Shimomura, and H. Ohno, Pressure-induced structure change of molten KCl, *High Pressure Res.*, *14*, 375-382, 1996.

Waseda, Y., and J. M. Toguri, The structure of molten FeO-SiO_2 system, *Met. Trans. B*, *9B*, 595-601, 1978.

S. Urakawa, Department of Earth Sciences, Okayama University, 3-1-1 Tsushima naka, Okayama 700, Japan.

N. Igawa, Department of Materials Science and Engineering, Japan Atomic Energy Research Institute, 2-4 Shirane, Tokai 319-11, Japan.

O. Shimomura, Photon Factory, National Laboratory for High Energy Physics, 1-1 Oho, Tsukuba 305, Japan.

H. Ohno, Department of Synchrotron Radiation Facilities Project, Japan Atomic Energy Research Institute, Kanaji, Kamigori 678-12, Japan.

Hydrogen in Molten Iron at High Pressure: The First Measurement

Takuo Okuchi and Eiichi Takahashi

Department of Earth and Planetary Sciences, Tokyo Institute of Technology, Tokyo, Japan

In order to evaluate hydrogen abundance in the Earth's core, high-pressure melting experiments of FeH*x* were carried out. The experiments were designed to simulate the iron-water reaction in the Earth's magma ocean so that FeH*x* was synthesized in hydrous silicate melts. Hydrogen concentration in liquid FeH*x* was measured for the first time with a new technique. The hydrogen concentration in the liquid FeH*x* was 20% higher than the coexisting solid FeH*x*. The hydrogen concentration in the liquid FeH*x* rapidly increased with increasing temperature. The accreted iron to the proto-Earth should have accumulated and have been in equilibrium with hydrous silicate melt at the bottom of the magma ocean. Hydrogen concentration in the core estimated from the result of our experiments was up to H/Fe = 0.69, which may reconcile most of the density deficit in the present outer core.

1. INTRODUCTION

The Earth's core has ~10% lower density than pure iron [*Jeanloz*, 1990; *Fukai*, 1992]. This deficit has been explained by the dissolution of light elements. Hydrogen is one important candidate, because theoretical studies predicted that iron-water reaction proceeds to form FeH*x* at high pressure due to the negative DV of the reaction [*Stevenson*, 1977; *Fukai*, 1984]. For solid FeH*x* coexisting with excess hydrogen, a series of in situ X ray observations confirmed large hydrogen concentration as H/Fe = 0.13 at 3 GPa to H/Fe ~ 1 above at 3.5 GPa [*Fukai et al.*, 1982; *Badding et al.*, 1991; *Yamakata et al.*, 1992]. For solid FeH*x* coexisting with solid silicate and water, the composition up to H/Fe = 0.3~0.4 was confirmed recently [*Yagi and Hishinuma*, 1995].

Because the FeH*x* is only stable at high pressure, in situ X ray diffraction measurements on lattice parameters has been essential for measuring hydrogen in it. However X ray measurement is not effective for liquid FeH*x*, which has no typical diffraction pattern. Hydrogen concentration in liquid FeH*x* has not been examined yet, though Earth's core is mostly liquid [*Usselman*, 1975; *Suzuki et al.*, 1984; *Urakawa et al.*, 1987; *Yagi and Hishinuma*, 1995; *Boehler*, 1996]. *Suzuki et al.* [1984] reported the formation of iron droplets in hydrous silicate melt, at temperature ~500 K lower than the melting point of pure iron. Dissolution of hydrogen was inferred from this observation. However hydrogen was not detected from the recovered droplets, because it escapes instantaneously from iron after decompression at room temperature [*Fukai and Akimoto*, 1983].

In this study we made the first measurement for hydrogen concentration in the molten FeH*x*, by quenching experiments at high pressure. The experiments were designed to simulate the iron-water reaction in the Earth's magma ocean. Efforts were made to retain hydrogen and to reduce the temperature gradient in the sample capsule during the experiments. By an instant decompression just after quenching for preventing the escape of hydrogen, we recovered the iron grains having numerous bubbles after

Properties of Earth and Planetary Materials
at High Pressure and Temperature
Geophysical Monograph 101

TABLE 1. Starting Compositions and Run Products

Run No.	Pressure	Temperature	Duration	Starting Composition	Run Products*
572	7.5GPa	1100°C	10 min	$MgSiO_3$ 1.0H_2O 2Fe	Fe, ol(Fo64), px, H_2O
499	7.5GPa	1200°C	5 min	$MgSiO_3$ 1.4H_2O 2Fe	Fe, ol(Fo73), px (En80), qX(Fo75-72, En75-37)
500	7.5GPa	1300°C	5 min	$MgSiO_3$ 1.6H_2O 2Fe	Fe, ol(Fo78-76), qX(Fo75-74 , En39-29)
508	7.5GPa	1400°C	1 min	$MgSiO_3$ 1.5H_2O 2Fe	Fe, ol(Fo78-75), qX(Fo78-49)
570	7.5GPa	1500°C	15 sec	$MgSiO_3$ 1.2H_2O 2Fe	Fe, qX(Fo80-30, En80-25)
616	5.0GPa	~1100°C	3 min	$MgSiO_3$ 1.1H_2O 2Fe	Fe, ol(Fo74-29), px(En37-29), H_2O

*Abbrevations; Fe:metallic iron, ol:olivine, px:orthopyroxene, qX: quench crystals, H_2O: water vapor

the synthesized FeH_x. Hydrogen in FeH_x was measured from the volume fraction of the bubbles.

2. EXPERIMENTAL

2.1. *Run Condition and Starting Materials*

Experiments were carried out between 1100°C and 1500°C at 7.5 GPa using a uniaxial multianvil apparatus SPI-1000 at Tokyo Institute of Technology. One additional experiment was conducted at 1100°C and 5.0 GPa. The detail of the apparatus, including pressure calibration procedure, was described previously [*Takahashi et al.*, 1993].

The starting materials were a mixture of iron, $Mg(OH)_2$, and $SiO_2 \cdot xH_2O$. The iron powder was a spec-pure (99.99%) reagent. We did not crush it prior to use, in order to recover large iron droplets after the experiments. The $Mg(OH)_2$ and $SiO_2 \cdot xH_2O$ were powdered reagents. These hydroxide reagents were sieved with a 380-mesh screen to facilitate chemical reaction. Their impurities were mostly adsorbed H_2O. This was expelled by holding at 110°C in an oven; then H_2O content of the reagents were measured from weight loss at 1000°C. Table 1 shows the run condition and starting compositions of the experiments.

2.2. *Sample Capsule*

The primary requirement for the capsule material is to retain the hydrogen resulting from the iron-water reaction throughout the experiments. The NaCl was reported as a good insulator for hydrogen [*Yamakata et al.*, 1992]. However, we found in preliminary experiments that an NaCl capsule dissolves instantly into water in the sample. In order to avoid the dissolution of NaCl we used Pt inner capsule in addition to the NaCl outer capsule. However the NaCl capsule easily deformed at the experiments so that the heater and the Pt capsule contacted and the Pt capsule melted due to local electric heating. Therefore we determined to use a single metallic capsule.

There are practically no metals inert for hydrogen at high hydrogen pressure and temperature [*Sugimoto and Fukai*, 1992]. Dissolution of hydrogen results in the drastic reduction of melting point of metals [*Fukai*, 1991]. We conducted a series of tests searching for a capsule material having a high enough melting temperature to conduct melting experiments of FeH_x samples. We tested Pt, $Au_{75}Pd_{25}$ alloy, and Au capsules and found Pt is most preferable. In order to buffer the hydrogen pressure, which inevitably escapes through the metallic capsule throughout the experiments, we used a very large (7-mm-long and 5-mm-diameter) Pt tube that can enclose sufficient amounts of water. Loss of the water at the sealing of the capsules was carefully measured, and was subtracted from initial composition of the samples. We confirmed that this large Pt capsule kept water longer than a smaller, 3-mm-diameter Pt capsule.

2.3. *Furnace Assembly*

In order to heat the larger capsule, more electric power was necessary, which usually causes severe damage to the tungsten-carbide anvils. Even more serious is the inhomogeneity of temperature distribution inside the capsule. We designed a furnace assembly to hold the 7 × 5 mm capsule, totally enclosed in a good thermal insulator, ZrO_2, in order to reduce the heat conduction to the WC anvils and, therefore, to reduce the temperature inhomogeneity. The assembly is shown in Figure 1. A thin, 0.05-mm-thick Mo ribbon was used for the electrode. A graphite tube was used for the heating element, which was 6.3 mm in outer diameter, 0.3 mm in thickness and 12.0

mm in length. After a number of testing experiments we found a graphite heater is more stable than metallic or oxide ($LaCrO_3$) heater. This may be because hydrogen cannot reduce graphite and cannot dissolve in graphite even at high pressure. The graphite heater was designed to be as thin as possible because graphite has high thermal conductivity; a thicker heater would produce a steeper temperature gradient inside the heater. The practical limit of the heater thickness, 0.30 mm, was determined by its stability against graphite-diamond transition throughout the experiments.

A pyrophyllite sleeve and spacer, baked at 1300°C for 1 h for dehydration, was located inside the heater to enclose the Pt capsule. This provided further thermal insulation at the nearest point of the capsule, instead of ZrO_2 insulator which easily reacts with hydrogen. We found the baked pyrophyllite changed its color from pale pink to gray if water leaked from the capsule. Accordingly, the water leakage during the experiments was monitored by the color of the recovered pyrophyllite spacers. We checked this color prior to further investigation of the run charges.

2.4. *Temperature Calibration*

Thermocouple metals such as Pt and W react with hydrogen and form metal hydrides at high hydrogen pressure and temperature [*Sugimoto and Fukai*, 1992] so that their emfs may be unreliable. In addition, the thermocouple inevitably transports the heat and perturbs the temperature distribution within the furnace. We found ~100°C lower temperature at the center of the furnace shown in Figure 1, when a thermocouple with an alumina sleeve was introduced instead of the Pt capsule. Because of these reasons, we did not use the thermocouple in the experiments with the large Pt capsules. Instead, the temperature distribution was determined by measuring the wollastonite component of coexisting two pyroxenes after the method of *Takahashi et al.* [1982]. A uniform glass, having composition of $MgSiO_3$: $CaSiO_3$: Al_2O_3 = 72.5 : 20.0 : 7.5 in molar %, was synthesized from mixtures of powder reagents, rapidly quenched in water after heating at 1600°C for 1 h. It was crushed, ground and sealed in the 7 × 5 mm Pt capsule. The same furnace (Figure 1) was used for the calibration, and we found the power input-temperature relation. Some of the calibration results are shown in Figure 2. We found surprisingly uniform temperature distribution; <100°C in difference throughout the whole capsule. In the experiments listed in Table 1, convection of hydrous silicate melt may occur in the capsule so that the temperature distribution may be even more uniform than shown in our Figure 2.

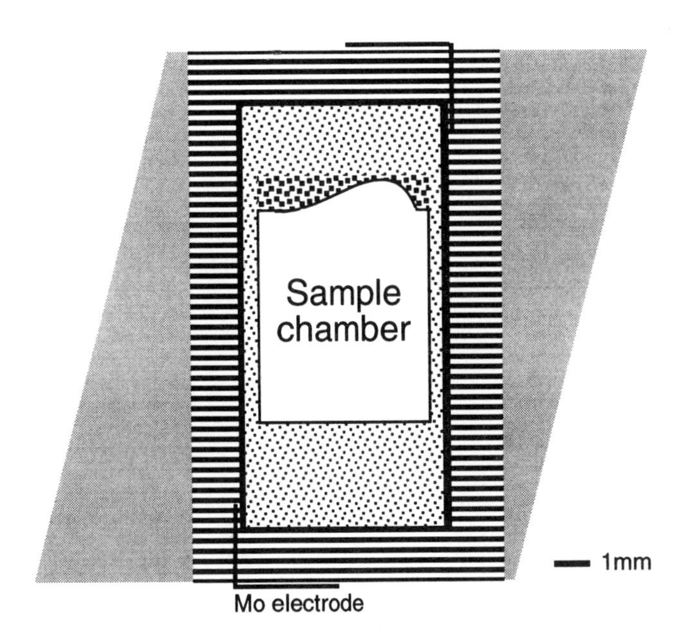

Figure 1. Furnace assembly.

2.5. *Run Duration*

The experiments should be finished before all the iron spherule sink to the bottom of the capsule. In one testing experiment conducted at 1500°C and 7.5 GPa for 30 sec, complete gravitational separation occurred and no iron spherules remained in the capsule. The hydrous silicate melts have extremely low viscosity in this case. With a number of such time studies, we managed to find a run duration just shorter than the sinking time of the iron spherules. Despite the short duration, only 15 sec at 1500°C as shown in Table 1, the iron spherules should have achieved chemical equilibrium for hydrogen with surrounding hydrous silicate melts. This is due to the fairly large diffusion coefficient of hydrogen in iron. The diffusion length $\sqrt{Dt}$ is ~700 mm at 1500 °C for t = 15 sec, where D is the hydrogen diffusion coefficient in iron [*Fukai and Sugimoto*, 1985]. This is much larger than the typical spherule size (see Figure 3b). All the experiments were terminated by shutting-off the electric power. The samples were quenched to ~200°C in a few seconds and to ~50°C in two minutes.

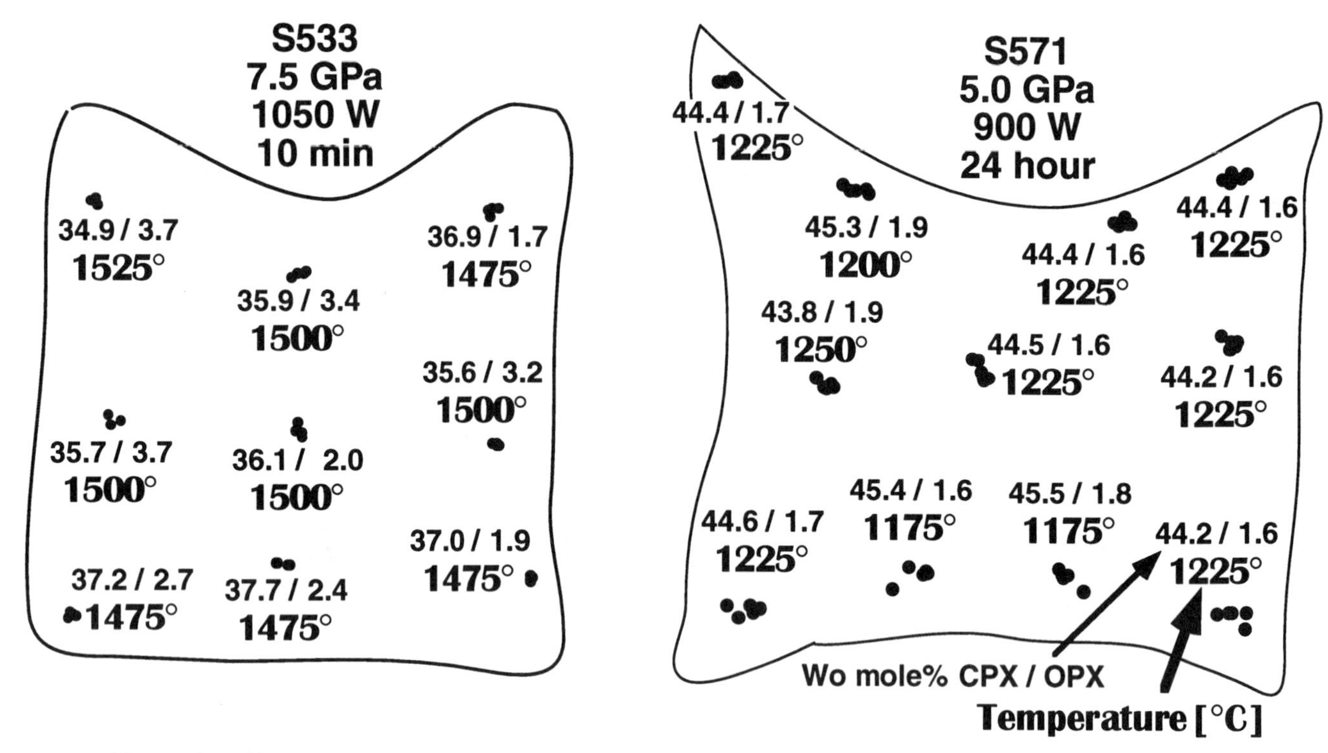

Figure 2. The results of temperature calibration. Molar percent of wollastonite component of coexisting clinopyroxene and orthopyroxene, and the corresponding temperature were shown at each measured point. The temperature was determined using the results of *Yamada and Takahashi* [1984].

Table 2 shows the results of all testing and calibration runs conducted for finding the preferable capsule material, furnace assembly and experimental conditions discussed in this chapter.

2.6. *Sample Analysis*

The JEOL-8800 electron probe microanalyzer (EPMA) at Tokyo Institute of Technology was used for all analysis. The accelerating voltage was 15 kV. The beam current was 12 nA for quantitative analysis of the silicates and 30 nA for qualitative analysis and taking back-scattered electron image of the iron grains. The melting of the silicate was identified by the occurrence of dendritic quench crystals (see Figure 7) with anomalously FeO-rich composition (see Table 1).

3. MEASURING HYDROGEN IN FeH_x

Fukai and Akimoto [1983] performed high-pressure experiments to synthesize FeH_x from iron and $Al(OH)_3$. They examined the reaction product by X ray diffraction after quenching and pressure release. Observed peaks were of a-iron. They interpreted the absence of FeH_x as due to its instantaneous decomposition as soon as the pressure was released. The FeH_x has never been recovered by conventional quenching experiments. Therefore the in situ measurements have been the only way to measure hydrogen in FeH_x, but is intrinsically ineffective for liquid FeH_x, as noted previously. Thus we developed a new technique to measure hydrogen in liquid (and solid) FeH_x. We made the decompression as fast as possible in order to prevent hydrogen escape from synthesized FeH_x grains. We confirmed that pressure can be released from 7.5 GPa to 1 atm in 5 sec without causing blowout when the furnace was sufficiently cold (~50°C) after waiting a few minutes after the quenching. We found numerous bubbles in the iron grains after such rapid decompression.

Figure 3 shows a back-scattered electron image of recovered iron grains after the experiment at 1200°C. We found two type of textures coexisted in this experiment. Figure 4 shows the result of qualitative analysis of the grains shown in Figure 3a, in which no peaks are seen other than iron and coated carbon. The bubbles are now empty but must have been filled with an element lighter than boron, and this must be hydrogen since there are no other candidates.

The occurrence of bubbles in the recovered iron grains clearly differs depending on whether solid or liquid FeH_x

Table 2. Results of the Testing Experiments

Run No.	Capsule		Heater		Iron	Hydrogen	Liquid Water	Pressure
	Material and Diameter [mm]	Melting Temperature	Material	Stability	Melting Temperature	Source Material	in Recovered Run Charges	Release Speed
262, 266, 286, 287, 368 375, 383, 385	Au3.0	1000~1100°C	$LaCrO_3$	OK	> 1100°C	H_2O + Fe	not found	normal
268, 269, 284, 285	Au3.0	1100°C	$LaCrO_3$	OK	1100°C	$LiAlH_4$ + FeO/Fe_2O_3	not found	normal
271, 274, 275, 276, 278 281, 282, 283, 429, 430	Pt3.0	>1500°C	$LaCrO_3$	OK	1200°C	H_2O + Fe	not found	normal
388	Au3.0	> 1100°C	$LaCrO_3$	OK	-	H_2O	found	normal
396	Au3.0 + AuPd2.3	AuPd 1200°C	$LaCrO_3$	OK	?	$LiAlH_4$	?	normal
401	NaCl4.5 + AuPd2.3	AuPd 1200°C	$LaCrO_3$	unstable	?	$LiAlH_4$	?	normal
404, 405, 410	NaCl4.5 + AuPd2.3	AuPd 1200°C	graphite	OK	> 1200°C	$LiAlH_4$	not found	normal
407, 408, 409	NaCl4.5	dissolved in H_2O	graphite	OK	?	H_2O + Fe	?	normal
411	NaCl4.5 + Pt1.6	Pt 1200°C	graphite	OK	?	$LiAlH_4$	?	normal
412, 414, 417, 418, 419	NaCl4.5	> 1200°C	graphite	OK	1200°C	$LiAlH_4$ + FeO/Fe_2O_3	not found	normal
416, 427, 428	NaCl4.5	> 1400°C	graphite	OK	1200°C	CaH_2 + H_2O	not found	normal
420, 425, 426, 431	Pt3.0	> 1400°C	graphite	OK	1400°C	H_2O + Fe	not found	normal
433, 439	Pt foil5.2	< 1000°C	Pt foil	melted	?	$LiAlH_4$ + H_2O	?	normal
436, 437, 438, 443, 445	Pt foil5.2 + Pt2.0-3.0	> 1200°C	graphite	unstable	?	$LiAlH_4$ + H_2O	?	normal
451, 452	NaCl4.5 + Pt3.0	< 1000°C	graphite	unstable	?	H_2O + Fe	?	normal
454, 456, 459, 465, 466 467, 468	Pt foil4.5-4.8	> 1500°C	graphite	OK	1300°C	H_2O + Fe	not found	normal
472, 474, 475, 476, 477 479, 480, 482 483	Pt foil4.5	> 1500°C	graphite	OK	1300°C	H_2O + Fe	not found	rapid
484, 495, 496, 498, 501 502, 504, 505, 509, 510	Pt5.0	> 1500°C	graphite	OK	1200°C	H_2O + Fe	found	rapid
511, 512, 515, 517, 519	Pt5.0	> 1500°C	graphite	OK	1200°C	H_2O + Fe	found	normal
448, 449, 450, 470	-		Temperature calibration using W-Re thermocouples					
533, 535, 536, 544, 546 567, 571	Pt5.0		Temperature calibration using the wollastonite content of pyroxenes					

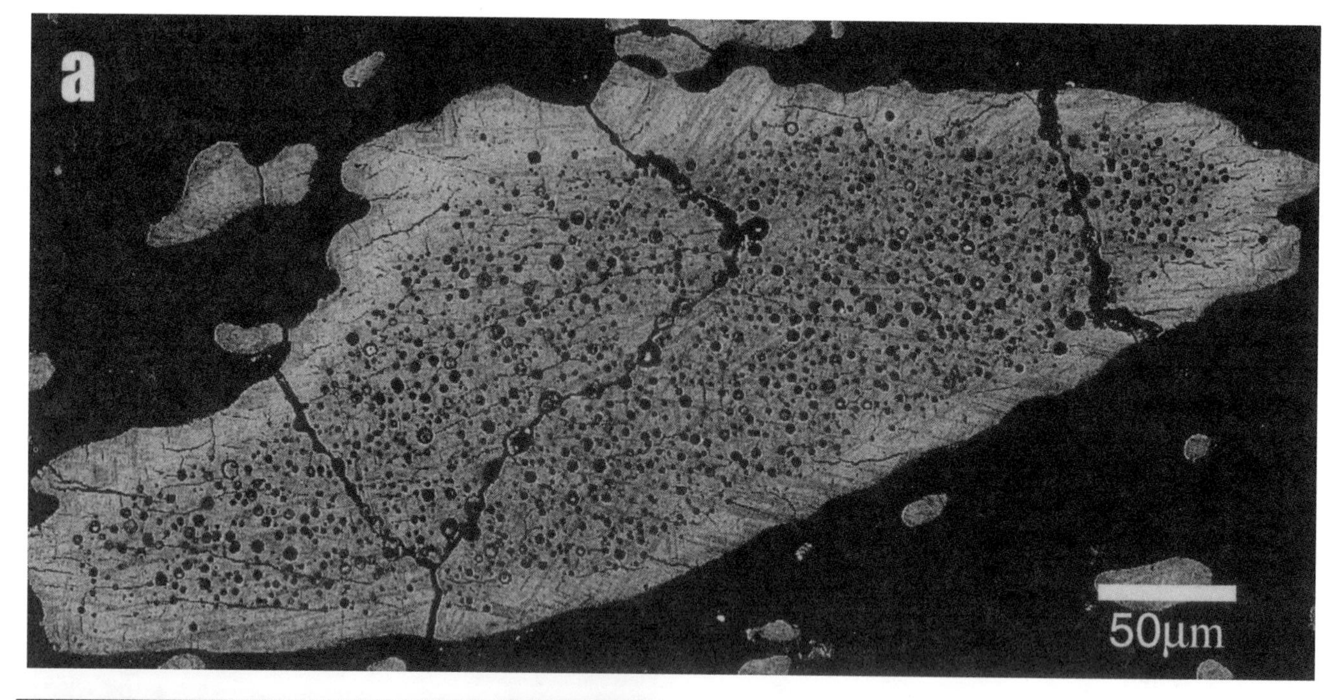

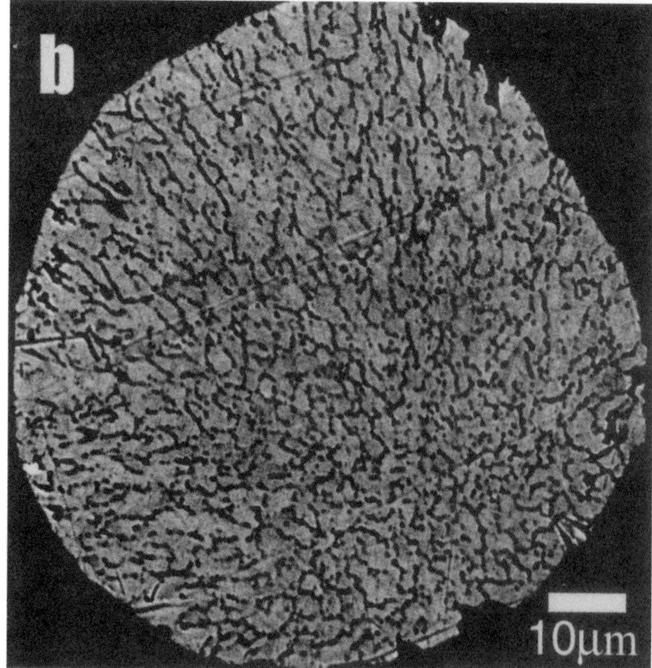

Figure 3. Back-scattered electron image of recovered iron grains in polished cross section of run at 1200°C. (a) Solid FeH*x*. Because of an instant decompression hydrogen became visible as numerous bubbles. (b) Liquid FeH*x*. Hydrogen bubbbles formed at the grain boundaries of quench crystals. These grains with different textures coexisted in the same run (#499). See discussion in the text.

was the host phase. Figure 5 shows schematic quench and decompression paths over the proposed phase diagram for the bcc metal and hydrogen system [*Fukai*, 1991]. Solid FeH*x*, recovered as irregular-shaped large crystals (Figure 3a), was quenched as a large crystal, and hydrogen bubbling took place homogeneously upon decompression. The width of the bubble-absent region at the edge of the grain may show the diffusion length of hydrogen during the decompression. On the other hand, liquid FeH*x*, recovered as frozen droplets (Figure 3b), was crystallized into aggregates of small polyhedrons upon quenching. Hydrogen diffused to the grain boundary of the polyhedral microcrystals and bubbling occurred there upon decompression.

Figure 6 shows the fine texture of frozen droplets after the liquid FeH*x*. We determined volume fraction of the bubbles from these images. If the system within the FeH*x* droplet was closed during the quenching and the decompression, the volume fraction of the bubbles should be proportional to dissolved hydrogen concentration. To determine the proportional factor, or to determine hydrogen density in the bubbles, we made one additional experiment at 5.0 GPa and about 1100°C (see Table 1). In this experiment we obtained the averaged bubble fraction in solid FeH*x* as 0.30 (30 vol. %). By in situ X ray diffraction measurements *Yagi and Hishinuma* [1995] reported the H/Fe molar ratio in solid FeH*x* as 0.40 at very similar condition (4.9 GPa and 1050°C). Using this ratio we determined the hydrogen density in the bubbles as 0.13

g/cm3. Using this density we calculated the hydrogen concentration in the solid and liquid FeH*x* at 7.5 GPa (see Figure 8).

4. RESULTS

Figure 7 shows the back-scattered electron image of polished cross sections of the run charges. The run products are shown in Table 1. No quench crystals were observed at 1100°C, while large spherical pores were present in the silicate mineral matrix (Figure 7a). The pores must have been filled with H_2O fluid since the temperature is above the decomposition point of hydrous minerals. The quench crystals are observed in the experiments above 1200°C (Figure 7b-e). The silicate melted almost totally with a small fraction of residual olivine above 1300°C (Figure 7c-e). In these charges the silicate melt is the host phase of H_2O coexisting with FeH*x*. *Inoue and Sawamoto* [1992] reported the solidus temperature of pyrolite-H_2O system as 1150°C at 7.7 GPa. We found the silicate solidus between 1100° and 1200°C for the Fe-$MgSiO_3$-H_2O system studied at 7.5 GPa, which is in good agreement with the pyrolite-H_2O system.

We can clearly recognize the temperature effect for hydrogen concentration in the liquid FeH*x* from Figure 6. The bubble fraction increased with increasing temperature. Figure 8 shows the bubble fraction (left scale) and the hydrogen concentration (right scale) against temperature. Table 3 shows averaged hydrogen concentration in the FeH*x*. The hydrogen concentration in the liquid FeH*x* rapidly increases with temperature, from 0.41 at 1200°C to 0.69 at 1500°C. At 1200°C the solid and the liquid FeH*x* coexisted. We found that hydrogen concentration in the liquid FeH*x* was 20% higher than the coexisting solid FeH*x*. This is consistent with the thermodynamic calculation which predicted a small increase of H/Fe after the melting of FeH*x* at high pressure [see the phase diagram in Figure 5; see also *Fukai*, 1992]. The distribution coefficient of hydrogen between liquid and solid FeH*x*, k = x_{liquid} / x_{solid} , was therefore found as 1.2. This value is somehow smaller than k = 1.87 at 1 atm [*Fukai*, 1991; 1992], which was measured at a very low hydrogen concentration in iron; H/Fe ~ 10^{-3}.

5. DISCUSSION

5.1. *Comparison with the Fe-H System*

Our new measurements showed that the hydrogen concentration in molten FeH*x* is smaller than H/Fe = 1 and rapidly increases with increasing temperature in the Fe-$MgSiO_3$-H_2O system at 7.5 GPa. *Yagi and Hishinuma* [1995] also showed that the hydrogen concentration in solid FeH*x* increases with temperature in the same system at 2.9 and 4.9 GPa. On the other hand, in the simple Fe-H system with excess hydrogen, it has been reported that the hydrogen concentration in solid FeH*x* decreases starting at H/Fe ~ 1 with increasing temperature at 6 GPa [*Yamakata et al.*, 1992]. They estimated the hydrogen concentration from the unit cell volume of solid FeH*x*. Their observation agrees with the result of thermodynamic calculation for the Fe-H system [*Fukai*, 1984]. The discrepancy of the temperature systematics between the Fe-silicate-H_2O and the Fe-H system is reconciled as follows:

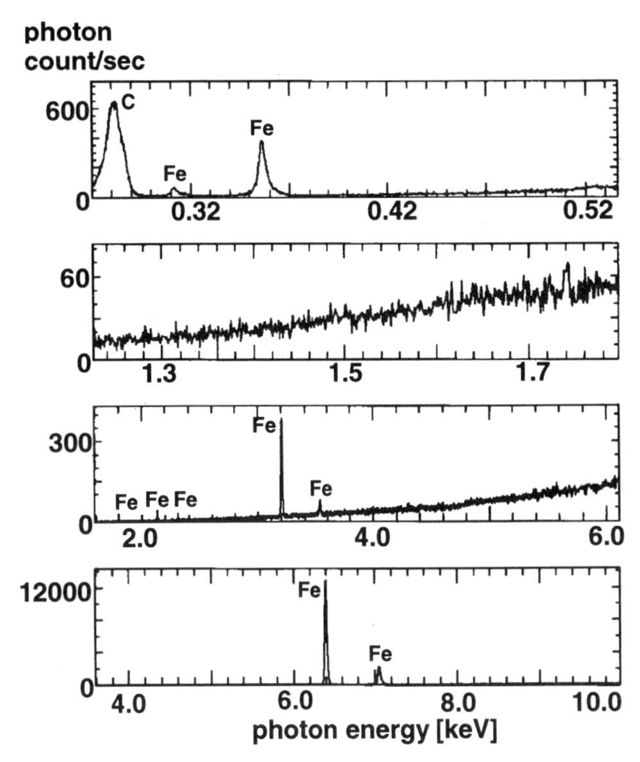

Figure 4. The result of qualitative analysis of the iron grain shown in Figure 3a. Counts of characteristic X ray photons emitted from the grain radiated by 30 nA of electron beam are shown against the energy of the photons. Only iron (Fe) and coated carbon (C) were detected from the grain through the analysis.

The iron reacts with water in the silicate melt in two steps:

$$Fe + H_2O \rightleftarrows FeO + H_2 \quad \text{(I)}$$
$$Fe + 1/2H_2 \rightleftarrows FeH \quad \text{(II)}$$

Reaction I defines hydrogen partial pressure (pH_2). Reaction II defines hydrogen concentration in FeH*x*. The

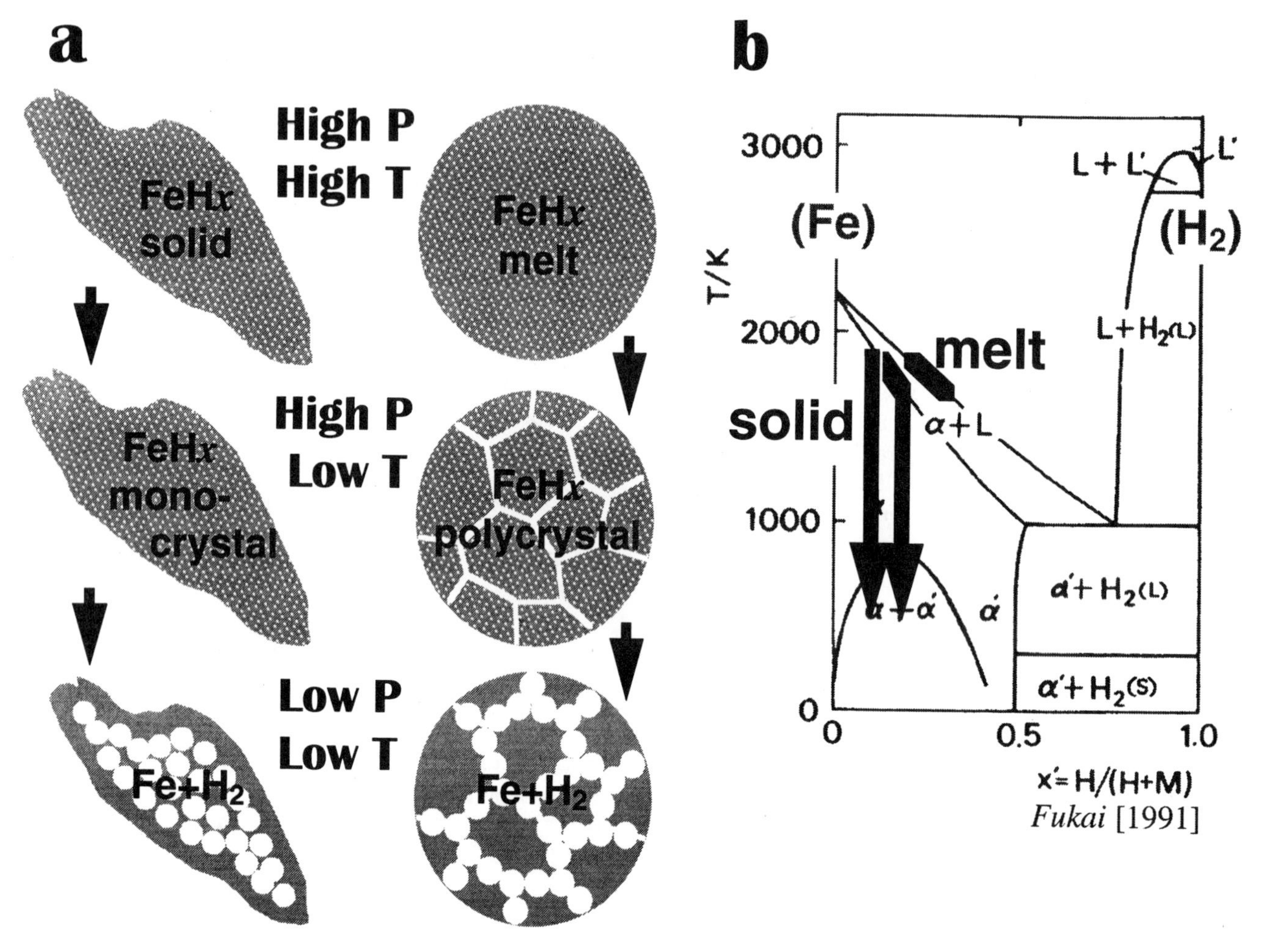

Figure 5. (a) Schematic diagram of the quench and decompression paths of synthesized FeH*x*. (b) The quench path of FeH*x* shown over the proposed bcc metal-hydrogen phase diagram [*Fukai*, 1991]. The phase diagram suggests small increase of hydrogen in FeH*x* at melting.

thermodynamic calculation on the Fe-H system showed that DH of reaction II becomes negative above at ~5 GPa and the H/Fe in solid FeH*x* approaches to unity if the system is saturated with hydrogen [*Fukai*, 1984]. Since the DH of reaction II is negative, the H/Fe in FeH*x* should decrease with increasing temperature. This was observed as a slight decrease of the H/Fe by *Yamakata et al.* [1992].

On the other hand, in the Fe-silicate-H_2O system, the H/Fe in FeH*x* is smaller than unity. The system is not saturated with hydrogen and reaction I controls the pH_2 in this case. Since ΔH of reaction I is positive, the pH_2 should increase with increasing temperature. This applies to our observation in which the H/Fe in FeH*x* rapidly increases with increasing temperature. We, therefore, conclude that the reaction I mainly controls the hydrogen concentration in FeH*x* in the Fe-silicate-H_2O system. To consider the iron-water reaction in the primordial Earth, the temperature systematics in Fe-silicate-H_2O system observed in our experiments is more realistic than that found in the Fe-H binary system previously studied.

5.2. *Hydrogen Abundance in the Core*

In the light of present experiments we can evaluate hydrogen abundance in the Earth's core, the problem that has been discussed repeatedly by many authors [e.g., *Stevenson*, 1977; *Fukai and Akimoto*, 1983; *Fukai*, 1984; *Suzuki et al.*, 1984; *Fukai and Suzuki*, 1986; *Badding et al.*, 1991; *Fukai*, 1992; *Yagi and Hishinuma*, 1995]. It is widely believed that the proto-Earth had accreted a few percent of water [*Ringwood*, 1977; *Wänke*, 1981;*Fukai*, 1984; *Fukai and Suzuki*, 1986; *Kuramoto and Matsui*,

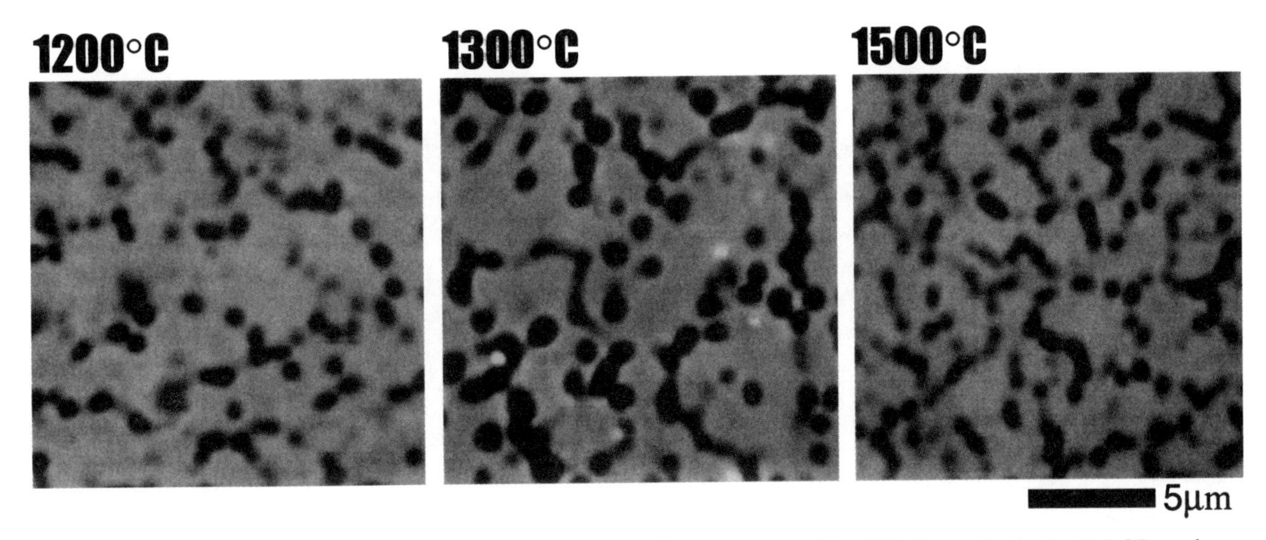

Figure 6. Fine textures of hydrogen bubbles formed in rapidly decompressed liquid FeH_x synthesized at 7.5 GPa and 1200°~1500°C. Bubble fraction apparently increases with increasing temperature.

1996]. The global magma ocean should have formed due to the release of gravitational energy and the blanketing effect of early steam atmosphere [*Safronov, 1978*; *Sasaki and Nakazawa*, 1986; *Abe and Matsui*, 1985, 1986]. The magma ocean should have been in equilibrium with the steam atmosphere and therefore may have dissolved most of the accreted water [*Abe and Matsui*, 1985, 1986] if the amount of impact-induced reduction of water to hydrogen was negligible [*Tyburczy et al.*, 1990]. The accreted iron had sunk in such a hydrous magma ocean. The iron should have accumulated at the bottom of the magma ocean because the large surface tension of molten iron, including FeH_x, prohibits its downward percolation through the solid silicate mantle [*Stevenson*, 1990; *Taylor*, 1992; *Hishinuma et al.*, 1994; also see *Urakawa et al.*, 1987].

The accumulated iron should have settled to the center of the proto-Earth by self-accelerating and catastrophic processes such as the Rayleigh-Taylor instability, since there are no other effective processes [*Stevenson*, 1981; *Ida et al.*, 1987]. During sucn processes there is no time to achieve equilibrium between the iron and the silicate. Therefore final chance for the sinking iron to react with the hydrous silicate is available at the bottom of the magma ocean, where the melt fraction of the silicate must be low enough to prevent the iron from percolating. This suggests that the silicate is just above at the wet solidus so that the silicate melt should contain water with its saturation level.

In the present experiments, we determined the hydrogen concentration in the FeH_x coexisting with the silicate melt containing saturation level of water (see the starting compositions in Table 1). Although the water concentration in our starting materials was much larger than in the material accreted to the proto-Earth, the silicate melt produced in our experiments may have adequately simulated the silicate melt in equilibrium with the accumulated iron at the bottom of the Earth's magma ocean. Accordingly our results may be directly applicable to determine the hydrogen concentration in molten iron at the bottom of the magma ocean. As discussed, the composition of the iron should be preserved until it had settled to the core.

The hydrogen abundance in the core at the completion of its formation therefore can be estimated using the value in Table 3, if 7.5 GPa was an adequate pressure at the bottom of the Earth's magma ocean. This corresponds to ~230 km depth of the magma ocean. Hydrogen in the core should increase with increasing depth of the Earth's magma ocean because of large negative DV of the reaction II [*Fukai*, 1992]. Table 3 shows that the hydrogen abundance in the core should increase rapidly with increasing temperature at the bottom of the magma ocean. The deeper the magma

TABLE 3. Averaged Hydrogen Concentration in FeH_x

Pressure	Temperature	State	H/Fe [mole]
7.5 GPa	1200°C	Solid	0.33
7.5 GPa	1200°C	Liquid	0.41
7.5 GPa	1300°C	Liquid	0.60
7.5 GPa	1400°C	Liquid	0.66
7.5 GPa	1500°C	Liquid	0.69

a. 1100°C

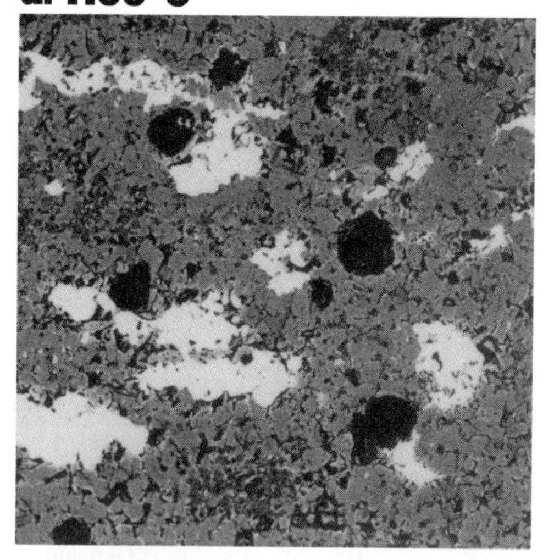

b. 1200°C

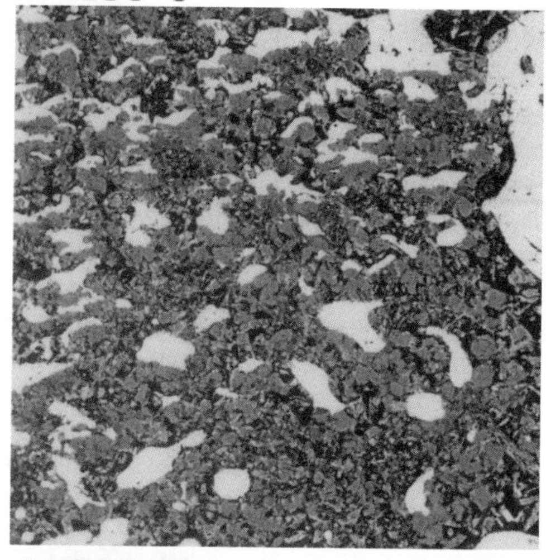

c. 1300°C

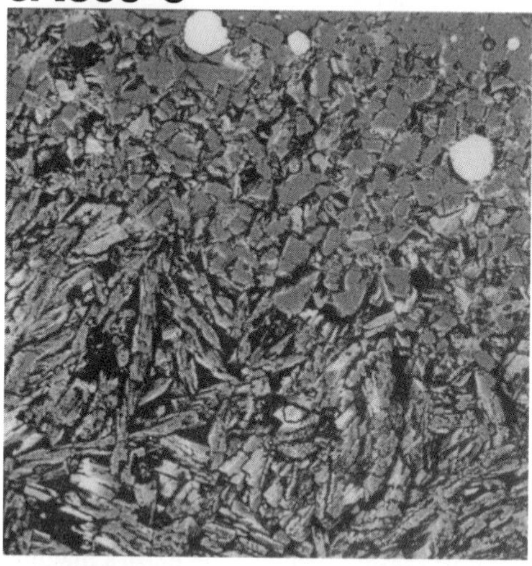

d. 1400°C

e. 1500°C

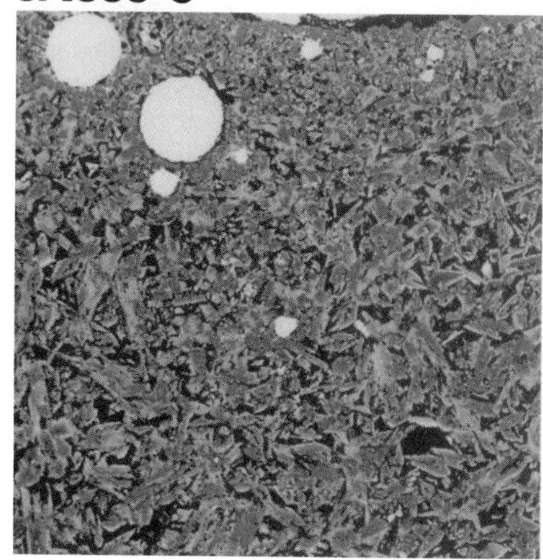

100μm

Figure 7. The back-scattered electron image of the run charges in polished cross section. Bright grains and droprets are iron with numerous bubbles.

ocean had been, the greater the pressure and the temperature at the bottom of the magma ocean, and the more hydrogen had dissolved in the core. A rather moderate depth of the magma ocean, ~230 km, corresponds to the hydrogen concentration in the core up to H/Fe = 0.69. This amount of hydrogen may reconcile most of the density deficit in the outer core observed now [*Fukai*, 1992].

5.3. The Role of Hydrogen in the Core for Mantle Dynamics

In the following discussion we assumed that the present outer core is in equiribrium with the lowermost mantle for hydrogen. The core has gradually cooled together with the cooling of the Earth. As discussed, reactions I and II should proceed leftward with the temperature reduction, so that hydrogen should have been exsolving from the outer core, reducing FeO in the lowermost mantle to metallic Fe and forming water. The metallic iron, actually the liquid FeH_x, may separate from the mantle and sink into the core. Accordingly the density of the lowermost mantle material may decrease. The water added to the mantle drastically reduces the viscosity and solidus temperature of the material [Karato et al., 1986; Inoue and Sawamoto, 1992]. These all are conditions yielding mantle instabilty. We believe hydrous melting and metal separation is taking place just above the core-mantle boundary. These may account for the dynamically unstable nature of the D" layer with heterogeneous thickness and negative dVp/dZ [*Young and Lay*, 1987; *Jeanloz*, 1990]. If the mantle plumes come from the D" layer [*Dziewonski*, 1984; *Fukao*, 1992], hydrogen in the core may produce them. More theoretical and experimental studies are required to examine this potential role of hydrogen.

6. CONCLUSION

Hydrogen concentration in the solid and liquid FeH_x was determined by high pressure quenching experiments at 7.5 GPa. Very large sample capsules were used for buffering hydrogen partial pressure. By careful designing of the furnace assembly with complete thermal insulation, the temperature gradient inside the capsule was reduced to <100°C. In order to measure hydrogen in the FeH_x, extremely high decompression speed was applied. Numerous hydrogen bubbles were formed in quenched iron grains after the decompression.

The hydrogen concentration was determined from the bubble fraction of iron grains using the tensional yield strength of pure iron. The hydrogen concentration in the liquid FeH_x was found as 20% higher than the coexisting solid FeH_x. The positive temperature dependence of the hydrogen concentration was clearly seen in the liquid FeH_x coexisting with the hydrous silicate melt. This observation shows the endothermic nature of the reaction of FeH_x formation in the hydrous silicate melt.

Hydrogen abundance in the core at the completion of its formation was estimated using the result of present experiments. The iron accreted to the proto-Earth should have been in equilibrium with a silicate melt saturated with water, at the bottom of the Earth's magma ocean. The FeH_x should have formed there and settled to the core without further chemical reaction with the deeper mantle because of the catastrophic transportation. The actual hydrogen abundance in the core depends on the depth of the magma ocean at the cessation of the Earth's accretion. The rather moderate depth of ~230 km corresponds to the hydrogen abundunce in the core up to H/Fe = 0.69. This reconciles most of the density deficit in the present outer core.

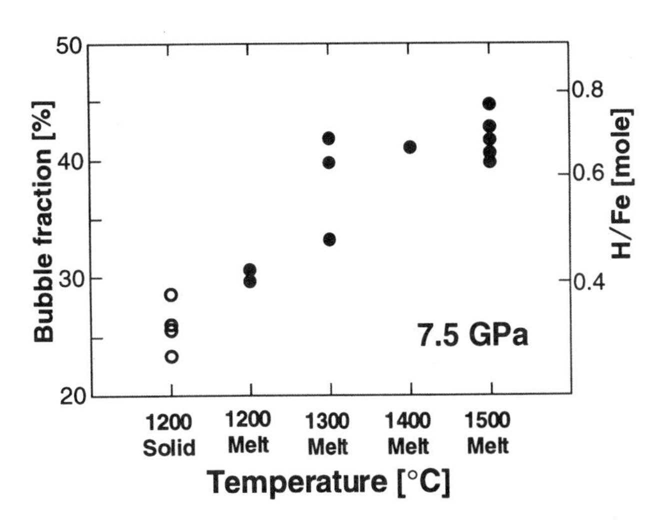

Figure 8. Change in bubble fraction of recovered iron grains (left ordinate) and hydrogen concentration in FeH_x (right ordinate) as a function of temperature (abscissa).

Acknowledgments. We thank D. J. Stevenson and one anonymous referee for their critical reviews and comments. We thank Y. Fukai, T. Yagi, K. Kawamura and H. Yurimoto for helpful discussions. This work was supported by the foundation for Research Fellowships of Japan Society for the Promotion of Science for Young Scientists.

REFERENCES

Abe, Y., and T. Matsui, The formation of an impact-generated H_2O atomosphere and its implications for the early thermal history of the earth, *Proc. Lunar Planet. Sci. Conf. 15th*, in *J. Geophys. Res.*, *90*, C545-C559, 1985.

Abe, Y., and T. Matsui, Early evolution of the earth: accretion, atmosphere formation and thermal history, *J. Geophys. Res.*, *91*, E291-302, 1986.

Boehler, R., Experimental constraints on melting conditions relevant to core formation, *Geochim. Cosmochim. Acta*, *60*, 1109-1112, 1996.

Badding, J. V., R. J. Hemley and H. K. Mao, High-pressure chemsitry of hydrogen in metals: In situ study of iron hydride, *Science*, *253*, 421-424, 1991.

Dziewonski, A. M., Mapping the lower mantle: determination of lateral heterogeneity in P velocity up to degree and order 6, *J. Geophys. Res.*, *89*, 5929-5952, 1984.

Fukai, Y., The iron-water reaction and the evolution of the Earth, *Nature*, *308*, 174-175, 1984.

Fukai, Y., From metal hydrides to the metal-hydrogen system, *J. Less-Common Metals*, *8*, 172-174, 1991.

Fukai, Y., Some properties of the Fe-H system at high pressures and temperatures, and their implications for the Earth's core., in *High Pressure Research: Application to Earth and Planetary Sciences*, *Geophys. Monogr. Ser.*, vol. 67, edited by Y. Syono and M. H. Manghnani, pp. 373-385, TERRAPUB/AGU, Tokyo/Washington D. C., 1992.

Fukai, Y., and S. Akimoto, Hydrogen in the Earth's core, *Proc. Japan Acad.*, *59*, 158-162, 1983.

Fukai, Y., A. Fukizawa, K. Watanabe and M. Amano, Hydrogen in iron - Its enhanced dissolution under pressure and stabilization of the g phase, *Jpn. J. Appl. Phys.*, *21*, L318-L320, 1982.

Fukai, Y., and H. Sugimoto, Diffusion of hydrogen in metals, *Adv. in Physics*, *34*, 263-326, 1985.

Fukai, Y., and T. Suzuki, Iron-water reaction under high pressure and its implication in the evolution of the Earth, *J. Geophys. Res.*, *91*, 9222-9230, 1986.

Fukao, Y., Seismic tomogram of the Earth's mantle: Geodynamic implications, *Science*, *258*, 625-630, 1992.

Hishinuma, T., T. Yagi and T. Uchuda, Surface tension of iron hydride formed by the reaction of iron-silicate-water under pressure, *Proc. Jpn. Acad.*, *70*, 71-76, 1994.

Ida, S., Y. Nakagawa and K. Nakazawa, The Earth's core formation due to the Rayleigh-Taylor instability, *Icarus*, *69*, 239-248, 1987.

Inoue, T., and H. Sawamoto, High pressure melting of pyrolite under hydrous condition and its geophysical implications, in *High-pressure research: Application to Earth and Planetary Sciences*, *Geophys. Monogr. Ser.*, vol. 67, edited by Y. Syono and M. H. Manghnani, pp. 323-331, TERRAPUB/AGU, Tokyo/Washington D. C., 1992.

Jeanloz, R., The nature of the Earth's core, *Ann. Rev. Earth Planet. Sci.*, *18*, 357-386, 1990.

Karato, S., M. S. Paterson and J. D. FitzGerald, Rheology of synthetic olivine aggregates: Influence of grain size and water, *J. Geophys. Res.*, *91*, 8151-8176, 1986.

Kuramoto, K., and T. Matsui, Partitioning of H and C between the mantle and core during the core formation in the Earth: its implications for the atmospheric evolution and redox state of early mantle, *J. Geophys. Res.*, *101*, *14909-14932*, 1996.

Ringwood, A. E., Composition of the core and implications for the origin of the Earth, *Geochem. J.*, 11, 111-135, 1977.

Safronov, V. S., The heating of the earth during its formation, *Icarus*, *33*, 3-12, 1978.

Sasaki, S., and K. Nakazawa, Metal-silicate fractionatin in the growing earth: energy source for the terrestrial magma ocean, *J. Geophys. Res.*, *91*, 9231-9238, 1986.

Stevenson, D. J., Hydrogen in the Earth's core, *Nature*, *268*, 130-131, 1977.

Stevenson, D. J., Models of the Earth's core, *Science*, *214*, 611-619, 1981.

Stevenson, D. J., Fluid dynamics of core formation, in *Origin of the Earth*, edited by H. E. Newsom and J. H. Jones, pp. 231-249, Oxford Univ. Press, New York, 1990.

Sugimoto, H., and Y. Fukai, Solubility of hydrogen in metals under high hydrogen pressures: thermodynamical calculations, *Acta Metall. Mater.*, *40*, 2327-2336, 1992.

Suzuki, T., S. Akimoto and Y. Fukai, The system iron-enstatite-water at high pressures and temperatures - formation of iron hydride and some geophysical implications, *Phys. Earth Planet. Inter.*, *36*, 135-144, 1984.

Takahashi, E., T. Shimazaki, Y. Tsuzaki and H. Yoshida, Melting study of a peridotite KLB-1 to 6.5 GPa, and the origin of basaltic magmas, *Phil. Trans. R. Soc. Lond.*, *A342*, 105-120, 1993.

Takahashi, E., H. Yamada and E. Ito, An ultrahigh-pressure furnace assembly to 100 kbar and 1500°C with minimum temperature uncertainty, *Geophys. Res. Lett.*, *9*, 805-807, 1982.

Taylor, G. J., Core formation in asteroids, *J. Geophys. Res.*, *97*, 14,717-14,726, 1992.

Tyburczy, J. A., R. V. Krishnamurthy, S. Epstein and T. J. Ahrens, Impact-induced devolatilization and hydrogen isotopic fractionation of serpentine: Implications for planetary accretion, *Earth Planet. Sci. Lett.*, *98*, 245-260, 1990.

Urakawa, S., M. Kato and M. Kumazawa, Experimental study on the phase relations in the system Fe-Ni-O-S up to 15 GPa, in *High-Pressure Research in Mineral Physics*, *Geophys. Monogr. Ser.*, vol. 39, edited by M. H. Manghnani and Y. Syono, pp. 95-111, TERRAPUB/AGU, Tokyo/Washington, D.C., 1987.

Usselman, T. M., Experimental approach to the state of the core, I, The liquidus relations of the Fe-rich portion of the Fe-Ni-S system from 30 to 100 kb, *Am. J. Sci.*, *275*, 278-290, 1975.

Wänke, H., Constitution of terrestrial planets, *Phil. Trans. R. Soc. Lond.*, *A303*, 287-302, 1981.

Yagi, T., and T. Hishinuma, Iron hydride formed by the reaction of iron, silicate, and water: Implications for the light element of the Earth's core, *Geophys. Res. Lett.*, *22*, 1933-1936, 1995.

Yamada, H., and E. Takahashi, Subsolidus phase relations between coexisting garnet and two pyroxenes at 50 to 100 kbar in the system $CaO-MgO-Al_2O_3-SiO_2$, in *Kimberlites. II: The Mantle and Crust-Mantle Relationships*, J. Kornprobst, Elsevier Science Publishers, Amsterdam, 247-255, 1984.

Yamakata, M., T. Yagi, W. Utsumi and Y. Fukai, In situ X-ray observation of iron hydride under high pressure and high temperature, *Proc. Jpn. Acad.*, *68B*, 172-176, 1992.

Young, C. J., and T. Lay, The core-mantle boundary, *Ann. Rev. Earth Planet. Sci.*, *15*, 25-46, 1987.

T. Okuchi and E. Takahashi, Department of Earth and Planetary Sciences, Tokyo Institute of Technology, 2-12-1 Ookayama, Meguro-ku, Tokyo 152, Japan

Element Partitioning Between Some Mantle Minerals and the Coexisting Silicate Melt Under High Pressure

Toshihiro Suzuki and Masaki Akaogi

Department of Chemistry, Gakushuin University, Mejiro, Tokyo 171, Japan

Element partitioning between minerals (garnet, magnesiowüstite, Ca-perovskite, merwinite, anhydrous phase-B) and silicate melt has been investigated at pressures 10–17 GPa, by using a 6-8 type multianvil high-pressure apparatus. The observed partition coefficients, D, were plotted on partition coefficient-ionic radius (PC-IR) diagrams. The D-profile in the PC-IR diagram of magnesiowüstite and anhydrous phase-B shows a peak at around 60–70 nm, which may correspond to the size of 6-coordination sites in their structure. Small differences were found among the D-profiles of majorite, Fe-rich garnet, and Ca-rich garnet, but their D-profiles resemble each other, and the fundamental characteristic of D-profile would be the same among these garnets. No obvious peak was found in the D-profiles of garnets and the D- values of most of the observed elements were 1–10, but La shows significantly small D-value compared with other elements. Partition coefficients of large ions, such as Ca and La, were large in Ca-perovskite and merwinite. A peak of D-profile was expected at around 90–100 pm in these minerals, and there is a bottom of D-profile at around 70–80 pm. The present experiments indicate that the observed D for Na is very small in Ca-perovskite, although large trivalent and divalent ions similar to Na in size are incorporated in it.

1. INTRODUCTION

The knowledge of partitioning behavior of elements between mineral and coexisting silicate melt is required as fundamental information to resolving the evolution process of the terrestrial planets. Since partition coefficients vary with pressure- temperature conditions or composition of the system, it would be desirable to have all the partition data for any possible conditions in the history of evolution of the terrestrial planets. However, it is impossible to accumulate such an enormous amount of information. Hence, we need to find some useful systematics to estimate the partition coefficients at desired conditions and compositions.

Onuma et al. [1968] pointed out that partition coefficients for elements with the same valence vary smoothly with ionic radius, and if we plot partition coefficients to ionic radius (PC-IR diagram), peaks of partition coefficients which correspond to the optimum size of the sites in the crystal structure appear in the PC-IR diagram. For example, the profile of partition coefficients of olivine/silicate melt system shows a peak at around 70 pm. This peak is regarded as the demonstration of size of M1 and M2 sites in the olivine structure, and the position of this peak will not change largely even if the composition or P-T condition is changed [e.g., *Suzuki and Akaogi*, 1995]. Therefore, if a mineral is equilibrated with

Properties of Earth and Planetary Materials
at High Pressure and Temperature
Geophysical Monograph 101

TABLE 1. Chemical Compositions of the Starting Materials (wt%)

	PL	FO	FA	WO1	WO2	San Carlos Olivine
SiO_2	40.18	33.51	28.56	43.81	42.64	40.51
MgO	35.71	38.32	4.45	2.80	0.93	50.32
CaO	2.30	1.41	1.00	38.02	34.04	0.08
Al_2O_3	2.44	1.00	0.79	0.75	0.78	0.01
FeO	7.15	6.65	56.24	1.49	1.14	8.72
MnO	0.71	1.28	1.08	1.11	1.06	0.06
NiO	2.48	2.67	0.87	1.30	1.21	0.32
V_2O_5	1.58	2.76	0.87	1.11	1.02	-
Sc_2O_3	0.86	1.50	0.82	0.89	0.82	-
In_2O_3	1.62	2.84	1.04	1.88	1.60	-
Cr_2O_3	0.96	1.55	0.69	0.92	0.43	-
Tm_2O_3	2.53	4.43	2.90	2.48	2.42	-
La_2O_3	0.84	1.38	0.69	2.63	1.56	-
Na_2O	0.43	-	-	0.83	10.37	-
Li_2O	0.25	0.57	-	-	-	

silicate melt, it is possible to estimate the relative abundance of elements with the same valence in a mineral from the profile of partition coefficients in the PC-IR diagram. Hence, as long as we employ the abundance ratio of elements which have similar geochemical characters, the "structure control" concept described above can be useful for estimating the partition behavior of elements.

Partition coefficients of elements between high pressure minerals and coexisting silicate melt had already been studied by several authors [e.g., *Kato et al.*, 1988, *Yurimoto and Ohtani*, 1992]. However, these data have not been generally analyzed by PC-IR diagram. As a basic information for the partition relation of elements, we have reported the systematics of partition coefficient of elements in olivine at high pressures using PC-IR diagram [*Suzuki and Akaogi*, 1995]. In this paper, we intend to observe PC-IR diagrams for some high-pressure minerals, such as majorite or Ca-perovskite.

2. EXPERIMENTAL PROCEDURE

In order to observe the partition behavior of various types of high-pressure minerals, we prepared 5 types of starting materials. The composition of prepared starting materials are listed in Table 1. The main components of these starting materials are pyrolite [*Ringwood*, 1975](PL), San Carlos natural olivine (FO), fayalite (FA), and $CaSiO_3$ (WO1 and WO2). In order to observe the partition coefficient profile in the PC-IR diagram, 1–3 wt% of Sc_2O_3, V_2O_3, Cr_2O_3, In_2O_3, Tm_2O_3, La_2O_3, MnO, NiO, and Na_2O components were added to the main components. Materials-PL and FO contain a small amount of the Li_2O component because these starting materials were prepared to observe the partitioning profile of olivine in our previous experiments [*Suzuki and Akaogi*, 1995]. In these experiments, addition of Li increased the trivalent ion contents in olivine, but did not affect the profile of the partition coefficient. Therefore we employ these starting materials again in the present experiments.

In order to prepare the starting materials listed in Table 1, the following substances were mixed in the desired proportions: San Carlos olivine, synthetic fayalite, wollastonite and Li_2SiO_3, and reagents of Al_2O_3, NiO, MnO, V_2O_5, Cr_2O_3, Sc_2O_3, In_2O_3, Tm_2O_3, La_2O_3, and Na_4SiO_4. Fayalite, wollastonite, and Li_2SiO_3 were synthesized by the procedure described in our previous paper [*Suzuki and Akaogi*, 1995]. Other components were high purity reagents with guaranteed quality.

The present experiments were performed by using a 6–8 type multianvil high-pressure apparatus. Pressure values were calibrated at 1000 °C by using the olivine-spinel transition of Fe_2SiO_4 at 5.2 GPa [*Yagi et al.*, 1987], the coesite-stishovite transition at 8.7 GPa [*Zhang et al.*, 1996] and the olivine–β spinel transition of Mg_2SiO_4 at 14.0 GPa [*Akaogi et al.*, 1989]. The tungsten carbide anvils used in the present experiments were truncated in triangular faces with 5-mm edge length. The sample assemblies used in the present experiments are illustrated in Figure 1. The hot junction of a Pt-Pt/13%Rh thermocouple was in contact with the outer surface of the Re heating element. Since the thermocouple could not be used above approximately 1700 °C, temperature was estimated from the applied electric power by extrapolating the temperature-electric power relation which was obtained below 1700 °C. The

TABLE 2. Experimental Conditions

Run No.	Pressure (GPa)	Temperature (°C)	Time (min)
PL-1	17	2600	30
FO-1	17	2500	30
FA-1	10	1900	30
WO1-1	15	2400	30
WO2-1	15	2400	30

specimen was enclosed in a graphite capsule, and the oxygen fugacity was fixed as a carbon coexisting condition.

These starting materials were held at 10–17 GPa and 1900–2600 °C for 30 minutes, and quenched isobarically. Recovered specimens were polished to sections to observe their texture and chemical composition. The crystal structure of the mineral in the recovered specimen was determined by the microfocused X ray diffractometer of the Institute for the Solid State Physics, the University of Tokyo. The Cr KƒÀ line, which was generated from a rotating anode type X ray source with 50 kV and 300 mA, was collimated to ~30 μm and focused on a polished specimen. The diffracted X ray was detected by a position-sensitive photon counter.

Chemical analysis was carried out with an automatic 4-channel EPMA system at the Ocean Research Institute, the University of Tokyo. The accelerating voltage was 25 kV and the beam current on the Faraday Cup was 20 nA. The standard for Si and Ca was wollastonite. Albite and hematite were used as the standard material for Na and Fe, respectively. Standards for other elements were their pure oxides. Chemical composition was calculated by using the

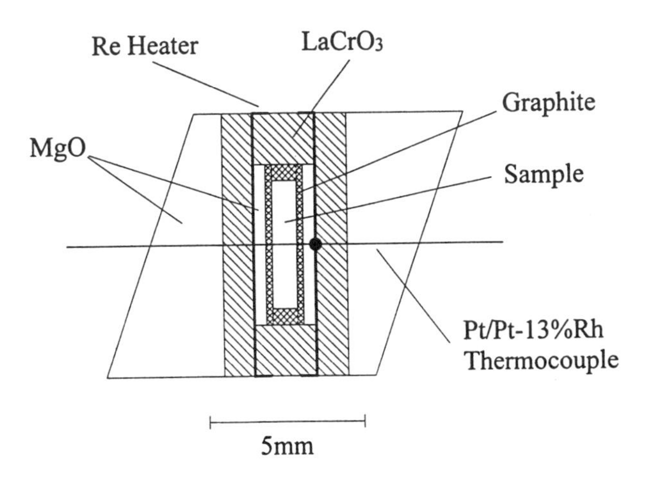

Figure 1. Sample assembly used in the present experiments. Semi-sintered magnesia was used as the pressure transmitting medium.

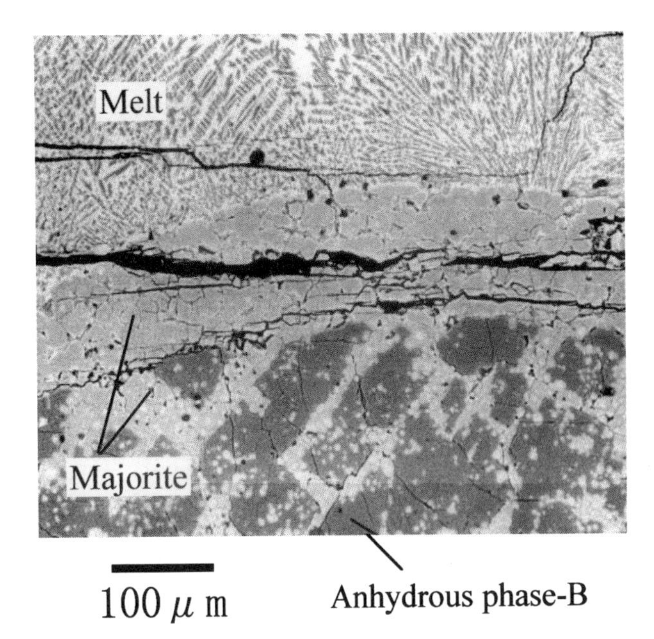

Figure 2. Back-scattered electron image of polished section of run PL-1. Liquidus phase was majorite, and anhydrous phase-B was found in the region where temperature is 20–30 degrees lower than liquidus.

ZAF correction method. Since a dendritic texture generally appeared in the quenched liquid, the composition of the liquid was obtained from the average of five analysis points of this region by using a defocused electron beam of EPMA with 20–30 μm diameter. The mineral which was in contact with this liquid was also analyzed 3 times using a focused electron beam to ~1 μm. In order to avoid a vaporization of Na from the sample, the electron beam was also expanded to ~5 μm in the measurement of mineral in the run product WO2 which contained ~10 wt% Na_2O. The averaged composition was compared with the result for the liquid region, and one set of partition data was thus acquired. Then 5–10 sets of partition data were obtained for each run product, and their averages were used as partition data of the sample.

3. RESULTS

Experimental conditions of the present experiments are listed in Table 2. A back-scattered electron image (BEI) of polished section of run product PL-1, whose main component was pyrolite, is shown in Figure 2. As shown in this figure, majorite was identified as the liquidus phase and partition relations were measured between majorite and silicate melt in PL-1.

In Figure 2, the center of the specimen is shown in the uppermost part of this figure. Since the temperature

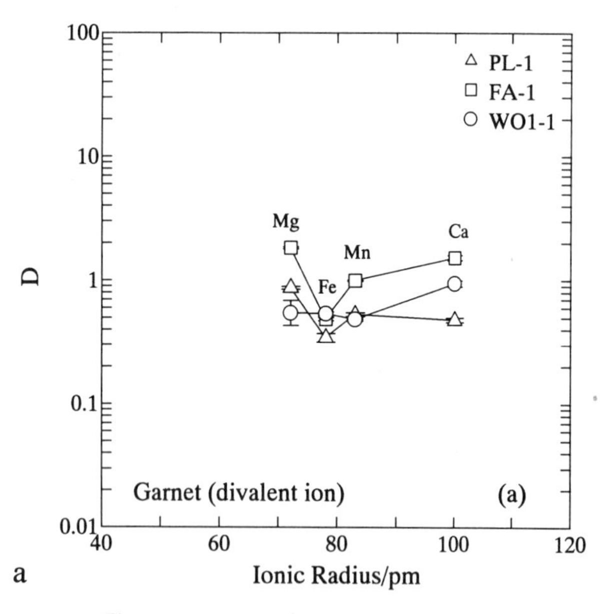

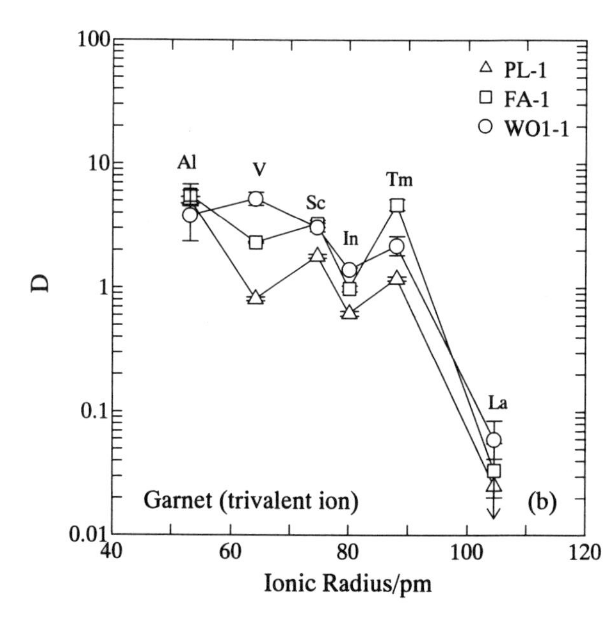

Figure 3. PC-IR diagram of garnet/melt system of divalent ions (a) and trivalent ions (b). Error bars show ±1σ standard deviation of the observed mean values. Since the La content of majorite of PL-1 was comparable to the detectable limit of EPMA, D_{La} of PL-1 exhibits the upper limit of the partition coefficient.

gradually decreased from the center to the end of the sample chamber, the upper part of Figure 2 was totally molten, whereas majorite existed as the liquidus phase in the lower part of this figure. The compositions of these majorite and coexisting silicate melts were measured at around the liquidus line by EPMA, and partition coefficients were calculated. In this paper, the partition coefficient, D, is defined as the mol ratio.

$$D = C_x^A \; / \; C_x^{melt} \qquad (1)$$

where C_x^A and C_x^{melt} represent the concentrations of an element X in mineral A and in silicate melt, respectively.

Although the ionic sizes vary with their coordination number, it is difficult to identify the occupying site in the structure and coordination number of each ion. Moreover, some elements may occupy a few kinds of sites of different coordinations. Therefore, we plotted the observed partition coefficients against ionic radius in 6-coordination number [*Shannon*, 1976], because the previous works on PC-IR diagrams have employed ionic radius in 6-coordination number in all elements [e.g., *Matsui et. al.*, 1977]. In this case, the position of the peak in the PC-IR diagram does not directly correspond to the size of the site in the crystal structure, but the relative size relation among ions are preserved, and the characteristics of the D-profile of the mineral are easily shown in the PC-IR diagram.

The acquired PC-IR diagrams of majorite/silicate melt are shown in Figure 3. Although the starting materials of the present study contain Ni and Cr, anomalous partitioning behaviors were found in these elements in our previous experiment [*Suzuki and Akaogi*, 1995]. Therefore, D-values of Cr and Ni are not shown in the PC-IR diagrams of the present experiments.

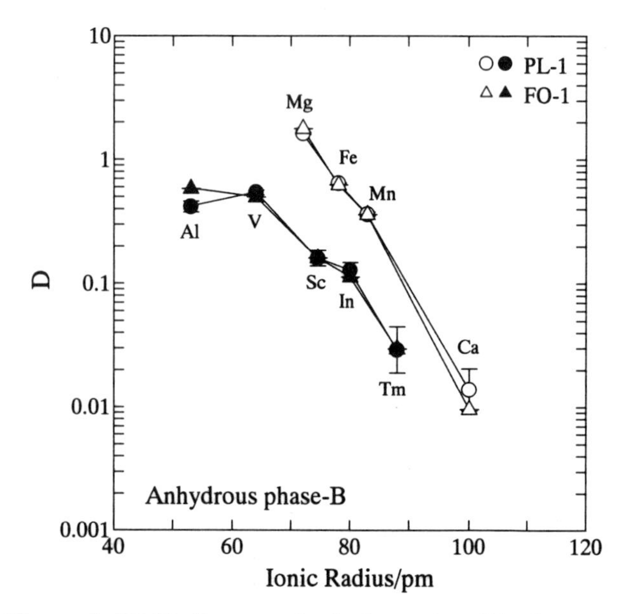

Figure 4. PC-IR diagram of anhydrous phase-B/melt system. Error bars show ±1σ standard deviation of the observed mean values. Open symbols represent the D values of divalent cations, and closed symbols represent trivalent cations.

The temperature gradient in the sample chamber of the present experiments was estimated from the temperature difference between the center and the end of the sample chamber. The difference was measured to be approximately 10% of the temperature at the center. Thus, estimated temperature gradient is approximately 10 degrees by 100 μm at around the liquidus line. As shown in Figure 2, anhydrous phase-B (anhy-B) was found where temperature is 10–20 degrees lower than the liquidus line. The observed melting relation is consistent with the previous high pressure melting experiments on dry peridotite [e.g., *Zhang and Herzberg*, 1994]. Although anhy-B grains contact with a very small amount of silicate melt which filled the interstitial regions of majorite grains, we cannot determine the chemical composition of such a small area. Hence we tentatively calculated D(anhy-B/melt) from the composition of the silicate melt at around liquidus, and the results shown in Figure 4.

The liquidus phase was magnesiowüstite in FO-1, and observed magnesiowüstite/silicate melt partition coefficients were plotted in the PC-IR diagram (Figure 5). Anhy-B was also found at a region where temperature is a few tens of degrees lower than the liquidus line. The observed melting relation of FO-1 is also consistent with the previous melting experiments of forsterite [e.g., *Presnall and Gasparik*, 1990]. The result of PL-1 and FO-1 suggests that the melting relations are not largely changed by the addition of "special" components of the present experiments, such as Sc_2O_3 or Tm_2O_3. The D(anhy-B/melt) was also calculated in the same way as the case of PL-1 and compared with that of PL-1 in Figure 4.

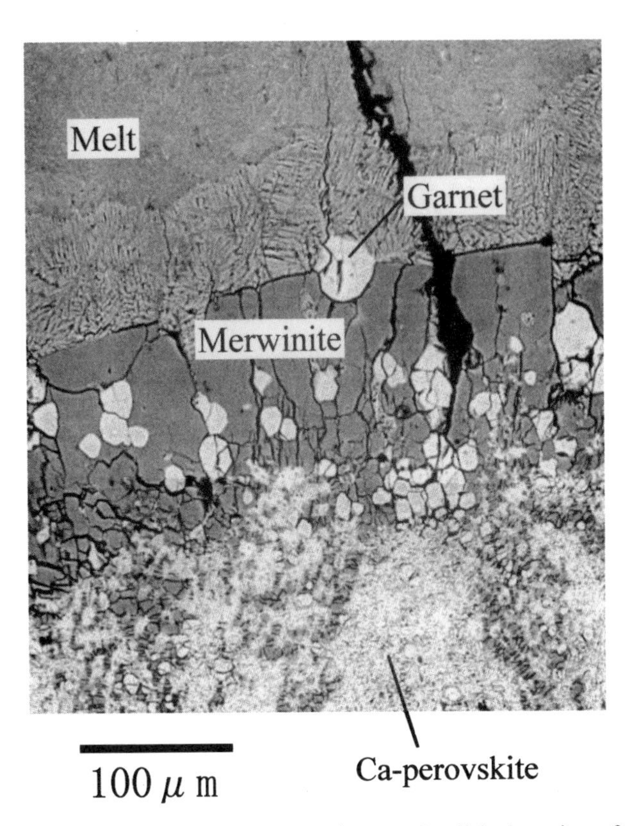

Figure 6. Back-scattered electron image of polished section of run WO1-1. Liquidus phase was merwinite, Ca-perovskite was found in the low temperature region of the sample.

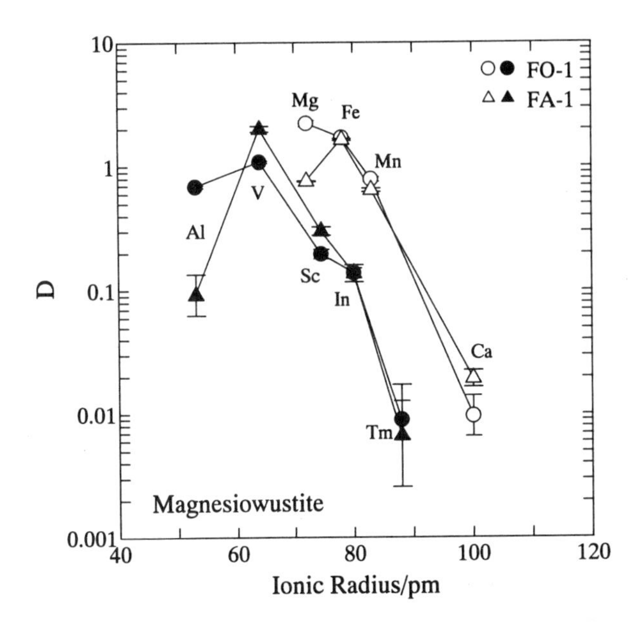

Figure 5. PC-IR diagram of magnesiowüstite/melt system. Error bars show ±1σ standard deviation of the observed mean values. Open symbols represent the D values of divalent cations, and closed symbols represent trivalent cations.

The liquidus phase of FA-1 was garnet, and the acquired PC-IR diagram of D(garnet/melt) is compared with D(majorite/melt) in Figure 3. Magnesiowüstite appeared in the region where temperature is a few tens of degrees lower than the liquidus line. The D(Mw/melt) was also estimated in the same way as D(anhy-B/melt), and results are shown in Figure 5.

The BEI of polished section of WO1-1 is shown in Figure 6. Merwinite ($Ca_3MgSi_2O_8$) was found as liquidus phase in WO1-1, and the observed D-values are shown in Figure 7. As shown in Figure 6, a small amount of garnet is also found at around liquidus. The D(garnet/melt) was also measured and compared with the D-profile of the other garnets (Figure 3).

As mentioned previously, the D-profile is dominantly determined by its crystal structure and will be not largely modified even if the composition of the system is changed. Hence, we tried to allow Ca-perovskite to be the liquidus phase by increasing Na_2O content in the starting material

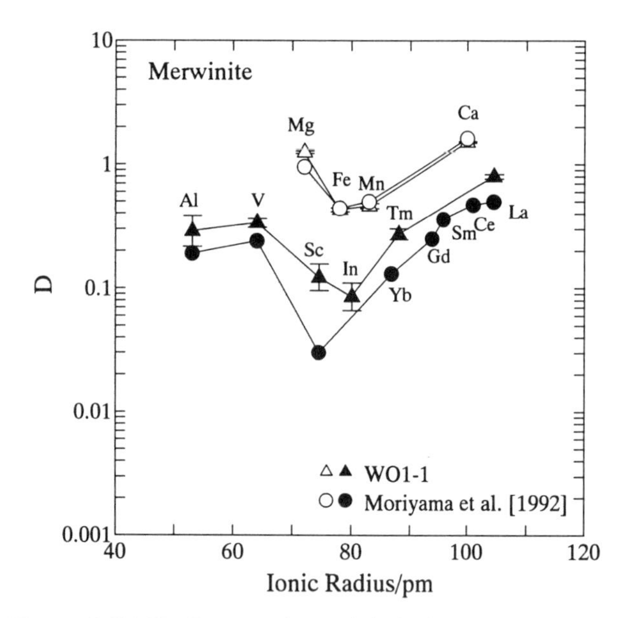

Figure 7. PC-IR diagram of merwinite/melt system. Error bars show $\pm 1\sigma$ standard deviation of the observed mean values. The results of *Moriyama et al.* [1992] are also shown in this figure. Open symbols represent the D values of divalent cations, and closed symbols represent trivalent cations.

(WO2-series). The recovered specimen of WO2-1 shows that Ca-perovskite existed at around the liquidus line (Figure 8), and acquired D(Ca-perovskite/melt) was shown in Figure 9.

4. DISCUSSION

4.1. *Effect of Temperature on Partitioning Behavior*

The effect of temperature on D is estimated by thermodynamic considerations [e.g., *Murthy*, 1992]. If we neglect non-ideality of the system, the equilibrium condition between crystal and silicate melt is expressed as

$$\mu_x^{L0} + RT\ln C_x^L = \mu_x^{S0} + RT\ln C_x^S \qquad (2)$$

where R is gas constant and T is temperature, and μ_x^{L0} and μ_x^{S0} represent standard chemical potentials of component x of the liquid phase and solid phase, respectively. Then the partition coefficient is expressed as

$$\ln D = (\mu_x^{L0} - \mu_x^{S0}) / RT. \qquad (3)$$

This equation predicts that every partition coefficient approaches 1 with an increase in temperature.

As mentioned previously, the temperature difference between the center and the end of the sample chamber was measured to be approximately 10% of the temperature at the center, and the estimated temperature variation of the region where D-values were obtained was 50 °C in maximum. The deviation in ln D values with a temperature variation of 50 degrees is estimated to be approximately 3%, which causes only a small error on the PC-IR diagram.

4.2. *D-Profile of Magnesiowüstite and Anhydrous Phase-B*

The crystal structure of magnesiowüstite is an NaCl-type structure, and cations occupy 6-coordination sites. This structure also has a 4-coordination site and is closely related to spinel structure. In the case of anhy-B ($Mg_{14}Si_5O_{24}$), Mg^{2+} occupies a 6-coordination site, and Si^{4+} occupies a 6- coordination site and a 4-coordination site [*Finger et. al.*, 1991]. As shown in Figures 4 and 5, D-profiles of these minerals show similar shapes in a PC-IR diagram. The D values of divalent ions of these minerals generally decrease with increase in ionic radius, and a peak was found at 60–70 pm in D-profiles of trivalent ions.

When we consider the MgO-FeO system as a simple binary system, D_{Mg}/D_{Fe}(Mw/melt) should be always larger than 1 and the MgO component is concentrated in magnesiowüstite, because the melting temperature of MgO is higher than that of FeO. Hence, it is very interesting that D_{Mg}/D_{Fe}(Mw/melt) >1 in FO-2, while D_{Mg}/D_{Fe}(Mw/melt) <1 in FA-1. Enrichment of the FeO component in magnesiowüstite was also found in a high-pressure melting experiment of carbonaceous chondrite [*Agee*, 1990].

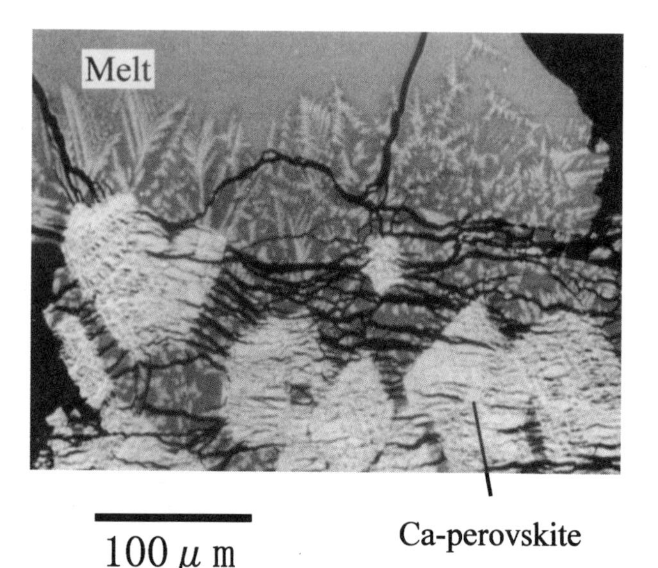

Figure 8. Back-scattered electron image of polished section of run WO2-1. Ca-perovskite was found as liquidus phase in this run product.

Although these observations may suggest the existence of a eutectic point in the MgO-FeO system, the starting material PL and FO or carbonaceous chondrite contain a large amount of the other components. Hence, the observed change of D_{Mg}/D_{Fe}(Mw/melt) may be ascribed to the difference in the composition of the system.

From the other point of view, although the D-profile of trivalent ions does not show any change in peak position, the observed change in D_{Mg}/D_{Fe}(Mw/melt) may suggest that the dimension of the site in the magnesiowüstite may change with its composition. It is not found in the other minerals that the peak position of the D-profile shifts with a change in the composition of the mineral. But the peak position of the silicate minerals may be fixed because these structures are mainly constructed by the Si-O framework; the dimension of the site of cations in silicate minerals may be determined by the structure of Si-O framework, and they are not allowed to change easily. But the cation site of the NaCl structure is constructed by packing of O^{2-} and may be simple enough to change the site dimension by the size of the occupying cation. More detailed experiments are required to resolve the partitioning relation of a magnesiowüstite/silicate melt system under high pressure.

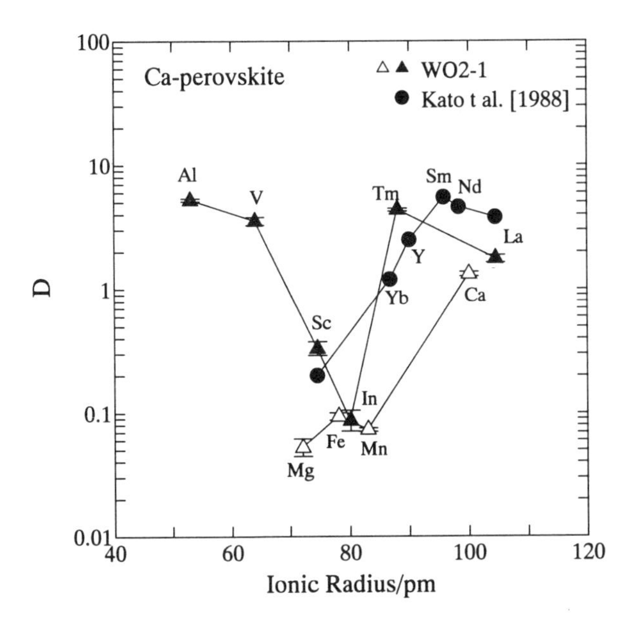

Figure 9. PC-IR diagram of Ca-perovskite/melt system. Error bars show $\pm 1\sigma$ standard deviation of the observed mean values. The result of *Kato et al.* [1988] are also shown in this figure. Open symbols represent the D values of divalent cations, and closed symbols represent trivalent cations.

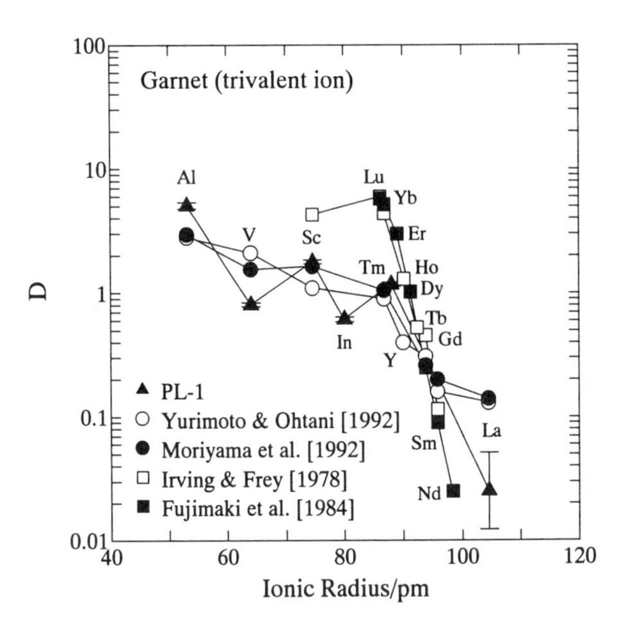

Figure 10. PC-IR diagram of trivalent ions of majorite/melt system. The results of the previous high-pressure experiments and the results of the pyrope/groundmass system are compared. Because the La content of majorite of PL-1 was comparable to the detectable limit of EPMA, the D_{La} of PL-1 exhibits the upper limit of the partition coefficient.

4.3. *D-Profile of Majorite*

Garnet structure has three different sites for cations: 4-, 6- and 8-coordination sites. Hence, it is expected that three peaks will appear in a PC-IR diagram. Unfortunately, the number of observed D values are too small to resolve such a complex D-profiie. The observed D(majorite/melt) of trivalent ions are compared with the previous studies in Figure 10. The observed D(pyrope/ groundmass) [*Irving and Frey*, 1978, *Fujimaki et al.*, 1984] are in agreement with the present results. Although small differences were found among the D-profiles of majorite, Fe-rich garnet and Ca-rich garnet, their D-profiles resemble each other and the fundamental characteristic of D-profile must be the same in the minerals with garnet structure (Figure 3). No obvious peak was found in the D-profiles of garnets, and D-values of most of the observed elements were 1–10, but La shows a significantly small D-value compared with other elements. Since the La content of majorite in PL-1 was comparable to the detection limit of La in EPMA analysis (Table 2), D_{La} of majorite shows an upper limit.

The observed D(majorite/melt) in the previous high-pressure melting experiments [*Yurimoto and Ohtani*, 1992, *Moriyama et al.*, 1992] shows similar values to those of the present experiments, but their D_{La}(majorite/melt) value is not consistent with the present result or with the pyrope/groundmass system. The observed D_{La}(majorite/melt) in the previous high-pressure experiments show significantly larger values than the present result. This difference may be derived from non-equilibrium partitioning of La in the previous studies, because the run duration times of the previous high pressure experiments were a few minutes, while those of the present experiments were 30 min. We selected 30 min for run duration time because our previous study [*Suzuki and Akaogi*, 1995] showed that the observed D values changed over 1 to 10 min, but further change in D values were not observed for run duration time longer than 10 min.

4.4. *D-Profile of Ca-Perovskite and Merwinite*

The observed D(merwinite/melt) is compared with the previous result [*Moriyama et al.*, 1992] in Figure 7. The present results are in agreement with the previous work. Merwinite structure has three distinct sites for Ca^{2+}; two of them have 8-coordination numbers, and one has a 9-coordination number [*Moore and Araki*, 1972]. The PC-IR diagram of merwinite suggests a broad peak at around 100 pm, which may represent these sites for Ca^{2+} in merwinite structure. The merwinite structure also suggests a peak at around 60 pm, which may correspond to the site for Si and Mg.

As shown in Figure 9, the observed D(Ca-perovskite/melt) values in the present study are in agreement with the results of *Kato et al.* [1988]. The PC-IR diagram of Ca-perovskite suggests two peaks in D-profile, and there is a bottom of D-value at around 70–80 pm. These peaks may be representing the two kinds of sites in the perovskite structure: a 12-coordinated site for Ca^{2+} and a 6-coordination site for Si^{4+}. When we take into account the results of *Kato et. al.*, [1988], the peak of D-value, which should be representing the dimension of the site of Ca^{2+}, may locate approximately at 100 pm.

This result suggests that large ions, whose ionic radii are 90–110 pm, may be easily incorporated into a Ca-perovskite structure. For example, if we compare the observed D_{La}, Ca- perovskite, and merwinite, they have D_{La}-values of 1–10, while those of other minerals are very small. Hence, if we consider the mineral abundance in the Earth's lower mantle, Ca-perovskite is the most probable candidate for the host mineral for large ions, such as rare earth elements.

Since the starting materials contain a considerable amount of rare earth elements, we can expect that two Ca^{2+} will be substituted by Na^+ and rare earth ions, such as La^{3+}.

$$2Ca^{2+} \rightleftharpoons Na^+ + La^{3+}. \quad (4)$$

Although La and Tm contents are large enough to incorporate considerable amount of Na^+ in Ca-perovskite structure, the amount of Na in the Ca-perovskite of WO2-1 was only 0.4 wt% (see Table 3) and the observed D_{Na} in Ca-perovskite is ~0.05. Therefore, rare earth elements are mainly incorporated by the substitution of (Ca^{2+} + Si^{4+}) by (rare earth ions + small trivalent ion), such as

$$Ca^{2+} + Si^{4+} \rightleftharpoons La^{3+} + Al^{3+}. \quad (5)$$

It may be possible that the observed small D_{Na} value in Ca-perovskite is caused by serious error in measurement of Na content in Ca-perovskite, because a considerable amount of Na in the specimen may be vaporized by the electron beam of EPMA. The observed total wt% of Ca-perovskite was 94–97%, but such a deficit of total amount has been always observed in the EPMA measurement of pure $CaSiO_3$ perovskite. It is generally believed that this deficit in total wt% is caused by the crack of the recovered "amorphous" phase which was converted from Ca-perovskite during release of pressure. Therefore, it is unlikely that this small D_{Na} was due to the Na loss in the EPMA measurement.

4.5. *Some Implications for the Geochemical Applications*

The origin and genesis of the rocks and magmas have been discussed from the abundance ratios of the elements which have similar geochemical characteristics. As mentioned previously, the shape of D-profile appeared in the PC-IR diagram is largely determined by the structure of the mineral. Therefore, relative abundance of the elements will be estimated from the PC-IR diagram. For example, the behavior of the rare earth elements (REE) in the partial melting process will be estimated from the observed PC-IR diagram.

Because the peaks of magnesiowüstite and anhydrous phase-B locate at around 70 pm, REE will concentrate more in the silicate melt than in the minerals. The concentration of the heavy rare earth elements (HREE) in the mineral will be slightly larger than the light rare earth elements (LREE), because ionic radius of HREE are smaller than that of LREE. The differentiation among REE will be clearly seen in the majorite fractionation. HREE will concentrate in majorite and LREE concentrate in silicate melt. In the case of Ca-perovskite or merwinite, the peak of the D-profile may locate in the middle of the ionic radius of the REE. Therefore, "middle" REE will concentrate in the mineral, and the "heavy" and "light" REE elements, such as La and Lu, may be concentrated in the silicate melt.

TABLE 3. Examples of Composition of Recovered Specimens

	PL-1			FO-1			FA-1			
	Garnet	Anhy-B	Liquid	Mw	Anhy-B	Liquid	Garnet	Mw	γ-Spinel	Liquid
SiO_2	47.32	32.30	39.87	0.22	31.84	38.29	34.03	0.78	30.20	29.29
MgO	29.31	55.04	33.60	67.78	53.75	32.58	8.79	3.49	7.96	4.74
CaO	1.35	0.03	2.80	0.03	0.03	2.75	2.12	0.02	0.02	1.35
Al_2O_3	9.03	0.74	1.73	0.99	0.74	1.27	4.89	0.13	0.08	0.71
Na_2O	0.03	0.05	0.29	-	-	-	-	-	-	-
FeO	2.80	5.21	8.30	10.93	4.06	6.46	25.92	82.89	58.33	54.28
MnO	0.47	0.33	0.89	1.41	0.63	1.57	1.03	0.65	0.42	1.03
NiO	0.23	2.32	1.03	9.94	3.47	1.90	0.19	1.44	1.09	0.43
V_2O_3	0.87	0.58	1.07	1.72	0.80	1.55	1.27	1.13	0.30	0.56
Sc_2O_3	1.19	0.11	0.66	0.27	0.23	1.21	2.31	0.22	0.07	0.66
In_2O_3	0.94	0.19	1.46	0.44	0.36	2.67	0.98	0.12	0.07	0.94
Cr_2O_3	2.13	1.17	0.81	4.41	1.93	1.28	3.63	1.43	0.33	0.44
La_2O_3	<0.01	<0.01	1.25	<0.01	<0.01	1.18	0.03	<0.01	<0.01	0.69
Tm_2O_3	2.82	0.04	2.42	0.10	0.18	5.02	14.65	0.07	<0.01	2.74
Total	98.50	98.15	96.19	98.35	98.12	98.21	99.86	92.42	98.90	97.91

TABLE 3. Continued

	WO1-1				WO2-1	
	Merwinite	Garnet	Ca-Perovskite	Liquid	Ca-perovskite	Liquid
SiO_2	36.58	38.49	47.63	39.93	40.51	47.86
MgO	11.73	5.86	0.23	9.31	0.09	1.22
CaO	45.01	27.70	42.77	31.19	38.46	28.57
Al_2O_3	0.34	4.08	0.61	0.70	2.83	0.51
Na_2O	0.77	<0.01	0.20	0.57	0.42	9.15
FeO	0.56	0.66	0.08	1.24	0.11	1.28
MnO	0.57	0.55	0.10	1.17	0.08	1.06
NiO	0.52	0.29	0.06	0.69	0.04	0.85
V_2O_3	0.21	2.86	0.24	0.59	1.76	0.51
Sc_2O_3	0.15	2.42	0.04	0.83	0.27	0.67
In_2O_3	0.27	2.93	0.05	2.08	0.12	1.14
Cr_2O_3	0.46	8.41	0.28	0.68	0.16	0.03
La_2O_3	1.94	0.20	2.01	2.38	2.28	1.15
Tm_2O_3	0.64	5.08	2.78	2.07	7.78	1.84
Total	99.74	99.53	97.07	93.42	94.89	95.83

5. CONCLUSIONS

D-profiles of magnesiowüstite and anhydrous phase-B show a peak at around 60–70 nm, which may correspond to the size of 6-coordination sites in their structure. The D-profile of majorite, Fe-rich garnet, and Ca-rich garnet resemble each other, and no obvious peak was found in the D-profiles of these garnets. However, D_{La}(garnet/melt) shows a significantly small D-value compared with other elements. Partition coefficients of large sized ions, such as Ca and La, were large in Ca-perovskite and merwinite. A peak of D-profile was expected at around 90–100 pm, and there is a minimum in the D-profile at around 70–80 pm in these minerals. Although the present experiments indicate that large trivalent and divalent ions are incorporated in Ca-perovskite structure, the observed D_{Na} is very small in Ca- perovskite.

Acknowledgments. We are grateful to T. Ishii for his helpful suggestions for chemical analysis. We are indebted to K. Kondo and T. Yagi for their help in measurements of the micro X-ray diffractometer. Critical readings by T. Katsura, C.B. Agee, and anonymous reviewer are acknowledged. This work was supported in part by funds from the Cooperative Program provided by the Ocean Research Institute, the University of Tokyo, and by Grants-in-Aid of the Ministry of Education, Science and Culture, Japan.

REFERENCES

Agee, C. B., A new look at differentiation of the Earth from melting experiments on the Allende meteorite, *Nature, 346,* 834-837, 1990.

Akaogi, M., E. Ito, and A. Navrotsky, Olivine-modified spinel-spinel transitions in the system Mg_2SiO_4-Fe_2SiO_4: calorimetric measurements, thermochemical calculation, and geophysical application, *J. Geophys. Res., 94,* 15,671-15,685, 1989.

Finger, L. W., R. M. Hazen, and C. T. Prewitt, Crystal structures of $Mg_{12}Si_4O_{19}(OH)_2$ (phase B) and $Mg_{14}Si_5O_{24}$ (phase AnhB), *Am. Mineral., 76,* 1-7, 1991.

Fujimaki, H., M. Tatsumoto, and K. Aoki, Partition coefficients of Hf, Zr, and REE between phenocrysts and groundmass, Proc. Lunar Planet. Sci. Conf. 14th, Part 2, *J. Geophys. Res., 89,* Suppl. B662- B672, 1984.

Irving, A. J., and F. A. Frey, Distribution of trace elements between garnet megacrysts and host volcanic liquids of kimberlitic to rhyolitic composition, *Geochim. Cosmochim. Acta*, 42, 771-787, 1978.

Kato, T., A. E. Ringwood, and T. Irifune, Experimental determination of element partitioning between silicate perovskites, garnets and liquids: constraints on early differentiation of the mantle, *Earth Planet. Sci. Lett.*, 89, 123-145, 1988.

Matsui, Y., N. Onuma, H. Nagasawa, H. Higuchi, and S. Banno, Crystal structure control in trace element partition between crystal and magma, *Bull. Soc. fr. Minéral. Cristallogr.*, 100, 315-324, 1977.

Moore, P. B., and T. Araki, Atomic arrangement of merwinite, $Ca_3Mg[SiO_4]_2$, an unusual dense-packed structure of geophysical interest, *Am. Mineral.*, 57, 1355-1374, 1972.

Moriyama, J., I. Kawabe, K. Fujino, and E. Ohtani, Experimental study of element partitioning between majorite, olivine, merwinite, diopside and silicate melts at 16 GPa and 2000 °C, *Geochem. J., 26,* 357-382, 1992.

Murthy, V. R., Geochemical evidence for an initially molten Earth, *Phys. Earth Planet. Inter., 71,* 46-51, 1992.

Onuma, N., H. Higuchi, H. Wakita, and H. Nagasawa, Trace element partition between two pyroxenes and the host lava, *Earth Planet. Sci. Lett., 5,* 47-51, 1968.

Presnall, D. C., and T. Gasparik, Melting of enstatite ($MgSiO_3$) from 10–16.5 GPa and the forsterite (Mg_2SiO_4)–majorite ($MgSiO_3$) eutectic at 16.5 GPa: Implications for the origin of the mantle, *J. Geophys. Res., 95,* 15,771-15,778, 1990.

Ringwood, A. E., Composition and Petrology of the Earth's mantle, McGraw-Hill, New York, 1975.

Shannon, R. D., Revised effective ionic radii and systematic studies of interatomic distances in halides and chalcogenides, *Acta Cryst., A32,* 751-767, 1976.

Suzuki, T., and M. Akaogi, Element partitioning between olivine and silicate melt under high pressure, *Phys. Chem. Minerals, 22,* 411- 418, 1995.

Yagi, T., M. Akaogi, O. Shimomura, T. Suzuki, and S. Akimoto, In situ observation of the olivine-spinel phase transformation in Fe_2SiO_4 using synchrotron radiation, *J. Geophys. Res., 92,* 6207-6213, 1987.

Yurimoto, H., and E. Ohtani, Element partitioning between majorite and liquid: a secondary ion mass spectrometric study, *Geophys. Res. Lett., 19,* 17-20, 1992.

Zhang, J., and C. Herzberg, Melting experiments on anhydrous peridotite KLB-1 from 5.0 to 22.5 GPa, *J. Geophys. Res., 99,* 17,729-17,742, 1994.

Zhang, J., B. Li, W. Utsumi, and R. C. Liebermann, In situ X ray observations of the coesite-stishovite phase transition: Revised phase boundary and kinetics, *Phys. Chem. Miner., 23,* 1-10, 1996.

M. Akaogi and T. Suzuki, Department of Chemistry, Gakushuin University, Mejiro, Tokyo 171, Japan

Thermodynamics of Iron Phases at High Pressures and Temperatures

S. K. Saxena and L. S. Dubrovinsky

Institute of Earth Sciences, Uppsala University, S-752 36 Uppsala.

The data on iron melting temperatures at high pressure in a laser-heated diamond-anvil cell, along with the previously existing static pressure data on phase equilibrium relations, have been used to obtain an internally consistent thermodynamic data base. With such a data base, it is not only possible to calculate phase equilibrium relations among the four well known phases (BCC, δ-BCC, FCC and HCP) but also to consider the effect of introducing additional possible phases yet to be verified. For the DHCP structure we have used a combination of recent X ray as well as laser-heating data to define an approximate field of stability. Similarly for explaining the disparity between the shock wave and static pressure data at pressures above 200 GPa, we have created a thermodynamically stable phase θ. This phase is stabilized because of high enthalpy and entropy; it melts at the core-mantle boundary at 6000 K or above which is much closer to the melting temperature from shock wave data than from the extrapolated static pressure data on melting of HCP (or DHCP).

INTRODUCTION

Iron is one of the most important elements in the solar system. It not only forms planetary cores but is also quite abundant in the mantles and the crusts of terrestrial planets. Earth has a large core and over 90% of it consists of iron. Therefore iron has always been a subject of high interest to geoscientists. This interest has been particularly intensified recently due to new developments in high pressure-temperature techniques. Iron is known to occur in four distinct polymorphs (BCC, δ-BCC, FCC and HCP). A thermodynamic data base is essential for phase equilibrium calculations involving iron and for predicting possible phase relations for designing new experiments. The thermodynamic data on all these four phases and melt were assessed by *Fernández Guillermot and Gustafson* [1985]. Since then we have amassed a significant amount of data both on the well characterized iron phases (α, γ, ε, and melt) as well as on a phase which appears to have a DHCP *structure* [*Saxena et al.*,1995]. Additionally in several publications, Anderson [e.g., *Anderson*, 1993, 1995] has emphasized the need of a high pressure phase θ to construct the iron phase diagrams at the core conditions. In this study, we will construct a thermodynamic data base by using all the available data (standard enthalpy, entropy and heat capacity at 1 bar) on the four phases and then determine the high pressure data on these phases from the high pressure experimental data which include measured melting temperatures, molar volumes, thermal expansions and compressibilities. We will also generate data on the two possible phases DHCP-β and θ for comparing the variations in Gibbs free energies as a function of pressure and temperature with those of the known phases. Such data demonstrate how a certain phase relation topology can be sustained without violating any thermodynamic laws.

Properties of Earth and Planetary Materials
at High Pressure and Temperature
Geophysical Monograph 101

THERMODYNAMICS

Magnetic Contribution

The magnetic contribution to the Gibbs free energy is described by the following expression:

$$\Delta G_m = RT \ln(\nu + 1) f(\tau) \quad (1)$$

where τ is defined as T/T_C. T_C is critical temperature for magnetic ordering, i.e., the Curie temperature for ferromagnetic ordering and the Neel temperature for antiferromagnetic ordering, and ν is a quantity related to the total magnetic entropy as

$$\Delta S_m(\iota) - \Delta S_m(0) = R \ln(\nu + 1) \quad (2)$$

where ι represents infinity. For $f(\tau)$, we use the expression from Hillert and Jarl [see *Sundman*, 1991]:

$$f(\tau) = 1-(1/A)\,(79\tau^{-1}/140+(474/497)\,((1/p)-1) \times (\tau^3/6+\tau^9/135+\tau^{15}/600)) \text{ for } \tau<1 \quad (3)$$

and

$$f(\tau) = -(1/A)\,(\tau^{-5}/10+\tau^{-15}/315+\tau^{-25}/1500) \text{ for } \tau>1 \quad (4)$$

where $A = (518/1125) + (11692/15975)\,((1/p) - 1)$ and p is a structural parameter.

Temperature and Pressure Dependence of the Gibbs Free Energy

The change in Gibbs free energy (ΔG) as a function of temperature (T) and pressure (P) can be obtained from these relations:

$$\left(\frac{\partial^2 \Delta G}{\partial T^2}\right)_P = -\frac{C_P(T)}{T} \quad (5)$$

and

$$\left(\frac{\partial \Delta G}{\partial P}\right)_T = V(P,T) \quad (6)$$

where $C_P(T)$ is heat capacity at constant P and at a temperature T and $V(P,T)$ is the molar volume at a temperature T and a pressure P. The free energy of a reaction is given by

$$\Delta G(P,T) = \Delta H_T^0 - T\Delta S_T^0 + \int_1^P \Delta V(P,T)\,dP \quad (7)$$

where ΔH^o_T and ΔS^o_T are the standard enthalpy and entropy of a reaction, respectively, at temperature T and 1 bar. They are given by

$$\Delta H_T^0 = \Delta H_{298}^0 + \int_{298}^{T} \Delta C_P dT \quad (8)$$

and

$$\Delta S_T^0 = \Delta S_{298}^0 + \int_{298}^{T} \frac{\Delta C_P}{T} dT \quad (9)$$

where ΔH^o_{298} and ΔS^o_{298} are the standard enthalpy and entropy of reaction at 298.15 K, ΔC_P is the heat capacity difference between products and reactants, and $\Delta V(P,T)$ is the volume change for the reaction.

The extrapolation of temperature and pressure dependence of the Gibbs free energy of a solid or a liquid phase to high temperature and pressure is complicated by the nature of the heat capacity, C_P, and the behavior of molar volume, $V(P,T)$. In order to obtain a suitable form of the Gibbs free energy for solids at high pressure and high temperature, it is necessary to discuss the heat capacity at high temperature and the equation of state in detail.

Heat Capacity at High Temperature

In general, C_P for a solid can be expressed as

$$C_P = C_V + \alpha^2 V K_T T + C_P' \quad (10)$$

where C_V is the heat capacity of a crystal at constant volume, α is the coefficient of thermal expansion, V is the molar volume, and K_T is the isothermal bulk modulus, all at temperature T, and C_P' is the contribution from cation disordering and anharmonicity (other than those incorporated in the $\alpha^2 V K_T T$ term).

The isobaric thermal expansion α and isothermal compressibility β_T, respectively, are given by

$$\alpha_P = \frac{1}{V_T}\left(\frac{\partial V}{\partial T}\right)_P \quad (11)$$

$$\beta_T = -\frac{1}{V_T}\left(\frac{\partial V}{\partial P}\right)_T \quad (12)$$

The inverse of β_T is the bulk modulus K_T. It is experimentally convenient to measure the heat capacity of solids at constant pressure (C_p) rather than at constant volume (C_v). Many different polynomial equations have been used for fitting the measured C_p data [*Saxena*, 1988 and *Saxena and Zhang*, 1989]. In this work, the following equations are used:

$$C_v = c_0 + c_1 T + c_2 T^{-1} + c_3 T^{-2} + \ldots \quad (13)$$

$$\alpha_p = a_0 + a_1 T + a_2 T^{-1} + a_3 T^{-2} + \quad (14$$

$$K_T = 1/(\beta_0 + \beta_1 T + \beta_2 T^2 + \beta_3 T^3 + \ldots) \quad (15)$$

where c_i are the coefficients of the heat capacity equation, a_i the coefficients in the expansion for isobaric thermal expansion, and β_i the coefficients of compressibility.

Molar Volume at High Pressure and High Temperature

The Birch-Murnaghan equation of state is given by

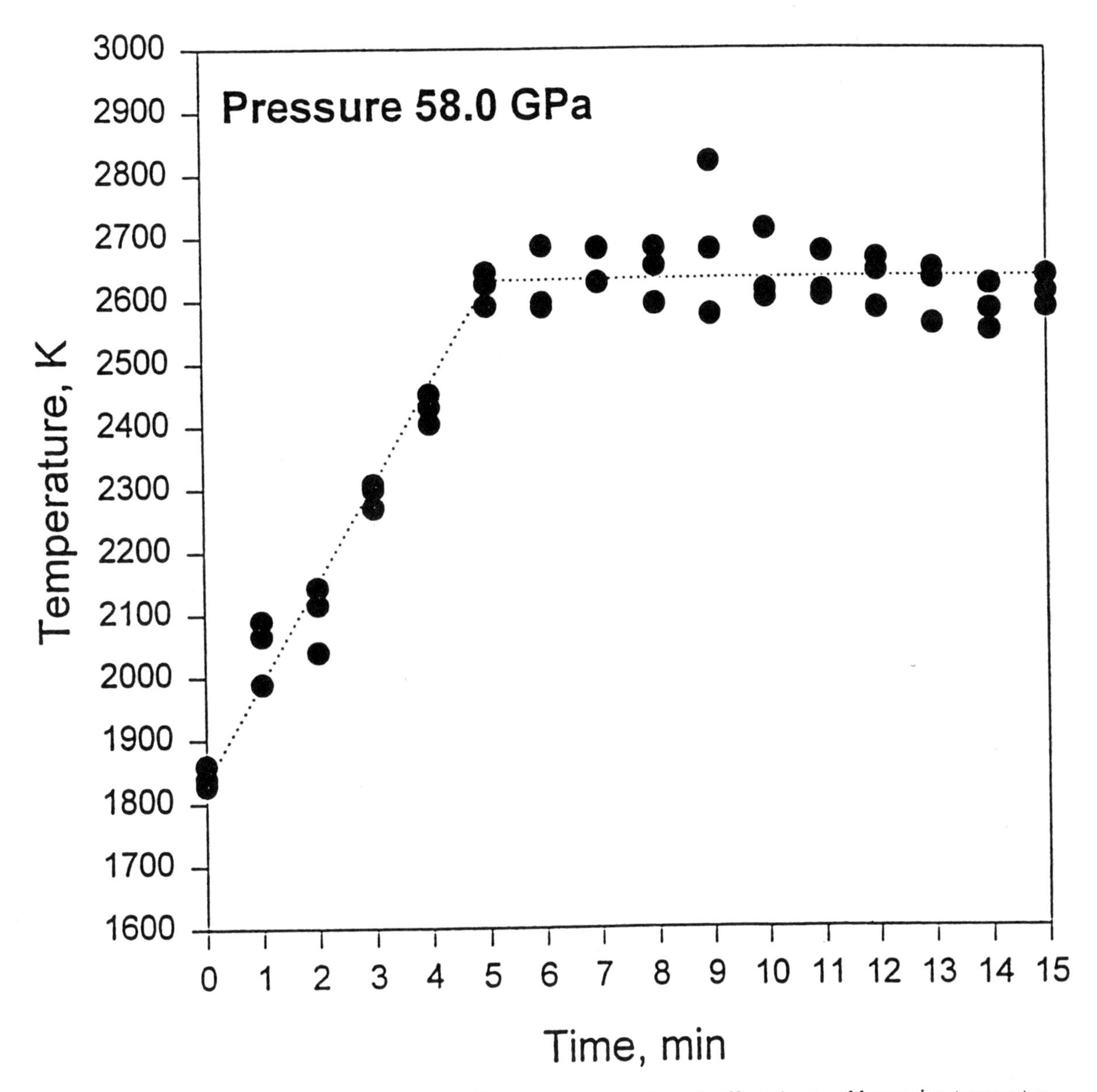

Figure 1. The recognition of melting by the oscillatory temperature (yo-yo) effect. A smoothly varying temperature-time (with a constant laser power) or a laser-power/temperature relation changes to an erratic pattern when the temperature oscillates around an average melting temperature. Iron at a pressure of 58 GPa.

$$P_{B\text{-}M} = \frac{3}{2}K_{T,0}\left[\left(\frac{V^0}{V}\right)^{7/3} - \left(\frac{V^0}{V}\right)^{5/3}\right]\left\{1 - \frac{3}{4}(4 - K'_{T,0})\left[\left(\frac{V^0}{V}\right)^{2/3} - 1\right] + \ldots\right\} \quad (16)$$

where $K_{T,o}$ and $K'_{P,o}(= [dK_T/d]_T)$ are the isothermal bulk modulus and its pressure derivative at 298 K, respectively.

With all the information on the temperature dependence of α and K available, one may proceed to apply the isothermal form of the BM equation [see *Saxena and Zhang*, 1989] at different temperatures as follows.

The VdP is calculated by adopting the third order Birch-Murnaghan equation of state, equation (16), where the temperature dependence of the isothermal bulk modulus is included and Vo/V is replaced by $V(1,T)/V(P,T)$. The temperature dependence of all variables, except of the pressure derivative K' is known from the data systematization at conditions of 1 bar and T as discussed in the previous section. By using the experimental data on in situ P-V-T determinations or with the help of the phase equilibrium experimental data we may determine the temperature dependence of the pressure derivative $K'p$ by expressing $(dKT/dP)T$ with an appropriate polynomial, e.g.

$$(\delta K_T/\delta P)_T = K'_{300} + K_4(T\text{-}300)\ \ln(T/300) \quad (17)$$

where K'_{300} is the pressure derivative in the Birch-Murnaghan equation and K_4 the temperature coefficient (not to be confused with K'' the second derivative of the bulk modulus). *Saxena et al.* [1993a] called this model the high-temperature Birch-Murnaghan (HTBM) model. For computational convenience PdV may be calculated from equation (5), instead of from VdP. The relation between PdV and VdP is given by

$$\int_1^P V dP = \int_{V(P,T)}^{V(1,T)} P dV + V(P-1) \tag{18}$$

where

$$\int_{V(P,T)}^{V(1,T)} P\,dV = \,)$$

$$\frac{3}{2}K_T V(1,T)\left[\frac{3}{4}(1+2x)(Y^{4/3}-1)-\frac{3}{2}(1+x)(Y^{2/3}-1)-\frac{1}{2}x(Y^2-1)\right] \tag{19}$$

where

$$x = \frac{3}{4}\left[4-\left(\frac{\partial K_T}{\partial P}\right)_T\right] \quad \text{and} \quad Y = \frac{V(1,T)}{V(P,T)}$$

Saxena et al. [1993a] have shown how internal consistency in the data set can be achieved and extrapolation of the data to core temperatures is possible.

DATA USED IN ASSESSMENT

Thermodynamic data

The thermochemical data on four iron phases BCC (δ-BCC), FCC, HCP, and melt at 1 bar were assessed by *Fernández Guillermot and Gustafson* [1985]. We used the physical property data [*Huang et al.*, 1987, *Manghnani et al.*, 1987, *Boehler et al.*, 1990] in addition to the data already used by *Fernández Guillermot and Gustafson* [1985].

Data on melting. Since there are many different sets of high pressure melting data, it is important to review our selection briefly. Before the use of X ray in determining melting of a solid [*Yoo et al.*, 1995], we used either visual observation and/or the method which depends on observing oscillation in temperature around an average temperature accepted as the melting temperature. This is illustrated in Figure 1 which shows the melting of iron. To study the reliability of our temperature measurements, we studied the melting of several substances using resistance wire heating. A small aperture in a platinum wire was filled with the sample and the wire was heated electrically with in situ X ray data collection. The temperature was determined from the equation of state for platinum. The entire electrical assembly was then transferred to the spectrometry site and this time the temperatures of melting were determined from the radiation thermometry using the Re calibration function. Figure 2 shows the result. It is obvious that we have achieved a high degree of precision in our method of temperature measurement.

We present new data on melting of iron at few different high pressures to test our present temperature calibration method and to compare the melting temperatures at 50 to 60 GPa with the X ray data of *Yoo et al.* [1995]. The new melting data are shown in Figure 3. At ~ 50 GPa, *Yoo et al.* [1995] found the melting at 2530 ± 50 K. At 58 GPa, we have a melting temperature of ~ 2600 ± 100 K (see Figure 3).

In addition to the data described above, we used all other available melting data at static pressures [*Liu and Bassett*, 1975, *Boehler*, 1986, *Boehler et al.*, 1990, *Boehler*, 1993, *Shen et al.*, 1993, *Saxena et al.*, 1993b, 1994] in our optimization.

The FCC-DHCP (or HCP) - Melt Triple-point. The triple-point was located by *Boehler* [1993] at ~ 100 GPa. This was based on the argument that there is the depression of the melting curve around 80–100 GPa [*Boehler et al.*, 1990]. *Boehler et al.* [1990] did state that this depression may be an

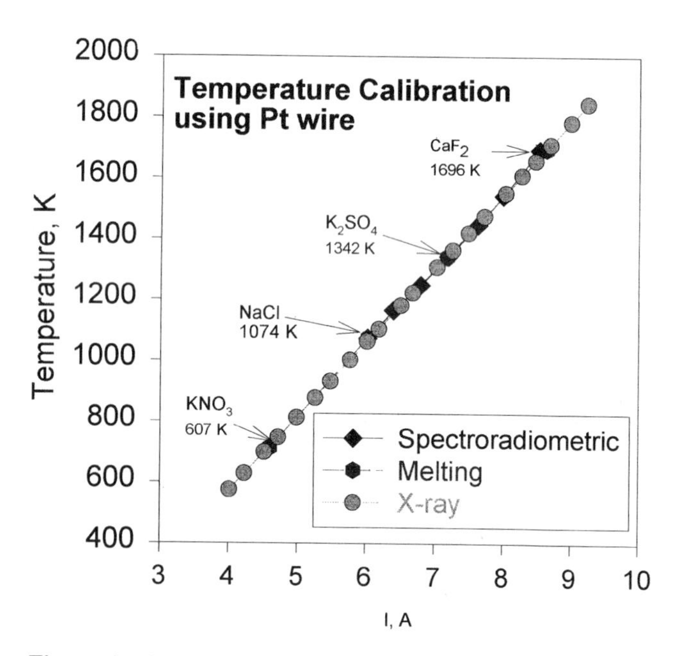

Figure 2. Comparison of the temperatures determined from spectroradiometry and hot resistance wire. Micrometer sized holes in a platinum wire are filled with differently melting compounds and the wire is heated electrically while in situ X ray data are collected for the sample and wire. The Pt equation of state is used to get the temperature of melting. The experiment is repeated with the same electrical set up and the temperature of the wire is determined by spectroradiometry

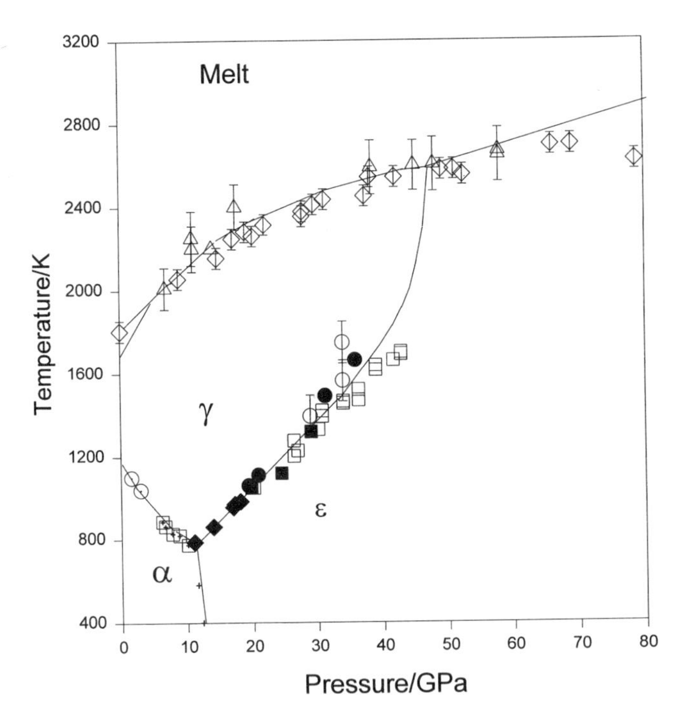

Figure 3. Iron phase diagram for the well-characterised phases, BCC (α and δ-BCC), FCC(γ), HCP(ε) and melt. New melting data from this study (triangles) are shown along with those of *Boehler* [1993](diamond). Several other data on melting are not plotted here, e.g., *Shen et al.* [1993]. The phase transition data are from *Boehler* [1986] (square), *Boehler et al.*[1987] (solid square), *Bundy* [1965] (plus in square, solid diamond, plus), *Mao et al.* [1987] (solid circle), *Mirwald and Kennedy* [1979] (dot in a circle), and this study (open circle).

artifact. Indeed in our melting experiments we were unable to confirm this depression.

Yoo et al. [1995] did not find the HCP-FCC phase transition all the way to melting at 50 GPa. On the basis of their arguments, it is possible to locate the triple-point between 40 and 60 GPa.

Phase Transformations. The topology of the phase relations among the four solid phases BCC (δ-BCC), FCC, and HCP to about 40 GPa as determined *by Fernández Guillermot and Gustafson* [1985] may be adopted here. Their work included the existing data up to 1984, and the additional data [*Boehler*, 1986, *Mao et al.*, 1987, *Boehler* at al., 1990, *Nasch and Manghnani*, 1994] continue to support the phase relations except for the triple point as discussed above. *Saxena et al.* [1995] found that at high pressures (>30 GPa) and at high temperatures (~1500 K), the high pressure HCP phase may change to a new DHCP structure. However, since the phase equilibrium field for this phase remains to be determined, we shall not include it in our optimization of the thermodynamic data. But we will separately determine the possible property of this and the θ phase for the purpose stated before.

ASSESSMENT OF DATA

The assessment procedure was discussed by *Saxena et al.* [1993a] and involves essentially the optimization of the available data of all types and estimating the missing data by fitting calculated equilibrium phase relations to experimental data. The optimization procedure is given in two thermodynamic program packages, the THERMOCALC [*Sundman et al.*, 1985] and ChemSage [*Eriksson and Hack*, 1991]. A complete thermodynamic data set on the iron phases is presented in Table 1. The data are presented in the form of Gibbs energy equations in temperature with a flexible polynomial format following the current practice in metallurgy. We note that the magnetic Gibbs free energy for the BCC and FCC phases is included but the electronic contributions are not. At high temperatures, one may use the electronic heat capacity as discussed by *Boness and Brown* [1990] to make the appropriate corrections to the heat capacity of HCP and Fe-melt. For the phase equilibrium calculations, we have checked that such corrections largely cancel out. An example of the match of the assessed pressure-volume-temperature data on HCP [*Huang et al.*, 1987] is shown in Figure 4.

The stability pressure-temperature field for β is not known with any certainty. The thermodynamic properties of the two phases HCP and DHCP with closely similar structure cannot be very different. Assuming that the combination of X ray and the laser-heating data [e.g., *Boehler*, 1993] define an approximate equilibrium curve (HCP-DHCP), we have

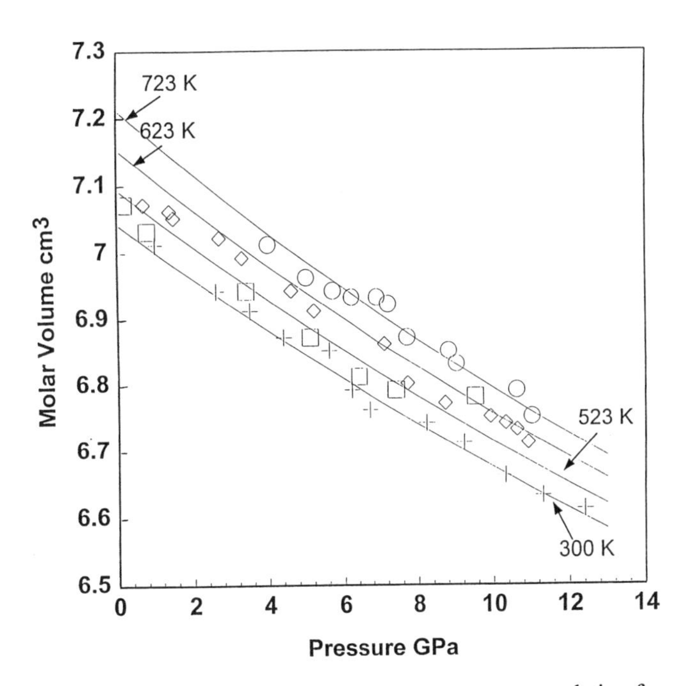

Figure 4. The modeled pressure-volume-temperature relation for the HCP phase using data of *Huang et al.* [1987]. Similar analysis is done using the data of *Manghnani et al.* [1987].

TABLE 1. Thermodynamic Data on Iron Phases (Gibbs Energy/J.mol^{-1} = $a + bT + cT\ln(T) + dT^2 + eT^3 + f/T$)

TMAX/K	a	bT	$cT\ln(T)$	dT^	eT^3	f/T
Fe-bcc						
1811.00	1.4000000E+03	1.2406000E+02	-2.3514300E+01	-4.3975200E-03	-5.8927000E-08	7.7359000E+04
	0.0000000E+00	EXTENSION TERMS:i*T^(0.0)				
6000.00	-2.5383581E+04	2.9931255E+02	-4.6000000E+01	0.0000000E+00	0.0000000E+00	0.0000000E+00
	2.2960300E+31	EXTENSION TERMS:i*T^(-9.0)				
Fe-fcc						
6000.00	1.6300921E+04	3.8147162E+02	-5.2275400E+01	1.7757750E-04	0.0000000E+00	-3.9535543E+05
	-2.4762800E+03	EXTENSION TERMS:i*T^(0.5)				
Fe-liq						
1811.00	-6.9564940E+03	-2.0136815E+02	1.2393000E+01	-1.0941500E-02	0.0000000E+00	6.5230000E+05
	3.0686000E+03	EXTENSION TERMS:i*T^(0.5)				
6000.00	-9.0073402E+03	2.9029866E+02	-4.6000000E+01	0.0000000E+00	0.0000000E+00	0.0000000E+00
	0.0000000E+00	EXTENSION TERMS:i*T*ln(T)				
Fe-hcp						
6000.00	1.6161040E+04	3.8286894E+02	-5.2275000E+01	1.7757735E-04	0.0000000E+00	-3.9535543E+05
	-2.4762800E+03	EXTENSION TERMS:i*T^(0.5)				
Fe-beta						
6000.00	1.8261040E+04	3.8106894E+02	-5.2275000E+01	1.7757735E-04	0.0000000E+00	-3.9535543E+05
	-2.4762800E+03	EXTENSION TERMS:i*T^(0.5)				
Fe-theta						
6000.00	2.9461040E+04	3.7906894E+02	-5.2275000E+01	1.7757735E-04	0.0000000E+00	-3.9535543E+05
	-2.4762800E+03	EXTENSION TERMS:i*T^(0.5)				

	Molar Volume		Thermal Expansion		
	cm^3	a	bT	f/T	g/T^2
Fe-bcc	7.0400	5.61240E-05	4.33730E-09	0.00000E+00	-4.11870E+00
Fe-fcc	6.8450	7.14780E-05	0.00000E+00	0.00000E+00	0.00000E+00
Fe-liq	7.3000	7.04840E-05	1.82470E-10	0.00000E+00	0.00000E+00
Fe-hcp	6.7300	7.04840E-05	1.82470E-10	0.00000E+00	0.00000E+00
Fe-beta	6.7500	7.04840E-05	1.82470E-10	0.00000E+00	0.00000E+00
Fe-theta	6.7300	7.04840E-05	1.82470E-10	0.00000E+00	0.00000E+00

	Compressibility (Mbar^{-1})			
	a	bT	dT^2	hT^3
Fe-bcc	4.90000E-01	1.40000E 04	2.50000E-08	0.00000E+00
Fe-fcc	5.00000E-01	4.00000E 04	3.50000E-08	0.00000E+00
Fe-liq	8.40670E-01	3.82500E-04	1.43970E-07	0.00000E+00
Fe-hcp	5.71400E-01	1.40000E-04	4.63774E-08	0.00000E+00
Fe-beta	5.71400E-01	1.40000E-04	4.63774E-08	0.00000E+00
Fe-theta	5.71400E-01	1.40000E-04	4.63774E-08	0.00000E+00

	K_0'	
	a	$cT\ln(T)$
Fe-bcc	5.40000E+00	0.00000E+00
Fe-fcc	8.60000E+00	0.00000E+00
Fe-liq	1.01000E+01	0.00000E+00
Fe-hcp	5.60000E+00	0.00000E+00
Fe-beta	5.61000E+00	0.00000E+00
Fe-theta	5.61000E+00	0.00000E+00

	Curie/Neel Temperature (K)	Magnetic Moment
Fe-bcc	1043.00	2.2200
Fe-fcc	-201.00	-2.1000

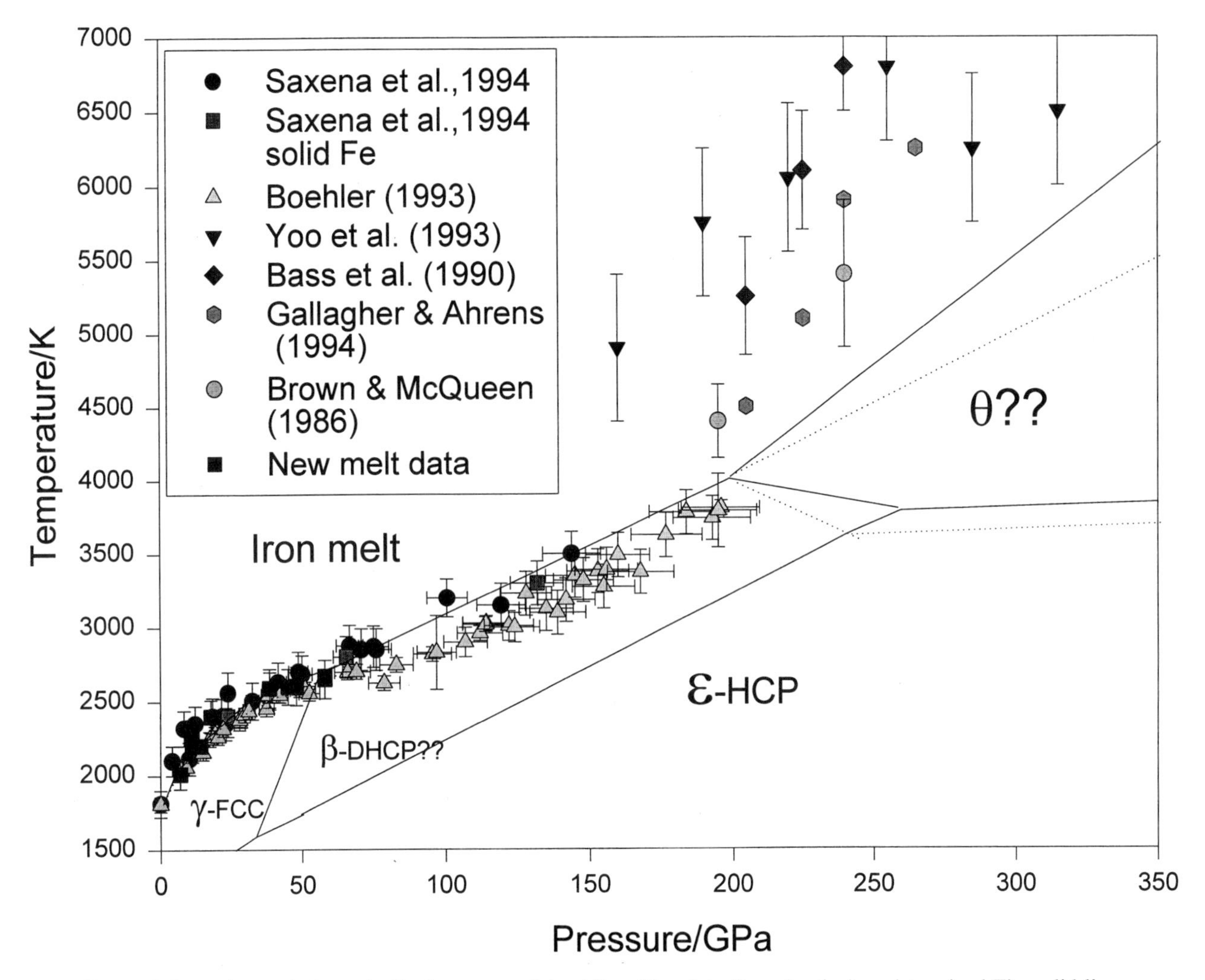

Figure 5. Iron phase relations. A. Static pressure data. All melting data (laser-heating) as determined The solid lines are calculated from the thermodynamic data listed in Table 1. Hugoniot data are shown for comparison with the static pressure melting data extrapolated with modeled thermodynamic data. The two data of *Brown and McQueen* [1986] signify a solid-solid and a solid-melt transition. We find that a high entropy-enthalpy θ phase can be stable with a triple-point θ-β-melt (or θ-ε-melt) between 190-200 GPa as suggested by *Anderson* [1993]. Such a phase could have a significantly high melting temperature; the higher the entropy the higher is the melting temperature. Two examples are shown; the dotted line is calculated with ΔH^{o}_{f} as 21 kJ/mol. and S^{o} as 38.2 and the solid line with ΔH^{o}_{f} as 33 kJ/mol. and S^{o} as 41.2.

estimated the thermodynamic data on the DHCP phase. Similarly for the purpose of exploring the possibility of a θ-iron phase, as postulated by *Anderson* (1993) based on *Boehler's* [1986] suggestion, we estimate the data on this phase. We accept *Anderson's* [1993] suggestion that a triple-point exists at about 190-200 GPa which permits the iron melting curve to change slope steeply and approach the from *Brown and McQueen* [1986] and *Gallaghar and Ahrens* data [1994] (see Figure 5). Here we have number of possibilities to consider how the θ-β-melt triple junction would appear. We have used the simplest possibility where the enthalpy of the θ-phase is an average of the enthalpies of the melt and the β-phase and the entropy is fixed by varying the slope of the β-θ equilibrium boundary to a position as shown in Figure 5 (dotted line). We emphasize that the derived thermodynamic data on the phases β and θ are for illustration only. The data on enthalpy, entropy and heat capacity for β do satisfy the topological aspects of the phase diagram as we know today (Figure 5). We have simplified the low pressure phase relations by assuming that the β phase [*Saxena et al.*, 1993b, 1995] is the same as the one described by *Boehler* [1993]. (Our own laser-power/temperature based transition data is closely similar to that of *Boehler* [1993)]in the high pressure range).

Figure 5 shows that the melting of the hypothetical phase θ may be modeled by assuming different combinations of enthalpy and entropy. Two such models are shown in Figure 5. For the average value of enthalpy (21 kJ/mol.) between melt and HCP, the calculated phase diagram is shown by the dotted curves. One may increase the melting temperature of θ by increasing the entropy and enthalpy as shown by the calculated solid curve (S°, 41.2 Gibbs/mol., $\Delta H°_f$, 33 kJ/mol.).

DISCUSSION AND CONCLUSIONS

We have presented a thermodynamic data base with which we may draw a phase diagram for iron up to the pressures and temperatures of the core. The topology of the relations among the well known phases are produced quite reliably. There is little doubt that there are iron polymorphs as yet unknown. While correctly reproducing the phase boundaries of the stable phases as we know to-day, this data base permits us to discuss how the phase relations would change if certain intermediate phases are found to be stable. A triple-point (FCC-HCP or DHCP-melt) could lie at lower pressures than that suggested by *Saxena* et al. [1994] or *Boehler* [1993]. In-situ X ray data on heated samples are needed to make a reliable determination of the triple-point.

The presence of θ-phase was postulated by *Anderson* [1993, 1995] to explain the large disparity between the temperatures of melting between shock wave data and the extrapolation from static pressure data. A formal thermodynamic treatment appears to suggest that a θ phase, with equilibrium boundaries against DHCP and HCP phases as shown in Figure 5, may melt at significantly higher temperatures than either DHCP or HCP phase. The θ-phase must be stabilized specially above 200 GPa by sudden change in a property, e.g., electronic entropy and heat capacity or a structural transformation different from the HCP poly types. If a theta phase could not be stable, the melting temperature of β or HCP phases at the core-mantle boundary may lie between 5000 to 5500 K; however if a theta phase could occur, it may be possible to consider iron melting at temperatures higher than 6000 K and still consider the melting data as consistent with both static pressure and shock wave experiments.

Acknowledgments. We thank O. Anderson for several inspiring discussions during his visit to Uppsala. The experiments at Brookhaven were possible with the active support of H. K. Mao and J. Hu. Financial support was provided by Swedish Natural Science Research Council and Wallenberg Foundation.

REFERENCES

Anderson, O. L., The phase diagram of iron and the temperature of the inner core, *J. Geomag. Geoelectr., 2*, 145-156, 1993.

Anderson, O. L., Mineral physics of iron and the core. *Rev. Geophys*. (U.S. National Report to International Union of Geodesy and Geophysics), 429-441, 1995.

Bass, J. D., T. J. Ahrens, J. R. Abelson, and T. Hua, Shock temperature measurements: New results for an iron alloy, *J. Geophys. Res. 95*, 21,757-21,777, 1990.

Boehler, R., The phase diagram of iron to 430 kbar, *Geophys. Res. Lett., 13*, 1153-1156, 1986.

Boehler, R., M. Nicol, and M. L. Johnson, Internally-heated diamond-anvil cell: Phase diagram and P-V-T of iron in *High-Pressure Research in Mineral Physics, Geophys. Monogr. Ser.*, vol. 39, edited by M. H. Manghnani and Y. Syono, pp. 173-176, TERRAPUB, Tokyo/AGU, Washington DC, 1987.

Boehler, R., Temperatures in the Earth's core from melting-point measurements of iron at high static pressures, *Nature, 363*, 534-536, 1993.

Boehler, R., N. von Bargen, and A. Chopelas, Melting, thermal expansion, and phase transition of iron at high pressures, *J. Geophys. Res., 95*, 21,731-21,736, 1990.

Boness, D. A., and J. M. Brown, The electronic band structures of iron, sulfur, and oxygen at high pressures and the Earth's core, *J. Geophys. Res., 95*, 21,721-21,730, 1990.

Brown, J. M., and R. G. McQueen, Phase transitions, Grüneisen parameters and elasticity for shocked iron between 77 GPa, *J. Geophys. Res., 91*, 7485-7494, 1986.

Bundy, F. P., Pressure-temperature phase diagram of iron to 200 kbar, 900° C, *J. Appl. Phys., 36*, 616-620, 1965.

Eriksson, G., and K. Hack, ChemSage. Handbook for ChemSage. GTT mbH, Aachen, Germany, 1991.

Fabrichnaya, O., Thermodynamic data for phases in the FeO-MgO-SiO_2 system and phase relations in the mantle transition zone, *Phys. Chem. Mineral. 22*, 323-332, 1995.

Fernández Guillermot, A., and P. Gustafson, An assessment of the thermodynamic properties and the (p,T) phase diagram of iron, *High Temp.-High Press., 16*, 591, 1985.

Gallagher, K. G. and T. J. Ahrens, First measurements of thermal conductivity of griceite and corundrum at ultra high pressures and the melting point of iron (abstract), *EOS Trans. AGU, 75*, 44, Fall Supplement, 653, 1994.

Huang, E., W. A. Bassett, and P. Tao, Pressure-temperature-volume relationship for hexagonal close packed iron determined by synchrotron radiation, *J. Geophys. Res., 92*, 8129-8135, 1987.

Liu, L., and W. A. Bassett, The melting of iron to 200 kbar, *J. Geophys. Res., 80*, 3777-3782, 1975.

Manghnani, M.H., L. C. Ming, and N. Nakagiri, Investigation of the α-Fe⇔ε-Fe phase transition by synchrotron radiation, in *High-Pressure Research in Mineral Physics, Geophys. Monogr. Ser.*, vol. 39, edited by M. H. Manghnani and Y. Syono, pp. 155-163, TERRAPUB, Tokyo/AGU, Washington DC, 1987.

Mirwald, P. W., and G. C. Kennedy, The Curie point and the α–γ transition of iron to 53 kbar - a reexamination. *J. Geophys. Res., 84*,656-658, 1979.

Mao, H. K, P. M. Bell, and C. Hadidiacos, Experimental phase relations of iron to 360 kbar, 1400°C, determined in an internally heated diamond anvil apparatus, in *High-Pressure*

Research in Mineral Physics, *Geophys .Monogr. Ser.*, vol. 39, edited by M. H. Manghnani and Y. Syono, pp. 135-145, TERRAPUB, Tokyo/AGU, Washington DC, 1987.

Ming, L. C., and W. A. Bassett, Laser heating in the diamond anvil press up to 2000° C sustained and 3000° C pulsed at pressures up to 260 kilobars, *Rev. Sci. Instrum.*, *9*, 1115, 1974.

Nasch, Ph. M,. and M. H. Manghanani, A modified ultrasonic interferometer for sound velocity measurements in molten metals and alloys, *Re. Sci. Instrum.*, *65*, 682-688, 1994.

Saxena, S. K., Assessment of thermal expansion, bulk modulus, and heat capacity of enstatite and forsterite, *J. Phys. Chem. Solid, 49,* 1233-1235, 1988.

Saxena, S. K., and J. Zhang, Assessed high-temperature thermochemical data on some solids, *J. Phys. Chem. Solid, 50,* 723-727, 1989.

Saxena, S. K., and J. Zhang, Thermochemical and pressure-volume-temperature systematics of data on solids, examples: tungsten and MgO, *Phys. Chem. Min.*, *17*, 45-51, 1990.

Saxena, S. K., N. Chatterjee, Y. Fei, and G. Shen, *Thermodynamic data on oxides and silicates,* 428 pp., Springer-Verlag, Heidelberg, 1993a.

Saxena, S. K., G. Shen, and P. Lazor, Experimental evidence for a new iron phase and implications for Earth's core, *Science*, *260*, 1312-1314, 1993b.

Saxena, S. K., G. Shen, and P. Lazor, Temperatures in Earth's core based on melting and phase transformation experiments on iron, *Science*, *264*, 405-407, 1994.

Saxena, S. K., L. S. Dubrovinsky, P. Häggqvist, Y. Cerenius, G. Shen, and H. K. Mao, Synchrotron X-ray study of iron at high pressure and temperature, *Science*, *269*, 1703-1704, 1995.

Shen, G., P. Lazor, and S. K. Saxena, Melting of wüstite and iron up to pressures of 600 kbar, *Phys. Chem. Minerals*, *20,* 91-96, 1993.

Sundman, B., An assessment of the Fe-O system. *J. Phase Equil. 12*, 127-140, 1991.

Sundman, B., B. Jansson, and J. O. Andersson, The Thermo-Calc Data bank system, *Calphad 9*, 153-190, 1985.

Yoo, C. S.,N. C. Holmes, and M. Ross, Shock temperatures and melting of iron at earth core conditions. *Phys. Rev. Lett. 70*, 3931-3934, 1993.

Yoo, C. S., J. Akella, A. J. Campbell, H. K. Mao, and R. J. Hemley, Phase diagram of iron by in situ X-ray diffraction: Implications for Earth's core, *Science*, *270*, 1473-1475, 1995.

Isothermal Compression Curve of Al_2SiO_5 Kyanite

T. Yagi, S. Inutsuka[1], and T. Kondo

Institute for Solid State Physics, University of Tokyo, Tokyo, Japan

Isothermal compression curve of Al_2SiO_5 kyanite was determined through a high-pressure in situ powder X ray diffraction study up to 10 GPa. Combination of synchrotron radiation and an imaging plate detector was adopted and the result was analyzed using conventional least squares calculation for the triclinic unit cell. Isothermal bulk modulus, K_{T0}, was determined to be 183±12 GPa when its pressure derivative, K'_{T0}, was assumed to be any value between 2 and 6. This bulk modulus is slightly larger than those of other low pressure polymorphs of Al_2SiO_5 and much smaller than the mixture of corundum plus stishovite.

1. INTRODUCTION

There are several polymorphs of Al_2SiO_5, such as andalusite, silimanite, and kyanite, and among them kyanite is a polymorph stable at relatively low temperature and high pressure [*Newton*, 1969; *Robie et al.*, 1984]. Kyanite is widely found in metamorphic rocks and is an important constituent of eclogites. Because of its importance for the petrology of the crust and upper mantle, many studies have been made to determine its stability under the condition of the Earth's interior. *Ringwood and Reid* [1969] reported that kyanite is stable up to 15 GPa and 1000°C. *Liu* [1974] observed that kyanite decomposes into a mixture of corundum plus stishovite at about 16 GPa using a laser-heated diamond anvil cell. Recently, *Irifune et al.* [1995] made a detailed study using multianvil apparatus and reported that the decomposition boundary can be expressed by an equation P=11.9+0.0008T (GPa). Furthermore, the existence of a new high pressure polymorph with V_3O_5-type structure is proposed above 40 GPa [*Ahmed-Zaid and Madon*, 1991].

[1]Now at Komatsu Soft Co.

Properties of Earth and Planetary Materials
at High Pressure and Temperature
Geophysical Monograph 101

In order to establish these phase relations and to argue kyanite's properties in the Earth's interior, it is important to know its elastic property such as bulk modulus. However, the low symmetry (triclinic) of its crystal structure made it difficult to determine the bulk modulus either by conventional ultrasonic method or by high pressure in situ X ray diffraction study. In the ultrasonic method, it is necessary to determine 21 independent elastic constants for triclinic symmetry. Kyanite has strong cleavage which makes it difficult to get single crystals large enough for the complete ultrasonic measurement. Resolution of a high-pressure in situ X ray diffraction measurement using the conventional powder X ray diffraction technique is not high enough to make meaningful analysis for such low symmetry materials. Accordingly, there have been no measurements of the bulk modulus of kyanite and it is estimated on the basis of various indirect methods. The purpose of the present study is to determine its bulk modulus using a high-resolution, powder X ray diffraction technique under hydrostatic condition.

2. EXPERIMENT

Natural kyanite from the Northern Territory, Australia, was kindly provided by T. Irifune who used it for their phase transformation study [*Irifune et al.*, 1995]. Chemical composition was analyzed by them using an

TABLE 1. Observed and Calculated *d*-Values of Kyanite at 10 GPa

h	k	l	d(obs) Å	d(cal) Å*	(do/dc)-1
0	1	1	4.2244	4.2267	-0.0006
-1	1	1	4.2244	4.2225	0.0004
0	2	0	3.7016	3.6977	0.0010
1	-1	1	3.7016	3.6981	0.0009
2	0	0	3.2927	3.2920	0.0002
0	2	-1	3.1305	3.1274	0.0010
1	1	1	3.1305	3.1311	-0.0002
0	2	1	2.9697	2.9658	0.0013
1	-2	1	2.9697	2.9678	0.0006
2	-2	0	2.8978	2.8977	0.0000
1	2	0	2.8978	2.8945	0.0012
0	1	-2	2.5641	2.5660	-0.0007
1	-1	-2	2.5641	2.5645	-0.0001
2	-1	-2	2.3135	2.3129	0.0003
2	-3	0	2.3135	2.3103	0.0014
0	2	-2	2.2315	2.2305	0.0005
-1	2	2	2.2315	2.2302	0.0006
1	1	2	2.1254	2.1285	-0.0015
2	1	-2	2.1254	2.1233	0.0010
1	-4	0	1.9289	1.9270	0.0010
3	-3	0	1.9289	1.9318	-0.0015
1	3	1	1.8975	1.8995	-0.0011
2	0	2	1.8975	1.8997	-0.0011
3	0	-2	1.8975	1.8943	0.0017
-3	3	1	1.8975	1.8992	-0.0009
0	2	-3	1.6462	1.6452	0.0006
1	-1	3	1.6462	1.6467	-0.0003
1	-4	2	1.5644	1.5646	-0.0002
-1	4	2	1.5644	1.5641	0.0002
0	5	0	1.4758	1.4791	-0.0022
-1	3	3	1.4758	1.4777	-0.0013
-1	5	1	1.4758	1.4744	0.0010
2	-1	3	1.4758	1.4735	0.0015
2	-5	1	1.4504	1.4481	0.0016
3	3	0	1.4504	1.4488	0.0011
1	3	-3	1.4504	1.4495	0.0006
1	-3	3	1.4504	1.4514	-0.0007
1	0	-4	1.3675	1.3674	0.0001
5	-2	0	1.3675	1.3684	-0.0006
4	2	0	1.3675	1.3673	0.0001
3	-5	1	1.3526	1.3537	-0.0008
-4	4	2	1.3526	1.3514	0.0009
4	0	-3	1.3526	1.3533	-0.0005
2	3	-3	1.3526	1.3536	-0.0008
1	4	2	1.3526	1.3522	0.0003

*calculated unit cell parameters:
$a = 6.9935(39)$, $b = 7.7081(34)$, $c = 5.4696(22)$
$\alpha = 90.01(4)$, $\beta = 101.12(4)$, $\gamma = 106.06(4)$
$V = 277.58(22)$

electron microprobe analyzer and was proved to be essentially Al_2SiO_5 with a very limited amount of Fe_2O_3 (~ 0.5 wt%). The kyanite specimen was finely ground in an agate mortar and was placed in a 300-μm hole of the stainless steel gasket. In order to maintain the hydrostatic condition, the sample chamber was filled with a 4 to 1 mixture of methanol and ethanol. Small chips of ruby were also placed as a pressure marker. The gasket was compressed between two diamond anvils with 600-μm culet using a modified Mao-Bell type diamond anvil cell [*Yagi and Akimoto*, 1982].

All the X ray experiments were made using synchrotron radiation at beam line 6-B of the Photon Factory, Tsukuba. In order to get high resolution of the diffraction, an angle dispersive technique was employed using an imaging plate (IP) detector. Incident X ray beam was monochromatized to a wave length of 0.6888Å and was directed on the sample through a collimator with a hole of 150 μm diameter. The distance between sample and the detector was about 150 mm, which was calibrated using diffraction of a silver standard placed on the table face of the anvil. Typical exposure time was 2 hours, when the synchrotron ring was operated at 2.5 GeV and around 250 mA. Pressure was measured before and after each exposure using the pressure scale by *Mao et al.* [1986]. There was no meaningful difference of pressure both within the sample chamber and before and after the exposure, and the averages of these values was used as pressure values. Two independent runs were made, and in both cases the pressure was first increased to the highest value and then the X ray measurements were made in the course of releasing pressure. In this way, we can avoid possible errors caused by the systematic change of the sample to detector distance, since most of the deformation of gasket occurs in the pressure increasing process, before the X ray measurements.

The two-dimensional record on the imaging plate was integrated along a polar axis which coincides with the Debye rings. The detail of this data processing was identical with the previous study [*Yagi et al.*, 1992]. Two different methods of analysis were adopted to calculate lattice parameters of the triclinic unit cell of kyanite. One is a whole powder pattern decomposition (WPPD) method [*Toraya*, 1986] which makes profile fits the whole powder pattern, using lattice parameters of the unit cell together with profile parameters as fitting parameters. The other is a conventional least-squares calculation of the lattice parameters using observed *d*-values of the diffractions. In the latter case, because of the low symmetry of the crystal and the limited resolution of the powder X ray diffraction, many diffraction lines are overlapped and it was difficult

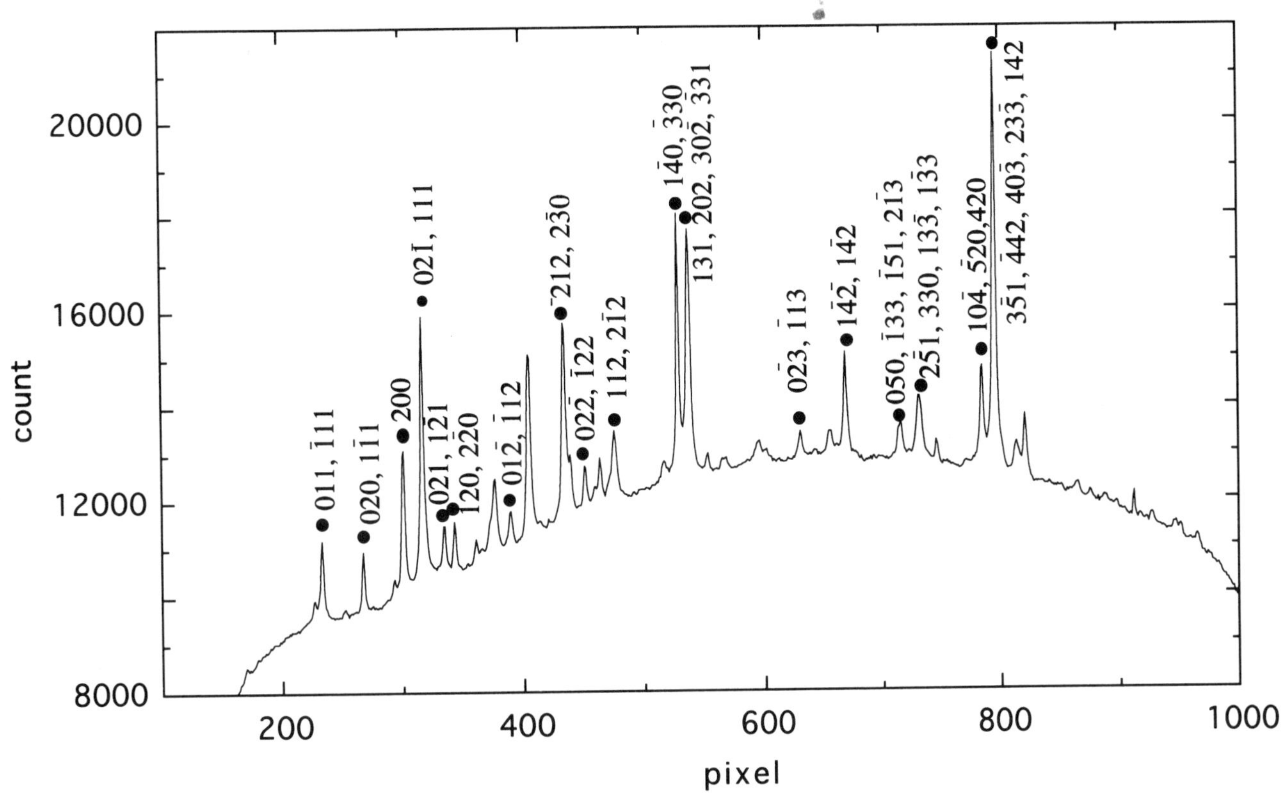

Figure 1. An example of X ray diffraction pattern of kyanite observed at 10.2 GPa. Vertical axis is an integrated intensity along the Debye ring and horizontal axis is a distance from the center of the Debye ring. One pixel is 100 μm on the imaging plate detector. Solid circles and numbers indicate the diffraction line and indexes used for the unit cell calculation.

to index all diffraction peaks unequivocally. In principle, the WPPD method is superior to the conventional least-squares calculation. In the case of kyanite, however, there are so many fitting parameters, as will be discussed later, that it was difficult to converge the calculation using this method. Thus, all the results in the present study were obtained by the conventional least squares calculation.

3. RESULTS AND DISCUSSION

An example of the diffraction profiles is shown in Figure 1. The lines used for the analysis are marked by circles with the indices given for each diffraction lines. All the data were analyzed using same set of diffraction lines and indices. An example of the unit cell calculation is shown in Table 1. Almost all the diffraction lines have more than one index and, in total, 45 different indices, which were selected based on the intensity of the JCPDS data, were given to 18 well-resolved diffraction lines. Each observed line is well explained by the overlap of several different diffraction lines. Unit cell parameters obtained at the ambient pressure, the average of two measurements after each run, are listed in Table 2, together with the unit cell parameters reported in the previous

TABLE 2. Unit Cell Parameters of Kyanite at Ambient Condition*

Author	*a* (Å)	*b* (Å)	*c* (Å)	α	β	β	V (Å^3)
Burnham [1963]	7.119(1)	7.847(1)	5.572(1)	89.98(1)	101.12(1)	106.01(1)	293.16
Skinner et al. [1961]	7.123(2)	7.844(2)	5.568(2)	89.54(5)	101.13(5)	105.59(5)	292.93
Winter and Ghose [1979]	7.126(1)	7.852(1)	5.572(1)	89.99(2)	101.11(2)	106.03(1)	293.60(9)
This study*	7.113(4)	7.838(4)	5.564(2)	89.97(3)	101.12(5)	106.05(4)	292.01(2)

* average of run 1 and run 2

TABLE 3. Summary of the Experimental Results

Pressure (GPa) run	0.001 #1	0.001 #2	3.53(2) #1	4.92(3) #1	5.08(2) #2	7.07(2) #1	8.50(2) #2	10.23(2) #1
a (Å)	7.114(4)	7.112(3)	7.068(4)	7.056(6)	7.057(4)	7.030(6)	7.014(4)	6.994(4)
b (Å)	7.837(4)	7.838(3)	7.784(4)	7.776(5)	7.773(4)	7.736(5)	7.730(3)	7.708(3)
c (Å)	5.564(2)	5.563(2)	5.526(2)	5.522(4)	5.517(2)	5.493(3)	5.485(2)	5.470(2)
α	89.98(4)	89.95(3)	90.01(4)	89.97(6)	90.00(4)	90.01(5)	90.01(4)	90.01(4)
β	101.10(5)	101.14(4)	101.12(5)	101.25(7)	101.13(5)	101.13(7)	101.12(4)	101.12(4)
γ	106.05(4)	106.05(3)	106.06(4)	105.94(6)	106.06(4)	106.04(6)	106.05(4)	106.06(4)
V (Å^3)	292.1(2)	292.0(2)	286.2(2)	285.3(3)	284.9(2)	281.2(3)	280.0(2)	277.6(2)

studies. The unit cell of kyanite determined in the present study is in agreement with those reported before. This result also indicates that the present method of analysis is reasonable for calculating the lattice parameters of a triclinic unit cell, although the wave length of X ray used for analysis is much shorter than that used in the previous studies. All the unit cells observed at various pressures are summarized in Table 3. Based on these results, relative compressions of the unit cell dimensions, a, b, c, and volume are calculated and summarized in Figures 2 and 3. It is clear from Figure 2 that the compressibilities of the three axes a, b, and c are very similar to each other and no meaningful differences were observed among them. Moreover, the angles between these axes, α, β, and γ, remain unchanged in the present pressure range, as shown in Table 3. In other words, this triclinic unit cell of kyanite was compressed isotropically.

The volume compression data was fitted by the Birch-Murnaghan equations of state. When two fitting parameters, bulk modulus K_{T0} and its pressure derivative K'_{T0}, are used, the least squares calculation gives the following values:

$K_{T0} = 189 \pm 14$ GPa
$K'_{T0} = 2.4 \pm 3.4$

However, the present pressure range is not large enough, compared to the large bulk modulus of the sample, to determine these two parameters simultaneously. In dense silicates, usually K'_{T0} possesses a value close to 4. Moreover, all the previous studies to estimate the bulk modulus of kyanite were made assuming $K'_{T0}=4$. Thus, the bulk moduli obtained by fixing $K'_{T0}=4\pm2$ is shown in Figure 4, together with the residuals of the least squares calculation. Variation of the residual is very small in this range and any combination of K_{T0} and K'_{T0} can explain the observed compression curve equally well. When K'_{T0} is fixed as 4, we obtain $K_{T0}=183\pm4$ GPa. The K_{T0} value varies from 191 GPa to 175 GPa when K'_{T0} is changed from 2 to 6. Considering these variations, we conclude that the bulk modulus of kyanite is 183±12GPa.

So far, there is no direct measurement of the compressibility of kyanite. *Brace et al.* [1969] estimated the isothermal bulk modulus of kyanite to be 140 GPa, based on the compressibility measurement of a mixture of kyanite and other material of known properties. This estimate, however, is sensitive to the assumption of the stress and strain across the grain boundary of the compressed samples. Although there is a preliminary trial to directly measure the stress and strain state across the grain boundary in NaCl [*Funamori et al.*, 1994], there is no information about kyanite, because it is very difficult to determine these quantities by experiment. Thus, a very large uncertainty is involved in the estimate of the bulk modulus by *Brace et al.* [1969]. *Irifune et al.* [1995] calculated the bulk modulus of kyanite based on its disproportionation boundary into corundum plus stishovite and the thermodynamic argument. The bulk modulus they obtained is 202±15 GPa, which is in reasonable agreement

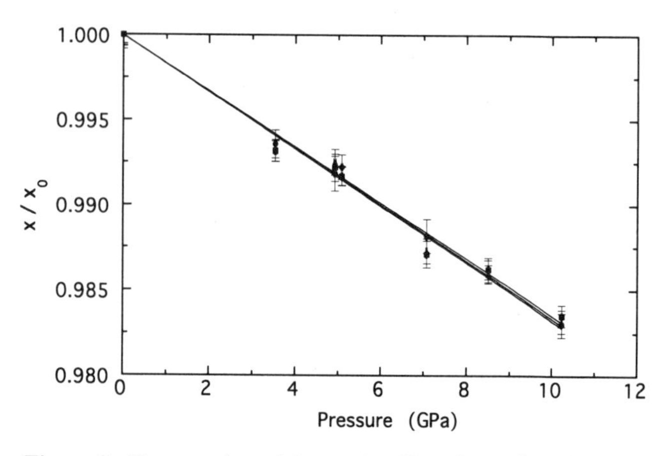

Figure 2. Compression of three axis of kyanite, a, b, and c, with pressure. Circles, squares, and triangles represent, respectively, a/a_0, b/b_0, and c/c_0. Three solid lines are the linear compressibilities of each axis fitted to each compression datum.

with the present observation, if we consider the uncertainties of both analyses. *Matsui* [1996] made a molecular dynamics calculation and obtained the bulk modulus of 197 GPa. Moreover, he predicted that the linear compressibility of each axis is similar (M. Matsui; private communication) which is in good agreement with the present observation.

The present result is quite reasonable compared to the bulk moduli of other polymorphs of Al_2SiO_5. The bulk moduli of andalusite and silimanite are, respectively, reported to be 159 GPa and 169 GPa [*Vaughan and Weidner*, 1978] or 158 GPa and 166 GPa [*Ralph et al.*, 1984]. The high-pressure polymorph has better packing of ions and has higher density compared to the low-pressure phases. As a result, in most cases high-pressure polymorphs have higher bulk moduli compared to those of their low-pressure phases. The bulk modulus of kyanite is clearly higher than these low-pressure phases and much smaller than that of the mixture of corundum plus stishovite (K~270 GPa) to which kyanite breaks down above about 13 GPa [*Irifune et al.*, 1995].

Present results demonstrated that it is possible to determine bulk modulus of materials with low symmetry, such as triclinic, using high-pressure, powder X ray diffraction techniques, when the combination of synchrotron radiation and imaging plate detector is adopted. For the analysis of the observed diffraction data, it is desirable to adopt the whole powder pattern decomposition method, using profile analysis of the whole diffraction. This method of analysis was quite successful in the case of $MgSiO_3$ tetragonal garnet which is only slightly distorted from cubic symmetry [*Yagi et al.*, 1992]. Using an identical experimental technique, however, it was unsuccessful in the present study. Because of the low symmetry of kyanite, almost all the observed diffraction lines were overlapped by several different lines and it was difficult to determine the profile parameters of a single diffraction line. In such cases, there are so many parameters in the calculation of the deconvolution of each profile that it is generally difficult to converge calculations. Although the conventional least-squares fitting of the lattice parameter is, in principle, not accurate in such calculations, it gives quite a reasonable result if the number of observed diffraction lines is large enough. For further accurate studies, single crystal analysis under pressure will be required.

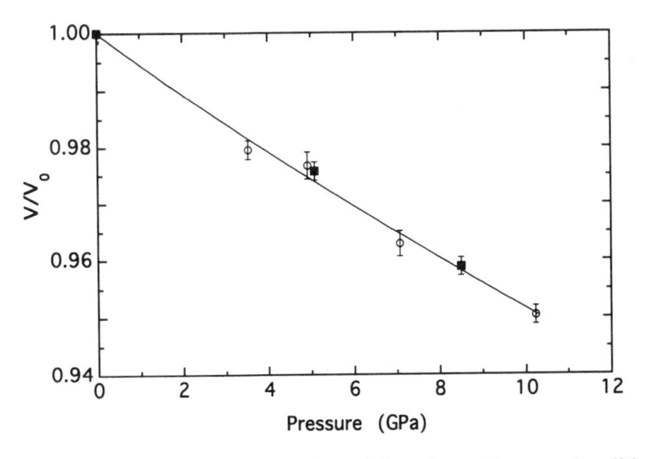

Figure 3. Volume compression of kyanite. Open and solid symbols represents results of run 1 and run 2, respectively. Solid line represents the Birch-Murnaghan equation of state fitted to these data (K_{T0}=189 GPa and K'_{T0} =2.4).

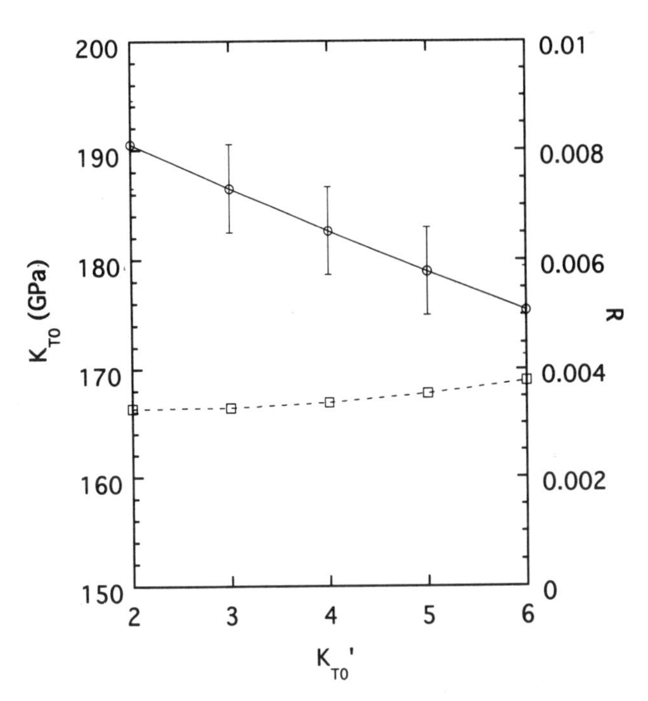

Figure 4. Solid line is a variation of the isothermal bulk modulus, K_{T0}, calculated by fixing K'_{T0} from 2 to 6. Dotted line represent the residuals of the least squares calculation.

Acknowledgments: The authors are grateful to T. Irifune for providing a natural kyanite sample used in the present study and for useful comments to the manuscript. Thanks are also due to M. Yamakata and Y. Fujihisa for their help in X ray diffraction study at the Photon Factory and to H. Sawa for the discussion of the WPPD analysis. This research was performed at the Photon Factory, KEK (Proposal number G95-318).

REFERENCES

Ahmed-Zaid, I., and M. Madon, A high-pressure form of Al_2SiO_5 as a possible host of aluminum in the lower mantle, *Nature, 353*, 426-428, 1991.

Brace, W. F., C. H. Scholz, and P. N. LaMori, Isothermal compressibility of kyanite, andalusite, and sillimanite from synthetic aggregates, *J. Geophys. Res., 74*, 2089-2098, 1969.

Burnham, C. W., Refinement of the crystal structure of kyanite, *Z. Kristallogr. 118*, 337-360, 1963.

Funamori, N., T. Yagi, and T. Uchida, Deviatoric stress measurement under uniaxial compression by a powder x-ray diffraction method, *J. Appl. Phys., 75*, 4327-4331, 1994.

Irifune, T., K. Kuroda, T. Minagawa, and M. Unemoto, Experimental study of the decomposition of kyanite at high pressure and high temperature, in *The Earth's Central Part: Its Structure and Dynamics*, edited by T. Yukutake, pp. 35-44, Terrapub, Tokyo, 1995.

Liu, L. G., Disproportionation of kyanite to corundum plus stishovite at high pressure and temperature, *Earth Planet. Sci. Lett., 24*, 224-228, 1974.

Mao, H. K., J. Xu, and P. M. Bell, Calibration of the ruby pressure gauge to 800 kbar under quasi-hydrostatic conditions, *J. Geophys. Res., 91*, 4673-4676, 1986.

Matsui, M., Molecular dynamics study of the structures and bulk moduli of crystals in the system CaO-MgO- Al_2O_3-SiO_2, *Phys. Chem. Minerals, 23*, 345-353, 1996.

Newton, R. C., Some high-pressure hydrothermal experiments on severely ground kyanite and sillimanite, *Am. J. Sci., 267*, 278-294, 1969.

Ralph, R. L., L. W. Finger, R. M. Hazen, and S. Ghose, Compressibility and crystal structure of andalusite at high pressure, *Am.. Mineral., 69*, 513 - 519, 1984.

Ringwood, A. E., and A. F. Reid, High-pressure phase transformation of spinel (1), *Earth Planet. Sci. Lett., 5*, 245-250, 1969.

Robie, R. A., and B. S. Hemingway, Entropies of kyanite, andalusite, and sillimanite : additional constraints on the pressure and temperature of the Al_2SiO_5 triple point, *Am. Mineral., 69*, 298-306, 1984.

Skinner, B. J., S. P. Clark Jr., and D. E. Appleman, Molar volumes and thermal expansion of andalusite, kyanite and sillimanite, *Am. J. Sci., 259*, 651-668, 1961.

Toraya, H., Whole-powder-pattern fitting without reference to a structural model: application to X-ray powder diffractometer data, *J. Appl. Crystallogr., 19*, 440-447, 1986.

Vaughan, M. T., and D. J. Weidner, The relationship of elasticity and crystal structure in andalusite and silimanite, *Phys. Chem. Miner.*, 3, 133-144, 1978.

Winter, J. K., and S. Ghose, Thermal expansion and high-temperature crystal chemistry of the Al_2SiO_5 polymorphs, *Am. Mineral., 64*, 573-586, 1979.

Yagi, T., and S. Akimoto, Rapid X-ray measurements to 100 GPa range and static compression of α-Fe_2O_3, pp. 81-90, Tokyo/Dordrecht: Center for Academic Publication Japan (CAPJ)/Reidel, 1982.

Yagi, T., Y. Uchiyama, M. Akaogi, and E. Ito, Isothermal compression curve of $MgSiO_3$ tetragonal garnet, *Phys. Earth Planet. Inter., 74*, 1-7, 1992.

S. Inutsuka, T. Kondo, and T. Yagi, Institute for Solid State Physics, University of Tokyo, Roppongi, Minato-ku, Tokyo 106, Japan.

Melting of Rare Gas Solids Ar, Kr, Xe at High Pressures and Fixed Points in the $P-T$ Plane

Andrew P. Jephcoat and Stanislav P. Besedin[1]

Department of Earth Sciences, University of Oxford, Parks Road, Oxford OX1 3PR, UK

We present melting temperatures determined at several pressures for solid argon, krypton, and xenon up to 47, 18, and 12 GPa respectively. At these pressures the observed melting temperatures rise to 2790±150 K for argon, 2175±150 K for krypton, and 2054±150 K for xenon. These data suggest that rare gas solids (RGS) melt at high temperatures relative to other classes of materials with increasing pressure. The argon melting temperatures are consistent with both empirical and theoretical predictions. Within experimental error, the melting curves for Ar and Kr are consistent with the Simon-Glatzel (SG) expression for the melting curve obtained at lower pressures. For xenon, the data indicate that melting occurs at lower temperatures than predicted by an SG relation based on melting data to 0.65 GPa, but above the melting curve expected from a corresponding-states scaling of the internally consistent melting curves of both Ne and Ar. In addition to the intersection of the argon and iron (Ar-Fe) melting curves in the range 42-47 GPa and 2750±150 K, the experiments suggest an intersection of the Kr-Fe melting curves near 23 GPa and 2300±150 K and of the Xe-Pt curves in the range 12-15 GPa and 2300±150 K.

1. INTRODUCTION

Observations of melting at high pressures in the laser-heated diamond-anvil cell (DAC) are perhaps one of the most difficult static high-pressure measurements to make because of constraints on the sample size and the proximity (within a few microns) to diamond. Apart from the problem with physical measurement of temperature, with proper account of gradients and the area sampled, the identification of the melting transition has to be made simultaneously. In spite of these difficulties, many determinations for melting of oxides and metals of interest to geophysics now exist, representing a major advance in knowledge of high-pressure, -temperature properties.

In this paper we examine the melting behavior of the rare gas solids at pressures in the 50 GPa range. These materials serve as model simple molecular solids and we compare the melting temperatures to those measured for other classes of materials. The pressure dependence of a melting curve may fit one of several types: It can continue to rise without limit, reach a maximum temperature where $\Delta V=0$, change slope at phase transitions in the liquid or solid, or change character from first to second order [*Bilgram*, 1987]. The description of the dynamics of melting and crystallization in addition pose a fundamental theoretical challenge and are

[1]Permanent address: Institute of Crystallography, Russian Academy of Sciences, Leninski pr.59, 117333 Moscow, Russia

Properties of Earth and Planetary Materials
at High Pressure and Temperature
Geophysical Monograph 101

TABLE 1. Summary of Experimental Data on the Melting of RGS

RG	A	B or P_0	c	T_0	Scale Factor	Function	Data Range[a]	Source
He	0.016067	0.0	1.565		10^{-4}	SG	0-24.0	1
	17.83518	31.86	1.54171		10^{-4}	SG	0-18.0	2
Ne	0.012062	1.478	1.4587		10^{-1}	SG	0-5.45	3
	15.70774	587.7	1.41852	11.685	10^{-4}	MSG	0-1.0	4
Ar	2.67348	2293.252	1.52299		10^{-4}	SG	0-1.1	5
	0.22636	0.0	1.5	83.2	1	SG2	0.23-1.78	6
Kr	0.304	0.0	1.4	116.1	1	SG2	0-1.2	6
	3.36253	1778.71	1.44084	38.096	10^{-4}	MSG	0-0.8	4
Xe	0.3445	0.0	1.31	161.5	1	SG2	0-0.7	6
	5.7106	1295.59	1.29515	95.445	10^{-4}	MSG	0-0.15	7

Sources: 1, *Vos et al.* [1990]; 2, *Loubeyre et al.* [1982]; 3, *Vos et al.* [1991]; 4, *Crawford and Daniels* [1971]; 5, *Hardy et al.* [1971]; 6, *Lahr and Eversole* [1962]; 7, *Michels and Prins* [1962]. The Kr and Xe fits show the largest divergence. The Scale Factor is required to obtain pressure in GPa after substitution of the parameters in the appropriate SG function.

[a]In GPa

relevant to a wide range of crystal-growth phenomena [*Oxtoby*, 1990; *Bilgram*, 1987].

Studies of melting at pressure in the RGS have been extensive and include the formulation and description in terms of empirical melting laws, [see e.g., *van der Putten and Schouten*, 1986 and *Crawford*, 1976 for review; *Stishov*, 1975], rationalization in terms of more fundamental descriptions of melting [*Ross*, 1969; 1973; *Vaidya*, 1984; *Kechin*, 1995] and studies of liquid and solid structure along the melting line [e.g., *Choi et al.*, 1993]. The melting curves of the RGS have been measured by a variety of high-pressure, low-temperature techniques up to a few kilobars [*Crawford* 1976]. In the DAC, the melting curve for helium has been determined to 24 GPa and 460 K [*Vos et al.*, 1990] and for neon to 5.45 GPa and 328 K [*Vos et al., 1991*] with quasi-isochoric methods. For the heavier RGS, where melting temperatures start to exceed the range accessible by conventional methods, DAC measurements until recently have only been made for argon to 6 GPa and 717 K with resistance-heating methods [*Zha et al.*, 1986]. The melting curves of the RGS belong to the (normal) class that increase with pressure [*Stishov*, 1975; *Bilgram*, 1987] and one objective here is to examine this behavior in the heavy RGS melting curves to higher temperatures and pressures with the laser-heated DAC. We cannot yet claim sufficient accuracy that allows a detailed comparison with melting curves measured in the several 100 K range, particularly with isochoric $P-T$ scanning methods, but the preliminary melting-curve trends do offer insight for further work.

A simple parameterization of melting curves that continue to rise with pressure is the empirical Simon-Glatzel (SG) equation

$$P = AT^c - B$$

where A, B, and c are fitted parameters.

At lower pressures and temperatures where P and T are near the triple point values of pressure and temperature, P_0 and T_0, the SG relation can be written

$$P = A[(T/T_0)^c - 1] + P_0$$

where P and T are the pressure and temperature at melting [*van der Putten and Schouten*, 1986]. (This form is labeled SG2 in Table 1.) A further 'modified SG' relation (MSG) has also been used by some authors to improve agreement over a particular data range:

$$P = A(T - T_0)^c - B$$

where T_0 is an additional (non-physical) parameter. The SG relation has been shown to be related to the Lindemann law under special conditions which are expected to break down at high compressions [*Ross*, 1969], but appear to remain valid in the case of the RGS.

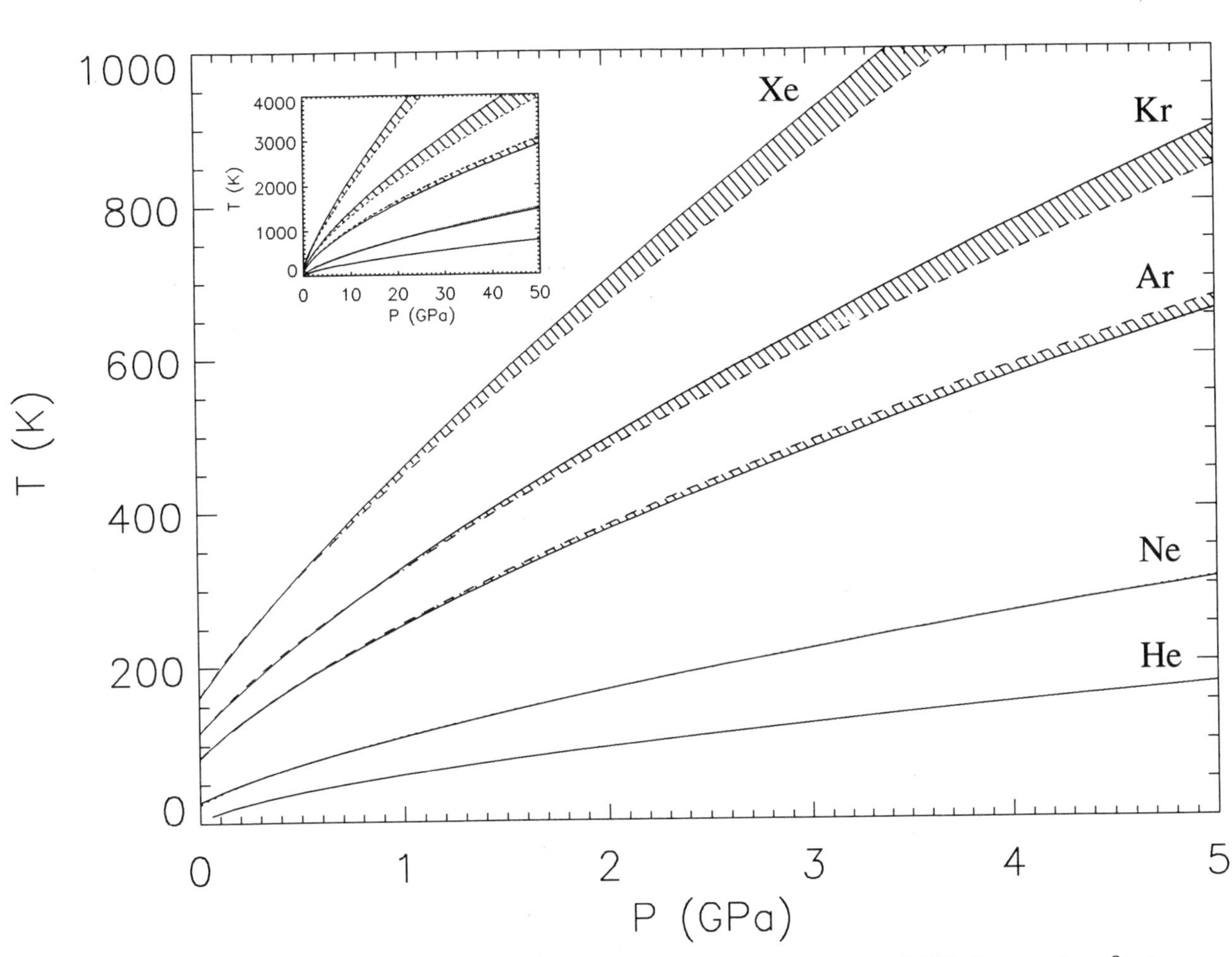

Figure 1. The melting curves of the RGS plotted over a pressure range up to 5 GPa from various fits to the SG relation. Over this pressure range, the melting curves are well documented for He, Ne, and Ar, but less certain for Kr and Xe. The shaded regions represent bounds between empirical fits in Table 1. The inset shows the melting curves extended to 50 GPa. Curves for He and Ne are likely well defined by the low pressure data. The SG relation for argon of *Hardy et al.*, [1971] agrees with the data obtained in this study to near 50 GPa. There is no fundamental reason to expect that the SG relation holds for Kr and Xe in this high pressure range, except that the trajectories of the melting curves are expected to be smooth and nonintersecting.

Despite this lack of a fundamental basis [e.g., *Stishov*, 1975], the SG or MSG relations have been observed to predict melting pressures within experimental error remarkably well when comparing measurements obtained over different density ranges for He, Ne and Ar: The ^{4}He melting curve, for example, is predicted well up to both 18 GPa [*Loubeyre et al.*, 1982] and to 24 GPa [*Vos et al.*, 1990] with the SG fit to data to 1 GPa [*Crawford and Daniels*, 1971]. The neon melting curve [*Vos et al.*, 1991] is in similar agreement with low pressure melting data [*Crawford and Daniels*, 1971], and the argon SG relation to 1.1 GPa [*Hardy et al.*, 1971] models the DAC data of *Zha et al.*, [1986] to 6 GPa.

Figure 1 shows the melting trends expected for the RGS based on published fits to experimental data of the empirical SG (or MSG) relation up to a maximum pressure of 5 GPa which is well within the data range for He, Ne, and Ar and in excess of the available data for Kr and Xe. The parameters used and functional form for all the RGS are listed in Table 1. The inset in Figure 1 shows extension of these empirical relations to 50 GPa where no observations of static melting for the RGS exist. Shaded regions are bounds of the empirical fits in Table 1; there is no *a priori* reason to expect actual melting to fall within these bounds. The various He, Ne, and Ar SG relations agree well to the highest pressure plotted: The melting point of argon predicted by the fits of *Lahr and Eversole* [1962] (to 1.82 GPa),

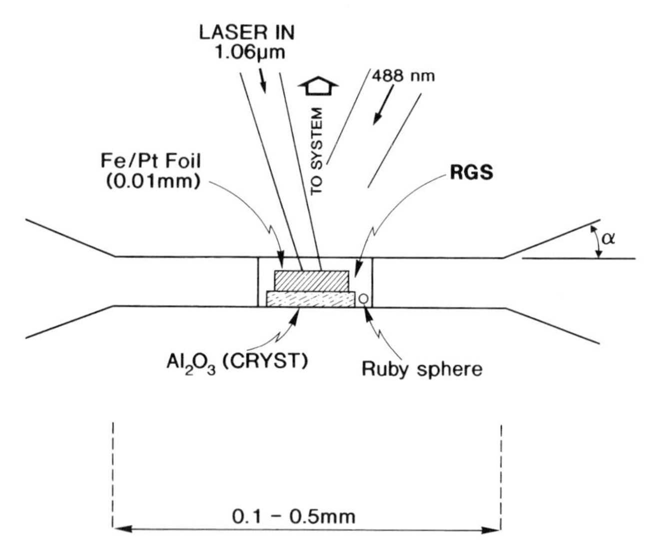

Figure 2. Schematic of now standard sample environment in the DAC for laser heating experiments. Metal foils (Pt, Fe) are embedded in RGS media that serve as both sample and insulating medium from the diamond. A thin wafer of Al_2O_3 can be included in the chamber to prevent contacting of lower diamond. Melting is detected at the surface of the foil; pressure measured from ruby sphere (0.005 mm diameter) before and after experiment.

Stishov and Fedisimov [1971] (to 1.53 GPa) differ by less than ~150 K at 50 GPa from that of *Hardy et al.* [1971] (to 1.1 GPa), well within error of determination of temperature in the DAC.

As noted by *Stishov* [1975], melting temperatures of rare-gas type solids may increase without limit under compression as both $\Delta S/R$ and $\Delta V/V_s$ tend to finite values with increasing temperature. On the assumption that the melting curves of the RGS do not cross with increasing density, the RGS melting points span a large region in pressure and temperature. Within the larger errors associated with the laser-heating technique, we present below results of preliminary experiments on the melting of solid Ar, Kr, and Xe at temperatures above the range of resistance-heating methods.

2. EXPERIMENT

As described previously [*Jephcoat and Besedin*, 1996], we have used an infra-red (IR), 1.064μm (Nd-YAG, Spectron SL-903) laser radiation source coupled with a DAC. In single-mode operation with a polarised TEM_{00} beam 25 W continuous wave (CW) power is available and up to 65 W (CW) power in unpolarised multimode operation. Thermal emitted light was collected through a spectrometer system using a fibre-optic to deliver the light to a 0.275-m focal length Czerny-Turner spectrometer entrance slit with diode-array detector. Control of incident laser power is achieved with counter-rotating pairs of quartz plates in single-mode operation and adjustment to the laser pump lamp current in multimode operation, with some loss of laser stability.

The heated spot diameter is larger when focused on metal foils in RGS media than on an insulating surface because of thermal conduction within the foil laterally away from the focus. Operation of the laser in multimode with a 50-mm focal-length lens also results in broader spot sizes (≥0.05 mm) and improves the spectral radiance measurements which we attribute to uniform temperatures across the spatial filtering aperture (equivalent to an ~0.003 mm square region area at the object). Illumination of the sample during heating is provided by an Ar^+ laser operating at 488 nm, and visual observation of the foil surface is made through an interference narrow-bandpass filter. A helium-neon laser illuminates the sample from below, through the lower diamond, and acts as a marker in sequential exposures: Hole formation in an opaque foil due to melting can be detected from the distinct appearance of the 632.8-nm line in the thermal emission spectrum. Figure 2 illustrates a typical sample configuration in the DAC. The main difficulty with soft RGS media is contacting of the foil with the upper diamond at high compressions. Pressures were obtained from a ruby sphere included in the sample chamber before and after heating; little relaxation of pressure was observed after initial heating.

Melting temperatures were measured in Fe-Kr, Pt-Xe, and Fe-Ar sample combinations with the metal serving as the absorber of the YAG laser radiation. The number of data points obtained for Xe and Kr are small because the melting temperature of the RGS approached that of the metal in the pressure range studied, making it difficult to reach stable higher temperatures within the limits of the laser power fluctuation. (For the experiments reported here, power stabilization was referenced from the laser output and not from sample brightness.) On melting of the metal foil a hole develops usually decoupling it from the laser beam. Surface tension of the molten metal is sufficient to cause droplet formation and loss of a flat reflecting surface. Higher P-T crossing points than those reported here could be reached with tungsten foils. Temperatures are obtained from a sequence of 0.1-s exposures at constant laser power, and we estimate the error based on the range from maximum to minimum temperatures in a 10-s period. This range is large (several hundred Kelvin without laser stabilization referenced to sample brightness) and means that our pressure points need to be quite widely separated to obtain temperature resolution of melting point. In addition, temperatures above ~1800 K are required in order to measure thermal emission profiles reliably.

3. MELTING CRITERIA

At low pressures, for example near 10 GPa, it is possible to observe a clear liquid-solid phase boundary in argon on heating with an absorber. A boundary forms between melt and solid at roughly constant distance from the foil edge [*Jephcoat and Besedin*, 1996]. A similar effect is well known near the melting point at room temperature and 1.2 GPa, where absorption of a few mW of visible laser light by ruby is sufficient to melt the solid. At higher pressures, no boundary between fluid and solid can be detected in this way [*Jephcoat and Besedin*, 1996]. In the experiments here we used a change in the interference (speckle) pattern [*Françon*, 1979] derived from the scattered, coherent 488 nm light of the Ar^+ laser used to illuminate the sample in the DAC. Under this illumination, a static speckle pattern from both gasket and sample metal surfaces is observed at constant T. It is local changes in the texture of the pattern under and in the vicinity of the IR laser spot during heating of an absorbing metal foil embedded in a rare-gas solid pressure medium that appears to provide a useful diagnostic of the melting transition in the 50 GPa range [see *Jephcoat and Besedin*, 1996].

The speckle pattern is sensitive to thermal expansion of either or both the metal foil and the overlying RGS solid: On first opening the laser shutter, the pattern expands abruptly but maintains a static appearance. This distortion in the pattern observed on heating is attributed to changes in refractive index contrast or internal reflection at the grain boundaries. That grain boundaries must exist in RGS at pressures well above the room temperature freezing point is supported by the fact that smooth powder X-ray patterns of RGS are well documented [e.g., *Jephcoat et al.*, 1987; *Hemley et al.*, 1989]. On raising the incident IR laser power, a continuous change in the speckle pattern structure with time can be observed above a threshold laser power exactly coincident with the IR spot. Increasing the power further above this threshold results in apparent vigorous motion (constant reworking of the speckle pattern) which on lowering the laser power, can be reduced to a small area coincident with the spatial filtering aperture. The onset of this continuous grain-boundary reconstruction is interpreted as melting of the RGS and is well correlated with any irregular shape of the hot parts of the absorbing foil. We see a sharp onset with increasing temperature supporting the use of this criterion as a diagnostic of melting. The perturbation in the speckle pattern obscures texture at the foil surface itself (i.e., in a pressure range where the foil melts at a higher temperature) forcing the need of a different method of melt determination in the metal (pinhole formation etc.). These effects were observed in argon well before iron was observed to melt and significant increase in power was required to melt the absorbing iron foil. As pressure increased, we observed the same changes in the speckle pattern, but with a smaller increment in power required to melt iron. We make the assumption that any pre-melting phenomenon in iron would not be expected to change with pressure such that it rises in temperature closer to the melting point and vanishes at a critical pressure. We therefore identify this change with melting in the RGS rather than the absorber. Further aspects of these observations at melting and related effects have been discussed elsewhere [*Jephcoat and Besedin*, 1996].

TABLE 2. $P-T$ Data and Melting Observations for Ar, Kr, Xe

Pressure	Temperature, K	
(GPa)	Min.	Max.
	Argon	
26.21	1783	1934
32.44	2158	2233
34.80	2161	2336
42.37	2492	2674
47.16	2558	2948
57.20	2640‡	
	Krypton	
18.44	2160	2235
23.54	2152	2279
	Xenon	
7.6	1770	1790
12.0	2054	

‡ No Melting observed

4. RESULTS

Table 2 lists the melting data for Ar, Kr, and Xe we obtained with the speckle contrast method described

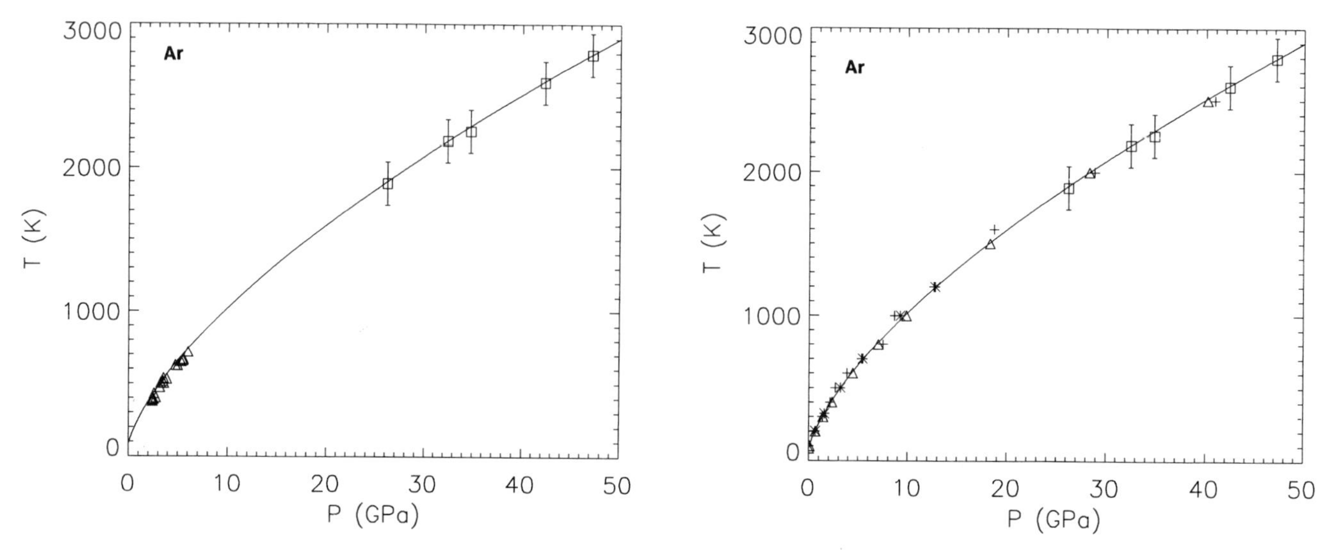

Figure 3. The melting curve of argon. (a) Experimental data to near 50 GPa. Open squares with temperature error bars are data from this study. Solid line is a SG parameterization of the melting curve determined by *Hardy et al.* [1971] to 1.14 GPa and is negligibly different from the present data within the large error on temperature. Data at lower pressures are from *Zha et al.* [1986]. (b) Comparison of various theoretical estimates of the argon melting curve with the data (squares) reported in this study. Solid line is the SG fit from *Hardy et al.* [1971]; + MD simulation of *Belonoshko* [1992] to 4000 K; △ from *Zha et al.* [1986]; * *Ross* [1973]. Temperatures were measured from an iron foil absorber.

above showing maximum and minimum temperatures obtained from a sequence of 0.1-s temperature measurements at nominally constant laser power. Maximum and minimum temperatures are not plotted separately, but the average chosen and a conservative error bar drawn based on our experience of temperature fluctuations observed. We emphasize that the laser was not stabilized from the sample brightness in these experiments. Melting temperatures for argon [*Jephcoat and Besedin*, 1996] are included here for completeness and plotted in Figure 3a. The melting curve of argon follows, within experimental error, an extrapolation of the SG equation based on data of *Hardy et al.* [1971] obtained up to 1.1 GPa, passing through the data of *Zha et al.* [1986] to 6 GPa and through the high pressure data of this study to 47 GPa. In Figure 3b we plot several theoretical determinations of the argon melting curve all of which agree with the new DAC data to 47 GPa as well as the extension of the SG relation of *Hardy et al.*, [1971].

Melting observations obtained in the DAC for krypton are plotted in Figure 4 and for xenon in Figure 5 (from Table 2). Superposed on the $P-T$ points are the melting curves predicted on the basis of the SG relations shown in Figure 1 and, in addition, the melting curve of the metal foil used as absorber in each experiment: For Kr in Figure 4, we plot the low-temperature melting curve of iron [*Boehler et al.*, 1990; *Boehler*, 1993; *Shen et al.*, 1993]; and for Xe in Figure 5, the platinum melting curve of *Kavner and Jeanloz* [1996]. Note that we observed clear melting in the one low pressure point at 18.4 GPa for solid krypton that is consistent with an SG extrapolation of lower pressure melting data based on the MSG fit of *Crawford and Daniels* [1971] reported to reproduce their melting curve most accurately for small extension above the data range. At the higher pressure we had difficulty in maintaining stable temperatures and observing the krypton melt in the speckle pattern. Inspection of the iron melting curve (Figure 4) suggests that these difficulties arose because we were close to the melting point of iron at this pressure and that fluctuation in absorbed laser power would lead to iron melting simultaneously within the krypton medium with consequent loss of incident power. In this case, the measured temperature is bounded by the melting point of the absorber.

In solid xenon, a similar effect was observed: At the lowest pressure datum we observe the onset of change in the speckle pattern indicating xenon melting occurring at a temperature in close agreement with an SG prediction based on lower pressure melting data [*Lahr and Eversole*, 1962]. At the second $P-T$ point, melting was also detected from change in the speckle pattern. At this pressure small pinholes appeared in the Pt foil at or near the onset of detectable melting in xenon. We note from Figure 5 that the melting curve of Pt [*Kavner and Jeanloz*, 1996] could cross the xenon melting curve in this region. The precise crossing point is uncertain because the platinum melting curve as shown is a guideline trend drawn through the original data with no representation of the underlying spread in melting temperatures measured. Given the error on the melting curve of Pt coupled with the fluctuation in the absorbed laser power in our experiment, temperatures probably rose into the liquid field of platinum embedded in solid xenon randomly at pressures near 12 GPa. At the lower pressure of 7.6 GPa, no evidence of Pt melting was observed and it appears therefore that the temperature difference between melting of solid Xe and Pt decreased with increasing pressure. The situation with regard to expected xenon melting temperatures is less clear, however, because we observe melting below the SG temperature at 12 GPa, but slightly above a corresponding states calculation (see below) that indicates the melting curve of xenon lies at lower temperatures compared to the SG extrapolation at all pressures. A further possibility that we cannot assess at the present time is that xenon may chemically interact with metals at these $P-T$ conditions.

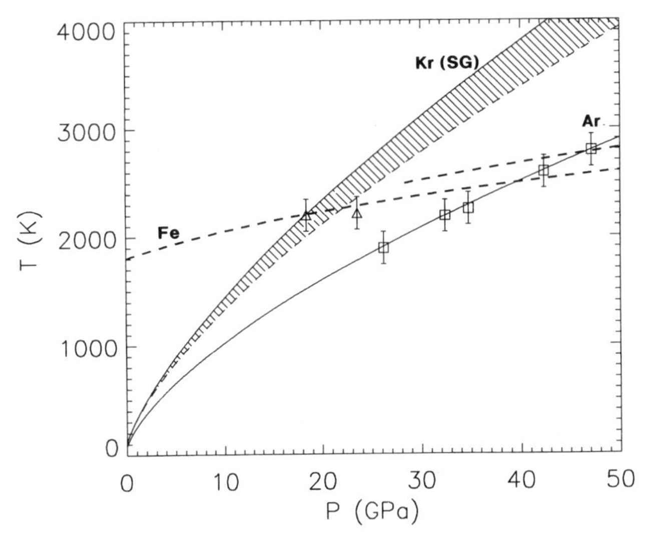

Figure 4. The melting points for krypton (△) obtained with an iron foil absorber. Also plotted are the SG relation extrapolations of *Lahr and Eversole* [1962] and *Crawford and Daniels* [1971], together with the melting curve of iron [*Boehler et al.*, 1990; *Boehler*, 1993; *Shen et al.*, 1993] (lower dashed line) and, for comparison, the melting data obtained for argon (□). The upper dashed line is the melting curve of iron inferred by *Jephcoat and Besedin* [1996] and agrees within the temperature measurement error with the earlier work cited above.

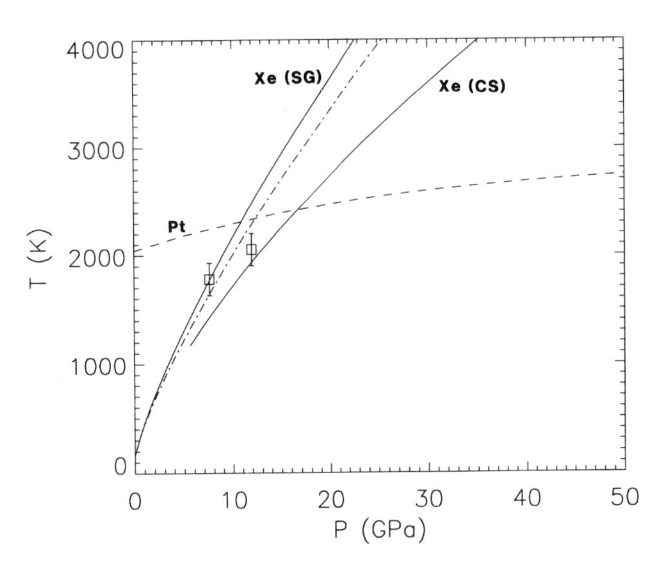

Figure 5. The melting points for xenon obtained with a platinum foil absorber (□). Also plotted are the melting curve extrapolated from *Lahr and Eversole* [1962] and *Michels and Prins* [1962] (SG). The melting curve of platinum as determined by *Kavner and Jeanloz* [1996] is shown (dashed line), and a corresponding states scaling (solid line, CS) of the expected xenon melting curve based on the melting data for neon of *Vos et al.*, [1991] and parameters of the exponential-6 effective intermolecular potential (see text). Errors in temperature determination are too large to show any perturbations to the melting curve of xenon due to the structural transition $fcc \rightarrow hcp$ that begins near 12 GPa [*Jephcoat et al.*, 1986].

In summary, for both Kr and Xe, these early results provide for the first time a direct indication of high melting temperatures in the RGS.

5. DISCUSSION

Theoretical calculations of the melting curve for argon are shown in Figure 3b by *Ross* [1973] to 12.8 GPa based on a 2-phase model intersection of Gibbs free energy isotherms (deduced with an intermolecular potential derived from shock compression data), that compared well to a single-phase (Lindemann) model of the solid. In *Zha et al.* [1986], calculations were made to 53.2 GPa with separate models of the liquid and solid again using a shock-derived intermolecular potential. *Belonoshko* [1992] has calculated the melting curve of argon to over 3500 K with a molecular dynamics method. All these theoretical techniques appear largely consistent with the experimental data up to the highest pressures near 50 GPa.

An alternative interpretation of the melting of RGS systems can be made through a corresponding states

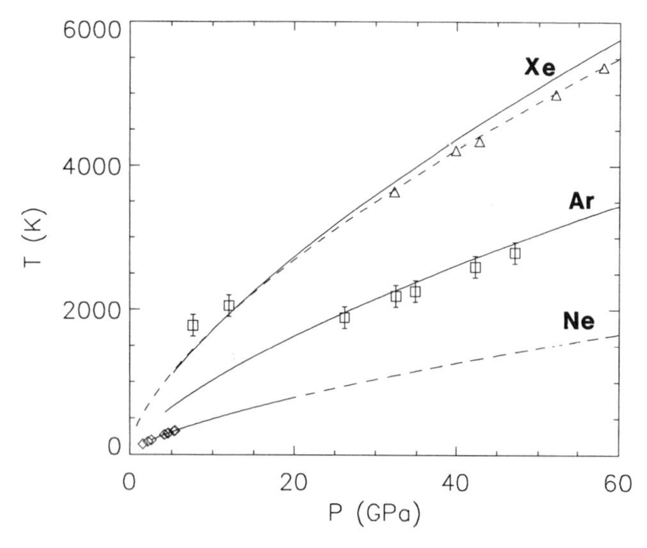

Figure 6. The melting points of the RGS expected from a corresponding states calculation based on data of *Vos et al.*, [1991] for Ne to 4.5 GPa (◇). The lower curve is the melting curve of neon shown extended with the SG relation of *Vos et al.* [1991]. The open squares with error are data from this study for Ar and Xe. The solid line passing above the argon melting data is the CS calculation based on Ne and Ar exp-6 parameters (see text). Triangles and dashed line represent the DAC Ar melting data scaled with CS to predict the Xe melting curve. The solid line passing above the Xe data is a CS scaling based on the Ne data of *Vos et al.*

(CS) scaling based on empirical potential parameters. *Vos et al.* [1990; 1991] reported melting of helium and neon to high pressure and these data can be used to predict the melting curves in other RGS through a corresponding states scaling on the basis of the exponential-6 effective intermolecular potential parameters stiffness, α, well depth ϵ, and potential minimum r_m. The scaling expands (or contracts) both the pressure and temperature axis such that the melting curve of Ne, for example, determined to 5.45 GPa and 328 K by *Vos et al.* [1991] maps the argon melting curve to 9.0 GPa and 953 K and the Xe melting curve to 11 GPa and 1835 K. The argon data obtained to 47 GPa here map onto the xenon melting curve to 58 GPa and 5300 K, assuming the potential parameters remain valid. Figure 6 shows melting curves predicted by CS for Ar and Xe based on the Ne data of *Vos et al.* [1991]. The values of α, ϵ, and r_m for neon and argon used were ([*Vos et al.*, 1991]), (Ne, Ar): α= (13.2, 13.2); ϵ/k= (42, 122 K); r_m=(3.18, 3.85 Å). For xenon, we used the effective potential parameters for isoelectronic CsI [*Ross and Rogers*, 1985; *Boehler et al.*, 1996]: α= 13.0; ϵ/k= 235 K; r_m=4.47Å.

The CS melting curves for He, Ne, and Ar are self-consistent as noted by *Vos et al.*, [1991] and the CS mappings He ⟶ Ne and Ar ⟶ Ne approximately overlap the experimental Ne data implying consistency of the potentials and accuracy of melting data. Figure 6 shows that the argon melting curve predicted by CS: Ne ⟶ Ar lies slightly above the data to 47 GPa. The CS calculation for the xenon melting curve obtained from the scaling Ne ⟶ Xe also lies above the scaling based on the argon melting data Ar ⟶ Xe, but these differences are small compared to the range over which the exp-6 parameters are valid. The results suggest that the low-pressure data for xenon melting (to 0.65 GPa [*Lahr and Eversole*, 1962]) cannot be used with their SG relation to predict melting at these higher pressures or that the assumption that the (unoptimized) potential parameters used for CsI are transferable to Xe is invalid. Both experimental data points for xenon melting fall marginally above the CS calculation (Figures 5, 6). It is well known that simple pair potentials can fail to describe the $P-V$ isotherm at 300 K at high densities [*Hemley et al.*, 1989] and a more complete test of the validity of the pair-potentials used for CS scaling here could be obtained from a comparison of the equations of state predicted with each potential and those measured for krypton to 130 GPa [*Jephcoat et al.*, 1991] or xenon to 137 GPa [*Jephcoat et al.*, 1987].

The relative position of melting points of different materials as a function of pressure offers the potential to constrain temperature independently of its absolute measurement. The assumption that there is no chemical effect such as melting point depression is strictly important, but may not actually be required to fix a melt-crossing point in practice because any perturbation to the true melting points of the isolated phases from that of the coexisting phases should be reproducible at given P and T. Melting of the RGS appears readily detected in the presence of a metal surface and the heavy RGS melting curves, in combination with selected metals, could then be used to provide fixed (reference) points in the P-T plane. In the sense that a fixed point is a reproducible state of an arbitrary standard system [e.g., *Zemansky*, 1968] and in the absence of means to measure pressure in DAC experiments that accurately reflect pressure conditions at the heated spot, a well characterized crossover in melting curves means that the uncertainty in pressure is removed. One advantage with RGS media, which are most likely quasihydrostatic at high homologous temperatures, is that strain effects at the melting interface are minimum.

In our previous experiments we observed the intersection of the argon melting curve with that of iron

TABLE 3. Estimated Melt-Crossing Points for the Heavy RGS and Fe and Pt Metals

RGS	Fe				Pt[c]	
	Low[a]		High[b]			
	P_m, GPa	T_m, K	P_m, GPa	T_m, K	P_m, GPa	T_m, K
Ar	40	2500	65	3500	60	3000
Kr	20	2300	20	2350	22	2500
Xe	10	2100	10	2100	12	2300

[a]Iron melting curve of *Boehler et al.* [1990]; *Boehler* [1993].
[b]Iron melting curve of *Williams et al.* [1987].
[c]From *Kavner and Jeanloz* [1996].

near 47 GPa [*Jephcoat and Besedin*, 1996], suggesting that the melting curve of iron is slightly above other determinations [*Boehler et al.*, 1990; *Boehler*, 1993; *Shen et al.*, 1993] (Figure 4). The difference is within the error of our measured temperatures, but indicates the applicability of defining a reference point. The experiments here suggest an equivalent intersection of the Kr-Fe melting curves occurs near 23 GPa and 2300±150 K, and the Xe-Pt curves in the range 12-15 GPa and 2300±150 K. Table 3 summarises melting-curve crossovers and lists approximate estimates of the intersection points for Fe and Pt based on the available melting curve data. These points are subject to large errors due to uncertainty in the measurements for one material or both. The low and high bounds on iron indicate the difference between reported upper and lower melting curves of pure iron [*Williams et al.*, 1987; *Boehler*, 1993; and *Shen et al.*, 1993].

Finally, it is worth noting that as the melting temperatures of the heavy RGS rise with pressure, above that of many metals and oxides, then on the basis of a correlation between melting point and cohesive energy [e.g., *Sutton*, 1993], the cohesive strength of the heavy rare-gases may exceed that of other materials above the melt-crossing pressure for each system. On the other hand, the solid phase of the RGS may be stabilized by entropy terms alone associated with packing geometry as has been predicted to occur in hard-sphere fluids [*Oxtoby*, 1990].

Acknowledgments. We thank anonymous reviewers for constructive comments. Work supported by The Royal Society (London) and NERC Grant Numbers GR3/07673A and GR9/1570A. SPB acknowledges The Royal Society for support under a FSU Research Fellowship.

REFERENCES

Belonoshko, A. B. Equation of state and melting transition of argon up to 8000 K and 4 Megabars: A molecular dynamics study, *High Press. Res.*, 10, 583-597, 1992.

Boehler, R., N. von Bargen, and A. Chopelas, Melting, thermal expansion and phase transitions of iron at high pressures, *J. Geophys. Res.*, 95, 21,731-21,736, 1990.

Boehler, R., Temperatures in the Earth's Core from melting-point measurements of iron at high static pressures, *Nature*, 363, 534-536, 1993.

Boehler, R., M. Ross, and D. B. Boercker, High-pressure melting of alkali halides, *Phys. Rev. B.*, 53, 556-563, 1996.

Choi, Y., T. Ree, and F. H. Ree, Crystal stability of heavy-rare-gas solids on the melting line, *Phys. Rev. B*, 48, 2988-2991, 1993.

Crawford, R. K. and W. B. Daniels, Experimental determination of the $P-T$ melting curves of Kr, Ne, and He, *J. Chem. Phys.*, 55, 5651, 1971.

Crawford, R.K. Melting, vaporization and sublimation in *Rare Gas Solids* Vol. 2, edited by M. L. Klein and J. A. Venables, pp. 663-728, Academic, London, 1976.

Françon, M., *Laser Speckle and Applications in Optics*, 158pp, Academic Press, New York, 1979.

Hardy, W.H., R. K. Crawford, and W. B. Daniels, Experimental determination of the P-T melting curve of argon. *J. Chem. Phys.*, 54, 1005-1010, 1971.

Hemley, R. J., C. S. Zha, A. P. Jephcoat, H. K. Mao, L. W. Finger, and D. E. Cox, X-ray diffraction and equation of state of solid neon to 110 GPa, *Phys. Rev. B*, 39, 11820-11827, 1989.

Jephcoat, A. P., L. W. Finger, and D. E. Cox, Compression of solid krypton to 130 GPa, *National Synchrotron Light Source Annual Report*, BNL 52317, 128, 1991.

Jephcoat, A. P., and S. P. Besedin, Melting and temperature determination in the laser-heated diamond-anvil cell, *Phil. Trans. Roy. Soc.*, A354, 1333-1360, 1996.

Jephcoat, A. P., H.-k. Mao, L. W. Finger, D. E. Cox, R. J. Hemley, and C.-s. Zha, Pressure-induced phase transitions in solid xenon, *Phys. Rev. Lett.*, 59, 2670-2673, 1987.

Kavner, A. and R. Jeanloz, High-Pressure Melting Experiments with the Laser-Heated Diamond Cell, in *Proc. 3rd NIRIM Int. Symp. on Advanced Materials*, pp. 143-147, Tokyo: International Communication Specialists, 1996

Kechin, V. V. Thermodynamically based melting-curve equation, *J. Phys.: Condens. Matt.*, 7, 531-535, 1995.

Lahr, P. H. and W. G. Eversole, Compression isotherms of argon, krypton, and xenon through the freezing zone, *J. Chem. Eng. Data*, 7, 42-47, 1962.

Loubeyre, P., J. M. Besson, J-P. Pinceaux, and J-P. Hansen, High-pressure melting curve of ^{4}He, *Phys. Rev. Lett.*, 49, 1172-1175, 1982.

Michels, A. and C. Prins, The melting lines of argon, krypton and xenon up to 1500 atm; representation of the results by a law of corresponding states, *Physica*, 28, 101-116, 1962.

Putten, L. van der, and J. A. Schouten, The thermodynamic discontinuities $\Delta S, \Delta H$, and ΔV along the melting curve. 1. Argon, mercury, nitrogen, *High Temp. - High Press.*, 18, 393-404, 1986.

Ross, M., Generalized Lindemann melting law. *Phys. Rev. B.*, 184, 233-242, 1969.

Ross, M., Shock compression and the melting curve for argon. *Phys. Rev. A*, 8, 1466-1474, 1973.

Ross, M. and F. J. Rogers, Structure of dense shock-melted alkali halides: Evidence for a continuous pressure-induced structural transition in the melt, *Phys. Rev. B*, 31, 1463-1468, 1985.

Shen, G.Y., P. Lazor, and S. K. Saxena, Melting of wüstite and iron up to pressures of 600 kbar. *Phys. Chem. Minerals*, 20, 91-96, 1993.

Stishov, S. M., and V. I. Fedosimov, Thermodynamics of the melting of argon, *ZhETF. Fiz. Pis. Red.*, 14, 326-329; *JETP Lett.*, 14, 217-220, 1971.

Stishov, S. M., The thermodynamics of melting of simple substances. *Sov. Phys.-Usp.*, 17, 625-643, 1975.

Sutton, A. P., Electronic structure of materials, 260pp., Oxford Clarendon Press, 1993.

Vaidya, S. N., Simple interpretation of Lindemann law of melting for rare-gas solids, *J. Phys. Chem. Sol.*, 45, 975-976, 1984.

Vos, W. L., M. G. E. van Hinsberg, and J. A. Schouten, High-pressure triple point in helium: The melting line of helium up to 240 kbar, *Phys. Rev. B*, 42, 6106-6109, 1990.

Vos, W. L., J. A. Schouten, D. A. Young, and M. Ross, The melting curve of neon at high pressure, *J. Chem. Phys.*, 94, 3835-3838, 1991.

Williams, Q., R. Jeanloz, J. Bass, B. Svendsen, and T. J. Ahrens, The melting curve of iron to 250 gigapascals: A constraint on the temperature at Earth's centre, *Science*, 236, 181-182, 1987.

Zemansky, M. W., *Heat and Thermodynamics*, 658pp., McGraw-Hill Kogakusha, Tokyo, 1968.

Zha, C-S., R. Boehler, D. A. Young, and M. Ross, The argon melting curve to very high pressures, *J. Chem. Phys.*, 85, 1034-1036, 1986.

S. P. Besedin and A. P. Jephcoat, University of Oxford, Department of Earth Sciences, Parks Road, Oxford, OX1 3PR, UK. (e-mail: stasb@earth.ox.ac.uk; andrew@earth.ox.ac.uk)

Structures and Phase Equilibria of FeS Under High Pressure and Temperature

Keiji Kusaba and Yasuhiko Syono

Institute for Materials Research, Tohoku University, Sendai, Japan

Takumi Kikegawa and Osamu Shimomura

Photon Factory, National Institute for High Energy Physics, Tsukaba, Japan

High-pressure behavior of FeS up to 16 GPa and 800°C is investigated by energy-dispersive type X ray powder diffraction method. Four phases (the troilite phase, MnP-related phase, NiAs type phase, and a high-pressure phase) are observed in this pressure and temperature range. The high-pressure phase at 10.30 GPa and 27°C, the structure of which has been unknown, can be indexed on a monoclinic system: a=8.044(3)Å, b=5.611(2)Å, c=6.433 (4)Å, β=93.11(4)°, and Z=12. The high pressure phase is calculated to be 12% denser than the troilite phase at ambient conditions. All these four phases are found to have an NiAs-type framework in their structures. Phase equilibria of FeS are determined by paying attention to diffraction lines with d-spacings larger than 3Å, due to a long-range order of the NiAs-type framework. The phase boundary between the MnP-related phase and the simple NiAs-type phase can be described by a straight line (P/GPa = 0.0212 x T/°C - 3.38) in the pressure and temperature range up to 11GPa and 700°C. The result shows that high-pressure behavior of the simple NiAs-type phase may play an important role in the interior of the terrestrial planets.

1. INTRODUCTION

High-pressure behavior of iron sulfides has been investigated in order to understand the interior of the terrestrial planets. For the investigation of iron sulfides, it is always important to note their stoichiometry, because iron sulfides have shown many crystal structures between FeS (troilite) and Fe_7S_8 (pyrrhotite) even at ambient conditions [*Morimoto*, 1976]. The differences are explained by an ordering of iron atoms or by defects in the simple NiAs-type structure. The ordering in the structure causes a variety of physical properties.

Stoichiometric FeS has been reported to have at least four structures in high-pressure and high-temperature conditions.

Properties of Earth and Planetary Materials
at High Pressure and Temperature
Geophysical Monograph 101

The ambient condition phase has the troilite structure, derivative of the NiAs-type structure. The unit cell of the troilite can be described by $a = \sqrt{3}A$ and $c = 2C$ using cell dimensions of the NiAs-type cell (A and C) [*King and Prewitt*, 1982]. At higher temperature than the troilite stable region, the simple NiAs-type phase is observed [*Morimoto*, 1976]. In the high-pressure condition, two phase transitions are observed at room temperature. The troilite phase transforms to the MnP-type with an orthorhombic cell ($a = C$, $b = A$ and $c = \sqrt{3}A$) which can be represented with a hexagonal cell ($a = 2A$ and $c = C$) of the NiAs-related structure [*King and Prewitt*, 1982]. The higher transition is observed above 7 GPa by X ray diffraction and Mössbauer spectroscopy [*King et al.*, 1978; *King and Prewitt*, 1982; *Kobayashi et al.*, in press]. The structure of the higher pressure phase has not been determined, since it was first observed by *Taylor and Mao* [1970]. From an X ray study using a single crystal, the high-pressure phase was suggested to have a NiAs-type related structure.

Recently, *Fei et al.* [1995] investigated the high-pressure phases of FeS using the X ray powder diffraction method with a synchrotron white X ray source. They proposed a phase diagram of FeS under high pressure and temperature, in which a curved phase boundary divided the MnP-related and the simple NiAs-type phases. The internal structure of the Mars was discussed based on their diagram.

Fei et al. [1995] also tried to determine the structure of the high-pressure phase existing above 7 GPa. The authors suggested a possible monoclinic cell with a=5.121(2)Å, b=5.577(2)Å, c=3.328(2)Å and β=95.95(4)° at 15 GPa. However, they also pointed out that diffraction patterns from several experiments were not consistent with each other even using the second-generation synchrotron X ray source. The nonreproducibility may be due to limitations of the diamond anvil apparatus, i.e., the diffracting volume of the sample in a diamond anvil cell is too small to collect reasonable powder X ray diffraction data, or to the limitation of the second-generation synchrotron source for X ray work about a low symmetry structure.

We can avoid these problems by using MAX80, installed at an AR-NE5 beam line in the accumulation ring (AR) of the National Laboratory for High Energy Physics (KEK) [*Shimomura et al.*, 1985]: The MAX80 system has a DIA-10 type vessel, by which a several thousand times larger volume can be compressed in comparison with a diamond anvil. The AR operates at 6.5 GeV, is a prototype of the third-generation synchrotron source, and the AR-NE5 bending beam line provides a wide range of white X rays. In fact, the combination of the MAX80 system and AR has made detailed structure work under high pressure possible on, for example, the baddeleyite type TiO_2 and the KOH type AgCl with monoclinic cells [*Sato et al.*, 1991; *Kusaba et al.*, 1995].

The aim of the present work on FeS is to propose a structure model of the high-pressure phase above 7 GPa and to ascertain the reported unusual phase boundary between the MnP-related and the simple NiAs-type phases, utilizing the advantage of the combination of the MAX80 system and AR.

2. EXPERIMENT

2.1. *Synthesis of Troilite*

A powdered specimen of FeS (troilite) was synthesized similar to *King and Prewitt* [1982]. A mixture of iron and sulfur with the ratio 1:1 was sealed in an evacuated glass tube and reacted at 690°C for two weeks to make stoichiometric FeS. According to Morimoto's phase diagram of FeS-Fe_7S_8 [*Morimoto*, 1976], the troilite type FeS is stable below 100°C. The reacted products annealed at 100°C for two weeks in the evacuated glass tube for atomic ordering and separation of nonstoichiometric components. Products were crushed to several-micrometer particles. Nonstoichiometric components were removed from the powdered products by a magnet, as nonstoichiometric $Fe_{1-x}S$ is a ferrimagnetic compound.

By the X ray powder diffraction method using PW1729 (Philips) with CuKα radiation, the product was confirmed to be a single phase with the troilite structure. The unit cell parameters for a hexagonal cell were a=5.966(1)Å and c=11.76(1)Å, which were in good agreement with the previous reports (ICDD card 37-477: a=5.9676(7)Å and c=11.761(3)Å, and *King and Prewitt*, [1982]: a=5.963(1)Å and c=11.754(1)Å). From a d_{114} value of 2.093Å, the chemical formula of the troilite specimen was determined to be $Fe_{0.998}S$ [*Yund and Hall*, 1969].

2.2. *High-Temperature Measurements*

The high-temperature behavior of FeS at ambient pressure was observed by the X ray powder diffraction method. The sample vessel for high-temperature, X ray observation was evacuated to a few kilopascal to protect FeS from oxidation. The X ray diffraction patterns were collected after holding at temperature for more than 20 minutes.

2.3. *High-Pressure Measurements*

In situ X ray observations were carried out using the large volume press MAX80. The MAX80 system can generate high-pressure and temperature conditions of more than 15 GPa and 1000°C with 3-mm sintered diamond anvils. The powdered troilite specimen and a mixture of NaCl and BN were separately encased in a boron-epoxy pressure medium, in order to take diffraction patterns of the specimen and the pressure marker separately. Two carbon disk heaters were also enclosed for elevating temperature to anneal the specimen and NaCl as a pressure marker. Temperature was measured by an R-type thermocouple, which was put between the pair of carbon disk heaters in the pressure medium. The X ray patterns even at room temperature were collected after heating to 300°C in order to reduce strain effects of the specimen and the NaCl. Pressure was calculated from the Decker scale for NaCl [*Decker*, 1971].

High-pressure behavior of FeS was observed by an energy-dispersive method using a Ge-solid state detector.

The incident white X ray beam size was 0.1 mm vertical x 0.3 mm horizontal. The thickness of the specimen was larger than 1 mm. Diffracting volume of the powdered specimen was larger than 0.03 mm^3. The optical system can exclude diffracted X rays from the pressure medium, and X rays diffracted only from FeS could be collected with 2θ angle higher than 3.0°.

A wide energy range between 30 and 130 keV was usable for the X ray observation and the most intensive range was between 40 and 80 keV. We utilized the advantage for phase identification and determination of the crystal system by choosing $2\theta = 3.0°$, so that diffraction lines larger than 3 Å could be observed in the most intensive energy range. Such diffraction lines clearly identify superstructures of the simple NiAs-type structure. In other words, we could easily identify the crystal structure even among very similar ones by focusing on these characteristic diffraction lines. For the determination of the crystal system of the high pressure phase above 7 GPa, X ray diffraction patterns were collected with several different 2θ values, to observe diffraction lines in a wide d-value range and to improve accuracy of the observed d-values.

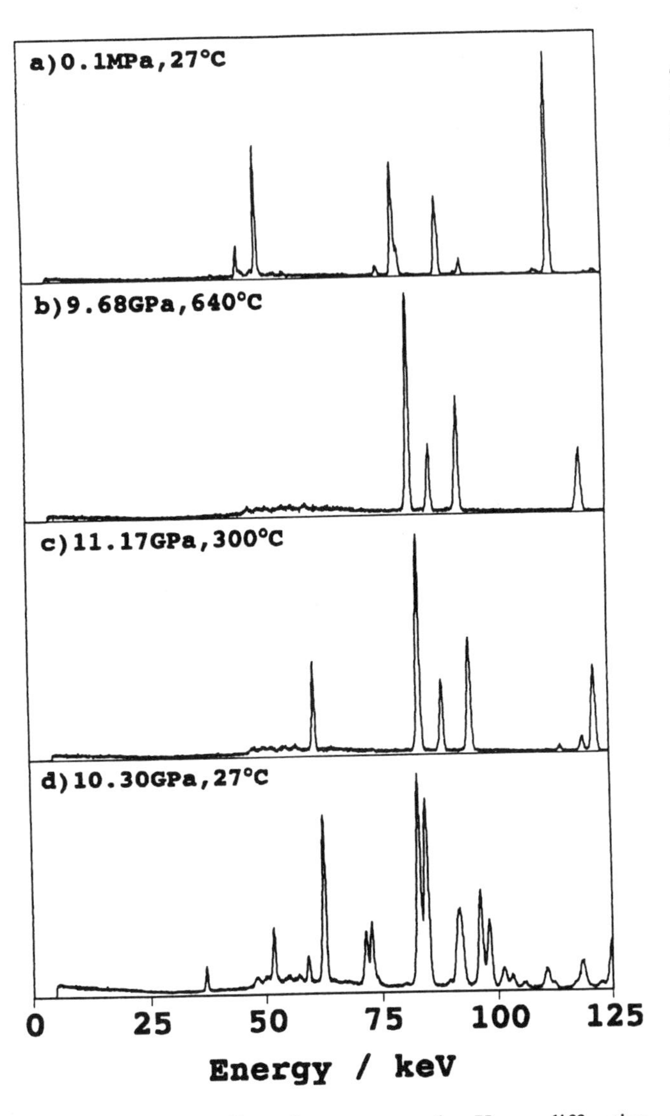

Figure 1. Energy-dispersive type powder X ray diffraction patterns of FeS taken at 2θ=3.00°: (a) the troilite type phase at 0.1 MPa and 27°C, (b) the simple NiAs-type phase at 9.68 GPa and 640°C, (c) a MnP-related phase at 11.17 GPa and 300°C, and (d) the high-pressure phase at 10.30 GPa and 27°C.

TABLE 1. Observed d-Values of High-Pressure Phase of FeS at 10.30 GPa and 27°C at Several 2θ Values for the Ge-Solid State Detector

$d_{obs.}$/Å[1]	$d_{obs.}$/Å[2]	$d_{obs.}$/Å[3]	$d_{obs.}$/Å[4]	$d_{ave.}$/Å[5]
6.43	6.43			6.43
4.60	4.60	4.60		4.60
4.01	4.02	4.02		4.02
3.80	3.80	3.80	3.79	3.80
3.32	3.32	3.33		3.32
3.26	3.27	3.27	3.28	3.27
2.860	2.861	2.862	2.862	2.861
2.805	2.806	2.805	2.804	2.805
2.584	2.586	2.585	2.587	2.586
2.466	2.465	2.465	2.464	2.465
2.414	2.418	2.417	2.417	2.417
2.341	2.342	2.341	2.341	2.341
2.300	2.300	2.299	2.296	2.299
	2.242	2.242	2.248	2.244
	2.144	2.144	2.144	2.144
	2.115	2.115	2.114	2.115
	2.029	2.027	2.027	2.028
	2.003	2.003	2.003	2.003
	1.936	1.936	1.935	1.936
	1.900	1.898	1.897	1.898
		1.790	1.788	1.789
		1.758	1.758	1.758
		1.700	1.700	1.700
		1.662	1.662	1.662
		1.631	1.631	1.631

[1-4] d-values were observed at the following 2θ angles of the detector ([1]; 2.50°, [2]; 3.00°, [3]; 4.49°, and [4]; 5.99°).
[5] Average d-values.

3. RESULTS AND DISCUSSION

3.1. *Four Phases of FeS Under High Pressure and Temperature*

In this study, we observed four phases of FeS in the pressure and temperature range up to 16 GPa and 800°C. Figure 1 shows typical X ray diffraction patterns of the observed four phases, taken at $2\theta = 3.0°$. Diffraction lines lower than 75keV were from superstructures of the simple NiAs-type structure. Phase identification is easily made by paying attention to the diffraction lines lower than 75 keV. The troilite type phase is identified by two diffraction lines lower than 50keV, the simple NiAs-type by no lines lower than 75 keV, and the MnP-related phase by a characteristic line around 60 keV. The high-pressure phase above 7 GPa has a very complicated X ray diffraction pattern.

The high-pressure phase with the MnP-related structure was observed on wide pressure and temperature conditions,

TABLE 2. Observed and Calculated *d*-Values of High Pressure Phase of FeS with the Monoclinic Cell at 10.30 GPa and 27°C

h	k	l	$d_{ave.}$/Å[1]	$d_{calc.}$/Å[2]
0	0	1	6.43	6.42
1	1	0	4.60	4.60
2	0	0	4.02	4.02
1	1	-1	3.80	3.80
2	0	1	3.32	3.33
2	1	0	3.27	3.27
2	1	1	2.861	2.861
0	2	0	2.805	2.806
1	1	2	2.586[3]	2.596
2	0	-2		2.577
1	2	-1	2.465	2.464
3	1	0	2.417	2.416
2	1	-2	2.341	2.342
2	2	0	2.299	2.300
3	1	-1		2.299
2	1	2	2.244	2.241
2	2	1	2.144	2.144
0	2	2	2.115	2.113
1	2	2	2.028	2.026
3	0	2	2.003	2.004
3	2	0	1.936	1.937
2	2	-2	1.898	1.898
4	1	-1	1.839	1.839
4	1	1	1.789	1.789
1	3	-1	1.758	1.758
0	2	3	1.700	1.702
4	0	2	1.662	1.663
3	2	2	1.631	1.631
2	3	1		1.630
3	0	3		1.630

[1]Average of observed *d*-values. From the average data, the high-pressure phase is suggested to have a monoclinic unit cell with $a = 8.044(3)$Å, $b = 5.611(2)$Å, $c = 6.433(4)$Å and $\beta = 93.11(4)°$.
[2]*d*-values calculated from the monoclinic cell.
[3]The diffraction line was so broadened that it was assigned by double indexes, as shown in Figure 2.

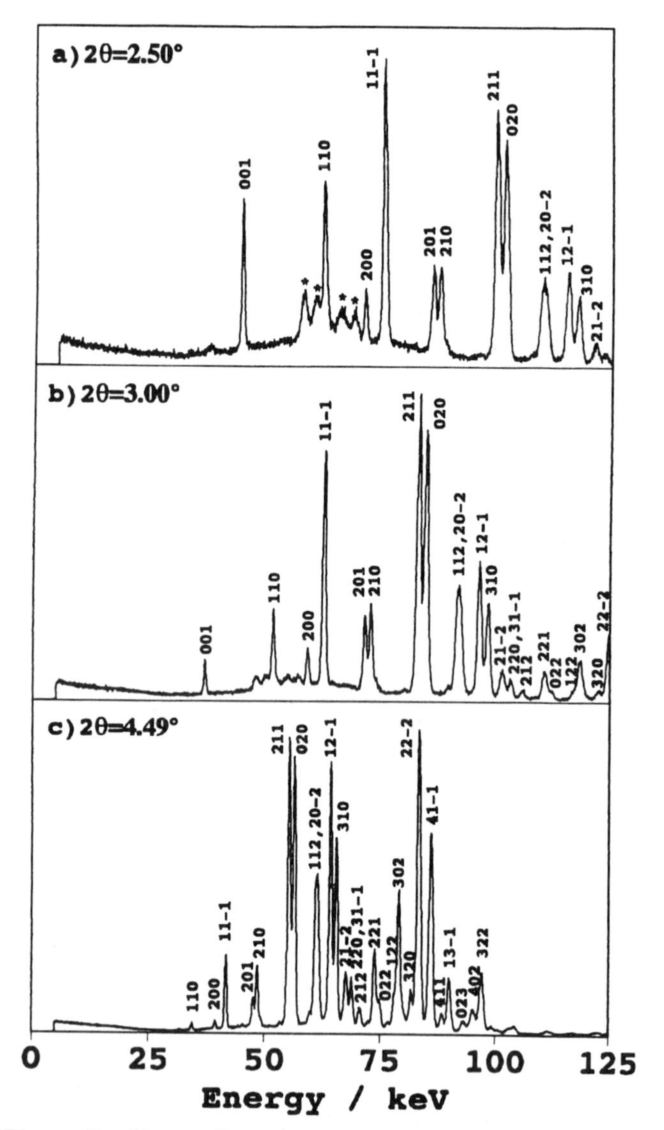

Figure 2. Energy-dispersive type powder X ray diffraction patterns of FeS with indexing on a monoclinic cell, collected at 10.30 GPa and 27°C: (a) taken at 2θ=2.50°, (b) taken at 2θ=3.00°, and (c) taken at 2θ=4.49°. Asterisks indicate diffraction lines from a boron-epoxy pressure medium.

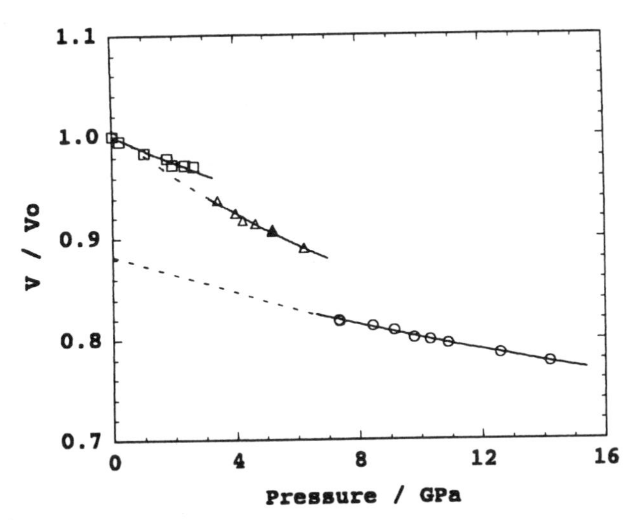

Figure 3. Compression behavior of FeS at 27°C. Open squares, triangles and circles show compression data of the troilite phase, the MnP-related phase, and the monoclinic high-pressure phase, respectively.

which are discussed in a following section on *Phase equilibria of FeS under high pressure and temperature.* The observed X ray diffraction lines could be explained by Fei's phase IV with a hexagonal cell [*Fei et al.*, 1995]. The cell parameters at 11.17 GPa and 300°C were determined to be a=6.549(1)Å, c=5.357(1)Å, and Z=8 from 26 d-value data. The determined cell dimensions were consistent with those reported by *Fei et al.*, [1995] (a=6.492(4)Å, c=5.301(2)Å, and Z=8 at 15 GPa and 350°C). Neither extra diffraction lines nor line splitting were observed in the wide pressure and temperature range. These facts suggest that the high-pressure phase does not have an orthorhombic cell (phase II by Fei's notation), but the high-pressure phase has a hexagonal cell, which was proposed to have a MnP-related structure (Phase IV) [*Fei et al.*, 1995].

3.2. *Structure of the High Pressure Phase of FeS Aabove 7 GPa*

A drastic change in the X ray diffraction patterns of FeS was observed around 7 GPa at 27°C, as shown in Figure 1c. Figure 2 shows the X ray diffraction patterns of the structure unknown high-pressure phase at 10.30 GPa and 27°C, taken at three different 2θ angles. The 26 diffraction lines were observed at 10.30 GPa, as listed in Table 1, except for broad diffraction lines from the boron-epoxy pressure medium. Almost all the diffraction lines were observed in different 2θ conditions, and each observed d-value showed small dispersion. Such excellent consistency is attributed to the large diffracting volume, which is one of the biggest advantages of a large-volume press apparatus for an in situ X ray observation.

An automatic indexing computer code, DICVOL [*Louër and Louër*, 1971] showed no possibility for an

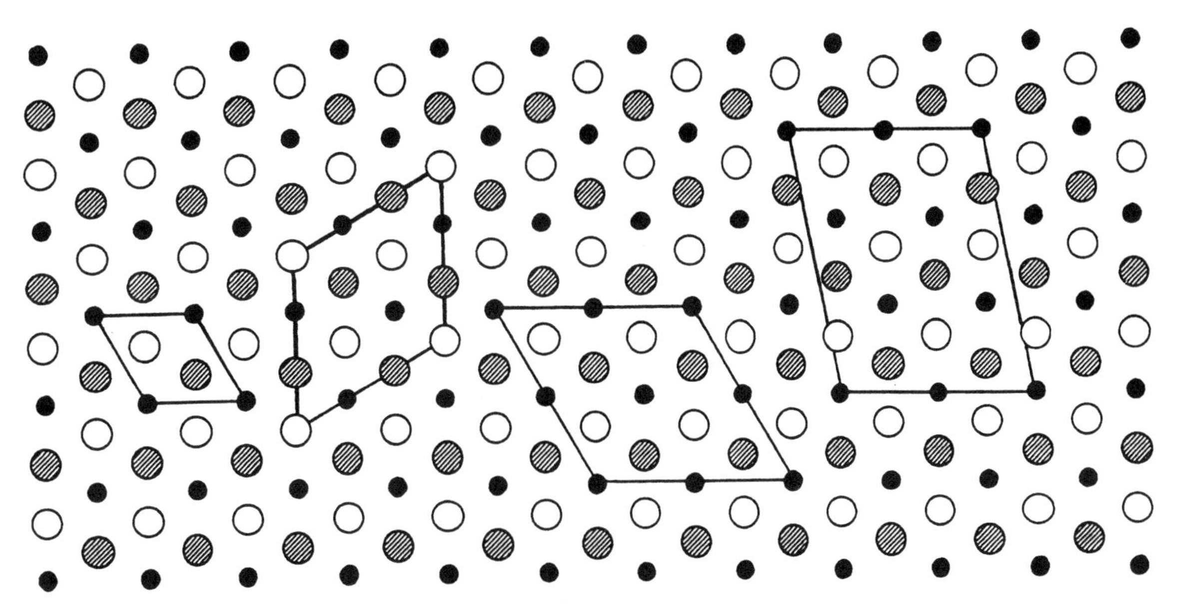

Figure 4. Schematic illustration of the simple NiAs-type structure, the troilite structure, the MnP-related structure, and the monoclinic structure (from left to right) in the NiAs-type structure projection along the c-axis. Solid small circles show Fe atoms at z=0 and z=1/2. Large circles show S atoms; hatched circles indicate z=1/4 and open circles indicate z=3/4.

orthorhombic or higher symmetry unit cell and only two possibilities for the monoclinic system in the calculation conditions: a, b, c < 15Å and 90° < β < 140°. One possibility is a_1=8.04Å, b_1=5.61Å, c_1=6.43Å, and β_1= 93.1°. The other possibility is a_2=10.57Å, b_2=5.61Å, c_2=6.43Å, and β_2=130.5°. However, the second possibility is presumed to be a mirage of the first one in the computer calculation process, because of $a_1\sin\beta_1=a_2\sin\beta_2$, $b_1=b_2$, and $c_1=c_2$. The unit cell parameters at 10.30 GPa and 27°C were refined to be a=8.044(3)Å, b=5.611(2)Å, c=6.433(4)Å, β=93.11(4)°, and V=290.0Å^3, as listed in Table 2 and shown in Figure 2. The determined unit cell parameters were different from those reported by *Fei et al.* [1995] (a=5.121(2)Å, b=5.577(2)Å, c=3.328(2)Å and β=95.95(4)° at 15 GPa). However, it is noteworthy to mention that several diffraction lines, in particular those with d-values larger than 4Å, were clearly observed in the present study and could not be explained by Fei's cell.

From the volume per chemical formula of the MnP-related structure (26.89Å^3 at 6.2 GPa and 27°C, and 24.87Å^3 at 11.17 GPa and 300°C), the monoclinic high-pressure phase was suggested to include 12 FeS in the monoclinic cell (V/Z=24.17Å^3 at 10.30 GPa and 27°C). All X ray diffraction patterns in the pressure range between 7 and 15 GPa could be explained by the monoclinic cell. This supports the accuracy of the indexing on the monoclinic cell for the higher pressure phase.

Compression data of four independent experimental runs are shown in Figure 3. From the present result, the monoclinic high-pressure phase was about 5% denser than the MnP-related phase at the phase transition pressure. By extrapolating the compression curve to 0.1 MPa, the volume of the high-pressure phase was estimated to be 12% smaller than that of the troilite phase.

According to the diffraction study using a single crystal [*King and Prewitt*, 1982], diffraction patterns of the monoclinic high-pressure phase were still based on the simple NiAs-type structure. On the basis of the results of the single crystal work and the present study, we constructed a structure model for the monoclinic high-pressure phase in the simple NiAs structure, as shown in Figure 4. The high pressure phase with the monoclinic cell can be considered as a superstructure of a distorted NiAs-type structure, similar to that of the MnP-related phase.

The phase transition from the MnP-related phase to the monoclinic high-pressure phase at 7 GPa just corresponds to the electronic state change of iron atoms, observed by ^{57}Fe Mössbauer spectroscopy; the hyperfine field suddenly disappeared above the phase transition [*King et al.*, 1978; *Kobayashi et al.*, 1996]. Furthermore, a positive temperature coefficient of electrical resistance had been

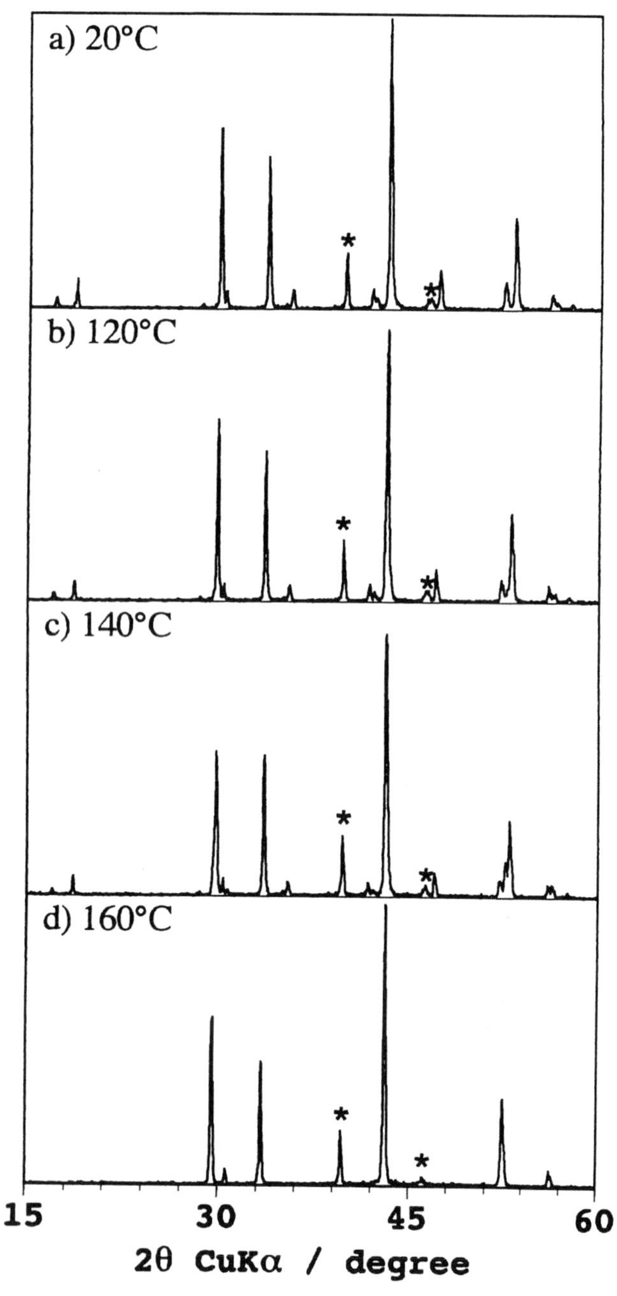

Figure 5. X-ray powder diffraction patterns of FeS at various temperature and ambient pressure conditions: (a) the troilite phase at 20°C, (b) the troilite phase at 120°C, (c) a mixture of the troilite and simple NiAs-type phases at 140°C, and (d) the simple NiAs-type phase at 160°C. Diffraction lines with asterisks are from the platinum sample holder.

observed in the pressure range for the monoclinic high-pressure phase [*Minomura and Drickamer*, 1963]. These previous studies show that the phase transition to the monoclinic phase occurs with a high spin - low spin transition of Fe^{2+} and/or oxidation of Fe^{2+} to Fe^{3+}. In either case, the electronic transition of iron atoms causes a drastic decrease of its atomic radius and explains a large volume decrease at the transition and also the distortion of the framework of the crystal structure from the simple NiAs-type structure. Such a pressure-induced transition to a low-symmetry phase with an electronic transition has been known in the case of some chalcogen compounds, for example ZnTe and CdTe [*Nelmes et al.*, 1995].

3.3. *Phase Equilibria of FeS Under High Pressure and Temperature*

Figure 5 shows X ray powder diffraction patterns of FeS at various temperature and ambient pressure conditions. The troilite structure was stable up to 120°C, whose characteristic diffraction lines were clearly observed at around 18° of 2θ CuKα. Above 160°C, diffraction lines from only the simple NiAs-type phase were observed. The X ray diffraction pattern at 140°C could be explained by a mixture of the troilite and simple NiAs-type phases. These results show that the troilite phase directly transforms to the simple NiAs-type phase at around 140°C, and that there is no stable region for the MnP-related phase with an extra diffraction line around 25° of 2θ CuKα. It is consistent with Morimoto's diagram at ambient pressure condition [*Morimoto*, 1976], but not consistent with Fei's phase diagram under high pressure and temperature [*Fei et al.*, 1995]. Furthermore, the fact that the simple NiAs-type phase (phase V by Fei's notation) was observed at 9.68 GPa and 640°C (Figure 1b), was contradictory to Fei's diagram.

In order to make clear the phase equilibria of FeS under high temperature and high pressure, we observed phase

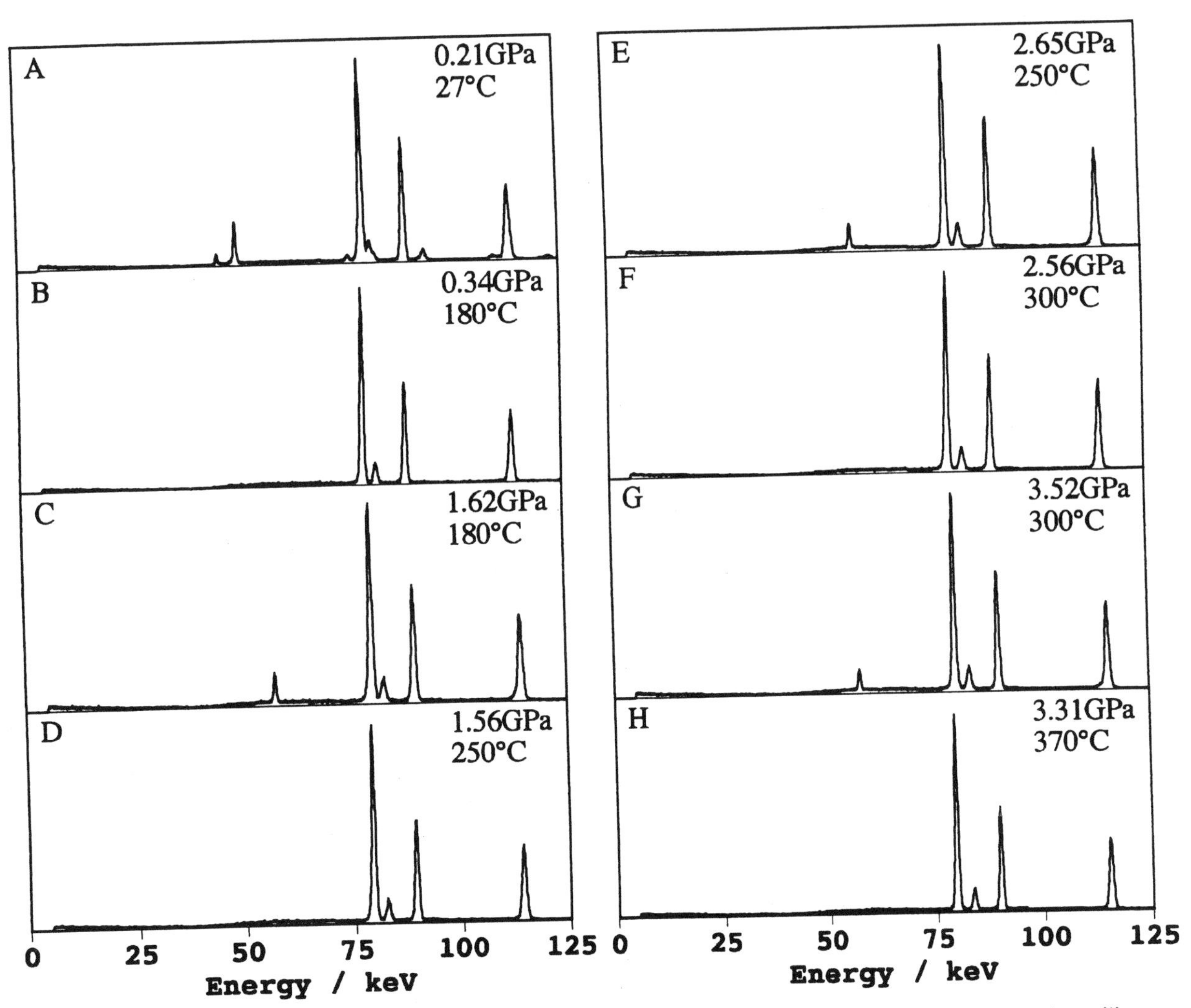

Figure 6. X ray diffraction patterns of FeS in the alternating temperature and pressure increasing process: A, the troilite phase. B, D, F, and H, the simple NiAs-type phase. C, E, and G, the MnP-related phase.

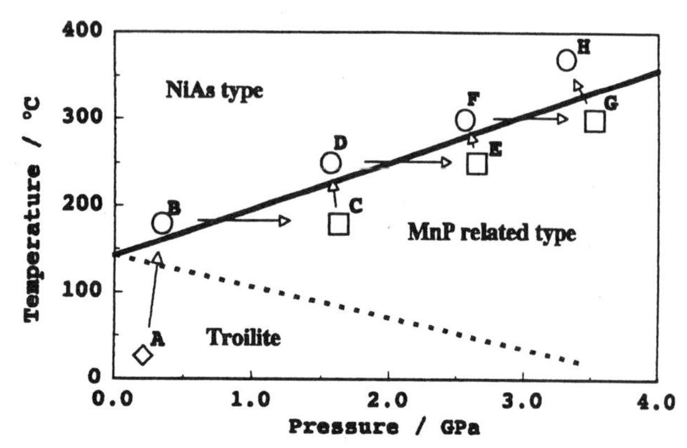

Figure 7. The alternating temperature and pressure increasing route along the phase boundary between the simple NiAs type and MnP related phases. Alphabetical notations correspond to those in Figure 6.

transitions by alternatively increasing temperature and pressure along the phase boundary. The X ray diffraction patterns in the alternative pass are shown in Figure 6, and the pressure and temperature route is drawn in Figure 7. Phase identification among the troilite (phase I by Fei's notation), MnP-related phase (phase IV), and simple NiAs-type phase (phase V) was easily made by taking note of the diffraction lines below 75keV in Figure 6, as mentioned previously in the section on *four phases of FeS under high pressure and temperature*. Figure 7 clearly indicates that the phase boundary between the MnP-related phase (phase IV) and simple NiAs-type phase (phase V) can be described by a straight line. The intersection of the straight line with the temperature axis is consistent with the result at ambient pressure condition.

All experimental results for phase equilibria of FeS up to 16 GPa and 800°C are summarized in Figure 8. The phase boundary between the MnP-related phase (phase IV by Fei's notation) and the simple NiAs-type phase (phase V) can be described by the simple straight line (P/GPa = 0.0212 x T/°C - 3.38) in the pressure and temperature range up to 11 GPa and 700°C.

Fei et al. [1995] discussed the interior of Mars using compression data of the MnP-related phase (phase IV by Fei's notation), on the basis of the unreasonable phase boundary between the MnP-related phase (phase IV) and simple NiAs-type phase (phase V). However, the simple NiAs-type phase (phase V) has been found to have a much wider stability region, particularly at high-pressure conditions. The present result shows that the high-pressure behavior of the simple NiAs-type phase (phase V) may play an important role in the interior of the terrestrial planets.

The phase boundary between the MnP-related phase and the monoclinic phase has not been determined yet, because the phase transition was found to have a hysteresis in the temperature cycle. The hysteresis may be due to a large volume change at the transition, as shown in Figure 4. Now, a tentative phase boundary between the MnP-related phase and the monoclinic phase is shown in Figure 8.

4. CONCLUSION

In the present work, we investigated the compression behavior of FeS up to 16 GPa and 800°C using a combination of a large-volume press (MAX80) and white X rays from a prototype of the third-generation synchrotron ring (KEK-AR). This combination provided new information: the structure of the high-pressure phase above 7GPa, which has been unknown, and more reliable phase equilibria under high pressure and temperature.

We still have a question about the compression behavior of the MnP-related phase. Figure 4 indicates that the MnP related phase is more compressible than the troilite phase at room temperature. *Minomura and Drickamer* [1963] observed a drastic electrical resistance decrease at the phase transition from the troilite phase to the MnP-related phase. The fact indicates that the phase transition is accompanied by an electronic state change. *Fei et al.*, [1995] also reported an interesting compression behavior of the MnP-related phase at high temperature, in which the *c/a* ratio had an anomaly around 5 GPa. They discussed the result as an electronic transition. Now we plan to measure electrical resistance of FeS simultaneously with X ray observation under high pressure.

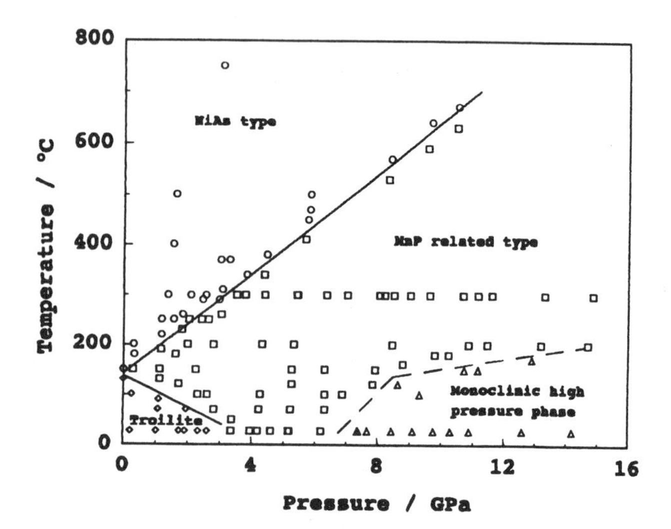

Figure 8. Phase equilibria of FeS under high pressure and temperature up to 16 GPa and 800°C. Symbol marks of ◊, ❒, ○, and Δ indicate the troilite phase, the MnP related phase, the simple NiAs type phase and the monoclinic high-pressure phase, respectively.

Acknowledgments. The authors thank Dr. H. Kobayashi, Department of Physics, Faculty of Science, Tohoku University, for his discussion about Mössbauer spectroscopy. This study has been performed under the approval of the Photon Factory Program Advisory Committee (Proposal No. 95G363). This work was also supported by a grant-in-aid for scientific research given by the Ministry of Education, Culture and Science, Japan.

REFERENCES

Decker D. L., High-pressure equation of state for NaCl, KCl and CsCl, *J. Appl. Phys.*, *42*, 3239-3244, 1971.

Fei Y., C. T. Prewitt, H. K. Mao and C. M. Bertka, Structure and density of FeS at high pressure and high temperature and the internal structure of Mars, *Science*, *268*, 1892-1894, 1995.

King H. E. and C. T. Prewitt, High-pressure and high-temperature polymorphism of iron sulfide (FeS), *Acta Cryst.*, *B38*, 1877-1887, 1982.

King H. E., D. Virgo and H. K. Mao, High-pressure phase transitions in FeS, using ^{57}Fe Mössbauer spectroscopy, in *Carnegie Inst. Washington Year Book*, *77*, pp. 830-835, 1978.

Kobayashi H., M. Sato, T. Kamimura, M. Sakai, H. Onodera, N. Kuroda and Y. Yamaguchi, Mössbauer study on Fe-S system with a NiAs-type structure under pressure, in *Proceedings of International Conference on the Applications of the Mössbauer Effect-1995*, Remini in Italy, *in press*.

Kusaba, K., Y. Syono, T. Kikegawa and O. Shimomura, A topological transition of B1-KOH-TlI-B2 type AgCl under high pressure, *J. Phys. Chem. Solids*, *56*, 751-757, 1995.

Louër D. and M. Louër, Méthode d'essais et erreurs pour l'indexation automatique des diagrammes de poudre, *J. Appl. Crystallogr.*, *5*, 271, 1971.

Minomura S. and H. G. Drickamer, Effect of pressure on the electrical resistance of some transition-metal oxides and sulfides, *J. Appl. Phys.*, *34*, 3043-3048, 1963.

Morimoto N., Nonstoichiometry and structure-disorder of pyrrhotite ($Fe_{1-x}S$), *J. Miner. Soc. Jpn.* (*in Japanese*), *12*, 326-338, 1976.

Nelmes, R. J., M. I. McMahon, N. G. Wright and D. R. Allan, Phase transitions in CdTe to 28 GPa, *Phys. Rev. B.*, *51*, 15723-15731, 1995.

Sato, H., S. Endo, M. Sugiyama, T. Kikegawa, O. Shimomura and K. Kusaba, Baddeleyite-type high-pressure phase of TiO_2, *Science*, *251*, 786-788, 1991.

Shimomura O., S. Yamaoka, T. Yagi, M. Wakatsuki, K. Tsuji, O. Fukunaga, H. Kawamura, K. Aoki and S. Akimoto, Multi-anvil type high-pressure apparatus for synchrotoron radiation, in *Solid State Physics Under High Pressure in Recent Advance with Anvil Devices*, edited by S. Minomura, pp. 351-356, KTK/Reidel, Tokyo/Dordrecht, 1985.

Taylor L. A. and H. K. Mao, A high-pressure polymorph of troilite, FeS, *Science*, *170*, 850-851, 1970.

Yund R. A. and H. T. Hall, Hexagonal and monoclinic pyrrhotites, *Econ. Geol.*, *64*, 420-423, 1969.

Keiji Kusaba, and Yasuhiko Syono, Institute for Materials Research, Tohoku University, Katahira, Aoba-ku, Sendai 980-77, Japan.

Takumi Kikiegawa, and Osamu Shimomura, Photon Factory, National Institute for High Energy Physics, Tsukuba 305, Japan.

Molar Volume, Thermal Expansion, and Bulk Modulus in Liquid Fe-Ni Alloys at 1 Bar: Evidence for Magnetic Anomalies?

P. M. Nasch and M. H. Manghnani

Hawaii Institute of Geophysics and Planetology, School of Ocean and Earth Science and Technology, University of Hawaii, Honolulu, Hawaii

New experimental data on the molar volume Ω, thermal expansion coefficient α, and ultrasonic sound velocity v_p in liquid Fe-Ni systems at temperatures between melting and 1975 K are reported. The molar volume and thermal expansion data were acquired using a penetrating γ radiation method; the sound velocity data were obtained by ultrasonic interferometry. In the temperature range of this study, the molar volume Ω increases and the sound velocity v_p decreases, both linearly with temperature. The adiabatic bulk modulus $K_S \propto v_p^2/\Omega$ of liquid Fe-Ni alloys is nearly independent of composition at Fe content greater than 65 wt%. The temperature derivative $\partial K/\partial T$ of both adiabatic and isothermal bulk modulus of pure liquid Fe decreases by approximately 50% upon being alloyed with 15 wt% Ni. The mixing behavior of thermodynamic and cohesive properties of liquid Fe-Ni is interpreted as resulting from the existence of disordered and localized magnetic states and correlations in the liquid state, i.e., well above the Curie temperature and extending from pure Fe into the Fe-Ni stability field. These magnetic contributions have strong mechanical effects on the structure in modifying the volume and elastic modulus by as much as 13% and 31%, respectively, in the case of pure liquid Fe. It is believed that the magnetic contribution, which is likely to be absent at core temperatures, should be removed from the measured 1-bar values of density and elastic moduli if these latter were to be used as precise anchoring points in high pressure-temperature EOS.

1. INTRODUCTION

The thermodynamic properties of binary liquid Fe-Ni systems are of fundamental geophysical importance. In particular, the molar volume Ω (or density, ρ), thermal expansion α, compressional wave velocity v_p, and their respective temperature dependences are all basic inputs for modeling the dynamics and rheology of the Earth's outer core. The 1-bar elastic properties of Fe-rich (i.e., with more than 75% Fe) liquids are critical in constraining deep Earth's interior models since any valid equation-of-state (EOS) must be anchored on the 1-bar data. As discussed recently [*Anderson*, 1996; *Cynn et al.*, this volume], the 1-bar thermal dependence of elastic properties can be valuable in inferring the pressure dependence of other physical properties of geophysical relevance.

The density and thermal expansion at 1 bar in liquid Fe-Ni find direct application in asteroidal core crystallization

Properties of Earth and Planetary Materials
at High Pressure and Temperature
Geophysical Monograph 101

modeling (less than a few kilobars core pressure) [*Haack and Scott*, 1992]. In relation to meteorites, in which a larger range of Ni content occurs, the overall equilibrium phase diagram of the Fe-Ni system at low pressure becomes of interest. In low acoustic attenuation liquids, such as Fe-Ni melts [*Nasch*, 1996], the shear modulus is ordinarily zero (i.e., the liquid does not resist change in shape). In this case the bulk sound velocity v_{Φ} equals the compressional (longitudinal) velocity v_p, and is related to the adiabatic bulk modulus K_S by the well known Laplace relationship

$$v_p = (K_S/\rho)^{1/2} \tag{1}$$

Here we present experimental results on molar volume Ω, thermal expansion α, and ultrasonic compressional wave velocity v_p in Fe-Ni melts. In the temperature range of this study (1700-1975 K), Ω increases and v_p decreases linearly with temperature. The adiabatic and isothermal bulk moduli are derived. Also, the Grüneisen first- and second-order parameters, thermal pressure, and specific heat ratio were computed using the present data, and their variation with Ni content is discussed in terms of possible magnetic contributions.

2. EXPERIMENTAL TECHNIQUES

2.1. *Gamma-Ray Densimetry*

A penetrating radiation technique has been used to measure the density and thermal expansion in molten Fe-Ni alloys. The density ρ of the material (or, conversely, the molar volume $\Omega = M/\rho$, where M is the molar mass) is determined from the attenuation which is produced by the passage of a collimated beam of monoenergetic γ radiation through the material, whose length, x, and attenuation coefficient, μ, are independently determined. The attenuation law of a beam of monoenergetic, well collimated radiation written in terms of γ-ray intensities is

$$I(x) = I_0 \exp\{-\mu \rho x\} \tag{2}$$

where I_0 and $I(x)$ are the measured γ-flux intensities of the incident beam and at the depth x inside the material, respectively. Corrections of $I(x)$ for instrumental losses (dead time) or nonideal geometry of collimation are of crucial importance. A detailed discussion of the experimental procedure, resulting errors, and results on pure molten metals (Fe, Ni and others) is given elsewhere [*Nasch and Steinemann*, 1995] and thus only a brief summary is given here.

Figure 1 depicts schematically the experimental arrangement for density measurement. Samples with nominal purities of at least 99.98% were first prepared by casting in order to assure a good starting homogeneity and to prevent the occurrence of exothermic mixing reactions during an actual high-temperature measurement cycle. The metal sample is contained in an Al_2O_3 ceramic crucible. A graphite sleeve surrounds the crucible and converts the electromagnetic energy, produced by a radio frequency induction generator, to Joulean heat. A ZrO_2 radiation shield serves for thermal insulation of the high-temperature sample assembly. The sample temperature is monitored by an alumina-sheathed Pt-Pt13%Rh thermocouple positioned in the melt. The measuring assembly is contained in a clear-fused quartz tube held by water-cooled flanges. The high-temperature chamber is evacuated with a high-vacuum system for the heating and melting cycles, then purified argon is introduced for the measurement cycle. The γ-rays (0.662 MeV) are produced by a Pb-shielded ^{137}Cs source of 350 mCi nominal activity. The γ beam is collimated both before entering and after exiting the sample-containing high-temperature assembly. The detection system and counting electronics consist of a temperature-controlled NaI(Tl) scintillator coupled to a photomultiplier tube, an amplifier with automatic gain stabilizer, a multichannel analyzer and a counter. Thermal control of the collimators and detector system ensures count-rate stability of the electronic counting system. A constant statistics was achieved by always measuring to 10^6 counts. All count rates are corrected for the dead time τ=10±1 μs of the electronic detection system. A detailed study of random and systematic errors has shown an upper limit of 0.75% as the highest possible error for the absolute density determination by means of the γ-ray attenuation technique [*Nasch and Steinemann*, 1995].

The mass attenuation coefficient μ is a function of sample atomic mass and γ-ray energy, and is independent of the physical state of the sample, i.e., μ is temperature independent. However, μ is geometry-dependent (because of nonideal collimation), and it is therefore necessary to measure μ in situ for each endmember. For this purpose, a solid sample for each pure endmember (Fe, Ni) with two parallel faces has been machined out of the same cast as for high-temperature run. For each material, the solid sample of known thickness x was then carefully mounted in the path of the γ beam to achieve normal incidence on one of the parallel faces. The transmitted amplitude $I(x)$ was measured at room temperature and the room-temperature product $\mu\rho$, i.e., the *linear* attenuation coefficient, was computed using Equation (2) and the measured I_0 obtained without sample. Then, the room-temperature density ρ of the solid sample was measured by X ray diffraction using a diffractometer calibrated with Si. The knowledge of ρ allows μ to be extracted from measured $\mu\rho$. In our geometry, the obtained mass attenuation coefficients μ for Fe and Ni are 0.0734 and 0.0755 $cm^2\ g^{-1}$, respectively. The propagation of random errors including those made in the solid metal X ray density measurement yields a resulting error on μ of 0.6% [*Nasch and Steinemann*, 1995].

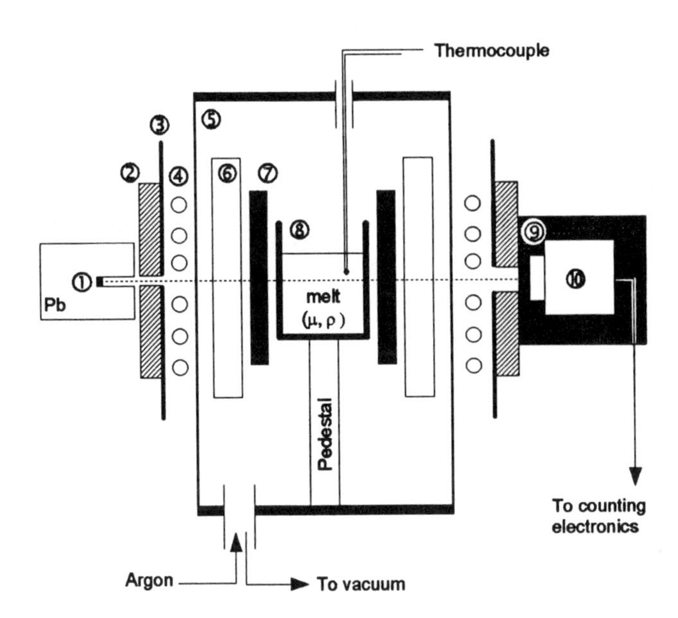

Figure 1. Schematic arrangement for γ-ray density measurement (not to scale): (1) γ radiation source in Pb vault, (2) collimator, (3) water-cooled heat shield, (4) rf inductor coil, (5) quartz tube, (6) ZrO_2 thermal insulator, (7) graphite susceptor, (8) Al_2O_3 crucible containing the melt, (9) water-cooled protective shield, and (10) NaI(Tl) scintillator and photomultiplier tube.

2.2. *Ultrasonic Interferometry*

The principle of our high-temperature ultrasonic interferometry is based on the system described by *Katahara et al.* [1981]. Briefly, a gated continuous wave is transmitted through a layer of melt by means of Al_2O_3 dense ceramic buffer rods. Acoustic impedance contrast between the rods and the melt generates multiple reflections within the melt layer that interfere with each other. A resonant, constructive standing wave is formed for melt thicknesses equal to integral multiples of half the wavelength ($n\lambda/2$), whereas antiresonant, destructive conditions occur when the melt thickness corresponds to $(n+1/2)\lambda/2$, where λ is the ultrasonic wavelength in the melt. When the upper buffer rod is moved up and away from the fixed lower buffer rod, the amplitude of the transmitted ultrasonic signal traces out an alternating constructive (maxima) and destructive (minima) interference pattern. Combining maxima separation measurements ($\lambda/2$) with the known carrier-wave frequency f, the velocity is computed by the relation $v_p = \lambda \cdot f$.

Experimental details are summarized in *Nasch et al.* [1994a]. However some recent changes (use of single-crystal sapphire buffer rod; use of heaters mounted on the buffer rod; improved vacuum system (no bell jar); alignment procedure; shorter rf impedance bus; and new rod displacement system, among others) have considerably ameliorated the performance of the apparatus for sound velocity measurements at high temperature [*Nasch*, 1996]. This is illustrated below with a recent run on pure (99.9+%) molten Fe. Figure 2 represents the latest mechanical arrangement of our interferometer, and details of the high temperature assembly are shown in Figure 3. The molten metal is contained in a Al_2O_3 crucible encapsulated in a graphite susceptor and insulated outside with a ZrO_2 radiation shield. The assembly is enclosed in a quartz tube. A controlled Ar atmosphere is introduced in the furnace above the melting point. As in the case of the γ-ray attenuation technique described above, a radio-frequency induction technique is used for the heating and melting. The sample temperature is monitored by an alumina-sheathed W-26%Re vs. W-5%Re thermocouple located in the melt. A second thermocouple feeds the temperature controller in turn connected to a 50-200 kHz, 15 kW rf induction heating unit. The achieved temperature stability is better than ±0.5 K at 1975 K. A clear fused quartz tube (11.5 cm in diameter) is placed around the high-temperature assembly and is sealed at both ends by water-cooled flanges, thus isolating the high-temperature load from the environment. This assembly allows a high-vacuum system (turbomolecular pump) to evacuate the quartz tube prior to the introduction of a pressurized atmosphere of argon. A 7-turn helical Cu coil couples the rf power to the load from outside the quartz tube. The upper buffer rod is held by a water-cooled housing mounted on a 32-cm-diameter aluminum plate. The entire plate moves up and down to allow for varying the melt thickness, and is guided by high-precision shafts fed through linear bearings attached to the plate which can be moved vertically by a motor driven threaded shafting. A counterweight attached to the plate via a pulley system (not shown in Figure 2) eliminates any torque on the plate and stress on the motor. A metal bellows makes the seal between the moving plate and the quartz top flange, allowing for the rod to move friction-free.

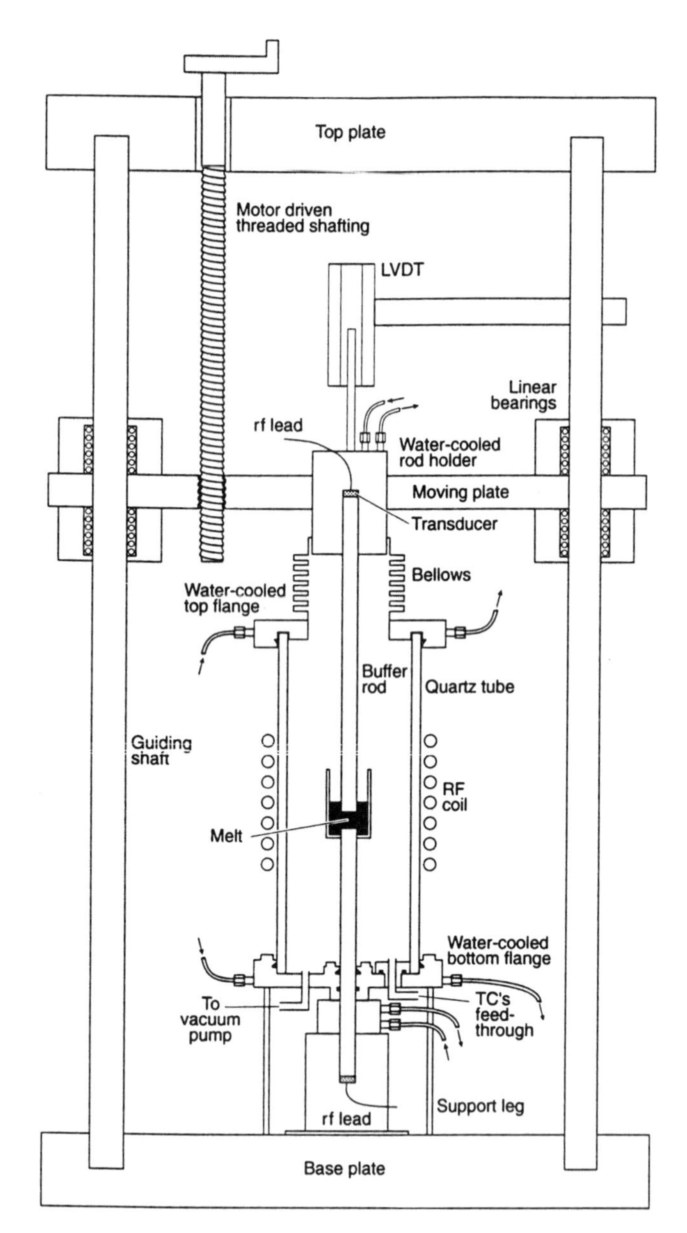

Figure 2. Schematic mechanical arrangement of the ultrasonic interferometer.

3. RESULTS

3.1. *Molar Volume and Thermal Expansion*

Density data points have been obtained for Fe_x-Ni_{1-x} alloys (x= 25, 50, 65, and 75 wt% Fe) in the temperature range from their respective melting points to a maximum temperature of 1975 K. The least-squares fits to the data of the density as a function of temperature over the range studied have shown that it is suitable to limit the polynomial development of the density to the first order. Thus, one has

$$\rho(T) = a + b(T - T_m) \tag{3}$$

where $a = \rho_m$ is the density at the melting point T_m and $b = (d\rho/dT)_P$. The coefficient of thermal expansion is defined by

$$\alpha \equiv \frac{1}{V}\left(\frac{\partial V}{\partial T}\right)_{P,N} = \frac{1}{\Omega}\left(\frac{\partial \Omega}{\partial T}\right)_P = -\frac{1}{\rho}\left(\frac{\partial \rho}{\partial T}\right)_P \tag{4}$$

and is thus calculated as the ratio $-b/a$. The errors on α are derived from least-squares statistical analysis and do not exceed 11%. We present in Table 1 our results of ρ_m, b and α, where pure endmember values (namely, Fe and Ni) are recalled from *Nasch and Steinemann* [1995] for convenience. Our density data are converted to molar volume by the relation $\Omega = M/\rho$ where $M = xM_{Fe} + (1-x)M_{Ni}$ is the molar mass of the alloy calculated from an additive relationship between the pure endmember molar mass values M_{Fe} and M_{Ni}, x being the Fe molar fraction. The compositional variation of the molar volume in liquid Fe-Ni from our study (solid circles) is compared in Figure 4 (bottom part) along an isotherm at 1823 K with the results

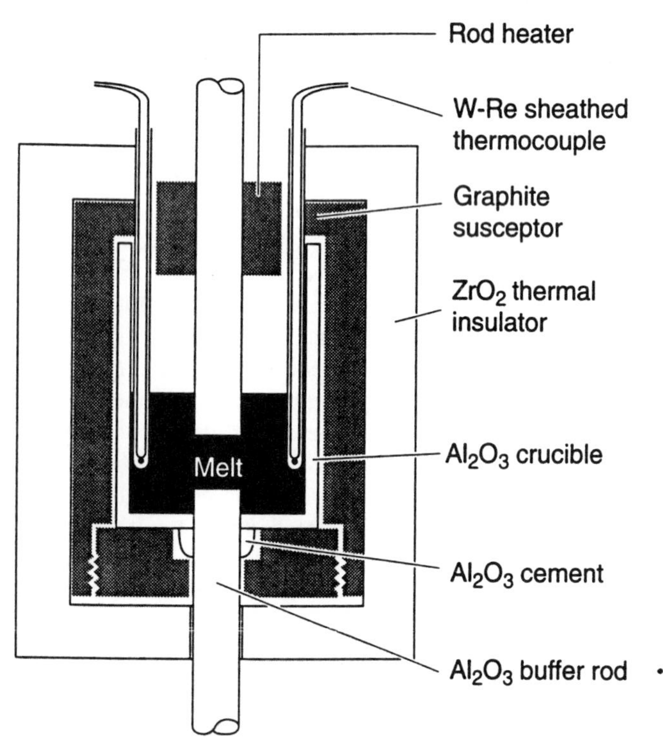

Figure 3. High-temperature sample assembly.

of *Popel' et al.* [1969] using the less accurate drop volume method (open circles), and with solid, low-temperature X ray lattice spacing data [*Pearson*, 1964] (open squares). Note the similarity in trend, with a smooth maximum occurring in the invar (Fe-35%Ni) region. Such a trend was also observed in solid fcc Fe-Ni alloys at high temperatures (>1000°C) [*Tanji et al.*, 1990]. Also shown in Figure 4 (top part) is the compositional variation of the volumetric thermal coefficient of expansion α. Note the broad minimum in the neighborhood of 75% Fe.

3.2. *Sound Velocity*

The compressional sound wave velocity v_p in liquid Fe, Fe-8%Ni, Fe-15%Ni, and Fe-40%Ni (which equals the bulk sound speed in these low-loss liquids owing to the absence of shear component) has been measured from the respective melting temperature to a maximum temperature of 1975 K. The temperature dependence of v_p is linear in the temperature range investigated, and, as in most metals and alloys, v_p decreases with increasing temperature. This is commonly attributed to continuous structural loosening (see *Iida and Guthrie* [1988]). When the temperature rises, changes in the coordination number and in the interatomic distance produce a net increase in the free-volume. Thus, the liquid becomes more compressible, which in turn lowers the sound velocity.

The run on pure Fe serves as a measure of the improvements made to the system since our preliminary report [*Nasch et al.*, 1994b]. Table 2 compares published values

TABLE 1. Density, ρ, at the Melting Point, Temperature Derivative, $\partial\rho/\partial T$, and Volumetric Thermal Expansion Coefficient, α in Liquid Fe_x - Ni_{1-x} at 1 bar Obtained by γ Ray Attenuation Densimetry

x	T range (K)	ρ, (g cm^{-3})	$\partial\rho/\partial T$ (10^{-4} g cm^{-3} K^{-1})	α (10^{-6} K^{-1})
0.00 [a]	1726-1973	7.81	- 7.26	93 ± 3
0.25	1703-1973	7.52	- 6.24	83 ± 3
0.50	1708-1973	7.32	- 5.71	78 ± 8
0.65	1718-1873	7.05	- 5.29	75 ± 8
0.75	1753-1873	7.12	- 4.98	70 ± 7
1.00 [a]	1809-1973	6.98	- 5.72	82 ± 8

[a] From *Nasch and Steinemann* [1995]

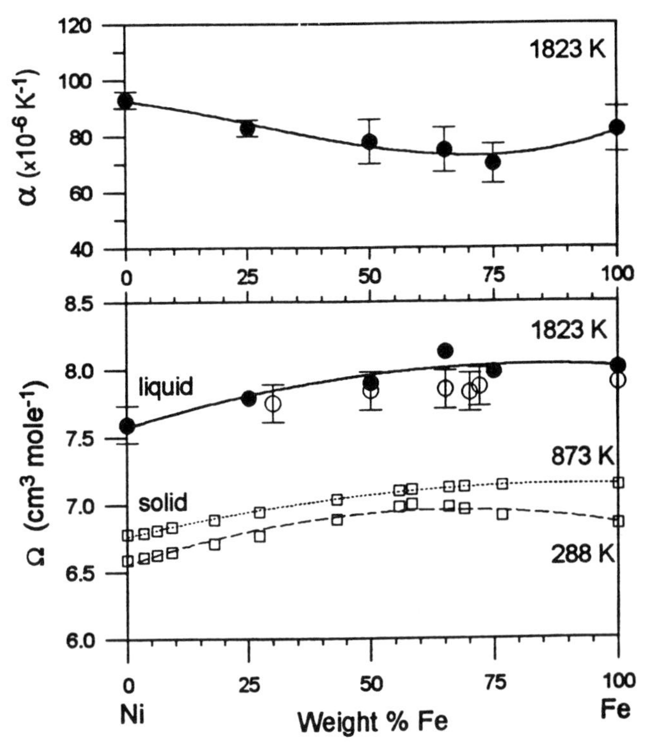

Figure 4. Compositional variation of the molar volume Ω (bottom) and volumetric thermal expansion α (top) in liquid Fe-Ni along a 1823 K isotherm. ●: this study; O: *Popel' et al.* [1969]; □ solid X ray lattice spacing data [*Pearson*, 1964].

for v_p and $\partial v_p/\partial T$ at the melting point (1809 K) of pure liquid Fe.

The 60% Fe datum, even though outside the composition of geophysical importance, serves as a connection with the data of *Nasch et al.* [unpublished] who employed a variable-path time-of-flight ultrasonic technique [*Seemann and Klein*, 1965; *Keita et al.*, 1981] for nickel-rich (>25% Ni) liquid alloys. The compositional variation of v_p in liquid Fe-Ni along an isotherm (1823 K) is depicted in Figure 5. The datum at 92% Fe is low and is due to a 2.4% temperature-independent systematic error caused by a decalibration of the melt thickness measuring device. Correction is not undertaken here, awaiting confirmation from a rerun.

With the exception of the datum at 92%Fe, the results from ultrasonic interferometry (solid circles) are nicely correlated with those obtained by means of the variable-path method (open squares), attesting the compatibility of the two techniques. The variation of v_p with composition shows a nonlinear departure from an ideal mixing behavior (dashed line). The ideal trend is based on the assumption of additivity of the compressibilities of Fe and Ni [*Van Dael*,

TABLE 2. Sound Velocity v_p at the Melting Point (1809 K) and Temperature Derivative $\partial v_p/\partial T$ in Liquid Fe at 1 bar

Method	Max. T (K)	v_p (m s^{-1})	$\partial v_p/\partial T$ (m s^{-1} K^{-1})	Error on v_p (%)	References
Pulse-Echo	1973	3930	-0.34	0.5	*Kurz and Lux* [1969]
Pulse-Phase	1923	3917	-0.42	0.3	*Kats et al.* [1978]
Variable-Path	1973	3912	-0.22	0.4	*Keita et al.* [1981]
Floating-Zone	1809	4052	------[a]	3.3	*Casas et al.* [1984]
Pulse Transmitting	1893	3983	-1.00	0.7	*Tsu et al.* [1985]
Ultrasonic Interferometry	1883	3820	-0.72	0.5	*Nasch et al.* [1994*b*]
Ultrasonic Interferometry	1950	3961	-0.86±0.15	0.6	This work

[a] Not applicable

1975]. Similar nonlinear trends have been reported for the bulk modulus at 1000°C in solid fcc Fe-Ni [*Renaud*, 1988; *Renaud and Steinemann*, 1990] and also for the viscosity in liquid Fe-Ni (e.g., *Iida and Guthrie*, [1988] and references therein).

The temperature coefficient of the sound velocity $\partial v_p/\partial T$ is plotted with respect to Fe concentration in Figure 6 and compared to data from *Nasch et al.* [unpublished]. Error bars are derived from statistical analysis. The compositional variation of $\partial v_p/\partial T$ also behaves nonmonotonically, and similar to the v_p trend, a pronounced dip is found near the 70% Fe composition.

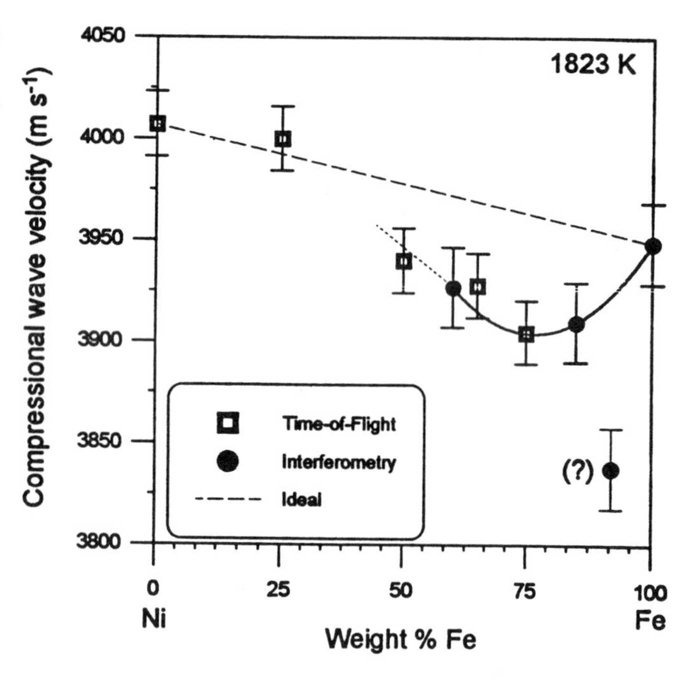

Figure 5. Compositional variation of the sound velocity v_p in liquid Fe-Ni along a 1823 K isotherm. The results from ultrasonic interferometry (●, this study) are compared to those obtained by means of the variable-path time-of-flight method (□) [*Nasch et al.*, unpublished]. The variation of v_p with composition shows a nonlinear mixing trend with a maximum departure from an ideal mixing behavior (dashed line) near 75% Fe.

4. DISCUSSION

Using Equation (1) together with the well known thermodynamic relation

$$\frac{K_S}{K_T} = 1 + \frac{T\alpha^2 \Omega K_S}{M C_P} \qquad (5)$$

where C_P is the specific heat at constant pressure, the adiabatic K_S and isothermal K_T bulk modulus for liquid Fe-Ni have been computed for every composition and are shown in Figure 7. No distinction is made in the following discussion between the present sound velocity data and those obtained by time-of-flight method. The molar volumes for 60, 85, and 92% Fe have been obtained by interpolation using a quadratic fit of Ω vs. x in Figure 4.

C_P values were taken from *Nasch and Steinemann* [1995] and references therein for the pure endmembers, and are assumed to obey Kopp's rule which states that in a binary mixture C_P varies linearly with concentration, (i.e., the activity coefficient is independent of temperature). Our derived K_S for liquid Fe at the melting point is 110±2 GPa and compares well with the reference isentrope value of 109.7±0.7 GPa recommended by *Anderson and Ahrens* [1994]. Figure 7 shows that, for a nickel content less than 35%, alloying nickel does not affect significantly the magnitude of K_S of liquid Fe. In the case of an ideal solution, we should have expected an almost linear mixing law. Note the softening in K_S with a deviation from the linear mixing maximum at about 65% Fe composition. This softening could indicate that some electronic effects still modify the elastic moduli. The temperature derivative of the two elastic moduli, $(\partial K_S/\partial T)_P$ and $(\partial K_T/\partial T)_P$, has been derived at the melting point by differentiating equations (1) and (5) with respect to temperature and using the known effects of temperature on molar volume and sound velocity. Further, in the case of $(\partial K_T/\partial T)_P$, we assumed a temperature-independent specific heat [e.g., *Barin et al.*, 1977]. The compositional variation of $(\partial K_S/\partial T)_P$ and $(\partial K_T/\partial T)_P$ is shown in Figure 8(a). Numerical values for these isobaric coefficients are

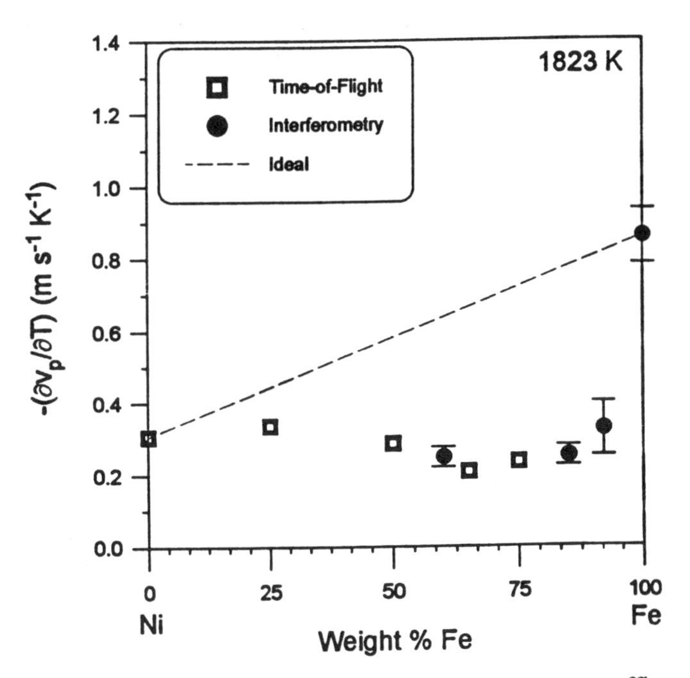

Figure 6. Compositional variation of the temperature coefficient of the sound velocity $\partial v_p/\partial T$ in Fe-Ni liquid alloys. Error bars are derived from statistical analysis. The coefficient $\partial v_p/\partial T$ behaves nonmonotonically, and a pronounced dip is found near 75% Fe.

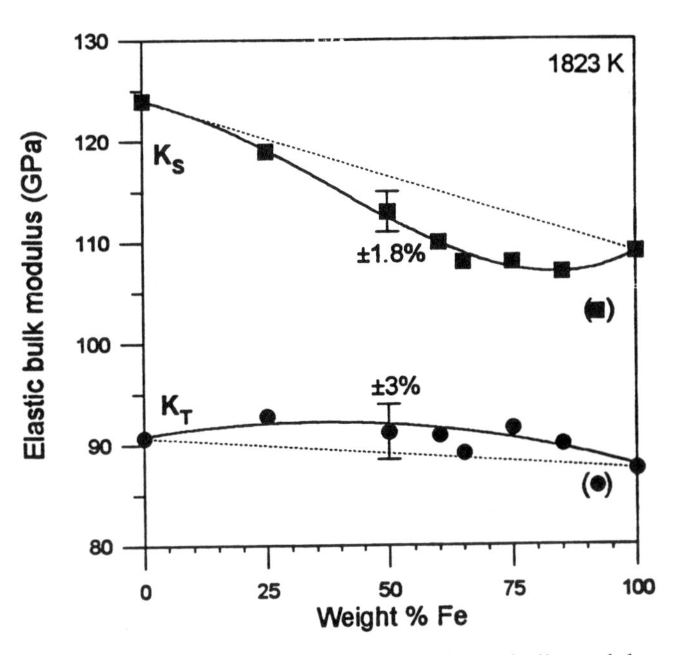

Figure 7. Adiabatic K_S and isothermal K_T bulk modulus in liquid Fe-Ni at 1823 K. Typical propagated errors are shown at the equiatomic composition. Note the softening in bulk moduli (especially prominent in K_S) which is most pronounced near 65-75% Fe.

summarized in Table 3 along with other various important thermodynamic properties of geophysical relevance:

Grüneisen parameter

$$\gamma_G = \frac{\Omega \alpha K_S}{M C_P} \tag{6}$$

Specific heat ratio

$$\frac{C_P}{C_V} = \frac{K_S}{K_T} = 1 + T \alpha \gamma_G \tag{7}$$

Thermal pressure coefficient

$$\left(\frac{\partial P}{\partial T}\right)_V = \alpha K_T \tag{8}$$

Second-order Grüneisen parameters

$$\delta_S = \frac{1}{\alpha K_S}\left|\left(\frac{\partial K_S}{\partial T}\right)_P\right| \tag{9a}$$

$$\delta_T = \frac{1}{\alpha K_T}\left|\left(\frac{\partial K_T}{\partial T}\right)_P\right| \tag{9b}$$

TABLE 3. Thermoelastic Properties of Liquid Fe-Ni Alloys at the Melting Point T_m and 1 bar

	Mol Fraction x in Fe_x - Ni_{1-x}								
Parameter	0.00	0.25	0.50	0.60	0.65	0.75	0.85	0.92	1.00
T_m (K)	1726	1703	1708	1715	1718	1753	1763	1783	1809
ρ (g cm^{-3})	7.81	7.52	7.32	*7.21* [a]	7.05	7.12	*7.05*	*7.02*	6.98
α ($\times 10^{-6}$ K^{-1})	93	83	78	*75*	75	70	*73*	*76*	82
v_P (m s^{-1})	4037	4040	3973	3953	3950	3922	3924	3850	3961
$\partial v_P / \partial T$ (m s^{-1} K^{-1})	-0.31	-0.34	-0.29	-0.25	-0.21	-0.24	-0.24	-0.35	-0.86
K_S (GPa)	127	123	116	113	110	110	109	104	110
$(\partial K_S/\partial T)_P$ (MPa K^{-1})	-31	-31	-26	-23	-20	-21	-21	-27	-57
K_T (GPa)	94	97	95	94	92	93	91	87	88
$(\partial K_T/\partial T)_P$ (MPa K^{-1})	-31	-31	-27	-25	-23	-23	-23	-27	-46
C_P/C_V	1.36	1.27	1.22	1.20	1.20	1.17	1.19	1.20	1.24
γ_G	2.2	1.9	1.7	1.5	1.6	1.4	1.5	1.4	1.6
αK_T (MPa K^{-1})	8.7	8.0	7.4	7.0	6.9	6.5	6.7	6.6	7.2
$(\partial\alpha/\partial P)_T$ ($\times 10^{-6}$ GPa^{-1} K^{-1})	-3.6	-3.3	-3.0	-2.8	-2.7	-2.7	-2.8	-3.6	-5.9
δ_S	2.6	2.9	2.8	2.6	2.4	2.7	2.6	3.3	6.2
δ_T	3.6	3.8	3.7	3.5	3.3	3.5	3.5	4.1	6.4
Φ (km^2 s^{-2})	16.3	16.3	15.8	15.6	15.6	15.4	15.4	14.8	15.7

[a] Values in italic have been interpolated

Pressure dependence of thermal expansion

$$\left(\frac{\partial\alpha}{\partial P}\right)_T = \frac{1}{K_T^2}\left(\frac{\partial K_T}{\partial T}\right)_P \tag{10}$$

Seismic parameter

$$\Phi = \frac{K_S}{\rho} = \left(\frac{\partial P}{\partial\rho}\right)_S = v_P^2 \tag{11}$$

Figure 8(b) depicts the compositional variations at the melting point of the first- and second-order Grüneisen parameters, and Figure 8(c) shows the thermal pressure coefficient (solid squares) and pressure derivative of thermal expansion (solid circles).

From a general inspection of Figures 4-8, it seems that all the properties display a change in trend upon mixing which is mostly prominent in the neighborhood of the invar Fe_3Ni stochiometry (i.e., in the range 65-85% Fe). A high-temperature invar "intrusion" seems unlikely since all invar-related effects should have completely disappeared above 1000°C [*Renaud and Steinemann*, 1990], excluding thus any ferromagnetic origin for these high-temperature "anomalies". Valence change [*Kleppa*, 1960; *Kleppa et al.*, 1961], charge transfer [*Waseda et al.*, 1984], and high spin-low spin transition, are among the possible effects that are modifying the interatomic association in liquid binary alloys. It is also conceivable a priori that the nonlinear mixing could result from an additional "pseudo-segregation" effect as if remnants of the metastable γ' (fcc) phase proposed by *Nakagawa et al.* [1979] and *Tanji et al.* [1990] were persisting in the liquid state. (γ' results from a hypothetical high-temperature metallurgical instability of a spinodal decomposition around the Fe_3Ni fcc phase). An immiscibility effect should be detectable in X ray diffraction experiments through measurements of the compositional-sensitive concentration-concentration fluctuation function. However, such study in liquid Fe-Ni [*Waseda and Jacob*, 1981] does not reveal any noticeable departure from perfect mixing near or at that stochiometry.

Our favored interpretation qualitatively compatible with

the observation is based on the existence of disordered local magnetic moments (DLM) [*Friedel et al.*, 1961; *Janak and Williams*, 1976; *Pettifor*, 1980] that are known to persist in some 3d transition metals well above the Curie temperature, and even in the liquid state [*Grimvall*, 1976; *Steinemann and Keita*, 1988]. In itinerant-electron theory of band magnetism [e.g., *Hasegawa and Pettifor*, 1983], DLM result from a band splitting due to a spin polarization and are accompanied by an increase in kinetic energy due to broadening of the band. DLM generate a repulsive magnetic interaction which produces a repulsive internal pressure (no ordering is required). DLM have thus strong mechanical effects on the structure through magnetic contribution to the volume and elastic modulus (Figure 9). The molar volume is "inflated" (compared to the nonmagnetic case) and the cohesive energy is reduced (bulk modulus is reduced). Iron is expected to carry strong DLM whereas nickel should have weak (or total lack of) DLM [*Janak and Williams*, 1976; *Steinemann and Keita*, 1988]. The softening in the

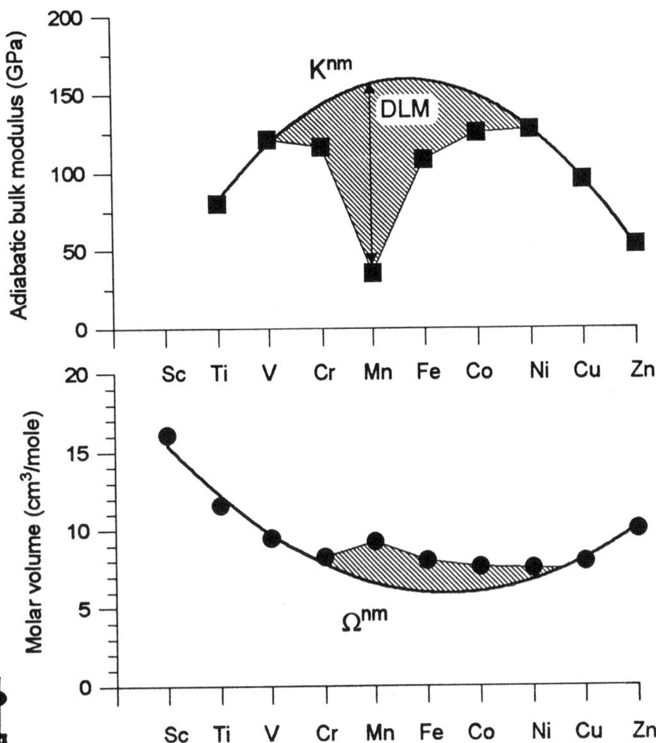

Figure 9. Adiabatic bulk modulus (top) and molar volume (bottom) of liquid 3d transition metals at the melting temperature. Data are taken from *Nasch and Steinemann* [1995] and references therein. In each case, the heavy solid line is a second-order polynomial fit on elements at left (Sc, Ti, V) and at right (Ni, Cu, Zn), and represents the nonmagnetic (nm) reference state. Deviations from the parabola (hatched areas) are interpreted as resulting from magnetic interactions (DLM) and are more pronounced at half-filled 3d band. In this picture, Fe carries strong DLM whereas Ni should have weak (or total lack of) DLM.

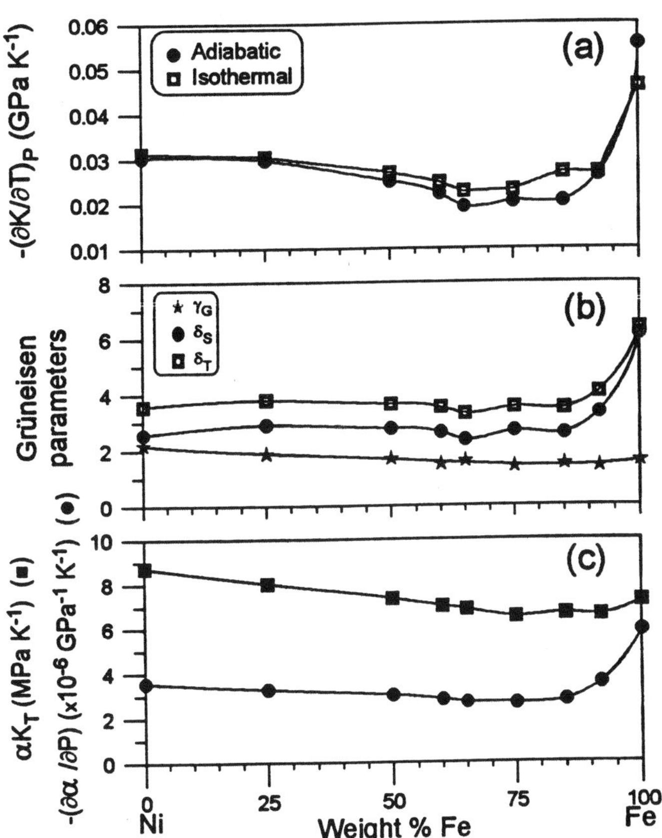

Figure 8. Thermoelastic properties of liquid Fe-Ni at the melting point. a) $\partial K_S/\partial T$ and $\partial K_T/\partial T$; b) first- (γ_G) and second-order (δ_S, δ_T) Grüneisen parameters; c) $\partial\alpha/\partial P$ and thermal pressure coefficient αK_T.

elastic modulus (Figure 7) and the small but positive excess molar volume (Figure 4) could be signs of the presence of DLM extending from pure Fe in the Fe-Ni stability field. Also, the temperature derivatives α, $\partial v_p/\partial T$, $\partial K/\partial T$, most sensitive to structural changes, show a marked drop from the pure iron values as the nickel content increases (Figures 4, 6, and 8(a), respectively). Such a magnetic contribution to the cohesion should have important consequences on the applicability of room-pressure data as anchoring points of EOS. For example, our values for the density ρ_o=6.98±0.05 (g cm^{-3}) and the bulk modulus K_{So}=110±2 (GPa) at 1 bar, 1809 K, can be corrected from DLM contributions following the procedure devised by *Steinemann and Keita* [1988]. According to them, the bulk modulus and molar volume for liquid 3d transition metals in absence of magnetic effects can be estimated using models for elastic constants of solid transition metals which predict a parabolic variation of K and Ω through the 3d transition series [see details in *Steinemann and Keita* 1988, and references therein]. A second-order polynomial fit on elements at left (Sc, Ti, V) and at right (Ni, Cu, Zn) of the series is used in this framework as the nonmagnetic reference state, and the excess quantities $K^{nm} - K$ and $\Omega - \Omega^{nm}$ (hatched areas in Figure 9) are interpreted as resulting from magnetic interactions. It follows that the nonmagnetic values for the bulk modulus and the density become K_{So}^{nm}=144 (GPa) and ρ_o^{nm}=7.89 (g cm^{-3}). Thus the DLM contribution is important in modifying the molar volume (13%) and the adiabatic bulk modulus (31%). It is not clear to the authors to what extent these corrections have on the interpretation of shock wave data. However, it is believed that the comparison, which is implicitly made in EOS fitting procedure, between high $P-T$ and normal conditions gains in justification [*Besson and Nicol*, 1990] if these latter are corrected from magnetic contributions no longer existing at very high $P-T$ conditions.

5. SUMMARY

The molar volume Ω, thermal expansion coefficient α, and compressional sound wave velocity v_p of molten Fe-Ni have been measured at several compositions from the melting point up to 1975 K. In the temperature range of this study (1700-1975 K), Ω increases and v_p decreases linearly with temperature. Many valuable physical parameters such as Grüneisen first- and second-order parameters, thermal pressure, specific heat ratio, and seismic parameter were computed using these data. Alloying Ni to pure Fe does not affect significantly the adiabatic bulk modulus K_S until the nickel content reaches the region of invar stochiometry (Fe-35%Ni). Near or at this composition, the compositional variations of Ω and K_S show a marked change in mixing trend, and α and $\partial v_p/\partial T$ are minimum in this region. Such a nonlinear mixing behavior is unlikely to be caused by ferromagnetic effects or immiscibility but can well be related to DLM-controlled liquid state properties. DLM have strong mechanical effects on the structure through magnetic contribution to the volume and elastic modulus. Since the fitting of EOS is constrained by the values of the density ρ_o and bulk modulus K_{So} at 1 bar, the magnetic contribution, which is absent at high temperature, should be removed from ρ_o and K_{So} if these were to be used as precise anchoring points in high pressure-temperature EOS.

Clearly, more work is needed to clarify the role of DLM on the 1-bar physical properties of Fe melts and their implications on shock-wave data interpretation and derived EOS for the core.

Acknowledgments. This work was funded by NSF (grants # EAR89-17531, EAR94-06790), W. M. Keck Foundation, University of Hawaii, and the Swiss National Science Foundation. S. G. Steinemann is gratefully acknowledged for initiating and supervising the density measurements in his metallurgical laboratory (University of Lausanne, Switzerland). Our deepest appreciation to J.-F. Jeanneret and W. Schneider of the University of Lausanne (Switzerland) for preparing the Fe-Ni samples. We also thank J. Balogh (University of Hawaii) for technical advice. P.M.N. further gratefully acknowledges financial support for graduate study from the following sponsors in Switzerland: Foundation L. & H. Mouttet (LEMO S.A.), Foundation Fern-Moffat (Société Académique Vaudoise), and Foundation Sunburst (Council of the Swiss Institutes of Technology). The authors are indebted to Thomas J. Ahrens, Don L. Anderson, Orson L. Anderson, and Hartmut Spetzler for making constructive criticism of the manuscript. School of Ocean and Earth Science and Technology contribution no. 4135.

REFERENCES

Anderson, O. L., Finding physical properties at high pressure using the high-temperature database (abstract), *U.S.-Japan Seminar on High Pressure-Temperature Research: Properties of Earth and Planetary Materials*, Maui, Hawaii, USA, January 22-26, 1996.

Anderson, W. W., and T. J. Ahrens, An equation of state for liquid iron and implications for the earth's core, *J. Geophys. Res., 99*, 4273-4284, 1994.

Barin, I., O. Knacke, and O. Kubaschewski, *Themochemical Properties of Inorganic Substances* (supplement), Springer-Verlag, Berlin, 1977.

Besson, J. M., and M. Nicol, An equation of state of g-Fe and some insights about magnetoelastic effects on measurements of the a-g-e triple point and other transitions, *J. Geophys. Res., 95*, 21,717-21,720, 1990.

Casas, J., N. M. Keita, and S. G. Steinemann, Sound velocity in liquid titanium, vanadium and chromium, *Phys. Chem. Liq., 14*, 155-158, 1984.

Friedel, J., G. Leman, and S. Olszewski, On the nature of the magnetic couplings in transitional metals, *Suppl. J. Appl. Phys., 32*, 325S-330S, 1961.

Grimvall, G., Polymorphism of metals. III. Theory of the temperature-pressure phase diagram of iron, *Physica Scripta, 13*, 59-64, 1976.

Haack, H., and E. R. D. Scott, Asteroid core crystallization by inward dentritic growth, *J. Geophys. Res., 97*, 14,727-14,734, 1992.

Hasegawa, H., and D. G. Pettifor, Microscopic theory of the temperature-pressure phase diagram of iron, *Phys. Rev., 50*, 130-133, 1983.

Iida, T., and R. I. L. Guthrie, *The Physical Properties of Liquid*

Metals, 288 pp., Oxford Science Publications, Clarendon Press, Oxford, 1988.
Janak, J. F., and A. R. Williams, Giant internal pressure and compressibility anomalies, *Phys. Rev. B, 14*, 4199-4204, 1976.
Katahara, K. W., C. S. Rai, M. H. Manghnani, and J. Balogh, An interferometric technique for measuring velocity and attenuation in molten rocks, *J. Geophys. Res., 86*, 11,779-11,786, 1981.
Kats, Ya. L., L. N. Sokolov, and G. N. Okorokov, Ultrasonic investigation of Fe-Cr melts, *Izv. Akad. Nauk. SSSR Met., 5*, 57, 1978.
Keita, N. M., H. Morita, and S. G. Steinemann, Cohesive properties of liquid transition metals, *Proc. 4th Int. Conf. Rapidly Quenched Metals*, 119-122, Sendai, Japan, 1981.
Kleppa, O. J., The volume change on mixing in liquid metallic solutions. I. Alloys of cadmium with indium, thallium, lead and bismuth, *J. Phys. Chem., 64*, 1542-1546, 1960.
Kleppa, O. J., M. Kaplan, and C. E. Thalmayer, The volume change on mixing in liquid metallic solutions. II. Some binary alloys involving mercury, zinc and bismuth, *J. Phys. Chem., 65*, 843-849, 1961.
Kurz, W., and B. Lux, Velocity of sound in iron and iron alloys in the solid and molten state (in German), *High Temp. - High Press., 1*, 387-399, 1969.
Nasch, P. M., *Elastic and Anelastic Properties of Liquid Iron and Iron Alloys: Applications to the Earth's Core*, Ph.D. thesis, 220 pp., University of Hawaii, USA, 1996.
Nasch, P. M., and S. G. Steinemann, Density and thermal expansion of molten manganese, iron, nickel, copper, aluminum and tin by means of the gamma-ray attenuation technique, *Phys. Chem. Liq., 29*, 43-58, 1995.
Nasch, P. M., M. H. Manghnani, and R. A. Secco, A modified ultrasonic interferometer for sound velocity measurements in molten metals and alloys, *Rev. Sci. Instrum., 65*, 682-688, 1994a.
Nasch, P. M., M. H. Manghnani, and R. A. Secco, Sound velocity measurements in liquid iron by ultrasonic interferometry, *J. Geophys. Res., 99*, 4285-4291, 1994b.
Pearson, W. B. (Ed.), *Handbook of Lattice Spacings and Structures of Metals*, Pergamon Press, Oxford, 1964.
Pettifor, D. G., Electronic structure calculations and magnetic properties, *J. Mag. Magn. Mat., 15-18*, 847-852, 1980.
Popel', S. I., L. M. Shergin, and B. V. Tsarevskii, Temperature variations of the densities and surface tensions of iron-nickel melts, *Russian J. Phys. Chem., 43*, 1325-1327, 1969.
Renaud, P., Propriétés magnétoélastiques des invars (in French), Ph.D. thesis, 106 pp., University of Lausanne, Switzerland, 1988.
Renaud, P., and S. G. Steinemann, High temperature elastic constants of Fe-Ni invar alloys, in *Physical Metallurgy of Controlled Expansion Invar-Type Alloys*, edited by K. C. Russell and D. F. Smith, pp. 225-234, The Minerals, Metals and Materials Society, 1990.
Seemann, H. J., and F. K. Klein, Schallgeschwindigkeit und Kompressibilität von Quecksilber und geschmolzenen Aluminum (in German), *Z. Angew. Phys., 19*, 368-374, 1965.
Steinemann, S. G., and N. M. Keita, Compressibility and internal pressure anomalies of liquid 3d transition metals, *Helv. Phys. Acta, 61*, 557-565, 1988.
Tanji, Y., Y. Nakagawa, S. G. Steinemann, and N. M. Keita, High temperature anomalies of sound velocities of Fe-Ni (fcc) alloys and their invar properties, in *Physical Metallurgy of Controlled Expansion Invar-Type Alloys*, edited by K. C. Russell and D. F. Smith, pp. 209-224, The Minerals, Metals and Materials Society, 1990.
Tsu, Y., K. Takano, and Y. Shiraishi, The velocities of ultrasound in molten iron, cobalt and nickel (in Japanese), *Res. Inst. Min. Dress. Met., 41*, report #942, 1-8, 1985.
Van Dael, W., *Thermodynamic Properties and Velocity of Sound*, Butterworth, London, 1975.
Waseda, Y., and K. T. Jacob, Computation of thermodynamic properties of liquid ferrous alloys from structural measurements, *Arch. Eisenhüttenwes., 52*, 131-136, 1981.
Waseda, Y., K. T. Jacob, and S. Tamaki, Current views on the microscopic thermodynamics of liquid alloys, *High Temp. Mat. Processes, 6*, 119-141, 1984.

P. M. Nasch and M. H. Manghnani, Hawaii Institute of Geophysics and Planetology, School of Ocean and Earth Science and Technology, University of Hawaii, 2525 Correa Road, Honolulu, HI 96822.

Pressure-Induced Phase Transition of B1 Oxides in Relation to Shock Compression Behavior of MnO

Yasuhiko Syono, Yuichi Noguchi, and Keiji Kusaba

Institute for Materials Research, Tohoku University, Sendai 980-77, Japan

Shock compression measurements of MnO have been carried out up to 120 GPa, and a phase transition with a volume decrease, $-\Delta V/V_0$, of about 6 percent is observed at 90 ± 3 GPa. Release adiabat measurements revealed that the high pressure phase was retained down to 37 GPa and the volume difference with respect to the rock salt (B1) phase was estimated to be about 23 percent. The observed large volume change in the phase transition of MnO excludes the possibility of the nickel arsenide (B8) structure for the high pressure phase, unlike the high pressure phase of $Fe_{1-x}O$. The cesium chloride (B2) structure is suggested to be a possible candidate from the systematics obtained for the B1-B2 transformation in alkaline earth monoxides, although the observed volume change would be too large. Furthermore, a possibility of change in the electronic configuration of transition metal ions at very high pressures, such as spin-pairing transition, is also inferred to assess the observed large volume change in MnO.

1. INTRODUCTION

The high pressure behavior of monoxides with the rock-salt (B1) structure has attracted the keen interest of earth scientists, because MgO and FeO are considered to be important constituents of the Earth's deep mantle. Phase transformations of alkaline earth monoxides of CaO, SrO, and BaO have been studied extensively at high pressures. Both CaO and SrO were found to transform directly from the B1 structure to the high-pressure phase with the cesium chloride (CsCl) type (B2) structure at 60 ± 2 and 36 ± 4 GPa, respectively [*Jeanloz et al.*, 1979; *Sato and Jeanloz*, 1981]. The phase transformation of CaO was also discovered under shock compression, in advance of the static work, at the pressure corresponding to the static compression, although temperature at the phase transition was estimated to be about 1000 °C [*Jeanloz et al.*, 1979]. BaO shows more complicated behavior at high pressure. *Liu and Bassett* [1972] first reported a high-pressure phase with the tetragonally distorted B2 structure above 14 GPa. Later *Weir et al.* [1986] confirmed this phase transformation, but found an additional intermediate phase with the nickel arsenide (NiAs) type (B8) structure for the pressure interval of 10-15 GPa. In contrast to these alkaline earth monoxides, no phase transition of MgO has ever been observed under static and shock compression [*Mao and Bell*, 1979; *Vassiliou and Ahrens*, 1981]. The most recent X ray diffraction measurements with the diamond anvil cell (DAC) have revealed no indication of phase change in MgO up to 230 GPa [*Duffy et al.*, 1995].

On the other hand, $Fe_{1-x}O$, among transition metal monoxides, behaves rather differently from alkaline earth monoxides. The phase transition pressure observed under

Properties of Earth and Planetary Materials
at High Pressure and Temperature
Geophysical Monograph 101

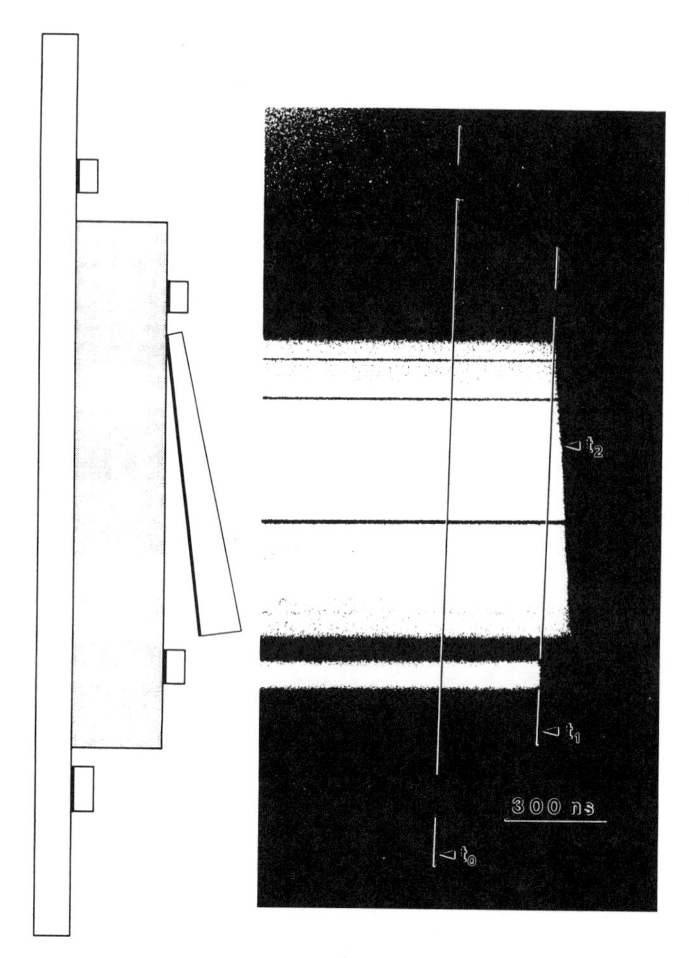

Figure 1. Schematic illustration of the inclined mirror assembly for the shock compression experiments and a streak photograph showing a kink in the inclined mirror trace due to the phase transition. Arrows t_0, t_1 and t_2 indicate incidence of the shock wave at the interface of the specimen and driver plate, arrival of the phase transition and the final deformational wave at the front surface of the specimen, respectively. This record represents a shot with impact velocity of 3.368 km/s.

shock loading is remarkably low and the volume change accompanied by the phase transition is also rather small [*Jeanloz and Ahrens,* 1980; *Yagi et al.*, 1988], if compared with those of alkaline earth monoxides. Much larger volume change of 14-20 percent was inferred from the reassessment of the original shock compression data of *Jeanloz and Ahrens* [1980], *Jackson and Ringwood* [1981], and *Jackson et al.* [1990]. Their conclusion was largely affected by omitting the 101 GPa datum from the high-pressure phase. However this omission may not be warranted, because it was found fairly close to the linear U_s-u_p relationship of the high-pressure phase.

No transition was induced for $Fe_{1-x}O$ at static pressure [*Yagi et al.*, 1985] unless the specimen was heated [*Fei and Mao*, 1994]. Metallic behavior of $Fe_{1-x}O$ has been observed at static pressure and elevated temperatures [*Knittle and Jeanloz*, 1986; 1991] and also in shock resistivity measurements [*Knittle et al.*, 1986] around the pressure range at which the phase change was observed under shock compression. Recent X ray diffraction measurements using DAC revealed the high-pressure phase of $Fe_{1-x}O$ possibly with the B8 structure above about 70 GPa at elevated temperatures [*Fei and Mao,* 1994], instead of the B2 structure, although its c/a ratio of 2.01 was considerably higher than those of most B8 compounds (1.2-1.7) [*Wyckoff*, 1963]. The volume change was estimated to be about 4 percent [*Mao et al.*, 1996], again consistent with the shock compression measurements [*Jeanloz and Ahrens* , 1980; *Yagi et al.*, 1988].

This finding stimulates the interest in the high-pressure behavior of other transition metal monoxides such as MnO, which is known to show no phase transition up to 60 GPa by static compression [*Jeanloz and Rudy,* 1987] and also no metallization up to 50 GPa by shock resistivity measurements [*Syono et al.*, 1975]. Very recently we have observed a new phase transition around 90 GPa by shock compression [*Noguchi et al.*, 1996] and suggested a possibility of the high-pressure phase to be of the B2 structure, similar to the alkaline earth monoxides. The aim of the present article is to discuss the high-pressure behavior of transition metal monoxides, in particular of MnO, and to compare with those of alkaline earth monoxides with the B1 structure. The article is also intended to envisage the pressure-induced phase transition of monoxides in a comprehensive fashion, giving special inference to the high-pressure behavior of $Fe_{1-x}O$ and MgO.

2. SHOCK COMPRESSION MEASUREMENTS OF MnO

The MnO single crystals used for the shock compression study were grown with the Verneuil method by Nakazumi Crystal Laboratory. The bulk density measured with the Archimedean method was 5.33 (2) g/cm^3, slightly smaller than the X ray density of 5.365 (4) g/cm^3 which was estimated from the unit cell dimension of 0.4445 (1) nm. This difference would be explained by the presence of a small amount of Mn_3O_4 with the density of 4.83 g/cm^3 [*Webb et al.,* 1988], as observed in the X ray powder diffraction pattern and also by SEM observation.

The (100) platelets of MnO single crystal with the thickness ranging from 2.1 to 2.6 mm were mounted on the high-density (19.2 g/cm^3) tungsten or 2024 Al driver plate (an exceptional shot was done for an approximately (110) platelet with 3.3-mm thickness.). To measure shock and free surface velocities, the conventional inclined mirror technique with streak photography was adopted [*Goto and Syono*, 1984]. A typical specimen assembly together with a measured streak record is shown in Figure 1. The specimen assembly was illuminated with an intense xenon flash lamp and the reflected light was introduced to a continuous-access, rotating-mirror type streak camera with a writing speed of 10 mm/μs.

Shock loading experiments were carried out by using a 20-mm bore, two-stage light gas gun and a 25-mm bore propellant gun for the velocity range of 1.63-3.93 km/s [*Goto and Syono,* 1984]. Shock Hugoniot parameters were determined by symmetrical impact with 1-mm tungsten or 2024 Al flyers and analyzed by both the free surface approximation and impedance match solution with the measured flyer velocity. The particle velocity of the final deformational state obtained by both methods generally showed good agreement, but the particle velocity determined by the impedance match solution was adopted because of higher accuracy.

The shock velocity (U_s) versus particle velocity (u_p) relation is summarized in Figure 2. We observed an indication of an elastic-plastic transition in the shot with the lowest impact velocity, although a large uncertainty in determining the Hugoniot elastic limit was inevitable. A linear relation of $U_s = 5.32 + 1.18\ u_p$ was obtained for the low pressure B1 phase of MnO. The C_0 value of 5.32 km/s is in good agreement with the bulk sound velocity measured with ultrasonic methods [*Oliver*, 1969; *Uchida and Saito*, 1972; *Sumino et al.*, 1980; *Webb et al.*, 1988; *Pacalo and Graham*, 1991].

A break of the linear U_s-u_p relation at U_s = 8.0 km/s was due to a phase transition which was observed as a kink in the inclined mirror image in the streak photograph as shown in Figure 1. The fact that single wave records were again observed for the runs with the shock velocity above 8.0 km/s lends support for this assignment. However, full description of the high-pressure phase could not be made in this experiment because of the limited capability of our two-stage light gas gun. Instead, we determined the release path from the shock-induced high pressure phase using a buffer technique [*Goto and Syono*, 1984]. Fused quartz glass and polymethylmetacrylate (PMMA) were used for the buffer material [*Marsh*, 1980].

The shocked state was computed from the mass and momentum conservation relations from the measured shock and particle velocities. The pressure-volume relation ob-

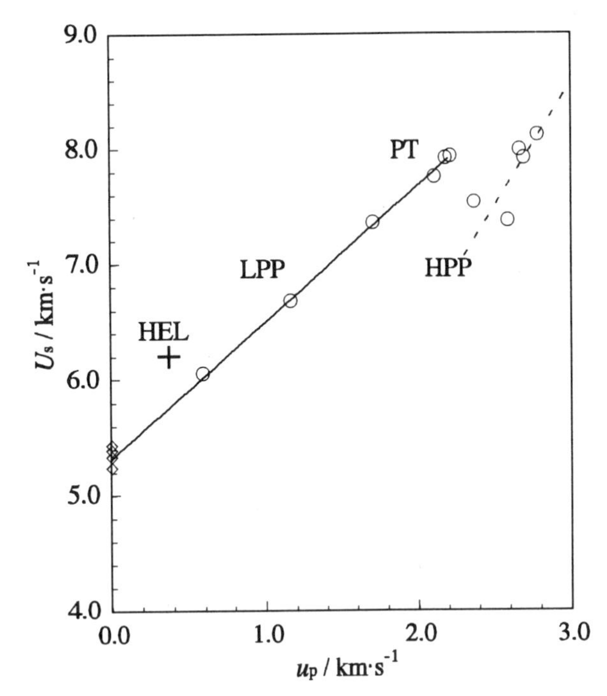

Figure 2. Shock velocity (U_s) versus particle velocity (u_p) relation of MnO. Solid line is the least squares fit for the low pressure B1 phase up to U_s = 8.0 km/s. Broken line is presumed for the high pressure phase. Diamonds on the U_s axis are bulk sound velocity determined by ultrasonic measurements (see text for the references.). HEL, PT, LPP and HPP are Hugoniot elastic limit, phase transition, low- and high-pressure phases, respectively.

tained is shown in Figure 3. The phase transition accompanied by a volume decrease, $-\Delta V/V_0$, of about 6 percent was observed at 90 ± 3 GPa. The measured larger volume change in MnO than in $Fe_{1-x}O$ is consistent with the observation of the kinked inclined-mirror record in MnO (Figure 1), which could not be detected in the phase transition in $Fe_{1-x}O$ [*Yagi et al.*, 1988]. The isotherm of the low pressure B1 phase of MnO was calculated with the aid of the Mie-Grüneisen equation of state. A Grüneisen constant of 1.51 (4) and Debye temperature of 534 K were used [*Anderson and Isaak*, 1995]. The temperature at the phase transition was estimated to be about 1000 °C. The bulk modulus, K_0, of the B1 phase MnO was determined to be 142 ± 1 GPa from Murnaghan-Birch fit, assuming K_0' = 4. The isotherm determined from the present Hugoniot measurements was slightly more compressible than that of the static compression using DAC [*Jeanloz and Rudy*, 1987].

The decompression behavior determined by the buffer technique is also shown in Figure 3. A simple extrapolation of the release path to zero pressure yields a volume

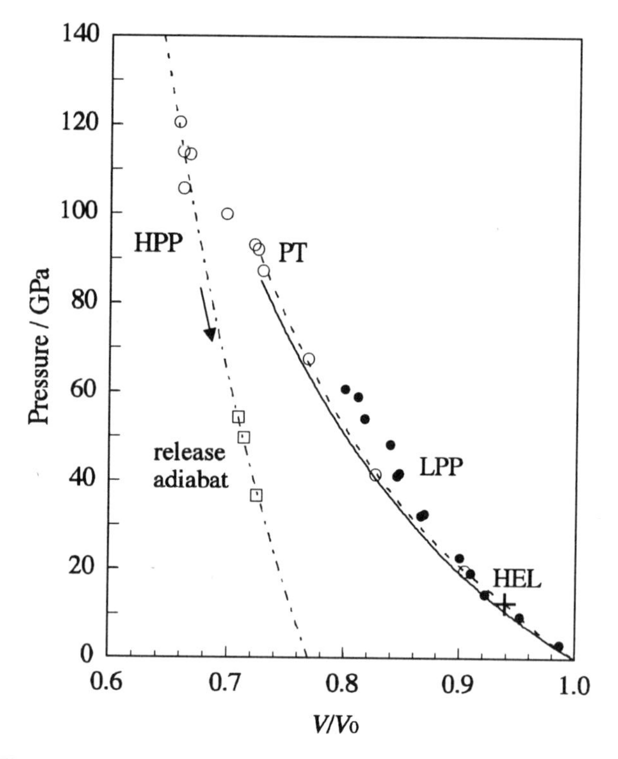

Figure 3. Compression curve of MnO. Open and solid circles are present shock experiments and static compression data by *Jeanloz and Rudy* [1987]. The solid line is a calculated isotherm. HEL, PT, LPP and HPP are Hugoniot elastic limit, phase transition, low- and high-pressure phases, respectively. The release adiabat indicated by rectangular was determined by the buffer technique using fused quartz glass and PMMA as the buffer materials.

change of about 23 percent. Details of experimental results have been [*Noguchi et al.*, 1996] and will be elucidated elsewhere.

3. PHASE TRANSITIONS OF B1 OXIDES AT HIGH PRESSURE

Phase transition pressures of alkaline earth monoxides [*Liu and Bassett*, 1972; *Jeanloz et al.*, 1979; *Sato and Jeanloz*, 1981; *Weir et al.*, 1986] and two transition metal oxides of $Fe_{1-x}O$ [*Jeanloz and Ahrens*, 1980; *Yagi et al.*, 1988] and MnO [*Noguchi et al.*, 1996] observed under static and shock compression are plotted against ionic radius ratio of $r_{M^{2+}}/r_{O^{2-}}$ in Figure 4, where M is divalent cations [*Shannon and Prewitt*, 1969; 1970]. A similar diagram has already been presented by *Jeanloz and Ahrens* [1980] and also *Sato and Jeanloz* [1981]. The volume change accompanied by the phase transition normalized by the volume at ambient conditions, V_0, is also shown as a function of $r_{M^{2+}}/r_{O^{2-}}$ in Figure 5. For BaO, we only plotted only the volume difference between BaO I with the B1 structure and BaO III with the tetragonal PH_4I (distorted B2) structure. The data points obtained for MnO in the present study apparently fall on the extrapolated curve (broken lines in Figure 4 and Figure 5 depicted as a guide to the eyes) for the B1-B2 transformation in alkaline earth monoxides, while those of $Fe_{1-x}O$ significantly deviate from these systematics, suggesting a possibility of the high pressure phase of MnO with the B2 or B2-related structure, similar to alkaline earth monoxides and different from $Fe_{1-x}O$ whose high-pressure phase was reported to have the B8 structure [*Fei and Mao*, 1994]. The possibility of the B1-B2 transition in MnO in the pressure range 50-90 GPa has previously been predicted by *Webb et al.* [1988] on the basis of shear mode softening.

The different high-pressure behaviors of MnO and $Fe_{1-x}O$, if any, can be understood from their electronic configurations, i.e. Mn^{2+} with the half-filled d^5 and Fe^{2+} with the high spin d^6. The Mn^{2+} ion generally shows ionic, highly localized, and crystal-field insensitive character among transition metal ions and is known to behave geochemically like Mg^{2+}, a typical element of alkaline earths.

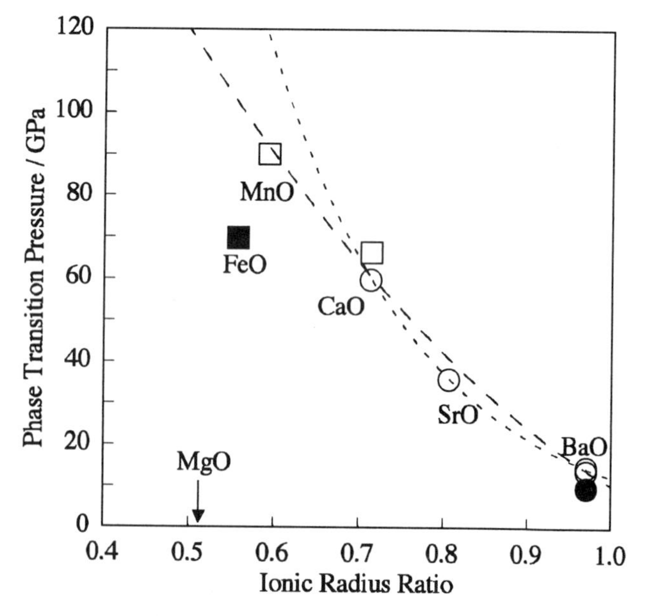

Figure 4. Phase transition pressure versus ionic radius ratio $r_{M^{2+}}/r_{O^{2-}}$ for alkaline earth and transition metal monoxides. The ionic radii of *Shannon and Prewitt* [1969; 1970] were used. Squares and circles are obtained by shock and static compression, respectively. Open (MnO excepted) and closed symbols indicate the transition to the B2 and B8 phase respectively. The broken and dotted lines are extrapolation of the systematic trend for the transition pressures to the B2 phase of the alkaline earth metal monoxides on the basis of a quadratic and exponential function of the ionic radius ratio respectively.

Even manganese monochalcogenides of MnS and MnSe crystallize as the B1 phase and show no metallic behavior, although MnSe shows a high-pressure phase transition to the B8 structure [*Cemic and Neuhaus*, 1972] and MnTe crystallizes with the B8 structure transforms to the B1 structure at high temperatures [*Panson and Johnston*, 1964]. On the other hand, the Fe^{2+} ion is more covalent and sensitive to ligand field. Actually FeS has a B8 (NiAs) derivative type structure at ambient pressure and becomes metallic above about 7 GPa [*Minomura and Drickamer*, 1963; *Kusaba et al.*, 1996]. *Wilson* [1972] has demonstrated this situation in a heuristic diagram, in which electrical and magnetic properties are strongly dependent on the electron configuration of transition metal ions. For example, ions with the d^5 configuration, such as Mn^{2+} as well as Fe^{3+}, form nonmetallic compounds, reflecting their highly localized character. On the other hand, FeS and also FeO situate close to the insulator-metal boundary line in the diagram, presumably being very susceptible to metallic behavior at high pressure.

Thermodynamical considerations on the relative stability of the B2 or B8 structures with respect to the B1 structure have been made extensively by *Navrotsky and Davies* [1981]. They estimated the transition pressure from the B1 structure to the B2 or the B8 structure by making use of thermochemical systematics obtained for many halides and oxides, which showed much lower transition pressure to the B8 phase (30 GPa) than that to the B2 phase (300 GPa) in $Fe_{1-x}O$, suggesting that the transition observed by *Jeanloz and Ahrens* [1980] near 70 GPa may be to the B8 structure. The estimation of the transition pressure of MnO with the B1 structure to the B2 or B8 structures in a similar manner using their systematics yields similar pressure values for the B1-B2 and B1-B8 transitions.

The large volume difference at zero pressure observed in MnO in the present study would be too large for the B1-B2 transition to compare with the values of 10-15 percent obtained for alkaline earth monoxides and alkaline halides [*Sato-Sorensen*, 1983; *Yagi et al.*, 1983; *Zhang and Bukowinski*, 1991], although it certainly excludes the possibility of the B1-B8 transition with no change in coordination number. The unit cell dimension of the hypothetical B2 type MnO was estimated to be 0.271 nm from the eight-coordinated ionic radii of Mn^{2+} and O^{2-} [*Shannon and Prewitt,* 1969], giving a unit cell volume of 0.020 nm^3. Thus the volume change in the B1-B2 transition would be only 10 percent, much smaller than the observed value. Furthermore the transition observed around 90 GPa in MnO might be remarkably low as a B1-B2 transition, if the trend of the B1-B2 transition pressures in alkaline earth oxides is extrapolated on the basis of an exponential

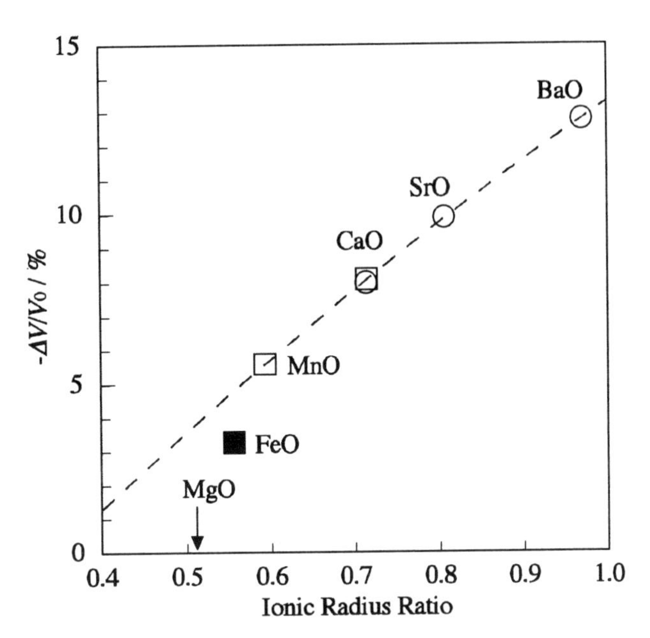

Figure 5. The volume change at the phase transition (normalized by V_0) of alkaline earth and transition metal monoxides is plotted against the ionic radius ratio. Squares and circles are the volume change obtained by shock and by static compression, respectively. The open symbol indicates the volume difference between B1 and B2 phase. For FeO, the volume difference estimated from shock compression measurements is shown by the solid symbol. The broken line is depicted as guide to eyes.

function of the ionic radius ratio (a dotted line in Figure 4). These considerations invoke that the electronic configuration of Mn^{2+} ions might be subjected to considerable change at very high pressures, similar to the observed spin-pairing transition in Fe_2O_3 above 50 GPa [*Goto et al.*, 1982]. The volume change for the high-spin-low-spin transition in MnO while keeping the B1 structure is calculated to be 20 percent from the listed value of ionic radii [*Shannon and Prewitt*, 1969; 1970]. Furthermore, the transition pressure range predicted by *Ohnishi* [1978], 70-130 GPa, is rather close to the observed transition pressure. Detailed arguments should await the determination of the structure of the high-pressure phase of MnO, which is now being studied by in situ X ray diffraction using DAC. Preliminary experiments have revealed a new phase with a metallic luster above about 90 GPa [*Kondo and Yagi*, personal communication].

In relation to the phase transition in MnO, the question remains why the B1 phase MgO is so stable and does not show the B1-B2 transition even to pressures as high as 230 GPa [*Duffy et al.*, 1995]. Because the B1-B2 transition is caused by a displacive mechanism [*Hyde and Andersson,*

1989; *Kusaba et al*, 1995], there would be no chance to fail in observing the stable B2 phase at room temperature compression or by shock compression. A simple explanation could be found in the fact that the volume change at the phase transition decreases with the increase in the transition pressure, hence yielding less contribution to Gibbs free energy in favor of the B2 phase. However, many first-principle theoretical calculations have been carried out [*Chang and Cohen*, 1984; *Causa et al.*, 1986; *Zhang and Bukowinski*, 1991; *Chizmeshya et al.*, 1994] and predicted much higher transition pressures than that inferred simply from the extrapolated transition pressure versus ionic radius ratio systematics. Actually an exponential fit for the trend of the B1-B2 transition pressures of alkaline earth monoxides (dotted line in Figure 4) could give a very high transition pressure (above 200 GPa) for MgO. Application of high pressure would probably make Mg-O bonding highly covalent, making the phase transition based on the ionic model unrealizable [*Matsui*, personal communication].

4. CONCLUDING REMARKS

In the present study, the electron configuration of the transition metal ions is suggested to be highly susceptible to high compression, seriously affecting the high-pressure behavior. The structural and electronic properties of transition metals at high pressures should be studied more systematically. Further considerations on the applicability of the B1-B2 transition model to the case of MnO would be necessary. Detailed experimental study of magnesiowüstite is also needed, because MgO and FeO apparently behave quite differently in the deep mantle, and solid solution systems should be very sensitive to pressure and composition.

Acknowledgments. Technical support in shock compression measurements by Kiyoto Fukuoka is gratefully acknowledged. The authors thank Tadashi Kondo and Takehiko Yagi for providing us X ray diffraction data of MnO at static high pressure prior to publication and also Yosiko Sato-Sorensen, Yoshito Matsui, Toshiyuki Atou and Ian Jackson for invaluable information and suggestions. They also indebted to helpful comments on the original manuscript by Tom Duffy and an anonymous reviewer.

REFERENCES

Anderson, O. L., and D. G. Isaak, Elastic constants of mantle minerals at high pressure, in *Mineral Physics and Crystallography: a Handbook of Physical Constants, AGU Reference Shelf 2,* edited by T. J. Ahrens, AGU, Washington, D. C., pp. 64-97, 1995.

Causa, M., R. Dovesi, C. Pisani, and C. Roetti, Electronic structure and stability of different crystal phases of magnesium oxide, *Phys. Rev. B 33,* 1308-1316, 1986.

Cemic, L., and A. Neuhaus, Über eine neue Hochdruckmodifikation des MnSe von NiAs-Typ und über die Mischbarkeit MnSe-MnTe, *High Temp.-High Pressures, 4,* 97-99, 1972.

Chang, L., and M. L. Cohen, High-pressure behavior of MgO: Structural and electronic properties, *Phys. Rev. B 30,* 4774-4781, 1984.

Chizmeshya, J. A., F. M. Zimmermann, R. A. La Violette, and G. H. Wolf, Variational charge relaxation in ionic crystals: An efficient treatment of statics and dynamics, *Phys. Rev. B 50,* 15,559-15,574, 1994.

Duffy, T. S., R. J. Hemley, and H. K. Mao, Equation of state and shear strength at multimegabar pressures: Magnesium oxide to 227 GPa, *Phys. Rev. Lett., 74,* 1371-1374, 1995.

Fei, Y., and H. K. Mao, In situ determination of the NiAs phase of FeO at high pressure and temperature, *Science, 266,* 1678-1680, 1994.

Goto, T., J. Sato, and Y. Syono, Shock-induced spin-pairing transition in Fe_2O_3 due to the pressure effect on the crystal field, in *High-Pressure Research in Geophysics,* edited by S. Akimoto and M. H. Manghnani, Center Acad. Publ. Japan /Reidel, Tokyo/Dordrecht, pp. 595-609, 1982.

Goto, T., and Y. Syono, Technical aspect of shock compression experiments using the gun method, in *Materials Science of the Earth's Interior*, edited by I. Sunagawa, pp. 605-619, Terra, Tokyo, 1984.

Hyde, B. G., and S. Andersson, *Inorganic Crystal Structures,* 430 pp., John Wiley, New York, 1989.

Jackson, I., and A. E. Ringwood, High-pressure polymorphism of the iron oxides, *Geophys. J. R. astr. Soc., 64,* 767-783, 1981.

Jackson, I., S. K. Khanna, A. Revcolevschi, and J. Berthon, Elasticity, shear mode softening and high-pressure polymorphism of wüstite ($Fe_{1-x}O$), *J. Geophys. Res., 95,* 21,671-21,685, 1990.

Jeanloz, R., T. J. Ahrens, H. K. Mao, and P. M. Bell, B1-B2 transition in calcium oxide from shock-wave and diamond-cell experiments, *Science, 206,* 829-830, 1979.

Jeanloz, R., and T. J. Ahrens, Equation of state of FeO and CaO, *Geophys. J. R. astr. Soc., 62,* 505-528, 1980.

Jeanloz, R., and A. Rudy, Static compression of MnO Manganosite to 60GPa, *J. Geophys. Res., 92,* 11,433-11,436, 1987.

Knittle, E., and R. Jeanloz, High-pressure metallization of FeO and implications for the earth's core, *Geophys. Res. Lett., 13,* 1541-1544, 1986.

Knittle, E., R. Jeanloz, A. C. Mitchell, and W. J. Nellis, Metallization of $Fe_{0.94}O$ at elevated pressures and temperatures observed by shock-wave electrical resistivity measurements, *Solid State Commun., 59,* 513-515, 1986.

Knittle, E., and R. Jeanloz, The high-pressure phase diagram of $Fe_{0.94}O$: A possible constituent of the Earth's core, *J. Geophys. Res., 96,* 16,169-16,180, 1991.

Kusaba, K., Y. Syono, T. Kikegawa, and O. Shimomura, A topological transition of B1-KOH-TlI-B2 type AgCl under high pressure, *J. Phys. Chem. Solids, 56*, 751-757, 1995.

Kusaba, K., Y. Syono, T. Kikegawa, and O. Shimomura, Structure of FeS under high pressure, *J. Phys. Chem. Solids*, 1996, in press, and also in this volume.

Liu, L., and W. A. Bassett, Effect of pressure on the crystal structure and the lattice parameters of BaO, *J. Geophys. Res.*, *77*, 4934-4937, 1972.

Mao, H. K., and P. M. Bell, Equations of state of MgO and εFe under static pressure conditions, *J. Geophys. Res., 84*, 4533-4536, 1979.

Mao, H. K., J. Shu, Y. Fei, J. Hu, and R. J. Hemley, The wüstite enigma, *Phys. Earth Planet. Inter., 96*, 135-145, 1996.

Marsh, S. P. (Ed.), *LASL Shock Hugoniot Data*, 658 pp., Univ. Calif. Press, Berkeley, 1980.

Minomura, S., and H. G. Drickamer, Effect of pressure on the electrical resistance of some transition metal oxides and sulphides, *J. Appl. Phys., 34*, 3043-3048, 1963.

Navrotsky, A., and P. K. Davies, Cesium chloride versus nickel arsenide as possible structure for (Mg, Fe)O in the lower mantle, *J. Geophys. Res., 86*, 3689-3694, 1981.

Noguchi, Y., K. Kusaba, K. Fukuoka, and Y. Syono, Shock-induced phase transition of MnO around 90 GPa, *Geophys. Res. Lett., 23*, 1469-1472, 1996.

Ohnishi, S., A theory of the pressure-induced high-spin-low-spin transition of transition-metal oxides, *Phys. Earth Planet. Inter., 17*, 130-139, 1978.

Oliver, D.W., The elastic moduli of MnO, *J. Appl. Phys.*, *40*, 893, 1969.

Pacalo, R. E., and E. K. Graham, Pressure and temperature dependence of the elastic properties of synthetic MnO, *Phys. Chem. Miner., 18*, 69-80, 1991.

Panson, A. J., and W. D. Johnston, The MnTe-MnSe system, *J. Inorg. Nucl. Chem., 26*, 701-703, 1964.

Sato, Y., and R. Jeanloz, Phase transition in SrO, *J. Geophys. Res.*, *86*, 11,773-11,778, 1981.

Sato-Sorensen, Y., Phase transition and equations of state for the sodium halides: NaF, NaCl, NaBr, and NaI, *J. Geophys. Res., 88*, 3543-3548, 1983.

Shannon, R. D., and C. T. Prewitt, Effective ionic radii in oxides and fluorides, *Acta Cryst.*, *B25*, 925-946, 1969.

Shannon, R. D., and C. T. Prewitt, Revised value of effective ionic radii, *Acta Cryst.*, *B26*, 1046-1048, 1970.

Sumino, Y., M. Kumazawa, O. Nishizawa, and W. Pluschkell, The elastic constants of single crystal $Fe_{1-x}O$, MnO and CoO, and the elasticity of stoichiometric magnesiowüstite, *J. Phys. Earth*, *28*, 475-495, 1980.

Syono, Y., T. Goto, J. Nakai, and Y. Nakagawa, Shock compression study of transition metal oxides, *Proc. 4th Intern'l Conf. on High Pressure, Kyoto, 1974, Spec. Issue of Rev. Phys. Chem. Jpn*, Phys.-Chem. Soc. Jpn, Kyoto, 466-472, 1975.

Uchida, N., and S. Saito, Elastic constants and acoustic absorption coefficients in MnO, CoO, and NiO single crystals at room temperature, *J. Acoust. Soc. Am.*, *51*, 1602-1605, 1972.

Vassiliou, M. S., and T. J. Ahrens, Hugoniot equation of state of periclase to 200 GPa, *Geophys. Res. Lett., 8*, 729-732, 1981.

Webb, S. L., I. Jackson, and J. D. Fitz Gerald, High-pressure elasticity, shear-mode softening and polymorphism in MnO, *Phys. Earth Planet. Inter.*, *52*, 117-131, 1988.

Weir, S. T., Y. K. Vohra, and A. L. Ruoff, High-pressure phase transitions and the equations of state of BaS and BaO, *Phys. Rev. B 33*, 4221-4226, 1986.

Wilson, J. A., Systematics of the breakdown of Mott insulation in binary transition metal compounds, *Advanc. Phys., 21*, 143-198, 1972.

Wyckoff, R. W. G., *Crystal Structures, Vol. 1*, John Wiley & Sons, New York, pp. 122-126, 1963.

Yagi, T., T. Suzuki, and S. Akimoto, New high-pressure polymorphs in some sodium halides, *J. Phys. Chem. Solids, 44*, 135-140, 1983.

Yagi, T., T. Suzuki, and S. Akimoto, Static compression of wüstite ($Fe_{0.98}O$) to 120 GPa, *J. Geophys. Res.*, *90*, 8784-8788, 1985.

Yagi, T., K. Fukuoka, H. Takei, and Y. Syono, Shock compression of wüstite, *Geophys. Res. Lett.*, *15*, 816-819, 1988.

Zhang, H., and M. S. T. Bukowinski, Modified potential-induced breathing model of potentials between closed-shell ions, *Phys. Rev. B 44*, 2495-2503, 1991.

Yasuhiko Syono, Yuichi Noguchi and Keiji Kusaba, Institute for Materials Research, Tohoku University, Katahira 2-1-1, Aoba-ku, Sendai 980-77, Japan

Volume Measurement of MgO at High Pressures and High Temperatures

Wataru Utsumi[1], Donald J. Weidner, and Robert C. Liebermann

Center for High Pressure Research and Department of Earth and Space Sciences, University at Stony Brook, Stony Brook, New York

The unit cell volume of periclase (MgO) has been measured under high pressures and high temperatures up to 10 GPa and 1400°C. The measurement was made by an in situ X ray diffraction technique with the DIA-type high multianvil high-pressure apparatus (SAM85) installed at the superconducting wiggler beamline at the National Synchrotron Light Source (NSLS, X17B). In order to avoid the effect of microscopic stress, a chunk of MgO sintered with a piston cylinder apparatus was used for the measurement. Special care was taken to obtain deviatoric-stress-free spectra; for the room temperature measurement, a liquid pressure medium was used, and in the high temperature experiments, the relaxation of deviatoric stress by the heating treatment was always checked. As a result, very precise unit cell volumes of MgO were determined using its six to eight diffraction peaks. MgO showed no structural distortion in the P-T conditions in this study. Fitting the experimental P-T-V data to the high-temperature Birch-Murnaghan equation of state yields K_0 = 153(3) GPa and $(\partial K_T/\partial T)_p$ = -0.034(3) GPa/K, assuming a constant value of 4 for $K'_{T,0}$. The high temperature equation of state of MgO can be used as an experimental pressure standard in high P-T condition where NaCl scale is useless.

1. INTRODUCTION

The purpose of this study is to measure the precise volumes of MgO under simultaneous high pressure and high temperature conditions and to obtain its accurate high temperature equation of state. The equation of state of MgO is an old and new issue in geophysics. The reason for its importance is that magnesiowustite (Mg,Fe)O is thought to be one of the most abundant minerals in the Earth's mantle, and therefore, for the calculation of chemical composition and density profiles of the lower mantle, its reliable equation of state at high pressure and high temperature is indispensable to compare laboratory measurements with seismic data. Moreover, MgO is expected to be a good high-pressure standard in high-pressure experiments, and thus a well established equation of state of MgO is required.

Many efforts have been previously made to determine the equation of state of MgO using various experimental techniques such as volume measurements under high pressure, sound velocity measurements, shock experiments, and so on. Most of them, however, are measurements at ambient temperature or pressure. In order to obtain an accurate equation of state, the experimental data under simultaneous high-pressure and high-temperature conditions are needed. Moreover, for the volume measurement under high pressure, recent

[1] Now at Japan Atomic Energy Research, SPring8, Kamigori, Hyogo 678-12, Japan

Properties of Earth and Planetary Materials at High Pressure and Temperature
Geophysical Monograph 101

X ray studies have suggested that non-hydrostaticity (deviatoric stress) may greatly affect to the volume measurement and cause unreliable results for the equations of state [*Duffy et al.*, 1995; *Wang et al.*, 1994, *Funamori et al.*, 1996]. High-pressure volume measurements of MgO at room temperature have been made repeatedly, but in the earlier experiments this deviatoric stress effect was ignored, resulting in some unreliability. Recent technical developments in high-pressure experiments, especially the use of the synchrotron radiation beam, have made it possible to carry out precise volume measurement under high pressures by taking the stress factors into consideration.

In the present study, we have performed in situ volume measurements of MgO in pressure-temperature ranges up to 10 GPa and to 1400°C using a multianvil high-pressure apparatus combined with synchrotron radiation. In the experiment, we paid attention to many factors which may cause an inaccurate result and tried to minimize these effects, such as non-hydrostatic stress, macroscopic stress, temperature gradient, and fluctuation of the pressure and temperature. New experimental P-T-V data sets are presented, and high temperature equations of state are calculated. We also discuss the possibility of the pressure and temperature determination in a high-pressure X ray experiment without thermocouple by using both NaCl and MgO as internal pressure standards.

2. EXPERIMENTAL

The experiment was carried out using a DIA type multianvil apparatus (SAM85), which is installed at the wiggler beamline (X17B) at the National Synchrotron Light Source in the Brookhaven National Laboratory [*Weidner et al.* 1992]. Sintered diamond anvils with 4×4 mm^2 truncation were used for the high pressure

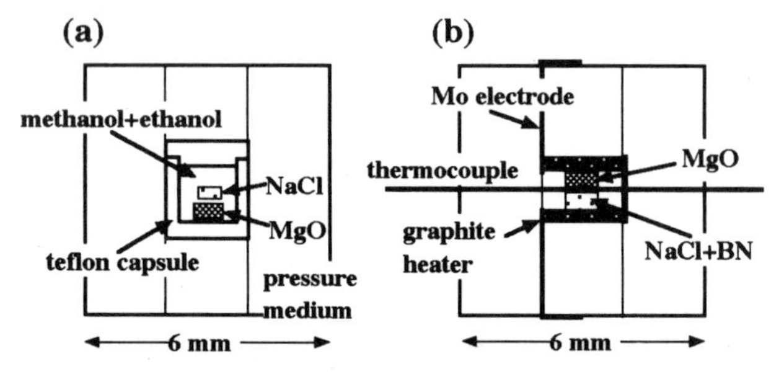

Figure 1. Cross section of the cell assembly. (a) Liquid cell for the room temperature compression. Teflon capsule is filled with mixture of methanol and ethanol. (b) High temperature cell with disk type graphite heaters. Sample and NaCl are placed separately between the graphite heaters.

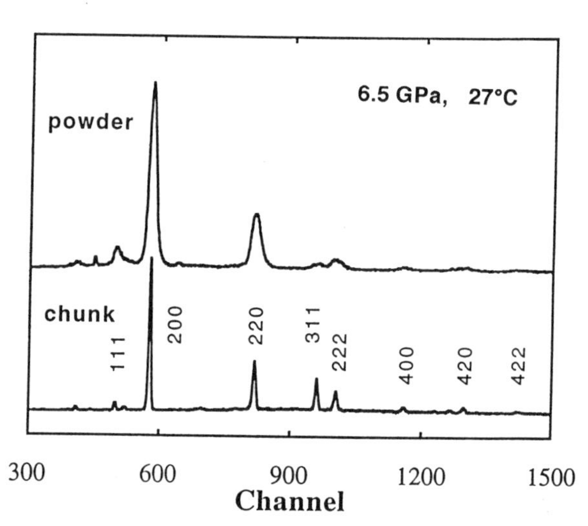

Figure 2. Comparison of the X ray diffraction profiles of the powder sample and chunk sample at 6.5 GPa and 27°C. For the powder sample, the linewidth of the diffraction peaks becomes large because of the microscopic stress effect.

generation. Two different cell assemblies were used as shown in Figure 1. For the room temperature experiment, a small teflon capsule was placed in a pressure medium (6-mm edge cube made of a mixture of boron and epoxy resin). This capsule was filled with a mixture of methanol and ethanol (4:1), so that a perfect hydrostatic condition was obtained. For the high temperature experiments, the MgO sample and pressure marker (mixture of NaCl and BN) were placed separately between two graphite disk heaters. Temperature was measured by a $W_{0.95}Re_{0.05}$-$W_{0.74}Re_{0.26}$ thermocouple, and pressure was calculated from the lattice constants of NaCl. The fluctuation of temperature indicated by the thermocouple was within ±5°C at 1400°C and less than ±2°C at lower temperatures.

X ray powder diffraction measurements were made by an energy dispersive method at constant 2-θ (10°) with a Ge solid state detector using a white synchrotron radiation beam. Incident X rays were collimated by slits of 0.1×0.3 mm^2 (horizontal-vertical) to restrict the investigated area in the cell assembly. The relation between d-spacing and channel number of the multichannel analyzer was calibrated at the beginning (and at the end) of the experiment with the fluorescence lines of molybdenum and the diffraction lines of Si, Al_2O_3, and MgO. The peak positions were determined by a least-squares fitting of Gaussian curve to the diffraction profiles, after subtracting the background.

A sintered chunk MgO was used for the present experiment. Figure 2 shows the difference between the X ray diffraction profiles of the powder sample and chunk sample at the same pressure. For the powder sample, the line width of each diffraction peak becomes large owing to the presence of microscopic stress at grain-to-grain level. This microscopic stress also causes the unsymmetrical shape of the diffraction peak. These broad and unsymmetrical diffraction peaks are not suitable for the precise volume measurement. On the other hand, the diffraction peaks of the chunk MgO remain sharp under high pressures so that the peak position can be determined precisely. The sintered chunk MgO was synthesized beforehand with a piston cylinder apparatus in addition to the 1% LiF. The synthesis conditions were 1.2 GPa, 700°, and 1 hour.

Three experimental runs were carried out. Run (a) was a room temperature compression using the liquid cell assembly where perfect hydrostatic condition was obtained. In run (b) and run (c), the sample was compressed in a solid pressure medium, and thus nonhydrostatic stress (deviatoric stress) may exist in the sample. This nonhydrostatic stress greatly affects determination of the peak position, giving rise to inconsistency in the cell parameter refinement. In order to make a precise volume measurement, we must reduce this non-hydrostatic stress. *Weidner et al.* [1994] reported that nonhydrostatic stress generally vanishes once the cell assembly has been heated above 600°C. In run (b), the sample was first compressed to a certain pressure at room temperature and then heated up to 800°C at the constant applied load. Volume measurements were made during the temperature decreasing process to room temperature at every 100°C interval. By heating the sample up to 800°C, nonhydrostatic stress is expected to vanish, and this stress effect can be avoided even at lower temperatures. In run (c), volume measurements were made during the isothermal decompression at 800°C, 1000°C, and 1400°C, where temperatures were high enough to release the nonhydrostatic stress.

3. RESULTS AND DISCUSSION

3.1. *Unit Cell Volumes*

MgO showed no structural distortion up to 9.52 GPa and 1400°C, the most severe condition in this study. Even at this high temperature condition, no remarkable grain growth of MgO occurred which was accompanied by an extraordinary change of the peak intensity, and six to eight diffraction peaks were clearly observed. Table 1 shows a representative refinement of unit cell parameter. In the present study, at least six diffraction peaks were always used for the calculation. As shown in the table, the agreement of the observed d-spacing values and the d

TABLE 1. A Representative Refinement of Unit Cell of MgO

hkl	d(obs.)	d(calc.)	d_{obs}/d_{cal}-1
111	2.4255	2.4267	-0.00048
200	2.1014	2.1015	-0.00007
220	1.4860	1.4860	-0.00001
311	1.2673	1.2674	0.00002
222	1.2134	1.2133	0.00006
400	1.0511	1.0508	0.00031
420	0.9400	0.9398	0.00017

P=7.9 GPa T=1400℃
a=4.2037±0.0003 (Å)
V=74.287±0.014 ($Å^3$)
($\Delta V/V=\pm 0.02$ %)

values calculated from the obtained unit cell are very good, and the unit cell volume was determined very precisely within a 0.02% volume uncertainty.

Unit cell volumes of MgO at various pressure-temperature conditions are listed in Table 2. For 1 bar volume data, we measured the sample volume in the high pressure cell after the high-pressure experiment. For run (c), however, the initial volume data was used, because we could not measure the final one owing to the blowout in the final decompression process. In order to avoid a systematic error in volume measurements, we used the normalized volume (V/V_0) for the calculation of equations of state.

All volume data are plotted in Figure 3. Data points obtained in different runs are represented by different symbols. The white circles are the room temperature compression data using a liquid pressure medium (run (a)). In run (b), temperature was varied at the constant press load, and volume measurements were made at every 100°C up to 800°C (black circles). The black squares represents volume data obtained in an isothermal decompression process in run (c). As shown in the figure, all the data obtained by the different runs (different P-T path and/or different sample assembly) are very consistent, e.g., room temperature data in run (b) are in perfect agreement with liquid compression data (run (a)). This means that nonhydrostatic stress in run (b) was released and the volume measurement was made precisely. Each data point has an error in volume (0.02%) which is much smaller than the length of the symbol. Precise error estimation of the pressure is rather difficult, but taking into account the temperature fluctuation and the uncertainty of NaCl volume measurement, we can conclude that in most P-T conditions the pressure uncertainty is within ±0.1 GPa, so that the error bar in

TABLE 2. Unit Cell Volumes of MgO at Various Pressures and Temperatures

P (GPa)	T (℃)	V (Å)	V/V_0	Volume Error
run (a)				
0.00	27	74.7130	1.00000	0.027%
1.13	27	74.1458	0.99241	0.030%
2.31	27	73.6454	0.98571	0.019%
3.42	27	73.1975	0.97972	0.015%
4.43	27	72.6787	0.97277	0.019%
5.38	27	72.2564	0.96712	0.023%
6.23	27	71.8971	0.96231	0.041%
7.07	27	71.6807	0.95941	0.026%
7.79	27	71.3265	0.95467	0.028%
run (b)				
0.00	27	74.7100	1.00000	0.013%
3.73	700	74.9067	1.00260	0.008%
3.56	600	74.6663	0.99942	0.006%
3.40	500	74.4244	0.99618	0.005%
3.24	400	74.1725	0.99281	0.013%
3.10	300	73.9800	0.99023	0.004%
2.94	200	73.7752	0.98749	0.009%
2.82	100	73.5857	0.98495	0.013%
2.73	27	73.4605	0.98328	0.010%
5.49	800	74.4206	0.99613	0.007%
5.25	700	74.1964	0.99313	0.013%
5.11	600	73.9711	0.99011	0.004%
4.95	500	73.7265	0.98684	0.009%
4.80	400	73.5269	0.98416	0.011%
4.66	300	73.3095	0.98125	0.009%
4.53	200	73.0829	0.97822	0.006%
4.40	100	72.9221	0.97607	0.007%
4.28	27	72.7600	0.97390	0.012%
7.75	800	73.1923	0.97969	0.011%
7.65	700	72.9845	0.97690	0.009%
7.50	600	72.7804	0.97417	0.008%
7.34	500	72.5752	0.97143	0.011%
7.05	400	72.4675	0.96998	0.011%
6.76	300	72.3231	0.96805	0.011%
6.47	200	72.1952	0.96634	0.015%
6.18	100	72.1054	0.96514	0.013%
5.97	27	72.0149	0.96393	0.016%
9.15	800	72.6191	0.97201	0.008%
9.01	700	72.4081	0.96919	0.014%
8.89	600	72.2025	0.96644	0.013%
8.72	500	72.0331	0.96417	0.014%
8.58	400	71.8399	0.96158	0.021%
8.42	300	71.6945	0.95964	0.014%
8.24	200	71.5189	0.95729	0.020%
8.10	100	71.3585	0.95514	0.022%
7.97	27	71.2892	0.95421	0.018%
run (c)				
0.00	27	74.7130	1.00000	0.015%
7.66	800	73.2858	0.98090	0.020%
6.67	800	73.7384	0.98696	0.017%
6.01	800	74.1349	0.99227	0.051%
4.82	800	74.6332	0.99894	0.027%
4.23	800	74.9186	1.00280	0.022%
3.53	400	74.0389	0.99098	0.034%
5.61	400	73.0346	0.97754	0.031%
7.82	400	72.0310	0.96411	0.020%
8.65	1000	73.3068	0.98118	0.019%
7.91	1000	73.5752	0.98478	0.007%
7.63	1000	73.6472	0.98574	0.024%
7.21	1000	73.8638	0.98864	0.017%
6.34	1000	74.3764	0.99550	0.017%
5.62	1000	74.6973	0.99980	0.022%
4.71	1000	75.1196	1.00540	0.025%
9.52	1400	73.4898	0.98363	0.029%
8.87	1400	73.8337	0.98824	0.034%
8.41	1400	74.0323	0.99089	0.039%
7.89	1400	74.2867	0.99430	0.019%
6.51	1400	74.9922	1.00370	0.038%
5.75	1400	75.5283	1.01090	0.029%
5.15	1400	75.7539	1.01390	0.013%

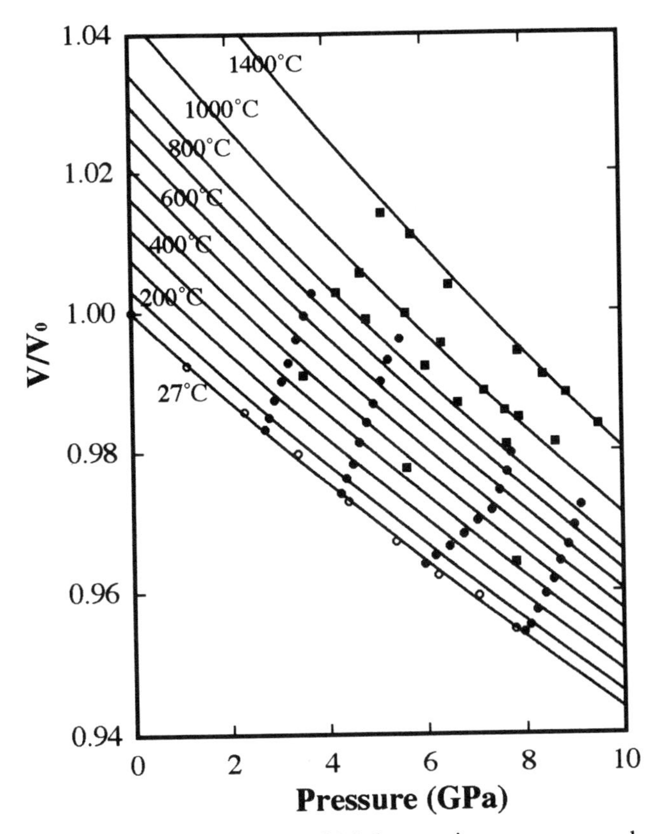

Figure 3. Volume change of MgO at various pressures and temperatures. The open circle, the solid circle, and the solid square represent the experimental data obtained in run (a), run (b), and run (c), respectively. Solid lines are the isothermal compression curves at various temperatures based on the best fit parameters obtained by this study (K_0=153 GPa, K'=4, and $(\partial K_T/\partial T)_p$= −0.034(3) GPa/$K$).

pressure is also smaller than the symbol. However, for the 1400°C data, the highest temperature in the present study, the pressure uncertainty might be larger, because of the instability of temperature and the grain growth of the NaCl.

3.2. *Equation of State at Room Temperature*

The room temperature data are fitted to the third order Birch-Murnaghan equation of state:

$$P = \frac{3}{2}K_0[(V_0/V)^{7/3} - (V_0/V)^{5/3}] \times \{1 - \frac{3}{4}(4 - K_0)[(V_0/V)^{2/3} - 1]\} \quad (1)$$

When only the P-V data obtained in run (a) (liquid cell) are used for the calculation, K_0 = 4(5) GPa, K_0' = 8(1.8) are obtained. While the calculation based on the run (b) data (27°C) leads to K_0 = 2(5) GPa and K_0' =.5(1.5). The results of these different runs agree well. Using data from both runs, the bulk modulus K_0 and its pressure derivative K_0' are determined as K_0=153(3) GPa and K_0'=4.1(1.1) . These values are in good agreement with the previous reported values by various techniques (162.3(2) GPa, 4.27(8) by *Chang and Barsch* [1969], 162.1(1.7) GPa, 3.11(11) by *Bonsczar and Graham* [1982], 162.5(2) GPa, 4.13(9) by *Jackson and Niesler* [1982], 156(11) GPa, 5.4(2) by *Mao and Bell* [1979], 152.9 GPa, 4.54 by *Carter et al.* [1971]).

3.3. *Equation of State at High Temperature*

In the high temperature equation of state, temperature derivative of the bulk modulus, $(\partial K_T/\partial T)_p$, is an important parameter. In order to calculate the $(\partial K_T/\partial T)_p$, we extend the Birch-Murnaghan equation (1) to the high temperature form

$$P = \frac{3}{2}K_{T,0}[(V_{T,0}/V)^{7/3} - (V_{T,0}/V)^{5/3}] \times \{1 - \frac{3}{4}(4 - K'_{T,0})[(V_{T,0}/V)^{2/3} - 1 \quad 2)$$

where $K_{T,0}$, $K'_{T,0}$, and $V_{T,0}$ are the isothermal bulk modulus, its pressure derivative, and the unit cell volume at temperature T and 1 bar, respectively. By neglecting second and higher order derivatives, the bulk modulus $K_{T,0}$ can be written as

$$K_{T,0} = K_0 + (\partial K_T/\partial_T)_p(T - 300) \quad (3)$$

The unit cell volume $V_{T,0}$ is given by the following expression

$$V_{T,0} = V_0 \exp\int \alpha_{T,0} dT \quad (4)$$

where $\alpha_{T,0}$ is the thermal expansivity at ambient pressure and can be assumed as

$$\alpha_{T,0} = a + bT \ (a,b = \text{constant}) \quad (5)$$

The $(\partial K_T/\partial T)_p$ value is calculated by fitting all the experimental P-T-V data in Table 2 to the high-temperature Birch-Murnaghan equation of state. Fixing the bulk modulus at room temperature K_0=153 (from the room temperature result) and assuming $K'_{T,0}$= 4 (const.), we obtained $(\partial K_T/\partial T)_p$ = −0.034(3) GPa/K. In this calculation, the thermal expansivity at ambient pressure is not fixed but treated as a variable parameter. Since

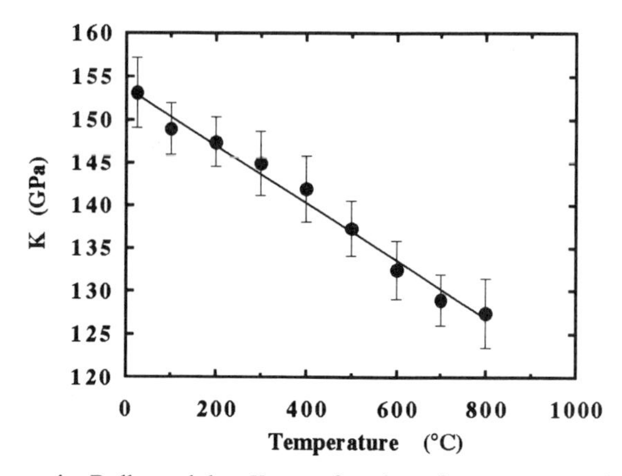

Figure 4. Bulk modulus K_T as a function of temperature. At each constant temperature, K_T can be calculated using isothermal volume data. Fitting the $K_T - T$ data to the linear function yields $(\partial K_T/\partial T)_p = -0.035(2)$ GPa/*K*.

$(\partial K_T/\partial T)_p$ is a second derivative of volume, it may change greatly depending on the selection of other parameters, particularly on the 1 bar thermal expansivity. Thus we re-calculated the $(\partial K_T/\partial T)_p$ value by fixing the 1 bar thermal expansivity at the previously reported value [*Suzuki* 1975]. It leads to $(\partial K_T/\partial T)_p = -0.0343(8)$ GPa/*K*, and thus the assumption of Suzuki's 1bar thermal expansivity does not affect the final conclusion at all.

There is another way to calculate $(\partial K_T/\partial T)_p$. Since we have obtained isothermal volume data, bulk modulus can be calculated at each constant temperature. In Figure 4, the bulk moduli K_T thus calculated are plotted as a function of temperature. From the slope of this temperature vs K_T graph, we can calculate $(\partial K_T/\partial T)_p$ directly. Using the K_T data at 27-800°C, $(\partial K_T/\partial T)_p$ = -0.035(2) GPa/*K* is obtained, which is very close to the previous number. K_T values at higher than 800°C are so scattered that they cannot be used for this $(\partial K_T/\partial T)_p$ calculation.

3.4. *MgO as a Pressure Standard*

In most of the high pressure X ray experiments using multianvil apparatus, NaCl has been commonly used as an internal pressure standard, because NaCl has a large compressibility and its equation of state is well established. However, as the scientific frontiers have been expanded, experimental pressure and temperature conditions have become more severe. In such a severe condition, the NaCl scale is useless because of the grain growth problem at high temperature and/or the phase transition at high pressure. Another pressure standard which can be used under much wider pressure and temperature ranges is now strongly required, and MgO is one of the candidate materials [*Jamieson et al.*, 1982]. Moreover, if a new pressure standard other than NaCl is well established, by measuring the volume of the two materials, we can determine the pressure and temperature simultaneously without a thermocouple. This possibility had been previously pointed out by *Jamieson et al.* [1982] and *Yagi et al.* [1985], but accuracy of the P-T determination was not high enough for the quantitative experiment because uncertainty of the volume measurement was rather large in those days. Recent experimental development has made it possible to carry out more precise volume measurement. Therefore, much more accurate P-T determination without a thermocouple can be expected.

In Figure 5, two isochore lines of NaCl and MgO are plotted in a P-T plane. From the cross point of these lines, pressure and temperature are determined. Each line has a certain width due to the volume uncertainty, thus pressure and temperature should be within a black region. Using the current high pressure X ray experimental technique combined with synchrotron radiation, the volumes of NaCl and MgO can be measured with a accuracy of 0.07% and 0.02%, respectively. As a result, in this case,

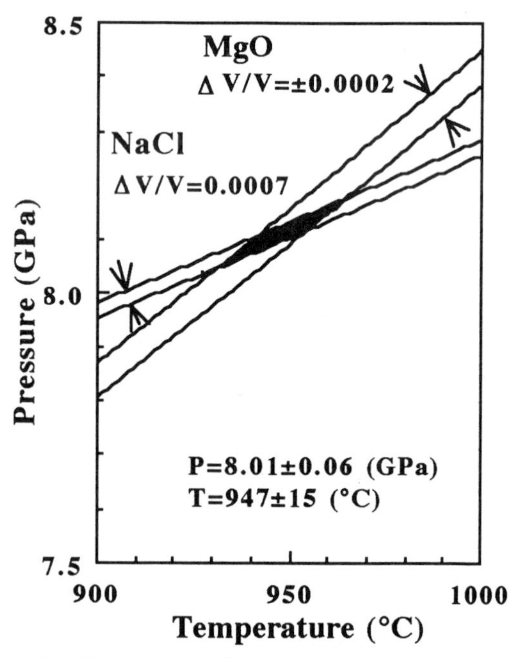

Figure 5. Pressure and Temperature determination by the double standards method. Solid lines are the isocore of NaCl (V/V_0 = 0.856±0.00007) and MgO (V/V_0 = 0.981±0.0002). Pressure and temperature can be determined as 8.01±0.06 GPa and 947±15°C.

pressure and temperature can be determined as 8.01±0.06 GPa and 947±15°C. This accuracy may be already good enough for some experiments. If the variation of the c-axis in hBN is well established and is used instead of the NaCl line, the temperature can be determined more precisely, because its slope is expected to be much smaller than that of NaCl.

This double standard method seems to be of practical use in a real experiment with synchrotron radiation beam. It should be pointed out, however, that powder mixture of MgO and NaCl is of no use for this purpose because of the "lame" effect. For the accurate pressure and temperature determination, MgO and NaCl should be placed separately in the pressure medium. This "lame" effect is described by *Wang et al.* in this volume.

Acknowledgments. The authors are grateful to M. Vaughn, K. Kusaba, B. Li, and Y. Wang for their support of the experiment and the helpful discussion.

REFERENCES

Anderson, O. L., and K. Zou, Formulation of the thermodynamic functions for mantle minerals, MgO as an example, *Phys. Chem. Minerals, 16*, 642-648, 1989.

Anderson, O. L., and D. G. Isaak, Elastic constants of mantle minerals at high temperature, in *A Handbook of Physical Constants: Mineral Physics and Crystallography, Reference Shelf Ser.*, vol. 2, edited by T. J. Ahrens , Washington, D.C., 1995.

Bonsczar, L. J., and E. K. Graham, The pressure and temperature dependence of the elastic properties of polycrystal magnesiowustite, *J. Geophys. Res., 87*, 1061-1078, 1982.

Carter, W. J., S. P. Marsh, J. N. Fritz, and R. G. McQueen, The equation of state of selected materials for high-pressure references, *Nat. Bur. Stand. Spec. Publ., 326*, 47-158, 1971.

Chang, Z. P., and G. R. Barsch, Pressure dependence of single crystalline magnesium oxide, *J. Geophys. Res., 74*, 3291-3294, 1969.

Duffy, T. S, R. J. Hemley, and H. K. Mao, Equation of state and shear strength at multimegabar pressures: Magnesium oxide to 227 GPa, *Phys. Rev. Lett. 74*, 371-1374, 1995.

Fei, Y., H. K. Mao, J. Shu, and J. Hu, P-V-T equation of state of Magnesiowustite ($Mg_{0.6}$,$Fe_{0.4}$)O, *Phys. Chem. Mineral,s 18*, 416-422, 1992.

Funamori, N., T. Yagi, W. Utsumi, T. Kondo, T. Uchida, and M. Funamori, Thermoelastic properties of $MgSiO_3$ perovskite determined by in situ x-ray observation up to 30 GPa and 2000K, *J. Geophys. Res, 101*, 8257-8269, 1996.

Isaak, D. G., R. Cohen, and M. J. Mehl, Calculated elastic and thermal properties of MgO at high pressures and temperatures, *J. Geophys. Res., 95*, 7055-7067, 1990.

Jackson, I., and H. Niesler, Elasticity of periclase to 3 GPa and some geophysical implications, in *High Pressure Research in Geophysics*, edited by S. Akimoto and M.H. Manghnani, pp. 93-113, Center for Academic Publ., Tokyo/Reidel, Dordrecht, 1982.

Jamieson, J. C., J. N. Fritz, and M. H. Manghnani, Pressure measurement at high temperature in x-ray diffraction studies: gold as a primary standard, in *High Pressure Research in Geophysics*, edited by S. Akimoto and M.H. Manghnani, pp. 27-48, Center for Academic Publ., Tokyo/Reidel, Dordrecht, 1982..

Manghnani, M. H., L. C. Ming, J. Balogh, S. B. Qudri, E .F. Skelton, and D. Schiferl, Equation of state and phase transition studies under in situ high P-T conditions using synchrotron radiation, in *Solid State Physics under Pressure: Recent Advance with Anvil Device*, edited by S. Minomura, pp. 343-350, Terra Scientific Publishing Company, Tokyo, 1985.

Mao, H. K., and P.M. Bell, Equation of state of MgO and Fe under static pressure conditions, *J. Geophys. Res., 84*, 4533-4536, 1979.

Spetzler, H., Equation of state of polycrystalline and single crystal MgO to 8 kilobars and 800K, *J. Geophys. Res., 75*, 2073-2087, 1970.

Suzuki, I., Thermal expansion of periclase and olivine and their anharmonic properties, *J. Phys. Earth, 23*, 145-159, 1975.

Wang, Y., D. J. Weidner, R. C. Liebermann, and Y. Zhao, P-V-T equation of state of (Mg,Fe)SiO_3 perovskite and constraints on composition of the lower mantle, *Phys. Earth Planet. Int., 83*, 13-40, 1994.

Weidner, D. J., M. Vaughan, J. Ko, Y. Wang, K. Leinenweber, X. Liu, A. Yeganeh-Haeri, R. Pacalo, and Y. Zhao, Large volume high pressure research using the wiggler port at NSLS, *High Pressure Research, 8*, 617-623, 1992.

Weidner, D. J., M. Vaughan, J. Ko, Y. Wang, X. Liu, A. Yeganeh-Haeri, R. Pacalo, and Y. Zhao, Characterization of stress, pressure, temperature in SAM-85, a DIA type high pressure apparatus, in *High Pressure Research: Applications to Earth and Planetary Sciences*, edited by Y. Syono and M. H. Manghnani, pp. 13-17, Terra Scientific, Tokyo/AGU, Washington, D.C., 1992.

Weidner, D. J., Y. Wang, and M. T. Vaughan, Yield strength at high pressure and temperature, *Geophys. Res. Lett., 21*, 753-756, 1994.

Yagi, T., O. Shimomura, S. Yamaoka, K. Takemura, and S. Akimoto, Precise measurement of compressibility of gold at room temperature and high temperatures, in *Solid State Physics under Pressure: Recent Advance with Anvil Device*, edited by S. Minomura, pp. 363-368, Terra Scientific Publishing Company, Tokyo, 1985.

Properties of LiF and Al_2O_3 to 240 GPa for Metal Shock Temperature Measurements

Kathleen G. Holland and Thomas J. Ahrens

Lindhurst Laboratory of Experimental Geophysics,
Seismological Laboratory,
California Institute of Technology, Pasadena, California

Shock temperature experiments employing a six-channel pyrometer were conducted on 200, 500, and 1000 Å thick films of Fe sandwiched between 3-mm-thick anvils of Al_2O_3 and LiF to measure the thermal diffusivity ratios Al_2O_3/Fe and LiF/Fe at high temperatures and pressures. Temperature decays of 3000 ± 800 K in 250 ns were observed at Fe pressures of 194 – 303 GPa, which reflect the conduction of heat from the thin metal films into the anvil material. These results were achieved via experiments employing LiF anvils at conditions of 164 – 165 GPa and 4190 – 4220 K and Al_2O_3 anvils at conditions of 156 – 304 GPa and 1290 – 2740 K. Thermal modeling of interface temperature versus time yields best fit thermal diffusivity ratios of 4 – 19 ± 1 (Fe/anvil) over the pressure and temperature range of the experiments. Calculated thermal conductivities for Fe, using electron gas theory, of 111 – 181 W/mK are used to calculate thermal conductivities for the anvil materials ranging from 2 to 13 W/mK. Debye theory predicts higher values of 8 to 35 W/mK. Data from previous experiments on thick ($\geq$ 100μm) films of Fe and stainless steel are combined with our present results from experiments on thin ($\leq$ 1000 Å) films to infer a 5860 ± 390 K Hugoniot temperature for the onset of melting of iron at 243 GPa. Our results address the question of whether radiation observed in shock temperature experiments on metals originates from the metal at the metal/anvil interface or from the shocked anvil. We conclude that the photon flux from the shocked assemblies recorded in all experiments originates from the metal. Within the uncertainties of the shock temperature data, the uncertainties in shock temperatures resulting from the radiation from the anvils is negligible. This is in direct disagreement with the conclusions of previous work by Kondo.

INTRODUCTION

The phase diagram and thermal properties of iron at ultra-high pressures are important because they provide vital information for understanding the Earth's core, and also because they add to the fundamental understanding of the behavior of materials at high pressure. However, experimental determination of these properties are difficult. Dynamic and static experiments each have their own challenges [*Gallagher et al.*, 1994; *Boehler*, 1994; *Williams et al.*, 1987; *Yoo et al.*, 1993] and the results in the 50 – 200 GPa range are disparate. Not only does the extrapolation of static results to higher pressure disagree with the higher pressure dynamic results, but static experiments conducted at dif-

Properties of Earth and Planetary Materials
at High Pressure and Temperature
Geophysical Monograph 101

ferent laboratories do not agree with each other. Shock temperature determinations on opaque materials such as iron are difficult because one cannot directly observe the interior of the shocked material and it is necessary to employ transparent anvil materials and observe the metal/anvil interface. This method has inherent difficulties [*Nellis et al.*, 1990], the three most important of which are addressed here. First, the shock-compressed anvil materials may have a sufficiently high opacity that they may be emitting enough light so that they interfere with the observation of the iron/anvil interface. Second, the data reduction requires knowledge of thermal parameters that are not easily measured at the conditions of the experiment. Third, imperfections at the iron-anvil interface can lead to local deposition of irreversible work and hence induce anomalously high temperatures.

When a shock wave propagates from one material to another that is both in ideal contact and has a similar shock impedance, most of the energy of a shock wave is transmitted, rather than reflected. The optical properties of Al_2O_3, which has a similar shock impedance to iron and is used as an anvil material in shock temperature experiments of metals, have been controversial. Initially *Grover and Urtiew* [1974] inferred that Al_2O_3 became opaque above 85 GPa. However both *Ahrens et al.* [1990] and *McQueen and Isaak* [1990] concluded that the Grover and Urtiew analysis was too simplified. A more detailed analysis demonstrated the transparency of Al_2O_3 to 200 GPa. *Ahrens et al.* [1990] and *Williams et al.* [1987] assumed that Al_2O_3 remained transparent while in the shocked state, implying that the anvil material is indeed a window material and that one can observe optical radiation from the metal at the anvil/metal interface. *Kondo* [1994] inferred that Al_2O_3 becomes opaque under shock loading and that observed radiation in shock temperature experiments originates from the anvil, not the metal. Thus *Kondo* claimed that shock temperature experiments on a metals may, in fact, be measuring the temperature of the anvil rather than the metal. How radiation from the metal can be distinguished from anvil radiation during a shock temperature experiment is germane, especially in view of the above results by *Kondo* [1994] with Ag films deposited on Al_2O_3. In the present paper we address this issue by studying the systematic differences in the radiation observed from sample assemblies, between experiments. For example, if observed interface temperature varies systematically with the type of metal sample used but not with the type of anvil, then the optical radiation is inferred to originate from the metal/anvil interface and therefore to reflect the properties of the metal.

One of the critical parameters that is used to calculate Hugoniot temperature from interface temperature is the thermal diffusivity ratio between the metal and the anvil, R. A series of experiments (discussed in the section on 'thin films') is carried out to measure this parameter. However, the 'thin film' experiments do not provide accurate shock temperature data for iron, so results from other experiments (discussed in the 'thick film' section) are analyzed using R values determined in 'thin film' experiments. Revised Hugoniot temperatures are reported. These results do not by themselves yield a pressure where the iron Hugoniot intersects the solid-liquid phase boundary. However, *Brown and McQueen* [1986] observed the onset of melting at 243 ± 3 GPa along the principal Hugoniot of iron by observing a decrease in sound velocity, so we have interpolated our temperatures results to infer melting at 5860 ± 390 K and 243 GPa.

In the present paper we use the lack of correlation of shock temperature with anvil material to address the issue of the origin of the optical radiation during shock temperature experiments on iron. We then discuss the data reduction of these experiments and later experiments to determine R. We also discuss issues related to effects and problems at the iron/anvil interfaces. Finally, we determine a revised phase diagram for iron, which infers a temperature for the onset of melting of an assumed ϵ phase along the principal Hugoniot.

RELEVANT EQUATIONS

Experimental observation of the temperatures of shocked metals is difficult because they are opaque. Thus, an anvil is used, and the temperature of the interface between the anvil and the metal T_i is observed. In the present case we assume the anvil has a slightly lower shock impedance than the metal sample [*Ahrens*, 1987]. We drive a shock wave into the sample inducing a Hugoniot pressure P_h, volume V_h, and temperature T_h. Upon reflection at the anvil, a release wave is reflected back into sample, resulting in a release-pressure P_r and release-density ρ_r in the metal. The temperature of the interior of the released sample [*Grover and Urtiew*, 1974] T_r is calculated from T_i via

$$T_i = T_r - \frac{T_r - T_a}{1 + \alpha} \tag{1}$$

where T_a is the internal shock temperature of the anvil, and α is defined by

$$\alpha = \sqrt{\frac{k_m \rho_m C_{p_m}}{k_a \rho_a C_{p_a}}} \tag{2}$$

where $C_p = C_v(1 + \gamma\alpha^{th}T)$ is the specific heat at constant pressure, ρ is the high-pressure density, k is the thermal conductivity, α^{th} is the coefficient of thermal expansion [*Duffy*, 1993], and $C_v = 3R_u/w$ is the specific heat at constant volume, where $R_u = 8.31441$ J/K·mol is the universal gas constant, and w is the mean atomic weight. The subscript 'a' denotes the anvil material and the subscript 'm' denotes the metal. Previous authors [*Ahrens et al.*, 1990] have used C_v instead of C_p in Equation 2 and this makes only a 0.4% difference in the resulting value for α. Bass also calculated k_a from

$$k_{T,0} = A + B/T \tag{3}$$

$$k_a = k_{T,P} = k_{T,0}\left(\frac{\rho_P}{\rho_o}\right)^{2\gamma+5/3} \tag{4}$$

where $A_{Al_2O_3} = -2.599$ W/mK, $B_{Al_2O_3} = 1.176 \times 10^4$ W/m, $A_{LiF} = -0.2$ W/mK, and $B_{LiF} = 3.7 \times 10^3$ W/m are measured at ambient pressure and high temperature and γ, ρ_o, and ρ_P are the Grüneisen parameter and the initial and compressed densities of the material. Other formulations [*Leibfried and Schlömann*, 1954; *Tang*, 1994] calculate $k_{T,0}$ analytically and produce similar results in the temperature range of interest. Equation 2 assumes $\gamma = \gamma_0$; without this assumption, a lower value of k_a would be predicted. However, we use the above formulation to facilitate a direct comparison of our T_h values with the results of Bass. Fourier conduction of heat from a high temperature metal film into a lower temperature anvil material is dependent upon the thermal diffusivity ratio, R.

$$R = \frac{\kappa_m}{\kappa_a} \tag{5}$$

where κ_m and κ_a are the thermal diffusivities of the metal and the anvil material respectively. Thermal diffusivity, κ, is related to thermal conductivity, k, via

$$\kappa = k/\rho C_p \tag{6}$$

where C_p is the specific heat at constant pressure and ρ is the high pressure density. The thermal conductivity of the metal, k_m, is calculated from the Wiedemann-Franz law:

$$k_m = L\sigma T \tag{7}$$

where $L = 2.45 \times 10^{-8}$ W $\cdot\,\Omega$/K^2 is the Lorenz number, σ is the electrical conductivity of the metal, and T the is temperature. This formulation assumes that the metal acts as a free-electron gas, and that phonon effects are negligible. This assumption is supported by the k_m(Fe) data of *Secco and Schloessin* [1989]. *Manga and Jeanloz* [1996] compared Secco's data to the σ(Fe) data of *Matassov* [1977] and concluded that the Wiedemann-Franz law was valid for Fe. Note that σ has a $1/T$ dependence and also a pressure P dependence, making Equation 7 more properly $k_m(P) = L\sigma(P, 1/T)T$. Thus when calculating k_m for a thin film experiment that has a different shock temperature than the Hugoniot temperature of Fe, one must use a calculated Hugoniot temperature in Equation 7 because that is the temperature at which the measurement of σ was made [*Manga and Jeanloz*, 1997].

To determine a Hugoniot temperature for an experiment employing a 'thick film' (≥ 1 μm) an R value is used to calculate k_m/k_a via Equation 5 and Equation 6. k_m/k_a is then used to calculate α via Equation 2, and α is used to calculate T_r via Equation 1. T_h is calculated from T_r via

$$T_h = T_r \exp\left[\gamma_o\rho_o\frac{(u_r - u_h)^2}{P_h - P_r}\right] \tag{8}$$

where u_r and u_h are the released and Hugoniot particle velocities.

Equation 1 does not apply for 'thin films', where the films cannot be approximated by an infinite half-space, so to determine a Hugoniot temperature for an experiment employing a 'thin film' (≤ 1000 Å), T_r is obtained by fitting the measured T_i and the calculated T_a to a finite element model [*Gallagher and Ahrens*, 1996] for symmetric heat flow from a thin film; Equation 8 is used to calculate T_h from T_r.

SHOCK TEMPERATURE EXPERIMENTS ON METALS

The shock temperature experiments on metals were performed on a 2-stage light-gas gun [*Ahrens*, 1987] via optical pyrometry [*Yang*, 1996]. There are two different types of sample configurations used, the 'thick film' setup, using 1 μm or thicker films and the 'thin film' experiments using 200, 500, or 1000 Å films. In both cases, the metal film is in contact with the anvil materials that serve to compress the metal at high pressure. The Al_2O_3 used for our experiments was obtained from the Adolph Mueller Company as spectral grade sapphire, and the LiF was obtained from Bicron Inc. as optical grade windows. The metal surface is intended to be viewed through one of the transparent anvils during the time the shock wave passes through the anvil.

To address the issue of interface quality, we grew films of 99.995% purity iron epitaxially on our anvil materials using argon ion sputtering in ultra high vacuum 5×10^{-9} torr. The 1000 Å film that was used for Shot #287 was deposited in 40 minutes on a 500°C preheated

LiF substrate, and a shallow angled electron scattering pattern (RHEED) was observed to show that the deposited film displayed limited long-range order [*Hashim et al.*, 1993]. An SEM (Figure 1) image was obtained to demonstrate that there was no micron scale porosity. The image shows no dark patches in the uppermost light grey area, which would imply that there was measurable porosity. Additionally the image shows clearly that the length-scale of surface roughness of the anvil is much larger than the thickness of the metal film, which is the thin pale line in the center of the figure. This leads to an effective porosity in the film, which results in 'thin films' having anomalously high shock temperatures.

Thick Film Experiments

Many shock temperature experiments on metals [*Williams et al.*, 1987; *Ahrens et al.*, 1990; *Yoo et al.*, 1993] have employed thick films or foils. In these cases the metal/anvil interface can be modeled as in infinite half-space. Specifically, a 'thick film' is one where the thickness is much greater than the thermal skin depth. One cannot determine from a single experiment which material is the source of the radiation, but one can determine the source from examining the data of many such experiments.

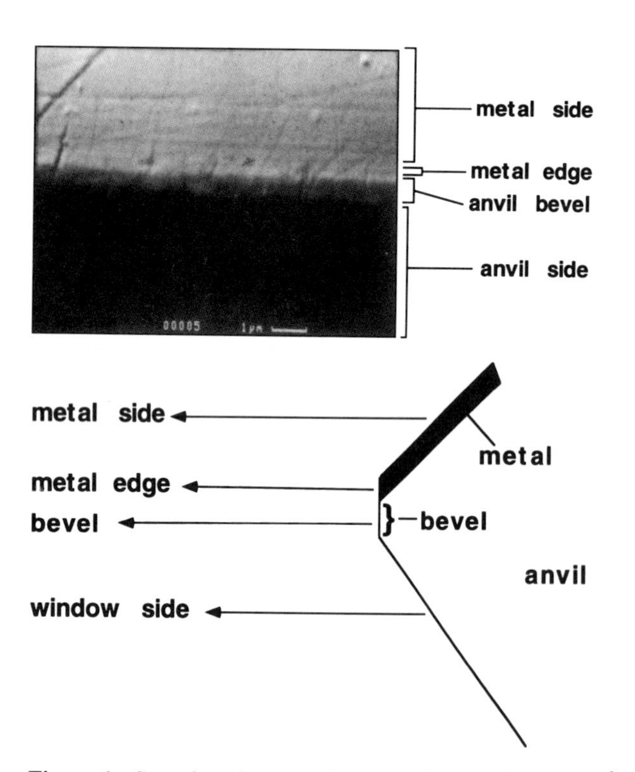

Figure 1. Scanning electron microscope image of the 1000 Å film used in Shot #297. The film is shown as deposited upon the Al_2O_3 anvil, tilted at a 45° angle to the camera. The dark lower section is the anvil, and the lighter section in the middle is the bevel at the edge of the anvil. The very thin pale line above the bevel is the edge of the Fe film, and the darker section at the top is the face of the Fe film, with the Al_2O_3 visible through the Fe.

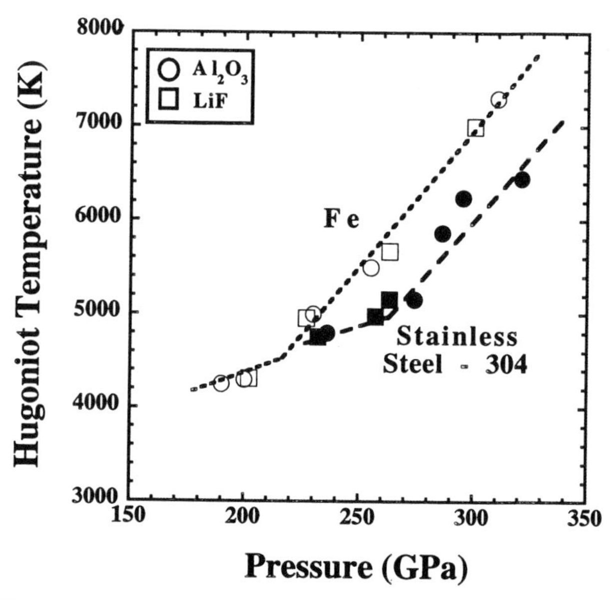

Figure 2. Comparison of Hugoniot temperatures for Fe (open symbols) and stainless steel, SS (solid symbols). Note the absence of coherent difference between temperatures for LiF (circles) and Al_2O_3 (squares).

Thick film shock temperatures show systematic temperature differences that depend on the film material used but not the anvil material. Figure 2 shows the results of experiments on Fe and stainless steel (SS), using both LiF and Al_2O_3 anvils. The Hugoniot temperatures for experiments on SS films are consistently higher than the ones for Fe films, as expected theoretically. There is no significant difference in the temperature achieved with different anvil materials. This result is difficult to reconcile if the radiation originates within the anvil, but it is consistent with the light originating in the metal. This is the first of several pieces of evidence supporting the transparency of Al_2O_3 during our experiments.

Thin Film Experiments

A series of shock temperature measurements was performed on thin iron films, sandwiched between two dielectric anvils. Shock temperatures are lower for dielectric anvils than for metal films. For the duration of the experiment, there is symmetric heat flow from the iron

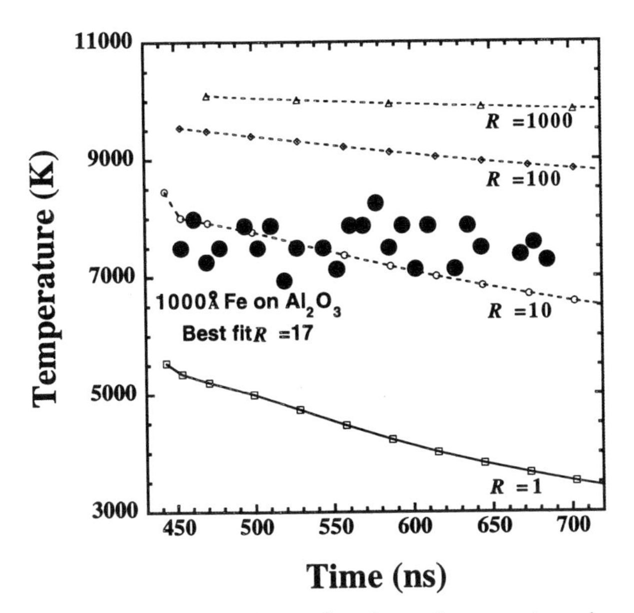

Figure 3. An experimentally observed temperature decay (Shot #287) from a 1000 Å film of Fe on an Al_2O_3 anvil at 164 GPa. The open symbols represent theoretical values for given values of *R*, the Fe/anvil thermal diffusivity ratio. The best fit *R* is 17 ± 2.

into the anvils. Because thickness is much greater than thermal skin depth, the interface temperature would not change even though heat flows across the interface [*Grover and Urtiew*, 1974]. When the film is thin, however, the interface temperature decays with time as heat flows across the boundary. We observed this decay and fit it to a finite element one-dimentional heat flow model [*King et al.*, 1989; *Gallagher and Ahrens*, 1996] to obtain thermal diffusivity ratios, R (defined by Equation 5), for the experimental materials (Figure 3). Since k_a has a $1/T$ dependence, heat conducts faster at earlier times. Thus the slope flattens out after about 40 ns of heat conduction; this correspond to times labeled 450 ns on Figure 3. Table 1 shows the best fit R values and corrects values given previously [*Gallagher and Ahrens*, 1996]. In the table, P_h is the Hugoniot pressure of the iron or of the anvil material during the experiment. T_h is the calculated Hugoniot temperature of the anvil material at the given P_h. T_i is the interface temperature measured during the experiment. ΔT is the RMS error in the grey body fit to temperature after time averaging. $k_{T,P}$ is the thermal conductivity of the anvil material; both theoretical values (Equation 4), and experimentally inferred values are listed. R is the thermal diffusivity ratio (Equation 5), and calculated values, as well as ones experimentally determined by the decay of T_i during the experiment, are listed. $\Delta k_{T,P}$ and ΔR are the RMS errors in experimentally determined $k_{T,P}$ and R values.

One of the inherent difficulties in our method of observing shock temperatures of opaque metals is the possibility that calculated conductivity values, used to calculate Hugoniot temperatures from interface temperatures via Equation 1, may be inaccurate. Our experiments provide data for R, the thermal diffusivity ratio, which is related via Equation 6 to the thermal conductivity ratio needed to calculate α (Equation 2). However it appears that the experimental values of R differ from previously calculated values by a less than factor of three for Al_2O_3 and a factor of five for LiF. The use of the 'fit' $k_{T,P}$ of Table 1 rather than values calculated via Equation 4 decreases the inferred Hugoniot temperature by about 350 K.

The thin film experiments were intended to constrain the thermal diffusivity ratio of the metal/anvil interface but can also be used to address the question of the source of the radiation. The systematics are consistent with the thick film experiments, in that no difference is observed between Hugoniot temperatures observed in experiments employing LiF and those employing Al_2O_3 anvils. Additionally, we observe a signal in which temperature decays with time, with the rate of decay depending on the thickness of the film. Thinner films (200 and 500 Å) decay faster than thicker films (1000 Å). Furthermore, since R for Fe/Al_2O_3 is lower than that of Fe/LiF, a 1000 Å film on Al_2O_3 should show a slower

Table 1. Experimental and Theoretical High Temperature and Pressure Thermal Conductivities of Corundum (Al_2O_3) and Griceite (LiF).

Shot #	Anvil	P_h (GPa) anvil	P_h (GPa) iron	T_h (K) anvil	T_i (K) interf.	ΔT_i (±K) interf.	$k_{T,P}$ ($\frac{W}{mK}$) calc.	$k_{T,P}$ ($\frac{W}{mK}$) fit	Δk ($\frac{W}{mK}$) fit	R Fe/anvil calc.	R Fe/anvil fit	ΔR (±) fit
285	Al_2O_3	244	303	2747	11000	120	16.5	6	2	70.5	19	12
286	LiF	166	261	4217	10200	490	8.57	2.3	1	29.1	11	13
287	LiF	164	269	4186	7550	520	12.6	3.4	1	11.4	4.3	11
296	Al_2O_3	165	197	1412	2×10^6	2×10^5	33.3	12	3	62.9	17	12
297	Al_2O_3	164	196	1406	6451	1030	33.4	12	3	62.7	17	13
304	Al_2O_3	156	194	1293	4813	677	35.3	13	3	58.0	16	12

decay than a film of the same thickness on LiF (Figure 4). If anvil material were the source of the observed radiation, then the decay time should not depend on film thickness.

In our thin film experiments the temperatures decay while emissivities remain approximately constant, giving an overall decrease in radiation with time. If the source of the radiation had been the anvil, then the effective observed emissivity should have increased with time because as more of the anvil material enters the shock state, more of it would be radiating. This is one of several pieces of evidence supporting the transparency of Al_2O_3 during our experiments.

Tang [1996] states that thermal contact resistance at the metal/anvil interface could be important in reducing shock temperature data. Because a contact resistance would allow a thermal boundary layer at the interface [*Swarts and Pohl*, 1989], Tang interprets the initial high intensity that is seen in some experiments as being a measure of the temperature in the interior of the metal. We disagree with this interpretation for two reasons. First, for many of our best sample assemblies we see no initial flash. If the flash was caused by an intrinsic thermal contact resistance then it would be seen in all experiments, not just some. For this reason we prefer to explain the initial intensity of some samples

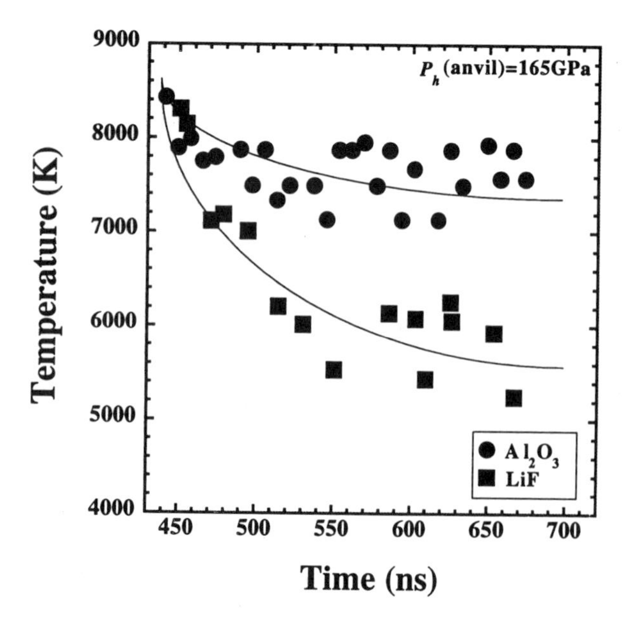

Figure 4. Comparison of conduction between Fe and (solid circles, Shot #287, P_h(Fe) = 269 GPa, P_h(LiF) = 164 GPa), and between Fe and Al_2O_3 (solid squares, Shot #297, P_h(Fe) = 196 GPa, P_h(Al_2O_3) = 164 GPa). Interface temperatures decrease as heat conducts out of the 1000 Å Fe films into the anvils.

as being a gap flash caused by an imperfect interface. Second, if Tang's interpretation is correct, we would expect the temperature of the initial flash to be consistent from experiment to experiment, but the observed grey body temperatures of the initial rises vary over a much wider range ($\pm$800 K) than the subsequent plateau temperatures ($\pm$250 K).

RADIATION FROM ANVIL MATERIALS

The reason the radiating anvil material is not observed is related to the large differences in T_h for the metal films and the relatively low shock temperatures of the anvils. There are two possible causes for radiation from the anvil, shear banding and grey-body emission from the continuum. If we were observing continuum anvil radiation, the amount of radiating material would increase with time (as the shock wave traverses the sample, it heats more and more of it), so the temperature would remain constant and the emissivity would increase with time. This would give an overall increase in photon flux with time. With a six channel pyrometer, we can resolve the emissivity time-dependence from the temperature time-dependence. In one experiment where the iron Hugoniot pressure was greater than 300 GPa (242 GPa in the Al_2O_3) this was observed; however, it was not the case for the other experiments, and we have never seen this behavior with LiF. Thus the radiation is not originating from the continuum of the anvil material for LiF and for Al_2O_3 below 242 GPa. If the radiation were caused by shear banding in the anvil material then, again, the amount of material involved in shear banding would increase with time, and the observed radiant intensity would increase with time. Additionally, shear banding dielectrics typically have very low emissivities $\epsilon \leq 10^{-2}$ [*Kondo and Ahrens*, 1983] whereas metals have emissivities in the range $0.1 \leq \epsilon \leq 1.0$. Our experiments show emissivities in the range 0.19 - 0.33. Therefore we conclude that for LiF and for Al_2O_3 below $\sim$ 240 GPa the radiation observed is from the iron and not the anvil. This is one of several pieces of evidence supporting the transparency of Al_2O_3 during our experiments.

As already stated, we did observe a single record that resembled *Kondo*'s [1994], where the Al_2O_3 Hugoniot pressure was $\sim$ 240 GPa. In this case, the radiation from the anvil material became brighter than the metal. Further we can distinguish between the two types of behaviors by examining the time dependence of emissivity. Kondo observes radiation from Al_2O_3 at Hugoniot pressures < 80 GPa, below our experimental range. Moreover he is observing radiation from Al_2O_3 against a background radiation from Ag films that are 400 K hotter than those of iron at 90 GPa.

DISCUSSION

One drawback of the thin film experiments is that in order for thermal decay to be observable on the $\sim$ 200 ns time-scale of the experiment, the film must be so thin as to be comparable to the surface roughness of the optically polished anvil materials and therefore comparable in thickness to the size of the gap between the driver anvil and the metal. This causes the metal to achieve shock temperatures much higher than for thick iron films, 10,000 K as compared to 6000 K. The higher temperatures are useful in that they expedite heat flow during the experiments, but they are not useful as a shock temperature measurement. Thus, only thick film experiments can give reliable interface temperatures. Reliable Hugoniot temperatures can be calculated from 'thick film' experiments by employing the thermal diffusivity ratios measured in the 'thin film' experiments. Figure 5 shows Hugoniot temperatures for iron and also phase boundaries obtained from static experiments. Solid Hugoniot temperatures are expected to show a decrease in slope where they intersect the solid-liquid phase boundary. The slope should increase again where the phase boundary intersects the liquid Hugoniot. We refer to this behavior as an 'offset' in the slope. If there is a phase boundary that crosses our data then the 'offset' is so small as to be unobservable with shock temperature experiments employing Al_2O_3 and LiF anvils. Therefore the present data do not agree in detail with the phase boundary inferred by *Yoo et al.* [1993], who observed this expected effect at $\approx$ 250 GPa with shock temperature experiments of iron using diamond anvils.

Sound speed measurements by *Brown and McQueen* [1986] detect what they interpreted as the melting of iron under Hugoniot conditions at 243 $\pm$ 2 GPa. These measurements are accurate for determining the pressure of melting, but the melting temperatures inferred from the experiments are calculated theoretically. Interpolation of our data to 243 GPa yields a temperature of 5860 $\pm$ 390 K. This is within the error bars of Brown and McQueen's theoretical calculation, and $\sim$ 13% (740 K) less than the melting temperature reported for dynamic experiments which use diamond as the anvil material [*Yoo et al.*, 1993], and 15% (900 K) greater than the melting temperature extrapolated from static compression data [*Boehler*, 1994]. Recent further exploration at high pressures and temperatures by *Yoo et al.* [1995, 1997]

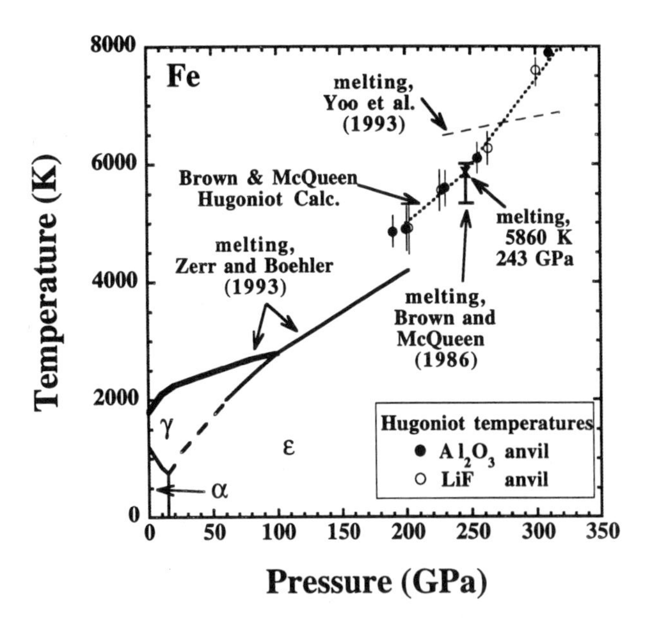

Figure 5. Hugoniot temperatures for Fe from thick film experiments using R values determined from thin film experiments. Both Al_2O_3 (solid circles) and LiF (open circles) anvils were used. Solid lines represent phase boundaries as reported by static measurements [*Boehler*, 1994]. The indicated uncertainty at 243 GPa represents melting determined via sound speed measurements [*Brown and McQueen*, 1986]. The 'X' represents our interpolated Hugoniot temperature of 5860 ± 390 K at 243 ± 2 GPa. The dotted line represents calculated Hugoniot temperatures from model b of the Brown and McQueen paper, which assumed γ = 1.34 and $\left.\frac{dE}{dP}\right|_v = 0.051\ m^3/Mg$.

suggest that our knowledge of the Fe phase diagram is incomplete.

CONCLUSIONS

For our experiments the anvil materials LiF and Al_2O_3 are shown to be transparent, using several lines of reasoning: First, there is a predictable systematic difference between Hugoniot temperatures of Fe and stainless steel. Second, there is no systematic dependence upon anvil material used for Hugoniot temperatures of Fe. Third, we observe a time dependence for emissivity, but no systematic time dependence for interface temperature during our 'thin film' experiments. Fourth, due to the high shock pressures of our Fe films, $P_h > 190$ GPa and due to the high temperatures caused by effective porosity in our 'thin films' experiments, our 'thin films' of Fe are expected to emit much more light than the 'thick films' of Ag reported by *Kondo* [1994] for lower pressures, < 80 GPa. Thus, light from our films more easily overwhelms any light from the continuum thermal emission of the anvil media.

We successfully conducted 'thin film' experiments to measure R. For Al_2O_3, values of 16 to 19 were obtained, which are up to a factor of 3 lower than calculated. For LiF, values of 4.3 to 11 were obtained, which are up to a factor of 5 lower than calculated.

Experimental R values were used to revise Hugoniot temperatures on Fe. The revised shock temperatures do not show the expected offset in slope as the Hugoniot intersects the fusion curve, so neither the shock pressure of the onset of melting nor that of the completion of melting is clearly obtained. However, we report here a Hugoniot temperature of 5860 $\pm$ 390 K at the 243 $\pm$ 2 GPa, the pressure where sound speed measurements detect the onset of melting of iron.

Acknowledgments. Research supported by NSF grant EAR 92-19906 (Ahrens), EAR 95-06377 (Ahrens) and DMR 92-02587 (Atwater). We thank Neil Holmes and Michael Manga for their prompt and helpful reviews. We thank Harry Atwater and Imran Hashim for use of the argon ion sputtering apparatus and for invaluable aid in making the iron coatings used for our 'thin film' experiments. We are grateful to Raymond Jeanloz and Michael Manga for technical discussions, to Stanley Love and Jessica Faust for helpful comments on the manuscript, as well as to M. Long, E. Gelle, A. Devora and Paul Carpenter for experimental assistance. Contribution No. 5847, Division of Geological and Planetary Sciences.

REFERENCES

Ahrens, T. J., Shock wave techniques for geophsyics and planetary physics, *Meth. of Exp. Phys.*, Vol. 24, edited by C. L. Luke, pp. 185–235, Academic Press, New York, 1987.

Ahrens, T. J., J. D. Bass, and J. R. Abelson, Shock temperatures in metals, in *Shock Compression of Condensed Matter - 1989*, edited by S. C. Schmidt, J. N. Johnson, L. W. Davidson, pp. 851–857, Elsevier, Amsterdam, 1990.

Boehler, R., The phase diagram of iron to 2 Mbar: New static measurements, in *High-Pressure Science and Technology - 1993*, edited by S. C. Schmidt, J. W. Shaner, G. A. Samara, M. Ross, pp. 919–922, AIP Press, New York, 1994.

Brown, J. M., and R. G. McQueen, Phase transitions, Grüneisen parameter, and elasticity for shocked iron between 77-GPA and 400-GPa, *J. Geophys. Res., 91*, 4785–4794, 1986.

Duffy, T., and T. J. Ahrens, Thermal expansion of mantle and core materials at very high pressures, *Geophys. Res. Lett., 20*, 1103–1106, 1993.

Gallagher, K. G., J. D. Bass, T. J. Ahrens, M. Fitzner, and J. R. Abelson, Shock temperature of stainless steel and a high pressure-high temperature constraint on thermal diffusivity of Al_2O_3, in *High-Pressure Science and Technology - 1993*, edited by S. C. Schmidt, J. W. Shaner, G. A. Samara and M. Ross, pp. 963–968, AIP Press, New York, 1994.

Gallagher, K. G., and T. J. Ahrens, Ultra-high-pressure thermal-conductivity measurements of griceite and corundum, in *Shock Waves, Vol. 2*, edited by B. Sturtevant, J. E. Shepherd and H. G. Hornung, pp. 1401–1406, World Scientific, Singapore, 1996.

Grover, R., and P. A. Urtiew, Thermal relaxation at interfaces following shock compression, *J. App. Phys., 45*, 146–152, 1974.

Hashim, I., B. Park, and H. Atwater, Epitaxial growth of Cu (001) on Si (001): Mechanisms of orientation development and defect morphology, *Appl. Phys. Lett., 63*, 2833–2835, 1993.

King, S. D., A. Raefsky, and B. H. Hager, ConMan: vectorizing a finite element code for incompressible two-dimensional convection in the earth's mantle, *Phys. Earth. Plan. Int., 59*, 195–207, 1989.

Kondo, K., and T. J. Ahrens, Heterogeneous shock-induced thermal radiation in minerals, *Phys. Chem. Min., 9*, 173–181, 1983.

Kondo, K.-I., Window problem and complementary method for shock-temperature measurements of iron, in *High-Pressure Science and Technology - 1993*, edited by S. C. Schmidt, J. W. Shaner, G. A. Samara and M. Ross, pp. 1555–1558, AIP Press, New York, 1994.

Leibfried, G., and E. Schlömann, Wärmeleitung in elektrisch isolierenden kristallen. *Nachr. Akad. Wiss. Göttingen, Math Phys. Klasse I A, 4*, 71–73, 1954.

Manga, M., and R. Jeanloz, Implications of a metal-bearing chemical boundary layer in D" for mantle dynamics, *Geophys. Res. Lett., 23*, 3091–3094, 1996.

Manga, M., and R. Jeanloz, Thermal conductivity of corundum and periclase and implications for the lower mantle, *J. Geophys. Res., 102*, 2999–3008, 1997.

Matassov, G., The electrical conductivity of iron-silicon alloys at high pressures and the Earth's core, Ph.D. Thesis, University of California, Livermore, Chapter 7, 1977.

McQueen, R. G., and D. G. Isaak, Characterizing windows for shock wave radiation studies, *J. Geophys. Res., 95*, 21752–21765, 1990.

Nellis, W. J., and C. S. Yoo, Issues concerning shock temperature measurements of iron and other metals, *J. Geophys. Res., 95*, 21749–21852, 1990.

Roufosse, M. C., and R. Jeanloz, Thermal conductivity of minerals at high pressure: the effect of phase transitions, *J. Geophys. Res., 88*, 7399–7409, 1983.

Secco, R. A., and H. H. Schloessin, The electrical-resistivity of solid and liquid Fe at pressures up to 7 GPa, *J. Geophys. Res., 94*, 5887–5894, 1989.

Schmitt, D. R., B. Svendsen, and T. J. Ahrens, Shock induced radiation from minerals, in *Shock Waves in Condensed Matter*, edited by Y. M. Gupta, pp. 261–265, Plenum Publishing Corporation, New York, 1986.

Schmitt, D. R., and T. J. Ahrens, Shock temperatures in silica glass: implications for modes of shock-induced deformation, phase transformation, and melting with pressure, *J. Geophys. Res., 94*, 5851–5871, 1989.

Swartz, E. T., and R. O. Pohl, Thermal boundary resistance, *Rev. Mod. Phys, 61*, 605–668, 1989.

Tan, H., and T. J. Ahrens, Shock temperature measurements for metals, *High Pressure Research, 2*, 159–182, 1990.

Tang, W., The pressure and temperature dependence of thermal conductivity for non-metal crystals, *Chinese J. High Press. Phys., 8*, 125–132, 1994.

Tang, W., F. Jing, R. Zhang, and J. Hu, Thermal relaxation phenomena across the metal/window interface and its significance to shock temperature measurements of metals, *J. Appl. Phys., 80*, 3248–3253, 1996.

Williams, Q., R. Jeanloz, J. D. Bass, B. Svendson, and T. J. Ahrens, Melting curve of iron to 250 GPa: a constraint on the temperature of the earth's center, *Science, 236*, 181–183, 1987.

Yang, W., Impact volaatilization of calcite and anhydrite and the effect on global climate from K/T impact crater at Chicxulub, Ph.D. thesis, California Institute of Technology, Chapter 4, Pasadena, CA, 1996.

Yoo, C. S., N. C. Holmes, M. Ross, D. J. Webb, and C. Pike, Shock temperatures and melting of iron at earth core conditions, *Phys. Rev. Lett., 70*, 3931–3934, 1993.

Yoo, C. S., J. Akella, A. J. Campbell, H. K. Mao, and R. J. Hemley, Phase-diagram of iron by in-situ x-ray-diffraction - implications for earth core, *Science, 270*, 1473–1475, 1995.

Yoo, C. S., A. J. Campbell, H. K. Mao, and R. J. Hemley, Detecting phases of iron - Response, *Science, 275*, 96–96, 1997.

T. J. Ahrens and K. G. Holland, Lindhurst Laboratory of Experimental Geophysics, Seismological Laboratory, California Institute of Technology, Pasadena, CA 91125. (e-mail: kathleen@gps.caltech.edu; tjacaltech.edu)

Elastic Properties of Forsterite at High Pressure Obtained From the High-Temperature Database

Hyunchae Cynn[1], Donald G. Isaak[2], and Orson L. Anderson

Center for Physics and Chemistry of Planets, Institute of Geophysics and Planetary Physics, University of California, Los Angeles, Los Angeles, California

We examine the possibility of computing the high pressure P elastic properties of forsterite using acoustic resonant frequencies measured at elevated temperature T (but at $P = 0$). The transformation into P space from T space requires three imposed conditions: (1) that the property αK_T be independent of volume V (α is the volume coefficient of thermal expansivity, and K_T is the isothermal bulk modulus); (2) that C_V, the heat capacity at constant volume, be quasiharmonic at high T; and (3) that the acoustic resonant mode frequencies be linear in P. We also compute the high-pressure elastic properties of NaCl and MgO to provide comparisons with the results for forsterite. We show that NaCl meets all three conditions, but that for MgO, the first condition is not met. For forsterite the second condition is not met.

1. INTRODUCTION

We examine the concept that the pressure dependencies of elastic properties can be computed from data on the temperature dependencies of acoustic resonant frequencies together with appropriate thermodynamic identities. Since the acoustic frequencies of normal mode vibrations are state variables as a function of P, T, and V, we believe this study is the first attempt to understand the anharmonic contribution of acoustic modes using acoustic vibrational frequencies. We focus on the case for forsterite, but also address the cases of NaCl and MgO. We begin by showing that in the quasiharmonic approximation, the temperature dependencies of normal mode frequencies are easily transformed into their respective pressure derivatives when reasonable assumptions are made. The vibrational part of the free energy of an ensemble of independent insulator oscillators obeying Bose-Einstein statistics is [*Landau and Lifshitz*, 1958]

$$\mathcal{F}_{VIB_i} = \left(\frac{\hbar\omega_i}{2kT}\right) \ell\text{n} \left(1 - e^{\hbar\omega_i/kT}\right), \tag{1}$$

where there are $3pN$ modal energies given by ω_1, ω_2, $\omega_3 \ldots 3pN$; N is Avogadro's number; p is the number of atoms in the formula unit; $\hbar$ is Planck's constant; and k is the Boltzmann constant.

An important approximation to (1) is provided by the assumption that all frequencies depend on V, but not explicitly on T. This is called the quasiharmonic approximation, defined by

$$\left(\frac{\partial\omega_i}{\partial V}\right)_T \neq 0, \tag{2}$$

[1]Now at Lawrence Livermore National Laboratory, Livermore, CA 94550.

[2]Also at Department of Math and Physics, Azusa Pacific University, Azusa, CA 91702.

Properties of Earth and Planetary Materials at High Pressure and Temperature
Geophysical Monograph 101

and

$$\left(\frac{\partial \omega_i}{\partial T}\right)_V = 0, \tag{3}$$

where $i = 1, 2, 3 \ldots 3pN$.

In resonant ultrasound spectroscopy (RUS), we deal with measured acoustic modal frequencies, which are at the center of the Brillouin zone ($\mathbf{k} = 0$). Since these acoustic vibrations are normal mode frequencies, (3) applies. There are, however, many acoustic normal frequencies at $\mathbf{k} = 0$, and in an RUS experiment, sometimes over 50 acoustic frequencies can be identified.

We assume that the acoustic normal modes, f_i, measured in RUS, follow the quasiharmonic approximation, that is

$$\left(\frac{\partial f_i}{\partial T}\right)_V = 0 \tag{4}$$

Expressing f_i as $f_i(P, T)$, we have from calculus

$$\left(\frac{\partial f_i}{\partial T}\right)_V = \left(\frac{\partial f_i}{\partial T}\right)_P + \left(\frac{\partial P}{\partial T}\right)_V \left(\frac{\partial f_i}{\partial P}\right)_T. \tag{5}$$

By means of a calculus equation involving P, V, and T, we have

$$\left(\frac{\partial P}{\partial T}\right)_V = \alpha K_T, \tag{6}$$

where α, the volume coefficient of thermal expansivity, is $(1/V)\,(\partial V/\partial T)_P$, and K_T, the isothermal bulk modulus, is $-V\,(\partial P/\partial V)_T$.

Therefore, within the quasiharmonic approximation, the pressure derivatives of frequencies are easily approximated from the temperature derivatives, being

$$\left(\frac{\partial f_i}{\partial P}\right)_T = -\frac{1}{\alpha K_T}\left(\frac{\partial f_i}{\partial T}\right)_P. \tag{7}$$

Then, assuming each f_i is linear in P, as was done in optic phonons, we can find

$$f_i(P)_T = f_i(P = 0,\, T) + \left(\frac{\partial f_i}{\partial P}\right)_T P.$$

The linear pressure dependencies of optic phonon frequency, including Raman and infrared vibration, are found in α- and β-olivine up to at least 14 GPa [*Chopelas*, 1990, 1991; *Hofmeister et al.*, 1989; *Cynn and Hofmeister*, 1994].

We also take the elastic moduli, $C_{ij}(P)_T$, to be linear in P, i.e.,

$$C_{ij}(P)_T = C_{ij}(P = 0, T) + \left(\frac{\partial C_{ij}}{\partial P}\right)_T P. \tag{8}$$

Equation (8) is not a fundamental assumption, since our calculations can be performed at a high density of points in P space. However, we compare our results with experiments that provide linear $C_{ij}(P)$ values over a finite P range and view any nonlinear effects as negligible in this comparison.

Application of (7) to find $f_i(P)_T$ at elevated compression depends on the assumption that $(\partial f_i/\partial P)_T$ is linear with $(\partial f_i/\partial T)_P$; that is, αK_T is independent of volume, a principle we shall investigate in some detail throughout this paper. (*Anderson* [1996] presented a review paper showing when one may expect αK_T to be independent of V in oxides and silicates.) Furthermore, (7) depends on the assumption that the higher-order pressure derivatives of f_i are negligible. From the measured $(\partial f_i/\partial T)_P$, $(\partial C_{ij}/\partial T)_P$ can be calculated using standard resonant ultrasound spectroscopy inversion techniques [*Ohno*, 1976; *Sumino et al.*, 1976]. C_{ij} versus T gives rise to an extensive high-T database. Examples are Tables 1–14 found in *Anderson and Isaak* [1995]. The values of αK_T, α, and density ρ can be found for many solids because of the high-T database as shown in Tables 15–42 of *Anderson and Isaak* [1995]. The relevant tables for forsterite used in our current study are Tables 9 and 18 of *Anderson and Isaak* [1995].

Before we can confidently use (7) and (8) to find elastic constants versus P, we must be sure of 3 conditions: (1) the thermal pressure must be independent of V; (2) equation (4) must be fulfilled; and (3) the mode frequencies must be linear in P. The need for condition 1

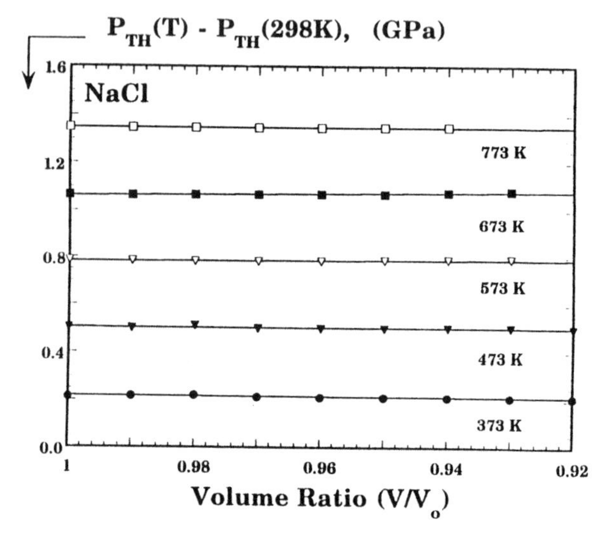

Figure 1. Isotherms of thermal pressure versus volume ratio ($\eta = V/V_0$) for NaCl. Below about 750 K, thermal pressure is independent of η.

arises from (7). In principle, αK_T could vary with T and P (or V), implying that the $(\partial f_i/\partial P)_T$ obtained from (7) is appropriate for only one coordinate of P, T space. Thus, the assumption that αK_T is independent of volume is necessary for condition 3 as well. We shall examine solids for which conditions 1 and 2 are fulfilled, or one or the other condition is fulfilled. The test of the first condition is whether αK_T is independent of V.

Equation (3) results in the classical specific heat, where C_V approaches the Dulong and Petit limit at very high T. If $(\partial \omega_i/\partial T)_V \neq 0$, then the derivative of the free energy for C_V has a T term, and C_V crosses the Dulong and Petit limit at high T. Thus violation of condition 2 may lead to anharmonicity. Conversely, if C_V is experimentally anharmonic, then we expect (3), and thus (4), to be violated, at least for some important modes. Therefore, a test of the second condition is whether C_V demonstrates significant anharmonicity below the melting temperature.

2. SODIUM CHLORIDE, NaCl

2.1. *Is αK_T Independent of V?*

The experiments of *Boehler and Kennedy* [1980] on NaCl show that up to 750 K, thermal pressure is independent of volume. This is illustrated in Figure 1 constructed from the *Boehler and Kennedy* [1980] experimental data. For NaCl, the thermal pressure at room temperature is independent of volume up to very high pressure, as indicated by the analysis of *Birch* [1986]. When P_{TH} is independent of V, αK_T is independent of V, as is now shown. The general form of the equation of state is

$$P(V,T) = P_1(V) + P_{TH}(V,T), \tag{9}$$

where the first term on the right, P_1, refers to the isothermal equation of state (EoS), and the second term on the right, P_{TH}, refers to the thermal pressure. Since P_1 is independent of T, $(\partial P/\partial T)_V = (\partial P_{TH}/\partial T)_V$, and therefore

$$P_{TH} = \int \alpha K_T \, dT \text{ at constant } V,$$

using (5).

Therefore, αK_T must vary with V in the same way P_{TH} varies with V. The thermodynamic identity relating αK_T and V is given by *Anderson* [1995]

$$\left(\frac{\partial \alpha K_T}{\partial V}\right)_T \equiv \left(\frac{\alpha K_T}{V}\right)(\delta_T - K'), \tag{10}$$

where δ_T is the Anderson-Grüneisen parameter, a dimensionless thermoelastic parameter defined as $-(1/\alpha K_T)$ $(\partial K_T/\partial T)_P$. Using the pressure data of *Boehler and Kennedy* [1980] and the shock-wave data of *Fritz et al.* [1971], *Birch* [1986] has shown that for NaCl, $\delta_T \approx K'$ out to large pressures. The data of δ_T and K' versus η (V/V_0) presented in Table 10 of *Birch* [1986] are reproduced in our Table 1. Through (10), we see that for NaCl, αK_T is independent of V over very large compressions at ambient temperature. The data showing that $\delta_T - K' = 0$ for a wide range of T at $P = 0$ are found in the tables of *Anderson and Isaak* [1995]. The condition that the thermal pressure is independent of V near ambient temperature is satisfied.

2.2. *Is There Anharmonicity in C_V at High T?*

The literature has several examples showing that for NaCl, $(\partial C_V/\partial T)_P = 0$ at high T. This property is demonstrated by Table 5 from *Anderson and Isaak* [1995], which gives C_V of NaCl up to 750 K ($T = 2.5\Theta$) with all the attendant properties that go into the calculation of C_V from the measured C_P.

2.3. *Computation of $C_{ij}(P)$*

The frequencies of normal modes of NaCl at ambient conditions were measured [*Yamamoto et al.*, 1987] on a sample with dimensions of $5.947 \times 4.529 \times 3.463$ mm^3 and a density of 2.160 g cm^{-3}. The C_{ij} at ambient conditions are (GPa) $C_{11} = 49.67(0.37)$; $C_{12} = 13.18(0.44)$; $C_{44} =$

TABLE 1. Dimensionless Thermoelastic Parameters for NaCl [after *Birch*, 1986] Using Experimental Data From *Boehler and Kennedy* [1980] and *Fritz et al.* [1971]

P (GPa)	V/V_0 (300 K)	K'_T	δ_T	γ
0	1.0000	5.5	5.3	1.62
1	.9627	5.0	4.9	1.55
2	.9324	4.8	4.7	1.51
3	.9067	4.6	4.5	1.46
4	.8845	4.4	4.3	1.43
5	.8649	4.2	4.2	1.40
10	.7910	3.8	3.8	1.27
15	.7397	3.4	3.5	1.19
20	.7004	3.2	3.1	1.12
25	.6685	3.0	2.9	1.07
30	.6416	2.8	2.7	1.03

12.68(0.02). The value of αK_T at ambient conditions is 2.856×10^{-3} GPa K^{-1} [*Yamamoto et al.*, 1987].

The measured frequencies for 4 of the 25 modes are listed in Table 2 for illustrative purposes. All 25 modes were used to find C_{ij} for NaCl at high pressure. The values of $(\partial f_i/\partial T)_P$ were found to three significant figures. Then $(\partial f_i/\partial P)_T$ was found by (7) and is shown in the fifth column of Table 2.

Using the pressure derivatives of the measured modes, we calculated the variation of C_{ij} with P using rectangular parallelepiped resonance (RPR) techniques of inversion [*Ohno*, 1976]. Results for C_{ij} versus P for NaCl are shown in Figure 2. Here we see that the measurements of K_S and the shear moduli at high pressure by *Spetzler et al.* [1972] agree fairly well with the calculated values arising from the high-temperature database. The calculated $(\partial K_S/\partial P)_T$ agrees with the measured value within 21%. The absolute difference between the measured and calculated K_S at 1 GPa is 0.78 GPa.

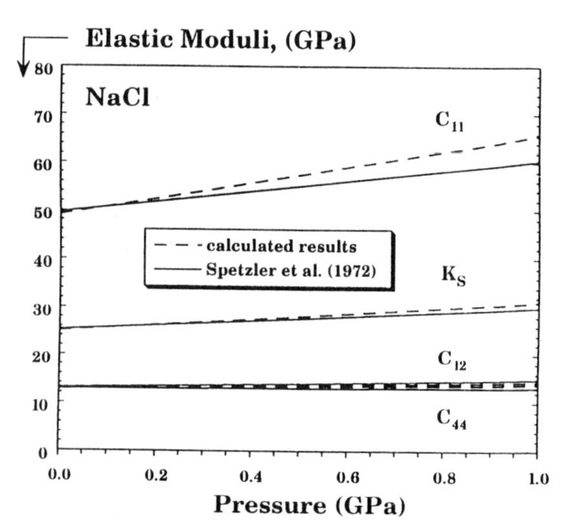

Figure 2. Elastic constants C_{ij} versus P computed from (7) for NaCl (dashed line) compared up to 1.0 GPa. Solid lines are from measurements of *Spetzler et al.* [1972].

3. MgO

3.1. *Is αK_T Independent of V?*

While we found that αK_T is independent of η for NaCl over a wide compression range at ambient conditions, this is not true for MgO. We know [*Anderson et al.* 1993; *Cynn et al.*, 1995] that at ambient conditions (and below 1000 K), αK_T in (7) depends on η (see Figure 3). Thus, we may expect that (7) will be somewhat inaccurate for MgO when extrapolating beyond ambient conditions.

3.2. *Is There Anharmonicity in C_V at High T?*

There is no evidence of anharmonicity in the specific heat of MgO. Figure 4 shows C_V and C_P of MgO. Curve a is the $C_P(T)$ at $P = 0$ [*Garvin et al.*, 1987]. Curve c is C_V computed from C_P, using the thermodynamic relationship between C_V and C_P:

$$\frac{C_V}{C_P} = 1 - \left(\frac{\alpha^2 K_T}{\rho C_P}\right) T, \qquad (11)$$

TABLE 2. RPR Modal Data for NaCl

Mode	294 K (kHz)	571 K (kHz)	$(\partial f_i/\partial T)$ (kHz/K)	$(\partial f_i/\partial P)$ (kHz/GPa)
Au–1 (torsion)	180.6	176.3	−0.0155	5.436
B1u–1 (flexure)	255.1	229.1	−0.0939	32.869
B1g–1 (shear)	297.3	286.9	−0.0375	13.147
Ag–1 (dilation)	370.6	320.2	−0.1819	63.714

where α is taken from *Anderson and Isaak's* [1995] tabulation. We see from Figure 4 that C_V never rises to the Dulong and Petit limit, even at $T = 2\Theta$. Curve b is computed from the classical equation combining the density of states and the quasiharmonic approximation:

$$C_V = 3Nk \int_0^{\omega_{\max}} \left\{\frac{e^x x^2}{(e^x - 1)^2}\right\} g(\omega)\, d\omega. \qquad (12)$$

The density of states used for $g(\omega)$ for MgO is from *Sangster et al.* [1970] and is shown in *Anderson* [1995], p. 122.

3.3. *Computation of $C_{ij}(P)$*

MgO is a solid in which αK_T is not independent of η except above about 2000 K (Figure 3), but the quasiharmonic postulate, our Condition 2, is obeyed.

Using (7) and (8) to calculate C_{ij} versus P for MgO, we should not expect the agreement with experiment we saw for NaCl. We calculated K_S and G versus P (up to 3 GPa) (Figure 5). We find that while our calculated K_S agrees with the measurement of K_S by *Jackson and Niesler* [1982] and that by *Spetzler* [1970], G does not agree as well with experiment.

We also calculate v_p and v_s at high pressure (up to 40 GPa) and compare with new experimental results from *Chopelas*, [1996] (Figure 6). The mismatch between the calculated and experimental results is due, in large part, to the use of linear extrapolation of our v_p and v_s at low pressure computed from (9) and the neglect of higher-order terms in C_{ij} with P. This comparison shows that

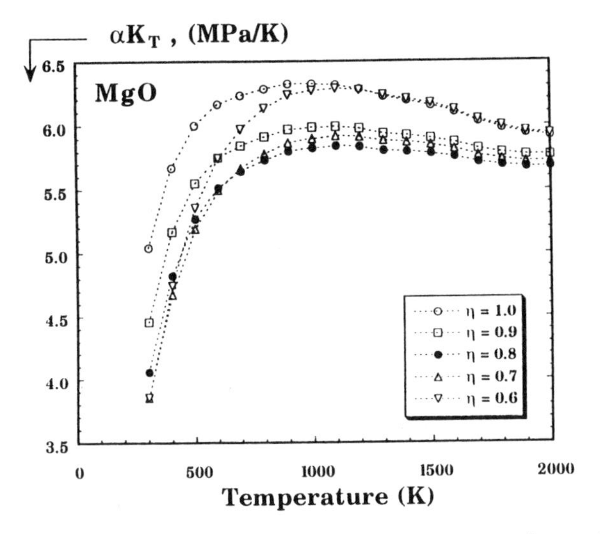

Figure 3. Isochores of αK_T versus T for MgO (from data in *Anderson et al.* [1993]). These show that αK_T is not independent of η.

a better understanding of higher-order pressure derivatives of resonant frequencies is needed to match with experiments at very high pressure.

4. Mg_2SiO_4

4.1. *Is αK_T Independent of V?*

For Mg_2SiO_4, there is considerable evidence that P_{TH} (and thus, αK_T) is independent of η over a wide range of η and T. Convincing experimental evidence comes from the fact that the value of P_{TH} versus T is the same at high compression (up to 7 GPa) as it is at room pressure. This is shown in Figure 7, which summarizes the results of *Guyot et al.* [1996], *Meng et al.* [1993], and *Isaak et al.* [1989].

One can see from (10) that when $\delta_T = K'$ over a large temperature range, αK_T is independent of η. The data from $\delta_T(T)$ at $P = 0$, taken from *Anderson and Isaak* [1995], are plotted in Figure 8. The value of K_0' at 300 K is taken from *Kumazawa and Anderson* [1969], and the value of $\partial^2 K_T/\partial T \partial P$ is taken from *Isaak* [1993], yielding $K_0' = 5.87$ at 1800 K. We see that the near coincidence of the δ_{T_0} and K_0' curves in Figure 8 yields $(\delta_T - K_0') = 0 \pm 0.2$ from 400 K to 1300 K. Therefore, by (10), αK_T is independent of η.

4.2. *Is There Anharmonicity in C_V at High T?*

If the quasiharmonic assumptions given by (2) and (3) are invoked, the equation for C_V converges to $3pR$ where $R = Nk$. When C_V approaches $3pR$ as a limit, as in the case of NaCl, the solid is said to be quasiharmonic in behavior.

Gillet et al. [1991] find that some optic modes of olivine and forsterite apparently violate (3). In their view, this temperature effect on the Raman modal frequencies is sufficient to add an extra term in C_V that is proportional to T, resulting in C_V crossing the Dulong and Petit limit at relatively low T. *Cynn et al.* [1996] tested the anharmonicity of forsterite by converting the measured C_P [*Barin and Knacke*, 1973] to C_V using the measured α. The results of this conversion are shown in Figure 9. In Figure 9, the horizontal line represents the Dulong and Petit limit. We see that C_V crosses the Dulong and Petit limit at 1500 K, when the α reported by Kajiyoshi [as quoted by *Isaak et al.*, 1989] is used in (11). Further, C_V crosses the Dulong and Petit limit at 1300 K when the new measurements of α up to 1800 K by *Bouhifd et al.* [1996] are used in (11). Thus, we see that a strong anharmonicity effect is found with a crossing of the Dulong and Petit limit at about $T \sim 1.6\Theta$. For purposes of comparison, we show (curve d) the result of using (12) where $g(\omega)$ is found from the calculations of *Rao et al.* [1988]. This demonstrates that the use of (12) will always guarantee the appearance of quasiharmonicity, which is not surprising since the formula was derived under the quasiharmonic postulate.

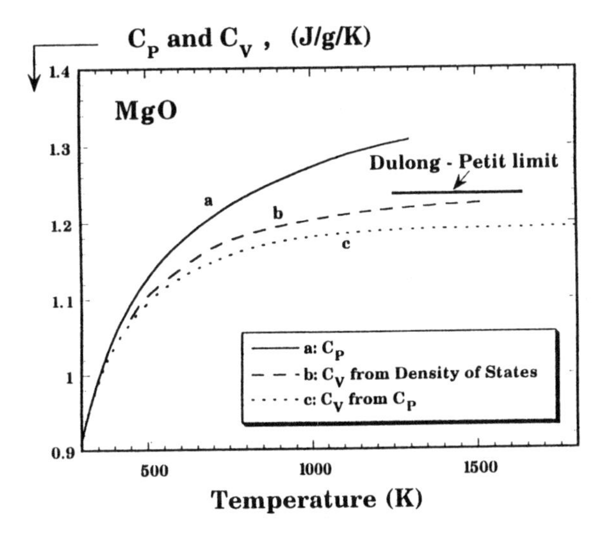

Figure 4. C_P and C_V for MgO at $P = 0$ (a) C_P from *Garvin et al.* [1987]. (b) the quasiharmonic C_V, calculated from (12) using the density of states from *Sangster et al.* [1970]. (c) C_V, determined from C_P using (11) and α data from *Anderson and Isaak* [1995].

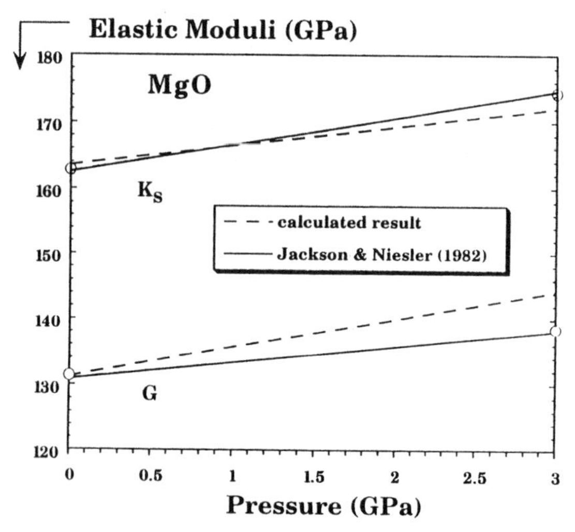

Figure 5. Elastic constants of MgO computed using (7) (dashed lines) and compared with data from *Jackson and Niesler* [1982] (solid lines) up to 3 GPa. *Spetzler*'s [1970] data, shown by open circles, are very close to those of *Jackson and Niesler*.

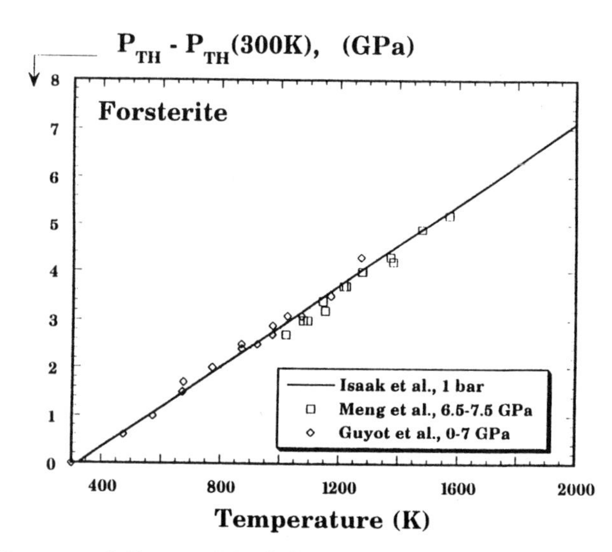

Figure 7. ΔP_{TH} for Mg_2SiO_4, showing that measurements at $P = 0$ coincide with those determined at high P ($P \sim$ 7 GPa). This shows that both P_{TH} and αK_T are independent of η.

4.3. *Computation of $C_{ij}(P)$*

As in the case of MgO, we find for forsterite the prospects for computing $(\partial C_{ij}/\partial P)_T$ from $(\partial C_{ij}/\partial T)_P$ weakened by the failure of one of the main imposed conditions that justify use of (7). Nevertheless, since the calculations of $C_{ij}(P)$ for MgO were reasonably accurate up to 3 GPa for K_S, we may also expect fair agreement with experiment for forsterite for K_S up to a few GPa pressure. However, even if the case for computing $(\partial C_{ij}/\partial P)_T$ is weak, we can use the comparison of the experimental data on $C_{ij}(P)$ with the predicted $C_{ij}(P)$ as a way of obtaining improved understanding on the shifts of the acoustic modes with pressure. This will be pursued in the next section. We show the calculated pressure dependence of each of the resonant modes for forsterite in Table 3.

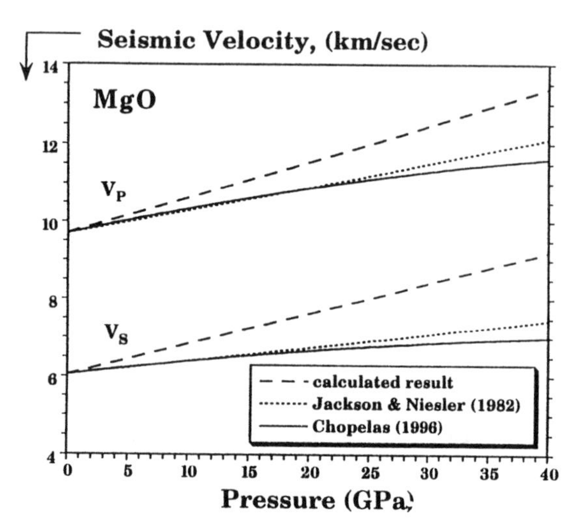

Figure 6. Comparison of v_p and v_s at high pressure for MgO. v_p and v_s from our calculation (dashed line) and from *Jackson and Niesler* [1982] (dotted line) are linearly extrapolated. *Chopelas*'s [1996] experimental data to 40 GPa are shown in the solid curve.

4.4. *The Shear and Longitudinal Velocities of Forsterite*

The nine $(\partial C_{ij}/\partial P)_T$ values for forsterite using the RPR-determined values of $(\partial f_i/\partial T)_P$ and (7) are shown in Table 4 (by first assuming (4) is valid — Column 7 — and then by substituting for (7), $(\partial f_i/\partial T)_V \neq 0$ — Column 8). The details of how changes from the assumption that $(\partial f_i/\partial T)_V = 0$ to $(\partial f_i/\partial T)_V \neq 0$ are done are provided below in the section entitled, "Estimation of $(\partial f_i/\partial T)_V$." These calculated $(\partial C_{ij}/\partial P)_T$ are compared with the results of measurements of the pressure derivatives by 5 different laboratories. We see that the calculated pressure dependencies of the shear moduli, C_{44}, C_{55}, and C_{66}, are high compared to experiment. However, the match between the calculated $K_S(P)$ and experiment is good. This match can be explained by a compensation between the three calculated values — $(\partial C_{11}/\partial P)_T$, $(\partial C_{22}/\partial P)_T$, and $(\partial C_{33}/\partial P)_T$ — which are high compared to the measured values of *Duffy et al.* [1995] and the three calculated values — $(\partial C_{23}/\partial P)_T$, $(\partial C_{13}/\partial P)_T$, and $(\partial C_{12}/\partial P)_T$ — which

TABLE 4. Pressure Dependence of the Elastic Constants of Forsterite

$\left(\frac{\partial C_{ij}}{\partial P}\right)_T$	Pulse Superposition[a]	Pulse Superposition[b]	Brillouin[c]	Pulse-Echo Overlap[d]	Brillouin[e]	This Study[f]	This Study[g]
C_{11}	8.47	8.32	9.83	7.08	5.45	10.50	3.62
C_{22}	6.56	5.93	7.03	5.09	5.17	8.11	3.88
C_{33}	6.57	6.21	7.55	5.29	4.72	8.68	3.70
C_{44}	2.12	2.12	1.72	1.80	1.67	4.09	2.63
C_{55}	1.66	1.65	1.50	1.40	1.41	4.15	2.39
C_{66}	2.37	2.32	2.20	2.09	1.76	4.84	3.07
C_{23}	4.11	3.53	—	3.17	3.45	1.73	0.24
C_{31}	4.84	4.23	—	3.40	3.04	2.91	1.46
C_{12}	4.67	4.30	—	3.56	3.66	3.19	1.77

[a] *Kumazawa and Anderson* [1969]; [b] *Graham and Barsch* [1969]; [c] *Shimizu et al.* [1982]; [d] *Yoneda and Morioka* [1992]; [e] *Duffy et al.* [1995]; [f] $(\partial f_i/\partial T)_V = 0$; [g] $(\partial f/_i\partial T)_V \neq 0$.

Our attempt to change $(\partial f_i/\partial T)_V$ to a non-zero value destroys agreement with the measured $d\Phi/dP$ (see also Figure 12). Any attempt to bring G into coincidence with the measured G by applying similar corrections for all the modes regarding the assumption $(\partial f_i/\partial T)_V = 0$ makes the agreement between our calculated K_S and the measured K_S worse, as shown in Figure 12. This shows that (7) is somewhat satisfactory for extrapolations of K_S, but (7) is not satisfactory for extrapolations of shear constants. This result indicates that modes with strong controls on the shear moduli are, indeed, strongly affected by anharmonicity effects. However, modes that correspond to dilation are not significantly affected by anharmonicity of the solid.

5. THE PROSPECTS OF MEASURING $(\partial f_i/\partial P)_V$

There is the prospect that the frequency of the individual RPR modes can be measured as a function of pressure, giving experimental values of $(\partial f_i/\partial P)_T$ to be used in (4). This measurement, coupled with $(\partial f_i/\partial T)_P$, will yield values of $(\partial f_i/\partial T)_V$ by (4). This is a difficult measurement, but the experiment is under way [*Isaak*, 1995]. We expect in the future to be able to distinguish the acoustic modes that are anharmonic and those modes that are quasiharmonic.

6. ESTIMATION OF $(\partial f_i/\partial T)_V$

We can find some idea of $(\partial f_i/\partial P)_T$ from the examination of the acoustic mode gamma [*Anderson*, 1988]. We have as the definition of the acoustic mode γ,

$$\gamma_{s,T} = \frac{1}{3} + \frac{K_T}{v_s}\left(\frac{\partial v_s}{\partial P}\right)_T \tag{13}$$

$$\gamma_{p,T} = \frac{1}{3} + \frac{K_T}{v_p}\left(\frac{\partial v_p}{\partial P}\right)_T, \tag{14}$$

where the sub s refers to shear mode propagation. This allows us to determine an effective $(\partial f_i/\partial P)_T$ associated with v_s and also v_p. We know the value of the acoustic mode gammas for forsterite at constant temper-

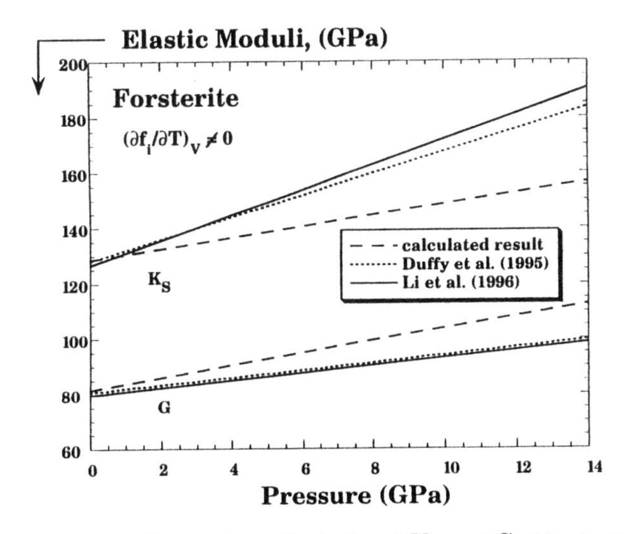

Figure 10. Comparison of calculated K_S and G of forsterite with data of *Duffy et al.* [1995] and *Li et al.* [1996] up to $P = 14$ GPa. This assumes $(\partial f_i/\partial T)_V = 0$.

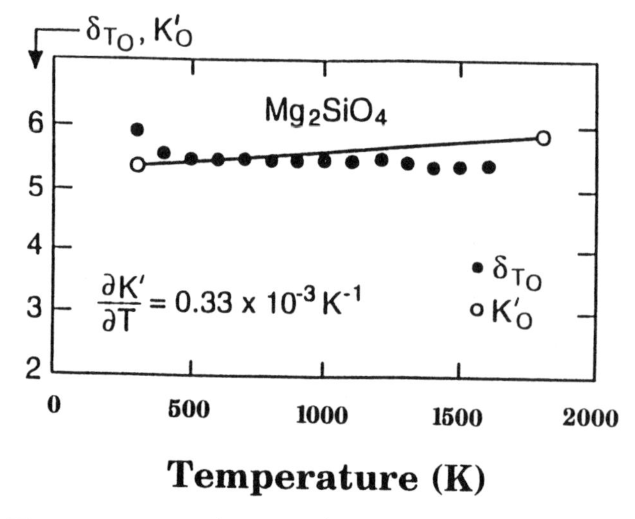

Figure 8. Plots of δ_{T_0} and K_0' versus T for Mg_2SiO_4 show that these two parameters are coincident up to temperatures on the order of 1300 K. Thus αK_T is independent of η over this same T interval. Data from *Anderson and Isaak* [1995], *Kumazawa and Anderson* [1969], and *Isaak* [1993].

are low compared to the measured values of *Duffy et al.* [1995].

Our results for forsterite indicate that, as in the case for MgO, the final calculation for K_S is fairly close to experiment, but the pressure derivatives for the shear moduli do not agree well with experimental values. This result is shown in Figure 10, where K_S and G are calculated out to 14 GPa from Table 4 and (7) and compared

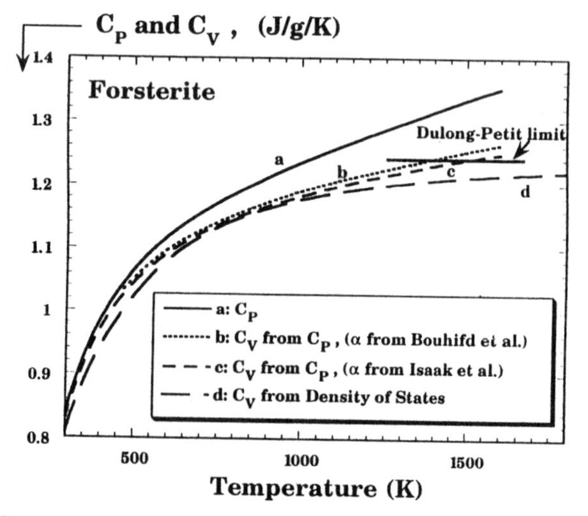

Figure 9. Specific heat, C_V, for Mg_2SiO_4 at $P = 0$. (a) C_P [*Barin and Knacke*, 1973]; (b) C_V calculated from C_P using (11) where α is from *Bouhifd et al.* [1996]; (c) C_V calculated from C_P using (11) where α is from *Isaak et al.* [1989]; (d) C_V calculated from (12) using the density of states found by *Rao et al.* [1988]. Figure modified from *Cynn et al.* [1996].

TABLE 3. Calculated RPR Frequencies for Forsterite at High Pressure

Mode	300 K kHz	$(\partial f_i/\partial T)$ kHz/GPa	$(\partial f_i/\partial P)$ kHz/GPa	1 GPa kHz
Au–1	647.43	−0.553	15.983	663.412
Au–2	771.50	−0.762	22.023	793.523
B2g–1	959.19	−0.787	22.746	981.936
B3g–1	965.26	−0.939	27.139	992.399
B2u–1	973.72	−0.639	18.454	992.174
B1g–1	1001.74	−0.952	27.500	1029.240
B1u–1	1004.25	−0.617	17.832	1022.082
B3u–1	1018.32	−0.808	23.353	1041.672
B2u–2	1061.10	−0.783	22.616	1083.715
B1u–2	1120.12	−0.884	25.549	1145.669
B1g–2	1194.37	−0.912	26.344	1220.713
B2g–2	1211.82	−0.930	26.864	1238.684
B3g–2	1219.49	−0.860	24.856	1244.345
B3u–2	1225.26	−0.859	28.427	1250.086
B3g–3	1302.69	−1.059	30.607	1333.296
Ag–4	1331.73	−0.962	27.803	1359.533
Au–3	1368.25	−1.231	35.578	1403.828
B1g–3	1398.06	−1.218	35.202	1433.262
B3u–3	1418.26	−1.047	30.246	1448.505
Ag–5	1432.11	−0.770	22.254	1454.364
B1u–3	1459.70	−1.047	30.260	1489.960
B2g–3	1468.40	−1.315	37.991	1506.391
B2u–3	1494.98	−1.021	29.494	1524.474
B1u–4	1564.84	−1.300	37.572	1602.412
B2u–4	1575.17	−1.350	39.003	1614.172
Ag–6	1636.52	−0.982	28.367	1664.887
B1u–5	1668.71	−1.101	31.806	1700.516
B2u–5	1688.22	−1.184	34.220	1722.439
B3u–5	1701.34	−1.084	31.329	1732.669
B2u–6	1707.46	−1.477	42.673	1750.133

with the measured results of *Li et al.* [1996] and *Duffy et al.* [1995]. Although the calculated K_S is close to that measured by *Li et al.* [1996], the departure of the extrapolated $G(P)$ from the measured $G(P)$ is large. We attribute this deviation to the failure of (4) and, consequently, the failure of (7). Lattice dynamics studies show that while K_S may be linear in P, the C_{ij} associated with the shear constants are markedly nonlinear [*Anderson and Liebermann*, 1970]. By assuming (4) and (7), the calculation of the seismic parameter, Φ, defined as K_S/ρ, shows that $\Phi(P)$ by our inversion agrees well with the *Li et al.* [1996] measurements (see Figure 11).

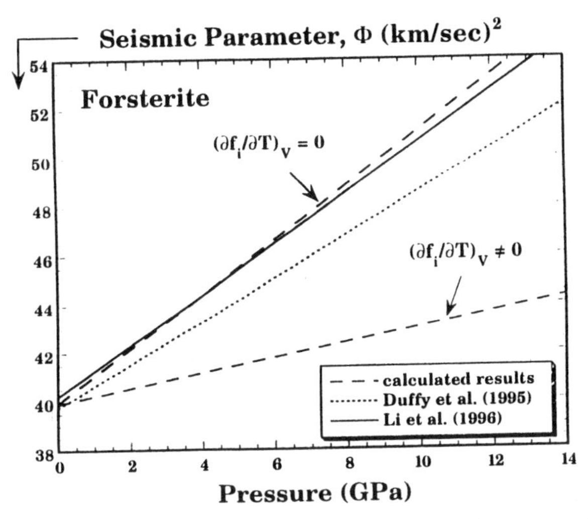

Figure 11. Comparison of calculated seismic parameter, Φ, with experimental data of *Duffy et al.* [1995] and *Li et al.* [1996], assuming that $(\partial f_i/\partial T)_V = 0$. The lower curve shows the result when correction is made for $(\partial f_i/\partial T) \neq 0$.

ature, so we can easily calculate $\gamma_{s,p}$ and $\gamma_{p,T}$ to find an averaged anharmonicity using acoustic data measured at high pressure and high temperature [*Kumazawa and Anderson*, 1969]. The averaged acoustic gamma at constant pressure, $\gamma_{ac,p} = 1/3\,(\gamma_{s,P} + 2\gamma_{p,P})$, is 2.76. This approach allows us to estimate $(\partial f_i/\partial P)_T$. This, coupled with $(\partial f_i/\partial T)_P$, gives us a value of $(\partial f_i/\partial T)_V$ through (5). We do not have $(\partial f_i/\partial P)_T$ for all the acoustic modes, so we assume $(\partial f_i/\partial P)_T$ is the same for

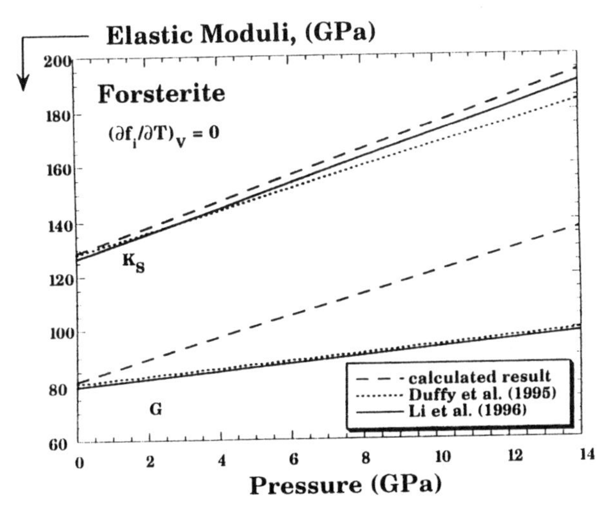

Figure 12. The effect of assuming $(\partial f_i/\partial T)_V \neq 0$ upon the calculated K_S and G. This assumption makes G closer to the experimental G, but farther away from the experimental K_S.

all modes. This gives us $(\partial f_i/\partial T)_V = -3.5\times10^{-5}$Hz/K as a correction to apply to the experimental values of $(\partial f_i/\partial T)_P$. This correction is uniformly applied to all the modal frequencies. Nevertheless, we obtain the effect shown in Figures 11 and 12 identified by the curves labelled $(\partial f_i/\partial T)_V \neq 0$. Comparing Figures 10 and 12, we see that $(\partial f_i/\partial T)_V \neq 0$, as calculated from the above equations lowers K_S and lowers G, and also from (11), lowers Φ. Thus, while the anharmonicity correction improves the comparison with the shear modulus, it makes the comparisons with the experimental bulk modulus worse. What is needed is a better theoretical idea of how to calculate the anharmonic effect of individual modes.

7. SUMMARY

We have found that for a solid (such as NaCl) in which αK_T is independent of V and C_V is quasiharmonic, the inversion of the measured $(\partial f_i/\partial T)_P$ yields reliable values of $(\partial f_i/\partial P)_T$ modes. This provides an accurate calculation of $K_S(P)$ and shear moduli up to 1.5 GPa.

We have found that for a solid (such as MgO), where C_V is quasiharmonic up to very high T but αK_T is not independent of V at ambient conditions, the inversion leading to $K_S(P)$ yields good results, but $G(P)$ departs from experiment.

We have found that for a solid (such as Mg_2SiO_4), where αK_T is independent of V but C_V is not quasiharmonic, the inversion leading to $K_S(P)$ and $\Phi(P)$ yields good results, but $G(P)$ departs from experiment.

These results show that caution is required when estimating elastic properties in P space from measurements at elevated T. The degree to which the quasiharmonic approximation, $(\partial f_i/\partial T)_V = 0$, is valid for acoustic resonant frequencies, f_i, appears to depend on the particular modal frequency. A better understanding of the degree to which each mode follows the approximation given by (4) would result in improvement of the fit between calculation of elastic properties and corresponding experiments. Our results are meant to stimulate further studies on how and why the anharmonicity varies from one mode to another.

Acknowledgements. The authors thank Moises Levy and Lars Stixrude for helpful discussions. HC especially thanks Hitoshi Oda and John D. Carnes for informative discussions. This work was supported by the Office of Naval Research and the National Science Foundation (EAR 94–05965). IGPP # 4626.

REFERENCES

Anderson, O. L., Simple solid-state equations for materials of terrestrial planet interiors, in *Physics of the Planets*, edited by S. K. Runcorn, pp. 27–60, John Wiley and Sons, Ltd., 1988.

Anderson, O. L., *Equations of State for Geophysics and Ceramic Science*, Oxford University Press, 1995.

Anderson, O. L., Volume dependence of thermal pressure in solids, *J. Phys. Chem. Solids*, submitted, 1996.

Anderson, O. L., and D. G. Isaak, Elastic constants of mantle minerals at high temperature, in *Mineral Physics and Crystallography: A Handbook of Physical Constants*, edited by T. Ahrens, pp. 64–97, AGU reference shelf 2, American Geophysical Union, Washington, DC, 1995.

Anderson, O. L., and R. Liebermann, Equations for the elastic constants and their pressure derivatives for three cubic lattices and some geophysical applications, *Phys. Earth Planet. Inter.*, *3*, 61–85, 1970.

Anderson, O. L., H. Oda, A. Chopelas, and D. G. Isaak, A thermodynamic theory of the Grüneisen ratio at extreme conditions: MgO as an example, *Phys. Chem. Miner.*, *19*, 369–380, 1993.

Barin, L., and O. Knacke, *Thermochemical Properties of Inorganic Substances*, Springer-Verlag, New York, 1973.

Birch, F., Equation of state and thermodynamic parameters of NaCl to 300 kbar in the high-temperature domain, *J. Geophys. Res.*, *91*, 4949–4954, 1986.

Boehler, R., and G. C. Kennedy, Equation of state of sodium chloride up to 32 kbar and 500°C, *J. Phys. Chem. Solids*, *41*, 517–523, 1980.

Bouhifd, M. A., D. Andrault, G. Fiquet, and P. Richet, Thermal expansion of forsterite up to the melting point, *Geophys. Res. Lett.*, *23*, 1143–1146, 1996.

Chopelas, A., Thermal properties of forsterite at mantle pressures derived from vibrational spectroscopy, *Phys. Chem. Miner.*, *17*, 149–156, 1990.

Chopelas, A., Thermal properties of β-Mg_2SiO_4 at mantle pressures derived from vibrational spectroscopy: implications for the mantle at 400 km depth, *J. Geophys. Res.*, *96*, 11817–11829, 1991.

Chopelas, A., Thermal expansivity of lower mantle phases MgO and $MgSiO_3$ perovskite at high pressure derived from vibrational spectroscopy, *Phys. Earth Planet. Inter.*, *98*, 3–15, 1996.

Cynn, H., O. L. Anderson, D. G. Isaak, and M. Nicol, Grüneisen ratios of MgO from the calculation of entropy, *J. Phys. Chem.*, *99*, 7813–7818, 1995.

Cynn, H., J. D. Carnes, and O. L. Anderson, Thermal properties of forsterite, including C_V, calculated from αK_T through the entropy, *J. Phys. Chem. Solids*, in press, 1996.

Cynn, H., and A. M. Hofmeister, High-pressure IR spectra of lattice modes and OH vibrations in Fe-bearing wadsleyite, *J. Geophys. Res.*, *99*, 17717–17727, 1994.

Demarest, H. H., Jr., Cube-resonance method to determine the elastic constants of solids, *J. Acoust. Soc. Am.*, *49*, 768–775, 1969.

Duffy, T. S., C.-S. Zha, R. T. Downs, H.-K. Mao, and R. J. Hemley, Elasticity of forsterite to 16 GPa and the composition of the upper mantle, *Nature*, *378*, 170–173, 1995.

Fritz, J. N., S. P. Marsh, W. J. Carter, and R. G. McQueen, The Hugoniot equation of state of sodium chloride in the sodium chloride structure, in *Accurate Characterization of the High Pressure Environment*, edited by E.C. Lloyd, *NBS Spec. Publ.*, *326*, 201–208, 1971.

Garvin, D., V. B. Parker, and H. J. White, Jr. (Eds.), *Codata Thermodynamic Tables: Selections for Some Compounds of Calcium and Related Mixtures: A Prototype Set of Tables*, 356 pp., Hemisphere Publishing Corporation, Washington, D.C., 1987.

Gillet, P., P. Richet, F. Guyot, and G. Fiquet, High temperature thermodynamic properties of forsterite, *J. Geophys. Res.*, *96*, 11,805–11,816, 1991.

Graham, E. K., and G. R. Barsch, Elastic constants of single-crystal forsterite as a function of temperature and pressure, *J. Geophys. Res.*, *74*, 5949–5960, 1969.

Guyot, F., Y. Wang, P. Gillet, Y. Ricard, Quasi-harmonic computations of thermodynamic parameters of olivines at high-pressure and high-temperature. A comparison with experiment data, *Phys. Earth Planet. Inter.*, *98*, 17–29, 1996.

Hofmeister, A. M., J. Xu, H.-K. Mao, P. M. Bell, and T. C. Hoering, Thermodynamics of Fe-Mg olivines at mantle pressures: Mid- and far-infrared spectroscopy at high pressure, *Amer. Mineral.*, *74*, 281–306, 1989.

Isaak, D. G., The mixed P, T derivatives of elastic moduli and implications on extrapolating throughout Earth's mantle, *Phys. Earth Planet. Inter.*, *80*, 37–48, 1993.

Isaak, D. G., New pressure measurement of elasticity using the rectangular parallelepiped resonance method, (abstract F 563) *Eos*, AGU, Fall Meeting, 1995.

Isaak, D. G., O. L. Anderson, T. Goto, and I. Suzuki, Elasticity of single crystal forsterite measured to 1700 K, *J. Geophys. Res.*, *94*, 5895–5906, 1989.

Jackson, I., and H. Niesler, The elasticity of periclase to 3 GPa and some geophysical implications, in *High-Pressure Research in Geophysics*, edited by S. Akimoto and M.H. Manghnani, Center for Academic Publications, Tokyo, 1982.

Kumazawa, M., and O. L. Anderson, Elastic moduli, pressure derivatives, and temperature derivatives of single-crystal olivine and single-crystal forsterite, *J. Geophys. Res.*, *74*, 5961–5972, 1969.

Landau, L. D., and E. M. Lifshitz, *Statistical Physics*, Pergamon Press, Ltd., London, 1958.

Li, B., I. Jackson, T. Gasparik, and R. C. Liebermann, Elastic wave velocity measurement in multi-anvil apparatus to 10 GPa using ultrasonic interferometry, *Phys. Earth Planet. Inter.*, *98*, 79–91, 1996.

Meng, Y., D. J. Weidner, G. D. Gwanmesia, R. C. Liebermann, M. T. Vaughan, Y. Wang, K. Leinenweber, R. E. Pacalo, A. Yeganeh-Haeri, and Y. Zhao, In-situ high P-T X-ray diffraction studies on three polymorphs (α, β, γ) of Mg_2SiO_4, *J. Geophys. Res.*, *98*, 22,199–22,207, 1993.

Ohno, I., Free vibration of a rectangular parallelepiped crystal and its applications to determination of elastic constants of orthorhombic crystals, *J. Phys. Earth*, *24*, 355–379, 1976.

Rao, K. R., S. L. Chaplot, N. Choudhury, S. Ghose, J. M. Hastings, L. M. Corliss, and D.L. Price, Lattice dynamics and inelastic neutron scattering from forsterite, Mg_2SiO_4:

phonon dispersion relation, density of states and specific heat, *Phys. Chem. Miner.*, *16*, 83–97, 1988.

Sangster, M. J. L., G. Peckham,, and D. H. Saunderson, Lattice dynamics of magnesium oxide, *J. Phys. Chem.*, *3*, 1026–1036, 1970.

E. M. Brody, Brillouin-scattering measurements of single-crystal forsterite to 40 kbar at room temperature, *J. Appl. Phys.*, *53*, 620–626, 1982.

Spetzler, H., The equation of polycrystalline and single crystal MgO to 8 kbars and 800 K, *J. Geophys. Res.*, *75*, 2073–2087, 1970.

Spetzler, H., C. G. Sammis, and R. J. O'Connell, Equation of state of NaCl: Ultrasonic measurements to 8 kbar and 800°C and static lattice theory, *J. Phys. Chem. Solids*, *33*, 1727–1750, 1972.

Sumino, Y., I. Ohno, T. Goto, and M. Kumazawa, Measurement of elastic constants and internal frictions on single-crystal MgO by rectangular parallelepiped resonance, *J. Phys. Earth*, *24*, 263–273, 1976.

Yamamoto, S., I. Ohno, and O. L. Anderson, High temperature elasticity of sodium chloride, *J. Phys. Chem. Solids*, *48*, 143–151, 1987.

Yoneda, A., and M. Morioka, Pressure derivatives of elastic constants of single crystal forsterite,in *High-Pressure Research: Application to Earth and Planetary Sciences*, edited by Y. Syono and M. H. Manghnani, pp. 207–214, Terra Scientific Publishing Company, Tokyo/ American Geophysical Union, Washington, DC, 1992.

H. Cynn, D. G. Isaak, and O. L. Anderson, Center for Physics and Chemistry of Planets, Institute of Geophysics and Planetary Physics, University of California, Los Angeles, CA 90095–1567

Hydrogen at High Pressures and Temperatures: Implications for Jupiter

W. J. Nellis, S. T. Weir, N. C. Holmes, M. Ross, and A. C. Mitchell

Lawrence Livermore National Laboratory, University of California, Livermore, California, USA

Electrical conductivities and shock temperatures were measured for shock-compressed liquid H_2 and D_2. Conductivities were measured at pressures of 93–180 GPa (0.93–1.8 Mbar). Calculated densities and temperatures were in the range 0.28–0.36 mol/cm^3 and 2000–4000 K. The resistivity data are interpreted in terms of a continuous transition from a semiconducting to metallic, primarily diatomic fluid at 140 GPa and 3000 K. Shock temperatures up to 5200 K were measured at pressures up to 83 GPa. These data are interpreted in terms of a continuous dissociative phase transition above 20 GPa. The continuous transition from a molecular to monatomic fluid means that Jupiter has no distinct core-mantle boundary. The dissociation model derived from the temperature measurements indicates a dissociation fraction of about 5% at 140 GPa and 3000 K. The isentrope of hydrogen was calculated starting from the surface temperature of Jupiter (165 K). At a metallization pressure of 140 GPa in Jupiter, the temperature is about 4000 K and about 10% of the hydrogen molecules are dissociated. The electrical conductivity was calculated along this isentrope by deriving a scaling relationship from the measured conductivities. The results indicate that hydrogen becomes metallic much closer to the surface of Jupiter than thought previously, a possible explanation of the very large magnetic field of Jupiter, but the metallic conductivity of the molecular fluid is two orders of magnitude lower than predicted for the monatomic fluid.

INTRODUCTION

Properties of hydrogen at high pressures and temperatures are needed to understand the nature of metallization and affects of electrical conductivity on the magnetic fields of giant planets, as well as to calculate the pressures, densities, and temperatures inside Jupiter and Saturn [*Hubbard*, 1980; *Stevenson*, 1982; *Zharkov and Gudkova*, 1992]· The interiors are at high pressures and high temperatures because of the planets' large mass and low thermal conductivity. Pressure and temperature at the center of Jupiter are about 4 TPa and 20,000 K [*Zharkov and Gudkova*, 1992]. Internal temperatures are well above the calculated melting curve of hydrogen [*Ross et al.*, 1981]. Properties of hydrogen also must be known to understand the interiors of new Jupiter-size planets now being discovered close to nearby stars [*Mayor and Queloz*, 1995]. These giant planets might also be composed of massive amounts of hydrogen, since hydrogen comprises 90% of all known atoms. For these reasons we have measured electrical conductivities and temperatures of fluid hydrogen and deuterium shock compressed to pressures up to 180 GPa. The conductivity experiments were performed with a reverberating shock wave to minimize the temperature and maximize pressure. The temperature measurements were performed at higher temperatures and lower pressures to investigate effects of

Properties of Earth and Planetary Materials
at High Pressure and Temperature
Geophysical Monograph 101

molecular dissociation. The theoretical model for dissociation derived from the temperature measurements shows that the conductivity experiments probed primarily the molecular phase.

EXPERIMENTS

High shock pressures were generated by impact of a hypervelocity impactor onto the front surface of a sample holder containing liquid hydrogen [*Nellis et al.*, 1983]. Al, Cu, or Ta impactors were accelerated to velocities up to 8 km/s with a two-stage light-gas gun. Both hydrogen and deuterium were used to obtain different shock-compressed densities and temperatures. In the conductivity experiments, hydrogen was contained between Al_2O_3 anvils which in turn were contained between Al disks [*Weir et al.*, 1996a]. Hydrogen shock pressure was determined by shock impedance matching the measured impactor velocity and known Hugoniot equations of state of the metal impactor, Al disk, and Al_2O_3 anvil. The pressure in hydrogen reverberates up to the first shock pressure in the Al_2O_3, independent of the equation of state of hydrogen [*Ogilvie and Duvall*, 1983]. Due to the large density mismatch between hydrogen and Al_2O_3, the first-shock pressure in hydrogen is a factor of ~25 lower than the first shock in Al_2O_3. The reverberation of the shock in hydrogen between the Al_2O_3 anvils compresses the hydrogen to the first shock pressure in the Al_2O_3 while maintaining a relatively low final hydrogen temperature. The configuration was illustrated previously [*Weir et al.*, 1994]. The electrical resistance was measured and electrical resistivity was obtained by calibration (conductivity is the reciprocal of resistivity). The relationship between resistance and resistivity was calculated by representing the sample by a cubic grid of 400-ohm resistors each 25 μm apart and solving Kirchoff's laws subject to the appropriate boundary conditions. The electrical resistivity of such a grid is 1 ohm-cm. The calculations were done as a function of sample thickness, which gives the electrical resistance as a function of sample thickness. The sample thickness achieved in each experiment was calculated by computational simulation. Three somewhat different electrical circuits were used, depending on the expected resistance to be measured. Two or four electrode probes were used depending on whether the expected conductivity was small or large compared to 10 $(\Omega\text{-cm})^{-1}$. Thus, the conductivity was measured using two isotopes and three circuits. In the temperature experiments hydrogen was contained between an Al disk and an Al_2O_3 or LiF window [*Holmes et al.*, 1995]. Pressure and density were obtained by shock impedance matching. The temperatures of the first shock and of this shock reflected off the window were determined by fitting the optical spectra of the emitted radiation to a greybody spectrum.

RESULTS

Although condensed molecular hydrogen is a wide bandgap insulator at ambient pressure (E_g=15 ev), at sufficiently high pressure the energy gap is expected to close to zero. Previous experiments have measured the electrical conductivity of hydrogen in the fluid phase at single-shock pressures up to 20 GPa and 4600 K [*Nellis et al.*, 1992]. In the present conductivity experiments, multiple-shock pressures of 93–180 GPa were achieved. Hydrogen is in the fluid phase because the calculated temperatures are well above the calculated melting temperatures of ~1000 K at 100 GPa pressures [*Ross et al.*, 1981]. The shock-compression technique is very well suited for measuring the electrical conductivity of hydrogen because (i) the high pressure reduces the energy gap, (ii) the reverberating shock maintains temperatures at a few 0.1 ev, ~10 times lower than the temperature which would be achieved by a single shock to the same pressure, (iii) the relatively low shock temperatures activate sufficient conduction electrons to produce measurable conductivities in the condensed phase, (iv) electrode dimensions and separations are of order mm's, which are straightforward to assemble, and (v) thermal equilibrium is obtained in a time less than the resolution (~1 ns) of the measurements. In these experiments the thickness of the hydrogen layer decreases from the initial value of 500 μm down to the compressed value of ~60 μm. Current flow reaches its equilibrium flow pattern in ~1 ns for a layer 60 μm thick with a minimum electrical resistivity of 5×10^{-4} Ω-cm. Supplemental experiments examining the electrical conductivity of shock-compressed Al_2O_3 [*Weir et al.*, 1996b] show that the conductivity of hydrogen at metallization (140 GPa) is 5 orders of magnitude greater than that of Al_2O_3. Thus, the conductivity measured is that of equilibrated hydrogen with negligible contribution from the shock-compressed Al_2O_3 holding the measurement electrodes.

The electrical resistivities decrease from about 1 Ω-cm at 93 GPa to 5×10^{-4} Ω-cm at 140 GPa and are constant at 5×10^{-4} Ω-cm at 155 and 180 GPa, as shown in Figure 1. The change in slope is indicative of the transition to the metallic state. Since our previous data [*Nellis et al.*, 1992] show that the electrical conductivity of hydrogen is thermally activated, we analyzed these new results in the range 93–135 GPa using our previous dependence for a fluid semiconductor:

$$\sigma = \sigma_0 \exp(-E_g(\rho)/2k_BT), \quad (1)$$

where σ is electrical conductivity, σ_0 depends on density ρ, $E_g(\rho)$ is the density-dependent mobility gap in the electronic density of states of the fluid, k_B is Boltzmann's constant, and T is temperature.

The density (ρ) and temperature (T) were calculated by simulating each experiment with a hydrodynamic computer code using a standard equation of state of hydrogen [*Kerley*, 1983]. Density and temperature must be calculated because these are normally determined by reflecting a laser beam from a surface subjected to multiple shock waves, in the case of density, and from the spectral dependence of emitted thermal radiation, in the case of temperature. Because of the high-rate deformations induced by the reverberating shock, the initially transparent single-crystal sapphire anvils turn into highly-defected, polycrystalline, opaque alumina. Thus, it is not possible at present for light to get in or out of the layer containg the hydrogen sample. At the conditions achieved in these experiments, this hydrogen equation of state is for the molecular fluid phase. Although a computational model for hydrogen introduces systematic uncertainties in the calculated densities and temperatures, the results are useful for understanding the slope change at 140 GPa. In all cases the maximum hydrogen pressure achieved in the simulation agreed to within 1% of the maximum reverberation pressure calculated by simple shock impedance matching, which does not depend on the equation of state of hydrogen. This agreement is a check on the simulations. The calculated densities and temperatures are plotted in *Weir et al.* [1996a].

For the fit to Equation 1, σ_0 was taken as a constant and $E_g(\rho)$ was assumed to vary linearly with density. Our least-squares fit of Equation 1 to the measured conductivities and calculated densities and temperatures gives $E_g(\rho) = 0.905 - (67.7)(\rho - 0.3)$, where $E_g(\rho)$ is in ev, ρ is in mol/cm^3 (0.28–0.31 mol/cm^3), and $\sigma_0 = 140$ (Ω-cm)$^{-1}$. A value of $\sigma_0 \approx 200$–300 (Ω-cm)$^{-1}$ is typical of liquid semiconductors [*Mott,* 1971]. The conductivity values calculated from the fit differ from the measured values by 23–53%, which is the range of uncertainties in the measurements. These uncertainties are modest because the conductivity varies by four orders of magnitude.

The E_g derived from this fitting procedure and k_BT are equal at a temperature of 0.3 ev and a density of 0.31 mol/cm^3. In this region of density and temperature the energy gap is smeared out thermally, activation of electronic carriers is complete, disorder is already saturated in the fluid, and conductivity is expected to be weakly sensitive to further increases in pressure and temperature. The latter is true because the conductivity of a disordered liquid metal is essentially constant when the number of carriers is essentially constant. The observation that the measured conductivity is constant within the error bars above 140 GPa demonstrates that carrier concentration is constant in this range. At 0.31 mol/cm^3 the calculated pressure is 120 GPa, which is close to the 140 GPa pressure at which the slope changes in the electrical resistivity (Figure 1). At higher pressures of 155 and 180 GPa the resistivity is essentially constant at 5×10^{-4} Ω or a conductivity of 2000 (Ω-cm)$^{-1}$, a value typical of the fluid alkali metals Cs and Rb at 2000 K [*Hensel and Edwards*, 1996]. Also, the minimum electrical conductivity of a metal is given by $\sigma = 2\pi e^2/3ha$, where e is the charge of an electron, h is Planck's constant, and a is the average distance between particles [*Mott and Davis*, 1971]. In this case $a = \rho_m^{-1/3}$ in proper units, where ρ_m is the density of hydrogen at metallization. The calculated value is 4000 (Ω-cm)$^{-1}$, which is in good agreement with the experimental value of 2000 (Ω-cm)$^{-1}$ when hydrogen becomes metallic. Thus, fluid hydrogen becomes metallic at about 140 GPa and 3000 K via a continuous transition from a semiconducting to metallic fluid.

About 5% of the hydrogen molecules are dissociated at metallization in the fluid at 3000 K [*Holmes et al.*, 1995].

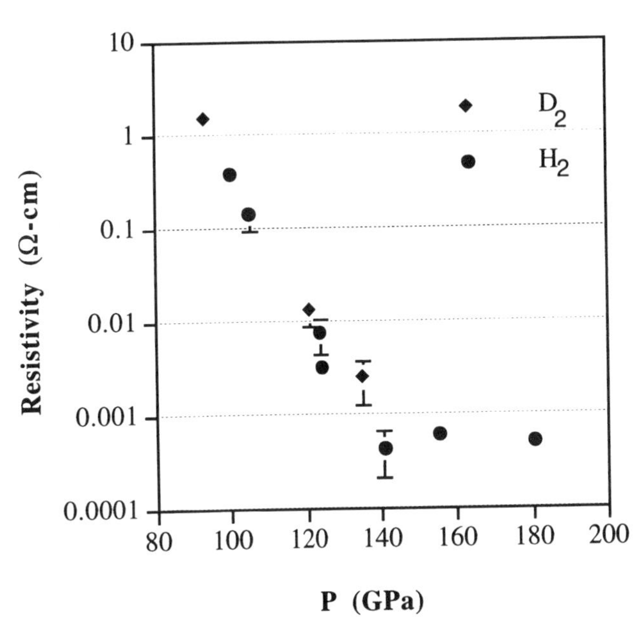

Figure 1. Electrical resistivity versus pressure for fluid hydrogen and deuterium. The saturation resistivity of 500 μΩ-cm above 140 GPa is that of the metallic fluid.

The sizes of the hydrogen molecule and atom and their initial electronic energy gaps are similar, which suggests that in the mixed-phase fluid at high temperatures the molecule and atom are mutually soluble and form a common mobility gap in their electronic density of states. Thus, the energy gap measured is that of the mixture. Additional experiments varying temperature and, thus, dissociation fraction are needed to determine the sensitivity of the band gap to the relative composition of molecules and atoms.

One question about the experiments is whether highly reactive hydrogen could be causing a fast reduction or some other reaction with Al_2O_3 at their interface, which is producing a new material which is causing the measured resistance. This is not possible because of the short time duration (~100 ns) of the experiment. Conservative estimates of the mass diffusion coefficient indicate that at most an interfacial layer 0.1 mm thick could be produced in 100 ns. For such a thin layer to produce the measured electrical resistance, its resistivity would have to be that of solid Cu at room temperature (1 μΩ-cm), which is unlikely in a disordered fluid mixture at 3000 K.

The metallization pressure of 140 GPa in the warm fluid is substantially lower than 300 GPa, the typical theoretical value for the crystalline solid at 0 K [*Ceperley and Alder*, 1987; *Ashcroft*, 1990]. The lower metallization pressure in the fluid is probably caused by the fact that hydrogen molecules approach closer to one another in the warm disordered fluid than in the cold crystal. Also, the electronic energy gap is known to be very structure dependent [*Chacham and Louie*, 1991; *Kaxiras et al.*, 1991] and the disorder allows the fluid to effectively sample many crystalline structures with lower band gaps. In addition several interactions occur in a lattice occupied by diatomic hydrogen molecules which have a high zero-point vibrational energy, orient their molecular axes relative to the crystal structure, and undergo charge transfer. These interactions in the crystal inhibit band-gap closure [*Ashcroft,* 1995] and are eliminated by melting, thus reducing the metallization pressure relative to that of the solid.

We have also measured temperatures of liquid deuterium and hydrogen shocked at pressures up to 83 GPa and 5200 K [*Holmes et al.*, 1995; *Nellis et al.*, 1995]. Temperatures were measured by fitting the optical radiation emitted from shock-compressed hydrogen to a greybody spectrum. The first-shock temperatures up to 20 GPa are in excellent agreement with predictions based on molecular hydrogen [*Ross et al.*, 1983]. The second-shock temperatures up to 83 GPa, obtained by reflection of the first shock off a window, are lower than predicted for the molecular phase. The lower measured temperatures are caused by a continuous dissociative phase transition above 20 GPa. This partial dissociation from the molecular to the monatomic phase absorbs energy, which causes lower temperatures than expected if hydrogen were to remain molecular. The theoretical model for molecular dissociation derived from the temperature data [*Holmes et al.*, 1995] was used to calculate the dissociation fraction in the conductivity experiments. At 140 GPa and 3000 K the dissociation fraction is calculated to be about 5%. Since only about 5% of the hydrogen molecules are dissociated at metallization, electronic conduction is probably caused by electrons delocalized from H_2^+ ions [*Ashcroft*, 1968]. This is a different phase than the monatomic one predicted initially by *Wigner and Huntington* [1935].

APPLICATION TO JUPITER

The shock temperature data were used to derive a model for molecular dissociation at high pressures and temperatures. Because hydrogen is in equilibrium in these experiments, this model was used to calculate the isentrope of hydrogen starting from the surface conditions of Jupiter, 165 K and 1 bar pressure [*Nellis et al.*, 1995]. Since Jupiter is ~90% hydrogen by number, this simple approach gives important general features of Jupiter. That is, the presence of He is neglected because the equation of state and electrical conductivity of Jupiter are dominated by hydrogen. This approach is generally sufficient, except for the situation of convection in the Jovian mantle, as discussed below.

When plotted as temperature versus pressure, the isentrope of hydrogen increases modestly and monotonically up to about 40 GPa, at which point the temperature reaches a broad plateau; the temperature of hydrogen might even decrease ~2% from 40 to 180 GPa, as shown in Figure 2. This weak temperature dependence is caused by the continual increase in molecular dissociation with increasing pressure. Thus, these calculations indicate that the interior of Jupiter is cooler and has much less temperature variation than believed previously. Of course, Jupiter contains about 10% He by number. Monatomic He is much hotter at a given density and pressure than is molecular hydrogen because He has no internal degrees of freedom to absorb energy in this regime. Thus, the warmer He probably causes the temperature of Jupiter to always increase monatomically with increasing pressure. In this case the volume coefficient of thermal expansion is also positive and a positive volume coefficient drives convection. Jupiter is known to be convective because it has a large external

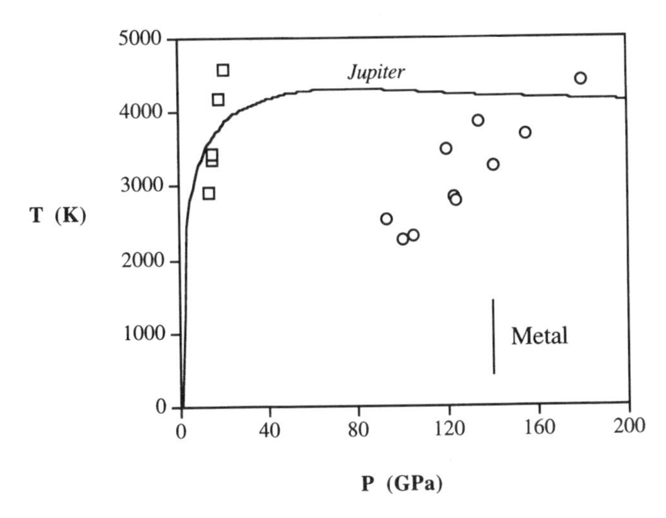

Figure 2. The solid curve is the isentrope of hydrogen calculated from surface conditions of Jupiter (165 K and 1 bar) plotted as temperature versus pressure. Open circles and squares are temperature and pressure at which electrical conductivities were measured. Metallization of hydrogen in Jupiter occurs at 140 GPa.

magnetic field which varies from 14 G at the north magnetic pole to 11 G at the south magnetic pole [*Smith et al.*, 1976], versus 0.5 G on Earth. The magnetic field is produced by the convective motion of electrically conducting hydrogen by dynamo action [*Stevenson*, 1983]. In addition, convective heat transfer to the surface is substantial and the reason why Jupiter radiates more internal energy than it receives from the sun [*Hubbard and Lampe*, 1969; *Hubbard*, 1968]. It is possible, however, that Jupiter is convectively quiescent over a radially thin region. A possible maximum in temperature versus pressure might induce an additional layer in the molecular region, as has been predicted to occur at 42 GPa by *Zharkov and Gudkova* [1992]. That is, a quiescent boundary layer over a long period of time might facilitate settling of ice and rock from hydrogen and cause an abrupt density change [*Nellis et al.*, 1995].

The continuous dissociative phase transition from the molecular to the monatomic phase means that there probably is no sharp boundary between the molecular mantle and the monatomic metallic-hydrogen core of Jupiter. Most recent models of Jupiter assume a sharp boundary between a molecular mantle and monatomic core at an internal pressure of 300 GPa.

A major issue about Jupiter is the region in which the magnetic field is produced. On the basis of calculations of electrical conductivities of dense fluid hydrogen [*Stevenson and Ashcroft*, 1974; *Stevenson and Salpeter*, 1977; *Hubbard and Lampe*, 1969; *Hubbard*, 1968; *Kirk and Stevenson*, 1987], the magnetic field has been thought to be produced primarily in the monatomic metallic core and in the molecular mantle, as well [*Kirk and Stevenson*, 1987; *Smoluchowski*, 1975; *Hide and Malin*, 1979]. However, the relative contributions of each region have been uncertain. Here we derive a scaling relationship for electrical conductivity and evaluate it as a function of pressure, density, and temperature along the isentrope of hydrogen starting from the surface conditions of Jupiter. To demonstrate that electrical conductivities were measured at pressures and temperatures representative of those in Jupiter, both our hydrogen isentrope and the temperatures and pressures achieved in the electrical conductivity measurements are plotted in Figure 2.

In a semiconducting fluid, electrical conductivities fit the relation in Equation (1) [*Nellis et al.*, 1996]. Energy gap E_g versus density was derived from the recently measured conductivities in two different density ranges [*Nellis et al.*, 1992; *Weir et al.*, 1996a]. The fit to these two data sets is

$$E_g = 20.3 - 64.7\rho, \tag{2}$$

in the range $0.13<\rho<0.3$ mol/cm^3 with E_g in ev. The prefactors σ_0 determined in the two sets of experiments differ by 10^4 at densities of 0.13 and 0.30 mol/cm^3 (the corresponding pressures differ by a factor of 10). We obtain values of σ_0 at intermediate densities by exponential interpolation of these data:

$$\sigma_0(\rho) = 3.4 \times 10^8 \exp(-44\rho). \tag{3}$$

Equation (3) fits the value of σ_0 determined at 0.13 mol/cm^3 (1.1×10^6 $(\Omega\text{-cm})^{-1}$), is close to the value obtained near 0.3 mol/cm^3 (1.4×10^2 $(\text{W-cm})^{-1}$), where hydrogen is undergoing a continuous transition from a semiconducting to metallic fluid, and σ extrapolates to the metallic value of 2000 $(\Omega\text{-cm})^{-1}$ (resistivity=5×10^{-4} Ω-cm) at 140 GPa.

The electrical conductivity of monatomic metallic hydrogen was predicted theoretically to be higher, 1–2 × 10^5 $(\Omega\text{-cm})^{-1}$ [*Stevenson and Salpeter,* 1977; *Hubbard and Lampe*, 1969], than measured for the molecular phase, 2 × 10^3 $(\Omega\text{-cm})^{-1}$, by *Weir et al.* [1996a] at the density and temperature at which metallization is observed (0.31 mol/cm^3 and 3000 K). The metallic hydrogen conductivity predicted theoretically is expected to be accurate at higher pressures for densities above 0.5 mol/cm^3, where the molecules are expected to be fully dissociated into a monatomic metallic fluid.

The electrical conductivity of hydrogen along the isentrope in Figure 2 was calculated with Equations (1)–

(3). In Figure 3 the resulting conductivity is plotted versus pressure along this isentrope. The conductivity was calculated up to 120 GPa, where the electronic energy gap approaches the temperature, and it was then extrapolated up to the metallic value of 2000 $(\Omega\text{-cm})^{-1}$ at 140 GPa.

For comparison, we also calculated electrical conductivities in the molecular envelope of Jupiter [*Kirk and Stevenson*, 1987] using Equation (1) with $\sigma_0=(\Omega\text{-cm})^{-1}$, a value typical of a liquid semiconductor [*Mott*, 1971] and different relations of $E_g(\rho)$ at 0 K for molecular hydrogen calculated by *Freidli and Ashcroft* [1977] and by *Min et al.* [1986], along an isentrope of hydrogen calculated by *Saumon et al.* [1995]. These two curves, labeled FA and M in Figure 3, can only be calculated up to 200 GPa where the temperature on the isentrope is equal to the bandgap of the molecular phase. As shown in Figure 3, for pressures below 20 GPa all three models approach a common value. For pressures above 40 GPa our calculated conductivities are 1 to 2 orders of magnitude greater than those calculated with FA and M. There is no physically meaningful way at present to interpolate from FA and M at 200 GPa in the molecular phase up to the calculated value of S for the monatomic phase at 300 GPa. Further theoretical results [*Stevenson and Salpeter,* 1977; *Hubbard and Lampe*, 1969] imply that the conductivity of monatomic metallic hydrogen at what was thought to be the core-mantle boundary, 300 GPa, 0.65 mol/cm^3, and 10,000 K, should be 10^5 $(\Omega\text{-cm})^{-1}$ (curve labeled S in Figure 3).

The metallization pressure of 140 GPa in the fluid observed by *Weir et al.* [1996a] is substantially lower than 300 GPa, the typical theoretical value for the solid at 0 K [*Ceperley and Alder*, 1987; *Ashcroft*, 1990]. This lower pressure implies that hydrogen in Jupiter becomes metallic at about 0.9 of the radius, as predicted by *Smoluchowski* [1975], rather than about 0.8 of the radius for 300 GPa. Because of the large volume of Jupiter this corresponds to about 50 more Earth masses of metallic hydrogen than thought previously. Also, an electrical conductivity as low as 20 $(\Omega\text{-cm})^{-1}$ might contribute to the magnetic field of Jupiter because a conductivity of this value is thought to be responsible for producing the magnetic fields of Uranus and Neptune [*Nellis et al.*, 1988]. As shown in Figure 3 a conductivity of 20 $(\Omega\text{-cm})^{-1}$ is achieved in Jupiter at 80 GPa. For these reasons, the external magnetic field of Jupiter would be produced in the molecular envelope substantially closer to the surface with a metallic conductivity about two orders of magnitude smaller than thought previously.

The above results imply that the magnetic field of Jupiter is produced in the region in which fluid molecular

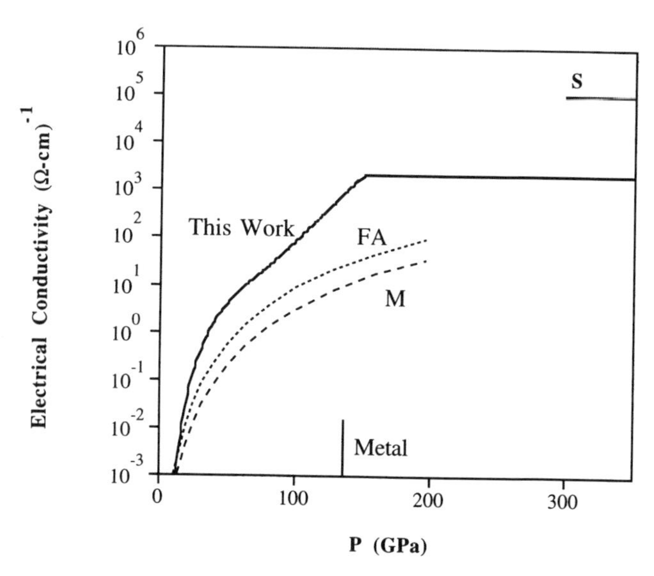

Figure 3. Electrical conductivities of hydrogen in Jupiter plotted versus pressure along isentropes of hydrogen. Curve labeled FA is from *Freidli and Ashcroft* [1977], M is from *Min et al.* [1986], and S is from results of *Stevenson and Salpeter* [1977] and *Hubbard and Lampe* [1969].

hydrogen undergoes a continuous dissociative phase transition. Using the theoretical model for molecular dissociation derived from shock temperature experiments, about 10% of the hydrogen molecules are dissociated at metallization in Jupiter (140 GPa and 4000 K). Once hydrogen metallizes by thermally smearing out the energy gap in this degenerate system, additional changes in electrical conductivity caused by the continuous molecular transition to complete dissociation at higher pressures are expected to be relatively small for an additional 2–3 fold increase in pressure. That is, in a fluid the increase in the electrical conductivity is dominated by the increase in the number of conduction electrons. Once they are all thermally excited to form a metal, further changes in conductivity are expected to be relatively small (~ factor of 2) as density and temperature increase somewhat with increase in pressure and depth. So that in the regime up to 2–3 times the observed metallization pressure, the electrical conductivity is about 2000 $(\Omega\text{-cm})^{-1}$. At higher pressures, temperatures, and depths at which molecular dissociation becomes complete, the pure monatomic phase is expected to form a two-component electron-proton metallic plasma, and the higher predicted conductivity [*Stevenson and Salpeter*, 1977; *Hubbard and Lampe*, 1969; *Abrikosov*, 1964] is expected to be the case. However, the external magnetic field is caused primarily by a conductivity of about 2000 $(\Omega\text{-cm})^{-1}$ in the outermost portion of Jupiter.

It is unlikely that small amounts of He, the ices, and rock mixed into the envelope contribute significantly to the electrical conductivity in Jupiter. The electrical conductivities of these materials are small compared to that of hydrogen. In addition, their concentrations are too small to affect the conductivity if they are in solution and would produce at most a tiny filamentary percolation path if they phase separate. These materials would be a small, electrically inert volume fraction within the hydrogen.

Acknowledgments. This work was performed under the auspices of the U. S. Department of Energy by the Lawrence Livermore National Laboratory under contract No. W-7405-ENG-48 with support from the LLNL Directed Research and Development Program and the U. S. National Aeronautics and Space Administration under grant W 16.180. We thank N. W. Ashcroft, W. B. Hubbard, D. J. Stevenson, A. A. Abrikosov, and D. A. Young for helpful discussions and D. Saumon for providing his isentrope of hydrogen. We thank E. See, P. McCandless, J. Crawford, K. Stickle, W. Brocious, S. Weaver, J. Ulrech, and R. Silva for their assistance in performing these experiments. We thank R. Kays of the Lockheed-Martin Co. for providing the liquid hydrogen storage dewar for liquid-H_2 coolant.

REFERENCES

Abrikosov, A. A., The conductivity of strongly compressed matter, *Sov. Phys.-JETP, 18*, 1399-1404, 1964.

Ashcroft, N. W., Metallic hydrogen: a high-temperature superconductor?, *Phys. Rev. Lett., 21*, 1748-1749, 1968.

Ashcroft, N. W., Pairing instabilities in dense hydrogen, *Phys. Rev. B, 41*, 10,963-10,971, 1990.

Ashcroft, N. W., Dense hydrogen: the reluctant alkali, *Phys. World*, *8*, 43-47, 1995.

Ceperley, D. M., and B. J. Alder, Ground state of solid hydrogen at high pressures, *Phys. Rev. B, 41*, 2092-2106, 1987.

Chacham, H., and S. G. Louie, Metallization of solid hydrogen at megabar pressures: A first-principles quasiparticle study, *Phys. Rev. Lett., 66*, 64-67, 1991.

Freidli, C., and N. W. Ashcroft, Combined representation method for use in band-structure calculations: Application to highly compressed hydrogen, *Phys. Rev. B, 16*, 662-672, 1977.

Hensel, F., and P. Edwards, The changing phase of liquid metals, *Phys. World, 9*, 43-46, 1996.

Hide, R., and S. R. C. Malin, The size of Jupiter's electrically conducting fluid core, *Nature, 280*, 42-43, 1979.

Holmes, N. C., M. Ross, and W. J. Nellis, Temperature measurements and dissociation of shock-compressed liquid deuterium and hydrogen, *Phys. Rev. B., 52*, 15,835-15,845, 1995.

Hubbard, W. B., Thermal structure of Jupiter, *Astrophys. J., 152*, 745-754, 1968.

Hubbard, W. B., and M. Lampe, Thermal conduction by electrons in stellar matter, *Astrophys. J. Suppl., 18*, 297-346, 1969.

Hubbard, W. B., Interiors of the giant planets, *Science, 214*, 145-149, 1980.

Kaxiras, E., J. Broughton, and R. J. Hemley, Onset of metallization and related transitions in solid hydrogen, *Phys. Rev. Lett., 67*, 1138-1141, 1991.

Kerley, G. I., A model for the calculation of thermodynamic properties of a fluid, in *Molecular-Based Study of Fluids*, edited by J. M. Haile and G. A. Mansoori, pp. 107-138, American Chemical Society, Washington, D. C., 1983.

Kirk, R. L., and D. J. Stevenson, Hydromagnetic constraints on deep zonal flow in the giant planets, *Astrophys. J., 316*, 836-846, 1987.

Mayor, M., and D. Queloz, A Jupiter-mass companion to a solar-type star, *Nature, 378*, 355-359, 1995.

Min, B. I., H. J. F. Jansen, and A. J. Freeman, Pressure-induced electronic and structural phase transitions in solid hydrogen, *Phys. Rev. B, 33*, 6383-6390, 1986.

Mott, N. F., Conduction in non-crystalline systems VI. Liquid semiconductors, *Phil. Mag., 24*, 1-18, 1971.

Mott, N. F., and E. A. Davis, *Electronic Processes in Non-Crystalline Materials*, p. 81, Oxford Press, London, 1971.

Nellis, W. J., A. C. Mitchell, M. van. Thiel, G. J. Devine, R. J. Trainor, and N. Brown, Equation-of-state data for molecular hydrogen and deuterium at shock pressures in the range 2-76 GPa (20-760 kbar), *J. Chem. Phys., 79*, 1480-1486, 1983.

Nellis, W. J., D. C. Hamilton, N. C. Holmes, H. B. Radousky, F. H. Ree, A. C. Mitchell, and M. Nicol, The nature of the interior of Uranus based on studies of planetary ices at high dynamic pressure, *Science, 240*, 779-781, 1988.

Nellis, W. J., A. C. Mitchell, P. C. McCandless, D. J. Erskine, and S. T. Weir, Electronic energy gap of molecular hydrogen from electrical conductivity measurements at high shock pressures, *Phys. Rev. Lett., 68*, 2937-2940, 1992.

Nellis, W. J., M. Ross, and N. C. Holmes, Temperature measurements of shock-compressed liquid hydrogen: Implications for the interior of Jupiter, Science, 269, 1249-1252, 1995.

Nellis, W. J., S. T. Weir, and A. C. Mitchell, Metallization and electrical conductivity of hydrogen in Jupiter, *Science, 273*, 936-938, 1996.

Ogilvie, K. M., and G. E. Duvall, Shock-induced changes in the electronic spectra of liquid CS_2, *J. Chem Phys., 78*, 1077-1087, 1983.

Ross, M., H. C. Graboske, and W. J. Nellis, Equation of state experiments and theory relevant to planetary modelling, *Phil. Trans. R. Soc. Lond., A303*, 303-313, 1981.

Ross, M., F. H. Ree, and D. A. Young, The equation of state of molecular hydrogen at very high density, *J. Chem. Phys., 79*, 1487-1494, 1983.

Saumon, D., G. Chabrier and H. M. Van Horn, An equation of state for low-mass stars and giant planets, *Astrophys. J. Suppl., 99*, 713-741, 1995.

Smith, E. J., L. Davis, and D. E. Jones, Jupiter's magnetic field and magnetosphere, in *Jupiter*, edited by T. Gehrels, pp. 788-829, University of Arizona Press, Tucson, 1976.

Smoluchowski, R., Jupiter's molecular hydrogen layer and the magnetic field, *Astrophys. J., 200*, L119-L121, 1975.

Stevenson, D. J., and N. W. Ashcroft, Conduction in fully ionized liquid metals, *Phys. Rev. A, 9*, 782-789, 1974.

Stevenson, D. J., and E. E. Salpeter, The phase diagram and transport properties for hydrogen-helium fluid planets, *Astrophys. J. Suppl., 35*, 221-237, 1977.

Stevenson, D. J., Interiors of the giant planets, *Ann. Rev. Earth Planet. Sci. 10*, 257-295 1982.

Stevenson, D. J., Planetary magnetic fields, *Rep. Prog. Phys., 46*, 555-620, 1983.

Weir, S. T., A. C. Mitchell, and W. J. Nellis, Electrical conductivity of hydrogen shocked to megabar pressures, in *High-Pressure Science and Technology-1993*, edited by S. C. Schmidt, J. W. Shaner, G. A. Samara, and M. Ross, pp. 881-883, American Institute of Physics, New York, 1994.

Weir, S. T., A. C. Mitchell, and W. J. Nellis, Metallization of fluid molecular hydrogen at 140 GPa (1.4 Mbar*), Phys. Rev. Lett., 76*, 1860-1863, 1996a.

Weir, S. T., A. C. Mitchell, and W. J. Nellis, Electrical resistivity of single-crystal Al_2O_3 shock-compressed in the pressure range 91-220 GPa (0.91-2.20 Mbar), *J. Appl. Phys., 80*, 1522-1525, 1996b.

Wigner, E., and H. B. Huntington, On the possibility of a metallic modification of hydrogen, *J. Chem. Phys. 3*, 764-770, 1935.

Zharkov, V. N., and T. V. Gudkova, Modern models of giant planets, in *High-Pressure Research: Application to Earth and Planetary Sciences*, *Geophys. Monogr. Ser*., vol. 67, edited by Y. Syono and M. H. Manghnani, pp. 393-401, Terra Scientific Publishing, Tokyo/AGU, Washington, D.C., 1992.

N. C. Holmes, A. C. Mitchell, W. J. Nellis, M. Ross, and S. T. Weir, Lawrence Livermore National Laboratory, University of California, Livermore, California, USA.

Advances in Equation-of-State Measurements in SAM-85

Yanbin Wang[1], Donald J. Weidner, and Yue Meng[2]

Center for High Pressure Research and Department of Earth and Space Sciences, University at Stony Brook, Stony Brook, New York

This paper describes the techniques we have developed over the past 5 years in measuring P-V-T relationships for mantle minerals at simultaneously high pressure and temperature, using a DIA type, cubic-anvil, high-pressure apparatus (SAM-85). Specific problems encountered in such experiments include (1) pressure and temperature generation and characterization, (2) nonhydrostatic stress, (3) pressure difference in a multiphase mixture, (4) overlapped diffraction lines caused by low symmetry, and (5) metastability issues. Solutions to each problem are discussed.

1. INTRODUCTION

Thermoelasticity of minerals plays a vital role in understanding mineralogy and dynamic processes of the Earth's mantle. One of the major experimental techniques in this field is to measure pressure-volume-temperature (P-V-T) relations. In most of the thermoelasticity studies to date temperatures have been limited to about 1000 K, which is about or below the Debye temperature of most high-pressure phases of silicates [e.g., see compilation of *Anderson*, 1995]. As thermoelastic properties are expected to behave differently above the Debye temperature, low-temperature data may not be directly applicable to mantle problems [e.g., *Anderson and Suzuki*, 1984; *Anderson et al.*, 1992].

[1] Now at Consortium for Advanced Radiation Sources, The University of Chicago, 5640 S. Ellis Ave., Chicago, IL 60637.
[2] Now at General Electric Company, 6825 Huntly Road, Worthington, OH 43085.

Properties of Earth and Planetary Materials at High Pressure and Temperature
Geophysical Monograph 101

Direct measurements of acoustic velocities are currently limited to either room temperature and high pressure or room pressure and high temperature. Brillouin scattering [*Meng and Weidner, 1994*; *Duffy et al.*, this volume], laser interferometry [*Zaug et al.*, 1993], and ultrasonic [*Gwanmesia et al.*, 1990] techniques are used to measure elastic constants and their pressure derivatives; all of these measurements are currently limited to room temperature. On the other hand, both ultrasonic [*Anderson et al.*, 1992] and Brillouin scattering [*Yeganeh-Haeri and Weidner*, 1990] techniques can provide direct measurements on elastic constants and their temperature derivatives, but these measurements are presently limited to zero or low pressures and may not be suitable for high-pressure phases as they are thermodynamically unstable under high temperatures at zero pressure. Therefore, P-V-T measurements provide important information on thermoelastic properties of high-pressure phases, and there is a need for collecting complete thermal elasticity data set at simultaneously high pressure and temperature for all of the major mineral components.

There are several critical issues in determining equations of state using P-V-T data. One must be extremely careful in the measurements and data reduction as progressively higher order derivatives are extracted from the data. Errors in the P-V-T measurements can lead to erroneous thermoelastic parameters. Possible sources of errors include (1) deviation from a perfect diffraction geometry (systematic errors due experimental setup), (2) nonhydrostatic stress, (3) pressure or temperature gradient in the sample, (4) pressure differential in a multiphase mixture, and (5) possible irreversible changes in high-pressure phases as some of the P-V-T measurements may be carried out under metastable conditions. Over the last few years, we have developed the state-of-the-art methodology for equation-of-state measurements. Details of the experimental setup have been discussed extensively in the past; we will focus on the rest of the issues related to accurate P-V-T measurements.

2. THE DIA APPARATUS

The DIA-type cubic-anvil, high-pressure apparatus (SAM-85) is installed on the superconducting wiggler beamline (X-17) of the National Synchrotron Light Source (NSLS). It can comfortably generate stable temperatures up to 1500 °C at 12 GPa; pressures up to 15 GPa have been reached at 1300 °C using sintered diamond anvils with 4x4 mm^2 truncation, a condition corresponding to depths within the transition zone. The system is described elsewhere by *Weidner et al.* [1992a, b], but we will emphasize a few salient features here.

The force of a uniaxial ram is transmitted to six anvils which simultaneously advance into the solid pressure medium that is cubic in shape. Gaps between the anvils of about 0.5 mm afford space for incident X rays to enter the sample chamber and diffracted X rays to exit. The cut-away view of the pressure medium is illustrated in Figure 1. The dimension of the sample volume is about 1 mm in diameter and 2 mm long. A thermocouple which passes through one of the anvil gaps monitors the temperature during the run. NaCl is used as the diffraction standard to define the pressure with the Decker equation of state for NaCl [*Decker*, 1971].

White radiation from the superconducting wiggler magnet provides an energy-dispersive analysis with energies between 15 and 100 keV. We typically use a 2θ angle of 5 to 10°. Data gathering time of less than 5 minutes from an X ray beam of dimensions of 100 × 100 μm provides robust diffraction patterns. The accuracy of the volume is generally in the order of 0.1% or less. With careful system calibration and minimization of nonhydrostatic stress, the accuracy can be maintained to within 0.05% for volume. Such accuracy is critical in determining the bulk modulus and its temperature and pressure derivatives.

By analyzing the lattice parameters of samples at several vertical positions within the sample chamber, we have determined that the temperature variation over the entire sample length is of the order of 20°C at 1200°C. Using multiple thermocouples, we find that the radial temperature gradient is less than 5°C at these conditions. Therefore, the maximum temperate variation within the scattering volume is about 5°C. We find no evidence of a pressure gradient at room temperature.

Recently, we have modified the conventional DIA anvils in order to improve their performance in terms of pressure capability and reliability. Details of the new anvil design are discussed by *Wang et al.* [this volume]. Several important improvements have been made: (1) A ~30% gain in maximum attainable pressure compared to the conventional anvil geometry; the tapered WC anvils give the pressure range comparable to that of straight sintered

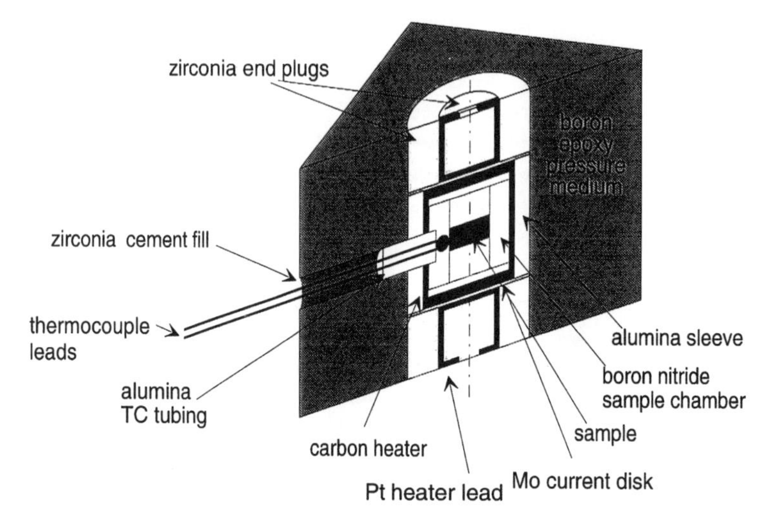

Figure 1. The cell assembly for SAM-85 using 4 mm straight or 3.5 mm tapered anvils.

diamond anvils. (2) The wedged anvil gaps act as pressure seals, thereby significantly reducing the number of blowouts, which earlier caused failures in the experiments. (3) The tapered anvil geometry greatly reduced hysteresis in the pressure-load relation, enabling us to obtain multiple heating cycles at various pressures and allowing complete P-V-T measurements within the entire P-T space with a single experiment. As an example, Figure 2 shows the P-T paths of one experiment for a sample of garnet solid solution with 38 mol% majorite and 62 mol% pyrope. (4) The tapered anvil geometry greatly increased survival rate of the thermocouples to close to 100%. Counting all the factors together, we have increased our efficiency in the P-V-T experiments by a factor about 5 with the tapered anvils.

3. UNDERSTANDING AND ELIMINATING NONHYDROSTATIC STRESS

One of the factors that affects both volume and pressure measurements is nonhydrostatic stress. We have characterized and measured non-hydrostatic stress in the sample [*Weidner et al.*, 1992a, b]. Both macroscopic and microscopic stresses will affect the determination of peak position, giving rise to inconsistency in the cell parameter refinement. As we use a diffraction standard to determine the pressure, these stresses are the main contributors in pressure uncertainty. *Weidner et al.* [1992a, b] have shown that the stress field in a DIA sample may be approximated as having a cylindrical symmetry, with the vertical (axial) component differing from the horizontal (girdle) components. A macroscopic deviatoric stress $\Delta\sigma = \sigma_1 - \sigma_3$ will make the apparent pressure differ from the real one by $(K/\mu)\Delta\sigma$, where K and μ are the aggregate bulk and shear

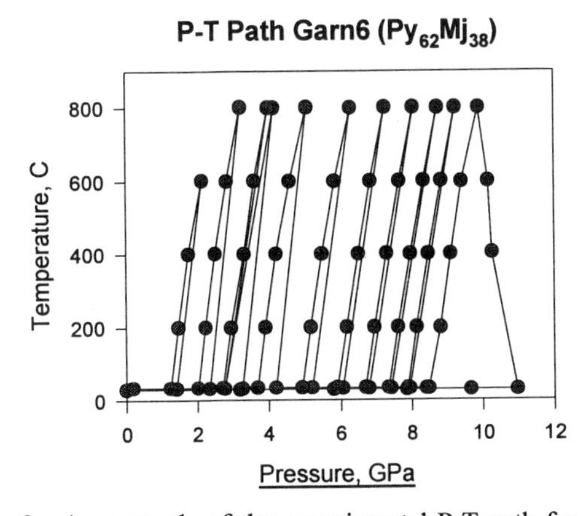

Figure 2. An example of the experimental P-T path for a garnet sample. The maximum temperature was restricted to 800°C because of the metastability of the sample. The entire P-T space offered by SAM-85 can be covered in a single run.

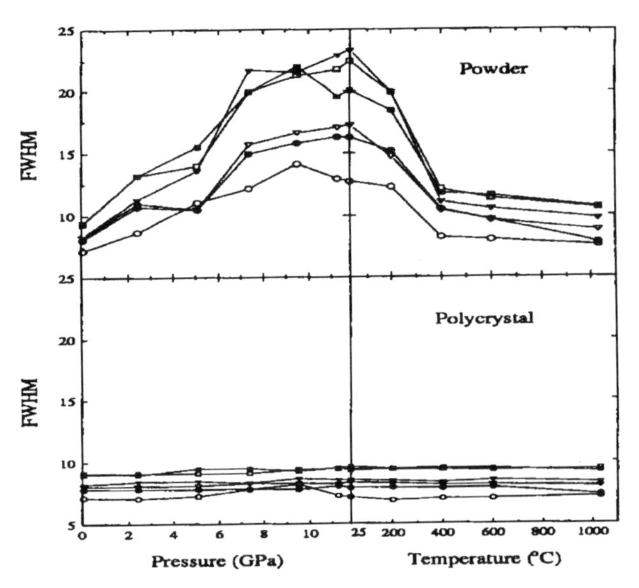

Figure 3. A comparison of full-width-at-half-maximum (FWHM) of various diffraction peaks of stishovite from a powdered (top two figures) and a sintered polycrystalline (bottom two figures) sample measured in a single experiment. The left two panels are data at room temperature and the right two panels are data at 11 GPa and high temperatures. Note the dramatic increase in FWHM for the powdered sample with increasing pressure at room temperature, whereas the sintered sample hardly show any broadening. Heating to above 500°C reduces the peak width to close to its original value.

moduli, respectively, of the pressure standard. The sign of this deviation depends on the orientation of the maximum compressive stress σ_1 with respect to the diffraction vector that is approximately vertical. While a correction may be applied to pressure once the macroscopic deviatoric stress is measured, similar corrections to sample volume are impractical.

As we use polycrystalline samples in the diffraction experiment, microscopic deviatoric stress is also present at grain to grain level due to elastic anisotropy, even when the boundary condition is perfectly hydrostatic. This microscopic stress field causes peak broadening and will also affect accuracy in volume measurement owing to larger errors in peak position determination.

Therefore, we prefer to use previously sintered and annealed polycrystalline samples with minimized internal stresses [*Gwanmesia and Liebermann*, 1992]. The advantages of this approach are clear from Figure 3, where two stishovite samples, one is a sintered pellet and the other loose powder were compressed and heated in the same cell assembly. In the case of sintered sample, microscopic deviatoric stress level remains low even at high pressures and the diffraction peaks remain sharp (Figure 3). For the powdered sample, a marked increase in peak width (more than a factor of 2 in FWHM) can be seen with increasing pressure at ambient temperature, significantly degrading the quality of the data for P-V-T use.

The stable high temperature capability of SAM-85 provides the most effective way to eliminate nonhydrostatic stress. Our previous studies on strength of NaCl show that above 500°C, the macroscopic deviatoric stress measured from peak shifts in NaCl becomes < 0.05 GPa, about the level of detectability (Figure 4). As the sample is embedded in NaCl, the stress boundary conditions on the sample would also be < 0.05 GPa. As strength of any material will decrease with increasing temperature, heating to certain temperature will eliminate peak broadening due to microscopic deviatoric stress in the sample. In the case of powdered stishovite, the FWHM of the diffraction lines returns almost to its ambient level at about 1000°C [*Li et al.*, 1995].

In a typical run, such as that shown in Figure 2, pressure is first increased to the maximum and then the sample is heated to the maximum temperature; P-V-T data collection begins at high temperature, where non-hydrostatic stress has been minimized. Further measurements are carried out at a constant load on decreasing temperature. During this process the sample pressure will decrease because of decrease in thermal pressure. At room temperature, the ram load is reduced by certain step and another temperature excursion starts again. Measurements again begin at the maximum temperature.

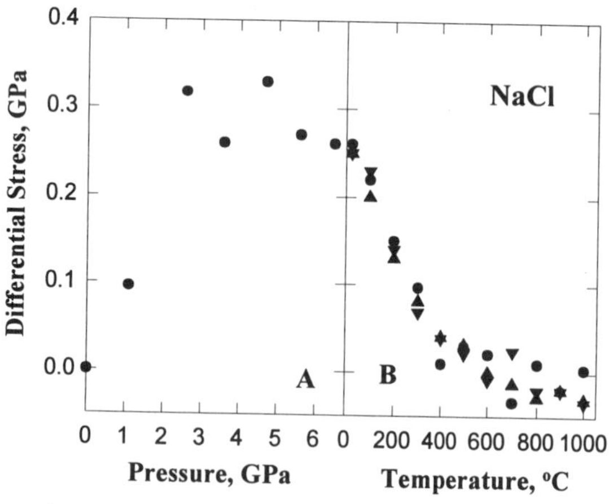

Figure 4. Strength of NaCl as a function of pressure and temperature as measured from relative shift in the diffraction lines [*Weidner et al.*, 1994]. (A) Room temperature data; differential stress builds up as pressure is increased and reaches the yield strength of NaCl at 3 GPa. (B) Nonhydrostatic stress in NaCl as a function of temperature at 6 GPa. By 500°C, the level of the non-hydrostatic stress is undetectable. Different symbols represent calculated stress at various positions in the sample chamber.

4. "LAMÉ EFFECT": PRESSURE DIFFERENCE IN A MULTIPHASE MIXTURE

It is a common practice in studies of equations of state to mix samples with pressure standard such as NaCl. It has been assumed that the pressure measured in the standard is the same as that in the sample and that the stress state in both is hydrostatic owing to the low yield strength of NaCl. This is, however, not always the case. To illustrate this, we examine a highly simplified, but still informative, example. Assume a spherical inclusion with radius a, in a spherical matrix with radius b ($b >> a$). The elastic bulk and shear moduli, K and μ, are denoted by subscripts i and m for inclusion and matrix, respectively. A hydrostatic pressure P_0 is applied at the outer boundary of the matrix ($r = b$).

Linear elastic solution of the spherically symmetric problem gives for the inclusion

$$P_i = P_0\,(1+4\mu_m/3K_m)/(1+4\mu_m/3K_i). \quad (1)$$

The result is independent of the size of the inclusion.

The pressure distribution in the matrix is different and given by

$$P_m = P_0 - 4\mu_m\,B_0\,(a/r)^3, \quad (2)$$

where

$$B_0 = (P_0/3K_m)\,(K_m/K_i-1)\,/\,(1+4\mu_m/3K_i),\ (\text{when } b >> a).$$

P_m approaches P_0 rapidly with increasing distance from the inclusion due to the r^{-3} dependence. Thus a strong pressure gradient exists in the matrix near the inclusion. The differential stress at any point in the matrix is

$$\sigma_{rr} - \sigma_{\theta\theta} = -6\mu_m B_0 (a/r)^3. \quad (3)$$

($\sigma_{rr} - \sigma_{\theta\theta}$) is the maximum at $r = a$, and decreases rapidly with increasing r so that at $r = 3a$, the differential stress is below 4% of the maximum.

For the case of dispersed small grains of second phase, the volume fraction of the matrix that is under high stress gradient is small. Experimentally observable quantity is thus the pressure difference, $P_i - P_0$ (hereafter termed "Lamé pressure"), which depends on the shear modulus of the matrix [for fluid $\mu_m = 0$, then $P_i = P_0$, according to (1)] and the contrast of bulk moduli between the inclusion and the matrix (if $K_i = K_0$, $P \equiv P_m \equiv P_0$). Suppose the matrix is NaCl (K_m = 24.9 GPa, μ_m = 14.4 GPa) and inclusion Au (K_i = 167 GPa), then elastic solution (1) predicts $P_i - P_0 = 0.59P_0$. In reality, however, the Lamé pressure will be limited by the yield strength of the matrix that supports the differential stress. According to (3), the maximum differential stress at the matrix-inclusion interface is $\sigma_{rr} - \sigma_{\theta\theta} = 0.29\ P_0$ for an NaCl/Au mixture; thus yielding is expected to initiate at the interface. Once the matrix yields ($\sigma_{rr} - \sigma_{\theta\theta} = \sigma_y$, where σ_y is the yield strength), the actual Lamé pressure will deviate from the linear function of P_0 predicted by the elastic solution presented above.

To examine the effects of Lamé pressure in actuality, we carried out the following experiment. A three-layered sample was used: one is a mixture of NaCl and Au (weight ratio 10:1), another a mixture of NaCl and BN (weight ratio 4:1), and the third pure NaCl. The sample assembly was compressed at room temperature to 8 GPa and then heated to 800°C at a constant ram load. Sample was temperature quenched and then the pressure was decreased to 5 GPa, where another temperature excursion was made up to 800°C, and quenched again.

Figure 5 shows the macroscopic deviatoric stresses calculated from relative peak shifts in both NaCl and gold in each layer, using the methodology described by *Weidner et al.* [1992a, b]. Results from both NaCl and gold show very similar behavior. At room temperature, nonhydrostatic stress builds up as pressure is increased and reaches the peak value (about 0.2 GPa) at about 160 bar ram load. The deviatoric stress level in the entire sample must be controlled by NaCl, as it is the dominant phase. The fact that the deviatoric stress measured here (0.2 GPa) is somewhat lower than the yield strength of 0.3 GPa at these pressures [*Weidner et al.*, 1994] may be due to the fact that the cell assembly is isotropic and that NaCl has not reached the yield point. Heating reduces the deviatoric stress very effectively; by 800°C, nonhydrostatic stresses are in the noise level. Temperature quench does not seem to have significant effects in this particular case. However, when the pressure is decreased at room temperature, deviatoric stress begins to build up again. Because of recrystallization in the pure NaCl sample, no reliable nonhydrostatic stress data can be obtained from peak shift, but the pressures calculated using available peaks in this sample are generally in agreement with those calculated from the NaCl+BN sample.

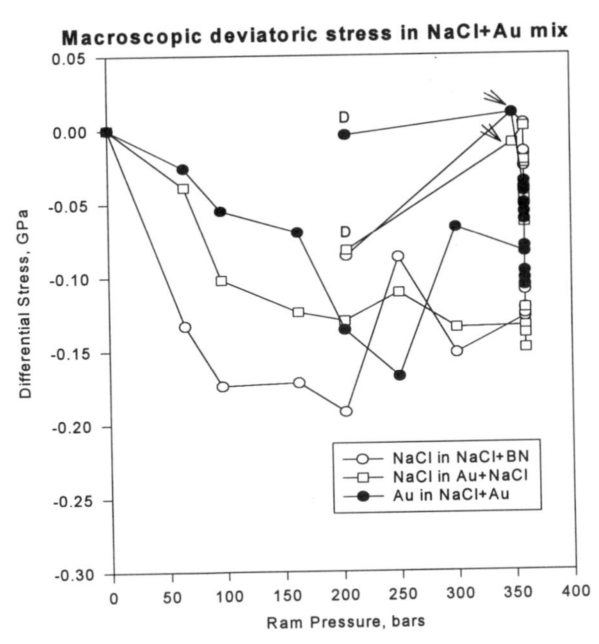

Figure 5. Macroscopic deviatoric stresses calculated using peak shifts in NaCl (open symbols) and Au (solid circles). The stresses increase with compression at room temperature (points from 0 to 360 bar) and decrease with increasing temperature (points at 360 bar). The quenched samples do not show any significant stress (arrowed points). Decompression causes the deviatoric stress to increase again (points labeled with D).

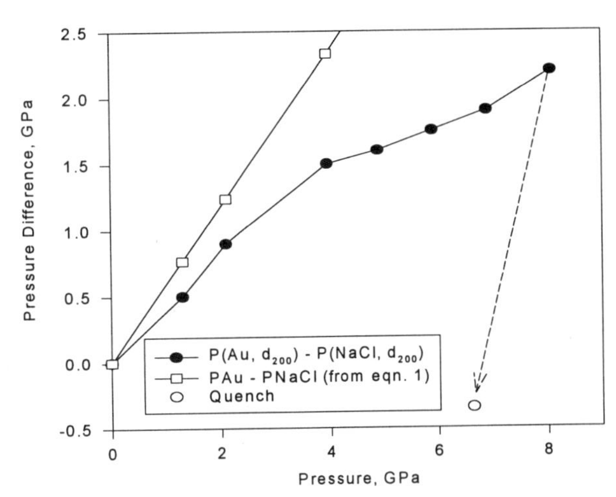

Figure 6. Measured (solid circles) and predicted pressure difference [equation (1); open squares] in an NaCl/Au mixture for compression at room temperature. The observed pressures are calculated from the *(200)* reflections of both materials. Open circle shows the pressure difference after heating to 800°C.

Figure 6 shows the difference in pressures calculated using the (*200*) peaks of NaCl and Au, the reference being the NaCl pressure. At room temperature, the Au pressure is significantly greater than that of NaCl as soon as compression starts and the difference increases with increasing pressure. The pressure difference (i.e., Lamé pressure) becomes much smaller after the sample has been heated to high temperatures, indicating that the difference at room temperature is not due to the particular Au equation of state used here. Also plotted in Figure 6 is the predicted Lamé pressure based on equation (1). It appears that the elastic Lamé pressure starts out with a slope quite similar to that observed, but it deviates from the observations, perhaps due to a combination of various effects, including (1) compacting of powdered NaCl matrix; (2) non-spherical shape of the Au grains; and, most importantly, (3) the matrix has yielded due to high differential stress.

The above results demonstrate the importance of "Lamé effect" in accurate equation-of-state measurements. For most silicates, the bulk moduli are of the order of 150 - 200 GPa, similar to that of gold. If a silicate sample is dispersed in NaCl for P-V-T measurements, similar pressure differences (order of 2 GPa at 10 GPa level as observed in the NaCl/Au mixture) may exist for room-temperature compression. This would result in a significant under estimate in bulk modulus for the silicate, because the actual pressure in the silicate is higher than that in the NaCl matrix. Similar caution should be applied to ruby pressure measurements, when the ruby chips are mixed with the sample. If the bulk modulus of the silicate is significantly lower than that of ruby, then the ruby pressure can be considerably higher than that in the silicate; thus resulting in an overestimate in the bulk modulus of the sample.

We choose not to use multiphase samples in P-V-T experiments to avoid this problem. Stacked monophase layers, with similar thickness, have a few interfaces terminating in soft BN liner in the sample chamber (Figure 1), which does not allow high stress to build up. More importantly, P-V-T measurements are carried out only after heating, when nonhydrostatic stress has been reduced to the noise level.

5. PROFILE FITTING PROCEDURES

Synchrotron radiation yields robust diffraction signals so that many diffraction lines can be extracted for determining cell parameters of the sample. With appropriately calibrated systems, uncertainties in volume determination are generally on the order of 0.1% or below. However, for minerals with low symmetry, a large number of overlapped peaks are unusable because of the uncertainties in their line positions. One solution to this is to use the entire diffraction profile. In essence, this is a Rietveld-type technique with either a Le Bail or an energy-dispersive fit with a prescribed, empirical, absorption profile, both of which optimize the cell parameters so that the entire diffraction profile is fitted.

Figure 7 is an example of a Le Bail fit to an observed diffraction profile of jadeite at 7 GPa and 1007°C [*Zhao et al.*, 1996]. This approach yields much more accurate cell parameters than the single peak fitting approach. A precision to well below 0.05% in volume can be achieved.

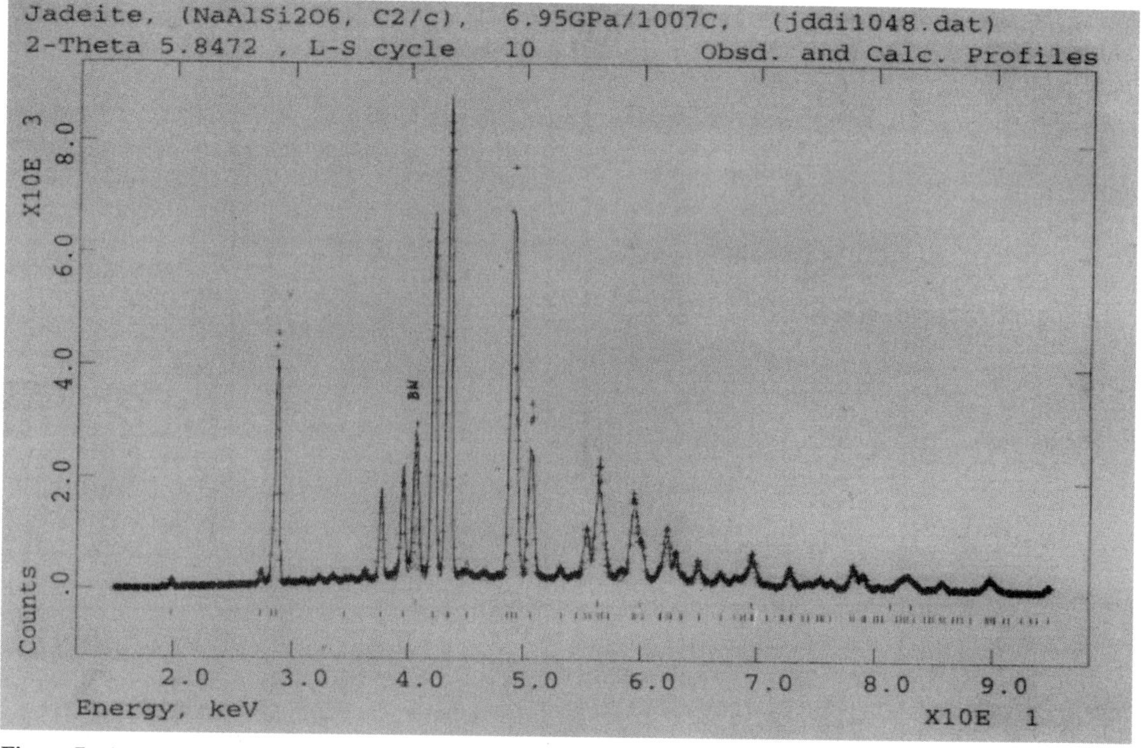

Figure 7. An example of profile fitting using GSAS on energy-dispersive data of jadeite at 7 GPa and 1000°C (*Zhao et al.*, 1996). Crosses are the observed data and the continuous line is the fit. Below the profile, the top tick marks are peak positions for BN and bottom ticks indicate peak positions of the sample.

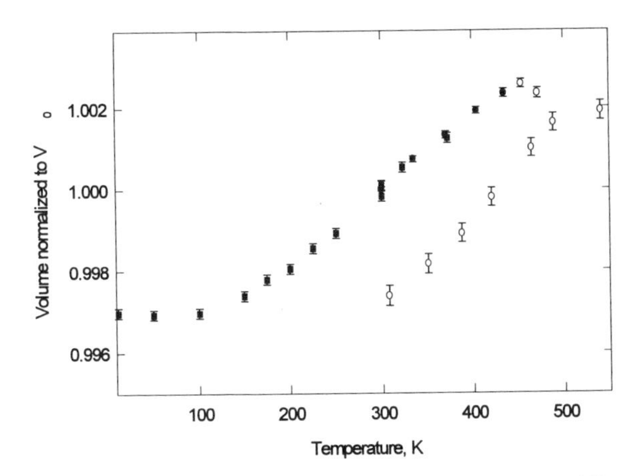

Figure 8. Normalized volume of $Mg_{0.9}Fe_{0.1}SiO_3$ perovskite as a function of temperature. The solid circles represent reversible measurements and open circles irreversible. Note that volume begins to decrease with increasing temperature above 430 K, and remains smaller as temperature is lowered from above 430 to 300 K. The irreversible volume at room temperature may be compared to the reversible volume at ~150 K.

Another advantage of this approach is that systematic zero-point shift can be estimated from the whole-pattern fit, thus providing a means to evaluate data points that may have been affected by changing the diffraction geometry due to human errors during the experiment.

6. MEASURING STATE FUNCTIONS FOR METASTABLE HIGH PRESSURE PHASES

Minerals that are metastable under the conditions of the P-V-T measurements deserve special care to be sure that no irreversible changes had taken place. In our previous studies, certain samples after the P-V-T experiment have been found by Raman spectroscopy, electron microprobe analysis, and transmission electron microscopy to undergo irreversible changes, which had significantly affected thermoelastic properties of the sample. One such example is $(Mg,Fe)SiO_3$ perovskite [*Wang et al.*, 1994]. At zero pressure, heating to above 430 K resulted in an irreversible volume-temperature behavior (Figure 8), and microprobe revealed that the sample has partly decomposed into oxides (Figure 9).

It may be interesting to compare the observed temperature induced irreversible changes in $(Mg,Fe)SiO_3$ perovskite with that induced by pressure in $CaSiO_3$ perovskite at room temperature. $CaSiO_3$ perovskite is unstable at ambient pressure and converts into an amorphous state. As the sample is decompressed from high pressures to below 1 GPa, a significant decrease in diffraction signal and an increase in the background are observed, due to the onset of amorphization. Volume is no longer a state function. Nevertheless, certain crystallinality is preserved and unit-cell volumes can be measured based on the remaining diffraction peaks. These unit-cell volumes decrease with decreasing pressure, resulting in an apparent negative bulk modulus if only the data below 1 GPa were considered; including these irreversible volume measurements will result in an erroneous equation of state [*Wang et al.*, 1996].

In most cases the change in P-V-T behavior could be very subtle and difficult to detect. However, these subtle changes often result in erroneous thermoelastic parameters. Thus, it has become a routine procedure in our equation-of-state studies to examine samples before and after the measurement using a variety of characterization techniques, such as microprobe, spectroscopy, and electron microscopy. Once irreversible change is detected, the P-V-T data cannot be regarded as representing state function.

7. CONCLUSIONS

We present here a state-of-the-art methodology for P-V-T equation-of-state measurements in large-volume apparatus. Accuracy is the key issue in such studies as the first-, second-, and even the third-order pressure and temperature derivatives are extracted from the measured volumes. Problems encountered in such experiments are discussed and solutions provided. The technical improvements have greatly enhanced our experimental capability.

One important issue not discussed here is the form of equation of state used in analyzing the data. All of the P-V-T data must eventually be put into a formulation in order to extract thermoelastic parameters. The parameters thus extracted depend on the form of the equation of state that is

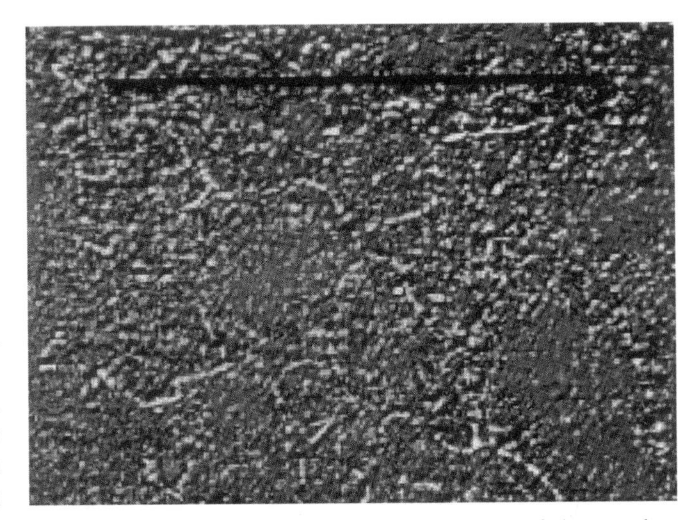

Figure 9. A back-scattered electron micrograph of the sample recovered from the V-T measurements shown in Figure 8. Bright areas are oxides enriched in Fe.

used to fit the data. Strictly speaking, the most widely used Birch-Murnaghan equation of state is applicable for nearly isotropic materials. Many minerals relevant to the mantle are highly anisotropic. For example, the linear compressibility of brucite along the *c* axis is 5 times that along the *a* axis [*Fei and Mao*, 1993]. Whether Birch-Murnaghan equation of state can give physically meaningful parameters is questionable.

Several types of thermal equations of state have been proposed and used in previous high-temperature studies; among them, the thermal pressure approach [*Anderson*, 1984] and the high-temperature Birch-Murnaghan equation of state [*Fei et al.*, 1992; *Wang et al.*, 1994] are the most commonly used ones in the literature. *Jackson and Rigden* [1996] proposed a Mie-Debye-Grüneisen equation of state, which combines the well-known Mie-Grüneisen thermal pressure equation of state with the Debye lattice vibrational energy model. This is an area where systematic theoretical as well as experimental studies are needed. One fundamental question is whether the previously determined equations of state based on data well below the Debye temperature can be applied to much higher temperatures relevant to the Earth. The answer to this question may await accurate experimental results at higher temperatures and pressures.

Acknowledgments. We thank M. T. Vaughan, K. Baldwin, and the users of SAM-85 for technical assistance and scientific discussions, and Y. Zhao and R. Jeanloz for helpful reviews. Work supported by NSF grants EAR 89-20239 to the Center for High Pressure Research and EAR95-26634. This is MPI contribution No. 189.

REFERENCES

Anderson, O. L., A universal equation of state, *J. Geodyn.*, *1*, 185-214, 1984.

Anderson, O. L., *Equations of State of Solids for Geophysics and Ceramic Science*, Oxford Monogr. Geol. Geophys. No. 31, Oxford University Press, New York, 1995.

Anderson, O. L., and I. Suzuki, Anharmonicity of three minerals at high temperature: Forsterite, fayalite, and periclase, *J. Geophys. Res.*, *88*, 3549-3556, 1984.

Anderson, O. L., D. Isaak, and H. Oda, High-temperature elastic constant data on minerals relevant to geophysics, *Rev. Geophys.*, *30*, 57-90, 1992.

Decker, D. L., High-pressure equation of state for NaCl, KCl, acd CsCl, *J. Appl. Phys.*, *42*, 3239-3244, 1971

Fei., Y., H. K. Mao, J. Shu, G. Parthasarathy, W. A. Bassett, and J. Ko, Simultaneous high P-T X-ray diffraction study of β-$(Mg,Fe)_2SiO_4$ to 26 GPa and 900 K, *J. Geophys. Res.*, *97*, 4489-4495, 1992.

Fei, Y. and H. K. Mao, Static compression of $Mg(OH)_2$ to 78 GPa at high temperature and constraints on the equation of state of fluid H_2O, *J. Geophys. Res.*, *98*,11, 875-11884

Gwanmesia, G. D., S. M. Rigden, I. Jackson, and R. C. Liebermann, Pressure dependence of elastic wave velocity for β-Mg_2SiO_4 and the composition of the Earth's mantle, *Science*, *250*, 794-797, 1990.

Gwanmesia, G. D. and R. C. Liebermann, Polycrystals at high pressure phases of mantle minerals: Hot-pressing and characterization of physical properties, in *High Pressure Research: Applications to Earth and Planetary Sciences*, *Geophys. Monogr. Ser.*, vol. 67, edited by Y. Syono and M. H. Manghnani, pp. 117-135, Terra Scientific Publishing, Tokyo/ AGU, Washington, D. C., 1992.

Jackson, I., and S. M. Rigden, Analysis of P-V-T data: Constraints on the thermoelastic properties of high-pressure minerals, *Phys. Earth Planet. Int.*, *96*, 85-112, 1996.

Li, B., J. Zhang, Y. Wang, D. J. Weidner, and R. C. Liebermann, Characterization of microscopic stress in a hot-pressed polycrystalline stishovite at high P and T, *Eos*, *Trans. Amer. Geophys. Union*, *75*, 346, 1994.

Meng, Y. and D. J. Weidner, High pressure elastic behavior of $MgAl_2O_4$ spinel, *Eos*, Trans. Amer. Geophys. Union, *75*, 347, 1994.

Wang, Y., D. J. Weidner, R.C. Liebermann, and Y. Zhao, P-V-T equation of state of $(Mg,Fe)SiO_3$ perovskite and constraints on composition of the lower mantle, *Phys. Earth Planet. Int.*, *83*, 13-40, 1994.

Wang, Y., D. J. Weidner, and F. Guyot, Thermal equation of state of $CaSiO_3$ perovskite, *J. Geophys. Res.*, *101*, 661-672, 1996.

Weidner, D. J., M. T. Vaughan, J. Ko, Y. Wang, X. Liu, A. Yeganeh-Haeri, R. E. Pacalo, and Y. Zhao, 1992a, Characterization of stress, pressure, temperature in SAM-85, a DIA type high pressure apparatus, in *High Pressure Research: Applications to Earth and Planetary Sciences, Geophys. Monogr. Ser.*, vol. 67, edited by Y. Syono and M. H. Manghnani, pp. 13-17, Terra Scientific Publishing, Tokyo/ AGU, Washington, D. C., 1992a.

Weidner, D. J., M. T. Vaughan, J. Ko, Y. Wang, K. Leinenweber, X. Liu, A. Yeganeh-Haeri, R. E. Pacalo, and Y. Zhao, Large volume high pressure research using the wiggler port at NSLS, *High Pres. Res.*, *8*, 617-623, 1992b.

Weidner, D. J., Y. Wang, and M. T. Vaughan, Yield strength at high pressure and temperature, *Geophys. Res. Lett.*, *21*, 753-756, 1994.

Yeganeh-Haeri, A., D. J. Weidner, Single crystal elastic property of $MgAl_2O_4$ up to 1200 K, *Eos*, Trans. Amer. Geophys. Union, *71*, 620, 1990.

Zaug, J., E. H. Abramson, J. M. Brown, and L. J. Slusky, Sound velocities in olivine at Earth mantle pressure, *Science*, *260*, 1487-1489, 1993.

Zhao, Y., R. B. von Dreele, T. J. Shankland, D. J. Weidner, J. Zhang, Y. Wang, and T. Gasparik, Thermoelastic equation of state of jadeite $NaAlSi_2O_6$: An energy-dispersive Rietveld refinement study for low symmetry and multiple phase diffraction, *Geophys. Res. Lett.*, in press, 1996.

Y. Wang, D. J. Weidner, and Y. Meng, Department of Earth and Space Sciences, University at Stony Brook, Stony Brook, NY 11794-2100.

Postspinel Transformations in the System Mg_2SiO_4–Fe_2SiO_4: Element Partitioning, Calorimetry, and Thermodynamic Calculation

M. Akaogi, H. Kojitani, K. Matsuzaka, and T. Suzuki

Department of Chemistry, Gakushuin University, Tokyo, Japan

E. Ito

Institute for Study of the Earth's Interior, Okayama University, Misasi, Japan

Compositions of coexisting spinel and magnesiowustite in the system Mg_2SiO_4–Fe_2SiO_4 have been experimentally determined at 18.5 and 20.4 GPa at 1873 K to constrain equilibrium boundaries of the postspinel transitions in relatively Fe_2SiO_4-rich composition. Calorimetric measurements of pyroxene and perovskite solid solutions in the system Mg_2SiO_3–Fe_2SiO_3 have been performed by a differential drop-solution method in a controlled atmosphere. Using the above data with published thermodynamic data on the high-pressure phases in the system MgO–FeO–SiO_2, phase relations of the postspinel transitions have been thermodynamically calculated. The calculated boundaries are generally consistent with the experiments by Ito and Takahashi. The calculated transition interval of spinel to perovskite + magnesiowustite is about 0.01–0.18 GPa for mantel spinel composition. The perovskite-magnesiowustite field expands to the Fe_2SiO_4-rich side with increasing temperature, in contrast to previous thermodynamic calculations. The present thermodynamic data show stability of an assemblage of magnesiowustite-stishovite in the Mg_2SiO_4-rich composition of the Mg_2SiO_4–Fe_2SiO_4 system between the fields of spinel and of perovskite + magnesiowustite at relatively low temperatures.

1. INTRODUCTION

Phase transformations in the system Mg_2SiO_4-Fe_2SiO_4 have been studied extensively to clarify mineralogy of the transition zone and the lower mantle. The postspinel transformations have been paid special attention to elucidate the nature of the seismic discontinuity observed at 660 km depth. *Yagi et al.* [1979] reported a reconnaissance study on the phase transitions of this system. *Ito and Takahashi* [1989] examined the detailed phase relations at about 22-26 GPa and 1373 and 1873 K in compositions of more than 40 mol % Mg_2SiO_4. They clarified the stability regions of spinel(sp), perovskite(pv) + magnesiowustite(mw), and others, and determined the very narrow pressure interval of the transition of spinel to perovskite + magnesiowustite. This narrow width is consistent with the sharpness of the 660-km discontinuity. The phase relations in the Fe_2SiO_4-rich compostions, however, have not yet been examined experimentally.

Thermodynamic modeling of the postspinel transitions of this

Properties of Earth and Planetary Materials
at High Pressure and Temperature
Geophysical Monograph 101

system was performed by *Fei et al.* [1991] using internally consistent thermodynamic data. *Stixrude and Bukowinski* [1993] also calculated the postspinel phase relations with thermodynamic potentials. Both of these studies reported some discrepancies in the phase relations between the thermodynamic calculations and the experimental determination by *Ito and Takahashi* [1989], e.g. the stability field of perovskite + magnesiowustite in the thermodynamic calculations shrunk to the Mg_2SiO_4-rich side with increasing temperature, while the reverse was observed in the experiments.

We have recently carried out precise calorimetric measurements of $MgSiO_3$ perovskite [*Akaogi and Ito*, 1993b] and stishovite(st) [*Akaogi et al.*, 1995], both of which are constituent minerals of the postspinel phase assemblages. These new thermodynamic data can be used to further constrain the postspinel phase relations by using the thermodynamic calculation.

In this study, we first performed Mg-Fe partitioning experiments between spinel and magnesiowustite in the system Mg_2SiO_4-Fe_2SiO_4 to determine accurately the postspinel transition boundaries in the Fe_2SiO_4-rich compositions and to constrain the thermodynamic data of the relevant phases. Second, enthalpies of the postspinel phases including $(Mg,Fe)SiO_3$ perovskite and MgO have been measured to construct an internally consistent thermodynamic data set together with our newly published data on $MgSiO_3$ perovskite and stishovite. Third, the postspinel transition boundaries in the whole compositional range have been calculated using our thermodynamic data set. Some implications of the calculated phase boundaries to the 660-km discontinuity and the mineralogy of the descending slabs are also discussed.

2. EXPERIMENTAL METHODS

2.1. *Synthesis and Characterization of Samples*

The starting materials for Mg-Fe partitioning experiments and the samples used for calorimetric measurements were synthesized from reagent-grade chemicals as follows. Forsterite was synthesized by heating a mixture of MgO and silicic acid (SiO_2 11.0 wt%H_2O) at 1723 K. Fayalite was made by heating a mixture of Fe_2O_3 and silicic acid at 1453 K in controlled oxygen fugacity using a mixture of H_2, CO_2, and Ar with volume ratios of 1:1:2. Olivine solid solutions with com-positions of $(Mg_{0.2},Fe_{0.8})_2SiO_4$ and $(Mg_{0.4},Fe_{0.6})_2SiO_4$ were synthesized using mixtures of the synthetic forsterite and fayalite by heating at 1453-1573 K in the same controlled atmosphere. Ferrosilite $FeSiO_3$ was made from a mixture of fayalite and anhydrous silica at 5 GPa and 1473 K, and $(Mg_{0.92},Fe_{0.08})SiO_3$ orthopyroxene from a mixture of forsterite, fayalite, and anhydrous silica at 5 GPa and 1273 K. Both of the pyroxenes were synthesized using a platinum capsule/heater by means of a split-cylinder type multi-anvil apparatus at Gakushuin University. All of the abovc samples were examined by powder X ray diffraction and polarizing microscopic observation, confirming single-phase materials. Microprobe analysis showed that the Mg/(Mg+Fe) ratios of the olivine and pyroxene solid solutions were within ± 1 mol % of their nominal values. Using a split-sphere type multi-anvil apparatus at the Institite for Study of the Earth's Interior, Okayama University, $(Mg_{0.92},Fe_{0.08})SiO_3$ perovskite was synthesized in a rhenium capsule/heater at 25 GPa and 1873-2073 K. Microfocused X ray diffraction, microscopic observation, and powder X ray diffraction showed that the sample was perovskite with a small amount (about 10 % in total) of magnesiowustite + stishovite. Metallic impurities such as rhenium were not observed in the sample.

2.2. *Mg-Fe Partitioning Experiments*

The partitioning experiments were carried out using the multi-anvil apparatus at Gakushuin University. Tungsten carbide anvils with a truncated edge length of 2.5 mm were used in combination with a semi-sintered magnesia octahedron of 7 mm on edge. Pressure was calibrated at room temperature using the pressure-fixed points of Bi I-II (2.55GPa), Ba I-II (5.5GPa), Bi III-V (7.7GPa), Ba II-III (12.6GPa), ZnS (15.5GPa), and GaAs (18.3GPa). Pressure calibration was also performed at 1873 K by using transitions in Mg_2SiO_4: α - β transition at 15.0 GPa and β - γ transition at 20.8 GPa [*Katsura and Ito*, 1989]. Two different types of starting materials were used to confirm equilibrium compositions. One was mechanical mixtures of forsterite and fayalite in molar ratios of 20:80 and 40:60. The other was olivine solid solutions with compositions of $(Mg_{0.2},Fe_{0.8})_2SiO_4$ and $(Mg_{0.4},Fe_{0.6})_2SiO_4$. These starting materials were put into a rhenium capsule/furnace in a semi-sintered magnesia octahedron and held at 18.5 and 20.4 GPa at 1873 K. The run duration was varied from 10 min to 5 hours to examine the change of compositions with time. The run temperature was measured on the outer surface of the central part of the furnace by a Pt/Pt-13%Rh thermocouple. No correction was made for pressure effect on thermoelectromotive force of the thermocouple. The quenched run products were confirmed to be the mixture of spinel, magnesiowustite, and stishovite by powder X ray diffraction of a part of each run product. The remaining run products were polished and analyzed by an electron microprobe at Ocean Research Institute, the University of Tokyo. The accererating voltage was 25 kV and the beam current on the Faraday cup was 20 nA. The beam size was about 1 μm in diameter. Standard materials were single crystals of wollastonite, periclase, and hematite for Si, Mg, and Fe, respectively. In all of the runs, typical grain sizes of spinel and magnesiowustite were about 20-50 and 5-20 μm, respectively. Compositions of spinel and magnesiowustite in the central part of each run product were determined by averaging about 10-20 analysis points. The compositions were calculated using the ZAF correction method.

2.3. *Calorimetry*

Enthalpies were measured using a twin Calvet-type micro-calorimeter at Gakushuin University. Molten $2PbO.B_2O_3$ was used as a solvent. Preparation method of the solvent was the same as that described by *Akaogi et al.* [1990]. Enthalpies of drop-solution for $FeSiO_3$ ferrosilite, $(Mg_{0.92},Fe_{0.08})SiO_3$ ortho-pyroxene, and $(Mg_{0.92},Fe_{0.08})SiO_3$ perovskite were measured by differential drop-solution calorimetry. This method is very similar to that by *Akaogi and Ito* [1993b], except for measuring the enthalpies in a controlled atmosphere. Two silica glass capsules of similar weights (about 3.5-4.5 mg), one containing the sample and a Pt wire (about 120-130 mg) and the other only the Pt wire, were dropped simultaneously from room temperature into the solvent in two chambers of the calorimeter kept at 978±1 K. The drop-solution runs were performed in purified Ar gas to prevent oxidation of the samples. The heat effects of the silica glass capsules and the Pt wires were almost cancelled in both of the sides. After small corrections for weight differences of the silica glass capsules and the Pt wires as well as for difference in the calibration factors in both sides of the calorimeter, the enthalpy of the drop-solution process which was the sum of heat content (H°_{978}-H°_{298}) and heat of solution ($\Delta H^\circ_{sol,978}$) of the sample was obtained. The powder samples were used for ferrosilite (5.75-7.57 mg in each run) and $(Mg_{0.92},Fe_{0.08})SiO_3$ ortho-pyroxene (4.57-5.76 mg in each run). Because it was very likely that the perovskite samples converted to an amorphous material by grinding, sintered pieces of the $(Mg_{0.92},Fe_{0.08})SiO_3$ perovskite (4.05-6.45 mg in each run) were used for the differential drop-solution calorimetry. All of the pyroxene and perovskite samples and the capsules were dissolved completely within about 50 min. The measured enthalpies of perovskite were corrected for the impurities (magnesiowustite + stishovite) of about 10%, using the measured enthalpy of stishovite [*Akaogi et al.*, 1995] and the enthalpy of magnesiowustite calculated from enthalpies of MgO and FeO with enthalpy of mixing of magnesiowustite solid solution shown below.

Enthalpy of solution of MgO (periclase) was also measured, using the same Calvet-type calorimeter. The calorimeter temperature of 983 K was slightly different, because the heat of solution mesurements were made in a different period. Synthetic single crystals of MgO (Nakazumi Crystal Co.) were crushed and mixed with α-quartz of about 35 wt%, and the mixed sample was dissolved in $2PbO.B_2O_3$ solvent by solution calorimetry method. The sample of each run contained 3.54-4.36 mg of MgO. Since α-quartz promoted dissolution, the mixed sample was completely dissolved in the solvent within 30 min. However, residual undissolved MgO was always observed when pure MgO was used as the sample for the solution runs. The enthalpy of solution of MgO was obtained by subtracting the heat of solution of quartz by *Akaogi et al.* [1995].

TABLE 1. Results of Mg-Fe partitioning between spinel and magnesiowustite in the system Mg_2SiO_4-Fe_2SiO_4 at 1873 K.

Run	Starting material[a] (Mg_2SiO_4 mol%)	Pressure (GPa)	Time (min)	Composition[b] Sp	Mw
2M-1	0.20Fo+0.80Fa	18.5	60	31.6(6)	8.6(6)
2M-2	0.20Fo+0.80Fa	18.5	300	30.8(10)	8.5(6)
2S-1	$Fo_{20}Fa_{80}$	18.5	180	30.3(10)	8.9(6)
4M-1	0.40Fo+0.60Fa	20.4	60	47.5(6)	15.1(7)
4M-2	0.40Fo+0.60Fa	20.4	300	48.0(5)	14.8(7)
4S-1	$Fo_{40}Fa_{60}$	20.4	10	45.2(9)	25.5(81)
4S-2	$Fo_{40}Fa_{60}$	20.4	60	48.6(8)	20.0(7)
4S-3	$Fo_{40}Fa_{60}$	20.4	180	48.1(6)	15.1(8)
4S-4	$Fo_{40}Fa_{60}$	20.4	300	47.9(4)	15.1(13)

[a] Starting material of xFo+(1-x)Fa indicates a mechanical mixture of Mg_2SiO_4 olivine and Fe_2SiO_4 olivine with molar ratio of x:(1-x). Fo_yFa_{100-y} represents olivine solid solution with y mol % Mg_2SiO_4.
[b] Standard deviation in parentheses is expressed in units of last digit.

3. RESULTS AND DISCUSSION

3.1. *Mg-Fe Partitioning between Spinel and Magnesiowustite*

The experimental results of Mg-Fe partitioning between spinel and magnesiowustite in the system Mg_2SiO_4-Fe_2SiO_4 are summarized in Table 1. The compositions of spinel and magnesiowustite in the runs at 1873 K, using the starting materials of olivine solid solutions (denoted ol s. s. runs) and mixtures of forsterite and fayalite (denoted mixture runs), are shown in Figures 1a and 1b, respectively. In the ol s. s. runs, spinel composition changed to Mg_2SiO_4-rich side and magnesiowustite composition to Fe_2SiO_4-rich side with increasing run duration within 3 hours, as shown in Figure 1a. However, no change in the compositions of either of the phases was observed between 3 and 5 hours. This shows that the compositions in the ol s. s. runs longer than 3 hours are close to equilibrium. In the mixture runs, the compositions of spinel and magnesiowustite are expected to change with time in opposite manner to that

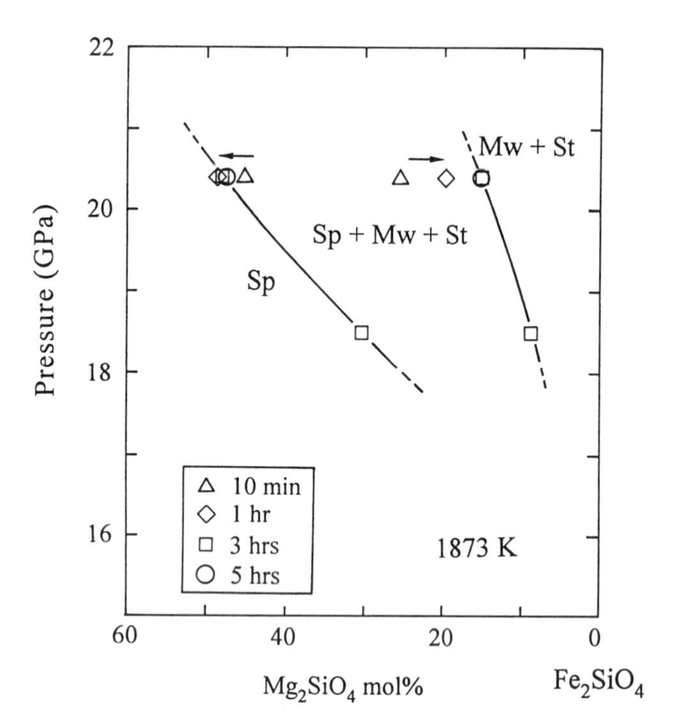

Figure 1a. Compositions of coexisting spinel and magnesiowustite in the system Mg_2SiO_4-Fe_2SiO_4 at 18.5 and 20.4 GPa at 1873 K, using the starting materials of olivine solid solutions. Arrows indicate directions of change of compositions of spinel and magnesiowustite with run time.

observed in the ol s. s. runs. Figure1b shows the results of the mixture runs in which the compositions of spinel and magnesiowustite in the runs of 1 hour agree with those of 5 hours. This indicates that 1 hour is enough to obtain the equilibrium compositions in this type of starting material. The compositions of spinel and magne-siowustite in the ol s. s. runs in Figure 1a are in good agreement with those of the mixture runs in Figure 1b within the errors. Therefore, equilibrium boundaries of the three-phase field of spinel + magnesiowustite + stishovite are tightly constrained at 18.5 and 20.4 GPa at 1873 K from the present results. The boundaries in Figures 1a and 1b are about 1.5-2 GPa higher than the thermodynamic calculation of *Fei et al.* [1991], and are generally consistent within about 5-10 mol % with the extenstion of *Ito and Takahashi's* [1989] boundaries which were experimetally determined above 21.5 GPa. The partition experiments between spinel and magne-siowustite at wider ranges of pressure and temperature are in progress.

3.2. *Calorimetric Data*

The enthalpies measured by differential drop-solution calori-metry of $FeSiO_3$ ferrosilite, $(Mg_{0.92},Fe_{0.08})SiO_3$ ortho-pyroxene, and $(Mg_{0.92},Fe_{0.08})SiO_3$ perovskite are listed in Table 2, together with the heat of solution of periclase. The enthalpy data of $MgSiO_3$ orthopyroxene and perovskite by *Akaogi and Ito* [1993b] and stishovite by *Akaogi et al.* [1995] measured at the same temperature using the identical method are also shown in Table 2. The heat of solution of periclase at 983 K in this study, shown in Table 2, is consistent with that measured at 970 K by *Charlu et al.* [1975] and that at 975 K by *Navrotsky et al.* [1994] within the errors. Using our enthalpy of solution data and the heat capacity equation of periclase by *Berman* [1988], enthalpy for drop-solution process of periclase from 298 K into the solvent at 978 K is calculated to be 37.09±0.66 kJ/mol, neglecting the effect of a small tem-perature difference on the heat of solution.

Figure 2 shows the enthalpies measured by differential drop-solution calorimetry of pyroxenes and perovskites shown in Table 2. The enthalpy of drop-solution of pyroxene dec-reases with the $FeSiO_3$ component. The enthalpy-composition curve for pyroxene solid solution in Figure 2 was obtained from the measured enthalpies for the three pyroxenes in Table 2 and the enthalpy of mixing (W_H) for the $MgSiO_3$-$FeSiO_3$ pyroxene solid solutions by *Chattilon-Colinet et al.* [1983]. The calculated enthalpy of drop-solution for an equimolar mixture of periclase + stishovite and that of wustite + stishovite are also shown in Figure 2. The former is the sum of the enthalpy of drop-solution

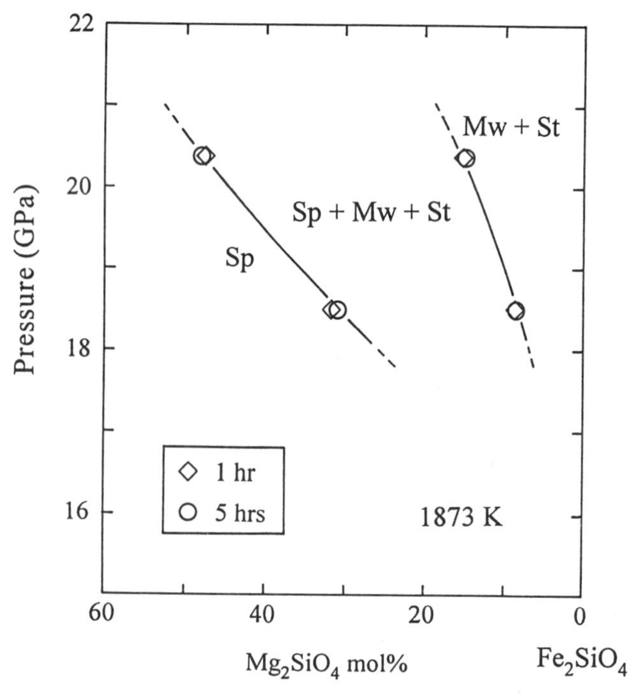

Figure 1b. Compositions of coexisting spinel and magnesiowustite in the system Mg_2SiO_4-Fe_2SiO_4 at 18.5 and 20.4 GPa at 1873 K, using the starting materials of mixtures of forsterite and fayalite.

of periclase and that of stishovite. The enthalpy of wustite + stishovite was calculated as 36.50±3.38 kJ/mol, using the enthalpy of formation of $FeSiO_3$ ferrosilite from wustite + α-quartz by *Saxena et al.* [1993], the enthalpy of α-quartz-stishovite transition by *Akaogi et al.* [1995], and the enthalpy of ferrosilite in Table 2. As shown in Fig. 2, enthalpy of drop-solution of $(Mg_{0.92},Fe_{0.08})SiO_3$ perovskite agrees with that of $MgSiO_3$ perovskite within the errors. This indicates almost no compositional dependence of enthalpy of drop-solution for the $MgSiO_3$-$FeSiO_3$ perovskite solid solutions. Therefore, enthalpy of hypothetical $FeSiO_3$ perovskite was estimated to be 11.06 kJ/mol, using these two measured enthalpy data with estimated W_H of 3.8 kJ/mol for the perovskite solid solutions by *Navrotsky* [1987]. From the difference between enthalpies of $FeSiO_3$ perovskite and of wustite + stishovite, enthalpy for the reaction of wustite + stishovite to $FeSiO_3$ perovskite at 298 K is calculated as 25.5 kJ/mol. We use this value in the following discussion because this value is similar to the ΔH°_{298} of 29.34±6.81 kJ/mol for the reaction of periclase + stishovite to $MgSiO_3$ perovskite. Because a large uncertainty accompanied the ΔH°_{298} for the reaction of wustite + stishovite to $FeSiO_3$ perovskite estimated above by extrapolation, we estimated the uncertainty of the ΔH°_{298} as follows. *Liu* [1976] reported that the assemblage of wustite and stishivite was stable up to at least 28 GPa at high temperature. This showed that the reaction boundary from wustite and stishovite to hypothetical $FeSiO_3$ perovskite should be placed at higher pressure than 28 GPa. Using the thermodynamic calculation method described below, the reaction boundary was calculated by varying the ΔH°_{298} leaving all of the other thermodynamic parameters as the values shown below. The result indicated that the lower bound of ΔH°_{298} was about 5.5 kJ/mol, which was smaller by 20 kJ/mol than the above estimated value. The upper bound of ΔH°_{298} was also estimated to be about 45.5 kJ/mol by calculation of the boundaries of the perovskite + magnesiowustite field which were consistent with the experimental determination. Therefore, the uncertainty of the ΔH°_{298} for the reaction of wustite + stishovite to $FeSiO_3$ perovskite was about ±20 kJ/mol.

TABLE 2. Enthalpies measured by differential drop-solution calorimetry of pyroxenes and perovskites in the system $MgSiO_3$-$FeSiO_3$ and stishovite, and by solution calorimetry of periclase.

Sample	ΔH° [a] (kJ/mol)	Number of runs	Ref.
$MgSiO_3$ opx	113.01±1.62	10	d
$(Mg_{0.92},Fe_{0.08})SiO_3$ opx	108.45±2.28	6	e
$FeSiO_3$ opx	89.05±3.23	8	e
$MgSiO_3$ pv	10.79±1.96	4	d
$(Mg_{0.92},Fe_{0.08})SiO_3$ pv	11.10±2.62[b]	4	e
SiO_2 st	3.04±0.91	6	f
MgO per	5.20±0.66[c]	4	e

[a] Error is two standard deviations of the mean.
[b] Corrected for the impurities of 10% in the perovskite samples to the observed value, 13.48±2.29 kJ/mol. See text.
[c] The ΔH° of periclase is enthalpy of solution. The ΔH° s for the other samples are enthalpies for differential drop-solution calorimetry.
[d] *Akaogi and Ito* [1993b]
[e] This study
[f] *Akaogi et al.* [1995]

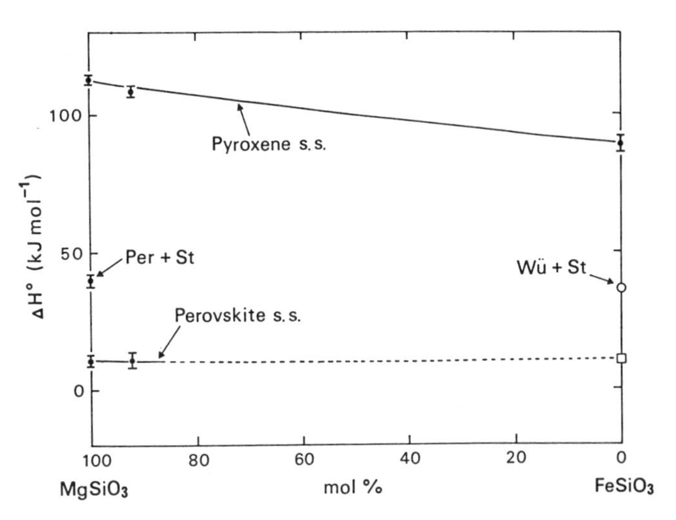

Figure 2. Relationships between enthalpy of drop-solution and composition for pyroxene and perovskite solid solutions in the system $MgSiO_3$-$FeSiO_3$, with enthalpies of drop-solution for the mixtures of MgO(periclase) + SiO_2(stishovite) and of FeO(wustite) + SiO_2 (stishovite).

3.3. *Method of Thermodynamic Calculation*

Phase relations of the post-spinel transitions in the system Mg_2SiO_4-Fe_2SiO_4 are calculated as follows. The calculation method is similar to that by *Akaogi et al.* [1989], but temperature dependence of the bulk moduli of the phases involved is included in this study. Using the thermodynamic data, the following phase relations are calculated :

$$sp = 2mw + st \quad (1)$$

$$2mw + st = pv + mw \quad (2)$$

$$sp = pv + mw \quad (3)$$

The equilibrium of (1) between spinel and magnesiowustite + stishovite, for example, is expressed as

$$Mg_2SiO_4(sp) = 2MgO\,(mw) + SiO_2\,(st) \quad (4)$$

for which we have at equilibrium

$$2\bar{\mu}^{mw}_{MgO} + \bar{\mu}^{st}_{SiO_2} = \bar{\mu}^{sp}_{Mg_2SiO_4} \quad (5)$$

where $\bar{\mu}^{A}_{i}$ is the partial molar free energy of the i component in phase A. For the other equilibrium

$$Fe_2SiO_4(sp) = 2FeO\,(mw) + SiO_2\,(st) \quad (6)$$

we have a similar equation

$$2\bar{\mu}^{mw}_{FeO} + \bar{\mu}^{st}_{SiO_2} = \bar{\mu}^{sp}_{Fe_2SiO_4} \quad (7)$$

In equations (5) and (7), $\bar{\mu}^{sp}_{i}$ (i=Mg_2SiO_4, Fe_2SiO_4) and $\bar{\mu}^{mw}_{j}$ (j=MgO, FeO) are expressed as

$$\bar{\mu}^{sp}_{i} = \mu^{sp}_{i} + RT\ln a^{sp}_{i} \quad (8)$$

$$\bar{\mu}^{mw}_{j} = \mu^{mw}_{j} + RT\ln a^{mw}_{j} \quad (9)$$

where a^{sp}_{i} is activity of the i component in the spinel phase, and μ^{sp}_{i} is the partial molar free energy of the pure i component with spinel structure. We take the standard state of all components to be pure phases at the pressure and tem-perature of interest. Then, activity of stishovite is equal to unity. From equations (5), (8) and (9), we have

$$\Delta H^{\circ}_{Mg}(T) - T\Delta S^{\circ}_{Mg}(T) + \int_{1atm}^{P}\Delta V_{Mg}(P,T)dP$$
$$+ RT\ln((a^{mw}_{MgO})^2/a^{sp}_{Mg_2SiO_4}) = 0 \quad (10)$$

where $\Delta H^{\circ}_{Mg}(T)$ and $\Delta S^{\circ}_{Mg}(T)$ are enthalpy and entropy changes at 1 atm and T for the reaction (4), and $\Delta V_{Mg}(P,T)$ is molar volume change at P and T for the reaction (4). Similarly, from the equations (7), (8) and (9), we have

$$\Delta H^{\circ}_{Fe}(T) - T\Delta S^{\circ}_{Fe}(T) + \int_{1atm}^{P}\Delta V_{Fe}(P,T)dP$$
$$+ RT\ln((a^{mw}_{FeO})^2/a^{sp}_{Fe_2SiO_4}) = 0 \quad (11)$$

The compositions of coexisting spinel and magnesiowustite can be calculated by solving the simultaneous equations (10) and (11), when activity-composition relationships in spinel and magnesiowutite are given. The relationships are expressed as

$$a^{sp}_{Mg_2SiO_4} = (X^{sp}_{Mg_2SiO_4}\cdot\gamma^{sp}_{Mg_2SiO_4})^2 \quad (12)$$

$$a^{mw}_{MgO} = X^{mw}_{MgO}\cdot\gamma^{mw}_{MgO} \quad (13)$$

where X^{A}_{i} and γ^{A}_{i} are mol fraction and activity coefficient, repectively, of the i component in phase A. We use the symmetric regular solution model to represent nonideal properties of spinel and magnesiowustite solid solutions. In this model, the activity coefficient is expressed as

$$RT\ln\gamma^{A}_{i} = W^{A}(1 - X^{A}_{i})^2 \quad (14)$$

where W^A is a nonideal parameter of phase A. The nonideal parameters are assumed to be independent of pressure and temperature. We adopt W^{sp} of 3.9 kJ/mol (for one site basis) and W^{mw} of 14.0 kJ/mol [*Hahn and Muan*, 1962; *Williams*, 1971; *Akaogi et al.*, 1989]. For W of perovskite, the estimated value of 3.8 kJ/mol by *Navrotsky* [1987] is used.

The enthalpy and entropy changes at T in equations (10) and (11) are given as

$$\Delta H^{\circ}(T) = \Delta H^{\circ}(T_0) + \int_{T_0}^{T}\Delta C_p(T')dT' \quad (15)$$

$$\Delta S^{\circ}(T) = \Delta S^{\circ}(T_0) + \int_{T_0}^{T}\Delta C_p(T')/T'dT' \quad (16)$$

where T_0 is a reference temperature and $\Delta C_p(T)$ is the heat capacity difference between low and high pressure assemblages at T and 1 atm. The effect of pressure on molar volume in equations (10) and (11) is calculated, using the third order Birch-Murnaghan equation with bulk moduli and their pressure and temperature derivatives. The effect of temperature on molar volume is calculated by thermal expansion coefficients. Table 3 shows molar volumes, thermal expansivities, bulk moduli, their pressure and temperature derivatives, and heat capacities of the phases in the MgO-FeO-SiO_2 system used for the phase boundary calculations. More detailed description of the calculation method is given by *Akaogi et al.* [1989, 1995]. The phase relations for the equations (2) and (3) are calculated in a similar manner to those of (1).

3.4. *Phase Diagrams of Postspinel Transitions*

The enthalpies and entropies used for calculation of the phase relations are shown in Table 4. These enthalpy data are obtained from the measured enthalpies in this and previous studies shown in Table 2 and Figure 2. The entropies in Table 4 were obtained as follows. Using the same method as that in *Akaogi and Ito* [1993b], the ΔS°_{298} for dissociation of Mg_2SiO_4(sp) into $MgSiO_3$(pv) + MgO(per) was calculated from the measured ΔH°_{298} [*Akaogi and Ito*, 1993b] and the dissociation pressure by *Ito and Takahashi* [1989], using new data on bulk modulus and its pressure and temperature derivatives of $MgSiO_3$ perovskite [*Funamori et al.*, 1996]. The ΔS°_{298} for dissociation of Fe_2SiO_4(sp) into 2FeO(wu) + SiO_2(st) was calculated from the ΔH°_{298} in Table 4 and the reversed

TABLE 3. Molar volumes, thermal expansivities, heat capacities, bulk moduli and their pressure and temperature derivatives of the phases in the system MgO-FeO-SiO_2.

Phase	V°_{298}	$\alpha = a + bT + cT^{-2}$ (K^{-1})			K_T	K_T'	$(\partial K_T/\partial T)_P$
	(cm^3/mol)	$a\times10^5$	$b\times10^9$	$c\times10$	(GPa)		(GPa/K)
Mg_2SiO_4 sp	39.487[a]	2.448	4.056	-6.029[b]	182.6[b]	5.0[c]	-0.0284[b]
Fe_2SiO_4 sp	42.040[d]	2.455	3.591	-3.703[e]	192.0[f]	5.0[g]	-0.0284[g]
$MgSiO_3$ pv	24.442[a]	1.982	8.180	-4.740[h]	261.0[h]	4.0[h]	-0.0280[h]
$FeSiO_3$ pv	25.395[i]	1.982	8.180	-4.740[j]	261.0[j]	4.0[j]	-0.0280[j]
MgO per	11.248[k]	3.753	7.941	-7.787[l]	160.3[m]	4.1[m]	-0.0272[n]
FeO wu	12.250[l]	1.688	2.040	0.190[l]	153.0[o]	4.9[o]	-0.0272[p]
SiO_2 st	14.014[k]	1.053	9.031	1.220[l]	314.0[q]	5.1[r]	-0.0470[e]

Phase	$C_p = A + BT^{-0.5} + CT^{-2} + DT^{-3}$ (J/mol.K)			
	$A\times10^{-2}$	$B\times10^{-3}$	$C\times10^{-6}$	$D\times10^{-8}$
Mg_2SiO_4 sp	2.178	-1.543	-0.569	-4.192[l]
Fe_2SiO_4 sp	2.813	-2.902	0.0	5.041[l]
$MgSiO_3$ pv	1.769	-1.565	0.0	-2.157[s]
$FeSiO_3$ pv	2.039	-2.246	1.664	1.563[t]
MgO per	0.586	-0.189	-1.664	0.234[k]
FeO wu	0.857	-0.871	0.0	3.954[l]
SiO_2 st	0.858	-0.346	-3.605	4.511[u]

[a] *Ito and Yamada* [1982], [b] *Meng et al.* [1993], [c] *Rigden et al.* [1991], [d] *Marumo et al.* [1977], [e] *Fei et al.* [1991], [f] *Sato* [1977] and *Bass et al.* [1981], [g] Same as Mg_2SiO_4 spinel, [h] *Funamori et al.* [1996], [i] Estimated using the data by *Kudoh et al.* [1990], [j] Same as $MgSiO_3$ perovskite, [k] *Robie et al.* [1978], [l] *Saxena et al.* [1993], [m] *Jackson and Niesler* [1982], [n] *Anderson et al.* [1992], [o] *Jackson et al.* [1990], [p] Same as MgO, [q] *Weidner et al.* [1982], [r] *Rigden et al.* [1994], [s] *Akaogi and Ito* [1993a], [t] Estimated using the relation of $C_p(FeSiO_3\ pv) = C_p(MgSiO_3\ pv) - C_p(per+st) + C_p(wu+st)$, [u] *Akaogi et al.* [1995].

boundary of this transition by Katsura et al. [1997]. When the boundaries for sp→ 2mw + st were preliminarily calculated using the ΔH°_{298} of 56.74 kJ/mol for the reaction (4), some discrepancies appeared between the calculated boundaries and the experimentally determined ones in Figure 1. Therefore, we increased the ΔH°_{298} by the standard deviation to partly remove the discrepancies. Using the increased value of 62.49 kJ/mol, the ΔS°_{298} for the reaction (4) was estimated as -3.6 J/mol.K to fit the calculated boundaries to those in Fig. 1. Combining the ΔH°_{298} of 25.5 kJ/mol for the reaction of FeO(wu) + SiO_2(st) = $FeSiO_3$(pv) with the free energy difference ΔG°_{298} for the same reaction obtained from the Mg-Fe partitioning data between perovskite and magnesiowustite [*Ito et al.*, 1984; *Ito and Takahashi*, 1989], the ΔS°_{298} was calculated to be -7.8 J/mol.K. Therefore, the ΔS°_{298} for Fe_2SiO_4(sp) = $FeSiO_3$(pv) + FeO(wu) was obtained as -13.4 J/mol.K. The errors of the ΔS°_{298} for both of the reactions were about ±2-3 J/mol.K.

The calculated boundaries for the postspinel transitions at 1373 K and 1873 K are shown in Figures 3a and 3b, respectively. The calculated boundaries are generally con-sistent with the experimental data by *Ito and Takahashi* [1989]. With increasing temperature, the boundaries shift to lower pressures, particularly in Mg-rich compositions, indicating the negative

TABLE 4. Enthalpy and entropy data used for calculation of phase relations of post-spinel transitions.

Transition	ΔH°_{298} (kJ/mol)	ΔS°_{298} (J/mol.K)
Mg_2SiO_4(sp)→$MgSiO_3$(pv)+MgO(per)	86.08±3.65	1.7±2.3
Mg_2SiO_4(sp)→2MgO(per)+SiO_2(st)	56.74±5.75	-4.6±1.0
	62.49[a]	-3.6[a]
MgO(per)+SiO_2(st)→$MgSiO_3$(pv)	29.34±6.81	6.3±2.5
Fe_2SiO_4(sp)→$FeSiO_3$(pv)+FeO(wu)	88.83[b]	-13.4[b]
Fe_2SiO_4(sp)→2FeO(wu)+SiO_2(st)	63.33±3.11	-5.6±2.4
FeO(wu)+SiO_2(st)→$FeSiO_3$(pv)	25.50[b]	-7.8[b]

[a] The values used for the calculation of the transition boundaries of sp → 2mw + st. See text.
[b] The estimated uncertainty is shown in text.

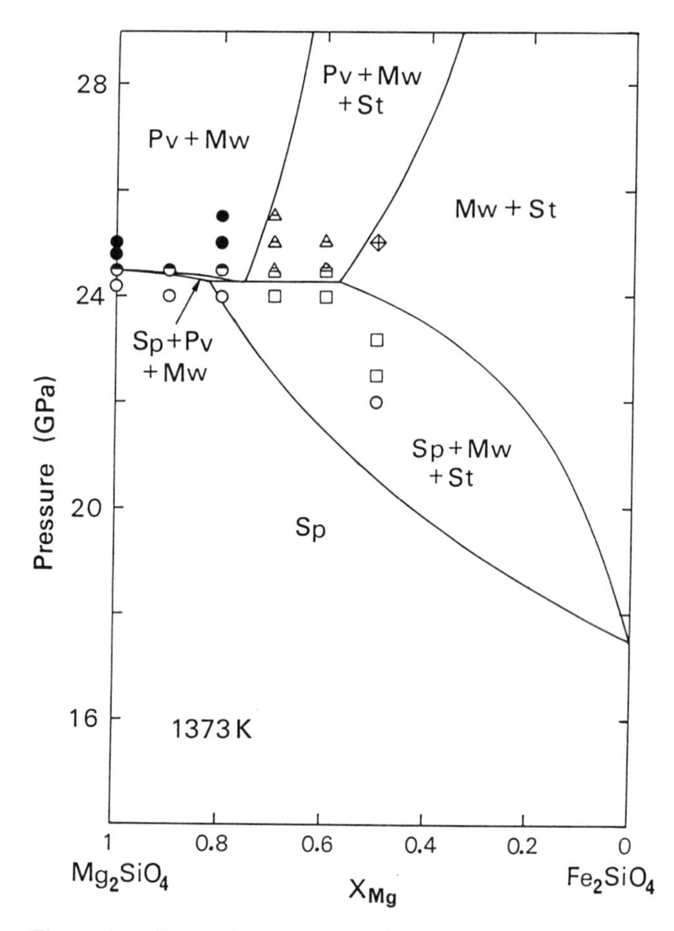

Figure 3a. Phase diagram of postspinel transitions in the system Mg_2SiO_4-Fe_2SiO_4 at 1373 K. All of the boundaries were calculated in this study. The data points at 22-25.5 GPa are from *Ito and Takahashi* [1989] (open circles, sp; solid circles, pv+mw; squares, sp+mw+st; triangles with bar, pv+mw+st; diamonds with cross, mw+st). Phase boundaries of the olivine-modified spinel-spinel tran-sitions which occur below about 20 GPa are not shown in the diagrams.

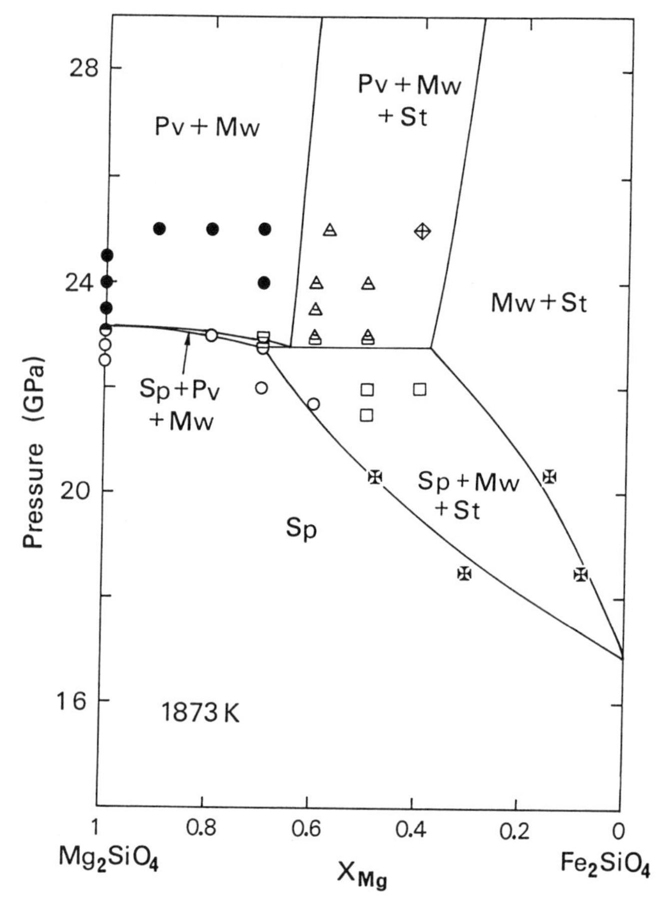

Figure 3b. Phase diagram of postspinel transitions in the system Mg_2SiO_4-Fe_2SiO_4 at 1873 K. All of the boundaries were calculated in this study. Four data points at 18.5 and 20.4 GPa represent compositions of coexisting spinel and magnesiowustite determined in this study (see Figure 1). The other data points at 21.5-25 GPa are from *Ito and Takahashi* [1989].

dP/dT slope of the postspinel transition boundaries. As shown in Figures 3a and 3b, the present thermodynamic data give the dP/dT slope of postspinel transition in Mg_2SiO_4 which is consistent with Ito and Takahashi's data. In Figures 3a and 3b, compositional ranges of the stability field of pv + mw at 1373 and 1873 K agree with those by Ito and Takahashi's experiments. Although some discrepancies between the calculated boundaries and the experimental data are observed in the fields of sp + mw + st at 1373 K and pv + mw + st at 1873 K, the experimenal boundaries are constrained only by the limited number of data points. The calculated transition loop of sp → pv + mw is very narrow and the transition pressures of the boundaries are almost independent of the composition. These results are consistent with Ito and Takahashi's experiments and thermodynamic calculations by *Wood* [1990], *Fei et al.* [1991], and *Stixrude and Bukowinski* [1993]. The calculated width of the sp → pv + mw transition for $(Mg_{0.8},Fe_{0.2})_2SiO_4$ is about 0.05 GPa at 1873 K in this study, and is consistent with the experimental determination of the interval of <0.15 GPa for the same composition and temperature by *Ito and Takahashi* [1989].

Figures 3a and 3b indicate that the calculated stability field of pv + mw expands to the Fe_2SiO_4-rich side with increasing temperature. This result is consistent with Ito and Takahashi's experiments. However, the thermodynamic calculation by *Fei et al.* [1991] with a similar method to ours showed that the pv + mw field shrank with increasing temperature. This difference is in part due to the fact that *Fei et al.* [1991] used the previous thermodynamic data on stishovite by *Akaogi and Navrotsky*

[1984] because preliminary calculation using the previous stishovite data showed the same behavior of the pv + mw region with temperature as calculated by Fei et al. When the revised data on stishovite by *Akaogi et al.* [1995] are used, the calculated boundaries of the pv + mw field are consistent with the experiments, as shown in Figures 3a and 3b. *Stixrude and Bukowinski's* [1993] thermodynamic calculation (global fit in their paper), which also showed a considerable contraction of the pv + mw field to the Mg-rich side with temperature, may be caused by using inadequate values of their thermodynamic potential parameters in the calculation.

The Fe/(Mg+Fe) in perovskite in equilibrium with magnesiowustite and stishovite in the system $MgSiO_3$-$FeSiO_3$ was calculated using the present thermodynamic data set and is shown in Figure 4. The calculated compositions at 25 GPa show maximum solubility of $FeSiO_3$ in perovskite and are consistent with the experimental data at 1373 and 1873 K at 25(±1) GPa by *Ito and Takahashi* [1989] and with those at 1473-1763 K at 26 GPa by *Fei et al.* [1996]. The maximum solubility of $FeSiO_3$ in perovskite clearly increases with temperature. As shown in Figure 4, the value also increases with pressure, but this tendency decreases with increasing temperature.

It has been proposed that the 660-km seismic discontinuity is caused by the transition of sp→pv + mw [e.g., *Ito and Takahashi*, 1989]. One of the crucial evidences is that recent seismological studies show that the 660-km discontinuity is very sharp: the width is less than 4-5 km [*Nakanishi*, 1988; *Benz and Vidale*, 1993; *Yamazaki and Hirahara*, 1994]. Generally an experimental phase equilibrium study is not very powerful to determine accurately the pressure interval of such a very sharp transition, because of the pressure resolution of about 0.2 GPa in the 20-25 GPa range. The thermodynamic calculation can tightly constrain the very sharp transition boundary. As shown in Figure 3, the calculated pressure interval in this study for the sp→pv + mw transition for mantle spinel composition with 10-15 mol % Fe_2SiO_4 is about 0.01-0.05 GPa at 1373-1873 K, which corresponds to a depth interval of about 0.3-1.2 km.

Because the pressure interval of the sp→pv + mw transition is of great geophysical importance as described above, we examine in more detail a possible range of the transition boundary width, considering the uncertainty of the thermodynamic data. The error of the measured ΔH°_{298} for transition of Mg_2SiO_4(sp) in Table 4 gave a negligible effect on the pressure interval. The uncertainty of ΔH°_{298} for the reaction of Fe_2SiO_4(sp) to $FeSiO_3$ (pv) + FeO(wu), calculated to be about ±20 kJ/mol from those of FeO(wu) + SiO_2(st) → $FeSiO_3$ (pv) and of the reaction (6), a little widened the pressure interval: the upper bound was about 0.18 GPa for the mantle spinel composition with 10-15 mol % Fe_2SiO_4. Therefore, considering

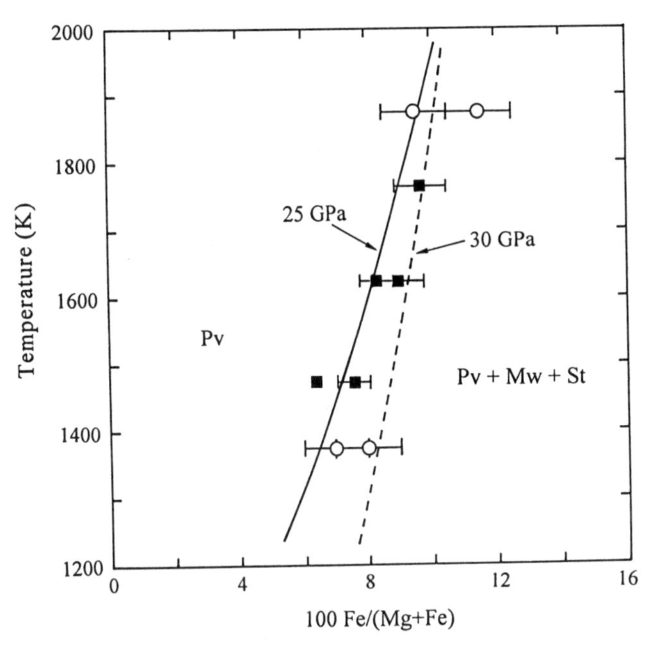

Figure 4. The Fe/(Mg+Fe) values in perovskite in equilibrium with magnesiowustite and stishovite in the system $MgSiO_3$-$FeSiO_3$, shown by the solid curve at 25 GPa and the dashed curve at 30 GPa calculated in this study. Open circles with errors are the experimental data at 25 GPa from *Ito and Takahashi* [1989], and solid squares with errors are the data at 26 GPa from *Fei et al.* [1996]. Pv and Pv+Mw+St represent the stability fields.

the uncertainties of the thermodynamic data, the possible pressure interval of the sp→pv + mw transition is about 0.01-0.18 GPa, which corresponds to depth interval of about 0.3- 4 km. This is consistent with the width (<4-5 km) of the seismically observed 660-km discontinuity.

Figures 3a and 3b show that the compositional range of the pv + mw field contracts to the Mg_2SiO_4-rich side with decreasing temperature. This suggests that spinel would not transform directly to perovskite + magnesiowustite in the system Mg_2SiO_4-Fe_2SiO_4 at very low temperature. This can be seen clearly in a P-T diagram for a fixed composition. Figure 5 shows transition boundaries in Mg_2SiO_4 calculated using the present thermodynamic data set. Above about 1000 K, Mg_2SiO_4(sp) dissociates into $MgSiO_3$(pv) + MgO(per) with increase of pressure. Below about 1000 K, however, spinel first transforms into an assemblage of 2MgO(per) + SiO_2(st), and further changes to $MgSiO_3$(pv) + MgO(per) at higher pressure. The stability field of 2MgO(per) + SiO_2(st) has not yet been experimentally observed in recent high pressure studies. This may be due to sluggish reaction at such low temperatures. However, the thermodynamic data in this study indicate that $MgSiO_3$ perovskite is energetically less stable than the

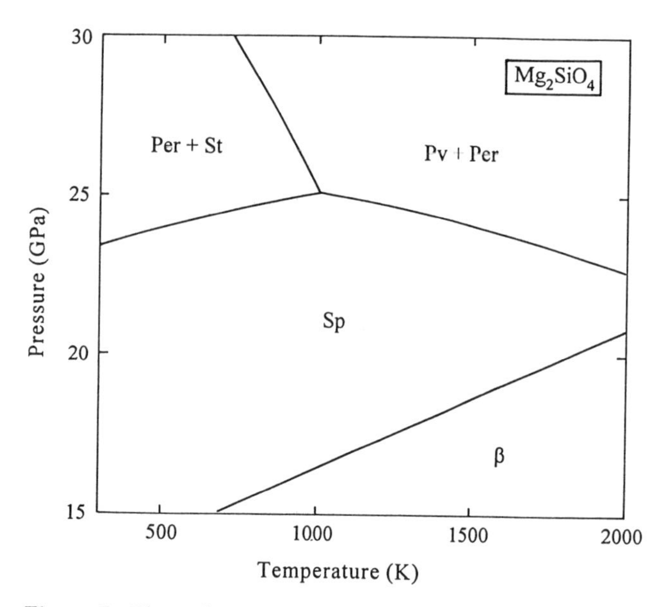

Figure 5. Phase diagram of Mg_2SiO_4. All of the boundaries were calculated in this study. β: modified spinel.

assemblage of MgO(per) + SiO_2(st) in the P-T region. Using the thermodynamic data, we examined possible stability of other assemblages which are denser than spinel and consist of the known phases in the system MgO-SiO_2 (e.g., $MgSiO_3$ ilmenite + MgO). However, it was confirmed that 2MgO(per) + SiO_2(st) is the most stable assemblage, unless a new, unexpected phase is stable in this P-T range. Recent quantum mechanical calculation by *Lacks and Gordon* [1995] shows that the most

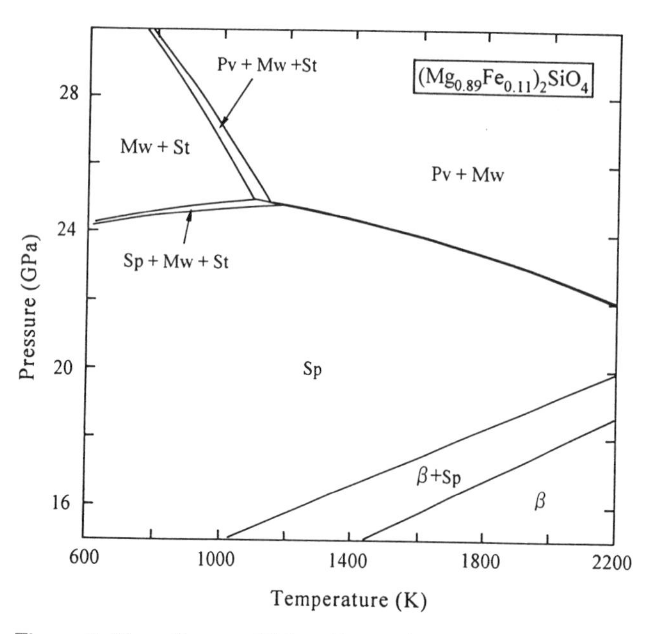

Figure 6. Phase diagram of $(Mg_{0.89}Fe_{0.11})_2SiO_4$. All of the boundaries were calculated in this study. β: modified spinel.

stable phase assemblage at 0 K is Mg_2SiO_4(sp) below 24.5 GPa, above which 2MgO(per) + SiO_2(st) is the most stable, and further above 31.5 GPa $MgSiO_3$(pv) + MgO follows. This result is generally consistent with the calculated boundaries in Figure 5 extrapolated to 0 K.

Figure 6 illustrates the calculated boundaries in a P-T space for $(Mg_{0.89}, Fe_{0.11})_2SiO_4$ which is a representative mantle spinel composition. Above about 1150 K, spinel directly transforms with pv + mw at about 22-24.5 GPa with an extremely narrow transition interval. Below that temperature, it is shown that spinel transforms to 2(Mg,Fe)O(mw) + SiO_2(st), passing though a narrow field of sp + mw + st. The mw + st assemblage further changes to pv + mw through an assemblage of pv + mw + st with a small pressure interval. However, these assemblages have not been experimentally observed. In the transformation experiments of *Wang et al.* [1995] at 26 GPa of San Carlos olivine (approximately $(Mg_{0.9}, Fe_{0.1})_2SiO_4$), they found that the olivine showed no change at 973 K after 19 hours, but at 1273 K olivine first transformed to spinel in the first 10 min and then the spinel began to dissociate into pv + mw after 1.5 hours. These observations suggest metastable nucleation and growth of spinel and sluggish reaction for the dissociation even at 1273 K. The mw + st field and the two narrow regions of the three-phase assemblages may not have been observed in previous experiments by the sluggish reaction kinetics. Further experimental studies are required to confirm whether these fields are really present. If the transition of spinel into mw + st is experimentally observed, spinel in subducted slabs may first transform into mw + st in the low temperature regime. This new type of transformation of mantle spinel will cause several new effects which are different from the transition to pv + mw in density, seismic velocities, and rheological properties of the slabs.

Acknowledgments. We are very grateful to H. Horiuchi for his help in microfocused X ray diffraction experiments, to T. Ishii for his help in microprobe analysis, and to A. Navrotsky for valuable discussion. We also thank A. Navrotsky and Y. Fei for their constructive review. This work was supported in part by Grants-in-Aid from Ministry of Education, Science, and Culture, Japan.

REFERENCES

Akaogi, M. and E. Ito, Heat capacity of $MgSiO_3$ perovskite, *Geophys. Res. Lett., 20*, 105-108, 1993a.

Akaogi, M. and E. Ito, Refinement of enthalpy measurement of $MgSiO_3$ perovskite and negative pressure-temperature slopes for perovskite-forming reactions, *Geophys. Res. Lett., 20*, 1839-1842, 1993b.

Akaogi, M., E. Ito, and A. Navrotsky, Olivine-modified spinel-spinel transitions in the system Mg_2SiO_4-Fe_2SiO_4: calorimetric measurements, thermochemical calculations, and geophysical applications, *J. Geophys. Res.*, *94*, 15,671-15,685, 1989.

Akaogi, M. and A. Navrotsky, The quartz-coesite-stishovite transformations: new calorimetric measurements and calculation of phase diagrams, *Phys. Earth Planet. Inter.*, *36*, 124-134, 1984.

Akaogi, M., H. Yusa, E. Ito, T. Yagi, K. Suito, and J. T. Iiyama, The $ZnSiO_3$ clinopyroxene-ilmenite transition: heat capacity, enthalpy of transition, and phase equilibria, *Phys. Chem. Miner.*, *17*, 17-23, 1990.

Akaogi, M., H. Yusa, K. Shiraishi, and T. Suzuki, Thermodynamic properties of α-quartz, coesite, and stishovite and equilibrium phase relations at high pressures and high temperatures, *J. Geophys. Res.*, *100*, 22337-22347, 1995.

Anderson, O. L., D. Isaak, and H. Oda, High temperature elastic constant data on minerals relevant to geophysics, *Rev. Geophys.*, *30*, 57-90, 1992.

Bass, J. D., R. C. Liebermann, D. J. Weidner, and S. J. Finch, Elastic properties from acoustic and volume compression experiments, *Phys. Earth Planet. Inter.*, *25*, 140-158, 1981.

Benz, H. M. and J. E. Vidale, Sharpness of upper-mantle discontinuities determined from high-frequency reflections, *Nature, 365*, 147-150, 1993.

Berman, R. G., Internally consistent thermodynamic data for minerals in the system Na_2O-K_2O-CaO-MgO-FeO-Fe_2O_3-Al_2O_3-SiO_2-TiO_2-H_2O-CO_2, *J. Petrol.*, *29*, 445-522, 1988.

Charlu, T. V., R. C. Newton, and O. J. Kleppa, Enthalpies of formation at 970 K of compounds in the system MgO-Al_2O_3-SiO_2 from high temperature solution calorimetry, *Geochim. Cosmochim. Acta*, *39*, 1487-1497, 1975.

Chatillon-Colinet, C., R. C. Newton, D. Perkins III, and O. J. Kleppa, Thermochemistry of $(Fe^{2+},Mg)SiO_3$ orthopyroxene, *Geochim. Cosmochim. Acta*, *47*, 1597-1603, 1983.

Fei, Y., H. K. Mao, and B. O. Mysen, Experimental determination of element partitioning and calculation of phase relations in the MgO-FeO-SiO_2 system at high pressure and high temperature, *J. Geophys. Res.*, *96*, 2157-2169, 1991.

Fei, Y., Y. Wang, and L. W. Finger, Maximum solubility of FeO in $(Mg,Fe)SiO_3$-perovskite as a function of temperature at 26 GPa: impication for FeO content in the lower mantle, *J. Geophys. Res.*, *101*, 11,525-11,530, 1996.

Funamori, N., T. Yagi, W. Utsumi, T. Kondo, T. Uchida, and M. Funamori, Thermoelastic properties of $MgSiO_3$ perovskite determined by in situ X ray observations up to 30 GPa and 2000 K, *J. Geophys. Res.*, *101*, 8257-8269, 1996.

Hahn, W. C. and A. Muan, Activity measurements in oxide solid solutions: the system "FeO"-MgO in the temperature range 1100 to 1300 °C, *Trans. Met. Soc. Am. Inst. Met. Eng.*, *224*, 416-420, 1962.

Ito E. and E. Takahashi, Postspinel transformations in the system Mg_2SiO_4-Fe_2SiO_4 and some geophysical implications, *J. Geophys. Res.*, *94*, 10,637-10,646, 1989.

Ito E., E. Takahashi, and Y Matsui, The mineralogy and chemistry of the lower mantle: an implication of the ultrahigh-pressure phase relations in the system MgO-FeO-SiO_2, *Earth Planet. Sci. Lett.*, *67*, 238-248, 1984.

Ito, E. and H. Yamada, Stability relations of silicate spinels, ilmenites and perovskites, in *High-Pressure Research in Geophysics*, edited by S.Akimoto and M. H. Manghnani, pp. 405-419, Center Acad. Publ. Japan, Tokyo, 1982.

Jackson, I. and H. Niesler, The elasticity of periclase to 3 GPa and some geophysical implications, in *High-Pressure Research in Geophysics*, edited by S.Akimoto and M. H. Manghnani, pp. 93-113, Center Acad. Publ. Japan, Tokyo, 1982.

Jackson, I., S. K. Khanna, A. Revcolevschi and J. Berthon, Elasticity, shear-mode softening and high-pressure polymorphism of wustite ($Fe_{1-x}O$), *J. Geophys. Res.*, *95*, 21,671-21,685, 1990.

Katsura, T. and E. Ito, The system Mg_2SiO_4-Fe_2SiO_4 at high pressures and temperatures: precise determination of stabilities of olivine, modified spinel, and spinel, *J. Geophys. Res.*, *94*, 15663-15670, 1989.

Katsura, T., A. Ueda, E. Ito, and K. Morooka, Postspinel transition in Fe_2SiO_4, in *High Pressure-Temperature Research: Properties of Earth and Planetary Materials*, edited by M. H. Manghnani and T. Yagi, Am. Geophys. Union, 1997, in press.

Kudoh, Y., C. T. Prewitt, L. W. Finger, A. Darovskikh, and E. Ito, Effect of iron on the crystal structure of $(Mg,Fe)SiO_3$ perovskite, *Geophys. Res. Lett.*, *17*, 1481-1484, 1990.

Lacks, D. J. and R. G. Gordon, Calculations of pressure-induced phase transitions in mantle minerals, *Phys. Chem. Miner.*, *22*, 145-150, 1995.

Liu, L. G., The high-pressure phases of $FeSiO_3$ with implications for Fe_2SiO_4 and FeO, *Earth Planet. Sci. Lett.*, *33*, 101-106, 1976.

Marumo, F., M. Isobe, S. Akimoto, Electron-density distribution in crystals of γ-Fe_2SiO_4 and γ-Co_2SiO_4, *Acta Crystallogr., Ser. B.*, *33*, 713-716, 1977.

Meng, Y., D. J. Weidner, G. D. Gwanmesia, R. C. Liebermann, M. T. Vaughan, Y. Wang, K. Leinenweber, R. E. Pacalo, A. Yeganeh-Haeri, and Y. Zhao, In situ high P-T X ray diffraction studies on three polymorphs (α,β,γ) of Mg_2SiO_4, *J. Geophys. Res.*, *98*, 22,199-22,207, 1993.

Nakanishi, I., Reflections of P'P' from upper mantle discontinuities beneath the Mid-Atlantic Ridge, *Geophys. J.*, *93*, 335-346, 1988.

Navrotsky, A., Models of crystalline solutions, in *Reviews in Mineralogy*, vol. 17, edited by I. S. E. Carmichael and H. P. Eugster, pp. 35-69, 1987.

Navrotsky, A., R. P. Rapp, E. Smelik, P. Burnley, S. Circone, L. Chai, K. Bose, H. R. Westrich, The behavior of H_2O and CO_2 in high-temperature lead borate solution calorimetry of volatile-bearing phases, *Am. Mineral.*, *79*, 1099-1109, 1994.

Rigden, S. M., G. D. Gwanmesia, J. D. Fitz Gerald, I. Jackson, and R. C. Liebermann, Spinel elasticity and seismic structure of the transition zone of the mantle, *Nature, 354*, 143-145, 1991.

Rigden, S. M., B. Li, and R. C. Liebermann, Elasticity of stishovite at high pressures, *Eos, Trans. Am. Geophys. Union, 75*, 596, 1994.

Robie, R. A., B. S. Hemingway, and J. R. Fisher, Thermodynamic properties of minerals and related substances at 298.15 K and

1 bar (10^5 pascals) pressure and higher temperatures, *U. S. Geol. Surv. Bull., 1452*, 456 pp., 1978.

Sato, Y., Equation of state of mantle minerals determined though high-pressure X ray study, in *High-Pressure Research: Applications in Geophysics*, edited by M. H. Manghnani and S. Akimoto, pp. 307-323, Acad. Press, New York, 1977.

Saxena, S. K., N. Chatterjee, Y. Fei, and G. Shen, *Thermodynamic Data on Oxides and Silicates*, 428 pp. Springer-Verlag, New York, 1993.

Stixrude, L. and M. S. T. Bukowinski, Thermodynamic analysis of the system MgO-FeO-SiO_2 at high pressure and the structure of the lowermost mantle, in *Evolution of the Earth and Planets, Geophys. Monogr. Ser.*, vol. 74, edited by E. Takahashi, R. Jeanloz, and D. Rubie, pp. 131-141, 1993.

Wang, Y., I. Martinez, F. Guyot, and R. C. Liebermann, The break down of olivine to perovskite + magnesiowustite, *Eos, Trans., Am. Geophys. Union, 76*, p. 618, 1995.

Weidner, D. J., J. D. Bass, A. E. Ringwood, and W. Sinclair, The single-crystal elastic moduli of stishovite, *J. Geophys. Res., 87*, 4740-4746, 1982.

Williams, R. J., Reaction constants in the system Fe-MgO-SiO_2-O_2 at 1 atm between 900° and 1300° C: experimental results, *Am. J. Sci., 270*, 334-360, 1971.

Wood, B. J., Postspinel transformations and the width of the 670-km discontinuity: a comment on "Postspinel transformations in the system Mg_2SiO_4-Fe_2SiO_4 and some geophysical implications" by E. Ito and E. Takahashi, *J. Geophys. Res., 95*, 12681-12685, 1990.

Yagi, T, P. M. Bell, and H. K. Mao, Phase relations in the system MgO-FeO-SiO_2 between 150 and 700 kbar at 1000° C, *Year Book Carnegie Inst. Washington, 78*, 614-618, 1979.

Yamazaki, A. and K. Hirahara, The thickness of upper mantle discontinuities, as inferred from short-period J-Array data, *Geophys. Res. Lett., 21*, 1811-1814, 1994.

M. Akaogi, H. Kojitani, K. Matsuzaka, T. Suzuki, Department of Chemistry, Gakushuin University, Toshima-ku, Tokyo 171, Japan.

E. Ito, Institute for Study of the Earth's Interior, Okayama University, Misasa, Tottori 682-01, Japan.

Pressure-Induced Structural Modifications of Metagermanates at Room Temperature : In Situ XRD and EXAFS Studies

T. Nagai, O. Ohtaka, A. Yoshiasa, T. Yamanaka

Department of Earth and Space Science, Osaka University, Toyonaka, Osaka, Japan

O. Shimomura[1]

Photon Factory, National Laboratory for High Energy Physics, Tsukuba, Japan

In situ X ray diffraction (XRD) and extended X ray absorption fine structure (EXAFS) measurements under high pressure and at room temperature were performed in order to verify the possibility of a pressure-induced amorphization of $MgGeO_3$-high-clinoenstatite and $CaGeO_3$-wollastonite. We found that the two metagermanates were transformed into new metastable phases without the amorphization. For $MgGeO_3$-high-clinoenstatite, a new reflection, which could not be assigned as its known high-pressure polymorphs, could be observed above 23 GPa. For $CaGeO_3$-wollastonite, EXAFS measurements under high pressure were conducted using MAX90. The observed pressure dependence of the Ge-O distance can suggest the existence of a metastable intermediate phase between wollastonite and perovskite. The intermediate phase corresponds to a rhodonite-like structure, which was proposed by a previous XRD experiments. These transitions, which were observed in the two metagermanates, suggest to be kinetically favored processes under room temperature compression. Such metastable pathways during phase transitions can be described as typical examples of the Ostwald step rule.

INTRODUCTION

When all the X ray diffraction lines of crystalline substances disappear under high pressure, we can judge that a pressure-induced amorphization occurs. Since the discovery of the amorphization of hexagonal ice [*Mishima et al.*, 1984], the structural behavior of crystalline materials under high pressure at kinetically low temperature has been a subject of great interest not only in the field of material sciences but also of geophysical science. The kinetically low temperature means that the temperature is low enough to inhibit the transformation into a thermodynamically stable phase under high pressure.

It is well known that quartz and feldspar convert to amorphous phases by shock waves at relatively low temperature [*Milton and De Carli*, 1963]. Several crystalline compounds, including geophysically significant materials, have recently been founded to undergo a vitrification by static compression at room temperature [cf. α-quartz structure compounds: *Hemley et al.*, 1988; *Itie et al.*, 1989; *Yamanaka et al.*, 1992b; *Wolf et al.*, 1992; olivine structure compounds: *Williams et al.*, 1990; *Richard and Richet*, 1990; *Goyot and Reynard*, 1992; *Andrault et al.*, 1995; anorthite: *Williams and Jeanloz*,

[1]Present address: SPring-8, Kanaji 1503-1, Kamigori, Ako, Hyogo 678-12

Properties of Earth and Planetary Materials
at High Pressure and Temperature
Geophysical Monograph 101

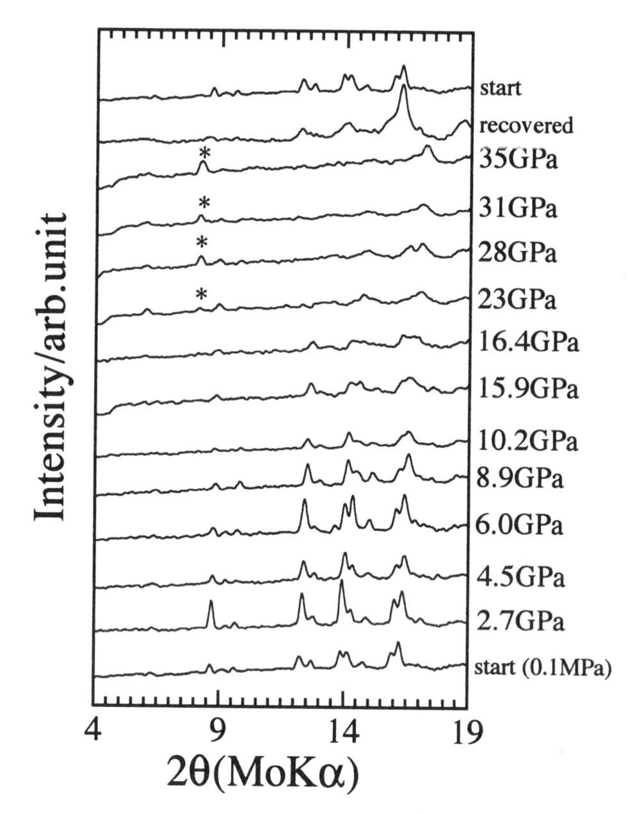

Figure 1. X ray diffraction patterns of $MgGeO_3$-high-clinoenstatite on compression up to 35 GPa and the diffraction pattern of a recovered sample. The reflection marked by * corresponds to the newly observed reflection above 23 GPa.

1989; serpentine: *Meade and Jeanloz*, 1991]. Although *Andrault et al.* [1996] have recently discussed the high-temperature behavior of germanate perovskite and pyroxenoids, we could find no information about the pressure-induced amorphization of the pyroxene related minerals, which have single tetrahedral chains. It is important to investigate their structural modification under room temperature compression, in order to clarify systematic understanding of the relation between the amorphization and the crystal structure.

Both $MgGeO_3$ and $CaGeO_3$ have the pyroxene related structure (so-called metagermanate), and the crystal structures of $MgGeO_3$ and $CaGeO_3$ at ambient conditions are considered to be analogous to $MgSiO_3$-high-clinoenstatite (*C2/c*) and $CaSiO_3$-wollastonite, respectively. In these compounds, *Andrault et al.* [1992] performed extended X ray absorption fine structure (EXAFS) studies under pressure. It was suggested that the pressure-induced coordination change of Ge atoms from fourfold to sixfold occurs from 8.5 GPa for $MgGeO_3$ and from 7 GPa for $CaGeO_3$ and that the coordination change might lead to the pressure-induced amorphization, because other works suggested that the amorphization of GeO_2 -quartz is closely related to the pressure-induced coordination change of Ge atoms from tetrahedral to octahedral coordination [*Itie et al.*, 1989, *Wolf et al.*, 1992, *Yamanaka et al.*, 1992b]. In situ X ray diffraction (XRD) measurements were performed up to 35 GPa for $MgGeO_3$-high-clinoenstatite and up to 27 GPa for $CaGeO_3$-wollastonite to investigate the possibility of the amorphization. In spite of the prediction from the EXAFS works of the amorphization, no pressure-induced, crystalline-to-amorphous transition was observed in either metagermanate. However, it was observed that newly observed metastable structural modifications did occur. Such metastable phase transitions can be understood by the Ostwald step rule. This principle states that a stepwise approach toward an equilibrium structure in a structural transformation is kinetically more favorable than a single-step approach.

We also conducted EXAFS studies on $CaGeO_3$ under high pressure using MAX90. *Andrault et al.* [1992] reported that the gradual coordination change of Ge atoms from fourfold to sixfold occurred in pressures between 7 and 12 GPa; however, the result of our experiments did not confirm such gradual change and suggested the appearance of the rhodonite-like phase, which was suggested by the XRD experiments [*Nagai and Yamanaka*, in press].

EXPERIMENTAL

Sample Preparation

The monoclinic phase of $MgGeO_3$ with space group *C2/c* ($MgGeO_3$-high-clinoenstatite) was synthesized as follows. At first an equimolar mixture of reagent-grade powders of MgO and GeO_2 was heated at 1473 K for 48 hours and then loaded at about 6 GPa and 727 K for an hour. High pressure syntheses were performed by a cubic-anvil type apparatus. The $CaGeO_3$-wollastonite was prepared from an equimolar mixture of reagent-grade powders of $CaCO_3$ and GeO_2 by heating at 1473 K for 72 hours. Each of the powder X ray diffraction patterns of the product showed a single phase of $MgGeO_3$-high-clinoenstatite or $CaGeO_3$-wollastonite.

Procedure for Diffraction Measurements

A diamond anvil cell [DAC] [*Yamaoka et al.*, 1979] was used to generate high pressures at room temperature. A powdered sample with a few ruby chips was loaded in the 0.3-mm-diameter hole of a stainless steel [SUS304] gasket, which was initially 0.3 mm thick and preindented to 0.15

mm thick. A 4:1 mixture of methanol:ethanol was used as a pressure transmitting medium, and the pressure was determined by a conventional ruby fluorescence measurement [*Piermarini et al.*, 1975].

In situ X ray diffraction measurements at high pressure using the DAC with the angular-dispersive method were conducted with a curved, position-sensitive detector [PSD, Inel CPS-120] and a molybdenum-target rotating anode X ray generator [Rigaku RU-200], operated at 50 kV and 180 mA. Incident X rays were monochromatized by pyrolytic graphite, and a 0.1-mmφ collimator was used. *Yamanaka et al.* [1992a] described this system in detail.

Procedure for EXAFS Measurements

EXAFS measurements using the MAX90 with synchrotron radiation were performed at BL-13B2, Photon Factory, National Institute for High Energy Physics, Tsukuba. Details of the MAX90 system are in *Shimomura et al.* [1992]. Sintered diamond anvils with a flat square of 3 × 3 mm^2 and a 6 × 6 × 6 mm^3 pressure medium made from a mixture of amorphous boron and epoxy resin were used to avoid a Bragg reflection which occurs at a certain energy. The well-mixed powder of the sample and BN filled half the volume of the sample chamber, and the other half was filled with the powder mixture of regent-grade Ge and BN. To estimate pressure in XRD experiments under high pressure, NaCl or Au is commonly used as a pressure calibrant, however, the present system could not measure XRD and EXAFS simultaneously. Therefore, the pressure values were estimated on the basis of the equation of state of the diamond type of Ge. The cell volumes were calculated by the first neighbor distance of Ge-Ge in the diamond type of Ge. It is well known that the diamond type of Ge is transformed into the β-Sn form at about 11 GPa [*Jamieson*, 1963]. Therefore, the pressure values over 11 GPa were estimated by extrapolating the relation between the oil pressures and the generated pressures up to 10 GPa. The accuracy of estimated pressures was not checked, however, judging from our previous XRD studies using the MAX90 system, the correlation between the loading oil pressures and the estimated pressures was reasonable and reproducible well. Intensities of the monochromatized incident X rays and those of the transmitted X rays were measured by ionization chambers.

RESULTS

$MgGeO_3$-High-Clinoenstatite

Diffraction patterns of $MgGeO_3$ were obtained by XRD measurements using the DAC up to 35 GPa at room temperature (Figure 1). The lattice parameters up to 16 GPa were derived from about 12 observed reflections by a least-squares method. Calculated volume compression data were fitted with the first order Birch-Murnaghan equation of state with $K_0' = 4$ (Figure 2). The K_0 value obtained is 127 (±8) GPa, which is consistent with the previously reported value of 131 GPa [*Liebermann*, 1974] obtained by measurements of elastic wave velocity under pressure. The agreement indicates that the present XRD measurements were reliable. The arrangement of the oxygen atoms of the pyroxene structure approximates a cubic closest packing (CCP) as reported by *Peacor* [1968]. In the structure composed of CCP oxygen, the ideal values of a/b and c/b are 1.1055 and 0.5774, respectively (a, b, and c denote lattice constants of $MgGeO_3$-high-clinoenstatite). The change of the observed values of a/b and c/b at various pressures (Figure 3) suggests that the oxygen atoms in the structure gradually depart from CCP as the pressure increases.

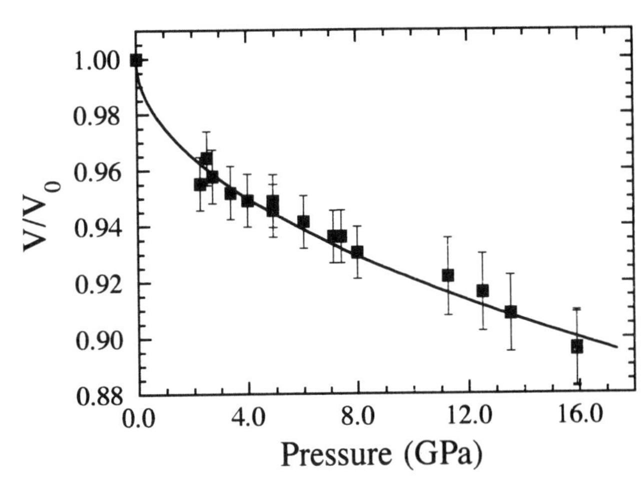

Figure 2. Variation of V/V_0 with increasing pressure on $MgGeO_3$-high-clinoenstatite. A solid curve was fitted by the Birch-Murnaghan equation of state with $K_0' = 4$.

No additional reflections were observed up to 16.4 GPa. However, diffraction profiles start broadening gradually above 10 GPa. The broadening of diffraction profiles obscures the peak separation, and some diffraction peaks are diminished and disappear with increasing pressure. It is a possible interpretation that the effect of non-hydrostatic pressure causes the profile broadening. However, the deviation from the hydrostatic condition seems to be smaller at around 10 GPa, even if the pressure medium consisting of the 4:1 mixture of methanol and ethanol is transformed into a glass state at pressure above 11.1 GPa [*Piermarini et al.*, 1973]. Indeed, the pressure distribution in the gasket hole was within 0.5 GPa at 15 GPa. On the

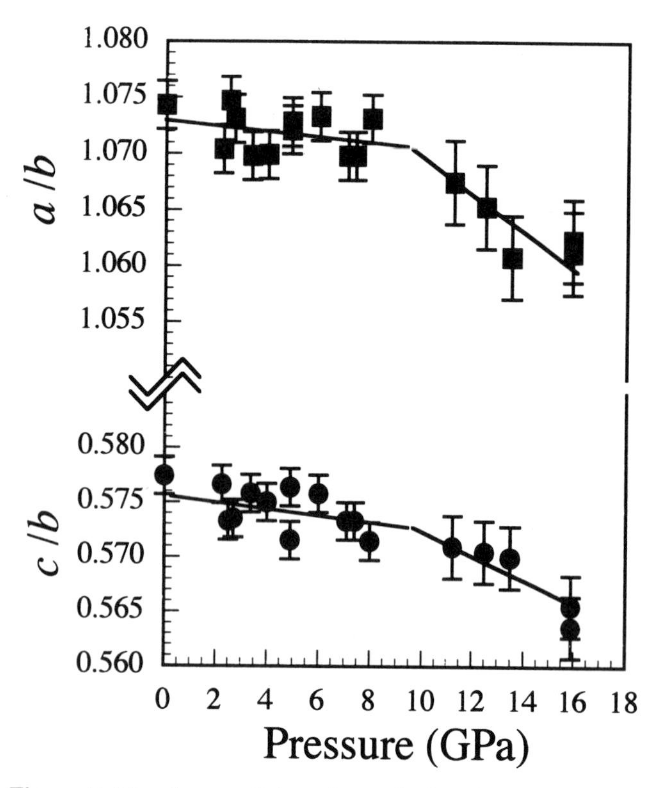

Figure 3. The change of the values of *a/b* and *c/b* ratio with increasing pressure.

other hand, an EXAFS study [*Andrault et al.*, 1992] suggested that the coordination change from fourfold to sixfold starts at about 8.5 GPa. It is another interpretation of the profile broadening that some structural modification starts gradually at around 10 GPa.

The drastic change of diffraction patterns occurs at around about 23 GPa. One new peak, $d = 4.96$ Å, appears and its profile becomes clearer with increasing pressure. The reflections from the high-clinoenstatite form decreased their intensities. At 35 GPa, only the new peak and a peak with $d = 2.38$ Å could be observed. Because of the peak broadening, it is unclear whether the $d = 2.38$ Å reflection corresponds to the 131 reflection of $MgGeO_3$-high-clinoenstatite or to a new peak. The $d = 4.96$ Å peak could not be assigned as a known high pressure polymorph of $MgGeO_3$-high-clinoenstatite (i.e., ilmenite, $LiNbO_3$ and perovskite form) and a ruby pressure maker. The diffraction patterns of a recovered sample from 35 GPa indicated that the new observed reflection disappears on decompression and reflections from $MgGeO_3$-high-clinoenstatite come out again, although the broadening of each profile takes place. This evidence suggests that the new reflection comes from $MgGeO_3$ and that a kinetically favored transition under room temperature compression takes place at around 23 GPa.

$CaGeO_3$-Wollastonite

Nagai and Yamanaka [in press] observed X ray diffraction patterns up to 27 GPa at room temperature. It was confirmed that $CaGeO_3$-wollastonite was first transformed into the metastable rhodonite-like structure at about 6 GPa and then converted to the perovskite structure at about 15 GPa. Totally amorphous state could not be observed up to 27 GPa. EXAFS studies [*Andrault et al.*, 1992] suggested that the Ge-O distance increases gradually in the pressure between 7 to 12 GPa. That evidence was interpreted by the coordination change of Ge atoms from fourfold to sixfold occurred; however, the result of the previous XRD experiments showed the existence of the rhodonite-like structure in the same pressure range. Because *Andrault et al.* [1992] determined the Ge-O distance at only three pressure points in the pressure range from 4 to 12 GPa, it was impossible to detect the detailed behavior of the Ge-O distance in the pressure range.

EXAFS studies were reexamined using the MAX90 system with synchrotron radiation X ray source. Figure 4 shows the EXAFS spectra at various pressures from room pressure to 12.4 GPa. There are some determinable parameters (e.g., the average distance of the nearest neighbor, the Debye-Waller factor, the coordination number, the mean free path of the electron) in the EXAFS analysis. And these parameters have strong correlation to each other. In this study, we focused mainly on

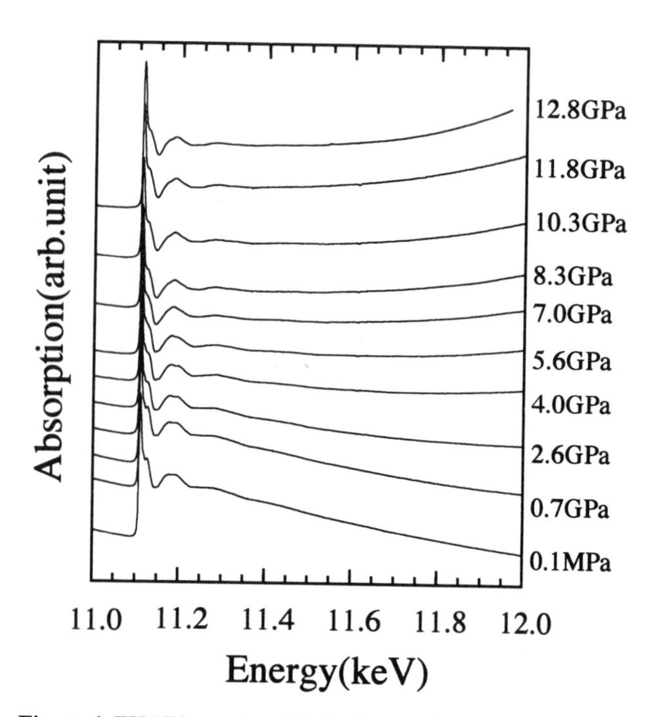

Figure 4. EXAFS spectra of $CaGeO_3$ at various pressures.

determining the distance from a Ge atom to the first nearest neighbor oxygen atom and the Debye-Waller factor. Therefore, the parameters except for the Ge-O distance and the Debye-Waller factor were estimated at ambient conditions and were fixed in the analysis of all other conditions.

The pressure dependence of the Ge-O distance is shown in Figure 5a. The Ge-O distance is compressed from room pressure to about 4 GPa and then seemingly starts to increase with pressures. Above 12 GPa, the values of the Ge-O distance represent the octahedral environment of Ge atoms. This result supports the appearance of the perovskite phase confirmed by the XRD experiments [*Nagai and Yamanaka*, in press]. *Andrault et al.* [1992] showed that the Ge-O distance gradually increases with pressures up to 15 GPa. However, this study indicates that the increase of the Ge-O distance is not gradual. The Ge-O distance decreases from 8 to 10 GPa and then increase with pressure up to 12 GPa again. If a gradual coordination change occurs from fourfold to sixfold, such a decrease of the Ge-O distance could not be observed. This decrease may support the existence not of an amorphous phase, which was predicted by *Andrault et al.* [1992], but of a crystalline phase, which was identified as a rhodonite-like phase by the XRD experiments [*Nagai and Yamanaka*, in press]. Figure 5b shows that the pressure dependence of Debye-Waller factor occurs in three regions. The first region is between room pressure and 4 GPa, the second region is between 8 GPa and 11 GPa, and the third region is above 12 GPa. This result also supports the wollastonite-to-rhodonite-to-perovskite transition under room temperature compression.

However, if the rhodonite phase appears, the increase of the Ge-O distance from 4 to 8 GPa, which suggests the increase of sixfolded Ge atoms, is a little strange, because Ge atoms in the rhodonite structure occupy only tetrahedral sites. The wollastonite-to-rhodonite transition is characterized in terms of the change of the repeat units of the GeO_4 chain and is not accompanied by the coordination change. Such change in a certain repeat unit of single tetrahedral chains is often founded in some pyroxenoid-to-pyroxenoid transition. *Aikawa* [1979] proposed that a tetrahedron composing atom (Si or Ge) moves from a tetrahedral site to another tetrahedral site through an adjacent vacant octahedral site in the process of a pyroxenoid-to-pyroxenoid transition. This transition mechanism suggests that part of the Ge atoms possibly remain at octahedral sites during the wollastonite-to-rhodonite transition under room temperature compression.

It has been pointed out that EXAFS measurements using DAC have a serious problem concerning Bragg reflections by single crystals of the diamond anvils [*Shimomura et al.*, 1992]. *Andrault et al.* [1992] used the EXAFS oscillations at the 300~400 eV energy region just above the Ge absorption edge to eliminate the diffraction caused by the diamond anvils. It is considered that the narrow energy range affects the accuracy of the determination of some parameters. On the other hand, in this study, the all obtained EXAFS oscillation for about 1000 eV above the absorption edge is free from the noise caused by Bragg reflections from anvils, because the sample was surrounded only by amorphous and polycrystalline

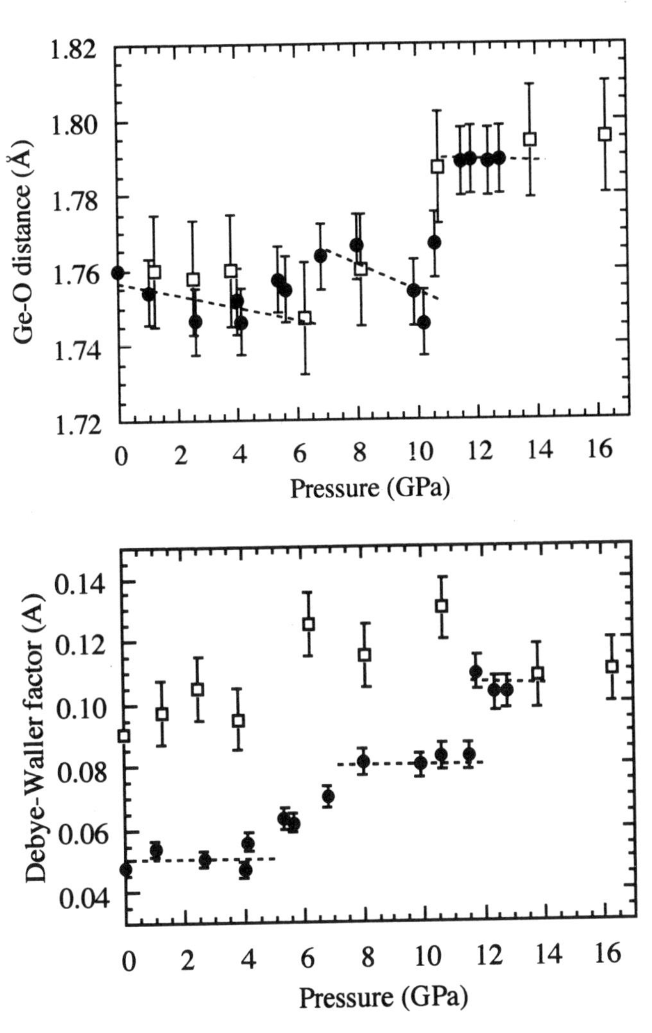

Figure 5. (top) Ge-O distances of $CaGeO_3$ at various pressures. Solid circles represent the data by the MAX90 system (this study) and open squares represent the data by the DAC [*Andrault et al.*, 1992]. The dashed lines are only guides. (bottom) The Debye-Waller factor for the Ge-O distance of $CaGeO_3$ at various pressures. Solid circles represent the data by the MAX90 system (this study) and open squares represent the data by the DAC [*Andrault et al.*, 1992]. The dashed lines are only guides.

materials. These advantages of EXAFS measurements using MAX90 made possible the precise determination of the Ge-O distance. Indeed, as shown in Figures 5a,b, the error bars in the present EXAFS results are much smaller than those in *Andrault et al.* [1992].

DISCUSSION

The phase relation of $MgGeO_3$ and $CaGeO_3$ has been well established in the literature. The ilmenite form, the $LiNbO_3$ form and the perovskite form are high pressure stable polymorphs of $MgGeO_3$ [*Ringwood and Seabrook*, 1962; *Kirfel and Neuhaus*, 1974; *Liu*, 1977; *Ito and Matsui*; 1979; *Ross and Navrotsky*, 1988; *Leinenweber et al.*, 1994]. The garnet form and the perovskite form are thermodynamically favorable polymorph of $CaGeO_3$ under high pressure [*Prewitt and Sleight*, 1969; *Sasaki et al.*, 1983; *Susaki et al.*, 1985; *Ross et al.*, 1986]. On the other hand, it is possible that the newly observed transitions of $MgGeO_3$ and $CaGeO_3$ are metastable and kinetically favorable. In the case of $MgGeO_3$, a newly observed phase above 23 GPa cannot be assigned and the measured reflection could be indexed as neither of the high-pressure thermodynamically stable polymorphs. These metastable transitions can be interpreted through the Ostwald step rule. The wollastonite-to-perovskite transition in $CaGeO_3$ involves a large activation energy barrier. However, when there is a metastable multiple-step reaction route, each step requires a relatively small activation energy. The wollastonite-to-perovskite transition via the rhodonite-like phase is possibly a kinetically favored pathway under room temperature compression.

Other germanates, for example, GeO_2-quartz and Mg_2GeO_4-olivine are transformed into the amorphous state under room temperature compression [*Wolf et al.*, 1992; *Yamanaka et al.*, 1992b; *Nagai et al.*, 1994] and the pressure-induced coordination change was reported at almost the same pressures as the amorphization occurrence [*Itie et al.*, 1989, *Reynard et al.*, 1994]. These evidences suggest that the coordination change becomes an important compression mechanism under room temperature pressurization. It is interesting that the metastable transition of $MgGeO_3$ can be observed without the amorphization, as well as the case of $CaGeO_3$, although the pressure-induced coordination change takes place. It is well known that the crystal structures of many oxides can be understood on the basis of the oxygen sublattice, because an oxygen atom is larger than the other constituent atoms. When the coordination change starts to occur locally without drastic reconstruction of the oxygen sublattice, the CCP oxygen sublattice is more suitable for accommodating many sixfold coordinated cations than the hexagonal closed packing (HCP) sublattice, because many face-shared octahedral sites are formed in the HCP sublattice and such a structure result in electrostatically unstable, according to the Pauling's third rule. The arrangement of oxygen atoms approximates to CCP in metagermanates. Therefore, when the pressure-induced coordination change of Ge atoms occurs locally in metagermanates, the electrical instability may not be caused seriously.

In terms of the oxygen packing, the quartz form is far from the closed packing; on the other hand, the rutile form, which is the high-pressure phase of the quartz form, is composed of almost closed packed oxygen. The olivine form is designed by HCP of oxygen atoms, and the spinel form, which is a high-pressure phase of the olivine form, has an almost CCP oxygen sublattice. The quartz-to-rutile and the olivine-to-spinel transition need the reconstruction of the oxygen sublattice and the reconstruction may lead to the amorphization. There may be some topological similarity of the oxygen packing between pyroxenoids and perovskite and the structural similarity may inhibit the amorphization, although weather the perovskite form is a goal of the pressure-induced transformation of $MgGeO_3$-high-clinoenstatite is a future problem.

Acknowledgments. We are grateful to T. Shimazu of Osaka University for his help at Photon Factory. EXAFS studies were performed under the approval the Photon Factory Advisory Committee (proposal no. 95-G333).

REFERENCES

Aikawa, N., Oriented intergrowth of rhodonite and pyroxemangite and their transformation mechanism, *Mineral. J., 9*, 255-269, 1979.

Andrault, D., M. Madon, J. P. Itie, and A. Fontaine, Compression and coordination changes in pyroxenoids: an EXAFS study of $MgGeO_3$ and $CaGeO_3$ wollastonite, *Phys. Chem. Minerals, 18*, 506-513, 1992.

Andrault, D., M. A. Bouhifd, J. P. Itie, and P. Richet, Compression and amorphization of $(Mg, Fe)_2SiO_4$ olivines: an x-ray diffraction study up to 70 GPa, *Phys. Chem. Minerals, 22*, 99-107, 1995

Andrault, D., J. P. Itie, and F. Farges, High-temperature structural study of germanate perovskite and pyroxenoids, *Am. Mineral., 81*, 822-832, 1996.

Guyot, F., and B. Reynard, Pressure-induced structural modifications and amorphization in olivine compounds, *Chem. Geol., 96*, 411-420, 1992.

Hemley, R. J., A. P. Jephcoat, H. K. Mao, L. C. Ming, and M. H. Manghnani, Pressure-induced amorphization of crystalline silica, *Nature, 334*, 52-54, 1988.

Iti,e J. P., A. Polian, G. Calas, J. Petiau, A. Fontaine, and H. Tolentino, Pressure-induced coordination changes in crystalline and vitreous GeO_2, *Phys. Rev. Lett., 63*, 398-401, 1989.

Ito, E., and Y. Matsui, High-pressure transformations in silicates, germanates, and titanates with ABO_3 stoichiometry, *Phys. Chem. Minerals, 4*, 265-273, 1979.

Jamieson, J. C., Crystal structure at high pressure of metallic modification of silicon and germanium, *Science, 139*, 762-764, 1963.

Kirfel, A., and A. Neuhaus, Zustandsverhalfen und elektrische Leit fahigkeit von $MgGeO_3$ bei Drucken bis 65kbar und Temperaturen bis 1300°C mit Folgerungen fur das Druckverhalten von $MgSiO_3$, *Z. Phys. Chem. Neue Folge 91*, 121-152, 1974.

Leinenweber, K., Y. Wang, T. Yagi, and H. Yusa, An unquenchable perovskite phase of $MgGeO_3$ and comparison with $MgSiO_3$ perovskite, *Am. Mineral., 79*, 197-199, 1994.

Liebermann, R. C., Elasticity of pyroxene-garnet and pyroxene-ilmenite phase transformations in germanates, *Phys. Earth Planet. Inter., 8*, 361-374, 1974.

Liu, L., Post-ilmenite phases of silicates and germanates, *Earth Planet. Sci. Lett., 35*, 161-168, 1977.

Meade, C., and R. Jeanloz, Deep-focus earthquakes and recycling of water into the earth's mantle, *Science, 252*, 68-72, 1991.

Milton, D. J., and P. S. De Carli, Maskelynite: formation by explosive shock, *Science, 140*, 670, 1963.

Mishima, O., L. D. Calvert, and E. Whalley , Melting ice' I at 77K and 10kbar: a new method of making amorphous solids, *Nature, 310*, 393-395, 1984.

Nagai, T., K. Yano, M. Dejima, and T. Yamanaka, Pressure-induced amorphization of Mg_2GeO_4-olivine, Mineral J., 17, 151-157, 1994.

Nagai, T., and T. Yamanaka, Structural modifications of $CaGeO_3$- wollastonite under room temperature compression, *Phys. Chem. Minerals*, in press, 1997.

Peacor, D. R., The crystal structure of $CoGeO_3$, *Z. Kristallogr., 138*, 258-273, 1968.

Piermarini, G. J., S. Block, and J. D. Barnett, Hydrostatic limits in liquids and solids to 100 kbar, *J. Appl. Phys., 44*, 5377-5380, 1973.

Piermarini, G. J., S. Block, J. D. Barnett, and R. A. Forman, Calibration of the pressure dependence of the R1 ruby fluorescence line to 195 kbar, *J. Appl. Phys., 46*, 2774-2780, 1975.

Prewitt C. T. and A. W. Sleight, Garnet-like structures of high-pressure cadmium germanate and calcium germanate, *Science, 163*, 386-387, 1969.

Reynard B., P. E. Petit, F. Guyot, and P. Gillet, Pressure-induced structural modifications in Mg_2GeO_4-olivine: a Raman spectroscopic study, *Phys. Chem. Mineral., 20*, 556-562, 1994.

Richard G., and P. Richet, Room-temperature amorphization of fayalite and high-pressure properties of Fe_2SiO_4 liquid, *Geophys. Res. Lett., 17*, 2093-2096, 1990.

Ringwood, A. E., and M. Seabrook, High-pressure transition of $MgGeO_3$ from pyroxene to corundum structure, *J. Geophys. Res., 67*, 1690-1691, 1962.

Ross, N.L., M. Akaogi, A. Navrotsky, J. Susaki, and P. McMillan, Phase transitions among the $CaGeO_3$ polymorphs [wollastonite, garnet, and perovskite structures]: Studies by high-pressure synthesis, high-temperature calorimetry, and vibrational spectroscopy and calculation, *J. Geophys. Res., 91*, 4685-4696, 1986.

Ross, N. L., and A. Navrotsky, Study of the MgGeO3 polymorphs [orthopyroxene, clinopyroxene, and ilmenite structures] by calorimetry, spectroscopy, and phase equilibria, *Am. Mineral., 73*, 1355-1365, 1988.

Sasaki, S., C. T. Prewitt, and R. C. Liebermann, The crystal structure of $CaGeO_3$ perovskite and the crystal chemistry of the $GdFeO_3$-type perovskites, *Am. Mineral., 68*, 1189-1198, 1983.

Shimomura, O., and T. Kawamura, EXAFS and XANES study under pressure, In High-Pressure Research in Mineral Physics, edited by Manghnani M.H. and Y. Syono, 187-194, TERRAPUB, Tokyo, 1987.

Shimomura, O., W. Utsumi, T. Taniguchi, T. Kikegawa, and T. Nagashima, A new high pressure and temperature apparatus with the sintered diamond anvils for synchrotron radiation use, in *High-Pressure Research: Application to Earth and Planet Sciences, Geophys. Monogr. Ser.*, vol. 67, edited by Syono Y., and M.H. Manghnani, 503-517, TERRAPUB, Tokyo/AGU, Washington, D.C., 1992.

Susaki J., M. Akaogi, S., Akimoto and O. Shimomura, Garnet-perovskite transformation in $CaGeO_3$: in-situ x-ray measurements using synchrotron radiation, *Geophys. Res. Lett., 12*, 729-732, 1985.

Williams, Q., and R. Jeanloz, Static amorphization of anorthite at 300K and comparison with diaplectic glass, *Nature, 338*, 413-415, 1989.

Williams, Q., E. Knittle, R. Reichlin, S. Martin, and R. Jeanloz, Structural and electronic properties of Fe_2SiO_4-fayalite at ultrahigh pressures: Amorphization and Gap Closure, *J. Geophys. Res., 95*, 21,549-21,563, 1990.

Wolf, G. H., S. Wang, C. A. Herbst, D. J. Durben, W. F. Oliver, Z. C. Kang, and K. Halvorson, Pressure induced collapse of the tetrahedral framework in crystalline and amorphous GeO_2, in *High- Pressure Research: Application to Earth and Planet Sciences, Geophys. Monogr. Ser.*, vol. 67, edited by Syono Y., and M.H. Manghnani, pp. 503-517, TERRAPUB, Tokyo/ AGU, Washington, D.C., 1992.

Yamanaka, T., S. Kawasaki, and T. Shibata, Time-resolved obser- vations of solid reactions and structure transitions using a PSD, an SSD and computer aided measurement and control, In Advances in X-ray Analysis, Vol. 35, edited by Barrett C.S. et al., pp. 415-423, Plenum Press, New York, 1992a.

Yamanaka, T., T. Shibata, S., Kawasaki and S. Kume, Pressure induced amorphization of hexagonal GeO2, in *High- Pressure Research: Application to Earth and Planet Sciences, Geophys. Monogr. Ser.*, vol. 67, edited by Syono Y., and M.H. Manghnani, pp. 493-501, TERRAPUB, Tokyo/AGU, Washington, D.C., 1992b.

Yamaoka, S., O. Fukunaga, O. Shimomura, and H. Nakazawa, Versatile type miniature diamond anvil high-pressure cell, *Rev. Sci. Instrum., 50*, 1163-1164, 1979.

The Stability of Almandine at High Pressures and Temperatures

Pamela G. Conrad

Geophysical Laboratory and Center for High Pressure Research
Carnegie Institution of Washington, Washington, D. C. 20015

We have characterized the subsolidus fate of almandine garnet at high pressure with in situ energy dispersive powder X ray diffraction. Using a new method of laser heating that reduces both axial and radial temperature gradients, we have conducted compression experiments on synthetic almandine in a diamond anvil cell. We confirm previous observations that a perovskite phase does not form from almandine ($Fe_3Al_2Si_3O_{12}$) and demonstrate that almandine decomposes to oxides FeO, SiO_2, and Al_2O_3 rather than transforming to a perovskite-structured phase at lower mantle pressures. This behavior differs from that of pyrope ($Mg_2Al_2Si_2O_{12}$), which transforms first to an ilmenite (or corundum) structure and then to perovskite at lower mantle conditions. We observe no other silicate products, such as a proposed Ca-ferrite structured $FeAl_2O_4$, produced from almandine. These results are in agreement with thermodynamic calculations.

1. INTRODUCTION

Whether or not garnet is stable to lower mantle pressures is important for both constraining the maximum depth of slab subduction and thus the structure and dynamics of not only the transition zone, but the entire mantle [*Irifune and Ringwood*, 1993]. Because garnets are so abundant in the upper mantle and transition zone, it is important to understand the limits of their stability at conditions approaching those of the lower mantle. The results of synthesis experiments in laser-heated diamond anvil cells vary with regard to the *P-T* stability limit of garnet. *O'Neill and Jeanloz* [1994] and *Kraft and Knittle* [1990] have observed Mg-Fe garnet at pressures as high as 50 GPa. *Kesson et al.* [1994, 1995] have not observed Mg-Fe garnet (either majoritic or aluminous) over 30 GPa, nor do they see the garnet persisting with the addition of Ca to the system.

Properties of Earth and Planetary Materials
at High Pressure and Temperature
Geophysical Monograph 101

These differences may result from factors such as chemical composition of starting materials and pressure or temperature gradients within the diamond anvil cell. Behavior of the pressure medium (or the lack of it) and choice of pressure calibrant as well as consideration of the thermal contribution to pressure may also lead to variations in pressure determination. Limitations of analytical techniques such as insufficient X ray brightness or lack of beam spatial resolution in conventional powder X ray diffraction could yield low signal/noise ratios in the diffraction patterns or other ambiguous results leading to differences in interpretation of data.

Liu [1974] observed the transition of pyrope garnet, first to intermediate high pressure structures then ultimately to an $MgSiO_3$ perovskite and corundum. In subsequent studies, *Liu* [1975] noted that the high pressure and temperature behaviors of pyrope and almandine differ, with the ultimate fate of almandine less certain than that of pyrope. He presented strong X ray diffraction evidence, however, for a disproportionation of almandine to a non-stoichiometric magnesiowüstite plus "unknown phase(s)."

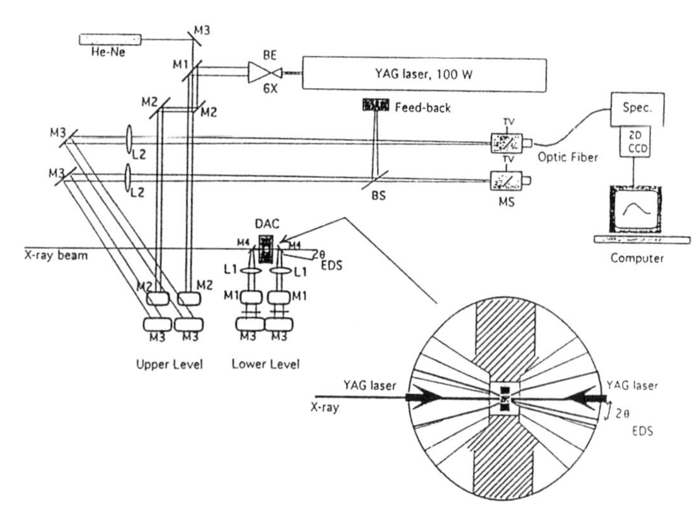

Figure 1. The double-sided multimode Nd:YAG laser layout illustrating path of the laser beams, thermal emission of sample for feed to the spectrometer for conversion to temperature measurement and X ray path. M=mirror, L=lens, BE= beam expander. The use of two levels of optic components was due to spatial constraints at the beamline. The inset is the detail of diamond cell gasket and sample. For further details, see *Shen et al.* [1996] and *Mao and Hemley* [1996].

Recent results on the post-garnet phase transitions of pyrope suggest that some iron must be present (Fe/Fe+Mg ~ 0.19) to achieve a single-phase perovskite field from pyrope, and then only at pressures exceeding about 33 GPa [*Fujino et al.*, this volume; *Kondo and Yagi*, this volume]. They note that the end-member pyrope transforms to an orthorhombic perovskite plus corundum-ilmenite phases even at pressures as high as 50 GPa.

The ultimate transformation of end-member almandine has been less well-studied. *Kesson et al.* [1995] offered two possible interpretations for the results of their high P-T synthesis experiments on almandine in a laser heated diamond anvil cell: one was the formation of a Ca-ferrite structured $FeAl_2O_4$ plus stishovite; the other, decomposition of almandine to constituent oxides. Both interpretations are thermodynamically permissible, although direct observation of the quench results with powder X ray diffraction was inconclusive.

We have carried out a study of the upper pressure limit of garnet stability with the almandine composition, $Fe_3Al_2Si_3O_{12}$. Refinements in experimental technique have made it possible to begin investigating the effect of chemical composition on the stability field of aluminous garnet. A series of laser heating experiments was conducted in a diamond anvil cell with a new method that significantly reduces temperature gradients within the cell [*Mao et al.*, this volume]. The reacted samples were probed in situ with polychromatic synchrotron radiation using a highly collimated X ray beam focused well within the reacted portion of the sample. Thermodynamic calculations for almandine stability are consistent with the observations to be presented.

2. EXPERIMENTAL METHODS

The starting material used for all experiments is a synthetic almandine prepared in a piston cylinder press at 3.5 GPa and 1273 K. The reactants, $Fe_2O_3 + Al_2O_3 + SiO_2$ oxides, were intimately mixed in stoichiometric proportions and subsequently heated at 1473 K in a reducing atmosphere of $CO + CO_2$. The log fO_2 of the oxide mixture was thus fixed at -10.5 before being welded into a gold capsule with an H_2O flux for 48 hours. Because Fe is readily soluble in Au, additional Fe was present in the capsule to prevent Fe loss from the starting oxide in the case of scavenging by the Au capsule. The X ray powder diffraction of the synthetic almandine gave a lattice parameter of a = 11.528 Å. According to *Woodland and Wood* [1989], this value of the cell is consistent with an almandine that is quite low in Fe^{3+} (about 3%). Mössbauer spectroscopy indicated the ferric iron content to be 3.2%.

The almandine crystals were finely ground (<5μm) and compressed into disks. A stainless steel gasket was pre-indented to about 25 GPa and drilled through, creating a 200-μm sample chamber. The gasket was then mounted on the piston anvil of a "mini" piston-cylinder type diamond anvil cell [*Mao and Hemley*, 1996]. This cell provides symmetrical optical access to the sample. The diamonds were mounted on stainless steel seats. A disk of the synthetic almandine with a diameter of roughly 100 μm was loaded into the gasket together with powdered NaCl that served as both a pressure medium and a thermal insulator against the diamond anvils, which are strong thermal conductors. A few grains of ruby were also loaded as pressure calibrant [*Mao et al.*, 1986]. The loaded sample was heated for at least twelve hours in a drying furnace at 423 K prior to being sealed in the diamond cell and compressed.

Samples were heated at high pressures by a 100 W multimode Nd:YAG laser (l = 1.064 mm). The laser beam can be split into two beams of equal power and intensity profile, each of which can be stabilized and directed through opposing sides of the diamond cell (Figure 1). The advantage of this laser heating configuration is minimization of the temperature gradient resulting from a decrease in heating efficiency with distance along the axis of the laser beam "beneath" the heating spot. The multi-mode beam has a much squarer beam intensity profile than a TEM_{00} beam, which has a Gaussian intensity distribution. Consequently, samples heated with the multimode Nd: YAG in the double-sided configuration have reduced temperature gradients, both axially and radially. Laser power stability is also clearly important if relatively constant temperatures are to be maintained. Our laser is stabilized by the feedback provided by photodiodes collecting thermal emission from the heating spots of both sides of the sample and applying an adjustment to the laser power. The technique is further described in *Shen et al.* [1996].

We probed the sample at simultaneous high pressure and temperature with this apparatus installed in the X ray hutch of beamline X17C of the National Synchrotron Light Source at Brookhaven National Laboratory. Temperatures were measured from both sides of the sample by an imaging spectroradiometric system with which a CCD detector records thermal emission spectra in profile across (radially) the laser heated spot [*Shen et al.*, 1996; *Heinz and Jeanloz*, 1987; *Heinz et al.*, 1991]. The width of this profile is ultimately determined by the size of the entrance slit to the spectrometer and the height of the profile can be defined by the pixel binning of the CCD detector.

The intensity of the thermal emission is proportional to the fourth power of the temperature, therefore to achieve an accurate transform of the emission spectrum to a temperature measurement, the intensity of emission must be distributed evenly across the laser heated spot. Again, the intensity distribution of the multimode laser beam is better suited for this purpose than that of a single-mode laser. The temperature is derived from a fit of the Planck radiation function to all of the thermal radiation wavelengths between 600 and 800 nm with the assumption of constant emissivity with wavelength. A conservative estimate of the error for a temperature measurement of 2000 is ± 200 K.

The X ray diffraction patterns were collected with a solid state Ge detector at a fixed 2θ angle of 11°. Angle calibration was based on the gold powder diffraction pattern; energy calibration was based on a series of known elements. Beam size at X17C was focused to about 5 × 16 μm with Kirkpatrick-Baez mirrors [*Mao and Hemley*, 1996], thus ensuring accurate collection from within the laser-heated spots only, the dimensions of which were about 50 μm. The size of the laser heated spots is larger than that which would result from a focused beam (35 μm); the defocusing was possible because almandine is a good enough absorber to allow sufficient heating without maximum power and we wished to obtain a large heating spot. The diffraction patterns were collected for five minutes per pattern on average, although some patterns were collected for as long as three hours.

TABLE 1. Results of Double-Sided Heating of Almandine

P (GPa)	*T* (K)	Result
10	1700	garnet
10	1950	garnet
10	2100	garnet
10	2350	garnet
12	2000	garnet
16	1773	garnet
17	2000	garnet
22	1773	oxides
28	2000	oxides
45	2000	oxides

Interplanar spacings (*d*) were calculated from the Bragg equation:

$$d(\mathring{A}) = \frac{6.1993}{E(\text{keV})\sin\theta}$$

After acquisition of powder diffraction data, pressure within the diamond anvil cell was measured again from the frequency shift of the ruby R_1 fluorescence line. We also tracked the pressure during the experiments by noting changes in peak positions of the NaCl medium, calculating pressure from the equations of state for both NaCl [*Decker*, 1971] and its higher-pressure, B2 structure [*Heinz and Jeanloz*, 1984].

3. HIGH *P-T* RESULTS

The experiments were conducted at a number of pressures and temperatures. We observed garnet at pressures as high as 46 GPa when heating with a conventional laser heating system (through one side of a DAC with a TEM_{00} Nd:YAG laser), whereas a complete transformation of almandine to oxides was found at a pressure of only 22 GPa with the double-sided/multimode laser heating technique. Results are summarized in Table 1. In seven successive heating events of the same spot, intensity of the oxide peaks grew stronger and garnet peaks diminished. Each time we heated the garnet for two minutes, we collected diffraction data at high temperature. Then diffraction data were collected on the quenched sample for an additional three minutes (Figure 2). Note that garnet diffraction peaks predominate at the same

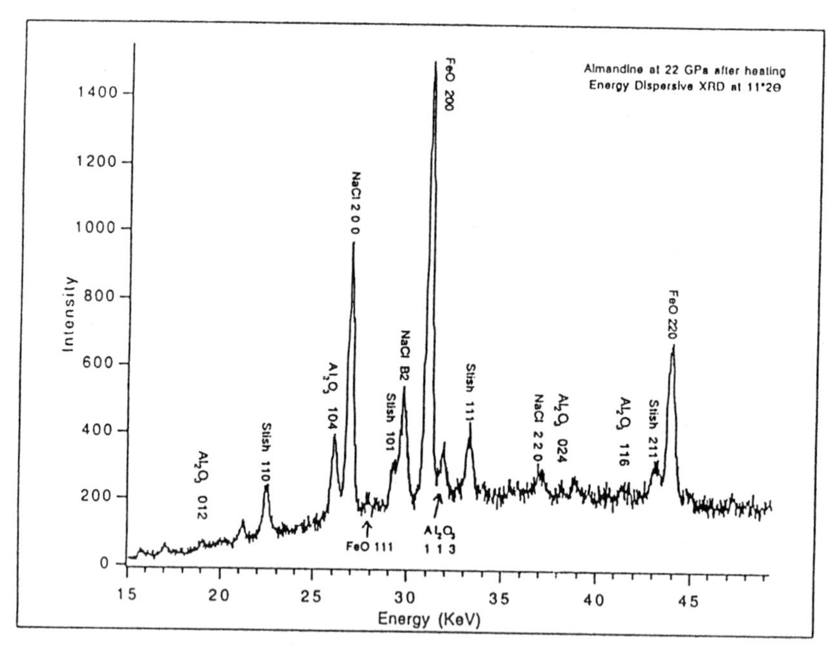

Figure 2. Energy-dispersive X ray powder pattern of sample after heating with the double hot plate Nd:YAG laser for 2 minutes. The 2θ angle is 11°.

experimental conditions (Figure 3) when heated only through one side of the diamond anvil cell. Because we did not observe these peaks either during or after heating with the double-sided method, we interpret the difference as insufficient heating of the starting material beneath the level of penetration by the laser with single-sided heating. We conclude that the double-sided heating technique is instrumental for ensuring complete transformation of the

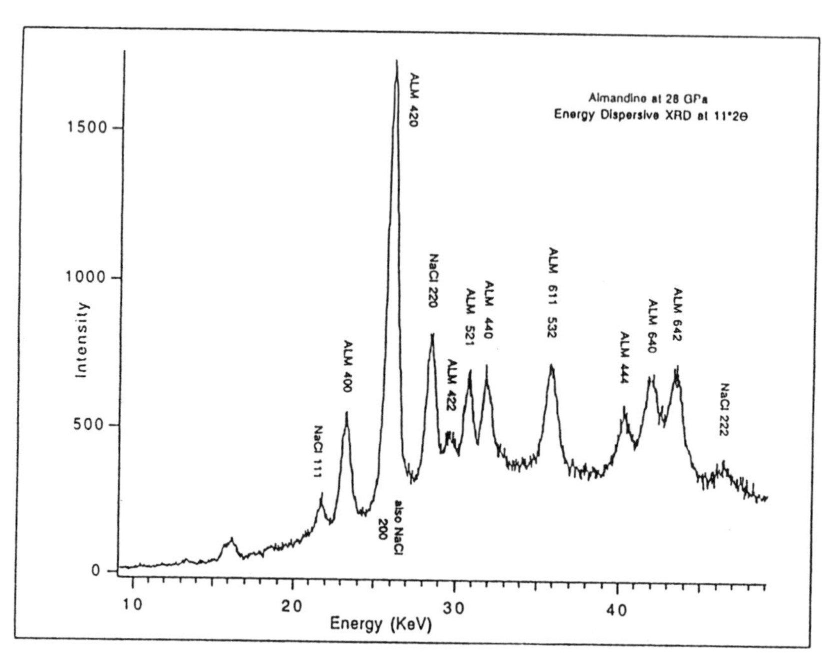

Figure 3. Energy-dispersive X ray powder pattern of sample after heating with a conventional TEM_{00} Nd:YAG laser for 2 minutes. The 2θ angle is 11°.

TABLE 2. Thermophysical Parameters for Almandine and Constituent Oxides: $\alpha = a_0 + a_1T + a_2T^{-2}$

Phase	V_0	K_{0T} (GPa)	K_0'	Refs.	a_0 (10^{-4})	a_1 (10^{-8})	a_2	Refs.
$Fe_3Al_2Si_3O_{12}$	115.28	175	4	[1]	0.1738	1.2443	-0.4524	[5]
FeO	12.250	146	4	[2]	0.1688	0.2040	0.0190	[5]
SiO_2	14.010	313	6	[3]	0.1023	1.3500	-0.0000	[6]
Al_2O_3	25.575	226	4	[4]	0.1871	0.9166	-0.0023	[7]

[1] *Sato et al.*, 1978
[2] *Fei*, 1996
[3] *Ross et al.*, 1990
[4] *d'Amour et al.*, 1978
[5] *Skinner*, 1966
[6] *Ito et al.*, 1974
[7] *Aldebert and Traverse, 1984*

starting material. The almandine will persist metastably at high pressures with insufficient heat to drive the decomposition reaction.

The NaCl (or other pressure medium) also serves as an important thermal insulator. Because of its high thermal conductivity, the diamond must be separated from the sample to achieve sufficiently high sample temperatures during heating. The iron content of almandine is such that it is a good absorber (as already mentioned) at the Nd:YAG wavelength whereas the NaCl is transparent to the laser, so the thermal insulation also helps in the reduction of the axial temperature gradient within the sample that would otherwise result. Because NaCl is transparent to the Nd:YAG laser radiation and much more thermally conductive than the almandine, a sample being heated from both sides maintains its internal temperature more efficiently than its surface temperature.

By using the laser power feedback method already described, we have avoided the melting that accompanies wide temperature fluctuations which may result from heating Fe-rich samples with an unstabilized Nd:YAG laser. With this method, we have obtained strong diffraction patterns from the oxide products of the almandine breakdown reaction, free of metallic iron or other melt products (refer once again to Figures 2 and 3).

4. THERMODYNAMIC CONSIDERATIONS

We now examine the thermodynamics of the observed reaction

$$Fe_3Al_2Si_3O_{12} \longrightarrow 3FeO + 3SiO_2 + Al_2O_3$$

For an isothermal process, the Gibbs free energy is

$$\Delta G = \Delta G^0 + \int_0^P dVdP$$

where ΔG^0 is the free energy of the reaction in the standard state (at 1 bar and 298 K). Measured *P-V* data are available for almandine and constituent oxides. Values for K_0 and K_0' (see Table 2) have been determined from Birch-Murnaghan fits of static compression data for almandine [*Sato et al.*, 1978] and oxides [*Fei*, 1996; *Ross et al.*, 1990; *d'Amour et al.*, 1978] and extrapolated over the range depicted in Figure 4. The integral was calculated over a range of pressures. At 29 GPa, $\Delta G = 0$, which therefore is the equilibrium thermodynamic transition pressure at room temperature. At this pressure, the $V_{tr}/V = 6\%$, whereas the same ratio at 1 bar is 10%. From the equations of state for both almandine and oxides, the transition is clearly density driven.

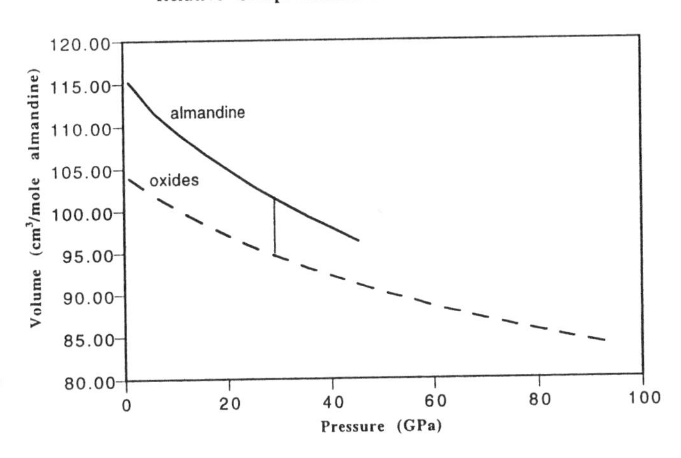

Figure 4. *P-V* relations for almandine and oxides per 12 oxygens. The decrease in volume with pressure is steeper with garnet (solid line) than with the same mole fraction of oxide products; that is, the difference between their respective volumes decreases with pressure, illustrating the density-driven nature of the decomposition reaction.

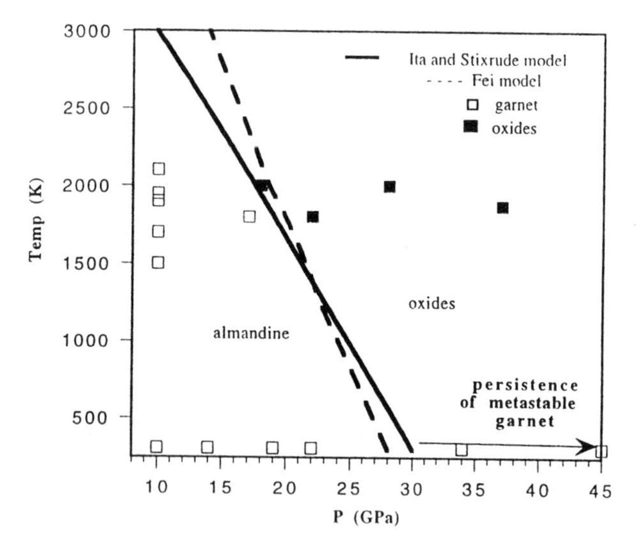

Figure 5. *P-T* curve for the decomposition of almandine to oxides. The solid line is a Debye thermodynamic model after *Ita and Stixrude* [1993], the dashed line a thermodynamic model after *Fei et al.* [1990]. Open squares are data points where sample remained garnet after heating, closed squares are sample that has decomposed to oxides. The arrow parallel to the *x* axis emphasizes the persistence of the unheated garnet to very high pressures.

In consideration of the effects of temperature, the Gibbs free energy of the reaction is given by

$$\Delta G(P,T) = \Delta H_T^0 - T\Delta S_T^0 + \int_0^P \Delta H_T^0(P,T)dP$$

where ΔH_T^0 is the enthalpy of the reaction (at a specific temperature T) and ΔS_T^0 is the entropy at that same T. Molar volume, isothermal bulk modulus, thermal expansivity, heat capacity and entropy can then be given for a specified P and T based upon measured parameters for the thermodynamic potential. The calculated *P-T* boundary for the almandine breakdown reaction is shown in Figure 5. The figure shows two thermodynamic calculations. The first is the model of *Ita and Stixrude* [1992], in which a Debye model is used to account for the thermal contribution to thermodynamic parameters. The thermodynamic data set is listed in *Stixrude and Bukowinski* [1990]. The second is the parameterization of *Fei et al.* [1990] with thermochemistry data listed in Table 3. For each phase $(dK_{T_0}/dT)_P$ = 0.02 GPa/Kelvin is assumed.

Both calculations are in good agreement with direct measurements and predict a negative *P-T* slope. They also predict a room temperature phase boundary at about 30 GPa for the Ita and Stixrude model and 26 GPa for the Fei model. The differences reflect not only different treatments for the effect of T on the bulk modulus, but also different parameterization of the thermodynamic potential and differences in the equations of state.

We did not observe the reaction at room temperature after compression to over 30 GPa, presumably because of kinetic barriers to the reaction as typically observed for silicates at these pressures [*Liu and Bassett*, 1986]. Complete heating is required to transform the entire sample because of the metastable persistence of the almandine at room temperature above its phase boundary at high pressures.

5. CONCLUSIONS

Almandine garnet decomposes to constituent oxides, FeO, SiO_2 and Al_2O_3 at pressures in excess of the mantle transition zone. It does not form a perovskite with garnet stoichiometry as does pyrope. The strongly temperature-dependent decomposition reaction will not proceed at room temperature, presumably because kinetic barriers cannot be

TABLE 3. Thermochemical Parameters for Almandine and Constituent Oxides: $C_p = a + bT + cT^{-2} + dT^{-3} + eT^{-1}$

Phase	ΔH°f	S°f298	*a*	*b*	*c*	*d*	*e*	Refs.
$Fe_3Al_2Si_3O_{12}$	-1758.49 kJ/mol	112.70 J/K·mol	171.640	8.6440 × 10^{-4}	5.2780 × 10^5	-1.1980 × 10^9	-8.6000 × 10^{-3}	[8]
FeO	-267.27 kJ/mol	57.59 J/K·mol	68.435	1.1940 × 10^{-3}	1.6970 × 10^6	1.3480 × 10^8	-1.1880 × 10^4	[9]
SiO_2	-858.818 kJ/mol	34.38 J/K·mol	65.535	9.5711 × 10^{-3}	-5.2702 × 10^6	7.7873 × 10^8	1.0674 × 10^3	[10]
Al_2O_3	-1675.71 kJ/mol	50.92 J/mol	141.810	2.0510 × 10^{-3}	1.4840 × 10^6	-3.4080 × 10^8	-2.0050 × 10^4	[11]

[8] *Chatillon-Colinet et al.*, 1983
[9] *Fei and Saxena*, 1986
[10] *Fei et al.*, 1990
[11] *Robie et al.*, 1978

overcome. Almandine is not stable to pressures greater than the limit at which pyrope has been observed.

The comparison between conventional and double-sided laser heating techniques illustrates the importance of reducing temperature gradients both axially and radially. Because the temperature measurements are an average of a radial cross section of the glowing sample, care must be taken to ensure minimization of radial temperature gradients in the heated sample. Thus, this new heating method not only achieves more complete transformation of the starting materials, it also seems to produce greater accuracy in temperature measurement.

The complete substitution of ferrous iron for magnesium in pyrope garnet does not extend its stability field to lower mantle pressures. A wider range of garnet compositions must now be explored before the issue of garnet stability at lower mantle conditions can be fully understood.

Acknowledgments. This work is supported by the National Science Foundation. I would like to thank H-k Mao and R. J. Hemley for both project guidance and assistance with manuscript preparation. Thanks are also due Yingei Fei and Joel Ita for assistance with thermodynamic models and to Guoyin Shen for assistance with the laser heating. A. Navrotsky and an anonymous reviewer made helpful suggestions for the manuscript.

REFERENCES

Aldebert. P., and J. P. Traverse, αAl_2O_3: A high temperature thermal expansion standard, *High Temperature-High Pressure, 16*, 127-135, 1984.

d'Amour, H., D. Schiferl, W. Denner, H. Schultz, and W. B. Holzapfel, High-pressure single crystal structure determinations of ruby up to 90 kbar using an automatic diffractometer, *J. Applied Phys, 49*, 4411-4416, 1978.

Chatillon-Collinet, C., O. J. Kleppa, R. C. Newton, and D. Perkins, Enthalpy of formation of $Fe_3Al_2Si_3O_{12}$ (almandine) by high temperature alkali borate solution calorimetry, *Geochim. Cosmochim. Acta , 47*, 439-444.

Decker, D. L., High Pressure Equation of state for NaCl, KCl and CsCl, *J. Applied Phys., 42*, 3239-3244, 1971.

Fei, Y., and S. K. Saxena, A thermochemical data base for phase equilibria in the system Fe-Mg-Si-O at high pressure and temperature, *Phys. Chem.. Miner., 13*, 311-324, 1986.

Fei, Y., S. K. Saxena, and A. Navrotsky, Internally Consistent Thermodynamic Data and Equilibrium Phase Relations for Compounds in the System MgO-SiO_2 at High Pressure and High Temperature, *J. Geophys. Res., 95*, 6915-6928, 1990.

Fei, Y., Crystal chemistry of FeO at high pressure and temperature, *Mineral Spectroscopy: A Tribute to Roger G. Burns*, Geochemical Society Special Publication No. 5, edited by M. D. Dyar, C. McCammon, and M. W. Schaefer, Houston, TX, 1996.

Heinz, D. L. and R. Jeanloz, Compression of the B2 high-pressure phase of NaCl, *Phys. Rev. B, 30*, 6045 - 6049, 1984.

Heinz, D. L. and R. Jeanloz, Temperature measurements in the laser-heated diamond cell, *High Pressure Research in Mineral Physics, Geophys. Monogr. Ser.*, vol. 39, edited by M. H. Manghnani and Y. Syono, pp. 113-127, AGU, Washington, 1987.

Heinz, D. L., J. S. Sweeney, and P. Miller, A laser heating system that stabilizes and controls the temperature: diamond anvil cell applications, *Rev. Sci. Instrum., 62*, 1568-1575, 1991.

Irifune, T., and A. E. Ringwood, Phase transformations in subducted oceanic crust and buoyancy relationships at depths of 600-800 km in the mantle, *Earth Planet. Sci. Lett., 117*, 101-110, 1993.

Ita, J. J., and L. Stixrude, Petrology, elasticity and composition of the mantle transition zone, *J. Geophys. Res.*, 97, 6849-6866, 1992.

Ita, J., and L. Stixrude, Density and elasticity of model upper mantle compositions and their implications for whole mantle structure, in *Evolution of the Earth and Planets, Geophys. Monogr. Ser.*, vol. 74, edited by E. Takahashi, R. Jeanloz, and D. C. Rubie, pp. 111-129, AGU, Washington, DC, 1993.

Ito, H., K. Kawada, and S. Akimoto, Thermal expansion of stishovite, *Phys. Earth Planet. Inter., 8*, 277-281, 1974.

Kesson, S. E., Mineral chemistry and density of subducted basaltic crust at lower-mantle pressures, *Nature, 372*, 767- 769, 1994.

Kesson, S. E., J. D. F. Gerald, J. M. G. Shelley, and R. L. Withers, Phase relations, structure and crystal chemistry of some aluminous silicate perovskites, *Earth Planet. Sci. Lett., 134*, 187-201, 1995.

Kraft, S. L., and E. Knittle, Observation of high pressure garnet at pressures in excess of 35 GPa, *EOS Trans. Am. Geophys. Union, 71*, 1666, 1990.

Liu, L.-G., High pressure reconnaissance investigation in the system $Mg_3Al_2Si_3O_{12}$-$Fe_3Al_2Si_3O_{12}$, *Earth Planet. Sci. Lett., 26*, 425-433, 1975.

Liu, L.-G. and W. A. Bassett, *Elements, Oxides Silicates: High Pressure Phases with Implications for the Earth's Interior*, 250 pp., Oxford, New York, 1986.

Mao, H.-k. and R. J. Hemley, Energy dispersive X ray diffraction of micro-crystals at ultrahigh pressures, *High Pressure Research, 14*, 257-267, 1996.

Mao, H. K., J. Xu, and P. M. Bell, Calibration of the ruby pressure gauge to 800 kbar under quasi-hydrostatic conditions, *J. Geophys. Res., 91*, 4673-4676, 1986.

O'Neill, B., and R. Jeanloz, $MgSiO_3$-$FeSiO_3$-Al_2O_3 in the Earth's lower mantle: perovskite and garnet at 1200 km depth, *J. Geophys. Res., 99*, 19,901-19,915, 1994.

Robie, R. A., B. S. Hemingway, and J. R. Fisher, Thermodynamic properties of minerals and related substances at 298.15 K and 1 bar (10^5 pascals) pressure and at higher temperatures, *U. S. Geol. Surv. Bull.*, 1452, 1978.

Ross, N. L., J. F. Shu, R. M. Hazen, and T. Gasparik, High-pressure crystal chemistry of stishovite, *Am. Mineral., 75*, 739-747, 1990.

Sato, Y., M. Akaogi, and S. Akimoto, Hydrostatic compression of the synthetic garnets pyrope and almandine, *J. Geophys. Res., 83*, 335-338, 1978.

Shen, G., H.-k. Mao, and R. J. Hemley, Laser-heated diamond

anvil cell technique: double-sided heating with multimode Nd:YAG laser, *Proceedings of ISAM '96*, Tskuba, Japan, 1996.

Skinner, B. J., Thermal expansion, in *Handbook of Physical Constants,* edited by S. P. Clark, Jr., pp. 75-95, Geological Society of America, Boulder, CO, 1966.

Stixrude, L., and M. S. T. Bukowinski, Fundamental thermodynamic relations and silicate melting with implications for the constitution of D", *J. Geophys. Res., 95*, 19311-19326, 1990.

Woodland, A. B., and B. J. Wood, Electrochemical measurement of the free energy of almandine ($Fe_3Al_2Si_3O_{12}$) garnet, *Geochim. Cosmochim. Acta*, *53*, 2277-2282, 1989.

P. G. Conrad, The Geophysical Laboratory, Carnegie Institution of Washington, 5251 Broad Branch Rd., N.W., Washington, DC, 20015.

A High-Pressure High-Temperature X Ray Study of Phase Relations and Polymorphism of HfO_2

J. Tang,[1] M. Kai, Y. Kobayashi, and S. Endo

Research Center for Materials Science at Extreme Conditions, Osaka University, Toyonaka, Japan

O. Shimomura and T. Kikegawa

National Laboratory for High Energy Physics, Tsukuba, Japan

T. Ashida

Department of Industrial Chemistry, Kinki University, Higashi-hiroshima, Japan

X ray diffraction experiments have been carried out to study the two pressure-induced phase transitions in HfO_2. Angle-dispersive experiments were performed at room temperature in a diamond anvil cell using synchrotron radiation and imaging plates. Experiments at high temperature were performed with a multianvil device using the energy dispersive technique and a rotating-anode generator with a solid state detector. It was found that the monoclinic-orthorhombic I (ortho I) transition occurred around 5 GPa with a volume reduction of 3.4% and the ortho I - ortho II transition occurred at 26 GPa with a volume reduction of 9.4%. The phase boundary between the ortho I and ortho II phases above 400°C is expressed approximately by a linear equation, T(°C)=4720-300P(GPa). The kinetic limit curves gradually towards the higher pressure side at lower temperatures because of the sluggishness of the transition. The ortho II phase was identified as the cotunnite-type structure for both at high pressure and quenched samples.

1. INTRODUCTION

Zirconia, ZrO_2, has been widely studied because of its interesting and valuable properties such as high values of hardness and toughness, which have made the material promising as mechanical parts used under severe conditions. High-pressure experiments have also been conducted to explore further its potentiality as a tough ceramics. On the other hand, high pressure phases have been investigated as a possible candidate for the post-stishovite structure which may exist in the lower mantle of the Earth. At ambient conditions hafnia, HfO_2, has the same crystal structure as ZrO_2, the monoclinic baddeleyite structure with space group $P2_1/c$ at ambient conditions. In this structure each zirconium or hafnium atom is coordinated with seven oxygen atoms. In addition, the sequence of transitions similar to ZrO_2 have been expected for HfO_2 with increasing pressure at room temperature.

After the first discovery by *Bendeliani et al.* [1967] of a phase transition around 6 GPa and 800K, *Bocquillon et al.* [1968] reported a transition in HfO_2 to a supposedly orthorhombic phase in samples quenched from high pressure and high temperature. Later *Ohtaka et al.* [1991] carried out Rietveld refinements of X ray and neutron diffraction data and concluded that the high pressure phase had an orthorhombic structure with the space group *Pbca*,

---

[1]Present address: National Research Institute for Metals, Tsukuba, Ibaraki 305, Japan.

Properties of Earth and Planetary Materials
at High Pressure and Temperature
Geophysical Monograph 101

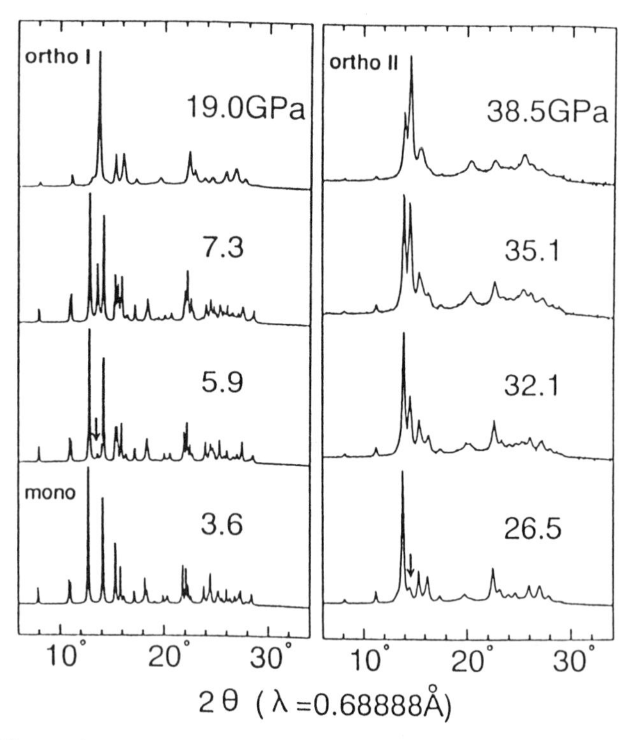

Figure 1. X ray diffraction patterns of HfO_2 obtained with monochramatized synchrotron radiation and an imaging plate under various pressures generated by a diamond anvil cell at room temperature .

which is the same as the orthorhombic I phase of ZrO_2. In this structure, each hafnium or zirconium atom is coordinated with seven oxygen atoms. They also determined the P - T diagram within the region of 3 - 6 GPa and 400 - 1100°C. *Arashi* [1992] determined by Raman spectroscopy that the critical pressure for the monoclinic - orthorhombic I (abbreviated as "ortho I" hereafter) transition was 4.3 GPa at room temperature.

A higher pressure phase was found by *Liu* [1980] in samples quenched from high pressure above 16 GPa and high temperature of about 1000°C generated in a diamond anvil cell (DAC) by laser heating. He determined that it had an orthorhombic cotunnite ($PbCl_2$)-type structure with space group *Pmnb*, in which each hafnium atom is coordinated with nine oxygen atoms.

Thereafter, several in situ experiments under high pressure have been performed at room temperature. *Adams et al.* [1991] reported that the monoclinic phase was still present, together with ortho I, even at 20 GPa. *Leger et al.* [1993] observed the monoclinic - ortho I transition at 10 GPa and the ortho I - orthorhombic II (ortho II) transition at 26 GPa, but they suggested that the high pressure phase above 26 GPa was not the cotunnite-type structure as proposed by *Liu* [1980], although it had an orthorhombic lattice. In addition, *Leger et al.* discovered the existence of another high-pressure phase with a tetragonal structure above 42 GPa. *Jayaraman et al.* [1993] carried out a Raman spectroscopic measurement and reported that the transition pressure for the monoclinic - ortho I transition was 4.3 GPa and that for the ortho I - ortho II transition was 12 GPa. They also suggested that the phase above 12 GPa has a different structure from the cotunnite-type. Very recently *Desgreniers and Larger* [1995] reported that the monoclinic - ortho I transition took place at 10 GPa and the ortho I - ortho II transition at 30 GPa, and proposed the cotunnite structure for the ortho II phase.

Summarizing these previous studies, it appears to us that there remain two unsettled important issues regarding the high pressure transitions in HfO_2: (i) the transition pressures for the monoclinic - ortho I and the ortho I - ortho II transitions at room temperature, and (ii) the crystal structure of the ortho II phase.

As to the first problem, since such a transition is often very sluggish at room temperature, excess pressure higher than the thermodynamical one is needed to induce the transition. In the present study, therefore, we have carried out X ray diffraction at high temperature using a multianvil device to determine the equilibrium phase boundary between the ortho I and the ortho II phases.

As to the second problem, we set out to get high quality X ray diffraction patterns for the ortho II phase at pressures as high as possible at room temperature and from samples quenched from high pressure and high temperature in the stable region of the ortho II phase.

2. EXPERIMENTAL

HfO_2 samples used in the present study were fine powders with grain size of ~100 nm and their purity was 98% with 2% of residual ZrO_2. They were single phase material with the monoclinic baddeleyite-type structure. Two experimental methods were used for acquiring high pressure diffraction data.

2.1. *X Ray Diffraction by a DAC at Room Temperature*

In the experiments using a DAC at room temperature, diffraction patterns were obtained from an angle-dispersive diffractometer using monochromatized synchrotron radiation (l=0.6888Å) and an imaging plate detector at the National Laboratory for High Energy Physics. The sample was filled with a pressure medium of 4:1 methanol-ethanol mixture in a hole of a gasket. Pressure calibration was made by the ruby fluorescence method.

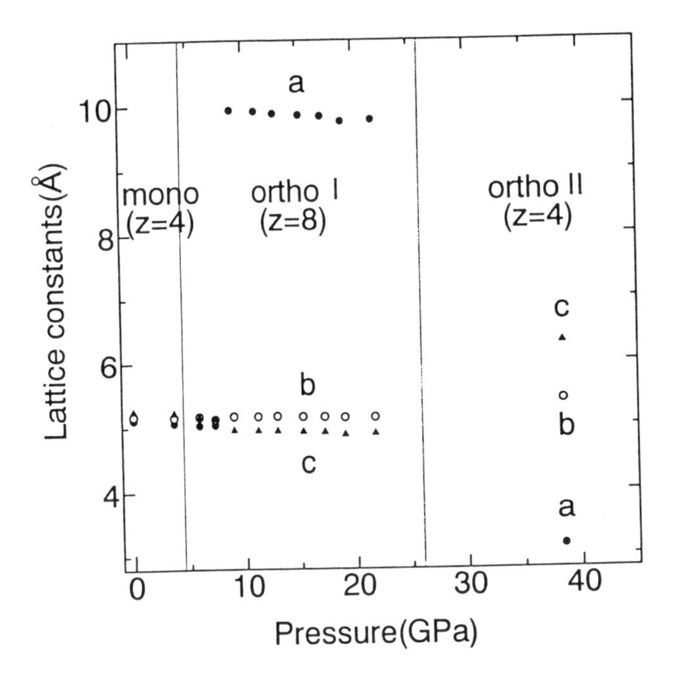

Figure 2. Lattice constants of the monoclinic, ortho I and ortho II phases plotted as a function of pressure at room temperature. The three symbols for the crystal axes are common to the three phases.

2.2. *X Ray Diffraction by a Multianvil at High Temperature*

A 6-8 type multianvil apparatus, driven by a hydraulic press and equipped with an X ray diffractometer, was used for high temperature experiments at Osaka University. Eight cubes made of WC-Co alloy with a front triangular face of 1.5 mm edge length were used as the second-stage anvils. The sample was sandwiched between two graphite disk heaters in a boron-epoxy octahedral pressure cell. The relation between the press load and the pressure generated in the cell at room temperature and the pressure deviation upon heating at various constant press loads were calibrated beforehand by measuring the lattice parameters of NaCl pressure calibrant. In some runs NaCl powders were mixed with HfO_2 to make in situ pressure measurements. Energy-dispersive X ray diffraction was performed using W radiation from a rotating anode type generator (Rigaku RU-300) and a solid state detector (SSD). The experimental details have been described in a previous publication [*Tang and Endo*, 1993].

In addition, careful X ray analyses using a micro-diffractometer were carried out to identify the possible phases present in the sample recovered from the final pressure- temperature condition in each run.

3. RESULTS AND DISCUSSION

3.1. *Transitions at Room Temperature*

Some of the diffraction patterns obtained in the compression process by a DAC are shown in Figure 1. A weak diffraction peak from the ortho I phase, indicated by an arrow, appeared at 5.9 GPa, and more peaks from the ortho I phase gradually emerged at higher pressures. The intensities of the ortho I peaks grew with increasing pressure, and an almost perfect single phase of ortho I was obtained at 19.0 GPa as indicated by the corresponding diffraction pattern at 19.0 GPa. In the pattern obtained at 26.5 GPa, a weak peak corresponding to the ortho II phase, indicated by another arrow, appeared, and an almost single phase of ortho II was obtained at 38.5 GPa, which is the maximum pressure used in the present study to avoid the complication that may arise from the formation of the tetragonal phase above 42 GPa [*Leger et al.*, 1993].

The lattice constants of the monoclinic, ortho I, and ortho II phases, were determined from the diffraction patterns. They are plotted as a function of pressure in Figure 2. The cotunnite-type structure for the ortho II phase has been

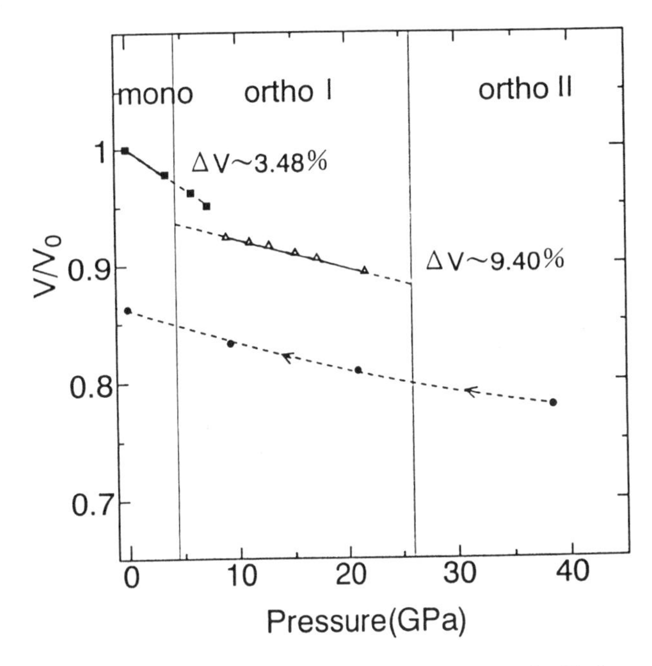

Figure 3. Volume of the monoclinic, ortho I and ortho II phases plotted as a function of pressure at room temperature. The value is normalized to the volume of the monoclinic phase at ambient pressure. The values of the monoclinoc and ortho II phases were obtained in the compression process and that of the ortho II phase was in the decompression process (see Text). Discontinuous volume reduction at the two transitions are indicated.

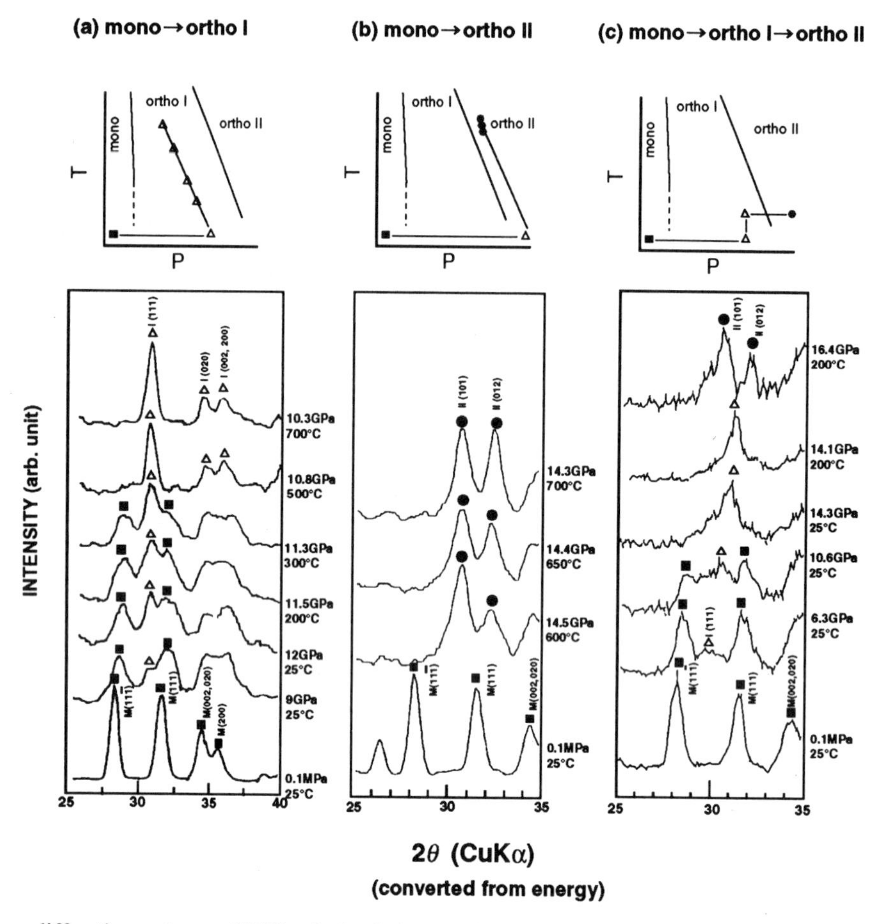

Figure 4. X ray diffraction patterns of HfO_2 obtained along various pressure-temperature paths, which are indicated in the upper panels, by a multianvil equipped with an energy-dispersive diffractometer using a rotating-anode type X ray generator and a solid state detector.

proposed as discussed later. The lattice constants were determined at 38.5 GPa only, because many of the diffraction lines from the ortho II phase overlapped with those from the ortho I phase at lower pressures. The lattice constants of the monoclinic phase agree well with those given by *Leger et al.* [1993] and the lattice constants of the ortho I phase agree well with those given *by Desgreniers and Lagarec* [1995].

The volumes of the three phases at high pressures are plotted in Figure 3, where the values have been normalized to the volume of the monoclinic phase at ambient pressure. The values for the ortho II phase were determined from the X ray patterns obtained in the decompression process. The volume change of the monoclinic - ortho I phase transition was measured as -3.48% and that of the ortho I - ortho II transition as -9.40%.

3.2. *Transitions at High Temperatures*

In order to determine the phase boundary between the ortho I and the ortho II phase, we carried out X ray diffraction experiments along various different paths in the P-T diagram. Three typical examples are shown in Figure 4. In Figure 4(a), the sample, in which the initial monoclinic phase was mixed with the ortho I phase at 12 GPa and room temperature, was heated under a constant press load. With increasing temperature, which resulted in a decrease in pressure, the amount of the monoclinic phase decreased until 700°C and 10.3 GPa, where a single phase of ortho I was obtained. However, in the case of a similar experiment at a little higher pressure around 600-700° and 14.5 GPa, only the ortho II phase was observed as shown in Figure 4(b). A direct transition from the ortho I to the ortho II phase was also observed by further compression of the ortho I phase from 14.1 to 16.4 GPa at a constant temperature of 200°C, as demonstrated by Figure 4(c).

3.3. *P - T Phase Diagram*

The results of X ray diffraction experiments in the present study are plotted in Figure 5, where the three phases are labelled by different symbols. The boundary between the monoclinic and the ortho I phases determined *by Ohtaka et al.* [1991] has also been included. The bars with letters at room temperature give the pressures reported by previous investigators for the monoclinic - ortho I and the ortho I - ortho II transitions. Our results for the monoclinic - ortho I transition agree well with those from Raman spectroscopy by *Arashi* [1992] and *Jayaraman et al.* [1993].

Our findings for the kinetic ortho I - ortho II transition at room temperature agree well with those given by *Leger et al.* [1993] and are close to those by *Desgreniers and Lagarec* [1995]. *Jayaraman et al.* [1993] suggested that this transition occurs at 12 GPa, but actually it is difficult to find any drastic change at that pressure in their Raman spectra. It seems to us that 28 GPa, the pressure under which they assumed the ortho II - tetragonal transition took place, in

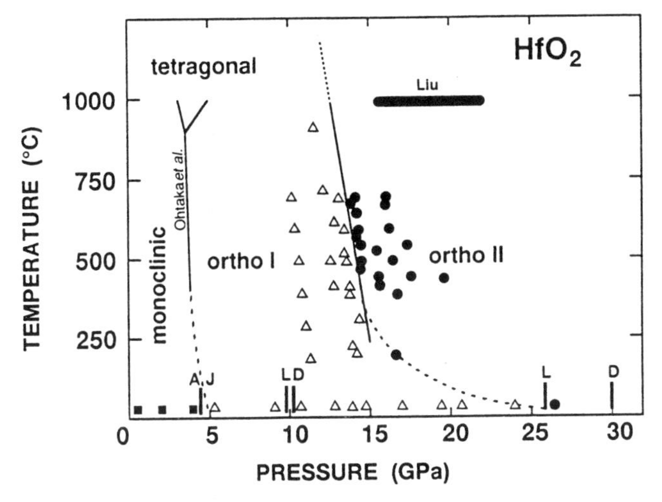

Figure 5. Pressure-temperature diagram of HfO_2 determined from the X ray diffraction measurements under various pressures and temperatures generated by a diamond cell and a multianvil. A closed square, open triangle and closed circle represent the monoclinic, ortho I and ortho II phase, respectively. The phase boundary between the monoclinic and the ortho I phases determined by *Ohtaka et al.* [1991] and the *P-T* region from where the ortho II phase was quenched by Liu are also plotted. The bars with letters at room temperature are the transition pressures for the monoclinic - ortho I and the ortho I - ortho II phase transitions, A: *Arashi* [1992], J: *Jayaraman et al.* [1993], L: *Leger et al.* [1993), and D: *Desgreniers and Lagarec* [1995].

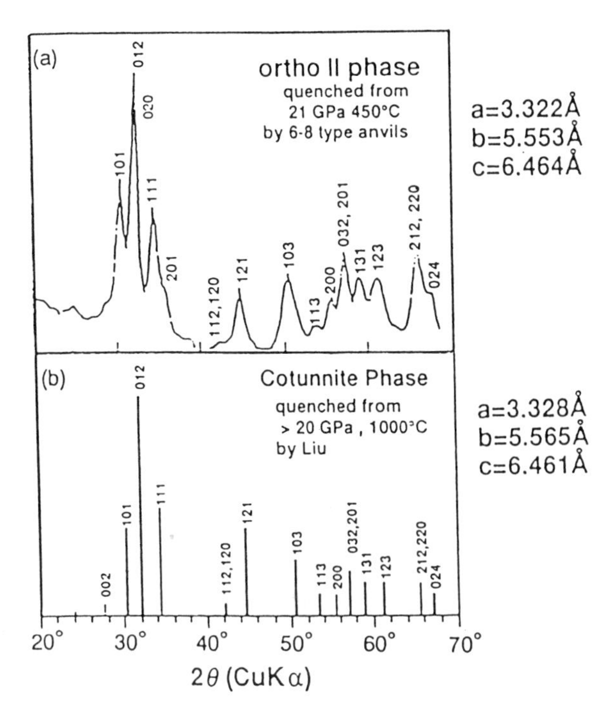

Figure 6. (a) X ray diffraction pattern for the ortho II phase in a sample quenched from 21 GPa and 450°C by a multianvil in the present study. Each peak was indexed as the cotunnite-type structure. (b) The diffraction data obtained by *Liu* from a sample quenched from pressure higher than 20 GPa and 1000°C generated with a diamond cell and laser heating.

effect corresponds to the ortho I - ortho II transition pressure.

At temperatures higher than about 400°C, the boundary between the ortho I and the ortho II phases, which we believe to represent the equilibrium phase boundary, can be expressed by a linear equation of T and P with a negative slope, T(°C)=4720-300P(GPa).

However, the boundary below 400°C, which actually represents the kinetic limit of the ortho I phase, must be curved so as to pass through the point 26 GPa at room temperature. This fact probably indicates that the ortho I - ortho II transition in HfO_2 becomes increasingly sluggish with decreasing temperature. The excess pressure of about 10 GPa is necessary to induce the transition at room temperature, considering the thermodynamical phase boundary is given by the above equation.

3.4. *Crystal Structure of the Ortho II Phase*

The structure of the ortho II phase was first proposed by *Liu* [1980] to be of the cotunnite-type for samples quenched from high pressures and high temperatures. But two contradicting papers were published later; one is a Raman

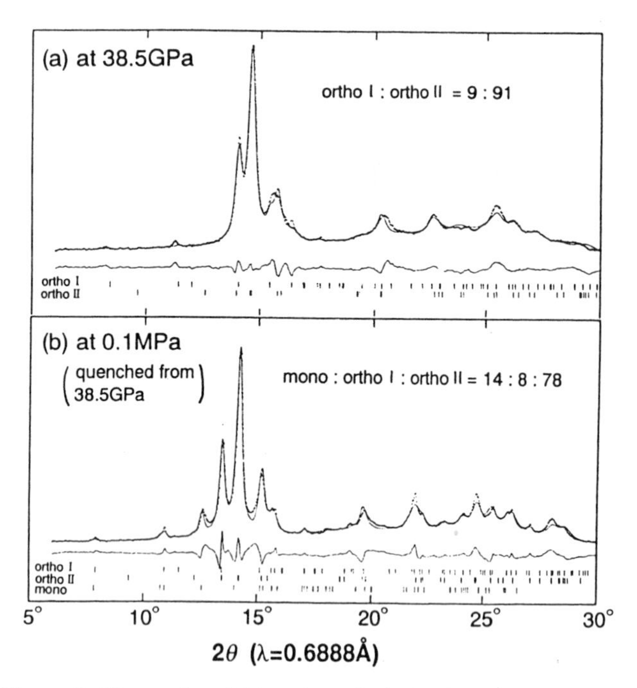

Figure 7. The results of Rietvelt analysis for the angle-dispersive X ray diffraction patterns of (a) a sample of HfO_2 at 38.5 GPa and (b) that quenched from 38.5 GPa at room temperature. Dots indicate the observed data and the curve the calculated pattern using the parameters of Table 1.

study and the other is an X ray diffraction study under high pressure at room temperature.

In the former *Jayaraman et al.* [1993] suggested that the ortho II phase appeared above 12 GPa and a higher pressure tetragonal phase existed above 28 GPa. They presented the ortho II phase as having a different structure from the cotunnite-type. Considering the other previous results and ours, we believe that their "ortho II" phase was not the true ortho II, but was actually the ortho I phase as described above. It is possible that they did not quench the ortho II phase from the pressure region where the ortho I phase was stable.

In the latter, *Leger et al.* [1993] discussed the propriety of the cotunnite structure for the ortho II phase based on an X ray diffraction pattern from a sample quenched from 47.5 GPa. It seems to us that the pattern was too complicated to deduce a reliable structure if, as proposed by them, all four phases including the highest pressure tetragonal phase coexist in it.

An X ray diffraction pattern of a sample quenched from 21 GPa and 450°C in the present experiment is shown in Figure 6(a). Compared with the diffraction data published by *Liu* [1980], shown in Figure 6(b), all peaks have been indexed by a cotunnite-type structure.

The question, however, is whether the ortho II phase has the cotunnite-type structure at high pressure also.

To clarify the situation, two X ray diffraction patterns obtained by a DAC are presented in Figure 7. The pattern at 38 GPa and room temperature is shown in Figure 7(a) and that from the sample quenched from that condition in Figure 7(b). Rietvelt analyses were made for both patterns. Pattern (a), taken at 38 GPa, was indexed as a mixture of the ortho I phase and the cotunnite-type ortho II phase with a mass ratio of 9:91, and pattern (b), quenched from 38 GPa, was found to be a mixture of three phases, namely, the monoclinic, the ortho I and the cotunnite-type ortho II phases with a ratio of 14:8:78. Therefore, the monoclinic phase reappeared in the quenched sample. The lattice constants and the atomic coordinates of oxygen and hafnium in a unit cell of the cotunnite-type structure were obtained from the two patterns and are summarized in Table 1.

Therefore, the cotunnite-type structure has been confirmed for the ortho II phase under high pressure at room temperature and for the quenched phase from high pressure.

4. CONCLUSIONS

Phase transitions in HfO_2 have been studied by X ray diffraction at high pressures at room temperature as well as

TABLE 1. Results of the structural parameters of the cotunnite-type ortho II phase determined by Rietvelt refinement of the X ray diffraction patterns of Figure 7.

	at 38.5 GPa	at 0.1 MPa quenched from 38.5 Gpa
a (Å)	3.175 (3)	3.302 (2)
b (Å)	5.412 (6)	5.569 (2)
c (Å)	6.283 (8)	6.453 (8)
Hf x	0.25	0.25
y	0.247 (7)	0.239 (6)
z	0.117 (3)	0.117 (5)
O1 x	0.25	0.25
y	0.39 (2)	0.38 (4)
z	0.38 (4)	0.37 (7)
O2 x	0.75	0.75
y	0.08 (2)	0.11 (4)
z	0.37 (3)	0.38 (8)
Rwp	4.42	5.45
Rp	3.22	4.07

at high temperatures. The following results have been obtained;

(i) The transition from the monoclinic phase to the ortho I phase at room temperature takes place around 5 GPa and is accompanied by a volume change of -3.4%, and that from the ortho I to the ortho II phase at 26 GPa with a volume change of -9.4%;

(ii) The phase boundary between the ortho I and ortho II phases determined by high-pressure experiments above 400°C can be expressed by a linear equation, T(°C)=4720-300P(GPa). But the actual transition below 400°C occurs at pressures much higher than the pressure extrapolated from the above equation. The difference can be explained by the kinetic effect that the ortho I - ortho II transition becomes increasingly sluggish with decreasing temperature;

(iii) The ortho II phase has the cotunnite-type structure as previously proposed by *Liu* [1980]. The coordination for hafnium atoms changes in sequence of 7-7-9 as the monoclinic - ortho I - ortho II transitions with increasing pressure.

REFERENCES

Adams, D. M., S. Leonard, D. R. Russel, and R. J. Cernik, X-ray diffraction study of hafnia under high pressure using synchrotron radiation, *J. Phys. Chem. Solids, 52,* 1181-1186, 1991.

Arashi, H., Pressure-induced phase transformation of HfO_2, *J. Am. Ceram. Soc.,75,* 844-847, 1992.

Bendeliani, N. A., S. V. Popova, and L. F. Vereshchagin, New high-pressure modifications of ZrO_2 and HfO_2, *Geokhimiya, 6,* 677-683, 1967.

Bocquillon, G., C. Susse, and B. Vodar, Allotropie de l'oxyde d'hafnium sous haute pression, *Rev. Int. Hautes Temp. Re fract., 5,* 247-251, 1968.

Desgreniers, S., and K. Lagarec, Quenched high density ZrO_2 and HfO_2, paper presented at Joint XV AIRAPT & XXXIII EHPRG Int. Conf., Warsaw, Poland, September 11-15, 1995.

Jayaraman, A., S. Y. Wang, and S. K. Sharma, Pressure-induced phase transformations in HfO_2 to 50 GPa studied by Raman spectroscopy, *Phys. Rev. B, 48,* 9205-9211, 1993.

Leger, J. M., A. Atouf, P. E. Tomaszewski, and A. S. Pereira, Pressure-induced phase transitions and volume changes in HfO_2 up to 50 GPa, *Phys. Rev. B., 48,* 93-98, 1993.

Liu, L., New high-pressure phases of ZrO_2 and HfO_2, *J. Phys. Chem. Solids, 41,* 331-334, 1980.

Ohtaka, O., T. Yamanaka, and S. Kume, Synthesis and X-ray structural analysis by the Rietveld method of orthorhombic hafnia, *Nippon Seramikkusu Kyokai Gakujutsu Ronbunshi, 99,* 826-827, 1991 (in English).

Tang, J., and S. Endo, P-T boundary of a-PbO_2 type and baddeleyite type high-pressure phases of titanium dioxide, *Am. Ceram. Soc., 76,* 796-798, 1993.

J. Tang, M. Kai, Y. Kobayashi, and S. Endo, Research Center for Materials Science at Extreme Conditions, Osaka University, Toyonaka, Osaka 560, Japan

O. Shimomura and T. Kikegawa, National Laboratory for High Energy Physics, Tsukuba, Ibaraki 305, Japan

T. Ashida, Department of Industrial Chemistry, Kinki University, Higashi-hiroshima, Hiroshima 739-21, Japan

Analytical Electron Microscopy of the Garnet-Perovskite Transformation in a Laser-Heated Diamond Anvil Cell

Kiyoshi Fujino and Nobuyoshi Miyajima

Division of Earth and Planetary Sciences, Graduate School of Science, Hokkaido University, Sapporo, Japan

Takehiko Yagi, Tadashi Kondo, and Nobumasa Funamori[1]

Institute for Solid State Physics, University of Tokyo, Roppongi, Minato-ku, Tokyo, Japan

Experimental analytical electron microscopy procedures for thin foils recovered from laser-heated diamond anvil cell experiments have been examined. The X ray microanalysis by an analytical electron microscope proved that spreading of the electron beam was necessary for perovskite to prevent the selective removal of elements during the electron beam irradiation. The calculated compositions are improved by using the experimentally determined K values calibrated for the foil's thickness. With this analytical procedure, recovered samples of natural pyrope garnets transformed at 30–50 GPa were examined. Iron-rich (Fe/(Fe+Mg) ~ 0.19) pyrope garnets are transformed into orthorhombic perovskite, an unknown Al-rich phase, and stishovite at ~30 GPa, whereas they are transformed wholly into orthorhombic perovskite at pressures ≥~33 GPa. However, iron-poor pyrope garnets decompose into perovskite, the unknown Al-rich phase, and stishovite even at 50 GPa. This difference may be attributed to the Fe and Al contents in starting material garnets. In an iron-rich sample, alternating lamellae of orthorhombic perovskite and lithium niobate phases with the same composition suggest the metastable transformation of perovskite to lithium niobate phase.

INTRODUCTION

Recent high-pressure and high-temperature experiments have clarified the constituent minerals of the upper mantle. However, those of the lower mantle are still unclear. To explore the phase relations of minerals under the lower mantle conditions, the method of laser-heated diamond anvil cell (DAC) experiments has the advantage, compared to multianvil cell (MAC) experiments, of generating higher pressures that correspond to lower mantle conditions. But the samples in DAC experiments are very small and very fine in grain size, in many cases submicrometer. To analyze those materials, an analytical electron microscope (AEM) is suitable because AEM can provide us with both diffraction patterns and precise compositions, simultaneously, of such fine grain samples. Therefore, a combination of DAC and AEM is a very powerful tool for the study of materials in the lower mantle conditions. However, special care must be taken to obtain the diffraction patterns and precise composition from samples

[1]Now at Department of Geology and Geophysics, University of California, Berkeley, CA 94720-4767.

Properties of Earth and Planetary Materials at High Pressure and Temperature
Geophysical Monograph 101

of the high-pressure phases at the lower mantle conditions. This is because such samples are very weak against an electron beam and they are subject to selective removal of the elements during ion-thinning and electron beam irradiation. In particular, special considerations are required for quantitative chemical analyses of the DAC samples by AEM.

In our AEM study of the DAC samples from the lower mantle conditions, we found some systematic deviations from the real values for compositions obtained by the usual procedure of the quantitative chemical analysis with AEM. The deviation was most significant in perovskite among the high-pressure phases. In an attempt to improve this deviation, we have examined the factors which would affect the quantitative analysis by AEM. Using the improved procedure by AEM, we have analyzed the recovered samples from DAC experiments corresponding to the lower mantle conditions.

Among the lower mantle mineral phases, the post-garnet phases and the host minerals for Al are actively debated. *Irifune et al.* [1996] report from their MAC high-pressure experiments that pyrope decomposes into perovskite and corundum phases at pressures greater than 26.5 GPa at 1500 °C, and the Al content in perovskite increases with pressure and temperature. They estimate that pyrope would transform into a single-phase perovskite at pressures ≥30~40 GPa. However, recent works by *Ahmed-Zaïd and Madon* [1995] and *Kesson et al.* [1995] on the transformations of pyrope garnet under lower mantle conditions give different results from each other, although they are using the same experimental method, a combination of DAC and AEM. *Ahmed-Zaïd and Madon* [1995] report from their experiments at 40 and 50 GPa and 1000–2500 K that aluminous garnet decomposes into perovskite and an Al-rich phase with the composition $(Ca,Mg)Al_2SiO_6$ (at moderate temperature), Al_2SiO_5 (at higher temperature) or $(Ca,Mg,Fe)Al_2Si_2O_8$ (in the case of Fe-rich garnet), and the Al content in perovskite decreases with pressure. On the other hand, *Kesson et al.* [1995] report that perovskite can contain as much as 30 mol % Al_2O_3 at pressures of 55–70 GPa; they further report that these aluminous perovskites are rhombohedral $R\bar{3}c$ and have slightly non-stoichiometric compositions. Meanwhile, *O'Neill and Jeanloz* [1994] report that iron-rich garnet, which coexists with perovskite, can survive at pressures up to 50 GPa, corresponding to ~1200 km depth, in their DAC experiments.

Our project was planned to settle these conflicting problems. In this paper, we present our improved procedure for quantitative chemical analysis of DAC samples by AEM and some preliminary results on the garnet-perovskite transformations in DAC experiments. The full results on the garnet-perovskite transformations will be reported on in a separate paper (N. Miyajima et al., unpublished information).

EXPERIMENTAL METHODS

Samples

Two sets of natural pyrope garnets with different iron contents were used as starting samples for DAC high-pressure experiments to examine the effect of different iron contents of the starting materials. One is pyrope garnet in garnet-lherzolite from Czechoslovakia (denoted as Pyc series); the other is pyrope in ultra-high pressure metamorphic rock from Dora Maira, Western Alps (denoted as Pyd series). In addition, perovskite and corundum phases synthesized by *Irifune et al.* [1996] in multianvil apparatus were also used to compare the effectiveness of the AEM quantitative analysis procedure. The chemical compositions of two natural samples were determined with an electron probe microanalyzer (EPMA) and are given in Table 1 together with the synthetic phases by *Irifune et al.* [1996].

High-Pressure Synthesis by a Laser-Heated Diamond Anvil Cell

Oriented single crystals (denoted as Pycc or Pydc) or powdered materials (denoted as Pycp or Pydp) of natural garnets in Table 1 were used for DAC experiments. Diamonds with culet diameters of 450 μm were used. Stainless steel or rhenium with a hole of ~150 μm in diameter was used as gasket material. Samples were loaded into the hole with ruby chips used to estimate pressures before and after heating. For single crystals, NaCl powder (for YAG laser) or Ar gas (for CO_2 laser) was also loaded, respectively, as the pressure medium. Most samples were heated by a YAG laser. Some samples were platinum-coated to accelerate the efficiency of laser-heating. Heating temperatures were not measured. Table 2 summarizes the experimental conditions.

X Ray Diffraction and Transmission Electron Microscopy

Recovered samples of powdered materials in DAC experiments were examined by X ray Debye-Scherrer or Gandolfi camera. Then, to prepare the thin foils for electron microscopy observation, all the recovered samples were Ar ion-thinned with a low-tension and low-current beam (2.5 keV and 0.3 mA) [*Wang et al.*, 1992], using a cooling stage to reduce the beam heating during thinning. Transmission electron microscopy examinations were

TABLE 1. Chemical Compositions of Natural Starting Materials for DAC Experiments and Synthetic Materials for Chemical Analysis by AEM

wt %	Garnet (Pyc)	Garnet (Pyd)	Perovskite (E841*)	Corundum (E841*)
SiO_2	42.28	43.80	48.20	18.47
TiO_2	0.55	—	—	—
Al_2O_3	22.00	25.10	16.92	70.14
Cr_2O_3	1.04	—	—	—
FeO	8.73	2.94	—	—
MnO	0.22	—	—	—
MgO	20.71	27.04	34.19	12.10
CaO	4.34	0.45	—	—
Total	99.85	99.32	99.31	100.71
No. of oxygens	12	12	3	3
Si	3.011	2.999	0.816	0.310
Ti	0.030	—	—	—
Al	1.847	2.026	0.337	1.386
Cr	0.059	—	—	—
Fe^{2+}	0.520	0.168	—	—
Mn	0.013	—	—	—
Mg	2.198	2.760	0.863	0.303
Ca	0.331	0.033	—	—
Sum	8.008	7.986	2.016	1.998

* Synthesized by *Irifune et al.* [1996].

carried out with a 200 kV JEOL JEM-2010 analytical electron microscope. Product phases in respective runs were identified partly by the X ray diffraction but mostly by the electron diffraction and the quantitative chemical analyses, described later.

Quantitative Chemical Analysis by AEM

Quantitative chemical analyses of the product phases were carried out with a NORAN Instruments/Voyager II energy-dispersive analytical system attached to the JEM-2010 electron microscope. Before obtaining the chemical compositions of the product phases of DAC experiments, the effectiveness of the usually adopted procedure of the quantitative chemical analysis by AEM for present high-pressure phases was examined using the several high-pressure samples with known compositions. The samples used for the examination are starting material garnet in Pycc7, perovskite (by DAC) in Pycc6, and corundum and perovskite (by MAC) in E841 by *Irifune et al.* [1996]. The compositions of respective samples except perovskite in Pycc6 are listed in Table 1. The composition of perovskite in Pycc6 could not be measured with EPMA because of too small grain sizes for EPMA, and it was assumed to be the same as that of the starting material garnet because perovskite was the only product phase in Pycc6. Two major points examined are the selective removal of the elements from the foils during the X ray intensity measurement with a bright electron beam [*Peacor*, 1992; *Kesson et al.*, 1995] and the effectiveness of the Cliff-Lorimer relation usually adopted to calculate the chemical compositions from the measured X ray intensities [*Cliff and Lorimer*, 1975].

We have checked the first point by using different spot sizes of the electron beam for the X ray measurement under the same beam current (20 pA/cm^2 for the smallest spot size on the small fluorescent screen). Figure 1 plots the measured X ray intensity ratios of M/Si (M=Mg, Al, Fe, Ca) versus the spot size of the electron beam for the four samples. The measured intensity ratios for garnet and corundum are nearly constant with the spot size up to 35 nm, while those for both perovskite phases decrease rapidly with decreasing the spot size of the electron beam beyond ~80 nm. This indicates that elements are selectively removed from the perovskite phases during the electron beam irradiation when the electron beam density is higher than some critical value. Therefore, for perovskite

TABLE 2. Experimental Conditions and Identified Phases of Run Products

Run No.	Starting material	Pressure (GPa)	Laser heating	Product phases
Pycp2	powder	30	YAG	perovskite, ual[a], stishovite, Fe metal*
Pycc3	crystal	33	YAG	perovskite, Fe metal*
Pycc5	crystal	35	YAG	perovskite, Fe metal*
Pycc7	crystal	35	CO_2	perovskite, lithium niobate**, Fe metal*
Pycc6	crystal	50	YAG	perovskite, Fe metal*
Pycp1	powder	50	YAG	perovskite, Fe metal*
Pydp1	powder	50	YAG	perovskite, ual[a], stishovite, Fe metal*

[a]Ual is the abbreviation of an unknown Al-rich phase whose structure is yet unknown.
*These are thought to be nonequilibrium phases.
**A lithium niobate phase is thought to have transformed metastably from a perovskite phase during decompression.

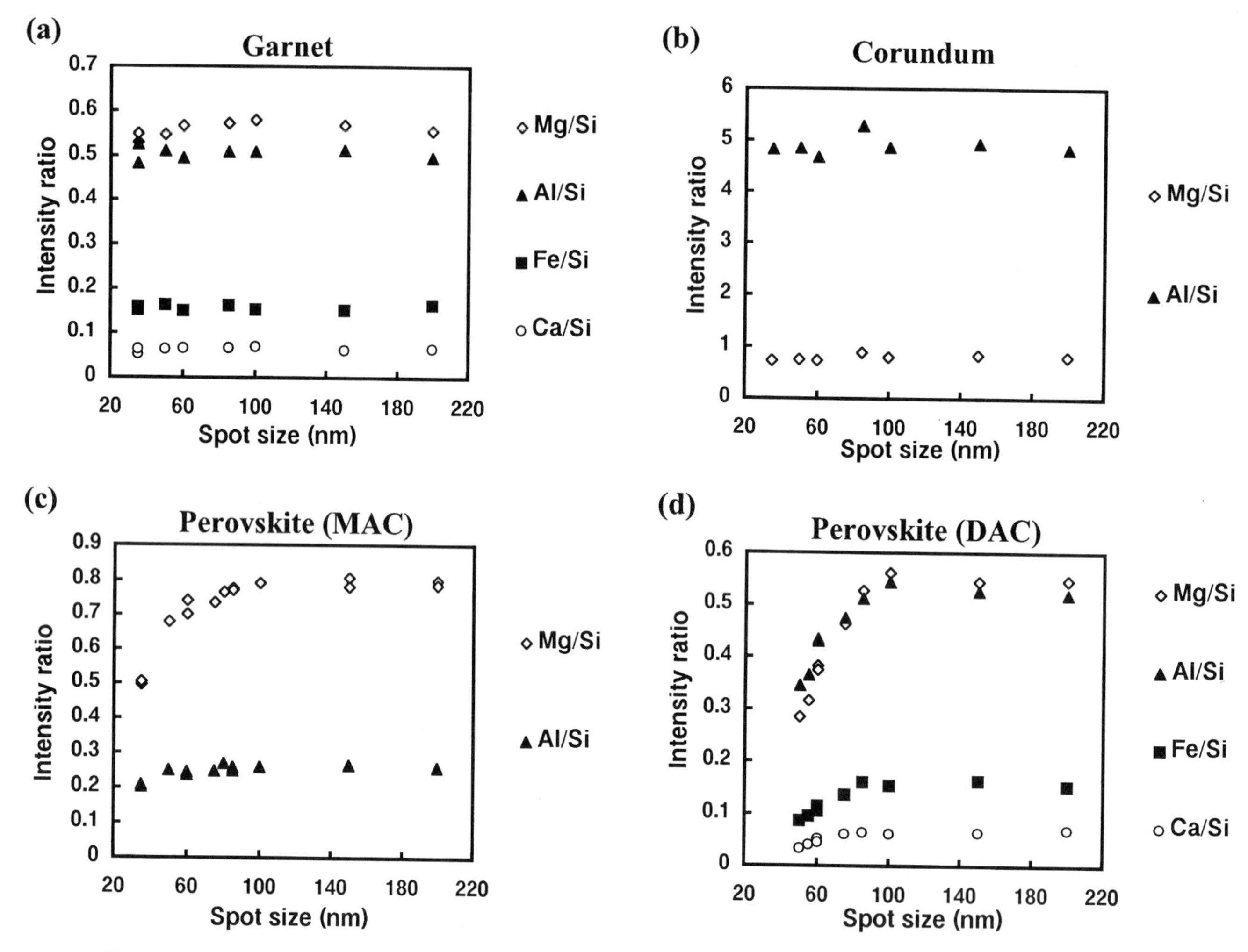

Figure 1. X ray intensity ratios versus different spot size of electron beam during the X ray measurement for high-pressure phases under the same electron beam current, 20 pA/cm^2 for the smallest spot size on the small fluorescent screen. (a) starting material garnet (pycc7), (b) corundum synthesized by MAC (E841 in *Irifune et al.* [1996]), (c) perovskite synthesized by MAC (E841 in *Irifune et al.* [1996]), (d) perovskite synthesized by DAC (pycc6).

phase, X ray intensities were measured with the electron beam of the size 85 nm, while for garnet and corundum phases, the measurement was made with the electron beam of the size 35 nm.

In the next step, we have examined the Cliff-Lorimer relation. In the usually adopted procedure of the quantitative chemical analysis of the thin foils by AEM, the compositions are calculated by the following formula [*Cliff and Lorimer*, 1975]:

$$C_i/C_j = K_{ij}(I_i/I_j) \qquad (1)$$

where C_i, C_j and I_i, I_j are the concentrations and X ray intensities of element i, j, respectively (in the present case, j=Si), and K_{ij} is a proportional coefficient called K value. Theoretically, in the case of foil specimens, X ray absorption and fluorescence can be neglected, and therefore the K values are independent of the foil's thickness. To check this point, the experimentally obtained K values from the measured X ray intensities and the concentrations with the four samples, were plotted against the total counts of all the X ray intensities which nearly correspond to the foil's thickness (Figure 2). In Figure 2, K values of Mg/Si and Al/Si are nearly constant or have small variations with the total X ray intensity for all the mineral phases, while K values of Fe/Si and Ca/Si vary significantly with the total intensity.

To overcome these variations of K values with the total X ray intensity, we adopted the following equation (2)

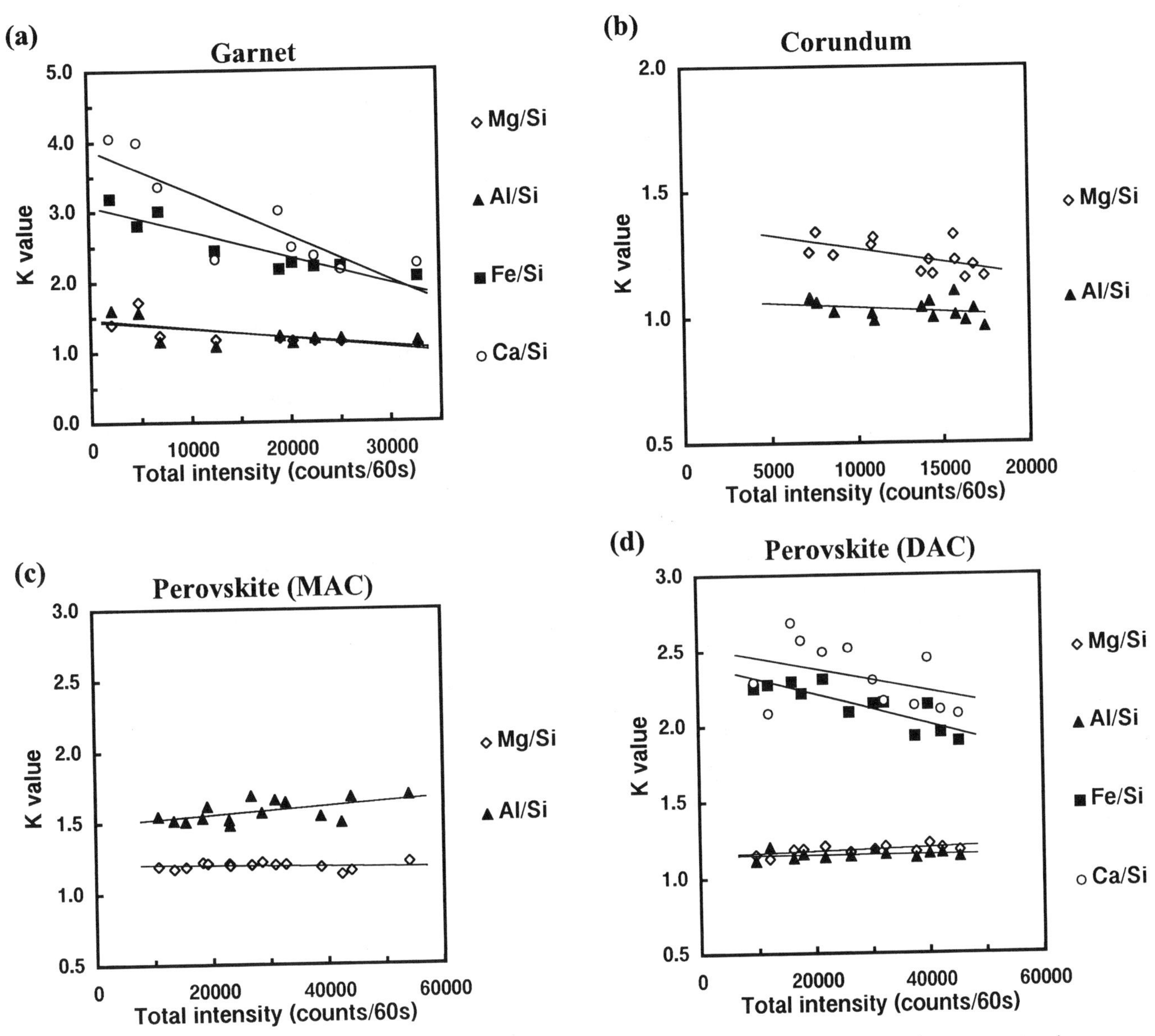

Figure 2. Experimentally determined K values for high-pressure phases under the same elecron beam current as in Figure 1. (a) garnet (pycc7), spot size 35 nm, (b) corundum (E841 in *Irifune et al.* [1996]), spot size 35 nm, (c) perovskite (E841 in *Irifune et al.* [1996]), spot size 85 nm, (d) perovskite (pycc6), spot size 85nm.

where K values vary linearly with the total X ray intensity T which corresponds to the foil's thickness,

$$C_i/C_j = (aT + b)\,(I_i/I_j) \qquad (2)$$

Using this equation, we have calibrated K values for the total X ray intensity T by the least squares fit for respective phases (solid lines in Figure 2). The compositions of the product phases were calculated using these K values calibrated as above and the measured X ray intensities with the proper spot size of the electron beam for respective phases.

RESULTS AND DISCUSSION

Phase Identification

The detailed analyses of the product phases of all the recovered samples in the present experiments are now underway, and the full report will be reported elsewhere

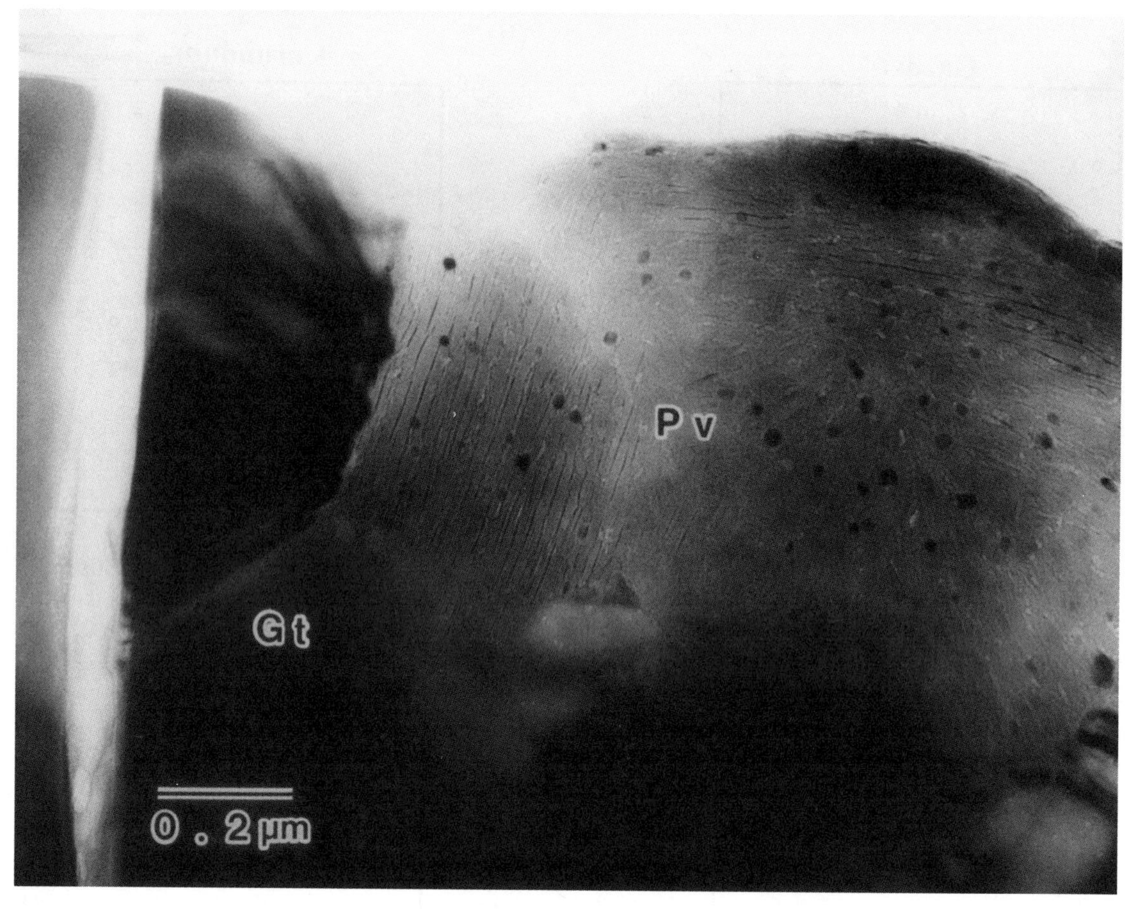

Figure 3. Electron micrograph of the laser-heated area in the recovered sample of pycc7. Lamella-like textures are often observed in the perovskite regeon. Small dark grains are Fe metals. Gt: starting material garnet, Pv: perovskite.

(N. Miyajima et al., unpublished information). The following is a preliminary report of the selected samples. In X ray examinations of the recovered samples of the powdered material, a diffraction pattern of orthorhombic perovskite was always detected. But no diffraction pattern of other product phases was recognized except those of an Al-rich phase with an unknown structure and stishovite in Pycp2.

In TEM examinations, most of the laser-heated areas in all the recovered samples became amorphous rapidly under the electron beam irradiation. This rapid amorphization, coupled with the chemical analysis, was used to identify a perovskite phase. Identified product phases, so far, for respective runs are summarized in Table 2. In the Pycc7 sample heated by a CO_2 laser, an electron diffraction pattern of orthorhombic perovskite could be obtained from the heated area. Also in this sample, lamella-like textures were often observed (Figure 3). These were identified as alternating lamellae of perovskite and lithium niobate phases by the electron diffraction patterns [N. Miyajima et al., unpublished information]. Ual in Table 2 is the abbreviation of an unknown Al-rich phase. This was initially identified as corundum phase, but it was proved that the diffraction pattern of this phase was different from that of corundum or ilmenite phase and its structure is still unknown. For all the samples, very small Fe-rich grains (~10 nm) were found (Figure 3). These grains were identified as Fe metals by the X ray microanalyses which show a strong iron peak but no corresponding oxygen peak. The appearance of similar Fe metals in a laser-heated DAC is also reported by *Kesson et al.* [1995]. At the present time, the reason for the formation of these Fe metals is not clear, and these Fe metals are thought to be a nonequilibrium phase.

Quantitative Chemical Analysis by AEM

As illustrated in Figure 2, K values are not always constant as expected from the Cliff-Lorimer relation, but vary with the specimen's thickness. In addition, some K values deviate significantly from those installed in the computer program MBTF (metallurgical and biological thin foil) where K values for Mg/Si, Al/Si, Fe/Si and Ca/Si are 1.1895, 1.0854, 1.0611 and 0.8957, respectively (for acceleration voltage of 200 kV, take-off angle of 25°, and ultra-thin window). We now think that the reason for the above diversities of K values can be attributed to the compositional change of the surficial part of the foil specimens caused by the ion-thinning illustrated in Figure 4. The surficial part of the foil specimen becomes amorphous by the ion bombardment and will have a chemistry different from the original part because of the selective removal of the elements.

In Figure 2 (c) and (d), K values of Al/Si slightly differ between perovskite (MAC) and perovskite (DAC), although they are both aluminous perovskites. This indicates that the selective removal of elements from the foil during ion-thinning is different depending on the method of sample synthesis, even if the materials have the same perovskite structure.

Table 3 demonstrates how the calculated compositions

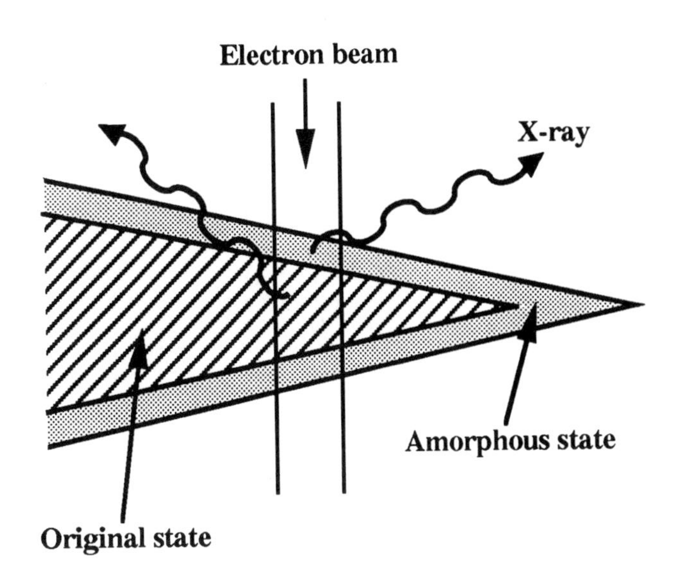

Figure 4. Schematic drawing of the ion-thinned foil specimen for AEM.

TABLE 3. Calculated Chemical Compositions of High-Pressure Phases in Pycp2, E841*, and Pycc7 Using Different K Values

wt %	Garnet (Pycp2)			Corundum (E841*)		Perovskite (Pycc7)		
	(EPMA) (AEM3[d])	(AEM1#)	(AEM2[a])	(AEM1#)	(AEM2[b])	(AEM1#)	(AEM2[c])	
SiO_2	42.28	45.49	42.63	17.65	17.77	45.28	41.46	42.10
TiO_2	0.55	0.74	0.71	—	—	0.57	0.52	0.67
Al_2O_3	22.00	21.67	21.27	71.66	70.32	21.85	28.21	21.52
Cr_2O_3	1.04	1.86	0.73	—	—	1.67	1.53	0.74
FeO	8.73	6.53	10.73	—	—	4.21	3.85	8.43
MnO	0.22	0.46	0.37	—	—	0.32	0.30	0.14
MgO	20.71	21.50	19.79	10.69	11.91	24.48	22.63	22.29
CaO	4.34	1.75	3.78	—	—	1.63	1.49	4.12
Total	99.85	100.00	100.00	100.00	100.00	100.00	100.00	100.00
No. of oxygens	12	12	12	3	3	3	3	3
Si	3.011	3.166	3.050	0.298	0.300	0.781	0.714	0.748
Ti	0.030	0.039	0.038	—	—	0.008	0.007	0.009
Al	1.847	1.778	1.795	1.424	1.400	0.444	0.573	0.450
Cr	0.059	0.102	0.041	—	—	0.023	0.021	0.010
Fe^{2+}	0.520	0.380	0.642	—	—	0.061	0.055	0.125
Mn	0.013	0.027	0.022	—	—	0.005	0.004	0.002
Mg	2.198	2.232	2.113	0.269	0.300	0.629	0.581	0.590
Ca	0.331	0.130	0.290	—	—	0.030	0.027	0.078
Sum	8.008	7.855	7.993	1.991	2.000	1.979	1.982	2.013

*Synthesized by *Irifune et al.* [1996].

#Calculated using K values installed in the computer program MBTF.

[a]Calculated using calibrated K values in Figure 2 (a).

[b]Calculated using calibrated K values in Figure 2 (b).

[c]Calculated using calibrated K values in Figure 2 (c).

[d] Calculated using calibrated K values in Figure 2 (d).

TABLE 4. Chemical Compositions of Product Phases in DAC Samples

Run No.	Pressure (GPa)	Product phases	Cation Numbers					Cation[a] Sum	
			Mg	Fe	Ca	Al	Si		
Pycp2	30	Perovskite	0.86	0.07	0.01	0.16	0.88	2.02	(O=3)
		Ual	0.60	0.01	0.03	0.95	0.43	2.07	(O=3)
		Stishovite[b]	0.00	0.00	0.00	0.11	0.91	1.02	(O=2)
Pycc7	35	Perovskite	0.56	0.13	0.08	0.47	0.74	2.01	(O=3)
Pycc6	50	Perovskite[c]	0.55	0.13	0.08	0.46	0.75	2.00	(O=3)
Pydp1	50	Perovskite	0.62	0.06	0.01	0.48	0.80	1.96	(O=3)
		Ual	0.54	0.00	0.01	0.99	0.49	2.02	(O=3)
		Stishovite[b]	0.04	0.00	0.00	0.15	0.87	1.06	(O=2)
Pyc[d]			0.55	0.13	0.08	0.46	0.75	2.00	(O=3)
Pyd[d]			0.69	0.05	0.01	0.51	0.76	2.00	(O=3)

[a]Including all the cations.
[b]Calculated using K values installed in the computer program MBTF.
[c]The composition of perovskite in Pycc6 was assumed to be the same as that of starting material garnet.
[d]Starting material garnets.

by AEM are improved by using the calibrated K values illustrated in Figure 2 (solid lines). Here, for the composition of perovskite in AEM2, calibrated K values of Mg/Si and Al/Si were taken from Figure 2 (c), but K values of other element pairs were taken from those installed in the computer program because perovskite in Figure 2 (c) contains only Mg, Al and Si.

In Table 3, the chemical composition of garnet in AEM1, calculated using the K values installed in the computer program, deviates from the stoichiometry of the chemical formula of garnet, while that of garnet in AEM2, calculated using the calibrated K values, satisfies the stoichiometry. On the other hand, for corundum the composition in AEM1 gives almost the same chemistry as that in AEM2. For perovskite, all the compositions in AEM1, AEM2, and AEM3 nearly satisfy the stoichiometry of perovskite. However, considering that garnet has wholly transformed into perovskite in this case, FeO in AEM1 is too low compared to that of starting material garnet, even if some iron content may have been removed as Fe metal from garnet. Also, Al_2O_3 in AEM2 is much higher than that of starting material garnet, indicating that K values calibrated with perovskite synthesized by MAC are still not appropriate for perovskite by DAC. Therefore, the compositions of perovskite phases of DAC samples were calculated based on the K values calibrated with perovskite by DAC in Figure 2 (d). Although ual is different from a corundum phase, the compositions of ual were calculated based on the K values calibrated with corundum in Figure 2 (b), because there are no calibrated K values with ual. The obtained compositions of product phases in selected runs in Table 2 are summarized in Table 4.

Al Content and Stability of Orthorhombic Perovskite

In Table 4, it is apparent that in iron-rich aluminous garnet (Pyc series), garnet decomposes into perovskite with the lower Al content, an unknown Al-rich phase, and stishovite at 30 GPa (Pycp2), while garnet wholly transforms into perovskite with nearly the same Al content at higher pressures. On the other hand, in iron-poor garnet (Pyd series), garnet seems to decompose into aluminous perovskite, the unknown Al-rich phase, and stishovite even at 50 GPa. This difference may depend on Fe and Al contents of the starting material garnets.

In Pycc7 where the alternating lamellae of perovskite and lithium niobate phases were often observed, the lamellar width was too narrow and we could not measure the compositions of respective lamellar domains. However, the composition of the mixture of alternating lamellae of both phases was nearly the same as that of orthorhombic perovskite, suggesting that the composition of the lithium niobate phase is the same as that of orthorhombic perovskite. This supports the recent report by *Funamori et al.* [1997] that the lithium niobate phase has metastably transformed from orthorhombic perovskite at high temperature and pressure during decompression, and there is no evidence to indicate that orthorhombic perovskite is not a stable phase in the upper part of the lower mantle. The rhombohedral perovskite which *Kesson et al.* [1995] reported would have been actually misidentified with this lithium niobate phase, and the slightly nonstoichiometric chemistry of their rhombohedral perovskite would be attributed to the improper K values or improper spot size of the electron beam in their quantitative analyses by AEM.

The compositions of two Ual phases in Table 4 are nearly the midpoint of Al_2O_3-$MgSiO_3$ join. However, their diffraction patterns differ from that of corundum or ilmenite phase. The fact that no iron-rich phase was observed, except nonequilibrium Fe metal, indicates that perovskite can accommodate the whole iron content in garnet as revealed in Table 4.

CONCLUSIONS

In quantitative chemical analyses of DAC samples by AEM, calculated compositions were significantly improved by using an appropriate spot size of the electron beam during the measurement of X ray intensities, and by introducing calibrated K values for the specimen's thickness into the calculation.

AEM examinations showed that iron-rich (Fe/(Mg+Fe) = ~ 0.19) aluminous garnet decomposes into perovskite, an unknown Al-rich phase, and stishovite at pressure around 30 GPa, while it wholly transforms into perovskite with the same composition at pressures greater than ~33 GPa, confirming the increase of the Al content in perovskite with pressure [*Irifune et al.*, 1996]. However, iron-poor aluminous garnet may decompose into perovskite, the unknown Al-rich phase, and stishovite even at ~50 GPa. This difference may be attributed to the Fe and Al contents in starting material garnets.

Alternating lamellae of orthorhombic perovskite and lithium niobate phases with the same composition, found in an iron-rich sample, suggest the metastable transformation of orthorhombic perovskite to the lithium niobate phase.

Acknowledgments. We thank the High Brilliance X-Ray Lab. of the Hokkaido University for high power X ray experiments, and T. Irifune and T. Usuki for the samples. This study is supported by the Grant-in-Aid for Research (#06402019) from the Ministry of Education, Science and Culture. N. Funamori and N. Miyajima have been supported by the Research Fellowships of the Japanese Society for the Promotion of Science for Young Scientists.

REFERENCES

Ahmed-Zaïd, I., and M. Madon, Electron microscopy of high-pressure phases synthesized from natural garnets in a diamond anvil cell: Implications for the mineralogy of the lower mantle, *Earth Planet. Sci. Lett.*, *129*, 233-247, 1995.

Cliff, G., and G.W. Lorimer, The quantitative analysis of thin specimens, *J. Microscopy, 103*, 203-207, 1975.

Funamori, N., T. Yagi, N. Miyajima, and K. Fujino, Transformation in garnet: from orthorhombic perovskite to $LiNbO_3$ phase on release of pressure, *Science, 275*, 513-515, 1997.

Irifune, T., T. Koizumi, and J. Ando, An experimental study of the garnet-perovskite transformation in the system $MgSiO_3$-$Mg_3Al_2Si_3O_{12}$, *Phys. Earth Planet. Inter.*, *96*, 147-157, 1996.

Kesson, S.E., J.D. Fitz Gerald, J.M.G. Shelley, and R.L. Withers, Phase relations, structure and crystal chemistry of some aluminous silicate perovskites, *Earth Planet. Sci. Lett.*, *134*, 187-201, 1995.

O'Neill B., and R. Jeanloz, $MgSiO_3$-$FeSiO_3$-Al_2O_3 in the Earth's lower mantle: Perovskite and garnet at 1200 km depth, *J. Geophys. Res.*, *99*, 19,901-19,915, 1994.

Peacor, D.R., Analytical electron microscopy: X-ray analysis, in *Reviews in Mineralogy*, Vol. 27, *Minerals and Reactions at the Atomic Scale: Transmission Electron Microscopy*, edited by P.R. Buseck, pp. 113-140, Mineralogical Society of America, Washington, D.C., 1992.

Wang, Y., F. Guyot, and R.C. Liebermann, Electron microscopy of $(Mg,Fe)SiO_3$ perovskite: Evidence for structural phase transitions and implications for the lower mantle, *J. Geophys. Res.*, *97*, 12,327-12,347, 1992.

Kiyoshi Fujino, and Nobuyoshi Miyajima, Division of Earth and Planetary Sciences, Graduate School of Science, Hokkaido University, Sapporo 060, Japan.

Takehiko Yagi, Tadashi Kondo, and Nobumasa Funamori*, Institute for Solid State Physics, University of Tokyo, Roppongi, Minato-ku, Tokyo 106, Japan.

Phase Transition of Pyrope Garnet under Lower Mantle Conditions

Tadashi Kondo and Takehiko Yagi

Institute for Solid State Physics, University of Tokyo, Tokyo 106, Japan

The high-pressure phase of pyrope ($Mg_3Al_2Si_3O_{12}$) was investigated using a laser heated diamond anvil cell in the pressure range of 20–60 GPa. The quenched samples were examined by X ray diffraction technique using synchrotron radiation and scanning electron microscopy with energy dispersive X ray analyzer (SEM-EDX). These analyses revealed that the dissociation of pyrope into a perovskite phase and a corundum structured phase starts at about 25 GPa and these two phases persist up to the highest pressure investigated. The single phase of perovskite isochemical with pyrope was not formed in the experimental conditions. The Al_2O_3 content of the perovskite structure was about 18 mol% at 45 GPa. Our study demonstrates that the post-garnet phase in the pyrope composition is a mixture of an aluminous magnesian perovskite with the orthorhombic symmetry and an aluminous phase with the corundum structure, which is consistent with the result of multianvil experiments reported by Irifune et al. However, the chemical compositions of two phases and their pressure dependencies are different from that of Irifune et al. and the transition pressure to a perovskite isochemical with pyrope is possibly much higher than that estimated by them.

1. INTRODUCTION

It is widely accepted that garnet is one of the major constituents of the upper mantle. Because of the importance of majorite garnet on the basis of the pyrolite mantle model [*Ringwood*, 1975] and wide range of the stability field of garnet in the transition zone, especially in its MORB composition in the subducting slab [*Irifune and Ringwood*, 1993], the precise phase relation of the system $MgSiO_3$-Al_2O_3 has been studied under upper mantle conditions by many authors [e.g., *Akaogi and Akimoto*, 1977; *Kanzaki*, 1987]. However, the behavior of this system under lower mantle conditions has not been clarified yet. The pioneering work in the system $MgSiO_3$-Al_2O_3 was made by *Liu* [1974,1977b], who suggested that pyrope transforms first to the ilmenite and then to the perovskite with increasing pressure. Recently, several groups have made experiments on natural or synthetic garnets under lower mantle conditions using the laser-heated diamond anvil cell (DAC). In these studies, the run products were examined either by X ray diffraction or by transmission electron microscopy (TEM) analysis. Nevertheless, the post-garnet phase are still controversial. *O'Neill and Jeanloz* [1994] reported that the garnet phase coexists with the perovskite phase up to 50 GPa, which corresponds to 1200 km depth in the mantle using natural and synthetic samples which contain a significant amount of iron. *Ahmed-Zäid and Madon* [1995] suggested the breakdown of natural pyrope garnet into the perovskite phase and the aluminum-rich phases and concluded that the solubility of Al_2O_3 into the perovskite structure

Properties of Earth and Planetary Materials
at High Pressure and Temperature
Geophysical Monograph 101

decreases with pressure. *Kesson et al.* [1995] synthesized an aluminous perovskite isochemical with the starting garnet from pure pyrope to pyrope$_{25}$-almandine$_{75}$ garnet at about 70 GPa. The perovskite phase belongs to the rhombohedral symmetry and is slightly nonstoichiometric. However, the phase relations at the lower pressures have not been reported. Very recently, Irifune et al. [1996] constructed a phase diagram in the system $MgSiO_3$-$Mg_3Al_2Si_3O_{12}$ up to 28 GPa using a multianvil apparatus. They found that the corundum-structured phase coexists with the aluminous perovskite phase up to 28 GPa. They reported that solubility of Al_2O_3 into the perovskite increases rapidly with pressure, which showed an opposite trend to the result of *Ahmed-Zäid and Madon* [1995]. Based on these results, *Irifune et al.* [1996] predicted that perovskite with pyrope composition would be stabilized at 30–40 GPa. As summarized here, these existing results differ from each other even when the same composition is used for the starting material, and the transformation of garnet in the lower mantle is still controversial. The reasons for these apparent discrepancies are not clear yet, but the phases formed and identified in each experiment seem to be sensitive to many factors such as starting material, experimental conditions, and analytical procedures.

At present, the DAC is the only apparatus to generate the wide range of pressure and temperature conditions corresponding to the lower mantle. The grain sizes of the phases transformed in the laser-heated DAC are usually so small that microanalysis using TEM has been adopted to characterize the recovered sample. However, the chemical analysis and phase identification using TEM has still many problems. In the present study, we have succeeded in growing large grains of the dissociated phases using laser-heated DAC, which made it possible to examine the recovered samples by X ray diffraction and SEM-EDX. As the behavior of garnet under lower mantle conditions seems to be sensitive to the chemical composition, we have selected the pyrope composition as the most basic system, and the purpose of this study is to clarify the high pressure phases of pyrope under the lower mantle conditions.

2. EXPERIMENTAL PROCEDURES

2.1. *Sample Preparation*

The starting material was a synthetic pyrope garnet which was crystallized from $Mg_3Al_2Si_3O_{12}$ glass at 4 GPa and 1100°C using the cubic anvil apparatus driven by a 700 ton hydraulic press. The crystalline sample was ground to about 1-μm powder in an agate mortar and was mixed with 3–4 wt% platinum black as an absorber of laser power. The sample was dried in an oven at 120°C for several days before loading to the DAC. The amount of platinum was much larger than in our usual experiments, because we found a large amount of unreacted starting material when the amount of platinum was smaller. Therefore, the sample was completely opaque even after it was compressed in the DAC. The $Mg_3Al_2Si_3O_{12}$ glass mixed with 1 wt% platinum black was also used as starting material to check the stability of garnet.

2.2. *High Pressure and Temperature Experiment*

High-pressure and -temperature experiments were performed using a laser-heated diamond anvil cell. The culet size was typically 0.45 mm. The diamond anvil with 0.35 mm culet was also used for the experiment above 50 GPa. Stainless steel with an initial thickness of 250 μm was pre-indented to 50 μm and used as a gasket to form the sample chamber. The powdered sample was loaded in a gasket hole of about 200 μm in diameter. A small ruby chip was put on the margin of the sample for pressure measurement. No pressure medium was used to obtain the maximum sample volume. The pressures were both before and after heating. The pressure difference from the center of the sample was measured by a separate run and was about 4 GPa at 40 GPa. The pressure decreased typically 3–4 GPa after heating, but in some cases it increased 1–2 GPa, probably because of the change in pressure distribution. Therefore, the average value before and after heating was adopted as the experimental pressure. The pressure range investigated was 20–60 GPa.

The sample in the DAC was heated by a single-mode Nd:YAG laser using the CW mode. The focused beam on the sample was about 20 μm in diameter. The laser power was monitored to keep a constant value of 10 W. In this study, we have tried to heat the sample as completely as possible in order to achieve phase equilibrium and to grow large grains of the transformed phases. When the laser irradiated one portion of the sample, it shined brightly at the beginning but the emission decreased gradually during heating. The estimated temperature range during heating was 1000–2000°C. The heated area became slightly more transparent, probably because of the grain growth of the high-pressure phases. Then the laser beam was moved to a new position and heated that in the same manner. The total duration time of heating was 40–50 minutes for each sample. The outermost rim of the sample was not heated in order to avoid any reaction or contamination of the sample from the gasket and the ruby.

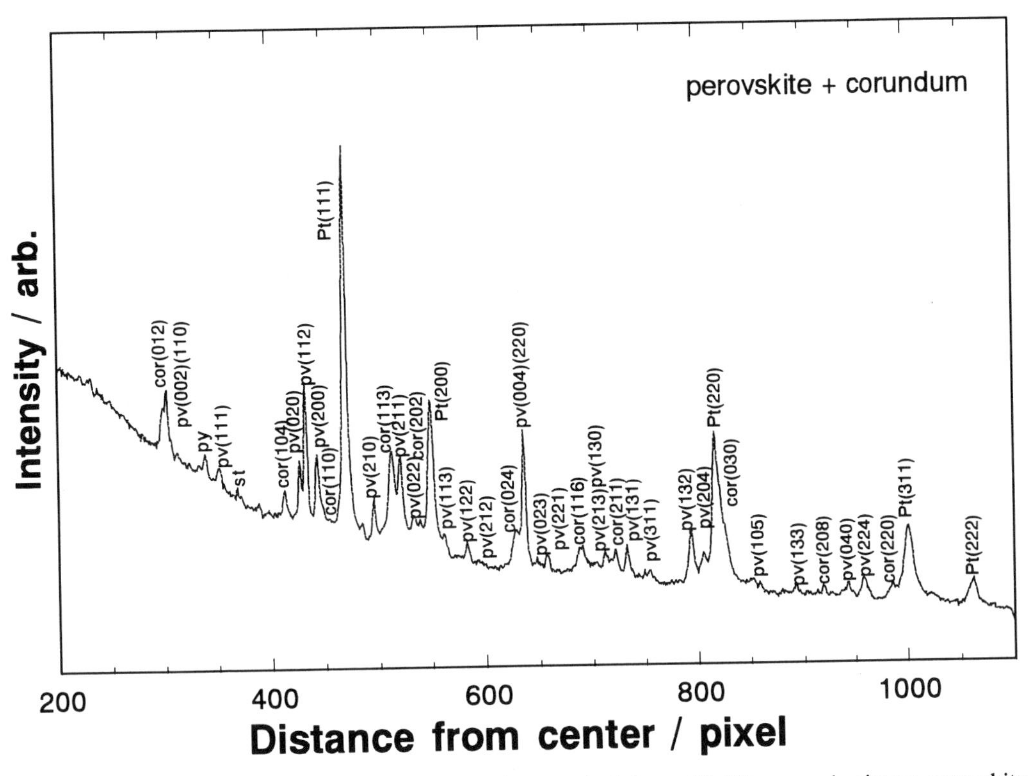

Figure 1. X-ray diffraction profile of recovered sample quenched from 45 GPa; pv, aluminous perovskite; cor, corundum; st, stishovite; 0, starting material of pyrope garnet.

2.3. *Characterization of the Samples*

After carefully removing the unreacted outer portion and the ruby, the surface layer of the sample, which was in contact with diamond and has a thickness of about 10 μm, was polished by diamond compounds. Then the recovered sample of about 25–30 μm in thickness was examined by X ray diffraction technique using a Debye-Scherrer camera combined with the Cr Kα tube X ray source for the preliminary determination of the stable phases. The quenched samples at above 30 GPa were also analyzed using an X ray diffraction system at the Photon Factory (BL-6B), National Laboratory for High Energy Physics, Tsukuba, using an imaging plate (IP) detector to determine the precise lattice parameters and to examine the existence of minor phases. The incident beam was monochromatized to the wavelength of the zirconium absorption edge (λ=0.6888Å) and collimated to 80 μm in diameter. The four diffraction lines from platinum, (111), (200), (311) and (222), were mixed with the sample as an internal standard to determine the sample-to-film distance.

The texture of the recovered sample was examined by a scanning electron microscope (JEOL, JSM-T220). The chemical compositions of grains in the heated region were also determined using an energy dispersive X ray analyzer (SEIKO EG&G, SED8600) attached to the SEM. The electron beam was focused to a few microns to obtain the maximum spatial resolution in analysis. The counting time was 120 seconds at 20 kV acceleration voltage and 0.7 nA beam current. A natural pyrope garnet or a small piece of glass of starting material was used as a standard material for measurement. Analytical error was about 0.5% in consecutive measurements at the same position. The measurement by scanning the beam over 50 μm square of the sample was carried out to determine the bulk composition and to check contamination during the preparation of thin sample.

3. RESULT

3.1. *X Ray Diffraction*

The X ray diffraction pattern of the sample quenched from 45 GPa is shown in Figure 1, which was obtained using IP and the synchrotron radiation. A perovskite-structured phase with the orthorhombic symmetry and a corundum-structured phase are clearly identified in this figure. The latter is not pure Al_2O_3 but we refer to this as

TABLE 1. Experimental Conditions and Identified Phases by X Ray Diffraction

Run No.	Experimental pressure GPa	Phases present* (before and after heating)
Crystalline starting material		
PYCR20	18 (20-16)	Ga + [St]
PYCR25	25 (24-26)	Ga + [Pv + Cor + St]
PYCR30	33 (36-30)	Ga + Pv + Cor + [St]
PYCR45	45 (48-41)	[Ga] + Pv + Cor + [St]
PYCR60	57 (62-52)	Pv+Cor
Glass starting material		
PYGL30	30 (33-27)	Pv+Cor

*Platinum was always observed. Ga, garnet; St, stishovite; Pv, aluminous magnesian perovskite; Cor, corundum. The phases in brackets exist only in trace amount.

"corundum" hereafter. The small amount of stishovite and the starting material of garnet are also observed. The high background is due to air scattering in the experimental hatch of the Photon Factory. In the quenched samples using the glass as a starting material, both the perovskite and corundum phases were also observed whereas stishovite and garnet phases were never observed.

The experimental conditions and identified phases are summarized in Table 1. The results indicate that the dissociation of pyrope garnet into perovskite and corundum phases starts at about 25 GPa, although the resolution of pressure is not so high because of the large pressure gradient in the sample chamber. The starting material of pyrope garnet mostly disappeared at about 30 GPa and the two-phase assemblage (plus stishovite) persists up to at least 57 GPa. Although the X ray diffraction pattern of the sample quenched at 57 GPa can be explained by these two phases, the calculation of precise lattice parameters was difficult because, in this sample, the diffraction from the high pressure phases was very weak relative to platinum. This may be related to the metastable transformation of unquenchable perovskite richer in Al_2O_3 to other structures such as glass or lithium niobate phase [*Funamori et al.*, 1997].

The X ray diffraction data for the sample quenched at 45 GPa are shown in Table 2. The lattice parameters of the perovskite and corundum phases of all the runs are summarized in Table 3 with errors which represent standard deviations in least squares calculation. The accuracy of lattice parameters of the corundum phase was generally lower than that of the perovskite because most peaks of the corundum phases overlapped with the diffraction lines of other phases. Moreover, the remaining starting material makes it difficult to determine the lattice parameters of other phases at 33 GPa. The absolute value of lattice parameters of the perovskite synthesized from the glass may have a large systematic error because the number of available diffraction lines from platinum as the internal standard is only two because of the low intensity of platinum peaks. However, it was clearly confirmed that only two phases, perovskite and corundum, are formed from pyrope composition at 30 GPa.

3.2. *SEM Observation and Chemical Analysis*

Figure 2 (a) shows a back-scattered electron image of the run product quenched from 45 GPa. The bright spots are platinum, while dark areas between grains are epoxy resin. Because the mean atomic numbers of the dissociated phases are close to each other, they cannot be distinguished clearly by the contrasts. The central portion of about 100 μm in diameter where the sample was well heated shows a crystalline texture consisting of relatively large grains. Although these grains seem to be single crystals, the smooth and continuous Debye-rings observed on the IP detector suggest that the true grain size would be sufficiently small. However, as will be discussed later, the result of chemical analysis suggests that each grain is chemically homogeneous and probably is polycrystalline, consisting of either perovskite or corundum. The texture of the heated area was quite different in the glass sample, see Figure 2(b), showing a uniform surface of the heated region without any large grains. The nature of laser heating seems to be quite complicated, as was discussed in detail in another paper [*Yagi et al.*, 1997].

The results of the chemical composition analysis are shown in Figures 3(a)–(d) for the sample heated at various pressures. The ideal composition of pyrope garnet is marked by a cross, and the open circles show the compositions analyzed at various points in the heated

TABLE 2. Observed (d_{obs}) and Calculated (d_{cal}) X ray Diffraction Data for the Perovskite and Corundum Phases Quenched at 45 GPa

h	*k*	*l*	d_{obs} (Å)	d_{cal} (Å)	d_{obs}/d_{cal} -1
Perovskite					
1	1	0	3.4387	3.4409	-0.0006
0	2	0	2.4746	2.4752	-0.0002
1	1	2	2.4449	2.4449	0.0000
2	0	0	2.3920	2.3930	-0.0004
1	2	0	2.1992	2.1986	0.0003
2	1	0	2.1549	2.1545	0.0002
2	1	1	2.0588	2.0579	0.0005
0	2	2	2.0167	2.0160	0.0004
1	1	3	1.9215	1.9215	0.0000
1	2	2	1.8582	1.8579	0.0002
2	1	2	1.8308	1.8311	-0.0001
2	2	0	1.7204	1.7204	-0.0000
0	2	3	1.6909	1.6913	-0.0002
2	2	1	1.6696	1.6700	-0.0003
2	1	3	1.5774	1.5776	-0.0001
1	3	0	1.5591	1.5600	-0.0006
1	3	1	1.5220	1.5221	-0.0001
3	1	1	1.4836	1.4834	0.0001
1	3	2	1.4229	1.4232	-0.0002
2	0	4	1.4065	1.4059	0.0004
1	3	3	1.2939	1.2939	-0.0000
0	4	0	1.2381	1.2376	0.0004
2	2	4	1.2222	1.2225	-0.0002
Corundum					
0	1	2	3.4827	3.4795	0.0009
1	0	4	2.5566	2.5568	-0.0001
1	1	0	2.3813	2.3741	0.0030
1	1	3	2.0903	2.0843	0.0029
0	2	4	1.7370	1.7398	-0.0016
1	1	6	1.6046	1.6043	0.0002
2	1	1	1.5424	1.5433	-0.0006

*The d_{obs} is calculated on the basis of the unit cell parameters of run PYCR45, in Table 3.

region of each sample. The composition of several aluminous silicates reported by *Ahmed-Zäid and Madon* [1991] and *Irifune et al.* [1991] are also shown by the open triangles and squares, respectively, in Figure 3(a). The bulk composition of the run product has deviated slightly toward the SiO_2-rich direction from the pyrope composition, probably because of a contamination from the agate mortar (see later). No elements other than Mg, Al, and Si have been recognized in the bulk analyses. The data points are concentrated in two regions, which suggest that at least two aluminous phases exist in the run product.

Figure 2. SEM photograph (back-scattered electron image) of recovered samples. Texture of polished surface of sample, (a) quenched from 45 GPa using crystalline starting material, and (b) quenched from 30 GPa using glass starting material.

They probably correspond to the aluminous perovskite phase and a corundum phase which were identified by the X ray diffraction. Although many other high-pressure aluminous phases have been reported to be stable under the present experimental conditions, such as $MgAl_2O_4$ with the calcium ferrite structure [*Irifune et al.*, 1991], Al_2SiO_5 with the V_3O_5 structure [*Ahmed-Zäid and Madon*, 1991] and $Mg_3Al_2Si_3O_{12}$ with the rhombohedral perovskite structure [*Fitz-Gerald and Ringwood*, 1991; *Kesson et al.*, 1995], we have never encountered grains with these compositions. The composition of the corundum phase observed in this study seems to be close to that of the unknown structured aluminous phase (open triangles) reported by *Ahmed-Zäid and Madon* [1995], which has the approximate compo-

TABLE 3. Lattice Parameters Calculated from the Observed Diffraction Lines.

Run No.	a Å)	b Å)	c (Å)	V(Å^3)
Perovskite				
PYCR30	4.789(1)	4.952(1)	6.939(1)	164.53(8)
PYCR45	4.786(1)	4.950(1)	6.949(1)	164.65(5)
PYCR60	4.77(1)	4.96(2)	6.97(1)	165.0(7)
PYGL30	4.790(1)	4.961(1)	6.962(3)	165.47(8)
Corundum				
PYCR30	4.735(6)		13.29(2)	258.1(8)
PYCR45	4.748(5)		13.06(3)	255.0(7)
PYCR60	4.755(2)		13.03(8)	255.1(3)
PYGL30	4.768(2)		13.11(1)	258.2(4)

sition $MgAl_2SiO_6$. However, the corundum phase reported by *Irifune et al.* [1996] is much richer in Al_2O_3 (open squares) than that of the present study. On the other hand, the compositions of perovskite phases are similar in all the studies and are slightly deficient in Al_2O_3 compared to the pyrope composition.

In the run product quenched at 45 GPa, the amount of remaining starting material was negligibly small and, accordingly, the measured compositions of perovskite and corundum phases shown in Figure 3(b) would be close to equilibrium. The perovskite phase contains about 18 mol% of Al_2O_3. Although the scatter of the measured composition is larger for the sample quenched at 57 GPa, the composition of perovskite gets closer to the starting composition, suggesting that the solubility of Al_2O_3 into perovskite was enhanced by pressure (Figure 3(c)).

On the other hand, in the glass starting material, all the measured compositions at various points, as well the bulk composition, are identical with the starting composition within the analytical error as shown in Figure 3(d). This may be caused by the very small grain size of the run products and the lack of contamination from an agate mortar, which is consistent with the fact that stishovite was not observed in the X ray chart.

4. DISCUSSION

The reason for the shift of bulk composition toward the SiO_2-rich side in the crystalline sample was first considered as the result of the volatilization of magnesium in the high-pressure phase [*Kesson and Fitz-Gerald*, 1991] or nonstoichiometric nature of the aluminous perovskite. However, stishovite observed in the crystalline sample requires the existence of a Mg-rich phase to satisfy the mass balance, but there was no evidence of a Mg-rich phase in our result. Although the same mineral assemblage as the crystalline sample is observed by X ray diffraction, the absence of stishovite in the glass sample and the agreement of the bulk composition of the high-pressure products with that of the starting composition suggest that the volatilization of magnesium seems unlikely. The crystalline sample has been intensively ground twice or more to get a micron-order grain size, while the glass sample was ground once with platinum. Therefore, this excess silica was possibly due to the contamination resulting from grinding the sample in an agate mortar and resulted in the formation of stishovite.

All the aluminous perovskites observed in the present study were identified to be orthorhombic and not to be the perovskites observed in the present study were identified to be orthorhombic and not to be the perovskite with rhombohedral symmetry as reported by *Kesson et al.* [1995]. The only phase with this symmetry was the corundum which coexisted with the perovskite. Most of the peaks for the the corundum phase can also be explained as the ilmenite structure. Characteristic peaks of ilmenite structure such as (003), (101), and (021) are expected to be fairly weak or they overlap with other peaks so that it is difficult to conclude that this phase really has a corundum structure (disordered) rather than an ilmenite structure (ordered) or intermediate phases. *Liu* [1974] reported the dissociation of pyrope garnet into perovskite and corundum, but he later changed this interpretation [*Liu*, 1977a, b]. Alternatively he suggested the stability field of ilmenite between those of pyrope garnet and perovskite. In the present study, corundum (or corundum-ilmenite solid solution according to the chemical composition) was observed in a wide range of pressure, and the composition of this phase is richer in Al_2O_3 than pyrope, indicating that the corundum found in the present study was not an ilmenite phase reported by *Liu* [1977a] and it exists stably to at least around 57 GPa. The unknown-structured

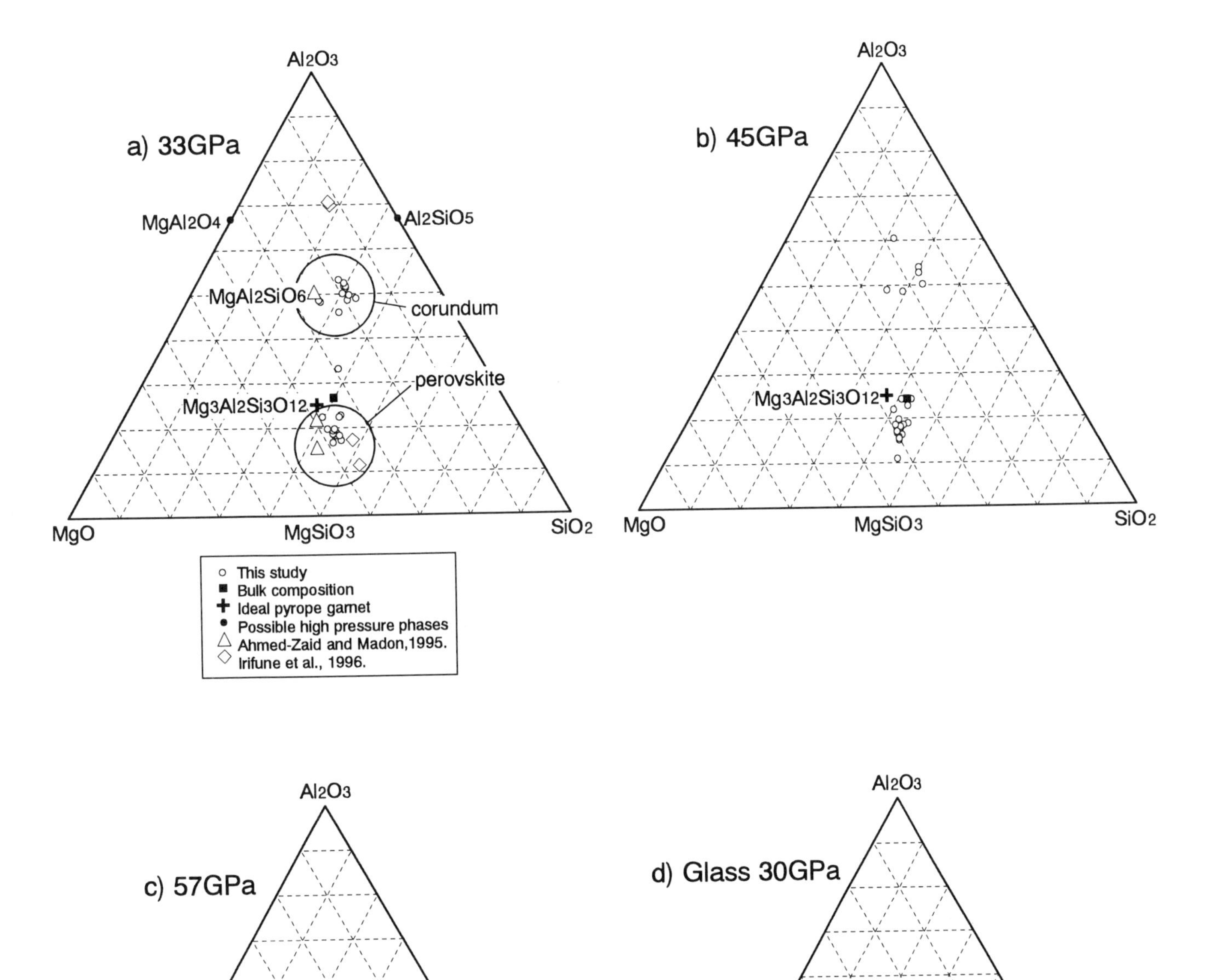

Figure 3. Chemical composition of various grains in the heated area. Solid circles represent the chemical composition of possible high-pressure phases in this system (see text). (a) 33 GPa, (b) 45 GPa, (c) 57 GPa, (d) 30 GPa (synthesized from glass starting material).

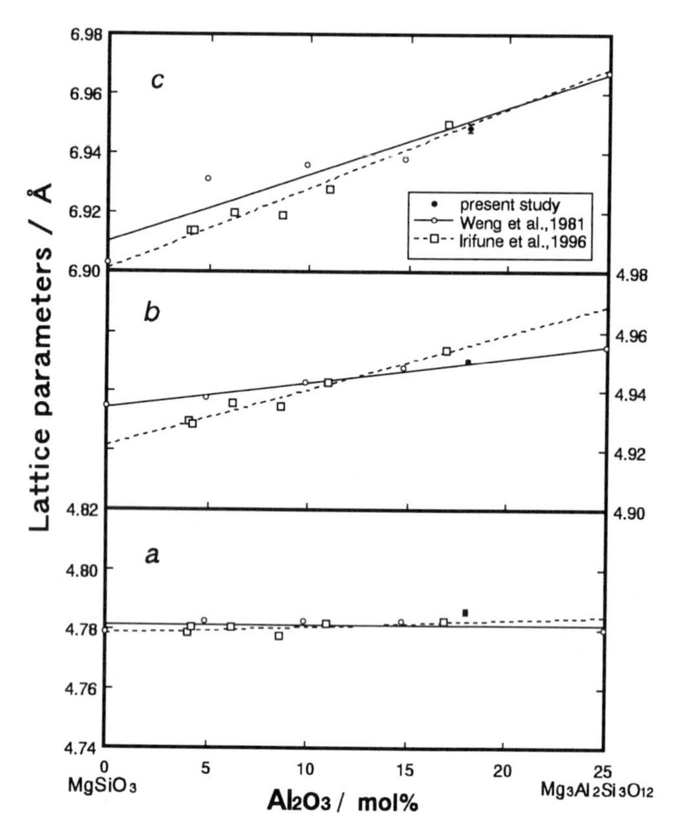

Figure 4. Relationship between lattice parameters and aluminum content in the magnesian perovskite. The result in the present study was plotted as a solid circle with error bar.

aluminous phase coexisting with perovskite phase reported by *Ahmed-Zäid and Madon* [1995] has the composition close to as those of corundums (see Figure 3(a)) and, hence, could be the corundum phase.

Consequently, the results of the present study up to 30 GPa are basically consistent with the phase relations proposed by *Irifune et al.* [1996]. However, contrary to their prediction that the single perovskite phase would be formed above 30–40 GPa, we have no evidence of the single phase of perovskite isochemical with pyrope even in the highest pressure experiment of 57 GPa. The lattice parameters of aluminous perovskite observed in this study are in good agreement with previous studies [*Weng et al.*, 1982; *Irifune et al.*, 1996] as in Figure 4. The unit-cell volume of perovskite increases slightly with increasing pressure for synthesis as shown in Figure 5 probably because of the increase of Al_2O_3 content, contrary to the result of *Ahmed-Zäid and Madon* [1995]. It was also smaller compared to the cell volume of perovskite with pyrope composition at 40 GPa reported by *Weng et al.* [1982], though they were not sure about the chemical composition of the run products. The increasing rate of the unit-cell volume with synthesis pressure is quite different from that of the lower pressure experiment by *Irifune et al.* [1996]. It is not clear whether this difference was caused by the temperature, because accurate temperature data were not available in both DAC experiments.

Kesson et al. [1995] reported that the single phase of perovskite ($Mg_{0.84}Al_{0.39}Si_{0.78}O_3$) was synthesized at 55 GPa or higher. However, the present experiments demonstrated that a corundum phase exists at 57 GPa in the same experimental procedure, though *Kesson et al.* [1995] did not mention a low-pressure product below the stable region of single phase perovskite. Our results suggest that the transition pressure of pyrope garnet into a single phase of perovskite is possibly much higher than the estimated pressure range of 30–40 GPa by *Irifune et al.* [1996], while the uncertainty of temperature might affect the phase relation.

5. CONCLUSION

Our results demonstrated that the post-garnet phase of pure pyrope is a mixture of aluminous perovskite and corundum-structured phases, and the assemblage is stable at least up to 57 GPa. It was essential for the identification of the phases present in the run products to perform both the high resolution X ray diffraction experiments of the bulk sample using synchrotron radiation and the chemical analysis of various points in the sample. Although the laser-heated DAC combined with TEM analyses has become a standard and strong tool to investigate the chemical and physical properties of the lower mantle region, it is dangerous to identify the stable phases by the

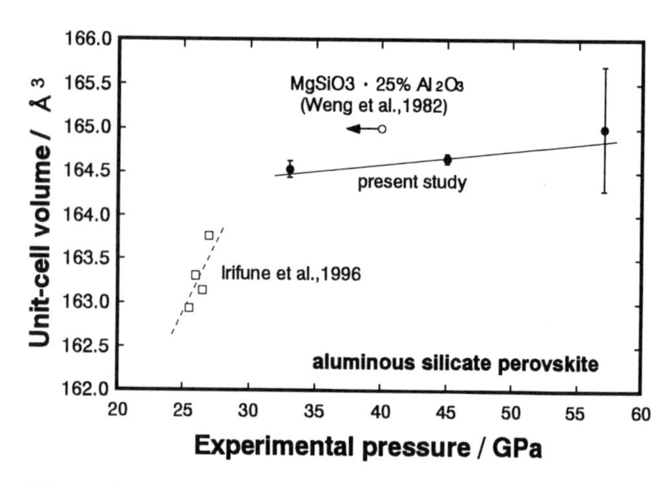

Figure 5. Variation of unit cell volume of recovered perovskite phase with experimental pressure. Solid circles with error bar represent the result in the present study.

local information obtained by TEM analysis alone. In this method it difficult to know the proportion of identified phases to the bulk sample and the results do not always reflect the stable phase in the run product of a DAC experiment, and it is also difficult to differentiate local contamination from the stable phases. Furthermore, prolonged laser heating was found to be effective to complete reaction of the starting material as much as possible. Consequently, very little starting material remained after laser-heating at 30 GPa.

Acknowledgments. We thank B.O'Neill in ISSP for helpful comments and discussion. We also thank T.Sakai for her kindly teaching in operation of SEM-EDX system in ISSP. We are grateful to T.Nobe, University of Tokyo, for her help in experiments, and to Professor O.Shimomura T.Kikegawa for their support of this study. This work was supported by a Grant-in-Aid for Scientific Research from the Ministry of Education, Science and Culture of the Japanese government.

REFERENCES

Ahmed-Zäid, I., and M. Madon, A high-pressure form of Al_2SiO_5 as a possible host of aluminum in the lower mantle, *Nature, 353*, 426-428, 1991.

Ahmed-Zäid, I., and M. Madon, Electron microscopy of high-pressure phases synthesized from natural garnets, kyanite and anorthite in diamond anvil cell, *Earth Planet. Sci. Lett.*, 129, 233-247, 1995.

Akaogi M., and S. Akimoto, Pyroxene-garnet solid-solution equilibria in the system $Mg_4Si_4O_{12}$-$Mg_3Al_2Si_3O_{12}$ and $Fe_4Si_4O_{12}$-$Fe_3Al_2Si_3O_{12}$ at high pressures and temperatures, *Phys.Earth Planet. Inter.*, 15, 90-106, 1977.

Fitz-Gerald, J. D., and A. E. Ringwood, High pressure rhombohedral perovskite phase $Ca_2AlSiO_{5.5}$, *Phys. Chem. Miner.*, *18*, 40-46, 1991.

Funamori, N., T. Yagi, N. Miyajima, and K. Fujino, Transformation in garnet from orthorhombic perovskite to $LiNbO_3$ phase on release of pressure, *Science*, *275*, 513-515, 1997.

Irifune, T., K. Fujino, and E. Ohtani, A new high pressure form of $MgAl_2O_4$, *Nature*, *349*, 409-411, 1991.

Irifune, T., and A. E. Ringwood, Phase transformations in subducted oceanic crust and buoyancy relationships at depths of 600-800 km in the mantle, *Earth Planet Sci. Lett.*, *117*, 101-110, 1993.

Irifune, T., T. Koizumi and Jun-ichi Ando, An experimental study of the garnet-perovskite transformation in the system $MgSiO_3$-$Mg_3Al_2Si_3O_{12}$, *Phys.Earth Planet. Inter.*, *96*, 147-157, 1996.

Kanzaki, M., Ultrahigh-pressure phase relations in the system $Mg_4Si_4O_{12}$-$Mg_3Al_2Si_3O_{12}$, *Phys. Earth Planet. Inter.*, *49*, 168-175,1987.

Kesson, S. E., and J. D. Fitz-Gerald, Partitioning of MgO, FeO, NiO, MnO and Cr_2O_3 between magnesian silicate perovskite and magnesiowustite: implications for the origin of inclusions in diamond and the composition of the lower mantle, *Earth Planet Sci. Lett.*, *111*, 229-240, 1991.

Kesson, S. E., J. D.Fitz-Gerald, J. M. G. Shelly and R. L. Withers, Phase relations, structure and crystal chemistry of some aluminous silicate perovskites, *Earth Planet Sci. Lett.*, *134*, 187-201, 1995.

Liu, L.-G., Silicate perovskite from phase transformations of pyrope-garnet at high pressure and temperature, *Geophys. Res. Lett.*, *1*, 277-280, 1974.

Liu, L.-G., Orthorhombic perovskite phases observed in olivine, pyroxene and garnet at high pressures and temperatures, *Phys. Earth Planet. Inter.*, *11*, 289-298, 1976.

Liu, L.-G., First occurrence of the garnet-ilmenite transition in silicates, *Science, 195*, 990-991, 1977a.

Liu, L.-G., The system enstatite-pyrope at high pressures and temperatures and the mineralogy of the Earth's mantle, *Earth Planet Sci. Lett.*, *36*, 237-245, 1977b.

O'Neill, B., and R. Jeanloz, $MgSiO_3$-$FeSiO_3$-Al_2O_3 in the Earth's lower mantle: Perovskite and garnet at 1200 km depth, *J. Geophys. Res.*, *99*, 19,901-19,915, 1994.

Ringwood, A. E., *Composition and Petrology of the Earth's Mantle*, McGraw-Hill, New York, 1975.

Weng, K., H. K. Mao, and P. M. Bell, Lattice parameters of the perovskite phase in the system $MgSiO_3$-$CaSiO_3$-Al_2O_3, *Carnegie Inst. Washington Yearb*., 81, 273-277, 1982.

Yagi, T., B. O'Neill, T. Kondo, N. Miyajima, and K. Fujino, Post garnet high pressure transition: Effect of heterogeneous laser heating and introduction of some new techniques, *Eur. J. Mineral.*, in press, 1997.

T. Kondo and T. Yagi, Institute for Solid State Physics, University of Tokyo, Roppongi, Minato-ku, Tokyo 106, Japan

In Situ Observation of ZrO_2 Phases at High Pressure and Temperature

Osamu Ohtaka, Takaya Nagai, and Takamitsu Yamanaka

Department of Earth and Space Science, Osaka University, Toyonaka, Osaka 560, Japan

Takehiko Yagi

Institute for Solid State Physics, University of Tokyo, Minato-ku, Tokyo 106, Japan

Osamu Shimomura

JAERI-RIKEN SPring-8 Project Team, Kamigori, Hyogo 678-12, Japan

Two kinds of in situ observation have been performed in order to reveal the high-pressure and -temperature polymorphism of ZrO_2 and to clarify the crystal structures for these polymorphs. By using a diamond anvil cell and a laser heating system, ZrO_2 was heated at about 1000°C under high pressures up to 30 GPa. Thermally quenched samples were investigated by X ray diffraction under high pressure. As a result, a hexagonal phase was formed above 20 GPa while a mixture of orthoI and orthoII was observed below 20 GPa. By using a cubic anvil type device and synchrotron radiation, diffraction data in the temperature range up to 900°C at 8 to 13 GPa were collected. The results indicated that ZrO_2 crystallizes in a distorted fluorite structure under these P-T conditions.

1. INTRODUCTION

Zirconia, ZrO_2, is one of the major components of modern ceramic materials. Recently, phase transition of ZrO_2 under high pressure has gathered much interest in both material and geophysical researches. Because the toughening mechanism of ZrO_2 ceramics is explained by a stress induced phase transition [*Gupta et al.*, 1978; *Kriven*, 1988], the phase transition at high pressure is to be studied from a microscopic viewpoint. The investigation of high pressure polymorphism of ZrO_2 is also important in geophysical implications as the polymorphic structure of ZrO_2 might be a possible candidate for the high pressure form of SiO_2 [*Liu*, 1980; *Liu and Bassett*, 1986].

At ambient pressure, pure ZrO_2 adopts three structures; monoclinic baddeleyite [*Smith and Newkirk*, 1965] up to 1170°C; a tetragonal structure [*Teufer*, 1962] from 1170 to 2370°C; and a cubic fluorite-type structure [*Smith and Cline*, 1962] stable to 2680°C. Both the monoclinic and tetragonal ZrO_2 are distorted fluorite structures and the coordination number of the zirconium ion is 7, 8, and 8 in the three forms, respectively. As schematically shown in Figure 1, four high pressure phases have been reported [*Bocquillon and Susse*, 1969; *Liu*, 1980; *Arashi et al.*, 1990; *Ohtaka et al.*, 1994]. The monoclinic-orthoI phase transition is given rise to by compression above 3 GPa at room temperature [*Bocquillon and Susse*, 1969; *Arashi and Ishigame*, 1982; *Ohtaka et al.*, 1991]. Based on a neutron powder diffraction experiment of orthoI quenched to ambient conditions, the space group of

Properties of Earth and Planetary Materials
at High Pressure and Temperature
Geophysical Monograph 101

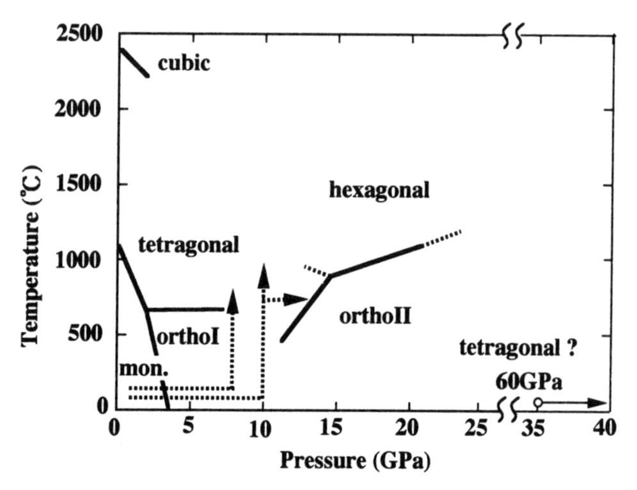

Figure 1. Pressure-temperature phase diagram of ZrO_2. Monoclinic-tetragonal (high temperature form) is after *Witney* [1965]; monoclinic-orthoI and orthoI-tetragonal are after *Block et al.* [1985]; orthoI-to-orthoII and orthoII-to-hexagonal are after *Ohtaka et al.* [1991]; tetragonal phase which appears above 35 GPa is after *Arashi et al.* [1990]. Dashed lines and allows indicate experimental paths along which X ray diffraction data were collected.

this phase was determined to be Pbca and structure refinement was performed [*Ohtaka et al.*, 1990]. OrthoI has a distorted fluorite structure similar to that of monoclinic ZrO_2 and the phase transition between them shows the characteristics of a martensitic transition. OrthoII is stable above 13 GPa at 800°C [*Ohtaka et al.*, 1991] and isostructural with cotunnite ($PbCl_2$) [*Liu*, 1980; *Devi et al.*, 1987; *Haines et al.*, 1995]. The orthoI-orthoII transition is accompanied by a nucleation and growth process [*Ohtaka et al.*, 1991]. Consequently, heating or excessive compression is required to promote this transition. On the other hand, once orthoII is formed under high pressure, it can be easily quenched to ambient conditions. Combining a diamond anvil cell and a laser heating system, *Ohtaka et al.* [1994] heated ZrO_2 powders higher than 1000°C at pressures above 20 GPa. Thermally quenched samples were investigated by using X ray diffraction under high pressure and revealed to have a hexagonal symmetry. This phase was retained in the course of releasing pressure down to 1 GPa but reverted to the monoclinic phase on the complete release of pressure. They also reported that its structure is derived from that of orthoII which is composed of pseudo-hexagonal packing of oxygen. *Arashi et al.* [1990] reported that the tetragonal phase, which is different from that of the ordinary high temperature form at ambient pressure, appears at pressures greater than 35 GPa. They showed that this phase is stable to at least 60 GPa and quenchable to ambient conditions. On the contrary, *Haines et al.* [1995] insisted that the diffraction data of this phase reported by *Arashi et al.* can successfully be indexed assuming orthoII structure.

Because of the experimental difficulties, these studies of ZrO_2 under high pressure have mainly been made at room temperature or, at most, a few hundred degrees. As shown in Figure 1, the phase relation at elevated temperature is still unknown. In order to reveal the phase relation and the crystal structure for the high-pressure and-temperature ZrO_2 phases, the present authors have attempted in situ observation under high pressure by means of two techniques: a laser heating system in a diamond anvil cell and the MAX80 system (a multi anvil type device for use with synchrotron radiation).

2. EXPERIMENTAL

2.1. *Laser Heating in a Diamond Anvil Cell*

The starting material was a fine powder of pure ZrO_2 (grain size is about 100 nm) provided by Tosoh Co. The powder has a purity of more than 99.9%. It was mixed with about 1wt% of platinum black to allow for absorbing laser radiation. A lever and spring type diamond anvil apparatus [*Yagi et al.*, 1985] was used. Thc anvils were one-third carat and Drukker standard-cut diamonds with a 0.5 mm culet. Stainless steel of 0.2 mm in thickness was used as a gasket. The sample with several pieces of ruby chips was placed in the gasket hole of 0.25 mm in diameter without any pressure transmitting medium. The generated pressure was measured by the ruby fluorescence method [*Mao et al.*, 1978].

The sample was heated by using a YAG laser under compression. The sample temperature was estimated to be around 1000°C from the radiation emitted from platinum black. A thin laser beam of 15 mm in diameter from the 12 W single mode YAG laser was scanned to heat the whole area of the sample and the total heating duration was about 1 hour. No reaction between sample powder and ruby chips was observed.

High pressure in situ X ray diffraction measurements were undertaken at room temperature with a high-power rotating-anode X ray generator operated at 55 kV and 160 mA. Filtered Mo Ka radiation was collimated with a 0.1 mm pinhole. X ray diffraction pattern was recorded on a Debye-Scherrer camera with a radius of 57.3 mm. The typical exposure time was 12 hours. Experimental conditions for individual runs were tabulated in Table 1. After the complete release of pressure, recovered samples were investigated by using a micro-focused X ray diffractometer.

2.2. *MAX80 System*

Powder samples of ZrO_2 and NaCl, as a pressure marker, were mixed with about 40 wt% BN or B-epoxy resin and

TABLE 1. Experimental Conditions and Results of Laser Heating Experiments

No. of specimen	pressure /GPa (before heating)	pressure /GPa (after heating)	phase (under pressure)	phase (recovered specimen)
1	28	21	hexagonal	monoclinic
2	18	15	orthoI and orthoII	monoclinic and orthoII
3	17	12	orthoI and orthoII	monoclinic and orthoII
4	7	6	orthoI	monoclinic

charged in B-epoxy resin pressure medium separately. Graphite disk heater was used and temperature in the pressure cell was monitored with a K-type thermocouple.

X ray observations under high pressure and temperature were carried out using the MAX80 system [*Yagi et al.*, 1987] with sintered diamond anvils of 3 mm truncation at AR-NE5C station of National Laboratory for High Energy Physics. The incident X ray beam was collimated to 0.1 mm in height and 0.3 mm in width, and diffracted X ray was measured with a pure germanium solid-state detector (SSD) by an energy dispersive method. The SSD was fixed at $2\theta = 6°$. The energy range used for the analysis was approximately 20–80 keV.

3. RESULTS AND DISCUSSION

Results of laser heating experiments are summarized in Table 1. The experiments were carried out at four different pressures. Observed *d* spacings and their hkl indices are shown in Table 2. The hexagonal phase reported by *Ohtaka et al.* [1994] was observed for the specimen 1 which was heated above 20 GPa. This phase was reverted to the monoclinic phase on the complete release of pressure. On the other hand, thermally quenched specimens 2 and 3 at lower pressures were found to be a mixture of orthoI and orthoII phases. It should be noted that hexagonal phase does not appear under these pressure range. Instead, orthoI or other phase which is not thermally quenched under pressure is likely to exist in this P-T condition. Coexisting orthoII is thought to be caused by the heterogeneous heating of laser beam; some region were left insufficiently heated. Recovered specimens 2 and 3 were a mixture of monoclinic and orthoII phases. Diffraction pattern of specimen 4, which was heated at the lowest pressure of 7 GPa, was indexed assuming orthoI or tetragonal structure and, after releasing pressure, monoclinic phase was recovered. Within the resolution limit of present X ray film method, we cannot discuss whether this phase is orthoI, tetragonal or other distorted fluorite-type structure closely related to both of the two.

TABLE 2. Representative d Spacings Observed under High Pressures and Their Indices

Specimen No. 1		Specimen No. 2.		Specimen No. 3		Specimen No. 4	
H,K,L	d obs.	H,K,L	d obs.	H,K,L	d obs.	H,K,L	d obs.
hex 111	3.03	oII 011	2.89	oII 011	2.90	oI 211	2.92
hex 200	2.88	oI 211	2.86	oI 211	2.88	oI 020	2.60
hex 201	2.66	oII 102	2.76	oII 102	2.78	oI 002	2.53
hex 103	2.08	oII 200	2.76	oII 200	2.78	oI 400	2.53
hex 113	1.85	oI 020	2.56	oI 020	2.57		
hex 220	1.66	oII 111	2.54	oII 111	2.55		
hex 213	1.56	oI 002	2.49	oI 002	2.50		
		oI 400	2.49	oI 400	2.50		

hex; hexagonal phase, oI; orthoI phase, and oII; orthoII phase.

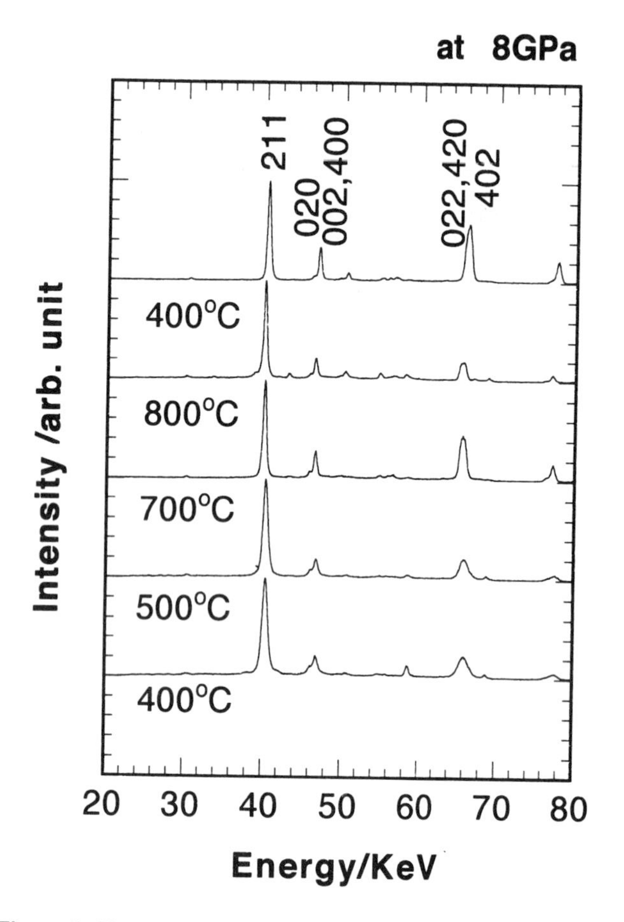

Figure 2. X ray diffraction patterns of ZrO_2 recorded at various temperature at 8 GPa. Representative diffraction lines are indexed assuming orthoI structure.

Experimental conditions for present X ray observations under high pressure and temperature using MAX80 system are indicated in Figure 1 with dashed lines and allows. X ray diffraction data were collected along these paths. Figure 2 shows X ray patterns recorded at 8 GPa in the temperature range from 400 to 800°C. As the temperature of specimen was elevated, diffraction lines become sharper, however, no other change was observed. As shown in Figure 3, the same results were observed for the run at 10 GPa. In this run, BN was mixed with specimen and diffraction from the BN overlapped with the main peak of specimen as indicated in the figure. Figure 4 shows diffraction patterns observed under various pressures at 700°C. These patterns in Figures 2, 3, and 4 can be indexed assuming orthoI structure. Throughout the present in situ observation under high pressure and temperature using MAX80 system, neither orthoII nor hexagonal phase was observed. It is difficult to distinguish orthoI from tetragonal phase, however, it can be concluded that distorted fluorite-type phase is stable at P-T conditions of this study. Because the X ray atomic scattering factor of oxygen is very small compared with that of zirconium, it is very difficult to determine the exact structure by X ray powder diffraction, particularly by an energy dispersive method. We are attempting a further investigation using an X ray angular dispersive method to collect diffraction data with higher resolution.

Acknowledgments. One of us (O.O) would like to thank T. Kondo and M. Yamakata for their kind help and supervision in the course of diamond anvil cell experiments at ISSP, University of Tokyo. We are grateful to Y. Matsumoto, T. Matsuda, M. Kikuchi, T. Hattori and A. Nakatuska of Osaka University for their help at Photon Factory. MAX80 experiments were conducted under the approval of the Photon Factory Advisory Committee (proposal no.

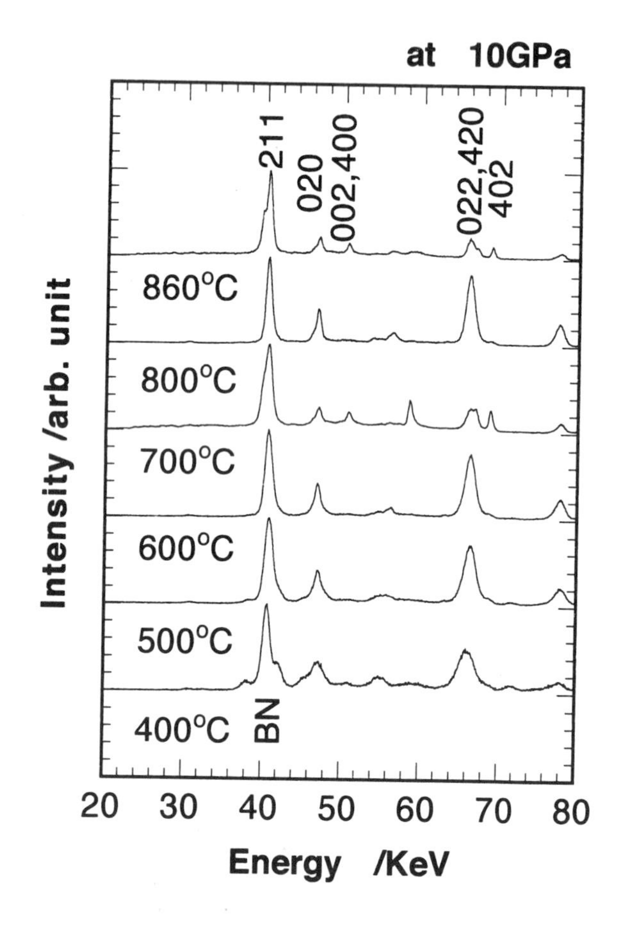

Figure 3. X ray diffraction patterns of ZrO_2 recorded at various temperature at 10 GPa. Representative diffraction lines are indexed assuming orthoI structure. Peak position of BN is also indicated.

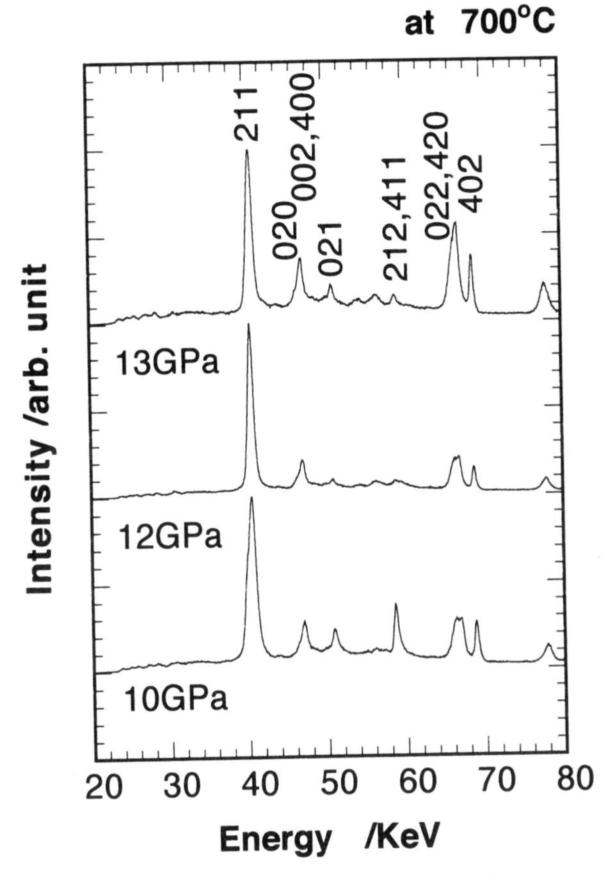

Figure 4. X ray diffraction patterns of ZrO_2 recorded at various pressure at 700°C. Representative diffraction lines are indexed assuming orthoI structure.

REFERENCES

Arashi, H., and M. Ishigame, Raman spectroscopic studies of the polymorphism in ZrO_2 at high pressures, *Phys. Status Solidi, 71*, 313-321, 1982.

Arashi, H., T. Yagi, S. Akimoto, and Y. Kudoh, New high-pressure phase of ZrO_2 above 35 GPa, *Phys. Rev. B, 41*, 4309-4313, 1990.

Block, S., J. A. da Jornada, and G. J. Piermarini, Pressure-temperature phase diagram of zirconia, *J. Am. Ceram. Soc., 68*, 487, 1985.

Bocquillon, G., and C. Susse, Diagramme de phase de la zircone sous pression, *Rev. Int. Hautes Temp. Refract., 6*, 263-266, 1969.

Devi, S. R. U., L. C. Ming, and M. H. Manghnani, Structural transformation in cubic zirconia, *J. Am. Ceram. Soc., 70*, C-218-219, 1987.

Gupta, T. K., F. F. Lange, and J. H. Bechtold, Effect of stress-induced transformation on the properties of polycrystalline zirconia containing metastable tetragonal phase, *J. Mater. Sci. 13*, 1464- 1470, 1978.

Haines, J., J. M. Leger, and A. Atouf, Crystal structure and equation of state of cotunnite-type zirconia, *J. Am. Ceram. Soc., 78*, 445-448, 1995.

Kriven, W. M., Possible alternative transformation tougheners to zirconia: Crystallographic aspects, *J. Am. Ceram. Soc., 711*, 1021-1030, 1988.

Liu, L. G. , New high-pressure phase of ZrO_2 and HfO_2, *J. Phys. Chem. Solids, 41*, 331-334, 1980.

Liu, L. G., and W. A. Bassett, *Elements, Oxides, and Silicates*, 250 pp., Oxford University Press, New York, 1986.

Mao, H. K., P. M. Bell, J. W. Shaner, and D. J. Steinberg, Specific volume measurements of Cu, Mo, Pd, and Ag and calibration of the ruby R1 fluorescence pressure gauge from 0.06 to1 Mbar, *J. Appl. Phys.* 49, 3276-3283, 1978.

Ohtaka, O., T. Yamanaka, S.Kume, N. Hara, H. Asano, and F. Izumi, Structural analysis of orthorhombic ZrO_2 by high resolution neutron powder diffraction, *Proc. Jpn. Acad., Ser. B, 66*, 193-196, 1990.

Ohtaka, O., T. Yamanaka, S.Kume, E. Ito and A. Navrostky, Stabilities of monoclinic and orthorhombic zirconia: Studies by high-pressure phase equilibria and calorimetry, *J. Am. Ceram. Soc., 74*, 505-509, 1991.

Ohtaka, O., T. Yamanaka, and T. Yagi, New high-pressure and temperature phase of ZrO_2 above 1000°C at 20 GPa, *Phys. Rev. B, 49*, 9295-9298, 1994.

Smith, D. K., and C. F. Cline, Verification of existence of cubic zirconia at high temperature, *J. Am. Ceram. Soc., 45*, 249-250, 1962.

Smith, D. K., and H. W. Newkirk, The crystal structure of baddeleyite (monoclinic ZrO_2) and its relation to the polymorphism of ZrO_2, *Acta Crystallogr., 18*, 983-991, 1965.

Teufer, G., The crystal structure of tetragonal ZrO_2, *Acta Crystallogr., 15*, 1187-1189, 1962.

Witney, E. D., Electrical resistivity and diffusionless phase transformations of zirconia at high temperatures and ultrahigh pressures, *J. Electrochem. Soc., 112*, 91-94, 1965.

Yagi, T., T. Suzuki, and S. Akimoto, Static compression of wustite ($Fe_{0.98}O$) to 120 GPa, *J. Geophys. Res. 90*, 8784-8788, 1985.

Yagi, T., M. Akaogi, O. Shimomura, H. Tamai, and S. Akimoto, High pressure and high temperature equations of state of majorite, in *High Pressure Research in Mineral Physics, Geophys. Monogr. Ser.*, vol. 39, edited by M.H.Manghnani and Y. Syono, pp. 141-147, Terra, Tokyo/AGU, Washington, D.C., 1987.

T. Nagai (takaya@setc.wani.osaka-u.ac.jp), O. Ohtaka, and T. Yamanaka, Department of Earth and Space Science, Osaka University, Toyonaka, Osaka 560, Japan

O. Shimomura, JAERI-RIKEN SPring-8 Project Team, Kamigori, Hyogo 678-12, Japan

T. Yagi, Institute for Solid State Physics, University of Tokyo, Minato-ku, Tokyo 106, Japan (yagi@kodama.issp.u-tokyo.ac.jp)

Post-Spinel Transition in Fe_2SiO_4

Tomoo Katsura, Atsushi Ueda[1], Eiji Ito, and Koh-ichi Morooka[2]

Institute for Study of the Earth's Interior, Okayama University, Misasa, Japan

The phase boundary of the dissociation of Fe_2SiO_4 spinel to wüstite+stishovite was investigated at temperatures of 1000-1600 K and pressures of 14-19 GPa using a 6-8 type multianvil apparatus. $FeCl_2$ was used as a catalyst to enhance the reaction rate, and complete the dissociation within 180 min above 1200 K. The catalyst also made it possible to conduct reversal runs. Excess wüstite and iron were added to the starting material in order to minimize the Fe^{3+} contents in spinel and wüstite. The dissociation reaction occurs at 17.3 GPa. No temperature dependence on the transition pressure was observed in the temperature range investigated.

1. INTRODUCTION

$(Mg,Fe)_2SiO_4$ olivine is considered to be the most abundant mineral in the upper mantle [*Ringwood*, 1979]. The 410-km and 660-km seismic discontinuities are usually accounted for by the olivine - modified spinel transition and the dissociation of spinel to perovskite+magnesiowüstite, respectively, in the system Mg_2SiO_4-Fe_2SiO_4. Hence, a knowledge of the phase relations in this system is indispensable for elucidating the constitution of the Earth's mantle. Accordingly, the phase relations in this system have been studied by many workers [*c. f.*, *Yagi et al.*, 1987; *Akaogi et al.*, 1989; *Ito and Takahashi*, 1989; *Katsura and Ito*, 1989; *Morishima et al.*, 1994]. The detailed phase relations of the olivine - modified spinel - spinel transitions have been clarified in the whole compositional range of the system. Those of the post-spinel transitions, however, have not yet been established for the Fe-rich part of the system.

[1] Now at Nikkato Co., Japan.

[2] Now at Mitsubishi Material Co., Japan.

Properties of Earth and Planetary Materials
at High Pressure and Temperature
Geophysical Monograph 101

More than 20 years ago, *Basset and Takahashi* [1970] and *Basset and Ming* [1972] reported that Fe_2SiO_4 spinel dissociates to wüstite (Fe_xO) and stishovite (SiO_2). The Fe/O ratio of wüstite is slightly smaller than unity [*Hazen and Jeanloz*, 1984]. Therefore the reaction should be written as

$$\underset{\text{spinel}}{Fe_2SiO_4} = \underset{\text{wüstite}}{2Fe_xO} + \underset{\text{stishovite}}{SiO_2} + \underset{\text{iron}}{2(1-x)Fe} \quad (1)$$

Kawada [1977] and *Ohtani* [1979] attempted to determine the phase boundary of this dissociation reaction. However, their results showed a striking disagreement with each other; *Kawada* [1977] proposed a positive (dP/dT) slope of the phase boundary whereas *Ohtani* [1979] proposed a negative one.

Morooka [1992] tried to determine the phase boundary more precisely by careful examination of the run products using scanning electron microscopy and microfocused X ray diffractometry. However, he observed the presence of three phase, that is, spinel+wüstite+stishovite over a 4 to 5 GPa pressure interval in the range of 15 to 20 GPa (Figure 1). Figure 2 is a typical example of the presence of three phases in a run product, indicating that the dissociation of spinel to wüstite+stishovite was not completed. Because of the sluggishness of the reaction, *Morooka* [1992] was not able to determine the phase boundary precisely.

One possible reason why the previous workers failed in

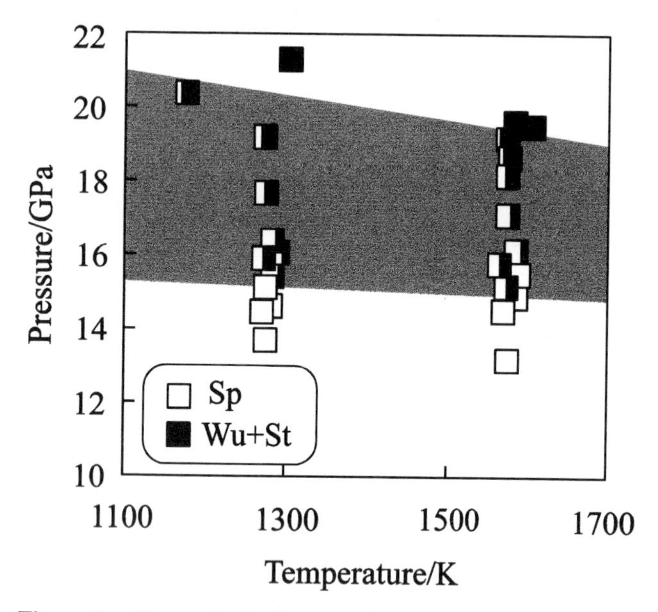

Figure 1. The synthesis diagram of Fe_2SiO_4 at pressures of 13-22 GPa and temperatures of 1200-1500 K from *Morooka* [1992]. The open and solid squares denote presence of spinel and wüstite+stishovite, respectively. A 5-GPa pressure interval shows the presence of three phases, that is, spinel+wüstite+stishovite.

determining the phase boundary is that the compositions of spinel and wüstite could have changed during experiments by oxidation. The Fe/O ratio in wüstite can vary, for example, from 0.96 to 0.85 at ambient pressure [*Hazen and Jeanloz*, 1984]. *Tobe* [1995] showed that Fe_2SiO_4 spinel forms a complete solid solution with Fe_2O_3 magnetite above 10 GPa. These facts suggest that when oxidation of the system occurs, iron will disappear from Equation (1), and instead, spinel+wüstite+stishovite will coexist over a certain pressure interval.

A second possible explanation is a slow reaction rate of the dissociation reaction, as implied by Figure 2. Stishovite has a very high melting point, for example, 3300 K at 17 GPa [*Zhang et al.*, 1993], whereas, the temperatures of the experiments made by the previous workers are below 1700 K. This temperature difference suggests that stishovite is very slow to react at temperatures well below its melting point.

The third possible source of uncertainty is a pressure drop while heating. Typically, in usual high-pressure experiments made in a multianvil apparatus, pressure is first applied to a desired value by increasing the oil pressure of the hydraulic press, and then temperature is raised to a desired value while the oil pressure is held. When the sample is heated, the pressure medium and gasket material are also heated and flow from the pressure cell more readily than at room temperature condition. Moreover, sintering and phase transitions within both the sample and pressure medium decrease the material volume in the pressure cell. Consequently, the sample pressure may decrease during heating, even if the oil pressure is held. If the pressure of the sample before heating is a little above the phase boundary, it will be lowered from the stability field of wüstite-stishovite to that of spinel on heating, resulting in cessation of the reaction.

In order to overcome these problems, we improved the experimental techniques as follows. First, we add excess metallic iron and wüstite to starting materials in order to keep the redox state of samples along the IW buffer, resulting in minimum amounts of Fe^{3+} in spinel and wüstite. Secondly, we add a small amount of catalyst to the staring material; the catalyst adopted here is mainly $FeCl_2$. The catalyst enhances and completes dissociation reaction, and also makes it possible to form spinel from wüstite+stishovite. Thirdly, we use anvils with larger truncated edge length than used in the previous studies. The large truncation makes the heating volume smaller relative to the pressure cell so that the pressure drop on heating will be smaller. The large truncation also enables more efficient thermal insulation, which minimizes the flow of gasket materials.

2. EXPERIMENTAL PROCEDURE

The starting material of 'normal' experiments was a mixture of fayalite, wüstite, iron, and catalyst with molar ratio of 100:20:10:5. The catalyst for the experiments

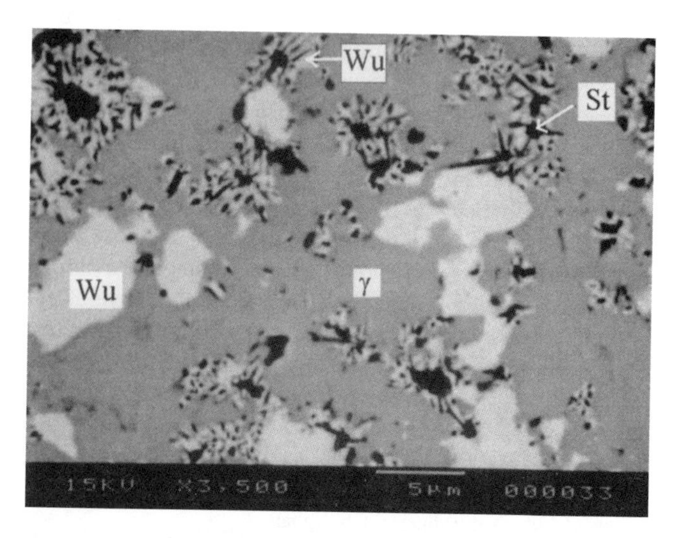

Figure 2. The back-scattered electron image of the run product from the spinel+wüstite mixture without a catalyst at 1300 K and 21 GPa. The solid bright and gray parts are wüstite and spinel, respectively. The fine grained mixture of white and black grains is an aggregate of wüstite+stishovite.

above 1200 K was $FeCl_2$, whereas, at 1000 K it was an even molar mixture of $FeCl_2$ and CsCl. The starting material of the 'reversal' experiments was a 100:105:120:15 mixture of hematite, iron, quartz, and $FeCl_2$. In these runs we used submicron powders of hematite and quartz which are expected to be very reactive. The hematite+iron and quartz, respectively, will initially become wüstite+iron and stishovite at high-temperature and high-pressure, and then will form spinel.

The high-pressure phase equilibrium experiments were carried out using a 6-8 type multianvil apparatus [*Ito et al.*, 1984]. The truncated edge length of the anvils and the edge length of the octahedral MgO+5%Cr_2O_3 pressure medium were 5.0 and 11.8 mm, respectively, whereas, those used by *Morooka* [1992] were 3.0 and 7.0 mm. The generated sample pressure was calibrated against the oil pressure of the hydraulic press using semiconductor-metal transitions in ZnS and GaAs at ambient temperature and the α–β transition in Mg_2SiO_4 at 1400 K. The results of the calibrations at ambient and high temperatures agree well, suggesting a very small temperature effect on pressure generation.

The sample assembly is shown schematically in Figure 3. The starting material is loaded in an Fe capsule. The heater is composed of $LaCrO_3$. The heater and capsule are electrically insulated by a MgO sleeve. The temperature outside the sample is monitored by a W_3Re_{97}-$W_{25}Re_{75}$ thermocouple with 0.05 mm diameter. The thermocouple and heater are electrically insulated by ZrO_2 sleeves.

Sample pressure was increased first by increasing the oil pressure, then the temperature was increased to the desired value at constant oil pressure. Experiments were made in the temperature range of 1000 to 1600 K, and experimental durations were 5-420 min. The sample was quenched by shutting off the supplied electric power, and sample pressure was decreased at a rate of 1 GPa/hour.

Recovered samples were mounted in epoxy resin and made into polished sections cut parallel to the longitudinal axis of the cylindrical heater so that the changes in phases and textures in the run products due to the temperature gradient could be observed. Phases present in the samples were identified by microfocused X ray diffractometry (MFXD). Phase textures and chemical compositions, respectively, were examined using scanning electron microscopy (SEM) and electron probe microanalysis (EPMA).

3. RESULTS AND DISCUSSION

Experimental conditions and results are summarized in Table 1.

At temperatures above 1200 K, the observed phase

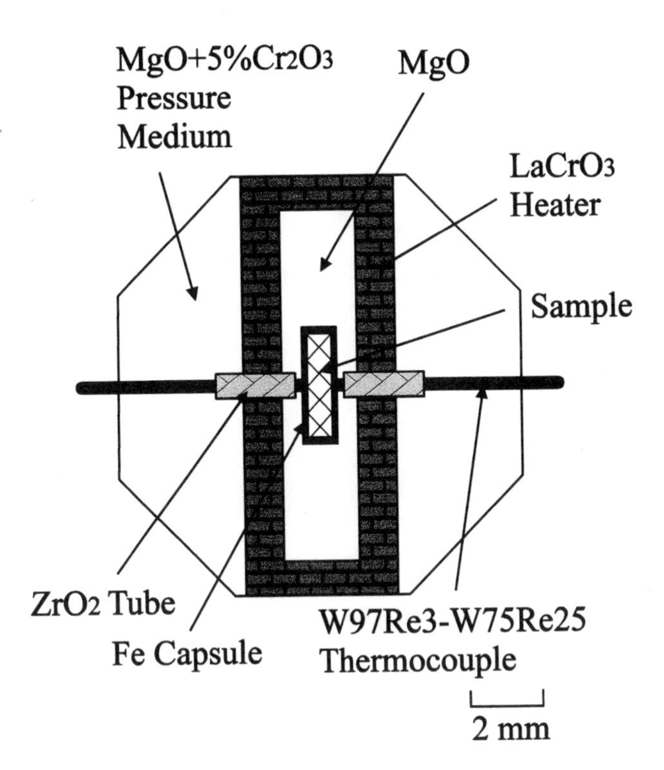

Figure 3. Schematic drawing of the furnace assembly.

assemblages of the normal runs were stishovite+wüstite+iron above 17.4 GPa (Figure 4a) and spinel+wüstite+iron below 17.2 GPa (Figure 4b). The texture observed by *Morooka* [1992], showing the incomplete dissociation reaction stops, was not seen in experiments made at conditions above 1200 K, suggesting that the $FeCl_2$ catalyst completes the dissociation reaction. No Cl-bearing phase could be found by SEM and EPMA observations. This is probably because most of the Cl-bearing phase was removed and/or dissolved in water while polishing. The area analysis by EPMA, however, showed a small amount of Cl present along the grain boundaries of the primary phases. It is noted that no variation in the phase assemblage associated with the temperature gradient was observed in any run product.

In contrast, the dissociation reaction was not completed in the experiments at 1000 K. The MFXD showed that the products at 17.7 and 18.4 GPa contain stishovite+spinel+ wüstite+iron whereas that at 17.2 GPa contains only spinel+ wüstite+iron. In these low temperature runs, the mixture of $FeCl_2$+CsCl was used for a catalyst instead of $FeCl_2$ for further enhancement of the reaction rate. Nevertheless, the reaction was not completed. The experimental temperature of 1000 K is too low to complete the dissociation reaction using the present method.

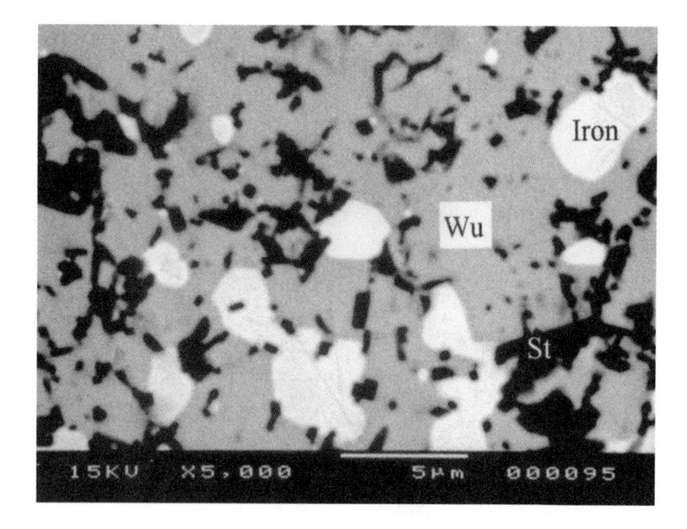

a

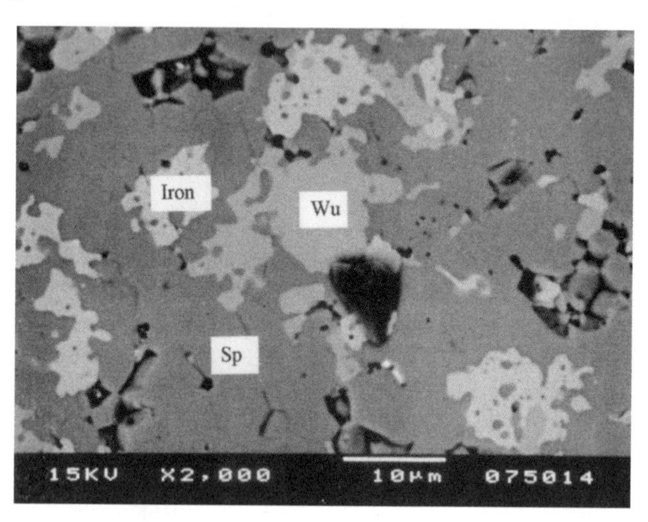

b

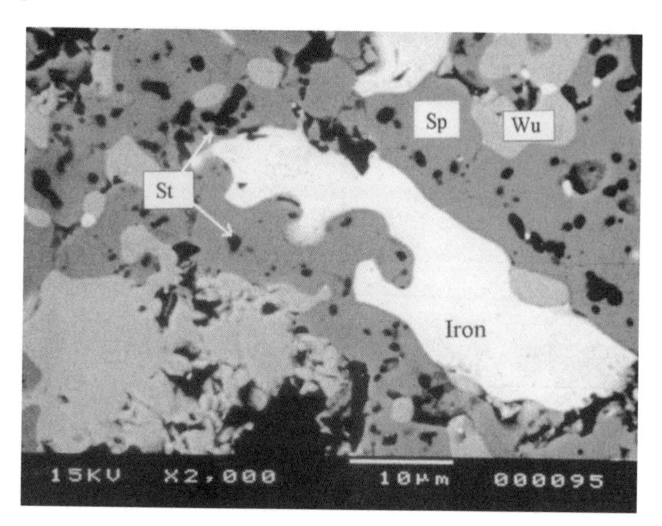

c

The reversal runs, using hematite+iron+quartz+$FeCl_2$ as the starting material, were conducted at 1200 K and 16.9 GPa, and at 1580 K and 16.7 GPa. At the former conditions, wüstite, iron, and a silica mineral were observed by SEM and EPMA analysis, but spinel was not found. However, a small amount of spinel was detected by MFXD. On the contrary, the run product at 1580 K shows that a large amount of spinel was formed, although a small amount of stishovite is still present (Figure 4c). Generally, reversal runs are very difficult to conduct for high-pressure dissociation reactions. One reason for this difficulty is that each of the dissociated phases (in this case, wüstite and stishovite) is stable in the stability field of the original phase (spinel). Another reason is that each of the dissociated phases usually has a higher melting point than the original phase if the dissociated phases have a eutectic relationship. In this study, the first successful reversal run for the post-spinel transition was achieved.

The phase diagram constructed on the basis of the observations described above (Figure 5) shows that the dissociation of Fe_2SiO_4 spinel to wüstite+stishovite occurs at 17.3 GPa with no temperature effect on the transition pressure. A very small or non-existent temperature dependence on the transition pressure is also suggested by the observation that there was no variation in phase assemblage due to the temperature gradient across the sample.

The present phase boundary is located at a lower pressure than those of *Kawada* [1977] and *Ohtani* [1979], and higher pressure than that of *Morooka* [1992]. One possible explanation for the discrepancy between this study and *Ohtani* [1979] is that *Ohtani* [1979] was not able to observe precisely the beginning of the dissociation reaction because of the sluggish kinetics. The kinetic effect should decrease with increasing temperature, and, therefore, the over-pressure needed to enhance the dissociation reaction should also decrease with increasing temperature. This could explain why *Ohtani* [1979] observed a negative pressure dependence.

The reason for the discrepancy between this study and

Figure 4. Back-scattered electron images of the run products with the $FeCl_2$ catalyst. (a) at 17.4 GPa and 1200 K from the mixture of fayalite, wüstite, and iron with $FeCl_2$. The run product is composed of stishovite, wüstite and iron with $FeCl_2$. (b) at 17.2 GPa and 1200 K from the mixture of fayalite, wüstite, and iron with $FeCl_2$. The run product is composed of spinel, wüstite, and iron. (c) at 16.2 GPa and 1575 K from the mixture of quartz, hematite, and iron with $FeCl_2$. The run product is composed mainly of spinel, wüstite, and iron, although a small amount of stishovite is found.

TABLE 1. Experimental Conditions and Results

Run#	Temperature (K)	Pressure (GPa)	Duration (min)	Phases Present
(Normal Run)				
64	1200	19.8	180	St-Wu-I
67	1200	14.8	180	Sp-Wu-I
68	1200	17.4	180	St-Wu-I
70	1200	16.2	180	Sp-Wu-I
71	1200	16.9	180	Sp-Wu-I
75	1400	17.2	20	Sp-Wu-I
77	1400	17.7	60	St-Wu-I
88	1000	18.4	18	St-Sp-Wu-I
89	1000	17.2	300	Sp-Wu-I
90	1000	17.7	300	St-Sp-Wu-I
95	1600	17.4	10	St-Wu-I
(Reversal Run)				
97	1200	16.9	420	St-Sp-Wu-I
98	1580	16.7	10	St-Sp-Wu-I

Sp, St, Wu, and I in the column 'phases present' denote spinel, stishovite, wüstite and iron, respectively.

Kawada [1977] could be that his samples were more oxidized than the IW buffer. Fe_2SiO_4 spinel can make a complete solid solution with Fe_3O_4 spinel, whereas, the maximum Fe^{3+} content in wüstite is much less than Fe_3O_4 spinel [*Hazen and Jeanloz*, 1984]. Therefore, once the system is oxidized, the Gibbs's free energy of spinel is decreased by $RT\ln(a^{Spinel}_{Fe2SiO4})$, whose absolute value is much larger than $RT\ln(a^{Wüstite}_{FeO})$, where R and T are the gas constant and absolute temperature, and a^x_y denotes the activity of component y phase x. Hence, the stability field of spinel increases to higher pressures with increasing oxidation of the system. This hypothesis also explains the positive pressure dependence of *Kawada* [1977], because the absolute value of $RT\ln(a^{Spinel}_{Fe2SiO4})$, increases with increasing temperature. In fact, the Fe capsules used in this study were found to be partially oxidized. Without excess iron, the spinel and wüstite in the present study would be more oxidized than those along the IW buffer as well.

The reason why the boundary obtained by *Morooka* [1992] is located at a lower pressure than the present study is probably because the pressures in *Morooka*'s [1992] samples were decreased while heating as a result of his smaller truncated edge length (3.0 mm) than that used in the present study (5.0 mm).

4. CONCLUDING REMARKS

1) The dissociation of Fe_2SiO_4 spinel to wüstite (Fe_xO)+stishovite (SiO_2) has a very slow reaction rate. In order to determine the phase boundary of this reaction at high pressure, use of a catalyst was required. In this study,

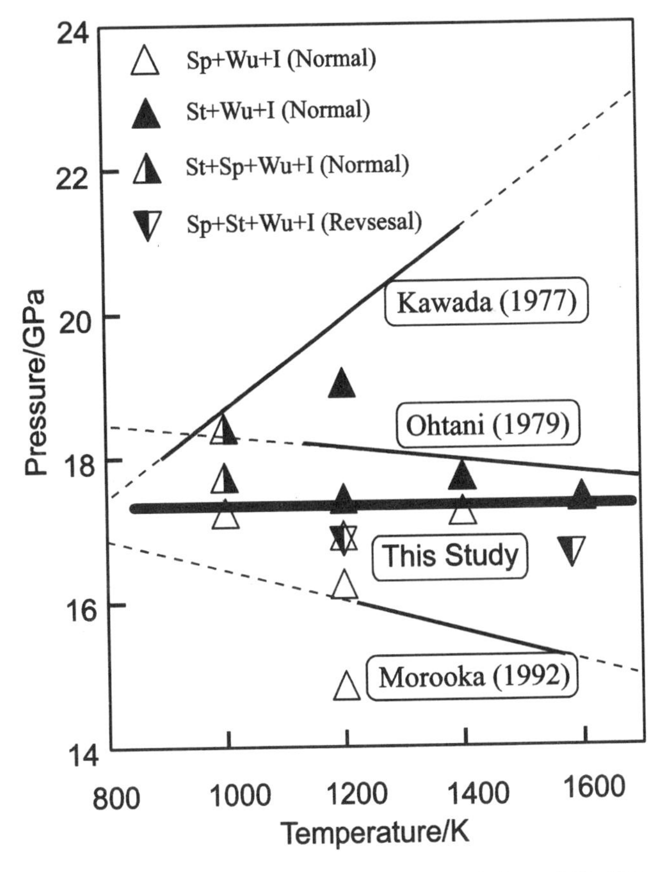

Figure 5. The phase diagram of the dissociation of Fe_2SiO_4 spinel to wüstite and stishovite. Open and solid symbols denote presence of spinel and stishovite in the run products, respectively. Upper and lower triangles denote the normal and reversal runs, respectively.

the $FeCl_2$ catalyst was used to enhance and complete the dissociation reaction above 1200 K, and also made it possible to conduct a reversal experiment at 1580 K. Generally, in high-pressure dissociation reactions, such as post-spinel transitions, accurate determination of the phase relations is very difficult because the dissociated phase are slow to react. Therefore, use of a catalyst is useful to study the phase relations of the post-spinel transitions [*c. f. Katsura and Ito*, 1996].

2) To study precisely the phase relations of Fe-rich phases, an oxygen buffer is necessary in order to keep redox condition. Without a buffer, samples are easily oxidized or reduced during experiments. Even though experimental temperatures in this study were relatively low for the multianvil experiment, oxidation of iron was clearly observed.

3) The experimental pressure of small pressure cells can become lowered by heating. To keep experimental pressure constant while heating, use of the largest possible pressure cells is desirable.

4) The transition pressure of the dissociation of Fe_2SiO_4 spinel is 17.3 GPa at temperatures of 1000-1600 K. Although the post-spinel transition in Mg_2SiO_4 has a negative (dP/dT) slope of the phase boundary, that in Fe_2SiO_4 has a flat (dP/dT=0) slope.

Acknowledgments. U. A. thanks M. Akaogi for his discussion and suggestions. T. K. thanks M. J. Walter for his correcting English.

REFERENCES

Akaogi, M., E. Ito, and A. Navrotsky, The olivine-spinel transformations in the system Mg_2SiO_4-Fe_2SiO_4; calorimetric measurements, thermochemical calculation, and geophysical application, *J. Geophys. Res.*, 94, 15671-15685, 1989.

Basset, W. A. and L. Ming, Disproportionation of Fe_2SiO_4 to $2FeO+SiO_2$ at pressures up to 250 kbar and temperatures up to 3000 °C, *Phys. Earth Planet. Int.*, 6, 154-160, 1972.

Basset, W. A., and T. Takahashi, Disproportionation of Fe_2SiO_4 to $2FeO+SiO_2$ at high pressure and temperature, *EOS Trans. AGU*, 51, 828, 1970.

Hazen, M. R. and R. Jeanloz, Wüstite ($Fe_{1-x}O$): a review of its defect structure and physical properties, *Rev. Geophys. Space. Phys.*, 122, 37-46, 1984.

Ito, E., and E. Takahashi, Y. Matsui, The mineralogy and chemistry of the lower mantle: an implication of the ultrahigh-pressure phase relations in the system MgO-FeO-SiO_2, *Earth Planet. Sci. Lett.*, 67, 238-248, 1984.

Ito, E. and E. Takahashi, Postspinel transitions in the system Mg_2SiO_4-Fe_2SiO_4 and some geophysical implications, *J. Geophys. Res.*, 94, 10,637-10,646, 1989.

Katsura, T., and E. Ito, The system Mg_2SiO_4-Fe_2SiO_4 at high pressures and temperatures: precise determination of stabilities of olivine, modified spinel, and spinel., *J. Geophys. Res.*, 94, 15663-15670, 1989.

Katsura, T. and E. Ito, Determination of Fe-Mg partitioning between perovskite and magnesiowüstite, *Geophys. Res. Let.*, 23, 2005-2008, 1996.

Kawada, K., The system Mg_2SiO_4-Fe_2SiO_4 at high pressures and temperatures and the earth's interior, *Ph. D. thesis of University of Tokyo*, pp. 187, 1977.

Morishima, H., T. Kato, M. Suto, E. Ohtani, S. Urakawa, W. Utsumi, O. Shimomura, and T. Kikegawa, The phase boundary between α-and β–Mg_2SiO_4 determined by in situ X-ray observation, *Science*, 265, 1202-1203, 1994.

Morooka, K., Dissociation of γ-Fe_2SiO_4 and the reaction between molten iron and silicate melts at high-pressure and high-temperature, *Master thesis of Okayama University*, pp. 57, 1992.

Ohtani, E., Melting relation of Fe_2SiO_4 up to about 200 kbar, *J. Phys. Earth.*, 27, 189-208, 1979.

Ringwood, A. E., *Origin of the Earth and Moon*, Springer-Verlag, New York, pp. 295, 1979

Tobe, H., *Master thesis of Osaka University*, pp.61, 1995. (in Japanese)

Yagi, T., M. Akaogi, O. Shimomura, T. Suzuki, and S.-I. Akimoto, In situ observation of the olivine-spinel transformation in Fe_2SiO_4 using synchrotron radiation, *J. Geophys. Res.*, 92, 6207-6213, 1987.

Zhang J., R. C. Liebermann, T. Gasparik, C. T. Herzberg, and Y. Fei, Melting and subsolidus relations in SiO_2 at 8-14 GPa, *J. Geophys. Res.*, 98, 19785-19793, 1993.

Tomoo Katsura, Atsushi Ueda, Eiji Ito, and Koh-ichi Morooka, Institute for Study of the Earth's Interior, Okayama University, Misasa, Tottori-ken, 682-01 Japan

An In Situ High Pressure X Ray Diffraction Study on Perovskite-Structured $PbZrO_3$ and $PbTiO_3$ to 57 GPa

L. C. Ming[1] and S.R. Shieh[2]

[1]*Hawaii Institute of Geophysics and Planetology*, [2]*Department of Geology and Geophysics*
University of Hawaii at Manoa, Honolulu, HI

Y. Kobayashi and S. Endo

Research Center for Materials Sciences at Extreme Conditions, Osaka University, Osaka, Japan

O. Shimomura and T. Kikegawa

National Laboratory for High Energy Physics, Tsukuba, Japan

An in situ X ray diffraction study has been carried out on two perovskite-structured compounds: the orthorhombic $PbZrO_3$ and the tetragonal $PbTiO_3$, to 51.3 and 57 GPa, respectively, at room temperature using a diamond anvil cell interfaced with synchrotron radiation. The orthorhombic phase of $PbZrO_3$ (I) transforms into a monoclinic phase (III) at ~35 GPa with no volume change, which persists to 51.2 GPa, the highest pressure in this study. In the case of $PbTiO_3$, the tetragonal phase (I) transforms to a cubic phase (II) at ~8 Gpa, which appears to be stable to 40 GPa. There seems to be no volume change at the transition, which is in accord with the conclusion from a high-pressure Raman study that the transition is second-order. Assuming the pressure derivative of the bulk modulus (K_0') to be 4.0, the zero-pressure bulk modulus (K_0) of $PbTiO_3$ (I) is calculated to be 82.4 GPa using the Birch-Murnaghan equation of state and is in good agreement with previous results. Similarly, we obtained K_0 = 154 GPa for $PbZrO_3$ (I).

1. INTRODUCTION

Among the ABO_3 perovskite family, $PbTiO_3$, $PbZrO_3$, and $PbHfO_3$ have received much attention because of their interesting dielectric and structural properties. At ambient conditions, $PbTiO_3$ is tetragonal and ferroelectric whereas $PbZrO_3$ and $PbHfO_3$ are orthorhombic and antiferroelectric. When heated at atmospheric pressure, the ferroelectric $PbTiO_3$ transforms to the cubic paraelectric phase at 500°C [e.g., *Samara,* 1969], while the antiferroelectric $PbZrO_3$ first transforms to a ferroelectric phase at 222°C and then to the cubic paraelectric phase at 235°C [*Pisarski and Ujima,* 1984]. The antiferroelectric $PbHfO_3$ transforms directly to a paraelectric phase at 210°C [*Samara,* 1969]. With pressure, the orthorhombic $PbHfO_3$ transforms to a tetragonal phase (II) at ~8 GPa and then into a different orthorhombic phase (III) at ~ 15 GPa, which transforms further to phase IV with an undetermined structure at ~45 GPa [*Ming et al.*, 1994; *Jayaraman et al.*, 1994). There seems to be no high-pressure data available for $PbZrO_3$. In view of the similarities in both structural and dielectric properties between $PbZrO_3$ and $PbHfO_3$, it is anticipated that the high pressure behavior of $PbZrO_3$ would be similar to that of $PbHfO_3$. In the case of $PbTiO_3$, it was found that the tetragonal phase (I)

Properties of Earth and Planetary Materials
at High Pressure and Temperature
Geophysical Monograph 101

transforms to a cubic phase (II) at pressures ~6 GPa and is a first order type according to *Ikeda* [1975]. However, according to *Sanjurjo et al.* [1983] and *Jayaraman et al.* [1984], such a transition occurs at ~12 GPa and is a second-order type. An optical absorption study also indicated a transition at 12 GPa and showed no further transformation at pressures to 35 GPa [*Zha et al.*, 1992]. What would be the nature of the transition? And why is there a large discrepancy in the transition pressure observed by different investigators?

We were thus motivated to carry out in situ high pressure X ray diffraction studies on both $PbTiO_3$ and $PbZrO_3$ in order to characterize their structural behaviors at high pressures. Studies of the perovskite-structured compounds with larger cations may be expected to serve as analogs for the structural behavior of $MgSiO_3$ perovskite at much higher pressure conditions.

2. EXPERIMENTAL METHODS

Fine-powder samples of $PbZrO_3$ (-325 mesh, 99.7% pure) and $PbTiO_3$ (Grade 1) used in this study were purchased from CERAC, Milwaukee, WI, and Johnson Matthey Chemicals Limited, NY, respectively. The X ray powder diffraction confirmed all diffraction lines expected for the orthorhombic $PbZrO_3$ with lattice parameters a = 5.880 (1) Å, b = 11.770 (2) Å, and c = 8.220 (1) Å. For $PbTiO_3$, a tetragonal unit cell with lattice parameters a = 3.900 (10) Å and c = 4.142 (11) Å was obtained. These results are in good agreement with the published values: a = 5.88 Å, b = 11.75 Å, and c = 8.21 Å for $PbZrO_3$ (ASTM# 20-608) and a = 3.899 Å, and c = 4.153 Å for $PbTiO_3$ (ASTM# 6-0452).

All the high-pressure X ray diffraction experiments were carried out in a gasketed diamond cell with synchrotron radiation. For $PbZrO_3$, both energy-dispersive and angular-dispersive X ray diffraction methods were employed, whereas for $PbTiO_3$ only the former method was used. The experimental setup for these two diffraction techniques and the diamond cell have been described elsewhere [e.g., *Ming et al.*, 1994; *Tang et al.*, this volume].

In the energy-dispersive runs, the fine-grained sample of $PbZrO_3$ or $PbTiO_3$ was mixed either with NaCl or with MgO or with very fine aluminium powder, which serves both as the pressure marker and the pressure medium. Pressure was determined from the (200) and (220) peaks of NaCl (B1), (200) of MgO, and (111) of Al along with their respective equation of state [*Decker*, 1971, for NaCl; *Jamieson et al.*, 1982, for MgO; and *Ming and Manghnani*, 1986, for Al]. At pressures higher than 27 GPa, NaCl (B1) transforms into NaCl (B2) and the pressure is estimated from the (110) peak and the equation of state of the B2 phase [*Heinz and Jeanloz*, 1984]. Experiments were carried out with an intrinsic Ge detector using white radiation of a bending magnet beam line (II-2) at Stanford Synchrotron Radiation Laboratory (SSRL). The 2θ angle between the direct beam and the detector for our energy dispersive experiments was set at ~14°. With the storage ring operating at 3 GeV and 90-60 mA, and a collimated incident beam 50 μ in diameter, each energy spectra was collected for 5 minutes in real time.

In the angular-dispersive runs, the pure sample powder was loaded together with a tiny ruby chip in a medium of 4:1 methanol-ethanol mixture. Sample pressure was determined from the well known ruby fluorescence technique [e.g., *Barnett et al.*, 1973]. Experiments were carried out with an imaging plate using a monochromatic radiation (λ =0.6888 Å) of a bending magnet beam line at the Photon Factory at KEK, Japan. The distance between the sample and the imaging plate is 164.48929 mm. With the storage ring operating at 2.5 GeV and 360 mA and a collimated beam of 75 μ in diameter, each energy spectra was collected for 80 minutes in real time.

3. RESULTS AND DISCUSSIONS

3.1 PbZrO3

Three separate runs were carried out to 14.5, 25, and 51.3 GPa, respectively, for this material. The first two were with the energy-dispersive mode using NaCl and Al, respectively, as the internal pressure calibrant, and the third run was in the angular-dispersive mode. X ray diffraction data obtained from the methods are generally in good agreement. However, we present in the following only the data obtained from the third run using an imaging plate. This is because of the much higher quality spectrum obtained and the much wider pressure range covered in this run. Nonetheless, it is to be anticipated that the bulk modulus derived from the first two runs (in a dry condition) will be higher than that from the third run under a hydrostatic/quasi-hydrostatic environment.

Figure 1 is a plot of d-spacing as a function of pressure. As can be seen, splitting appears at ~6 GPa for several diffraction lines (e.g., at d = ~2.9 Å, ~2.35 Å, ~1.83 Å, ~1.65 Å and 1.45 Å) and also at pressures above 37 GPa. The splitting of a diffraction line at a given pressure usually indicates the occurrence of a phase transition at that pressure. A closer examination, however, has shown that it is not the case for $PbZrO_3$ at ~6 GPa. This is because the original orthorhombic unit cell can fit X ray diffraction data reasonably well to ~33 GPa, implying that the splitting of those lines simply resulted from the different compressibility along the three different axes. In

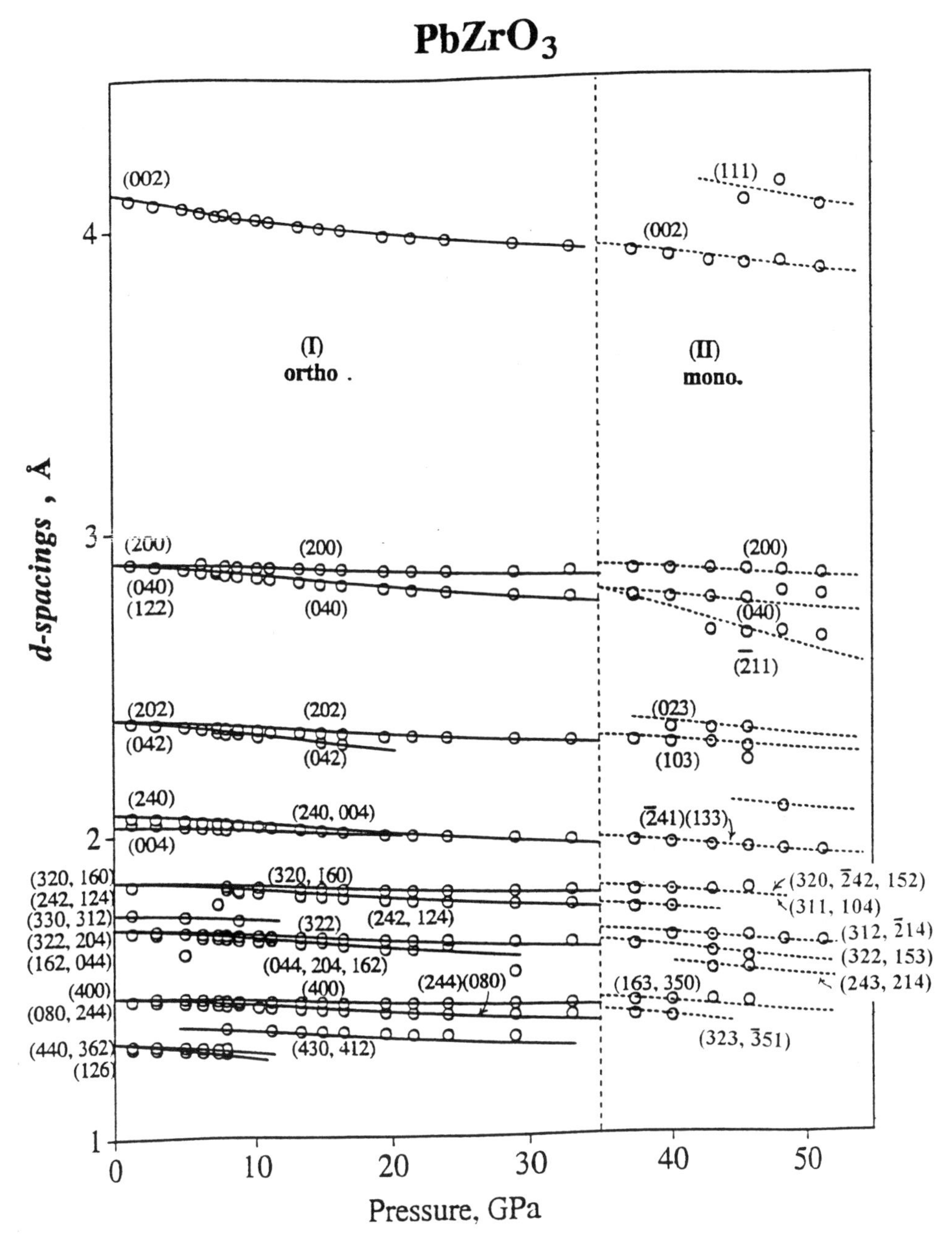

Figure 1. The effect of pressure on the *d*-spacings of $PbZrO_3$. The dotted vertical line indicates the orthorhombic $PbZrO_3$ (I) transforms to the monoclinic $PbZrO_3$ (II) at ~35 GPa.

determining the unit cell of the high-pressure phase of $PbZrO_3$ at pressure above 37 GPa, we have set two criteria: (1) it provides a good fit to the diffraction data, and (2) it yields only a slightly smaller molar volume than that of its low-pressure counterpart. The first criteria is necessary but not sufficient because the fitting process usually generates many solutions that fit equally well for a given set of *d*-spacings. The second criteria comes from crystal chemistry considerations for the perovskite structure, viz. the most efficiently packed crystal structure such as perovskite is unlikely to have phase transformations associated with a large volume change. These two criteria led us to a monoclinic unit cell for the $PbZrO_3$ (II) above 37 GPa. It is therefore suggested that the orthorhombic phase (I) transforms into the monoclinic phase (II) at about 35 GPa and persists to 51.3 GPa, the highest pressure in this study.

TABLE 1. X Ray Diffraction Data of $PbZrO_3$ (I) at 10.2 GPa and $PbZrO_3$ (II) at 43 GPa

	Phase (I) 10.2 GPa			Phase (II) 43.0 GPa		
(I/Io)[a]	*d*(obs.)	*d*(cal.)[b]	(hkl)	*d*(obs.)	*d*(cal.)[c]	(hkl)
M	4.035	4.030	(002)	3.878	3.869	(002)
VS	2.884	2.886	(200)	2.863	2.854	(200)
		2.859	(004)			
		2.861	(122)	2.766	2.766	(040)
					2.764	(210)
M				2.657	2.661	(-211)
M	2.345	2.346	(202)	2.332	32.338	(023)
M	2.325	2.322	(042)	2.284	2.291	(103)
S	2.028	2.031	(240)	1.946	1.947	(-241)
		2.105	(004)		1.946	(133)
W	1.826	1.823	(320)			
		1.810	(160)			
W	1.807	1.814	(242)	1.799	1.803	(-242)
		1.805	(142)		1.802	(152)
					1.799	(322)
					1.794	(311)
						(104)
M	1.663	1.661	(322)	1.647	1.644	(312)
					1.638	(-214)
	1.648	1.652	(162)			
		1.647	(044)	1.597	1.592	(322)
		1.652	(204)			(153)
VW				1.541	1.538	(243)
					1.537	(214)
VW	1.447	1.443	(400)	1.437	1.443	(350)
					1.437	(163)
M	1.429	1.430	(080)		1.432	(323)
		1.431	(244)		1.432	(-351)

[a] VS stands for very strong; S, strong; M, medium strong; W, weak; VW, very weak.
[b] Calculated on the basis of an orthorhombic unit cell with $a = 5.774$ (4) Å, $b = 11.440$ (9) Å, and $c = 8.059$ (6) Å
[c] Calculated on the basis of a monoclinic unit cell with $a = 5.722$ Å, $b = 11.064$ Å, $c = 7.756$ Å, and $\beta = 94°$.

Typical X ray diffraction spectra for the orthorhombic phase (I) and the monoclinic phase (II) are shown in Figure 2, and the corresponding *d*-spacing data are given Table 1. The lattice parameters and the molar volumes of each phase thus calculated at each given pressure condition are tabulated in Table 2 and plotted in Figure 3. It can be seen that lattice parameters (*a*, *b*, and *c*) of the phases (I) and (II) are very similar to one another and there is no volume change at the transition pressure. These features indicate the close structural relationship between phase I and phase II and are in accord with the general structural behavior in the ABO_3 perovskite materials that phase transformations are usually introduced by a small distortion due to the tilt and/or rotation of the corner-linked octahedrons [e.g., *Liu et al.*, 1988; *Liu and Liebermann*, 1993].

It is worth noting that in both $PbZrO_3$ and $PbHfO_3$, the most intense peak at $d = \sim 2.9$ Å behaves very similarly in that it becomes a doublet at some high pressure and then a

Table 2. Lattice Parameters (a, b, c, in Å and β in °) and the Molar Volume (V in cm^3/mole and V/V_0) of $PbZrO_3$ (I) and $PbZrO_3$ (II) as a Function of Pressure.

P	a	b	c	β	V	V/V_0
Orthorhombic $PbZrO_3$ (I)						
0.0001	5.880	11.772	8.220	90	42.837	1.000
1.30	5.830	11.660	8.190	90	41.915	0.978
2.97	5.830	11.617	8.162	90	41.618	0.972
4.94	5.807	11.569	8.127	90	41.106	0.960
6.16	5.791	11.540	8.100	90	40.754	0.951
7.88	5.785	11.523	8.080	90	40.551	0.947
8.78	5.768	11.474	8.086	90	40.290	0.941
10.2	5.773	11.440	8.059	90	40.071	0.935
11.1	5.768	11.411	8.068	90	39.980	0.933
13.2	5.764	11.391	8.014	90	39.601	0.924
14.8	5.758	11.373	8.016	90	39.521	0.923
16.4	5.756	11.343	7.971	90	39.182	0.915
19.5	5.739	11.244	7.924	90	38.497	0.899
21.5	5.733	11.221	7.909	90	38.305	0.894
24.0	5.728	11.214	7.929	90	38.345	0.895
29.0	5.715	11.146	7.866	90	37.723	0.881
32.9	5.704	11.112	7.837	90	37.398	0.873
Monoclinic $PbZrO_3$ (II)						
37.4	5.730	11.112	7.832	94.2	37.443	0.874
40.0	5.724	11.076	7.798	94.1	37.125	0.867
43.0	5.722	11.064	7.756	94.0	36.877	0.861
45.7	5.722	11.048	7.746	93.8	36.773	0.858
48.3	5.705	11.050	7.720	93.0	36.589	0.858
51.3	5.700	11.020	7.710	94.0	36.373	0.849

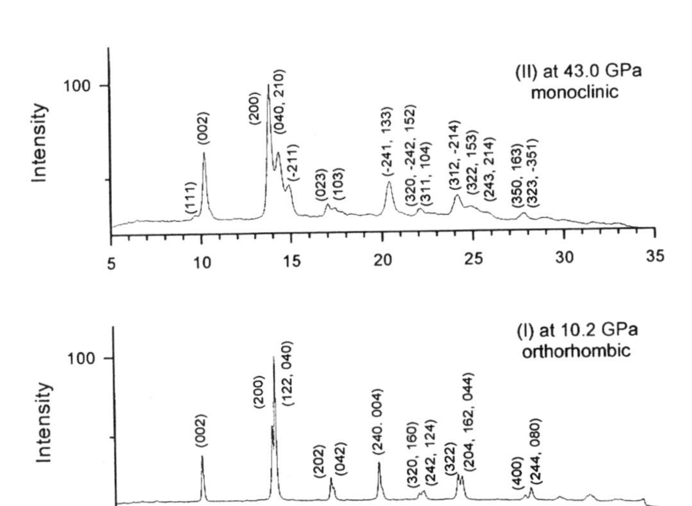

Figure 2. Selected spectrum of $PbZrO_3$ at 10.2 and 43 GPa.

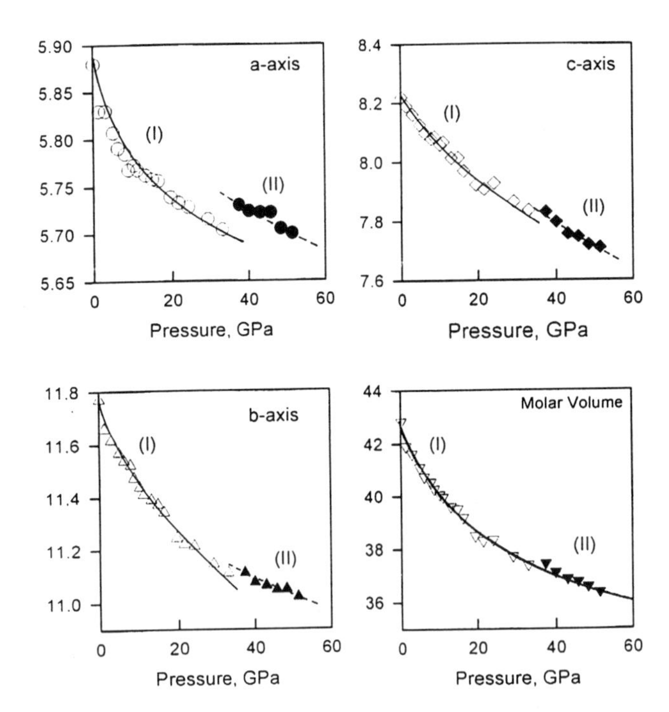

Figure 3. The effect of pressure on the lattice parameters (a, b, and c) and the molar volumes of $PbZrO_3$ (I) and (II).

triplet at much higher pressures. This suggests that these antiferroelectric perovskites assume a lower symmetry as pressure increases. However, the phase transformation sequence between them is different. For example, there is no tetragonal phase in $PbZrO_3$ as observed in $PbHfO_3$ (II) and no monoclinic phase in $PbHfO_3$ as found in $PbZrO_3$ (II) [*Ming et al.*, 1994].

The bulk modulus K_0 and its pressure derivative K_0' can be calculated by fitting the P-V data to the second-order Birch-Murnaghan equation:

$$P = 1.5\, K_0[X^{-7/3} - X^{-5/3}][1 - \xi(X^{-2/3} - 1],$$

where $X = V/V_0$, $\xi = 0.75(4 - K_0')$ and $K_0' = (d\,K_0/dP)_T$.

Fitting the P-V data of $PbZrO_3$ to 32.7 GPa and assuming $K_0' = 4$, we obtained a value of K_0 to be 154 GPa for $PbZrO_3$ (I). This value is slightly larger than $K_0 = 142$ GPa of $PbHfO_3$ (I) [*Ming et al.*, 1994]. This is not consistent with the empirical relationship such that $K_0\ V_0$ = constant proposed for the isostructural materials [*Anderson*, 1972], because $PbZrO_3$, which has a slightly larger molar volume (V_0 = 42.837 cm^3/mol) than that of $PbHfO_3$ (V_0 = 42.801 cm^3/mol), should have a slightly smaller bulk modulus.

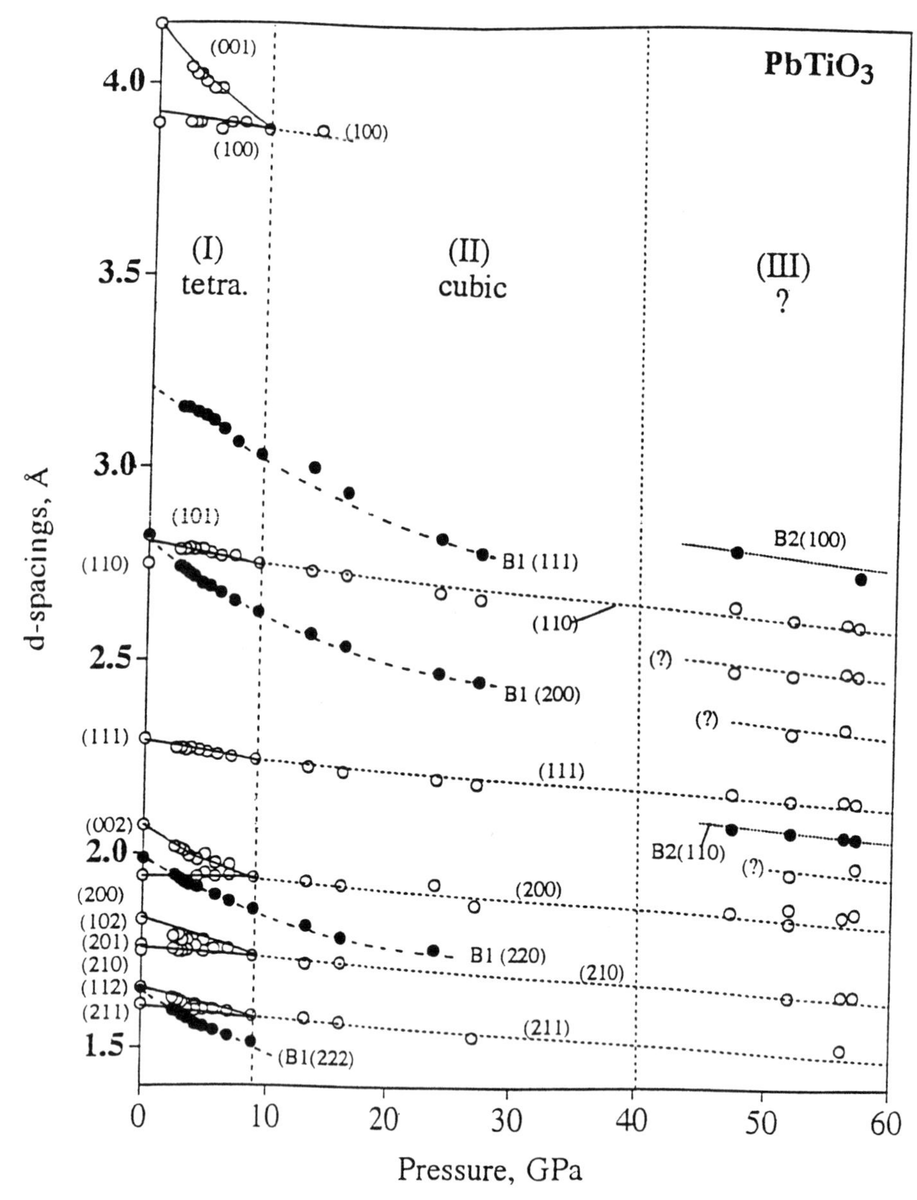

Figure 4. The effect of pressure on the *d*-spacings of $PbTiO_3$ + NaCl where open and solid symbols stand for $PbTiO_3$ and NaCl respectively. The dotted vertical lines indicate the transition pressures at 8 and ~40 GPa, respectively for the tetragonal $PbTiO_3$ (I) to the cubic $PbTiO_3$ (II), and for $PbTiO_3$ (II) to $PbTiO_3$ (III?). B1 and B2 represent the low-pressure (NaCl str.) and high-pressure (CsCl str.) of NaCl, respectively. For clarity, fluorescence lines and the possible diffraction lines from the gasket material are not plotted in the figure. The disappearance of the (100) of $PbTiO_3$ (II) at pressures above 14 GPa is because it moves into the fluorescence peak of Pb.

3.2 $PbTiO_3$

Only one energy-dispersive X ray diffraction run was carried out for $PbTiO_3$ to 57 GPa. Figure 4 is a plot of *d*-spacings of $PbTiO_3$ as a function of pressure showing the transformation of the tetragonal phase to the cubic phase. This occurs when several pairs of peaks such as (001)/(100), (200)/(002), (102)/(201), and (112)/(211) merge into one at ~8 GPa. The cubic phase II appears to persist to 57 GPa. However, above 47 GPa, besides those lines from the cubic $PbTiO_3$ and the NaCl (B-2), there are three very weak lines (see Table 3). It may be possible that the cubic phase II further changes to a denser phase at higher pressures. Further studies are needed to confirm such a speculation.

TABLE 3. X Ray Diffraction Data of $PbTiO_3$ at 56 GPa

I/Io	d(obs.)	d(cal.)[a]	Phase PbTiO3 (II)	NaCl (B2)
W	2.926	2.903		(100)
W	2.758?			
S	2.607	2.605	(101)	
W	2.479?			
W	2.331?			
M	2.146	2.126	(111)	
S	2.053	2.053		(110)
M	1.846	1.842	(200)	
W	1.642	1.647	(210)	
W	1.504	1.504	(211)	

[a] Calculated on the basis of a cubic unit cell with $a = 3.684$ Å and a = 2.903 Å for $PbTiO_3$ (II) and NaCl (B2), respectively.

TABLE 4. The Effect of Pressure (in GPa) on Lattice Parameters (a and c in Å), the Axial Ratio (c/a), Molar Volume (V_m in cm^3/mol), and the Volume Ratio (V/V_0) of $PbTiO_3$

P	a	c	c/a	Vm	V/V0
		Tetragonal PbTiO3 (I)			
0.0001	3.900	4.142	1.062	37.945	1.000
2.50	3.884	4.027	1.037	36.598	0.965
2.80	3.874	4.020	1.038	36.337	0.958
3.20	3.876	3.999	1.032	36.185	0.954
3.60	3.904	3.983	1.020	36.563	0.964
4.20	3.902	3.965	1.016	36.361	0.958
4.80	3.870	3.991	1.031	36.001	0.949
5.60	3.874	3.945	1.018	35.724	0.942
6.90	3.881	3.931	1.014	35.716	0.941
		Cubic PbTiO3 (II)			
8.70	3.876			35.072	0.924
13.1	3.857			34.559	0.910
15.9	3.842			34.157	0.900
23.8	3.814			33.416	0.881
27.0	3.776			32.427	0.855
47.2	3.731			31.282	0.824
51.8	3.698			30.459	0.803
56.0	3.684			30.114	0.794
57.0	3.676			29.918	0.789

The transition pressure of ~8 GPa observed in this study is higher than the previously reported value of 5.8 GPa [*Ikeda*, 1975] based on the dielectric and X ray diffraction measurements but is much less than 12 GPa based on the high-pressure Raman measurements [*Sanjurjo et al.*, 1983; *Jayaraman et al.*, 1984] and the 11.5 GPa based on the optical absorption measurement [*Zha et al.*, 1992]. The large discrepancy in the reported transition pressures is probably due to the different degree of non-hydrostaticity of the sample environment. Since previous Raman and the optical absorption measurements were carried out in a diamond-anvil cell with alcohol mixture as the pressure medium, and our experiment was carried out in a diamond-anvil cell with NaCl as the pressure medium, it is not unreasonable that those measurements under truly hydrostatic condition would yield a much higher transition pressure than our measurements under quasi-hydrostatic conditions, provided that the transition is highly sensitive to the shear stress. It is also possible that both the sample type (single crystal vs. polycrystalline) used and the microstructure of the sample (i.e., level of defects, impurities) have some effects on the transition pressure. Since no detailed descriptions on the sample and its pressure medium were given in *Ikeda's* report, it is not possible for us to offer any comments on his transition pressure being so low. Based on the transition pressure (i.e., 8 GPa) at room temperature and the transition temperature of 490°C, [*Samara*, 1970] at 1 bar pressure, we calculated the slope of the phase boundary between the tetragonal and the cubic phase as -58.4 °C /GPa.

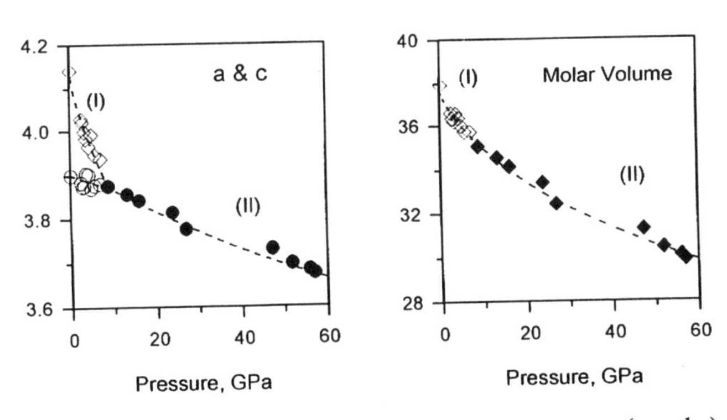

Figure 5. The effect of pressure on the lattice parameters (a and c) and molar volumes of $PbTiO_3$ (I) and (II).

Lattice parameters (a, c, and c/a) and molar volume (V and V_0) thus calculated are tabulated in Table 4 and plotted in Figure 5. One important question is concerned with the nature of the tetragonal-cubic transition in $PbTiO_3$: Is it a first-order type or a second-order type? Based on the X ray diffraction measurements to 8 GPa, *Ikeda* [1975] suggested that it is a first-order type transition because a volume change of 0.25 $Å^3$ per unit cell was observed at the transition. However, high-pressure Raman measurements have suggested it be a second-order type because the coalescence of pairs of modes (i.e., A1(2TO) + E(2TO) and A1(3TO) + E(3TO)) occurs at the transition [e.g., *Sanjurjo et al.*, 1983]. Our *P-V* data shown in Figure 5 can be fitted reasonably well by a smooth curve. It appears to be unnecessary to introduce a discontinuity of $\Delta V = 0.25$ $Å^3$ at the transition pressure (8 GPa) as suggested by *Ikeda* [1975]. Our X ray data seem to be consistent with the suggestion that the transition is second-order. It should be remembered that X ray diffraction data alone cannot provide a conclusive remark on this issue (i.e, first-order vs. second-order). This is because the precision of the *d*-spacing in the energy-dispersive X ray diffraction is, at best, of 0.1%, yielding 0.3% uncertainty for the volume. Any volume change within 0.3% would be difficult to evaluate.

It is evident that the *c*-axis is much more compressible than the *a*-axis, thus resulting in a decrease of the *c/a* ratio as pressure increases. The linear compressibility of *a*-and *c*-axes can be represented approximately by linear equations:

$$a/a_0 = 1 - 6.14 \times 10^{-4} \times P \text{ (GPa)},$$
$$c/c_0 = 1 - 6.26 \times 10^{-3} \times P \text{ (GPa)},$$

thus resulting in the compressibility along the *c*-axis as $\beta_c = 6.26 \times 10^{-3}$ GPa^{-1}, which is an order of magnitude higher than that along the *a*-axis ($\beta_a = 6.14 \times 10^{-4}$ GPa^{-1}). These results are in good agreement with the single crystal measurements to 6.9 GPa [*Nelmes and Katrusiak*, 1986].

The decrease of the tetragonality, as indicated by the *c/a* , can be represented as

$$c/a = 1.055 - 0.0069 \times P \text{ (GPa)}.$$

The value of $- 6.9 \times 10^{-3}$ for $d(c/a)/dP$ is in good agreement with 0.76×10^{-3} reported by *Ikeda* [1975].

Fitting the *P-V* data for $PbTiO_3$ (I) from the data given in Table 4 using the Birch-Murnaghnan equation, we obtain a value of K_0 = 82.4 GPa when K_0' is assumed to be 4.0. This K_0 value is in fairly good agreement with K_0 = 85.7 GPa by earlier X ray diffraction measurements [*Kabalkina and Vereshchagin*, 1962] and with K_0 = 88 GPa obtained very recently using Brillouin light scattering [*Bass and Kalinchev*, 1994]. It is interesting to note that $PbTiO_3$ has the smallest molar volume (V_0 = 37.945 cm^3/mol) among the three perovskites discussed in this paper (i.e., V_0 = 42.837 cm^3/mol for $PbZrO_3$ and V_0 = 42.801 cm^3/mol for $PbHfO_3$), and yet it also has a much smaller K_0 value in comparison with the other two (i.e., 154 GPa for $PbZrO_3$, and 142 GPa for $PbHfO_3$). This is not in accord with $K_0 V_0$ = constant systematics, and it would be understood from the fact that $PbTiO_3$ is basically different from $PbZrO_3$ and $PbHfO_3$ both in structural (tetragonal vs. orthorhombic) and in dielectric (ferroelectric vs. antiferroelectric) properties.

Acknowledgment. This work was partially done at SSRL, which is operated by the Department of Energy, Office of Basic Energy Science. We would like to thank A. Jayaraman for reading the manuscript and giving us many valuable suggestions and Y.H. Kim for helping collecting synchrotron data at SSRL. We also thank W. Utsumi, G. Samara, and an anonymous reviewer for their valuable suggestions and comments on the manuscript. This work was partially supported by a NSF grant (DMR94-02443). School of Ocean and Earth Science and Technology No. 4189.

REFERENCES

Anderson, O. L., Patterns in elastic constants of minerals important to geophysics, in *The Nature of the Solid Earth*, edited by E. Robertson, pp. 575-613, McGraw-Hill, New York, 1973.

Barnett, J. D., S. Block, and G. J. Piermarini, An optical fluorescence system for quantitative pressure measurement in the diamond-anvil cell, *Rev. Sci. Intrum.*, *44*, 1-9, 1973.

Bass, J. D., and A. G. Kalinchev, Elastic and piezoelastic properties of $PbTiO_3$ tetragonal single crystal by Brillouion scattering, *Eos, Trans. Am. Geophys. Union, Spring Meeting* (abs.), 187, 1994.

Decker, D. L., High pressure equation state of NaCl, KCl, and CsCl, *J. Appl. Phys.*, *42*, 3239-3247, 1971.

Hazen, R. M., L. W. Finger, Bulk modului and high-pressure crystal structures of rutile-type compounds, *J. Phys. Chem. 42*, 143-151, 1981.

Heinz, D. L., and R. Jeanloz, Compression of the B2 phase of NaCl, *Phys. Rev. B*, *30*, 6045, 1984.

Ikeda, T., Effect of hydrostatic pressure on the phase transition of ferroelectric $PbTiO_3$, *Solid State Comm*, *16*, 103-104, 1975.

Jamieson, J. C., J. N. Fritz, and M. H. Manghnani, Pressure measurement at high temperature in X-ray diffraction studies: gold as a primary standard, in *High Pressure Research in Geophysics*, edited by S. Akimoto and M.H. Manghnani, pp. 27-48, Center for Academic Publishing, 1982.

Jayaraman, A., S. K. Sharma, L. C. Ming, and S. Y. Wang, High pressure Raman studies on ABO_3 perovskite type structure compound $PbHfO_3$ to 40 GPa: Pressure-induced phase transitions, *J. Phys. Chem. Solids*, *55*, 1207-1212, 1994.

Jayaraman, A., J. P. Remeika, and R. S. Katiyar, A high pressure Raman study of ferroelectric to paraelectric transition in $BaTiO_3$ and $PbTiO_3$, *Mat. Res. Soc. Sym. Proc.*, *22*, edited by C. Homan, R. K. MaeCrone, and E. Whalley, pp. 165-168, Elsevier, New York, 1984.

Kabalkina, S. A., and L. F. Vereshchagin, An X ray study of the effect of hydrostatic pressure on the structure of lead titanate, *Soveit Phys.- Doklady*, *7*, 310-312, 1962.

Liu, X., Y. Wang, and R. C. Liebermann, Orthorhombic-tetragonal phase transition in $CaGeO_3$ perovskite at high temperature, *Geophys. Res. Lett.*, *15*, 1231-1234, 1988.

Liu, X. and R. C. Liebermann, X ray powder diffraction study of $CaTiO_3$ perovskite at high temperatures, *Phys. Chem, Minerals*, *20*, 171-175, 1993.

Ming, L. C., D. Xiong, and M. H. Manghnani, Isothermal compression of Au and Al to 20 GPa, *Physica*, *139 & 140B*, 174-176, 1986.

Ming, L. C., A. Jayaraman, S. Shieh, Y. H. Kim, and M. H. Manghnani, An in-situ high pressure X-ray diffraction study of $PbHfO_3$ to 52.5 GPa at room temperature and pressure-induced phase transformation, *J. Phys. Chem. Solids*, *55*, 1213-1219, 1994.

Nelmes, R. J., and A. Katrusiak, Evidence for anomalous pressure dependence of the spontaneous strain in $PbTiO_3$, *J. Phys. C: Solid State Phys.*, *19*, L25-L30, 1986.

Pisarski, M., and Z. Ujima, The influence of hydrostatic pressure and defects in the Pb and O sub-lattice of the parameters of phase transitions in $PbZrO_3$, *Mat. Res. Soc. Sym. Proc,. 22*, *Mat. Res. Soc. Sym. Proc., 22*, edited by C. Homan, R. K. MaeCrone, and E. Whalley, pp. 169-174, Elsevier, New York, 1984.

Samara, G. A., The effects of hydrostatic pressure on ferroelectric properties, in *Advances in High Pressure Research*, edited by R. S. Bradley, pp. 155-239, Academic Press, New York, 1969.

Sanjurjo, J. A., E. Lopez-Cruz, and G. Burns, High pressure Raman study of zone-center photons in $PbTiO_3$, *Phys. Rev. B*, *28*, 7260-7267, 1983.

Zha, C. S., A. G. Kalinchev, J. D. Bass, C. T. A. Suchicital, and D. A. Payne, Pressure dependence of optical absorption in $PbTiO_3$ to 35 GPa: observation of tetragonal to cubic phase transition, *J. Appl. Phys.*, *72*, 3705-3707, 1992.

S. Endo and Y. Kobayashi, Research Center for Materials Sciences at Extreme Conditions, Osaka University, Osaka, 560, Japan.

T. Kikegawa and O. Shimomura, National Laboratory for High Energy Physics, Tsukuba, 305, Japan.

L. C. Ming, Hawaii Institute of Geophysics and Planetology, University of Hawaii at Manoa, Honolulu, HI 96822 (e-mail: ming@soest.hawaii.edu).

S.R. Shieh, Department of Geology and Geophysics, University of Hawaii at Manoa, Honolulu, HI 96822.

Phase Relations and Physical Properties of Fe_2SiO_4-Fe_3O_4 Solid Solution Under Pressures up to 12 GPa

T. Yamanaka , H. Tobe, T. Shimazu, A. Nakatsuka, Y. Dobuchi, O. Ohtaka, and T. Nagai

Department of Earth and Space Science, Graduate School of Science, Osaka University, Osaka, Japan

E. Ito

Institute for Study of the Earth's Interior, Okayama University, Misasa, Japan

Phase study of the Fe_2SiO_4-Fe_3O_4 solid solution system has been carried out under high pressures up to 12 GPa at 1200°C by multianvil apparatus. A complete spinel solid solution between Fe_3O_4 and γ-Fe_2SiO_4 has been found at pressures over 10 GPa. γ-Fe_2SiO_4 with a normal spinel structure is stable at pressures above 7 GPa. A spinelloid structure similar to aluminosilicate V (Pmma) in the $NiAl_2O_4$-Ni_2SiO_4 system is found in a wide intermediate compositional range x=0.37 to 0.73 in $Fe_{3-x}Si_xO_4$ at pressures between 3 and 9 GPa. X ray single-crystal structure analyses of several samples of $Fe_{3-x}Si_xO_4$ spinel indicate the site occupancy of $(Fe^{3+}_{1-x+y}\ Si^{4+}_{x-y})\ [Fe^{2+}_{1+x}Fe^{3+}_{1-x-y}Si^{4+}_{y}]\ O_4$. This cation distribution affects the electrical conductivity mainly because of the electron hopping between Fe^{3+} and Fe^{2+} in the octahedral site. Measurement of the electrical conductivity of the spinel solid solution has been made at low temperatures in the range 80 to 300 K. The transition temperature of the Verwey order between Fe^{3+} and Fe^{2+} decreases with the Si content in $Fe_{3-x}Si_xO_4$ and the energy gap becomes smaller with Si.

1. INTRODUCTION

Phase relations in the system Fe_2SiO_4-Fe_3O_4 are important to understand the transition zone and the spinel lherzolite. Substitution of Ti^{4+} in Fe_3O_4 makes titaniferrous magnetite, which is found in basic igneous rock and forms a continuous solid solution between magnetite Fe_3O_4 and ulvospinel Fe_2TiO_4. But Si^{4+} substitution in magnetite is rarely found in natural minerals. However, the recent discovery of silicious magnetite in scarn minerals [*Shiga*, 1988, 1989] stimulated the investigation of the solid solution between magnetite and iron-silicate.

The olivine-spinel transformation commands attention in order to comprehend the phase transition in Earth's subduction zones; to that end, many high pressure studies of various silicate spinels have been carried out. The Fe_2SiO_4 has the olivine structure (α-Fe_2SiO_4) at ambient condition and transforms into spinel structure (γ-Fe_2SiO_4) under high pressure [*Akimoto et al.*, 1965, 1967; *Ito and Yamada*, 1982]. The β-spinel phase or spinelloid phases found in Mg_2SiO_4 and Co_2SiO_4 have never been known in the natural compound.

Ion substitution with different ionic radii and charges causes a tropochemical superstructure and provides spinelloid polymorphs in the intermediate solid solution of

Properties of Earth and Planetary Materials
at High Pressure and Temperature
Geophysical Monograph 101

TABLE 1. Experimental Conditions, Obtained Products and Composition of Each Run in $Fe_{3-x}Si_xO_4$

Run No.	Press. (GPa)	Starting ratio Fa : Mat	Composition of product (x)	Phase of product	Apparatus
1	6	5 : 5	0.48	spl	cubic
2	6	3 : 7	0.37, 0.20	spl + sp	cubic
3	6	7 : 3	0.65	spl	cubic
4	6	1 : 9	0.05	sp	cubic
5	6	9 : 1	0.73, 0.90	spl + ol	cubic
6	10	5 : 5	0.49	sp	6 - 8
7	10	3 : 7	0.28	sp	6 - 8
8	10	7 : 3	0.70	sp	6 - 8
9	3	5 : 5	0.34, 1.00	spl + ol	cubic
10	3	3 : 7	0.34, 0.02	spl + sp	cubic
11	3	7 : 3	0.34, 1.00	spl + ol	cubic
12	4	7 : 3	0.36, 1.00	spl + ol	cubic
13	5	9 : 1	0.60, 1.00	spl + ol	cubic
14	5	7 : 3	0.60, 1.00	spl + ol	cubic
15	6	Fa	1.00	sp	cubic
16	9	5 : 5	0.49	sp>>spl	6 - 8
17	10	Fa	1.00	sp	6 - 8
18	12	Fa	1.00	sp	6 - 8

Starting ratio (Fa: fayalite, Mat: magnetite)
Product (sp: spinel, spl: spinelloid, ol: olivine)
Apparatus (cubic: cubic anvil, 6 - 8: split sphere multianvil apparatus).

various systems, such as $NiAl_2O_4$-Ni_2SiO_4 [*Ma*, 1974], $MgGa_2O_4$-Mg_2GeO_4 [*Barbier and Hyde*, 1986; *Barbier*, 1989], $MgFe_2O_4$-Mg_2GeO_4 and $NiGa_2O_4$-Ni_2GeO_4 [*Hammond and Barbier*, 1991]. No intermediate phase has been proposed in the system $FeAl_2O_4$-Fe_2SiO_4, $CoAl_2O_4$-Co_2SiO_4, $FeGa_2O_4$-Fe_2GeO_4, and $CoGa_2O_4$-Co_2GeO_4.

Structures of five aluminosilicate polytypes phases I…V have been refined by *Ma et al.* [1975], *Ma and Tillmans* [1975], *Ma and Sahl* [1975], and *Horioka et al.* [1981]. Discussion of the thermodynamical stability of these phases has been investigated by calorimetric study [*Akaogi and Navrotsky*, 1984]. All five phases are spinel derivative structures and are designated as spinelloid [*Horioka et al.*, 1981]. In the course of these studies, spinelloid structure has been found in the system Fe_3O_4-Fe_2SiO_4 [*Canil et al.*, 1990] and the crystal structure of $Fe_{2.2}Si_{0.8}O_4$ was reported as an isostructure with nickel aluminosilicate V [*Ross et al.*, 1992]. However, the phase diagram of this binary system has never been proposed.

In this paper we present phase relations in this system and clarify the pressure and compositional stability range of both spinel and spinelloid solid solution. We have also investigated the electrical conductivity and crystallography of Fe_2SiO_4-Fe_3O_4 solid solution by X ray crystal structure refinement and direct current (DC) resistivity measurements.

2. EXPERIMENTAL

2.1. *High Pressure Study*

In the course of this study the phase diagram of the Fe_2SiO_4-Fe_3O_4 solid solution was made to clarify the stability region of spinel and spinelloid at high pressure and temperature. Mixtures of reagent grade Fe_3O_4 and pure α-Fe_2SiO_4 were used for high pressure synthesis. The latter was prepared from FeOOH and SiO_2 under controlled oxygen pressures [*Nitsan*, 1974] using CO_2 and H_2 gases. These starting materials were loaded in gold capsules 3.0 mm in diameter and sealed. A cubic-anvil type device with WC anvils of 10 mm edge length and pressure media of 12.5 mm cube was used below 7 GPa. For the experiments above 7 GPa , a split sphere multi-anvil type apparatus was applied using WC cubes with an 8-mm truncated edge length and 14-mm edge octahedral pressure media. The cell assemblies of both types are illustrated in Figure 1. The run temperatures were measured with a Pt/Pt13Rh thermocouple inserted into the center of the cell. The samples were held at 1200°C for 5 h and then quenched under pressure.

The phase identification of the recovered samples was carried out by a conventional X ray powder diffractometer with monochromatic CoKα radiation of λ=1.78965Å.

Their compositional analyses were undertaken with an electron probe microanalyzer (EPMA) energy-dispersive method, with Fe_2O_3 (hematite) and pure α-Fe_2SiO_4 used as standards. The composition of each phase was determined by applying the Bence-Albee calibration method and the homogeneity of the composition was also confirmed. The texture and grain size were examined by SEM image. The experimental conditions and products of individual runs are summarized in Table 1. In this paper, the composition of samples is indicated by x, referring to $Fe_{3-x}Si_xO_4$.

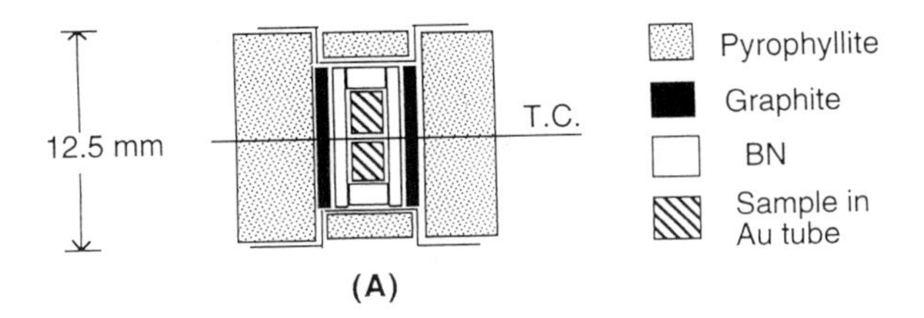

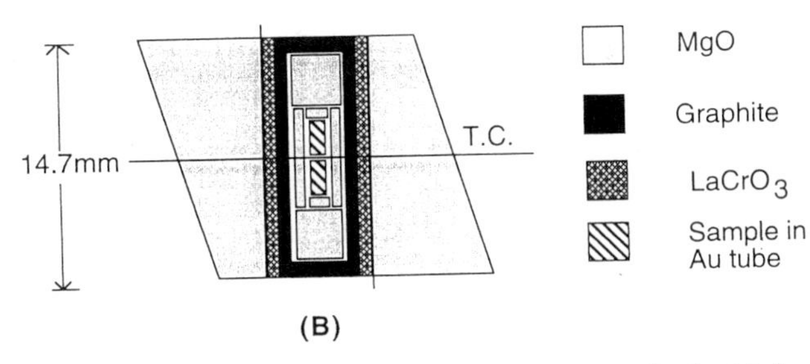

Figure 1. Cell assemblages for high pressure synthesis. (a) for the cubic type used below 7 GPa and (b) for 6-8 split sphere type of multianvil.

2.2. *Single-Crystal Structure Analysis*

Structure analyses of the spinel solid solution and spinelloid have been executed by single-crystal diffraction study and Mössbauer spectra measurement. The variation of the cation site occupancies of $Fe_{3-x}Si_xO_4$ was also examined in the spinel structure. The single-crystal was grown by keeping the sample for 5~6 hours under pressure. A specimen with ~100 mm in diameter in each recovered sample was selected for the structure analysis.

Diffraction intensities were collected by four-circle diffractometer Rigaku AFC-6 with MoKα radiation (λ=0.71069Å) monochromated by pyrolytic graphite and emitted from a rotated anode X ray generator with 12 kW. The ω-2θ scanning mode and scan speed of 2°/min in 2θ were applied for the intensity measurement. Reflections with intensities Fo>3σFo and within the range $0.1Å^{-1} < \sin\theta/\lambda < 1.32Å^{-1}$ were used for the structure refinement, where Fo is the observed structure factor and σ represents a standard deviation of Fo. The intensities were corrected for Lorentz-polarization effects and X ray absorption in the mode of spherical shape. The lattice constants of $Fe_{3-x}SiO_4$ various solid solutions were determined by the least-squares procedure applying to the observed 2θ values of 25 reflections.

The structure refinement using each data set was conducted by the full matrix least-squares program RFIN-IV written by *Finger* [1975] and RADY by *Sasaki and Tsukimura* [1987]. Converged values of the atomic coordinates, anisotropic temperature factors, site occupancy parameters and isotropic extinction parameters are presented in Table 2. The initial structure parameters were from *Yamanaka* [1986]. The reliability factor of the least-squares refinement is $Rw(F) = \Sigma w(|F_{calc}| - |F_{obs}|)^2/\Sigma w|Fo|^2$, where F_{calc} and F_{obs} are the calculated and observed structure factors, respectively; w is a weight factor.

2.3. *Electrical Resistivity Measurement*

It has been known that a high electrical conductivity in magnetite at room temperature is induced mainly from electron hopping between Fe^{2+} and Fe^{3+} in the octahedral site of the solid solution. At low temperatures below 120 K the hopping ceases by ordering these cations, which is called the Verwey order [*Verwey et al.*, 1947]. This transition temperature varies with the composition of the $Fe_{3-x}Si_xO_4$ spinel. Measurement of the electrical resistivity of the solid solution was made at temperatures in the range 80–300 K under N_2 dry gas-flow condition by a direct current (DC) four-probe method. Dimensions of the recovered sample were ~1.0 × 0.8 × 2.0 mm. A constant DC current of 0.120 mA was applied. The electrical resistivity was calculated by $\rho = (V/I)(S/l)$, where V is a voltage across the two electrodes, I is a constant DC current, S and l are a cross section of the sample and distance between the two electrodes, respectively. The Si compositional dependency of $Fe_{3-x}Si_xO_4$ on the Verwey transition temperature Tv and the electrical conductivity gap at the transition temperature have also been elucidated.

3. RESULT

3.1. *Phase Relation of the Fe_3O_4-Fe_2SiO_4 System*

The run products of the high pressure synthesis were first observed by back scattered electron imaging. Table 1

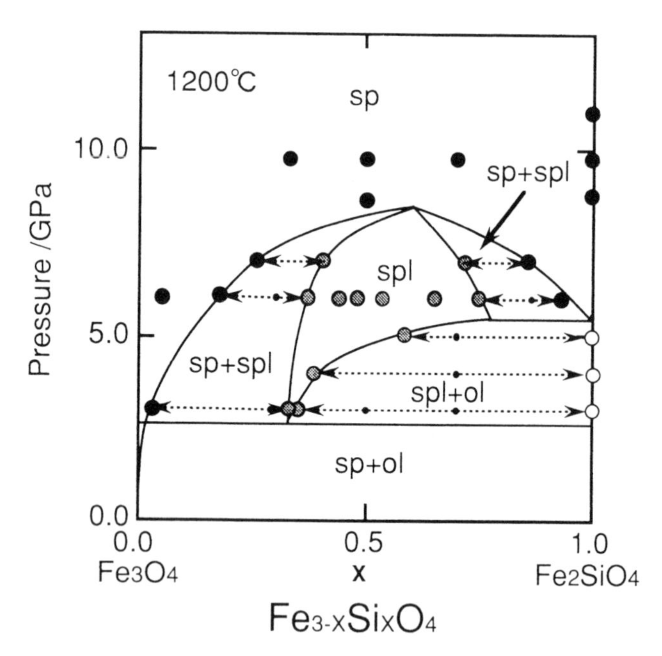

Figure 2. Phase diagram of Fe_2SiO_4-Fe_3O_4 system at 1200°C. The solid, open, and shaded circles represent spinel, olivine and spinelloid, respectively.

shows the results of phase identification with X ray diffraction (XRD) examination. The abbreviations SP, SPL, and OL indicate spinel, spinelloid, and olivine structure, respectively. The single phase of the spinel structure was obtained by the syntheses at 10 GPa. The phase diagram of the pseudo-binary system Fe_3O_4-Fe_2SiO_4 at 1200°C is presented in Figure 2 in which the stable regions of spinel, spinelloid, and olivine and two phase mixtures among these three phases are determined.

A complete solid solution between Fe_3O_4 and γ-Fe_2SiO_4 has been found at pressures over 10 GPa and 1200°C. It has been known that the former has an inverse spinel structure $(Fe^{3+})[Fe^{3+}, Fe^{2+}]O^{2-}_4$ and that the latter has a normal spinel structure of $(Si^{4+})[Fe^{2+}, Fe^{2+}]O^{2-}_4$ stable at high pressure above 7 GPa. The Si^{4+} substitution Fe^{3+} in a tetrahedral site of Fe_3O_4 causes the intracrystalline cation exchange and charge transfer from Fe^{3+} to Fe^{2+} in an octahedral site by the following ideal reaction:

$$(1-x)(Fe^{3+})[Fe^{3+}, Fe^{2+}O_4 + x(Si^{4+})[Fe^{2+}, Fe^{2+}]O_4 = (Fe^{3+}_{1-x}Si^{4+}_{x}))[Fe^{3+}_{1-x}, Fe^{2+}_{1+x}]O_4 \quad (1)$$

A spinelloid is stable in a wide intermediate compositional range from x=0.37 to 0.73 in $Fe_{3-x}Si_xO_4$ at pressures from 3 to 9 GPa. *Ross et al.* [1992] synthesized a single-crystal of $Fe_{5.2}Si_{0.8}O_8$ at 7 GPa and 1200°C with a spinelloid structure, and they confirmed that this phase was isostructural with nickel aluminosilicate V (Pmma) in the $NiAl_2O_4$-Ni_2SiO_4 system, referred to spinelloid V; [*Horioka et al.*, 1981]. The present X ray diffraction (XRD) patterns of spinelloid phases with x=0.48, 0.54, 0.65 synthesized at 6 GPa and 1200°C were successfully indexed with the aid of Rietveld synthesis in reference to structure analysis by Ross. It has been confirmed that no other polytype than aluminosilicate V type is stable in the Fe_3O_4-Fe_2SiO_4 system. Indices of reflections in the powder diffraction patterns of x=0.48, 0.54, and 0.65 are listed in Table 3, in which the lattice constant and unit cell volume of each sample are also presented. Figure 3 shows the XRD pattern of the spinelloid sample. The spinelloid phase that appeared in the present study was spinelloid V. The unit cell volume of the spinelloid phase indicates an almost linear dependence on composition.

The complete solid solution of spinel along the entire Fe_3O_4-Fe_2SiO_4 join occurs at 9 GPa, and at 7 GPa the range between x=0–15 and x=0.75–1.0 is interrupted by the presence of spinelloid phase in the range x=0.28~0.6 (indicated in Figure 2). Present results support those by *O'Neill and Canil* [1992]. Below 5 GPa, the two-phase region of spinelloid and olivine (fayalite) becomes larger with decreasing pressure.

3.2. *Structure of $Fe_{3-x}Si_xO_4$ Spinel*

The measured unit cell volumes of spinel phases are plotted against composition (x) and shown in Figure 4 Within the experimental precision, the unit cell volume shows an almost linear relation with composition. The lattice constant is not necessarily followed by the Vegard rule in the whole range of the solid solution, because there are two sites in the spinel structure. Single-crystal structure

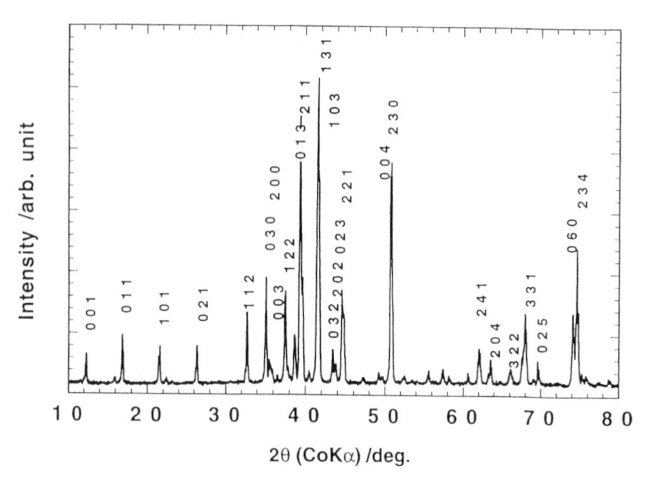

Figure 3. Powder diffraction pattern of $Fe_{2.52}Si_{0.48}O_4$ spinelloid synthesized at 6GPa and 1200°C. Present reflection indexing is referred to *Ross et al.* [1992].

TABLE 2. Structure Parameters of $Fe_{3-x}Si_xO_4$ [a]

Sample(x)	0.0[b]	0.09	0.28	0.75	0.92	1.0
a(Å)	8.3940	8.392(2)	8.374(2)	8.286(1)	8.256(1)	8.2374(9)
u	0.3797	0.3792(3)	0.3769(1)	0.3700(2)	0.3666(1)	0.3658(2)
R		0.019	0.021	0.030	0.025	0.021
wR		0.020	0.020	0.030	0.024	0.021
site occupancy						
Oct (x 2)						
Ai (Fe)	1.0	0.998	0.975	0.963	0.999	1.0
Ai (Si)	0.0	0.002	0.025	0.037	0.001	0.0
Tetr (x 1)						
Ai (Fe)	1.0	0.914(7)	0.769(4)	0.324(7)	0.082(5)	0.0
Ai (Si)	0.0	0.086	0.231	0.676	0.918	1.0
$^{VI}Si/Si_{total}$	0.0	0.044	0.179	0.099	0.021	0.0
Temp. factor						
$\beta_{11}(A) \times 10^{-5}$		188(4)	163(4)	185(8)	131(7)	119(8)
$\beta_{11}(B)$		250(3)	235(3)	274(5)	193(3)	121(8)
$\beta_{12}(B)$		21(4)	26(3)	25(5)	0(3)	-5(4)
$\beta_{11}(Oxy)$		285(7)	302(7)	327(13)	204(7)	137(9)
$\beta_{12}(Oxy)$		21(13)	67(25)	79(22)	3(9)	-5(12)
Bond distance						
A-O(Å)	1.886	1.878(2)	1.840(1)	1.722(2)	1.667(1)	1.653(1)
B-O(Å)	2.059	2.063(2)	2.078(1)	2.114(2)	2.136(1)	2.138(1)
(B-O)/(A-O)	1.092	1.099	1.129	1.228	1.281	1.293
V_A(Å^3)	3.441	3.399(2)	3.200(1)	2.621(2)	2.379(1)	2.314(2)
V_B(Å^3)	11.627	11.692(2)	11.955(1)	12.563(2)	12.904(1)	12.92(2)
V_B(Å^3)/V_A(Å^3)	3.379	3.440	3.736	4.793	5.424	5.58

[a] Yamanaka (1986)
[b] Shull, Wollan, and Kochler (1951)

analyses of spinel solid solution with x=0.0, 0.1, 0.28, and 1.0 were undertaken. Spinel structure has a space group Fd3m and z=8. Cations occupy 8 tetrahedral and 16 octahedral interstices in the oxygen lattice. Table 2 shows that a large percentage of Si^{4+} ion replaces Fe^{3+} in the octahedral site. An X ray single-crystal structure analysis of $Fe_{2.72}Si_{0.28}O_4$ reveals the cation distribution of $(Fe^{3+}_{0.769}Si^{4+}_{0.231})[Fe^{2+}_{1.280}Fe^{3+}_{0.671}Si^{4+}_{0.049}]O_4$.where cations in the parenthesis and bracket represent tetrahedral and octahedral cations, respectively. Si^{4+} substitutes for Fe^{3+} in the octahedral site and occupies 3% of the octahedral cations. It is a significant feature that not less than 17% of total silicon occupies the octahedral site. *Marumo et al.* [1974, 1977] first proposed the disorder of Si ion in both cation sites of silicate spinels. Consequently the spinel solid solution can be expressed by the following cation distribution:

$$(Fe^{3+}_{1-x+y}Si^{4+}_{x-y})[Fe^{2+}_{1+x}Fe^{3+}_{1-x-y}Si^{4+}_{y}]O_4 \qquad (2)$$

The disorder results in γ-Fe_2SiO_4 (x= 1.0, y = 0.023 ±

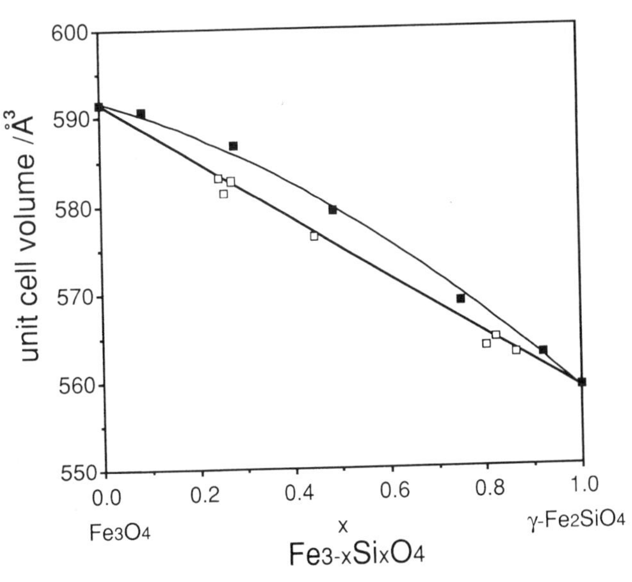

Figure 4. Unit cell volume of spinel solid solution $Fe_{3-x}Si_xO_4$. Solid squares are the present data and open squares are from *O'Neill et al.* [1992]

TABLE 3. Observed and Calculated d Spacings of Synthesized Spinelloid Phase for $Fe_{3-x}Si_xO_4$.

hkl	x=0.48 d_{obs}	x=0.48 d_{cal}	x=0.54 d_{obs}	x=0.54 d_{cal}	x=0.65 d_{obs}	x=0.65 d_{cal}
001	8.340	8.355	8.363	8.354	8.336	8.320
011	6.094	6.101	6.087	6.088	6.065	6.053
101	4.817	4.793	4.800	4.795	4.786	4.784
021	3.920	3.938	3.923	3.925	3.905	3.898
112	3.175	3.177	3.177	3.176	3.160	3.164
030	2.964	2.977	2.964	2.964	2.941	2.941
003	2.782	2.784	2.785	2.785	2.778	2.773
122	2.701	2.705	2.701	2.701	2.690	2.688
013	2.656	2.658	2.657	2.658	2.652	2.645
211	2.633	2.638	2.639	2.639	2.633	2.633
131	2.518	2.529	2.520	2.521	2.510	2.506
103	2.508	2.514				
202	2.416	2.396	2.399	2.398	2.396	2.392
221	2.358	2.348	2.347	2.347	2.343	2.340
230	2.089	2.087	2.084	2.083	2.074	2.074
241	1.739	1.736	1.732	1.732	1.725	1.723
204	1.698	1.699	1.700	1.700	1.697	1.695
322	1.640	1.643	1.643	1.643	1.640	1.639
331	1.600	1.601	1.600	1.600	1.596	1.595
025	1.566	1.565	1.564	1.564	1.558	1.557
251	1.500	1.500				
234	1.477	1.476	1.475	1.475	1.469	1.468
	a=5.848(3)Å b=8.932(5)Å c=8.355(3)Å V=438.4 Å^3		a=5.857(2)Å b=8.891(3)Å c=8.354(2)Å V=435.0Å^3		a=5.848(3)Å b=8.826(3)Å c=8.319(3)Å V=429.4Å^3	

0.01), γ-Co_2SiO_4 (x = 1.0, y = 0.034 ± 0.008), and γ-Ni_2SiO_4 (x = 1.0, y = 0.005 ± 0.012).

Previous and present experimental results suggest that a small amount of Si ions intrinsically occupies the octahedral site. This is probably because mixing entropy of the disorder reduces the free energy.

Interatomic A-O and B-O distances and site volumes of AO_4 and BO_6 are listed in Table 2, where A and B represent tetrahedral and octahedral cations, respectively. These distances are obtained from the lattice constant (a) and u-parameter, which is an atomic coordinate of oxygen position (uuu), as shown by

$$\text{A-O} = \sqrt{3}a(u-1/4)$$
$$\text{B-O} = a[(u-5/8)^2 + 2(u-3/8)^2] \quad (3)$$

As shown in Figure 5, A-O distance decreases with γ-Fe_2SiO_4 content. The decrease in A-O mainly results from the substitution of a smaller cation of Si^{4+} for a larger of Fe^{3+} in the tetrahedral site. Conversely, B-O distance increases with the γ-Fe_2SiO_4 content. This distance changes because Fe^{2+} replaces Fe^{3+} in the octahedral site. The A-O distance change is more effective to the lattice constant than the B-O. These bond lengths are functions of Si partition to tetrahedral and octahedral sites.

Not only ionic radii but also mixing entropy may control this cation distribution. More data on structure analyses of different compositions are needed to discuss the cation disorder. Mössbauer resonance spectra of the solid solution are in accord with the above model of the cation distribution. The detailed results will be reported elsewhere. This mode of the cation distribution considers the magnetic spin moment of the solid solution as a function of Bohr magneton (mB) in the fashion of ferrimagnetism. If the Si^{4+} ion occupies only a tetrahedral site, the variation of its magnetization M would be expressed as $4(1\text{-}x)\mu B$ which is shown as follows:

$$(Fe^{3+}{}_{1-x}Si^{4+}{}_{x})\,[Fe^{2+}{}_{1}\,Fe^{3+}{}_{1-x}\,Fe^{2+}{}_{x}]O_4$$
$$\underset{(1\text{-}x)\mu B}{\rightarrow} \qquad \underset{4\mu B}{\rightarrow}\ \underset{5(1\text{-}x)\mu B}{\leftarrow}\ \underset{4x\mu B}{\leftarrow} \quad (4)$$

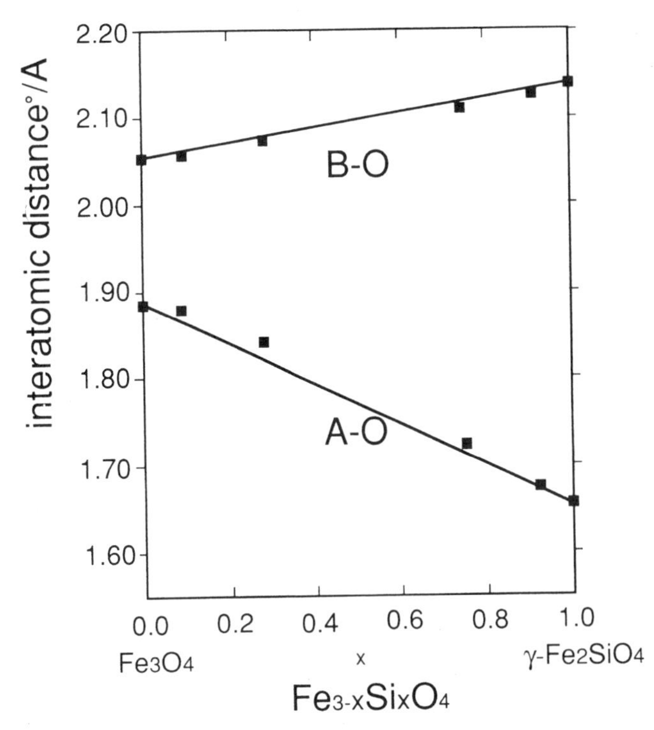

Figure 5. Interatomic distances of A-O and B-O of spinel solid solution as a function of composition, where A and B indicate tetrahedral and octahedral cations.

where arrows indicate the orientation of magnetic spin of Fe^{2+} and Fe^{3+} ions.

But the cation distribution shown in (2) ideally brings M=4(1-x) +10y (μB). The measurement of the magnetic susceptibility of these samples has been made with SQUID. The experimental results will be presented together with the Mössbauer study.

3.3. *Temperature Dependency of Electrical Conductivity of $Fe_{3-x}Si_xO_4$*

The cation distribution indicated by formula (2) controls both the electrical conductivity and magnetic properties of the solid solution. It is well known that magnetite Fe_3O_4 (x=0) has an inverse spinel structure and ferrimagnetic property below 848 K in the mode of $Fe^{3+}(3d^5)$ in the tetrahedral and octahedral site and $Fe^{2+}(3d^6)$ in the octahedral site. On the other hand γ-Fe_2SiO_4 (x=1.0) having a normal spinel structure is characterized as para-magnetism at room temperature and antiferromagnetism at temperature 16 K. The observed Neel point T_N agrees well accords with *Suito et al.*, [1984] who reported T_N =12 K. The present magnetic measurement with a thermal scan technique indicates that T_N varies with the compositional change. The electrical conductivity also changes with the composition as well as with magnetic properties. A fairly high electrical conductivity of Fe_3O_4 (x=0.0) (2.64 × 10^2 $\Omega^{-1}m^{-1}$) and its temperature change can define a semiconductive character. An electron hopping between Fe^{3+} and Fe^{2+} in the octahedral site results in conductivity at room temperature. The conductivity is induced mainly from electron transfer, and it can be detected in the wide range of solid solution. Electrical conductivity varies from semiconductor to insulator with increasing Si content. The γ-Fe_2SiO_4 (x=1.0) is an insulator (5.0 × 10^{-6} $\Omega^{-1}m^{-1}$), because two Fe^{2+} cations are located in the octahedral site, resulting in no electron hopping mechanism. The electrical conductivity in the intermediate composition of $Fe_{3-x}Si_xO_4$ is shown in Figure 6. Table 4 indicates its drastic drop with Si content. At low temperature the electron hopping between ferrous and ferric cations comes to a halt. The electron localization results in the ordered cation distribution and causes the lattice deformation, resulting in the transition from cubic to monoclinic structure, which is known as the Verwey transition [*Verwey* et al., 1941]. The Verwey transition temperature of $Fe_{3-\sigma}O_4$ including cation defects is lowered from about 120 to 90 K with increasing nonstoichiometric σ in magnetite [*Tamura*, 1990]. Then the conductivity is controlled by the hole or vacancy in the lattice as well as electron transfer. The samples used in this experiment are probably good stoichiometric compounds, because of their synthesis under reducing conditions. The present study reveals that the transition temperature is decreasing with γ-Fe_2SiO_4 content in $Fe_{3-x}Si_xO_4$ from 124 K at x=0 (Fe_3O_4) to 102.2 K at x=0.28 as shown in Table 4 and also clarifies that the electrical conductivity gap at the transition temperature becomes smaller. These

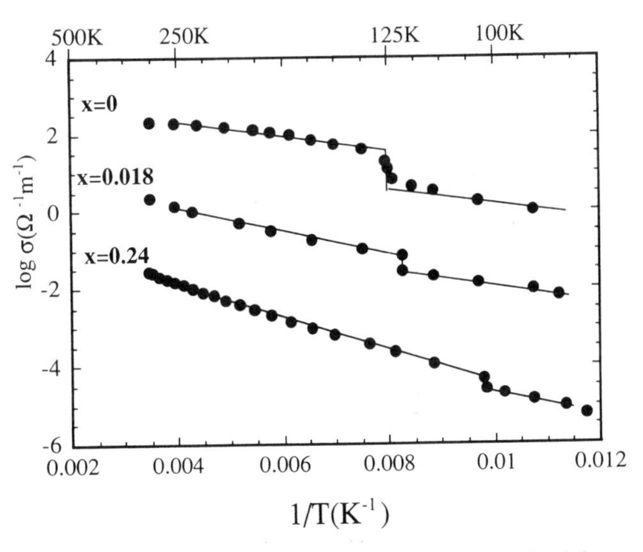

Figure 6. Temperature dependence of electrical conductivity σ in $Fe_{3-x}Si_xO_4$. Verwey transition temperature is lowered with x.

TABLE 4. Temperature Dependence of Electrical Conductivity (s) $\Omega^{-1}m^{-1}$

$Fe_{3-x}Si_xO_4$	288 K	93 K
x=0	2.643×10^2	9.872×10^1
x=0.018	2.315×10^0	8.915×10^{-3}
x=0.28	2.790×10^{-2}	1.327×10^{-5}

$Fe_{3-x}Si_xO_4$	Verwey transition temperature (K)	electrical conductivity gap (Dlogs)
x=0.0	124.0	0.853
x=0.018	120.9	0.425
x=0.28	102.2	0.299

experimental results can be explained by the probability of electron super exchange in the Fe^{2+} and Fe^{3+} cation pair and by the lattice deformation.

4. SUMMARY

The psuedo-binary Fe_2SiO_4-Fe_3O_4 system is important to study processes in the Earth's upper mantle. In the present study, the phase relations of the Fe_2SiO_4-Fe_3O_4 solid solution system have been investigated under high pressures up to 15 GPa at 1200°C by multianvil apparatus. A complete solid solution between Fe_3O_4 and γ-Fe_2SiO_4 has been found at pressure 10 GPa. A spinelloid structure similar to aluminosilicate V is stable in a wide intermediate compositional range x=0.37 to 0.73 in $Fe_{3-x}Si_xO_4$ at pressures from 3 GPa to 9 GPa. The X ray single-crystal structure analysis indicates that spinel solid solution has the following cation distribution: $(Fe^{3+}_{1-x+y}Si^{4+}_{x-y})[Fe^{2+}_{1+x}Fe^{3+}_{1-x-y}Si^{4+}_y]O_4$. The cation distribution regulates the electrical conductivity and magnetic properties.

REFERENCES

Akaogi, M., and A. Navrotsky, Calorimetric study of the stability of spinelloids in the system $NiAl_2O_4$-Ni_2SiO_4, *Phys. Chem. Mineral., 10,* 166-172, 1984.

Akimoto, S., H. Fujisawa, and T. Katsura,Demonstration of the electrical conductivity jump produced by the olivine-spinel transition, *J. Geophys. Res., 70*, 443-449, 1965.

Akimoto, S., E. Komada, and I. Kushiro, Effect of pressure on the melting of olivine and spinel polymorph of Fe_2SiO_4, *J. Geophys. Res., 72*, 679-686, 1967.

Barbier, J., New spinelloid phases in the $MgGa_2O_4$-Mg_2GeO_4 and $MgFe_2O_4$-Mg_2GeO_4 system, *Eur. J. Mineral, 1*, 39-46, 1989.

Barbier, J. and B. G. Hyde, Spinelloid phases in the system $MgGa_2O_4$-Mg_2GeO_4, *Phys. Chem. Minerals, 13*, 382-392, 1986.

Canil, D., H. O'Neil, and C. R. Ross II, A preliminary look at phase relations in the system Fe_3O_4-γFe_2SiO_4 at 7GPa (abstract), *Terra Abstracts, 3*, 65, 1990.

Finger, L., A system of FORTRAN IV computer programs for crystal structure computation, NBS technical note, 854, 1975.

Hammond, R., and J. Barbier, Spinelloid phases in the nickel gallosilicate system, *Phys. Chem. Minerals, 18*, 184-190, 1991.

Horioka, K., M. Nishiguchi, N. Morimoto, H. Horiuchi, M. Akaogi, and S. Akimoto, Structure of aluminosilicate (phase V): a high-pressure phase related to spinel, *Acta Cryst., B37*, 638-640, 1981.

Horioka, K., K. Takahashi, N. Morimoto, H. Horiuchi, M. Akaogi, and S. Akimoto, Structure of aluminosilicate (phase IV): a high-pressure phase related to spinel, *Acta Cryst., B37*, 635-638, 1981.

Ito, E., and H. Yamada, Stability relations of silicate spinels, ilmenites and perovskites, in *High-Pressure Research in Geophysics*, edited by S. Akimoto and M. H. Manghnani, pp. 405-419, Center for Academic Publications, Tokyo/Reidel, Boston, 1982.

Ma, C. B., New orthorhombic phases on the join $NiAl_2O_4$-Ni_2SiO_4, Stability and implications to mantle mineralogy, *Contrib Mineral Petrol, 5*, 257-279, 1974.

Ma, C. B., K. Sahl, and E. Tillmans, Nickel aluminosilicate, phase I, *Acta Crystallogr., B31*, 2137-2139, 1975.

Ma, C. B., and E. Tillmans, Nickel aluminosilicate, phase II, *Acta Crystallogr., B31*, 2139-2141,1975.

Ma, C. B., and K. Sahl, Nickel aluminosilicate, phase III, *Acta Crystallogr., B31*, 2142-2143, 1975.

Marumo, F., M. Isobe, Y. Saito, T. Tagi, and S. Akimoto, Electron density distribution in crystals of γ-Ni_2SiO_4, *Acta Crystallogr., B30*, 1904-1906, 1974.

Marumo, F., M. Isobe, and S. Akimoto, Electron density distribution in crystals of γ-Fe_2SiO_4 and γ-Co_2SiO_4, *Acta Crystallogr., B33*, 713-716, 1977.

Nitsan, U., Stability field of olivine with respect to oxidation and reduction, *J. Geophys. Res., 79*, 706-711, 1974.

O'Neill, H. S. C., and D. Canil, Phase relations along the join Fe_2SiO_4-Fe_3O_4 at 7 and 9 GPa and the oxidation state of the transition zone (abstract), *Eos, Trans. AGU, 73*, Spring Meeting Suppl., 297, 1992.

Ross, C. R., II, T. Armbruster, and D. Canil, Crystal structure refinement of a spinelloid in the system Fe_3O_4-Fe_2SiO_4, *Am. Mineral., 77*, 507-511, 1992.

Sasaki, S., and K. Tsukimura, Atomic positions of K-shell electron in crystals, *J. Phys. Soc. Jpn, 56*, 437-440, 1987.

Shiga, Y., Silican magnetite from the Kamaishi mine, Japan, *Min. Geol., 38*, 437-440, 1988.

Shiga, Y., Further study on silican magnetite, *Min. Geol., 39*, 305-309, 1989.

Shull, C. G., O. Wollan, and E. Kochler, Neutron-deuteron scattering amplitudes, *Phys. Rev., 84*, 833-92, 1951.

Suito, K., Y. Tsutsui, S. Nusu, A. Onodera, and F. E. Fujuta, Mössbauer effect study of the γ-form Fe_2SiO_4, *Mat. Res. Soc. Proc., 22*, 295-298, 1984

Tamura, S., Pressure dependence of the Verwey temperature of $Fe_{3-\sigma}0_4$, *J. Phys. Soc. Jpn, 59,* 4462-4465, 1990.

Verwey, E. J., P. W. Haayman, and F. C. Romeijn, Physical Properties and Cation Arrangement of Oxides with Spinel Structures, *J. Chem. Phys., 15,* 81-187, 1947.

Yamanaka, T., Crystal structure of Ni_2SiO_4 and Fe_2SiO_4 as a function of temperature and heating duration, *Phys. Chem. Mineral., 13,* 227-232, 1986.

Y. Dobuchi, T. Nagai, A. Nakatsuka, O. Ohtaka, H. Shimazu, K. Tobe, and T. Yamanaka, Department of Earth and Space Science, Graduate School of Science, Osaka University, 1-16, Machikaneyama, Toyonaka, Osaka 560

E. Ito, Institute for Study of the Earth's Interior, Okayama University, Misasa, Tottori-ken 682-01, Japan

Plastic Flow of Mn_2GeO_4 I: Toward a Rheological Model of the Earth's Transition Zone

Quan Bai and Harry W. Green II

Institute of Geophysics and Planetary Physics and Department of Earth Sciences
University of California, Riverside, California

Knowledge of the rheological properties of the major mineral phases in the Earth's mantle transition zone between 410 and 660 km depths is lacking because quantitative high-T deformation experiments at P = 13-25 GPa are technologically impossible. To provide a quantitative estimate of the rheology of the upper portion of the transition zone, we have initiated a study of the rheology of α- and β-Mn_2GeO_4. The $\alpha \leftrightarrow \beta$ phase transition in Mn_2GeO_4 occurs at 3.5-5.0 GPa, within the P range of our modified Griggs-type apparatus. These two phases are structural analogues of α- and β-$(Mg_{0.9}Fe_{0.1})_2SiO_4$, the most abundant minerals residing on either side of the 410-km seismic discontinuity. Our working hypothesis is that the change in rheological behavior upon the $\alpha \leftrightarrow \beta$-olivine phase transition in Earth's mantle can be estimated quantitatively from the creep data of α- and β-Mn_2GeO_4 through normalization of mechanical data against the respective solidus T. Here, we report for the first time rheological data for α-Mn_2GeO_4 olivine. We also examine the systematics of dislocation creep rheological properties of olivines with respect to homologous T. Polycrystals of α-Mn_2GeO_4 olivine were deformed at strain rates ($\dot{\epsilon}$) of 10^{-5}-10^{-4} s^{-1}, T of 1000-1200 K and P of 0.4-3.8 GPa under both dry and wet conditions using both Griggs apparatus and gas-medium deformation apparatus. As observed in mantle olivine, water significantly weakens α-Mn_2GeO_4. At similar homologous T and f_{O_2}, the strength of α-Mn_2GeO_4 agrees with those of the other olivines within a factor of ~4, suggesting that dislocation creep of olivines is determined to a large extent by common bonding and structural characteristics. Therefore, combining the new data on α-Mn_2GeO_4 with the well-documented data for α-$(Mg_{0.9}Fe_{0.1})_2SiO_4$ olivine should allow calculation of the rheology across the upper mantle-transition zone boundary when rheological data on β-Mn_2GeO_4 become available from our ongoing studies.

1. INTRODUCTION

The silicate olivine, α-$(Mg_{0.9}Fe_{0.1})_2SiO_4$, is the most abundant mineral in the upper mantle above ~410 km depth. It undergoes polymorphic phase transitions with increasing depths to wadsleyite (β-phase) with a modified spinel structure and then to the γ-phase with a true spinel structure.

Properties of Earth and Planetary Materials
at High Pressure and Temperature
Geophysical Monograph 101

The $\alpha \leftrightarrow \beta$ transition has widely been considered to be responsible for the prominent seismic discontinuity occurring at ~410 km depth, which marks the upper boundary of the mantle transition zone. Therefore, β-olivine is probably the major mineral in the upper half of the transition zone and to a large extent its properties should control those of this portion of the Earth. Compared to the recent advances in understanding of the phase relations and the elastic and thermal properties of mantle minerals, knowledge of the rheological behavior of deep mantle materials lags far behind. Continued progress in realistic computer modelling of mantle processes is calling for experimental rheological data. Because apparatus capable of high-temperature deformation experiments with stress control and measurement at transition-zone pressures does not exist, the only method currently available for determination of transition-zone rheology is through the study of analogue materials which undergo the same phase transformation(s) as does mantle olivine but at pressures low enough to allow quantitative laboratory measurement. According to published phase relations for the germanate and silicate analogues of $(Mg,Fe)_2SiO_4$ olivine, only α-Mn_2GeO_4 and α-Co_2SiO_4 transform to the β-polymorph. β-Co_2SiO_4 is stable only at pressures greater than 7 GPa but α-Mn_2GeO_4 transforms to the β-phase at pressures $P \gtrsim 3.7$ GPa and temperatures $T >$ 1000 K [*Akimoto and Sato*, 1968; *Morimoto et al.*, 1969]. These *P-T* conditions are within the capability of our unique modified Griggs-type deformation rig, making Mn_2GeO_4 the only available analogue system for high-pressure quantitative study of the changes in mechanical properties associated with the $\alpha \leftrightarrow \beta$ transition.

Dislocation creep and diffusion creep are the two end member deformation mechanisms that operate in mantle flow processes. In this study, the behavior of dislocation creep is investigated. To describe the dislocation creep processes in a ternary system such as α- and β-Mn_2GeO_4, the most convenient independent thermodynamic parameters to control experimentally are temperature (T), pressure (P), oxygen fugacity (fo_2) and activity of a componential oxide (a_{ox}). In a hydrothermal environment, water fugacity (f_{H_2O}) is another thermodynamic parameter controlling the concentration of hydroxyl species and possibly the rate of creep [e.g., *Bai and Kohlstedt*, 1992, 1993]. Therefore, the steady-state strain rate ($\dot{\epsilon}$) of olivine can generally be expressed as a function of T, P, fo_2, a_{ox} and f_{H_2O} as well as applied stress (σ):

$$\dot{\epsilon} = f(\sigma, T, P, f_{O_2}, a_{ox}, f_{H_2O}) \tag{1}$$

In this first study in the attempt toward a rheological model of the Earth's transition zone, we performed high-pressure high-temperature deformation at fixed values of fo_2 and a_{ox} to determine the flow strength of α-Mn_2GeO_4 as a function of strain rate ($\dot{\epsilon}$), temperature (T), total pressure (P), and water fugacity (f_{H_2O}). The rheological data for α-Mn_2GeO_4 were then compared to available data for five silicate olivines at similar homologous temperatures and the general characteristics of olivine rheology are discussed. The next step will be to collect data on β-Mn_2GeO_4 and calculate the rheology of mantle wadsleyite.

2. EXPERIMENTAL TECHNIQUES

2.1. *Sample Material*

Polycrystalline Mn_2GeO_4 samples were fabricated from MnO and GeO_2 in stoichiometric proportions for Mn_2GeO_4 with slight excess GeO_2. Well-mixed powders were cold pressed in small quantity and fired at high temperatures (~1300 K) at controlled oxygen fugacity to form Mn_2GeO_4 by solid-state reaction. This process was repeated until x-ray powder diffraction analysis showed peaks only for α-Mn_2GeO_4. The resultant material was reground into powder, cold pressed again and packed into a Mo-lined steel container, which was then flushed with Ar gas and heated under vacuum to remove gases adsorbed and trapped in pore space. The container was then welded shut under vacuum and the sample material was sintered via hot isostatic pressing at a temperature of 1100 K and a pressure of 0.3 GPa for 8 h. Optical and scanning electron microscope observations on petrographic thin sections indicate that a near homogeneous texture was obtained with an average grain size of ~80 μm. The specific density was measured to be very near 100%. A few volume percent of brownish-red grains of GeO_2 provided a solid-state buffer to control the activity of this component at unity. Infrared spectroscopy observation showed that the hot-pressed sample materials contained no water within the resolution (~10 H/10^6Ge) of the infrared spectrometer. Specimens for deformation experiments were cored from this material. Cylindrical samples ~3 mm diameter and ~7 mm long were prepared for experiments in the Griggs apparatus; a sample 6.0 mm diameter and 13.7 mm long was prepared for the experiment using a gas-apparatus.

2.2. *Deformation Apparatus*

The deformation experiments were mainly performed with two modified Griggs-type deformation rigs capable of 2 and 5 GPa maximum confining pressures, respectively. Both apparatus utilize a molten-salt cell so that stress resolution of a few MPa can be achieved. The 5-GPa rig includes a servo-controlled hydraulic drive and the servo-controller is interfaced to a computer to automate and program deformation cycles. The 2-GPa rig uses a conventional gear-train drive system coupled with a computer-controlled stepping motor. Data acquisition of pressure, load, temperature, and displacement are made through computers. Modifications were made to the sample assembly and graphite furnace that in most cases reduced the longitudinal temperature gradient in the sample to within 10 K at experimental temperatures as measured by two type B thermocouples (Pt-30%Rh vs Pt-6%Rh) located near the two ends of the sample. Detailed descriptions of these machines have been reported by *Green and Borch* [1989] and *Tingle et al.* [1993]. Deformation experiments at constant displacement rate were performed using molten alkali halide as confining medium at pressures <1.5 GPa and solid of NaCl+KCl at higher P because we have not yet developed a salt that remains molten at $P > 1.5$ GPa at $T < 1100$ K.

A Paterson gas-medium apparatus [*Paterson*, 1990], suitable for high-temperature creep experiments at pressures up to 0.7 GPa, was used to perform a single constant-stress (creep) experiment at a pressure of 0.4 GPa and temperatures of 1100-1200 K. The gas-medium rig is equipped with an internal load cell for in-situ stress measurement while load is applied with a servo-controlled electro-mechanical actuator. A resistance-heated furnace constructed with three independent windings provides a hot zone (±3 K) of 5 cm and allows deforming large samples.

2.3. *Sample Assembly*

Samples for the Griggs apparatus were inserted into Pt capsules of 250 μm wall thickness lined with Ni foil (including Ni caps at both ends) 30 μm thick. After being welded shut, this sample assembly was then enclosed in the salt confining medium. The high ductility of Pt prevents tearing of sample capsules and chemical attack of samples by the surrounding salt. The Ni layer minimizes loss of Mn from the sample to the Pt capsule, and partial oxidation of Ni during the experiments controlled the oxygen fugacity of the sample at Ni/NiO buffer. No correction was made for the strength of the wall of the Pt capsule or the pressure effect on thermocouple emf.

For dry experiments, to remove water adsorbed on sample surfaces during coring and polishing, the samples were heated to ~500 K in vacuum for >24 h before embedding in the sample assembly which, including the salt cell, was then kept in a vacuum oven maintained at ~500 K until being used for deformation runs. For experiments under hydrothermal conditions, drops of distilled water were added to the sample before welding the Pt capsule closed. In addition, ~15 vol% talc powder was mixed with the salt pressure-medium; the decomposition of talc under experimental conditions released water and minimized loss of water from the sample to the surrounding salt by hydrogen diffusion through the Ni and Pt walls. The water fugacity was fixed by both the total pressure and the oxygen fugacity. The water fugacities at different pressures were calculated from the results of *Pitzer and Sterner* [1994].

The sample for the creep test performed using the gas-medium apparatus was jacketed in a thick-walled (1.8 mm thick) Ni sleeve with Ni discs of 0.15 mm thick placed at each end. This sample can, together with the Al_2O_3 and ZrO_2 spacers and the hardened steel deformation pistons, was then inserted into an Fe jacket which separated the sample from the Ar confining gas. The contribution to the measured stress resulting from the finite strength of the Ni sleeve was corrected based on the published data for Ni [*Frost and Ashby*, 1982].

2.4. *Data Analysis*

Theoretical analyses for a variety of deformation mechanisms yield a power-law rheology when a single dislocation creep mechanism dominates [e.g., *Poirier*, 1985]. In this paper, a power-law form for Eq (1) was used to analyze the mechanical data. For deformation at given values of oxygen fugacity and oxide activity, this equation takes the form

$$\dot{\epsilon} = A\,\sigma^n\,(1+f_{H_2O})^h \exp\left[-\frac{\Delta E^* + \Delta V^* \bullet P}{RT}\right] \qquad (2)$$

where A is a preexponential factor, n is the stress exponent, h is the water fugacity exponent for creep, and R is the gas constant. ΔE^* and ΔV^* are the activation internal energy for creep (called activation energy hereafter) and activation volume for creep, respectively. For deformation at dry conditions, $f_{H_2O} = 0$.

3. EXPERIMENTAL RESULTS

To determine the solidus temperature for the Mn_2GeO_4-GeO_2 system needed for calculating homologous temperature, six static annealing experiments were performed in an Ar-gas atmosphere at a pressure of 0.1 MPa and temperatures ranging from 1500 to 1650 K. A solidus temperature T_m of 1580±20 K was determined, at which a melt fraction of <1 vol% was produced. Almost complete melting occurred at $T = 1630$ K.

Twelve deformation experiments were performed at total pressures of 0.4 - 3.8 GPa and temperatures of 1000-1200 K

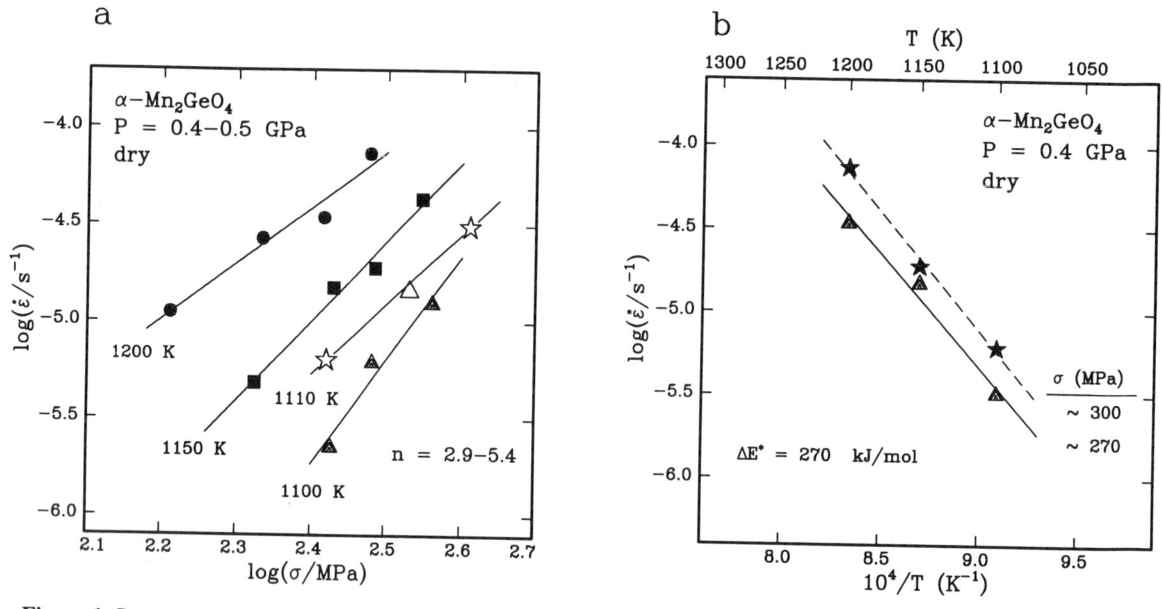

Figure 1. Stress- and temperature-dependence of creep of Mn_2GeO_4 olivine polycrystals. (a) Log-log plot of strain rate vs stress for steady-state flow. Solid symbols are for run PI-297 performed in a Paterson gas apparatus; open symbols are for deformation runs GL-629 (stars) and GL-631 (triangle) performed in a Griggs apparatus with molten-salt confining medium. (b) Semi-log plot of strain rate vs inverse temperature. Data were obtained at stresses of 270 and 300 MPa from run PI-297, yielding an activation energy, ΔE^*= 270 kJ/mol.

under both dry and hydrothermal conditions. The experimental conditions and mechanical results for the deformation runs are reported in Table 1.

3.1. *Deformation at P = 0.4 - 1.0 GPa*

One creep experiment (run PI-297) was carried out at a total pressure of 0.4 GPa and temperatures of 1100, 1150 and 1200 K under dry conditions, using the gas-medium deformation rig. At each of the temperatures, deformation cycles were performed at 3 or 4 different stress values to determine the stress exponent (n) for creep, Figure 1. Least squares fitting of the $\dot{\epsilon}-\sigma$ data for each of the temperatures to the power-law relation, Eq (2), yielded a stress exponent n = 2.9 - 5.4 and indicated that the dominant deformation mechanism is dislocation creep. The fit of each data set to Eq (2) was shown as a solid line in Figure 1a. The creep-rate data of run PI-297 are plotted against inverse temperature in Figure 1b for two differential stresses, ~270 and ~300 MPa. Fitted to Eq (2), the two data sets yielded an activation energy for creep $\Delta E^* = 270 \pm 50$ kJ/mol.

Two constant-displacement-rate deformation experiments (runs GL-629 and GL-631) were also performed using a molten-salt medium in the Griggs apparatus at P = 0.4 and 0.5 GPa and T = 1100 and 1110 K, similar to the P-T conditions for run PI-297 using the gas-medium apparatus. As shown in Table 1 and Figure 1a, the mechanical data of runs GL-631 and PI-297 agree to a factor of 1.07 at the same temperature of 1100 K. In run GL-631, two deformation tests were also performed at a higher pressure of 1.0 GPa at 1100 K. At P = 1.0 GPa, the sample strength was larger by a factor of ~1.2 than that at P = 0.4 GPa.

3.2. *Deformation at P = 1.5 - 3.8 GPa*

A typical load-displacement curve for deformation at P = 1.5 GPa using the molten-salt medium high-pressure apparatus is shown in Figure 2a, illustrating that friction on the movable piston is essentially constant during advance of the piston. The "hit" point (indicated by an arrow) was normally sharp and thus uncertainty in the contribution of friction to sample strength measurement was generally very small. In general, steady-state flow, characterized by a quasi-constant flow stress at a given strain rate, was achieved after compression at constant displacement rate by 4 - 8% strain, depending on strain rate, for a given set of temperature and pressure.

Deformation experiments at pressures of 1.5 - 3.8 GPa were carried out using the salt-medium high-pressure apparatuses (Table 1). Typical stress-strain curves for specified temperatures and strain rates are plotted in Figure 2b the solid and the dashed curves represent those at P = 1.5 and

TABLE 1. Summary of Experimental Conditions and Mechanical Results

Run #	*P* (GPa)	*T* (K)	$\dot{\epsilon}$ (s^{-1})	σ (MPa)	dry/wet	*n*
GL-629.1	0.5	1110	3.09×10^{-5}	408	dry	3.6
.2	"	"	6.31×10^{-6}	263	"	
GL-631.1	0.4	1100	1.45×10^{-5}	340	dry	-
.2	1.0	"	1.55×10^{-5}	400	"	2.0
.3	"	"	3.39×10^{-5}	590	"	
PI-297.1	0.4	1100	3.25×10^{-6}	267	dry	5.4
.2	"	"	6.16×10^{-6}	303	"	
.3	"	"	1.25×10^{-5}	365	"	
.4	"	1150	4.82×10^{-6}	211	"	4.1
.5	"	"	1.49×10^{-5}	269	"	
.6	"	"	4.31×10^{-5}	353	"	
.7	"	"	1.89×10^{-5}	305	"	
.8	"	1200	1.12×10^{-5}	163	"	2.9
.9	"	"	2.67×10^{-5}	216	"	
.10	"	"	7.37×10^{-5}	299	"	
.11	"	"	3.43×10^{-5}	260	"	
GL-591.1	1.5	1100	1.17×10^{-4}	880	dry	3.8
.2	"	"	1.15×10^{-5}	475	"	
GL-596	1.5	1000	8.13×10^{-6}	>672	dry	-
GB-102.1	1.5	1100	9.55×10^{-5}	962	dry	3.3
.2	"	"	3.31×10^{-5}	589	"	
.3	"	"	1.07×10^{-5}	504	"	
GL-620.1	1.5	1100	3.09×10^{-5}	416	wet	4.3
.2	"	"	1.00×10^{-4}	547	"	
GB-104.1	2.5	1100	1.10×10^{-4}	1225	dry	3.0
.2	"	"	4.07×10^{-5}	835	"	
.3	"	"	1.38×10^{-5}	613	"	
.4	"	"	5.37×10^{-5}	907	"	
GB-117.1	2.5	1100	4.90×10^{-5}	441	wet	3.8
.2	"	"	1.86×10^{-4}	630	"	
.3	"	"	1.32×10^{-4}	560	"	
GB-135	3.4	1100	4.79×10^{-5}	500	wet	-
GB-128.1	3.3	1100	1.58×10^{-5}	800	dry	-
.2	3.7	"	1.14×10^{-5}	1000	"	-
GB-130.1	3.4	1100	1.22×10^{-5}	1000	dry	-
.2	3.8	"	1.31×10^{-5}	1200	"	-
.3	"	"	1.41×10^{-5}	1100	"	-

2.5 GPa, respectively. Under dry conditions, sample strength increases steadily with increasing total pressure. However, under wet conditions, sample strength was approximately independent of pressure. This observation indicated that the hardening effect of total pressure was offset by the softening effect of water under hydrothermal conditions; samples deformed under wet conditions were weaker by a factor of ~1.5 to ~2.7 than those deformed under dry conditions. Detailed data are given below.

Deformation at 1.5 GPa. Three dry experiments (GL-591,

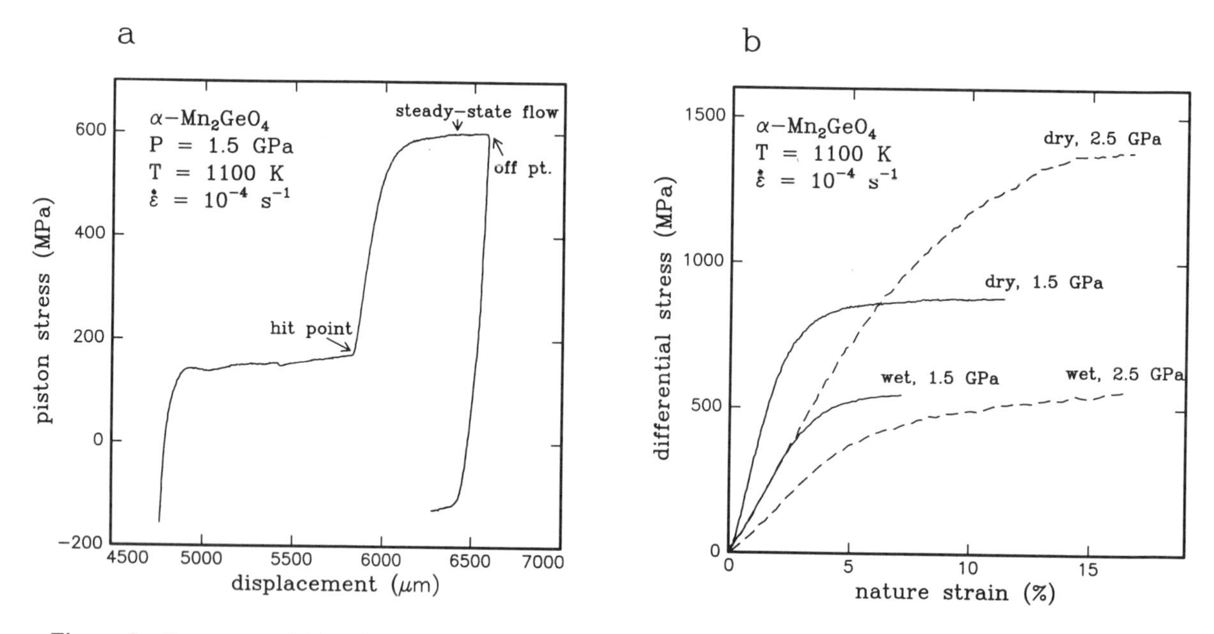

Figure 2. Force record (a) and stress-strain curves (b) from representative high pressure experiments. (a) Force (expressed as piston stress) vs displacement for experiment GL-591 performed with a molten-salt confining medium. Arrows indicate point at which deformation started (hit point), stage of steady-state flow, and point at which deformation ceased (off point). The point at which the piston lifted off the sample and molten salt intruded between sample capsule and piston is represented by the "knee" at the lower right. (b) Stress-strain curves for representative experiments. Solid curves represent 1.5 GPa experiments under dry (GL-591) and hydrothermal (GL-620) conditions in molten salt. Dashed curves represent 2.5 GPa experiments in solid salt under dry (GB-104) and wet (GB-117) conditions.

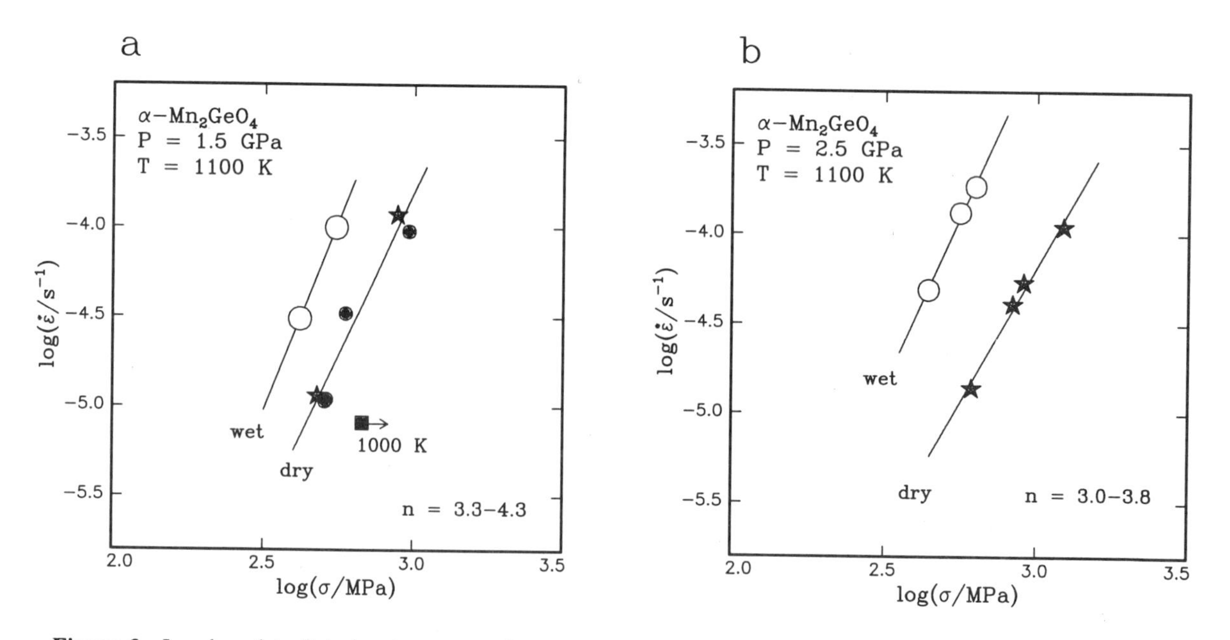

Figure 3. Log-log plot of strain rate vs stress for runs at P = 1.5 GPa (a) and 2.5 GPa (b). Experiments at 1100 K temperature were performed under both dry (solid stars and dots) and wet conditions (open circles). Data from both types of samples are fit to power-law relations. Similar stress exponents were obtained for both types of samples. At a given strain rate, dry samples were stronger than wet samples by a factor of 1.6 at 1.5 GPa and 2.1 at 2.5 GPa. The run at 1000 K yielded only a lower limit of sample strength due to intrusion of molten-salt into the capsule.

TABLE 2. List of Data Sources for the Rheological Strengths and Liquidus Temperatures Used in Figures 5a and 5b for the Comparison of Rheology Between Different Olivines

Olivines	T_m (K)[a]	Data source: creep data	Data source: T_m data	Comments[b]
Mn_2GeO_4	1580	this study	this study	polycrystals, P = 0.4 GPa
Mn_2SiO_4	1524	Bai et al. (ms submitted, 1996)	Bai et al. (ms submitted, 1996)	single crystals compressed along $[101]_C$, P = 0.1 MPa
Mg_2SiO_4	1830	Gleason & Green (1996)	Grieg (1927)	polycrystals, P = 1.1 - 1.2 GPa
$(Mg_{0.9}Fe_{0.1})_2SiO_4$	1720	single crys.: Bai et al. (1991) polycrys.: Hirth & Kohlstedt (1995)	Grieg (1927) & Bowen & Schairer (1935)	single crystals: compressed along $[101]_C$, P = 0.1 MPa; polycrystals: P = 0.3 GPa
Fe_2SiO_4	1450	Ricoult and Kohlstedt (1985)	Osborn & Muan (1960)	single crystals buffered against wüstite and compressed along $[110]_C$ at P = 0.1 MPa
Co_2SiO_4	1655	Ricoult & Kohlstedt (1985)	Asanti & Kohlmeyer (1951)	single crystals compressed along $[110]_C$ at P = 0.1 MPa

[a] T_m here refers to solidus of olivine plus enstatite system except for the faylite which was buffered against FeO.
[b] The general grain size for the polycrystalline silicate and germanate olivines are in the range of 30 - 60 μm.

GL-596 and GB-102) were performed at P = 1.5 GPa and T = 1000 and 1100 K (Table 2; Figure 3a). Least squares fits of the two 1100-K data sets to a power-law relation yielded stress exponents of 3.3 and 3.8, indicating that dislocation creep was the dominant deformation mechanism. The solid line labeled "dry" in Figure 3a represents a fit to Eq (2) of both data sets.

The single data point from experiment GL-596 at 1000 K provided only a lower bound (672 MPa) to the creep strength at a strain rate of 8.13×10^{-6} s^{-1} (Figure 3a) because observation of the capsule after the experiment showed a tear and leakage of some molten salt into the Pt capsule which may have weakened the sample. Nonetheless, the difference in strength between the samples deformed at the two T's indicated that the activation energy for creep is >180 kJ/mol, consistent with the result of run PI-297.

Experiment GL-620 was performed under hydrothermal conditions at T = 1100 K and P = 1.5 GPa, yielding a stress exponent of 4.3 (Table 1). Comparing the data to those of dry samples, Figure 3a, a reduction of sample strength by a factor of ~1.6 was observed due to the hydrothermal environment. Thus, as observed for mantle α-olivine [*Chopra and Paterson*, 1984; *Mackwell et al.*, 1985], water can significantly weaken α-Mn_2GeO_4.

Deformation at 2.5-3.8 GPa. One dry experiment (GB-104) was performed at P = 2.5 GPa and T = 1100 K. The stress-strain rate data, Figure 3b, yielded a stress exponent of 3.0. The sample strength at 2.5 GPa was larger than that at a pressure of 1.5 GPa by a factor of ~1.3. A further increase in creep strength was observed in runs GB-128 and GB-130 at higher P's of 3.4 - 3.8 GPa (Table 1 and Figure 4a).

One wet experiment (GB-117) was also performed at P = 2.5 GPa and T = 1100 K, yielding a stress exponent of 3.8 (Table 1; Figure 3b), similar to those of the dry sample. However, an absolute comparison of the strengths between dry and wet samples showed a decrease of a factor of ~2.1 due to the hydrothermal environment. Thus, the water weakening effect at P = 2.5 GPa was greater than that at P = 1.5 GPa. Still greater water weakening was observed in experiments (comparing runs GB-135 with GB-128) at ~3.3 GPa pressure (Table 1 and Figure 4a).

3.3. Effects of P and f_{H_2O} on Flow Strength

To quantify the effect of total pressure on flow strength, the mechanical data of dry samples at 1100 K were analyzed by plotting the flow strength against total pressure (Figure 4a). The strength data were normalized to a common strain rate of 10^{-5} s^{-1} using a stress exponent of 3.5 -- the average n value for all runs. Least squares fitting of these data to a power-law relation, Eq (2), yielded an activation volume (ΔV^*) of 10.6 ±0.5 cm^3/mol.

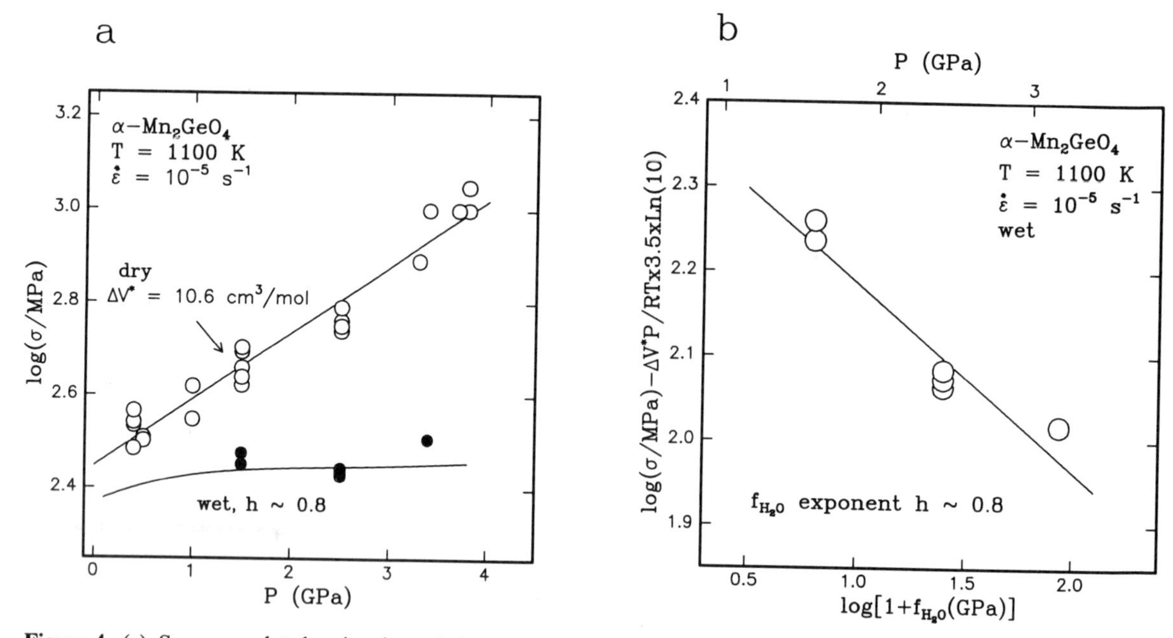

Figure 4. (a) Summary plot showing the variation of the strength of olivine with total pressure under dry (open circles) and hydrothermal (closed circles) conditions. The two data sets were separately fit to the power-law creep equation; the small disagreement at the extrapolated value of P = 0 could represent either data scatter or small differences in wet and dry flow mechanisms. (b) Dependence of sample strength on water fugacity. The data were from wet deformation runs GL-620 (P = 1.5 GPa), GB-117 (P = 2.5 GPa) and GB-135 (P = 3.4 GPa). On the vertical axis, the log values of sample strength were corrected for the increase of sample strength with total pressure due to the activation volume ΔV^* = 10.6 cm^3/mol. On the lower abscissa, the values of water fugacity were calculated from the values of pressure given on the upper abscissa, using the thermodynamic results of *Pitzer and Sterner* [1994].

If it is assumed that the creep mechanism is the same for deformation at dry and wet conditions, the water fugacity exponent for creep of Mn_2GeO_4 can be estimated from Eq (3) as follows: the mechanical data for the wet samples were first corrected for the effect of total pressure due to the activation volume term, $\exp(-P \bullet \Delta V^*/RT)$, assuming the same activation volume for both dry and wet samples. The data were also reduced to a common strain rate of 10^{-5} s^{-1} using a stress exponent of 3.5. The sample strength corrected in this way is plotted in Figure 4b against water fugacity, quantifying the decrease of sample strength with increasing f_{H_2O} after the hardening effect of pressure was accounted for. Fitting the data to Eq (3) yielded a f_{H_2O} exponent h=0.8±0.1.

4. DISCUSSION

4.1. Constitutive Equation

As shown in the previous section, two deformation experiments using the molten-salt medium, Griggs-type apparatus were performed at a pressure of ~0.4 GPa, the same pressure as that of the creep experiment using the gas-medium creep rig. Although the two groups of experiments employed different pressure media and different deformation modes (i.e., constant-stress vs constant-strain rate), the mechanical data were in very good agreement (Figure 1a). The small difference in sample strength by a factor of ~1.07 may be due to uncertainties associated with correcting the strength of the Ni sleeve for the gas-rig experiment, experiment-to-experiment scatter, or less precision in the Griggs rig which uses specimens of only 25% the cross section of the gas rig. The agreement demonstrates the consistency of mechanical data between the two major experimental approaches in the field of rock deformation. This is the third successful comparison between these two types of apparatus (see also *Meagher et al.* [1992] and *Gleason and Tullis*, [1993]). The data at the low pressures of 0.4 and 0.5 GPa also compared well with those at higher pressures 1.0 - 2.5 GPa in the context of power-law flow. These results and the constant ΔV^* over a pressure range of 3.4 GPa further indicate the robustness of our mechanical results.

The effects of strain rate, temperature, total pressure and water fugacity on the flow strength of α-Mn_2GeO_4 were well described in terms of power-law relations, Equation (2). Except for two experimental cycles giving rise to n = 2.0 and 5.4, the other eight runs yielded a range of 2.9 to 4.3 for the stress exponent, regardless of experimental conditions (Table 1). No systematic change of stress exponent was observed with varying experimental conditions. Thus, similar dislocation processes are likely to operate in the experimental pressure range for both dry and wet samples.

To determine the full constitutive equation describing the dislocation creep behavior for Mn_2GeO_4, the average value for the stress exponent (n = 3.5) and the measured values for the activation energy (ΔE^* = 270 kJ/mol) and the activation volume (ΔV^* = 10.6 cm^3/mol), as well as the water fugacity exponent for creep (h = 0.8) were used to determine the preexponential parameters (the A's) for deformation under both dry and wet conditions. The constitutive equations were obtained for an oxygen fugacity controlled at the N/NiO buffer, which is ~10^{-12} and 10^{-6} atm for P = 0.4 GPa and T = 1100 and 1600 K, respectively, and GeO_2 activity fixed at unity. For deformation at dry conditions,

$$\dot{\epsilon} = (1.8 \pm 0.2)\times 10^{-1}\ \sigma^{3.5\pm 1.0} \exp\left[\frac{\Delta E^* + \Delta V^* \times P}{R\,T}\right] \quad (3)$$

If the creep mechanism is the same for wet samples,

$$= (2.5 \pm 1.0)\times 10^{-1}\ \sigma^{3.5\pm 1.0}\,(1 + f_{H_2O})^{0.8\pm 0.1} \exp\left[\frac{\Delta E^* + \Delta V^* \times P}{R\,T}\right] \quad (4)$$

In both equations, ΔE^* = 270±30 kJ/mol and ΔV^* = 10.6±0.5 cm^3/mol, $\dot{\epsilon}$ is in s^{-1}, σ is in MPa, P is in GPa, T is in K, and f_{H_2O} is in GPa.

The deformation behavior as described by Equations (3) and (4) are illustrated in Figure 4a. The wet and dry data sets were analyzed separately (except for assuming that ΔV^* = 10.6 cm^3/mol applies for the wet samples). The offset of strength for the dry and wet Mn_2GeO_4 at P = 0 (where they should be identical) most likely reflects experimental scatter rather than a significant rheological difference. The increase of water fugacity with total pressure offsets the hardening effect of total pressure, resulting in a negligible variation of strength with pressure in a hydrous environment.

4.2. *Comparison with Other Olivines*

The results for the creep behavior of α-Mn_2GeO_4 provides a baseline for a comparison of the rheologies between the α- and β-phases. The use of this analogue material to mantle olivine to provide quantitative constraint on the possible change in flow strength upon the $\alpha \leftrightarrow \beta$ phase transition in mantle olivine relies on the demonstration of an analogy in rheological properties between this germanate olivine and mantle silicate olivine. To delineate the general mechanical behavior for the olivine family, the results of this study are compared in Figures 5a and 5b with those available for a variety of silicate olivines deformed in the dislocation creep regime under dry conditions.

Table 2 provides data for five other olivines to compare with Mn_2GeO_4: $(Mg_{0.9}Fe_{0.1})_2SiO_4$, Mg_2SiO_4, Mn_2SiO_4, Fe_2SiO_4 and Co_2SiO_4. The data were all normalized to a strain rate of $10^{-5}\ s^{-1}$ and oxygen fugacity at Ni/NiO buffer, using the stress exponents and the oxygen fugacity exponents determined by the individual studies. If available, low-pressure creep data for polycrystals — conditions most similar to those for Mn_2GeO_4 olivine -- were used, with experimental pressures in the range of 0.3-0.5 GPa for $(Mg_{0.9}Fe_{0.1})_2SiO_4$ and Mn_2GeO_4 and 1.1-1.2 GPa for Mg_2SiO_4. For single crystals of $(Mg_{0.9}Fe_{0.1})_2SiO_4$, Mn_2SiO_4, Fe_2SiO_4 and Co_2SiO_4 — all deformed at 0.1 MPa pressure -- the creep strengths corresponding to those of $[101]_C$[1] loading direction were used for the comparison, because they generally lie closest to those of polycrystals for a given olivine. Creep data for Fe_2SiO_4 and Co_2SiO_4 are only available for single crystals compressed along the $[110]_C$ orientation, with the former buffered against FeO while the latter buffered against SiO_2. The sample strength data for these two olivines were increased by a factor of ~1.6 to approach those of $[101]_C$ according to the plastic anisotropy result for mantle olivine [*Bai et al.*, 1991] and Mn_2SiO_4 olivine [*Bai et al.*, Manganese olivine II: high-temperature creep, ms submitted, 1996]. Further, except for those of fayalite, the mechanical data for the other olivine crystals correspond to that at the SiO_2 or GeO_2 solid-state buffer. No normalization was made for total pressure since the correction for the hardening effect of pressure on sample strength is small for pressures less than 1 GPa (e.g., within a factor of 1.2 for Mn_2GeO_4).

Steady-state flow stresses for the six olivines are plotted in Figure 5a against inverse T. Except for Mn_2GeO_4, the other olivines were deformed in a similar temperature range (between 1300 and 1600 K). The flow strengths of these olivines fall into a range within a factor of ~30 at $T \approx$ 1400 K, with $(Mg_{0.9}Fe_{0.1})_2SiO_4$ and Mg_2SiO_4 being the strongest and Fe_2SiO_4 being the weakest. Extrapolating for <200 K to a common temperature, the creep strength of Mn_2GeO_4 would agree with that of Mn_2SiO_4.

The creep strengths of the olivines correlate well with their

---

[1] The symbols $[110]_C$, $[101]_C$ and $[011]_C$ refer to the three 45° orientations along which compressive stress was applied. For example, $[101]_C$ is the direction at 45° to the [100] and [001] axes and at 90° to the [010] axis.

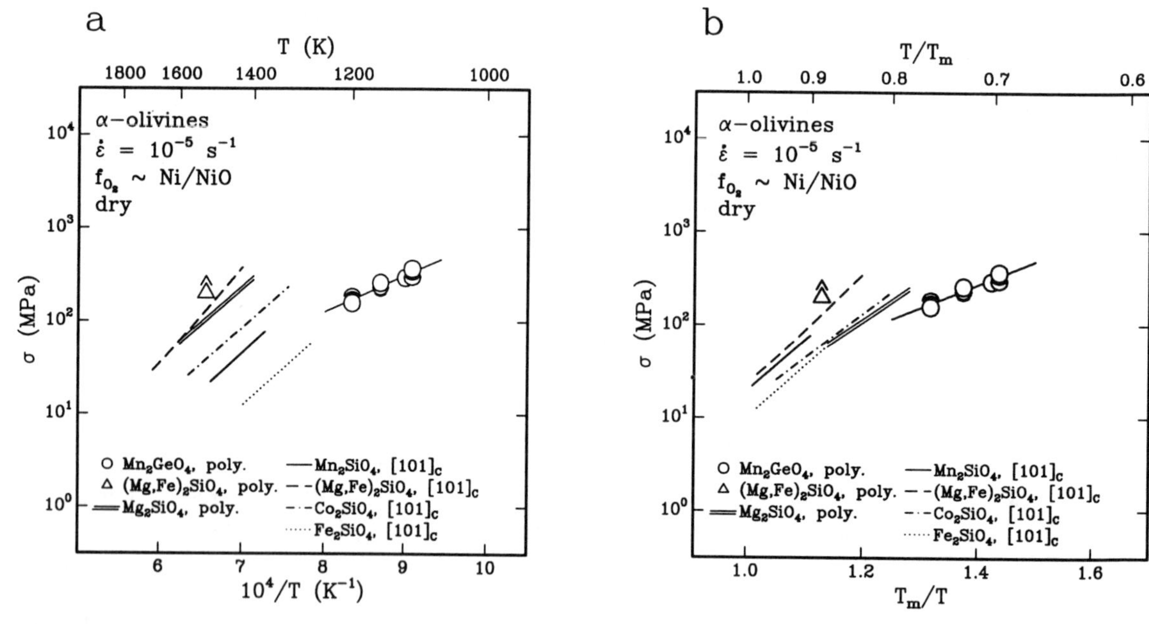

Figure 5. Compilation of dry creep data for six silicate and germanate Mn_2GeO_4 olivines. Sources for mechanical data and melting temperatures are listed in Table 2. Sample strengths are plotted against inverse temperature $10^4/T$ in (a) and against inverse homologous temperature T_m/T in (b). The T_m value used for a given olivine is the one-atmosphere solidus temperature for the olivine coexisting with excess SiO_2 or GeO_2, except for fayalite, for which the solidus with a wüstite buffer is used. For a clearer comparison, the data were all normalized to a common strain rate of 10^{-5} s^{-1} and an oxygen fugacity level at the Ni/NiO buffer, using the stress exponents and the oxygen fugacity exponents determined by the individual studies. Creep data for Fe_2SiO_4 and Co_2SiO_4 are only available for single crystals compressed along the $[101]_c$ orientation; the strength data for these two olivines were increased by a factor of 1.6 to approach those of $[101]_C$ according to the plastic anisotropy result for mantle olivine [*Bai et al.*, 1991] and Mn_2SiO_4 olivine [Bai et al., ms. submitted, 1996). No correction was made for temperature or total pressure for all data.

solidus temperatures, T_m. The sequence with decreasing creep strength and decreasing solidus is as follows: (1) $(Mg_{0.9}Fe_{0.1})_2SiO_4$ and Mg_2SiO_4 with T_m = 1720-1830 K, (2) Co_2SiO_4 with T_m = 1655 K, (3) Mn_2SiO_4 and Mn_2GeO_4 with T_m = 1524-1580 K, and (4) Fe_2SiO_4 with T_m = 1450 K. These systematics of rheological strength for the silicate and germanate olivines are further demonstrated in Figure 5b. Following *Karato* [1989], we plot the sample strengths against their inverse homologous temperatures T_m/T, based upon T_m measured at ambient pressure. After normalization by their respective solidus temperatures, the creep strengths of the olivines merged to a narrow range so that they agree, for example, within a factor of <4 at T/T_m = 0.8. The systematics of plasticity upon normalization by solidus temperature for the silicate and germanate olivines suggests that solid-state flow strength of olivines are largely determined by the bonding and the structural characteristics common to the family of silicate and germanate olivines, despite the variation in the identities of the tetrahedral and octahedral cations.

The activation energy (~300 kJ/mol) for the germanate olivine is smaller than those (~500-800 kJ/mol) for the five silicate olivines as shown by the data in Figure 5a. However, the germanate olivine was deformed at lower homologous temperatures than the silicate olivines, consistent with the lower temperature dependence. Further experiments need to be performed at higher homologous temperatures to provide mechanical data that can be directly compared with the silicate olivines.

4.3. Activation Volume

Previously reported measurements of the activation volume for dislocation creep in olivine have yielded a wide range of values. The pressure dependence of dislocation creep strength of mantle olivine was previously measured by *Ross et al.* [1979] using a solid silicate pressure medium in a Griggs apparatus at P < 1.5 GPa and by *Borch and Green* [1989] using a similar apparatus but with a molten salt medium at P < 2.5 GPa, yielding activation volumes of 13.4 and 27 cm^3/mol, respectively. *Gleason* et al. [1994] reported values of 20-33 and 26-42 cm^3/mol for the activation volume for forsterite polycrystals based on high-quality creep data at T = 1600 K, P = 1-2 GPa, and $\dot{\epsilon}$ = 10^{-4} and 10^{-5} s^{-1}. Recently, *Bussod and Sharp* [1994] estimated an average value of 6 cm^3/mol for the activation volume based on measurement of grain sizes in multianvil experiments at pressures up to 12 GPa. In addition, activation volumes of 19, 14 and 5 cm^3/mol for dislocation climb were measured through three dislocation recovery studies at pressures up to 0.5, 2, and 10 GPa, respectively [*Kohlstedt et al.*, 1980; *Karato and Ogawa*, 1982; *Karato et al.*, 1993].

Considerable controversy has swirled around these disparate results, much of which has involved questions concerning the accuracy of stress measurement in the Griggs high-pressure apparatus. The results of this study now allow us to address several aspects of this controversy with much greater certainty than before. The results of *Ross et al.* [1979] can be discounted because the silicate confining media used at that time have been shown to greatly overestimate sample strength [e.g., *Green and Borch*, 1989; *Gleason and Tullis*, 1993]. That leaves a contrast between the large values (~30 cm^3/mole) obtained directly in the Griggs apparatus [*Borch and Green*, 1989; *Gleason et al*, 1994] and the much smaller values inferred from measurements of dislocation climb [*Kohlstedt et al.*, 1980; *Karato and Ogawa,* 1982; *Karato et al.*, 1993] or estimated from comparison of low pressure results with inferred stress from multianvil experiments for which there are no direct strength measurements [*Bussod and Sharp*, 1994]. *Green* [1988] pointed out that very large values of activation volume are incompatible with mantle flow. However, *Borch and Green* [1987; 1989] argued that activation volume for creep must fall off dramatically at high pressure, just as the effect of pressure on melting of mantle rock decreases markedly. Our new results on Mn_2GeO_4 now show that the Griggs apparatus does not impose large and erroneous values of activation volume on the data, and the agreement at 0.4 GPa with the gas apparatus shows that there is no systematic offset between the two types of machinery. Thus, there is no apparatus-related reason to remain suspect of the large values determined for silicate olivines in the Griggs machine. Moreover, the markedly lower activation volume and activation energy determined here in the germanate system are consistent with prediction of the homologous temperature hypothesis and the argument that germanates proxy for the high pressure properties of silicates. We suggest therefore that the activation volume for creep of mantle olivine exceeds 25 cm^3/mol at low pressure and falls to much lower values by 10 GPa.

The experimental results of this study and the comparison of rheology between the olivines argue strongly for good crystal structure-plasticity systematics for silicate and germanate olivines upon normalization to the respective solidus temperatures. These rheological systematics provide a sound basis for our philosophy of using the analogue system α- and β-Mn_2GeO_4 to infer the rheology of mantle peridotite across the upper mantle-transition zone boundary. These mechanical data on α-Mn_2GeO_4 olivine will now form the basis against which we will compare results of deformation experiments to be performed in the stability field of β-Mn_2GeO_4 to pursue a realistic rheological model of the Earth's transition zone.

5. CONCLUSION

1. The $\dot{\epsilon}$ dependence of the strength of α-Mn_2GeO_4 in the dislocation creep regime can be described by a power-law relation with a stress exponent of 3.5±1.0 for the range of experimental P's and T's under both dry and wet conditions. The activation energy for creep of dry α-Mn_2GeO_4 is 270±30 kJ/mol while the activation volume for creep is 10.6±0.5 cm^3/mol.

2. Relative to the dry samples, sample strength measured under hydrothermal conditions at T = 1100 K and P = 1.5 and 2.5-3.5 GPa was reduced by a factor of ~1.6 and ~2.1, respectively. After correction for the effect of total pressure, strain rate becomes proportional to $(1+f_{H_2O})$ to the (0.8±0.1)th power. Thus, as observed in mantle α-olivine, water can significantly weaken α-Mn_2GeO_4.

3. The hardening effect of total pressure can be offset by the weakening effect of water fugacity, resulting in a negligible variation of olivine strength with increasing total pressure under hydrothermal conditions.

4. After normalization by respective solidus temperatures, the strengths of different silicate and germanate olivines merged from a scatter of a factor of ~30 to a narrow range within a factor of ~4. Hence, olivine rheology exhibits good systematics with respect to homologous temperature, consistent with our working hypothesis. Therefore, the results presented here should provide a good basis for comparison of the rheology of Mn_2GeO_4 polymorphs and calculation of the rheological properties of the Earth's mantle transition zone.

Acknowledgments. Z.-M. Jin, Su Wang, Gayle Gleason and Eric Riggs offered much help in the lab; numerous discussions with them about high pressure research were insightful and constructive. Frank Forgit did an excellent job in fabricating the sample assembly components used in the Griggs-type deformation apparatus and in keeping the deformation machines running. The deformation run PI-297 was performed using a gas-medium creep rig in David Kohlstedt's laboratory; the authors are indebted for his kindness, support and many suggestions. Mark Zimmermann helped perform experiment PI-297 and reduce data from this experiment. Z.-C Wang is thanked for help in determining the melting behavior of α-

Mn_2GeO_4. This research was supported by the National Science Foundation through grant #EAR-9508132.

REFERENCES

Akimoto, S. and Y. Sato, High-pressure transformation in Co_2SiO_4 olivine and some geophysical implications, *Phys. Earth Planet. Inter.*, *1*, 498-504, 1968.

Asanti, P. and E.J. Kohlmeyer, *Z. anorg. Chem.*, *265*, 96, 1951.

Bai, Q., S.J. Mackwell and D.L. Kohlstedt, High-temperature creep of olivine single crystals: 1. mechanical results for buffered samples, *J. Geophys. Res.*, *96*, 2441-2463, 1991.

Bai, Q. and D. L. Kohlstedt, Substantial hydrogen solubility in olivine and implications for water storage in the mantle, *Nature*, *357*, 672-674, 1992.

Bai, Q. and D. L. Kohlstedt, Effects of chemical environment on the solubility of hydrogen in olivine and incorporation mechanisms, *Phys. Chem. Miner.*, *19*, 460-471, 1993.

Borch, R. S. and H. W. Green, II, Dependence of creep in olivine on homologous temperature and its implications for flow in the mantle, *Nature*, *330*, 345-348, 1987.

Borch, R. S. and H. W. Green, II, Deformation of peridotite at high pressure in a new molten salt cell: Comparison of traditional and homologous temperature treatment, *Phys. Earth Planet. Int.*, *55*, 269-276, 1989.

Bowen, N. L. and J. F. Schairer, *Am. J. Sci.*, 5th Ser., *29*, 158, 1935.

Bussod, G. Y. and T. G. Sharp, Experimental evidence for the pressure dependence of creep in polycrystalline olivine at 13 Gpa (abstract), *Eos Trans. AGU*, *75*, 586, 1994.

Chopra, P. N. and M. S. Paterson, The role of water in the deformation of dunite, *J. Geophys. Res.*, *89*, 7861-7876, 1984.

Frost, H. J., and M. F. Ashby, *Deformation-Mechanism Maps*, Pergamon Press, 1982.

Gleason, G., H. W. Green, II, and Y. Zhou, The rheology of periclase- and enstatite-buffered forsterite polycrystals (abstract), *Eos Trans. AGU*, *75*, 586, 1994.

Gleason, G. and J. Tullis, Improving flow laws and piezometers for quartz and feldspar aggregates, *Geophys. Res. Lett.*, *20*, 2111-2114, 1993.

Green, H. W., II, Dissolution of volatiles in mantle olivine: Rheological implications, in *Geophysics and Petrology of the Deep Crust and Upper Mantle*, *Circ. 956*, edited by J. S. Noller, S. Kirby, and J. E. Nielson-Pike, p. 77-79, U.S. Geological Survey, Washington, DC, 1988.

Green, H. W., II, and R. S Borch, A new molten salt cell for precision stress measurement at high pressure, *Eur. J. Mineral.*, *1*, 213-219, 1989.

Grieg, J. W., *ibid*, **13**, 133-154, 1927.

Hirth, G. and D. L. Kohlstedt, Experimental constraints on the dynamics of the partially molten upper mantle, 2, Deformation in the dislocation creep regime, *J. Geophys. Res.*, *100*, 15,441-15,449, 1995.

Karato, S., Plasticity-crystal structures systematics in dense oxides and its implications for the creep strength of the Earth's deep interior: A preliminary results, *Phys. Earth Planet. Inter.*, *55*, 234-240, 1989.

Karato, S. and M. Ogawa, High-pressure recovery of olivine: Implications for creep mechanisms and creep activation volume, *Phys. Earth Planet. Inter.*, *28*, 102-117, 1982.

Karato, S.-I, D. C. Rubie, and H. Yan, Dislocation recovery in olivine under deep upper mantle conditions: Implication for creep and diffusion, *J. Geophys. Res.*, *98*, 9761-9768, 1993.

Kohlstedt, D. L., H. P. K. Nichols, and P. Hornack, The effect of pressure on the rate of dislocation recovery in olivine, *J. Geophys. Res.*, *85*, 3122-3130, 1980.

Meagher, S., R. S. Borch, J. Groza, A. K. Mukherjee and H. W. Green, II, Activation parameters of high-temperature creep in polycrystalline nickel at ambient and high pressures, *Acta Metall. Mater.*, *40*, 159-166, 1992.

Mackwell, S. J., D. L. Kohlstedt and M. S. Paterson, The role of water in the deformation of olivine single crystals, *J. Geophys. Res.*, *90*, 11,319-11,333, 1985.

Osborn, E. F. and A. Muan, *Phase equilibrium diagrams of oxide systems*, Plate 6, Published by the American Ceramic Society and Edward Orton, Jr., Ceramic Foundation, 1960.

Paterson, M. S., Rock deformation experimentation, in *The Brittle-Ductile Transition in Rocks*, *Geophys. Monogr. Ser.*, vol. 56, edited by A. G. Duba, W. B. Durham, J. W. Handin, and H. F. Wang, pp. 187-194, AGU, Washington, D.C., 1990.

Pitzer, K. S., and S. M. Sterner, Equations of sate valid continuously from zero to extreme pressures for H_2O and CO_2, *J. Chem. Phys.*, *101*, 3111-3116, 1994.

Poirier, J.-P., *Creep of Crystals: High-Temperature Deformation Processes in Metals, Ceramics and Minerals*, Cambridge Univ. Press, Cambridge, 1985.

Ross, J. V., H. G. AveLallemant and N. L. Carter, Activation volume for creep in the upper mantle, Science, *203*, 261-263, 1979.

Ricoult, D. L., and D. L. Kohlstedt, Creep of Co_2SiO_4 and Fe_2SiO_4 crystals in a controlled thermodynamic environment, *Philos. Mag.*, Part A, 51, 79-93, 1985.

Tingle, T. N., H. W. Green, T. E. Young and T. A. Koczynski, Improvement to Griggs-type apparatus for mechanical testing at high pressures and temperatures, *PAGEOPH*, *141*, 523-543, 1993.

Q. Bai and H. W. Green, II, Institute of Geophysics and Planetary Physics and Department of Earth Sciences, University of California, Riverside, CA 92521

Rheology Measurements at High Pressure and Temperature

Donald J. Weidner, Yanbin Wang[1], Ganglin Chen, Junichi Ando, and Michael T. Vaughan

Center for High Pressure Research and Department of Earth and Space Sciences, State University of New York at Stony Brook

Measurement of rheological properties of Earth materials in a multianvil device is accomplished with the aid of synchrotron X rays. Deviatoric stresses are a result of the interaction between individual grains, generated by compressing a polycrystalline powder. The stress is quantified by measurement of the broadening of the diffraction peaks. Generally, the magnitude of the deviatoric stress will exceed the nominal pressure unless the sample fails or flows. The change of the peak width with time defines the strain rate of the sample. Taken together, these data provide constraints on the rheological properties for these conditions. In situ X ray diffraction measurements of $MgAl_2O_4$ spinel at 10 GPa confining pressure and temperatures up to 1100°C provide an example of such data. Strain rates of the order of $10^{-7}s^{-1}$ were measured at 600°C with deviatoric stresses of the order of a few GPa. The total plastic strain in the sample amounted to a few per cent. The stress-strain rate-temperature conditions quantitatively agree with previous studies and correspond to the plasticity-power law creep boundary for this material. These results are supported by the microstructural observations of recovered samples with transmission electron microscopy. The current limits of this approach are defined by the resolution of X ray energy. Future improvements will come using monochromatic sources with high-resolution detection systems. This will enable lower stress determinations and hence higher temperatures.

1. INTRODUCTION

The rheological behavior of mantle minerals at high pressure and temperature is a fundamental property which has a bearing on the dynamic processes within the Earth's mantle. Yield strength has a direct application to phenomena which occur on a short time scale such as the maximum shear stresses associated with deep focus earthquakes and the maximum shear stress in the vicinity of a grain which undergoes a volume-reducing phase transformation thereby defining the mechanical constraints on the kinetics associated with these phase transformations [*Rubie et al.*, 1992]. The flow behavior of mantle minerals controls the style of convection within the mantle and the rate of glacial rebound. Knowledge of these rheological properties can provide insights into these phenomena. The most useful experiments would be single-crystal creep experiments at mantle strain rates, pressures, and temperatures. Short of this goal, it is still very useful to measure the rheological properties of the various high-pressure phases at elevated pressure and temperature. Such data can help constrain the role of phase transformations on the viscosity structure of

[1]now at Consortium for Advanced Radiation Sources, The University of Chicago 5640 S. Ellis Avenue, Chicago, IL, 60637

Properties of Earth and Planetary Materials at High Pressure and Temperature
Geophysical Monograph 101

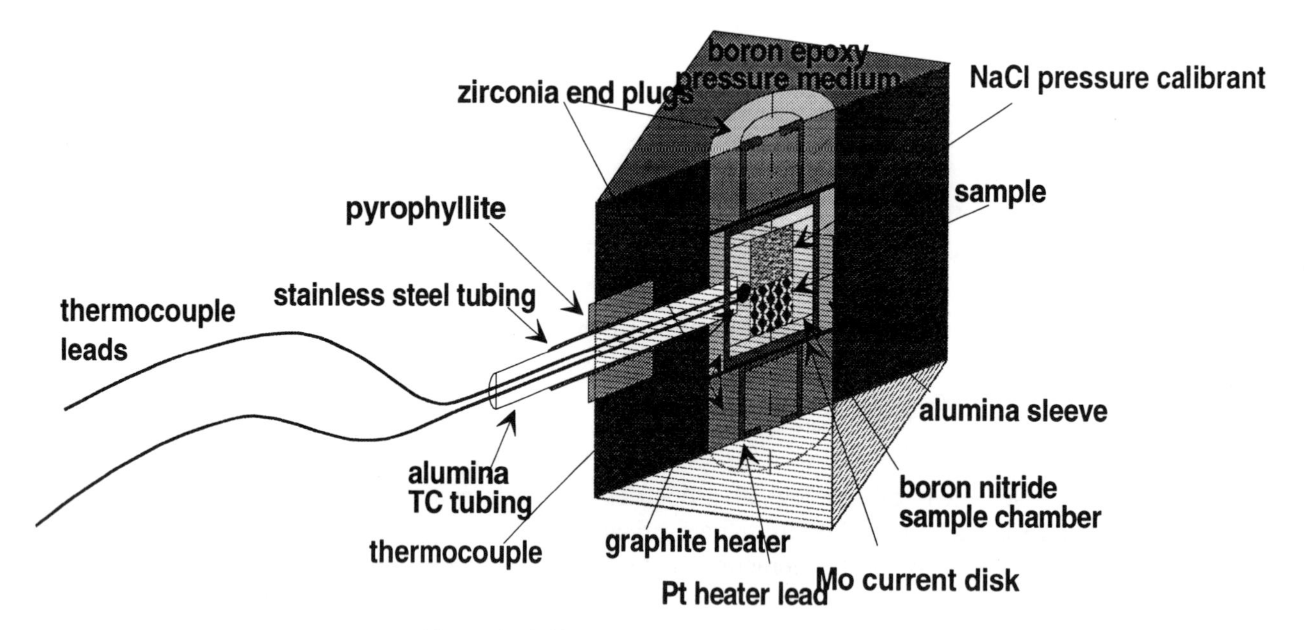

Figure 1. Cell assembly for SAM85 apparatus.

the mantle and limits on the magnitude of the viscosity profile. While the rheological properties of olivine have been well defined to pressures of a few kbar and temperatures close to 2000°C, very little is known about these properties for high-pressure phases. *Meade and Jeanloz* [1990] have reported strength measurements for several important materials at high-pressure and room temperature. However, strength is very temperature dependent and there are not any determinations of yield strength at high temperatures for high-pressure phases at this time. A crucial piece of information needed in understanding viscosity profile of the mantle is flow law of the high-pressure phases within their thermodynamic stability fields. This cannot be measured with conventional deformation apparatus

In this paper we report a methodology for measuring rheological properties of minerals at pressure of 10 GPa and temperature in excess of 1000°C. We measure yield strength as a function of pressure and temperature, as well as relaxation behavior to extract information on rheology. Results for $MgAl_2O_4$ spinel agree very well with the literature data. The methodology at this time is limited to rather large deviatoric stress and small plastic strain.

2. EXPERIMENTAL METHOD

In this study, we use a DIA-type high-pressure apparatus to create deviatoric stress in the sample at elevated pressure and X rays to measure the strain which will be inverted to define the stress that the sample is supporting. Temperature is increased during the experiment. A grain-to-grain deviatoric stress (microscopic stress field) is generated as a result of heterogeneous elastic properties in a polyphase aggregate and/or elastic anisotropy of the individual grains of the sample. This stress field varies from point to point within the sample and is expressed as line broadening of the diffracted X ray lines. The stress is monitored as a function of time at different temperatures. The relaxation rate of the stress contains information about the effective power law of the particular flow process that is limiting the stress in the sample.

2.1. *The DIA Apparatus*

The DIA-type cubic-anvil high-pressure apparatus (SAM-85) on the superconducting beam line (X-17) of NSLS has been developed and is operated by CHiPR (Center for High Pressure Research). The system is described elsewhere by *Weidner et al.* [1992a,b], but we will describe a few of the salient features here.

The force of a uniaxial ram is transmitted to six anvils which simultaneously advance into the cube-shaped solid pressure medium. Gaps between the anvils of about 0.5 mm afford space for X rays to enter the sample chamber and diffracted X rays to exit. The cross section of the pressure medium is illustrated in Figure 1. A cylindrical hole in the sample chamber contains a graphite furnace and the sample, which is isolated from the heater by a BN liner. The dimension of the sample volume is about 1 mm in diameter and 1.5 mm long. A thermocouple, which passes through one of the anvil gaps and is directly in contact with the sample, monitors the temperature during the run. A diffraction standard, with a well known equation of state, is used to define the pressure.

White radiation from the superconducting wiggler magnet yields an energy dispersive analysis with energies between 15 and 100 keV. In this experiment we used a two theta angle of 7.5°. Data gathering times of less than 1 minute from an X ray beam of dimensions of 100 × 200 microns provide robust X ray patterns that can be analyzed for position and width.

By analyzing the lattice parameters of samples at several vertical positions within the sample chamber, we have determined that the temperature variation over the entire sample length is of the order of 20°C at 1200°C. Using multiple thermocouples, we find that the radial temperature gradient is less than 5°C at these conditions. We find no evidence of a vertical pressure gradient at room temperature.

2.2. *Microscopic Stress*

While the loading system can generate a uniform stress field throughout the sample, heterogeneities within the sample, along with elastic anisotropy, will be responsible for generating stresses which vary from grain to grain. Theories regarding stress induced by thermal expansion in a polyphase aggregate made of anisotropic grains is applicable to the current problem. Studies such as those by *Walsh* [1973], *Kreher* [1990], *Tvergaard and Hutchinson* [1988], and many others allow one to estimate properties of the microscopic deviatoric stress for particular material properties. These studies reveal that the microscopic stress varies within the individual grains, but as *Evans and Clarke* [1980] point out, the grain size should not affect the amplitude of the microscopic stress field. In this study, a loose powder specimen is used, and the principal source of the deviatoric stress arises from the heterogeneity associated with the pore space.

We can model the manner that the diffraction profile will reflect this stress field. Each diffraction peak is the sum of diffraction from only a subset of grains within the specimen. These grains have the particular orientation which aligns the specific set of lattice planes with the diffraction vector. Figure 2 illustrates this picture. The solid grains represent the grains that are in diffraction condition. Each nondegenerate diffraction peak will sample a different subset of grains. The small arrows around each grain represent the deviatoric stress field at the position of these grains. The effect of this microscopic deviatoric stress field will be the broadening of X ray diffraction

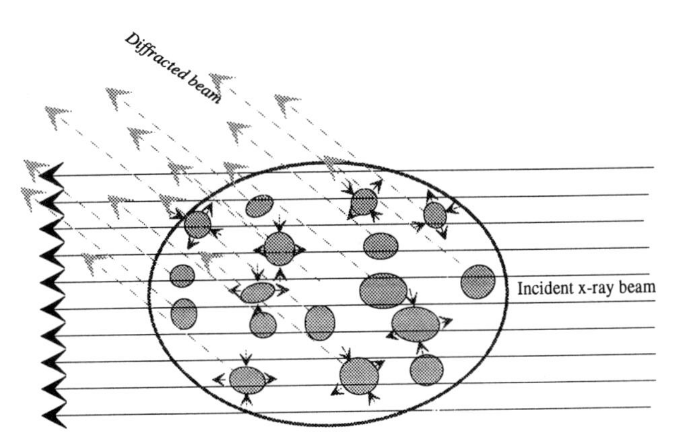

Figure 2. Schematic of X ray diffraction from grains in a randomly varying stress field. Each grain will provide a peak with a slightly different d spacing, thus altering the shape of the diffracted peak.

lines, and the amount of line broadening is determined by the distribution of longitudinal strain parallel to the diffraction vector. The profile of the diffracted peak is thus determined by the magnitude distribution of the principal stresses and the orientation distribution of the stress field. Figure 3 demonstrates the strain broadening for various principal stresses which are randomly oriented in space. The curves were generated assuming that the principal stress of each grain was the same and the d spacing of a particular peak was calculated for orientations that were uniformly distributed over space. A specific set of elastic moduli were assumed for these calculations. The stresses in the diagrams are normalized by S_{11}, the elastic compliance along the [100] direction. Different principal stress systems give quite different peak shapes. If the stress is $(\sigma,0,0)$, or uniaxial tension (tension is positive), then the peak is asymmetric, with the peak actually occurring at smaller values of d than an unstressed sample. The anticorrelation between stress and peak position is due to a Poisson's effect. For a randomly oriented stress field of this type, there will be many more stress vectors perpendicular to the diffraction vector than parallel to it. Tension will create compression perpendicular to the tension direction. Such asymmetric peaks are often observed during loading at low pressure.

Symmetric peaks result from symmetric stress distributions as illustrated by the second two model peaks. If all of the grains have reached yielding conditions, then we might expect the stress to be $(\sigma,-\sigma,x)$ with $-\sigma < x < \sigma$, where σ is the yield strength of the material. This stress field has a nonzero trace ($x/3$). Peaks that have been calculated for this stress field with the trace removed have a similar shape but are more nearly Gaussian.

Thus, the broadening of the X ray peak at the half-height

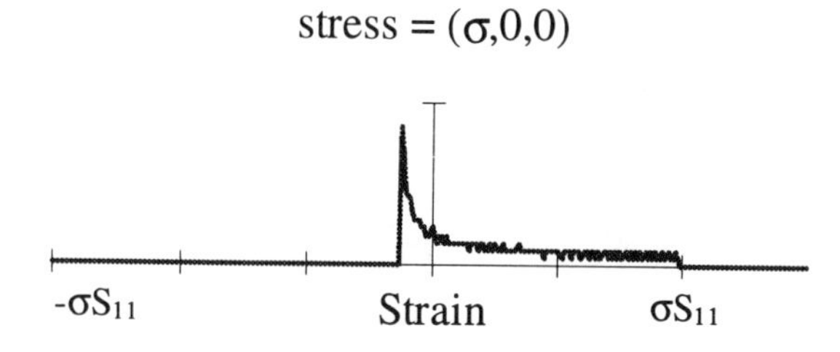

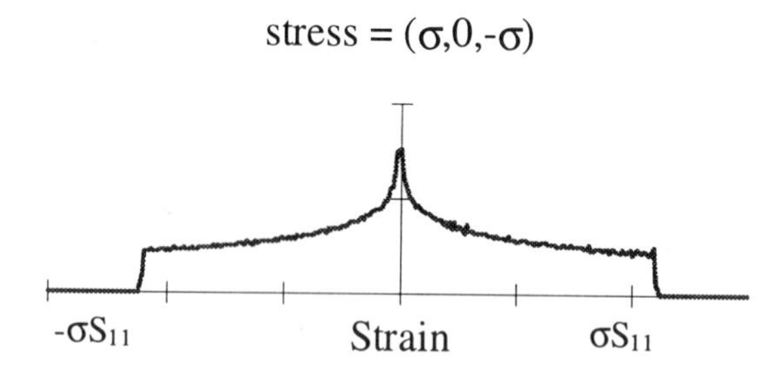

stress = (σ,-1,σ) -σ<x<σ

-σS11 Strain σS11

Figure 3. Calculated contribution of strain to the shape of the diffraction peak for different principal stress vectors that are randomly oriented in the sample.

level reflects the strain of most of the grains in the sample. If only a small percentage of the grains were actually experiencing deviatoric stress, then the base of the peak would broaden, but not the peak at half-height. Furthermore, S_{11} appears to be an appropriate elastic moduli to convert strain broadening to deviatoric stress. In a polycrystalline aggregate this is equivalent to $1/E$, where E is Young's modulus.

Measured spectra for the $MgAl_2O_4$ spinel sample at room pressure and at 10 GPa are illustrated in Figure 4. As can be seen, the centers of the peaks shift to higher energies (smaller *d* spacings) with increasing pressure and the peaks broaden considerably. In fact, more strain energy is partitioned into the deviatoric stress field than into the hydrostatic component.

To determine the magnitude of the differential strain ($\varepsilon_3 - \varepsilon_1$), we must deconvolve the instrument response from the measured spectra. We use the ambient (stress-free) diffraction signal as representing the instrument response. Since the convolution of two Gaussian functions yields a Gaussian whose variance is the sum of the individual variances, we define the peak broadening, W_P, by

$$W_P = (W_O^2 - W_I^2)^{1/2}$$

where W represents the full width at half maximum (FWHM), subscripts O represents the observed width, and I the initial width.

Peak broadening can result not only from strain heterogeneity associated with deviatoric stress but also from small grain size. Thus, it is conceivable that the peak broadening with increasing pressure and narrowing with increasing temperature reflects changes in grain size. *Gerward et al.* [1976] show that line broadening due to strain, W_e, is proportional to energy, while that due to particle size, W_d, is independent of energy for energy dispersive data. Hence the observed peak broadening, W_P, is given by

$$W_P^2 = (\varepsilon E)^2 + W_d^2$$

where ε represents the magnitude of the strain. *Gerward et al.* [1976] suggest plotting W_P^2 from different diffraction peaks which occur at different energies, E, *vs.* E^2. Then the slope of a straight line is related to the strain and the intercept is the contribution from the small grain size. Figure 5 illustrates this relationship for the spinel peaks at 11 GPa before any heating. This condition should be most affected by grain size reduction since the sample has been compressed cold with no heating that would allow

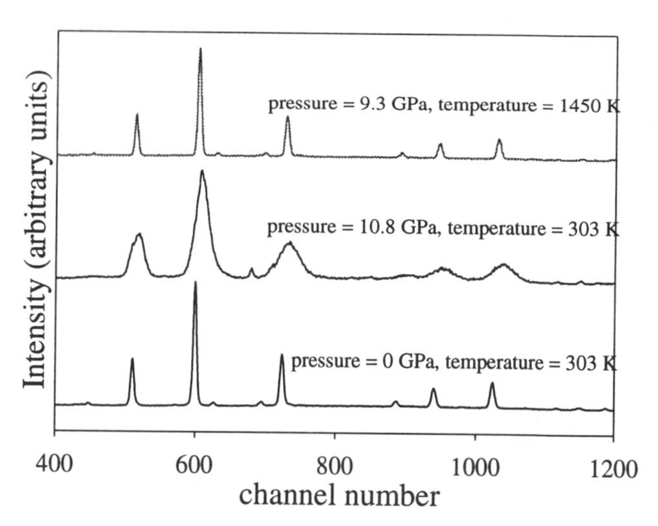

Figure 4. Spectra of spinel at room pressure at 11 GPa prior to heating and at 9 GPa and 1400K.

recrystallization. As seen in the figure, the intercept is zero, confirming that there is no contribution to the broadening from grain size changes during the experiment.

Peak broadening will also occur for an elastically anisotropic material in a macroscopically uniform deviatoric stress field. Each grain that contributes to a particular diffraction peak is oriented with parallel diffraction vectors. However, they may be randomly oriented with respect to rotation about the diffraction vector. Since Poisson's ratio is direction dependent in an anisotropic grain, the *d* spacing along the diffraction vector will vary as the grains are rotated in an anisotropic stress field. However, for the DIA geometry, this affect should not contribute to the peak width since the stress field perpendicular to the diffraction vector is generally uniform [*Weidner et al.*, 1992a,b].

Thus, the measured strain results from the elastic deviatoric stress that is being supported by the sample and thus reflects the deviatoric stress field. By analogy with the model studies, multiplying W_e by Young's modulus yields the magnitude of $(\sigma_3 - \sigma_1)$. If the magnitude of this stress exceeds the strength of the sample, the sample responds by flow and the stress reduces. Monitoring peak width as a function of time at different temperatures provides data relevant to a stress relaxation experiment at these temperatures.

To test whether or not the loading system has applied a sufficiently large deviatoric stress to reach the yield strength of the sample, we establish the following criteria:

1. Saturation of deviatoric stress. If the differential stress in the sample increases linearly with the applied load throughout the entire range of the load, then we would conclude that the sample has not yet reached its yield point. If, however, the differential stress in the sample saturates at some load, then this indicates that the yield point of the sample has been reached. Only in our study of diamond [*Weidner, et al.*, 1994] did the differential stress remain linear over a 10 GPa pressure interval. In fact, diamond did not exhibit yield until the temperature reached 1000°C. This example for a very strong sample demonstrates that the measure of differential stress is valid as is this criteria for yield. Figure 6 illustrates the observed strain for different diffraction peaks as a function of loading pressure for spinel. The departure for loading above about 5 GPa suggests yield is occurring at these pressures at room temperature.

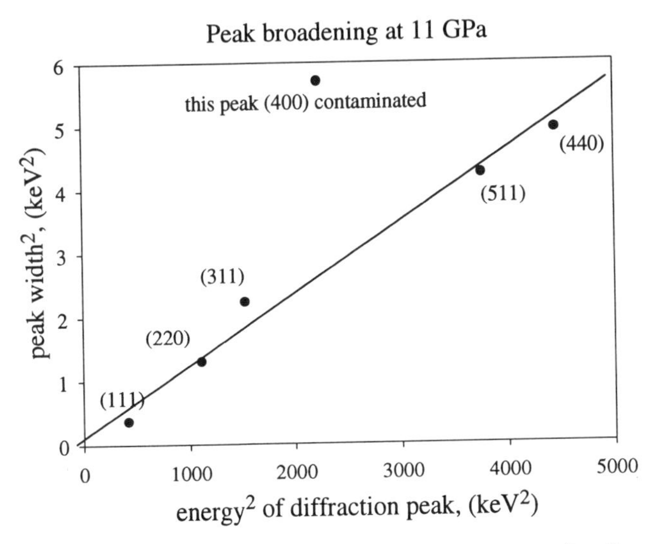

Figure 5. The square of the peak broadening, Wp, as a function of the square of the diffraction energy. The slope of the straight line is proportional to the square of the strain; the intercept is the contribution of grain size. There is no observable contribution of grain size broadening on this data.

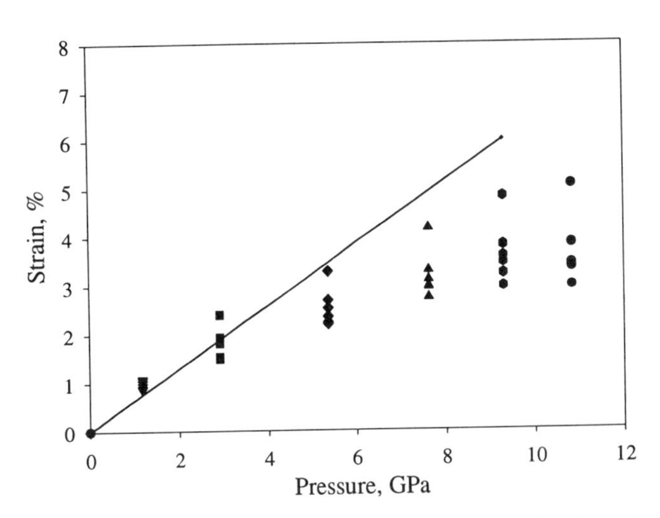

Figure 6. Strain for several diffraction peaks as a function of nominal pressure on loading. A purely elastic response would be a straight line. Any kind of stress relaxation brings the points beneath the elastic curve.

2. Temperature dependence. The hottest portion of the sample assembly is the sample itself. Thus, increasing temperature will mostly lower the sample strength. If the differential stress in the sample does not decrease with temperature, we conclude that the yield strength of the sample has not been obtained.

3. Time dependence. If plastic flow is occurring in the sample, then the differential stress will reduce with time. The data for spinel demonstrate a considerable time dependence at 400°C and higher.

Above we suggest several sources of microscopic deviatoric stress in the sample. Figure 7 shows peak broadening as a function of applied load at room temperature and as a function of temperature at elevated pressure for two samples of stishovite in the same experiment [*Li et*

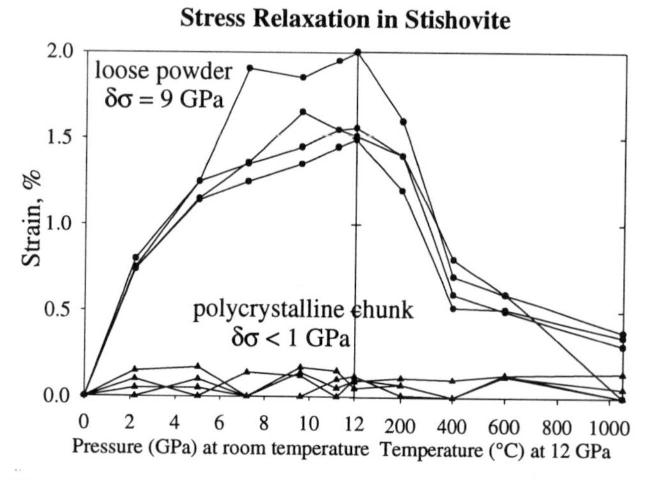

Figure 7. Strain in two stishovite samples as a function of cold loading followed by heating at elevated pressure. The polycrystal (chunk) is an as-sintered sample while the powder was ground from a portion of the chunk.

al., 1996]. Both stishovite samples were previously created at high pressure in a single well sintered chunk. A part of the chunk was separated off and ground to a polycrystalline powder forming the 'powder' sample. The remainder portion of the chunk served as the 'polycrystal'. As seen in this diagram, almost no change in peak width is observed for the sintered chunk as both pressure and temperature change. Furthermore, the initial peak width is similar to that of the powder, i.e., instrument limited. The powder, on the other hand, is considerably broadened with applied load and narrowed with heating, thus demonstrating a considerable magnitude of microscopic deviatoric stress. We conclude that the heterogeneity generated by the void space in a polycrystalline sample is the main driving force for microscopic deviatoric stress. Even though stishovite is extremely anisotropic, the sintered chunk did not show appreciable microscopic deviatoric stress when the sample was loaded to over 10 GPa.

The strains observed with X rays are elastic strains. Nonrecoverable strains associated with fracture and flow are not reflected in the spacing of diffraction planes. Thus, observed strains as related to elastic moduli through Hooke's law can define the stress state of the material. This model assumed that the material behaves elasto-plastically (*i.e.*, no hardening, for which case our model overestimates the yield strength). This, however, should not significantly affect the following analysis for relaxation measurements because only time derivatives are considered.

Typical experiments include loading by hydrostatic (or nearly hydrostatic) pressure and then heating with diffraction data taken at each pressure and temperature. The total strain of the sample is controlled by the magnitude of the hydrostatic load, which essentially defines the sample volume. Then

$$\varepsilon_{tot} = \varepsilon_{pl} + \varepsilon_{el}$$

where ε_{pl} represents the total nonelastic portion of strain and ε_{el} is the elastic portion. As the sample temperature is changed and the strength lowered, then ε_{tot} remains essentially constant and the plastic component of strain increases as the observed elastic portion decreases. Thus, we can estimate the total plastic strain as the deficit between elastic and total strain. This generally amounts to about 1-5% of plastic strain during these measurements. We can estimate the strain rate of the deformation process. Since the total strain is fixed,

$$\partial\varepsilon_{pl}/\partial t = -\varepsilon_{el}/\partial t.$$

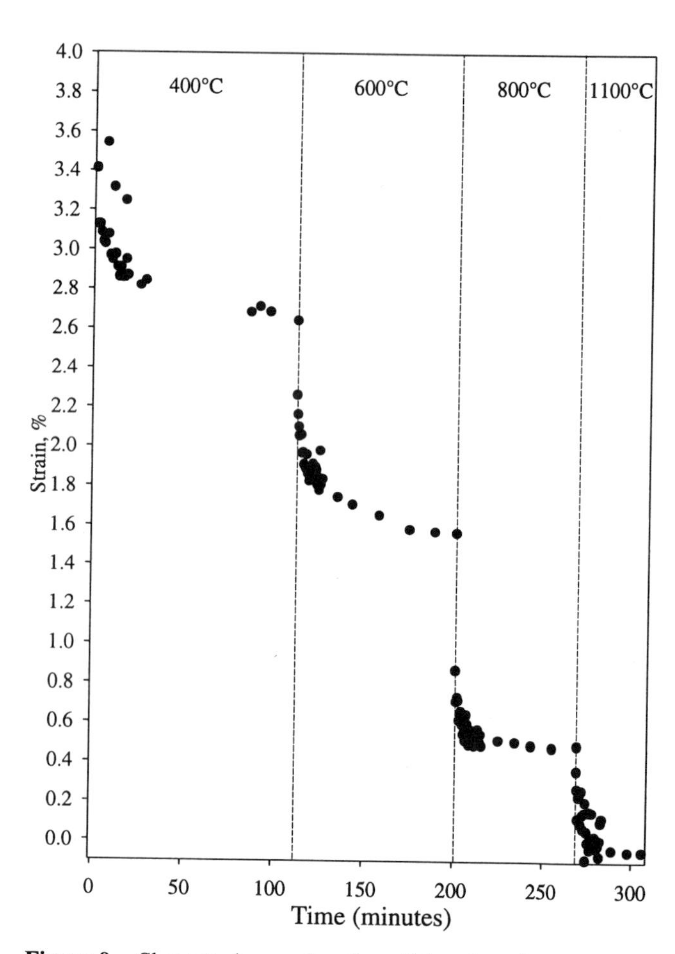

Figure 8. Shear strain as a function of time for spinel at different temperatures and constant pressure (10 GPa). The data points refer to the average of 6 different diffraction lines.

This relation will be true only when the deformation is uniform throughout the sample. If there is a variation of strain rate throughout the sample and flow is controlled by the faster strain rate, then this average serves as a lower bound on the flow strain rate in the sample.

The time dependence of the differential strain can be defined in terms of parameters of the governing equations. If the material follows a power law relationship, then

$$\partial\varepsilon_{pl}/\partial t = k\sigma^n$$

or, since

$$\partial\varepsilon_{pl}/\partial t = -\partial\varepsilon_{el}/\partial t \propto -\partial\sigma/\partial t$$

$$\partial\sigma/\partial t = k'\sigma^n$$

yielding

$$d\sigma/\sigma^n = k'dt$$

Thus,

$$(1-n)\ln(\sigma) = \ln(t) + k''.$$

As a result, stress (or elastic strain) *vs.* time will yield a slope of $(1-n)$ on a log-log plot.

3. RESULTS FOR $MgAl_2O_4$ SPINEL

A powder sample of spinel was loaded into the pressure chamber and compressed to 11 GPa in SAM85. Then the sample was heated to 400°C in a few seconds. Diffraction data were collected continuously for a period of several

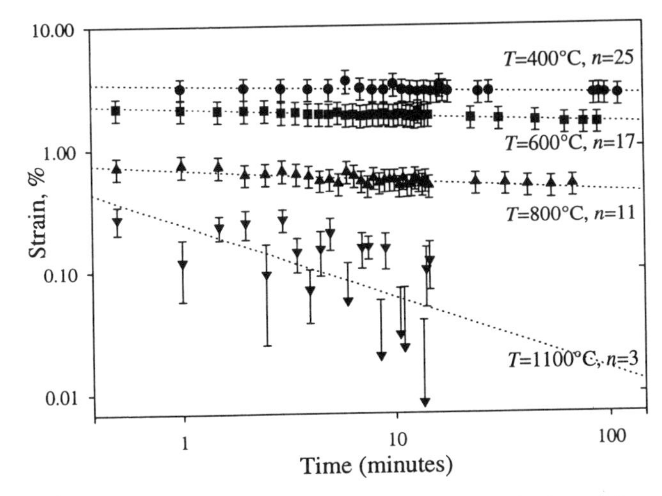

Figure 9. Log of strain for spinel as a function of the log of time for run 1 (same data as in Figure 8, with error bars added). At each temperature, the data define a straight line, except data at 1100°C, where the scattering is due to instrumentation resolution limit. Lines indicate the slope for different values of the stress exponent, *n*.

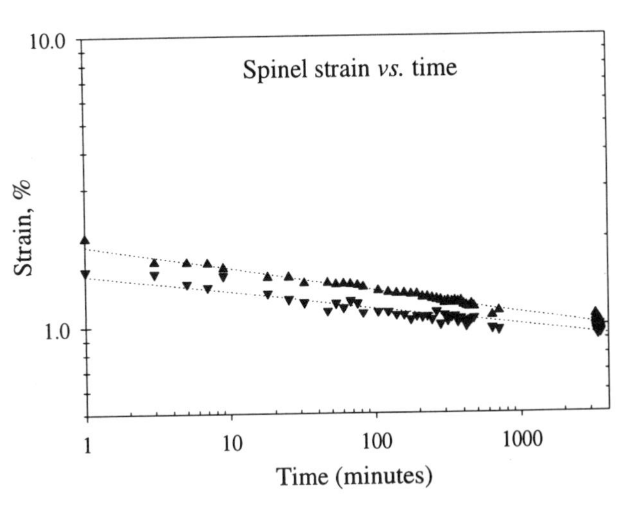

Figure 10. Log of strain as a function of log of time for run 2, a three-day experiment at 600°C. Data for two different diffraction peaks are shown.

minutes and each spectrum was collected for 30 seconds. Data were collected for about 2 hours at this temperature. Then the temperature was raised to 600°C in 1 second and the same procedure was used to collect several spectra. This was repeated at 800°C and 1100°C. A second sample was run in a similar manner. The difference is that the temperature was raised to 600°C immediately from room temperature and the temperature and pressure were maintained for three days. After that, we observed the microstructures of recovered samples with a transmission electron microscopy to identify the deformation mechanisms.

3.1. *Mechanical Data*

Time dependence of differential stress is illustrated in Figures 8 and 9 for the different temperatures of the first data set. Each data point is the average of the strain measured from six different diffraction lines; the error bars in Figure 9 represent the corresponding standard deviation. Here the shear stress ($\tau = [\sigma_3 - \sigma_1]/2$) is given as a function of time in minutes. At each temperature there is an initial fast relaxation that slows down with time. The strain rate derived from the slope of the equivalent strain *vs.* time curve ranges from 10^{-5} to 10^{-7}/s. The data given in terms of strain are plotted as a function of time on a log-log plot in Figure 9. The lines are straight, suggesting a well defined value for n at each temperature. Lines are drawn on the figure with slopes corresponding to different values of *n*. The lower temperature data have large values of *n*, but by 1100°C the value of *n* decreases to about 3, a value that is consistent with a power law creep mechanism. Figure 10 shows a 3-day relaxation experiment at 600°C.

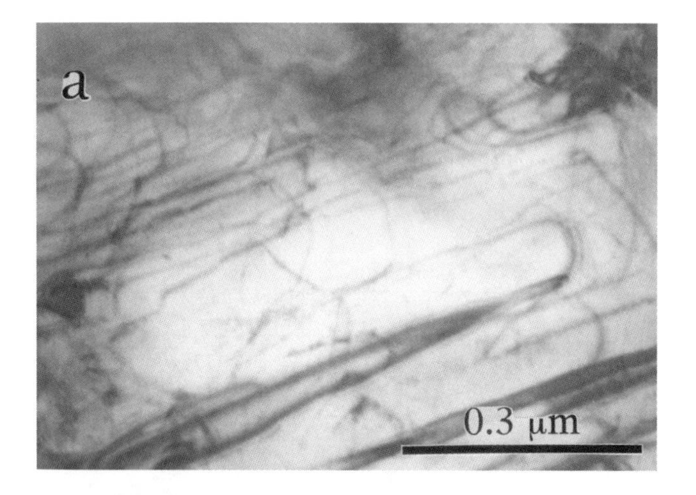

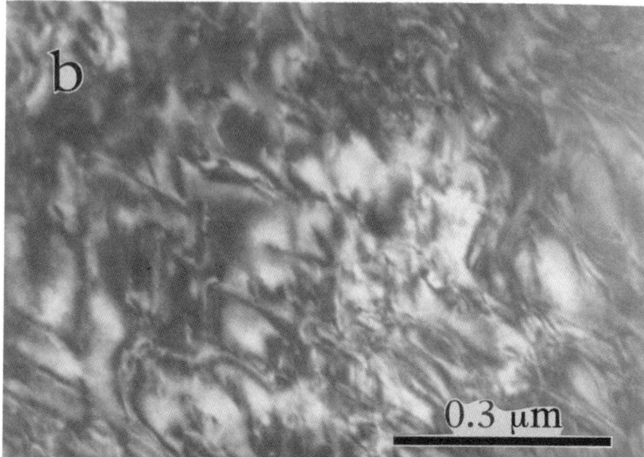

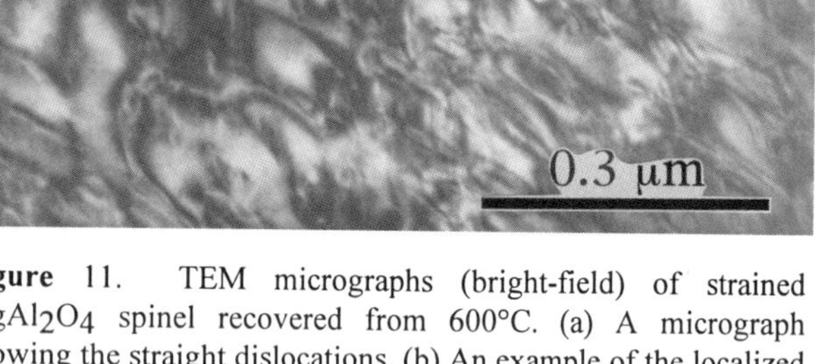

Figure 11. TEM micrographs (bright-field) of strained $MgAl_2O_4$ spinel recovered from 600°C. (a) A micrograph showing the straight dislocations. (b) An example of the localized regions of dislocation tangles complicatedly where are commonly observed in this sample.

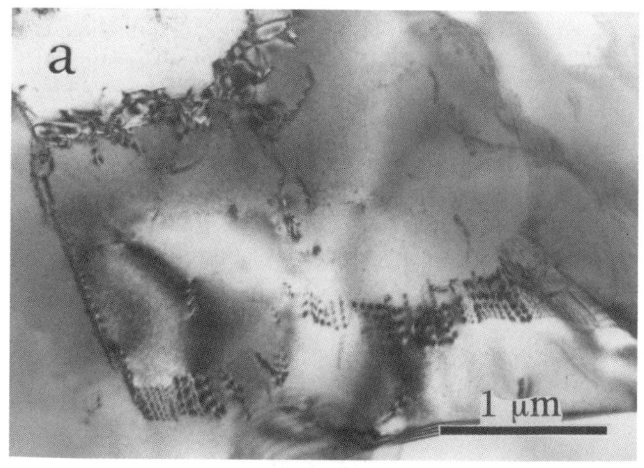

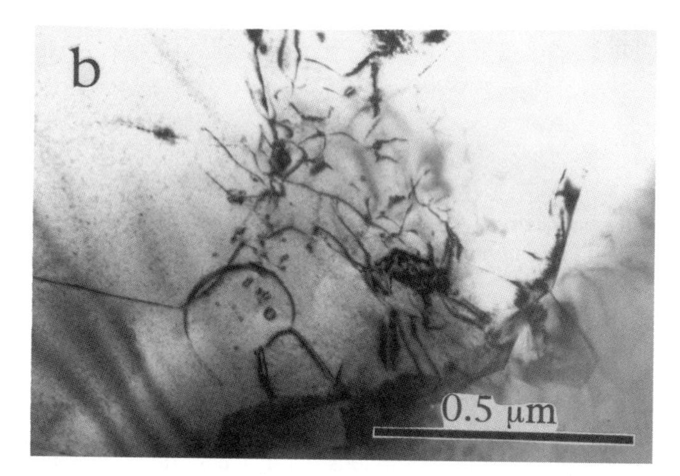

Figure 12. TEM micrographs (bright-field) of strained $MgAl_2O_4$ spinel recovered from 1100°C. (a) A subgrain with a comparatively low density of free dislocation. (b) A micrograph showing curved dislocations and dislocation loops.

Again the data define a straight line in a log-log representation. The slope and magnitude are consistent for the two different experiments.

3.2. *Microstructures*

In both the samples recovered from 600°C and 1100°C, many dislocations exist within the spinel grains. The density of free dislocations decreases with increasing temperature. The dislocation microstructures of these samples are entirely different. At low temperature, many dislocations are straight and tangled complicatedly (Figure 11). While at high temperature, subgrain boundary, curved dislocation and dislocation loop are conspicuously developed (Figure 12). The former microstructures indicate that the dislocations glided only on the slip planes without overcoming the obstacles such as dislocation tangling. This means the absence of dislocation climb which is stimulated by the enhanced lattice diffusion. On the other hand, the latter microstructures are typical characteristics responsible for the dislocation climb. Therefore the main deformation mechanisms of these samples can be identified dislocation glide (or plasticity) and dislocation creep (or power law creep) at low and high temperature, respectively.

The latter sample was heated at 400°C, 600°C and 800°C before 1100°C. The difference of microstructures between two recovered samples indicates that the microstructures created at lower temperatures are not preserved after higher temperature heating in the latter sample, and that the microstructures change with temperature and deformation mechanism.

The change of dislocation density with temperature is related to the change of deformation mechanism and/or the decrease of the driving stress of plastic strain. If the applied stress is constant through the temperature, the decrease of dislocation density is caused only by the high temperature annihilation of dislocations in the power law creep regime. In this case, it is expected that the strain rate increases in the higher temperature. On the other hand, the dislocation density is proportional to applied stress. Therefore it can be thought that the decrease of dislocation density is caused only by the decrease of applied stress.

Spinel has been studied in several different experiments in the past. *Frost and Ashby* [1982] summarize the rheological properties of spinel with a deformation mechanism map as illustrated in Figure 13. Low temperature gives rise to a 'plastic' regime characterized by high stress and very high values of the effective n. As temperature increases, the strain rate lines fan out, consistent with lower values of n and lower values of stress. This is termed the 'power law' region. We plot our data for shear stress as a function of temperature on the Frost and Ashby deformation map. Indeed our data follow the deformation region corresponding to our inferred strain rate very precisely. Our effective n values are quite consistent with those deduced from this diagram. The fit of our data to the deformation map is much better than the uncertainty of the deformation map. Moreover, these plots indicate that the deformation mechanisms of our experiments changed plasticity to power law creep between 600°C and 800°C. These results are supported by the microstructural observations. The microstructure also indicates that the applied stress promoting plastic strain decreased at higher temperature. Our data are for different stress-temperature regions than were the previous data. Furthermore, all of the previous data are for lower confining pressure than used in the present experiment.

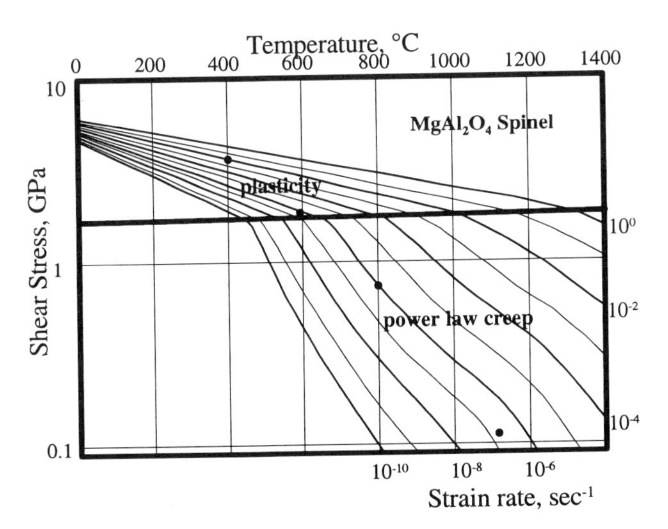

Figure 13. Observed shear strength of spinel (dots) plotted on the *Frost and Ashby* [1982] deformation map.

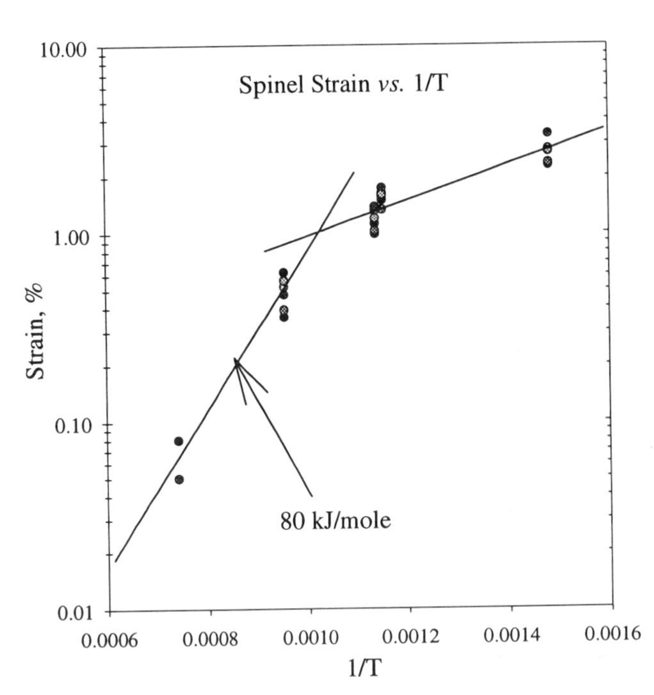

Figure 14. Log of strain as a function of inverse temperature. Slopes of lines on this diagram connect points corresponding to the same deformation process represent activation energies for these processes. Breaks in slope suggest changes of deformation mechanism such as plasticity to power law creep.

Rheological mechanisms are generally considered as activated processes. The activation energy can be derived from the slope of the data in a log strain *vs.* $1/T$ plot. Our data are illustrated in such a plot in Figure 14, where the strain corresponds to that after several 10s of minutes at a particular temperature. The low strain data exhibit scatter as we are close to the resolution of the technique for small strain. Nonetheless, there is a clear break in the curve consistent with a change in mechanism corresponding to the *Frost and Ashby* [1982] plastic to power law transition. Furthermore, the slope of the curve indicates an activation energy of 60 kJ/mol, a value consistent with previous studies.

4. CONCLUSIONS

Rheological properties of high-pressure phases, along with the pressure and temperature dependence of these properties, elude the standard techniques for defining these properties. Here we describe the application of a new technique to the characterization of the rheology of $MgAl_2O_4$ spinel, a material with a substantial data base. We find quantitative agreement with this study and the extant data. These results suggest that we can define these properties at pressures exceeding 10 GPa at temperatures to around 1000°C. Higher temperatures, corresponding to weaker samples, are limited by the resolution of the method.

The method rests on using a polycrystalline sample prepared as a loose powder. The stress field that is created in closing the holes at high pressure is the stress field monitored through an analysis of the broadening of the diffraction peaks. Since more energy is actually partitioned into this deviatoric stress field, the magnitude of this deviatoric stress saturates throughout the sample. Every grain is probably pushed to its deformation limit. By the time that the sample is weakened at the highest temperature, it has undergone an average plastic strain of several per cent.

This technique clearly has limitations. Steady-state creep generally requires much greater plastic strains. Thus, this method is probably not sampling the rheological properties at steady state. As strains become small, accuracy is poorer. Energy-dispersive radiation is inherently low resolution. Higher resolution will require monochromatic radiation with high-resolution detection. Probably an order of magnitude improvement in peak width resolution is possible with such strategies. This will allow us to measure lower shear stresses and hence maintain resolution to higher temperatures.

Acknowledgment. We thank Yujin Wu for assistance with the data analysis. This work was supported by NSF grant EAR8920239 to the Center for High Pressure Research. MPI contribution number 180.

REFERENCES

Evans, A. G., and D. R. Clarke, Residual stresses and microcracking induced by thermal contraction inhomogeneity, in *Thermal Stresses in Severe Environments*, edited by Hasselman, D. P. H. and R. A. Heller, Plenum Press, pp. 629-649, New York, 1980.

Frost, H. J., and M. F. Ashby, *Deformation-Mechanism Maps, The Plasticity and Creep of Metals and Ceramics,* 166 pp., Pergamon Press, 1982.

Gerward, L., S. Morup, and H. Topsoe, Particle size and strain broadening in energy-dispersive X ray powder patterns, *J. App. Phys., 47,* 822-825, 1976.

Kreher, W., Residual stresses and stored elastic energy of composites and polycrystals, *J. Mech. Phys. Solids, 38,* 115-128, 1990.

Li, B., S. M. Rigden and R. C. Liebermann, Elasticity of stishovite at high pressure, *Phys Earth Planet. Inter.* (Ringwood Volume, edited by T. Irifune and S. Kesson), *96,* 113-127, 1996.

Meade, C. , and R. Jeanloz, The strength of mantle silicates at high pressure and room temperature: implications for the viscosity of the mantle, *Nature, 348,* 533-535, 1990.

Rubie, D. C., C. A. Stewart and H. Schmeling, The effect of olivine rheology on kinetics of the olivine-spinel transformation (abstract), *Eos Trans. AGU, Spring Meeting Suppl. 73,* 311, 1992.

Tvergaard, V., and J. W. Hutchinson, Microcracking in ceramics induced by thermal expansion or elastic anisotropy, *J. Am. Ceram. Soc. 71,* 157-166, 1988.

Walsh, J. B., Theoretical bounds for thermal expansion, specific heat, and strain energy due to internal stress, *J. Geophys. Res. 78,* 7637-7646, 1973.

Weidner, D. J., M. T. Vaughan, J. Ko, Y. Wang, K. Leinenweber, X. Liu, A. Yeganeh-Haeri, R. E. Pacalo and Y. Zhao, Large volume high pressure research using the wiggler port at NSLS. *High Pres. Res., 8,* 617-623, 1992a.

Weidner, D. J., M. T. Vaughan, J. Ko, Y. Wang, X. Liu, A. Yeganeh-Haeri, R. E. Pacalo and Y. Zhao, Characterization of stress, pressure, and temperature in SAM85, A DIA type high pressure apparatus, in *High-Pressure Research: Application to Earth and Planetary Sciences*, *Geophys. Monogr. Ser.*, vol. 67, edited by Y. Syono and M. H. Manghnani., pp. 13-18, Terra Scientific Publishing Company Tokyo/AGU, Washington, D.C., 1992b.

Weidner, D. J., Y. Wang and M. T. Vaughan, Strength of diamond, *Science, 266,* 419-422, 1994.

J. Ando, G. Chen, M. T. Vaughan, Y. Wang, and D. J. Weidner, Center for High Pressure Research and Department of Earth and Space Sciences, State University of New York, Stony Brook NY 11794

Yield Strength, Slip Systems and Deformation Induced Phase Transition of San Carlos Olivine up to the Transition Zone Pressure at Room Temperature

Mu Chai and J. Michael Brown

Geophysics Program, University of Washington, Seattle, Washington

Yanbin Wang*

Center for High Pressure Research and Department of Earth and Space Sciences, University at Stony Brook, Stony Brook, New York

Controlled deformation experiments on single-crystal San Carlos olivine have been conducted in the diamond anvil cell to a confining pressure of 16 GPa at room temperature. The tensorial stress states of the samples were determined using ruby spectroscopy, and yield strengths of San Carlos olivine have been determined to 20 GPa. Results appear independent of the load directions, and are in general agreement with the previous measurements within the experimental uncertainties. Transmission electron microscopy on the recovered samples indicates that plastic deformation is associated with high dislocation density, twinning, and stacking disorder. At room temperature, samples were partially transformed to a high-pressure phase, whose diffraction patterns can be indexed as the b-phase. The transformation is proposed to be induced by the plastic deformation. The result may have implications on the stability of cold olivine within subducted slabs.

INTRODUCTION

Rheological investigations of mantle minerals at high pressures include studies of (1) yield strength, (2) the deformation mechanism, and (3) the effect of the deviatoric stress on phase transformations. Most deformation experiments have been carried out in a piston cylinder type apparatus where pressures are generally below 5 GPa. Recently, progress has been reported in using multianvil presses for rheological studies [*Weidner et al.*, 1995] and some effort has been directed at conducting deformation experiments in the diamond anvil cell (DAC) [*Kinsland and Bassett*, 1977; *Meade and Jeanloz*, 1988]. In general, accurate stress calibrations remain problematic in the high pressure regime.

Recently, *Chai and Brown* [1996] developed a technique to measure single-crystal yield strength in the diamond anvil cell under high confining pressure. Using this technique, we have deformed San Carlos olivine single crystals at room temperature under confining pressures of

*Now at Consortium for Advanced Radiation Source, The University of Chicago, 5640 S. Ellis Ave., Chicago, IL 60637

Properties of Earth and Planetary Materials
at High Pressure and Temperature
Geophysical Monograph 101

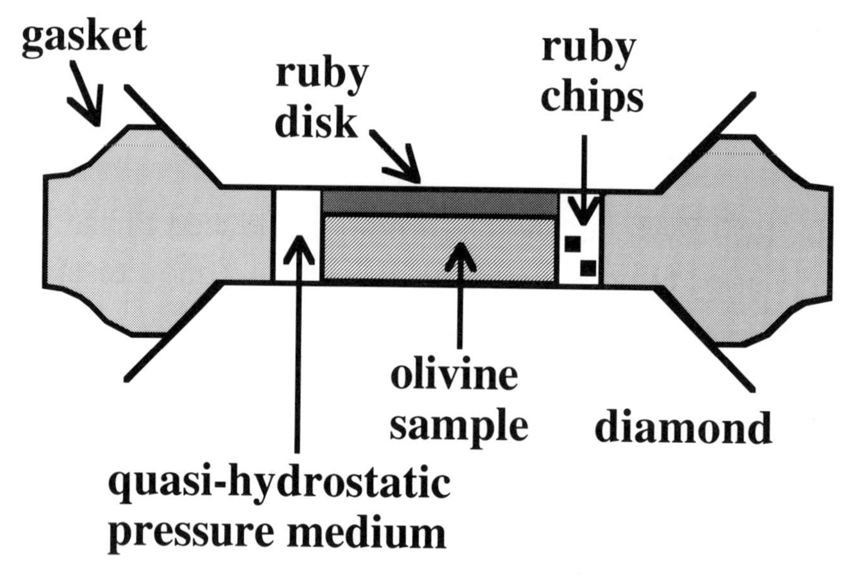

Figure 1. Experimental arrangement in the diamond anvil cell.

up to 16 GPa, with measured differential stresses of up to 15 GPa. The yield strength has been measured under those conditions. Transmission electron microscopy (TEM) analysis on the recovered samples indicates (1) that the deformation is completely plastic and (2) that deformation has induced phase transformation even at room temperature.

Olivine is believed to be the most abundant mineral within the lithosphere [*Ringwood*, 1970]. As the thermal boundary layer of a convecting mantle, the lithosphere is expected to be much colder and stronger than the surrounding mantle. For a slab subducting at a rate of 8 to 10 cm per year, the interior could be as cold as 600°C at a depth of 650 km [*Kirby et al.*, 1991]. At such pressures and temperatures, deformation through low temperature plasticity is expected [*Frost and Ashby*, 1982]. The study of low temperature plasticity of olivine is therefore directly relevant in the understanding of physical processes associated with the dynamics of the mantle.

EXPERIMENTAL TECHNIQUE

The experiment arrangement in the current study, schematically shown in Figure 1, is similar to that used by *Chai and Brown* [1996]. An oriented single crystal disk of San Carlos olivine, 200 μm in diameter and 35 to 75 μm thick, was placed in a metal gasket chamber between two diamonds. A single crystal ruby disk of the same diameter and about 20 μm thick was sandwiched between the sample and one of the diamonds. The diamond anvil cell was cryogenically loaded using argon as a quasi-hydrostatic pressure medium with micron-sized ruby chips distributed in it. The sample sandwich was first hydrostatically compressed to a desired pressure before the diamonds touched the sample. Deviatoric stresses developed as the diamonds were forced into direct contact with the sandwiched sample. Different confining pressures were achieved by varying either the initial thickness of the pre-indented gasket or the gasket material (pre-hardened stainless steel, air-hardened steel, Inconel, or rhenium). Clearance between the sample and the gasket chamber sidewalls ensured that the sample was supported by the pressure-transmitting medium instead of the more rigid gasket materials.

The geometry of the current experiments (axial loading of cylindrical samples by nearly incompressible diamond endplates) is likely to generate a stress state characterized by two principal stresses. The load stress (σ_1) is directed normal to the diamond faces while at the pressure medium-sample interface, the lateral stress (σ_3) is equal to the confining pressure. Fluorescence of ruby was excited using a 10 mW He-Cd laser focused to a spot size of approximately 10 microns. Spectra were collected on a grid of points separated by 20 microns. Data were fit as described in *Munro et al.* [1993] and are assumed to represent the vertically averaged properties of the stressed ruby at each point. Pressures in the pressure medium and in sample stacks were determined from the R_2 line of ruby. Following the work of *Shen and Gupta* [1993] *and Chai and Brown* [1996], we note that in nonhydrostatic experiments, the shift of the more intense R_1 line of ruby gives a systematically biased estimate of pressure while the R_2 line is apparently insensitive to deviatoric stresses and thus gives a better determination of pressure. The areal

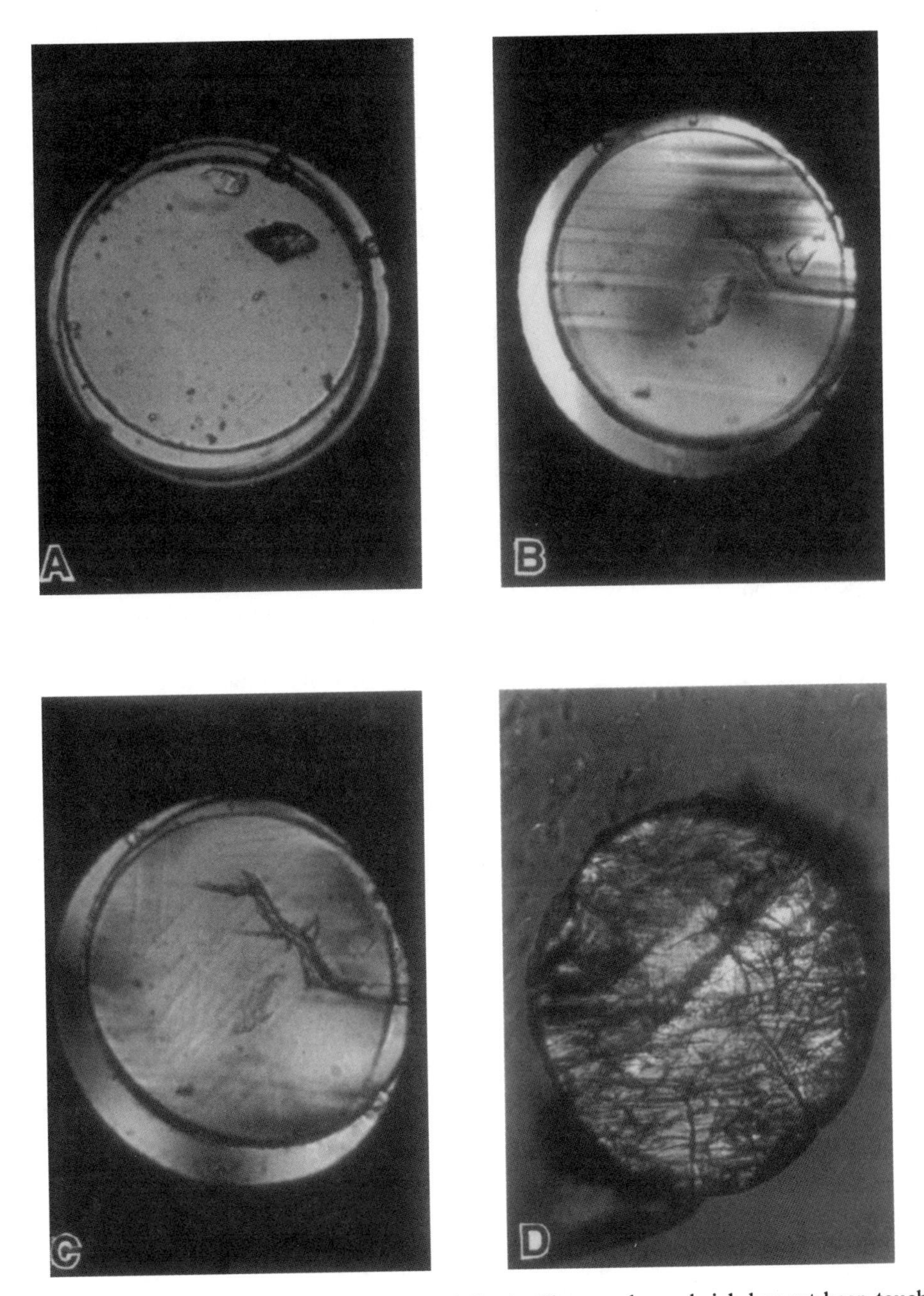

Figure 2. Top view of the sandwiched sample in the DAC. A: The sample sandwich has not been touched by the diamonds yet. The small ruby chips on the side in the pressure medium are used to measure the confining pressure, the pressure in the cell was at 9 GPa. Two ruby chips drifted onto the top of the sandwiched sample. B: Plastic lamellae first occurred as the diamond squeezed the sample. The crack was probably caused by local high stress due to the ruby chips on the top. C: As the sample was further squeezed, more slip systems were activated as indicated by the lineations. D: A recovered sample.

distribution of pressures and R_1-R_2 splittings of the ruby fluorescence in the olivine-ruby stack were used to characterize the state of stress in the olivine. The load stress (σ_1) was inferred on the basis of the pressure measurements. Further details are given below.

Plastic deformation was optically monitored through a microscope as shown in Figure 2. Figure 2A shows the top views of the sandwiched sample in the DAC. Deformation of the sample is associated with the occurrence of linear plastic lamellae (Figure 2B and C). The stress state of the

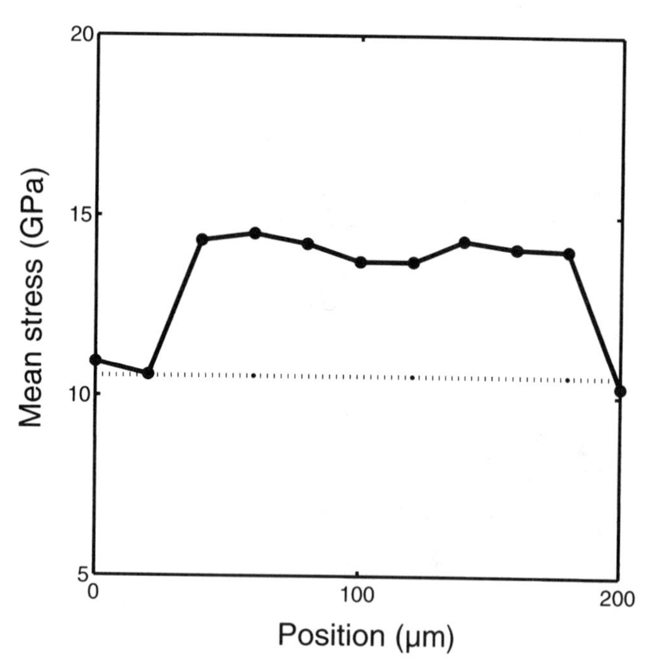

Figure 3. One profile of the mean stress distribution across the sandwiched sample.

sample was measured at onset of plastic deformation. This provided an estimate of the yield strength of the sample. Figure 2D shows a recovered sample. Radial strain was measured optically to a precision of a few microns. Recovered samples were transferred onto copper grids, Ar-ion thinned, and carbon coated for TEM studies.

RESULTS AND DISCUSSIONS

Stress Determination

Figure 3 shows one profile of the mean stress ((σ_1 + $2\sigma_3$)/3) distribution across an axially loaded sandwiched sample. The confining pressure is indicated by the horizontal straight line. All profiles were characterized by a nearly discontinuous jump from the confining medium pressure (10.5 GPa in Figure 3) to a nearly constant mean stress across the sample stack (14 GPa in Figure 3). The kinks in the mean stress profile appeared to be associated with the deformation lamellae of the sample.

The tensor stress state within an elastically compressed sample depends on the external loads and boundary conditions. With free slip between the sample and the diamonds, stresses would be uniform within the stack (the lateral stress σ_3 everywhere equal to the confining pressure and the axial stress σ_1 equal to the load stress)

Stresses were calculated for a elastically loaded isotropic cylinder using an analytical model [*Balla*, 1960]. In this model, a cylinder is compressed between two infinitely rigid end plates (in reality the rigidity of a diamond is finite). A friction factor between the sample and the end plates varied from 0 to 1, corresponding to free-slip and no-slip boundaries, respectively. The parameters needed in the calculation are aspect ratio (height to diameter), Poisson's ratio, the average load stress, and the lateral stress on the sides of the cylinder. The parameters of the model can be directly measured in the experiments, with the exception of the average load stress. In the calculations shown in Figure 4, an isotropically average Poisson's ratio of 0.25 was used for olivine. The aspect ratio was 1:4 and the lateral stress at the cylinder surface was set equal to the confining pressure. Calculated stresses were integrated vertically over the top half of the cylinder. The average load stress was determined by trial and error such that the calculated mean stress best matches the observed mean stress.

In Figure 4, stress profiles are plotted including (1) the vertically averaged stresses across the cylinder with a no-slip boundary condition, (2) the stresses expected for free-slip boundaries, and (3) the mean stress profile from Figure 3. In contrast to the free-slip situation and the experimental data, the axial (σ_1) and lateral stresses (σ_3) in the no-slip situation vary from the center to the edge by

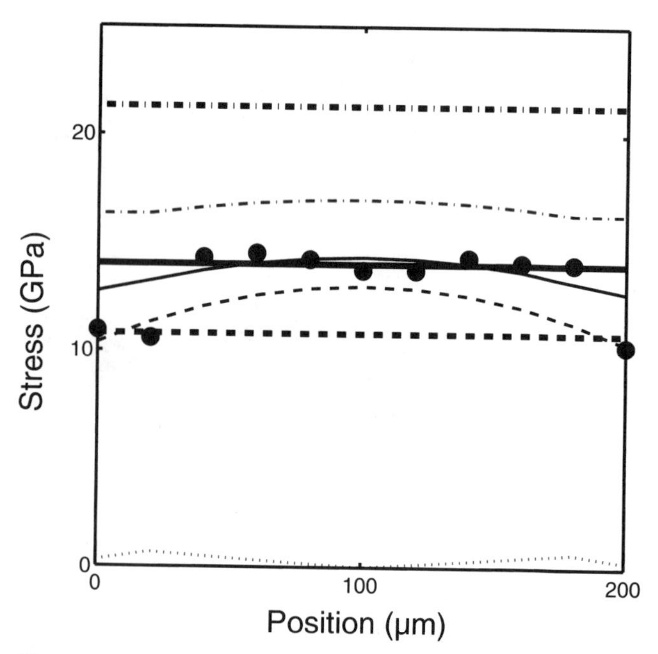

Figure 4. Calculated stress profiles for a loaded cylindrical plate. The thick lines are for the free-slip condition and the thin lines for the no-slip condition, respectively. Solid lines are mean stresses; dot-dashed lines, σ_1; dashed lines, σ_3; and the dotted line is σ_{13}. Distribution of mean stress from Figure 3 is plotted as the filled circles.

almost 30%. Although friction between the diamonds and sample would be expected to inhibit slip, it is evident that the mean stress distribution of the sample is more consistent with free slip condition (uniform stress distribution). The free-slip boundary condition gives the highest estimated differential stress.

The free-slip approximation was further validated by *Chai and Brown* [1996] through measurements of the R_1-R_2 splitting in ruby as a function of differential stress. The R-line separation increases when loaded along the *a*-axis and decreases when loaded along the *c*-axis. Using the free-slip approximation, observed R_1-R_2 splittings are plotted as a function of calculated differential stress (σ_1-σ_3) in Figure 5. The splitting calculated from equation (9) of *Shen and Gupta* [1993], based on shock wave data, plotted as the solid lines is in accord with the static data.

Finally, as discussed below, the lack of a dependence of (σ_1-σ_3) at yield on sample thickness is an additional test of the free-slip approximation in our experiments. The no-slip analysis of *Meade and Jeanloz* [1990] was based on the idea that the radial stress gradient in an axially loaded sample is balanced by the shear traction σ_{13}. Yield strength is proportional to $h(\partial p/\partial r)$, where h is the thickness of the sample and $\partial p/\partial r$ is the radial pressure gradient in the sample. The absence of a radial stress

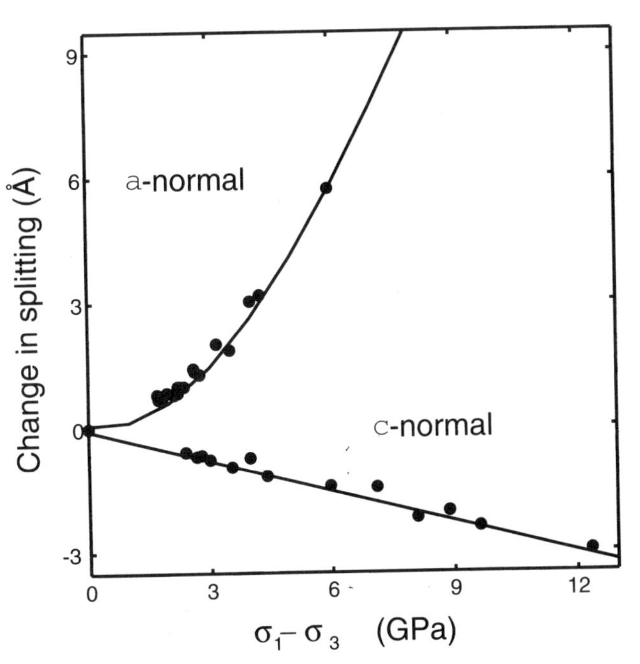

Figure 5. The observed R_1-R_2 line splittings for axially loaded ruby single crystals using differential stresses derived based on the free-slip boundary approximation (from *Chai and Brown* [1996]). The solid lines are predicted on the basis of the calibration reported by *Shen and Gupta* [1993].

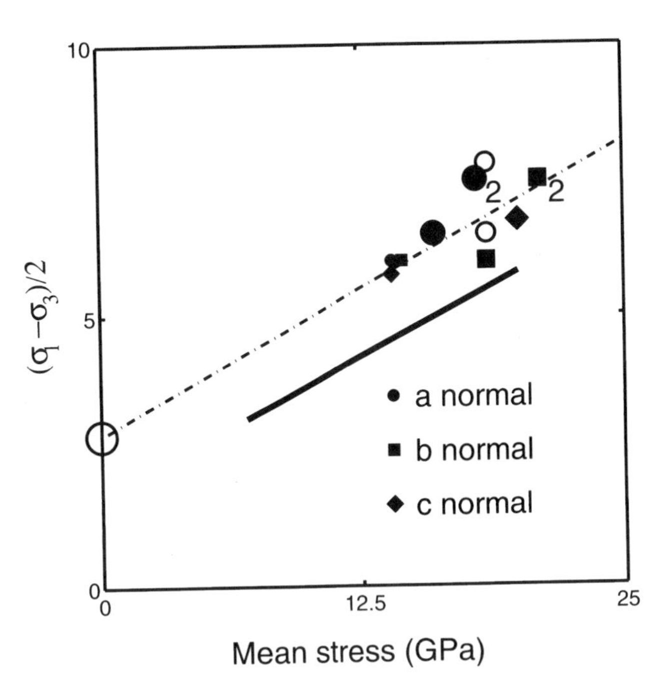

Figure 6. The yield strength of olivine as a function of mean stress. Different symbols represent different orientations of the olivine; symbols with numerical subscript 2 indicate observations made after compression beyond the onset of deformation. Larger and small solid symbols represent 75 micron and 35 micron olivine samples, respectively. The large circle is the ambient pressure yield strength of *Evans and Goetze* [1979]; the heavy line is from *Meade and Jeanloz* [1990]; the small circles are from *Furnish et al.* [1986].

approximate uncertainty of about 1 GPa is estimated on the basis of fits to the primary data as shown in Figure 4. Also plotted in Figure 6 are the ambient pressure data of *Evans and Goetze* [1979] and a line giving the results *of Meade and Jeanloz* [1990].

The yield strengths of some samples were recorded again after further deformation. These points (denoted with numerical subscript 2 in the figure) may have higher gradient and no thickness dependence for differential stress at the yield point in the current study argue against this stress balance.

Yield Strength

Yield strength is taken to be equal to the maximum resolved shear stress (σ_1-σ_3)/2 at visual onset of deformation. Three different crystallographic orientations (with crystal axes parallel to load axes) and two different thicknesses (35 and 70 μm) were investigated. Based on the current data, strength measurements on olivine at high pressure and room temperature are plotted in Figure 6. An strength as a result of strain hardening [*Nicolas and*

TABLE 1. Experimental Conditions for San Carlos Olivine Single Crystals

Sample	Orientation	Confining Pressure σ_3	Mean Stress $(\sigma_1+2\sigma_3)/3$	Differential Stress $\sigma_1 - \sigma_3$	Maximum Radial Strain
B1	[010]	14.0 GPa	19 GPa	15.0 GPa	15%
B2	[010]	16.5 GPa	21 GPa	13.5 GPa	30%
C1	[001]	15.5 GPa	20 GPa	13.5 GPa	10%

An average value of the mean stress and differential stresses is listed. Stresses varied by less than 1 GPa across the sample disk. Total strain is based on the change in radius of the sample disks as measured before and after deformation. We follow the mineralogical convention using the space group P*bnm* for olivine, with edge lengths *b*>*c*>*a*. In this assignment, the "six-fold" axis of the pseudo-hexagonal close-packed oxygen sublattice is parallel to the [100] axis.

Poirier, 1976]. Sample thickness had no discernible effect on the determination of the yield strength and no strong orientational dependence is apparent. *Evans and Goetze* [1979] also reported no orientation dependence. *Furnish et* al. [1986] measured the single crystal yield strength of olivine (the Hugoniot elastic limit) at 18.5 GPa. The strength of olivine shocked along (100) was 6.5 GPa and for a direction midway between (100) and (013) the strength was 7.8 GPa .

A dashed line, drawn by visual inspection, through the zero-pressure point of *Evans and Goetze* [1979] and our data is shown in Figure 6. The *Meade and Jeanloz* [1990] result, plotted in the figure as the solid line, has a similar slope with the current data being offset upward by a little more than 1 GPa. The yield strength in a single crystal normally represents that of a particular slip system. As discussed in the next section, the orientations of our samples did not favor the easy slip systems. A lower strength may be expected in polycrystals with randomly oriented grains. Thus, we consider the results to be in generally good agreement.

Low-Temperature Plastic Deformation Mechanisms

Recovered samples from the deformation experiment (Table 1 gives the experimental conditions) were analyzed with TEM. All of the examined samples show high dislocation densities throughout the crystals. An accurate determination of the dislocation density is difficult; a lower bound is estimated to be 10^{11} cm^{-2}. Each originally single crystal now contains numerous domains. In sample B1, the domains are generally rectangular and the domain boundaries are approximately parallel to the {100}, {010} and {001} planes, with the elongation direction parallel to 100*. The samples B2 and C1 contain extremely fine lamellar domains whose boundaries are again along the 100*. Selected-area electron diffraction (SAED) patterns taken from these areas show that the domains are related by twinning (Figure 7A). The most commonly observed twinning relation is rotation with respect to the [100] axis, with the (002) plane of one twin variant parallel to the (012) plane of the other. Other twinning relations are also observed, which all share the common (100) plane, with the perpendicular planes (011), (020), (021), (031), and so forth, parallel to each other across the domains (see SAED pattern in Figure 7B). These complicated twinnings contrast with the slip systems observed in experiments at low pressure and low temperature which are dominantly [001](100) [*Raleigh*, 1968; *Young*, 1969; *Carter and AveLallemant*, 1970]. The orientations of our samples do not favor these easy slip systems which are either parallel or perpendicular to loading axis, therefore extensive twinnings are expected under such high shear stresses.

These twinning relations and the observed microstructures can be interpreted as a result of significant increase in stacking faults in directions perpendicular to the 100* of the olivine crystals. In some areas, superlattice reflections are observed. In Figure 8A, taken from sample B1, the superlattice reflections indicate a twofold repeat in the $0\bar{1}2$* direction, with a corresponding *d*-spacing about 5.75 Å. Figure 8B shows another example where the superlattice reflections reveal a fourfold repeat in the 021* direction, corresponding to a *d*-spacing as large as 15.5 Å. The superlattice reflections, forbidden in an ordinary olivine structure, indicate the loss of symmetry elements due to the presence of large number of stacking faults and may be an indication of onset of structural polymorphism. In Figure 7A, an additional set of diffraction spots are present (indicated by the white arrows), which cannot be explained by superlattice reflection in olivine.

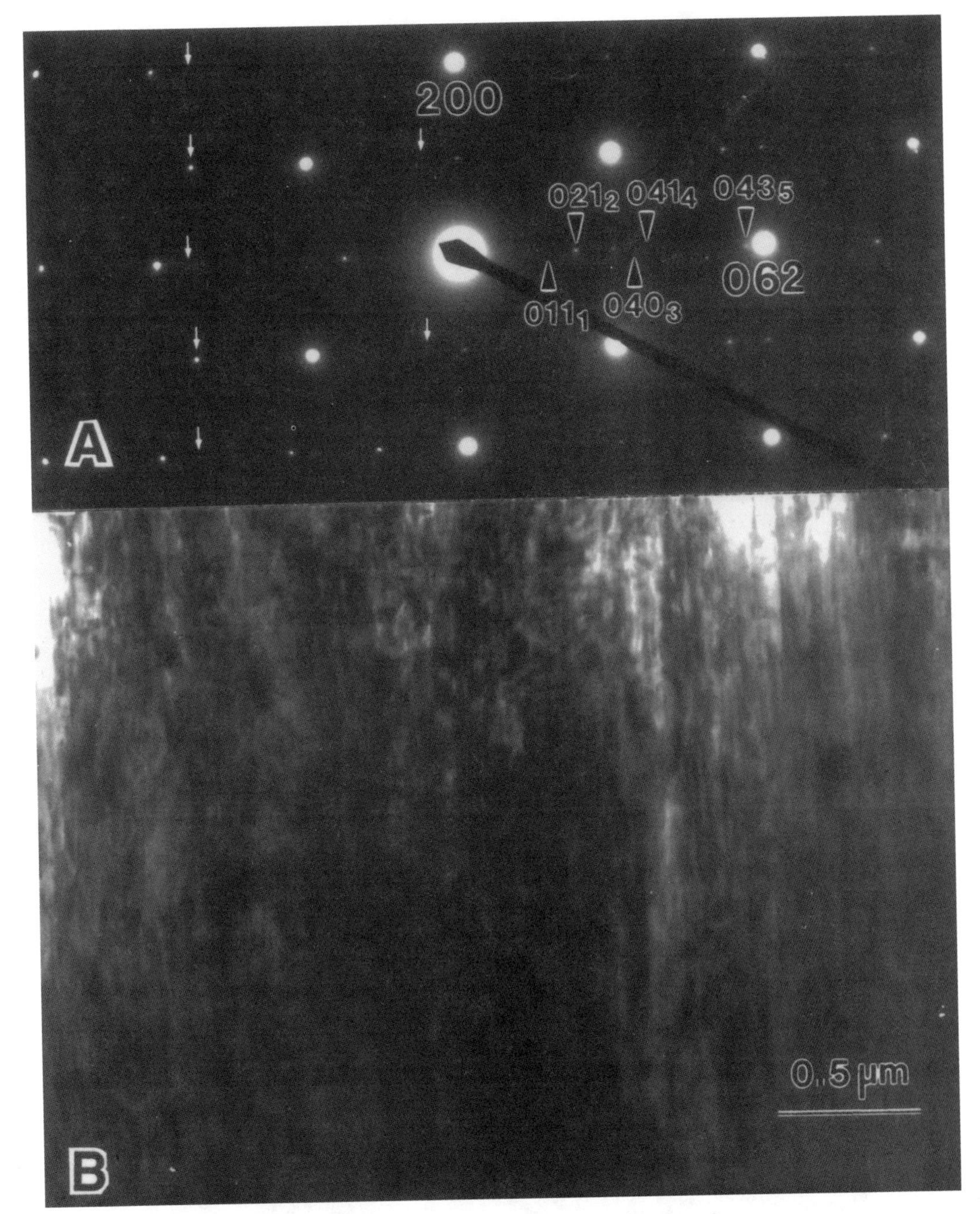

Figure 7. Typical deformation microstructure (sample C1). Selected-area electron diffraction (SAED) pattern (A) from a homogeneous microstructure (B) with lineations whose traces are parallel to the 100* direction. The SAED pattern indicates complex twinning relations. Spots 200 and 062 are from the olivine matrix. Spots due to five different twinning orientation relations are labeled by numerical subscripts. Additional superlattice spots are also present (e.g., vertical white arrows), which cannot be indexed as from the P*bnm* olivine structure.

Additionally, nearly all of the electron diffraction patterns indicate plasticity is the dominant deformation mechanism in the 4% shortening in the *c*-axis; the original 5.98 Å edge length now is about 5.7–5.75 Å. The other axes, on the other hand, do not show such large change in dimension. These observations indicate that the samples have undergone considerable structural modifications under the experimental conditions.

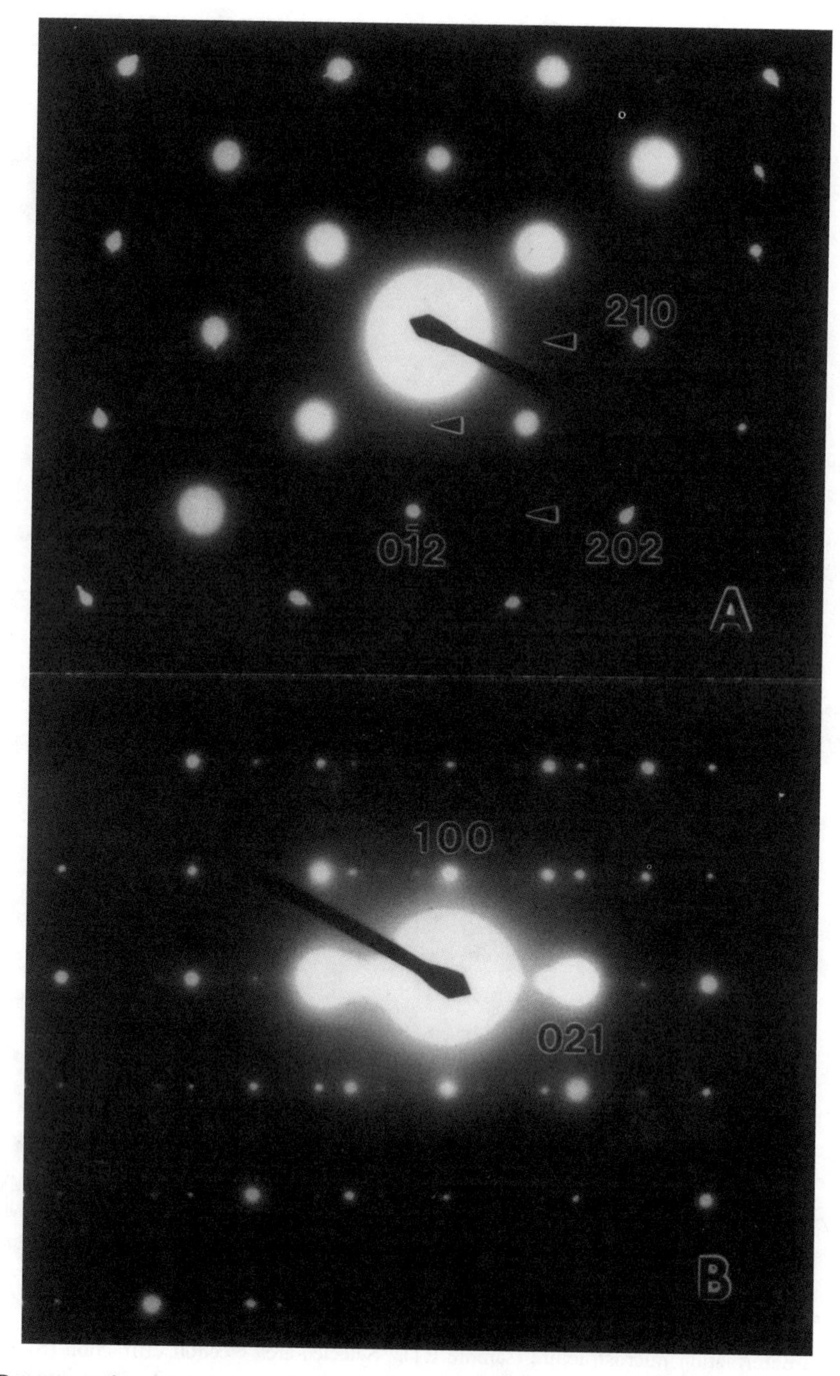

Figure 8. SAED patterns showing structure disorder in the deformed samples. A: SAED pattern, taken from sample B1, arrowheads indicate twofold repeat along the 0$\bar{1}$2* and 120* directions. B: A pattern from sample C1 showing a fourfold repeat along the 021* direction, resulting in a superstructure with a *d*-spacing of 15.5 Å.

Deformation Induced Phase Transformation

Upon close examination we found that all three samples contain domains whose SAED patterns can be indexed on the basis of the β-phase, which has close topotaxial relationships with olivine. Unfortunately, the small domain sizes did not allow a complete structure identification. Figure 9 is a bright-field electron micrograph taken from a 0.5-μm-wide zone in sample B1 which contains very thin layers of high-pressure phase.

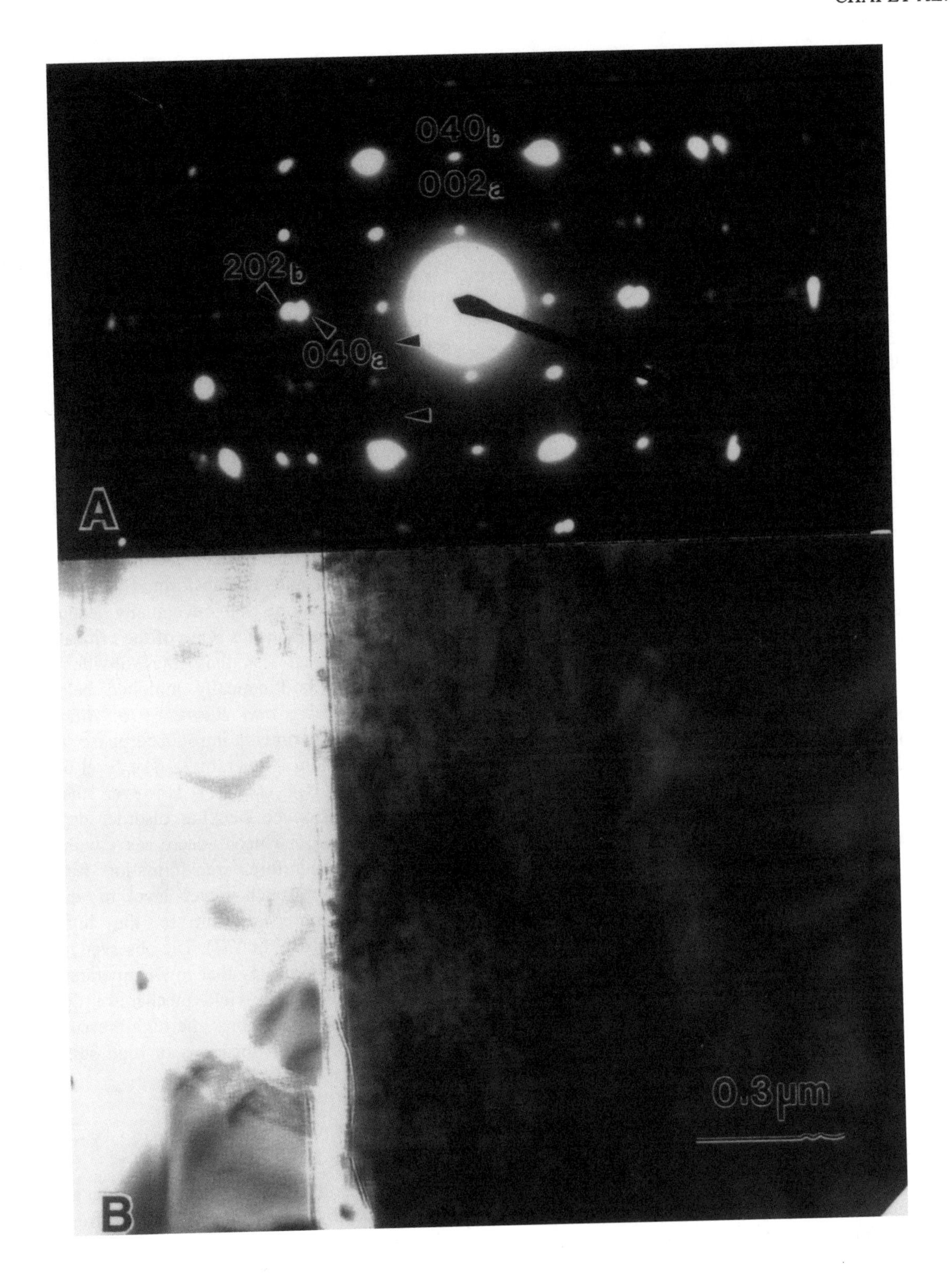

Figure 9. Formation of β in the deformed olivine (sample B1). The SAED pattern (A), taken from a linear, highly deformed band (vertical band in the middle of B), can be indexed as a combination of olivine (denoted by subscript a) and β-phase (denoted by subscript b).

TABLE 2. Zero-Pressure Lattice Misfit Between Adjacent Olivine and β-Phase Atomic Planes

Epitaxial Relation	Lattice Spacing, Å	Misfit (%)
$(100)_\alpha//(10\bar{1})_\beta$	$d_{100}{}^{\alpha}$= 4.755, $d_{10\bar{1}}{}^{\beta}$=4.67	1.8
$(031)_\alpha//(040)_\beta$	$d_{031}{}^{\alpha}$= 2.955, $d_{040}{}^{\beta}$=2.85	3.6
$(100)_\alpha// 10\bar{1}$	$d_{100}{}^{\alpha}$= 4.755, $d_{10\bar{1}}{}^{\beta}$=4.67	1.8
$(021)_\alpha//(040)_\beta$	$d_{021}{}^{\alpha}$= 3.477, $d_{040}{}^{\beta}$=2.85	18
$(001)_\alpha//(020)_\beta$	$d_{001}{}^{\alpha}$= 5.982, $d_{020}{}^{\beta}$=5.79	3.2
$(020)_\alpha//(101)_\beta$	$d_{020}{}^{\alpha}$= 5.102, $d_{101}{}^{\beta}$=4.67	8.5

*We adapt the *Imma* space group for the β-phase (a=8.248 Å, b=11.444 Å, and c= 5.696 Å).

The (002) and (040) diffractions of olivine and (040) and (202) diffractions of β -phase are labeled in Figure 9 A. Diffraction spots (040) of olivine and (202) of β-phase are well separated, while (002) of olivine and (040) of β-phase overlap due to the permanent shortening of olivine along c-axis. Diffraction (101) of β-phase (not labeled in the figure) is weak and connected with the (020) diffraction of olivine. Overall, three topotaxial relations between α and β have been observed:

$$(100)_\alpha//(10\bar{1})_\beta \text{ and } (031)_\alpha//(040)_\beta,$$
$$(100)_\alpha//(10\bar{1})_\beta \text{ and } (021)_\alpha//(040)_\beta,$$
$$\text{and } (100)_\alpha//(020)_\beta \text{ and } (020)_\alpha//(101)_\beta.$$

The subscripts α and β denote olivine and the β -phase, respectively. They are all different from those proposed by *Madon and Poirier* [1983] and *Poirier* [1981] for a martinsitic-like transformation. These orientations have relatively small strain mismatch between the two crystal lattices. Table 2 summarizes d-spacings for the related directions, which exhibit strain mismatch of less than 4%, except for $(021)_\alpha$ and $(040)_\beta$ for which the strain mismatch is as large as 18%.

We do not believe that the phase transformation observed in our samples occurs via the so-called martinsitic-like mechanism for the following reasons: (1) samples were deformed with various orientations relative to the maximum compressive principal stress (σ_1), and yet there is no unique topotaxy that relates the two structures; (2) microstructurally, the majority of the areas do not show clearly defined interfaces between the two phases; and (3) no macroscopic faulting was observed at strains as large as 30%.

Based on the microstructure, the twinning relations, the superlattice reflections in olivine, as well as the various orientational relations between olivine and the β-phase, we suggest a mechanism of structural polymorphism via the generation of high density of stacking faults. An intermediate highly disordered phase provides a low activation energy path for the reaction, thereby enhancing the rate of transformation. The intermediate disordered step is "frozen in" at laboratory time scales by low temperature; a moderate increase in temperature would likely lead to rapid transformation. Indeed, *Guyot et al.* [1991] proposed a two-step mechanism for the α-β transformation in forsterite.

Because brittle behavior (faulting) of rocks is generally suppressed by high pressure, there have been a number of proposed source mechanisms for earthquakes below a depth of 350 km. In the transformational faulting mechanism [*Kirby*, 1987; *Green and Burnley*, 1989; *Burnley et al.*, 1991; *Green et al.*, 1990, 1992], because the olivine in the subducting slab is substantially colder than the surrounding mantle, it cannot transform into β-phase as it passes through the 410-km discontinuity into the pressure stability field of the β-phase. But with heating at a greater depth, it starts to transform. The transformed phase tends to weaken the slab, eventually leading to failure which causes earthquakes. One of the critical assumptions in this proposal is that the olivine to β-phase transformation is kinetically inhibited below a certain temperature [*Sung and Burns*, 1976; *Rubie and Ross*, 1994]. Our experiments introduced phase transformation in olivine at room temperature. The level of shear stress was high ($(\sigma_1-\sigma_3)/\ \sigma_3 \approx 1$). However, the stress level required to introduce stacking disorder depends on the yield strength, which decreases with increasing temperature. Therefore, transformation by deformation may occur at a much lower level of shear stress at temperatures more relevant to the interior of the subducting slab (e.g., 600°C). The observed microstructure in our samples indicates that low-temperature laboratory. As long as the same yield mechanism operates under subduction zone conditions, the high-pressure phase may be quickly formed in locations of high stress within the downgoing slab.

CONCLUSIONS

We have conducted rheological studies at transition zone pressures with controlled deviatoric stresses. Direct evidence of plastic deformation of single crystals under such extreme differential stresses and confining pressures is provided by our data. Our results support the notion that confining pressure generally suppresses fracture. The yield strength measured in this study is in agreement with the previous results. The observation of the phase transformation in olivine at room temperature, which is proposed to be deformationally induced, opens questions concerning the stability of olivine in the subducted slabs. Further experiments are suggested at moderate temperature (<600°C) to investigate the formation of the β-phase in deforming samples. The current technique lends a powerful tool to investigating the role of shear stresses in the kinetics of phase transformation in general.

Acknowledgments. This work was partially supported by the National Science Foundation and the Institute of Geophysics and Planetary Physics, Los Alamos. We acknowledge continued encouragement of this project from Stephen Kirby. Discussions with David Mao and Harry Green are also acknowledged. We thank the reviewers for their careful reading of this paper.

REFERENCES

Balla, A., Stress conditions in triaxial compression, *J. Soil Mechanics and Foundation Division, SM 6*, 57-84, 1960.

Burnley, P. C., H. W. Green, and D. J. Prior, Faulting associated with the olivine to spinel transformation in Mg2GeO4 and its implications for deep-focus earthquakes, *J. Geophys. Res.*, *96*, 425-443, 1991.

Carter, N. L., and H. G. AveLallemant, High temperature flow of dunite and peridotite, *Geol. Soc. Amer. Bull*, *81*, 2181-2202, 1970.

Chai, M., and J. M. Brown, Effects of static non-hydrostatic stress on the R lines of ruby single crystals, *Geophys. Res. Lett.*, *23*, 3539-3542, 1996.

Evans, B., and C. Goetze, The temperature variation of hardness of olivine and its implication for polycrystaline yield stress, *J. Geophy. Res.*, *84*, 5505-5524, 1979.

Frost, H. J., and M. F. Ashby, *Deformation-Mechanism Maps-the Plasticity and Creep of Metal and Ceramics*, 166 pp., Pergamon Press, 1982.

Furnish, M. D., D. Grady, and J. M. Brown, Analysis of shock wave structure in single-crystal olivine using VISAR, *in Shock Waves in Condensed Matter*, edited by Y. Gupta, pp. 595-600, Plenum, New York, 1986.

Green, H. W., and P. C. Burnley, A new self-organizing mechanism for deep-focus earthquakes, *Nature*, *341*, 733-737, 1989.

Green, H. W., T. E. Young, D. Walker, and C. H. Scholz, Anticrack-associated faulting at very high pressure in nature olivine, *Nature, 348*, 720-722, 1990.

Green, H. W., T. E. Young, D. Walker, and C. H. Scholz, The effect of nonhydrostatic stress on the α - β and α - γ olivine phase transformations, in *High-Pressure Research Application to Earth and Planetary Sciences, Geophys. Monog. Ser.*, vol. 67, edited by Y. Syono and M. H. Manghnani, pp. 229-235, Terra Scientific Publishing Company, Tokyo/AGU, Washington, D.C., 1992.

Guyot, F., G. D. Gwanmesia, and R. C. Liebermann, An olivine to beta phase transformation mechanism in Mg_2SiO_4, *Geophys. Res. Lett., 18*, 89-92, 1991.

Kinsland, G. L., and W. A. Bassett, Strength of MgO and NaCl polycrystals to confining pressures of 250 kbar at 25 C, *J. Appl. Phys., 48*, 978-985, 1977.

Kirby, S. H., Localized polymorphic phase transformation in high-pressure faults and applications to the physical mechanism of deep earthquakes, *J. Geophys. Res., 92*, 13789-13800, 1987.

Kirby, S. H., W. B. Durham, and L. A. Stern, Mantle phase changes and deep-earthquake faulting in subducting lithosphere, *Science, 252*, 216-225, 1991.

Madon, M., and J. P. Poirier, Transmission electron microscope observation of a, b and g (Mg,Fe)2SiO4 in shocked meteorites: planar defects and polymorphic transitions, *Phys. Earth Planet. Int., 33*, 31-44, 1983.

Meade, C., and R. Jeanloz, Yield strength of B1 and B2 phases of NaCl, *J. Geophys. Res., 93*, 3270-3274, 1988.

Meade, C., and R. Jeanloz, The strength of mantle silicates at high pressures and room temperature: implications for viscosity of the mantle, *Nature, 348*, 533, 1990.

Munro, R. G., G. J. Piermarini, and S. Block, Model line-shape analysis for the ruby lines used for pressure measurement, *J. Appl. Phys., 57*, 165-169, 1993.

Poirier, J. P., On the kinetics of olivine-spinel transition, *Phys. Earth Planet. Int.*, 26, 179-187, 1981.

Raleigh, C. B., Mechanisms of plastic deformation in olivine, *J. Geophys. Res., 73*, 5391-5406, 1968.

Ringwood, A. E., Phase transformations and the constitution of the mantle, *Phys. Earth Planet. Int. 8*, 109-155, 1970.

Rubie, D. C., and C. R. Ross, II, Kinetics of the olivine-spinel transformation in subducting lithosphere: experimental constraints and implications for deep slab processes, *Phys. Earth Planet. Int., 86*, 223-241, 1994.

Shen, X. A., and Y. M. Gupta, Effect of crystal orientation on ruby R-line shifts under shock compression and tension, *Phys. Rev. B, 48* , 2929-2940, 1993.

Sung, C., and R. G. Burns, Kinetics of high-pressure phase transformation: implications to the evolution of the olivine-spinel transformation in the downgoing lithosphere and its consequences on the dynamics of the mantle, *Tectonophysics, 31*, 1-32, 1976.

Weidner, D. J., Y. Wang, G. Chen, and J. Ando, Rheology measurements at high pressure and temperature (abstract), *EOS Trans. AGU, Fall Meeting Suppl., 76*, f584, 1995.

Young, C., Dislocations in the deformation of olivine, *Am. J. Sci., 267*, 841-852, 1969.

J. M. Brown and M. Chai, Geophysics Program, University of Washington, Seattle, Washington, 98195

Y. Wang, Center for High Pressure Research and Department of Earth and Space Sciences, University at Stony Brook, Stony Brook, New York, 11794-2100

High Pressure Viscosity of an Fe-S Liquid : Experimentally Derived Estimate of the Viscosity of Earth's Outer Core

R. A. Secco, G. E. LeBlanc[1], H. Yang, and J. N. Seibel

Department of Earth Sciences, University of Western Ontario, London, Ontario

The viscosity of $Fe_{73}S_{27}$ (wt%) has been measured at 1100-1300°C and 2.0-5.0 GPa. Initial measurements were made using the traditional high pressure Stokes viscometry (quench and probe) technique. Because of the low viscosities, a new method (electro-detection) was developed for in situ sphere speedometry measurement. The method is based on the detection of an electrical resistance anomaly produced by an insulating sphere passing through a pair of electrodes located in the conducting liquid sample. The method was first bench tested and potential field Modeling was carried out to determine the electrical size of the sphere. Application of the electrodetection technique at high pressure yields viscosities in the range ~44 Pa-s at 3.9 GPa and 1100°C to ~2 Pa-s at 3.0 GPa and 1300°C. The high viscosities of $Fe_{73}S_{27}$, relative to the 1 atm viscosities of pure Fe, are attributed to the S content which controls the liquid structure to the extent that it conditions the formation of large viscous flow units. Using viscous activation energy and volume values determined in this study, extrapolation of the viscosity of $Fe_{73}S_{27}$ to core-mantle boundary pressure and temperature conditions gives a range of 10^{-4}-10^{4} Pa-s. This range of viscosity corresponds to the lower end of outer core viscosity estimates and is consistent with the theoretical estimates of the molecular viscosity.

1. INTRODUCTION

It is widely held that the motion of outer core fluid is responsible for the generation and maintenance of the geomagnetic field [*Jacobs*, 1987]. While the process of motional induction in the geodynamo is phenomenologically well understood, the velocity field of the fluid must be prescribed in order to compute the magnetic field using kinematic dynamo theory [*Braginsky*, 1991]. The choice of velocity field is often motivated by geophysical constraints but large uncertainties in some physical properties of the fluid make the choice, to some extent, arbitrary. The inverse problem of deriving core fluid motion from downward-continued, time-varying geomagnetic field models [e.g., *Voorhies*, 1986, 1987; *Bloxham and Jackson*, 1991] requires an assumption about the fluid viscosity, usually taken to be inviscid. In spite of the fact that some models have incorporated higher viscosities than expected for the outer core without consequence since the Lorentz and Coriolis forces are much larger than viscous forces [*Glatzmaier and Roberts*, 1995], this conclusion can be no more reliable than the assumption on which it is predicated. Viscosity is one of the important physical properties of the outer core fluid.

A recent compilation of 39 estimates of outer core viscosity [*Secco*, 1995], derived from theoretical studies as well as from observables in seismology, geodesy and geomagnetism, has shown that outer core viscosity is ill-constrained with estimates ranging over 14 orders of magnitude. In general,

[1]Now at Department of Geology, McMaster University, Hamilton, Ontario, Canada L8S 4M1.

Properties of Earth and Planetary Materials
at High Pressure and Temperature
Geophysical Monograph 101

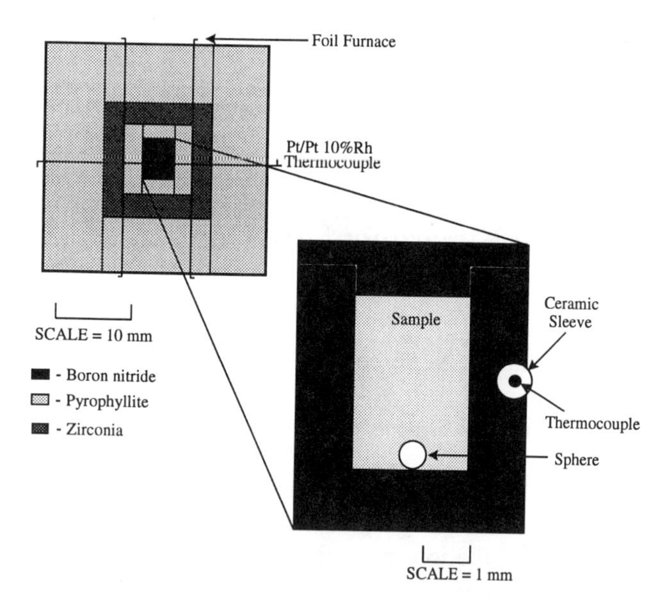

Figure 1. Cross section of a pressure cell designed for a quench and probe viscosity experiment.

the theoretical estimates occupy the lower end of the range since they account for only the intrinsic or molecular viscosity. The actual viscosity of the outer core must have incorporated an eddy viscosity component due to the effects of fluid motion. The seismological estimates are based on the absence of shear wave energy propagation and thus represent upper bounds [*Sato and Espinosa*, 1967]. Geodetic estimates of viscosity from length of day and polar motion studies suffer from long observation times which can introduce large uncertainties from additional energy sources and/or sinks. There is no estimate based on high pressure viscosity of liquids of outer core composition because only recently [*Secco*, 1994; *LeBlanc*, 1995] have experimental data become available. We report on an experimental investigation of the viscosity of a $Fe_{73}S_{27}$ (wt%) liquid at pressures in the range 2-5 GPa and at temperatures in the range 1100-1300°C. The resulting temperature and pressure dependences of viscosity, activation energies and volumes respectively, are used in a large extrapolation to outer core pressures and temperatures to give a range of estimates for the molecular viscosity of the outer core.

2. EXPERIMENTAL METHODS

Samples of $Fe_{73}S_{27}$ were prepared by mixing ultrapure (99.999% purity) fine powders of Fe and S (Aldrich Chemical Co.). The high pressure viscosity experiments were carried out in the 1000 ton cubic anvil press at the University of Western Ontario using the rising sphere Stokes viscometry method. Two types of techniques were used to measure sphere velocity: (a) post-mortem analysis of sphere position in the quenched and sectioned sample (hereafter referred to as the 'quench and probe' technique) and (b) a new in situ sphere speedometry technique (hereafter referred to as the 'electro-detection' technique) was developed using the electrical conductivity contrast of the insulating sphere and metallic liquid sample as the sensing property. In both methods, the choice of sphere material is based on chemical inertness with the sample, its ability to withstand the *P*, *T* conditions of the experiment, and its density contrast with the melt. For the electro-detection method, there is an additional requirement of sufficient electrical conductivity contrast with the melt to ensure a measurable signal. For those experiments ruby was used, but for the quench and probe method, both BN and ruby spheres were used. The typical diameter of the spheres was in the range 0.4-0.6 mm. BN spheres were ground to sphericities of <0.0001 from BN chunks in an air-driven grinder based on the design of Bond [1951]. Synthetic ruby spheres, with sphericities of <0.00001, were obtained commercially (Small Parts Inc., Florida; and Swiss Jewel Co., Philadelphia).

2.1. *High Pressure Quench and Probe Method*

A cross section of the pressure cell designed for a quench and probe experiment is shown in Figure 1. In this method, one or more spheres started at one end of the sample while the temperature was raised at constant pressure. Above the sample liquidus temperature, the sphere rose for a prescribed time after which the sample was quenched. The retrieved sample was sectioned and ground to locate the sphere, and the distance the sphere moved was measured under a microscope.

The high pressure application of the quench and probe method is fraught with difficulties and limitations. The total time of sphere motion is generally not well known since the time of sphere mobilization is poorly constrained. If a rising sphere traverses the entire height of the container and is found at the top, the velocity calculated is a minimum value since its arrival time is undetermined. In order to trap a sphere in the middle of the sample, large sample volumes are required, which poses challenges in maintaining sample temperature homogeneity. Finally, the quench and probe method cannot be reliably used for liquid viscosities below approximately 1 Pa-s since the time duration required for ascent is exceedingly short (<< 1 min) for cell geometries and materials typically used. This is especially true for pure liquid metals whose viscosities are typically of the order of 10^{-3} Pa-s at 1 atm [*Iida and Guthrie*, 1988]. To address these problems, a new method was developed for in situ sphere speedometry.

2.2. *Electro-Detection Method at 1 atm*

2.2.1. *Bench Tests.* The method is based on the in situ detection of sphere motion by measuring the electrical resistance anomaly as the sphere passes through a plane of sensing electrodes. The method was first tested on a bench model viscometer (scaled 10×) using two sets of electrodes mounted on the inside of a plexiglas cylinder as shown in Figure 2. A steel ball bearing, glued to the end of a thread, was lowered through a 0.95 molal aqueous solution of KCl. The conductivity of the solution (10^{-1} S cm^{-1}) was chosen to give a conductivity (σ) contrast with the steel sphere ($\sigma_{sphere}/\sigma_{KCl}$) of approximately 10^6, which is of similar magnitude to the conductivity contrast of the $Fe_{73}S_{27}$ liquid and a ruby sphere. Electrical resistance measurements were made either in static mode as a function of sphere depth, or in dynamic mode as a function of time with the sphere descending with a known velocity. The sphere descent velocity was controlled by unwinding a thread-wrapped wheel attached to a motor shaft as shown in Figure 2.

2.2.2. *Potential Field Modeling.* Two-dimensional models (scaled 60×) of the high pressure cell sample container were constructed in order to determine the electrical size of the sphere. Electromagnetic theory provides an analytical solution for the potential in the space surrounding a conducting space in a uniform electric field [*Jefimenko*, 1966] but the detection of a sphere in a practical situation depends on the noise level of the resistance meter and of the system. For typical sample/electrode geometries used in this study, the fluctuation in sample resistance is on the order of 0.1%. We have modelled this by potential field plotting (using a Graduate Field Plotter; Sensitized Coatings Ltd., Great Britain) on conductive paper as shown schematically in Figure 3. Electrodes of Cu foil were taped to the conductive paper ($\sigma_{paper} = 1.5 \times 10^{-2}$ S cm^{-1}) and a disk of Cu was used to model the sphere. Because it is much easier to increase the local conductivity of the paper (e.g., by using a metallic object) than to decrease the conductivity, the model experiments were configured to model a conducting sphere in an insulating medium. The potential applied across the two Cu electrodes was typically 1 V. Potentials at locations within the model container boundaries were determined using a field plotter probe. When held within ~1 mm of the conductive paper, the probe tip discharges and a spark marks the paper at the location where the probe potential, which is variable and can be set with the probe potential set dial, is less than or equal to the paper potential at that point. Accurate location of an equipotential line requires a probe approach from the low field side. The record of spark-defined potential field lines (each potential line was mapped with at least 80 points and as many as 400 points) was digitized for subsequent analysis and examples of recorded data are shown later.

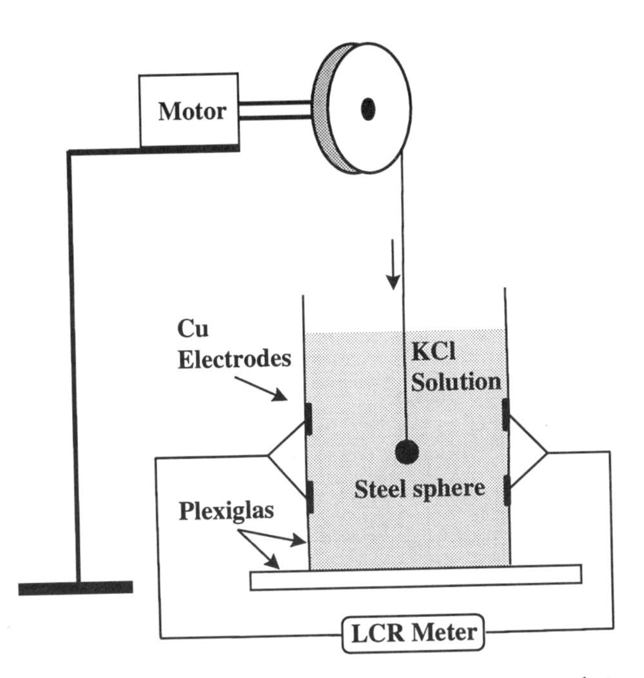

Figure 2. Schematic diagram of benchtop viscometer to test electro-detection method of in situ sphere speedometry.

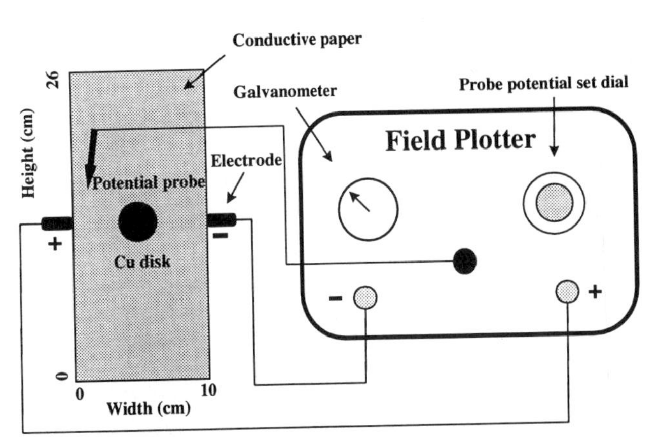

Figure 3. Schematic diagram of potential field model to determine electrical size of disk.

2.3. *Electro-Detection Method at High Pressure*

The electro-detection method was calibrated at high pressure using liquid jadeite, for which the viscosity is known up to 2.4 GPa [*Kushiro*, 1976]. A detailed description of the calibration is given elsewhere [*LeBlanc and Secco*, 1995]. A cross section of the pressure cell used for the Fe-S viscosity experiments is shown in Figure 4. The high pressure method is essentially the same as used in the 1 atm

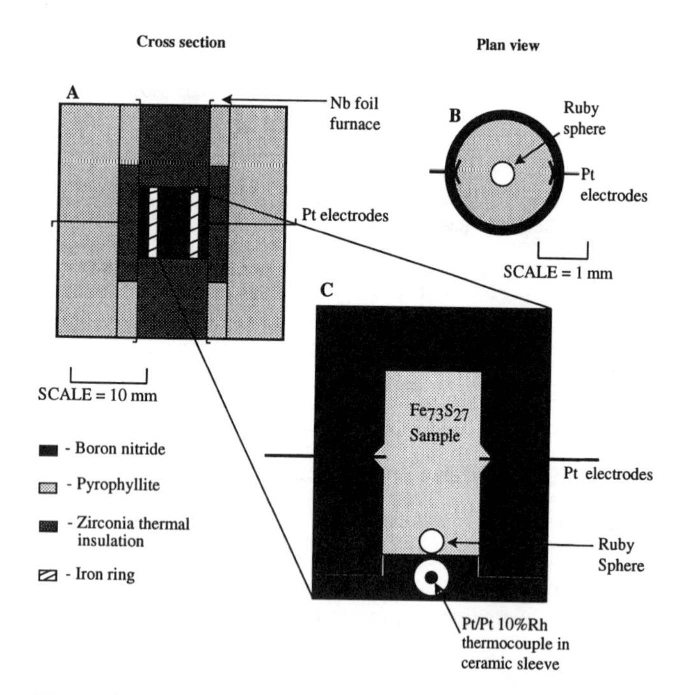

Figure 4. Cross ssection of a pressure cell designed for an electro-detection viscosity experiment.

bench model except that only one set of electrodes was used. This reduced the complexity of cell construction but, more importantly, it provided a means to determine the sphere velocity at the center of the sample where the sphere is least influenced by container end effects. The electrodes were recessed in an internal groove machined in the inside wall of the BN container. To prevent the BN container walls from collapsing on initial loading, a cylindrical shell of steel surrounded the BN container. Several experiments failed as a result of the sphere being anchored to the bottom of the BN container. This usually resulted from the packing of sample powder on top of the sphere which was placed at the bottom of the cup during preparation of the cell. This problem was simply and very effectively solved by placing the sphere on the top of the packed sample powder and then inverting the container (with the cover on the bottom as shown in Figure 4) on assembly of the cell.

3. RESULTS AND DISCUSSION

The average sphere velocity, v, was calculated from the known time and distance of travel and used in the Stokes equation below, to calculate the viscosity:

$$\eta^* = 2r_S^2 \Delta\rho g / 9v \qquad (1)$$

where η^* is the uncorrected shear dynamic viscosity, r_S is the radius of the sphere, $\Delta\rho$ is the difference in density between the sphere and the melt, and g is the gravitational acceleration. Equation (1) applies to a Newtonian fluid, for which there are two connecting coefficients of viscosity given in the stress tensor, η and λ, where λ is the dilatational viscosity and is associated with fluid compressibility [*Ryan and Blevins*, 1987]. Since a rising sphere shears the fluid through which it moves Stokes method measures only η.

Corrections were applied for the effects of the container boundaries on the sphere motion. *Bacon* [1936] *and Shartsis and Spinner* [1951] empirically determined that the *Faxen* [1923] corrections for wall effects were the most accurate but were restricted to radius ratios of the container (r_C) to the sphere of between 0.02 and 0.32. *Kingery* [1959] incorporated the correction for container end effects (E) determined by *Ladenburg* [1907] with the corrections for wall effects (W) to give the modified Faxen corrected Stokes equation:

$$\eta = \eta^* W/E \qquad (2)$$

where

$$W = 1 - 2.104\,(r_S/r_C) + 2.09\,(r_S/r_C)^3 - 0.95\,(r_S/r_C)^5 \qquad (3)$$

and

$$E = 1 + 3.3\,(r_S/h_C) \qquad (4)$$

where h_C is the height of the container. For the density of the melt, we have used the high pressure data of *Usselman* [1975a]. At 1 atm, the densities of BN (Grade A, Carborundum Co., New York) and ruby are 2.2×10^3 kg m^{-3} and 4.07×10^3 kg m^{-3}, respectively. The sphere density was corrected for the run pressure using the known values of the

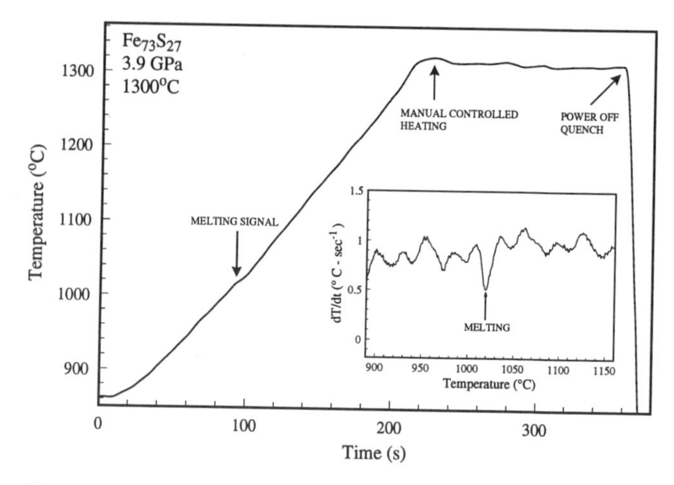

Figure 5. Example of the temperature history of a typical experiment. Melting is indicated by the latent heat absorption at 1020°C which is amplified in the insert on the plot of heating rate vs. temperature.

isothermal bulk modulus, K_T, of 254 GPa and its pressure derivative, K_T', of 4.3 for ruby [*Jephcoat et al.*, 1988] and K_T of 43.7 GPa and K_T' of 4.2 for BN [*Lynch and Drickamer*, 1966] in the 3rd order Birch-Murnaghan equation of state. Density corrections for thermal expansion, α, of ruby [$\alpha_{ruby} = 1.6 \times 10^{-5}$ K^{-1}; *Clark*, 1966] and BN ($\alpha_{BN} = 8 \times 10^{-6}$ K^{-1}; supplier technical data) at the run temperatures were also applied.

An example of the temperature history of an experiment at 3.9 GPa and 1300°C is shown in Figure 5. The expected melting temperature [*Usselman*, 1975b] of 1020°C is indicated by the latent heat absorption and consequent change in heating rate as shown in the inset of Figure 5.

3.1. *Quench and Probe Results*

An example of a sectioned cell showing a ruby sphere in the middle of a quenched sample is shown in Figure 6. As many as 13 separate experiments using as many as 3 spheres in one sample were carried out at a single set of *P*, *T* conditions at different time durations to check reproducibility. Results for 3.9 GPa and 1300°C, are shown in Figure 7. Due to the high order correction terms in Equation (2), which have generally been ignored in most previous high pressure (quench and probe) viscosity studies, where the sphere radius appears to

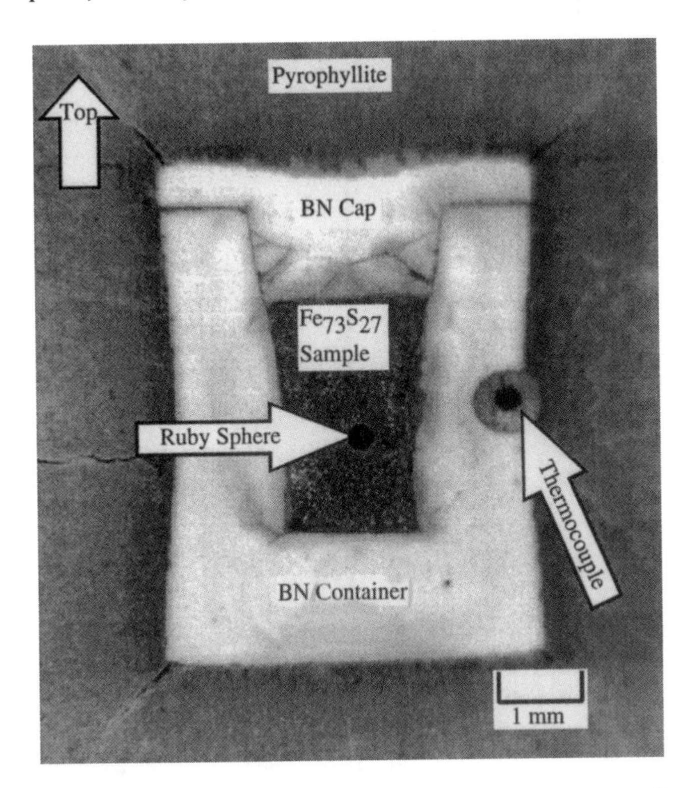

Figure 6. Example of sectioned quench and probe pressure cell showing the ruby sphere located in the middle of the sample.

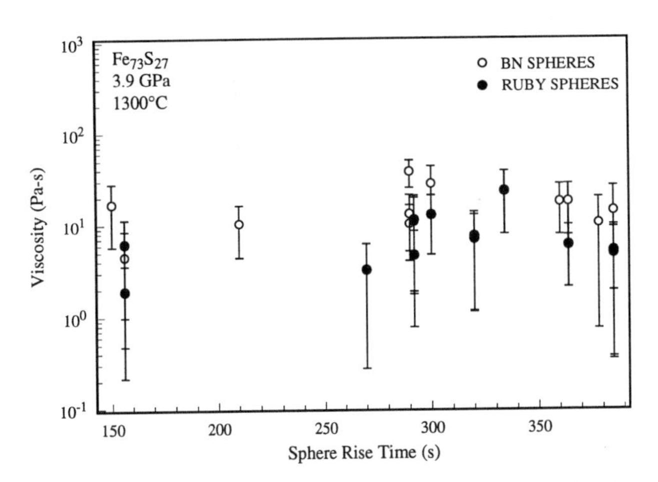

Figure 7. Viscosity results from thirteen separate quench and probe experiments with as many as three spheres in a single sample plotted versus rise time. These viscosity values are maximum values because in all experiments but two, the spheres were found at the top of the sample.

the third and fifth powers, the relative errors are 14-29%. All viscosity values in Figure 7 are maximum values since the quench and probe method gives no information about time at which the sphere begins to rise. In addition, in all experiments except two, the spheres were found at the top of the sample. For both reasons, the calculated velocity is a minimum and the viscosity is therefore a maximum. Despite the large variation in measured viscosity, there is no apparent dependence of viscosity on sphere rise time which indicates that the spheres reached terminal velocity very early in their ascent. Clearly, however, the scatter in the data attest to the need for a more accurate means of measuring the velocity of the sphere.

3.2. *Results of Bench Tests of Electro-Detection Method*

Measurements of ac resistance at varying frequency from 1 kHz to 100 kHz, with varying sphere sizes from 2.4 mm to 6.6 mm diameter, and at varying sphere descent velocities through the KCl solution were carried out using the benchtop viscometer shown in Figure 2. Two sets of data are plotted in Figure 8 corresponding to dynamic (Figures 8a-c) and static (Figure 8d) mode. In both cases, resistance anomalies are pronounced when the sphere passes each electrode plane. In some tests, as shown in Figures 8b and 8c, the resistance increased when the high-conductivity sphere passed the electrode plane because the sphere was sometimes coated with an insulating layer of glue that was used to secure the thread. The results of these bench tests showed that the sphere velocity could be determined by combining the known separation of electrodes and by measuring the time duration of the resistance anomaly.

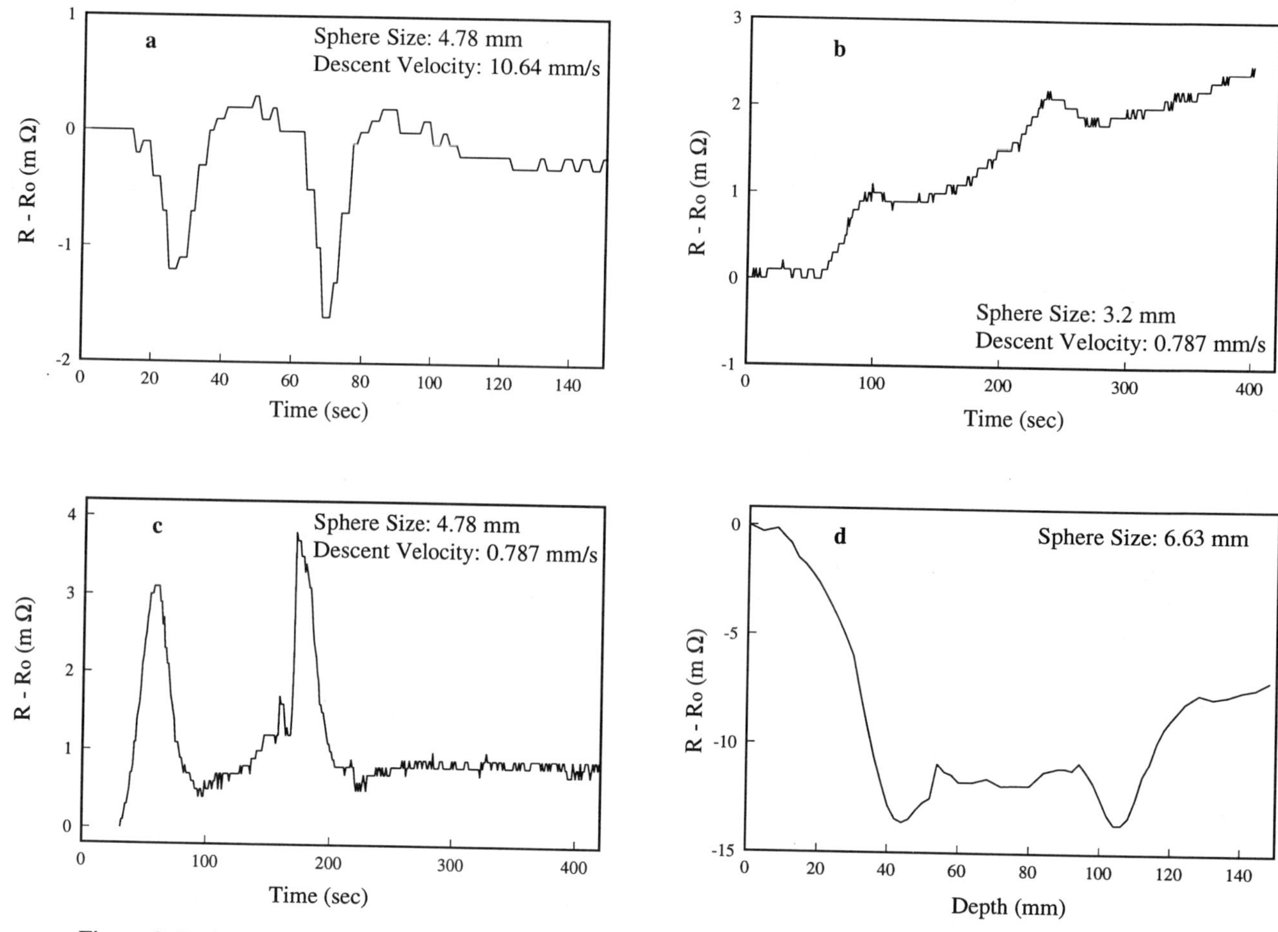

Figure 8. Resistance versus time (a-c) and depth (d) at various sphere descent velocities showing the anomaly as the sphere passes the two sets of electrodes in the benchtop viscometer.

3.3. *Potential Field Modeling Results*

Plots of digitized potential field lines for different fixed positions of the Cu disk are shown in Figures 9b-d and are compared with the 'background' potential field without the disk as shown in Figure 9a. The potential field distortion surrounding the disk is sensed by the probing electrodes before the disk crosses the center line. That is, the electrical size is larger than the physical size of the disk. To estimate the electrical size of the disk, which we define as the radius of the field distortion caused by the disk, potential measurements at a fixed location on the center line were made first without the disk to define a reference potential. The disk was then introduced between the electrodes and potential measurements were made at a fixed location (i.e., corresponding to one reference potential determined in the absence of the disk) for each disk position as it was displaced from the center line. The results from three different experiments are plotted as potential (a percentage of the total potential applied across the electrodes which was typically on the order of 1V) versus disk position as shown in Figure 10 for three different reference potentials. All three sets of data clearly indicate that the electrical size, r_e, is more than twice the physical size, r_p. From these results, we take the average of the ratio of electrical size to physical size to be 2.63. This value is later used for the calculation of the true sphere velocity.

3.4. *Results of Electro-Detection Method at High Pressure*

An example of the resistance record of an experiment at 4.0 GPa and 1200°C shows the anomalously high resistance as the sphere passes through the plane of electrodes (Figure 11). With only one set of electrodes, the width of the resistance anomaly signals the entrance and exit of the sphere from the electrode plane. The temperature was held constant within ±2°C as shown by the lower trace in Figure 11. The background resistance value of the liquid sample increased

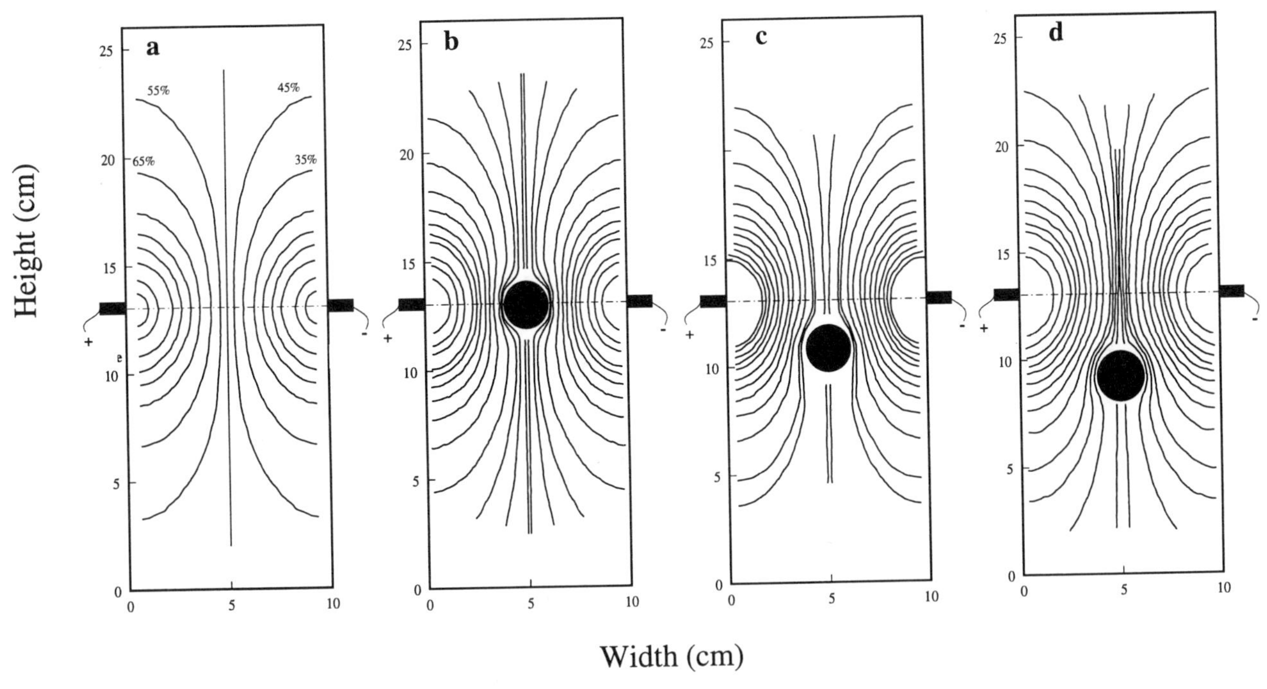

Figure 9. Potential field lines plotted on conductive paper for (a) no Cu disk, (b) Cu disk in center, and (c) and (d) with Cu disk displaced from model center to determine the electrical size of the disk.

by <3%. While we do not know the reason for this, we speculate that even slight changes to the shape of the sample container, specifically in the vicinity of the electrodes, could have magnified effects on the measured resistance of the sample especially if the electrode exposure to the liquid is reduced.

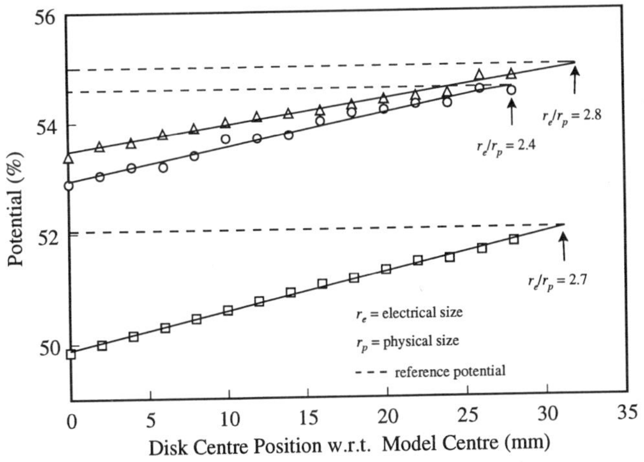

Figure 10. Potential at a fixed location plotted versus displacement of Cu disk from model center for three different reference potentials. The ratio r_e/r_p is the displacement of the disk centre at the reference potential crossing (or electrical size, r_e) to the radius of the disk (or physical size, $r_p = 5.8$ mm).

The viscosity data for $Fe_{73}S_{27}$ are shown on an Arrhenius plot in Figure 12 for pressures 2.0-5.0 GPa and temperatures 1100-1300°C and are compared with 1 atm viscosity values of pure liquid Fe determined in six previous studies. In Figure 13, the same viscosity data are plotted versus pressure. The data in both figures have been corrected following the new determination of electrical size of the sphere and thus

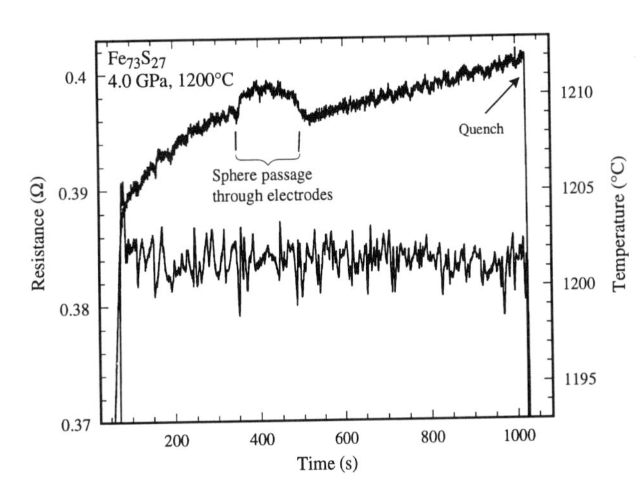

Figure 11. Example of the resistance record (upper trace) showing the anomaly resulting from the ruby sphere passing through the electrodes in a 4.0 GPa and 1200°C experiment. The lower trace is the temperature record.

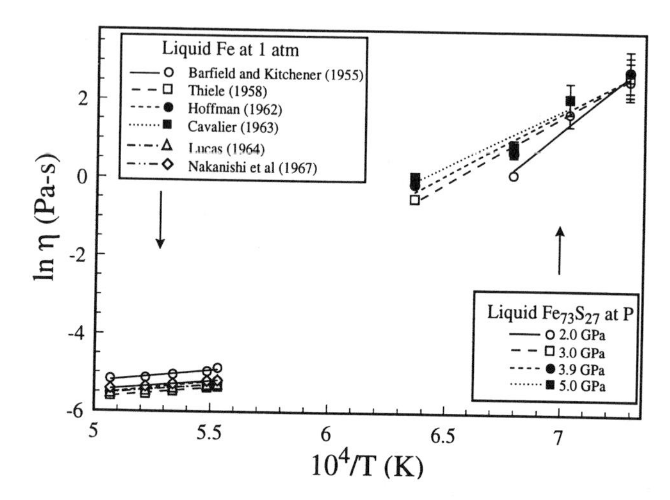

Figure 12. Arrhenius plot of viscosity of liquid $Fe_{73}S_{27}$ at pressures of 2.0, 3.0, 3.9, and 5.0 GPa compared with 1 atm viscosity data of pure liquid Fe from six previous studies.

they are a factor of 2.63 lower than the viscosity data reported in *LeBlanc and Secco* [1996]. The comparison in Figure 12 clearly shows that the viscosities measured in this study are significantly higher than what would have been otherwise expected from a normal volume dependence of the viscosity [*Bridgman*, 1958] of the pure liquid of the major Fe component. We have therefore considered in detail the possibilities of experimental influences on this result. These are only briefly discussed here and a more detailed discussion can be found in LeBlanc and Secco [1996]. The possibility of convection was tested in a separate experiment by measuring the maximum (i.e., axial) temperature gradient in the liquid sample at 4.0 GPa. The measured axial temperature gradients varied from 2.2°C mm^{-1} at 1100°C to 6.7°C mm^{-1} at 1500°C making the temperature difference between the hottest and coldest parts of the sample between approximately 4°C and 10°C, respectively. From the measured temperature gradients, and using values for thermal expansion of 8.2×10^{-5} K^{-1} [*Nasch and Steinemann*, 1995] and 7.8×10^{-6} m^2 s^{-1} for diffusivity [*Touloukian*, 1970] for liquid Fe, a Rayleigh number of 0.07 is calculated. This is 5 orders of magnitude lower than the critical Rayleigh number for the onset of convection [*Charlson and Sani*, 1970]. A Prandtl number of 77 was calculated, which also indicates that convection is not a concern in these experiments.

The quenched sample was analyzed with electron microprobe and found to have a bulk composition of $Fe_{71.5}S_{28.5}$. Backscattered electron images of a 3.9 GPa and 1300°C run product are shown in Figure 14. The dark dendrites are $Fe_{66.3}S_{33.7}$ and the grey matrix is $Fe_{77.1}S_{22.9}$. The bright halos around the dendrites are Fe which most likely formed by exsolution during quenching. X ray diffraction patterns of the quenched products also showed FeS and pure Fe peaks, both consistent with the microprobe data. The variation between start and end compositions of the sample cannot account for such large viscosities measured in this study.

The interaction between the magnetic field, induced by the high ac current in the Nb furnace, and the Fe liquid alloy sample was examined. Fluid motion resulting from the movement of the sphere could give rise to a Lorentz force but since the magnetic field is also an ac field, the net Lorentz force effect on the fluid is zero.

$Fe_{73}S_{27}$ is the eutectic composition at 3.0 GPa [*Usselman*, 1975b]. Eutectic compositions can form atomic associations in the liquid that lead to higher viscosities than expected from the regular monotonic trend describing the variation of viscosity from one end member to the other [*Chadwick*, 1965; *Iida and Guthrie*, 1988]. In addition, liquid binary systems at compositions where there is intermetallic compound formation in the solid can also show anomalously high viscosities. However, these viscosity anomalies are small (i.e., of the same order of magnitude as the viscosity of neighbouring compositions) compared to the difference in viscosity between liquid Fe and the viscosity of $Fe_{73}S_{27}$ measured in this study.

Finally, we examined the influence of S on the viscosity of Fe-S liquids. It is well known that S has several different liquid structures [*Brazhkin et al.*, 1992]. A rapid increase in viscosity with temperature, reaching a maximum at 187°, is associated with the well-understood polymerization reaction where long chains of octatomic (S_8) ring molecules form which bind the liquid [*Touro and Wiewiorowski*, 1966; *Eisenberg*, 1969; *Chang and Jhon*, 1982; *Cates*, 1987]. With increasing temperature, the S_8 chains break up and the viscosity decreases as for a Newtonian fluid. Recently, Heath

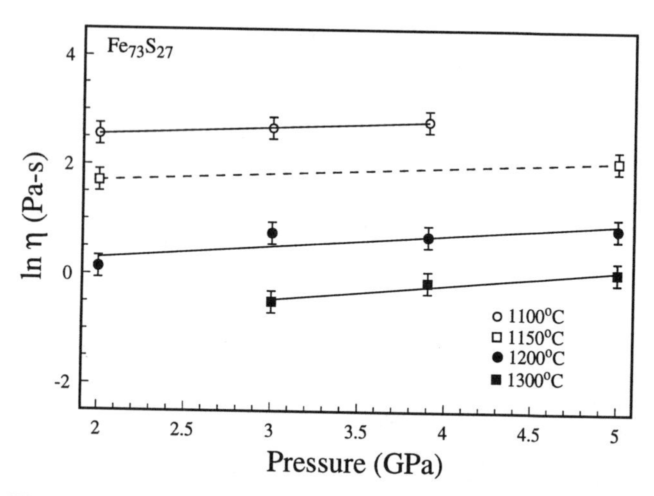

Figure 13. Pressure dependence of the viscosity of liquid $Fe_{73}S_{27}$ at 1100, 1150, 1200, and 1300°C between 2 and 5GPa.

[1995] measured the viscosity of pure liquid S at pressures of 1.0 and 4.0 GPa. The 4.0 GPa data show that even at temperatures up to 700°C, the viscosity of liquid S is in the range 1-6 Pa-s. A 1 atm study of the viscosity of Fe-S liquids [*Vostryakov et al.*, 1964] with S contents of 20-30 wt% showed an increase in the viscosity with increasing S. Taking into account the complex nature of the structural configuration of pure S liquid at high pressure, we postulate that the presence of S in $Fe_{73}S_{27}$ liquid in the experiments reported here controls the liquid structure to the extent that it conditions the formation of large viscous flow units.

4. EXTRAPOLATION TO OUTER CORE CONDITIONS

The estimate of any core property based on experimental data at pressures of 5GPa and temperatures of 1300°C involves a large extrapolation. The extrapolation was performed using the temperature and pressure dependences, or activation energies (Q_η) and volumes (ΔV_η) respectively, of the viscosity. Q_η decreases with increasing pressures from 405 ± 77 kJ mol^{-1} at 2.0 GPa to 240 ± 72 kJ mol^{-1} at 5.0 GPa. ΔV_η increases with increasing T, from $(1.51 \pm 0.01) \times 10^{-6}$ m^3 mol^{-1} at 1100°C to $(3.61 \pm 0.19) \times 10^{-6}$ m^3 mol^{-1} at 1300°C [*LeBlanc and Secco*, 1996]. *Backus* [1968] used the Q_η and ΔV_η data of *Natchrieb and Petit* [1956] on liquid Hg below 1 GPa and 400 K in the following equation:

$$\ln(\nu/\nu_0) = -(3\gamma + 2)\ [1 - (\rho_0/\rho)^{1/3}] + (Q_\eta + P\Delta V_\eta)/RT - (Q_\eta + P_0\Delta V_\eta)/RT_0 \quad (5)$$

where γ is Gruneisen's parameter, ν is kinematic viscosity ($\nu = \eta/\rho$), and all parameters subscripted zero are for a reference state. The reference state was chosen to be 1100°C and 2 GPa, that is, within the P, T conditions investigated since the extrapolation was carried out using Q_η and ΔV_η data from this study. For the extrapolation of viscosity to the core-mantle boundary, the following parameter values were used: $\nu_0 = 2.6 \times 10^{-3}$ m^2 s^{-1}, $\gamma = 1.6$ (for liquid Fe [*Nasch and Steinemann*, 1995]), $\rho_0 = 5257$ kg m^{-3}, $\rho = 9903$ kg m^{-3}, $P_0 = 2 \times 10^9$ Pa, $P = 1.36 \times 10^{11}$ Pa, $T_0 = 1373$ K, $T = 2750$ K [*Zerr and Boehler*, 1993]. The extrapolated viscosity is very sensitive to both ΔV_η and Q_η. The maximum calculated viscosity is 4×10^4 Pa-s using the minimum Q_η value (at 5 GPa) of 240 kJ mol^{-1} and the maximum ΔV_η value (at 1300°C) of 3.61×10^{-6} m^3 mol^{-1}. The minimum calculated viscosity is 2×10^{-4} Pa-s using the maximum Q_η value (at 2 GPa) of 405 kJ mol^{-1} and the minimum ΔV_η value (at 1100°C) of 1.51×10^{-6} m^3 mol^{-1}. This range of viscosity values is plotted in Figure 15 as the filled rectangle and is compared with the many other estimates that have been derived from other methods. Despite the extrapolation over two orders of magnitude in pressure, the calculated values span a range similar to the theoretically based estimates. This confirms the expectation discussed earlier that both the liquid metal theories used in the estimates shown in Figure 15 and the experiments reported here pertain only to the molecular, or intrinsic, viscosity component. It is interesting to note that the back-extrapolated value of temperature-dependent ΔV_η values to the melting temperature of 0.59 cm^3 mol^{-1} is very similar to the ΔV_η of 0.59-0.62 cm^3 mol^{-1} for Hg, the only other liquid metal for which an activation volume has been determined [*Nachtrieb and Petit*, 1956; *Bridgman*, 1958].

A)

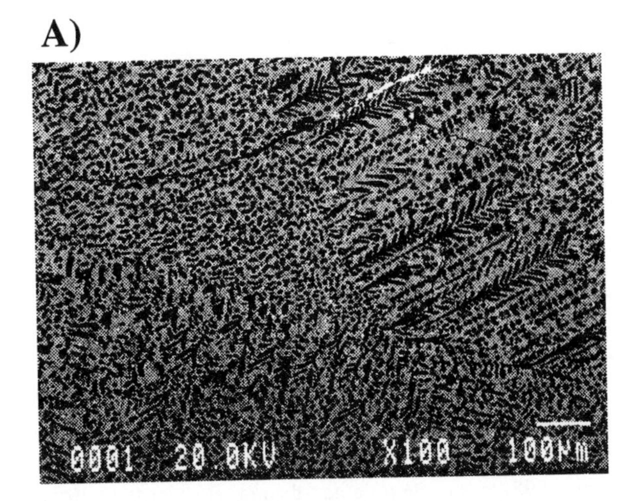

B)

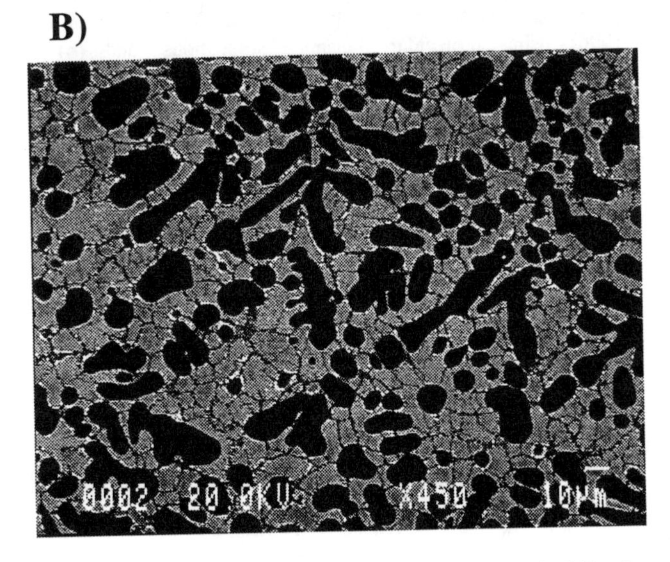

Figure 14. Backscattered electron images of (a) quenched $Fe_{73}S_{27}$ liquid showing dark FeS dendrites in a lighter colored, Fe rich matrix and (b) showing pure Fe (white) rims around dendrites.

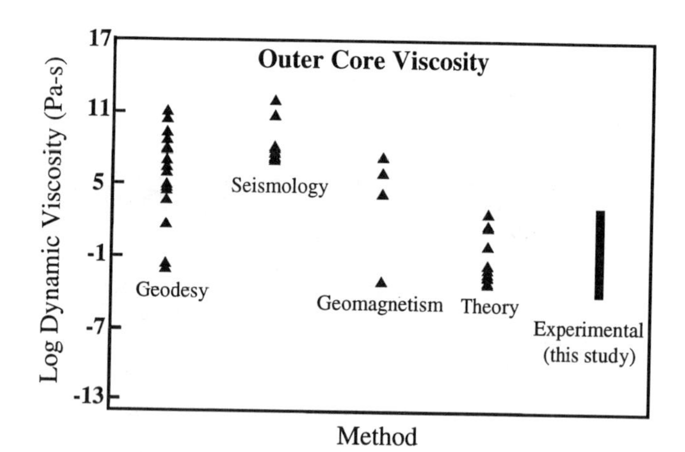

Figure 15. Viscosity estimates of Earth's outer core (modified from *Secco*, 1995) to include (a) a new theoretical estimate of 10^{-3} Pa-s for the viscosity at the top of the outer core by *Wasserman et al.* (1996) using molecular dynamics simulations, and (b) the experimentally derived estimate based on the high-pressure viscosity data in this study.

Acknowledgments. We thank R. Tucker and M. Thrasher for technical assistance. This work was supported by an NSERC research grant (RAS) as well as an NSERC student summer scholarship (JNS).

REFERENCES

Backus, G. E., Kinematics of geomagnetic secular variation in a perfectly conducting core, *Phil. Trans. Roy. Soc. Lond., A-263*, 239-266, 1968.

Bacon, L. R., Measurement of absolute viscosity by the falling sphere method, *J. Frankel Inst., 221*, 251-273, 1936.

Barfield, R. N., and J. A. Kitchener, The viscosity of liquid iron and iron-carbon alloys, *J. Iron Steel Inst., 180*, 324-329, 1955.

Bloxham, J., and A. Jackson, Fluid flow near the surface of Earth's outer core, *Rev. Geophys., 29*, 97-120, 1991.

Bond, W. L., Making of small spheres, *Rev. Sci. Instr., 22*, 344-345, 1951.

Braginsky, S. I., Towards a realistic theory of the geodynamo, *Geophys. Astrophys. Fluid Dynamics, 60*, 89-134, 1991.

Brazhkin, V. V., R. N. Voloshin, S. V. Popova, and A. G. Umnov, Metallization of the melts of Se, S, I_2 under highpressure, *High-Press. Res., 10*, 454-456, 1992.

Bridgman, P. W., *The Physics of High Pressure*, 445 pp., G. Bell and Sons, Ltd., London, 1958.

Cates, M., Theory of the viscosity of polymeric liquid sulfur, *Europhys. Lett, 4*(4), 497-502, 1987.

Cavalier, G., Mésure de la viscosité du fer, du cobalt at du nickel, *C.R. Acad. Sci. 256*, 1308-1311, 1963.

Chadwick, G. A., Eutectic solidification, pp. 327-352, in *Liquids, Structure, Properties, Solid Interactions.*, edited by T. J. Hughel, Elsevier, New York, 1965.

Chang, M. C., and M. S. Jhon, Viscosity and thermodynamic properties of liquid sulfur, *Bull. Korean Chem. Soc., 3*, 133-139, 1982.

Charlson, G. S., and R. L. Sani, Thermoconvective instability in a bounded cylindrical fluid layer, *Int. J. Heat Mass Transfer, 13*, 1479-1496, 1970.

Clark, S. P., *Handbook of Physical Constants*, 687 pp., The Geological Society of America, New York, 97, 1966.

Eisenberg, A., The viscosity of liquid sulfur. A mechanistic reinterpretation, *Macromolecules, 2*(1), 44-48, 1969.

Faxën, H. Die bewegung einer starren kugel langs der achse eines mit zaher flussigkeit gefullten rohres, *Arkiv fur Matematik, Astronomi och Fysik, 17*, 1-28, 1923 (as cited in Bacon, 1936).

Glatzmaeir, G. A., and P. H. Roberts, A three-dimensional self-consistent computer simulation of a geomagnetic field reversal, *Nature, 377*, 203-209, 1995.

Heath, D. R., Viscosity and density of liquid sulfur at high pressures and temperatures: Implications for the Earth's outer core and volcanism on Io, BSc. thesis, 73 pp., Univ. of Western Ontario, London, 1995.

Hoffmann, K., Ph.D. thesis, Saarbrucken, 1962 (cited in Lucas, 1964).

Iida, T. and R. I. L. Guthrie, *The Physical Properties of Liquid Metals*, Clarendon Press, Oxford, 1988.

Jacobs, J. A., *The Earth's Core*, 416 pp., Academic Press, Toronto, 1987.

Jefimenko, O. D., *Electricity and Magnetism: An Introduction to the Theory of Electric and Magnetic Fields*, 155 pp., Meredith Pub. Co., New York, 1966.

Jephcoat, A. P., R. J. Hemley, and H. K. Mao, X-ray diffraction of ruby (Al_2O_3:Cr^{3+}) to 175 GPa, *Physica B, 150*, 115-121, 1988.

Kingery, W. D., Viscosity in property measurements at high temperatures, John Wiley and Sons Inc., New York, 1959.

Kushiro, I., Changes in viscosity and structure of melt of $NaAlSi_2O_6$ composition at high pressure, *J. Geophys. Res., 81*, 6347-6350, 1976.

Ladenburg, R., Uber der innere reibung zaher flussigkeiten, *Ann. Physik, 22*, 287, 1907 (as cited in Kingery, 1959).

LeBlanc, G. E., High pressure viscosity of an Fe-S liquid: Experimental approach to outer core viscosity, MSc. thesis, 127 pp., Univ. Western Ontario, London, 1995.

LeBlanc, G. E. and R. A. Secco, High pressure Stokes' viscometry: A new *in-situ* technique for sphere velocity determination, *Rev. Sci. Instr., 66* (10), 5015-5018, 1995.

LeBlanc, G. E. and R. A. Secco, Viscosity of an Fe-S Liquid up to 1300°C and 5GPa, *Geophys. Res. Lett., 23*, 213-216, 1996.

Lucas, L-D., Viscosité du fer pur et du system Fe-C jusqu'a 4,8% C en poids, *C.R. Acad. Sci. Paris, 259*, 3760-3767, 1964.

Lynch, R. W., and H. G. Drickamer, Effect of high pressure on the lattice parameters of diamond, graphite, and hexagonal boron nitride, *J. Chem. Phys., 44*, 181-184, 1966.

Nachtrieb, N. H., and J. Petit, Self-diffusion in liquid mercury, *J. Chem. Phys. 24*, 746-750, 1956.

Nakanishi, K., T. Saito, and Y. Shiraishi, On the viscosity of molten iron and its dilute binary alloys of aluminum, silicon, and oxygen, *Jap. Inst. Metals J. 37*(7), 881-887, 1967.

Nasch, P. M., and S. G. Steinemann, Density and thermal expansion of molten manganese, iron, nickel, copper, aluminum and tin by means of the gamma-ray attenuation technique, *Phys. Chem. Liq.*, *29*, 43-58, 1995.

Ryan, M. P. and J. Y. Blevins, *The Viscosity of Synthetic and Natural Silicate Melts and Glasses at High Temperatures and 1 bar (10^5 Pascals) Pressure and at Higher Pressures*, U.S. Geological Survey Bull. 1764, Denver, 1-29, 1987.

Sato, R. and A. F. Espinosa, Dissipation in the Earth's mantle and rigidity and viscosity in the Earth's core determined from waves multiply reflected from the core-mantle boundary, *Bull. Seis. Soc. Amer. 57*, 829-857, 1967.

Secco, R. A., High pressure measurements of viscosities of Fe-S liquids, 947-949, in *High Pressure Science and Technology - 1993*, edited by S. C. Schmidt, J. W. Shaner, G. A. Samara, and M. Ross, American Institute of Physics Press, New York, AIP Proc. #309, vol. 1, 1994.

Secco, R. A., Viscosity of the Outer Core, in *Mineral Physics and Crystallography*, A Handbook of Physical Constants, AGU Reference Shelf Series, vol. 2, edited by T. J. Ahrens, pp. 218-226, Washington, D.C., 1995.

Shartsis, L. and S. Spinner, Viscosity and density of molten optical glasses, *J. Res. NBS, 46*, 176-194, 1951.

Thiele, M., Ph. D. thesis, Berlin, 1958 (cited in Lucas, 1964).

Touloukian, Y. S., *Thermal diffusivity: Thermophysical Properties of Matter*, *10*, 83-100, IFI/Plenum, New York, 1970.

Touro, F. J. and T. K. Wiewiorowski, Viscosity-chain length relationship in molten sulfur systems, *J. Phys. Chem.*, *70*(1), 239-241, 1966.

Usselman, T. M., Experimental approach to the state of the core: Part II. Composition and thermal regime, *Amer. J. Sci. 275*, 291-303, 1975a.

Usselman, T. M., Experimental approach to the state of the core: Part I. The liquidus relations or the Fe-rich portion of the Fe-Ni-S system from 30 to 100 kb, *Amer. J. Sci.*, *275*, 278-290, 1975b.

Voorhies, C. V., Steady flows at the top of Earth's core derived from geomagnetic field models, *J. Geophys. Res.*, *91, B12*, 12, 444-12466, 1986.

Voorhies, C. V., The time-varying geomagnetic field, *Rev. Geophys.*, *25*, 929-938, 1987.

Vostryakov, A. A., N. A. Vatolin, and O. A. Yesin, Viscosity and electrical resistivity of molten alloys of iron with phosphorus and sulfur, *Phys. Metals Metallogr.*, *18*, 167-169, 1964.

Zerr, A. and R. Boehler, Melting of (Mg,Fe)SiO_3-perovskite to 625 kilobars: Indication of a high melting temperature in the lower mantle, *Science, 262*, 553-555, 1993.

R. A. Secco, G. E. LeBlanc, H. Yang, and J. N. Seibel, Department of Earth Sciences, University of Western Ontario, London, Ontario, Canada N6A 5B7.

Stability of Hydration States and Hysteresis of Rehydration in Montmorillonites as a Function of Temperature, H_2O Pressure, and Interlayer Cations

W. A. Bassett and T.-C. Wu[1]

Department of Geological Sciences, Cornell University, Ithaca, New York 14853

The hydration states of Ca- and Mg-exchanged montmorillonite from Wyoming bentonite have been studied as a function of temperature and pressure in a hydrothermal diamond anvil cell (HDAC) by X ray diffraction at the Cornell High Energy Synchrotron Source (CHESS). Runs were made at constant volume allowing pressure to be calculated from the equation of state of water and the homogenization temperature for each sample studied. Observations were made at pressures from the liquid-vapor coexistence boundary up to approximately 1 GPa. Conversion from the 19 to 15Å hydration state was observed to take place in the range 260-350°C for Ca-montmorillonite and in the range 200-250°C for Mg-montmorillonite, with a slight increase with increasing pressure. The rehydration from the 19 to 15Å state occurred at the same temperature (no hysteresis) at pressures just above the liquid-vapor coexistence boundary but decreased to 75°C below dehydration at 0.6 GPa for Ca-montmorillonite and 75°C below dehydration at 0.25 GPa for Mg-montmorillonite. There is evidence that the dehydration is the equilibrium boundary and that the rehydration hysteresis is caused by kinetics. The dehydration boundary is very nearly parallel to the pressure axis (dT/dP = 0). Thus, according to the Clapeyron equation, the volume change is essentially zero. From this we conclude that the molar volume or density of the water inside and outside of the montmorillonite is the same. If this is true, then it follows that entropy increases during dehydration. We attribute this to the change in the structure of the water inside and outside of the montmorillonite. When the conversion temperatures for Ca- and Mg-montmorillonites are combined with our earlier measurements on Na-montmorillonite, systematic trends can be observed in the effect of pressure and interlayer cation species on the dehydration temperature and rehydration hysteresis.

1. INTRODUCTION

Montmorillonites are sheet silicates that have low charges on the silicate (2:1) layers. As a result of the low charges, these clay minerals are characterized by interlayers consisting of a relatively small number of cations

[1]Now at Applied-Komatsu Technology, 3101 Scott Blvd., M/S 9156, Santa Clara, CA 95054

Properties of Earth and Planetary Materials
at High Pressure and Temperature
Geophysical Monograph 101

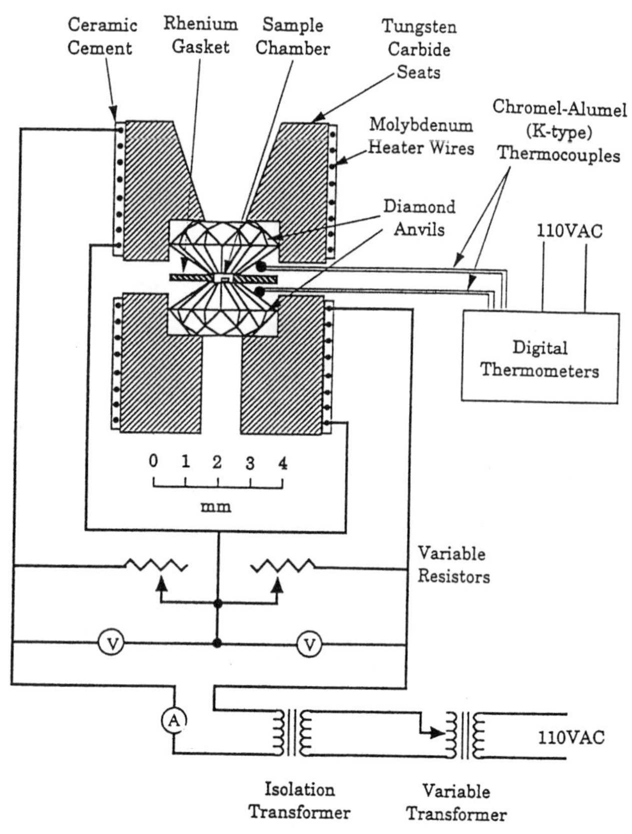

Figure 1. Schematic diagram showing the arrangement of the sample, diamond anvils, heaters, and electric circuits for heating, controlling and measuring the temperature of the sample.

(compared with other sheet silicates such as micas) and an abundance of water of hydration in the form of H_2O molecules associated with the cations. When the cations are divalent there are three common configurations of the cations and H_2O molecules leading to three discrete hydration states with d_{001}-spacings of ~19Å, ~15Å, and ~12.5Å. These hydration states are called 3-layer, 2-layer, and 1-layer, respectively, even though the distribution of H_2O molecules is more complex than simple layers [*Chang et al.*, 1995; *McBride*, 1994]. For this reason, we refer to the hydration states by their d-spacings rather than the number of layers of water.

The hydration of clays and the conditions for retention and release of their water is of interest to geologists because it affects the origin of mobile fluids in deeply buried sediments and possibly even in subducted sediments. Furthermore, hydrous phases in general have very different rheologic properties from anhydrous phases and can therefore influence the ductility of sedimentary rocks and their ability to flow and form traps for hydrocarbons.

Using the hydrothermal diamond anvil cell (HDAC) and synchrotron radiation [*Wu et al.*, in press], we have been able to make more detailed observations and significantly extend the range of temperatures and pressures of dehydration and rehydration studies of montmorillonites than in earlier studies [e.g., *Stone and Rowland*, 1955; *Koster van Groos and Guggenheim*, 1984, 1986, 1987].

2. EXPERIMENTAL METHODS AND PROCEDURES

The HDAC has proved to be a valuable instrument for studying the dehydration of hydrous minerals as a function of temperatures up to 1200°C and H_2O pressures up to 2 GPa [*Bassett et al.,* 1993]. The HDAC is a diamond anvil cell in which a sample consisting of phases of solid, liquid, and vapor can be loaded into a sample chamber consisting of a hole 500 μm in diameter drilled in a 125-μm-thick Re foil. The sample chamber is sealed and the sample is encapsulated when the diamond anvils are pressed against the open ends of the hole.

The temperature is raised by electric resistance heaters consisting of Mo wire wound around the tungsten carbide seats that support the diamond anvils (Figure 1). Chromel-alumel thermocouples attached to the outside surfaces of the diamond anvils measure the temperatures of the anvils with an accuracy of ±2°C. Variable resistors serve as shunts that can be used to balance the temperatures of the upper and lower diamonds to within 1°C. The diamonds are driven together when the upper and lower platens are pulled together by the three driver screws shown in Figure 2. The cell has been designed so that the driver screws experience little change in temperature and therefore cause little or no change in pressure in the sample. A slightly reducing gas consisting of 1% hydrogen and 99% argon surrounds the diamond anvils and heaters to prevent oxidation of the diamonds and the Mo wire at high temperatures.

In most experiments, there is no attempt to apply pressure to the sample by driving the anvils together. Instead, pressure is applied to the gasket to form a secure seal, and sample pressure is allowed to build up in the sample chamber as a result of increased temperature. The Re gasket and diamond anvils have been found to maintain a remarkably constant volume throughout a run consisting of increasing and decreasing the temperature hundreds of degrees and the pressure hundreds of MPa.

Pressure is determined by means of the equation of state of the H_2O in the sample [*Wagner and Pruss,* 1993; *Wagner et al.,* 1994; *Saul and Wagner,* 1989]. As long as both liquid and gas coexist, the sample in the chamber follows the liquid-vapor coexistence curve (L-V) in the phase diagram (Figure 3); therefore, pressure can be determined from the measured temperature. If as the temperature increases, the bubble of gas disappears, the pressure in the sample chamber increases along an isochore from the P-T point (homogenization point, T_h) at which it disappears. If the volume of the sample chamber remains constant, the pressure can be determined from this curve when the temperature has been measured. The volume of the solid sample is typically 25% of the sample chamber or less, and any changes such as solution or dehydration have a negligible effect on the pressure. The volume of the sample chamber is monitored by observing the diameter of the image of the hole and by the interference of laser light reflected off the upper and lower anvil faces defining the top and bottom of the sample chamber. Once the gasket has been cycled a few times through heating and cooling, the change in sample chamber volume during a run is less than 2% on heating and less than 1% on cooling. Pressure can be determined with an accuracy within ± 5%.

We carried out in situ X ray diffraction studies using an intense white synchrotron radiation beam in B Hutch at the Cornell High Energy Synchrotron Source (CHESS). Energy dispersive X ray diffraction patterns were collected by setting the intrinsic germanium detector at a fixed 2θ angle of 3° and plotting intensity versus energy of photons scattered at that angle. This combination of techniques allowed us to collect a complete diffraction pattern every one or two minutes. This, in turn, made it possible for us to follow the response of the sample to a change in pressure and temperature until the change in the sample had ceased. It also allowed us to determine if a feature such as peak broadening was a transitory or long-term phenomenon.

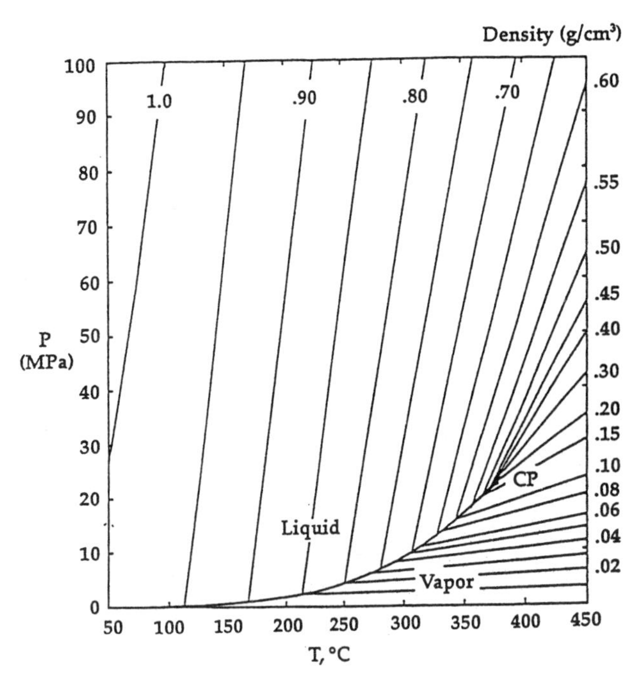

Figure 3. Pressure-temperature plot of the equation of state of H_2O. The curved line represents the liquid-vapor coexistence curve and CP represents the critical point. The isochores are labeled by density in g/cm^3.

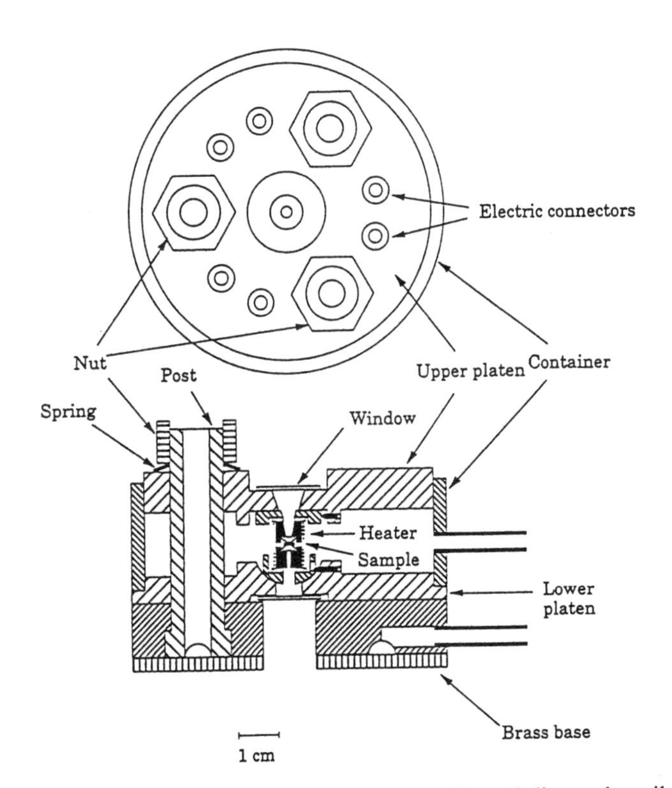

Figure 2. Plan and elevation of the hydrothermal diamond anvil cell (HDAC). Force is applied to the diamond anvils when the nuts are tightened. A reducing gas mixture surrounds the anvils and heater to prevent oxidation. Air can be forced through the posts to keep them cool and prevent expansion.

Samples of Ca- and Mg-exchanged montmorillonites derived from Clay Mineral Source Clay SWy-1 (Wyoming bentonite) were studied in this investigation. We will refer to these as Ca-montmorillonite and Mg-montmorillonite. Details of the ion exchange techniques used to produce these clays can be found in *Koster van Groos and Guggenheim* [1987]. In each case these samples were in the third hydration state with d_{001} = 19Å when they were loaded into the sample chamber of the HDAC with distilled, deionized water.

3. RESULTS

3.1. *Ca-montmorillonite*

Figure 4 shows a typical series of diffraction patterns of Ca-montmorillonite during heating and cooling along the 1.053 g/cm^3 isochore. The 001 peak shifted to smaller d-

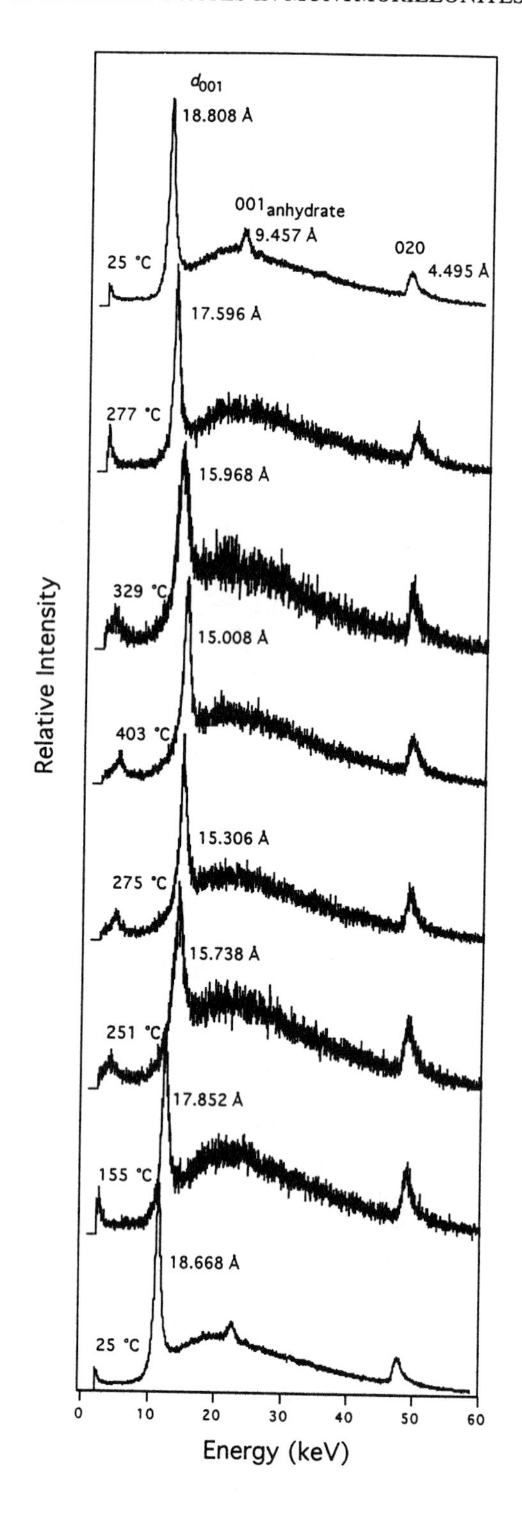

Figure 4. A series of energy-dispersive X ray diffraction patterns of Ca-montmorillonite during a heating-cooling cycle with isochoric density of 1.053 g/cm^3, showing the dehydration process starting from the top and the rehydration process ending at the bottom.

spacings as water of hydration was lost from the interlayer sites. The 020 peak remained essentially unchanged as it reflects the high degree of ordering that remains unchanged within the silicate layers.

Complete heating and cooling cycles were carried out along six isochoric paths with densities of 1.048 g/cm^3, 1.053 g/cm^3, 0.877 g/cm^3, 0.895 g/cm^3, 0.704 g/cm^3, and 0.720 g/cm^3. These six runs fall into three groups, 1.05 g/cm^3, 0.9 g/cm^3, and 0.7 g/cm^3. Data collected showed excellent agreement between the two runs of each density group. Figures 5a, b, and c show three plots of d_{001}-spacing versus temperature for the three isochore groups. Figures 5d, e, and f show the variation of peak width defined as the full width at half maximum (FWHM) during the 19-15Å dehydration and rehydration.

Several features of the plots shown in Figure 5 deserve comment. In every isochoric path the d_{001}-spacing decreases gradually from ~19 to ~17.5Å before it starts rapidly decreasing. The temperature range for the ~17.5 to 15Å dehydration varies slightly from isochore group to isochore group. It is 300 to 350°C along isochores 1.05 g/cm^3, 275 to 340°C along 0.9 g/cm^3, and 260 to 310°C along 0.7 g/cm^3.

The difference between the dehydration and rehydration temperature is essentially zero for the runs at low density (hence low pressure), i.e., the variation of d_{001}-spacing during rehydration followed the same path as during dehydration. At higher densities (higher pressures) rehydration occurs at significantly lower temperatures than dehydration. This rehydration hysteresis, appearing as an opening loop in Figures 5 a, b, and c, increases with increasing density (pressure).

In Figures 5d, e, and f d_{001}-spacing is plotted against the peak width (FWHM). The FWHM variation along each dehydration and rehydration path, determined by Gaussian peak fitting, shows a perfect correlation with the d-spacing regardless of the hysteresis.

A run was made along the 0.625 g/cm^3 isochore. The dehydration range for this isochore was identical with that of the 0.704 g/cm^3 and 0.720 g/cm^3 isochores. This was expected because the dehydration occurred while the P-T path lay along the L-V curve in all three cases.

Another run was made with much less liquid water than vapor. In this run homogenization occurred with the disappearance of the liquid phase indicating a path along an isochore in the vapor phase below the L-V curve. It was impossible to make an accurate determination of this homogenization temperature owing to the difficulty of seeing when the last of the liquid phase disappeared. In this run the 19 to 15Å dehydration occurred in the temperature range from 240 to 255°C, considerably lower than the same

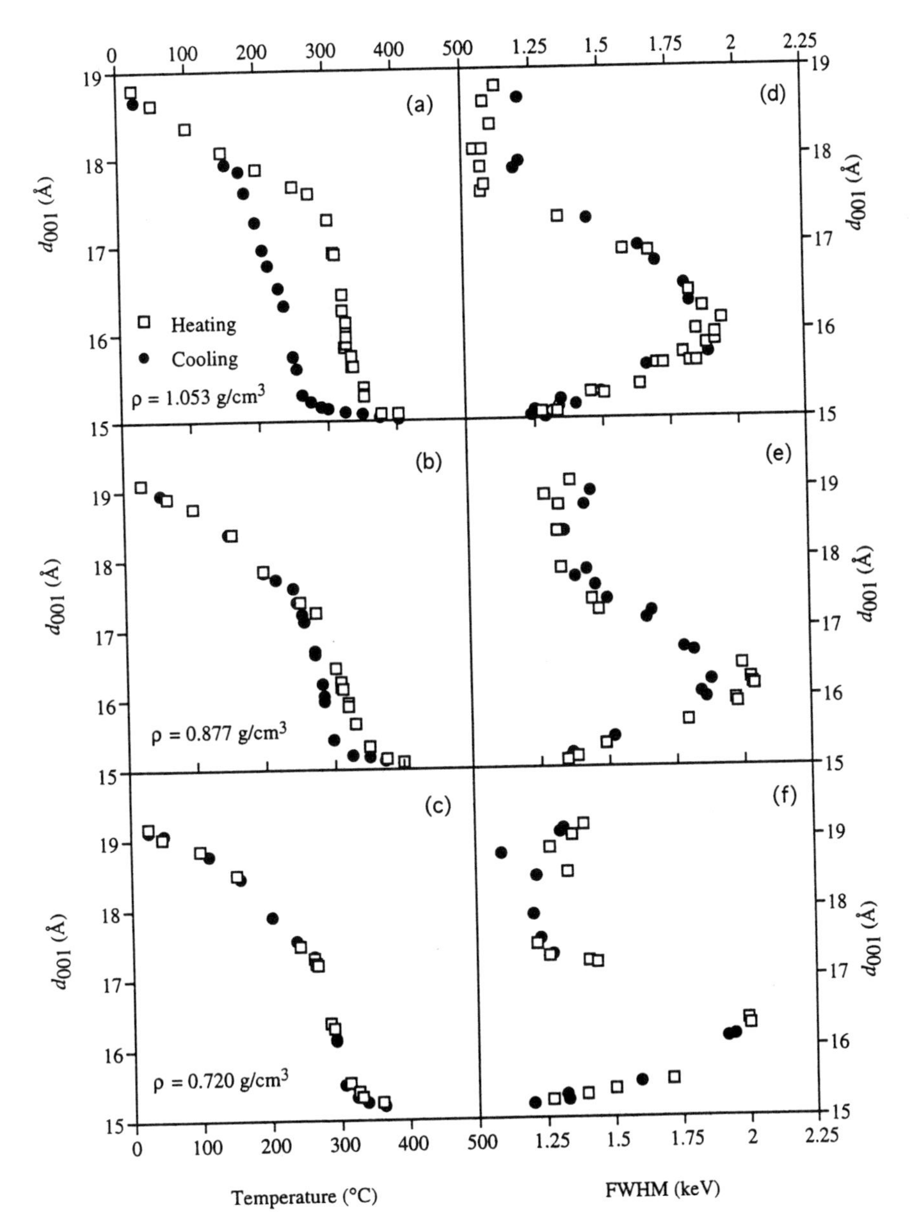

Figure 5 a, b, c. Plots of the 001 d-spacing of Ca- montmorillonite e versus temperature; **d, e, f.** plots of the 001 d-spacing of Ca-montmorillonite versus the full width half maximum (FWHM) of the 001 peak. The isochoric densities are given in the diagrams.

dehydration reaction along the L-V curve in the runs described above. Figure 6 is a pressure-temperature plot for the 19-15Å conversion in Ca-montmorillonite.

3.2. *Mg-montmorillonite*

Five runs were made on Mg-montmorillonite. Four of these started with the 19Å hydration state and one, the lowest density, started with the 15Å state. For three distinctly different isochores, 1.024 g/cm^3, 0.95 g/cm^3, and 0.75 g/cm^3, 19 to 15Å dehydration temperatures were found to be 215-250°C, 200-245°C, and 200-245°C, respectively. Complete heating and cooling cycles were carried out on the 0.95 g/cm^3 and 0.75 g/cm^3 isochores. Figure 7 shows d_{001}-spacing plotted against both temperature and peak width (FWHM). The hysteresis is found to increase from nearly zero along the 0.75 g/cm^3 isochore and increase to ~50°C along the 0.95 g/cm^3 isochore. Other than a small difference in peak width during heating and cooling along the 0.95 g/cm^3 isochore, the behavior was very similar to

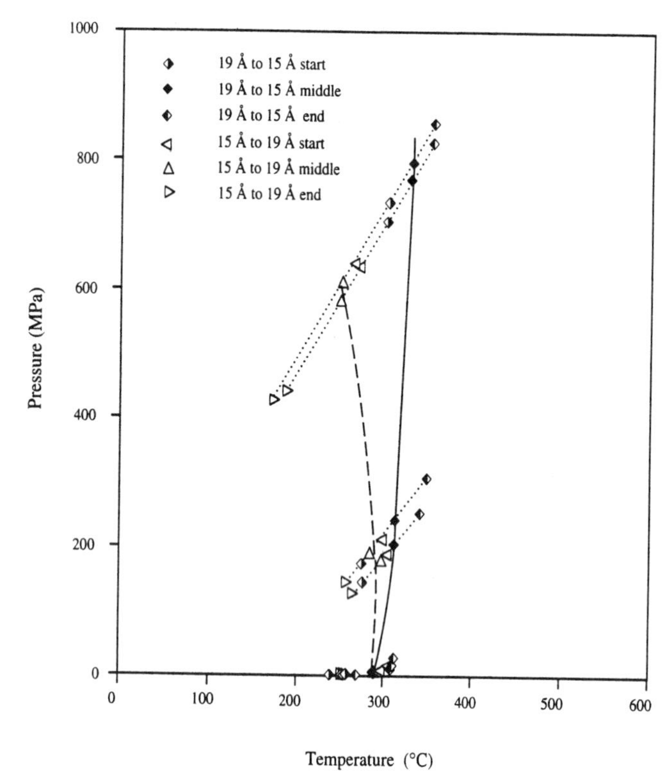

Figure 6. Pressure-temperature diagram of Ca-montmorillonite showing the start, finish, and midpoint of dehydration and rehydration for the 19-15Å conversion. The solid line connects midpoints of the dehydration process and the dashed line connects midpoints of the rehydration process. The separation of these two lines indicates the size of the rehydration hysteresis. Points collected along the same isochore are connected by dotted lines.

that of Ca-montmorillonite. Figure 8a is a pressure-temperature plot of the 19-15Å conversion in Mg-montmorillonite. It shows a pattern of dehydration and rehydration similar to Ca-montmorillonite but with temperatures shifted somewhat lower.

The 15 to 12.5Å dehydration was measured along three heating paths, 1.024 g/cm^3 isochore, 0.75 g/cm^3, and an isochore in the vapor phase below the L-V curve. The first two, which were continuous from the 19Å hydrate, showed 15 to 12.5Å dehydration at 600-605°C and 590-600°C, respectively. The third one carried out in the vapor phase started from the 15Å hydrate and showed dehydration to 12.5Å at 400 to 500°C.

The 12.5 to 15Å rehydration for the 0.75 g/cm^3 isochore started at ~150°C and showed a hysteresis of ~450°C. The 12.5 to 15Å rehydration for the 1.024 g/cm^3 isochore started at ~50°C and showed a hysteresis of ~550°C. The final products for those two runs showed the presence of two hydration states, 12.5Å and 15Å, indicating that the rehydration process did not go to completion. Figure 8b is a pressure-temperature plot for the 15-12.5Å conversion in Mg-montmorillonite.

The run carried out in the vapor field started with the 15Å hydrate and was estimated to have a density of ~0.10 g/cm^3. The dehydration from 15 to 12.5Å was complete at 500°C. During the cooling the rehydration started at ~430°C and was complete at 350°C, showing a rehydration hysteresis of 80°C. The lower hydration temperature and smaller rehydration hysteresis in this lower density run verifies that the trend of pressure dependence in the 15 to 12.5Å dehydration-rehydration cycle is the same as the trend observed for the 19 to 15Å cycles.

4. DISCUSSION

4.1. *Dehydration*

In the plots of d_{001} versus temperature (Figures 5,7) we define the starting and ending points of dehydration and rehydration where the most rapid change in slope occurs and we define the midpoint where the maximum FWHM occurs. Figure 6 shows a pressure-temperature plot of the 19-15Å dehydration and rehydration reactions in Ca-montmorillonite. The effect of pressure on the hysteresis can be seen in the separation of the solid line connecting midpoints of the dehydration reaction and the dashed line connecting the midpoints of the rehydration reactions. Figure 9 shows our 19 to 15Å dehydration data for Ca-montmorillonite plotted as dotted lines and the next stage of dehydration from 15 to 12.5Å [*Koster van Groos and Guggenheim,* 1987] as open circles. These experimental data are compared with the 50% dehydration curve for smectites in sedimentary basins predicted by the thermodynamic model of *Ransom and Helgeson* [1995]. Plots of proposed basin geotherms [*Hower et al.,* 1976; *Burst,* 1969] in Figure 9 allow comparisons between the various sets of data and the pressure-temperature conditions during burial.

Our measurements as well as those of *Koster van Groos and Guggenheim* [1987] show that dehydration temperatures are very pressure sensitive (increase with pressure) below the liquid-vapor curve (i.e., in the vapor phase of water) and are quite insensitive to pressure above the liquid-vapor curve. The slope, dT/dP, of the 19-15Å dehydration changes abruptly as it crosses the L-V curve due to the large change in density of water across the L-V curve. The slope of the 15-12.5Å dehydration, however, changes more gradually because it passes beyond the critical point where the change in density of water is more gradual. Nonetheless, the pressure dependence of the

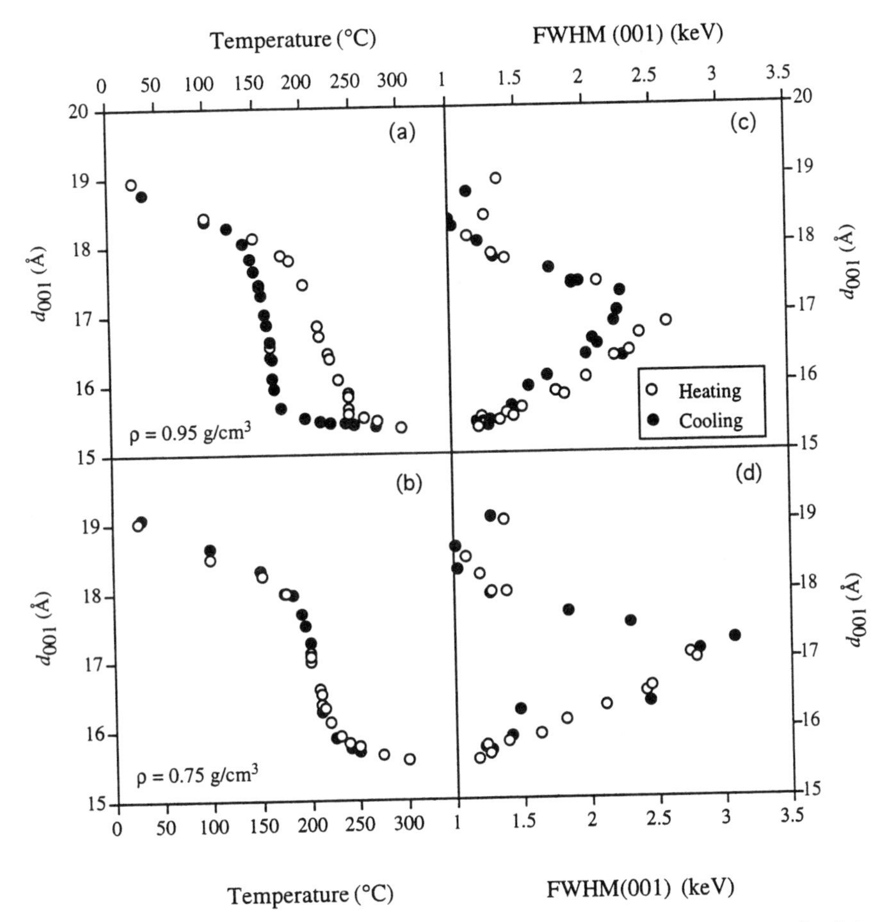

Figure 7 a, b. Plots of the 001 d-spacing of Mg- montmorillonite e versus temperature; **c, d.** plots of the 001 d-spacing of Mg-montmorillonite versus the full width half maximum (FWHM) of the 001 peak. The isochoric densities are given in the diagrams.

dehydration temperature is large at low pressures and small at high pressures.

According to the Clapeyron equation ($dT/dP = \Delta V/\Delta S$, where ΔV is change in molar volume and ΔS is change in entropy) the very small pressure dependence of the dehydration temperature indicates a very small change in density (ΔV) across the dehydration boundary. This, in turn, indicates a small difference in density between the interlayer water and the external water, whether liquid or supercritical fluid. Because the dehydration boundaries cross isochores, this relationship must apply over a range of water densities. Therefore, density of the interlayer water for a given hydrate such as the 19Å hydrate or the 15Å hydrate must vary in response to the density of the external water. It follows as well that the thickness of the hydration layer is not a direct function of the density since the 19-15Å conversion takes place over a range of densities. If ΔV is zero across the phase boundary, then according to the Clapeyron equation the entropy, S, must undergo a finite increase during the conversion from the 19 to 15Å hydrate. This is consistent with the transfer of water from the bonded state in the interlayer position to the disordered state of the external fluid.

Huang et al. [1994] proposed that the interlayer water in Na-montmorillonite has a higher density than the fluid water surrounding the montmorillonite. We did not attempt to test that observation in the present study. However, the measurements reported in the present study were made with a great deal more care and were repeated several times. We believe, therefore, that the new results are much better constrained than those of Huang et al. and that if the measurements on Na-montmorillonite were as well constrained, they too would show a dT/dP close to zero.

4.2. *Pressure Dependence of Rehydration Hysteresis*

Although pressure has little effect on the dehydration temperatures in both Ca- and Mg-montmorillonites at

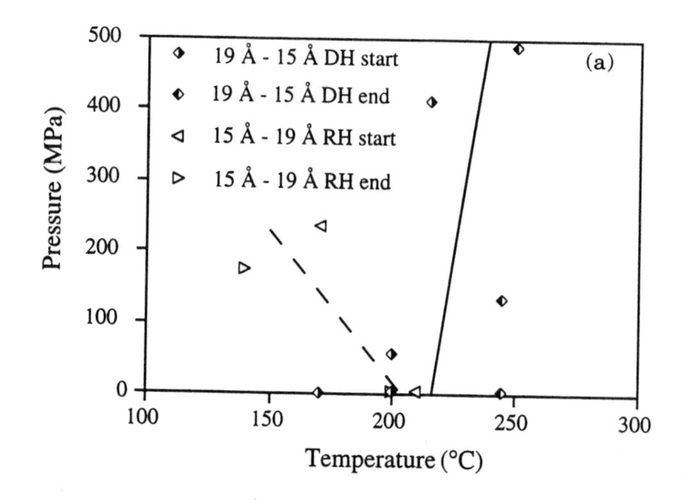

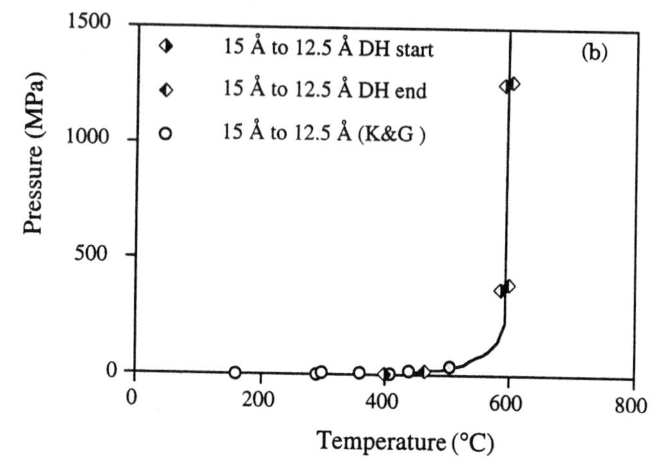

Figure 8. Pressure-temperature diagram of Mg- montmorillonite showing (a) the start, finish, and midpoint of dehydration and rehydration for the 19-15Å conversion. The solid line connects midpoints of the dehydration process and the dashed line connects midpoints of the rehydration process. The separation of these two lines indicates the size of the rehydration hysteresis. (b) the dehydration data for the 15 to 12.5Å conversion. Our data are shown as diamonds and those of *Koster van Groos and Guggenheim* [1987] are shown as open circles.

pressures above the L-V curve of water, it does have an effect on the rehydration temperatures (Figures 5, 6, 7, 8). An explanation for this observed hysteresis may lie in a two-stage mechanism: (1) change in d-spacing by reconfiguring of interlayer water, (2) exchange of water molecules between the interlayer position and the surrounding fluid to complete the balance of internal and external water densities. During increasing temperature, the internal and external densities would be balanced due to rapid diffusion when the interlayer dimension is large. During rehydration, the diffusion would be impeded by the smaller interlayer dimension resulting in a metastable 15Å phase until the reconfiguration took place to yield the 19Å phase. Reconfiguration without exchange would have a positive ΔV and would therefore be inhibited by pressure. We conclude, therefore, that the dehydration curve is closer to the equilibrium boundary because there is no metastable phase during dehydration.

4.3. *Interlayer Cations*

The size and charge of interlayer cations is known to play an important role in the hydration behavior of smectites [*McBride,* 1994]. In Figure 10 we compare the dehydration temperatures of Ca- and Mg-montmorillonites with those for Na-montmorillonite reported by *Huang et al.* [1994]. Wyoming bentonite was the starting material for all three samples and the techniques and pressure-temperature ranges used were very similar. In Figure 10 we plot the dehydration temperatures versus the ionic potential defined as the ratio of ionic charge/radius. The 19 to 15Å dehydration temperature shows a decrease with increasing ionic

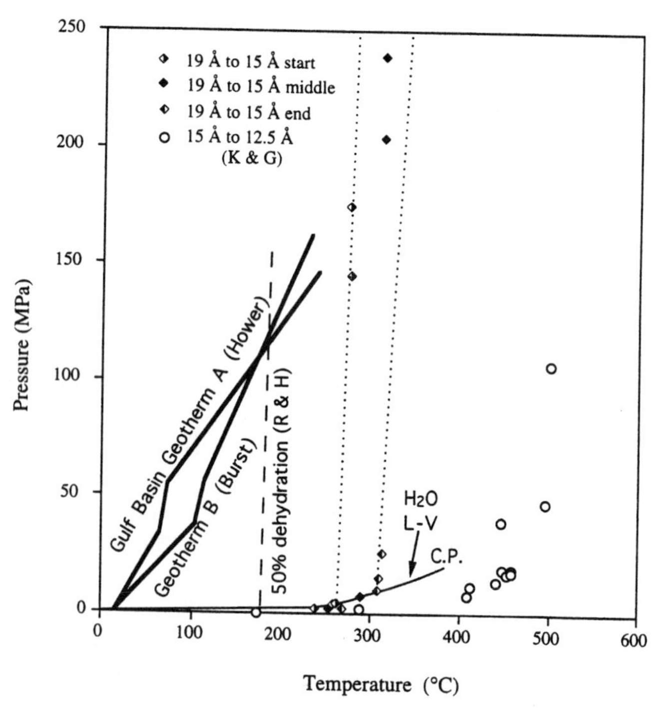

Figure 9. The pressure-temperature diagram for Ca- montmorillonite showing the 19-15Å dehydration at pressures up to 250 GPa as dotted lines representing the start and finish of dehydration. The 15-12.5Å dehydration data of *Koster van Groos and Guggenheim* [1987] are shown as open circles. The 50% dehydration of the 15Å hydrate based on a thermodynamic model of *Ransom and Helgeson* [1995] is shown as a dashed line. Two proposed geotherms for Tertiary sedimentary basins in the U.S. Gulf Coast [*Hower et al.,* 1976; *Burst,* 1969] are shown as heavy solid lines.

potential, whereas the 15 to 12.5Å dehydration temperature shows an increase with increasing ionic potential.

The long-range electrostatic attractive force between the negatively charged silicate sheets and the positively charged cations is the dominant force binding the layers together. The cations, in addition to providing the cohesive force, affect the configuration of the interlayer water molecules. The Mg ion, which has a smaller size than the Ca ion and stronger charge than the Na ion, more strongly favors the sixfold coordination of water molecules just as it favors sixfold sites in most oxides and silicates. The 15Å hydrate owes its spacing to the six-fold coordination group. It seems reasonable, therefore, that the Mg-montmorillonite should have a 15Å hydrate that is stable over a larger temperature range than the Ca- and Na-montmorillonites. This is consistent with the pattern seen in Figure 10.

5. GEOLOGICAL IMPLICATIONS

Figures 5-10 clearly show that the 19Å hydrate persists to temperatures in the range 200 to 385°C at H_2O pressures greater than the liquid-vapor curve. Likewise the 15Å hydrate persists to temperatures in the range 450 to 600°C. Our ability to reverse these reactions indicates that the hydrates, including the 19Å hydrate, are equilibrium phases.

In their model based on thermodynamic calculations

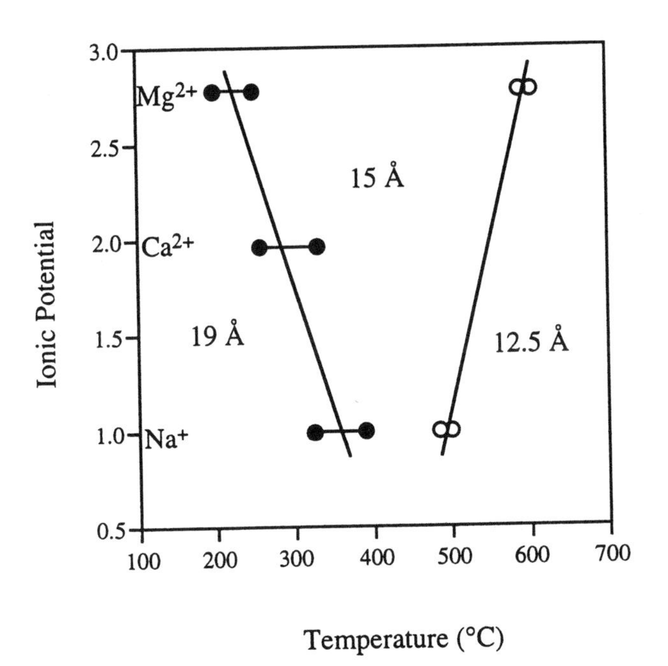

Figure 10. Dehydration temperatures of Na-, Ca-, and Mg- montmorillonites plotted against ionic potential (charge/radius) for those ions. Pairs of connected dots represent the ranges over which dehydration occurs.

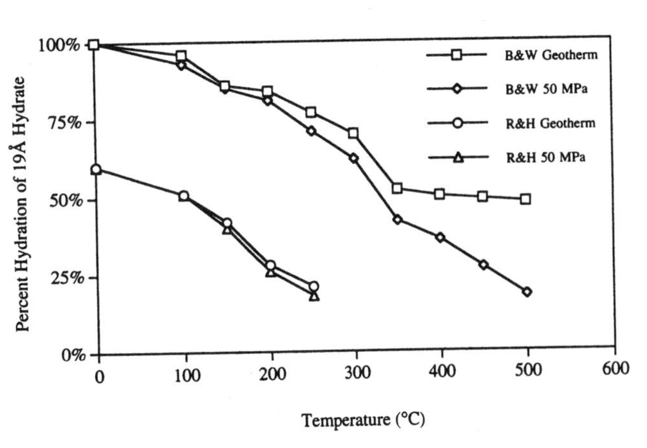

Figure 11. The percent hydration of Ca- montmorillonite as a function of temperature at 50 MPa and along an average geotherm based on the models of *Burst* [1969] and *Hower et al.* [1976]. In this diagram 100% is based on the water content of the 19Å hydrate. B&W refers to the results of *Bassett and Wu* reported in this paper. R&H refers to values reported by *Ransom and Helgeson* [1995].

Ransom and Helgeson [1995] treated the 15Å hydrate and the anhydrate as end members of a solid solution series. They chose to disregard the 19Å hydrate because they assumed the H_2O in these higher hydrates has thermodynamic properties about the same as those of bulk water. Therefore, in thermodynamic calculations any effect from this H_2O would cancel across the reaction. They calculated the extent of continuous dehydration from 15Å state as a function of pressure and temperature based on these assumptions. In Figure 9 the dashed line just over 200°C represents their 50% loss of water from the 15Å hydrate. This is at odds with our observations that montmorillonites retain significantly more water to significantly higher temperatures when the H_2O pressure is above the liquid-vapor curve. In Figure 11 we show the percent of water retained by Ca-montmorillonite as the temperature is increased at 50 MPa and along an average geotherm based on the models of *Burst* [1969] and *Hower et al.* [1976]. These curves were calculated on the basis of both the thickness of interlayer water and the change in density of the water. Our data are labeled B&W and those of *Ransom and Helgeson* are labeled R&H Our data indicate a retention of about 80% of the water at 200°C. That is 80% of the water in the 19Å hydrate. The 19Å starting material, in turn, contained approximately 20% more water than there is in the 15Å hydrate assumed by *Ransom and Helgeson* [1995] at ambient temperature and atmospheric pressure.

The persistence of the 19Å hydrate to higher temperatures in the presence of H_2O pressure has important implications for rheologic properties and fluid dynamics in buried sediments. Unaltered smectites would remain very ductile and might be good candidates as seals for deep oil reservoirs. If, however, illitization of smectites takes place by reaction with other cations such as K and Al, then they may release considerably more water than was previously thought to be the case.

The persistence of the 15Å hydrate to even higher temperatures in the presence of H_2O pressure may have implications for the upper mantle. If smectites survive alteration by reactions with other ions, they may serve as an effective conveyor of large quantities of water down subduction zones. Such water is believed to play an important role in plate boundary volcanism and earthquakes.

Acknowledgments. This work was funded by grants from the National Science Foundation and Exxon Production Research.

REFERENCES

Bassett, W. A., A. H. Shen, M. J. Bucknum, and I.-M. Chou, A new diamond anvil cell for hydrothermal studies to 2.5 GPa and from -190 to 1200°C,. Rev. Sci. Instrum., 64, 2340-2345, 1993.

Burst, J. F., Diagenesis of Gulf Coast clayey sediments and its possible relation to petroleum migration, *Am. Assoc. Pet. Geol. Bull., 53,* 73-93, 1969.

Chang, F. R. C., N. T. Skipper, and G. Sposito, Computer simulation of interlayer molecular structure in sodium montmorillonite hydrates, *J. Am. Chem. Soc., 11,* 2734-2741, 1995.

Hower, J. E., M. E. Hower, and E. A. Perry, Mechanism of burial metamorphism of argillaceous sediments I: mineralogical and chemical evidence, *Geol. Soc. Am. Bull., 87,* 725-737, 1976.

Huang, W.-L., W. A. Bassett, and T.-C. Wu, Dehydration and hydration of montmorillonite at elevated temperatures and pressures monitored using synchrotron radiation, *Am. Mineral., 79,* 683-691, 1994.

Koster van Groos, A., and S. Guggenheim, The effect of pressure on the dehydration reaction of interlayer water in Na-montmorillonite (SWy-1), *Am. Mineral., 69,* 872-879, 1984.

Koster van Groos, A., and S. Guggenheim, Dehydration of a K-exchanged montmorillonite (SWy-1) at elevated pressure, *Clays Clay Miner., 34,* 281-286, 1986.

Koster van Groos, A., and S. Guggenheim, Dehydration of a Ca- and Mg-exchanged montmorillonite (SWy-1) at elevated pressure, *Am. Mineral., 72,* 292-298, 1987.

McBride, M. B., *Environmental Chemistry of Soils,* 406 pp., Oxford, New York, 1994.

Ransom, B., and H. Helgeson, A chemical and thermodynamic model of dioctahedral 2:1 layer clay minerals in diagenetic processes: dehydration of dioctahedral aluminous smectite as a function of temperature and depth in sedimentary basins, *Am. J. Sci., 295,* 245-281, 1995.

Saul, A., and W. Wagner, A fundamental equation for water covering the range from the melting line to 1237 K at pressure up to 25000 MPa, *J. Phys. Chem. Ref. Data, 18,* 1537-1564, 1989.

Stone, R. L., and R. A. Rowland, DTA of kaolinite and montmorillonite under H_2O vapor pressure up to six atmospheres. Third National Conference of Clay Minerals Society, Houston, Texas, NAS-NRC 39, 103-116, 1955.

Wagner, W., and A. Pruss, International equations for the saturation properties of ordinary water substance. Revised according to the international temperature scale of 1990. Addendum to Journal of Physical and Chemical Reference Data 16, 893 (1987), *J. Phys. Chem. Ref. Data, 22,* 783-787, 1993.

Wagner, W., A. Saul, and A. Pruss, International equation for the pressure along the melting and the sublimation of ordinary water substance. *J. Phys. Chem. Ref. Data, 23,* 515-527, 1994.

Wu, T.-C., W. A. Bassett, W.-L. Huang, S. Guggenheim, and A. F. Koster van Groos, Montmorillonite under high pressures: stability of hydrate phases, rehydration hysteresis, and the effect of interlayer cations. *Am. Mineral.,* in press.

W.A. Bassett, Department of Geological Sciences, Snee Hall, Cornell University, Ithaca, NY 14853

Effect of Pressure on the Crystal Structure of Hydrous Wadsleyite, $Mg_{1.75}SiH_{0.5}O_4$

Yasuhiro Kudoh

Institute of Mineralogy, Petrology, and Economic Geology, Tohoku University, Sendai, Japan

Toru Inoue[1]

Department of Earth Sciences, Ehime University, Matuyama, Japan

The effect of pressure on the crystal structure of hydrous wadsleyite has been investigated by means of single-crystal four-circle X-ray diffractometry using a miniature diamond-anvil pressure cell. The linear compressibilities and bulk modulus are β_a =1.66(10), β_b =1.75(15), β_c =2.75(14) (10^{-3} /GPa) and K_0 =153(9) GPa (K_0'=4). The structural results at 3.7 GPa show that the O1-O edge lengths of the Mg3O6 octahedron, on which the hydrogen atoms are expected to be located when the Mg3 site is vacant, are insensitive to pressure.

1. INTRODUCTION

Wadsleyite (β-Mg_2SiO_4) is presumed to be the most abundant mineral in the mantle transition zone at depths of 400-550 km. Since the presence of water or hydroxide has a crucial influence on the seismic and rheological properties of the earth's mantle, the possibility that hydrogen is present in wadsleyite has been studied extensively [*Smyth*, 1987, 1994; *Downs*, 1989; *McMillan et al.*, 1991; *Gasparik*, 1993]. *Inoue* [1994] synthesized wadsleyite with a significant amount of H_2O, 3.1 wt% as confirmed by secondary ion mass spectrometry (SIMS) [*Inoue et al.*, 1995]. The crystal structure of hydrous wadsleyite with 3.1 wt% H_2O revealed that the hydrogen content is coupled to the existence of a vacancy at the Mg3 site, giving a structural constraint of 3.3 wt% for the maximum H_2O content [*Kudoh et al.*, 1996].

In this paper, we report the results of a single-crystal X-ray diffraction study undertaken up to 6.8 GPa concerning the effect of pressure on the crystal structure of hydrous wadsleyite.

2. EXPERIMENTAL

The specimen used in this study was a single crystal of hydrous wadsleyite,$Mg_{1.75}SiH_{0.5}O_4$ synthesized by *Inoue* [1994] using an MA8-type apparatus at conditions of 1300°C and 15.5 GPa. An optically clear single crystal with dimensions of 35x37x71 μm as hand-picked under an optical microscope, analyzed at room pressure by *Kudoh et al.* [1996], and mounted in the diamond anvil cell, along with a piece of ruby crystal about 10 μm in diameter used as a pressure indicator for the high pressure measurement. A 0.25-m-thick Inconel 750X plate was used for the gasket. A hole, 200 μm in diameter, was drilled in the center of the plate to contain the single crystal of hydrous wadsleyite and the ruby. The fluid pressure medium was a 4:1 mixture of methanol:ethanol [*Piermarini et al.*, 1973]. The pressure was calibrated with the ruby fluorescence method [*Barnet et al.*, 1973].

---

[1]Also at Center for High Pressure Research and Department of Earth and Space Sciences, State University of New York at Stony Brook, Stony Brook, NY 11791-2100, USA.

Properties of Earth and Planetary Materials
at High Pressure and Temperature
Geophysical Monograph 101

TABLE 1. Unit Cell Parameters of Hydrous Wadsleyite, $Mg_{1.75}SiH_{0.5}O_4$. Uncertainties in last digit are given in parentheses.

P(GPa)	a(Å)	b(Å)	c(Å)	V(Å^3)
0.00*	5.663(1)	11.546(2)	8.247(4)	539.2(5)
3.7	5.632(2)	11.481(4)	8.163(2)	527.9(2)
5.2	5.613(3)	11.443(5)	8.119(2)	521.6(4)
6.3	5.601(2)	11.422(6)	8.111(3)	519.0(4)
6.8	5.599(3)	11.394(8)	8.097(4)	516.5(5)

* *Kudoh et al.* [1996]

The lattice constants were measured with a Rigaku AFC-7S four-circle diffractometer using MoKα (λ = 0.70926 Å) radiation, and operating at 40 kV and 16 mA. The lattice parameters determined from 13 to 17 reflections with 2θ between 13° and 42° are listed in Table 1. X-ray diffraction intensities were measured at 3.7 GPa up to sin θ/λ = 0.7. After Lorentz and polarization corrections, the intensities of symmetrically equivalent reflections were averaged and 74 independent structure factors (Fo) were obtained. Final agreement factors (R) for 63 Fo's used (Fo>3σ Fo) were R=8.7% and Rw=9.3%, applying the weight $1/\sigma_{hkl}$ (Fo) for each reflection. For the structural refinement at high pressure, the atomic coordinates at room pressure of *Kudoh et al.* [1996] were used as starting coordinates. The occupancies of the Mg1, Mg2, and Mg3 sites were fixed to those at room pressure and were not refined. All calculations were performed using the *teXsan* [1992] crystallographic software package of the Molecular Structure Corporation. Neutral atom scattering factors were taken from *Cromer and Waber* [1974].

3. RESULTS AND DISCUSSION

3.1. *Linear compressibilities and bulk modulus*

Using the data listed in Table 1, the linear compressibilities were calculated using the the least-squares method, and the zero-pressure, isothermal bulk modulus was calculated using the method of finite-strain analysis [*Jeanloz and Hazen,* 1991], assuming K'_{0T} =4 (dK_{0T}/dP) (Table 2). The bulk modulus thus obtained was K_{0T} = 153(9) GPa. The values of linear compressibilities and bulk modulus are consistent with those of *Yusa and Inoue* [1994].

The value of bulk modulus of hydrous wadsleyite with 3.1 wt% H_2O is 9~12% smaller than the isothermal bulk moduli values, 165 Gpa [*Hazen et al.*, 1990], 171 GPa [*Jeanloz and Hazen*, 1991], and 174 GPa [*Fei et al.*, 1992] for -Mg_2SiO_4. These bulk moduli values for hydrous and anhydrous wadsleyite are nearly propotional to the densities. The 3.31 g/cm^3 density of hydrous wadsleyite with 3.1 wt% H_2O [*Kudoh et al.*, 1996] is 9.5% smaller than the 3.474 g/cm^3 value of anhydrous wadsleyite [*Horiuchi and Sawamoto,* 1981]. This fact supports the view that when the mean atomic weight [*Birch,* 1961] is nearly constant (19.9 for hydrous wadsleyite with 3.1 wt% H_2O and 20.0 for anhydrous wadsleyite), the bulk modulus is simply correlated to the density (*Anderson and Nafe,* 1965).

The linear compressibility of the *c* axis is around 2.5 × 10^{-3} GPa^{-1} (2.75 × 10^{-3} GPa^{-1} in this study and 2.32 × 10^{-3} GPa^{-1} in *Yusa and Inoue* [1994]), which is close to the 2.2 × 10^{-3} GPa^{-1} value of the Mg-O bond [*Kudoh et al.*, 1992]. The linear compressibilities of the *a* and *b* axes are around 1.7 × 10^{-3} GPa^{-1}, which is almost equal to the average of the 1.3 × 10^{-3} GPa^{-1} and 2.2 × 10^{-3} GPa^{-1} values (1.75 × 10^{-3} GPa^{-1}) of the Si-O and Mg-O bonds [*Kudoh et al.*, 1992]. This fact can be interpreted by the fact that the crystal structure of wadsleyite is based on the cubic closest packing of oxygens along the [201] direction [*Kudoh et al.*, 1996] and one-third of the Mg-O bonds are parallel to the *c* axis. The contribution of the Si-O and Mg-O bonds to the axial compressions are almost equal for the *a* and *b* axes, but that of the Mg-O bond is most significant for the *c* axis.

3.2. *Crystal structure at 3.7 GPa*

Table 3 lists the positional parameters and isotropic temperature factors of hydrous wadsleyite at 3.7 GPa. Table 4 compares the bond lengths at room pressure and those at 3.7 GPa. From Table 4, it is obvious that the effect of pressure is most significant on the Mg2-O bonds. The Mg3-Mg3' separation is insensitive to pressure (2.997 Å at room pressure and 3.00 Å at 3.7 GPa), which can be interpreted as follows. The $Mg3O_6$ octahedra make a double column parallel to the a axis. The $Mg1O_6$ octahedra

TABLE 2. Linear Compressibilities (10^{-3}/GPa) and Bulk Modulus K_0 (GPa) for Hydrous (H-beta) and Anhydrous (Beta) Wadsleyite. Uncertainties in last digit are given in parentheses.

	a	b	c	K_0	K_0'	Ref.
H-beta	1.66(10)	1.75(15)	2.75(14)	153(9)	4	a
H-beta	1.67(3)	1.87(3)	2.32(4)	155(2)	4.3	b
Beta	1.73(5)	1.62(6)	2.27(4)	167(5)	4.3	c
Beta	1.73	1.63	2.34	174		d*

a. This study; b. *Yusa and Inoue* [1994]; c. *Tanaka et al.* [1987]; d. *Sawamoto et al.* [1984]; *adiabatic moduli

connect each double column of the $Mg3O_6$ octahedra, making a three dimensional net work. On the other hand, the $Mg2O_6$ octahedra attach to the double columns of the $Mg3O_6$ octahedra without connecting any other octahedra. This implies that the $Mg2O_6$ octahedron has more steric freedom for compression than the $Mg1O_6$ and $Mg3O_6$ octahedra. The crystal structure of hydrous wadsleyite is shown in Figure 1.

All edge lengths of the $Mg2O_6$ octahedra decrease significantly (Table 5), while the edges of the $Mg1O_6$ and $Mg3O_6$ octahedra do not change significantly. As a result, the O1-O edge lengths of $Mg3O_6$, on which the hydrogen atoms are expected to be located when the Mg3 site is vacant [*Kudoh et al.*, 1996], are insensitive to pressure at 3.7 GPa. The significant decrease of the O1-O4 edge length may be attributed to the fact that hydrogen may not be located on this edge at 3.7 GPa: the O1-O4 length of 2.830 Å at room pressure is reduced to 2.740 Å at 3.7 GPa, which is nearly equal to the 2.788 Å value of the O1-O4 edge length of anhydrous wadsleyite [*Horiuchi and Sawamoto,* 1981]. This implies that the $Mg3O_6$ octahedron with more Mg vacancies is less compressible than the $Mg2O_6$ octahedron with less Mg vacancies. This situation is similar to the case of spinel and wustite, where the bulk modulus increases with increasing cation vacancy [*Sumino et al.*, 1980].

The valence sums at 3.7 GPa were calculated using the refined bond lengths and the method of *Brown* [1981]. The detailed procedure of calculation is given in *Kudoh et al.* [1996]. The valence sum values for oxygen atoms at 3.7 GPa thus obtained are listed in Table 4, together with those at room pressure [*Kudoh et al.* 1996] for comparison. There is no significant difference between the valence sum values at 3.7 GPa for each oxygen atom and those at room pressure, this being consistent with the fact that the O-O lengths for O-H·O are insensitive to high pressure in wadsleyite at 3.7 GPa.

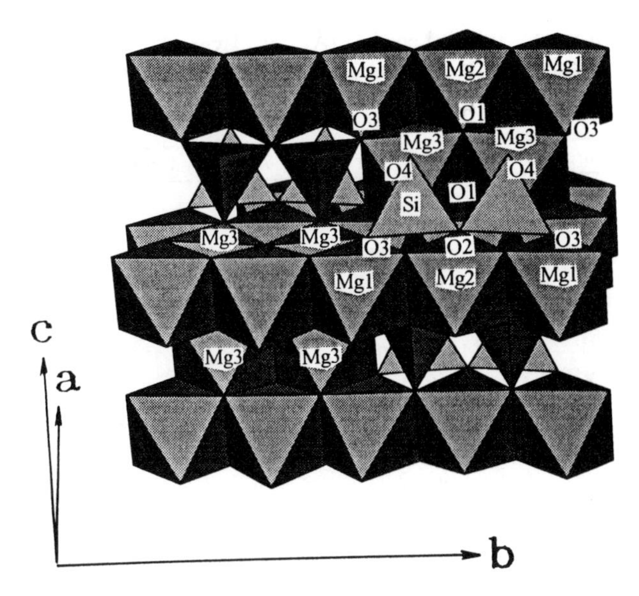

Figure 1. View of the hydrous wadsleyite structure at room pressure (drawn with the atomic parameters of *Kudoh et al.* [1996]).

TABLE 3. Atomic Coordinates, B_{iso} (Å^2) and Valence Sum** at 3.7 GPa. Uncertainties in last digit are given in parentheses

atom*	x	y	z	B_{iso}	VS_{37}	VS_0
Si	0.0000	0.1202(15)	0.6196(14)	1.0(3)		
Mg1	0.0000	0.0000	0.0000	0.0(6)		
Mg2	0.0000	0.2500	0.9687(26)	1.2(7)		
Mg3	0.2500	0.1175(23)	0.2500	0.7(4)		
O1	0.0000	0.2500	0.2246(44)	0.0(9)	1.4	1.3
O2	0.0000	0.2500	0.7270(46)	0.2(9)	2.0	2.0
O3	0.0000	0.9864(25)	0.2571(45)	1.4(9)	1.9	2.1
O4	0.2566(26)	0.1309(46)	0.9941(18)	0.0(4)	1.9	1.9

* Site occupancies for Si, Mg1, Mg2 and Mg3 are 100%Si, 97%Mg, 94%Mg and 78%Mg, respectively.
** VS is valence sum of Brown [1981]. VS_{37} is VS at 3.7 GPa. VS_0 is VS at room pressure.

TABLE 4. Bond Lengths (Å) of Hydrous Wadsleyite, $Mg_{1.75}SiH_{0.5}O_4$. Uncertainties in last digit are given in parentheses.

Atom	Mg1 (97%Mg)	Mg2 (94%Mg)	Mg3 (78%Mg)	Si (100%Si)
$^{v}O1$		2.14(1)	2.065(4)	
		2.09(4)	2.08(2)	
$^{iii}O2$		2.067(9)		1.718(5)
		1.97(4)		1.73(3)
$^{iv}O3$	2.127(8)		2.107(6)	1.617(7)
	2.10(4)		2.06(3)	1.58(4)
$^{iv}O4$	2.058(5)	2.082(5)	2.105(4)	1.633(4)
	2.09(4)	2.00(4)	2.10(2)	1.66(2)
Average	2.081(6)	2.089(7)	2.092(5)	1.650(5)
	2.09(4)	2.01(4)	2.08(2)	1.66(3)

First row: at 1 bar [Kudoh et al., 1996]
Second row: at 3.7 GPa [This study]

TABLE 5. Edge Lengths (Å) of Hydrous Wadsleyite, $Mg_{1.75}SiH_{0.5}O_4$. Uncertainties in last digit are given in parentheses

Pressure		1 bar*	3.7 GPa**	Chemistry***
Mg(1)O6				
	O4 - O4 [2]	2.87(1)	3.0(1)	
	O3 - O4 [4]	2.874(8)	2.85(4)	
	O4 - O4 [2]e	2.953(7)	2.89(3)	
	O3 - O4 [4]e	3.068(8)	3.07(4)	
Mg(2)O6				
	O1 - O4 [4]e	2.830(8)	2.74(4)	(O-H--O)
	O4 - O4 [2]	2.91(1)	2.7(1)	
	O4 - O4 [2]e	2.953(7)	2.89(3)	
	O2 - O4 [4]	3.079(8)	2.95(4)	
Mg(3)O6				
	O1 - O4 [2]e	2.830(8)	2.74(4)	(O-H--O)
	O3 - O3 [1]	2.834(1)	2.818(3)	
	O1 - O1 [1]e	2.853(2)	2.85(1)	O-H--O
	O3 - O4 [2]	2.929(8)	2.96(4)	
	O1 - O3 [2]e	3.061(6)	3.04(3)	O-H--O
	O3 - O4 [2]e	3.068(8)	3.07(4)	
	O1 - O4 [2]	3.023(8)	3.00(4)	O-H--O

Kudoh et al.*, [1996]; ** This study; * Expected partial occupation of hydrogen

REFERENCES

Anderson, O. L., and J. E. Nafe, The bulk modulus-volume relationship for oxide compounds and related geophysical problems, *J. Geophys. Res.*, *70*, 3951-3963, 1965.

Barnet, J. D., Block, S., and G. J. Piermarini, An optical fluorescence system for quantitative pressure measurement in diamond-anvil cell, *Rev .Sci. Instr.*, *44*, 1-9, 1973.

Birch, F., The velocity of compressional waves in rocks to10 kilobars, part2, *J. Geophys. Res.*, *66*, 2199-2224, 1961.

Brown, I. D., The bond-valence method: an empirical approach to chemical structure and bonding, in *Structure and Bonding in Crystals, vol II,* edited by M. O'Keeffe and A. Navrotsky, p. 1-30, Academic Press, New York, 1981.

Cromer, D. T., and J. T. Waber, Mean atomic scattering factors in electrons for free atoms and chemically significant ions, *International Tables for X-ray Crystallography, vol IV, Table2.2A*, Birmingham, England, 1974.

Downs, J. W., Possible sites for protonation in-Mg_2SiO_4 from an experimentally derived electrostatic potential, *Am. Mineral.*, *74*, 1124-1129, 1989.

Fei, Y., H.. K. Mao, J. Shu, G. Parthasarathy, W. A. Bassett, and J. Ko, Simultaneous high-P, high-T X ray diffraction study of β-$(Mg,Fe)_2SiO_4$ to 26 GPa and 900 K, *J. Geophys. Res.*, *97*, 4489-4495, 1992.

Gasparik, T., The role of volatiles in the transition zone, *J. Geophys. Res.*, *98*, 4287-4299, 1993.

Hazen, R. M., J. Zhang, and J. Ko, Effects of Fe/Mg on the compressibility of synthetic wadsleyite: β-$(Mg_{1-x}Fe_x)_2SiO_4$ ($x<0.25$), *Phys. Chem. Minerals*, *17*, 416-419, 1990.

Horiuchi, H., and H. Sawamoto, β-Mg_2SiO_4: single-crystal X-ray diffraction study, *Am. Mineral.*, *66*, 568-575, 1981.

Inoue, T., Effect of water on melting phase relations and melt composition in the system Mg_2SiO_4-$MgSiO_3$-H_2O up to 15 GPa, *Phys. Earth Planet. Inter.*, *85*, 237-263, 1994.

Inoue, T., H. Yurimoto, and Y. Kudoh, Hydrous modified spinel, $Mg_{1.75}SiH_{0.5}O_4$: a new water reservoir in the mantle transition region, *Geophys. Res. Lett.*, *22*, 117-120, 1995.

Jeanloz, R., and R. M. Hazen, Finite-strain analysis of relative compressibilities: application to the high-pressure wadsleyite phase as an illustration, *Am. Mineral.*, *76*, 1765-1768, 1991.

Kudoh, Y., C. T. Prewitt, L. W. Finger, and E. Ito, Ionic radius-bond strength systematics, ionic compressibilities, and an application to $(Mg,Fe)SiO_3$ perovskites, in *High-Pressure Research: Application to Earth and Planetary Sciences*, Geophys. Monogr. Ser., vol. 67, edited by Y. Syono and M. H. Manghnani, pp.215-218, Terra Scientific Publishing Company, Tokyo/AGU, Washington, D.C., 1992.

Kudoh, Y., T. Inoue, and H. Arashi, Structure and crystal chemistry of hydrous wadsleyite, $Mg_{1.75}SiH_{0.5}O_4$: possible hydrous magnesium silicate in the mantle transition zone, *Phys. Chem. Minerals*, *23*, 461-469, 1996.

McMillan, P. F., M. Akaogi, R. K. Sato, B. Poe, and J. Foley, Hydroxyl groups in-Mg_2SiO_4, *Am. Mineral.*, *76*, 354-360, 1991.

Piermarini, G. J., Block, S., and J. D. Barnet, Hydrostatic limits in liquids and solids to 100 kbar, *J. Appl. Phys.*, *44*, 5377-5382, 1973.

Sawamoto, H., D. J. Weidner, S. Sasaki. S., and M. Kumazawa, Single-crystal elastic properties of modified spinel (beta) phase of magnesium orthosilicate, *Science*, *224*, 749-751, 1984.

Smyth, J. R., β-Mg_2SiO_4: a potential host for water in the mantle?, *Am. Mineral.*, *72*, 1051-1055. 1987.

Smyth, J. R., A crystallographic model for hydrous wadsleyite (β-Mg_2SiO_4): an ocean in the Earth's interior?, *Am. Mineral.*, *79*, 1021-1024, 1994.

Sumino, Y., M. Kumazawa, O. Nishizawa, and W. Pluschkell, The elastic constants of single crystal $Fe_{1-x}O$, MnO and CoO, and the elasticity of stoichiometric magnesiowustite, *J. Phys. Earth*, *28*, 475-495, 1980.

Tanaka, S., H. Sawamoto, A. Fujimura, T. Akamatu, H. Hashizume, and O. Shimomura, Fine measurement of compressibility of β-Mg_2SiO_4 using synchrotron radiation (abstract), 28th High Pressure Conference of Japan, Kobe, Nov. 4-6, pp. 18-19, 1987.

teXsan, Crystal Structure Analysis Package, Molecular Structure Corporation, 1992.

Yusa, H., and T. Inoue, Compressibility measurements of hydrous β-phase by X-ray diffraction study under high pressure (abstract), 35th High Pressure Conference of Japan, Hiroshima, Nov. 14-16, p. 20, 1994.

Y. Kudoh, Institute of Mineralogy, Petrology, and Economic Geology, Faculty of Science, Tohoku University, Sendai 980, Japan

T.Inoue, Department of Earth Sciences, Faculty of Science, Ehime University, Matuyama 790, Japan

Raman Spectra of Hydrous β-Mg_2SiO_4 at Various Pressures and Temperatures

Lin-gun Liu[1,2], T. P. Mernagh[3], C.-C. Lin[1], Ji-an Xu[1], and T. Inoue[4]

Variations of Raman spectra of the hydrous β-phase of Mg_2SiO_4 were investigated up to about 210 kbar at room temperature and in the range 108-723 K at atmospheric pressure. Unlike the anhydrous β-phase, which remains intact up to at least 875 K, the hydrous β-phase converted to a defective forsterite at 723 K in about 5 minutes. The increase of H_2O content in the β-phase appears to shift most of the Raman modes toward low frequencies. With the exception of the two Raman bands which correspond to the OH groups, the Raman frequencies of the hydrous β-phase increase linearly with increasing pressure but decrease linearly with increasing temperature within the experimental uncertainties and the range investigated. The two OH Raman bands possess the opposite trend. A comparison of these data with those of the anhydrous β-phase suggests that the hydrous β-phase may be slightly less compressible than the anhydrous β-phase, and that the volume thermal expansivity for hydrous β-phase is also slightly less than that for anhydrous β-phase.

INTRODUCTION

It is widely accepted that the β-phase of Mg_2SiO_4 is probably the most abundant mineral in the transition zone of Earth's mantle. *Liu* [1987] estimated that the Earth's mantle might contain more than 5 times her present water content in the near-surface geochemical reservoirs (hydrosphere + crust). Although there are many dense hydrous magnesium silicates (e.g., phases A, B, C, ...), which accommodate H_2O and are known to be stable at the pressure and temperature conditions relevant to the Earth's mantle, it would be rather significant if the β-phase could also accommodate H_2O. Considering the electrostatic potential and the polyhedral coordination, *Smyth* [1987] suggested that the O(1) sites in the β-phase structure may be a host for hydroxyl. If all of the O(1) sites are assumed to be replaced by OH group, the H_2O contents in the hypothetical hydrous β-phase would be 3.3 wt% [*Inoue et al.*, 1995]. On the basis of the deficits in microprobe analysis weight totals, *Gasparik* [1993] estimated the maximum possible amount of H_2O in the β-phase is 7 wt%. However, earlier experimental studies have found approximately 0.06 wt% [*McMillan et al.*, 1991] and 0.32 wt% [*Young et al.*, 1993] of H_2O in the β-phase. Depending upon the starting materials and experimental P-T conditions, *Inoue* [1994] and *Inoue et al.* [1995] have recently successfully synthesized several hydrous β-phases which contain up to 3 wt% H_2O.

One of the samples synthesized by *Inoue et al.* [1995] was employed in the present study. We have investigated the Raman spectra of this hydrous β-phase at high pressures and room temperature, and at various temperatures at atmospheric pressure. The data thus obtained were also compared with those of the anhydrous β-phase to see the possible effect of H_2O on Raman shift at various pressures and temperatures.

[1] Institute of Earth Sciences, Academia Sinica, Nankang, Taipei, Taiwan, ROC

[2] Research School of Earth Sciences, Australian National University, Canberra, A.C.T., 2601 Australia.

[3] Australian Geological Survey Organisation, Canberra, A.C.T., 2601 Australia.

[4] Department of Earth Sciences, Ehime University, Matsuyama 790, Japan

Properties of Earth and Planetary Materials
at High Pressure and Temperature
Geophysical Monograph 101

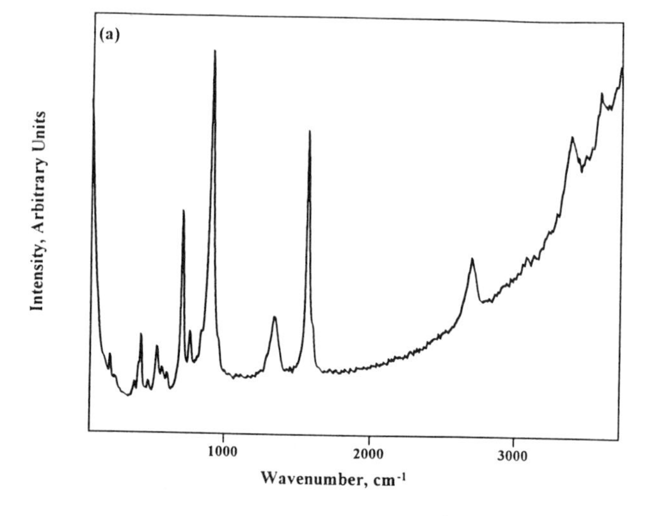

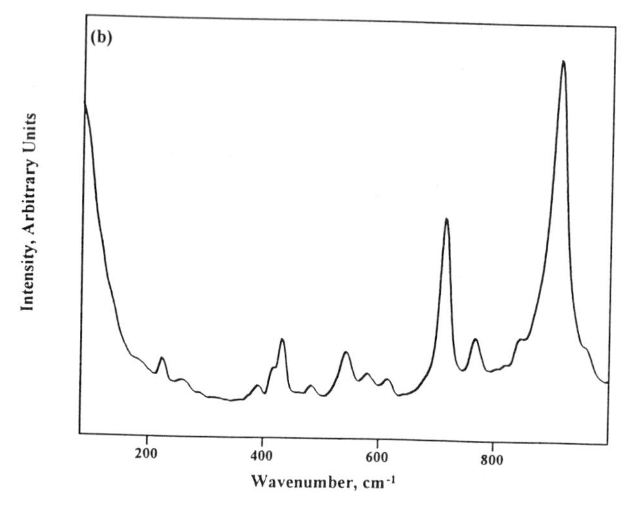

Figure 1. The ambient Raman spectrum of the hydrous β-phase used in the present study. The spectra were obtained using 514.5-nm excitation at 30 mW power. (a) In the range 100~3700 cm^{-1} and (b) 100~1000 cm^{-1}.

EXPERIMENTAL PROCEDURE

Single crystals of the hydrous β-phase, a few tens of μm size, used in the present study were synthesized at 155 kbar and 1200 °C for 22 minutes by *Inoue et al.* [1995]. The electron microprobe analysis of the present sample yielded the Mg/Si = 1.809 ± 0.049, and the secondary ion mass spectrometry study gave H_2O = 2.6 ± 0.3 wt%. One such sample having a composition (Mg/Si = 1.803 ± 0.013 and H_2O = 2.5 ± 0.3 wt%) very close to the present sample was confirmed to have the β-phase structure by *Inoue et al.* [1995], using the single-crystal X ray diffraction method. Thus, the chemical formula of the present sample should be very close to $Mg_{1.75}SiH_{0.5}O_4$ as assigned by *Inoue et al.* [1995].

The ambient Raman spectrum of the present sample obtained in this study is shown in Figure 1, of which the three peaks between 1000 and 3000 cm^{-1} correspond to the remnant graphite coated on the surface for earlier analysis. The two peaks near 3355 ± 3 and 3584 ± 5 cm^{1} correspond to the OH groups. All the Raman bands below 1000 cm^{-1} correspond closely to those of the anhydrous β-phase. More discussion on the ambient Raman spectrum of hydrous β-phase is given in the next section.

In the high-pressure experiments, one or two small chips of the sample, approximately 20~30 μm in size, were placed inside the hole (100~150 μm in diameter and 60~80 μm in depth) in a hardened stainless steel gasket in a standard diamond anvil cell. The anvil faces are approximately 600 μm in diameter. Four separate experimental runs were conducted. In the first experiment, a very thin chip of ruby (~5 μm in thickness) was placed inside the hole, and an ethanol-methanol (1:4) mixture was used as a pressure transmitting medium. The whole assembly was then sealed by compressing the two diamond anvils. The samples were set on top of the thin ruby chip, and the Raman spectra of the samples and the ruby fluorescence spectrum were measured at the same spot. This procedure tends to reduce the errors caused by the pressure gradient across the samples. Pressures were measured using the ruby-fluorescence technique. Because both the ambient 914 cm^{-1} band and the OH Raman bands of the sample were interfered with by the intense Raman bands of the ethanol-methanol mixture at pressures greater than about 140 kbar, the pressure investigation was terminated at this pressure. This pressure is not very much beyond the solidification point (~110 kbar at room temperature) of the ethanol-methanol mixture, therefore possible problems which may arise from a non-hydrostatic condition should not be too serious in the present study.

In the second set of high-pressure experiment, ruby powders were used as the pressure indicator and water was used as a pressure transmitting medium. Although H_2O freezes at 9.3 kbar at room temperature, the hydrostatic behavior of ices VI and VII is about the same as the ethanol-methanol mixture at pressures below 100 kbar and is superior to both solid CO_2 and the ethanol-methanol mixture in the pressure region 100~500 kbar or greater [*Liu*, 1982; 1984]. In this experiment, the pressure variation was investigated up to 210 kbar, but only the Raman bands of the hydrous β-phase below 1000 cm^{-1} were studied. At pressures greater than 210 kbar, the quality of the Raman spectrum is too poor to be measured.

The experimental conditions of the third and fourth sets are the same as those in the second set except paraffin and paraffin oil were respectively used as the pressure media. Thc hydrostaticity of paraffin oil is superior to that of paraffin, and both are similar to that of H_2O in the two hundred kilobar range, as judged from the quality of the peaks of ruby fluorescence. In these two experiments, only the OH band near 3355 cm^{-1} was investigated up 90 kbar, beyond which the peak signal is too weak to be reliably measured. The weak OH band near 3584 cm^{-1} was not detected at high pressures in these two experiments.

In the temperature variation experiment, sample chips of approximately 20~30 μm in size were placed at the center of a 16-mm-diameter crucible which has a sapphire window as the base. The crucible was then placed directly on a small silver block in a Linkam THM 600 heating/freezing stage which consists of a double-walled, anodized aluminum thermal cell. The cell can be purged with dry nitrogen to exclude moisture and air. Low temperatures are obtained by pumping liquid nitrogen through an annulus in the silver block, and a resistance heater opposes the cooling effect of the nitrogen to yield the desired temperature. In the heating mode only the resistance heater is used along with water cooling of the cell. The temperature is monitored by a platinum resistance thermometer attached to the heater. The temperature control unit is completely automatic and can be programmed to maintain any desired temperature or to change temperature at a constant rate.

In both modes, the sample temperature is controlled by thermal conduction between the sample and the silver block. The stage has been calibrated at both high and low temperatures by observing the phase changes in synthetic fluid inclusions placed in the center of the crucible. A more detailed description of the experimental procedures is given by *Liu and Mernagh* [1994]. Horizontal thermal gradients may lead to errors of up to 1% in temperature measurement. After varying the temperature, the samples were kept at the new temperature for at least 5 minutes (except the run carried out at 723 K) before recording the Raman spectrum, in order to allow the samples to reach thermal equilibrium.

Laser Raman spectra for the hydrous β-phase at ambient pressure and various temperatures were recorded from 100 to 4000 cm^{-1} on a Microdil-28 Raman microprobe using a spectral bandpass of approximately 3 cm^{-1}. These spectra were acquired with a 50X ULWD Olympus microscope objective and 15 accumulations with 15 sec integration time. In the temperature variation study, Raman spectra were recorded in three separate regions (between 75 and 560 cm^{-1}, between 505 and 960 cm^{-1}, and between 3265 and 3525 cm^{-1}) corresponding to the spectral coverage on the detector produced by the 1800 g/mm gratings of the spectrometer. Thus, the gratings were kept stationary while the temperature was varied. This should provide the most accurate means for directly comparing the variation of frequency with temperature. All spectra were obtained with the 514.5 nm line from a Spectra Physics model 2020 argon ion laser using 30 mW power at the sample. The pressure variation study was carried out on a Renishaw-2000 Raman microprobe, and the excitation source was a Coherent INNOVA 2W argon ion laser. The spectra were recorded with a 25X UM Leitz microscope objective and 3 accumulations at 1500~2000 sec integration time with 80 mW power on the sample. The focused laser spot on the sample was estimated to be approximately 2~3 μm in diameter in the temperature experiment and about 5 μm in diameter in the pressure experiment. Wavenumbers are accurate to ± 1 cm^{-1} as determined from plasma emission lines. The frequency of each Raman band was obtained by Lorentzian curve fitting using the Jandel Scientific Peakfit computer software.

AMBIENT RAMAN SPECTRUM OF THE SAMPLE

The ambient Raman spectra of the hydrous β-phase are shown in Figure 1. One of the spectra (Figure 1a) shows all the Raman modes observed in the range 100 to 3700 cm^{-1}, and the other (Figure 1b) gives the details below 1000 cm^{-1}. As pointed out earlier, the three bands between 1000 and 3000 cm^{-1} are caused by the graphite coating on the sample. The band near 415 cm^{-1} corresponds to the most intense Raman band of the sapphire window in the heating stage.

McMillan et al. [1991] observed a band at 3322 ± 3 cm^{-1} and perhaps a weak feature near 3580 cm^{-1}, suggesting the presence of OH groups in the β-phase. Two rather strong bands at 3355 ± 3 and 3584 ± 5 cm^{-1} are obvious in Figure 1. The frequency of the first band is significantly different from that reported by *McMillan et al.* [1991], and the better signal of the present spectrum is likely due to the higher H_2O content of the present sample.

The frequencies of the Raman bands of the present sample below 1000 cm^{-1} are compared with those of the anhydrous β-phase [*Liu et al.*, 1994] and the hydrous β-phase [*McMillan et al.*, 1991] in Table 1. There were many other weak bands in the spectrum shown by *McMillan et al.* [1991] but, because the frequencies for these weak bands were not given, we are not able to include them in Table 1. However, on the basis of the available data reported by *McMillan et al.* [1991], it is clear that the frequencies of the Raman bands of the hydrous β-phase observed by *McMillan et al.* [1991] are nearly identical to the anhydrous β-phase. In other words, the presence of OH groups in the β-phase structure does not appear to affect its Raman frequencies. This is very likely to be true, if one considers that there is only 0.06 wt% H_2O in the sample. On the other hand, the Raman frequencies of the present sample are rather different from those of the anhydrous β-phase. With a

TABLE 1. Comparison of the Ambient Raman Frequencies Between the Anhydrous β-phase and Hydrous β-phase and the Values of $-(\partial\nu_i/\partial T)_P$ and $(\partial\nu_i/\partial P)_T$ Determined for the Various Raman-Active Modes of the Hydrous β-phase.

Anhydrous β-phase	Hydrous β-phase			
Liu et al.	McMillan et al.	This work		
$\Delta\nu_i$ cm^{-1}	$\Delta\nu_i$ cm^{-1}	$\Delta\nu_{i,amb}$[a] cm^{-1}	$-(\partial\nu_i/\partial T)_P$ cm^{-1}/deg	$(\partial\nu_i/\partial P)_T$ $cm^{-1}/kbar$
	3580?	3584	− 0.01	−1.3
	3322	3355	− 0.14	−1.11
918	918	914	0.0137 (0.0159)[b]	0.37 (0.547)[c]
881sh		847sh	0.01 (0.022)	(0.52)
842	842			(0.45)
779	778	771	0.022 (0.020)	0.7 (0.52)
723	723	716	0.0146 (0.0164)	0.328 (0.443)
620	620	616	0.008 (0.019)	(0.34)
583	584	583	0.009 (0.015)	0.5?
551	551	546	0.018 (0.019)	0.38 (0.54)
445	443	483	0.006 (0.026)	0.4 (0.31)
427		432	0.024 (0.019)	
343		390	0.011 (0.030)	
300				
(268)		262	0.01	
233		225	0.012 (0.009)	

[a]$\Delta\nu_{i,amb}$ was calculated from the equation $\Delta\nu_i = a_i + b_i T$ with T = 298 K. sh stands for shoulder.
[b]From *Liu and Mernagh* [1994] for the anhydrous β-phase.
[c]From *Liu et al.* [1994] for the anhydrous β-phase.

few exceptions, most of its Raman bands shift significantly towards lower frequencies, in comparison with those of the anhydrous β-phase. The reasons for the changes in the Raman spectra are discussed in detail in *Mernagh and Liu* [1996].

PRESSURE DEPENDENCE OF RAMAN SPECTRA

In addition to two OH bands, up to about 5 Raman bands may be reliably measured in the present sample at high pressures. The spectra became progressively weaker with increasing pressure, in particular for the 3584 cm^{-1} OH band and other weak bands. Regardless of these drawbacks, the vibrational frequencies for the two OH bands were observed to decrease linearly with increasing pressure and for all other bands increased linearly with pressure within the experimental uncertainties and the range investigated. Because the 3584 cm^{-1} OH band is rather weak and both the 3355 cm^{-1} OH band and the most intense 914 cm^{-1} band were interfered with by the intense Raman bands of the ethanol-methanol mixture at pressures greater than about 140 kbar, it was decided that the investigation should be terminated at this pressure in the first experiment. In the second experiment, the pressure was extended to 210 kbar, but variation for only three Raman bands was investigated. The frequencies of the two most intense Raman bands of the hydrous β-phase (914 and 716 cm^{-1}) in the second run appear to increase nonlinearly with increasing pressure. However, when the data obtained in the two experiments were combined (see Figure 2), only a linear variation is warranted for these

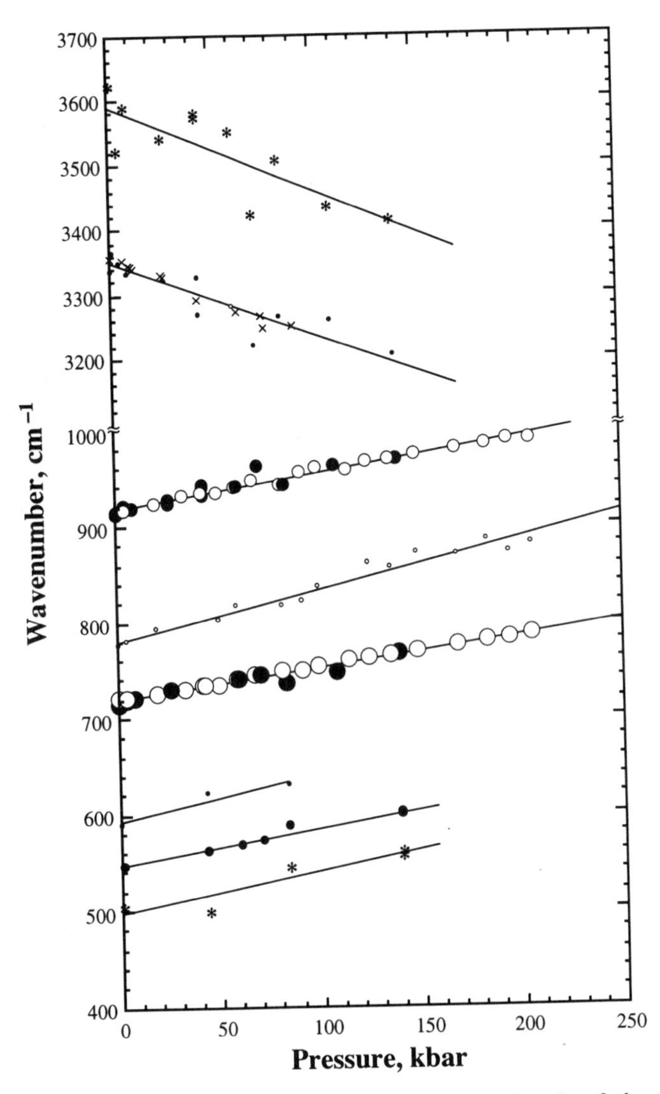

Figure 2. Pressure dependence of the Raman bands of the hydrous β-phase at room temperature. The solid dots represent data obtained in ethanol-methanol, the open dots in H_2O, and the crosses in paraffin pressure media, respectively. Except asterisk used for two very weak bands, the size of the symbols roughly indicates the relative intensities of the bands. The lines are linear regressions of the data.

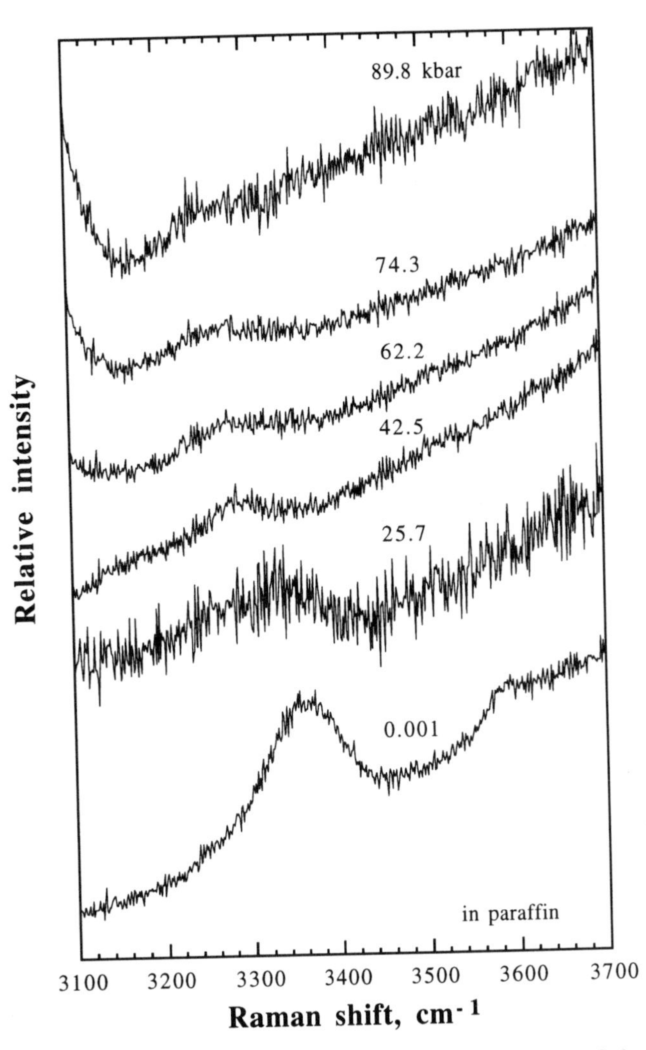

Figure 3. Selected Raman spectra for the OH bands of the hydrous β-phase as a function of pressure at 25 °C. The spectra were obtained using 514.5-nm excitation at 80 mW power.

data. Only the OH bands were investigated in the third and fourth experiments. Selected Raman spectra of these OH bands at various pressures and room temperature are shown in Figure 3. The 3584 cm^{-1} OH band was not observable at high pressure, and the spectra became progressively weaker with increasing pressure. Thus, the experiment was terminated at ~90 kbar.

The variation of the vibrational frequencies for all the Raman bands of the hydrous β-phase with pressure is shown in Figure 2 (the solid dots for the first, the open dots for the second, and the crosses for the third and fourth runs), and the values of $(\partial \nu_i/\partial P)_T$ derived by linear regressions of the various Raman bands are listed in Table 1. The size of the symbols in Figure 2 roughly indicates the relative intensities of these modes. Note that the relative intensities for the ambient 914 and 716 cm^{-1} bands of the hydrous β-phase were reversed in all the high pressure runs. Thus, it appears that the relative intensities of these two Raman bands are orientation-dependent and the pressure tends to control the orientation of these crystals. It is not fully understood how the hydrostatic pressure controls the orientation of the crystals. However, the single crystals used in the present experiment were picked up from a flake of thin-section which may always settle in the same orientation in the pressure chamber.

In considering the quality of the data, only variations for the 3355 cm^{-1} OH band and the other two strong

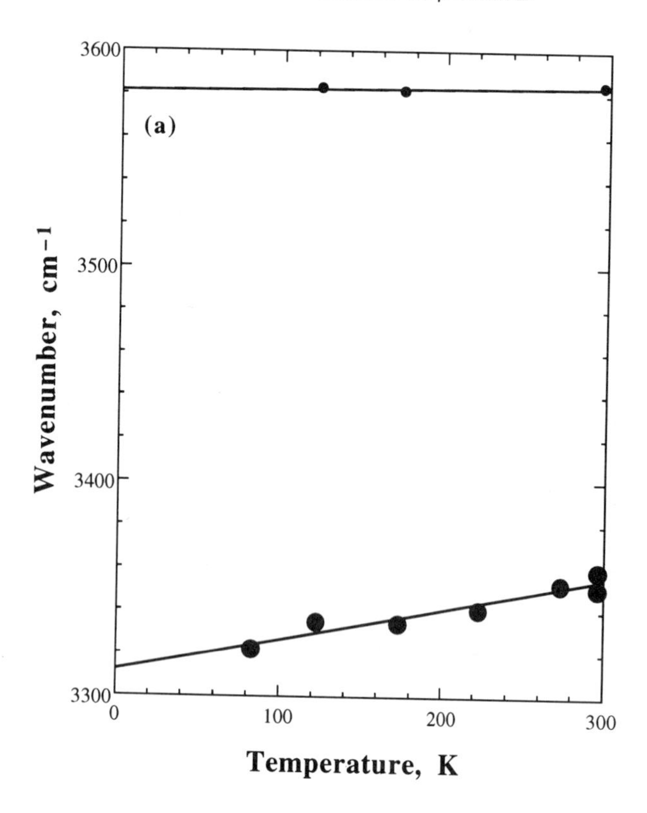

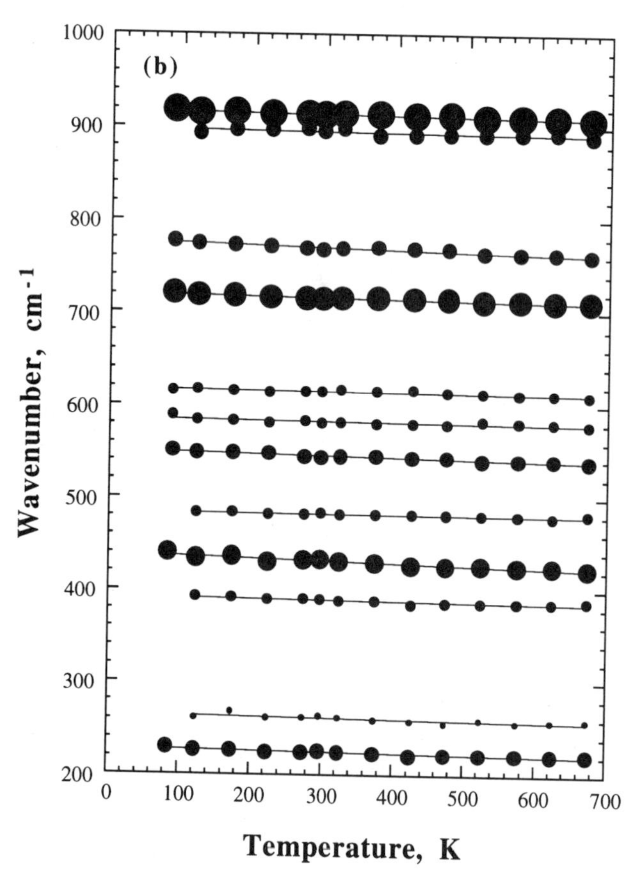

Figure 4. Temperature dependence of the Raman modes of hydrous β-phase at atmospheric pressure. The size of the symbols roughly indicates the relative intensities of the bands. The lines represent the linear regression of the data.

bands at 914 and 716 cm^{-1} are well determined. The rate of variation of the OH band is about a factor of 2 to 3 greater than that of the ordinary bands and is in the opposite direction. The decrease in frequency with pressure of the OH bands parallels those of ice VII reported by *Pruzan et al.* [1990], and it has been commonly observed in many hydroxides.

The values of $(\partial \nu_i/\partial P)_T$ for the 914 and 716 cm^{-1} bands were also compared with those for their corresponding Raman bands in anhydrous β-phase in Table 1. These values for the hydrous phase are about 30% smaller than those for the anhydrous one. By assuming that the mode vibrational frequencies of crystals depend mostly on the variation of volume, the high-pressure Raman data given in Table 1 suggest that the anhydrous β-phase may be slightly more compressible than the hydrous β-phase. It may appear to be surprising that an anhydrous phase is more compressible than its corresponding hydrous phase. However, hydrogen bonding may also strengthen the bonds and make the hydrous phase less compressible. In addition, there is no other pair of anhydrous and hydrous phases which may be used for a comparison. Fortunately, the data of temperature dependences of these two phases are consistent with those of pressure dependences (see next section).

The present conclusion concerning the compressibility between the hydrous and anhydrous β-phases, however, is opposite to that suggested by the tentative compression data for the hydrous β-phase reported by *Yusa and Inoue* [1994], who found that the hydrous β-phase is more compressible than the anhydrous β-phase.

TEMPERATURE DEPENDENCE OF RAMAN SPECTRA

Raman spectra of the hydrous β-phase were recorded over the temperature range 108 - 723 K at atmospheric pressure in the present study. The lower temperature is limited by the liquid N_2 used to cool the sample stage, and the high temperature is limited by the stability of the sample.

The variations of all Raman modes of hydrous β-phase with temperature are shown in Figure 4, and the values of $(\partial \nu_i/\partial T)_P$ derived by linear regressions of these modes are given in Table 1. The size of the symbols in Figure 4 roughly indicates the relative intensities of these modes. Within the experimental uncertainties and the range investigated, the frequency of all the Raman bands decreases linearly with increasing temperature, except for the two OH bands which possess the opposite trend. The linear behavior, however, may be due to the very limited temperature region which was investigated in this study.

In general, the resolution of the various Raman bands of the hydrous β-phase observed in the temperature variation study is superior to those observed in the high-pressure experiment, except for the two OH bands. This is also true in our earlier studies of the anhydrous β-phase [*Liu et al.*, 1994; *Liu and Mernagh*, 1994]. Thus, more Raman bands for the hydrous β-phase were measured in the temperature experiment. The weak OH band at ambient 3584 cm^{-1} was often unobservable in the lower temperatures and totally unobservable above room temperature. Thus, the value of $(\partial v_i/\partial T)_P$ determined for this band is unreliable. The strong OH band at ambient 3355 cm^{-1} was reliably measured at lower temperatures, but the band became so broad above room temperature that any attempt to include the data at higher temperatures would spoil the quality of the data obtained at lower temperatures. Thus, only the data obtained at temperatures below 298 K were shown in Figure 4a and were included in data analysis. Like the high-pressure experimental data, the rate of change for the 3355 cm^{-1} OH band is much greater than those for other Raman bands of the hydrous β-phase to be discussed next.

The rates of change, $(\partial v_i/\partial T)_P$, determined for all other Raman bands of the hydrous β-phase were also compared with those of the anhydrous β-phase in Table 1. Note that the temperature dependences (or the values of $(\partial v_i/\partial T)_P$ in Table 1) of Raman modes of the hydrous β-phase are in general smaller than those of their corresponding Raman modes of anhydrous β-phase. By adopting the same assumption that the mode vibrational frequencies of crystals depend mostly on the variation of volume, the temperature variation data of Table 1 suggest that the volume thermal expansivity for the hydrous β-phase is slightly smaller than that for the anhydrous β-phase. The latter feature parallels the results of our studies of these two phases at high pressures described above.

CONVERSION OF THE HYDROUS β-PHASE AT HIGH TEMPERATURE

Several Raman spectra were recorded as the temperature approached 723 K as we had already observed that the hydrous β-phase lost its water content and converted to a defective forsterite above this temperature. Figure 5 displays a sequence of these spectra recorded over the range 500 ~ 950 cm^{-1}. The spectrum recorded at 718 K gave those Raman bands corresponding to only the hydrous β-phase for an indefinite time. Once the temperature was raised to 723 K, the spectrum started to change within one minute.

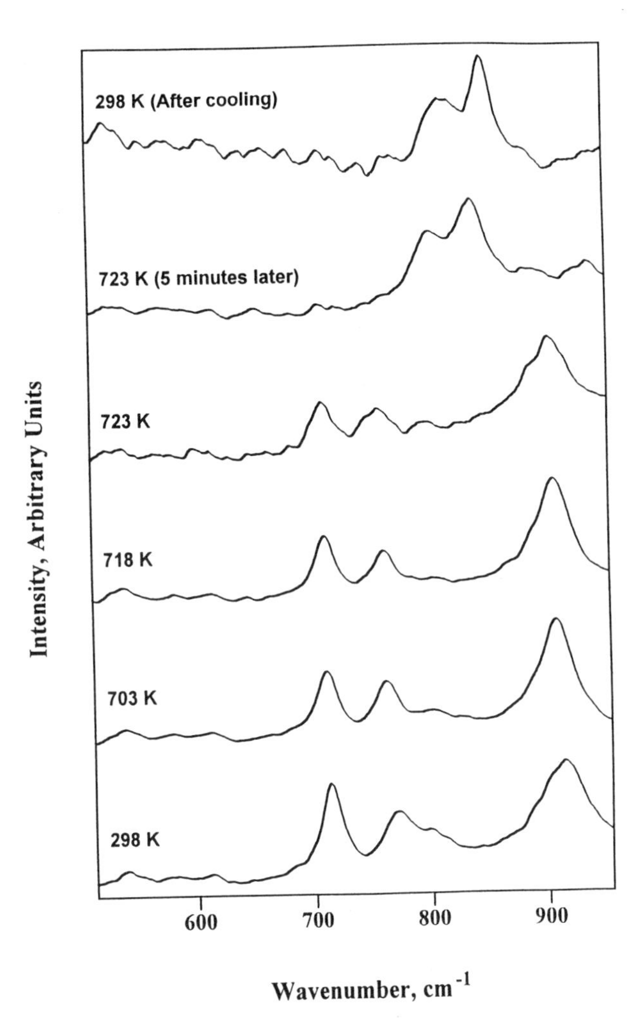

Figure 5. Raman spectra of the hydrous β-phase at temperatures near 723 K and before and after heating at atmospheric pressure. The spectra were obtained using 514.5-nm excitation at powers of 30 mW.

The spectrum at 723 K shown in Figure 5 was recorded in that time period. After about 5 minutes, the Raman bands corresponding to the hydrous β-phase all disappeared, and a totally new spectrum was recorded (see Figure 5). No further change, except for the shift in frequencies due to temperature variation, was observed when temperature was further raised to 873 K. After cooling to room temperature, the recorded Raman spectrum of the sample is similar to those obtained at temperatures greater than 723 K, except for a small shift in frequencies due to the temperature difference. The Raman frequencies of the sample cooled to room temperature correspond very closely to those of forsterite. Thus, it is concluded that the hydrous β-phase dehydrated near 723 K and then converted to forsterite. Because the Mg/Si ratio of the present hydrous β-phase sample is significantly smaller than that of a stoichiometric olivine, the forsterite observed in the recovered sample should be defective.

McMillan et al. [1991] have also examined the Raman spectra of their hydrous β-phase before and after heating it to 853 K, and have found that their sample had converted to forsterite after heating for 5 minutes. However, on the basis of the information given, one cannot judge whether 853 K is the actual dehydration temperature of their sample because these authors did not provide more information regarding the dehydration temperature. On the other hand, *Liu and Mernagh* [1994] found that the anhydrous β-phase remains intact at 873 K for an indefinite time. This may suggest that the dehydration temperature of the hydrous β-phase studied by *McMillan et al.* [1991] is in the range 723~853 K, because their hydrous β-phase contains only about 0.06 wt% of H_2O. In other words, a β-phase containing more H_2O tends to be kinetically less stable when it is heated to high temperature at atmospheric pressure.

Reynard et al. [1996] reported that in some cases of their experiments there appeared an intermediate phase of defective spinelloid structure when anhydrous β-phase converted to forsterite between 800 and 1000 K at atmospheric pressure. This conversion temperature is much higher than the dehydration temperature observed in the hydrous β-phase and there is no evidence for the existence of an intermediate phase during dehydration in the present study.

Acknowledgments. TPM publishes with the permission of the Executive Director of AGSO.

REFERENCES

Gasparik, T., The role of volatiles in the transition zone, *J. Geophys. Res.*, *98*, 4287-4299, 1993.

Inoue, T., Effect of water on melting phase relations and melt composition in the system Mg_2SiO_4-$MgSiO_3$-H_2O up to 15 GPa, *Phys. Earth Planet. Inter.*, *85*, 237-263, 1994.

Inoue, T., H. Yurimoto, and Y. Kudoh, Hydrous modified spinel, $Mg_{1.75}SiH_{0.5}O_4$: a new water reservoir in the mantle transition region, *Geophys. Res. Lett.*, *22*, 117-120, 1995.

Liu, L., Compression of ice VII to 500 kbar, *Earth Planet. Sci. Lett.*, *61*, 359-364, 1982.

Liu, L., Compression and phase behavior of solid CO_2 to half a megabar, *Earth Planet. Sci. Lett.*, *71*, 104-110, 1984.

Liu, L., Effect of H_2O on the phase behavior of the forsterite-enstatite system at high pressures and temperatures and implications for the Earth, *Phys. Earth Planet. Inter.*, *49*, 142-167, 1987.

Liu, L., and T. P. Mernagh, Raman spectra of high-pressure polymorphs of Mg_2SiO_4 at various temperatures, *High Temp-High Press, 26*, 631-637, 1994.

Liu, L., T. P. Mernagh, and T. Irifune, High pressure Raman spectra of β-Mg_2SiO_4, γ-Mg_2SiO_4, $MgSiO_3$-ilmenite and $MgSiO_3$-perovskite, *J. Phys. Chem. Solids, 55*, 185-193, 1994.

McMillan, P. F., M. Akaogi, R. K. Sato, B. Poe, and J. Foley, Hydroxyl groups in β-Mg_2SiO_4, *Am. Mineral., 76*, 354-360, 1991.

Mernagh, P. T., and L. Liu, Raman and infrared spectra of hydrous β-Mg_2SiO_4, *Can. Mineral*. in press, 1996.

Pruzan, Ph., J. C. Chervin, and M. Gauthier, Raman spectroscopy investigation of ice VII and deuterated ice VII to 40 GPa. disorder in ice VII, *Europhys. Lett., 13*, 81-87, 1991.

Reynard, B., F. Takir, F. Guyot, G. D. Gwanmesia, R. C. Liebermann, and Ph. Gillet, High-temperature Raman spectroscopic and X ray diffraction study of β-Mg_2SiO_4: insights into its high-temperature thermodynamic properties and the β- to α-phase-transformation mechanism and kinetics, *Am. Mineral., 81*, 585-594, 1996.

Smyth, J. R., β-Mg_2SiO_4. : A potential host for water in the mantle? *Am. Mineral., 72*, 1051-1055, 1987.

Young, T. E., H. W. Green, II, A. M. Hofmeister, and D. Walker, Infrared spectroscopic investigation of hydroxyl in β-$(Mg,Fe)_2SiO_4$ and coexisting olivine: implications for mantle evolution and dynamics, *Phys. Chem. Minerals, 19*, 409-422, 1993.

Yusa, H., and T. Inoue, Compressibility measurements of hydrous β-phase by X ray diffraction study under high pressure, Program and Abstract, 35th High Pressure Conference of Japan, p. 20, 1994.

High-Pressure Infrared Spectra of Feldspars: Constraints on Compressional Behavior, Amorphization, and Diaplectic Glass Formation

Q. Williams

Department of Earth Sciences and Institute of Tectonics, University of California, Santa Cruz, Santa Cruz, California, USA

The infrared spectra of $KAlSi_3O_8$-microcline and $NaAlSi_3O_8$-albite are presented to pressures of 30 GPa. Three separate deformational regimes of the aluminosilicate framework are observed on compression of these feldspars: between 0 and about 11 GPa, the infrared spectra document that compression occurs via a combination of Si-O-Si(Al) angle bending coupled with some tetrahedral compression. Compression of tetrahedra appears to be greater in microcline than in albite. At 11–12 GPa, the appearance of a new tetrahedral stretching band near 980 cm^{-1} indicates that some depolymerization of SiO_4 tetrahedra commences in these phases: this depolymerization is most plausibly associated with a severe distortion or coordination change of some of the AlO_4 tetrahedra in the structure and appears to be reversible on decompression. Above ~20 GPa, a major loss in intensity in the SiO_4 stretching bands occurs, and a spectral region associated with octahedral coordination of silicon increases in intensity: this intensity shift occurs to a greater extent in albite than in microcline. Previous spectra of $CaAl_2Si_2O_8$-anorthite show reasonably similar behavior to that of albite, although two pressure-induced crystallographic transformations complicate the behavior of anorthite. Spectra on compression indicate that albite converts to a disordered, diaplectic glass-like material at 24 (±4) GPa, while microcline does not convert to a glass-like phase to 30 GPa. This is consistent with the observation that microcline forms diaplectic glass less readily than either of the plagioclase end-members. Moreover, the pressure at which apparent AlO_4 coordination changes commence is coincident with the onset pressure of the "mixed phase" regime present under shock-loading, implying that the volumetric changes which occur along the Hugoniot are produced by the extreme shifts in local structure observed under static compression.

Properties of Earth and Planetary Materials
at High Pressure and Temperature
Geophysical Monograph 101

INTRODUCTION

The processes which generate amorphization of materials under pressure remain enigmatic. Specific aspects of these transitions which are uncertain include (1) the microscopic mechanisms of such transitions; (2) their thermochemistry; (3) the role of chemical composition and local structure in determining whether (or at what pressures) such transitions occur; (4) the spatial scales at which amorphization occurs, and the length-scale(s) of correlated order within amorphizing samples; (5) the degree to which the amorphized material is related to isochemical fusion-formed glasses; and (6) even the appropriate probe with which to constrain the pressure at which these transitions occur [*Mishima et al.*, 1984; *Hemley,* 1987; *Klug et al.*, 1989; *Williams and Jeanloz,* 1989; *Williams et al.*, 1993; *Kingma et al.*, 1993a,b; *Sciortino et al.*, 1995; *Redfern,* 1996]. The majority of compounds which have been documented to undergo pressure-induced amorphization contain tetrahedrally coordinated cations. In many cases, this transition appears to be associated with an instability of the tetrahedra relative to higher coordination [*Williams and Jeanloz,* 1989; *Itie et al.*, 1989; *Williams et al.*, 1990; 1993; *Tse and Klug,* 1991; *Wolf et al.*, 1992; *Serghiou and Hammack,* 1993; *Binggeli et al.*, 1994]. Within silicates, relatively little attention has focussed on the role of differing charge-balancing cations in modulating the pressure (or volume) at which amorphization occurs. For example, Fe_2SiO_4-fayalite has been demonstrated to amorphize at about 40 GPa when compressed in both an alcohol medium and with no pressure medium [*Williams et al.*, 1990; *Richard and Richet,* 1990], while isostructural Mg_2SiO_4 compressed with no pressure medium does not amorphize until pressures near 70 GPa [*Guyot and Reynard,* 1992]: the underlying reasons for this difference remain obscure.

Here, I present infrared spectra under pressure of $KAlSi_3O_8$-microcline and $NaAlSi_3O_8$-albite to pressures of 30 GPa. These data not only provide information on the pressure-induced amorphization of these compounds, but also constrain the structural changes which take place both under compression and through the amorphization transition of feldspar compounds. Previous published static measurements on microcline and albite have been limited to a pressure of 5 GPa [*Angel et al.*, 1988; *Couty and Velde,* 1986; *Downs et al.*, 1994], although an abstract indicates that data exist on microcline to 7 GPa [*Allan and Angel,* 1996]. The present data to 30 GPa thus dramatically expand the pressure range at which the structures of albite and microcline have been probed. In order to constrain the role of different cations in producing the amorphization transition, I examine the similarities between the albite and microcline data and previous measurements on $CaAl_2Si_2O_8$-anorthite [*Williams and Jeanloz,* 1989]. The results on albite and microcline demonstrate that profound structural changes occur at high pressures in these feldspar compounds; in the case of albite, these changes result in irreversible pressure-induced amorphization at about 24 GPa. In microcline, however, it appears that irreversible amorphization does not occur at pressures to 30 GPa. This result is in accord with the difficulty in shock-synthesizing diaplectic $KAlSi_3O_8$ glass relative to either $NaAlSi_3O_8$, $CaAl_2Si_2O_8$, or intermediate plagioclase compositions. The results on albite confirm that the instability of tetrahedral coordination of silicon and aluminum is of primary importance in producing amorphization of feldspars. These data thus yield insight into the interpretation of the Hugoniots of different feldspars and provide a crystal-chemical explanation for the pressure range and mechanism of diaplectic feldspar glass formation.

EXPERIMENTAL

The infrared measurements on albite and microcline were conducted using techniques described elsewhere [*Williams and Knittle,* 1996]. Briefly, a Bruker IFS-66v FTIR equipped with an attached infrared microscope, a KBr beamsplitter, and a liquid-nitrogen cooled MCT detector was used to collect all spectra of albite and microcline. Pressures were generated using modified Merrill-Bassett diamond anvil cells and were determined using the ruby fluorescence technique [*Mao et al.*, 1978]. Three to five separate ruby grains were monitored in all experiments to characterize pressure gradients. Type II diamonds with 500-mm-diameter culets were utilized; characteristic sample diameters were between 120 and 160 mm. Samples for the infrared measurements were powdered with an average grain size of about 2 mm and mixed at about the 10 wt% level with KBr, which served as both a pressure medium and infrared window. The frequencies of overlapping vibrational bands were constrained by peak-fitting of the spectra using the Levenberg-Marquardt algorithm.

The albite used was a natural intermediate albite from Bancroft, Ontario, while the microcline was from Keystone, South Dakota. Stoichiometries were determined using an ARL-SERL electron microprobe using a 2-mm spot size and ten analyses of different spots in the sample to approximately determine chemical homogeneity. The compositions corresponded to 91.4 (±0.2)% albite, 6.1 (±0.4)% anorthite, and 2.4 (±0.6)% orthoclase, and 96.1 (±0.1)% orthoclase, 3.4 (±0.5)% albite, and 0.5 (±0.4)% anorthite components for the albite and microcline, respectively. Ambient pressure polycrystalline X ray diffraction yielded values of $t_1o - \langle t_1m \rangle$ of 0.30 (±

0.05) for the albite and 0.98 (±0.02) for the microcline using the methods (and estimated errors) outlined in *Kroll* [1983] and *Kroll and Ribbe* [1983]. The present infrared

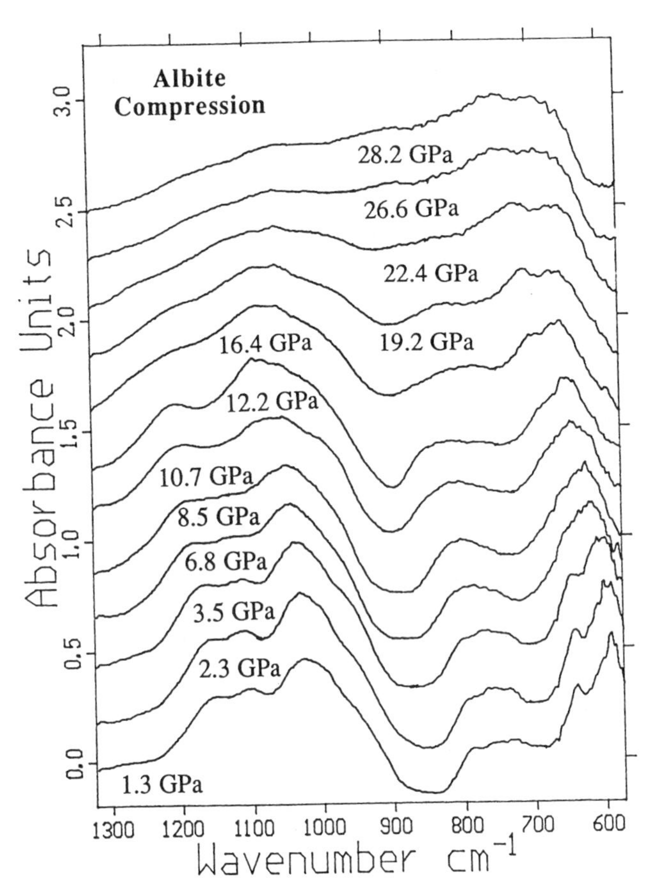

Figure 1. Representative infrared spectra of albite in situ on compression at 300 K; spectra are offset from one another for clarity.

spectra are in excellent accord with previous spectra of feldspars with these levels of disorder [*Salje et al.,* 1989; *Harris et al.,* 1989]. Because of the short characteristic length-scale of infrared spectroscopy, the presence or absence of microstructures in the starting materials should have little effect on the spectra, as long as their abundance is volumetrically small [e.g., *Salje,* 1994]. The anorthite infrared spectra were collected using techniques and sample material described elsewhere [*Williams and Jeanloz,* 1989].

One of the primary interests in compressing feldspars to pressures above 10 GPa lies in constraining the conditions and processes which produce amorphization of these materials; such amorphization processes may depend on the amount of shear stress present in the sample [e.g., *Kingma et al.,* 1993a,b]. However, it is important to recognize that there is no pressure medium which is truly hydrostatic at pressures above ~15 GPa. As such, experiments above this pressure will always contain some non-hydrostatic stress component. Yet, while non-hydrostatic stress conditions may alter the precise conditions (or range of conditions) at which metastable phases amorphize, they do not appear to alter the underlying structural changes which occur prior to and during amorphization [*Kingma et al.,* 1993b]: in short, the structural changes which this study is designed to explore. As an incidental aside, the pressure range over which amorphization of quartz and coesite occurs within a KBr medium is in good accord with the range observed in rare gas media with smaller pressure gradients [*Hemley,* 1987; *Williams et al.,* 1993; *Kingma et al.,* 1993a], thus indicating that non-hydrostatic stresses within KBr do not notably alter the conditions of amorphization of the silica polymorphs.

RESULTS

Major Structural Changes

The spectra of albite under compression are shown in Figure 1, with mode shifts shown in Figure 2; microcline spectra are shown in Figure 3, with mode shifts in Figure 4. The low pressure spectra of each material are in good agreement with previous measurements on disordered albite [e.g., *Salje et al.,* 1989] and microcline [*Iiishi et al.,* 1971a; *Harris et al.,* 1989]. Qualitatively, it is apparent that a profound shift in spectral amplitude takes place in

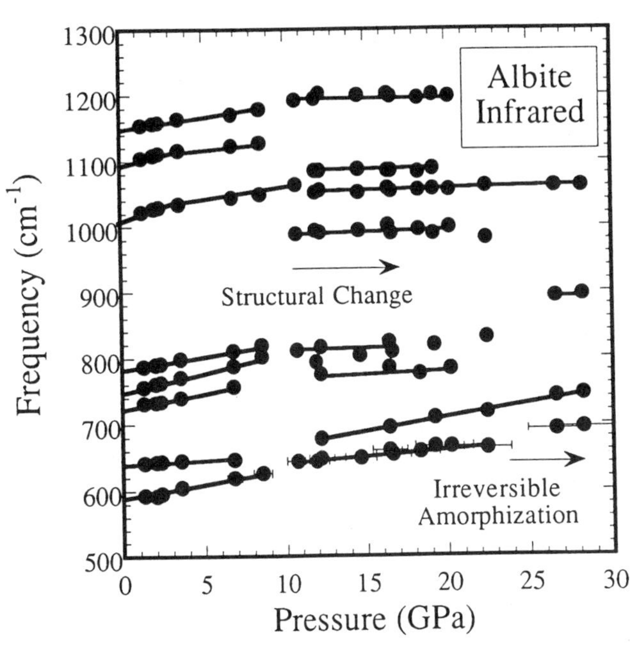

Figure 2. Mode shifts of albite at 300 K on compression. Error bars in pressure derived from measurements of pressure gradients are shown on the lowest frequency mode.

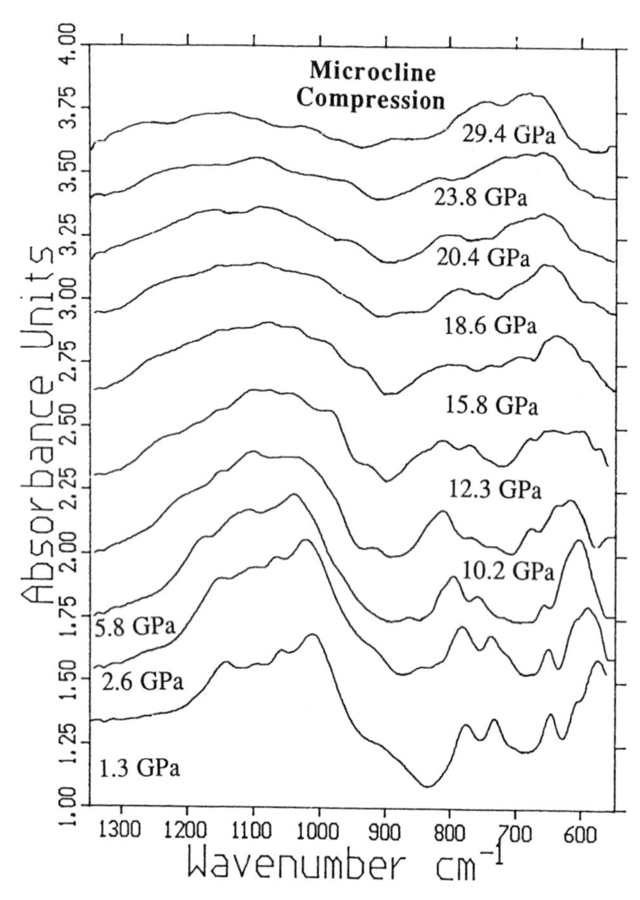

Figure 3. Representative infrared spectra of microcline in situ on compression at 300 K; spectra are offset from one another for clarity.

each material above about 16 GPa, with the peaks associated with the stretching vibrations of the Si(Al)-O tetrahedra between ~1000 and 1200 cm^{-1} [*Iiishi et al.,* 1971a] becoming progressively broader and lower in amplitude relative to the spectral region between 600 and 900 cm^{-1}. Within microcline, this effect is less severe than in albite: discrete $Si(Al)O_4$ tetrahedral stretching vibrations remain clearly present to the highest pressures of these measurements (29.4 GPa), while in albite at similar pressures the tetrahedral stretching peaks are comparatively low amplitude features superposed on an extremely broad absorption continuum, which steadily increases in amplitude to low frequencies. For comparison, *Williams and Jeanloz* [1989] demonstrated the profound effect of pressure in decreasing the amplitude of the tetrahedral stretching bands of anorthite, with an associated increase in spectral amplitude between 700 and 900 cm^{-1}; a selection of spectra of anorthite under pressure is presented in Figure 5, with mode shifts in Figure 6. These spectra more closely resemble those of albite than microcline at pressures in excess of 23 GPa: relatively small features lying over a broad absorption extending and increasing progressively in amplitude from near 1200 cm^{-1} to ~650 cm^{-1}.

Notably, such decreases in intensity of infrared tetrahedral stretching vibrations have been observed previously on compression of Fe_2SiO_4-fayalite, SiO_2-quartz and coesite, $CaAl_2Si_2O_8$-anorthite, GeO_2-quartz and a range of silicate glasses and have been attributed to a progressive shift in coordination from four-fold towards six-fold coordination of the tetrahedrally coordinated cations in each of these materials [*Williams et al.,* 1990; *Williams et al.,* 1993; *Williams and Jeanloz,* 1988; 1989; *Madon et al.,* 1991]. The interpretation of this spectral shift in intensity as being associated with increased cation coordination has been extensively discussed, and it has been confirmed by X ray scattering techniques on silica glass and EXAFS measurements on both glassy and quartz-structured polymorphs of germania [*Meade et al.,* 1992; *Itie et al.,* 1989].

Nevertheless, it appears from the differences between the highest pressure spectra of microcline and those of albite and anorthite that differing degrees of nearest neighbor ordering are present within these three phases at high pressures. Microcline retains a significant quantity of discrete Si tetrahedra at high pressures, while comparatively few discrete Si environments are present within albite or anorthite. Further insight into the response of

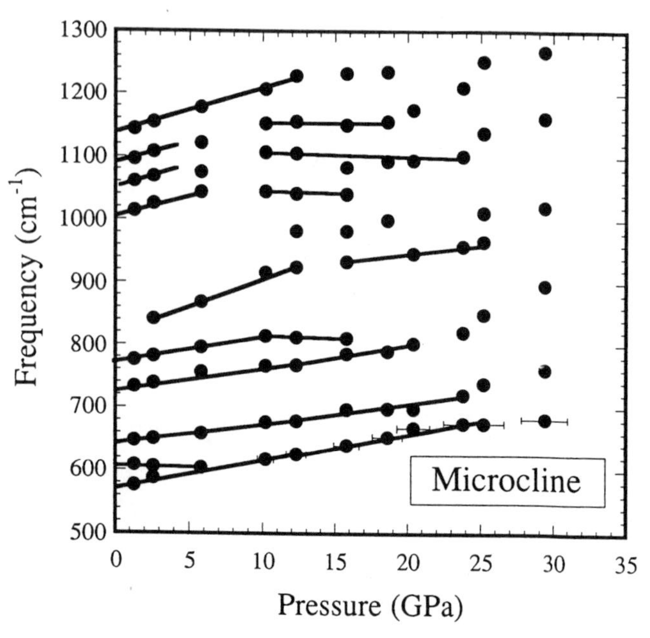

Figure 4. Mode shifts of microcline at 300 K on compression. Error bars in pressure are shown on the lowest frequency mode.

these materials to compression can be gained from spectral changes occurring at lower pressure conditions. In particular, as is apparent from both the spectra and the observed mode frequencies at pressure (Figures 1, 3, and 5), in each feldspar a broadening of the Si(Al)-O stretching vibrations occurs to lower frequency as pressure is increased above about 10–14 GPa. Within microcline, this broadening manifests itself as a discrete peak which

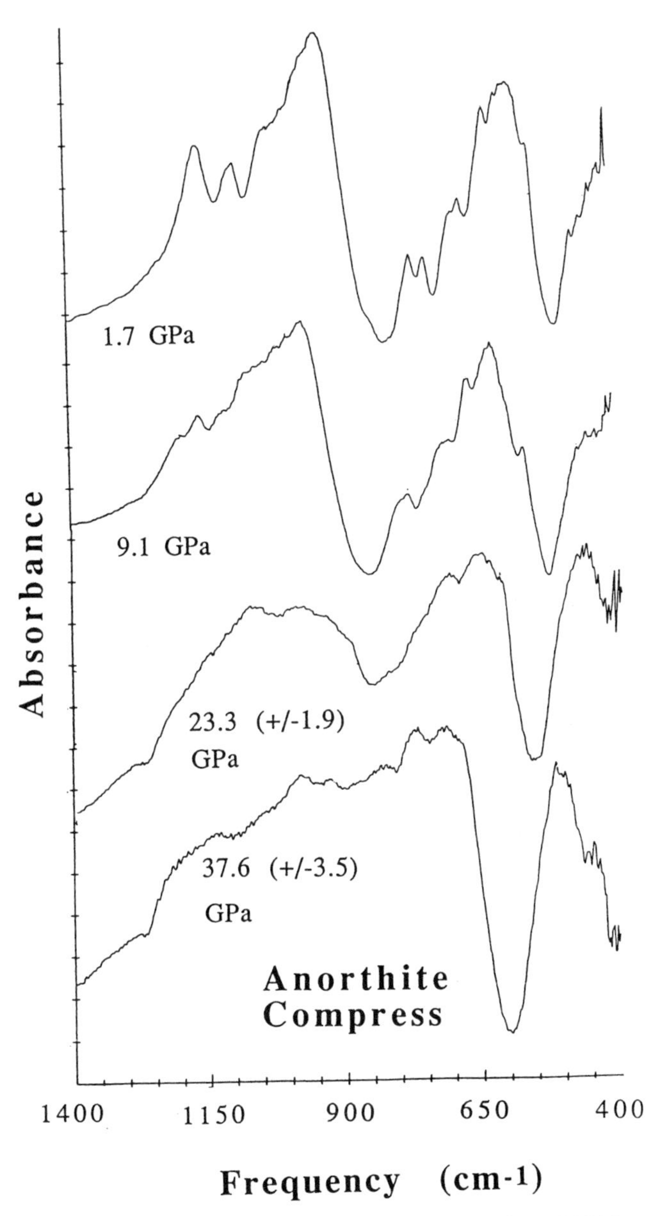

Figure 5. Representative infrared spectra of anorthite at 300 K on compression; the spectrum at 1.7 GPa is given in *Williams and Jeanloz* [1989], as are spectra of anorthite at 33.4 GPa and following decompression.

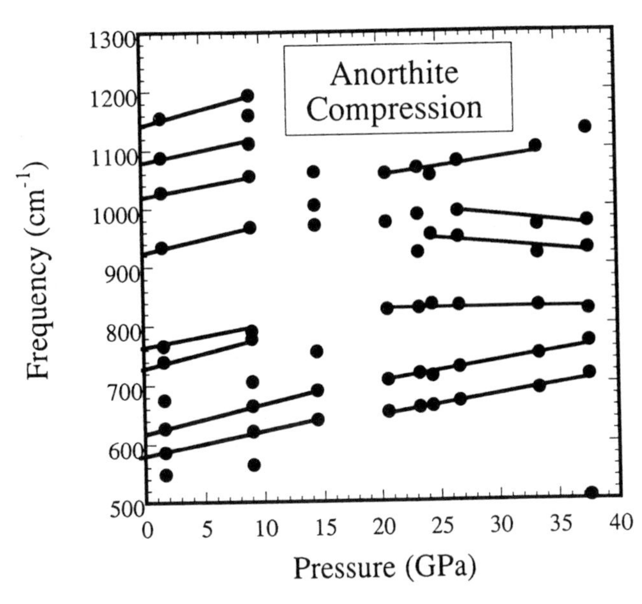

Figure 6. Mode shifts of anorthite at 300 K on compression. Some original spectra are given in *Williams and Jeanloz* [1989]; others in Figure 5.

appears near 980 cm^{-1} at ~12 GPa (Figure 3), while in albite a shoulder appears near this position at about 11 GPa (Figure 1). In anorthite, the onset in pressure of this broadening is only approximately bracketed between 9 and 15 GPa.

These new pressure-induced bands, with frequencies near 980 cm^{-1}, are certainly associated with stretching vibrations of tetrahedra. However, this frequency is lower than that associated with the stretching vibrations of fully polymerized silicate tetrahedra; instead, this band location is more similar to those observed in phases which have only partially polymerized silica tetrahedra, such as pyroxenes (particularly when the effect of pressure in producing a generally positive shift of such vibrations is taken into account) [e.g. *Tarte*, 1965; *Lazarev,* 1972]. This decrease in frequency of the tetrahedral stretching vibrations with smaller degrees of polymerization is simply generated by the reduction in average strength of the bonds surrounding a given tetrahedra, as is produced by reducing the number of adjoining tetrahedra.

Alternatively, vibrations which primarily involve stretching motions of AlO_4 tetrahedra could occur in this frequency range. However, within both microcline and albite, such vibrations are coupled with those of the adjoining silicate tetrahedra and fall at higher frequencies [*Iiishi et al.,* 1971a, b; *Zhang et al.,* 1996]. Because of the nature of the normal mode displacements in K-feldspar [*Iiishi et al.,* 1971b], decoupling of the AlO_4 stretching vibrations from the SiO_4 groups would probably require a complete breakdown of the four-membered rings of tetrahedra within the structure. Such a major shift in structure is unlikely below 15 GPa, as this would likely produce spectral changes considerably larger than are observed in this pressure range. Therefore, the appearance of the ~980 cm^{-1} feature is attributed to partial depolymerization of some silicate tetrahedra within the structure.

The appearance of this new band in the microcline is associated with a dramatic decrease in amplitude of the highest amplitude Si(Al)-O stretching vibration, with a zero pressure frequency near 1010 cm^{-1}, and of the smaller band with a zero pressure frequency near 1045 cm^{-1}. It is these two bands which *Iiishi et al.* [1971a,b] associate with coupled vibrations of silicon and aluminum tetrahedra based on isochemical substitution of iron for aluminum in microcline; bands above ~1100 cm^{-1} are associated with nearly pure Si-O stretching vibrations. Therefore, it appears that the most straightforward interpretation of these observations is that the aluminum tetrahedra within microcline begin to distort to a more highly coordinated state between 10 and 12 GPa: this eliminates the Si(Al)-O coupled tetrahedral vibrations, and gives rise to a vibration of depolymerized Si-O tetrahedra near 980 cm^{-1}.

The interpretation that a shift in aluminum environment occurs between 10 and 12 GPa is generally consistent with the behavior of the lower frequency bands of microcline, as well. The band at 770 cm^{-1} is an Si-O-Si bending vibration (such displacements are described as "Si-Si [or Si-Al(Si)] stretching" by *Iiishi et al.* [1971a]: however, the energy of this vibration is primarily sensitive to changes in the Si-O-Si(Al) angle), and is almost unchanged in frequency by chemical substitution of the aluminum, while that near 730 cm^{-1} is assigned to an Si-O-Al bending vibration. The former band persists at nearly constant amplitude to 10 GPa, while the latter band decreases in amplitude: near 12 GPa, the spectrum becomes considerably more complex. Similarly, the strong band near 584 cm^{-1}, which is probably associated with an O-Si(Al)-O bending vibration, appears to split and markedly decrease in intensity at the pressure at which the 980 cm^{-1} band appears in the spectrum. Additionally, the pressure shifts of most of the bands appears to shift markedly near 10 GPa. In particular, the shifts of the Si-O-Si(Al) bending vibrations which lie between 700 and 800 cm^{-1} decrease markedly near 10 GPa. The near constancy of the frequency of these vibrations at pressures above 10 GPa implies that the rate of change of the Si-O-Si(Al) angles with pressure has dramatically decreased. As such angle changes are the primary low-pressure mechanism of compaction of the feldspar framework [*Angel et al.,* 1988; *Downs et al.,* 1994], it appears that a change in compressional mechanism of the crystal is likely to occur near this pressure. Therefore, both the appearance and disappearance of bands and changes in the pressure shifts of the Si-O-Si(Al) vibrations each are consistent with a distortional change in aluminum coordination commencing with an onset pressure near 10 GPa.

There are several fairly minor changes which occur below 6 GPa in the spectrum of microcline. A weak feature appears near 840 cm^{-1} between 1.3 and 2.6 GPa that may be correlated with a peak present near 940 cm^{-1} at pressures in excess of 20 GPa: the origin of this feature is unclear, and it has no analogue in the spectra of either albite or anorthite. Also, the Si(Al)-O stretching vibration with a zero pressure frequency of 1052 cm^{-1} essentially disappears by 5.8 GPa. These changes may be associated with a minor shift in the compressional mechanism of this phase which has been proposed to occur between 3 and 4 GPa: this shift has only been cursorily described, but does not appear to involve a change in space group [*Angel,* 1994; *Allan and Angel,* 1996].

The evolution of the albite spectra with pressure is broadly similar to that of microcline. In this case, the onset of a moderate shoulder at 980 cm^{-1} is accompanied by the disappearance of the SiO_4 stretching band with a zero pressure frequency of 1095 cm^{-1}; a weak splitting of the band with an initial frequency of 1015 cm^{-1} may also occur near this pressure. As with microcline, there is a decrease in the pressure shifts of the Si-O-Si(Al) bending vibrations near 12 GPa. It is near this pressure that the bands associated with O-Si(Al)-O bending vibrations (between 580 and 650 cm^{-1}) begin to increase in frequency more rapidly with compression: such an effect may also be present in microcline. This change in pressure shifts is consistent with a shift in compressional mechanism from predominantly Si-O-Si(Al) angle changes to tetrahedral distortions occurring near this pressure. A comparable change in pressure shifts is seen in silica glass under compression [*Williams et al.,* 1993] and is associated with tetrahedral distortion becoming a major mechanism of compaction prior to the onset of a gradual shift in cation coordination.

In apparent contrast to the behavior of albite and microcline, anorthite undergoes at least two crystalline transitions under compression: the first of these occurs near ~2.7 GPa and is associated with an increase in symmetry across the $P\bar{1} \rightarrow I\bar{1}$ transition [*Angel,* 1988; *Angel et al.,* 1988; *Hackwell and Angel,* 1995]. The manifestations of this transition are distinct, yet fairly subtle within the infrared spectra: a general broadening of the spectra and a shift in the shape of the highest frequency

TABLE 1. Mode Shifts of Albite

Pressures below 9 GPa			
Frequency (cm^{-1})	Mode Shift (cm^{-1}/GPa)	Grüneisen Parameter	Assignment*
1150	3.6 (±0.3)	0.18	Si-O Stretching
1095	4.1 (±0.3)	0.21	Si-O Stretching
1017	4.1 (±0.3)	0.23	Si(Al)-O Stretching
785	4.5 (±0.3)	0.33	Si-O-Si Bending
755	6.3 (±0.4)	0.48	Si-O-Si Bending
726	4.5 (±0.4)	0.35	Si-O-Al(Si) Bending
642	1.2 (±0.2)	0.11	O-Si(Al)-O Bending
587	5.1 (±0.3)	0.50	O-Si(Al)-O Bending

*From *Iiishi et al.* [1971a,b]

tetrahedral stretching bands (Figure 5). Indeed, the spectrum at 9.1 GPa agrees well in its overall morphology with the spectra of $I\bar{1}$-anorthite taken at high temperatures by *Redfern and Salje* [1992], as does the Raman spectrum of this phase [*Daniel et al.,* 1995]. At pressures near 10 GPa, *Daniel et al.* [1995] have shown that anorthite appears to convert to a different crystal structure based on Raman spectroscopy. The present spectra are consistent with such a change, as the number of vibrational bands decreases near this pressure: in particular, the three highest frequency tetrahedral stretching vibrations disappear (Figures 5 and 6). Such a decrease in the number of bands is consistent with an increase in symmetry of the crystal structure of anorthite. *Daniel et al.* [1995] have indicated that the new phase may have significantly contracted T-O bond distances. That the highest frequency infrared stretching bands disappear through this transition is not consistent with such a T-O bond contraction; rather, it appears that this transition is more plausibly associated with a shift in the bond angles of the four membered tetrahedral rings of the structure. Additionally, relatively profound changes in peak intensities take place within anorthite: the strongest tetrahedral stretching feature near 920 cm^{-1}, which is present at such low frequencies because of the high Al content of anorthite relative to albite and microcline, decreases markedly in intensity above 10 GPa. This decrease in intensity could be indicative of a decrease in tetrahedral aluminum content in the compressed anorthite. Moreover, the disappearance of the highest frequency bands might be associated with depolymerization of the silica tetrahedra. However, no discrete new bands appear in the anorthite spectra between 10 and 22 GPa: rather, the 750-900 cm^{-1} region simply increases in absorbance. It is likely that this increase in absorbance is associated with a shift in coordination of a portion of the cations within the structure, but the separate behavior of aluminum and silica tetrahedra cannot easily be distinguished in this material.

At pressures of 30 GPa and above, anorthite does continue to have some absorbance in the region between 950 and 1150 cm^{-1} which is likely to be associated with tetrahedral coordination of silicon (taking into account the positive pressure shifts of most of the vibrational bands of anorthite). As with albite and microcline, the highest amplitude features in the spectra lie between 650 and 800 cm^{-1}: this region is known to absorb strongly in disordered silicates (either fusion-formed glasses or pressure-amorphized materials) at high pressures, and this spectral region is thus probably associated with octahedral silicon and aluminum [e.g. *Williams and Jeanloz,* 1988; *Williams et al.,* 1990; 1993].

Mode Shifts and Mode Grüneisen Parameters

Mode shifts of albite and microcline are given in Tables 1 and 2; these shifts are calculated to 9 GPa in albite and 11 GPa in microcline, as at higher pressures structural changes both make some modes difficult or impossible to track, and may produce discontinuities in mode frequencies. Also, at high pressures some shifts become strongly nonlinear. In most cases, a simple linear fit is used to calculate the mode shifts, as the amount of curvature in the ~10 GPa range of these data is generally fairly small (an exception is the lowest frequency tetrahedral stretching band of microcline). The present mode shifts are in general accord with the more limited data set on the shifts of the infrared modes of albite to 4.5 GPa of *Couty and Velde* [1986], which showed frequency shifts of between 3 and 5 cm^{-1}/GPa for a similar set of infrared bands.

Mode Grüneisen parameters were calculated from $K_0(d\omega/dP)/\omega_0$, where K_0 is the bulk modulus of the feldspar and ω_0 is its initial frequency. Ultrasonically

TABLE 2. Mode Shifts of Microcline

(Pressures below 11 GPa)			
Frequency (cm^{-1})	Mode Shift (cm^{-1}/GPa)	Grüneisen Parameter	Assignment*
1136	7.4 (±0.5)	0.36	Si-O Stretching
1092	6.2 (±0.4)	0.31	Si-O Stretching
1052	5.3 (±0.5)	0.28	Si(Al)-O Stretching
1010	7.2 (±0.8)	0.39	Si(Al)-O Stretching
768	4.3 (±0.3)	0.31	Si-O-Si Bending
728	4.3 (±0.3)	0.32	Si-O-Al(Si) Bending
645	3.0 (±0.3)	0.26	O-Si(Al)-O Bending
608	~-0.4	-0.04	O-Si(Al)-O Bending
579	5.4 (± 0.5)	0.51	O-Si(Al)-O Bending

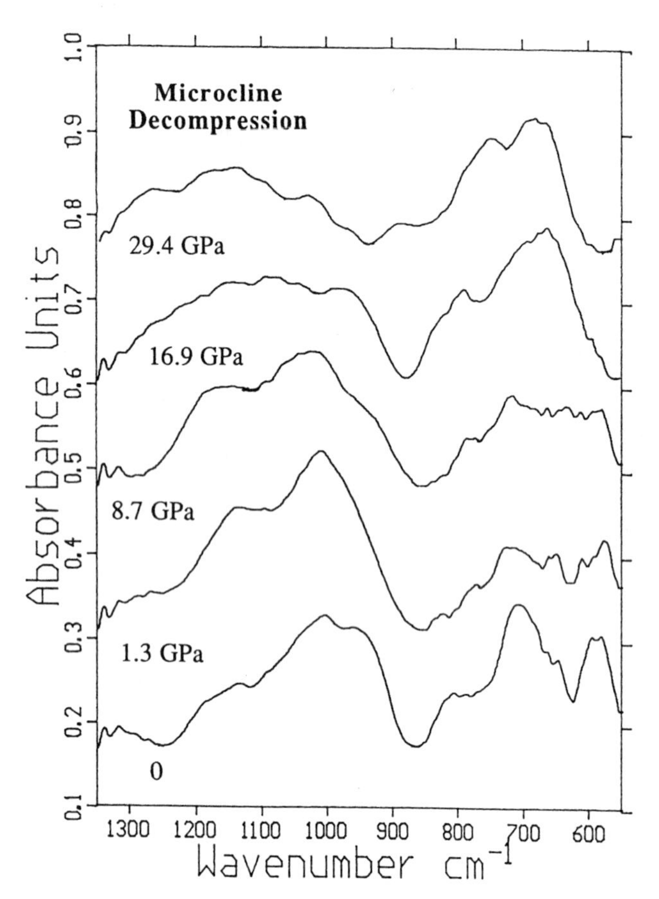

Figure 7. Representative spectra of microcline on decompression from 29.2 GPa.

determined bulk moduli of 57 and 55 GPa are used for albite and microcline, respectively [*Ryzhova,* 1964; *Ryzhova and Alexandrov,* 1965; *Bass,* 1995]. These values are compatible with those determined from Murnaghan fits to static compression data (58 and 63 GPa: *Angel* [1994]), but not with earlier linear pressure-volume fits [*Angel et al.,* 1988]. Estimated errors on the mode Grüneisen parameters are controlled primarily by the errors in the mode shifts and thus are generally about 10%. The calculated Grüneisen parameters (Tables 1 and 2) indicate that, in spite of the anisotropic compressibility of these feldspars [*Angel et al.,* 1988; *Downs et al.,* 1994; *Bass,* 1995], the vibrational modes behave generally similarly in both these phases under compression. Only the O-Si(Al)-O bending modes with zero pressure frequencies of 608 cm^{-1} in microcline and 645 cm^{-1} in albite have Grüneisen parameters which fall outside of the 0.15–0.55 range. It is notable that the mode shifts (and Grüneisen parameters) of the vibrations assigned to Si-O-Si bending and Si-O-Al(Si) bending vibrations are quite similar in each phase. Yet, structural determinations at pressures to 4 GPa indicate that Si-O-Al angles in low albite contract significantly more rapidly than Si-O-Si angles [*Downs et al.,* 1994]. However, the calculated normal modes of these differing vibrations incorporate both significant aluminum and silicon displacements [*Iiishi et al.,* 1971a,b], implying that these modes are of significantly mixed character and thus are not simply correlated with Si-O-Si or Si-O-Al bond angles within the structure.

While many of the mode Grüneisen parameters are similar between albite and microcline, those of the tetrahedral stretching bands are significantly larger in microcline, indicating that the tetrahedra in microcline may undergo more compaction than in albite. Also, the Grüneisen parameters of microcline's tetrahedral stretching bands are comparable to those of the Si-O-Si(Al) bending bands. Thus, it is possible that the presence of the relatively large (and compressible) potassium ion within the feldspar structure allows a greater degree of compaction of the tetrahedra in this phase. The comparatively large values of the Grüneisen parameters of the tetrahedral stretching vibrations in both albite and microcline relative to (for example) quartz, in which mode Grüneisen parameters of the tetrahedral stretches lie between -0.10 and 0.05 [*Hemley,* 1987; *Williams et al.,* 1993], also indicate that more compression of the tetrahedra takes place in feldspars relative to quartz. This difference in behavior between feldspars and quartz, which has been viewed as anomalous [*Zhang et al.,* 1996], simply indicates that the mechanism of volume change in quartz is not particularly analogous to that of the feldspars. Coesite appears to be a more accurate analogue, as it has a four-membered tetrahedral ring framework similar to that of the feldspar structure [e.g., *Jackson and Gibbs,* 1988], and also has tetrahedral stretching and bending mode Grüneisen parameters similar to those of the feldspars [*Williams et al.,* 1993]. Notably, the mode Grüneisen parameters of Tables 1 and 2 are between 25% and a factor of eight smaller than the Grüneisen parameters determined from the temperature shifts of these vibrational modes [*Zhang et al.,* 1996], with the discrepancy being largest for the tetrahedral stretching bands. This difference illustrates the extremely large role that volume-independent anharmonic effects play in the temperature shifts of the infrared bands of the feldspars.

Behavior on Decompression

On decompression from 29 GPa, the tetrahedral stretching vibrations of microcline generally increase in amplitude relative to those observed at high pressure (Figure 7); however, the spectrum of the quenched material is notably different from the starting material (Figures 3 and 7). Indeed, a feature near 950 cm^{-1} is prominent in the quenched material, and the spectrum below 850 cm^{-1} bears little resemblance to that of the starting material. The quenched spectrum retains relatively discrete peaks relative to those observed within fused glasses of similar chemistry [e.g., *Stöffler,* 1974], implying that it is not disordered at the level of either a melt-quenched or diaplectic glass. The quenched material is clearly no longer single-phase, crystalline microcline; nor does it particularly resemble the material present at high pressures, and thus it is probably produced on decompression. Notably, microcline is unique among the feldspars in having had unidentified crystalline phases observed in samples following shock compression above about 19 GPa [*Robertson,* 1975]. This unidentified phase was attributed to a disordered (but crystalline) phase inversion product of the compressed state, and it is thus possible that the material generated on quench following static compression is similar to that generated under shock.

Within albite, the onset of irreversible amorphization is bracketed by the present study as occurring between 20 and 28 GPa from quenched samples (Figure 8). The sample decompressed from 20 GPa has a generally similar, although broader and less well-resolved spectrum to the starting material. This similarity implies that its local structure (at the unit-cell length scale) closely resembles that of crystalline albite. The sample decompressed from 28 GPa has tetrahedral stretching vibrations which resemble those of albite glasses formed by both fusion and shock [*Velde et al.,* 1987]: the intermediate Si-O-Si(Al) bending vibrations lie at somewhat higher frequency (770 cm^{-1}) then those of fusion quenched glasses, implying that a contraction of the Si-O-Si(Al) angles has occurred. Also, the pressure-quenched material has distinct features present in the spectrum between 550 and 700 cm^{-1}: diaplectic feldspar glasses have discrete peaks in this region, while fusion quenched glasses do not [e.g., *Ostertag,* 1983; *Velde et al.,* 1987].

We can thus constrain the onset of irreversible amorphization to occur at 24 (±4) GPa in intermediate albite compressed in a KBr matrix. It is notable that there is a discontinuity in mode shifts near precisely this pressure (Figure 2), and it is probable that this shift indicates the pressure at which the lattice becomes irretrievably distorted. A representative portion of our spectra taken on decompression of albite is shown in Figure 9, for a sample decompressed from 20 GPa. It is of interest that the primary change in the spectrum takes place below 14 GPa on decompression: the feature which we have attributed to depolymerized SiO_4 vibrations near 980 cm^{-1} is discretely resolved on decompression, and disappears between 14 and 8 GPa. Accordingly, it appears that the structural changes which produce the broadening to lower frequency of the tetrahedral stretching vibrations at 10–12 GPa are reversible in albite on decompression; structural changes occurring above about 24 GPa result in irreversible amorphization.

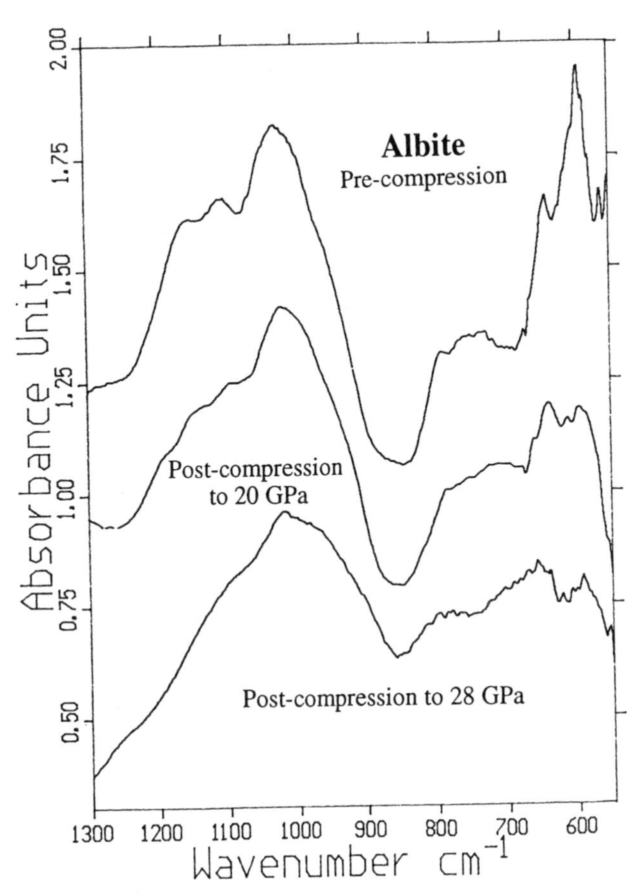

Figure 8. Infrared spectra of albite at 1.3 GPa prior to compression at 300 K, compared to spectra of albite at 0.4 GPa following compression to 20.2 (±1.3) GPa, and at 0.2 GPa following compression to 28.2 (±1.9) GPa.

Finally, the spatial distribution of amorphized material within these samples is a topic of considerable interest [*Redfern,* 1996]. It has been clearly demonstrated that amorphization of crystals can proceed in a spatially heterogeneous manner [*Williams and Jeanloz,* 1989; *Kingma et al.,* 1993b], and the length-scale(s) over which correlated order is observed to persist may depend on the nature of the probe [*Redfern,* 1996]. Therefore, we wish to assess whether the higher pressure infrared results represent spectra of an entirely amorphous sample or a mixture of amorphous and crystalline material. For albite,

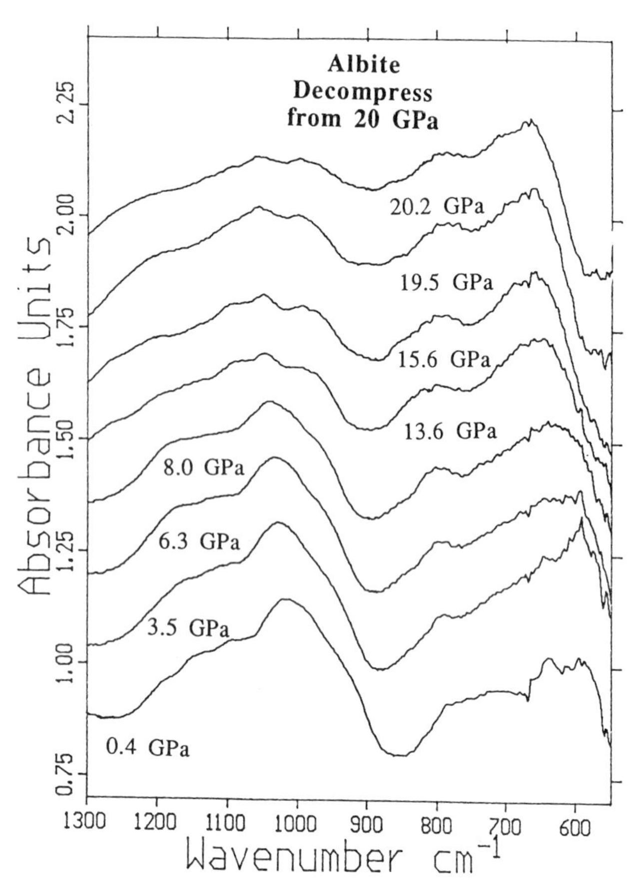

Figure 9. Infrared spectra of albite on decompression from 20 GPa at 300 K.

the highest pressure spectra bear essentially no resemblance to the spectrum of crystalline albite, and no signature of crystalline albite remains in our spectra of samples quenched from these conditions (Figure 8). Therefore, it appears that the highest pressure spectra represent those of fully amorphized albite: our spectra in the 12–24 GPa pressure range could, however, involve a mixture of amorphous and crystalline albite. Such a mixture of amorphous and crystalline phases is observed during the amorphization of quartz and GeO_2 [*Itie et al.*, 1989; *Kingma et al.*, 1993a]. Thus, it is possible that the apparent breakdown of tetrahedral aluminum between 12 and 24 GPa may be spatially localized in albite. Nevertheless, the pressure interval over which amorphization proceeds in albite may be reasonably narrow (in anorthite it is less than about 6 GPa [*Williams and Jeanloz*, 1989; *Redfern*, 1996]). For microcline, the situation is more complex: the structural state accessed in our highest pressure spectra is clearly more disordered than the starting material, but could still retain (probably highly distorted) microcline-like domains.

Implications for Diaplectic Glass Formation and Feldspar Hugoniots

A number of studies have examined the reponse to dynamic compression of various feldspars [*Ahrens et al.*, 1969; *Ahrens and Liu*, 1974; *Simakov et al.*, 1974; *Jeanloz and Ahrens*, 1980], as well as the physical properties of diaplectic feldspar glasses quenched from high dynamic pressures [e.g., *Stöffler*, 1984; *Ostertag*, 1983; *Syono et al.*, 1977; *Robertson*, 1975]. Previous results have shown that the pressure at which diaplectic glass is generated in $KAlSi_3O_8$ is strongly dependent on the nature of the starting material: while *Kleeman* [1971] observed large amounts of diaplectic glass formation in porous samples shocked above 24 GPa, *Robertson* [1975] shocked single crystals of microcline to peak pressures of 41.7 GPa and found only minor evidence of glass formation. *Ostertag* [1983] observed weak birefringence in single crystal microcline which had been shocked to 45 GPa, and his infrared spectra indicate that many of the peaks of the crystalline phase persist in samples shocked to at least 34 GPa. Thus, it is likely that the glasses generated by *Kleeman* [1971] from porous material may have been fused, while non-porous single crystals probably do not undergo such melting at pressures to at least 42 GPa. Furthermore, *Stöffler* [1974] noted that shocked potassium feldspars are likely to have a larger degree of long-range order than plagioclase feldspars shocked to comparable conditions, as even apparently amorphized $KAlSi_3O_8$ samples have unusually high indices of refraction and a considerably more complex infrared spectrum in the 600-800 cm^{-1} region than plagioclase feldspars.

For comparison, single crystals of albite-rich plagioclases (typically oligoclase of composition near Ab78) appear to quench to diaplectic glasses at shock pressures above ~28 GPa [*Ostertag*, 1983; *Stöffler*, 1984]. This pressure is essentially coincident with the estimate of the amorphization pressure of albite derived from the present infrared spectra. More anorthite-rich samples may convert to diaplectic glass at slightly lower pressures, near 25 GPa [*Ostertag*, 1983; *Williams and Jeanloz*, 1989]. Thus, microcline forms diaplectic glass less readily than either albite or microcline: the results of Figures 3 and 7 indicate that it is also the most difficult of the common feldspars to amorphize under static conditions. Moreover, its high pressure infrared spectra are consistent with microcline retaining relatively discrete tetrahedral and (probably) octahedral environments at high pressures, while both albite and anorthite appear to contain a broad continuum of structural environments at pressures above about 25 GPa. Therefore, the presence of the large potassium ion within the feldspar lattice apparently imposes a degree of order on the deformation of the aluminosilicate framework which inhibits the formation of diaplectic glass. The precise nature of the structural effect of potassium on amorphization is difficult to constrain; one possibility is simply that a relatively larger amount of the compaction of the microcline structure is produced by compression of the potassium ion, while less is taken up by deformation and coordination changes of the aluminosilicate network. Alternatively (and speculatively), the pressure-induced collapse of the aluminosilicate framework may produce a considerably more disordered structure around the smaller calcium and sodium ions than around the larger potassium. This qualitative explanation for the difference in amorphization and diaplectic glass-forming behavior of these feldspars has features in common with the suggestion of *Tse et al.* [1994] that discrete "guest molecules" may provide templates on which crystalline structures may reform on decompression. In the case of microcline, it may be that the comparatively large potassium ions provide an imperfect template which yields a different crystalline phase on decompression.

The Hugoniot of end-member albite is not well-constrained; however, reasonably extensive data exist on oligoclase of composition $Al_{75}Ab_{19}Or_{5.5}$ [*Ahrens et al.*, 1969]. The lower pressure onset of the "mixed-phase" regime in this material is not tightly constrained, but was proposed to lie near 14 GPa [*Ahrens et al.*, 1969]. This mixed-phase region, defined as a pressure range of anomalous sample compressibility, terminates near 40 GPa. Fewer data exist for microcline, with the onset of the mixed phase regime in this material being only loosely constrained to lie between 11 and 30 GPa [*Ahrens et al.*, 1969]. In contrast, the Hugoniot of isochemical orthoclase is quite well-characterized, with the onset of the mixed-phase regime occurring at 11.5 (±1.0) GPa [*Ahrens and Liu,* 1974]. Notably, this pressure is in outstanding agreement with the pressure range in which the infrared spectra of microcline change character: the association of this change with the start of a shift in coordination of aluminum under static loading provides a natural explanation for the anomalous compressibility commencing at these conditions under shock loading, as well. The Hugoniot temperature of feldspars at 11 GPa is less than 350 K [*Ahrens et al.*, 1969], implying that extensive diffusional reorganization/recrystallization of samples is unlikely to occur over the limited time-scale of shock loading. Rather, it appears that a structural transition as occurs at 300 K in the infrared spectra of statically compressed feldspars probably produces the anomalous compressibility observed on the Hugoniot at a similar onset pressure.

CONCLUSIONS

The infrared spectra of statically compressed feldspars demonstrate not only that profound structural changes take place on metastable compression of these materials, but also that unexpected differences in the response to compression (and decompression) exist between microcline, albite, and anorthite. Specifically, as with the low pressure polymorphs of SiO_2, shifts in the coordination of tetrahedral cations take place at pressures above 10 GPa in these feldspars. These coordination changes appear to initiate with the AlO_4 tetrahedra, with SiO_4 tetrahedra beginning to convert to five- or six-fold coordination at pressures in excess of 20 GPa. These increases in cation coordination proceed (at comparable pressures) to a greater extent in albite or anorthite than in microcline, with microcline being significantly less disordered at high pressures. Additionally, apparently unlike albite or microcline, anorthite undergoes at least two crystalline phase transitions prior to amorphizing.

The behavior of feldspars during and following shock-loading shows considerable similarities to the present observations of structural changes during and following static compression. The onset of the "mixed-phase" regime, in which feldspars undergo anomalous densification with shock pressure, is essentially identical to the pressure at which we infer coordination changes to commence in the feldspars. Also, that diaplectic microcline glass is far less readily produced than diaplectic plagioclase glasses is probably directly correlated with the greater difficulty of pressure-amorphizing microcline under static conditions.

Acknowledgments. I thank J. P. Itie, E. Knittle, E. Morris and an anonymous reviewer for helpful comments on the manuscript, and D. Sampson for electron microprobe assistance. Anorthite spectra were collected at U.C. Berkeley, for which I thank R. Jeanloz. Work supported by the U.S. National Science Foundation and the W. M. Keck Foundation.

REFERENCES

Ahrens, T. J., C. F. Petersen, and J. T. Rosenberg, Shock compression of feldspars, *J. Geophys. Res., 74,* 2727-2746, 1969.

Ahrens, T. J., and H. P. Liu, A shock-induced phase change in orthoclase, *J. Geophys. Res., 78,* 1274-1278, 1974.

Angel, R.J., High pressure structure of anorthite, *Am. Mineral., 73,* 1114-1119, 1988.

Angel, R. J., Feldspars at high pressure, in *Feldspars and Their Reactions,* I. Parson, Ed., pp. 271-312, Kluwer Academic, Amsterdam, 1994.

Angel, R. J., R. M. Hazen, T. C. McCormick, C. T. Prewitt, and

J. R. Smyth, Comparative compressibility of end-member feldspars, *Phys. Chem. Minerals, 15,* 313-318, 1988.

Allan, D. R., and R. J. Angel, A high-pressure structural study of microcline ($KAlSi_3O_8$) to 7 GPa (abstract), Sixth International Symposium on Experimental Mineralogy, Petrology, and Geochemistry, Bayreuth, April, *Terra Abstracts, 8, Suppl. 2,* p. 2, 1996.

Bass, J. D., Elasticity of minerals, glasses and melts, in A *Handbook of Physical Constants, Vol. 2,* edited by T. J. Ahrens, pp. 45-63, AGU, Washington, D.C., 1995.

Binggeli, N., J. R. Chelikowsky, and R. M. Wentzcovitch, Simulating the amorphization of α-quartz under pressure, *Phys. Rev., B49,* 9336-9340, 1994.

Couty, R. and B. Velde, Pressure-induced band splitting in infrared spectra of sanidine and albite, *Am. Mineral., 71,* 99-104, 1986.

Daniel, I., P. Gillet, and S. Ghose, A new high-pressure phase transition in anorthite ($CaAl_2Si_2O_8$) revealed by Raman spectroscopy, *Am. Mineral., 80,* 645-648, 1995.

Downs, R. T., R. M. Hazen, and L.W. Finger, The high-pressure crystal chemistry of low albite and the origin of the pressure dependency of Al-Si ordering, *Am. Mineral., 79,* 1042-1052, 1994.

Guyot, F. and B. Reynard, Pressure-induced structural modifications and amorphization in olivine compounds, *Chem. Geol., 96,* 411-420, 1992.

Hackwell, T. P. and R. J. Angel, Reversed brackets for the $P\bar{1} \rightarrow I\bar{1}$ transition in anorthite at high pressures and temperatures, *Am. Mineral., 80,* 239-246, 1995.

Harris, M. J., E. K. H. Salje, G. K. Guttler, M. A. Carpenter, Structural states of natural potassium feldspar: An infrared spectroscopic study, *Phys. Chem. Minerals, 16,* 649-658, 1989.

Hemley, R. J., Pressure dependence of Raman spectra of SiO_2 polymorphs: α-quartz, coesite, and stishovite, in *High-Pressure Research in Mineral Physics, Geophysical Monogr. Ser.,* vol. 39, edited by M. H. Manghnani and Y. Syono, pp. 347-360, Terra Scientific Publishing Co., Tokyo/AGU, Washington, D.C., 1987.

Iiishi, K., T. Tomisaka, T. Kato and Y. Umegaki, Isomorphous substitution and infrared and far infrared spectra of the feldspar group, *N. Jb. Miner. Abh., 115,* 98-119, 1971a.

Iiishi, K., T. Tomisaka, T. Kato and Y. Umegaki, The force field of K feldspar, *Zeit. Krist., 134,* 213-229, 1971b.

Itie, J. P., A. Polian, G. Calas, J. Petiau, A. Fontaine, and H. Tolentino, Pressure-induced coordination changes in crystalline and vitreous GeO_2, *Phys. Rev. Lett., 63,* 398-401, 1989.

Jackson, M. D. and G. V. Gibbs, A modeling of the coesite and feldspar framework structure types of silica as a function of pressure using modified electron gas methods, *J. Phys. Chem., 92,* 540-545, 1988.

Jeanloz, R. and T. J. Ahrens, Anorthite: Thermal equation of state to high pressures, *Geophys. J. R. Astr. Soc., 62,* 529-549, 1980.

Kingma, K. J., R. J. Hemley, H. K. Mao, and D. R. Veblen, New high-pressure transformation in α-quartz, *Phys. Rev. Lett., 70,* 3927-3930, 1993a.

Kingma, K. J., C. Meade, R. J. Hemley, H. K. Mao, and D. R. Veblen, Microstructural observations of α-quartz amorphization, *Science, 259,* 666-669, 1993b.

Kleeman, J. D., Formation of diaplectic glass by experimental shock loading of orthoclase, *J. Geophys. Res., 76,* 5499-5503, 1971.

Klug, D. D., Y. P. Handa, J.S. Tse, and E. Whalley, Transformation of ice VIII to amorphous ice by "melting" at low temperature, *J. Chem. Phys., 90,* 2390-2392, 1989.

Kroll, H., Lattice parameters and determinative methods for plagioclase and ternary feldspars, *Rev. Mineral., 2 (2nd Edition),* 101-120, 1983.

Kroll, H. and P.H . Ribbe, Lattice parameters, composition and Al, Si order in alkali feldspars, *Rev. Mineral., 2 (2nd Edition),* 57-99, 1983.

Lazarev, A. N., *Vibrational spectra and structure of silicates,* 307 pp., Consultants Bureau, New York, 1972.

Madon, M., Ph. Gillet, Ch. Julien, and G. D. Price, A vibrational study of phase transitions among the GeO_2 polymorphs, *Phys. Chem. Minerals, 18,* 7-18, 1991.

Mao, H. K., P. M. Bell, J. W. Shaner, and D. J. Steinberg, Specific volume measurements of Cu, Mo, Pd and Ag and calibrationof the ruby fluorescence gauge from 0.06 to 1 Mbar, *J. Appl. Phys., 49,* 3276-3283, 1978.

Meade, C., R. J. Hemley, and H. K. Mao, High-pressure x-ray diffraction of SiO_2 glass. *Phys. Rev. Lett., 69,* 1387-1390, 1992.

Mishima, O., L. D. Calvert, and E. Whalley, "Melting ice" I at 77 K and 10 kbar: A new method of making amorphous solids, *Nature, 310,* 393-395, 1984.

Ostertag, R., Shock experiments on feldspar crystals, *J. Geophys. Res. Suppl.,* B364-B376, 1983.

Redfern, S. A. T., Length scale dependence of high-pressure amorphization: The static amorphization of anorthite, *Mineral. Mag., 60,* 493-498, 1996.

Redfern, S. A. T. and E. Salje, Microscopic dynamic and macroscopic thermodynamic character of the $I\bar{1} \rightarrow P\bar{1}$ phase transition in anorthite, *Phys. Chem. Minerals, 18,* 526-533, 1992.

Richard, G. and P. Richet, Room-temperature amorphization of fayalite and high-pressure properties of Fe_2SiO_4 liquid, *Geophys. Res. Lett., 17,* 2093-2096, 1990.

Robertson, P. B., Experimental shock metamorphism of maximum microcline, *J. Geophys. Res., 80,* 1903-1910, 1975.

Ryzhova, T. V., Elastic properties of plagioclase, *Bull. Acad. Sci. USSR Geophys. Ser., 7,* 633-635, 1964.

Ryzhova, T. V. and K. S. Alexandrov, The elastic properties of potassium-sodium feldspars, *Bull. Acad. Sci. USSR Phys. Sol. Earth,* 53-56, 1965.

Salje, E., Phase transitions and vibrational spectroscopy in feldspars, in *Feldspars and Their Reactions,* edited by I. Parson, pp. 103-160, Kluwer Academic, Amsterdam, 1994.

Salje, E., B. Guttler, and C. Ormerod, Determination of the degree of Al, Si order Q_{od} in kinetically disordered albite using hard mode infrared spectroscopy, *Phys. Chem. Minerals, 16,* 576-581, 1989.

Sciortino, F., U. Essmann, H. E. Stanley, M. Hemmati, J. Shao,

G. H. Wolf, and C. A. Angell, Crystal stability limits at positive and negative pressures, and crystal-to-glass transitions, *Phys. Rev. E, 52,* 6484-6491, 1995.

Serghiou, G. C. and W. S. Hammack, Pressure-induced amorphization of wollastonite ($CaSiO_3$) at room-temperature, *J. Chem. Phys., 98,* 9830-9834, 1993.

Simakov, G. V., M. N. Pavlovskiy, N. G. Kalashnikov, and R. F. Trunin, Shock compressibility of twelve minerals, *Izv. Earth Phys., 10,* 488-492, 1974.

Stöffler, D., Glasses formed by hypervelocity impact, *J. Non-Cryst. Solids, 67,* 465-502, 1984.

Stöffler, D., Deformation and transformation of rock-forming minerals by natural and experimental shock processes. II. Physical properties of shocked minerals, *Fortschr. Miner., 51,* 256-289, 1974.

Syono, Y., T. Goto, Y. Nakagawa, and M. Kitamura, Formation of diaplectic glass in anorthite by shock-loading experiments, in *High Pressure Research: Applications in Geophysics,* edited by M. H. Manghnani and S. Akimoto, pp. 477-489, Academic Press, New York, 1977.

Tarte, P., Etude eperimentale et interpretation du spectre infrarouge des silicates et des germanates, *Mem. Acad. R. Belg., 35 (4a,b),* 1-260, 1-139, 1965.

Tse, J. S. and D. D. Klug, Mechanical instability of alpha-quartz-A molecular dynamics study, *Phys. Rev. Lett., 67,* 3559-3562, 1991.

Tse, J. S., D. D. Klug, J. A. Ripmeester, S. Desgreniers, and K. Lagarec, The role of non-deformable units in pressure-induced reversible amorphization of clathrasils, *Nature, 369,* 724-727, 1994.

Velde, B., Y. Syono, R. Couty, and M. Kikuchi, High pressure infrared spectra of diaplectic anorthite glass, *Phys. Chem. Minerals, 14,* 345-349, 1987.

Williams, Q., R. J. Hemley, M. B. Kruger, and R. Jeanloz, High pressure vibrational spectra of α-quartz, coesite, stishovite, and amorphous silica, *J. Geophys. Res., 98,* 22,157-22,170, 1993.

Williams, Q. and R. Jeanloz, Static amorphization of anorthite at 300 K and comparison with diaplectic glass, *Nature, 338,* 413-415, 1989.

Williams, Q. and R. Jeanloz, Spectroscopic evidence for pressure-induced coordination changes in silicate glasses and melts, *Science, 239,* 902-905, 1988.

Williams, Q. and E. Knittle, Infrared and Raman spectra of $Ca_5(PO_4)_3F$-fluorapatite at high pressures: Compression-induced changes in phosphate site and Davydov splittings, *J. Phys. Chem. Solids, 57,* 417-422, 1996.

Williams, Q., E. Knittle, R. Reichlin, S. Martin, and R. Jeanloz, Structural and electronic properties of fayalite at ultrahigh pressures: Amorphization and gap closure, *J. Geophys. Res., 95,* 21,549-21,563, 1990.

Wolf, G. H., S. Wang, C. A. Herbst, D. J. Durben, W. F. Oliver, Z. C. Kang, and K. Halvorson, Pressure induced collapse of the tetrahedral framework in crystalline and amorphous GeO_2, in *High-Pressure Research in Mineral Physics, Geophysical Monogr. Ser.,* vol. 39, edited by M. H. Manghnani and Y. Syono, pp. 503-517, Terra Scientific Publishing Co., Tokyo/AGU, Washington, D.C., 1992.

Zhang, M., B. Wruck, A. G. Barber, E. K. H. Salje, and M. A. Carpenter, Phonon spectra of alkali feldspars: Phase transitions and solid solutions, *Am. Mineral., 81,* 92-104, 1996.

Q. Williams, Department of Earth Sciences and Institute of Tectonics, University of California, Santa Cruz, Santa Cruz, CA 95064 USA

Observation of Phase Transformations in Serpentine at High Pressure and High Temperature by In Situ X ray Diffraction Measurements

Koji Kuroda and Tetsuo Irifune

Department of Earth Sciences, Ehime University, Matsuyama 790, Japan

In situ X ray diffraction measurements on natural serpentine have been performed at high pressure and high temperature using a combination of a double-stage multianvil system and synchrotron radiation. No clear evidence for the amorphization of serpentine was obtained upon compression to 28 GPa at room temperature, in contrast to an earlier study using a diamond anvil cell and rotating anode X ray source. However, when the temperature was increased under pressure, amorphization was observed in a limited interval between about 200 and 400°C, at pressures greater than 14 GPa. At higher temperatures, X ray diffraction peaks became apparent quickly, suggesting rapid crystal growth of high pressure phases. In runs at pressures greater than 20 GPa, the final run products consisted of a phase assemblage with an unknown hydrous phase. The chemical composition of this phase is close to $Mg_2Si_3O_6(OH)_4$, which is nearly identical to that of phase F. However, its X ray diffraction pattern is different from phase F and is rather similar to the pattern of phase D reported by Liu. It was demonstrated that serpentine completely dehydrate to form forsterite + enstatite + water at relatively low temperatures less than 700–800°C and pressures less than 10 GPa, while it transforms to a phase assemblage including the hydrous phases, phase D and superhydrous phase B, at pressures greater than 20 GPa. This assemblage with dense hydrous magnesium silicate persists to high temperatures of at least 1200°C at pressures corresponding to the lower parts of the mantle transition region and the uppermost lower mantle.

INTRODUCTION

Amorphization of serpentine, which is the major hydrous phase constituting upper parts of the oceanic lithosphere [i.e., *Cannat et al.*, 1995] was reported at pressures above 10 GPa and at room temperature, using a combination of a diamond anvil cell and X ray diffraction method with a rotating anode X ray source [*Meade and Jeanloz*, 1991], as well as Raman spectroscopic measurements [*Meade et al.*, 1992]. Acoustic emissions were also observed upon compression, which was considered to be associated with the dehydration and/or the amorphization of serpentine. Such observations led these authors to conclude that some of the deep-focus earthquakes, which originate deeper than 300 km in the mantle, may be induced by the pressure-induced amorphization of serpentine under these pressures. However, the amorphization of serpentine was observed only at the ambient temperatures, and it is not clear whether this phenomenon is actually related to the acoustic emissions observed at high temperatures to about 600°C [*Meade and Jeanloz*, 1991].

Properties of Earth and Planetary Materials
at High Pressure and Temperature
Geophysical Monograph 101

On the other hand, in spite of the prime importance of serpentine as a possible carrier of water into deep mantle, only a few experimental studies on its stability in the mantle conditions have been carried out [*Yamamoto and Akimoto*, 1977; *Liu*, 1986; *Inoue*, 1994; *Ulmer and Trommsdorff*, 1995]. Among these studies, only *Liu* [1986] made experiments under the conditions corresponding to the depths of the mantle transition region and those of the lower mantle. Liu performed a series of high-pressure runs on natural serpentine using a diamond anvil cell and the X ray powder diffraction method, and reported the phase relations at various pressures to 28 GPa at temperatures near 1000°C. However, as no chemical analyses were performed in Liu's experiments and the temperature of his run was not well defined, this study should be regarded as of reconnaissance nature. Moreover, the presence of a few new dense hydrous magnesium silicates (DHMSs) have been reported since Liu's study [*Gasparik*, 1990, 1993; *Kanzaki*, 1991], some of which are difficult to identify solely by the X ray powder diffraction measurements.

We have developed an experimental technique that enables in-situ X ray measurements of phase transformations at pressures to 28 GPa and temperatures exceeding 1500°C on a routine basis, using sintered diamond (advanced diamond composite, ADC) anvils and synchrotron radiation as reported in this volume [*Irifune et al.*, this volume]. As the first experiment using this technique, we examined the nature of the amorphization of serpentine, because this had been observed only at room temperature using a diamond anvil cell. The preliminary experimental results on serpentine and its implications for the mechanism of deep-focus earthquakes have been given elsewhere [*Irifune et al.*, 1996], and we describe here further details of our experiments, as well as the results of some additional runs. Possible phase relations of serpentine in the deep mantle are also discussed on the basis of both the present in situ X ray diffraction measurements and the quench experiments.

STARTING MATERIALS AND EXPERIMENTAL PROCEDURE

Natural antigorite with a composition of $(Mg_{0.95}, Fe_{0.05})_3Si_2O_5(OH)_4$ was used as a starting material for our high pressure runs. In one run (ME-3), a natural lizardite sample, which has a similar composition but contains a certain amount of Al_2O_3 (~ 3 wt%), was also used to study the effect of crystallographic form upon the amorphization of serpentine. The natural specimen was inspected under the microscope, and the homogeneous part of the sample was carefully selected and ground under acetone in a hardened ceramic mortar. Thus prepared fine powders of the sample were intimately mixed with Au powder at 100:1 by volume, the latter having been used for pressure estimation, and enclosed in a capsule of cemented amorphous boron.

We used TiC sheet heaters and the temperature was measured by a $W_{95}Re_5$-$W_{74}Re_{26}$ thermocouple, whose hot junction was placed in the middle of the boron capsule. The hybrid anvil system using four ADC and four WC anvils (truncated edge length, TEL = 1.5 mm) was used for the second-stage anvil set. Pressure medium was made of sintered magnesia and preformed gaskets of pyrophyllite were used. Further details of the cell assembly are described in *Irifune et al.* [this volume].

In our in situ X ray diffraction experiments, pressure was applied first to the target value over 10–20 hours. In the course of increasing pressure, X ray diffraction data were taken every 20–30 tons (the maximum load used in

TABLE 1. Experimental Conditions and the Results of In Situ X Ray Measurements for Serpentine

Run No.	P_{max} (GPa)	T_{max} (°C)	Conditions for amorphization	Phases at T_{max}
ME-4	13	700	-	α + En + (W)
ME-6	17	1200	(11 GPa, 450 - 470°C)	α + En + (W)
ME-2	20	1050	18 GPa, 300 - 400°C	γ + St + (W)
ME-1	27	1200	22 GPa, 150 - 300°C	PhD + ShB*+ (W)
ME-3	28	1500**	26 GPa, 250 - 400°C	PhD + ShB* + (W)

P_{max} and T_{max} are the maximum pressure and temperature attained in each run, while the conditions for amorphization refer to the temperature interval between the temperature of the onset of amorphization and that of crystallization of high-pressure phases at the given pressure.
* confirmed in the quenched sample; ** may be lower than the nominal temperature (see, text).
-, no amorphization was observed; ( ), complete amorphization was not observed (see, text).
α = forsterite; En = enstatite; W = water; γ = Mg_2SiO_4 spinel; St = stishovite; PhD = phase D; ShB = superhydrous phase B.

the present series of runs was 300 tons, for ME-1 and ME-3). The pressures at these press loads were estimated on the basis of the unit-cell volume changes of Au, using an equation of state by *Anderson et al.* [1989]. The temperature was then gradually increased at the target pressure, keeping the press load constant. The X ray diffraction pattern was taken every 50–200°C for typically 500 seconds at each temperature. Temperature was increased up to 700–1500°C, normally over several hours. The nominal pressure dropped with increasing temperature at the fixed load, probably due to stress relaxation both in the sample and the surrounding pressure medium-gasket system. However, the pressure approached the initial value with further increasing temperature, probably because the thermal pressure in the sample overcame such effects. Further details of the present in situ X ray diffraction measurements are given in *Irifune et al.* [this volume].

The temperature was kept constant for ~30 minutes at the maximum temperature of each run, and the run was quenched by terminating the supply of electrical power. Then the sample was unloaded over 5–10 hours, depending on the maximum pressure of the run. The recovered sample was polished and examined by an electron microprobe analyzer and a microfocus X ray diffractometer. As there was significant temperature gradient across the sample (~100–200°C/mm) and the end parts of the sample were sometimes contaminated from the adjacent TiC heaters, only those near the thermocouple were examined by the above techniques.

In order to investigate the nature of phase D, which appeared as a high pressure phase in the present in situ X ray measurements above 18 GPa, a synthesis run was performed using the chemical composition of this phase. As this phase possesses a composition close to $Mg_2Si_3O_6(OH)_4$, we prepared a starting material of a mixture of $Mg(OH)_2$ and amorphous SiO_2 in the appropriate proportions to form this composition. An additional synthesis run was also made using antigorite starting material to confirm the chemical compositions and the X ray diffraction data of the final product from the in situ X ray diffraction experiment. These two synthesis runs were carried out using the standard cell assembly and the similar procedure as described in *Irifune et al.* [1992].

EXPERIMENTAL RESULTS AND DISCUSSION

In Situ Observations of the Phase Transformations and Amorphization of Serpentine at High Pressure and High Temperature

Experimental conditions and the results of the present high pressure runs are summarized in Table 1. In the first run (ME-1), pressure was slowly increased to about 27 GPa at room temperature over 14 hours (Figure 1). In the course of this process, the X ray diffraction pattern was

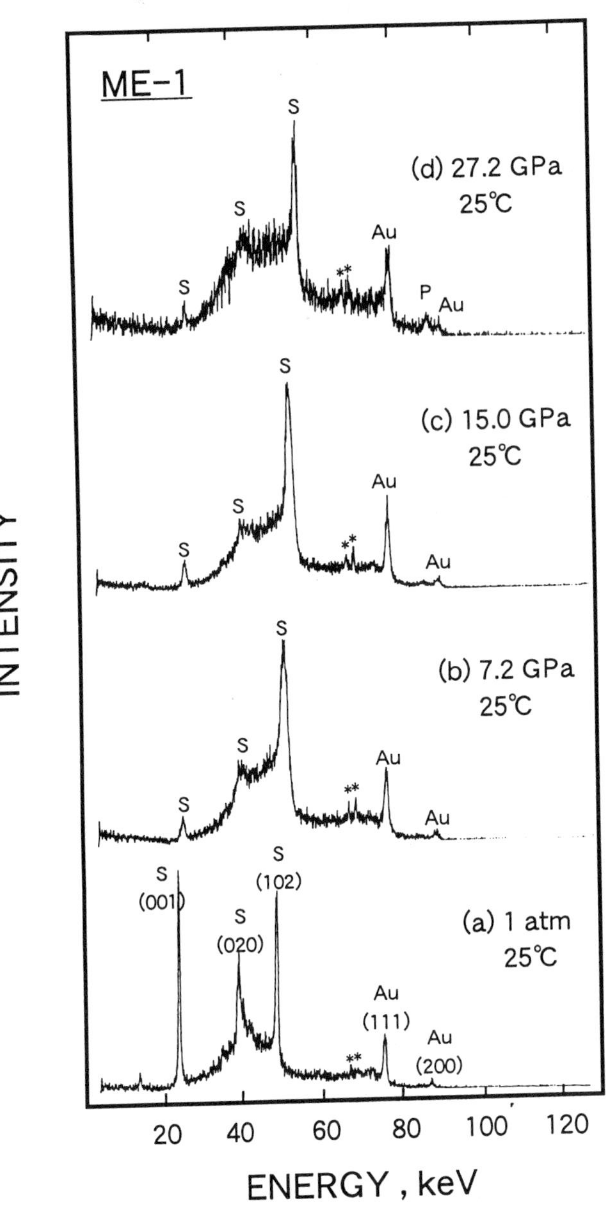

Figure 1. Variation of the X ray diffraction pattern in ME-1 with increasing pressure at a room temperature. The data were collected using an energy dispersive method, and the diffracted X ray beam was detected at 2 θ = 4° with a collimator and via vertical and horizontal slits [see *Irifune et al.*, this volume]. No amorphization of serpentine was observed at pressures to 27 GPa. Relatively high backgrounds around 40 keV are due to amorphous boron used as the sample capsule. Similarly, a diffraction peak from the surrounding MgO pressure medium (P) is seen at 27 GPa. S = serpentine (antigorite); Au = gold;
= characteristic line of Au.

taken for typically 10 minutes at every 2–3 GPa. The intensities of the diffraction lines corresponding to antigorite were carefully compared with those from the mixed Au powder. The relative intensity of the (001) reflection peak at 1 atm (Figure 1a) became notably small when pressure was applied to a few GPa (Figure 1b), while essentially no changes in the intensities of antigorite relative to Au were observed at higher pressures to 27 GPa, as shown in Figures 1c,d. Although we see a notable X ray background inherent to amorphous phases in a photon energy interval of about 30–60 keV, this presumably originated from the amorphous boron capsule surrounding the sample. Likewise a minor peak (P) from the MgO pressure medium was also detected particularly at pressures greater than 20 GPa (Figure 1d). Careful adjusting of the positions of slits and collimators enabled us to reduce such effects in the subsequent runs.

When we increased temperature slowly at a fixed load (300 ton) in ME-1, the diffraction peaks of antigorite became weaker at temperatures of 100–150°C, as compared to those at room temperature (Figure 2a). Then we collected X ray diffraction data at 200°C for 10 minutes and found that most of the peaks of serpentine disappeared quickly and an amorphous halo became obvious (Figure 2b). We then collected the data every 50–100°C, and found that new diffraction peaks appeared and grew quickly at temperatures above 300°C (Figure 2c). At the same time, the amorphous halo was reduced, and the diffraction peaks of the high pressure phases became sharper at temperatures higher than about 600°C (Figures 2d,e).

Examination of the quenched products demonstrated that the coexisting phases at the maximum temperature (1200°C) and at about 24 GPa were an unknown phase + superhydrous phase B (+ H_2O). The X ray diffraction pattern of the former phase was quite similar to that of phase D reported by *Liu* [1987], and we tentatively refer to this phase as phase D according to Liu. Note that the phase D used by *Yamamoto and Akimoto* [1977] was later reassigned to a new phase discovered in the product from serpentine at 22 GPa [*Liu*, 1986, 1987], as the original phase D was in fact the natural mineral chondrodite with a purer composition [*Yamamoto and Akimoto*, 1977; *Liu*, 1987]. Further investigation of the present new phase was made, the results of which will be given in the following section.

In the second run (ME-2), we increased temperature to 200°C at every 2-5 GPa above 10 GPa to find the minimum pressure where amorphization of antigorite would occur. However, no amorphization was observed to the maximum press load of 200 ton (cf. Figure 3a),

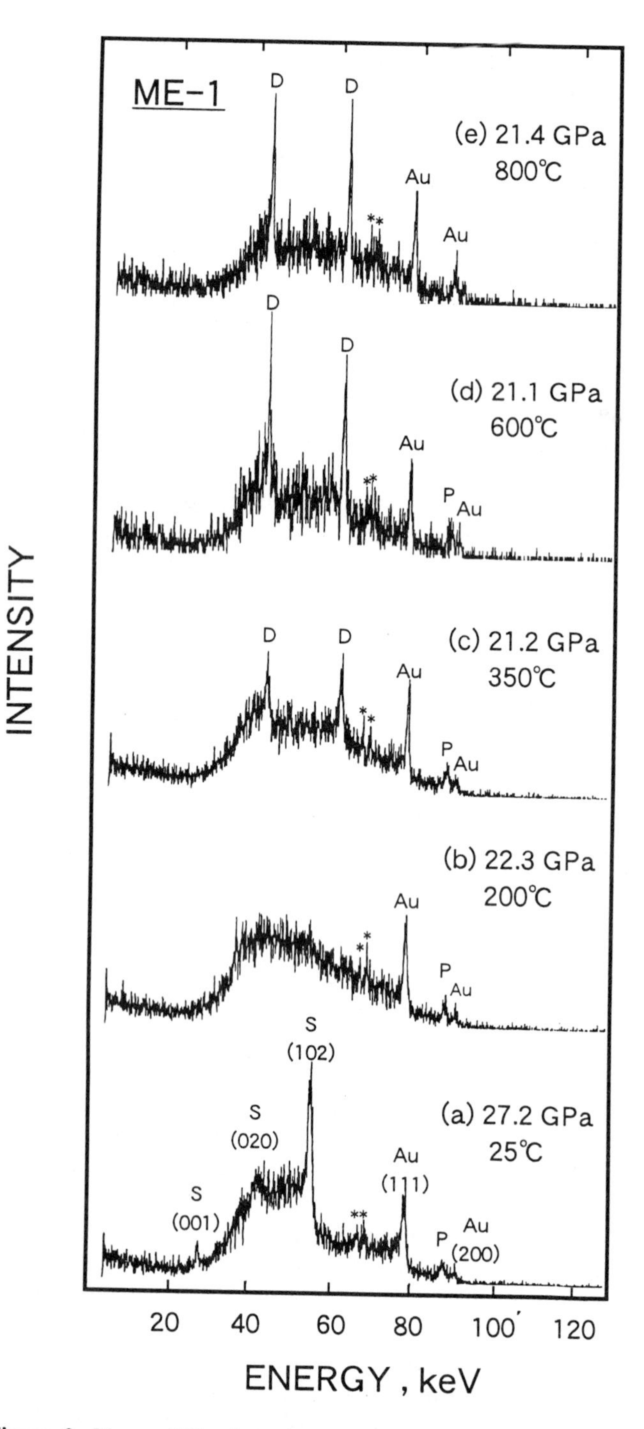

Figure 2. X ray diffraction pattern in ME-1 as a function of temperature at a fixed load of 300 tons. The diffraction peaks of serpentine (a) disappeared at temperatures of 150–300°C and an amorphous halo was clearly seen (b). High pressure phases, including phase D, crystallized rapidly at temperatures higher than 300°C (c-e), and the final product was identified as an assemblage of phase D + superhydrous phase B (+ water) at 1200°C and 24 GPa.

corresponding to a pressure of about 20 GPa. Then, at this pressure, we increased the temperature and found that antigorite became amorphous at a temperature greater than 300°C (Figure 3b), which is slightly higher than that expected from the observations in ME-1. At temperatures greater than 400°C, new diffraction peaks, such as those

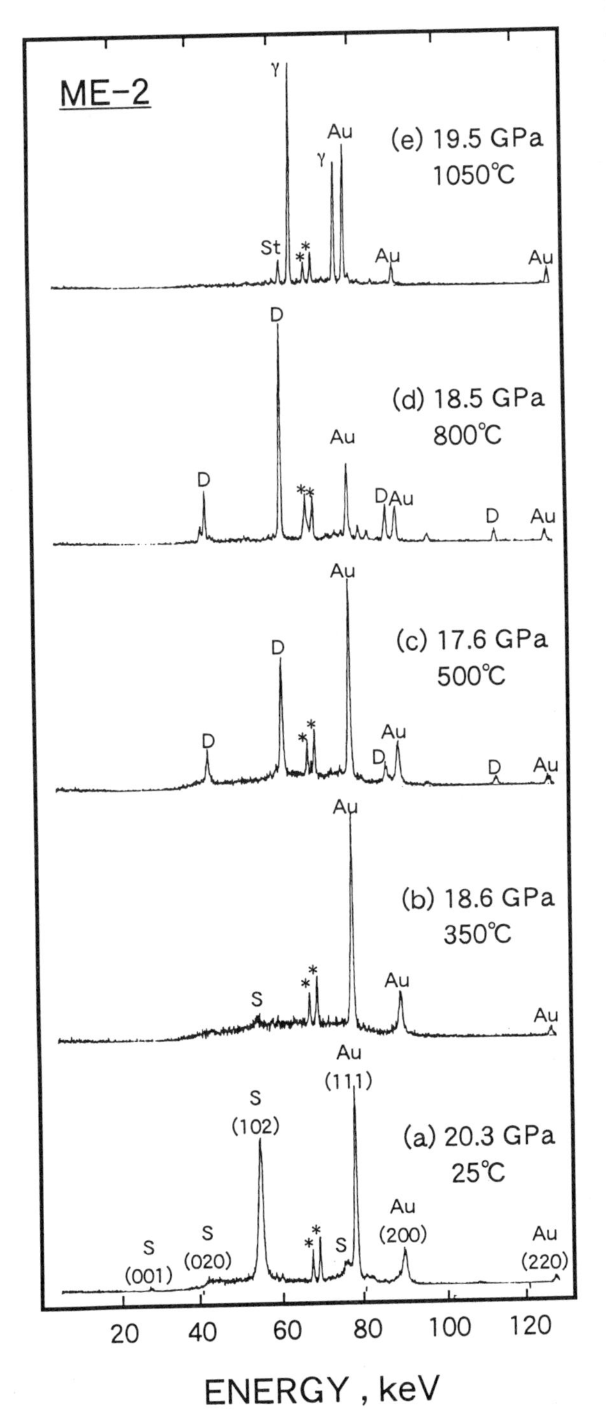

Figure 3. Some examples of the X ray diffraction patterns in ME-2. No amorphization of serpentine occurred at pressures up to about 20 GPa at a room temperature (a), as is the case of ME-1. Then the amorphization was observed in a temperature interval of 200–400°C (b), slightly higher than that in ME-1. At higher temperatures, rapid crystal growth of high pressure phases was observed (c and d). The final product at 19.5 GPa and 1050°C (e) consisted of γ- Mg_2SiO_4 and stishovite. γ = γ-Mg_2SiO_4; St = stishovite.

corresponding to phase D, appeared quickly as observed in run ME-1 (Figures 3c,d). Formation of γ-Mg_2SiO_4 + stishovite was confirmed at temperatures near 1000°C (Figure 3e), suggesting that complete dehydration occurred at these temperatures at pressures about 19 GPa. The final quench product obtained at 20 GPa and at 1200°C was in fact identified as an assemblage of γ-Mg_2SiO_4 + stishovite (+ H_2O) by both powder X ray diffraction and electron microprobe analyses.

We then further explored the minimum pressure of amorphization in runs ME-4 and ME-6. In ME-6, the sample was pressurized to about 16 GPa, and then temperature was increased gradually. We did not see any evidence of amorphization of antigorite at pressures 16–12 GPa with increasing temperature to 400°C at a fixed load of 70 ton (Figures 4a, b). However, we noticed that the intensity of the (102) reflection became slightly smaller as compared to the reference peak of Au after being held for 30 minutes at 450°C and at about 10 GPa (Figure 4b). The diffraction peaks of high pressure phases, such as phase A, were manifested and the intensities of the peaks of serpentine greatly reduced at a slightly higher temperature of 470°C (Figure 4c). Thus no obvious sign of amorphization was obtained in this run, but this observation suggests that amorphization could have happened near these pressure/temperature conditions if the run was held for longer period of time. When the temperature was further increased, the complete dehydration to form forsterite + enstatite was observed near 800°C (Figure 4d).

In run ME-4, performed at the lowest press load of 50 ton, antigorite persisted to about 450°C and about 7 GPa, and the diffraction peaks of high pressure phases appeared at this temperature at the expense of those of antigorite. Thus no amorphization was observed at such pressures. The complete dehydration to forsterite + enstatite occurred at a temperature near 700°C, at about 6 GPa.

One run (ME-3) was made using lizardite starting material under the conditions similar to those of ME-1, and we obtained virtually the same results from these two runs with different forms of serpentine; amorphization of lizardite was observed at temperatures above 250°C, at

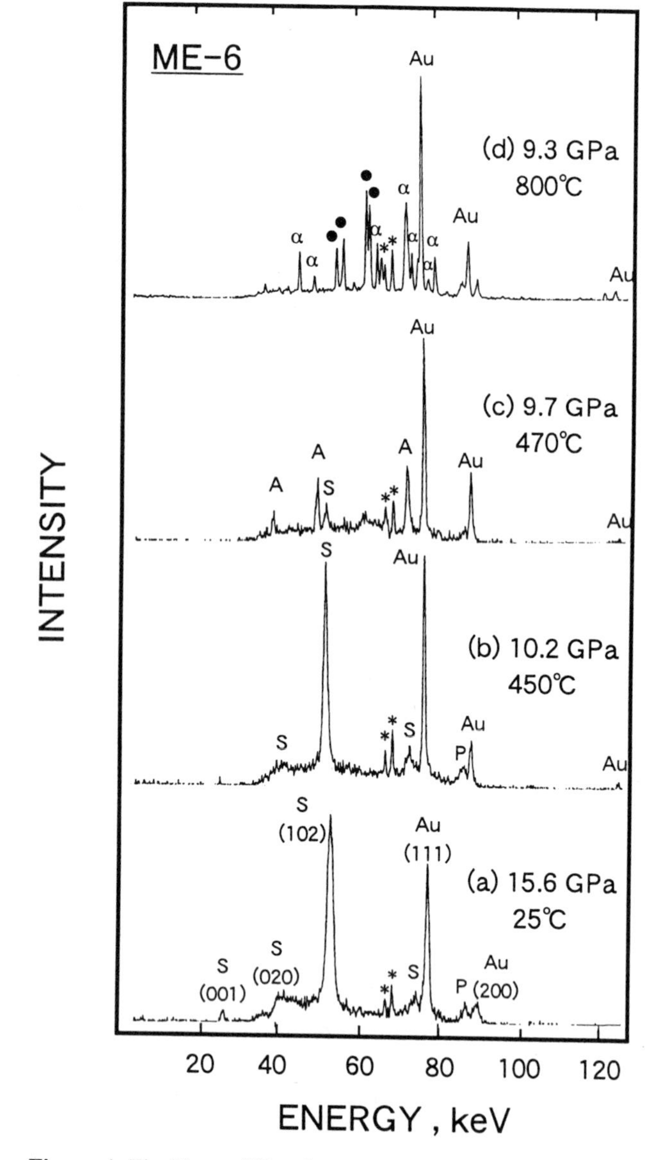

Figure 4. The X ray diffraction pattern of serpentine (a) did not change significantly with increasing temperature to 450°C at pressures about 10 GPa (b) in ME-6. At a slightly higher temperature of 470°C, the diffraction peaks of serpentine were appreciably reduced, and crystallization of the high pressure phases, such as phase A, was observed (c). The sample transformed to an assemblage of α-Mg_2SiO_4 and enstatite near 800°C, at about 9 GPa, suggesting the complete dehydration of serpentine. A = phase A; α= α-Mg_2SiO_4 ; Åú = enstatite.

pressures about 25–27 GPa, while the high pressure phases crystallized rapidly at temperatures above 400°C. We observed persistence of phase D and superhydrous phase B at temperatures as high as 1500°C at a pressure near 26 GPa in this run. However, it should be noted that this temperature may be somewhat overestimated due to the possible reaction of the thermocouple with the boron capsule at 1000–1200°C [*Irifune et al.*, this volume].

In Figure 5, we summarize the region where we observed amorphization of serpentine. The amorphization took place near the lower temperature boundary (~200°C) of this region, while the upper limit (~00°C) represents the temperature where the high pressure phases started to crystallize. These temperatures appear to become slightly lower as pressure increases, but this may be within the uncertainties of the temperature measurements. Both of these temperatures would become significantly lower if we could run for longer duration, when we take kinetic effects on the transformations into account. Thus these temperatures should be regarded as the highest bounds. On the other hand, we were unable to constrain the minimum pressure for the amorphization of serpentine. However, as seen in Figure 5, this phenomenon may be observed only at pressures greater than about 14 GPa, for the temperature interval of 200–400°C.

The present results are quite different from those reported by *Meade and Jeanloz* [1991] and *Meade et al.* [1992], in which gradual amorphization of serpentine was observed at pressures above 10 GPa at room temperature. Nevertheless, these and our present results may be reconciled when we take the effects of stress and kinetics upon the amorphization into account, because these authors used a diamond anvil cell in which stress environments are very different from those in our multi-anvil experiments: it is likely that the larger differential stress in diamond anvil cell promotes amorphization of serpentine at lower tempera-tures. Moreover, *Meade and Jeanloz* [1991] describe that their runs were made over a month, which is far longer than ours. We observed that rapid crystal growth of high pressure phases occurs at temperatures of 300–400°C in our run durations, which suggests that the amorphous state of serpentine is no more stable at temperatures above such critical temperatures. These observations led to a conclusion that amorphization of serpentine is unrelated to the occurrence of deep-focus earthquakes in subducting slabs, because the temperatures of the slabs are generally higher than these temperatures [*Irifune et al.*, 1996].

Characteristics of an Unknown Hydrous Phase

One quench run was carried out at the pressure and temperature condition similar to that of ME-1 (26 GPa, 1200°C), in order to further examine the run product. In this run, a platinum capsule was used to avoid contamination of the sample from surrounding boron capsule and TiC heaters. Tables 2 and 3 show the results of electron microprobe analyses of the run product and the

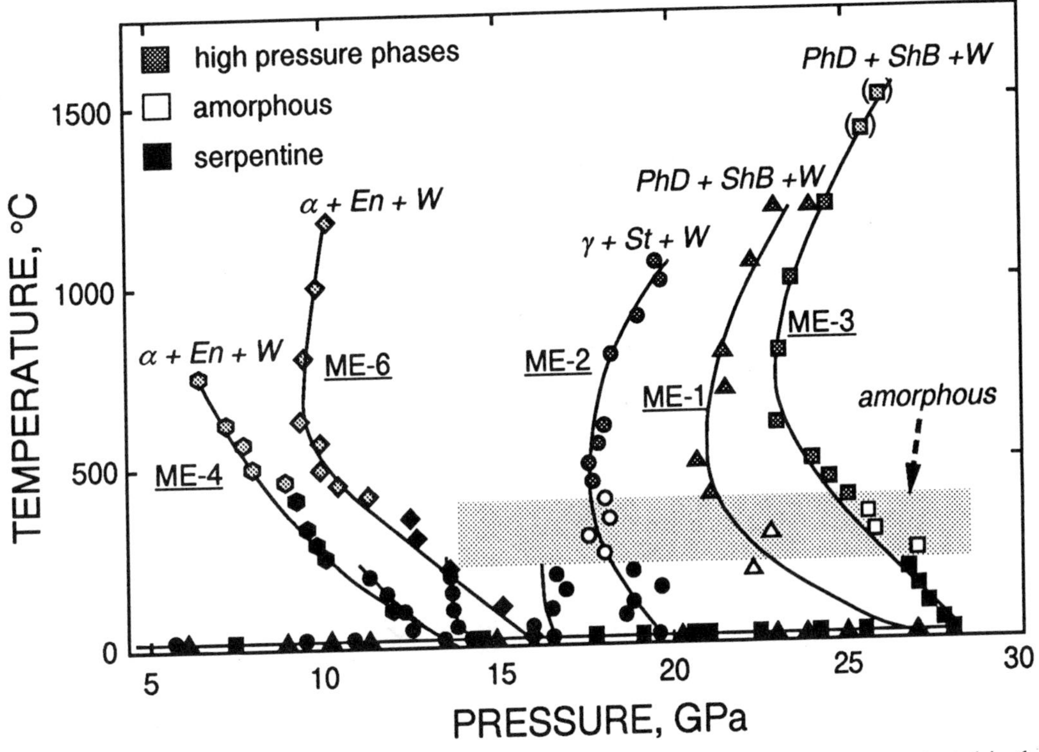

Figure 5. The pressure and temperature conditions where the X ray diffraction measurements were conducted in the present experiments. The region of the amorphization of serpentine is indicated by the shadow area. The phase assemblages under the maximum temperatures of these runs, determined on the basis of both in situ measurements and quenching method, are also shown on the top parts of the run trajectories. The temperatures above 1200°C in ME-3 may be somewhat overestimated [see text and *Irifune et al.*, this volume].

TABLE 2. Electron Microprobe Analyses of the Starting Material (Antigorite) and the Phases Present in the Sample Produced at 26 GPa and at 1200°C.

	Antigorite	Phase D	Superhydrous phase B
SiO_2	41.68	58.81	29.58
Al_2O_3	1.31	2.46	1.01
FeO	3.69	0.74	2.49
MgO	37.57	25.65	62.64
Total*	84.25	87.66	95.72
No. of oxygens	7	8	16
Si	2.015	2.918	3.010
Al	0.075	0.145	0.120
Fe	0.149	0.185	0.210
Mg	2.709	1.895	9.505
Sum	4.948	5.143	12.845
Mg/(Mg+Fe)	94.8	91.1	97.8
(Mg+Fe)/Si	1.42	0.71	3.23

* The deficits of the totals are supposed to be due to the presence of H_2O in these phases.

TABLE 3. Comparison of the X Ray Powder Diffraction Data for the Antigorite Sample Quenched from 26 GPa, 1200°C and Those of Phase D (Synthesized at 18 GPa, 1000°C) and Superhydrous Phase B [*Kanzaki*, 1993].

Run product at 26 GPa, 1200°C		Phase D		Superhydrous phase B*		
I/I_{100}	d_{obs}	d_{obs}	I/I_{100}	d_{obs}	I/I_{100}	(hkl)
5	4.69					
20	4.32	4.31	20			
5	4.14	4.11	15	4.11	15	(120)
10	3.45			3.44	12	(130)
100	3.01	2.996	100	2.99	44	(122)
20	2.539			2.545	12	(200)
20	2.409	2.391	10	2.401	25	(142)
20	2.234			2.233	36	(230)
50	2.107	2.098	90			
40	2.057			2.057	60	(161)
				2.043	100	(143)
5	1.949					
40	1.873	1.867	25	1.871	43	(153)
40	1.608	1.608	60			
		1.499	10			
30	1.472	1.472	20	1.464	99	(303)
10	1.385	1.380	30	1.376	14	(352)
20	1.367	1.368	10	1.372	12	(235)
40	1.315	1.316	5			
5	1.240	1.238	10			
10	1.186	1.186	10			
10	1.086	1.087	5			
		1.063	10			
		1.047	10			
		1.006	5			

*Some of the calculated diffraction peaks that do not correspond to any of the run products were omitted from this table for simplicity.

X ray powder diffraction data, respectively. Both these data and the electron microscopic observations (Figure 6) demonstrate the presence of a superhydrous phase B + an unidentified phase (+ H_2O). The latter phase has a composition close to $Mg_2Si_3O_6(OH)_4$, which is similar to that reported for phase F [*Kanzaki*, 1991; *Ohtani et al.*, 1995; *Kudoh et al.*, 1995]. However, the X ray powder diffraction pattern of the present unknown phase was very different from those reported for phase F [*Kudoh et al.*, 1995].

We then made further experiments using a starting material possessing a composition of the above unknown phase. The run was carried out at 18 GPa and 1000°C, and we obtained a single-phase crystalline product of the above composition. An X ray powder diffraction measurement was conducted on this sample, and the results are also listed in Table 3. The obtained data are in good agreement with those for the unknown phase observed in our serpentine run at 26 GPa and at 1200°C. We found the X ray diffraction pattern of this phase to be quite similar to that of phase D (+ periclase), reported by *Liu* [1986, 1987], except for a few minor diffraction peaks including one at 9.5 Å. Liu observed phase D as the transformation product from serpentine at pressures above 22 GPa and at about 1000°C in a diamond cell. As some of the diffraction peaks were close to those of periclase, *Liu* [1987] assumed that serpentine decomposed into an assemblage of periclase + phase D and suggested that phase D has a composition of $MgSiH_2O_4$. We confirmed, however, that serpentine transforms to an assemblage of superhydrous B + phase D (which has a composition close to $Mg_2Si_3O_6(OH)_4$) + H_2O under the conditions comparable to those of Liu's experiment.

The relation of our phases D and F is a matter of question. These phases have virtually the same composition, and their stability conditions also appear to be

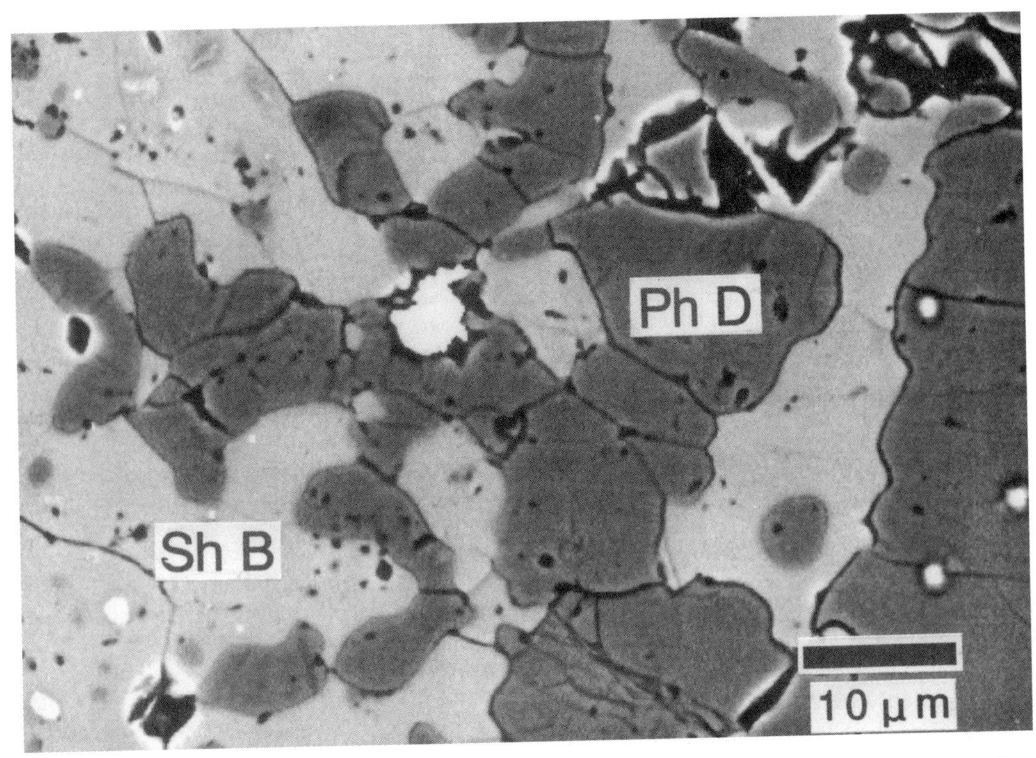

Figure 6. An SEM (back scattered electron) image of the run product quenched from 26 GP and 1200°C. Serpentine crystallized to an assemblage of phase D (PhD, dark grains) and superhydrous phase B (ShB, brighter grains). The presence of a number of small pores in the sample suggests the coexistence of dehydrated water (vapor) under these conditions. Bar = 10 mm.

overlapped according to *Kanzaki* [1991], *Ohtani et al.* [1995], and our present study. *Kanzaki* [1991] reported the similarity of the X ray pattern between phase F and Liu's phase D in his first report of the synthesis of phase F. However, the subsequent crystallographic analyses using single crystals of this phase yielded a completely different powder diffraction pattern [*Kudoh et al.*, 1995]. As the calculated density for phase F based on the crystallographic data is quite low (~ 2.83 g/cm^3) and the recalculated X ray powder pattern was almost identical to that of superhydrous phase B, we suppose there is something wrong with the crystallographic analyses of the single crystal phase F. Unfortunately, we were unable to index the present X ray powder diffraction data of phase D on the basis of any known crystallographic structures, and further experimental study is required to address this issue.

Phase Relations of Serpentine at High Pressure and Some Implications

Although the determination of the phase relations of serpentine at high pressure and high temperature is not the main purpose of the present study, some information about the nature of the phase transformations is available from our in situ X ray diffraction measurements. It is generally difficult to identify the stable phases in the course of increasing temperature via in situ X ray diffraction measurements, as run duration is very limited and usually insufficient for chemical equilibrium of the coexisting phases to be achieved. Moreover, the chemical compositions of the phases are not determined during the run. However, we can identify the phases present under the maximum pressure and temperature conditions by analyzing the quenched samples, and we can also impose some constraints on the P-T stability fields of these phases. Nevertheless, it should be noted that there is a possibility that some of the relased water upon dehydration of serpentine may escape from the capsule in the present cell assembly and the phase relations observed in the present study should be regarded as of reconnaissance nature.

At pressures between 6 and 10 GPa, the products at the maximum temperatures of 900–1200°C (ME-4 and ME-6) were forsterite and enstatite (Figure 5), which is consistent with the experimental results of the earlier studies using the quenching method [*Yamamoto and Akimoto*, 1977; *Liu*, 1986; *Ulmer and Trommsdorff*, 1995]. Although we were unable to identify the phase assemblages at lower temperatures unambiguously, our in situ X ray diffraction measurements demonstrated that the complete dehydration of such low temperature phase assemblages, including phase A, occurred at temperatures near 700°C at 6 GPa (ME-4) and near 800°C at 10 GPa (ME-6). These temperatures are again consistent with those determined from quench experiments by *Yamamoto and Akimoto* [1977] and *Ulmer and Trommsdorff* [1995].

Our experiments also showed that serpentine completely dehydrates to form γ-Mg_2SiO_4 + stishovite at temperatures higher than about 1000°C and pressure near 20 GPa (Figure 5) via a phase assemblage including phase D at the lower temperatures. In contrast, the phase assemblages of phase D + superhydrous phase B + water was found to persist at temperatures of at least 1200°C in two runs at pressures above 24 GPa (ME-1 and ME-3).

The present study suggests that some dense hydrous magnesium silicates (DHMSs) such as phase A, phase D, and superhydrous phase B may be present in subducting slabs, if temperatures of the upper parts of the slabs are significantly (~ 500–800°C) lower than the surrounding mantle. Such temperature distributions may be realized in relatively thick and old slabs, and thus a certain amount of water could be transported deep into the mantle transition region and the uppermost parts of the lower mantle. Further investigations of the phase relations in serpentine and the related DHMSs are needed to explore this process and to discuss the inventory and the distribution of water throughout the mantle of the Earth.

Acknowledgments. We thank the following persons for help and encouragement at various stages of the present study: T. Inoue, N. Funamori, T. Uchida, T. Yagi, N. Miyajima, K. Fujino, S. Urakawa, T. Kikegawa, and O. Shimomura. We thank L.-G. Liu and C. Meade for the benefit of discussion and T. Gasparik, P. Ulmer and T. Kato for review. Assistance by M. Mizobuchi, M. Miyashita, N. Kubo, M. Isshiki, N. Nishiyama, Y. Yamasaki during the experiments at KEK is also acknowledged. The present study is supported by a grant-in-aid for scientific research from the Ministry of Education, Science and Culture of Japan, Inoue Science Foundation, and also by Japan Society for Promotion of Sciences (JSPS).

REFERENCES

Anderson, O.L., D. G. Isaak, and S. Yamamoto, Anharmonicity and the equation of state for gold, *J. Appl. Phys., 65*, 1534-1543, 1989.

Cannat, M., C. Mevel, M. Maia, C. Deplus, C. Durand, P. Gente, P. Agrinier, A. Belarouchi, G. Dubuisson, E. Humler, and J. Reynolds, Thin crust, ultramafic exposures, and rugged faulting patterns at the Mid-Atlantic Ridge (22°–24°N), *Geology, 23*, 49-52, 1995.

Gasparik, T., Phase relations in the transition zone, *J. Geophys. Res., 95*, 15751-15769, 1990.

Gasparik, T. The role of volatiles in the transition zone, *J. Geophys. Res., 98*, 4287-4299, 1993.

Inoue, T., Effect of water on melting phase relations and melt compositions in the system Mg_2SiO_4-$MgSiO_3$-H_2O up to 15 GPa, *Phys. Earth Planet. Inter., 85*, 237-263, 1994.

Irifune, T., Y. Adachi, K. Fujino, E. Ohtani, A. Yoneda, and H. Sawamoto, A performance test for WC anvils for multianvil apparatus and phase transformations in some aluminous minerals up to 28 GPa, in *High-Pressure Research: Application to Earth and Planetary Sciences, Geophys. Monogr. Ser.*, vol. 67, edited by Y.Syono and M. H. Manghnani, pp. 43-50, TERRAPUB/AGU, Tokyo/ Washington, D.C., 1992.

Irifune, T., K. Kuroda, N. Funamori, T. Uchida, T. Yagi, T. Inoue, and N. Miyajima, Amorphization of serpentine at high pressure and high temperature, *Science, 272*, 1468-1470, 1996.

Kanzaki, M., Stability of hydrous magnesium silicates in the mantle transition zone, *Phys. Earth Planet. Inter.*, 66, 307-312, 1991.

Kanzaki, M., Calculated powder X ray patterns of phase B, anhydrous B and superhydrous B: re-assessment of previous studies, *Mineral. J.*, 16, 278-285, 1993.

Kudoh, Y., T. Nagase, S. Sasaki, M. Tanaka and M. Kanzaki, Phase F, a new hydrous magnesium silicate synthesized at 1000°C and 17 GPa: Crystal structure and estimated bulk modulus, *Phys. Chem. Minerals, 22*, 295-299, 1995.

Liu, L.-G., Phase transformations in serpentine at high pressures and temperatures and implications for subducting lithosphere, *Phys. Earth Planet. Inter., 42*, 255-262, 1986.

Liu, L.-G., Effect of H_2O on the phase behavior of the forsterite-enstatite system at high pressures and temperatures and implications for the Earth, *Phys. Earth Planet. Inter., 49*, 142-167, 1987.

Meade, C. and R. Jeanloz, Deep focus earthquakes and the recycling of water into the Earth's mantle, *Science, 252*, 68-72, 1991.

Meade, C., R. Jeanloz, and R. J. Hemley, Spectroscopic and X ray diffraction studies of metastable crystalline-amorphous transitions in $Ca(OH)_2$ and serpentine, in *High-Pressure Research: Application to Earth and Planetary Sciences, Geophys. Monogr. Ser.*, vol. 67, edited by Y. Syono and M. H. Manghnani, pp. 485-492, TERRAPUB/AGU, Tokyo/ Washington, D.C., 1992.

Ohtani, E., T. Shibata, T. Kubo, and T. Kato, Stability of hydrous phases in the transition zone and the upper most part of the lower mantle, *Geophys. Res. Lett., 22*, 2553-2556, 1995.

Ulmer, P. and V. Trommsdorff, V., Serpentine stability to mantle depths and subduction-related magmatism, *Science, 268*, 858-861, 1995.

Yamamoto, K. and S. Akimoto, The system MgO-SiO_2-H_2O at high pressures and high temperatures - stability field for hydroxyl-chondrodite, hydroxyl-clinohumite and 10 Å phase, *Am. J. Sci., 277*, 288-312, 1977.

K. Kuroda and T. Irifune, Department of Earth Sciences, Ehime University, 2-5 Bunkyocho, Matsuyama, Ehime 790, Japan

SUBJECT INDEX

AUTHOR INDEX